Primrose McConnell's

The Agricultural Notebook

20th Edition

Edited by

Richard J. Soffe

Seale-Hayne
University of Plymouth
UK

Blackwell
Science

Editorial Offices:
Osney Mead, Oxford OX2 0EL, UK
Tel: +44 (0)1865 206206
Blackwell Science, Inc., 350 Main Street, Malden, MA 02148-5018, USA
Tel: +1 781 388 8250
Iowa State Press, a Blackwell Publishing Company, 2121 State Avenue, Ames, Iowa 50014-8300, USA
Tel: +1 515 292 0140
Blackwell Publishing Asia Pty Ltd, 550 Swanston Street, Carlton South, Victoria 3053, Australia
Tel: +61 (0)3 9347 0300
Blackwell Wissenschafts Verlag, Kurfürstendamm 57, 10707 Berlin, Germany
Tel: +49 (0)30 32 79 060

First edition published 1883
18th edition published 1988
Reissued in paperback 1992
Reprinted 1994
19th edition published 1995
Reprinted 1996, 1997, 1998, 1999, 2000
20th edition published 2003

Library of Congress
Cataloging-in-Publication Data is available

ISBN 0-632-05829-3

A catalogue record for this title is available from the British Library

Set in 9 on 11pt Ehrhardt
by SNP Best-set Typesetter Ltd., Hong Kong
Printed and bound in Great Britain by Ashford Colour Press Ltd, Gosport, Hants.

For further information on
Blackwell Science, visit our website:
www.blackwell-science.com

Contents

Preface to the 20th Edition

This new edition takes *The Agricultural Notebook* into its third century. The first edition appeared in 1883. It was compiled by Primrose McConnell, a tenant farmer of Ongar Park Hall, Essex. As a student of Professor Wilson at Edinburgh University, he found 'a great want of a book containing all the data associated with the business of farming'.

The editor is very grateful to the team that has contributed to the new edition; also to a wider group of colleagues and friends who have helped guide the content of this edition. The challenge of the changing rural scene is evident in the range of conferences currently on offer considering the future of rural areas. The editor must take a balanced view regarding which sections should be added, and which should be replaced. Thirty contributors across the rural spectrum from Scotland to Devon and Norfolk to Wales are included. We hope this reflects developments within the rural sector. New sections added include (1) a marketing perspective on diversification, (2) organic farming and (3) farming and wildlife.

Our primary aim has been to meet the needs of students studying a range of agricultural, food and rural subjects. We have received many useful comments from universities and schools across the world who have been using *The Agricultural Notebook* to teach geography and a host of related subjects – please keep in contact with us and send in your suggestions. We additionally hope it will be of value to farmers, landowners and advisors in their many roles.

Primrose McConnell: a brief biographical sketch

P.W. Brassley

'I wish I had not been born for a hundred years to come, for there will be so many things found out after I am done with . . .', wrote Primrose McConnell in the spring of 1906. It was typical of him: always fascinated by the latest discoveries and inventions, but concerned to test them against his own extensive knowledge and experience.

McConnell was born at Lesnessock Farm, near Ochiltree in Ayrshire, on 11 April 1856, the son and grandson of tenant farmers. After leaving Ayr Academy he was apprenticed to a Glasgow engineering firm, but subsequently, in the 1870s, went to the University of Edinburgh, which did not then offer degrees in agriculture but prepared students for the diploma examinations of the Highland and Agricultural Society. McConnell passed in 1878, and a little later also passed the certificate examinations of the Royal Agricultural Society of England. When Edinburgh eventually introduced a degree, he returned, and was the second student to be awarded the BSc in Agriculture in 1886.

The *Notebook* is his lasting memorial, but his writing was based on the foundation of his other activities, as farmer, scientist, engineer and inventor, traveller, lecturer and all-round man of agricultural affairs. Farming formed the foundation of his life. He first rented the 636-acre Ongar Park Hall Farm, about 20 miles from London, near Epping in Essex, initially in partnership with his father. When he began farming, half of the land was in arable, but this was in 1883, when cheap grain from the new world was beginning to make life difficult for corn producers on stiff London clays. He therefore grassed down about 200 acres and based his farming upon dairying (he had about 60 cows in milk at any one time) and feeding bullocks, heifers and sheep. In 1905 he moved. Why? 'For the very good reason that I was losing more money than I could afford', he wrote soon afterwards, 'but also for various other reasons', which in fact centred on a dispute with his landlord over how he was to be compensated for capital invested in buildings. The case went to court, and McConnell had the better of the legal arguments, but he was left with a jaundiced view of landlords. He was the owner-occupier of his new farm, North Wycke, 500 acres of land 'as flat as a table' near Southminster in Essex, between the Crouch and Blackwater estuaries, with 'nothing higher than a tree or a house between me and the Ural Mountains'. When he began to farm there, it was half arable and half grass, and he kept 80 cows, nine work horses, a pony, two dogs, three tomcats and '135 head of poultry of all breeds under the sun, including those that do not lay in winter'. He was understocked, he knew, but he was too short of capital to buy any more. He continued to farm there until he died in 1931.

McConnell farmed to make money, but not only to make money. Fascinated by the scientific and technical problems involved, he attempted to deal with them professionally, as befitted an agricultural graduate. He tried machine milking, found that it resulted in decreased yields, and so went back to hand milking. Then, after considering his experiences for a year or so, he wrote a detailed article for the *Agricultural Gazette* setting out the costs, technical details, yield changes and probable explanations, before concluding that 'It is rather a dangerous thing to prophesy as to future inventions, and we do not know what mankind may accomplish in another generation. We may, therefore, still see a successful milking machine, but it has not arrived yet.' He was an early advocate of milk recording and kept a Gerber fat testing machine in his dairy. He experimented with silage and he designed his own elevator. The string-binder, he thought, was the greatest invention of the nineteenth century. He had 'an outfit of every possible kind of tool in my workshop on the farm that is likely to be of use', and was 'never . . . happier than when at the bench or vice'. When he wanted to try out a new plough, he would use it himself for a day, with a dynamometer between the horses and the plough. Not surprisingly, the shortcomings of farm machinery provoked some of his more vitriolic comments.

McConnell also led a busy life away from the farm. He lectured, at various times, at the Glasgow Veterinary College (where he was appointed Professor of Agriculture at the age of 24) and Oxford and Edinburgh universities and the Essex Winter School of Agriculture (the forerunner of Writtle College), and examined at Reading, Wye and the Royal Agricultural College. He was a Fellow of the Geological Society. He was on the Council of the Dairy Show and a milking judge there, and involved with the British Dairy Farmers' Association and the Eastern Counties Co-operative Dairy Farmers' Association. He visited farms in Holland in 1899 and made at least two trips to North America, in 1890 and 1893, on one occasion meeting some of the Sioux who had taken part in the Custer massacre in 1876. He was also one of the pioneers in the migration of farmers from Scotland to Essex in the late nineteenth century. No sooner had he found his farm at Ongar Park Hall than he was writing articles about the potential of Essex farms for the *North British Agriculturalist*.

And it is as a writer that he is now best remembered. 'I began to write to the farm papers at the age of eighteen, when first learning to hold the plough', he recalled, and he produced eleven editions of the *Notebook* between 1883, when he was 27, and 1930, the year before his death at the age of 75. He also wrote *The Elements of Farming* (1896), an elementary textbook, *The Elements of Agricultural Geology* (1902), *The Diary of a Working Farmer* (1906) and *The Complete Farmer* (1908), in addition to articles in the journals of the Royal Agricultural Society of England and the Bath and West Society. Later he spent many years as dairy editor of the *Agricultural Gazette* and editor of *Farm Life*.

Thus he was academically successful, and he clearly enjoyed writing. But it is also evident from his diary that he enjoyed physical work too: he writes enthusiastically of making his own cheese, digging, ploughing and broadcasting, and stooking even though the sheaves are drawing blood from his forearms. He was a teetotaller and a dissenting churchman (his wife Katherine was the daughter of a Free Church minister) who took a five-day study tour with the British Dairy Farmers' Association as his annual holidays. He took an unsentimental attitude to landscape: 'in a level district you get a great wide sky, and the sun shines longer'. But if this suggests the stereotypical dour Scot, then his irascibility, his sense of humour and his benevolent interest in the world around him keep breaking through, as does his pride in his family, when he mentions that his daughter Ann is an accredited dairymaid, and prints a photograph of his sons Archibald and Primrose (who was to be killed in the last days of the First World War). Farming, and thinking and writing about farming, provided him with stimulation and satisfaction. Fortunately, there was always something new to learn: 'Agriculture is a very wide subject, and no one can master it all within the limits of an ordinary lifetime'.

Sources

Most of the material used here is taken from McConnell's *The Diary of a Working Farmer* (1906) and from his obituary, published in the [Essex] *Weekly News* of Friday, 10 July 1931. For the latter, an enormous amount of other biographical material on McConnell, and many perceptive editorial comments, the author is indebted to Elizabeth Sellers of Chelmsford.

Contributors

Paul Brassley, BSc (Hons), BLitt., PhD.

Present appointment: Senior Lecturer Agricultural Economics, Faculty of Land, Food and Leisure, Seale-Hayne, University of Plymouth.

After reading agriculture and agricultural economics at the University of Newcastle-upon-Tyne went on to research in agricultural history at Oxford. Since his appointment to his present post at Seale-Hayne has taught agricultural economics and policy and researched agricultural history from the seventeenth to the twentieth centuries.

Peter Brooks, BSc (Hons), PhD.

Present appointment: Professor of Animal Production and Head of Research, Faculty of Land, Food and Leisure, Seale-Hayne, University of Plymouth

Obtained degree in Animal Production and PhD from Nottingham University. Was Head of Agriculture at Seale-Hayne College (subsequently University of Plymouth) for 18 years before becoming the faculty's Head of Research. Has researched and published extensively on the production, behaviour, management and nutrition of pigs. Current research focussed on the development of liquid feeding systems for pigs.

Adam Carter, BSc (Hons), MSc.

Previously: Lecturer in Rural Resource Management, Faculty of Land, Food and Leisure, Seale-Hayne, University of Plymouth.

Graduated in Rural Environment Studies at Wye College, University of London, before completing an MSc in Forestry at the Oxford Forestry Institute, University of Oxford. Previous experience in British silviculture and forestry extension work in Northern Sudan with Voluntary Services Overseas. Current research interests in forest ecology, silviculture and medicinal uses of forest products.

Richard Coates, FRICS, MRAC.

Present appointments: Buildings officer to the Anglican Church Schools of Papua New Guinea
Was Sole Principal of Rural Design Consultancy, Lemprice Farm, Budleigh Salterton, Devon. (now rtd)

Chartered Building Surveyor and Land Agent with a special interest in design in the countryside – both in agricultural, construction and other rural building projects. Forty years' experience in field including 18 years as Resident Building Surveyor to Clinton Devon Estates. Only designer to have won the coveted CLA Farm Buildings Award four times. Principal contributor of articles in *Countryside Building*, the journal of the Rural Design and Building Association (Chairman 2000–2001).

R.A. Cooper, CDA (Hons), NDA, MSc, PhD.

Present appointment: Principal Lecturer, Animal Production, Faculty of Land, Food and Leisure, Seale-Hayne, University of Plymouth.

Taught animal production at Shropshire Farm Institute before moving to University of Malawi as lecturer in animal husbandry and assistant farms director. Joined Seale-Hayne in 1974 following MSc at Reading University and completed PhD on interactions between growth promoters and reproductive physiology in ewe lambs in 1982. Main research interests in aspects of goat production, particularly in milking does, and in water intake studies.

John Eddison, BSc, PhD, CBiol, MIBiol.

Present appointment: Principal Lecturer, Applied Ethology, Faculty of Land, Food and Leisure, Seale-Hayne, University of Plymouth.

Graduated in Zoology from Leeds University and then conducted PhD research at Aberdeen University into the effect of environmental variation on ecological communities. Post-doctoral research followed at the Edinburgh School of Agriculture investigating the grazing behaviour and ecology of hill sheep. Took up his position in 1983 and now has a number of research projects on the behaviour and welfare of farm animals.

Tim Felton, LLb (Hons) of the Middle Temple, Barrister at Law.

Present appointment: Senior Lecturer, Law and Business Management, Faculty of Land, Food and Leisure, Seale-Hayne, University of Plymouth.

After reading law at Leeds University was called to the Bar of the Middle Temple. Following a period of legal practice he took up a career in practical farming and obtained a Diploma in Farm Management from Seale-Hayne. Prior to taking his present position he was share farming a mixed dairy, arable and beef farm at Tiverton, Devon of 185 hectares. Particular interests include access to the countryside and employment law issues in the rural environment.

David Fuller

Present appointment: Agriculture Colleges Liaison Officer, Royal Society for the Protection of Birds (RSPB), Sandy, Bedfordshire.

Has held the above post since its creation in 1997. Prior to working for the RSPB he spent over 33 years in the Ministry of Agriculture, Fisheries and Food (MAFF), administering many of the grant and subsidy schemes in East Anglia, Yorkshire and Lancashire. He also spent several years in Whitehall on policy work and in particular the development of the Environmentally Sensitive Areas (ESAs) scheme. He is a keen ornithologist and licenced bird ringer and carries out bird survey work on farmland in his local area of Norfolk.

Michael P. Fuller, BSc, PhD.

Present appointment: Reader in Crop Improvement and Acting Head of Department, Department of Agriculture and Food Studies, Faculty of Land, Food and Leisure, Seale-Hayne, University of Plymouth.

Graduated from Leicester University and then completed a PhD at the Welsh Plant Breeding Station, Aberystwyth (now IGER) on Frost Resistance in Grasses. Previous appointments: Leeds University and Sports Turf Research Institute; was visiting Professor to Angers University and is a Visiting Fellow of Exeter University. Current research interests include ice nucleation and resistance to frost; growth and development of cauliflowers and plant tissue culture.

Anita J. Jellings, BSc (Hons), PhD (Cantab).

Present appointment: Principal Lecturer, Faculty of Land, Food and Leisure, Seale-Hayne, University of Plymouth.

Read Applied Biology at the University of Bath and completed a PhD at the University of Cambridge before taking up a research fellowship at the University of York investigating constraints on photosynthetic capacity. Current interests are centred on sustainable agricultural systems.

J.A. Kirk, BSc, PhD.

Present appointment: Principal Lecturer, Animal Production, Faculty of Land, Food and Leisure, Seale-Hayne, University of Plymouth.

Spent nine years farming before reading agriculture with agricultural economics at the University College of North Wales, Bangor. Postgraduate research on the growth and development of Welsh Mountain lambs led to a PhD. Current research interests are the growth and development of animals and the improvement of meat quality.

Nicolas H. Lampkin, BSc (Hons), PhD.

Present appointment: Senior lecturer, Agricultural Economics, Institute of Rural Studies, University of Wales, Aberystwyth.

Graduated in agricultural economics from University of Wales, Aberystwyth, followed by PhD research on the economics of conversion to organic farming. Subsequent research has focused on the financial performance of organic farming and the role of organic farming in the development of the Common Agricultural Policy. Currently co-ordinator of organic agriculture teaching and research at IRS and director of Organic Centre Wales. Author/editor of *Organic Farming* (Farming Press) and the *Organic Farm Management Handbook* (UWA/EFRC).

A. Langley, BSc, MSc, HNC (Building), NDAgrE, MIAgrE.

Present appointment: Lecturer, SAC, Edinburgh.

After two years involvement with potato cultivation investigations at the National Institute of Agricultural Engineering (Scottish Station), joined SAC in 1974 as a Mechanisation Adviser, covering all aspects of mechanisation and specialising in cultivations, irrigation and crop storage. Focused on lecturing from 1990 and covered a range of topics including crop storage and processing, field production equipment and waste management. Also involved in organising and co-ordinating national machinery field demonstrations.

Tony G. Marangos, BSc (Hons), PhD, CBiol, MIBiol, RNutr.

Present appointment: Principal, Nutrition Solutions, Consultant Nutritionist, Burnham House, Fairfield Road, Shawford, Winchester, Hampshire, SO21 2DA

Educated at Reading and London Universities where he gained a PhD. Since then has held senior technical (nutrition) and marketing positions in the animal feed and supply industry. In 2000, after 25 years, began his own independent nutrition consultancy, *Nutrition Solutions*, and now advises national and international organisations.

D.J. Mattey, MIAgrE, MIOSH, MWeldI.

Present appointment: Consultant, International Labour Organisation (ILO) Training Centre (Turin); and The International Social Security Association (Geneva) Former HM Chief Inspector of Agriculture for HSE, and Regional Director (Midlands)

Robert E.L. Naylor, BSc (Hons), PhD, DSc, CBiol, FIBiol.

Present appointment: Partner in Trelareg Consultants specializing in crops and environment and Honorary Professor in Crop Science and Protection at the University of Aberdeen.

Spent 30 years at Aberdeen University teaching crop physiology, weed science, seed science and rural biodiversity and researching in specialist areas of these topics. On Board of Editors of Seed Science and Technology and Crop Protection. Senior Editor for *Journal of Agricultural Science*. Edited a new edition of the *Weed Management Handbook* for the British Crop Protection Council. Author of over 100 research papers and book chapters.

R.M. Orr, BSc, PhD, MIBiol.

Present appointment: Senior Lecturer, Faculty of Land, Food and Leisure, Seale-Hayne, University of Plymouth.

Read agriculture with animal science at the University of Aberdeen. Postgraduate research at Edinburgh University on appetite regulation led to a PhD. Teaching and research interests include food quality and nutrition.

H.E. Palmer

Formerly Farm Woodland Consultant, Scottish Agricultural College

Graduated in Agriculture from the University of Reading and completed an MSc in Forestry and its relation to land use at the Oxford Forestry Institute, University of Oxford. Wide experience in temperate forestry and farm woodland management, and has worked as a woodlands adviser and consultant on farms and estates in Scotland for 12 years. Member of the Institute of Chartered Foresters.

Robert Parkinson, BSc (Hons), MSc, PhD, MILT.

Present appointment: Principal Lecturer, Soil Science, Faculty of Land, Food and Leisure, Seale-Hayne, University of Plymouth.

Studied at Leeds and Reading Universities before taking up a post as Research Officer at Birkbeck College, University of London, working on soil water dynamics and agricultural drainage. This research subsequently led to the award of PhD. Current research interests include nutrient management in agricultural systems, biowaste recycling and composting, and soil/hydrological controls on the reconstruction of species-rich grassland.

John I. Portsmouth, NCP, NDP, DipPoult, NDR, FPH.

Present appointment: Managing Director, JP Enterprises, Consultant Nutritionist, Tremaen, Maenporth Hill, Falmouth, Cornwall.

Began his own private consulting service in 1991 following some 30 years in animal nutrition and animal health industry. Author of several books and many technical articles on poultry nutrition and management. Specialist subjects are calcium metabolism in laying hens and micro-nutrient requirements of poultry. In 1988 given a distinguished service award for services in nutrition to the UK poultry industry.

David Sainsbury, MA, PhD, BSc, MRCVS, FRSH, CBiol, FIBiol.

Present appointment: Director of The Cambridge Centre for Animal Health and Welfare, 101 Madingley Road, Cambridge.

After a period of research and lecturing at the Department of Veterinary Hygiene, Royal Veterinary College, London, until 1993 worked at the Department of Clinical Veterinary Medicine, University of Cambridge in the Division of Animal Health. Main interests have been concerned with the health and well-being of farm livestock especially under intensively managed conditions. Author of five textbooks and some 130 scientific papers.

R.W. Slee, MA, PhD, Dip Land Economy.

Present appointment: Senior Lecturer Rural Economics, Department of Agriculture, University of Aberdeen.

Before moving to Aberdeen in 1989, spent over 10 years as a lecturer at Seal-Hayne Faculty, University of Plymouth. Has written widely on rural economic change, including contributions to *The Changing Countryside* (eds J. Blunden and N. Curry, Croom Helm, 1985) and *A Future for Our Countryside* (eds J. Blunden and N. Curry, Blackwells, 1988) Author of *Alternative Farm Enterprises* (2nd edn, Farming Press, 1989). Current research interests include agrienvironmental policy questions and the economic impact of rural tourism.

Richard J. Soffe, MPhil, MCIM, M Inst M, ARAgS.

Present appointment: Senior Lecturer, Farm Business Management, Faculty of Land, Food and Leisure, Seale-Hayne, University of Plymouth.

Previous appointments have included a lectureship at Sparsholt College, Winchester, and assistant farm manager of a large estate in Hampshire. Research interests include rural leadership management and marketing. Course Director of the innovative and well respected Challenge of Rural Leadership at the University of Plymouth. He was co-editor of the eighteenth edition, editor of the nineteenth edition of *The Agricultural Notebook* and is editor of the current edition.

Mark A.H. Stone BA (Hons), MSc (Econ), MCIPD.

Present appointment: University of Plymouth Teaching Fellow and Principal Lecturer in People Management and Electronically Supported Open and Distance Learning.

After reading economics and politics at the University of Central Lancashire he went on to a masters degree in Employment Studies at University College Cardiff. He spent five years working in industry, during which time he achieved membership status within the Chartered Institute of Personnel and Development. He is involved in People Management and Information Technology teaching. He along with Dr Neil Witt formed the Communication and Learning Technology Group in 1997 to facilitate cross-faculty research and development in this area. He is currently leading an HEFCE TDTL Project into Student Progression and Transfer.

M.A. Varley, BSc, PhD, MIBiol, CBiol.

Present appointment: Lecturer, Department of Animal Physiology and Nutrition, University of Leeds.

Before his present position, spent 4 years as a lecturer in animal science at Seale-Hayne Faculty, University of Plymouth, and then moved to the Rowett Research Institute where he was a Senior Scientific officer for 6 years. Research interests include reproductive physiology of the sow, neonatal immunology and animal behaviour and welfare.

Martyn Warren, BSc, MSc, FIAgM, FRSA MILT.

Present appointment: Head of Land Use and Rural Management Department, Faculty of Land, Food and Leisure, Seale-Hayne, University of Plymouth.

Martyn Warren was educated at Newcastle and Reading Universities, and since 1994 has been Head of Land Use and Rural Management at the University of Plymouth, at its Seale-Hayne campus. He is author of *Financial Management for Farmers and Rural Managers*, in its 4th edition after nearly 20 years in print, and is Editor of *Farm Management*, the journal of the Institute of Agricultural Management. His main research interest is in the role of information and communication technologies in rural businesses and communities.

R.J. Wilkins, BSc, PhD, DSc, FIBiol, CBiol, FRAgrS.

Present appointment: Visiting Professor, Faculty of Land, Food and Leisure, Seale-Hayne, University of Plymouth; Visiting Professor, Department of Agriculture, University of Reading; Research Associate, Institute of Grassland and Environmental Research; Academician, Russian Academy of Agricultural Science.

After completing a PhD at University of New England, Australia, he was appointed to the Grassland Research Institute, Hurley, in 1966. Held a series of posts in that Institute and its successor, the Institute of Grassland and Environmental Research, until retirement in 2000 from positions including Institute Deputy Director, Head of the North Wyke Research Station, Devon, and Head of the Soils and Agroecology Department. Research interests include the production and utilisation of grassland and the environmental implications of grassland farming. Recipient of the RASE Research Medal and the British Grassland Society Award; President of British Grassland Society, 1986–7 and 1995–6.

Part 1

Crops

1

Soil management and crop nutrition

R.J. Parkinson

Introduction

Fertile soil is essential for sustained agricultural production, and has been exploited as a growing medium since animals and plants were domesticated. Cultivation of soils in order to create improved conditions for the sowing of crops can be dated back to at least 6000 BC (Goudie, 2000). Primitive farmers exploited the natural buildup of soil fertility by growing crops on land that was allowed to remain uncultivated for several years after the previous crop was harvested. The intensification that followed the enclosure of land and the adoption of sequential cropping patterns in rotation has ultimately lead to greater demands being placed on the soil to supply the needs of the growing crop and to allow machinery access over much of the cropping year.

Today, it is possible to grow some crops continuously on many soils, with the addition of fertilizers for plant nutrition and agrochemicals to control weeds, pests and diseases. Over the centuries, crop yields have increased dramatically, but our understanding of basic soil properties has not always increased proportionately. During the 1970s and 1980s evidence began to accumulate which indicated that soil fertility could decline under intensive agricultural use, mainly due to modifications of physical properties that had not previously been considered important. These changes were highlighted by the Houghton Report on *The Sustainable Use of Soil* (Cm 3165, 1996). Soil erosion and enhanced leaching losses of nutrients became more common on arable land, particularly under continuous winter cereal cropping. The challenge for farmers in the twenty-first century is to manage the soil to optimize yields while reducing unnecessary applications of fertilizers and other agrochemicals. Managing soil in a sustainable manner can only be done with a full appreciation of soil physical, chemical and biological properties. In this chapter the principles of soil management and crop nutrition are discussed.

Soil physical properties

Soil components

Soils are a complex mixture of mineral materials and organic matter that evolved over long periods of time, in some cases thousands of years, as a result of interactions between parent materials, soil organisms and climate. Analysis and understanding of the physical properties of soils and the way they respond when cultivated demands a detailed knowledge of the basic soil components, and how these components are arranged. A fertile soil will contain a mixture of mineral particles, organic matter and void spaces. These voids may be occupied by air or water. In Fig. 1.1 the proportion of these components is given for a well-managed topsoil. The properties of most agricultural soils are dominated by weathered mineral material. The primary components comprise the *texture* of a soil. Some soils, such as peats, are composed mostly of organic matter, in which case a texture description, based on the mineral fraction, is not appropriate. The combination of primary components with organic matter forms the *structure* of soils. In this section the agricultural significance of texture and structure will be described, followed by other important soil physical properties, most of which are controlled by texture or structure, and may be modified by agricultural practices.

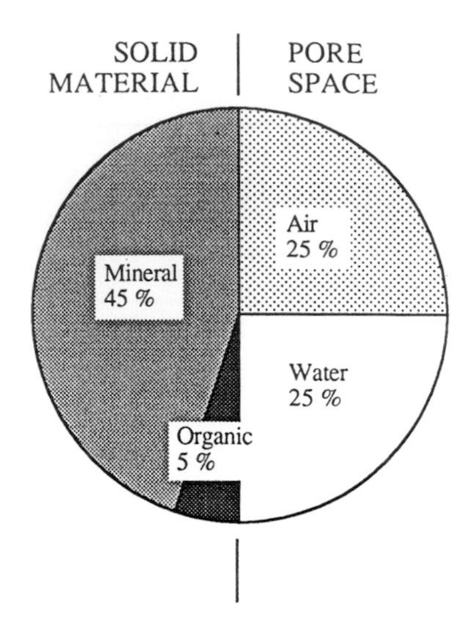

Figure 1.1 Components of a well-managed topsoil, expressed on a volume basis.

Texture

Mineral particles in soil range widely in size from large stones to minute clay fragments. The proportion of various sized particles has a major impact on soil physical, chemical and biological properties. Hence the description and characterization of soil texture is of primary importance to farmers and growers. Texture determination can be carried out approximately, but rapidly, by hand texturing in the field, or quantitatively by laboratory analysis.

Field determination

Hand texturing is a rapid but relatively crude method of determining soil texture. A soil sample is moistened in the fingers and rubbed in order to break down any natural structures that exist. Coarse sand grains can be seen with the naked eye, and will make the sample feel 'gritty'. Fine sand and silt feel smooth, and can make the sample 'slippery'. Clay binds soils together, and imparts a sticky feel. A description of these different size components is given in Table 1.1. More detailed description of the determination of soil texture in the field can be found in Rowell (1994). Accurate hand texturing requires technical skill and years of practice. Care must be taken when hand texturing, as some organic matter fractions feel slippery or soapy, similar to silt-sized particles. Mineral soils with high organic matter levels can feel finer in texture than is actually the case.

Table 1.1 Range of particle size and properties for mineral components of soil

Particle	*Diameter (mm)*	*Characteristic properties*
Gravel or stones	>2.0	—
Coarse sand	0.2–2.0	Coarse builder's sand or beach sand, particles clearly visible
Fine sand	0.06–0.2	Egg timer sand, just visible with naked eye
Silt	0.002–0.06	Flour, visible with hand lens
Clay	<0.002	Plasticine or putty, visible using electron microscope

Laboratory determination

Laboratory determination of texture or particle size distribution by a process known as mechanical analysis is time consuming and is only carried out when detailed information is required. The procedure is based upon sieving and sedimentation in water after thorough disaggregation. MAFF (1986) describes a three-phase process. Firstly, the soil sample is passed through a 2-mm sieve to remove stones. The analysis continues on the fine earth fraction that passes through the sieve. Secondly, the soil is dispersed using hydrogen peroxide (which destroys organic matter) and sodium hydroxide (which separates individual particles). Finally, the suspension of soil and water is passed through a series of sieves or a process of controlled sedimentation is carried out, so that the precise proportion of different sized mineral particles can be determined following dry weighing. The equivalent diameters of sand, silt and clay particles are shown in Table 1.1. This is the system employed in the UK (see Rowell, 1994); in other countries different size classes are used. The results of a particle size analysis can be plotted on a graph that has been subdivided into named texture classes (Fig. 1.2). A plot of percentage clay versus sand includes the silt component by difference, as the total must add up to 100%. For example, a soil containing 40% sand and 30% clay (and hence 30% silt) would be described as a clay loam. Having defined texture in a quantitative manner, it is possible to predict more accurately the behaviour of the soil in specific management situations.

Mineral materials

The mineral components of soils derive primarily from the underlying parent material. Hence the character of a soil will depend intimately on the type of rock from which the soil has been formed. In simple terms, soils composed of coarse particles derived from rocks such as

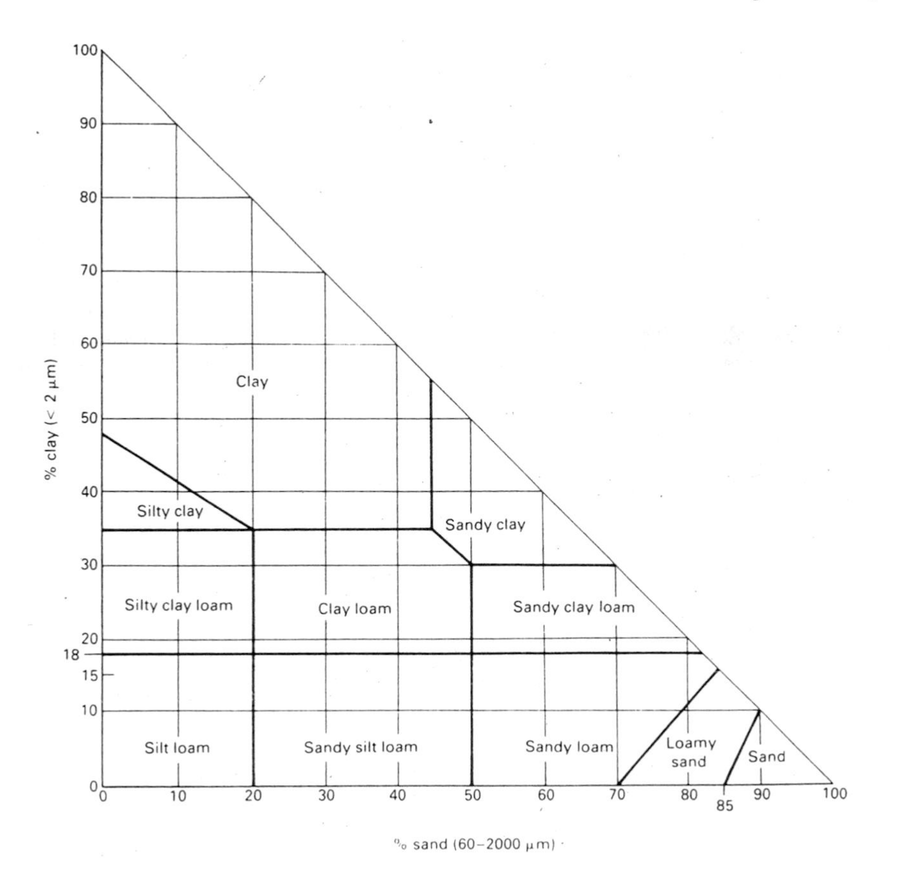

Figure 1.2 Soil texture classification as used in the UK (after Mullins, 1991).

sandstone and granite will produce coarse-textured, sandy soils. Mudstones and slates will produce fine-textured soils. The chemical character of a soil is also related to the size and origin of the mineral components. Sand and silt-sized particles tend to be dominated by only a few primary, rock-forming minerals, of which quartz (silicon dioxide) is the most common. Quartz is chemically unreactive, and therefore plays no part in nutrient retention by soils. In contrast, the clay fraction is dominated by clay minerals that have a varied chemical composition and can be chemically and physically very reactive. The potential for soil mineral particles to be physically and chemically reactive is partially explained by the increasing surface area with decreasing particle size. Table 1.2 shows the theoretical relationship between particle diameter and surface area. Clay minerals that have an expanding lattice structure, such as montmorillonite, exhibit large surface areas that allow water and nutrients to be retained, while sandy soils possess a small surface area and tend to be chemically inert. Silt is intermediate in behaviour, and does not have the large surface area possessed by clay, nor does it have the often beneficial properties attributed to sand (see next section for more details). In practice, soils are a mixture of all three mineral components, although one may dominate and hence control soil behaviour in the field.

Table 1.2 Relationship between particle size and surface area

	Surface area ($m^2 g^{-1}$)
Fine sand	0.1
Silt	1.0
Kaolinite clay	15–20
Montmorillonite clay	700–800

Texture and soil management

Texture exerts a profound influence on soil management. Ultimately the choice and flexibility of cropping, as well as potential yields, all depend on soil texture. Detailed knowledge of texture variations within and between fields can help farmers to manage soils and inputs such as fertilizers and manures effectively. The most important impacts of texture are given below. Many of these relationships are discussed further in other sections of this chapter, and in more detail by Davies *et al.* (1993).

Drainage status

Clay-rich soils retain water against gravity and therefore have high water contents during the winter months. At such a time a clay loam might hold 50–60% water, in comparison with a sandy loam which may hold only 25–30%. High water contents lead to reduced oxygen levels and poor plant growth as well as risk of soil damage by vehicles and livestock (see 'Agricultural land drainage' for more discussion of the consequences of poor drainage).

Water availability to plants

While finer textured soils retain more water than coarser soils, much of it is held so strongly (by capillary forces) that crop plants cannot extract the water. Consequently, silt soils tend to have the highest reserves of plant-available water, and are the most drought tolerant. Sandy soils tend to be very droughty due to the small volume of stored water.

Workability and trafficability

Access to land in the critical autumn and spring months can be controlled by soil texture. Sandy, well-drained soils have few access restrictions, making them suitable for the growth of a wide range of arable crops. Clay textured soils are often difficult to cultivate, being too hard when dry and too soft when wet, and frequently cannot support the weight of agricultural machinery during the winter. Cultivation operations must be timed very carefully in order to minimize the risk of soil damage due to cultivation when too wet.

Nutrient retention

Clay-sized particles retain nutrients very effectively, while sandy soils are often described as 'hungry', that is nutrients need to be added frequently but in small doses in order to supply the requirements of crop plants (see 'Nutrient retention – cation and anion exchange' for further discussion).

Soil texture describes the fine earth fraction (particles <2 mm). Stones and gravel are therefore excluded from this discussion, but quite clearly the presence of significant proportions of stone-sized material can have a major influence on the growth of root crops and cultivation operations. Many of the glacially derived soils of northern Britain and the flinty chalk-derived soils of southern Britain contain significant quantities of stones that limit the choice of crops. No hard and fast rule can be given as to the amount of stones that will restrict crop choice or influence cultivations, as stone type and distribution are equally important. Soil organic matter can influence many of the properties listed above. High levels of organic matter can increase nutrient retention, increase water retention and make soils more workable (see 'Soil organic matter and the carbon cycle' for further discussion).

Structure

Formation of soil structure

As soils develop, mineral particles of sand, silt and clay are mixed with organic matter by soil organisms. This mixing process creates stable aggregates and hence the soil structures. Immature soils, such as might be found on a river flood plain or sand dune, tend to show little evidence of structure development, but most soils in agricultural use are well structured. Stable soil structure develops over long periods of time, as soils shrink and swell throughout the year, and as organic residues are combined with mineral particles, often through the action of soil organisms. A simplistic representation of soil structural components is given in Fig. 1.3, which helps to explain the important process of structure

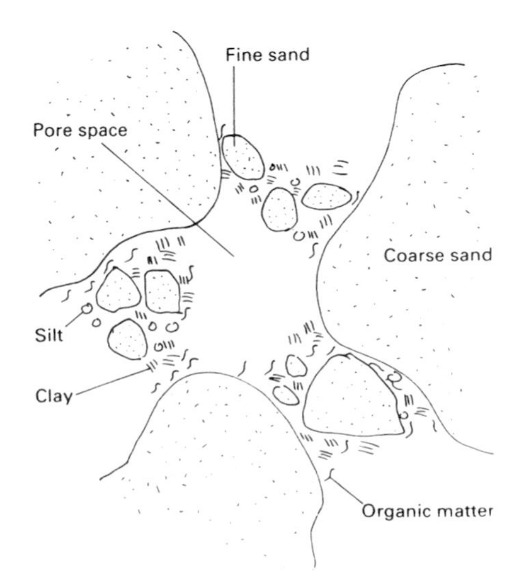

Figure 1.3 Detailed representation of soil structure formation.

formation by the intermixing of soil mineral and organic components. Certain soil characteristics aid structure development, namely organic matter (most important), clay, calcium carbonate and iron oxide (least important). A well-structured soil will display fine, even-sized structures in the topsoil, with progressively larger aggregates in the lower horizons of the soil. Figure 1.4 shows several example soil profiles broken down into structure types.

Importance of soil structure

Well-structure soils display high porosity, low density, adequate water storage, free drainage and movement of air within the soil profile. Plant roots are able to exploit the whole soil volume, and will display a fine fibrous root system. The fine stable aggregates near the soil surface will be resistant to collapse and therefore will allow free passage of water and air through the surface layers. Roots will extract water and nutrients from within the aggregates, while excess water will drain away through gaps between the aggregates, resulting in no prolonged periods of waterlogging. The importance of structure cannot be underestimated in agricultural production systems, although assessment of soil structure is a difficult task. Experience can be gained by frequent field soil examination, particularly of soil profile pit faces that have been allowed to weather naturally for a few days, after which time natural structure patterns become more visible.

Modification of soil structure

Many agricultural practices modify soil structure and the stability of soil aggregates; the creation of a seedbed by ploughing and secondary cultivation, for example, is direct modification of soil structure designed to suit the needs of the seed to be sown. Unlike texture, structure is not permanent. Soil aggregate stability can be changed by the cultivation and intensive use of soils. Serious structure breakdown can lead to soil compaction and soil erosion. Many soils in the UK have suffered from problems relating to modification of structure by farming practices. The intensification of agricultural activity in the latter part of the twentieth century has lead to some deterioration of soil structure. Most notably, arable cropping has caused organic matter levels in some soils to decline, due to the removal of crop products, such as grain and straw. This reduction has become critical for many sandy soils that tend to have naturally low organic matter contents. Rates of organic matter decline are variable, but soils with organic matter levels of less than 5% are susceptible to compaction and soil erosion.

Tilth is the term applied to the finely structured surface soil that has been worked down by cultivation implements to create ideal conditions for the germination and growth of crops. Unfortunately the repeated cultivation of soils destroys natural, stable aggregates, resulting in weaker, finer structures that ultimately may

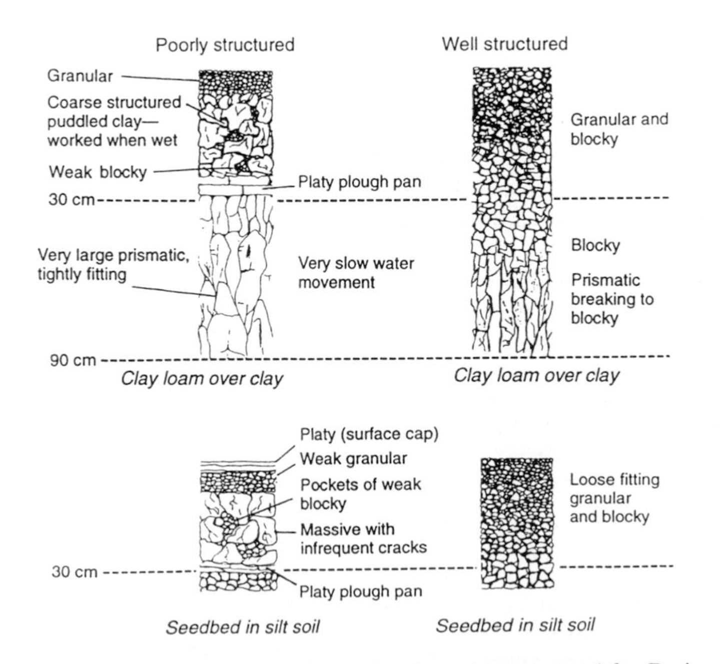

Figure 1.4 Some examples of soil structure types under good and poor management (after Davies *et al.*, 1993).

collapse, hence leading to deterioration in the soil physical environment. It is therefore important that soil is not overcultivated, particularly for autumn-sown crops where soils are exposed to the full force of the winter weather with minimal protection from a growing crop.

Soil structure can be improved by a number of agricultural practices: use of (long) grass leys in rotations, adding manures and other organic materials, adding lime, and the use of deep-rooting green manure crops. All these will help to stabilize the soil and maintain fertility.

Soil density

Figure 1.1 gave a typical breakdown of soil components for a well-structured topsoil, which is made up of 50% solids and 50% pore space. As the structural properties of a soil change under agricultural management, so will the density and pore space. As roots need to access water and air held in these pore spaces, an understanding of such changes is important.

Bulk density

Bulk density is defined as the mass of oven-dry soil per unit volume, and depends on the densities of the constituent soil particles and, most importantly, how these constituents are packed together. Bulk density is usually determined by extracting a soil core of known volume, oven drying at 105°C for 24 hours to remove the soil water and then weighing the core. Values of bulk density range widely, as shown in Table 1.3. In general, root access to soil pore spaces becomes difficult above a bulk density of 1.5–1.6 $t\,m^{-3}$. Soil compaction results in an increase in bulk density. In Fig. 1.5 some example bulk density profiles are given for a soil that has suffered surface compaction due to excessive animal grazing (Profile B) and a soil that has been compacted at plough depth due to repeated cultivation when soil conditions were too wet, resulting in smearing and structure destruction (Profile C).

Table 1.3 Bulk density and total pore space for agricultural soils

Bulk density ($t\,m^{-3}$)	*Pore space* (%)	*Description*
0.5–0.8	>70	Loose, uncompacted topsoils. Peats and organic soils
~1.0	60–65	Permanent pasture, woodland soils, well structured
~1.5	45	Compacted, root penetration difficult
~2.0	25	Dense, no root growth

Under good management, bulk density will tend to reduce until an equilibrium value is reached for a given soil and cropping situation. Bulk density is simply an indication of the general status of the soil in physical terms – to understand the effect on crop plants, it is necessary to describe pore space changes associated with increases in bulk density.

Given bulk density, it is possible to calculate the mass of soil in a given area. For example, assuming a bulk density of 1.0 $t\,m^{-3}$, 1 ha of soil down to plough depth (200 mm) will have a mass of 2000 t. This represents a considerable mass of material that is moved every time the soil is ploughed.

Particle density

It is necessary to know the density of mineral particles (or when unweathered, solid rock) in order to calculate the pore space in a soil. Particle density is defined as the mass per unit volume of mineral particles in soil, and does not include air spaces or water. Typical values for particle density range from 2.50–2.70 $t\,m^{-3}$. For example, a 1 m^3 solid granite block weighs approximately 2.65 t. A value of 2.65 $t\,m^{-3}$ can be assumed for quartz-dominated mineral material in British soils. Organic matter weighs considerably less than mineral material; values of 1.20–1.50 $t\,m^{-3}$ are common for organic material. Given this difference, it is important to state the

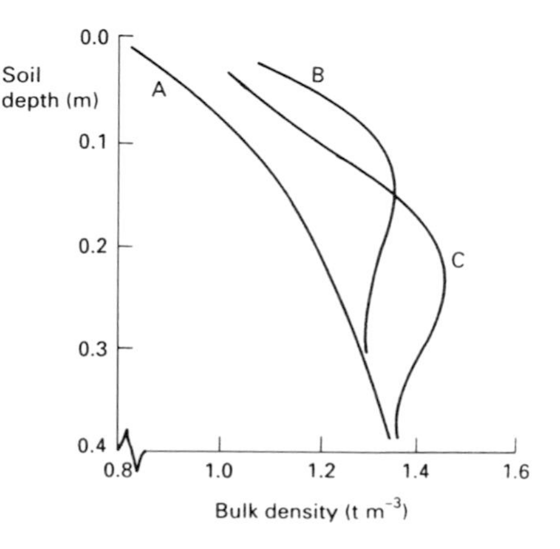

Figure 1.5 Example bulk density profiles. *A* Uncompacted soil; *B* heavily grazed soil showing surface compaction; *C* cultivated soil showing compaction at plough depth.

organic matter content of a mineral soil when describing pore space of soils.

Pore space

The size, shape and arrangement of soil aggregates control not only the density but also the total porosity of a soil. When density changes as a result of soil management practices, porosity also changes. Total pore space is defined as the volume of pores expressed as a fraction of the total soil volume, and is usually determined by measuring the bulk density and assuming a particle density of $2.65\,\text{t}\,\text{m}^{-3}$.

$$\%\ \text{Pore space} = [1 - (\text{bulk density}/\text{particle density})] \times 100$$

For the example given in Fig. 1.1, the total pore space is 50%. By substitution into the equation above, this equates to a bulk density of $1.32\,\text{t}\,\text{m}^{-3}$. A well-structured topsoil under permanent grass with a bulk density of $1.0\,\text{t}\,\text{m}^{-3}$ would have a total pore space of 62%; i.e. greater than half the soil is pore space. Increases in bulk density, which are associated with compaction, lead to reductions in pore space. Further examples of increasing bulk density and decreasing pore space are given in Table 1.3.

Total porosity does not provide any direct information about the size of the individual pores, or their function. It is simply an expression of the total volume of a soil that may act as a store of air or water.

Water retention

Water occupies pore space in soil. The mechanisms by which water is held in soils, and then released to plants or allowed to drain out of the profile, depend upon the size of the pores. Pores can be classed as having various functions according to their approximate diameter. The larger pores (termed macropores, >60 µm in diameter), for example drying cracks and earthworm burrows, allow excess water to drain out of the soil, as gravitational forces exceed the low capillary forces in such large pores. These voids are very important during the winter months when heavy textured soils are prone to waterlogging. Macropores tend to be the first to be lost when soils are compacted, hence leading to drainage problems. In addition, these pores allow air to enter the soil profile.

The smaller pores (<60 µm in diameter) store water against gravity, due to capillary forces. These forces become stronger the smaller the pore diameter. Eventually a point is reached where the pore is so small that plants cannot overcome the capillary force, and so any water contained in that pore is considered to be unavailable. Figure 1.6 displays the relationship between soil texture and the quantity of water, expressed as a percentage of the soil volume, that is available (held in mesopores between 0.2 and 60 µm in diameter) or unavailable (held in micropores <0.2 µm in diameter). These differences are of vital importance in cropping systems; sandy soils may have only 5–10% available water, while at the other end of the range, silty soils may contain more than 20% available water. In a dry summer these differences may lead to crop failure or poor yields on sandy soils, while heavier textured soils may be able to maintain crop growth throughout a drought period.

As well as soil texture, the depth of a soil exerts a controlling influence over the total amount of water that is available to plants in the soil profile. The total available water that is stored in three example soils is given in Table 1.4. Soil A is coarse textured, but deep, while soils

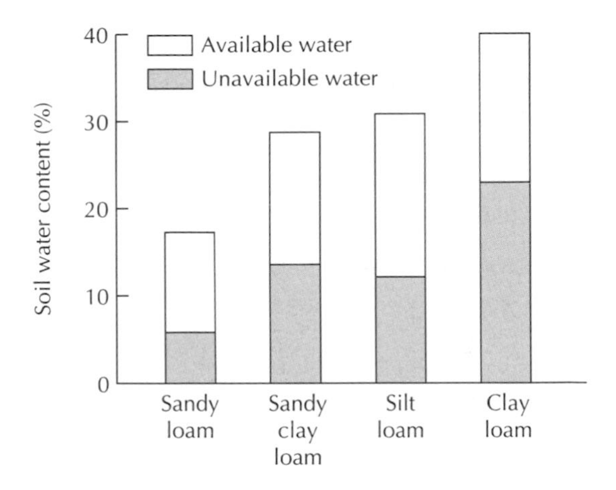

Figure 1.6 Soil texture and available water capacity.

Table 1.4 Total plant-available water for three example soil types

Soil depth (mm)	*Available water (mm)*		
	A *Sandy loam*	*B* *Silt loam*	*C* *Silt loam (0.5 m deep)*
0–200	40	80	80
200–500	21	45	45
500–1000	25	50	—
Total	86	175	125

B and C are fine textured, although soil C is only 0.5 m deep, below which weathered rock is encountered. The impact that texture and soil depth have on the amount of water available to a plant is clearly seen in this example. The need for irrigation is dependent upon soil texture and hence available water capacity, as well as the climate and weather patterns in a given area.

Aeration

Soil air differs from the free atmosphere above the soil surface in being enriched in carbon dioxide and sometimes depleted in oxygen. A steady supply of oxygen is needed in soils for root respiration and to support aerobic bacteria that carry out important functions in the soil (see 'Micro-organisms'). If the volume of macropores that allow drainage and aeration falls below 10%, it is likely that there will be insufficient air for root respiration. Any practice that increases the volume of air-filled pore space, such as soil loosening, is to be encouraged.

The consequences of poor aeration are that soil conditions change to the detriment of most crop plants, although the susceptibility to poor drainage varies from crop to crop. Breakdown of organic matter in anaerobic conditions can lead to the production of organic acids and ethylene (C_2H_4), both of which inhibit root extension. Nitrate may be denitrified to gaseous nitrous oxide (N_2O) or dinitrogen (N_2), and lost from the soil system.

Soil strength and cultivation

Careful management of soils will lead to continued high productivity and few physical problems. However, it is not always possible to cultivate under optimum conditions, and it is sometimes necessary to keep animals out on the land when high soil water contents will lead to damage occurring. The ability of a soil to resist deformation and damage depends upon the relationship between water content and strength. This relationship is unique for each soil type. In general, high soil water contents lead to low strength, and hence greater susceptibility to potential damage under applied load, for example a tractor tyre or an animal hoof. Figure 1.7 displays this general relationship for a well-structured soil. Optimum conditions for cultivation occur when the soil will fail in a brittle, friable manner, as indicated by the asterisked arrow in Fig. 1.7.

Various problems can ensue if trafficking or animal trampling ('poaching') occurs when the soil is at a high water content such that deformation occurs in a plastic

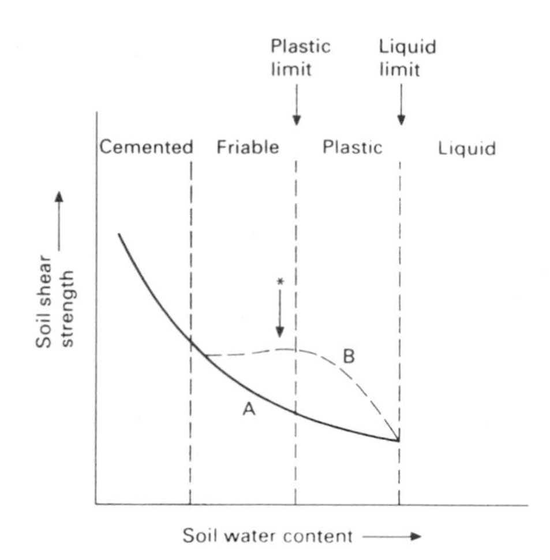

Figure 1.7 Relationship between soil strength and water content. *A* Soil aggregate; *B* whole soil, if well structured.

rather than a brittle manner. Some of these problems are described below, and are discussed in more detail in Davies *et al.* (2001).

Cultivation pans

A cultivation pan or plough layer will form following repeated cultivation under conditions that are too wet. Structural aggregates will be destroyed and a compacted smeared layer will be formed (see Fig. 1.5). Such pans can occur in soils of any texture, but they are most common in heavy textured soils that tend to have higher water contents. The presence of a pan will lead to temporary waterlogging, even in better drained soils, as well as higher bulk density, both of which will restrict root development. Alleviation of the problem is usually by soil loosening to below the depth of the pan under dry conditions, when brittle failure will occur.

Surface compaction

Surface compaction may be due to machinery access or livestock grazing of land that is too wet, resulting in plastic deformation, smearing and compaction of the surface layer (see Fig. 1.5). In consequence, aeration will be restricted and surface run-off may occur. Given the opportunity, soils will naturally 'restructure' at the surface, as a result of wetting and drying cycles that occur in all soils during the year. Removing the cause of the initial damage and allowing the soil to recover naturally is often all that is needed in such circumstances.

'Puffy' seedbeds

Overcultivation of soils with high sand and silt contents can lead to structure breakdown with very light, puffy

Table 1.5 Machinery work days for three example soils in south-west England (after Findlay *et al.*, 1984)

Location	*Soil association and texture*	*Machinery work days*	
		Autumn (1 Sept–31 Dec)	*Spring (1 Mar–30 Apr)*
Bridgnorth	Loamy sand	86	29
Frilsham	Sandy clay loam	73	18
Hodnet	Silty loam/clay loam	32	3

seedbeds. In such cases rolling may be necessary to ensure good soil : seed contact following drilling. Failure to roll may result in uneven germination and poor establishment.

Machinery work days

Autumn and spring are the critical periods of the year when soil damage is possible due to high water contents at times when essential cultivations need to be carried out. The concept of machinery work days describes the potential cultivation opportunities that exist for given soil types. These opportunities are measured during the main autumn and spring periods, and describe those occasions when harvesting, tillage and drilling operations can be conducted without risk of structural damage to the soil. Soil Survey regional bulletins (see 'Further reading') contain a detailed description of the system employed in England and Wales. These descriptions form a basis for the discrimination of soils according to their flexibility of management at those critical times of the year when access to the land is necessary in order to be able to establish and then harvest arable crops. Table 1.5 lists the machinery work days for three soil types that occur in south-west England. Texture can be seen to exert a strong influence on the number of machinery work days.

Soil erosion

Erosion is a natural process, but changes in soil properties caused by intensive use of land can lead to accelerated rates of soil loss. On a global scale, there are many examples of severe erosion, often associated with the removal of protective vegetation cover in areas subjected to intense rainfall or high winds. The 'Dust-Bowl' of the mid-west United States and severe gullying in the footslopes of the Himalayas demonstrate the importance of protecting vulnerable soils. In the UK, accelerated rates of erosion are associated most frequently with winter cereals, and to a lesser extent potatoes, sugar beet and oilseed rape (MAFF, 1999a). The Houghton Report on *The Sustainable Use of Soil* (Cm 3165, 1996) noted that up to 15% of arable land was at risk of erosion in any one year. Erosion not only results in the loss of valuable topsoil, hence leading to potential yield reductions, but also causes significant pollution of water courses. Examples include sediment clogging gravel-bottomed rivers and nutrients (such as nitrogen and phosphorus) contributing to eutrophication. Erosion can also occur in grassland and moorland, mainly due to overgrazing which in upland areas has been found to contribute to accelerated erosion.

Water erosion is the most common process of soil loss in the UK. First, rainsplash detaches soil particles. Various factors, summarized in Table 1.6, increase the risk of splash detachment occurring, such as soil aggregate stability and the soil surface protection or cover. Declining organic matter content and the increased area of winter cereals are the main factors leading to water erosion in the UK. Losses in excess of 10 t ha^{-1} $year^{-1}$ are becoming more common. Strategies to control water erosion must be based on awareness of the problem. MAFF (1999a) details steps that should be taken; the key elements are summarized in Table 1.6, with maintenance of a rough seedbed with some stubble or other surface cover being essential. The principles of Integrated Crop Management include many of these strategies. If these are unsuccessful, then it will be necessary to revert to more grass in the rotation to allow the soil to recover naturally.

Erosion by wind is a problem in some eastern counties of England, notably the Fenland and the Vale of York, where vulnerable soils (peats and fine sands/silts) are intensively cropped with sugar beet and potatoes. Fine seedbeds produced in the spring can be vulnerable to 'windblow' if dry weather follows drilling. The establishment of windbreaks at field boundaries and use of non-cultivated strips within the field can help to reduce the risk of soil and crop loss by wind erosion.

Table 1.6 Soil erosion: high-risk situations and control strategies

Factors leading to high risk of erosion	*Erosion control strategies*
Water	
Sandy or fine silty texture	Survey farm to identify vulnerable soils and sites where erosion has occurred in the past
Valley features which concentrate run-off	Use field margins as buffer zones to detain run-off
Steep slopes (10%)	Replace or add field boundaries, cultivate across slope
Long unbroken slopes	Reduce frequency of cultivation, add organic manures
Low organic matter content	Adjust timing and intensity of cultivation to avoid compaction and leave seedbed rough
Compacted soils	Leave stubble on surface; use non-inversion tillage
Fine seedbeds	
Lack of surface protection, e.g. stubble	
Wind	
Silty texture or peaty soils	Identify soils at risk
Fine seedbeds, particularly in late spring	Avoid overcultivation, leave seedbed rough
Large, flat fields, few hedges	Plant windbreaks, subdivide field, plant different crops

Agricultural land drainage

Drainage needs and benefits

Why drain?

The removal of excess water from soil by an artificial drainage system can reduce soil management problems and increase crop yields. The need for drainage in the UK arises either from heavy textured soils retaining excess winter water, or from a high water table, for example on a river floodplain. Heavy textured soils have an ability to retain nutrients and provide the water needed to satisfy the requirements of demanding crops, and hence are capable of producing high yields, particularly of cereals and grass. For example, the majority of $10\,t\,ha^{-1}$ wheat crops have been obtained from clay loam or clay textured soils, but only with efficient drainage systems to remove excess winter water from the soil. The move in recent years to lower-input farming systems puts more emphasis on the need to ensure that soil physical conditions are optimal; in addition, the increased variability in weather patterns associated with global warming is predicted to result in increased rainfall in north-west Europe. Both these factors demonstrate the need to ensure that drainage systems function effectively.

Annual patterns of water loss and gain are very variable across the UK. As a result of higher rainfall and less sunshine, the soils in western Britain tend to pose more drainage problems than do those in the east. This does not mean that the arable soils in eastern Britain do not need drainage. The greater demands placed on soils in arable cropping systems have lead to the need for increased flexibility of soil management in arable situations, and hence many soils in eastern Britain have been drained prior to being cropped intensively.

The first drainage systems date from Roman times, but it was only after the large-scale enclosure of common lands in the eighteenth century that drainage became widespread. Techniques used varied widely, with most systems relying on stones, straw or other bulky material to form and stabilize a channel to remove excess water. The mechanical production of clay tiles in the mid-nineteenth century led to dramatic improvements in the quality of drainpipes, and 100-year-old systems can still be found that are working well today. Estimates of areas drained in the nineteenth and early twentieth centuries suggest that up to 50% of the agricultural land in southern England was drained, all by hand. Mechanical drain installation became the norm by the middle part of the twentieth century, when 50 000–100 000 ha were drained annually. The current rate of drainage is very much less, in the region of 10 000 ha $year^{-1}$, due to the progressive reduction in grants for drainage schemes and changes in the economics of arable crop production in the 1980s and 1990s. In addition, the conservation value of wetland habitats is now more appreciated than formerly, such that decisions to drain areas of land are no longer based solely on considerations of likely yield increases after drainage. The conservation value of species-rich grasslands, for example, may depend on land remaining undrained. In some cases grants are now available to encourage farmers to choose this option (e.g. Countryside Stewardship Scheme). The following discussion refers only to land of little conservation value where drainage is needed to improve soil management and raise yields.

Drainage benefits

Lowering soil water contents by drainage results in several changes in the crop rooting environment, all of which are beneficial to the crop and soil management (Parkinson, 1988). These benefits are summarized below.

Duration of waterlogging

Autumn-sown crops rely on efficient water table control to allow development of a root system. Failure to control waterlogging in the late autumn/winter/early spring period will result in a stunted root system unable to exploit deeper-seated water and nutrient reserves during the following summer.

Soil workability and trafficability

A reduction in water content will increase soil strength. Successful cultivation usually depends on the soil water content being below the lower plastic limit. Drainage lowers the water content of the topsoil, resulting in increased cultivation opportunities during the critical autumn and spring months. In grazing systems, the benefits from drainage will come from the increased number of days that livestock can graze a sward without risk of damage to the soil structure.

Soil temperature

Drainage reduces soil specific heat capacity and therefore can lead to higher temperatures. Castle *et al.* (1984) noted a 2°C elevation in spring soil temperatures on a clay soil in eastern England compared to when underdrained. Such an increase may lead to the more rapid germination and emergence of spring-sown crops, and may accelerate the development of winter-sown crops. It must be noted, however, that soil temperature does not always rise following drainage. In some field experiments no benefits have been found.

Efficiency of fertilizer use

More aerobic, warmer soil conditions will lead to more efficient use of applied fertilizers, particularly nitrogen top dressings. Growing roots will absorb nutrients more readily and less will be lost by leaching or denitrification.

Arable crop yield benefits

The benefit obtained as a result of installing a drainage system can be most easily measured in terms of crop yield, but the variability of the British climate often produces a wide range of yield benefits from year to year. The yield advantage for most crops when comparing drained and undrained soils is 10–25%. For example, average winter wheat yields can be increased by up to $1.0\,t\,ha^{-1}$. In wet years, however, efficient drainage can make the difference between crop failure and success.

Grassland/livestock benefits

In the wetter regions of the UK, drainage is essential in order to maximize grass utilization. The benefits can be expressed in terms of dry matter yield or liveweight gain. For example, Tyson *et al.* (1992) reported that drainage of a clay soil in Devon resulted in a 5-day longer growing season and an increase of 11% in beef cattle liveweight gain compared to the undrained soil. Drainage can alter the composition of the sward, increasing the proportion of productive grasses at the expense of weed species, as well as increasing the response to nitrogen. In addition to liveweight gain, other benefits in a livestock production system can include a reduction in the incidence of liver fluke and foot problems.

Drainage systems

Land can be drained by a system of ditches or pipes laid in the soil, or a combination of both. In the case of pipe drainage, additional short-lived measures, such as mole draining, can be carried out to increase the effectiveness of the drainage system. Both permanent and temporary systems serve specific purposes and must be installed following recommended guidelines (see Castle *et al.*, 1984). The principles and some of the practical points are outlined here.

Open ditch drainage

Most drainage systems find outlets into an open ditch that usually leads to a larger watercourse. These ditches are a vital component of a drainage system, and in some cases may be the sole method of water removal. Careful design, construction and maintenance are therefore very important. Ditches allow direct access for water, have a large capacity to carry storm flows and are easily maintained. However, they can hinder cultivations, they need fencing to exclude livestock and are susceptible to wall collapse and blockage by vegetation.

Ditch specifications

Design standards require that ditches must be of sufficient capacity for the catchment area drained. Theoretical capacity can be calculated given catchment size and design rainfall rate (see, for example, Farr and Henderson, 1986, p. 131). Ditch width and depth will depend upon soil and geology. The more stable the soil, the steeper the permissible slope. Some examples are given in Table 1.7. Ditches dug in sandy materials must have

Table 1.7 Ditch channel side slope ratios (horizontal : vertical) and soil type (after Castle *et al.*, 1984)

	Channel side slope ratio	
	Channel <1.3 m deep	*Channel >1.3 m deep*
Fen peat	Vertical	0.5 : 1
Heavy clay	0.5 : 1	.1 : 1
Clay loam or silt loam	1 : 1	1.5 : 1
Sandy loam	1.5 : 1	2 : 1
Sand	2 : 1	3 : 1

lower slopes than those in more stable finer-textured soil. Ditch floor gradient will in practice depend upon local topography. The gradient should be uniform and not too steep (leading to channel erosion) or too shallow (leading to silting). A gradient of 0.5–1.0% is generally considered to be adequate. Ditch sidewall collapse can be minimized by guarding against water erosion and livestock damage. Pipe drain outfalls should be fitted with splash plates, ditches should be fenced and, in areas with highly erodible soils, the sidewalls should be grassed down. Ditches must be piped under roads and farm tracks. The pipes used, normally concrete, must be large enough to carry anticipated peak flows. The design flow can be calculated (see MAFF, 1982), but the minimum recommended pipe diameter is 225 mm. This size of pipe will serve a catchment area of up to approximately 12 ha.

Pipe drainage

Pipe drains remove excess water without reducing the area of land cropped or interfering with field operations. Installing a permanent system of clay tiles or plastic pipes to carry water below plough depth can solve drainage problems efficiently and can be a worthwhile investment. It is particularly important that the principles of operation, design, installation and maintenance are all understood and followed, as a buried drainage scheme is difficult to inspect and maintain once installed. The successful operation of a drainage pipe system depends upon the flow of water from the soil surface into the drain itself.

Water flow in soil depends upon the hydraulic conductivity. In a uniformly permeable material, such as a sandy textured soil that is underdrained because of a high regional water table, as, for example, would be found on a river floodplain near sea level, water movement takes place through the whole soil (see Fig. 1.8a). In heavy textured soils with low rates of water movement, particularly in the subsoil, water tends to move in the topsoil and into the trench created by the drain-laying operation (see Fig. 1.8b). The bulk of the subsoil plays little part in the disposal of excess water. Rates of water movement depend on soil texture. Table 1.8 gives examples of water flow rates, class descriptions and soil types that are used in drainage design (MAFF, 1982). Water flow rates in coarse textured soils can be 100 or 1000 times more rapid than in clay soils, hence reinforcing the need for drainage of many heavy soils.

Drain materials

Several choices need to be made when installing a drainage system. Drains may be clay or plastic. Clay tiles are normally cylindrical and 250 mm in length. Tiles are supplied in a palletized form, which allows for easy mechanical handling. The most common sizes are 75, 100 or 150 mm internal diameter. Good-quality tiles are essential, as the failure of one tile can cause the failure of a complete lateral. The tiles are laid in the trench to butt up to one another, but with a gap of 0.5–2.0 mm resulting from the uneven ends of successive pipes. Plastic pipes have largely superseded clay tiles. The former have several advantages, being easier and lighter to handle, and more suited for mechanical laying. In addition, disjointed drain runs are unlikely. However, the rough interior surface of a plastic pipe results in a lower hydraulic carrying capacity, and the material is inherently weaker. Slots run the length of plastic pipes so water entry is relatively unrestricted. Standard joints and junctions are available for both plastic and clay pipes, so that laterals can be led into main drains. Plastic pipes are available prewrapped with a filter, which may be necessary in silty soils to prevent blockage of the slots by sediment.

Some form of gravel is placed over pipes in about half of field drainage systems installed in England and Wales. As permeable fill may account for 50% of total installation costs, it is important to justify its use. Permeable fill acts as a connector to allow water movement from the topsoil to the drainpipe. In addition, backfill forms a permeable surround, to improve water entry to the pipe, and acts as a filter to prevent soil particles entering the drain. Washed gravel with a mean particle diameter of 20–50 mm is the most suitable permeable fill material. The use of permeable fill is not recommended for medium and coarse textured soils.

Where a pipe discharges directly into a ditch, the first 1.5 m of the pipe should be sealed, rigid and frost resistant. The pipe should be able to discharge freely – therefore the invert of the pipe should be at least 150 mm above the normal ditch water level. Where the ditch sides are unstable the pipe should be supported by a con-

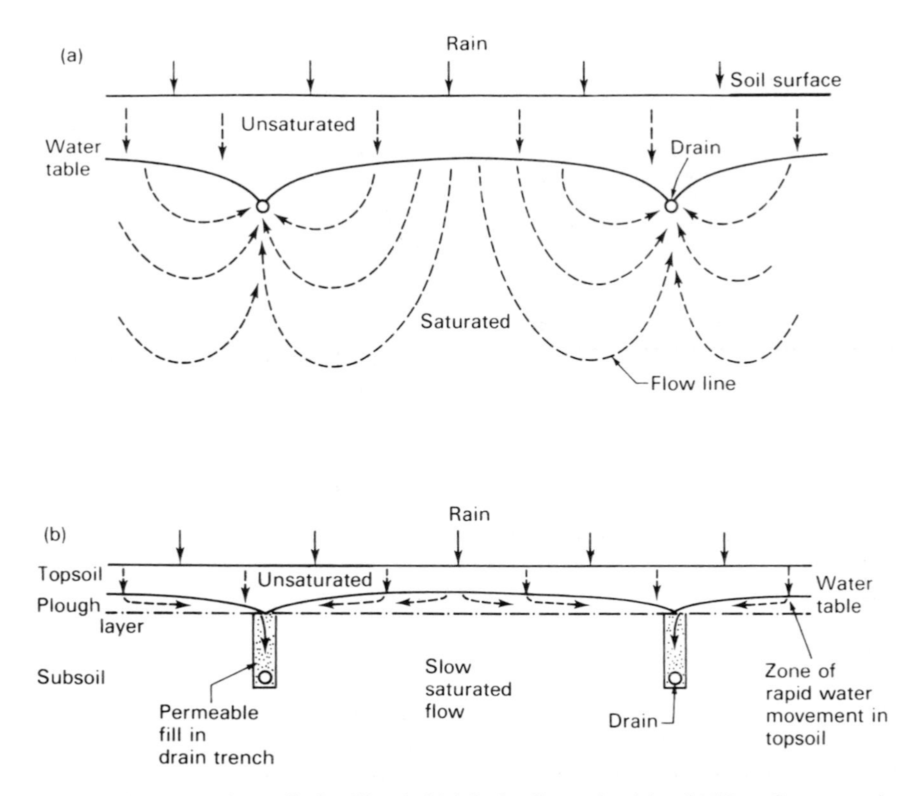

Figure 1.8 (a) Water flow routes in a soil of uniformly high hydraulic conductivity. (b) Water flow routes in a soil with a low subsoil hydraulic conductivity.

Table 1.8 Saturated hydraulic conductivity classification and soil type

Saturated hydraulic conductivity (m day^{-1})	*Class*	*Example soil types*
<0.01	Very slow	Clay, silty clay[1]
0.01–0.1	Slow	Clay loam[1]
0.1–03	Moderately slow	Silty clay[2]
0.3–1.0	Moderately rapid	Sandy clay loam, clay loam[2]
1.0–10	Rapid	Loamy sand
>10	Very rapid	Gravel

[1] Poor structure.
[2] Good structure.

crete headwall. Failing this, the ditch side should be grassed down.

Installation

Drainage work should be carried out when the soil is dry and hence strength is high. Drainage through a crop can be justified in some cases if disruption to farming operations can be kept to a minimum. Under dry conditions compaction of the soil surface and smearing of the drain trench sides are unlikely. Pipes must be laid on a smooth firm bed, shaped to support the pipe. Permanent pipes must be laid at least 600 mm deep to avoid damage by moling or subsoiling operations. Clay tiles must be firmly butted together, and located on one side of the drain trench to ensure water flow from one pipe to the next. Care must be taken with plastic pipes when installing them in air temperatures below 5°C, as they become brittle. It is essential to level accurately during the drainage operation so that a continuity of flow occurs. Laser levelling techniques allow grades of 0.1% to be attained, although for practical purposes 0.2% is the limit to which most machines work. Slopes of greater than 4.0% are likely to lead to erosion problems in the drain trench.

Trenched drainage

A trench, normally 150–300 mm wide, is excavated using a vertical endless chain with blades attached to its links. Spoil is brought up by the chain and pushed out

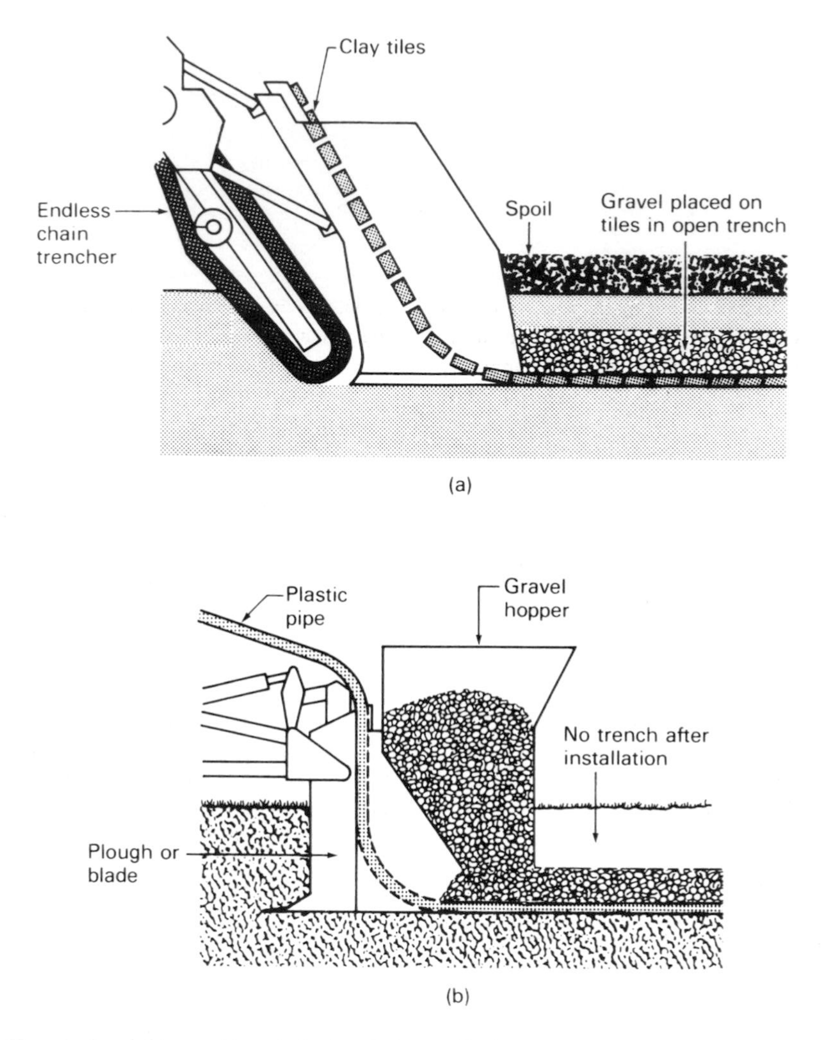

Figure 1.9 (a) Trenched and (b) trenchless drain installation methods.

to either side of the trench by augers. Figure 1.9 illustrates trenched and trenchless drain installation. Work rates of up to 30 m $hour^{-1}$ are possible, although in weakly structured or stony soils progress may be considerably slower. Several important features favour the trenched systems of drain laying:

Advantages:

- Pipes laid in trench can be inspected before backfilling.
- Old drainage systems can be found and tied in to the new system.
- Permeable fill can be placed in the ditch from a hopper to a uniform depth.
- Grade of the trench can be easily checked.

Disadvantages:

- In unstable soils, trench wall collapse may occur.
- Open trench requires more permeable fill than trenchless drainage.
- Slower workrate than trenchless drainage; hence more expensive.

Trenchless drainage

The drain-laying machine features a large plough which cuts a slit to a predetermined depth, and drops a plastic pipe down a narrow box behind the plough. If required, permeable fill is supplied from a hopper mounted above the pipe chute (Fig. 1.9). Approximately 25% of drain installation is by the trenchless method, which is usually employed to lay close-spaced systems in grassland with small plastic pipes (<75 mm). Surface ridging caused by the passage of the drain-laying machine will need to be rolled before normal farming operations can continue.

Secondary drainage

Moling

Heavy textured soils often require, in theory, a drain spacing of 5 m or less in order to dispose of excess winter rainfall. While it may not be economic to lay permanent drains at such a close spacing, temporary channels (secondary drainage) may be employed. Mole drainage is a process whereby a system of 'moles' is pulled at 2 to 3-m spacing across a network of permanent drains that may be spaced 30–40 m apart. Moling must be carried out under specific conditions and is only appropriate to certain soil types.

A mole plough is essentially a circular bullet followed by an expander which leaves a channel connected to the soil surface by a series of cracks (see Fig. 1.10a). Moling can only be carried out successfully in soils with a substantial clay content; 30% clay is commonly quoted as the minimum allowable (Castle *et al.*, 1984). Clays that expand and contract in response to changes in water content are more likely to provide stable long-lasting channels than non-expanding clays. Moling should be carried out at a depth of 0.5–0.6 m, at an angle approaching 90° to the field drain (Fig. 1.10b), with a minimum channel slope of 0.5%. These unlined channels are prone to collapse at angles <0.5% due to standing water.

The timing of mole installation is crucial to successful secondary drainage. Figure 1.11 demonstrates the effect of the passage of a mole plough through a soil with a dry surface layer but a moist subsoil. Such conditions will normally occur in early or mid-summer in the UK. If a stable channel is formed at depth, it may continue to carry water to the permanent drain for at least 5 years.

Subsoiling or soil loosening is not specifically a drainage operation, but can be described as any form of soil cultivation that is intended to shatter the soil beneath normal cultivation depth. When carried out as a drainage operation, the intention is to work the soil when dry, so that extensive shattering occurs, thus creating a system of artificial cracks that will conduct water to the drain or at least to lower horizons. Subsoiling is most suitable for those soils that will not hold a mole channel, generally soils with a high silt or stone content. The winged subsoil shoe 'heaves' the soil, allowing it to fail under tension and so produce an extensive network of cracks. This operation is normally carried out after harvest on arable land.

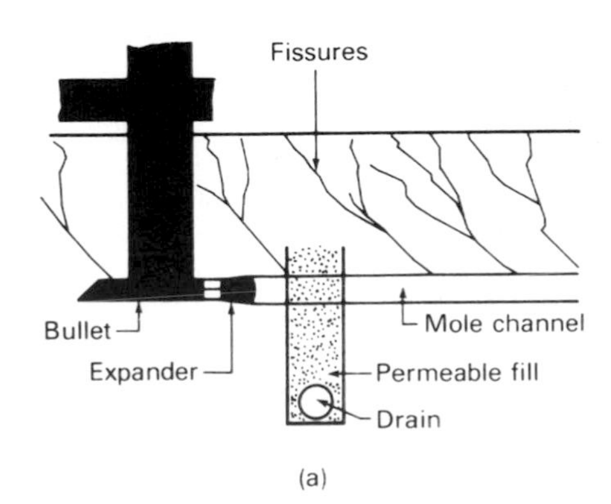

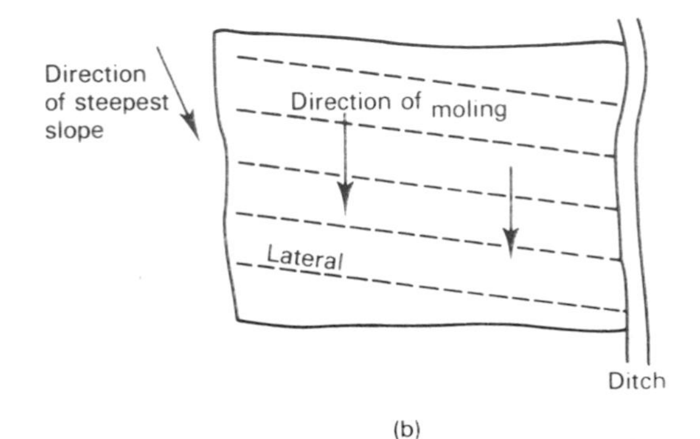

Figure 1.10 Pulling a mole channel. (a) Section through soil illustrating connection with permeable fill. (b) Field plan of moling direction relative to laterals.

Drainage system design

Site investigation

Before carrying out the installation of a single drain or a complete drain network, it is important to identify the

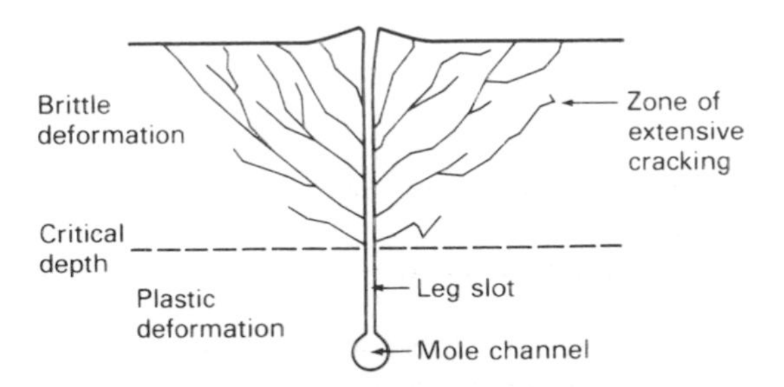

Figure 1.11 Ideal mole channel configuration.

causes of the problem. Examination of the soil profile can yield much useful information. The presence of compacted pans, ochreous mottles and grey colours all indicate problems. A drainage contractor or independent drainage advisor will prepare a detailed solution to a particular problem, with a plan, but it is useful for a farmer to be familiar with the initial cause and extent of the problem.

Drainage systems

Regularly spaced systems are used for draining soils with low hydraulic conductivity or uniformly sloping land, with drains spaced according to the rate of water movement through the soil. In heavy textured soils such as clays and clay loams, which can hold a mole, permanent drains may be installed at 30 to 40-m spacing, with a secondary treatment superimposed upon that system (see Fig. 1.10b). A regularly spaced system can be designed to use the slope of the land to collect water, in a herringbone design.

Irregular/random systems

Complicated geology or undulating relief may necessitate the design of a random system, which may tap a spring line or drain a wet valley bottom (see Fig. 1.12). The resulting design may leave much of an area undrained, only removing water from natural collecting zones.

System design

Recommendations for the design of field drainage pipe systems are given in detail in MAFF (1982). Design criteria are based upon determining the size of pipe needed to conduct water away so that damage to crops is minimized. Pipe sizes for laterals, main drains and ditches can be calculated knowing the characteristics of the site.

- *Land use:* design rates assume that horticultural crops are more valuable and more susceptible to damage than arable crops, with grass being most tolerant of waterlogging.
- *Soil type:* soil texture and structure control the relative rate of water movement such that, for example, a well-structured clay loam will drain more rapidly than a poorly structured silty clay loam.
- *Rainfall probability:* the probability of a certain amount of rain falling in one day varies across the UK. Drainage design is based upon keeping the water table below a certain depth in the soil, so that a given system will be able to dispose of a specific quantity of rainfall in one day.
- *Ground and pipe gradient:* on steeper slopes, water disposal is more rapid. However, pipes are often laid across the slope, so it is important to know these gradients when designing a system.

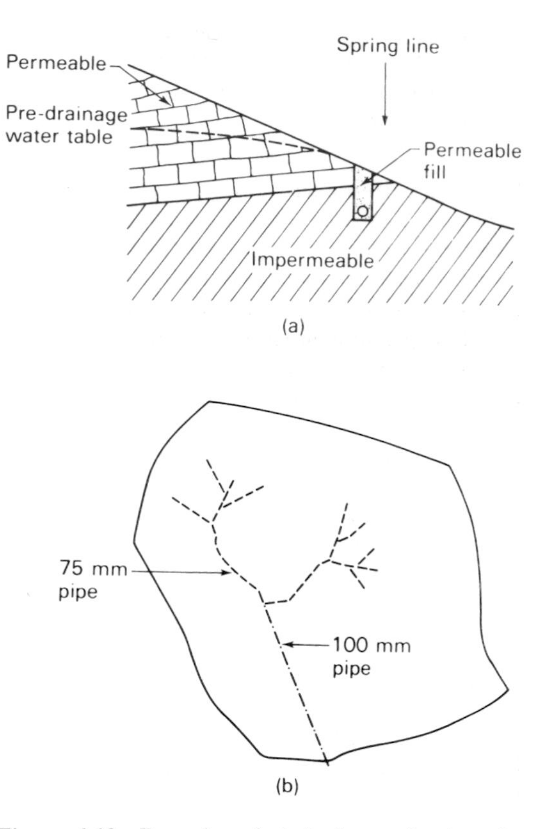

Figure 1.12 Examples of drain layout for irregular random systems. (a) Spring-line interception, in section. (b) Random layout, in plan.

Given this information, and assuming a certain pipe spacing and drainage system to be installed, i.e. 40-m spacing with permeable fill and moles, the pipe size needed can be calculated. This pipe size will allow the system to deal with most (but not all) of the rainfall that can, on balance of probability, be expected in any year.

Drain maintenance

A drain maintenance programme is essential in order to ensure efficient operation. Annual inspection will reveal any problems that have arisen due to soil instability, vegetation growth or animal disturbance. Ditches should be kept clear of excessive vegetation that will restrict water movement and might lead to silting. Fences should be kept in good order to prevent animal damage of ditch sides. Pipe drain outfalls should be kept clear of obstructions. Blockages can be cleared by rodding if necessary. Moles should be redrawn every 5 years, while subsoiling should be carried out as conditions and time allow and require.

Drainage and conservation

Wildlife habitats may be affected adversely by land drainage operations. While the principal aim of field drainage is to reduce water contents to enhance crop growth, there are many established methods of reducing the impact of drainage on the flora and fauna of an area. Most important is to plan for conservation so that, for example, an old pond in the corner of a field can remain undisturbed during and after a drainage operation. Ditches can be designed with wildlife in mind, for example with a wide shallow section to carry stormflow. Advice on conservation measures can be obtained from local Farming and Wildlife Advisory Groups (see 'Useful websites').

Economics of drainage

In recent years, there has been a steady reduction in the level of government grants available for drainage work, from 65% in the mid-1970s to 15% in 1994. Grant aid is no longer available for drainage work. The cost will depend upon lateral spacing, field size, materials used and soil type. Typical installation costs (Nix, 2000) for excavating a trench, supplying and laying an 80-mm plastic pipe are £1.15–1.40 m^{-1}. Adding permeable fill over the pipe to within 375 mm of the ground surface adds £1.50–2.00 m^{-1} to costs. An intensive regular spaced system, say 20-m spaced laterals with permeable fill, may cost between £1400 and £1600 ha^{-1}. Such a system could be expected to have a useful life of at least 25 years. Mole draining costs £45–60 ha^{-1}. Digging a new ditch (1.8-m top width, 0.9-m depth) costs £1.50–2.00 m^{-1}. Trenchless systems are cheaper to install, but quality control is not as good, so the life of such a system may be less than for a trenched installation.

Current estimates of land drained or redrained each year (10 000 ha $year^{-1}$) are much less than that needed to be drained in order to replace old drains that have failed (40 000 ha $year^{-1}$). In the current economic climate, and given recent changes in land-use policies and reduction of land values, it is unlikely that the level of drainage activity will increase to match the rate of system decay. As a consequence, soil management problems may become more severe on land farmed intensively.

Soil biological properties

Soil is a living entity. Soil fertility depends upon the activities of a vast range and number of organisms that affect soil physical conditions (for example earthworm channels and structure formation) and chemical conditions (for example nitrogen availability to plants). Soil is distinguished from inert geological material by the incorporation of degraded organic matter. Soil biology is extremely complex, involving a huge variety of organisms.

Micro-organisms

A discussion of soil properties would not be complete without a consideration of the role of soil organisms in the maintenance of soil fertility. In many situations, particularly extensive land management, biologically driven processes maintain nutrient supply and optimum soil physical conditions. Biological processes are carried out by organisms, which constitute the *soil biomass*. This living powerhouse of the soil may make up 2–3% of all organic matter, and hence within the topsoil (300 mm) may weigh up to 5 t ha^{-1}. Once dead, soil organisms become part of the organic matter fraction. When fully decomposed, plant and animal remains are indistinguishable, and become part of the humus fraction. All aspects of soil biology are important in agricultural situations. For a detailed discussion, see Lynch (1983).

Bacteria

Bacteria are minute soil organisms, often being <5 μm in length. They live in water films around soil particles and under ideal conditions can duplicate themselves within 24 hours. Hence the number of these organisms in soil is enormous, but it is difficult to separate bacteria from the rest of the organic matter to calculate total weight or numbers. The commonly accepted range is in the region of 10^7–10^9 organisms g^{-1}, equivalent to a live weight of about 1–2 t ha^{-1}. This equates to about 1% of the soil organic matter.

There are many different species of bacteria, which exhibit a wide variety of metabolism and ability to decompose diverse substrates. Detailed descriptions of bacterial and other soil organisms can be found in Lynch (1983). Most bacteria are classified as heterotrophs, which are organisms that require complex organic molecules for growth. Bacteria secrete enzymes which breakdown complex molecules into relatively simple, soluble compounds, such as glucose, which can be easily absorbed. Some bacteria are autotrophic, that is they can synthesize their cell constituents from simple inorganic molecules given a supply of energy, either from sunlight or chemical oxidation reactions. In addition to the heterotrophic/autotrophic subdivision, it is possible to classify bacteria according to their oxygen requirements. Some are aerobic, requiring oxygen for respiration, while others are anaerobic.

Examples of some important bacterially moderated processes include:

- degradation of organic matter to release many nutrients as by-products: heterotrophic aerobes
- $NH_4^+ \rightarrow NO_2^- \rightarrow NO_3^-$: *Nitrosomonas* and *Nitrobacter*
- nitrogen fixation: *Azotobacter*, *Rhizobium* spp.
- sulphate oxidation: *Thiobacillus*

Actinomycetes

These organisms are classified as bacteria, but tend to be larger in size but fewer in number. In morphological and physiological terms they represent a transition between bacteria and fungi. Numbers of actinomycetes may range from 10^5 to $10^8\,g^{-1}$, most of which will be found in the topsoil. Actinomycetes are important degraders of organic matter, particularly more resistant fractions such as lignin and complex organic molecules. As with bacteria, these organisms do not favour acid or waterlogged soils. *Streptomyces* are responsible for the characteristic smell of freshly turned soil.

Fungi

Fungal colonies tend to be much less numerous than bacteria ($10^5\,g^{-1}$), but due to the larger size of fungi, these organisms can be equivalent to half the bacterial biomass. They occur in all soils, and are commonly associated with the initial breakdown of organic debris. They can be highly efficient at converting organic matter into fungal tissue, particularly in acidic woodland soils where bacterial numbers may be low. Fungi also tolerate variations in soil water contents better than bacteria, and they are able to degrade resistant materials such as lignin. Mycorrhizal fungi are of particular importance in phosphate (P) deficient soils, where symbiotic associations exist between the fungi, which enhance P absorption, and the host plant, commonly a coniferous tree, which supplies a source of carbon for the fungi. The root systems of saplings which are to be planted can be inoculated with mycorrhizal fungi, hence reducing the need to fertilize the soil. Mycorrhizae are not important in most agricultural soils, as the use of fertilizers suppresses their activity.

Macro-organisms

There is a wide range of types and numbers of macro-organisms in agricultural soils. These range from the small arthropods, such as springtails and beetles, to moles and rabbits, all of which may have an important function in disturbing soil and incorporating organic matter. Only one group of macro-organisms will be discussed here – the earthworms, Darwin's 'natural ploughs of the soil'. Earthworms consume more plant litter material than all the other invertebrates in total. The total biomass varies widely according to soil conditions, from $<20\,kg\,ha^{-1}$ under intensive, organic-matter-poor arable land, to $>500\,kg\,ha^{-1}$ under neutral-pH permanent grassland. Earthworms feed on dead organic matter and incorporate that material into the soil, at the same time mixing organic with mineral material. Earthworms need calcium for their metabolism, and are therefore not common in acid soils ($pH<5.0$). Large populations of worms, such as *Allolobophora* and *Lumbricus* spp., may ingest and excrete $50–100\,t\,ha^{-1}$ annually. During that process, large pores will be created which may aid drainage. These organisms are found to be most important in direct-drilled soils, where numbers can be much greater under regular ploughing. It appears that earthworms are sensitive to disturbance, and so many intensively managed soils with low organic matter contents have become depleted in earthworms in recent years. As a result the benefits of litter incorporation, soil mixing and aeration have been lost.

Soil organic matter and the carbon cycle

Soil organic matter consists of plant material and animal remains and excreta. Ultimately all these materials will decompose in the soil to form humus, which contributes to many aspects of soil fertility. Some farming systems, notably the more extensive and the 'organic' systems, rely heavily on this organic matter fraction to supply many of the plant nutrients as well as the physical benefits that derive from high humus levels in the soil. Some intensively farmed soils have in recent years become difficult to manage; undoubtedly falling organic matter levels have played an important role in this decline of fertility.

Organic matter decomposition

The degradation of plant and animal remains is a complex process, carried out by a variety of organisms, which have been described above. Initial stages of decomposition are dominated by macro-organisms which incorporate and break down material, and in the process increase the surface area and allow bacterial inoculation of this fresh substrate. The soluble and rapidly decomposed fractions, such as simple sugars and starches, may well break down within a matter or days after incorporation. The more resistant fats, waxes and lignins may take months or even years to decay. Soil conditions are important in this decay process, as described above. Most rapid decomposition will occur in near-neutral, damp, warm soils. The result of these decomposition processes is the production of a suite of complex organic chemicals known in simple terms as humus, an amorphous, dark material in which the structure of the original constituents is not distinguishable. There are various humus fractions, discriminated on the basis of molecular weight and chemistry. For further details see White (1997). Ultimately, all organic matter in soil will decompose to its constituent components, carbon dioxide, water and minerals.

In natural soils, types and degree of stratification of humus reflect the nature of the decomposition processes. Neutral or weakly acid brown earths rarely show an accumulation of leaf litter at the soil surface. Humus is well decomposed and intimately mixed with the mineral material. Acidic soils are characterized by a buildup of slowly decomposing organic matter at the surface. Poor humification leads to the development of layers of organic material in various stages of decomposition, as would typically be found in a moorland situation. High water contents can also restrict humification, leading to the development of peat. The higher the rainfall and the lower the temperature, the slower the decomposition processes, so that soils in the wetter and cooler parts of the UK tend to have higher organic matter contents.

Benefits of organic matter

- *Physical* – maintenance of stable soil structure, raising water-holding capacity and creating better conditions for cultivation through reduction of cohesion.
- *Chemical* – retention and release of nutrients and increase in cation exchange capacity.
- *Biological* – the supply of a substrate to various beneficial soil organisms, such as bacteria and earthworms.

These benefits are most important for soils with low clay contents. Improvements in fertility of sandy soils following the addition of organic matter, such as manure or crop residues, can be very marked.

The carbon cycle

The transformation of carbon from gaseous carbon dioxide in the atmosphere to plant, animal and then soil organic matter forms part of the carbon cycle. This cycle is completed when organic matter is released from decomposing humus back to the atmosphere (Fig. 1.13). In agricultural terms the magnitude of the fluxes is important, as carbon accumulates in plant material due to photosynthesis, and is lost from soils following organic matter decomposition. In global terms, the influence of industrial activity, notably the burning of fossil fuels, is changing the balance of carbon within the cycle. Fossil fuels represent carbon that has been

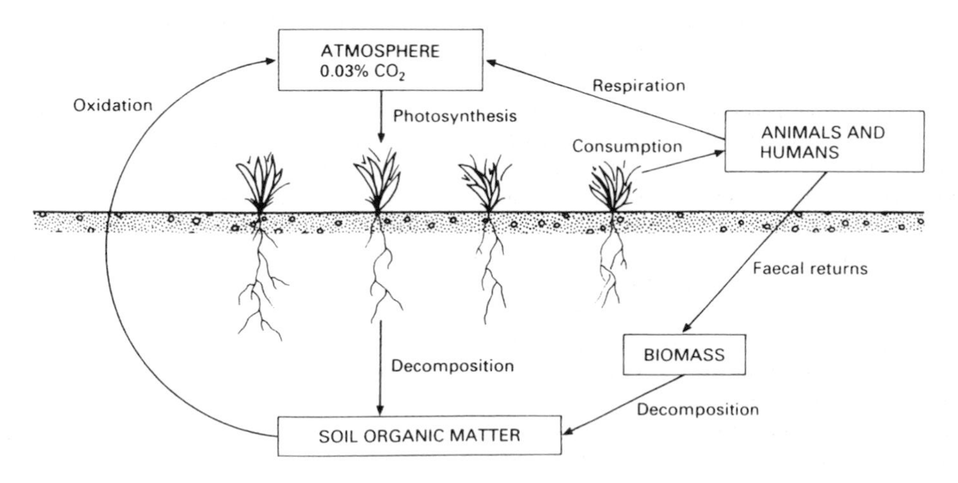

Figure 1.13 The carbon cycle.

trapped in rocks by geological processes, and stored for millions of years. Release by combustion is elevating 'natural' levels (Goudie, 2000), and having a major impact on the global carbon cycle. The turnover of carbon in agricultural systems can be measured in years, rather than millions of years. Changes in the input and loss processes within agricultural soils have lead to changes in this cycle, which are summarized below.

Impact of agricultural practices on soil organic matter

In general, soils under arable crops have lower organic matter contents than grassland. Table 1.9 shows the range of typical organic matter contents observed under selected cropping systems. Note that some literature sources refer to organic carbon, which constitutes approximately 60–65% of organic matter. The values in Table 1.9 represent the balance between inputs and losses. The major inputs and loss processes are:

- *Inputs* – crop residues (e.g. straw, roots, sugar beet tops), green manures (crops sown and ploughed into the soil, not harvested), animal manures (farmyard manure and slurry).
- *Losses* – frequent cultivation, drainage, liming and fertilizing may all lead to greater losses of organic matter by oxidation.

Intensively managed soils show the greatest changes in terms of organic matter turnover, relative to uncultivated soils. Only permanent pasture and arable soils receiving minimal cultivations escape these changes. Any change in organic matter status reflects the balance between inputs and outputs. Grassland soils often have a stable organic matter content, while taking a soil out of grass may lead to a decline in organic matter. Figure 1.14 displays two example situations where cropping practice has had an impact on organic matter levels. In both cases the changes occur over a number of years, with the changes being most rapid soon after the change in cropping practice. The fen soils of East Anglia demonstrate the consequences of prolonged intensive cultivation following drainage several centuries ago. Some fen soils have lost more than 0.5 m of peaty topsoil, such that the underlying clay is being ploughed up to the surface, causing major soil management problems.

The carbon : nitrogen (C : N) ratio describes the rate of decomposition of organic materials added to the soil. For well-decomposed humus the ratio is usually between 10 and 12 : 1. Bacterial biomass has a C : N ratio of 6 : 1, while most plant and animal remains contain much more carbon and much less nitrogen. In consequence, the rate of breakdown of materials with wide C : N ratios, such as cereal straw, will be slow unless there is a source of nitrogen that the bacteria can use. Following straw incorporation there is often a flush in bacterial activity as decomposition begins. This will lead to nitrate-N being removed from the plant-available pool. This nitrogen is immobilized in bacterial biomass, although it will eventually be released as the organic

Table 1.9 Typical organic matter contents of agricultural soils. Note: peats and organic soils (>30% organic matter) are excluded from this table

	Organic matter (%)
Arable, straw/residues removed	3–5
Arable, straw/residues incorporated	4–7
Grass/arable rotation	5–10
Permanent grass	10–20
Woodland	15–30

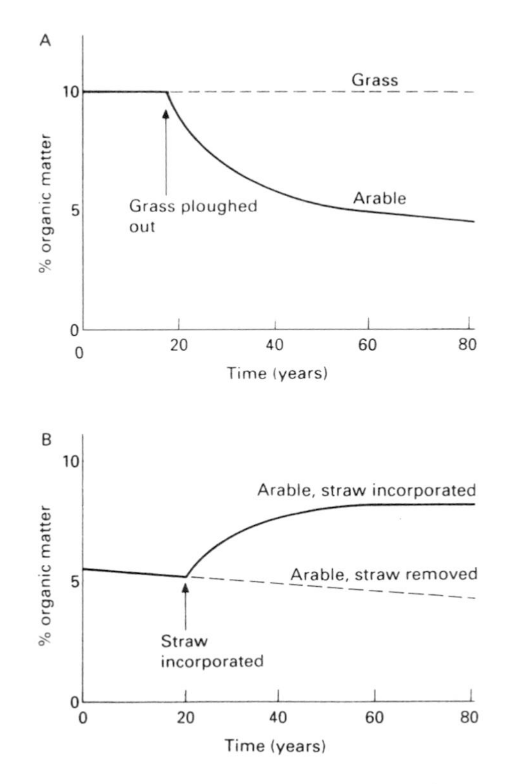

Figure 1.14 Effect of management on organic matter content. *A* Following the conversion of grassland to arable cropping; *B* following the change from straw burning/removal to incorporation.

matter decays in years to come. Materials with narrow C:N ratios can be decomposed readily, and hence make more suitable forms of organic addition if rapid benefits are needed.

Several general rules should be followed where possible in order to maintain organic matter status of soils:

- Return as much of the crop residue as possible.
- Minimize disturbance by cultivation, which should be shallow and infrequent.
- Use green manures.
- Maximize crop cover.
- Use longer grass/clover leys.
- Incorporate manures quickly to minimize oxidation losses.
- Maintain pH levels at optimum for soil organisms.

Adherence to these guidelines will raise soil fertility and reduce the dependence upon fertilizers, while making the soil easier to manage.

Impact of pesticides on soil organisms and organic matter

A wide variety of pesticides are used in agriculture. Many are short-lived and degrade rapidly in the soil, while others are persistent, or may leach out of field soils into adjacent watercourses. Pesticide efficiency may be limited in certain soil types due to adsorption. Under these circumstances it may be necessary to increase application rates to achieve effective control. The environmental consequences of such action may be significant; hence it is important to match pesticide use to soil type as well as to cropping situation.

Pesticide adsorption

The greater the clay and organic matter content of a soil, the stronger the adsorption of residual herbicides. Reduction in the effectiveness of residual herbicides in direct-drilled soils with high surface organic matter and ash contents has been noted in some situations. This adsorption is at its most extreme in peat soils, while on some sandy soils with low organic matter levels, certain soil-acting herbicides cannot be used because of the risk of damage to the crop. The bipyridyl herbicides, paraquat and diquat, are strongly and irreversibly adsorbed by clay minerals and hence are rendered inactive on contact with the soil. However, adsorption onto organic matter is reversible, so these herbicides adsorbed by peaty soils can be taken up, and may damage crops. Losses of pesticides by leaching may occur from sandy soils, which do not exhibit strong adsorption properties. In addition, surface erosion of soil has been found to lead to substantial losses of applied agrochemicals, such as atrazine applied to maize crops and monolinuron applied as a herbicide to winter cereal crops (Environment Agency, 2001).

Degradation

The degradation rate, or alternatively the persistence of pesticides, depends upon soil type. Most pesticides in common use are broken down by microbial action or degrade chemically following application to the soil. The rate of breakdown may be rapid, just a matter of days, as shown, for example, by the synthetic pyrethroids, or may take many months, as in the case of other organophosphorus pesticides. The more rapidly a material breaks down, the smaller the risk of long-term soil organism damage and losses by leaching (Merrington *et al.*, 2002).

Soil classification

Soil types and classification

Soil formation

Soils develop as a result of interactions between a number of soil-forming factors, namely parent material, climate, vegetation, organisms, topography and time. As each of these factors varies, so too will the resulting soil characteristics. In agricultural situations, some soil formation processes are modified – for example the relationship between soil and vegetation – while a basic knowledge of the different soil types created by these processes will help to give an insight into reasons for variations in soil fertility between fields and between regions (Merrington *et al.*, 2002).

The main characteristics of significance to land users derive from the parent material. The geology of an area will be very closely related to the pattern of soil types. In this case geology includes not only the 'hard' rocks, but also the superficial materials, such as river floodplain deposits, from which a soil might be derived. Soils form over long periods of time. Most of the soils in the UK have developed since the end of the last ice age, approximately 15 000–20 000 years ago. Processes of soil evolution are slow, whereas accelerated soil erosion from arable fields can be very rapid, leading in extreme cases to a significant reduction in soil depth. Soils deepen by weathering at a rate of between 0.01 and 0.1 mm year^{-1}. Any losses greater than this (0.1 mm is equivalent to approximately 1 t ha^{-1}) will result in the soil becoming less deep and hence less able to support crop plants.

Soil types

Brown earths

Well-drained soils of varying texture, moderately acid to alkaline pH, warm brown in colour, usually well structured, no marked organic matter accumulation at the surface. Brown earths are often uninteresting to look at, due to lack of marked horizonation, but this shows that the soil is fertile, with a large organism population actively mixing the soil and creating a favourable rooting environment. These soil are capable of supporting most crop types, site and climate permitting.

Gleys

Poorly drained soils with mottled yellow/grey colours at depth, reflecting waterlogging during some time of the year. Waterlogging may be due to heavy texture or high regional watertable, i.e. on a floodplain. The grey colours develop when iron is changed from an oxidized (Fe^{3+}, red/brown) to a reduced (Fe^{2+}, grey/blue) state. With drainage and careful management, these soils can support some arable crops, notably cereals. These soils possess the advantage of high available water contents, and hence are drought resistant.

Podzols

Acid, sandy soils found on heathlands, moorlands and coniferous plantations. These soils show marked stratification of the soil profile, with layers of undecomposed organic matter at the surface, and possible iron pan at depth. Deep cultivation and the addition of lime and organic matter can raise the fertility of these soils, but they will continue to be draughty and develop patchy acidity under intensive management, for example horticultural cropping.

Peats

Peat soils are unusual in being organic not mineral based. Peats may accumulate in two environments. Lowland peats (e.g. Somerset Levels, Cambridgeshire Fens, Lancashire Moss) develop over thousands of years in wet, marshy situations at or close to sea level. The soils can exhibit a neutral or alkaline pH, and can be very productive when drained. However, rapid soil erosion and conservation pressures have virtually stopped further drainage and agricultural improvement taking place. Upland peats (e.g. much of the Pennines, Dartmoor, Cambrian Mountains, southern uplands) form where high rainfall prevents organic matter decomposition. These soils are invariably acid and incapable of much improvement due to the dominant influence of climate.

Soil classification

In order to understand differences in soil type and management characteristics across the farmed landscape, it is necessary to classify and then map the soils. Soils in Britain have been described and classified at a variety of scales by the National Soil Resources Institute (England and Wales) and the Macaulay Institute (Scotland). Complete mapping of British soils was completed in the early 1980s at a scale of 1:250 000. More detailed maps are available for selected areas, usually at scales of 1:25 000 or 1:63 630. Further details and sources of information are given at the end of this chapter. As scale of mapping changes, so does the level of detail contained within the map. For example, the National Soil Map of England and Wales, at 1:250 000, provides information that gives a guide to soil types in an area, and is therefore useful in general terms, but cannot be used for detailed farm management purposes. Soils in England and Wales are grouped into associations, which are combinations of soil series, defined by parent material, texture and mineral composition. Each association is described on the basis of parent material, main soil and site characteristics, cropping and land use.

Land classification and crop suitability maps

The agricultural potential of land, as opposed to just soil, depends on site and climate. The Agricultural Land Classification (ALC) system employed by MAFF (MAFF, 1988) subdivides land into five grades in terms of limitation, typical cropping range and the expected level of consistency of yield. Many of the limitations are soil based, such as drainage status, stone content or soil depth, but can also include other factors, such as ground slope or climatic regime. The description of grades is given in Table 1.10. Agricultural land in England and Wales has been classified using this system, at an original mapping scale of 1:63 360. This information integrates all those physical factors that might affect crop yield or control choice of cropping, and hence forms a useful basis for the assessment of land potential. However, the grades themselves do not indicate the cause of any restricted potential. Only detailed site and soil analysis can yield that information.

Land can also be classified in terms of potential to support specified crop types, such as arable, grassland

Table 1.10 Agricultural land classification grades (after MAFF, 1988)

Grade 1 – Excellent quality agricultural land
Land with no or very minor limitations to agricultural use. A very wide range of agricultural and horticultural crops can be grown and commonly includes top fruit, soft fruit, salad crops and winter-harvested vegetables. Yields are high and less variable than on land of lower quality.

Grade 2 – Very good quality agricultural land
Land with minor limitations which affect crop yield, cultivations or harvesting. A wide range of agricultural and horticultural crops can usually be grown but on some land in the grade there may be reduced flexibility due to difficulties with the production of the more demanding crops such as winter-harvested vegetables and arable root crops. The level of yield is generally high but may be lower or more variable than Grade 1.

Grade 3 – Good to moderate quality agricultural land
Land with moderate limitations which affect the choice of crops, timing and type of cultivation, harvesting or the level of yield. Where more demanding crops are grown, yields are generally lower or more variable than on land in Grades 1 and 2.

Subgrade 3a – Good quality agricultural land: land capable of consistently producing moderate to high yields of a narrow range of arable crops, especially cereals, or moderate yields of a wide range of crops including cereals, grass, oilseed rape, potatoes, sugar beet and the less demanding horticultural crops.
Subgrade 3b – Moderate quality agricultural land: land capable of producing moderate yields of a narrow range of crops, principally cereals and grass or lower yields of a wider range of crops or high yields of grass which can be grazed or harvested over most of the year.

Grade 4 – Poor quality agricultural land
Land with severe limitations which significantly restrict the range of crops and/or level of yields. It is mainly suited to grass with occasional arable crops (e.g. cereals and forage crops) the yields of which are variable. In moist climates, yields of grass may be moderate to high but there may be difficulties in utilization. The grade also includes very droughty arable land.

Grade 5 – Very poor quality agricultural land
Land with very severe limitations which restrict use to permanent pasture or rough grazing, except for occasional pioneer forage crops.

or forestry. These crop-suitability assessment systems require detailed soil and site information, and can only be compiled from soil survey information. These capability classifications are more robust than the ALC system for that reason. Areas of land can be classified not only in terms of their potential to support given crops, but also in terms of land management practices. For example, classifications of land exist that describe nitrate leaching risk, and suitability for straw incorporation. Further information can be obtained from the Soil Survey Regional Bulletins.

Soil chemical properties and crop nutrition

Crop nutrient requirements and uptake

Agricultural crops require a wide range of essential nutrients for growth. With the exception of carbon, which is obtained from atmospheric carbon dioxide, these are all stored and released from the soil. All the essential nutrients required by green plants are taken up in inorganic form, unlike animals where some organic compounds are also needed. Some of these nutrients are applied commonly from fertilizer sources, such as nitrogen, phosphorus and potassium, while others such as iron, copper and manganese derive from natural soil sources.

There are 16–19 elements known to be essential for plant growth. These are listed in Table 1.11 with a summary of their main functions in the plant, the form and typical quantity taken up by the plant and example deficiency symptoms. An essential nutrient is defined as one that is required for the normal growth of the plant, is directly involved in the nutrition of the plant and cannot be substituted by another nutrient. Not all plants have the same essential requirements, e.g. sodium (Na), silicon (Si) and cobalt (Co) are needed by some plant species only. Sometimes essential plant elements are divided into macronutrients (the first 11 elements in Table 1.11) and micronutrients. This division is based on the concentration of each element in the plant, but plants differ in their nutrient composition so such a distinction is not always meaningful. Also the importance of a nutrient is not related to the quantity contained in the plant tissues, and an insufficient quantity of any nutrient can lead to a restriction in crop growth and/or quality. More detailed information on the uptake and role of plant nutrients can be found in Marschner (1995). Plant tissue analysis will be needed in most cases to confirm suspected deficiency symptoms.

Nitrogen is taken up by most crop plants in the largest quantities, closely followed by potassium. Example crop uptakes quoted in Table 1.12 for the macronutrients are given in kilograms per hectare, while removals of

Table 1.11 Essential plant nutrient form, uptake, function and deficiency symptoms

Element	*Form of uptake by plant (from soil solution unless stated)*	*Major function in plant*	*Plant deficiency symptoms*
C	CO_2 from atmosphere	Major constituent of organic material; can account for 40% of dry weight of plant	
H	Soil water and leaf	Linked with C in all organic compounds	Lack of leaf turgor, wilting
O	CO_2 and H_2O; O_2 during root respiration	With C essential in carboxylic groups; with H in oxidation–reduction processes	
N	NO_3^- and NH_4^+	Essential constituent of cell compounds, e.g. proteins, chlorophyll	Older leaves become senile prematurely, yellow/brown colour. Spindly plants that lack vigour are dwarfed, pale coloured and tillerless. Can be rapidly corrected by fertilizer N
P	HPO_4^{2-} and $H_2PO_4^-$	Vital constituent of living cells, associated with storage and transfer of energy and protein metabolism	Difficult to identify. Plant colour initially dark green/blue, reddish stem in severe cases. Stunted, poor root growth and seed formation. Difficult to correct as symptoms appear late. Use water-soluble P fertilizer
K	K^+	Essential for efficient water control (osmosis) and translocation of carbohydrate; found in some enzymes	Loss of plant turgor, reduced growth rate. Older leaves yellow/brown spots or brown edges. Potatoes and brassicas show marginal necrosis of leaves. Use muriate of potash
S	SO_4^{2-}; SO_2 absorbed from atmosphere	Constituent of some essential amino acids, e.g. cysteine and methionine; also in enzyme proteins	S nutrition linked to N; hence S deficiency indicated by lack of response to N. Brassicas, grass and legumes commonly deficient
Ca	Ca^{2+}	Essential role in cell walls and biological membranes	Generally rare. Poor translocation leads to fruit problems, e.g. bitter pit in apples, blossom end rot in tomatoes. Spray fruit with calcium nitrate
Mg	Mg^{2+}	Essential constituent of chlorophyll, some enzymes and some organic acids	Older leaves show intervenal chlorosis, with leaf veins dark green. Sugar beet – black necrotic areas, potatoes – purple tint. Foliar epsom salts. Mg deficiency can be induced by luxury uptake of K
Na[1]	Na^+	Essential in some plants only (e.g. of marine origin); partially interchangeable with K^+ in most plants	Sugar beet, fodder beet, mangolds, spinach require Na as an essential nutrient; deficiency indicated by apparent water stress. Use Kainit or Na-enriched compound fertilizer

Si[1]	Probably $Si(OH)_4$	Used in cellulose framework; interacts with P in plant (mechanism uncertain)	Lack of cell rigidity
B	H_3BO_3 or BO_3^{3-}	Assists in carbohydrate synthesis, uptake of Ca^{2+} and absorption of NO_3^-	Common in root crops (sugar beet and swedes), brassicas and legumes. Death of apical growing point, leading to crown rot in beet, terminal leaves yellow or red. Use boronated fertilizer or foliar spray
Mn	Mn^{2+}	Associated with chlorophyll formation and some enzyme systems	Chlorosis of younger leaves (grey speck in oats, 'speckled yellows' in sugar beet). Brown spots and streaks in barley, white intervenal streaks in wheat, stunted leaves in potatoes. Apply manganese sulphate to soil (125–250 kg ha^{-1}) or 5–10 kg ha^{-1} as foliar spray
Cu	Cu^{2+} or copper chelates	Small quantities needed in chloroplasts and for enzyme systems converting NO_3^- to protein	Spiralling stunted leaves with yellow tips and poor grainfill in cereals. Deficiency in grass ('teart' pasture) due to excess Mo. Apply 60 kg ha^{-1} copper sulphate to soil, or 2–3 kg ha^{-1} chelated copper as foliar spray
Zn	Zn^{2+}	Assists with starch formation and some enzyme systems	Fruit trees, maize, linseed and field beans are sensitive. Apply 5–10 kg ha^{-1} zinc sulphate to soil, or zinc chelate as foliar spray
Mo	MoO_2^{4-}	Small quantity essential for enzymes controlling N nutrition (also for N-fixing bacteria)	Lack of heart formation in cauliflowers ('whiptail'), reduced by liming or foliar spray of 0.25–0.50 kg ha^{-1} sodium molybdate
Fe	Fe^{2+} or Fe^{3+} or Fe chelates	Essential in chlorophyll and enzyme activities; associated with enzymes in photosynthesis	Chlorotic marking on younger leaves (unlike Mn which affects all leaves equally). Fruit trees also affected. Apply iron chelate as foliar spray
Cl	Cl^-	Involved in evolution of O_2 during photosynthesis (excess Cl more common than a deficiency)	Required in small quantities, hence deficiency uncommon
Co[1]	Co^{2+} or Co chelates[2]	Not essential for most species; essential for N fixation by bacteria and could be used in N nutrition in plant	Required in small quantities, hence deficiency uncommon

[1] These three elements are not essential in all plant species.
[2] Chelates are organic-metal complexes that can maintain the availability of some metal ions over a wide range of pH.

Table 1.12 Removal of nutrients (kg ha^{-1} year^{-1}) by some crops (after Simpson, 1986)[1]

Crop	*Yield (t ha^{-1})*	*Dry matter yield (t ha^{-1})*	*N*	*P_2O_5*	*K_2O*	*Ca*	*Mg*	*S*
Cereal	6 t grain 3.5 t straw	8	120	50	70	15	10	30
Sugar beet	40 t roots 25 t tops	12	200	45	240	70	25	30
Potatoes (tubers only)	50 t	10	180	50	240	10	15	20
Grass silage	30 t	—	160	40	160	45	15	15
Hay	—	8	100	30	120	30	10	10
Clover	—	5	180	25	120	100	15	15
Kale	50 t	10	200	60	220	250	20	100
Natural grass heath	2 t	0.4	10	3	10	2	1	2

Trace elements (g ha^{-1}) removed annually in crops:

Iron (Fe)	600–2000	Zinc (Zn)	100–400
Manganese (Mn)	300–1000	Copper (Cu)	30–100
Boron (B)	50–300	Molybdenum (Mo)	5–20

micronutrients, such as copper or molybdenum, are several orders of magnitude smaller. Nevertheless, these micronutrients are essential, and soils must be able to supply them to crop plants. In the absence of fertilizer additions crops will still grow, indicating that natural soil processes will supply nutrients in the ratios required by plants, albeit at lower levels than the more demanding crops need to yield well. Nutrient availability depends on soil type, organic matter content and environmental conditions. Matching crop requirements to soil conditions is a difficult task, but one that must be attempted in order to utilize applied fertilizer nutrients efficiently and to minimize losses. Central to good soil nutrient management is the understanding of processes of nutrient retention and release.

Nutrient uptake processes are complicated, and depend on the intimate contact between root hairs and nutrients in soil water. Nutrient uptake takes place by mass flow and diffusion. Mass flow is brought about by transpiration, for the water taken up by a root to meet the transpirative demand contains ions dissolved in it. Diffusion occurs when ions are being taken up faster than they are carried to the surface by mass flow, for this sets up a concentration gradient between the root surface and the body of the soil down which the ion will diffuse. Essential to effective uptake of nutrients is a soil that can be exploited fully by developing root systems. Hence a well drained and structured soil is a vital requirement, without which applied fertilizer or manure-derived nutrients will not be used effectively by crop plants.

Nutrient retention – cation and anion exchange

The retention and release of nutrients to plants depends on a variety of soil and environmental factors, of which the chemistry of soil clay and organic matter is the most important. Soils are not inert media – they interact with applied agrochemicals in a very dynamic manner. Before individual nutrients are discussed in detail, the ability of soils to retain and release nutrient ions will be described. These processes not only control the supply of plant-available nutrients, but also act to modify the whole soil chemical environment, with important consequences for crop management.

Charges on soil surfaces

Soil is not an inert material. Most importantly, the clay fraction, iron oxides and organic matter can act in a chemically interactive manner, due to electrical charges at the surfaces of these particles. The origin of these charges depends upon the chemical structure of the material. Charges on clay particles are associated either with defects in clay lattice structures, or with unsatisfied broken bonds at the edges of clay platelets. In both cases, the charge imbalance tends to be negative in British soils, resulting in an attraction for ions carrying the opposite, positive charge. Organic matter is likewise negatively charged. Only iron and aluminium oxides show a reversed pattern, tending to attract negatively charged ions due to these components having a positive charge. For a more detailed discussion of this topic see White (1997).

Ion exchange processes

Nutrient ions are available for root uptake in solution – that is from water stored in soil pores. This water will be in a quasi-equilibrium with the chemistry of ions held on charged surfaces described above. As roots extract certain nutrients, the equilibrium is upset, resulting in ions leaving the charged surfaces to re-establish the balance. This process is known as ion exchange. Exchange reactions can work in the other direction, so that fertilizer nutrients, for example, can be retained by soils until needed by plants.

Cation exchange

A cation is any ion with a positive electrical charge, such as the hydrogen ion H^+ or the sodium ion Na^+. Both H^+ and Na^+ are monovalent cations – that is, they possess only one charge. A cation may have more than one positive charge, such as magnesium Mg^{2+} (divalent) or iron Fe^{3+} (trivalent). The ability of a soil to exchange cations depends primarily on the quantity and type of clay and organic matter present. Exchange reactions in soils can be complex, but they can be simply visualized in terms of 'swops' of equivalent charged ions. Imagine that a clay particle has one calcium ion (Ca^{2+}) attached to its surface (see equation below). Rainwater percolating down through the soil usually contains dissolved hydrogen ions. This rainwater will wash around the clay particle, so that an exchange reaction may occur, whereby two hydrogen ions will replace one calcium ion. As a result, charge balance is maintained and the calcium ion will be washed further down through the soil profile and may be lost in drainage water. The reaction may go in either direction, as is indicated by the double-headed arrow. The reaction might be expected to move from right to left when lime is added to soil to counteract soil acidity.

$$Ca^{2+}[clay] + 2H^+ \leftrightarrow H^+[clay]H^+ + Ca^{2+}$$

In acid soils, H^+ and Al^{3+} are the dominant cations, while in calcareous soils Ca^{2+} and Mg^{2+} are dominant. For short periods after fertilizer application potassium (K^+) or ammonium (NH_4^+) may be dominant.

Soils with a large cation exchange capacity are chemically stable and nutrient retentive. Cation exchange capacities of some example soils are given in Table 1.13. Cation exchange is expressed in a variety of units; those quoted here are in terms of milliequivalents of charge per 100 g of soil. It is the magnitude of the differences between soil types as shown in this table that are important. Clay textured soils have an ability to hold at least five times as many cations on exchange sites as sandy textured soils. In this latter category, the importance of organic matter becomes clear; well humified (decomposed) organic matter can have a cation exchange capacity in excess of 200 meq $(100\,g)^{-1}$.

Table 1.13 Cation exchange capacities of mineral and organic soils

	Cation exchange capacity (milliequivalents (100 g soil)$^{-1}$)
Sandy loam	5–15
Sandy clay loam	10–20
Clay loam	20–40
Clay	30–50
Lowland peat	150–200
Upland peat	40–60

Anion exchange

Anion exchange occurs in soil to a much more limited extent, and is much less important in terms of any influence on soil chemical reactions, although for individual nutrient anions, such as NO_3^- and $H_2PO_4^-$, it can be locally important. Iron and aluminium oxide surfaces can carry positive charge sites that attract anions, and exchange reactions can occur, but this process is subsidiary to cation exchange.

pH and nutrient availability

The acidity or alkalinity of a soil exerts a strong influence on nutrient availability and uptake, and hence on crop growth. Acidity or alkalinity is the balance between hydrogen (H^+) and hydroxyl (OH^-) ions in a solution. The hydrogen ion concentration in soil solution reflects the soil chemical environment, and hence is used as an indicator of chemical conditions. pH can be defined as follows:

$$pH = -\log[H^+]$$

Expressed in words, pH is the negative logarithm of the hydrogen ion concentration, usually measured in a mixture of 1 part soil to 2.5 parts distilled water. The pH range of some natural and agricultural soils is given in Table 1.14. Most soils in the UK, with the exception of those derived from limestone or calcareous boulder clay, tend to be acid, due to the acidifying effect of rainfall, which is naturally acidic, even when unpolluted by oxides of sulphur or nitrogen. Soils with a low cation exchange capacity tend to become acidic most rapidly –

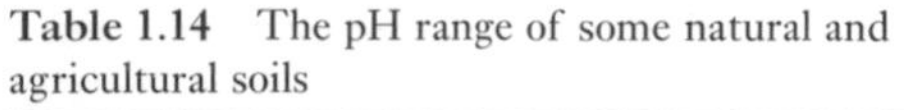

Table 1.14 The pH range of some natural and agricultural soils

	pH range
Sandy heathland	3.5–5.0
Calcareous brown soil	6.5–8.0
Upland peat	3.5–4.5
Cultivated soil, non-calcareous	5.0–7.0
Cultivated soil, calcareous	7.0–8.0
Permanent pasture, lowland	5.0–6.0
Permanent pasture, upland	4.5–5.5
Lowland peat	4.0–7.0

Figure 1.15 Relationship between pH, nutrient availability and organism activity (after Brady, 1999).

that is coarse textured soils. Organic soil such as peats can be either acidic or neutral/alkaline depending upon the origin of the peat.

Soil pH affects nutrient availability in a very dramatic manner, as is shown in Fig. 1.15. Although individual crops differ in their response to soil acidity, certain general principles hold true regarding nutrient availability to plants. Those nutrients that are stored in organic sources in soils, such as nitrogen, are released when soil organisms break down the organic material, a process known as mineralization. Figure 1.15 shows that the activity of macro-organisms such as earthworms and micro-organisms such as bacteria is pH dependent, with maximum breakdown activity occurring at pH 6–7. Other nutrients such as potassium will be held on exchange sites providing the soil is not swamped with hydrogen ions – hence the restricted availability at low pH, due to the high hydrogen ion concentration. Other nutrients, such as phosphorus, display more complex patterns. With the exception of certain trace elements, such as Fe, Cu and Zn, maximum nutrient availability occurs in the range pH 6–7.

Correcting acidity by liming

The soil benefits of the addition of lime are well known. In addition to reducing soil acidity, lime will increase soil organism activity (see Fig. 1.15) and enhance structure stability, thus improving physical conditions. These benefits were known to the Romans, but the mode of action of liming materials was only fully appreciated when principles of soil chemistry, and cation exchange in particular, were elucidated. Soil becomes acid as calcium, magnesium, potassium and sodium are displaced from exchange sites by hydrogen ions and then leached from the profile. Natural weathering processes will replace the leached cations, but only at a slow rate. A recent survey (Chambers and Garwood, 1998) demonstrated that the proportion of soils with pH values below the recommended levels for arable and grass crops has increased in the last two decades. The reasons for soil becoming progressively more acid with time are complex, but the following factors are important:

- acid deposition from polluted rainfall, including ammonia deposition
- use of acidifying nitrogen fertilizers
- heavy cropping which removes calcium, magnesium, etc.
- oxidation of organic matter which generates free H^+ ions in the soil

Clearly the soil type and parent material are important in supplying the dominant basic or 'acid resisting' ions, calcium and magnesium. Soils with a large cation exchange capacity will acidify more slowly than sandy, organic-poor soils. Hence, for any given pH, a soil with a high clay and organic matter content will have a greater lime requirement than a sandy, organic-matter-deficient soil. For a detailed description of the determination of the lime recommendation for a given soil and cropping situation, see 'Liming'.

Macronutrients

Nutrients are an integral part of the soil : plant system that 'flow' from one part of the system to another according to a variety of complex processes that interact together. Modification of one part of the system will affect another. For example the effect of applying a fertilizer on crop yield cannot be considered fully without taking into account other sources of nutrients, such as nutrients held in soil organic matter. There are three main stores of nutrients within the soil:

- the inorganic store, mainly in the soil
- the biomass store, mostly living organisms, including plant material
- the humus store, the dead and decaying remains of plant and animal material

These stores and the relationships between them are represented diagrammatically in Fig. 1.16. The relative importance of each store varies from nutrient to nutrient; these are discussed individually below. For any given soil : crop situation, the application of fertilizer or manure may only contribute a small proportion of the total soil reserve of a given nutrient. It is therefore important to appreciate that even intensively fertilized soils can supply significant quantities of nutrients from non-fertilizer sources.

Nitrogen

Nitrogen (N) is the most important macronutrient in terms of the quantities taken up by crops, but efficient use of nitrogen in fertilizer and manurial sources is problematic, due to the rapid losses of the main plant-available form of N, the nitrate anion (NO_3^-), either by leaching or by conversion to gaseous forms of N. Minimizing these losses is best achieved by a full understanding of the N transformation processes in the soil : plant : atmosphere system.

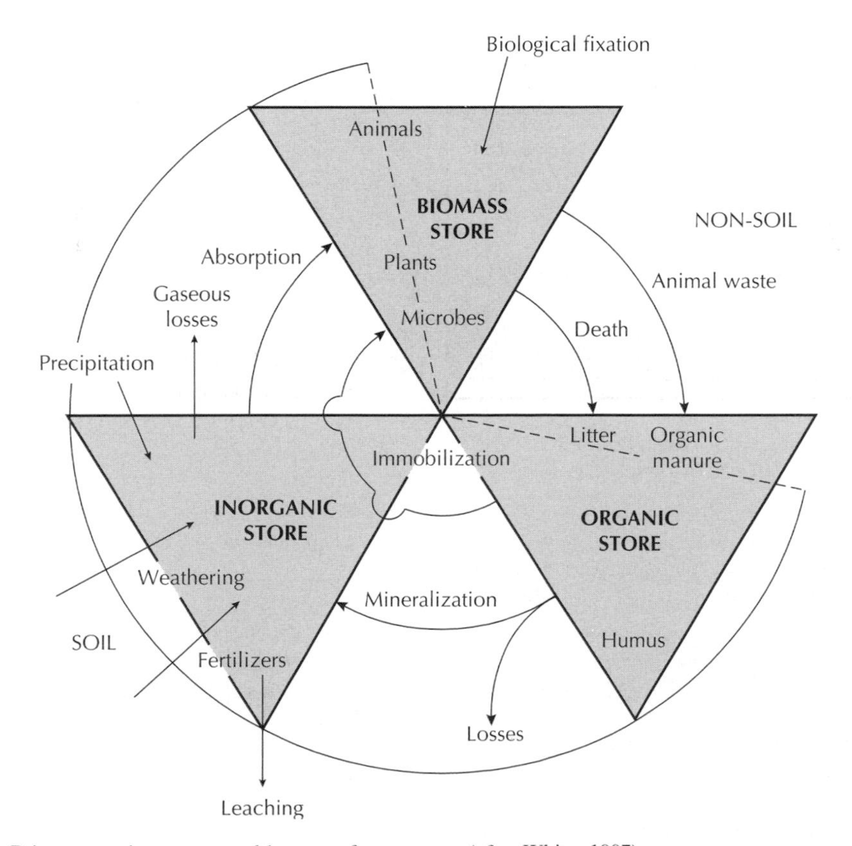

Figure 1.16 Primary nutrient stores and key transfer processes (after White, 1997).

The nitrogen cycle

Nitrogen compounds in soil may be either organic, as components of organic matter, or inorganic, as ammonium (NH_4^+) or nitrate (NO_3^-) ions. The main stores and transfers between those stores are given in Fig. 1.17. It can be seen from this figure that there are no significant rock or parent material sources of nitrogen. All nitrogen in soils is ultimately derived from the atmosphere, which contains 79% N (as the gas dinitrogen, N_2). Rainfall may add 5–20 kg N ha^{-1} $year^{-1}$. These additions are observed to be high where intensive livestock husbandry leads to ammonia volatilization from manure and slurry. During the long process of soil formation, N is fixed into the soil through the action of free-living algae, or bacteria, such as *Azotobacter*, or symbiotic fixation, primarily carried out by legumes. These organisms will raise the level of organic N such that a mature soil might contain 2000–5000 kg N ha^{-1}.

Once a substantial pool of organic N has accumulated, the process of mineralization is carried out by bacteria in the soil. Mineralization leads to the production of nitrate, but several steps are involved in the process.

Organic N	$\rightarrow$	NH_4^+	
NH_4^+	$\rightarrow$ *Nitrosomonas*	NO_2^-	(nitrite)
NO_2^-	$\rightarrow$ *Nitrobacter*	NO_3^-	

All these reactions are controlled by soil and environmental conditions. They can only occur under aerobic conditions, hence waterlogged soils are not a suitable environment for mineralization. However, some water is needed, as mineralization will be restricted in very dry soils. The process is temperature dependent, such that mineralization will be limited below a soil temperature of 4–5°C. Soil pH is also important, with mineralization occurring most rapidly at neutral pH values. Since the production of nitrate is potentially able to proceed quicker than the production of nitrite, the latter is not normally found in significant quantities in soils. The quantity of mineral N present in a soil at any time will be very variable, but might fall within the range 10–50 kg N ha^{-1}.

The products of mineralization, NH_4^+ and NO_3^- (mineral N), react in different ways in the soil. NH_4^+ is

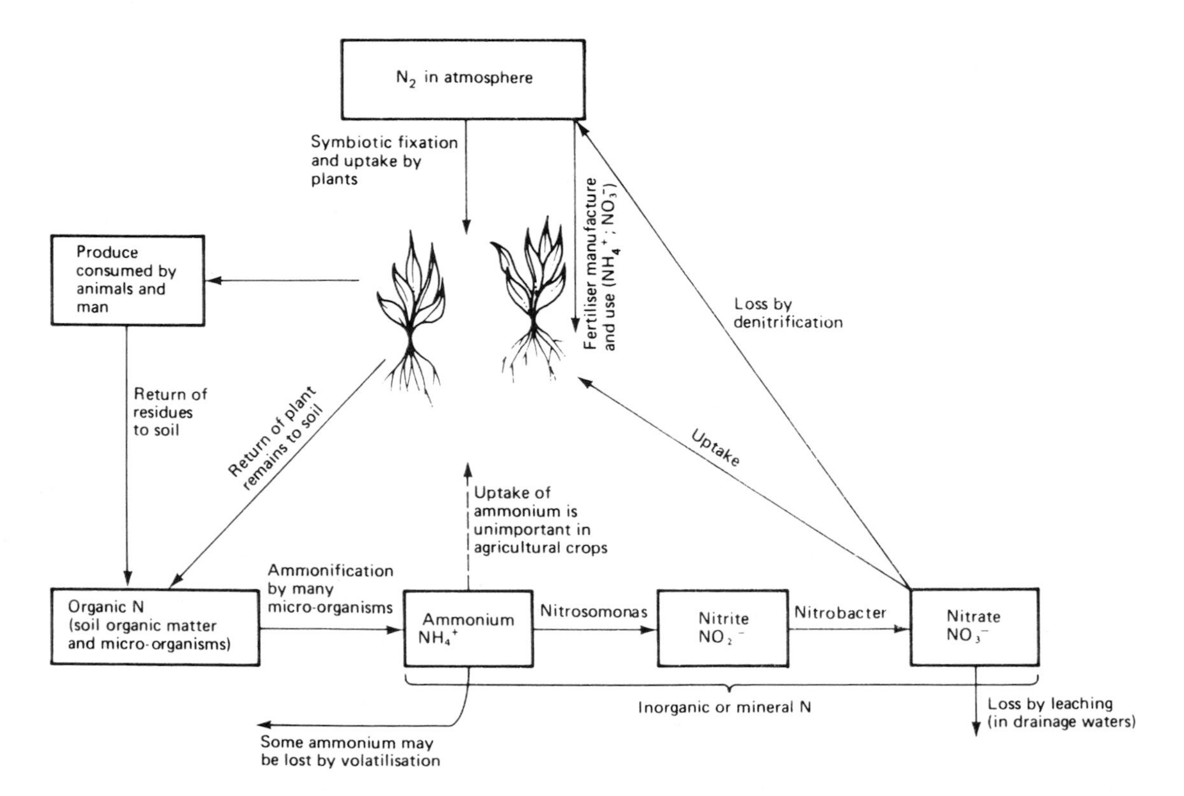

Figure 1.17 The nitrogen cycle.

attracted to cation exchange sites, and so is less likely to be leached. In contrast, NO_3^- is very mobile in soils, and vulnerable to leaching losses during periods of drainage. The fate of mineral N depends once again on soil and environmental conditions. Plant uptake may occur, mineral N might be leached or converted into gaseous forms (the process of denitrification, which takes place in wet soils), or bacteria might convert the 'available N' into biomass and hence add to the organic reserves. All these processes are summarized in Fig. 1.17.

N additions

N additions to the soil : plant system may come from a variety of sources. Fertilizer N is frequently applied to agricultural soils. The common forms of fertilizer N, such as ammonium nitrate (NH_4NO_3), are readily soluble and plant available. However, experiments have shown that, on average, only half of most fertilizer N is taken up by the crop to which it is applied. The other half may be lost, for example by leaching, or be incorporated into soil organic matter through the action of bacteria. Addiscott *et al.* (1991) give a full description of the fate of fertilizer N. Organic manures and crop residues are important sources of N, but the availability of the variable quantities of N contained within these materials (see 'Nutrient sources: manures') depends on those processes described above that control the N cycle. Only when bacteria have broken down the crop residues and animal wastes does the N become available to crop plants.

Phosphorus

Sources

Phosphorus (P) is unusual amongst the major nutrient elements in that concentrations in soil solution are frequently extremely low (equivalent to less than 5 kg P ha^{-1} in most situations), but plant requirements (see Table 1.12) can be significant. The main sources of P are from soil parent materials and fertilizer/manure additions. In Fig. 1.18 a simplified P cycle is represented. Release of P from insoluble mineral sources is slow, such that the pool of available P is always small. In mature soils organic matter forms a major reserve of P which will become available to plants as mineralization occurs. Organic matter sources of P become important in sandy soils and peats that tend to have low levels of mineral P. Most agricultural soils have sufficient reserves of P, built up over many decades. Fertilizer P (as rock phosphate, basic slag, superphosphate or bonemeal) has been added to soils since the mid-nineteenth century. Rainfall additions of P are insignificant. Total P contents range from 500–2500 kg ha^{-1}.

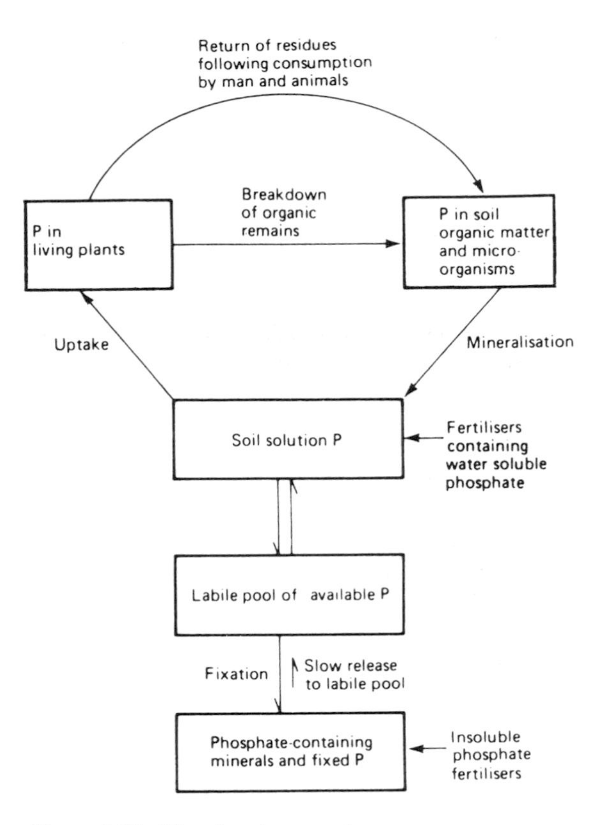

Figure 1.18 The phosphorus cycle.

Soil reactions and availability

P is very insoluble, so that a large proportion of total P, in excess of 95%, is unavailable to plants at any time. Release of P from insoluble mineral sources is controlled by chemical equilibrium processes. Plant uptake of orthophosphate ($H_2PO_4^-$ and HPO_4^{2-}) removes P from soil solution, leading to more of the labile pool moving into solution (Fig. 1.18). Plant requirements of P can be up to 3 kg P ha^{-1} day^{-1}, which, given the quantity in solution at any one time, indicates the importance of these chemical reactions. Orthophosphate being an anion does not take part in cation exchange reactions, but may be absorbed on the smaller number of anion exchange sites in soils. P fixation is a process whereby soil solution/labile P is rendered insoluble by reactions with calcium (in alkaline soils), iron or aluminium (in acid soils). Fixation leads to added P becoming unavailable to plants over short time periods, and can result in fertilizer P uptake being less than 25% in P-deficient soils. Fixation is least strong at pH 7. In most agricultural soils in the UK there has now been a sufficient history of fertilizer P use and application of manures that the most powerful sites of P fixation have been satisfied, such that newly applied P is fixed less strongly than was the case in previous years. However, it is still important to place fertilizer P close to the developing seed if crop demands are high.

Losses
Losses tend to be small, due to P insolubility. Leaching through the soil is not a major loss in agricultural terms (<10 kg P ha^{-1} $year^{-1}$), but can make a significant contribution to nutrient enrichment in watercourses. There are no gaseous loss processes. Surface soil erosion is the main route of P loss. Phosphorus losses can be important from overgrazed compacted soils, where surface run-off will carry P bound to sediment and in organic forms down slopes to adjacent watercourses. Arable soil erosion will also lead to P losses. The main 'loss' is the fixation of P, which reduces the plant-available supply but adds to total soil reserves. This fixed P may eventually become available to plants over a period of years.

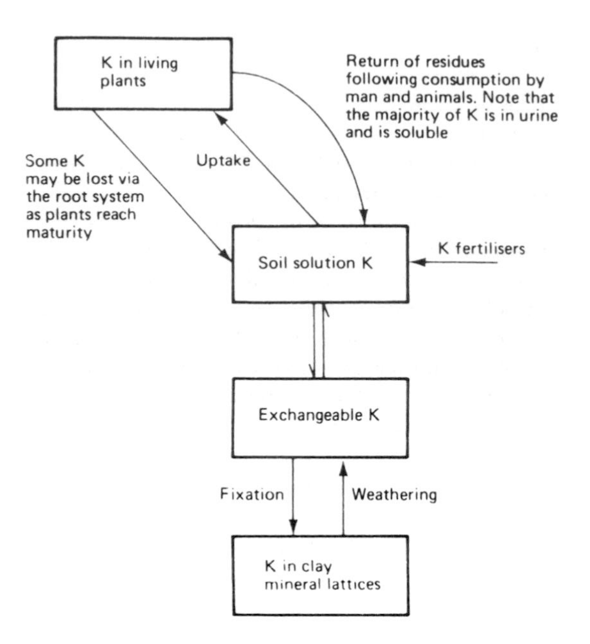

Figure 1.19 The potassium cycle.

Potassium

Sources
Potassium (K) is derived from the weathering of clay minerals in the parent material; hence any clay-rich soil will be well supplied with K. Many rock-forming minerals, such as feldspars, are rich in K, so that as these minerals break down due to weathering, K is released. Organic matter is a poor source of K, as plant material contains K in the fluid phase. Hence most K will be lost at an early stage from decaying plant material. Rainfall may add 1–10 kg K ha^{-1} $year^{-1}$. Total K contents may be very large (>10 000 kg ha^{-1}), while exchangeable K may range from 500–1000 kg ha^{-1}.

Soil reactions and availability
The chemistry of soil K is much less complicated than that of either N or P. K exists in exchangeable and solution forms as the simple cation K^+ (see Fig. 1.19). Plants' uptake of K^+ can be equal to or greater than that of N (see Table 1.12). Soils with a large cation exchange capacity will be able to retain a large amount of K in a relatively plant-available form. Some fixation of available K may occur, but this is a relatively unimportant process in most soils.

Losses
K is subject to leaching losses, particularly from sandy, organic-poor soils, but this loss is rarely considered to be important in agricultural systems, due to the ease of K availability from soil and fertilizer sources. Ion balance dictates that drainage water must contain an equal proportion of anions and cations. Hence losses of NO_3^-, for example, must be balanced by losses of a cation, such as K^+.

Calcium and magnesium

Sources
The weathering of minerals forms the main supply of plant-available calcium (Ca^{2+}) and magnesium (Mg^{2+}). These two macronutrients behave in a similar manner to K in soils. Total soil reserves may be very large, particularly in limestone soils. Exchangeable Ca^{2+} and Mg^{2+} may range between 1000 and 5000 kg ha^{-1}. Soil Mg levels in coastal areas tend to be elevated due to sea salt spray additions.

Soil reactions, availability and losses
As with K, these macronutrients are held on the exchange complex and released into solution as plant uptake occurs. Ca and Mg play an important role in controlling soil pH, as discussed previously. Fertilizer additions accumulate on the exchange complex, and may be subject to some temporary fixation, but again this process is not of major importance. Large leaching losses of Ca and Mg will occur from agricultural soils (e.g. 100–200 kg Ca ha^{-1} $year^{-1}$), but these are small in relation to total reserves, and do not cause a major environmental concern.

Sulphur

Sources
Soil sulphur (S) is derived originally from sulphide minerals, which are oxidized to sulphate (SO_4^{2-}) in weathering. Clays and shales frequently contain large quantities of S. In mature soils, most S is found in the topsoil in the organic form. Additions in rainfall range from 10–30 kg S ha^{-1} year^{-1}, and are found to be towards the upper end of this range in areas close to industrial sources of atmospheric S such as coal-fired power stations. Sulphur emissions from this source declined dramatically during the 1990s, such that some crops in eastern England are beginning to show signs of sulphur deficiency. Sea spray is another source of S. These two sources are sufficient in most cases to maintain adequate S levels in most cropping situations. Total levels may range from 200–2000 kg S ha^{-1}.

Soil reactions, availability and losses
Organic S is released by microbial decomposition, to give the plant-available form, SO_4^{2-}. In many respects, S behaves like P in soils, but is not subject to the problems of fixation that restrict P availability. Rainfall inputs and dry deposition (dust) from industrial sources account for the majority of crop requirements in many areas of the UK. Leaching losses do occur, but are balanced by natural inputs.

Micronutrients

Micronutrients are essential to plants, but found in small concentrations, often less than 100 mg kg^{-1} dry matter in plant material. A full list of micronutrients is given in Table 1.11. For a detailed discussion of micronutrient availability, the reader is referred to one of the standard texts on soils and plant nutrition, such as Wild (1993) or Marschner (1995). Trace elements include plant micronutrients as well as other elements that are not essential for plant growth, such as cadmium, nickel and lead. These will not be discussed further. Median total micronutrient concentrations of soils in England and Wales are given in Table 1.15. Values range widely according to the element, from <100 mg kg^{-1} dry soil (Zn, Cu, Co) to >10 000 mg kg^{-1} dry soil (Fe).

Deficiencies of micronutrients may occur in agricultural crops either due to an absolute lack of a particular micronutrient, or due to inadequate availability, which may, for example, be caused by overliming (see Fig. 1.15). In general, sandy soils and peats are most susceptible to micronutrient deficiencies. Deficiencies have become more common in recent years due to more intensive cropping and fertilizers becoming 'purer', that is containing fewer impurities. Major deficiency symptoms are described in Table 1.11. Further details relating to individual micronutrients, together with common problems are given below.

Table 1.15 Median micronutrient concentrations for all soils, England and Wales (after McGrath and Loveland, 1992)

Element	*mg (kg dry soil)$^{-1}$*
Cobalt	10
Copper	18
Iron	26 786
Manganese	577
Sodium	242
Zinc	82

Boron (B)
Boron is present in the rock mineral tourmaline, from which it is weathered to form plant-available borate (BO_3^{3-}). B accumulates in organic matter, and is freely available to plants in all except alkaline soils. Overliming can induce boron deficiency. Boronated fertilizers can be used to overcome these problems when growing B-demanding crops such as sugar beet. B is toxic to some plants at a level only slightly above that needed for normal growth. Leaf tips become brown and rapid necrosis of the whole leaf can follow. The problem is worse in dry areas and is exacerbated where irrigation water high in B is used. Sensitive crops include runner beans and grapes; semi-sensitive crops are barley, potatoes, tomatoes and legumes.

Copper (Cu)
Copper is released from rock minerals by weathering, and held on the exchange complex. Cu availability decreases with increasing pH, hence alkaline, sandy or mineral-poor organic soils may be Cu deficient. Cu deficiency may be overcome by a foliar application of copper sulphate. When present in excess, Cu^{2+} replaces other metal ions, particularly Fe^{2+}; root growth is restricted and chlorosis occurs.

Iron (Fe)
Iron is very abundant in soils, mainly in rock minerals, hydrated and non-hydrated iron oxides and as plant-available Fe^{2+}, which only occurs in small amounts in soils as this is the reduced form of iron. The oxidized form, Fe^{3+}, is insoluble and immobile. Iron deficiencies are uncommon, but can occur in fruit trees grown on well-drained, alkaline soils.

Manganese (Mn)

Manganese is abundant in most soils, being weathered from rock minerals to form the plant-available Mn^{2+} ion, which is retained on the exchange complex. Alkaline and freely drained soils can be deficient in plant-available Mn, while very acid soils and those that are poorly drained can contain high concentrations of Mn that may be toxic to plants. Excess Mn causes brown spots and uneven chlorophyll on older leaves. Silicon (Si) can minimize the harmful effects of excess Mn.

Chlorine (Cl)

As large amounts of Cl are applied with K in muriate of potash, the effects of excess Cl are common, particularly on soils that have been affected by salt. Cl toxicity is seen as burning of leaf tips, bronzing and premature yellowing of leaves. Sugar beet, barley, maize and tomatoes are tolerant, but potatoes, lettuce and many legumes are sensitive to excess Cl. Where there is a risk of a sensitive crop being affected, sulphate of potash should be applied in place of muriate.

Potentially toxic elements (PTEs)

Some elements such as iodine, fluorine, aluminium, nickel, chromium, selenium, lead and cadmium are not essential for plant growth and their presence can lead to toxicity. Sewage sludge (biosolids) contains a variety of PTEs, such as nickel, cadmium and lead, that can accumulate in soils and plants. Application of these materials is controlled by the Safe Sludge Matrix (ADAS, 2000).

Soil sampling and analysis

Soil sampling and description

Soil sampling and description are important procedures that must be carried out with care and precision. Fertilizer recommendations are often based on soil analyses, so it is important to ensure that samples taken are representative of the whole of a given field. Soils are naturally occurring materials. Chemical properties can vary widely from point to point within a field, even under 'uniform' agricultural management. Preparations for taking soil samples for analysis include an assessment of a field to be sampled. If there are apparent, systematic variations across the field, such as a break in slope or an obvious poorly drained area, then sample these areas separately.

Soil sampling is usually confined to the topsoil, which may be less deep in grassland than in arable land. For arable land, the top 15 cm should be sampled, whereas in grassland 10 cm is adequate. Sampling is carried out with a hollow circular corer, so that an uncontaminated sample can be collected. Sampling pattern is important. Figure 1.20 demonstrates a suggested sampling pattern to collect a number of subsamples that can be bulked together to derive one homogeneous sample representative of the whole field. In this example, 25 subsamples have been collected. In general, the more subsamples the better. Samples should be placed in clean dry bags that can be effectively sealed to prevent contamination.

Soil sampling strategy will be different if the objective is to determine the extent of variation across an area in order to provide information to help explain causes of yield variation. Such approaches are becoming more common with the wider adoption of precision farming techniques. In such situations, soil samples are usually taken every hectare, with sample locations chosen to reflect contrasts within large fields.

Soil profile description

In addition to sampling for laboratory analysis, field soil profile description is an important part of good soil management practice. Excavation of a profile face, if possible into the subsoil and ideally down to the soil parent material, will reveal much information regarding root proliferation, stone content, compacted layers, straw and other residue decomposition, drainage problems and so on. This physical evaluation is most important when considering soil loosening or a drainage operation.

Soil analysis

Soil samples are normally air-dried, crushed (to separate out, but not break the stones) and passed through a 2-mm sieve prior to analysis, which is carried out on the 'fine-earth' fraction. It is analysed to determine a variety of parameters, such as 'plant-available' nutrient concentrations, pH, organic matter content and soil texture. Routine soil analysis will cost £10–15 ha^{-1}, and should be conducted every 3–4 years.

Available or extractable nutrients

As described previously, the total amount of a given plant nutrient may be very large, and bear little relationship to the quantity to which a crop plant has easy access. Hence a variety of methods have been tried and tested to estimate the quantity of a given nutrient that is likely to become available to a crop during the next year. Methods used have been found by trial and error to predict reasonably closely, for most soil types, the

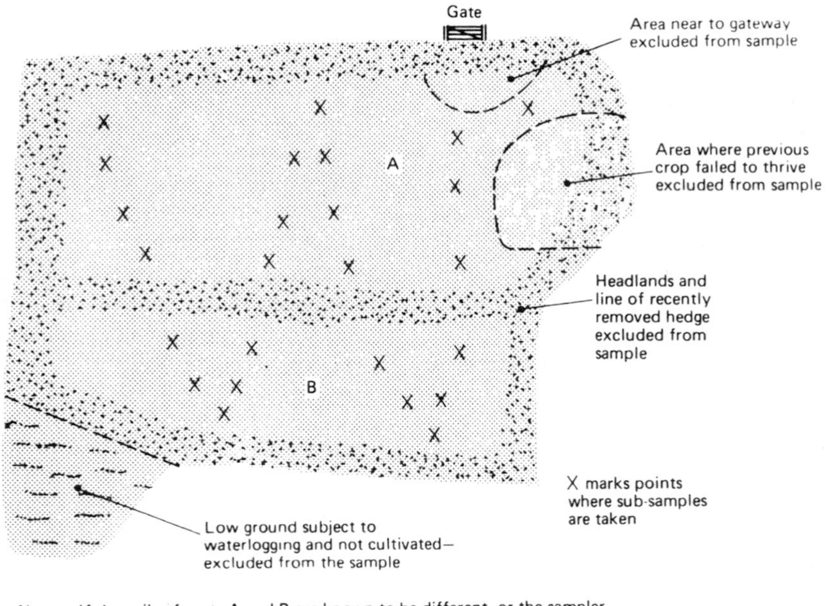

Figure 1.20 Example field soil sampling pattern.

nutrient release and hence the shortfall that is needed from fertilizer sources. Details of methods used by ADAS are given in MAFF (1986). The concentrations of available macronutrients, such as P (extracted using sodium bicarbonate), K and Mg (extracted using ammonium acetate) are usually expressed as an index. Determination of N availability based on soil analysis is subject to more variation due to soil mineralization/immobilization processes (see 'Nitrogen'), but recent advances in analytical techniques allow more confidence to be placed in estimates of soil mineral nitrogen (SMN) concentrations which are now used more often to predict N fertilizer requirements. The use of indices based on soil analysis is discussed in 'Fertilizer recommendations'.

Nitrogen

A substantial proportion of plant-available nitrate is released by mineralization of organic matter, which is difficult to predict as mineralization rate depends upon environmental conditions, notably soil temperature and water content, rather than release from purely chemically mediated processes as is mainly the case for P, K and Mg. Mineral nitrogen (nitrate and ammonium nitrogen, referred to as Nmin) concentration in soil can be measured, but the total amount of nitrate that might become available to a crop over a season cannot be so easily determined. It is becoming more common to measure mineral nitrogen in the soil profile in spring as an aid to planning N fertilizer strategy for high-value crops. In this case, soil samples are collected from each soil horizon down to rooting depth, usually 0–30, 30–60 and 60–90 cm. The total mineral N is calculated following analysis using moist soil samples from which the mineral N has been extracted by shaking with potassium chloride solution. This procedure will cost in the order of £20–30 ha^{-1} depending upon number of samples taken. Where analysis of soil mineral nitrogen is not carried out, the N status of the soil is derived based on previous cropping, including fertilizer and manure use, soil type and winter rainfall (MAFF, 2000a). See 'Fertilizer recommendations' for full details.

Trace elements

Trace elements are rarely determined in standard agricultural situations. Exceptionally, micronutrient concentrations are determined in soil material to ascertain deficiency problems. Such analyses are expensive and only warranted when analysing problems in high-value cropping systems. More usually, plant tissue analysis will be conducted when a micronutrient deficiency is suspected.

Nutrient sources: manures

The need for effective manure management

Approximately 90 million tonnes of farm manure is applied to agricultural land annually in the UK. This is in addition to direct returns from grazing animals. These animal manures have some value as they contain plant nutrients such as N, P, K and S. The controlled application of these manures to farmland can represent both a relatively safe and a money-saving method of manure use. The estimated fertilizer N replacement value of farm manures handled in the UK is approximately £60 million per annum. However, livestock manures represent a significant environmental hazard when handled incorrectly. There are two main environmental hazards associated with poor manure utilization:

- Point source water pollution incidents attributed to the agriculture industry continue to be dominated by failure to utilize solid manures and liquid slurries effectively. Slurries and manures with high biochemical oxygen demand due to soluble nutrients present cause oxygen depletion in water courses which may have an impact on aquatic life. Many of these problems result from inadequate storage or field spreading procedures.
- Farm buildings and field spreading operations are significant sources of ammonia and odour pollution. Ammonia released from manures to the atmosphere can cause acidification and nitrogen enrichment in adjacent semi-natural habitats, leading to a decline in species diversity. Odour pollution represents a serious nuisance factor.

During the 1980s and 1990s the livestock industry, with support from government grants, invested in improved handling facilities. Furthermore, MAFF has published the *Code of Good Agricultural Practice for the Protection of Water and of Soil* (MAFF 1998a, 1998c) which contains detailed guidance for avoiding pollution from farm manures. All livestock farmers are encouraged to produce a farm waste management plan for their farm, to aid effective utilization of the nutrients in manures (see 'Management aspects of use of livestock manures' for more details).

Factors affecting manure composition and efficiency of utilization

Animal type and diet

Before outlining the probable nutrient value of animal manures (including slurry) it must be stressed that all such materials can be extremely variable in composition according to the animal type, diet and storage of the manure. Full details are given in MAFF's *Fertiliser Recommendations for Agricultural and Horticultural Crops*, 7th edition (2000a), which is an essential reference for planning manure and fertilizer utilization programmes. Table 1.16 lists the approximate amount of excreta (undiluted) produced by livestock. Animal type and diet affect nutrient content. For example, animals such as dairy cows on high-protein diets produce manures enriched in N in comparison with fattening cattle. Poultry in particular produce manure with a very high N content, which can lead to ammonia loss problems from poultry units.

Table 1.16 Estimated quantities of excreta and nutrients produced by livestock during the housing period (from MAFF, 2000a)

Type of livestock	*Output during housing period*					
	Body weight (kg)	*Housing period (% of year)*	*Undiluted excreta (t or m^3)*	*Nitrogen (N) (kg)*	*Phosphate (P_2O_5) (kg)*	*Potash (K_2O) (kg)*
Cattle – dairy	550	50	9.6	48	19	48
Cattle – beef	400	66	6.2	31	12	31
Pig – sow + litter	130–225	100	4.0	20	20	16
Pig – bacon, meal fed	35–105	90	1.5	11	8	6
Poultry – 1000 laying hens	2200	97	41	660	545	360
Sheep – adult ewe	65	8	0.1	0.8	0.2	0.4

Composition changes following collection and storage

Nutrients are present in manures in a variety of forms, some that are plant available and some that are bound to organic matter. For reliable fertilizer planning it is important to know the manure nutrient content and availability. Important factors to take into account are:

- the dilution of the excreta, for example by parlour washings and rainfall onto yard surfaces
- the amount and composition of straw or other absorbent material used for bedding
- the duration and method of storage.

Gaseous loss of N as ammonia

The quantity lost will vary with the conditions of storage and temperature. Loss of 10% N is average, but loose-stacked farmyard manure (FYM) that is turned prior to spreading can lose up to 40%. Where slurry is stored for long periods, 10–20% of N can be lost and agitation before removal will aggravate the loss. Leaching losses from FYM stored in the open can be very variable, but are typically in the range 10–30% of N, 5–10% of P_2O_5 and 20–40% of K_2O: the flatter the heap the greater the likely loss. Seepage losses from slurry can result in 20–30% loss of N, a little loss of P_2O_5 and 20–30% loss of K_2O.

The above sources of loss can be additive, so that material that has been stored outdoors for a long period will have a greatly reduced manurial value due to gaseous, leaching and seepage losses. Table 1.17 states the estimated composition and likely nutrient value of livestock manures, based on MAFF (2000a). Both FYM and slurry tend to be very variable in dry matter and composition.

Other points to note are the relatively high K value associated with cattle manures and the high N value of poultry manure. An alternative to using standard estimates shown in Table 1.17 is to obtain an analysis of the manure or slurry. The availability of nutrients in manures and slurries varies according to manure type, time and method of application and soil type. Hence, calculating the fertilizer replacement value of manures is not straightforward, but is important for effective nutrient utilisation.

Crop availability of manure nutrients

Table 1.17 shows that the availability of some nutrients is less than the total applied in manure. For P and K, availability can be predicted with some reliability. It is usually assumed that 60% P and 90% K in manures will

Table 1.17 Typical nutrient content and availability (kg t^{-1} or kg m^{-3}) of solid manure (FYM) and slurry, on a fresh weight basis (based on MAFF, 2000a)

	Dry matter (%)	*Total N*	*Available N*		*Total P_2O_5*	*Available[2] P_2O_5*	*Total K_2O*	*Available[2] K_2O*	*Total S*	*Total Mg*
			Spring[1]	*Autumn[1]*						
FYM										
Cattle	25	6.0	1.2–1.5	0.3–0.6	3.5	2.1	8.0	7.2	1.8	0.7
Pig	25	7.0	1.4–1.7	0.3–0.7	7.0	4.2	5.0	4.5	1.8	0.7
Sheep	25	6.0	1.2–1.5	0.3–0.6	2.0	1.2	3.0	2.7	—	—
Poultry layer manure	30	16.0	5.6	1.6–3.2	13	8	9	8.1	3.8	2.2
Poultry broiler litter	60	30.0	9.0	3.0–6.0	25	15	18	16	8.3	4.2
Slurry										
Dairy	6	3.0	1.1	0.2–0.4	1.2	0.6	3.5	3.2	0.8	0.7
Beef	6	2.3	0.8	0.1–0.3	1.2	0.6	2.7	2.4	0.8	0.7
Pig	4	4.0	2.0	0.2–0.8	2.0	1.0	2.5	2.3	0.7	0.4
Dirty water										
Dairy	1	0.3	0.2	0.1	—	—	0.3	0.3	—	—

[1] Surface-applied manure of slurry; N availability is increased by incorporation (arable) or injection (grass).
Note: 1 kg P_2O_5 = 0.44 kg P; 1 kg K_2O = 0.83 kg K.
[2] Available refers to amount of a nutrient likely to be available to the next crop. The remaining nutrient may be released to crops in subsequent years.

become available to the next crop independent of time or method of application. The plant availability of N in manures spread in the field depends upon:

- time of application – autumn-applied manures are less effective due to winter leaching
- application technique – incorporation tends to result in more effective utilization of N because of reduced NH_3 losses
- soil type – losses are greater from sandy soils due to the poorer retention of nutrients and vulnerability to leaching

Management aspects of use of livestock manures

Farm waste management planning

Sound manure management relies upon detailed evaluation of manure production and nutrient value, which should be matched to land available on the farm to utilize the manure. The Water Code (MAFF, 1998a) gives guidance on how to prepare a farm waste management plan. Following the guidance will minimize the risks of water pollution by nutrients or pathogens. The ADAS MANNER (*MAN*ure *N*itrogen *E*valuation *R*outine) programme, available free from ADAS (www.adas.co.uk) provides specific information for manure nitrogen budgeting).

The key stages in completing a farm waste management plan are:

- Identify zones where manures should not be spread, such as within 10 m of a ditch or watercourse, very steep slopes, SSSIs.
- Identify areas where manures should not be spread under certain conditions, such as fields that flood, or sloping impermeable soils.
- Calculate how much land is available for spreading manures.
- Using stock numbers and standard areas needed to spread manure (for example 0.032 ha cow^{-1}) determine the minimum land area needed to spread the manure.
- Assuming the land available is greater than the land needed, then manure can be allocated according to cropping requirements. Total application of manure should not exceed the MAFF recommended limit of 250 kg N ha^{-1} $year^{-1}$, which is equivalent to approximately 40 m^3 ha^{-1} cattle FYM or 80 m^3 ha^{-1} cattle slurry (see Table 1.17).

Animal health risks from application of livestock manures

Hypomagnesaemia in grazing stock

Farmyard manure and slurry from ruminant stock are rich in potash and where high rates of such manures are applied to grassland in late winter and spring, there will be an increase in the potash content of the grass. Spring herbage is naturally low in Mg, and as a high plant content of K depresses Mg content, then there is an enhanced risk of hypomagnesaemia in animals grazing grass that has received cattle manures. Where such manures have to be applied to grass that is allocated to spring grazing, then it should be applied well before Christmas.

Animal disease

The main hazard that exists is from the infection of pasture with bacteria of the *Salmonella* group – normally as a result of applying infected cattle or poultry manures and slurries. In order to minimize the risk:

- Store the manure for 2–3 weeks before application, as most problems arise where fresh material is applied to grass.
- Allow rain to wash the manure from the herbage before grazing.
- Ensure watercourses are kept free from contamination.

Toxic elements

The main risk is from manure obtained from fattening pigs, as this can contain high levels of Cu and Zn, both derived from feed additives. Cu and Zn build up slowly in the soil and frequent application of pig manure to the same field can lead to potential toxicity problems. Particular risk arises on grazed grassland, where evidence suggests that slurry contamination on the herbage is more at fault than high Cu and Zn levels in the herbage itself: so physical contamination should be allowed to disappear prior to grazing. Specifically, sheep should never be allowed to graze grass contaminated with Cu as extensive liver damage can occur.

Manure management in specific farming situations

Mixed grass/arable farms

Where suitable arable crops are grown, the most beneficial crops for high application rates are potatoes, maize and cereals. Following a manure use survey, Chalmers *et al.* (1998) noted that the area of crops receiving manure annually in the UK was as follows:

	%
Winter wheat	12
Winter barley	13
Oilseed rape	9
Sugar beet	24
All tilled land	19
Grassland	44

Manures applied to arable land will supply not only nutrients but also organic matter, which can help to enhance soil fertility (see 'Soil biological properties').

All-grass farms

All-grass farms produce more manure per total farm hectare than mixed grass/arable farms and have fewer suitable crop situations for its use. If storage does not exist for the whole winter's production, then the following should be considered:

- In November/December apply to next year's spring grazing area;
- In January–March apply to areas for spring conservation cuts;
- In May/June apply thinly to areas cut for conservation in spring.
- High rates applied to grass can cause physical shading of the grass and loss of tillers, particularly if the grass is not growing rapidly at the time.
- Cattle manures are high in potash and their use in spring can promote hypomagnesaemia.
- Most manures cause taint and subsequent refusal problems by grazing stock: some farmers find sheep graze behind cattle slurry better than cattle.
- There is a *Salmonella* risk where fresh cattle slurry is applied to grass regrazed by cattle.
- Pig slurry is often high in Cu and Zn: grass should be free from slurry contamination before grazing, particularly with sheep.

Pig and poultry units

Pig and poultry units are always 'exporters' of manure and this manure requires disposal at regular intervals throughout the year as the stock are nearly always housed. A mixed farm can more easily accommodate and efficiently utilize pig and poultry manure than either an all-grass or an all-arable farm. Poultry manure has a higher N content than other manures and is valuable for application to grassland in the March–September period.

Other organic manures

Straw

Straw can be incorporated into the soil, preferably after it has been chopped, and this will provide some phosphate and potash, but, as noted above, it may cause a deficiency of N that will need correction in the first year or so. Where manure contains much fresh, unrotted straw, it is possible that the soil bacteria breaking down the straw need more N than is contained in the manure: thus the material can deplete N status in the short term. To correct this, additional fertilizer N may be used at the rate of 10 kg of N for every 1 t of fresh straw used. Note that this N should be applied to the next crop and not to the manure or the soil at the time of manure application. However, in the long term, regular annual incorporation of straw will stimulate the N cycle in the soil so that 20–30 kg ha^{-1} of extra available N per year can be anticipated.

Composts

Composts can be made from straw alone, straw mixed with another crop waste, or from other wastes alone. Stacking and turning solid FYM accelerates the rate of breakdown and stabilization by aerobic composting. During the process micro-organisms decompose the manure, generating temperatures up to 70°C which will suppress pathogens and render weed seeds non-viable. The resulting composted manure will be considerably reduced in mass, resulting in less material to spread. Composted manure contains more nutrients on a fresh weight basis than fresh manure, but the total nutrient content will be less due to NH_3 volatilization and NO_3 and K leaching losses.

Greenwaste compost from domestic sources (hedge and grass clippings, for example) is being increasingly used in agriculture. As a general rule all such materials tend to be woody and hence lack available N for crops. Example analyses for composts, sewage sludge and other organic materials are given in Table 1.18. For adequate

Table 1.18 Composition of some composts and organic materials

	Dry matter (%)	*Composition as spread (kg t^{-1} or kg m^3)*		
		N	*P_2O_5*	*K_2O*
Greenwaste compost	70	12.0	7.0	6.0
Fresh seaweed	20	2.0	1.0	12.0
Digested liquid sludge	4	2.0	1.5	Trace
Thermally dried sludge	95	35.0	45.0	Trace

bacterial decomposition additional soluble N, such as from ammonium nitrate, at a rate of up to 20 kg fertilizer per tonne of composting material might be needed to prevent N starvation in the crop. Greenwaste composts have been reported to stimulate the soil microflora and suppress some crop diseases such as take-all in wheat and club root in brassicas. In general, greenwaste composts should not be used just for their nutrient value, but also for the beneficial effects these materials can have on soil fertility and disease suppression.

Sewage sludge is often available, either from cesspits on the farm or from urban treatment works. Depending on the treatment process different types are produced: liquid digested sludge (material taken straight from sedimentation tanks) is not recommended for use in agriculture according to *The Safe Sludge Matrix* (ADAS, 2000). Sludges can contain significant amounts of N and P (see Table 1.18), but the availability, particularly for N, can be difficult to predict (MAFF, 2000a). Aerobic or anaerobic digestion to remove oils and reduce pathogens produces digested liquid sludge. The above materials may be dewatered and may also be conditioned with lime or other chemicals. Domestic sewage sludge often contains high levels of potentially toxic elements, notably Zn. Sludges arising from industrial plants should be analysed as one or more of some 12 high-risk elements may be present.

Fresh seaweed should not be confused with the liming material calcified seaweed. Seaweed is rich in potash (and sodium) as shown in Table 1.18 and a dressing of 25 t ha^{-1} would provide some 300 kg ha^{-1} of K_2O. Sometimes seaweed is used to make composts, but, like so many organic manures, it is bulky material and expensive to transport. Thus seaweed is used only in close proximity to the coast and is valuable for some horticultural crops and potatoes. Also it is used to provide humus on some light soils.

Nutrient sources: inorganic fertilizers

Nitrogen fertilizers

Ammonium nitrate

This is a widely used fertilizer that is supplied as a straight fertilizer or as a component in compound fertilizers. This material is very soluble in water and also hygroscopic, so it must be stored in a dry place in sealed bags. Half the N is in ammonium form and half in nitrate form; in this way it has only half the acidifying effect of ammonium sulphate for any given amount of N applied. Practically, pure ammonium nitrate is sold as a straight fertilizer containing 33.5–34.5% N. The composition of ammonium nitrate in comparison with other N fertilizers is given in Table 1.19. Ammonium nitrate

Table 1.19 Composition of major N, P and K fertilizers

	Formula	*N(%)*	*P_2O_5(%)*	*K_2O(%)*
Nitrogen fertilizers				
Ammonium nitrate	NH_4NO_3	33.5–34.5	—	—
Urea	$CO(NH_2)_2$	46	—	—
Ammonium sulphate	NH_4SO_4	21	—	—
Calcium ammonium nitrate	$NH_4NO_3 + CaCO_3$	26–28	—	—
Anhydrous ammonia	NH_3	82	—	—
Aqueous ammonia	$NH_3 + H_2O$	18–30	—	—
Phosphate fertilizers				
Superphosphate	$Ca(H_2PO_4)_2.CaSO_4$	—	18–20	—
Triple superphosphate	$Ca(H_2PO_4)_2$	—	47	—
Basic slag	Complex CaFe phosphate	—	8–10	—
Rock phosphate	$Ca_5(PO_4)_3F$	—	30–35	—
Mono-ammonium phosphate	$NH_4H_2PO_4$	12	52	—
Di-ammonium phosphate	$(NH_4)_2HPO_4$	18	46	—
Potash fertilizers				
Muriate of potash	KCl	—	—	60
Sulphate of potash	K_2SO_4	—	—	50
Kainite	KCl, $MgSO_4$ + other salts	—	—	12
Silvinite	KCl + other salts	—	—	21

is a very powerful oxidizing agent and if subjected to heat or flame it can explode. Therefore straight ammonium nitrate and compounds containing a high proportion of ammonium nitrate should not be stored in barns with hay or straw and never stored in bulk, as the risk of explosion is thus enhanced. Being soluble in water, ammonium nitrate is often used as a major N source in liquid fertilizers.

Urea

Urea is the most concentrated solid source of N that is currently available on any wide scale. It is very soluble in water and hygroscopic. Sometimes its granules are softer than those of ammonium nitrate and do not spread so well. When urea is applied to most soils it breaks down rapidly to NH_4^+. However, gaseous ammonia (NH_3) can be formed if the soil contains free Ca, or is very dry, resulting in loss to the atmosphere. If the ammonia is lost to the atmosphere, then the fertilizer is less effective as a source of N for crop nutrition; this loss of ammonia is most likely when the soil is alkaline and if the urea is applied to the soil surface in dry weather. In seedbeds the high concentration of ammonia near germinating seeds can cause toxicity and serious loss of plant stand. On poorly structured soils liable to surface capping the released ammonia can be trapped in the soil and not only kill germinating seedlings but also severely check the growth of roots of established plants.

Urea is best used on acid-to-neutral soils, surface-applied to established crops during periods of frequent rainfall. Thus, in the UK, urea has a place on grassland and for top dressing cereals in spring. Compared with ammonium nitrate, the cost per unit of N in urea varies widely. Each tonne of urea contains 33% more N than ammonium nitrate. It follows that even if it is assumed that N in urea is 15% less effective than N in ammonium nitrate, urea can be up to 20% more expensive but still a better buy. Urea is widely used as a major N source in liquid fertilizers, as liquid application overcomes some of the physical problems of urea noted above, and when mixed with phosphoric acid to make a compound fertilizer, the acid nature of the material further reduces the risk of ammonia loss.

Ammonium sulphate

Ammonium sulphate was once the most important source of inorganic N; it is soluble in water and although all the N is in NH_4^+ form, it is quick-acting under field conditions. However, because of its high ammonium ion content it has an acidifying action in the soil. While the N concentration is limited to about 21%, it does contain S (60% as SO_3). Ammonium sulphate has been replaced by ammonium nitrate in the UK and some other temperate countries and by urea in tropical and subtropical countries and some high-rainfall temperate areas. The reduction in S deposition from atmospheric sources during the 1990s has lead to an increased incidence of S deficiency in cereal crops, leading to more interest in ammonium sulphate as a combined N and S fertilizer.

Calcium ammonium nitrate

This is a mixture of ammonium nitrate with calcium carbonate and is sometimes called nitrochalk. It has a higher N content than ammonium sulphate and has little acidifying action in the soil. Also the hygroscopic and explosive properties of ammonium nitrate are nullified by the presence of chalk, making it easy and safe to store, even in bulk. However, calcium ammonium nitrate tends to be more expensive per unit of N purchased than ammonium nitrate.

Anhydrous ammonia

Anhydrous ammonia is the most concentrated source of N available, being pure ammonia. At normal temperature and pressure this material is a gas, but when stored at high pressure it is a liquid containing the equivalent of 82% N. At this concentration it is very economical to transport on a weight-to-value basis, but the whole process of storage, transport and application needs special equipment, both to maintain up to the point of application the high pressure needed and to prevent hazards from uncontrolled loss of ammonia. Anhydrous ammonia is applied below the surface of the soil using special injection equipment, and because of the high cost of this operation it is economical to apply only fairly high rates of N (at least $100\,kg\,N\,ha^{-1}$) at each application. Also it is essential that loss of ammonia to the atmosphere be kept at a minimum during application by careful sealing of the slits. For these reasons the use of anhydrous ammonia is best restricted to row crops that need high individual N dressings. The very high concentration of ammonia released in the soil in the vicinity of the slits causes partial sterilization of the soil and has the effect of retarding the action of the bacteria that convert NH_4^+ to NO_3^- so that anhydrous ammonia can act as a slow-release fertilizer. Although widely used in some countries where suitable crops are grown in stone-free, friable soils, anhydrous ammonia has not established itself in the UK, mainly because it was not found suitable for grassland due to sward damage caused by injection equipment and loss of ammonia from the slits in the soil.

Aqueous ammonia

Aqueous ammonia is a solution of ammonia in water, a solution at normal pressure containing 18–30% N.

Aqueous ammonia has most of the advantages of the anhydrous form except high N concentration, but to offset that it does not need such specialized equipment. In crop situations where a liquid straight N fertilizer is required, then aqueous ammonia has an important place. Its concentration can be increased by: (1) partial pressurization, where storage and application under only a modest pressure can enable concentration to increase to 40% N; and (2) mixture with urea and/or ammonium nitrate to give 'no pressure' solutions of about 30% N. To prevent gaseous loss of ammonia all these materials should be either injected into the soil or worked in immediately after application: they should not be used on calcareous soils.

Potassium nitrate, sodium nitrate and calcium cyanamide

Potassium nitrate (14% N) and sodium nitrate (16% N) are both very quick-acting and highly soluble forms of N which are expensive but valuable in some horticultural crops and in particular for use in foliar feeds. Calcium cyanamide (21% N) is not important in the UK but is a fertilizer containing N in both amide and cyanide forms. It is soluble in water, and in the soil it is converted to urea: during this conversion some toxic products are released which can kill germinating weeds and slow the rate of nitrification. Breakdown requires water and so the product is not effective in dry conditions.

Slow-release N fertilizers

Most of the commonly used and cheaper forms of N are rapidly available to the growing plant and yet many crops would gain greatest benefit from applied N if it were available over a period covering most of the vegetative growth of the plant. This is probably the most convincing reason for considering organic N sources. When inorganic sources are used there are three ways in which this objective can be met:

(1) Apply frequent small dressings of conventional N fertilizers during vegetative growth, for example as with spring N applications on some winter wheat crops and on grassland.
(2) Use a conventional fertilizer material that has received a coating to reduce its rate of breakdown into plant-available forms, for example sulphur-coated urea.
(3) Apply complex compounds of N, which require considerable chemical change in the soil before they are available to the plant.

There is much research to find suitable materials in the last of the above categories. To be successful such a compound must be economic in price and supply available N at a rate required by the crop. The materials on the market tend to be expensive, make available to the crop a relatively low proportion of the total N they contain and have rates of release controlled by soil temperature and moisture so that during humid weather most of the N is released too rapidly for full crop benefit. Among synthetic slow-release fertilizers are:

- urea formaldehyde (ureaform, 40% N): this is made by reacting urea with formaldehyde. The rate of breakdown to release available N is controlled by soil bacteria;
- isobutylidene di-urea (IBDU, 32% N): rate of release of available N depends on differential particle size and soil moisture status;
- sulphur-coated urea (36% N) and resin-coated ammonium nitrate (AN) (26% N): both depend on differential times for the coatings to break down before releasing available N.

Urea and IBDU are used in commercial glasshouse production because not only is the slow rate of N release of particular value to some of the crops grown, but also the rate of breakdown can be controlled to a large extent by adjustment to the management of the house.

Phosphorus fertilizers

Superphosphate

Superphosphate is made by treating ground rock phosphate with sulphuric acid and producing water-soluble monocalcium phosphate and calcium sulphate. Usually some rock phosphate remains in the product as the quantity of sulphuric acid is restricted to prevent the final product containing free acid. Superphosphate was once the most widely used form of water-soluble phosphate and formed the basis of most compound fertilizers. Because of its relatively low P_2O_5 content (see Table 1.19) it limited the concentration of phosphate in compounds, but because of the $CaSO_4$ present it did apply useful quantities of sulphur.

Triple superphosphate

This is made by treating rock phosphate with phosphoric acid. This produces mainly water-soluble monocalcium phosphate with no calcium sulphate. Thus, triple superphosphate contains 47% P_2O_5 but no sulphur. Because of its high concentration, triple superphosphate is used widely in the production of high-analysis compound fertilizers.

Ammonium phosphate

This is made by adding ammonia to phosphoric acid, producing both mono- and di-ammonium phosphates. These compounds are very soluble in water and quick-acting in the soil, and supply both N and P in a highly available chemical combination. They are not used as straights to any extent because of the few crop situations that require a low N:high P ratio, but they do form an important base for many compound fertilizers.

Basic slag

Basic slag is a by-product of the steel industry and contains phosphates, lime and trace elements. In the smelting process, P contained in iron ore is held in the furnace bound to CaO, becoming a solid 'slag' as the material cools. The phosphate is in complex chemical combination with calcium and has to be ground before it can be used as a fertilizer. Although the phosphate in slag is not water soluble, it has been found that it can release phosphate slowly over a period of years on soils of pH range 4–7, although the rate of release is greater on acid soils. The trace element content of basic slag is also important, particularly Mg, Fe, Zn, Si and Cu.

Modern steel plants do not produce slags containing much P, although some older plants (mainly outside the EU) can provide slag with 8–10% P_2O_5. None of the 'slag substitutes' on the market seems able to emulate all four of the main attributes of traditional slag, namely:

(1) steady release of available P over several years on acid, neutral and alkaline soils
(2) supply of valuable trace elements
(3) some liming value
(4) cost per unit of P about half the cost of water-soluble P.

Rock phosphates

These are now being offered to farmers as a source of phosphate where rapid release of P is not essential but soil P status needs to be maintained. Rock phosphates will only release P under acid conditions, although some experiments have shown that soil pH from conventional analysis is not necessarily a true indication of the pH that might exist around root hairs, so that at times rock phosphate has shown some value on neutral soils. Rock phosphates vary considerably in their composition and potential agronomic value. In all cases they must be ground finely to have any value at all, current legislation being that 90% of the material should pass a 100-mesh sieve. Some of the hard crystalline apatites have very little fertilizer value even when crushed finely and used under acid conditions. On the other hand, soft rock phosphates such as gafsa do have value and are regarded as the best of untreated rock phosphates. Rhenania phosphate is produced by disintegrating rock phosphate with sodium carbonate and silica in a rotary kiln: this 'sintered' product contains much of the P in the form of calcium sodium phosphate and as such the P is rendered a little more available than in the original rock. Senegal rock phosphate contains aluminium phosphate, and when this is heated in a kiln and then allowed to cool in a humid atmosphere, an expanded type of rough granule is produced that greatly enhances the surface area of the material in contact with the soil after application. This calcined Senegal rock is claimed to be an effective P source, even on alkaline soils.

No rock phosphate has a rapid release of available P, so that these materials should not be used in situations where the need of the crop for P is rapid, for example at crop establishment or on crops that have a big demand for P, such as potatoes. Also, the crop recovery of P from rock phosphates is less than the recovery from water-soluble sources, so that the amount of useful P in rock phosphate is less than in the water-soluble types: this fact should be remembered when the costs of rival products are considered. As a guide, the cost of each kilogram of P_2O_5 in a water-insoluble phosphate should be less than half that in a water-soluble product before it becomes a better buy for most uses.

Others

In addition to the above, which can all be used in solid forms, there are a number of liquid products based on polyphosphates. The basic component of these solutions is superphosphoric acid, made from orthophosphoric acid and one of a number of polyphosphoric acids. An ammonium compound is used to neutralize the appropriate acid, and N:P solutions of ratios in the order 11 : 37 can be obtained. These products are expensive but do provide a highly concentrated source of available P in liquid fertilizers. There can be problems when potassium salts are mixed with polyphosphates, as they can cause crystallization and precipitation.

Potassium fertilizers

Muriate of potash (potassium chloride)

This is mined and is sold in either a powdered form or a fragmented (granular) form. It is the source of K in nearly all compound fertilizers and its use as a straight fertilizer in the UK is very limited. Where crops require substantial amounts of K it should be remembered that equal quantities of Cl are also applied if muriate of potash is used. Hence in a crop such as potatoes it is sometimes advisable to use potassium sulphate instead of muriate (see Table 1.19).

Sulphate of potash and other K fertilizers

This is made by treating muriate with sulphuric acid and so is a more expensive source of K unless S is required (contains 45% SO_3). However, for Cl-sensitive crops the extra cost of the sulphate salt is recommended. Some manufacturers make a series of compound fertilizers containing potassium sulphate and these are widely used by horticulturists, not only because of the fact that some crops are sensitive but also because of the risk of a buildup of Cl^- ions in the soil where regular heavy manuring is carried out. See Table 1.19 for composition of K fertilizers. Other K fertilizers include:

- Potassium nitrate (saltpetre) – this is expensive and its main fertilizer use is in foliar applications on fruit trees and some horticultural crops.
- Potassium metaphosphate – this compound is a water-insoluble source of K (unlike all the other sources mentioned which are soluble in water), and it can be used where it may be necessary to keep ionic concentrations at a low level in the root vicinity.
- Potassium magnesium sulphate, magnesium kainite and silvinite – these are sometimes used where some Mg is required in addition to K; also kainite is often used for sugar beet because of the Na content.
- Agricultural salt (mainly NaCl) – this can replace KCl in some situations, for example where sugar beet or mangolds are grown. Sodium chloride should not be applied to poorly structured heavy soils as it may promote deflocculation, and it should not be applied less than 3 weeks before sowing lest it affect germination: in fact it is often applied in autumn and ploughed in. Application of Na reduces the need for K.

Nutrient sources: organic fertilizers

Nutrient availability from slow-release fertilizers

Organic fertilizers are derived from either plant or animal materials. Not all the nutrients contained in such materials are in organic form and those that are in organic form are not readily or completely available to plants. Complex organic compounds will become part of the soil organic cycle and could perhaps have an eventual nutrient value, depending on the activity of the soil biomass.

Compared with inorganic sources of nutrients, organic sources have the following features:

- They are not immediately soluble in water and so not readily leached.
- Because they have to break down to become partially soluble, they can act as a slow-release source of plant nutrient.
- They can be applied at heavy rates without risk of injury to roots or germinating seeds as they have little ionic activity.
- They can stimulate microbial activity.
- They are much more costly per unit of plant food (unless by-products).
- The recovery of nutrients contained in the materials is low.

Because of the above features, organic fertilizers have restricted use in cropping systems that use mineral fertilizers, but they do have a place in organic production systems, particularly market gardening and horticulture where the slow-release characteristics have application for some of the high-value crops that are grown. See Table 1.20 for a summary of the nutrient sources permitted by the Soil Association *Standards for Organic Food and Farming* (1999). Note that most of the 'more available' nutrient sources in this table are restricted; that is use is only permitted after prior approval.

A number of regulations have come into force since the BSE crisis that restrict the use of animal by-products in organic fertilizers. Current regulations restrict the use of meat and bone meal products to protected cropping systems in order to minimize the risk of transfer of BSE to livestock. The use of blood products is under review and legislation may change in the near future.

Organic N, P and K fertilizers

- *Hoof and horn meals, hoof meal, horn meal* – these contain 12–14% N and can be obtained either coarsely or finely ground: fine materials release N more quickly. They should be worked into the soil before planting or sowing.
- *Dried blood* – this contains 12–13% N. It is very expensive but of great value in glasshouse crops where it is quick-acting.
- *Shoddy (wool waste)* – analysis varies from 3–12% N, depending on the proportion of wool contained

Table 1.20 Nutrient sources permitted in organic production systems in the UK (after Soil Association, 1999)

Nutrient	*Source*	*Relative availability and utilization*
N	Blood meal; hoof and horn meal	12–15% N; work into soil before sowing. Restricted
	Composted manures	Slow release N, some P, S; from organic sources
	Plant waste materials and by-products e.g. greenwaste compost	Slow release, may suppress plant diseases
P	Meat and bone meals	20–25% P_2O_5 (and 3–5% N). Insoluble. Restricted
	Rock phosphate	Very slow acting, variable composition
	Basic slag	Variable composition. Restricted
K	Rock potash, adularian shale	Use if it has low Cl content and low solubility
	Sylvinite	More soluble, also Mg source. Restricted
	Wood ash	Variable composition depending on storage
Ca and Mg	Epsom salts	Soluble, rapid acting source of Mg
	Calcareous magnesium rock; gypsum; ground chalk and limestone; calcified seaweed	Speed of action depends on particle size
S	Flowers of sulphur	Fast acting, apply to sandy S-deficient soils. Restricted
	Potassium sulphate	If K index below 2 and sandy soils. Restricted
Trace elements	Liquid and dried seaweed	Use based on evidence of deficiency, e.g. soil analysis. Foliar sprays restricted

amongst other wastes. Shoddy is a very slow-release material, should be worked into the soil before planting and should be analysed before purchase.

- *Bone meals* – these contain 20–24% P_2O_5 (insoluble in water) together with 3–4% N. The phosphate acts very slowly and is of most value on acid soils. Use is restricted to protected crops.
- *Steamed bone meals and flours* – these contain 26–29% P_2O_5 (insoluble in water) together with about 1% N. These materials are made from bones that are steamed to obtain glue-making substances and a good deal of the N is removed in the process. The bones are generally ground after extraction and the phosphate acts more quickly than in ordinary bone meal. Use is restricted to protected crops.
- *Meat and bone meal (also meat guano or tankage)* – these contain 9–16% P_2O_5 (insoluble in water) together with 3–7% N. They are made from meat and bone wastes and analysis varies; the phosphate has slow availability. Use is restricted to protected crops.
- *Fish and meals and manures (also fish guano)* – these contain 9–16% P_2O_5 (insoluble in water) together with 7–14% N. They are waste products from fish processing. Use is restricted to protected crops.
- *Organic potash fertilizers* – potassium does not occur in chemically organic form, but it does occur in organic materials (such as livestock manures and bird guano). Also potassium occurs in some 'natural' materials (such as wood ash, mica and adularian shale).

Forms of fertilizer and application methods

Solid fertilizers

Most fertilizers are applied in solid forms, and normally manufacturers go to great lengths to ensure that the material stores without becoming compacted and spreads evenly and freely at the time of application. Water-insoluble materials, such as some types of phosphate, have to be in solid form, but most other materials could be sold in liquid form if necessary. The main advantages of solid types are:

- high concentration of nutrients in the material, reducing transport and storage costs and leading to a high rate of work at spreading;
- when sold in 50 kg plastic bags, very cheap storage is possible and the material will keep in good storage condition for many months. One tonne open sacks can be easily handled by tractor-mounted forks, although storage life is shorter as bags are open to the atmosphere so hygroscopic materials may absorb water;
- when sold in bulk, relatively simple modifications to existing buildings will enable successful storage and make possible spreading systems that have low labour requirements;
- the farmer can buy large quantities at times when market conditions are favourable;

- the farmer can have a range of fertilizer types available, can use only that quantity which is needed at a given time, and can judge the application rate easily by counting the number of bags used.

Where water-soluble ingredients are used in solid fertilizers, it normally takes very little soil moisture to render these ingredients available. Similarly, where the same ingredients are applied to a dry soil, the quantity of water applied with the fertilizer is so minute that it has no irrigation effect. So it can be assumed that once brought into contact with the soil, liquids behave in the same way as comparable solid fertilizers, and generally no differences are observed in relation to growth and crop yields. The Fertilizer Manufacturers' Association (2000a,b) gives industry recommendations regarding quality and use of fertilizers.

Essential to the effective use of fertilizers is the calibration of field spreaders. Problems of uneven crops, lodging, reduced yield and nutrient losses to watercourses at field margins can be attributed to lack of maintenance of machinery and poor spreading practice. Regular calibration and appropriate care when spreading are vital in order to minimize losses and maximize beneficial uptake of nutrients from fertilizer sources.

Solid materials are sold as powders, granules, prills or fragments.

Powders

Powders can vary in degree of fineness and for some materials (e.g. water-insoluble phosphates) powders are essential to ensure activity of the material in the soil; powders can be difficult to spread evenly and this is critical if relatively low rates per hectare are needed; under these conditions full-width spreaders are advised.

Granules

Granules are made during manufacture by passing the fertilizer powder through a rotary drum, often with an inert granulating and coating agent; sometimes several fertilizer materials are mixed at the same time, so that each granule contains a mixture of materials. All granules made in this way are designed to break down quickly in the soil and release their nutrients, but sometimes the inert carrier remains visible in the soil for some days. The size and texture of the granules will have a marked effect on the sowing rate and spreading width of fertilizer distributors, and both these granule characteristics are specific to the type of material being granulated and the conditions under which the granulator is working: it follows that granules will vary in spreader performance and each consignment should be checked when being used.

Prills

Prills are made by the rapid cooling of a hot fertilizer liquid, are homogeneous in character and usually have a very smooth surface. Like granules they can vary in size, and great care is necessary at spreading time as prills usually flow more freely than granules and spreader-setting is critical to get correct rates and even spread. Prills normally consist of one fertilizer material only but this may contain more than one nutrient, for example a prilled ammonium phosphate would contain both N and P.

Fragments

Fragments are made by very coarse grinding of a solid material, usually followed by a screening process to produce material of a certain size range, for example some samples of muriate of potash are made in this way. Fragments behave in a similar way to very uneven granules (but note that granules are always spherical in shape because of the way they are made), but fragments and granules should not be mixed and spread together as marked segregation will occur both in the spreader and during the spread itself. Some fragmented material has a very abrasive action on working parts in the spreader and can cause rapid wear.

Most manufacturers take great trouble to ensure compatibility within each fertilizer type, but mixing of different fertilizer types on the farm prior to or during application can lead to poor spreading performance.

Liquids and gases

Some materials must be used in liquid form, either because they do not exist as solids at normal temperature and pressure (such as polyphosphates) or because their intended use precludes solid form (e.g. foliar nutrient application).

The main features of liquid fertilizers are:

- easy handling on the farm, as they can be pumped in and out of store and spread by sprayer fitted with non-corrosive working parts and nozzles;
- relatively low concentration of nutrients, as some chemicals will crystallize out during storage and at low temperatures if concentration is raised;
- storage tanks required, which can be expensive; this can mean that storage capacity is restricted so that an annual requirement for fertilizers cannot be purchased at the lowest price;
- the possibility that herbicides can be mixed with fertilizer to further reduce spreading costs.

In the UK liquid fertilizers hold a fairly small share of the total fertilizer market, unlike the situation in some other countries. There is no simple guide as to whether a farmer should use liquids or solids as the decision rests on several aspects of general farm management, such as fertilizer purchasing policy, availability of existing stores for solid fertilizer storage, contacts for contra-trading with merchants selling either solid or liquid fertilizers, types of crop grown and labour and/or machine availability for spreading. It is likely that the use of liquid fertilizers will continue to increase, particularly as a result of changing farm management practices, such as contracting out fertilizer application as a means of both saving costs and obtaining access to up-to-date technology in terms of application equipment and products.

As mentioned earlier, anhydrous ammonia in gaseous form can be used as a source of N. However, because of the high pressure needed for storage and the special injection equipment required, this is a contractor-based service. It is valuable for relatively few crops and soils in the UK.

Compound fertilizers

When fertilizers are sold with a single main chemical ingredient, they are known as straights: examples are ammonium nitrate, superphosphate and muriate of potash. Sometimes straights can supply more than one nutrient, for example ammonium phosphate supplies N and P and potassium nitrate supplies N and K. Occasionally straights may comprise a mixture of ingredients, but if they still supply only one main nutrient they are still known as straights. For example some phosphatic fertilizers contain a mixture of water-soluble and water-insoluble phosphate, but still supply only P.

Many crops need more than one of the three major nutrients N, P and K, and in most of these situations it is convenient to apply all the nutrients at the same time. For this purpose, manufacturers have produced a wide range of compound fertilizers, that is fertilizers that contain more than one ingredient and supply more than one nutrient. Compound fertilizers are more expensive than straights, but many farmers find the extra cost is justified because if several straights were used the cost of application would be increased and it is difficult to mix and effectively spread straights on the farm. As a guide, the extra cost of nutrients in a compound compared with straights should reach 15% before the compound looks expensive. The composition of a compound fertilizer is stated in terms of the ratio of $N:P_2O_5:K_2O$. For example, a common general-purpose compound fertilizer includes these three macronutrients in a ratio 20:10:10. A compound for use on grass aftermath, a situation that demands high N and K input, might contain 25:0:15. A compound designed to supply mainly P and K in a cereal seedbed might contain 0:15:15.

Fertilizer placement

Band placement

Most fertilizers are spread evenly over the surface of the soil or crop, and are used at the period when the crop is most in need of the nutrients contained. For example, PK fertilizers are worked into a seedbed prior to sowing and N fertilizers are top-dressed on to some growing crops, particularly winter cereals in the spring and on to grassland.

In some row-crops, placing fertilizer in a band near to the zone of the soil in which the majority of the plant roots will feed can both reduce the total quantity of fertilizer needed and check weed growth outside the crop rows. Fertilizer placement in this way tends to reduce the speed of fertilizer application and if carried out at sowing time by combine-drilling will also slow down the rate at which drilling can be carried out. Also, if the fertilizer rate is too high or too close to the seed or plant root, the material can cause scorch and this can reduce plant survival and check growth in established plants. Band placement of fertilizer is most advantageous where:

- soil nutrient status is very low, for here responses to placed fertilizer can be greater than to very high rates evenly spread over the field;
- crops need very high rates of nutrient application and are grown in discrete rows set well apart. Here, placement not only reduces the overall amount of fertilizer needed but also ensures the roots come into contact with sufficient material. A good example of this is the main-crop potato grown on normal mineral soils;
- the applied nutrient has limited mobility in the soil: this applies particularly to P that scarcely moves in the soil even when applied in water-soluble form.

Rotational placement

Some soils are reasonably well supplied with nutrients, but further fertilizer application is recommended to maintain the adequate level. This situation applies particularly to P and K. It is thus suggested that it may not be necessary to use P and K fertilizers for every crop but adopt a rotational fertilizer programme where PK fertilizers are used periodically and used immediately before a PK-sensitive crop is grown. This approach can save money but needs careful monitoring to ensure that soil nutrient levels do not decline: also it should be noted that PK fertilizers will be needed on some fields on the farm each year as the rotation dictates.

Deep placement

In dry weather the growth of crops slows down, mainly due to lack of water for growth and transpiration but partly because the roots cannot absorb the soil nutrients which are located near the surface. Experiments have shown that deep placement of fertilizers can maintain a faster growth rate in crops in dry weather than where the same nutrients are placed on the surface in a conventional way. It should be noted that, for this deep placement to work, the nutrients should be at least 150 mm below the surface and this is deeper than normal injection of anhydrous or aqueous ammonia.

Table 1.21 Crop tolerance to soil pH. Values indicate thresholds below which yield might be restricted

Agricultural crops	*pH*	*Horticultural crops*
—	6.6	Mint
—	6.3	Celery
Sainfoin, lucerne	6.2	—
Trefoil	6.1	Lettuce
Beans, peas, barley, sugar beet, vetches	5.9	Asparagus, beetroot
Mangolds	5.8	Spinach
—	5.7	Carrot, onion, brussels sprouts
Wheat, oil seed rape, maize, white clover	5.6	Cauliflower
—	5.5	Cucumber, sweet corn
Kale, linseed, mustard, swede, turnip	5.4	Cabbage, rhubarb, parsnip
Cocksfoot, oats, timothy	5.3	—
—	5.1	Chicory, parsley, tomato
Potato, rye	4.9	—
Wild white clover, fescue grasses, ryegrass	4.7	—

Liming

Crop response to pH and the need for lime

Adequate lime status is essential for the following reasons:

- Lime reduces soil acidity, and most plants will not thrive under acid conditions. Table 1.21 gives pH guidelines for major crops.
- It increases the availability of certain plant nutrients in the soil.
- Adequate soil pH encourages soil biological activity, thus enhancing the organic matter cycle in the soil, often releasing available N.
- It improves the physical condition of the soil by increasing the stability of crumb structures.

As explained in 'Nutrient retention – cation and anion exchange', there is a constant loss of lime from the soil, mainly due to crop removal and leaching. Thus it is necessary to have a regular liming policy on all soils other than those with a naturally high Ca content. The frequency of liming will depend on soil type, rate of leaching (controlled by soil texture, rainfall, fertilizer and N use) and sensitivity of crops grown. Soil analysis gives a good indication of soil pH and lime measurement, and each field should be sampled on a regular basis, at least every 4 years. Optimum pH values can be identified for general cropping situations on mineral soils:

- Grassland – soil pH should not be allowed to fall below 5.5 and soil should be limed to bring the pH up to 6.0. This will include grassland with the occasional wheat crop. Grassland with the occasional barley crop should be limed to 6.2.
- Arable – soil pH must be kept above the level required by the most sensitive crop grown. General advice is to apply lime when pH falls to 6.0 and bring the pH up to 6.5.
- Where grass is grown in rotation, it is essential that soil pH is kept up in preparation for succeeding arable crops, as it is difficult to restore rapidly a low pH. If acid grassland is ploughed for arable crops, not only must some lime be worked into the soil, but a sensitive crop such as barley should not be grown for several years.
- Peat soils can be allowed to become more acid without affecting yield. In general, optimum values are 0.7 pH units lower than for mineral soils, for example arable peat soils should be limed to an optimum of 5.8.

Liming policy can affect two diseases – 'club root' of brassicas (*Plasmodiophera brassicae*) and 'common scab' of potato (*Streptomyces scabies*). Club root is less severe on well-limed soils whereas common scab is more prevalent where lime has been applied recently. Thus, in a rotation where brassicas and potatoes are grown, lime that may be required should be applied before the brassica crop, leaving at least 2 years before potatoes are grown.

Lime recommendation

It is possible to calculate the amount of lime needed to raise the pH of a soil to a value that is deemed suitable for given crop plants – this is known as the lime recommendation (MAFF, 2000a). In order to calculate the lime recommendation it is necessary to measure or estimate the size of the cation exchange complex, because soil pH measurements only give an indication of 'active acidity', rather than a true measure of the quantity of hydrogen ions that might be released into soil solution over a period of several years. Most advisors will use a 'look-up' table, such as given in Table 1.22. Having measured the soil pH and assessed the soil texture, the lime recommendation can be determined.

It is assumed that the amount of lime required will in practice be slightly greater than that needed to attain the actual soil pH due to variations in soil pH across any field. The speed of action of a liming material will depend upon when it is applied and the particle size of the material. Lime applied prior to cultivation in the autumn or spring will be rapidly absorbed onto the exchange complex. If a slower release is required, as in an upland situation, then a coarse grade of material can be selected. This will dissolve slowly and therefore counteract any trend towards acidity for a number of years.

Overliming can cause problems and arises when either a regular liming policy is carried out without regard to high soil pH or when the soil is so acid that substantial quantities of lime are needed and mistakenly applied in one heavy dressing. In the latter case, it is essential that some lime is worked into the soil, so that extreme acidity can be alleviated and no part of the soil is overlimed, and the remainder of the requirement applied some months later. A good rule is never to apply more than $3\,t\,ha^{-1}$ of CaO equivalent at one dressing. Overliming can cause the following deficiencies: Mn, causing grey speck in oats, speckled yellows in sugar beet and marsh spot in peas; B, causing heart rot in sugar beet and crown rot in swedes; Cu, causing severe stunting, particularly in cereals; Fe, causing chlorosis in many crops, particularly fruit-bearing species. Overliming also encourages scab in potatoes.

Liming materials

Liming materials occur naturally over a wide area of the UK and they are mostly Ca- and Mg-rich sedimentary rocks such as chalk, magnesian limestone and Carboniferous limestone. The effectiveness of liming materials depends upon:

- fineness of grinding
- neutralizing value
- permissible variation (normally ± 5%)

Fineness of grinding affects the speed of action of the lime and the total quantity of the lime applied that may become useful: the finer the material the more rapid its action and some very coarse materials may never break down sufficiently to be of value. Not only is fineness important, but also the hardness of the material; it is more essential for hard material such as limestone to be ground fine than a softer form such as chalk. The physical and chemical characteristics of a range of common liming materials are given in Table 1.23.

Neutralizing value (NV) is the standard basis for comparing liming materials. NV can be determined by laboratory analysis and is expressed as a percentage of the effect that would be obtained if pure calcium oxide (CaO) had been used. For example, if a sample of ground limestone has NV = 55, then 100 kg of this material would have the same neutralizing value as 55 kg of CaO.

Table 1.22 Lime recommendations for arable and grass crops ($t\ CaCO_3\ ha^{-1}$) (after MAFF, 2000a)

Soil pH	*Sands and loamy sands*		*Sandy loams and silt loams*		*Clay loams and clays*		*Organic soils*		*Peaty soils*	
	Arable	*Grass*	*Arable*	*Grass*	*Arable*	*Grass*	*Arable*	*Grass*	*Arable*	*Grass*
6.2	3	0	4	0	4	0	0	0	0	0
6.0	4	0	5	0	6	0	4	0	0	0
5.5	7	3	8	4	10	4	9	3	8	0
5.0	10	5	12	6	14	7	14	7	16	6

Table 1.23 Liming materials

	Composition	*Source*	*Neutralizing value (NV) and screen grade*	*Recommended use*
Ground limestone	$CaCO_3$ (<15% MgO)	Limestone quarries (including challk)	NV 50 (max. 56); 100% must pass 5-mm sieve; 40% 150-μm sieve	Medium to slow release, arable and grass
Screened limeston (limestone dust)	$CaCO_3$ (<15% MgO)	By-product of industrial limestone production, e.g. road stone	NV48; 100% must pass 5-mm sieve; 30% 150-μm sieve	As above
Coarse screened limestone	$CaCO_3$ (<15% MgO)	By-product of industrial limestone production, e.g. road stone	NV 48; 100% must pass 5-mm sieve; 15% 150-μm sieve	Slow release, not rapid effect; apply at 120% normal rate
Ground magnesian limestone	$CaMg(CO_3)_2$ (>15% MgO)	Magnesian limestone quarries (north-east England)		Useful for soils with long-term Mg deficiency
Ground chalk		Chalk (Cretaceous) quarries in south-east and east England	NV 50; 98% must pass through 6.3-mm sieve	
Calcareous sand (shell sand)	Shell sand $CaCO_3$	Beaches, e.g. Comwall and Scotland, shell	NV 25–40, variable; all the material must pass 6.3-mm sieve	Slow release, can contain some trace elements
Calcified seaweed	Marine sea sand $CaCO_3$	Latin name *Lithothamnion calcareum* dredged from sea floor	NV 40–50, variable	Source of trace elements; expensive as a liming material
Sugar beet factory sludge	$CaCO_3$ and impurities	By-product of sugar beet processing	NV 15–20 when wet; up to 50% when dry	Some N, P and K in residues
Lime sludge	$CaCO_3$ and $Ca(OH)_2$	By-product of sewage treatment	NV variable, especially when wet	

For example, a field has a lime requirement of 1.0 t ha^{-1} of CaO and two materials are available, ground limestone (NV 50) at £12 t^{-1} delivered and hydrated lime (NV 70) at £36 t^{-1} delivered. The two materials can be compared by calculating the cost of NV:

$$\text{Ground limestone} = 1200/50$$
$$= 24.0\ \text{p kg}^{-1}\ \text{CaO equivalent}$$
$$\text{Hydrated lime} = 3600/70$$
$$= 51.4\ \text{p kg}^{-1}\ \text{CaO equivalent}$$

Thus the ground limestone is the cheaper buy, except that the hydrated lime may be easier to apply (if not being applied by a contractor) and is quicker acting, factors that may override the price difference in some circumstances.

Some liming materials contain magnesium as well as calcium; where the addition of Mg is considered beneficial, then the use of such material may be justified even though the cost of each unit of NV is greater than a straight Ca lime. In addition to the common liming materials given in Table 1.23, other materials include waste from soap works, paper works and bleach works. Many of these materials are wet, but can be spread by contractor and they will dry out to form a fine powder. Some have a slight caustic quality and should be applied to the soil rather than to a crop.

Historically other liming materials have been used. For example, when fuel was cheap it was common practice to burn limestone or chalk in kilns to produce burnt lime (CaO). With an NV of 100 this material was the cheapest form of lime to transport, but it had to be kept free from contact with water. When water is added to burnt lime it forms hydrated (or slaked) lime [$Ca(OH)_2$] and the chemical reaction can evolve considerable heat if allowed to proceed without control. Hydrated lime has an NV of about 70, and although more expensive than the liming materials mentioned earlier, it can be used in cases of emergency as it is extremely fine and quick acting.

Fertilizer recommendations

Crop response

The response of individual crops, and even individual crop varieties, is obtained from field experiments, where the shape of the crop response curve to increasing levels of nutrient input is studied. Figure 1.21 gives a theoretical crop response curve, assuming that there are no limitations to crop yield other than the nutrient being applied. Although the exact shape of this curve will vary with crop, soil type and climate, the following principles hold good:

- Under field conditions, some yield is obtained where no fertilizer nutrient is used, the yield A depending on the level of available nutrient in the soil. Fertilizer recommendations take this into account by allowing for soil index and making specific recommendations for each crop.
- There is a maximum yield (B) which is obtained from a certain level of nutrient input (M): using more nutrient than M will not give a greater yield – it may give the same yield as M or less depending on the nutrient and the crop.
- The shape of the response curve is such that the yield return per each extra kilogram of nutrient declines. For example, it can be seen from Fig. 1.21 that the first half of the nutrient dressing M gives a much greater yield increase than the second half of the rate concerned. Thus maximum yield is reached at a point where yield return per unit of extra nutrient is so low that it would not pay for the extra nutrient used.

Where a new crop situation develops, either by the extension of an existing crop into a new soil type or climate area or by the introduction of a new crop into the country, then soil scientists and agronomists soon conduct experiments to establish agronomic criteria for economic optima to be calculated. Good examples in recent years have been the work on forage maize and oilseed rape, where recommendations were updated regularly for several years and now the fertilizer recommendations for these crops in the UK are different from those suggested in other countries where soil, climate and cropping systems are not the same as in the UK.

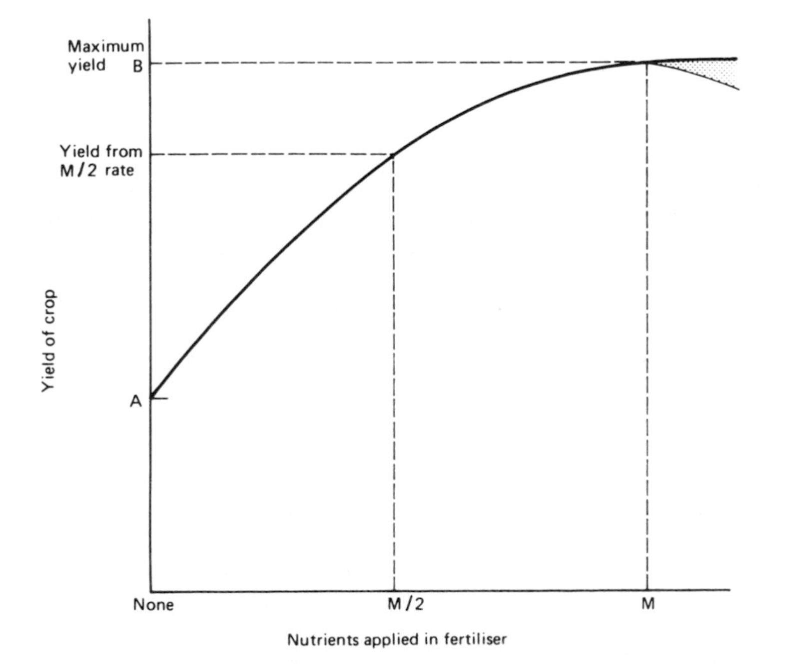

Figure 1.21 Theoretical crop yield response to increasing rate of fertilizer nutrient applied, assuming no other limitations to yield.

Financial return and fertilizer costs

Financial return is considered in establishing fertilizer recommendations by giving monetary value to the crop yield and nutrient costs. Figure 1.22 is based on the data shown in Fig. 1.21 and shows a curve for 'profit from fertilizer' which is taken as value of crop minus nutrient cost. The point O is very important as it is the highest rate of nutrient application that will show an increasing financial return. It is called the optimum rate and is below the maximum rate (M). Optimum rate depends on the ratio of crop value : nutrient cost, as well as on the shape of the agronomic response curve. Optimum rates are not very sensitive to changes in nutrient costs or value of crop output. For most crops in the UK there is a large bank of experimental data that gives reliable information on nutrient responses over a wide range of soil and climatic conditions. Thus, ADAS and manufacturers' recommendations are sound guides to probable optimum rates of application. The periodic update of recommendations [MAFF's *Fertiliser Recommendations RB209* (2000a) is the 7th edition] takes steady trends into account.

The easiest way to compare the relative costs of two compound fertilizers is to relate each of them to the cost of straights, using the following steps:

(1) Establish the cost of 1 kg of each nutrient in straights.
(2) Calculate the nutrient value of the compound, costing all nutrients in the compound as if they were straights.
(3) Express the extra price of the nutrients in the compound as a percentage of the cost as straights.

For example:

Assume ammonium nitrate (34.5% N) costs £130 t^{-1}, then 345 kg N costs 13 000 p = 37.7 p kg^{-1}.

Triple-superphosphate (45% P_2O_5) costs £130 t^{-1}, then 450 kg P_2O_5 costs 13 000 p = 28.9 p kg^{-1}.

Muriate of potash (60% K_2O) costs £110 t^{-1}, then 600 kg K_2O costs 11 000 p = 18.3 p kg^{-1}.

Compare the value of 20:10:10 @ £107 t^{-1} with 25:5:5 @ £102 t^{-1}

	20:10:10	25:5:5
N	200 × 0.377	250 × 0.377
	= £ 75.40	= £ 94.25
P_2O_5	100 × 0.289	50 × 0.289
	= £ 28.90	= £ 14.45
K_2O	100 × 0.183	50 × 0.183
	= £ 18.30	= £ 9.15
Value	= £ 122.60	= £ 117.85

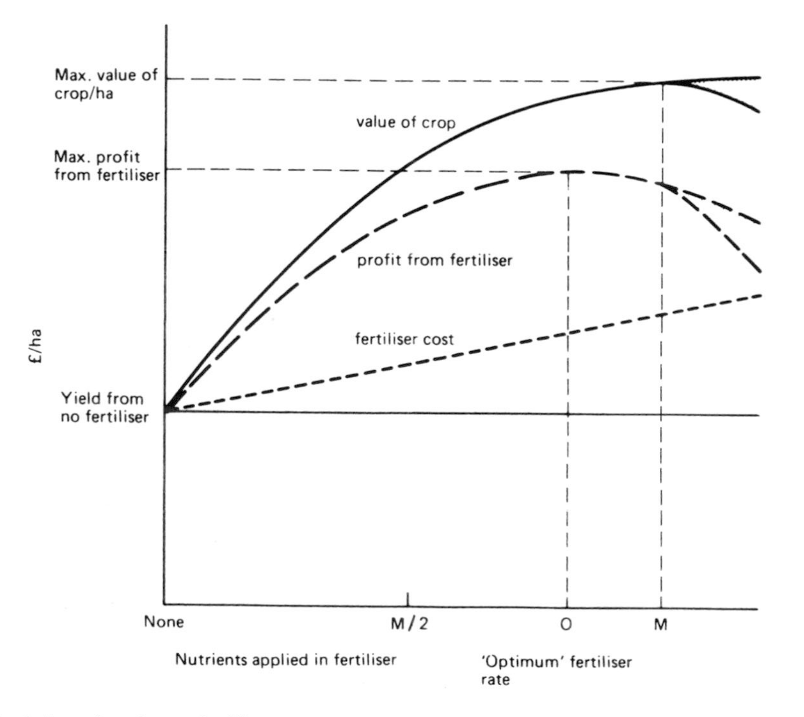

Figure 1.22 Calculation of optimum fertilizer rate.

Cost	= £ 107.00	= £ 102.00
Difference	= £ 15.60	= £ 15.85

In this example, as is common with many compound fertilizers, there is little to choose between different compounds, but both are cheaper than equivalent straights. Note that prices of fertilizer, particularly straight N fertilizer, have become more volatile, and have risen steeply in recent years.

Fertilizer recommendations and soil nutrient index system

Soil management and soil nutrient supply

An essential prerequisite to the assessment of the fertilizer need of a crop is to plan a soil, manure and fertilizer policy. Optimum efficiency of applied fertilizer nutrients depends on taking account of the factors that affect crop nutrient uptake. Even then, certain factors, such as the weather, are beyond the control of the farmer. Crops take up nutrients through root systems; soil physical conditions must allow root systems to develop unchecked to maximize the opportunity for growth and development. Hence it is assumed that prior to construction of a detailed fertilizer plan, soil management has been such that crops will be able to take advantage of the applied nutrients.

Soils will supply nutrients to crop plants independently of the addition of manures or fertilizers. Mineralization or breakdown of organic matter and the weathering of soil minerals release macro-nutrients and trace elements at a rate determined by climatic and soil factors (see 'Soil chemical properties and crop nutrition' for more details). Clay and organic-matter-rich soils are naturally more fertile than sandy soils. Manures and fertilizers supply additional nutrients to raise crop yields. Limitation of supply of nutrients not contained in fertilizers is the main reason for yield response curves such as that shown in Fig. 1.21 reaching a plateau. Soil nutrient supply can be predicted from soil analysis or consideration of the history of previous cropping, which gives an indication of likely nutrient-rich residues left in the soil. Given information relating to the soil nutrient supply and the response of a specific crop to fertilizer nutrients, then a fertilizer programme can be devised. The exact approach taken for N is different to that for P, K and Mg, because the release of N from organic sources is difficult to predict. In recent years considerable advances have been made in the prediction of N supply to crops, resulting in more detailed predictions being available (MAFF, 2000a). In consequence, fertilizer N use (expressed on a per hectare basis) has remained static, and for some crops has declined since the mid-1980s, while the response to nutrients has continued to rise. The result is more efficient use of fertilizers and reduced losses.

P and K fertilizer policy

In order to determine the requirement for P and K in a specific situation, it is necessary to know:

- The soil analyses for these nutrients, which should be conducted at least every 4 years. Soil analysis techniques are described earlier in this chapter. Care must be taken to ensure the sample is representative of the whole field. Table 1.24 gives the range of soil nutrient concentrations that correspond to each index value.
- The target nutrient index for the crop rotation. For arable, forage and grassland crops, the target index is P 2 and K 2–. For vegetables that are more demanding, the target index in P 3, K 2+. Where the soil index is at the target level, the fertilizer application rate should be sufficient to replace the nutrients removed by the crop.
- The need to build up or run down the P and K status. Where the soil P or K index is below the target level for the rotation, the fertilizer policy should be to raise the index gradually to the target level. Manures can be used to aid this process. Building up P levels is a gradual process, so simply applying more P than is recommended will not rectify a deficiency situation, i.e. a P index of 0 or 1. As K is more soluble, it is possible to raise levels more quickly using muriate of potash. Reducing nutrient status similarly takes time, and depends

Table 1.24 Soil nutrient concentrations and soil index values for P (sodium bicarbonate extract), K and Mg (ammomium nitrate extract) (after MAFF, 2000a)

Index	*Phosphorus* ($mg\,l^{-1}$)	*Potassium* ($mg\,l^{-1}$)	*Magnesium* ($mg\,l^{-1}$)
0	0–9	0–60	0–25
1	10–15	61–120	26–50
2	16–25	121–180 (2–); 181–240 (2+)	51–100
3	26–45	245–400	101–175
4	46–70	401–600	176–250
5	71–100	601–900	251–350
6	101–140	901–1500	351–600
7	141–200	1501–2400	601–1000
8	201–280	2401–3600	1001–1500
9	>280	>3600	>1500

upon crop offtake and other losses. K levels in sandy soils can fall rapidly, but P levels will fall more slowly due to smaller crop demands and reduced availability/solubility of soil P. When conducting a policy of running down soil nutrient status, regular soil sampling is important to ensure that nutrient supply does not become limiting to crop yield.

- The responsiveness of the crop grown to fertilizer and the likely crop offtake of nutrients. This will depend upon crop yield and variety, and the type of fertilizer used. See MAFF (2000a) for more information.
- The quantity of nutrients supplied by organic manure applications – see 'Nutrient Sources: manures'.
- The residues of P and K from previous fertilizer and manure applications.

Once these factors are known (or approximated in some cases) then fertilizer recommendation tables such as are given in MAFF (2000a) can be used to determine application rates. Regular crop monitoring and soil sampling are the keys to effective P and K fertilizer application strategy.

Nitrogen supply to crops

The supply of N to crop plants is much more difficult to predict than P or K, yet N is essential to many plant functions (see Table 1.11), is taken up in substantial quantities by crops (see Table 1.12) and has a larger effect on crop growth and quality than any other nutrient. However, too much N can reduce yield, lead to lodging in cereals, poor silage fermentation in conserved grass and may lead to nitrate pollution of water.

The supply of N to crops is controlled by:

- soil mineral nitrogen (SMN), which is dominated by nitrate-N (NO_3-N) and ammonium-N (NH_4-N). Recent cropping, soil type and management and environmental factors control the amount of SMN present in the root zone at any one time.
- potential for N mineralization (release from organic sources) also is dependent on recent soil and crop management and environmental factors, which will influence the quantity of N that will be released to a crop during the next growing season.
- organic manures contain large amounts of total N (see Table 1.17), a declining proportion of which will be released to add to the SMN supply in the years subsequent to application.
- atmospheric N that is deposited in rainfall may come from natural sources, pollution from vehicle exhaust emissions and NH_3 from livestock enterprises. Such contributions are variable, but may add 10–40 kg N ha^{-1} $year^{-1}$ depending upon location.
- fertilizer additions which should be applied to crops to make up any projected shortfall in the crop requirements for N.

This supply is balanced by losses, which include:

- leaching of water-soluble NO_3-N. Figure 1.23 shows the range of average losses that have been observed under different crops. These average figures hide a wide variation from year to year and soil to soil. In general, the more freely draining sandy soils are more prone to leaching loss, and the amount of excess autumn and winter rainfall is important. Ammonium-N is more tightly held to clay and organic matter in the soil, as it has a positive charge, but on mineralization to NO_3 is easily leached.
- denitrification, the gaseous loss of N (see 'Nitrogen'), occurs in poorly drained soils, particularly when warm. For example, losses can be high (2–5 kg N ha^{-1} day^{-1}) in late spring grassland after a top dressing of N fertilizer.
- ammonia volatilization from livestock manures applied to fields but not incorporated can be important sources of loss, particularly to livestock farmers. Losses are very variable according to manure/slurry composition and conditions at time of spreading. Surface-applied manures rich in NH_4-

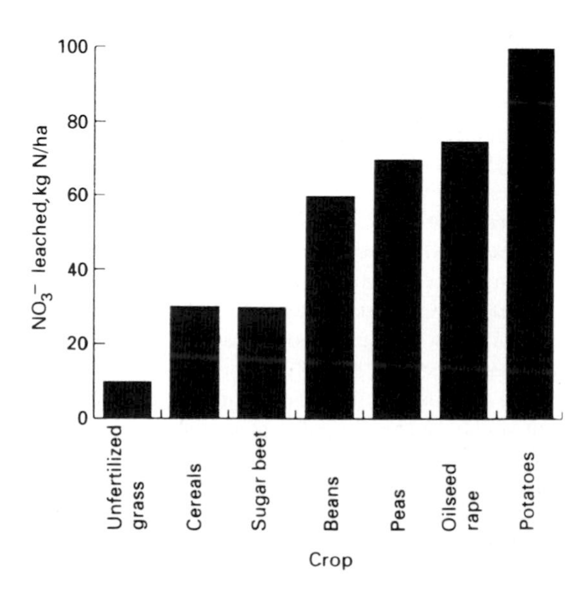

Figure 1.23 Nitrate losses from selected crops without the use of manures grown on a sandy soil (from MAFF, 1999b).

N spread on a warm, windy day can lose up to 80% of the available N within 24 hours.

Nitrogen index and N fertilizer policy

For a given farm economic situation, it is possible to determine the N requirement of a crop. Given the factors that affect N supply and losses, it is necessary to determine:

- the crop requirement for N
- the supply of N from the soil
- the cost of the fertilizer and the likely value of the crop

The crop requirement for N can be determined from tables in MAFF (2000a). These tables specify that for most common crops the N requirement is based on knowledge of the Soil N Supply (SNS) index and the soil type. In Table 1.25 an example of the determination of the SNS index for a high-rainfall area is given. This high-rainfall area includes much of southern and south-western England (except East Anglia, the area around London and the Midlands), Wales and northern England. There are other tables that state the SNS index for low and moderate rainfall situations. The system now recommended by MAFF includes seven index levels, in contrast to three in the former system. This demonstrates our clearer understanding of the complexities of N supply to crop plants. The shaded area of Table 1.25 indicates situations that are rarely likely to occur in farming practice. The asterisks indicate situations where high levels of SMN may have accumulated, in which case it is recommended that SMN levels be determined by soil analysis.

Nitrogen index values in Table 1.25 apply primarily to arable situations; where grassland has been ploughed out, then different index values apply. These are shown in Table 1.26, and should be used for 3 years after a ley has been ploughed out. In this table, high N refers to amounts in excess of 250 kg N ha^{-1} $year^{-1}$ and available manure nitrogen used on average in the last 2 years, or high clover content, or lucerne. These indices assume that little or no organic manures have been applied. The index should be raised by unity if regular heavy dressings of manures have been applied. The index values given refer to ley–arable situations. If a permanent grass ley is ploughed out, then the situation given in Table 1.26 for a 3- to 5-year ley, high N, grazed (i.e. 3) should be increased by unity to 4.

Sulphur

Sulphur (S) is needed by plants in similar quantities to P, and until recently most crop requirements were met by the soil from atmospheric sources or fertilizers that contain S as a by-product. However, during the 1990s there was a marked decline in the levels of sulphur dioxide emissions from power stations burning fossil fuels. It is likely that this trend will continue in the future. For example, much of East Anglia now only receives 10–15 kg S ha^{-1} $year^{-1}$ from atmospheric sources. Some crops, notably oil seed rape, cereals and grass, now show yield benefits to added S. Soil analysis is not as reliable as plant tissue analysis, but results from the latter may be available too late to correct a deficiency in the current crop. Ammonium sulphate can be used on soils where S deficiency is diagnosed. Manures typically contain 1–2 kg m^{-3} total S, expressed as SO_3, although poultry manures can contain more (see Table 1.17).

Calculating fertilizer application rates

By law every bag or consignment of fertilizer must have on it a statement of the plant food value of its contents, and under EU regulations the phosphate value must be as P_2O; and the potash value as K_2O. Thus a compound fertilizer might be called 20 : 10 : 10 and state on the bags that its contents were:

20% N 10% P_2O_5 10% K_2O

If the farmer applied 300 kg/ha of this 20 : 10 : 10 fertilizer to a field, then the application of N would be 60 kg ha^{-1} (20% of 300), P_2O_5 would be 30 kg ha^{-1} (10% of 300) and K_2O would also be 30 kg ha^{-1}. Recommendations are given as the requirements for N, P_2O_5 and K_2O and so the quantity of product to apply needs to be calculated. The rate of fertilizer to be applied in a specific situation can be calculated as follows:

$$\text{Fertilizer application rate}\left(\text{kg ha}^{-1}\right) = \frac{\text{Nutrient application rate}\left(\text{kg ha}^{-1}\right)\times 100}{\text{Percentage of nutrient in fertilizer}}$$

For example:

For first-cut silage it is recommended that a specific crop receives 125 kg ha^{-1} of N, 30 kg ha^{-1} of P_2O_5 and 30 kg ha^{-1} of K_2O, with the phosphate and potash and some of the N in late February and the rest of the N in mid-March. The farmer has in stock a 16 : 16 : 16 compound fertilizer and ammonium nitrate (34.5% N). How much of each fertilizer should he apply?

In February:

$$\frac{30\times 100}{16} = 188 \text{ kg ha}^{-1} \text{ of 16:16:16 fertilizer}$$

Table 1.25 Soil N supply (SNS) index for high-rainfall areas (>700 mm annual rainfall, or >250 mm excess winter rainfall), based on last crop grown (after MAFF, 2000a).

	Soil nitrogen supply (SNS) index						
	0	*1*	*2*	*3*	*4*	*5*	*6*
	Measured SNS (kg ha^{-1}), where SNS = SMN (0–90 cm depth) + crop N + estimate of N mineralization						
	<60	*61–80*	*81–100*	*101–120*	*121–160*	*161–240*	*>240*
Light sands or shallow soils over sandstone	Cereals, potatoes, peas, beans, oil seed rape, forage crops (cut), low/ medium/high N veg, rotational set-aside	Sugar beet	[1]	[1]	[1]	[1]	[1]
Medium soils or shallow soils (not over sandstone)		Cereals, sugar beet, forage crops (cut), low/medium N veg, peas, beans, potatoes, oil seed rape	High N veg, rotational set-aside	[1]	[1]	[1]	[1]
Deep clay soils		Cereals, sugar beet, fodder crops (cut), low/medium N veg	Peas, oil seed rape, beans,potatoes, high N veg,rotational set-aside	[1]	[1]	[1]	[1]
Deep fertile silty soils		Cereals, sugar beet, low N veg, forage crops (cut)	Medium N veg	Oil seed rape, potatoes, peas, beans, high N veg, rotational set-aside	[1]	[1]	[1]
Organic soils				All crops	All crops	All crops	All crops
Peat soils						All crops	All crops

[1] These indices can occur where there has been a history of grassland or frequent application of manures; it is recommended to carry out an SMN analysis.

Table 1.26 Soil nitrogen supply (SNS) index values following ploughing out of grass leys on medium textured soils in all rainfall areas of England and Wales (after MAFF, 2000a)

	Year 1	*Year 2*	*Year 3*
All leys, 2 or more cuts annually receiving little or no manure 1–2 year leys, low N	1	1	1
1–2 year leys, 1 or more cuts 3–5 year leys, low N, 1 or more cuts 1–2 year leys, high N, grazed 3–5 year leys, low N, grazed	2	2	1
3–5 year leys, high N, 1 cut then grazed 3–5 year leys, high N, grazed	3	3	2

In March:
The compound applied 30 kg ha^{-1} of each nutrient, including N, so N rate to apply in March is 125 − 30 = 95.

$$\frac{95 \times 100}{34.5} = 275 \text{ kg ha}^{-1} \text{ of ammonium nitrate}$$

Some fertilizer recommendations, particularly from non-UK European countries, give phosphate and potash recommendations in terms of P and K and not P_2O_5 and K_2O. Great care should be taken when reading literature to ensure confusion does not occur.

- to convert P to P_2O_5 multiply by 2.29 (P_2O_5 to P multiply by 0.44).
- to convert K to K_2O multiply by 1.20 (K_2O to K multiply by 0.83).

Table 1.27 contains the fertilizer recommendation for maincrop potatoes grown on land that has received 50 t ha^{-1} fresh farmyard manure. When manure nutrient value is taken into account, using the procedure described in MAFF (2000a), there is no requirement for fertilizer K. The precise N requirement will depend upon the variety grown and the market requirements. This need could be met by using an N:P compound or using straights. If triple superphosphate was chosen to supply the P, then the requirement would be:

$$\frac{75 \times 100}{47} = 160 \text{ kg ha}^{-1} \text{ of triple superphosphate.}$$

Table 1.28 gives example fertilizer recommendations for an 8 t ha^{-1} winter wheat crop grown on a clay soil, with the straw ploughed in and no manure used since the harvest of the previous crop. The SNS index will vary for this situation depending upon the previous crop (see Table 1.25). The N requirement declines progressively as the N supply from soil increases. Practice has shown that SNS index 0 situations are unlikely to occur with this soil type, hence there is no recommendation for Index 0. The P_2O_5 and K_2O recommendations for Index 2 feature the recommendation for a maintenance dressing of fertilizer (M). Where soil is at the target level for a particular rotation, sufficient fertilizer should be applied to replace the quantity of nutrients removed from the field by the harvested crop. This maintenance (M) dressing can be calculated knowing P type, yield and nutrient content of the crop material removed from the field.

For example, winter wheat (grain only) contains 7.8 kg P_2O_5 t^{-1} of fresh material. Therefore an 8-t ha^{-1} crop will remove 62 kg P_2O_5 ha^{-1} (hence the recommendation of 60 M for P_2O_5 in Table 1.28), while a 6-t ha^{-1}

Table 1.27 Fertilizer recommendations for maincrop potatoes, 90–120 days growing season, sandy clay loam soil, variety group 3 (long haulm longevity), likely tuber yield 50 t ha^{-1}, with 50 t ha^{-1} fresh farmyard manure directly incorporated into the soil in spring prior to planting. SNS index = 1, P index 2, K index = 1 (after MAFF, 2000a)

	fertilizer (kg ha^{-1} to apply)		
	N	*P_2O_5*	*K_2O*
Total required	120–180	180	325
From 50 t ha^{-1} FYM	75	105	360
From fertilizer	45–105	75	0

Table 1.28 Example fertilizer recommendations for winter wheat, for medium and deep clay soils, expected yield approximately 8 t ha^{-1}, straw ploughed in, no manure applied since harvest of the previous crop (after MAFF, 2000a). M = Maintenance dressing

SNS index	*In spring N*	*Soil index*	*In autumn P_2O_5*	*K_2O*[1]
0	—	0	110	95
1	220	1	85	70
2	180	2	60 M	45 M (2−), 20 (2+)
3	150	3	20	0

[1]Reduce potash recommendation by 25 kg ha^{-1} on sand-textured soils where the soil-available K is 100 mg l^{-1} or higher.

crop would remove only 47 kg P_2O_5 ha^{-1}. If the crop was anticipated to be less than 8 t ha^{-1}, then the maintenance fertilizer application would be reduced accordingly. The same applies to situations where yields considerably in excess of 8 t are expected. Maintenance dressings for non-responsive crops can be reduced provided that the input balances offtake over the whole rotation. It is possible to adjust inputs in this way to minimize risk of uncontrolled losses, for example P in surface run-off where bare soil persists during the winter period.

Soil management, crop nutrition and sustainable production

Managing soil, and particularly nutrients, within agricultural situations is a complex and difficult task, due to the interaction between natural and 'artificial' processes. For example, the efficiency of fertilizer utilization by crops is typically 50% (N), 25% (P) and 50% (K), although actual values will vary according to soil, weather and crop management. The nitrogen cycle (Fig. 1.17) demonstrates the complex range of transformations that occur in soils. Fertilizer N is only one source of potential leaching losses – natural processes can also lead to significant removal of N and other nutrients from soils. In recent years concerns have focused on those nutrients that have a major impact on the wider environment, notably nitrate and phosphate in water and nitrogen gases in the atmosphere. Efficient nutrient management based on an understanding of soil properties will lead to a conservation of nutrients within the soil, reduced reliance on artificial fertilizers and reduced environmental impact.

Principles of good soil management are embodied in the *Code of Good Agricultural Practice for the Protection of Soil* (MAFF, 1998b). This guide, when read with the partner volumes on the protocion of water and air (MAFF, 1998a,c), provide a framework for the more sustainable use of soil. Many examples can be used to illustrate the consequences of failure to consider the wider impacts of increased inputs into agricultural systems, but perhaps the most well studied is the nitrate problem. Significant nitrate losses from arable and grassland soils have been observed for at least two decades now. Figure 1.23 taken from MAFF (1999b) plots the losses that typically occur from a range of crops, although individual losses will vary widely according to climate, location, soil type and management regime. Principles of good soil, fertilizer and manure management to reduce these losses have been described in the *Code of Good Agricultural Practice for the Protection of Water* (MAFF, 1998a). Adherence to the code will benefit the farmer through efficient nutrient utilization. Some example recommendations are given below:

- Plough up grassland only when necessary, and reseed or drill next crop as quickly as possible.
- Apply organic manures to growing crops, and not in the autumn.
- Split fertilizer dressings, and do not apply too early in spring.
- Maximize crop cover in the autumn.
- Incorporate crop residues to fix nitrogen.
- Delay autumn cultivations.
- Reduce grassland grazing intensity in the autumn.

Adherence to these and other recommendations is most important in areas vulnerable to nitrate leaching, such as regions where groundwater accumulates in aquifers overlain by permeable soils, e.g. sandy or limestone derived materials. The designation of nitrate vulnerable zones (NVZs) where many of the practices described above are mandatory will help to reduce nitrate concentrations in drinking water to levels below 50 mg NO_3 l^{-1}, the maximum specified by European Community Directive 80/778/EEC (see www.defra.gov.uk for latest information on NVZs). Nutrient enrichment of natural waters, eutrophication, is caused not only by nitrate, but also by phosphate, which can be lost from agricultural systems by surface run-off and soil erosion. Livestock systems have been observed to cause major losses when farm wastes are applied under inappropriate conditions or when surface compaction by animals has enhanced surface water run-off. Again, adherence to the codes of good practice for soil and water protection will minimize such losses (Merrington *et al.*, 2002).

Indicators used to assess progress towards a more sustainable agriculture (for example, MAFF, 2000b) describe other aspects of soil and nutrient management that can have a significant impact on the long-term productivity of soils as well as environmental quality. For example, the organic matter content of agricultural topsoils in England and Wales was found to have generally decreased by 0.5% over 15 years up to 1995. Losses of organic matter lead directly to lower inherent fertility. Similarly, the incidence of soil erosion by water carrying P-rich sediments into ecologically sensitive water bodies has increased in recent years. Many of these losses have been attributed to the increased area of winter cereals and maize being grown in the UK (Environment Agency, 2001).

There are a range of production approaches employed by the agriculture industry today that take note of these threats to continued sustained production. For example, many of the principles outlined above are embodied in the methods adopted by organic farmers. These are discussed further by Lampkin (2002) in Chapter 11 on Organic Farming. Maintenance of long-term soil productivity can be ensured by practising methods that enhance natural soil fertility and so lead to a reduced reliance on agrochemicals. Reduction in the use of artificial fertilizers and other imported materials is central to the concept of sustainability, as is the aim of minimizing uncontrolled losses from agricultural systems (Merrington *et al.*, 2002). Integrated crop management principles are based on careful consideration of natural resource exploitation in crop production, and the use of careful management based on rotations and sound soil/crop/pest management. The move towards less-intensive systems with more attention paid to the soil and environmental controls on production will lead to greater efficiency of nutrient use and fewer problems of declining soil fertility. Enlightened soil management based on a sound knowledge of soil properties is central to continued productive agriculture.

Acknowledgement

Previous editions of *The Agricultural Notebook* contained a separate chapter on crop nutrition authored by Dr John Brockman, who died in 1998. In this edition aspects of crop nutrition have been incorporated into this revised chapter. I would like to acknowledge the contribution that John made to these sections of this chapter, as well as his wider contribution to the study of grassland and soil management.

References

ADAS (2000) *The Safe Sludge Matrix*, http://www.adas.co.uk

Addiscott, T., Whitmore, A. & Powlson, D. (1991) *Farming, Fertilizers and the Nitrate Problem*. CAB International, Wallingford.

Brady, L.C. (1999) *The Nature and Properties of Soils*, 12th edn. Prentice Hall, New Jersey.

Castle, D.A., McCunnal, J. & Tring, I.M. (1984) *Field Drainage, Principles and Practices.* Batsford, London.

Chalmers, A., Burnhill, P., Fairgrieve, J. & Owen, L. (1998) *The British Survey of Fertilizer Practice. Fertilizer Use on Farm Crops 1997*. FMA, Peterborough.

Chambers B.J. & Garwood, T. (1998) Lime loss rates from arable and grassland soils. *Journal of Agricultural Sciences (Cambridge)*, **131**, 455–64.

Cm 3165 (1996) *Royal Commission on Environmental Pollution, 19th Report: The Sustainable Use of Soil.* The Stationery Office, London.

Davies, B.D., Eagle, D.J. & Finney, J.B. (2001) *Resource Management–Soil.* Farming Press, Ipswich.

Environment Agency (2001) *State of the Environment Report.* www.environment-agency.gov.uk/state_of_enviro

Farr, E. & Henderson, W.C. (1986) *Land Drainage.* Longmans, London.

Fertilizer Manufacturers' Association (2000a) *Best Fertilizer Quality.* FMA, Peterborough.

Fertilizer Manufacturers' Association (2000b) *Fertilizer Spreaders – Choosing, Maintaining and Using*. FMA, Peterborough.

Findlay, D.C., Colbourne, G.J.N., Cope, D.W., Harrod, T.R., Hogan, D.V. & Staines, S.J. (1984) *Soils and Their Use in South West England.* Soil Survey of England and Wales Bulletin No. 14, Harpenden.

Goudie, A. (2000) *The Human Impact on the Natural Environment*, 5th edn. Blackwell Science, Oxford.

Lampkin, N. (2002) Organic farming. In: *The Agricultural Notebook*, 20th edn. pp. 288–303. Blackwell Science, Oxford.

Lynch, J.M. (1983) *Soil Biotechnology: Microbiological Factors in Crop Productivity*. Blackwell Science, Oxford.

McGrath, S.P. & Loveland, P.J. (1992) *The Soil Geochemical Atlas of England and Wales.* SSLRC, Cranfield.

MAFF (1982) *The Design of Field Drainage Pipe Systems.* Reference Booklet No. 345. HMSO, London.

MAFF (1986) *The Analysis of Agricultural Materials.* Reference Book 427. MAFF Publications, London.

MAFF (1988) Agricultural Land Classification of England and Wales. MAFF Publications, London.

MAFF (1998a) *Code of Good Agricultural Practice for the Protection of Water.* MAFF Publications, London.

MAFF (1998b) *Code of Good Agricultural Practice for the Protection of Soil.* MAFF Publications, London.

MAFF (1998c) *Code of Good Agricultural Practice for the Protection of Air.* MAFF Publications, London.

MAFF (1999a) *Controlling Soil Erosion. A Manual for the Assessment and Management of Agricultural Land at Risk of Water Erosion in Lowland England.* MAFF Publications, London.

MAFF (1999b) *Tackling Nitrate from Agriculture.* MAFF Publications, London.

MAFF (2000a) *Fertiliser Recommendations for Agricultural and Horticultural Crops (RB209)*, 7th edn. MAFF Publications, London.

MAFF (2000b) *Towards Sustainable Agriculture*. MAFF Publications, London.

Marschner, H. (1995) *Mineral Nutrition of Higher Plants*, 2nd edn. Academic Press, London.

Merrington, G., Winder, L., Parkinson, R. & Redman, M. (2002) *Agricultural Pollution–Problems and Practical Solutions*. Spon Press, London.

Mullins, C. (1991) Physical properties of soils in urban areas. In: Bullock, P. & Gregory, P.J. (eds) *Soils in the Urban Environment*, pp 87–118. Blackwell Science, Oxford.

Nix, J. (2000) *Farm Management Pocketbook*, 31st edn. Wye College Press, Ashford.

Parkinson, R.J. (1988) Field drainage and the soil environment. *Outlook on Agriculture*, **17**, 140–5.

Rowell, D.L. (1994) *Soil science – Methods and Applications*. Longmans, Harlow.

Simpson, K. (1983) *Soil*. Longmans, Harlow.

Simpson, K. (1986) *Fertilizers and Manures*. Longmans, Harlow.

Soil Association (1999) *Standards for Organic Food and Farming*. Soil Association, Bristol.

Tyson, K.C., Garwood, E.A., Armstrong, A. & Scholefield, D. (1992) Effect of field drainage on the growth of herbage and liveweight gain of grazing beef cattle. *Grass and Forage Science*, **47**, 290–301.

White, R.E. (1997) *Principles and Practice of Soil Science. The Soil as a Natural Resource*, 3rd edn. Blackwell Science, Oxford.

Wild, A. (1993) *Soils and the Environment: an Introduction*. Cambridge University Press, Cambridge.

Further reading: regional Soil Survey memoirs

Hodge, C.A.H., Burton, R.G.O., Corbett, W.M., Evans, R. & Seale, R.S. (1984) *Soils and Their Use in Eastern England*. Soil Survey of England and Wales Bulletin No. 13, Harpenden.

Jarvis, M.G., Allen, R.H., Fordham, S.J., Hazelden, J., Moffat, A.J. & Study, R.G. (1984) *Soils and Their Use in South East England*. Soil Survey of England and Wales Bulletin No. 15, Harpenden.

Jarvis R.A., *et al.* (1984) *Soil and Their Use in Northern England*. Soil Survey of England and Wales Bulletin No. 10, Harpenden.

Ragg, J.M., *et al.* (1984) *Soils and Their Use in Midland and Western England*. Soil Survey of England and Wales Bulletin No. 12, Harpenden.

Rudeforth, C.C., Hartnup, R., Lea, J.W., Thompson, T.R.E & Wright, P.S. (1984) *Soils and Their Use in Wales*. Soil Survey of England and Wales Bulletin No. 11, Harpenden.

Soil maps may be obtained from the National Soil Resources Institute, Silsoe Campus, Bedford, MK45 4DT (England and Wales), or Macaulay Institute Institute, Craigiebuckler, Aberdeen, AB9 2QJ (see website addresses for both these organizations below).

Useful websites

www.bsss.bangor.ac.uk/ – British Society of Soil Science homepage, contains many links to websites with information on soil use and management.

www.efma.org/ – European Fertilizer Manufacturers Association.

www.environment-agency.gov.uk/state_of_enviro/ – Environment Agency State of the Environment report, includes discussion of soil, fertilizer and manure use in agriculture.

www.fma.org.uk/ – Fertilizer Manufacturers' Association.

www.fwag.org.uk/ – Farming and Wildlife Advisory Group.

www. defra.gov.uk – Department for Environment, Food and Rural Affairs, includes publication details for Codes of Good Agricultural Practice for the Protection of Soil, Air and Water.

www.maculay.ac.uk/ – Macaulay Land Use Research Institute, Aberdeen.

www.silsoe.cranfield.ac.uk/nsri/ – National Soil Resources Institute, Silsoe Campus, Cranfield University.

2

Crop physiology

M.P. Fuller & A.J. Jellings

Fundamental physiological processes

Analysis of green plant material shows that about 60–90% of fresh weight is made up of water. Of the remainder, the dry matter (DM), about 45% is carbon, 45% oxygen, 5% hydrogen, 2–3% nitrogen and 2–3% other elements such as potassium, phosphorus, calcium, magnesium and sulphur.

The crop obtains the carbon, oxygen and hydrogen by the process of photosynthesis, and water, nitrogen, potassium and other mineral elements by uptake from the soil. These two processes are fundamental to crop physiology, and thus to efficient crop production.

Water and nutrient uptake

Water enters the roots near the root tips by diffusion across the cell wall. It must reach the xylem vessels embedded in the root by passing through or around the cells of the root to be delivered to the pores in the leaf surface (stomata) via the leaf veins (vascular bundles). It is the evaporation of the water from the stomata (transpiration) that creates the main driving force for water uptake from the soil. The moving column of water from the soil to the atmosphere through the plant is the transpiration stream. The creation of the transpiration stream is largely a physical not a biological process, in which the plant can be thought of as a tube for the flow of water.

The pathway of water from soil to atmosphere thus involves four key steps: (1) through the soil to the roots, (2) in and out of cells across membranes, (3) along the xylem and (4) out of the leaves into the air. Water moves along the pathway down gradients of water potential (water potential may be defined as the concentration of water, and is thus always a negative value since pure water is said to have a water potential of zero), encountering resistance to flow in the soil, at membranes and cell walls and at the stomata. Water uptake will be at a maximum when the stomata are fully open in conditions of warm dry moving air, with high soil water availability. Water uptake will be reduced by low soil water availability, frozen soil conditions, low soil temperatures, low light levels, high humidity and low air temperatures. A reduction in water uptake will eventually result in wilting and death.

Of the water taken up from the soil, 95% is not used directly by the plant but moves through and out of the plant (a plant may transpire its own weight of water in 1 hour). Approximately 5% of the water taken up is used by the plant to give turgidity to the cells, to provide a medium for metabolism and to participate in the reactions of the cell.

Water uptake is the means by which plants gather mineral ions from the soil (or other root medium). Mineral ions are taken up by the roots in solution and thus the ions travel by the same pathway in the plant to be delivered to the cells of those leaves that are actively transpiring. Mineral ions in excess of the requirement of transpiring tissues are passed into the phloem to be retranslocated to non-transpiring plant tissues.

Root cells can be selective in their uptake of mineral ions by making use of the selective permeability of cell membranes, i.e. the relative concentrations of ions in different parts of the plant will be different from each other, and different from that of the soil solution. In this way plants can absorb elements in relative proportions different from that of the soil mineral composition.

Active uptake of minerals requires energy from respiration and a plant under stress or whose roots are in anaerobic conditions will be unable to acquire an adequate mineral supply. In addition, mineral uptake is reduced by the same factors that reduce water uptake.

Photosynthesis and respiration

Ninety-five per cent of the dry matter of plants (carbon-based material) is created from the gaseous environment, by the process of photosynthesis. The molecules of the green pigment of plants, chlorophyll, absorb radiation in the visible light range 400–700 nm (photosynthetically active radiation or PAR), and by means of the electron transport chain are used to make ATP (adenosine triphosphate) and NADPH (reduced nicotinamide adenine dinucleotide phosphate). The energy thus conserved in these compounds can be used to drive the reactions of the carbon fixation cycle which fixes atmospheric carbon dioxide by converting it to sugars. Thus photosynthesis stores radiant energy by converting it into a chemical form. Agriculture manages the process so that human society can utilize its products as food and fuel.

The summary equation for both parts of photosynthesis is:

$$6CO_2 + 6H_2O \xrightarrow{\text{light}} C_6H_{12}O_6 + 6O_2$$

All the reactions of photosynthesis occur in the chloroplasts. The key enzyme for the fixation of carbon dioxide is ribulose bisphosphate carboxylase/oxygenase ('Rubisco'). This is the most fundamentally important enzyme to life on earth, and it makes up a large part of the soluble protein fraction in the global plant biomass; in crop production it is one of the main beneficiaries of added nitrogen.

The rate of photosynthesis depends on both internal and external factors. Internally the amount and activity of the participating enzymes, the amount of chlorophyll and the anatomy of the leaf all dictate the photosynthetic capacity. The rate achieved at any one time will be affected by several environmental factors, of which the most important are light intensity, temperature and carbon dioxide concentration, the rate always being determined by the factor in shortest supply. Figure 2.1 demonstrates the general relationship between these three main factors.

About half of the sugars produced by photosynthesis are used by the plant to provide energy through the process of respiration which occurs in the mitochondria and is essentially the reverse of photosynthesis. The other half is assimilated into DM. Photosynthesis and respiration are thus the means by which the plant utilizes light energy (which cannot itself be stored) to provide a continuous and translocatable energy source.

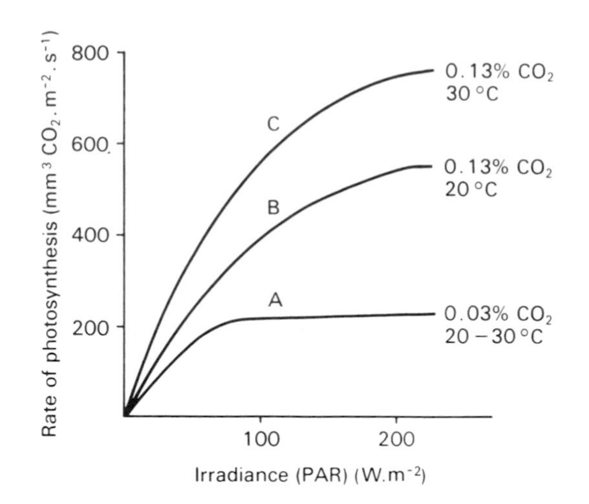

Figure 2.1 General relationship of photosynthetic rate with light intensity, carbon dioxide concentration and temperature in leaves of temperate crops. Curve A demonstrates that light saturation occurs at very moderate light levels under atmospheric CO_2 concentrations. Curve B shows that, if CO_2 concentration can be raised above atmospheric levels, photosynthetic rate can be stimulated, thus CO_2 concentration is the limiting factor to photosynthetic rate under normal conditions (curve A). In curve B temperature becomes limiting, as at higher temperature (curve C) there is a further increase in photosynthetic rate; this latter case is only applicable in controlled environments where CO_2 concentration may be artificially raised, e.g. glasshouse lettuce production.

In most common temperate crops (which utilize C3 metabolism) at average and above average levels of light intensity, carbon dioxide concentration (currently at 350 ppmv in the atmosphere) is the most common limiting factor for the rate of photosynthesis as light saturation is achieved at less than the light intensity experienced on most UK summer days. In addition, photorespiration, which takes up oxygen and releases carbon dioxide, can be severe under good summer conditions of high light and temperature, making carbon dioxide compensation points high. Currently atmospheric levels of CO_2 are rising, due to the burning of fossil fuels, and will lead to a CO_2 fertilization effect in C3 photosynthetic plants.

Many tropical crops, e.g. sugar cane and sorghum, which use a modified (C4) photosynthetic pathway with little or no accompanying photorespiration, are not limited in their photosynthetic rate by carbon dioxide concentration until much higher light intensities. The only UK crop with this tropically adapted metabolism is maize, which, on hot summer days, may gain the advantage of higher photosynthetic rates and faster DM accumulation than its neighbouring temperate crops.

The internal leaf carbon dioxide concentration can be controlled to some extent by the opening and closing of the stomata. The stomata may open passively in response to changes in the gradient of water potential, but are mostly controlled by light levels, leaf temperature and internal carbon dioxide levels, which cause active, energy-requiring, stomatal opening. Photosynthesis will cease if water stress causes stomatal closure.

The carbon fixed by photosynthesis is moved around the plant as sugars in the phloem. The extent and direction of translocation is very important in crop production since it determines the partitioning of DM within the plant. A tissue actively importing sucrose is termed a sink, and one that is exporting sucrose is termed a source. Rate of growth and metabolic demand affect the strength of a sink, and positive feedback between sources and sinks allows increased demand by the sinks to stimulate increased rate of assimilation by the source tissues. Thus, for example, as tubers form within a potato crop the photosynthetic rate of the leaves tends to increase (under standard conditions), but falls again if tubers are removed. The rate of translocation, determined by the size of the phloem vessels, also has some control over assimilate partition.

For temperate crop plants photosynthesis is only about 1% energy efficient (i.e. 1% of sunlight is converted to crop DM). This low efficiency occurs for a number of intrinsic reasons: the conversion of light energy into chemical energy is about 25% efficient; about half of the total radiation is unusable by the plant; about half of the PAR is lost by reflection and transmission, and through respiration and photorespiration. This gives a possible maximum photosynthetic efficiency of about 5% which is further reduced to 1% by imperfect crop production conditions, e.g. incomplete crop canopy, stress from pathogens, climatic factors, etc.

Growth

Growth is strictly the accumulation of dry matter (DM), though the term is often used loosely to describe any increase in size. An overall increase in size will include a certain amount of water accumulation although this is not true growth.

Accumulation of DM occurs fundamentally at the cellular level. DM fixed by photosynthesis is laid down as new compounds within the cell so that individual cells tend to increase their mass with time. At the same time cells may divide to form new cells thereby redistributing DM. Thus, when an individual leaf increases in DM it is through a combination of both an increase in cell size and an increase in cell number. The way in which the cells are organized will determine whether the increase in DM is visible as an increase in leaf area, in leaf thickness or in leaf density. Sites of growth, e.g. cereal grains, are often distant from photosynthetic tissues and require the DM to be translocated to them.

In order to analyse crop growth it is usual to determine dry weight and leaf area at known time intervals. These data can be used to obtain the mean crop growth rate (Fig. 2.2) between successive sampling dates (t_x and t_y) and mean leaf area index. [Thus, growth analysis, which measures the DM accumulation, estimates *net* photosynthesis (to measure *gross* photosynthesis more sophisticated techniques must be employed, such as gas exchange analysis under varying conditions, or the incorporation of radioactive carbon dioxide).]

Mean crop growth rate $\left(\overline{\text{CGR}}\right) = \left(W_y - W_x\right)/P$ $\left(\text{between } t_x \text{ and } t_y\right)$

Mean leaf area index $\left(\overline{\text{LAI}}\right) = \left(A_x + A_y\right)/2P$ $\left(\text{between } t_x \text{ and } t_y\right)$

where W = dry weight, A = leaf area and P = cropping area.

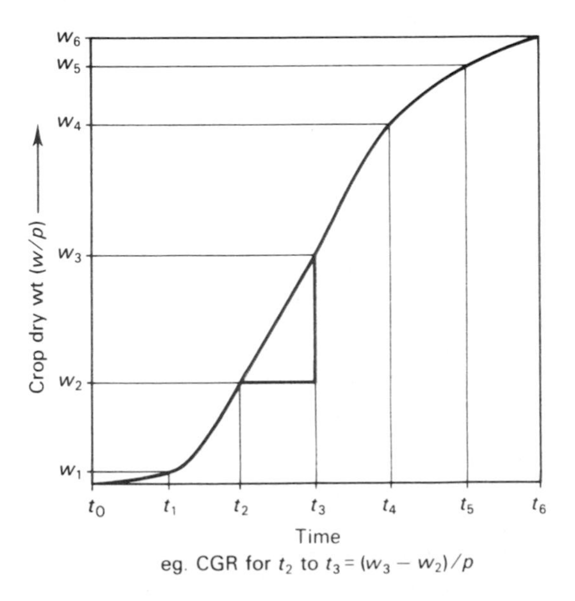

Figure 2.2 Typical crop dry matter growth curve showing calculation of mean crop growth rate ($\overline{\text{CGR}}$).

From the measurements of dry weight and leaf area an estimation of crop photosynthetic efficiency (net assimilation rate) can be made:

$$\text{Mean net assimilation rate}\left(\text{NAR}\right) = \frac{\left(\log_e A_y - \log_e A_x\right)\left(W_y - W_x\right)}{\left(A_y - A_x\right)}$$

The relationship between these three parameters of growth is:

$$\overline{\text{CGR}} = \overline{\text{LAI}} \times \overline{\text{NAR}}$$

There are a number of other growth analysis formulae which are used to analyse growth in more depth and these have been the foundation of mathematical modelling of crop growth.

Potential yield

The total DM of a plant is termed its biomass. Potential biomass is determined largely by photosynthetic capacity. The proportion of the potential biomass that is economic yield is determined by the extent to which DM is partitioned into useful plant parts; this gives rise to the expression harvest index.

$$\text{Harvest index} = \frac{\text{Economic yield}}{\text{Total biomass}}$$

Both photosynthetic capacity and harvest index are very largely genetically determined so that potential yield for any crop can be theoretically calculated for any given climate.

Many factors may limit photosynthesis to less than capacity, and may influence the partitioning of DM to give a less favourable harvest index, so that *actual* yields rarely reach *potential* yields. This shortfall is termed the yield gap. The practice of efficient crop production tries to close the yield gap by understanding the relationships outlined in this chapter, and acting upon them as described in other chapters of this book.

Crop development cycle

From germination to maturation, crops go through a development cycle which involves an increase in complexity usually accompanied by an increase in biomass. The cycle is a result of the interaction of the genotype (see 'Plant breeding' at the end of this chapter) of the plant and the environment it is grown in. The environment provides not only the basic needs of the crop for growth, i.e. light, nutrients, water and heat, but also stimuli to trigger developmental sequences.

During its development the crop exists in several 'states' and passes from one state to another in an organized manner via a number of 'processes' (Fig. 2.3).

Germination

The germination of a non-dormant seed will be initiated given the presence of water, an aerobic medium and sufficient warmth. (For most temperate crops the temperature must be above 0–5°C for germination, with an optimum of about 20–25°C.) A few cold-sensitive crops will not germinate below a higher temperature, e.g. forage maize requires a minimum temperature of 8–10°C for germination.

Water is imbibed through the testa or through the micropyle and returns the existing cells to full turgor. When a seed enters germination it passes from a quiescent state with extremely low metabolism and a high resistance to drought and frost stress into a state of high metabolic activity and extreme susceptibility to stress. Initially water content of the plant rises from 5–10% to 30–40% as water is taken up by imbibition. At this point most crop species progress directly into active metabolism, but in those that show dormancy mechanisms germination may cease until dormancy has been broken. Thus, many weed seeds exist in the soil in an ungerminated but imbibed state. In the absence of dormancy or when dormancy has been broken, germination can continue and metabolic rate rises dramatically, with an increase in protein and nucleic acid synthesis. Food reserves are mobilized, often as a result of hormone stimulation of enzyme synthesis. For example, in barley, gibberellic acid stimulates the mobilization of protease and the synthesis of α-amylase, which catalyses the breakdown of starch in the endosperm for translocation as sugars.

The expansion of the cells of the embryo and its development into the seedling root (radicle) and shoot (plumule) rupture the testa, allowing access to oxygen for respiration, and further expansion of the seedling. From this time the seedling ceases reliance on its stored food reserves and begins to harvest the resources of its environment. Initially, a greater proportion of the seed reserves are directed to root growth in most species to ensure an adequate supply of water and minerals before the photosynthetic machinery becomes fully operational.

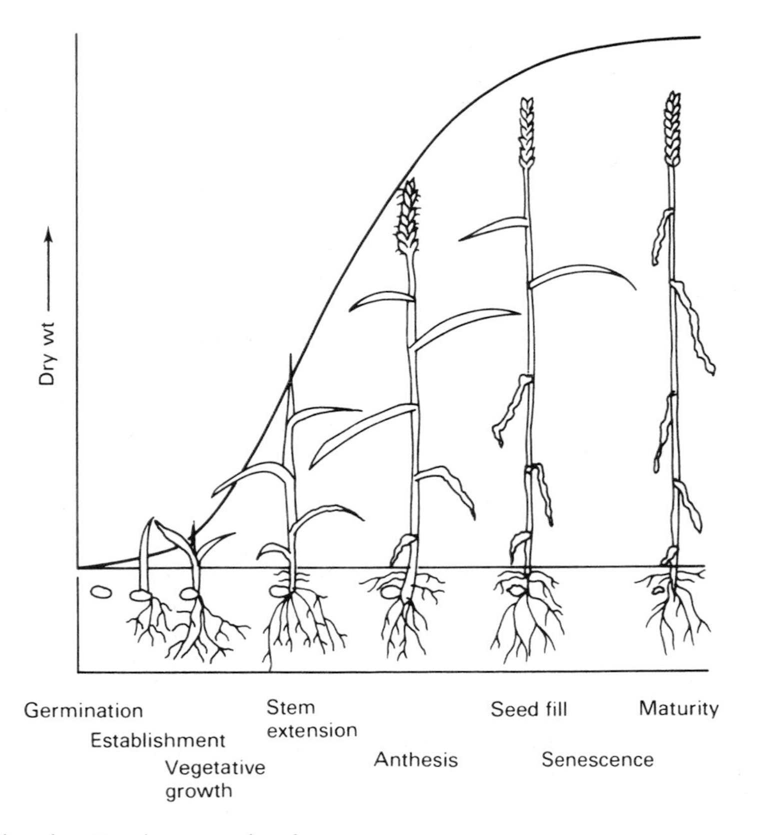

Figure 2.3 Growth cycle pattern in an annual seed crop.

The size of the seed, which tends to be related to the volume of food reserves, dictates the length of time for which the developing embryo can delay becoming fully independent, e.g. germinating field beans can survive on seed reserves for much longer than germinating oilseed rape. This factor has considerable influence on the optimum depth of sowing for each crop species. In order to achieve early photosynthesis to supplement the seed reserves, the cotyledons are frequently expanded above ground (epigeal germination) in small-seeded crops.

For effective establishment, commercial crop seed lots should exhibit a rapid rate of germination once imbibed, and a high germination capacity (indicated by the germination percentage). High vigour seeds are particularly required when field conditions are not optimal for germination.

Vegetative development

During vegetative development, the shoot apex develops primordia which subsequently develop into leaves. A leaf first appears as a bump (primordium) on the side of the apex and then undergoes cell division and differentiation. A midrib consisting of protoxylem and protophloem develops and establishes contact with the vascular system of the stem so that it can obtain assimilates and water for its continued development. As the leaf emerges into the light, cell components differentiate, proplastids develop into chloroplasts and the leaf appears green. Photosynthesis can proceed as soon as the chloroplasts and stomata are functional.

The point of attachment of the leaf on to the stem is important as it is here that the vascular connections are made, allowing water to pass to the leaf for transpiration and for assimilates to pass from the leaf for relocation to other growing parts or storage organs. This connection of leaf and stem is called a node, and the stem between two nodes an internode. In many plants, particularly cereals and oilseed rape, there is a stage of development when the internodes are not expanded and the plant takes on a prostrate or rosette appearance. This adaptation keeps the apex of the plant close to the ground where it is more protected from frost and is a useful feature of overwintered crops. In the spring, such plants demonstrate internode expansion and the apex is pushed up into the air as the crop approaches flowering. Internode expansion is primarily expansion growth of stem cells and is followed by the strengthening of the cell walls to give the stem stiffness. At any one time usually only one internode is expanding rapidly, although expansion phases of successive internodes frequently overlap. In most plants, this internode expansion is under the control of gibberellin in the plant, and chemicals that interfere with its synthesis or action are commercially available as stem shorteners.

The final number of leaves that a plant bears (assuming it does not branch) is dependent on the duration of the vegetative phase of the stem apex. At some point the stem apex will receive a stimulus which will trigger a change from producing leaf primordia to producing floral primordia. This change in function of the apex is invisible to the naked eye and usually does not alter the outside appearance of the plant for some considerable time. For example, in winter barley, floral induction often occurs in November or December, but the ear does not appear until the following May. Systematic analysis of developing crops, including the microscopical examinations of stem apices, has resulted in the availability of development keys for many species which have become a valuable tool for crop management, e.g. nitrogen timing to winter wheat. The stimuli that trigger changes in the development of the stem apex include photoperiod and low temperature (vernalization) and, in some species, growth for a set number of days of thermal time.

At the same time as shoot development is occurring, root development is also progressing. Root apices branch to provide a root network specialized for the uptake of water and soil minerals. Apices are located behind a protective root cap and produce new epidermal cells as root hairs which increase the root surface area and therefore the absorption rate of water and minerals.

Normally in a crop the root system develops in a progressively deeper and more extensive manner, maintaining a balance with the shoot size (root : shoot ratio); however, as rapid spring growth occurs root growth often declines. It becomes important therefore to establish a well-rooted crop prior to rapid spring growth.

Reproductive development

Reproductive development starts with the induction of flowering. Externally the plant will continue to expand leaves which have been initiated up to the point of floral induction. By the time the floral structures are visible to the naked eye they are already fully developed, with stamens bearing immature pollen and ovaries with developed ovules. The actual process of flowering involves the opening of the flowers, rupture of the stamens and release of pollen (anthesis), pollination of the stigma, growth of the pollen tube and, finally, fertilization of the ovule. In relation to the whole period of floral development, flowering is only a brief phase.

Some crop plants show remarkable synchrony of flowering, e.g. cereals, with all florets flowering within a few days (determinate flowering). Others, however, have a protracted flowering period lasting weeks (indeterminate flowering), e.g. field beans and oilseed rape. Determinate flowering is a desirable feature in a crop plant since it leads to an evenness in ripening.

Seed development

Following fertilization, cell proliferation to create embryo and storage cells occurs. The embryo tissues develop into a polarized structure which defines the sites of the stem apex and the root apex. Often the shoot apex undergoes further development and will produce leaf primordia, e.g. in wheat the first three to four leaves are already initiated in the embryo before the seed is shed from the plant. The mass of storage tissue cells becomes filled with storage compounds – starch, lipids, protein.

The duration of seed filling is temperature dependent, lasting longer at cooler temperatures. If the supply of assimilates during seedfill is limited, the seed may not fill to its maximum capacity and will appear shrivelled or may abort. In a crop plant, therefore, greater seed-filling capacity will occur in a cool environment where the length of seedfill will be protracted, provided the crop is not stressed by drought and provided photosynthesis is not impaired. This contributes to the higher yield potential of winter wheat in Scotland compared with southern England. As seed filling progresses, the moisture content of seed falls. At the end of seed filling, the process of maturation begins, with continued loss of moisture from the seed, the cessation of development of the embryo and frequently the induction of dormancy. Vascular connections between the seed and the mother tissues are discontinued and the seed becomes an independent body. Some further seed development can occur after seed shed but is usually completed when moisture content declines to 5–10%.

Agriculturally, seeds are of major importance as the main food source for animals and man (e.g. wheat, barley, rice, maize, beans), as well as providing many beverages (e.g. coffee), medicines (e.g. evening primrose), spices (e.g. mustard) and industrial oils (e.g. linseed).

Senescence

Senescence is occurring throughout the life of a crop plant as old leaves are being replaced by young leaves. However, 'crop senescence' is a term that is usually applied to a crop where the rate of senescence is greater than the rate of growth and the net effect is that the crop is dying. Crop senescence is usually noticeable from about mid-seedfill onwards.

Senescence is a two-phase process, the first phase of which is an organized breakdown of the tissues, where many of the products of the breakdown can be remobilized to other parts of the plant. In this way certain nutrients can be reused for new growth. The initial stages of senescence are not visible to the naked eye but involve a depression of photosynthetic activity followed by a dismantling of the photosynthetic apparatus, leading to a breakdown of chlorophyll. The second phase of senescence is the rapid self-destruction of cellular integrity by powerful hydrolytic enzymes.

Following senescence, leaves are often shed from the plant after the formation of an abscission zone in the stem at the point of attachment of the leaf. The cells in this zone become filled with inert substances which effectively act as a barrier to pathogens at the scar left when the leaf is shed. Abscission layer formation is a feature of dicotyledonous plants; monocotyledonous plants, such as cereals, do not form such layers.

Environmental influences on growth and development

Temperature

Temperature and growth rate

The growth rate of most plants responds to changes in temperature (Fig. 2.4). The maximum temperature for growth (T_{max}), the optimum (T_{opt}) and the base temperature (T_{base}) vary between species, but for UK crops typically fall between 0 and 35°C. For crops native or well adapted to the UK, e.g. perennial ryegrass and winter wheat, T_{base} is around 0–1°C. However, for crops originating from warmer climates, e.g. maize and sugar beet, T_{base} is around 6–8°C. As a consequence, winter wheat and ryegrass are able to grow early in the spring (March and April) while maize and sugar beet do not commence rapid growth until much later (June/July). In a temperate climate this has important consequences with respect to canopy development (see 'Light capture by crop canopies'), radiation capture and yield potential.

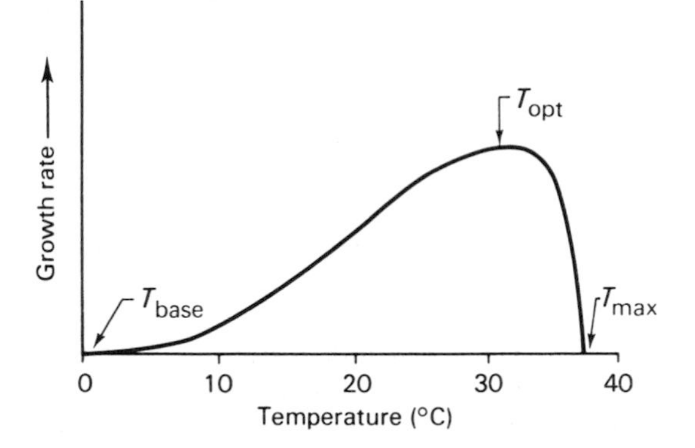

Figure 2.4 Growth rate response of plants to temperature.

Temperature and development

The progression of a plant through a sequence of development is closely related to the temperature it experiences over a period of time. Temperature varies both diurnally (within a day) and on successive days, and plants integrate their response to temperature over time. Thus, the amount of heat a plant experiences in a day can be called its daily heat unit and a set number of heat units are required between developmental stages. Some heat unit systems are simply the sum of the average daily air temperatures, while others use the daily maximum and minimum air temperatures in a more complex formula.

Heat units may be used in crop production in at least two important ways. First, with reference to past meteorological data, areas can be mapped for suitability for growing particular crops. For example, forage maize can only be grown successfully in the UK in areas that experience over 1250 heat units over 6°C during the months of May–October inclusive (see 'Forage maize' in Chapter 3). Secondly, by using up-to-date meteorological data, the stage of development of a crop in the field can be pinpointed and its subsequent development over the forthcoming week or two can be predicted. Where husbandry inputs are very time responsive, such prediction can be of great benefit in helping to achieve optimum use of inputs. Such prediction schemes are available for nitrogen and growth regulator timing to cereals.

Heat units may be modified by photoperiod, which can effectively delay or hasten the thermal response.

Temperature can also affect development by acting as a trigger for a change in the development sequence. In particular, low temperature can trigger flowering (vernalization) and dormancy.

Vernalization

Many biennials (e.g. sugar beet), winter annuals (e.g. winter cereals) and perennials (e.g. perennial ryegrass) require a period of vernalization over winter before flowering is possible the next spring. In the UK this is normally achieved quite adequately during the winter, but if winter cereals are sown very late (after February–March) then flowering and yield may be seriously impaired. Within winter wheat there exist varietal differences in the vernalization requirement and care should be taken not to sow varieties with low vernalization requirement early in the autumn and, conversely, not to sow varieties with a high requirement too late.

Maincrop sugar beet production exploits only the first year of the biennial cycle of this plant, and flowering in the first year (bolting) reduces crop yield, interferes with machinery and can introduce a weed beet problem. The vernalization requirement of the crop usually prevents flowering in the first year, but if the crop is sown early in the spring or if unusually low temperatures are experienced, flowering of a proportion of the crop may occur. Plant breeders have attempted to produce varieties of high vernalization requirement so that early sowing can be practised.

Dormancy

Low-temperature-induced dormancy can be regarded as a survival strategy in certain plants. True dormancy is common in woody perennials and bulbs but is rare in UK agricultural crops. Perennial ryegrass exhibits a winter quiescence which appears similar to dormancy, but upon close study it is found that new leaves are continually being produced at the same time as the death of old leaves occurs. Low temperatures in winter and the lack of net growth are often mistakenly referred to as dormant periods. Close inspection would reveal that development has progressed whereas in true dormancy development ceases.

Dormancy is also apparent in seeds and helps prevent germination in unfavourable environments. Thus weeds that have little or no frost resistance in the vegetative state require a cold shock to the seed (obtained over winter) to break their dormancy before germinating in the following spring. In this way the weed avoids frost which may kill off the species.

Other dormancy-breaking stimuli include extremes of heat, exposure to light, exposure to high concentrations of specific ions, degradation of seed coverings, washing away of inhibitors and the passage of time to allow seeds to fully mature or ripen. Seed dormancy is common in weed species and this, coupled with the prolificity of seed set of many weeds, makes it virtually impossible to eliminate a weed species completely from a field. In contrast, crop species have had dormancy mechanisms bred out of them to ensure rapid and even germination after sowing. Even so, cereals retain some dormancy that helps prevent sprouting in the ear before harvest, which is extremely important both for breadmaking wheat and malting barley.

Temperature-induced stress

Temperature stress to plants can be caused by extreme heat or cold, although heat stress is not normally encountered in UK crop production. Cold stress, however, particularly frost damage, commonly occurs and can lead to complete plant death or to a degree of damage. Some crop plants have little or no inherent frost tolerance, e.g. maize and potato, and for this reason are restricted in their cropping to the frost-free months (May–October) in lowland areas. Other crop species have varying degrees of frost tolerance and there is frequently varietal variation within a species.

The obvious advantage of frost resistance in a crop species is that it allows the crop to be planted in the autumn, which generally raises the yield potential. Species of plants that are capable of withstanding frosts in the winter are not equally frost-hardy throughout their life cycle. During the declining temperatures and daylengths of autumn they undergo a conditioning process known as hardening (or acclimation). This raises the frost-hardiness in 'anticipation' of the ensuing winter. In the following spring as temperatures and daylength increase, this hardiness declines and the crop may become susceptible to damage by late frosts. Hardening temperatures are typically in the range 0–8°C.

Some plants escape damage by frost by a frost-avoidance mechanism. The extreme of this is for the plant to exist in a very resistant form, e.g. a seed or a bulb. Another avoidance mechanism is undercooling (or supercooling) where the tissues attain a temperature below their freezing point without the formation of damaging ice crystals. This may only be effective for mild frosts (−1 to −2°C) of short duration and can be dependent upon the absence of ice nucleators (e.g. certain bacteria). Other plants protect their sensitive tissues with layers of hardy tissue which delay and therefore prevent the penetration of damaging frosts, e.g. the wrapper leaves on winter cauliflowers and the bud scales on woody deciduous perennials. With some high-value crops it can be worthwhile investing in frost protection measures such as the use of polythene sheeting, e.g. for early potatoes, or of fogging or misting techniques, e.g. to provide blossom protection in orchards.

Light

Photoperiod

Plants have adapted to use photoperiod in many ways because it is the only truly predictable parameter of the environment detectable by the plant. Notable examples occur with flowering and germination which help a species to synchronize critical phases of development. All photoperiodic responses involve the compound phytochrome as the detector compound at the cellular level.

In respect of flowering, plants can either be photoperiodic responsive or day-neutral. Day-neutral plants flower after a set time from germination or dormancy breaking, e.g. annual meadow grass. Photoperiodic-responsive plants are either 'long-day' or 'short-day' types requiring exposure to the correct daylength to either start or accelerate flowering. Temperate plants are frequently long-day types, which helps ensure flowering in the summer months.

Light capture by crop canopies

The crop canopy is defined as the structure of leaves, stems and flowers found above ground. At full development the depth of the canopy can vary from 30 cm (grass) to more than 30 m (agroforestry).

It is important that for a crop to achieve high growth rates and high yield potential it must (1) fully intercept the incoming solar radiation, (2) make efficient use of that radiation within the canopy and (3) produce its canopy at a time when incoming radiation is at its maximum.

It is found that the growth rate and the DM production of many crops are linearly related to the quantity of radiation intercepted during the life of the crop. Incoming radiation and therefore cropping potential vary considerably throughout the year in temperate regions with average daily radiation totals of 16 MJ m^2 day^{-1} in June/July falling to only 2 MJ m^2 day^{-1} in December/January.

Within a day, actual radiation receipts (the daily insolation) depend on the time of year, which dictates the daylength and maximum radiation intensity, and the degree of cloud cover. Cloud cover may reduce the actual radiation reaching the crop by as much as 75% compared with a clear sunny day. In this way excessive cloud cover can reduce cropping potential, whilst in years of high radiation receipt (little cloud) cropping potential is raised. This was certainly a major contributory factor in the record-breaking yields obtained from cereals in 1984.

Measurements of crop canopies have revealed two important expressions which are useful: leaf area index (LAI) and leaf area duration (LAD). LAI is a measure of the surface area of the photosynthetic area of a crop per area of ground occupied, whilst LAD is the integral of LAI with time (LAI × time). Thus, on a graph of LAI versus time, LAD is the area under the LAI curve. The value of LAI that has to be achieved by a crop before efficient interception of radiation occurs varies with the crop and the plant spacing but commonly needs to exceed 3–4. Some crops that are well adapted to improving radiation conditions in the spring, e.g. cereals, are able to adapt their canopy structure and raise their LAI to more than 10 and achieve high growth rates and high physiological efficiency.

Factors affecting light interception by crops

Early sowing Advancing the sowing date of most crops leads to a better chance of early ground cover and a better timing of the peak LAI closer to mid-summer when radiation levels are higher. This effect is maximized if the crop is frost resistant and sowing can be done the previous autumn, e.g. cereals, oilseed rape. Such early sowing usually also prolongs the growing period, allowing more radiation to be intercepted by the improved LAD. This can be illustrated with reference to wheat (Fig. 2.5) where autumn sowing leads to

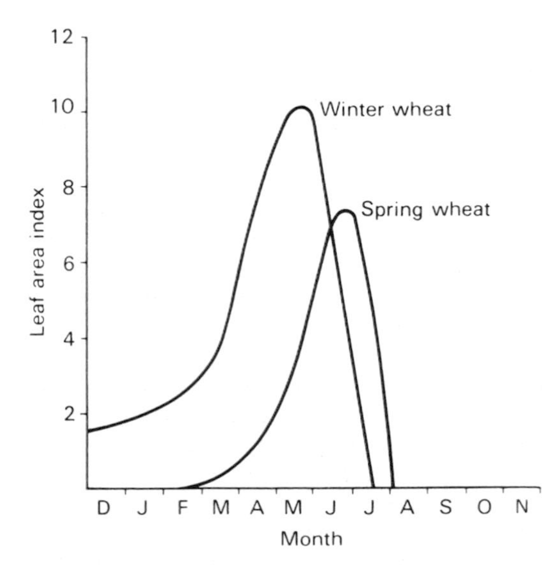

Figure 2.5 Leaf area index development in winter and spring wheat.

increased yields through improved canopy timing, higher LAI and increased growing period leading to greater light interception over the life of the crop. Many crops are, however, limited in the degree to which the sowing date can be advanced by temperature limitations, e.g. sugar beet, potatoes, maize, peas. Here, the advantages of early canopy development are offset by the risks of frost damage or poor, slow germination. In some high-value cash crops temperature protection can be supplied by plastic film which acts as a 'mini-greenhouse', e.g. for first early potatoes, early vegetables and sweetcorn.

Seed rate Increasing seed rate can have the advantage of producing an early rise in LAI with ground cover occurring early. This may be of particular importance in crops with short growing periods, e.g. spring and summer sown crops. Indeed, in spring cereals seed rate becomes more and more critical with later and later sowing. For crops with long growing periods compensatory growth (side-branching, tillering) frequently diminishes the advantages of high seed rate. Using high seed rates can produce disadvantages to the crop by leading to excessively high competition between plants. This has the effect of lowering the size of important structures, e.g. grain or root size, and can lead to lodging (canopy collapse).

Nitrogen The application of nitrogen to most arable crops raises yield potential. This is achieved through a stimulation of cell growth, particularly in the developing leaves, which increases leaf size and thus LAI. Also, leaf longevity is improved, i.e. senescence is delayed, which helps to improve LAD. Nitrogen also improves photosynthetic capacity by increasing levels of the CO_2 harvesting protein (Rubisco) and increased chlorophyll-binding protein leading to higher chlorophyll levels (greener leaves).

If nitrogen is available in very high amounts, then overstimulation of growth can occur which can be counterproductive. Leaf area index may be increased to such an extent that lower leaves in the canopy are severely shaded and fall below the compensation point (respiration greater than photosynthesis) and such leaves effectively become 'parasitic' on the plant. Also, excess nitrogen may cause a weakening of stems leading to complete canopy collapse (lodging). Furthermore, stimulation of vegetative growth by nitrogen can be at the expense of the growth of important structures, e.g. tubers in potatoes, roots in sugar beet.

Canopy structure Photosynthesis in the leaves of most crop plants is saturated at radiation intensities of about 300 $W\,m^{-1}$ (or 150 $W\,m^{-1}$ of photosynthetically active radiation). This means that, on many days in spring and summer, leaves at the top of a canopy are saturated in radiation for maximum photosynthesis. Unfortunately, such leaves will continue to absorb radiation but will be unable to utilize it and the radiation is therefore wasted because it is not available to the rest of the canopy. There is benefit to crop photosynthesis if the extra radiation can be allowed to penetrate the canopy and be intercepted and used by lower leaves. Canopy structures that allow such penetration have been shown to improve crop growth rate and yield (Fig. 2.6). It is perhaps not surprising to find that crops such as cereals which are well adapted to the UK climate show the best type of canopy structure.

Common problems depressing canopy effectiveness

Lodging The collapse of the canopy caused by wet and/or windy weather conditions and/or the weakening of the stem and/or roots will drastically reduce crop growth rate. Lodging is most serious in deep well-structured canopies with high yield potential, e.g. cereals. As well as depressing yield, lodging may also lead to a spoiling of the economic product by soil contamination or rotting diseases, and can lead to a surge of weed growth up through the crop. Lodging may not be serious if the crop is nearly ripe but will frequently slow down and delay harvest.

Lodging commonly results from weakened stems through high nitrogen use, high seed rates, stem-based diseases or weak-stemmed varieties and from poor supportive root systems occurring in unconsolidated seedbeds.

Foliar diseases Foliar diseases depress crop yield by depressing the effective leaf area and thus lowering the LAI. Many diseases are actually affecting a greater leaf area than is apparent from their visual symptoms, and this means that they have to be controlled at a relatively early stage in their infection. Infection on lower leaves in the canopy is generally insignificant since these leaves are contributing least to the gross production of the crop. Similarly, in autumn-sown crops autumn disease infection is frequently not worth controlling since crop growth rate is slow and protection of leaf area is not as necessary as during the main growth period in spring.

Unfortunately, well-structured crop canopies are frequently ideal environments for disease multiplication and spread, and in attempting to achieve high yields from good canopies, disease control will normally be necessary.

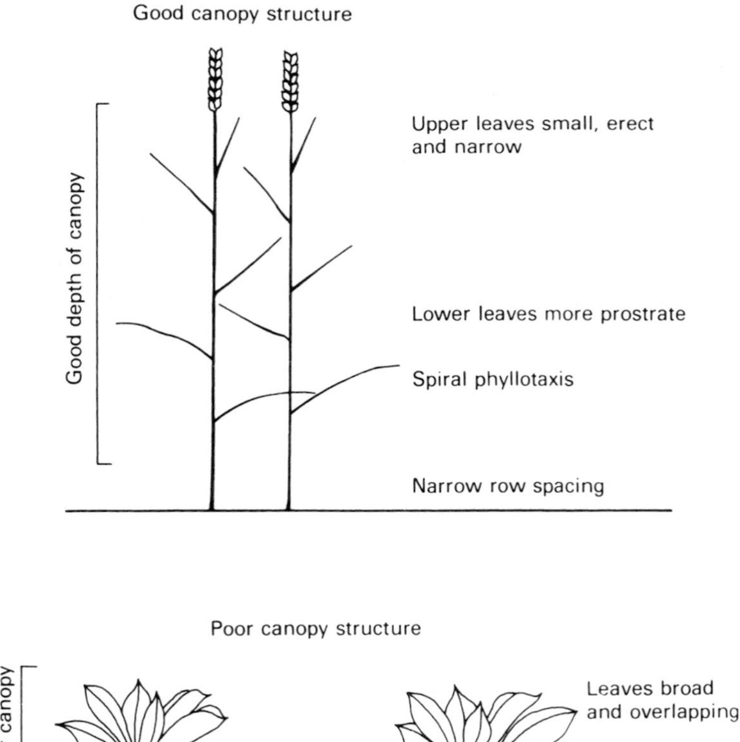

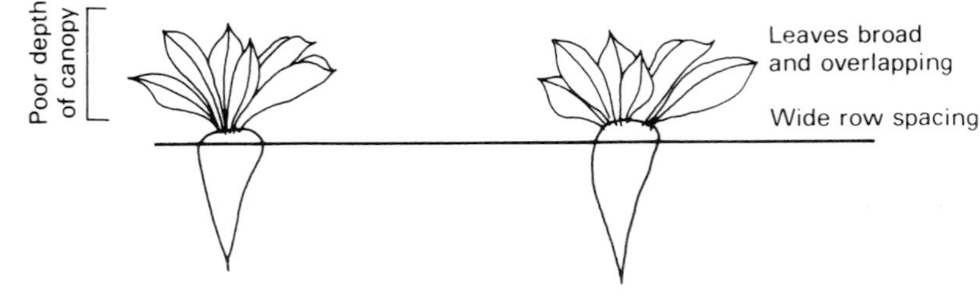

Figure 2.6 Crop canopy structures.

Drought Most UK crops can withstand a good deal of water stress without injury. However, even mild stress can provoke a drought response which lowers productivity of the canopy. Stomata close to conserve moisture under drought stress and at the same time restrict carbon dioxide uptake into the leaf and depress photosynthesis. If drought stress is severe or prolonged, it will have an adverse effect on canopy structure since older leaves will senesce and die whilst younger leaves may wilt or roll up. New leaves produced during drought stress will be smaller and less photosynthetically responsive.

Since the risk of drought stress is frequently towards the end of a growth cycle, it is likely to affect the crucial partitioning of assimilates to the economic portions of the crop and can drastically affect economic yield.

Competition

Competition occurs when two or more plants are growing in an environment and the combined demands of the plants exceed the supply of one or more of the limiting factors for growth and development. These factors include water, soil nutrients, soil oxygen, carbon dioxide and light. Space is frequently referred to as a limiting factor, but in reality space embraces two or more of the factors already listed.

Within a crop, competition occurs between plants of the same species and is termed intraspecific competition. Between plants of different species it is termed interspecific competition.

Intraspecific competition (plant populations)

In the extreme case of a crop plant growing in complete isolation, i.e. in the absence of competition, its individual yield will give an indication of the maximum yield possible per plant. As the population is increased, competition between plants will increase and individual yield per plant will decrease. However, yield per unit area (crop yield) will increase (Fig. 2.7). This is because, despite each plant not fulfilling its full potential production, a more efficient use of the limiting factors for growth is being made by the community of plants (the crop). This is particularly the case with respect to the efficiency of utilization of light and the ability of the crop canopy to capture more light at higher populations. As population increases, the yield response starts to diminish until a plateau is reached and no further response to plant population can be achieved (asymptotic response curve – Fig. 2.8). In biological terms the optimum plant population in such a case is at the point where the plateau starts whereas in practice the optimum is lower than this when seed costs are taken into account. A dense community of plants also has the advantage of synchronizing and accelerating ripening.

Plant density affects the partitioning of assimilates within the plant, with high densities tending to lead to a greater vegetative component and a lower reproductive or storage component per plant. If the adverse effect of partitioning away from an economic portion is not outweighed by the advantage of increased plants per unit area, then economic yield will decline beyond a certain point (parabolic response curve – Fig. 2.8). Often an asymptotic yield response is associated with total biomass production and crops grown for complete utilization, e.g. grass and other fodder crops. In contrast, the parabolic yield response is associated with crops grown for economic yield, e.g. cereals, oilseed rape, potatoes, sugar beet, peas.

When considering crops showing a parabolic yield response, obtaining the optimum plant population is very important. However, if a suboptimal population is established, then compensatory growth may occur as individual plants can have a greater share of the limiting resources and attain a larger size. In crops able to show a high degree of compensatory growth, e.g. winter cereals (by way of tillering), the parabolic curve becomes a wide flat-topped response curve.

In many crops, especially those that show a high degree of compensatory growth, the unit of population may not always be simple to define. For example, in potatoes it is the stem population and in cereals the tiller population that is of the greatest value. Competitive demands in the crop are not constant but change with time and may lead to alterations in the density of the unit of population (Fig. 2.9). In the case of cereals, the mortality of stems in the spring is a result of competitive demands within the plant for assimilates and can be greatly offset by applying nitrogen to improve leaf area and thus assimilate supply.

Increasing plant populations can also lead to higher mortality rates within a crop (thinning response).

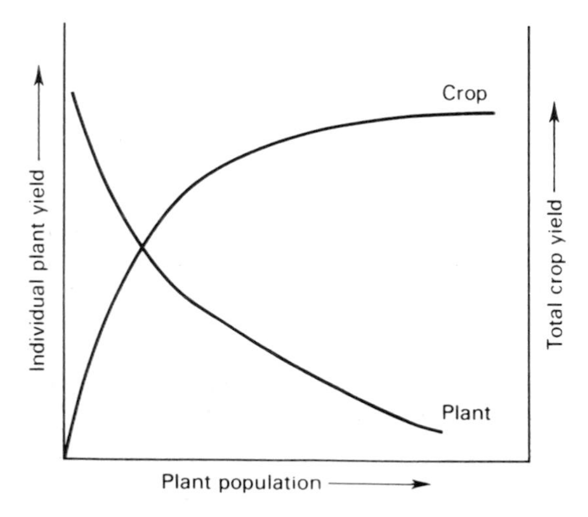

Figure 2.7 Plant and crop yield in response to increasing plant population.

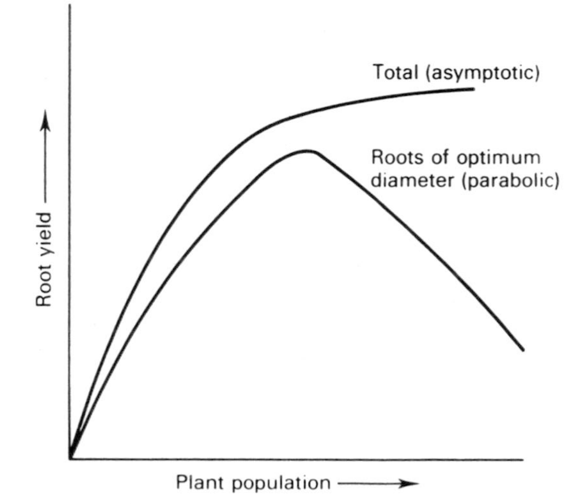

Figure 2.8 Response of root crops to increasing plant population.

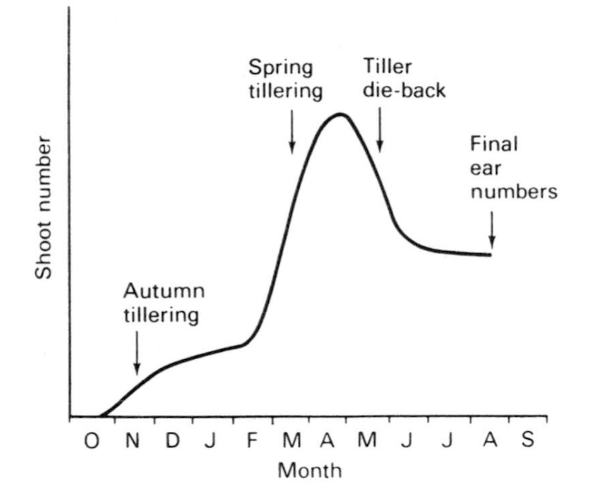

Figure 2.9 Variation with time of the unit of density in a winter cereal crop.

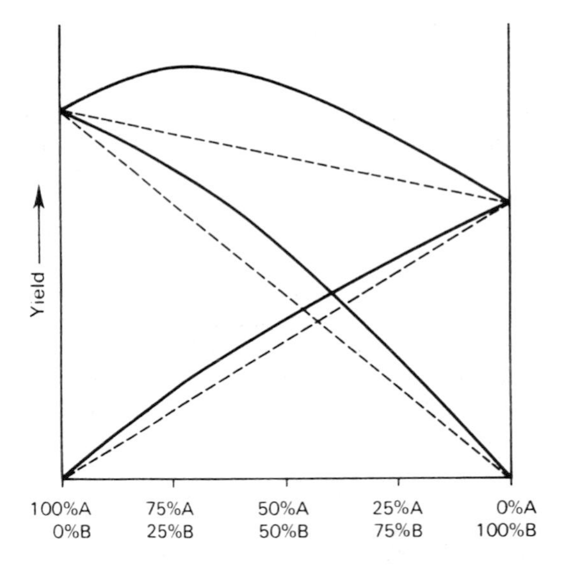

Figure 2.10 Theoretical replacement series diagram to analyse the performance of a mixture of species A and B. Solid lines indicate observed yields; broken lines indicate expected yields (extrapolated from monoculture yields) if no competition were to occur. Diagram demonstrates mutual cooperation with a maximum mixture yield benefit from 70%A and 30%B.

Usually, in a highly competitive situation, 'weaker' genotypes or individuals located in an unfavourable position will be disadvantaged and may well die. This could be considered to be a waste of seed but may be a necessary loss to achieve the desired plant population.

Interspecific competition

Competition for resources occurs between species in crop mixtures, e.g. grass/clover swards, cereal/legume arable silage crops, where two or more crops are harvested (or grazed) together, and between a crop and its weeds. (Intercropping, where two or more crops are grown together but harvested separately, e.g. maize and beans, is very little used in the UK but is common in other countries. The physiological principles of intercropping are identical to those of mixtures.)

The total biomass produced per unit area of ground when more than one species is present will frequently be greater than that produced by one of the species grown alone at the same density. This is because species of different habit are able to utilize the available resources more efficiently, chiefly by better spatial distribution within both the aerial and soil environments, e.g. by forming a more efficient leaf canopy, or by tapping a greater depth of soil for water and nutrients. In general, the greater the difference in habit of the component species the greater will be the yield advantage of the mixture, and conversely the less will be the degree of competitive pressure between the component species. This yield advantage is exploited by crop mixtures, but may be disadvantageous in crop/weed situations, depending on the degree of weed competition.

Quantitative analysis of the degree of competition and its effects is important to determine (1) the optimum mixture proportion, and (2) the economic threshold for weed control. Many forms of analysis have been devised appropriate to each type of interspecific competition situation. One of the simplest and most widely useful is the replacement series analysis for two-species competition. Yield data are collected from the monoculture of each species, and from a range of mixture proportions (it is usual to employ the same plant density throughout the experiment though this may lead to bias in the case of two species with widely differing single plant sizes), and analysed using a replacement series diagram (Fig. 2.10). This form of analysis can indicate the optimum mixture proportion in yield terms, the extent of the yield advantage and the behaviour of the component species within the mixture. Three cases are possible:

(1) mutual cooperation, in which the yield of both components is enhanced, giving an overall yield advantage;

(2) mutual inhibition, in which the yield of both components is depressed, giving an overall yield reduction;
(3) compensation, in which one component is enhanced and the other depressed; this case may lead to a yield advantage, a yield depression or a balanced result, depending on the relative effects on each component.

Analysis of this type often demonstrates that one of the component species has competitive superiority. Competitive superiority is conferred by general spatial superiority, e.g. greater height, a broader leaf canopy, better root distribution, all enabling the species to gain a disproportionately high share of the light, water and nutrients, by shading its neighbours, achieving a higher photosynthetic rate, increasing its transpiration rate and thus its rate of mineral uptake, resulting in higher growth rates and even greater competitive superiority. Competitive advantage may also be achieved through temporal means, e.g. weeds emerging prior to crop emergence will tend to dominate, and vice versa, though this relationship will be mediated by intrinsic, and temperature-influenced, growth rate and habit differences. Yield advantage from competition may be enhanced by symbiotic relationships, e.g. legumes donate atmospherically fixed nitrogen from their decaying nodules to their non-legume partners. Competitive advantage is gained by some species through allelopathy, the exudation of toxic substances from their roots.

Spatial and temporal distribution interact and are profoundly influenced both by the environment and the overlying plant density, making physiological interpretation and general prediction of competitive situations extremely difficult, particularly when more than two species are involved.

Modification of crops by growth regulators and breeding

Plant growth regulators

Regulatory compounds known as plant hormones influence many of the processes of plant growth and development. These compounds act both independently and interactively at several levels of concentration to achieve control. The internal (endogenous) levels of some of these hormones can be manipulated either by applying further hormone to the plant (exogenous application) or by applying synthetic regulators which alter the endogenous hormone levels.

Plant growth regulators may be regarded as a short-term answer to a plant breeding problem though they may in effect be a more cost-effective solution. Discovery and characterization of plant growth regulators can be extremely difficult since effects on plants are frequently transitory and compensated by subsequent growth phases.

Plant breeding

The form of any organism (phenotype) is fundamentally determined by its genes (genotype). The degree to which any character is expressed is modified by the environment. Crossing of individuals or within populations, followed by a lengthy programme of selection, enables the plant breeder to change the genotype of an organism and so to modify its phenotype, thus creating a new cultivar.

The objectives of any breeding programme will be numerous but will usually include those leading to an improvement in yield, e.g. improved climatic adaptation, pest and disease resistance, partition of biomass, and those leading to an improvement in quality of product. A considerable proportion of the post-war increase in UK crop production is attributable to the plant breeders. For example, it has been estimated that more than 60% of the yield increase for winter wheat has been due to improved cultivars.

Plant breeding has been accelerated in recent years by the advances made in plant biotechnology. In particular, plant tissue culture and molecular biology are improving the speed of cultivar improvement and breaking down some of the barriers in species hybridization. Transfer of individual genes between unrelated organisms is now possible. Genetic fingerprinting also allows for more precise crossing and selection programmes.

Further reading

Forbes, J.C. & Watson, R.D. (1992) *Plants in Agriculture*. Cambridge University Press, Cambridge.
Hay, R.K.M. & Walker, A. J. (1989) *An Introduction to the Physiology of Crop Yield*. Longmans, Harlow.
Sinclair, T.R. & Gardner, F.P. (eds) 1998 *Principles of Ecology in Plant Production*. CAB International, Wallingford.

3

Arable cropping

A.J. Jellings & M.P. Fuller

Introduction

All crop production seeks to utilize the natural, free resources of the biosphere (energy as light and heat, oxygen, soil, water, mineral nutrients) to obtain maximum productivity from plants useful to man as food, fuel, fibre, flavourings, and medicines. The biological and physical principles upon which this exploitation is based are outlined in Chapters 1 and 2, and these principles underlie crop production everywhere.

Commercial crop production in the UK operates the fundamental principles of productivity from natural resources in a particular economic, political, and social context; this context changes with time. Thus, artificial inputs and energy can be added to the available natural resources in order to increase productivity, but only if profitability can be achieved, and compliance with the appropriate policy and legislation framework can be maintained.

This chapter will consider arable cropping in the context of the UK at the present time, by presenting firstly the principles that are generally applicable, and secondly the particular husbandry of the major arable crops. The reader should recognize that husbandry practice changes with the introduction of new varieties, improved understanding of biological and agronomic principles, and the development of more efficient techniques and machinery. Cropping practice is also influenced by consumer and societal requirements regarding the quality of food and care for the environment, and by economic and political trends that change the relative value of inputs and outputs.

Arable crop planning and production practice

Choice of arable crop

The choice of an appropriate arable crop for any given situation depends on many factors. Potential choices must be evaluated against these factors, using sound information about the crop and the situation in which it will be grown. These factors can be grouped around three key questions:

- Can it be grown at this site?
- Can it be sold (or used) profitably?
- What effect will it have on the farm system and business as a whole?

There is no substitute for local knowledge and experience, and observation of the type of crops grown in the locality will give an instant indication of what can be grown. This commonsense method does not indicate what alternatives might be grown (or what cannot be grown), however, as the local producers may be conservative in their choice of crop enterprises. An analysis using basic principles can be very valuable.

Climatic constraints

The climate will impose restrictions on the crop types and varieties that can be grown in a particular location. A good knowledge of the normal weather pattern for the site is essential; accurate historic average data can be found in *The Agricultural Climate of England and Wales* (MAFF, 1984), or from www.metoffice.com. Key climatic features are the number of days of frost per year, the average monthly temperatures, and the rainfall pattern. These should be considered in relation to the crop features of: minimum and optimum temperature for germination; base temperature for growth; accumulated temperature (thermal time) requirement from sowing to harvest maturity; sensitivity to frost; sensitivity to drought; and any particular requirements in relation to development, for example for vernalization, pollination, or ripening (see Chapter 2).

The greatest flexibility for cropping, with respect to climate, is found in areas with mild frost-free winters, relatively even rainfall distribution, and summers that supply more than 750 day degrees above 10°C between May and October.

Soil constraints

The soil will restrict the crop types that can be grown in a particular location. A good knowledge of the soil characteristics, and their interaction with rainfall pattern, for the site is essential (see Chapter 1). Crops vary in their requirements with respect to soil depth, soil texture and optimum pH, and in their sensitivity to compacted or stony soils, drought, and periods of waterlogging. In general, root crops have the most demanding soil requirements.

The soil and soil water characteristics not only affect the crop directly but also influence the timeliness of mechanised operations, particularly workability at time of establishment, and trafficability at times of major inputs and harvest. Soils offering the greatest flexibility for cropping are deep and medium textured with few stones, in areas with evenly distributed rainfall.

Potential productivity

The potential yield from a particular crop at a particular site is of considerable significance in relation to choice of crop and cropping system. For most crops the value of the output will be strongly correlated with yield per unit area, and this will be one of the determinants of whether it will be profitable to grow the crop. The potential yield will also be a strong determinant of the level of inputs, and thus variable costs, to be devoted to the crop; in this way potential yield also influences the cropping system employed.

As was seen in Chapter 2 dry matter production is fundamentally determined by the amount of radiation intercepted by the crop. The amount of radiation potentially available is determined by latitude, and the amount potentially intercepted is determined by the genotype of the crop. Thus it is possible to calculate the theoretical maximum dry matter production for any geographical location and any crop. In the UK this theoretical dry matter production is about 35 t/ha for winter wheat; other annual crops have lower theoretical limits due to inferior phenotypes especially with respect to canopy structure. Not all of the dry matter produced becomes yield, much is partitioned into parts of the plant with less commercial value (for example for winter wheat the theoretical maximum grain yield in the UK is approx. 17 t/ha); canopy structure and partitioning can be improved by plant breeding.

Much more significant to the producer is the effect of environment and management on dry matter production, and thus yield. The theoretical dry matter yield is likely to be much reduced by the effects of site and season, and the management of the crop within the geographical and genetic constraints will become critical for profitability. Site effects on potential yield must be assessed carefully to inform crop choice decisions, and crop management planning. These effects are related to climate and soil factors and their interaction, for example a site with low summer rainfall and shallow soil will suffer a yield penalty because dry matter production will be limited by water stress, even with high radiation interception (see Chapter 2). Prediction of potential yield for a particular site is an inexact science; the site class system for grass production (Chapter 4) is one predictive system, and the Agricultural Land Classification of England and Wales (Chapter 1) can also be used in this way. Seasonal weather effects cannot be predicted with accuracy and will only be known as they develop; day-to-day crop management must monitor these effects and adapt accordingly.

It should always be remembered that good crop management will only ensure that the potential yield is approached, it can never raise the potential yield; input decisions must reflect this crucial fact, within an appropriate financial management framework (Chapter 12). The crop enterprises section of this chapter indicates the range of average commercial yields for each crop; this does not imply that yields cannot be outside this range in very good, or in poor, conditions.

Crop marketing

It is fundamentally important that the output from a crop must be saleable (or useable on farm). It is vital that the producer is aware of the range of market outlets available, the quality of product required by each outlet, and any special arrangements, such as quotas or contracts, associated with each outlet.

The basic marketing chain is similar for all crop products and is shown below.

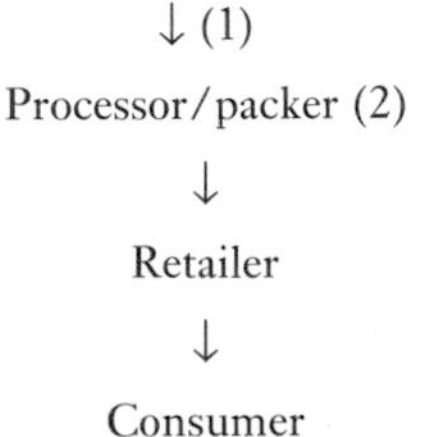

(1) The producer sells either directly to the processor/packer, or to a merchant or wholesaler who sells on to the processor/packer. In some situations it is possible for the producer to take on the roles of processor/packer and retailer and to sell direct to the consumer through a farm shop, farm-gate sales, pick-your-own enterprise, or delivery service, e.g. 'veg box' scheme.
(2) All crop products require some form of processing and/or packing before retail to the consumer, though the sophistication of this step varies widely from, for example, the superficial trimming of the outside leaves of a cauliflower for farm-gate sale, to the high technology of sugar extraction from sugar beet.

The importance of knowing the market options for any crop product, and their associated quality requirements cannot be overemphasized, as this is crucial for sound strategic planning. Generally, the greater the sophistication of the processing step, the fewer marketing options are available to the producer (e.g. see 'Sugar beet' in this chapter), and the more likelihood there is of needing a contract for production of the crop. Where there is a limit to demand, a contract to grow will reduce the risk to an acceptable level (sale of the product is secure provided the quality requirements are met), and there may be a legislative requirement where quotas exist, or in relation to crops grown on set-aside land. It is rare for a 'new' producer to obtain a contract for the highly profitable crops requiring sophisticated processing and with a distinct limit to demand (e.g. evening primrose) as existing contract holders are unlikely to go out of the crop. Where limited outlets exist for a crop product, location will be a strong influence on the suitability of a crop for a specific site, as transport costs can reduce financial margins considerably.

For crops with a range of market outlets it is essential to know the quality requirements of each, and to make sensible judgements regarding the relative profitability of contrasting outlets. It is generally true that the more stringent quality requirements have a higher cost of production, though this may be balanced by a premium price, i.e. the value of the output may be higher. Crops grown under contract must meet the quality specifications, or the output may be worthless. Crops grown for on-farm use, e.g. fodder crops, should also be evaluated in terms of quality requirements, and cost of production, particularly in comparison with bought-in feed.

The value of the output will depend on the price per tonne and the yield; the price at any one time will vary with quality. Both price and yield are unstable (to a greater or lesser extent depending on the crop): the price will vary over time according to supply and demand, and yield will depend on site and season. A predicted value can be estimated for financial planning purposes using typical seasonal patterns in conjunction with recent price trends; John Nix's *Farm Management Pocketbook* is useful in this regard, as are the weekly prices published in trade journals such as *Farmers Weekly*. The income from the value of the output may be supplemented by an Arable Area Payment for those crops included in the Scheme. Net income will depend on the costs of production (see Chapter 12), which will be influenced by the cropping system.

It is rare for a single crop enterprise to be managed in isolation, and usually potential crops must be evaluated in relation to a profile of enterprises being managed on a single (farm) unit. Labour planning, capital investment in machinery and buildings, and business objectives will also need to be considered as part of the strategic planning process (see Chapter 12), and will in turn influence the cropping system employed.

Choice of cropping system

Political influences

The Arable Area Payments Scheme partially compensates producers for the lower (world) prices for those crops whose production was previously subsidized. This change in the price structure of the industry places a greater emphasis on maximization of production from natural resources together with considered evaluation of the benefits of each input, in order to reduce costs and optimize gross margins.

The increased emphasis on more effective use of freely available natural resources, in combination with consumer pressure against perceived risks to human health and environment from 'intensive' farming methods, have increased the need to develop and adopt more sustainable cropping systems. Sustainable systems seek to make best possible use of natural resources whilst minimizing both the use of external inputs and the polluting effects of waste products. Thus efficient cropping systems minimize the need for artificial nutrients and pesticides by utilizing sound rotations, making use of manures and slurries, increasing biodiversity, employing cultural control methods whenever possible, and by ensuring that artificial inputs are applied safely and efficiently.

Integrated crop management

Integrated crop management (ICM) systems integrate care for the environment into safe, efficient food production by the use of farming practices that balance the economic production of crops through the application of rotations, cultivations, choice of variety and judicious use of inputs, with measures that preserve and protect the environment. They have been developed as an extension of the well-known integrated pest management techniques (see Chapter 6) which rely on a combination of cultural, genetic, biological, and chemical control to minimize losses from weeds, diseases, and pests. ICM systems appropriate to the UK have been developed and demonstrated by the LEAF (Linking Environment and Farming) organization set up by the Royal Agricultural Society of England; a joint initiative between the NFU and several of the major supermarket retailers has also developed ICM protocols for vegetable production.

Organic systems

The most well known of the specialist systems of production is organic farming, for which producers conform to particular production 'rules' based on the almost complete elimination of artificial inputs to production, and for which a premium is obtainable for some products, depending on market demand. For detailed information the reader is referred to Chapter 11.

Specialized systems

There may be special features of the geographical location which will influence the cropping system employed, for instance if the holding is within a National Park, an Environmentally Sensitive Area, or a Nitrogen Vulnerable Zone, or is entered in the Countryside or Arable Stewardship Scheme. Management agreements may be optional or obligatory in such situations; appropriate advice should be sought from the relevant authority in these cases.

Crop rotations

Most crop production systems depend on an effective and appropriate crop rotation.

A crop rotation is a sequence of different crops on the same piece of land. For example, a simple six-course rotation might consist of two courses of winter wheat

Table 3.1 Example of a simple crop rotation, shown temporally and spatially

Year	*Field A*	*Field B*	*Field C*	*Field D*	*Field E*	*Field F*
1	Wheat	Wheat	Beans	Wheat	Barley	OSR
2	Wheat	Beans	Wheat	Barley	OSR	Wheat
3	Beans	Wheat	Barley	OSR	Wheat	Wheat
4	Wheat	Barley	OSR	Wheat	Wheat	Beans
5	Barley	OSR	Wheat	Wheat	Beans	Wheat
6	OSR	Wheat	Wheat	Beans	Wheat	Barley

followed by winter beans, a further course of winter wheat, winter barley, and oilseed rape (OSR). On a farm scale the temporal arrangement of the rotation in one field will be reflected spatially each year across fields (Table 3.1).

There are many workable crop rotations, reflecting the diversity of crop enterprises and their combinations within production units across the UK, and it is impossible to give all the possibilities here. On mixed farms a 2- or 3-year ley will form the backbone of the rotation with forage maize or other fodder crop also in the sequence. (Rotational set-aside may also be included if Arable Area Payments are being claimed.)

Crop rotations are used to:

- maintain soil fertility, and obtain maximum benefit from natural nutrient cycling;
- give opportunity for cultural control of weeds and/or cheaper chemical control of weeds;
- control, or minimize effect of, soil-borne pests and diseases;
- encourage biodiversity within the farm system;
- make optimum use of natural resources;
- spread risk and maximize financial returns (rotational gross margin).

Historically, the chief importance of rotations was in relation to soil fertility. Before the invention of artificial bagged fertilizer, methods for maintaining soil fertility were essential to ensure that yields did not decline as the land became exhausted. There was great reliance on manure and legumes for restoring the fertility after exhaustive crops. Rotating different crops around the same piece of land helped this process.

For example the Romans often used a three-course rotation of two cereals followed by a fallow year. The fallow was allowed to naturally regenerate with weeds and spilt cereal seed and this was used for grazing. Grazing animals dunged on the regenerating fallow thus replenishing the nutrients to some extent. The famous Norfolk Four-Course rotation (developed by Turnip Townsend in the eighteenth century) used wheat, followed by a root crop, followed by barley undersown with clover, followed by grass and clover. The clover added to the nitrogen status, and the ley and the root crop allowed nutrients from manure to be added back to the soil during grazing.

Although artificial fertilizers reduce our dependence on rotations for the maintenance of soil fertility, good rotational design can reduce the amount needed. Rotations ensure that all fields benefit from restorative crops such as leys, clover and beans, applications of manure, and deep cultivation. Soil structure will benefit and residues from legumes and manure will provide nitrogen and other nutrients, and thus cut fertilizer costs. In some situations rotations assume considerable importance, e.g. organic systems, Nitrate Sensitive Areas. The effect of each crop on the soil nutrient indices (see Chapter 1) should be noted to aid rotational design in this respect.

Many soil- and residue-borne pests and diseases are specific to a crop or group of crops and related weeds, which must be present for the population to multiply. Growing one crop too frequently allows the associated pest/disease to build up to infestation levels, so that subsequent crops of the same type fail completely or produce uneconomic yields. The pest/disease must then be starved out by growing crops that are not susceptible. Unfortunately some pests and diseases survive in the ground without a host plant for many years and starving out becomes a long-term commitment; growing resistant varieties (when available) is of value in some cases. Hence it is cheaper and safer to adopt a sound rotation, in conjunction with other measures such as use of clean seed, destruction of trash and volunteers, provision of good crop growth conditions, to prevent pest/disease buildup. See Chapter 6 for further information on the use of rotations in maintaining crop health.

Rotations have to work practically. One of the most basic considerations is that the previous crop must be out of the ground and the soil prepared for sowing the next crop at the optimum time. When the gap between harvest and sowing is too long, potential leaching problems are created, and the empty field will represent a lost opportunity for profit. High-value crops will take precedence, sometimes requiring lower-value crops to 'fit' round them, for example wheat is often sown later than the optimal time when following sugar beet or a vegetable crop. It should also be remembered that few rotations are 'perfect' and fulfil all the ideal requirements.

Green manures and set-aside cover crops

The function of a green manure crop is to occupy ground that would otherwise be fallow, to reduce soil erosion, and to take up available nutrients from the soil, primarily to avoid leaching and loss of nutrients. The nutrients thus 'stored' in the plants are returned to the soil prior to planting the true crop by incorporating the green manure into the soil by appropriate means. Any plant may be used as a green manure crop, but ideally the plant type should be cheap to establish, tolerant, and should establish and grow quickly thus ensuring rapid ground cover and enabling rapid uptake of available nutrients, even in less than ideal climatic and soil conditions. The choice of green manure should be considered carefully in relation to the practised rotation; it is very important that it should not present potential weed, volunteer, disease or pest problems. There is some potential for additional benefit by considering the green manure as a short break crop, i.e. in relation to soil-borne pests and diseases, and, if leguminous, it may contribute additional nitrogen to the system. The use of green manures as 'trap' crops for troublesome pests, e.g. nematodes, requires skill and careful monitoring of pest development stage but can be useful in organic systems. Typical green manures are winter rye, mustard, *Phacelia*, rape, and vetches.

Cover crops for set-aside are essentially green manures, having the same function, but with restrictions on choice and specified management guidelines. The *Arable Area Payments. Explanatory Guide* (DEFRA) should be consulted for detailed guidance, as the regulations change from time to time.

Enterprise planning

Each enterprise must be planned within the overall strategic business plan in order to predict labour, equipment, and input requirements; this plan will be modified on a day-to-day basis according to the season and other unpredictable events. Effective and efficient recording of field features and inputs is vital for effective enterprise management, and to provide the traceability required for compliance with quality assurance schemes, codes of practice, and legislation, e.g. LERAPS (see Chapters 1 and 6).

Field constraints

Individual fields should be assessed in relation to the proposed cropping; an effective field record system is essential for this. Soil maps (texture, pH, nutrient indices), previous cropping and associated inputs, and historical 'problems', e.g. poorly drained areas or weed patches, must be taken into account during enterprise planning. Particular features of the site should also be accounted for, such as access, slope, topography, exposure, footpaths, archaeological sites, and boundary features such as woodland, hedges, and field margins. The effects of existing features and past cropping must be incorporated into the enterprise plan for effective crop and field margin management. Precision farming utilizes recent advances in technology [global positioning systems (GPS), geographical information systems (GIS), and decision support systems (DSS)] to plan inputs at a sub-field scale and achieve increased efficiency.

Varietal choice

Many varieties are available for each crop type, each variety exhibiting a unique profile of characteristics. The characteristics of importance will vary for each crop type but will usually be related to yield, quality, resistance to disease, time to maturity, and height. Varieties should be carefully chosen to suit the market/end use, the site of production, and the farming system. Recommended lists of varieties are published by the National Institute of Agricultural Botany (NIAB) and other organizations, and these should always be consulted to effect appropriate choices.

Crop establishment

It is false economy to sow anything but seed of high germination and vigour, whether purchased or home-saved. Only licensed seed merchants can legally sell seed, and are required to state the germination rate. The statutory germination test is carried out under ideal laboratory conditions and may not indicate the field germination rate if the seed is of low vigour, particularly in cold, poor soil conditions. The vigour of a seed lot is attributable to many factors including age, condition of mother plant, size, weight, mechanical damage, and presence of pathogens. There are various methods of testing vigour;

high vigour seed should always be used in less than perfect field conditions. If home-saved seed is to be used the germination should always be tested, and if suitable for sowing the seed should be cleaned and dressed.

Seed treatments are used to promote good seedling establishment, minimize yield loss, improve quality, and avoid the spread of harmful organisms. Successful practical seed treatments must be consistently effective under varied conditions, be safe to handlers and operators, be safe to wildlife, and have a wide safety margin between the dose that controls harmful organisms and the dose that harms the plant. The harmful organisms associated with the establishment phase may be seed-borne, soil-borne, or air-borne pathogens that attack the emerging and developing seedling. Seed treatments can be a cheap, easy, and effective way to protect crops in the early stages of growth, and to avoid the need for less well-placed agrochemical inputs.

Crop seed rate recommendations should be used with care as varieties, and seed lots of the same variety, can vary significantly in their thousand seed weight (TSW), and allowance must be made for losses due to the field conditions pertaining. It is the final plant population that is important for optimal yield and quality; therefore the appropriate seed rate should be calculated for each situation depending on TSW and field conditions. The following formula should be used:

$$\text{Seed rate}\,(\text{kg/ha}) = \frac{\text{Target plant population}\,(\text{no./m}^2) \times \text{TSW}\,(\text{g})}{\text{Germination}\,(\%) \times \text{field factor}}$$

[The inclusion of a field factor in the seed rate calculation attempts to allow for field losses due to adverse seedbed conditions such as temperature, moisture or condition. Such field losses can only be estimated and may vary from 0.5 (50% establishment of the seeds capable of germinating) for poor conditions to 0.9 (90% establishment) for good conditions.]

Example calculation: Crop with target plant population of 250 plants/m^2, TSW of 50 g, germination of 80%, and estimated establishment of 60%, i.e. field factor of 0.6.

$$\text{Seed rate} = (250 \times 50)/(80 \times 0.6) = 260.42\,\text{kg/ha}$$

Sowing ungerminated seed directly into open field

Seedbed quality is very important to successful establishment, especially for small seeds (e.g. oilseed rape) and sensitive seeds (e.g. peas). Seed needs to be surrounded by moist aerated soil of adequate temperature for germination. Degree of fineness of tilth should reflect seed size in order to ensure contact between seed and soil and therefore moisture. Very fine seedbeds can be liable to capping on some soils, so some small surface clod is desirable in appropriate conditions. Rough seedbeds cause uneven germination, slow establishment, and plants lacking vigour; in addition residual herbicides do not work satisfactorily, and precision drills sow inaccurately. Seedbeds should be neither compacted nor loose, but reasonably firm.

Depth of sowing depends on seed size and availability of soil moisture. In general sow no deeper than necessary to ensure good coverage as quick emergence ensures a vigorous plant, but recognize that seed needs adequate moisture to germinate and availability will decrease towards soil surface. Smaller seed must be sown shallow because of limited food reserves, larger seed can be sown deeper without risk.

Broadcasting seed is quick and cheap, and can be performed using a suitable fertilizer distributor for cereals, grass, and forage brassicas. Its disadvantages are uneven depth of sowing (seed is often harrowed in) and uneven spatial distribution, making it suitable only in low yield potential situations or as a last resort, e.g. when sowing has been seriously delayed by poor weather. A special case is winter field beans which are best broadcast and ploughed under, often giving a yield benefit compared to drilled crops.

Precision drills are used where accurate seed placing and precise plant populations are required. Singling and thinning by hand are now rarely necessary due to drilling to a stand using graded or pelleted seed, with allowance for non-germinating seeds. Success depends on a high quality seedbed, precise drilling, effective protection against seedling pests and diseases, and a successful weed control programme.

Direct drilling, using special drills with zero or minimal cultivations following the use of a total herbicide (paraquat or glyphosate), can be used for cereals, grass, oilseed rape, and forage brassicas, but well-structured soils are needed, and if used sequentially certain weeds may become a problem.

Bed systems are used for root and vegetable crops to eliminate fanging, give quick vigorous establishment, and reduce reliance on good ground conditions for timely machine operations (Fig. 3.1). Crops are grown in strips or beds which the tractor straddles, thus restricting wheelings and compaction to the areas between the beds; all equipment widths must be matched to the bed width (or multiples of the bed width) to use this system effectively. Bed width is usually determined by harvester design.

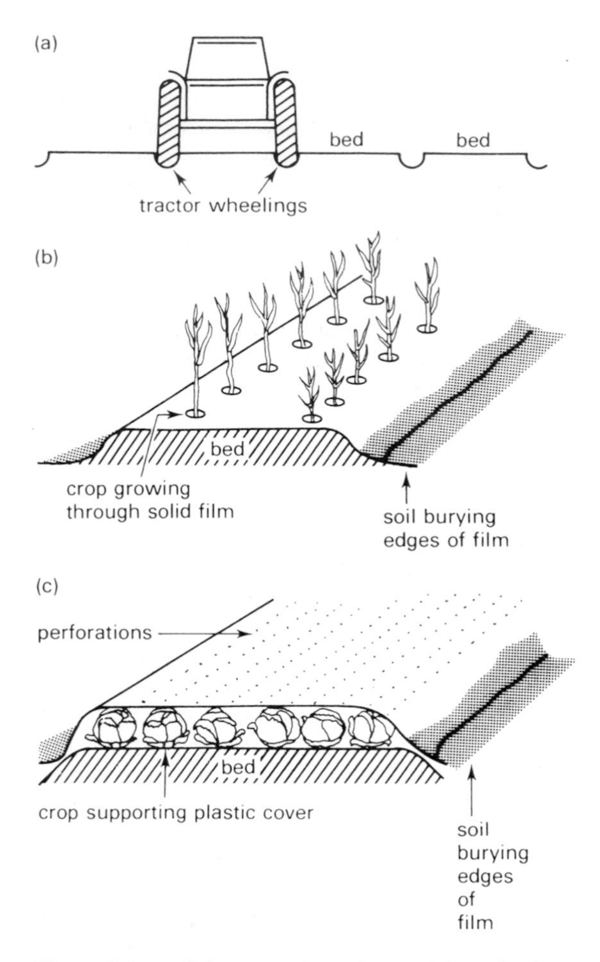

Figure 3.1 **a** Bed system; **b** surface mulch; **c** floating plastic film.

Plastic covers for protection during establishment in the field

Plastic covers can be used to raise the soil temperature and offer protection from cold weather, in order to achieve earlier emergence and growth and advance maturity in comparison to crops established without protection. The use of covers is only economic for high value crops, particularly field vegetables and early potatoes. Covers may be used as surface mulches or floating films (Fig. 3.1). For surface mulching solid plastic film is laid in contact with the soil surface, held down by soil at bed edges, and the crop grows through holes pierced at appropriate positions. Photodegradable film is used so that it disintegrates after a certain time (dependent on grade of plastic), from 1 to 4 months after laying. Floating films are perforated (polythene or woven material) to allow exchange of gases and water/water vapour as the crop grows beneath the film, giving a 'cloche' effect. Different grades of film or net are available with different perforation, and therefore ventilation, characteristics. Timing of floating film removal can be critical as plants may not be hardened to exposed conditions, and temperatures below the film can become excessive as ambient temperature increases. It should also be remembered that light levels under the film are less than those 'outside'.

Transplants

The high value and demanding market quality specifications of some vegetable crops can make the increased costs of transplanting (as opposed to sowing seed directly into the field position) very worthwhile. The use of transplants can give a higher yield of the highest quality product, reduce the need for stringent seedbed preparation, reduce weed control costs, avoid unfavourable field conditions, and release field space for other crops. Transplants may be 'bare rooted' (lifted from raising beds with little soil attached to the roots) or raised in individual blocks, modules, or cells in growing medium which is transferred to the field with minimal root disturbance. Raising transplants is a specialized business requiring careful hygiene and critical management of nutrition and environment during establishment; most growers purchase transplants from nurseries/suppliers.

Inputs and their timing

Efficient use of inputs and their timing depends on close monitoring of the crop and growing conditions, together with the use of reliable product information. Site and season have very strong effects on crop growth and development and there can be no foolproof 'recipe' for the production of a particular crop. Valuable aids to decision making are the development keys available for the major, and some of the minor, crops, e.g. the Zadoks key for cereals (Table 3.3). Inputs should always be applied according to the state of the crop (or its associated weeds, diseases or pests) and the weather, *not* by calendar date. More detailed advice is given for particular crops in the following section.

Arable crop enterprises

The annual DEFRA publication *Agriculture in the United Kingdom* should be consulted for current detailed crop production figures. The 2000 census returns gave the following production areas for the UK:

	ha
Total cropped area	4 665 000
of which:	
Cereals	3 348 000
Oilseed rape	332 000
Linseed	71 000
Field beans and peas	208 000
Sugar beet	173 000
Potatoes	166 000
Vegetables	119 000
Bare fallow area	37 000

Combinable crops

Combinable crops are those grown for the grain or seed, harvested at relatively low moisture contents. Grain/seed ripening requires warm dry conditions and thus all the UK combinable crops are grown for harvest in the summer/early autumn period in order to take advantage of the seasonal climate; products are then stored to supply the market all year round. The majority of the UK arable cropping hectarage is used for combinable crops, and thus the cropping system on the majority of UK arable farms is based on the combinable crop season and requirements.

Most UK combinable crops are grown under the Assured Combinable Crops Scheme (ACCS) which ensures that the product is produced using good agricultural practice (GAP) and is traceable.

Cereals

Markets

Cereals are grown for animal feed and human consumption both for use in the UK and as exports. The animal feed grain market accepts wheat, barley, oats, and triticale, and the price obtained reflects the extractable starch or energy content of the grain; thus wheat commands a higher price than barley and barley a higher price than oats. Oats can be sold into the specialist horse feed market at higher prices. Quality standards in the UK are appearing in the animal feed grain sector and the price can be reduced if grain is below a target specific weight. Generally though the main quality requirement is sound grain of below 15% moisture content with a contaminants content below 5%. The feed price is generally the lowest price in the market.

Grains sold for human consumption include the major markets of wheat for bread and biscuits, and barley for brewing. Other minor/specialist markets include wheat for cake flours, barley for other domestic uses, wheat and oats for breakfast cereals, wheat for starch and gluten extraction, and rye for crispbread. The human consumption markets command premiums over feed grains but demand stricter quality requirements which are assessed in the laboratories at the intake site. Examples of the quality requirements for bread wheat and malting barley are given in Table 3.2. Export markets have been growing in recent years and specific restrictions apply for each destination country.

Intervention markets also exist for wheat and barley. This is where the EU buys surplus grain on the market, stores it, and resells it onto the market at a later time. All intervention wheat grain must now meet quality standards similar to those for bread wheat in the UK yet command a price often lower than the feed wheat price. Price support for the cereal sector has altered dramatically with the disappearance of support prices, leading to cereal prices dropping to world cereal prices. Instead, Arable Area Payments (AAPs) are made on cereals grown on IACS (Integrated Administration and Control System) registered land, provided growers comply with the statutory degree of set-aside.

Growth cycle

Cereals are monocotyledonous plants belonging to the Poaceae (previously Gramineae) family (which includes all the grasses). They are all annual or winter annuals completing their growth cycle in less than 1 year. Autumn-sown varieties (winter types) differ from spring-sown varieties in that they require a vernalization period (exposure to cold) before they will flower. Thus winter varieties are in the ground for longer than spring varieties and this gives them a potential yield advantage. For wheat, approximately 97% of the UK crop is winter wheat whereas for barley approximately 55% is winter barley.

Each seed produces a single plant which during its vegetative phase initiates leaves and side shoots (tillers). Each tiller has the ability to produce one ear, but normally a plant overproduces tiller numbers and many die off in spring leaving only the strongest to produce ears.

Table 3.2 Quality standards for bread wheat and malting barley in the UK

Standards	*Notes*
Bread wheat	
Named varieties	Only HGCA Group 1 varieties are accepted (breadmaking types with the right combination of grain proteins)
Hard endosperm	Not essential but preferred by the miller
Hagberg Falling Number over 250	Assessment of alpha amylase activity (high levels, indicated by low HFN, ruin the crumb structure of the loaf)
Protein content over 11%	Indication of high gluten content necessary to give strength to the dough
Specific weight above 76 kg/hl	Indicates well-filled grains and good flour extraction rate
Absence of ergots	Poisonous fungal fruiting bodies
Contaminants below 1%	
Good colour and free of smells	
Malting barley	
Named varieties	Varieties bred for malting quality and approved by the Institute of Brewers
Germination above 96%	Indicates good maltose extraction potential
Low nitrogen content	Normally below 1.6%; can cause cloudiness in beer
Pale lemon colour	
Free from odours	
Contaminants below 3%	

The stimulus for the apex of the plant to produce a flowering head (ear) is made early in the growth cycle of the plant, e.g. 6 to 8 weeks after sowing in a winter barley crop and at the 3- to 5-leaf stage in spring barley. The embryonic ear develops inside the sheaths of the developing leaves and appears (ear emergence) several months later.

After ear emergence pollination takes place (anthesis) and then grain filling. Most of the grain sites of wheat and barley are self-pollinated as the flower opens. The grain is filled with starch produced by photosynthesis of the top two leaves in the canopy during the grain filling period (80%) and from the stored starch in the stem and photosynthesis in lower leaves (up to 20%).

At the end of grain filling the grain loses moisture and matures in the field before it is ready to combine. Harvesting is best carried out when grain moisture is about 15%, but in wet summers harvesting up to 20% may be practised and grain then artificially dried to less than 15%.

In order to harmonize the chemical inputs to a cereal crop, a key has been produced for the classification of the development of cereal crops (Table 3.3). All recommendations of agricultural chemicals and fertilizers to cereals are referenced to this growth and development key.

The components of yield of a cereal crop are:

$$\text{Ears/m}^2 \times \text{grains/ear} \times \text{seed weight}$$

The size of these components varies with the developmental stages of the crop (Fig. 3.2). For ears per square metre and grains per ear there is an overproduction and a dying off phase as the plant adjusts its potential grain sinks to the supply of carbohydrate coming from the leaves (source). The grower has most influence over the ears per square metre component of yield through the manipulation of seed rate and nitrogen application.

Soil types and rooting

Cereals can be grown on a wide variety of soil types provided a suitable seedbed can be cultivated. The yield potential of the crop (particularly wheat) is, however, related to the moisture-holding capacity of the soil. This is particularly the case in a dry summer. Cereals yielding about 6 t/ha will transpire about 100 to 125 mm of water during grain filling, and whilst some of this will come from rainfall most must come from the soil's storage capacity. Early maturing crops should be grown on soils that are particularly low in available water-holding capacity.

The grain filling periods of the major UK cereals in the south of England are:

Winter barley – June
Winter wheat – July to early August
Spring barley – July to early August
Spring wheat – mid-July to mid-August

Rye and triticale have good tolerance of low soil moisture-holding capacity and can outyield wheat on droughty soils in dry years.

Table 3.3 Zadok's decimal code for the growth stages of cereals

Code		
0	GERMINATION	
00	Dry seed	
01	Start of imbibition	
02	—	
03	Imbibition complete	
04	—	
05	Radicle emerged from seed coat	
06	—	
07	Coleoptile emerged from seed coat	
08	—	
09	Leaf just at coleoptile tip	
1	SEEDLING GROWTH	
10	First leaf through coleoptile	
11	First leaf unfolded	
12	2 leaves unfolded	
13	3 leaves unfolded	
14	4 leaves unfolded	
15	5 leaves unfolded	
16	6 leaves unfolded	
17	7 leaves unfolded	
18	8 leaves unfolded	
19	9 or more leaves unfolded	
2	TILLERING	
20	Main shoot only	
21	Main shoot and 1 tillers	
22	Main shoot and 2 tillers	
23	Main shoot and 3 tillers	
24	Main shoot and 4 tillers	
25	Main shoot and 5 tillers	
26	Main shoot and 6 tillers	
27	Main shoot and 7 tillers	
28	Main shoot and 8 tillers	
29	Main shoot and 9 or more tillers	
3	STEM ELONGATION	
30	Pseudostem erection (winter cereals only) or stem elongation	
31	1st node detectable	
32	2nd node detectable	
33	3rd node detectable	
34	4th node detectable	
35	5th node detectable	
36	6th node detectable	
37	Flag leaf just visible	
38	—	
39	Flat leaf ligule just visible	
4	BOOTING	
40	—	
41	Flag leaf sheath extending	
42	—	
43	Boot just visibly swollen	
44	—	
45	Boot swollen	
46	—	
47	Flat leaf sheath opening	
48	—	
49	First awns visible	
5	EAR EMERGENCE	
50	—	
51	First spikelet of ear just visible	
52	—	
53	$\frac{1}{4}$ of ear emerged	
54	—	
55	$\frac{1}{2}$ of ear emerged	
56	—	
57	$\frac{3}{4}$ of ear emerged	
58	—	
59	Emergence of ear completed	
6	FLOWERING	
60	—	
61	Beginning of flowering (not easily detectable in barley)	
62	—	
63	—	
64	—	
65	Flowering half-way	
66	—	
67	—	
68	—	
69	Flowering complete	
7	MILK DEVELOPMENT	
70	—	
71	Seed coat water ripe	
72	—	
73	Early milk	
74	—	
75	Medium milk	Increase in solids of liquid endosperm visible when crushing the seed between fingers (75–79)
76	—	
77	Late milk	
78	—	
79	—	
8	DOUGH DEVELOPMENT	
80	—	
81	—	
82	—	
83	Early dough	
84	—	
85	Soft dough (finger-nail impression not held)	
86	—	
87	Hard dough (finger-nail impression held, head losing chlorophyll)	
88	—	
89	—	
9	RIPENING	
90	—	
91	Seed coat hard (difficult to divide by thumb-nail)	
92	Seed coat hard (can no longer be dented by thumb-nail)	
93	Seed coat loosening in daytime	
94	Over-ripe, straw dead and collapsing	
95	Seed dormant	
96	Viable seed giving 50% germination	
97	Seed not dormant	
98	Secondary dormancy induced	
99	Secondary dormancy lost	

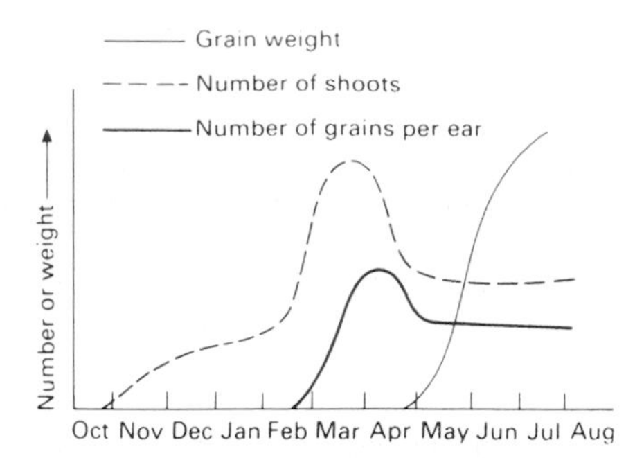

Figure 3.2 How the components of yield of a winter cereal crop vary with time.

Cereals can root down to 2 m, but good crops are frequently produced on soil depths of 1 m. Winter-sown crops root more deeply than spring crops because they have longer to expand their roots; root growth practically ceases when rapid shoot growth commences in spring. Cereals have two root systems: the seminal root system which originates from the seed and is the deep penetrating root system, and the nodal or adventitious root system that originates from the base of the crown of the plant and is the root system that proliferates in the topsoil and takes up the majority of the spring-applied nitrogen and summer rainfall.

Seedbed

Traditionally, winter and spring seedbeds differ, with spring seedbeds being much finer than winter seedbeds. Fine winter seedbeds are sometimes prepared and have been favoured where soil-residual herbicides were used. The most common winter seedbed has a fine tilth at drilling depth (2 to 5 cm) with a mixture of sizes of clods on top, the biggest being about 5 to 10 cm in diameter. Overly deep seedbeds should be avoided or drilling will be too deep, and loose seedbeds should be avoided or problems with lodging could arise. This does not mean that winter seedbeds should be rolled following drilling as this runs the risk of breaking the surface clods which are necessary to help prevent soil capping and erosion during the heavy winter rains (particulary important on silty soils). If rolling is necessary then a ring or Cambridge roll should be used. Spring seedbeds can be much finer than winter seedbeds and rolling helps to consolidate the drying soil and gives better germination.

The use of no-tilth or mini-tilth seedbeds was popular when straw burning was permitted but declined with the straw burning ban in 1992. More cultivations are needed now to help dispose of unwanted straw and stubble. On suitable soil types (well-structured, calcareous loams) direct drilling into stubble can still be beneficial. On heavy soils, reduced cultivations are still necessary to avoid the problems of large sun-baked clods following ploughing. Recently, with the decline in cereal profitability, there has been renewed interest in reducing cultivations and this has been encouraged by some new engineering innovations, particularly in combination cultivators and drills. Minimal cultivations can play a large role in reducing soil erosion.

Plant population and seed rate

An adequate plant population is essential to achieving a good, evenly ripened crop. Given ideal conditions then, a winter cereal will tiller sufficiently to give an adequate crop density from a plant population of about 100 plants/m^2. However, higher populations of 250 to 300 are recommended because they give a safety margin for losses, a more uniform crop, and greater early weed suppression. Populations above 350 are wasteful of seed and can lead to lodging (commonly seen on headlands where double drilling has occurred). Seed costs are a relatively small component of the variable costs of growing cereals and therefore a high seed rate is frequently considered to be a good insurance policy. Spring cereals have less time to produce strong tillers and as a consequence are established at higher plant populations to compensate (spring barley – 350 to 400 plants/m^2, spring wheat – 400 to 500 plants/m^2).

The seedrate for a field is determined from the following:

- the desired plant population
- the seed weight (thousand grain weight)
- the expected establishment percentage (field factor).

Of these factors, the field factor is of course unknown and must be guessed in relation to previous experience. As a guide, good drilling conditions at the optimum drilling date should give an establishment of about 85%, moderate conditions about 70%, and poor conditions only about 55%. Conditions, particularly soil temperature, deteriorate quite rapidly after mid-October and it is recommended that seed rate is raised substantially for November sowings. Early spring sowings are frequently made into poor germination conditions (wet and cold) as are late sowings (dry and poor seedbeds). Guideline seed rates are given in Table 3.4. There is variation in seed size between varieties especially in winter wheat and a knowledge of thousand grain weight can influence the seed rate markedly.

Poorly established winter wheat crops should not be ploughed up and planted with spring barley unless there are substantial areas of the field at less than 50 plants/m^2.

Table 3.4 Guideline seed rates for cereals in England and Wales

Crop	*Sowing date*	*Normal seed rate (kg/ha)*
Winter wheat	Late Sept–mid Oct	140–160
	Nov–Dec	180–220
Winter barley	Mid Sept–early Oct	140–150
	Mid Oct–end Oct	150–190
Winter oats	Early Oct	150–160
	Late Oct	160–180
Rye	Late Sept–mid Oct	150–180
Triticale	Early Oct–mid Oct	120–150
Durum wheat	Oct–early Nov	180–210
	Feb–mid March	190–220
Spring wheat	Late Feb–early March	190–220
	Mid March	220–230
	Nov–Dec	200–220
Spring barley	Feb–March	125–150
	April	160–170
Spring oats	Feb–early March	170–190
	Late March	190–200

Stitching in some spring crop into poorly established areas of a winter crop can be effective but will delay the harvest of the field whilst the spring crop ripens.

Sowing date

Cereals are remarkably tolerant of a wide variety of sowing dates but perform best over a relatively short window of sowings. Winter wheat, for example, can be sown from early September to March but performs best if sown at the end of September to early October. Winter varieties of course need to experience a period of cold in order to flower and produce grain, but the cold requirement is met even if the crop is sown as late as February/March. Wheat has a greater requirement for cold than does barley which in turn requires more than oats.

The windows and optimal sowing dates are as follows:

Crop	*Normal window*	*Optimal*
Winter barley	Early Sept to early Oct	Last 2 weeks of Sept
Winter oats	Late Sept to mid Oct	First week of Oct
Winter wheat	Late Sept to end Oct	First 2 weeks of Oct
Spring wheat	Late Nov to end March	February
Spring barley	Late Feb to end Apr	March
Spring oats	March to end Apr	March

For winter cereals, earlier sowing has the advantages of a higher yield potential, higher germination rate, and greater tillering capacity, but it carries with it a greater risk of disease infection, pest infestation (particularly aphids carrying barley yellow dwarf virus – BYDV), and greater weed germination. Many growers deliberately delay the sowing of winter cereals in order to reduce the need for autumn pesticides. This is particularly important for growers of organic cereals, who will commonly sow winter wheat in November at very high seed rates and avoid the use of herbicides and reduce the damage from BYDV. Growers of large cereal hectarages will, however, need to begin sowing as early as possible in the autumn in order to complete before winter rains bring the soil to field capacity.

The priorities of sowing a large winter acreage are many and varied, but good husbandry should take account of the following:

- First wheats can be sown early with little disease carryover risk, thus maximizing yield potential (if possible this should be exploited with a milling wheat).
- Fields at risk of take-all (3rd to 4th cereal after a 2-year break or 2nd cereal after a 1-year break) should be sown towards the end of sowing.
- Heavy soils should receive priority.
- Ploughed out grassland should always be left at least 2 to 3 weeks before sowing to avoid frit fly and aphid carry over.

Varieties of winter wheat show some variability in their ability to tolerate different sowing dates. Generally, varieties that are disease susceptible and/or have a low vernalization requirement are less suited to early sowing.

There is less pressure on spring sowings because there is less area to establish on most farms. Sowing is dictated more by soil conditions than anything else and early sowing is only possible on light, easily drained soils. Germination percentage is likely to be low in cold spring soils and seed rates need to be higher to compensate. With late spring sowings seed rate also needs to be increased as less time is available for tillering.

Sowing depth

Cereals have relatively large seeds with a good carbohydrate store and therefore can germinate from quite deep sowings. It should be remembered though that the seed stores are used for early leaf production as well as for emergence growth. Thus, deeply planted cereals will show low early vigour. Plants that have come from seed sown too deeply often have a hooped appearance often referred to as 'hockey sock' appearance; such appearance is common on volunteer plants which have germinated following ploughing. Shallow sowings can suffer from an inadequately developed root system which is susceptible to frost heave over the winter. Frost heave is where ice forms under the crown of the plant and pushes the plant upwards, snapping the roots. The normal depth of sowing is 2 to 2.5 cm and is adequate for both winter and spring varieties. Under warm germination conditions it can be advantageous to sow slightly deeper than 2.5 cm especially if the seedbed is dry. Sowing depth is controlled by drill settings and influenced by the level of consolidation of the seedbed. Slow drilling leads to the even placement of the seed and a better control of depth. The practice of broadcasting and cultivating the seed into the seedbed results in the poorest control of sowing depth and inevitably results in greater establishment losses. When broadcasting is used, seed rates are increased by 5 to 10% to account for this.

Seed dressings

Most cereal seed is dressed to protect it against seedbed pests and germination and establishment diseases. Fungicides are available for seed dressing but are expensive. These fungicides have a systemic action which gives some early disease protection to the establishing seedling and can be very effective in spring cereal production where early mildew infection can be significant. The seed dressings are applied to the seed by the seed supplier or by a mobile seed cleaning firm if home-saved seed is used. Home-saved seed now attracts a Breeders Rights payment for newer varieties which is collected by a self-declaration method.

Establishment pests

Cereal crops can fail in some instances following the attack by certain pests during establishment. A more complete description is given in Chapter 6, but a brief consideration of the most important pests is given here.

Slugs

Damage caused by slugs is most serious when they graze on the germinating seed underground causing grain hollowing. Effective control can be obtained by mixing slug pellets with seed before sowing in high-risk situations. The post-sowing broadcasting of slug pellets is less effective since the application is frequently made too late, i.e. after damage has been caused. High-risk situations include: heavy soil types, high presence of trash, and where the previous crop favoured slug multiplication in summer, e.g. potatoes, grass, OSR, wet summer and autumn, direct drilling.

Fly larvae

Various fly larvae cause damage to cereals known as 'dead heart'. This is where the larvae have penetrated the shoot and eaten away the apical meristem. These larvae include frit fly, wheat bulb fly, and yellow cereal fly. Frit fly are common following grass and can be largely controlled by a delay in the sowing of the cereal of at least 14 to 21 days after sward destruction. Wheat bulb fly is more localized in its distribution, only being found in the east of England and the East Midlands. The fly lays its eggs on bare soil in July and August and therefore is commonly found after second early potatoes. Post-emergence chemical control of fly larvae is possible after diagnosis of infestation.

Wireworm

This small soil-borne pest is also frequently found after grass. Control by the routine use of a gamma HCH seed dressing is effective for low populations. Where there is a risk of wireworms the seed rate should be increased to compensate for plant losses.

Establishment diseases

Diseases at establishment can be divided into two groups: firstly those that cause germination failures and secondly those that lead to early plant infection which has the potential to cause epidemic disease at a later stage. Germination failure caused by disease is rare but can occur in poor germination conditions with seed of low vigour. On the whole, certified cereal seed will have a good germination capacity (EU minimum of 85%) and be vigorous enough to establish satisfactorily.

Early plant-infection organisms include the seed-borne diseases bunt (*Tilletia caries*), leaf stripe (*Pyrenophora graminea*) *Fusarium* spp., net blotch (*Pyrenophora teres*) and loose smut (*Ustilago nuda*), and the foliar diseases mildew (*Erysiphe graminis*), *Septoria* spp., and *Rhynchosporium*. Special mention is warranted of the virus diseases cereal yellow dwarf virus (commonly known as barley yellow dwarf virus or BYDV) and barley yellow mosaic virus (BYMV). BYMV is soil-borne and is serious enough to force some growers out of barley on infected land, but some tolerant varieties are now being developed and used successfully. BYDV is more widespread and is spread by an aphid vector. Aphids carrying the virus can infect a newly established winter cereal either by migrating from ploughed-under grass or grassy stubble and/or by flying in from the air-borne population which exists during autumn. The incidence of BYDV is more common in the south of England and in particular in the south-west, where the aphid population continues to fly later in the autumn thus putting more sowings at risk. Control of the aphids is the key to control of BYDV. Where the cereal follows a ley or grassy stubble then it is important to completely destroy the sward and this in turn will destroy the aphids. Chemical destruction of the sward using paraquat leads to a fast sward destruction, and sowing can follow 2 weeks later; if glyphosate is used the 3- to 4-week delay is necessary as sward destruction is slower. If only ploughing is used then at least 4 weeks' delay is necessary. Flying aphids need to be controlled using insecticides sprayed in the first week of November when flying has ceased. In high risk areas (south-west and south) and high risk years (high percentage of infectivity) early sown crops of barley may need to be sprayed twice, once in mid October and again in early November. Only crops that emerge by mid October are at risk of BYDV, which includes most winter barley and early sown winter wheat and winter oats. Spring sowings do not usually suffer yield loss by BYDV since infection occurs late in the life cycle of the crop when aphids begin flying again in summer. Current research has identified varietal resistance to BYDV, and breeding programmes are aimed at producing resistant varieties.

Weed control

Weed control in cereal production is an integral part of modern growing systems. During the expansion days of cereals in the UK (late 1970s and early 1980s) profitability from cereals was good and many herbicides were used with up to five different chemicals per crop. In the 1990s the use of herbicides declined due to a combination of environmental pressures and most importantly a decline in profitability. A summary of weed control is given below, but details of herbicide choice are given in Chapter 6.

Weed control options include the following:

(1) Pre-sowing – applied to the soil and incorporated before sowing of the crop; only used for soil-acting granules to control wild oats.
(2) Pre-emergence – applied after sowing and before emergence; chemicals are soil acting (residual); good approach for the control of germinating grassy weeds (blackgrass, sterile brome).
(3) Post-emergence – applied after crop and weed emergence; chemicals may be both soil acting and contact or systemic acting.
    (a) Autumn applied – controls seedling weeds germinating with the crop.
    (b) Spring applied – controls larger weeds present in the crop and seedling weeds germinating in the spring.

Winter cereal weed control is also aided by a high seed rate, late sowing, and a vigorously growing crop.

Each region and each field has its own particular weed problems, but there is a list of common cereal weeds. The competitive ability of these species varies from species to species and from year to year (Table 3.5). The key to cost-effective chemical weed control is the identification of weed species and the ability to assess high-risk infestation levels.

Chemical weed control is now subject to the government regulations contained in the Food and Environmental Protection Acts (FEPA) and all sprayer operators must pay due regard to the regulations. For new operators, a competence certificate must be held for the safe use and handling of crop chemicals (see Chapter 15). In several parts of the country herbicide-resistant weeds are appearing, e.g. blackgrass and wild oats, and growers must use herbicides with alternative modes of action to control these populations.

Nutrients

Most soils contain all the necessary micro-nutrients for continuous cereal production but will frequently be deficient in one or more of the three macro-nutrients nitrogen (N), phosphate (P) and potassium (K). Also over time arable cropping leads to the removal of calcium and magnesium, resulting in the acidification of the soil which needs to be corrected by the application of lime.

Phosphate and potassium

The removal of phosphate and potassium by a cereal crop is summarized in Table 3.6. Grain removes virtually equal amounts of P and K whilst straw removes

Table 3.5 Competitive ability of weeds in cereals

Very competitive	Wild oats Cleavers
Moderately competitive	Blackgrass Poppy Rough stalked meadow grass Chickweed Brome grass Mayweeds Forget-me-not
Poorly competitive	Speedwells Red deadnettle Field pansy

Table 3.6 Phosphate and potash removed by cereal crops (adapted from MAFF, 2000)

	kg/t (of grain)		*As an 8-t/ha crop*	
	P_2O_5	K_2O	P_2O_5	K_2O
Grain only	7.8	5.6	62.4	44.8
Grain and straw				
Winter wheat/barley	8.6	11.8	68.8	94.4
Spring wheat/barley	8.8	13.7	70.4	109.6
Winter/spring oats	8.8	17.3	70.4	138.4

disproportionate levels of K. Application levels of P and K to the soil for a cereal crop therefore depend on the expected grain yield, whether the straw is to be removed from the field or not, and the level of P and K reserves already in the soil. Soil reserves are indicated by the soil index (see Chapter 1), with a high index indicating a good reserve. Good husbandry aims to maintain the P and K indices at about 2 or above, but a cereal crop only demands an index of about 1. If the index is a low 1 or a 0 then P and K must be applied to the seedbed of the crop for that crop. If the index is a 1 or low 2 then the P and K are also applied to the seedbed but may be used by the succeeding crop. If the index is a 2 or higher then the P and K can be applied as a top dressing at any time during the life of the crop since it is only for use by the succeeding crop(s). Some soils, particularly heavy clays, have naturally high levels of K and may not need any application of K to the soil. Recommended levels of application of P and K are given in Table 3.7.

Lime

Details of lime requirement are given in Chapter 1. For cereals, a soil pH of about 6–6.5 is required. If the pH drops below 6 then lime is required. Among the cereals, the tolerance of low pH is as follows:

Crop	*Critical pH* (pH below which yield is restricted)
Rye	4.9
Triticale	5.2 (estimated)
Wheat	5.5
Barley	5.8

It can be seen that barley is the most sensitive to acidic soils and liming should take place before the barley crop in a rotation.

Nitrogen

Nitrogen fertilizer is still the most cost-effective input into a cereal crop. Its present low cost and high yield response means that the economic optimum is virtually the same as the optimum for maximum yield. Where soil N reserves are low the response to the application of a kilogram of N can be 30 kg of grain and averages 10–20 kg of grain (Fig 3.3). With soils that are more fertile the response is lower but still cost effective. Nitrogen is

Table 3.7 Phosphate and potash recommendations (kg/ha) for cereals [adapted from MAFF 2000]. Reduce potash by 50 kg/ha on clay soils, except carboniferous days. WW = winter wheat, WB = winter barley, SB = spring barley, WO = winter oats, SO = spring oats

	Soil index				
	0	*1*	*2*	*3*	*4+*
Straw incorporated					
WW, WB (8 t/ha)					
Phosphate	110	85	60	20	0
Potash	95	70	45[1]	0	0
SW, SB, rye, triticale (6 t/ha)					
Phosphate	95	70	45	0	0
Potash	85	60	35[1]	0	0
WO, SO (6 t/ha)					
Phosphate	95	70	45	0	0
Potash	85	60	35[1]	0	0
Straw removed					
WW, WB (8 t/ha)					
Phosphate	120	95	70	20	0
Potash	145	120	95[1]	25	0
SW, SB, rye, triticale (6 t/ha)					
Phosphate	105	80	55	0	0
Potash	130	105	80[1]	20	0
WO, SO (6 t/ha)					
Phosphate	105	80	55	0	0
Potash	155	130	105[1]	35	0

[1] Where the potash index is a high 2 (i.e. 2+) then reduce K applied by 15 to 25 kg/ha.

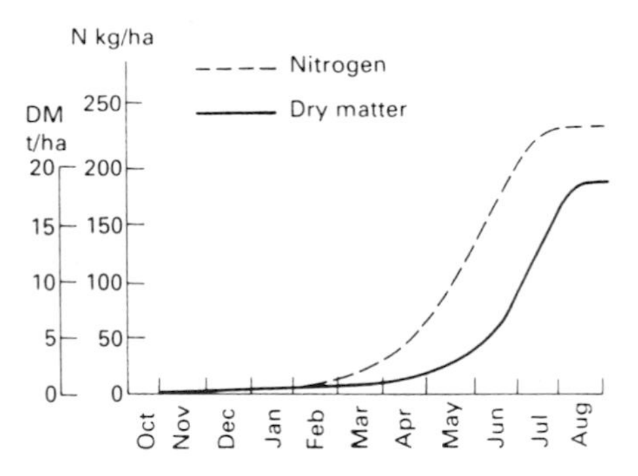

Figure 3.3 Nitrogen uptake and dry matter accumulation of winter wheat.

the only input to the established crop that actually lifts yield potential; all other inputs (fungicides, herbicides, insecticides, growth regulators) merely protect the yield potential that is established.

Nitrogen is available to the crop mostly in the form of nitrate, although ammonium can also be taken up and can be converted to nitrate in the soil (see Chapter 1). Both are soluble forms of nitrogen and as such can be lost from the soil by leaching if rainfall is high, leading to drainage flow. The most commonly used nitrogen fertilizer for cereals is ammonium nitrate (usually 35% N), although urea is also available and can be cheaper. However, some N can be lost by volatilization if urea is applied in warm conditions. The majority of nitrogen is taken up by the roots, but small amounts can be taken up by the leaves if urea solutions are sprayed onto the foliage. Soil leaching of N is most likely to occur when the soil levels of nitrate are high but the crop demand is low, a situation that frequently occurs in autumn. Leaching can also occur if fertilizer is applied in early spring when the soil is near field capacity and the application is followed by heavy rainfall.

Amounts of nitrogen

Judging the amount of nitrogen to apply to a cereal crop is at best an estimate. Responsible farming aims to not over-apply nitrogen so as to cause leaching problems. The soil nitrogen supply (SNS) index is estimated from the soil type, the previous crop and the annual rainfall (low, medium, high) (see Chapter 1). Recommendations of the amount of nitrogen to apply are then given in compliance with *MAFF RB209* (2000) (Table 3.8) based on many years of measuring crop yield response and nitrate leaching. Over-applying nitrogen may lead to a grower transgressing ACCS guidelines and a loss of quality assurance.

Some regions of the UK where the drinking water is collected in underground reservoirs (aquifers) have been designated Nitrogen Sensitive Areas (NSAs) and growers have management agreements to apply reduced levels of nitrogen.

Timing of nitrogen

The timing of nitrogen application is very important in cereal growing. Uptake precedes the growth response (Fig. 3.3) and it is important to have soluble nitrogen available at the time of uptake. Unfortunately, the release of the soil nitrogen is slow in the spring because the soil temperature is low and it important that the grower applies fertilizer to meet the early nitrogen demands of the crop.

The main response timing to applied nitrogen is at the beginning of stem extension or growth stage 31 (first node detectable). For winter wheat where large amounts of N are being applied (over 120 kg/ha), a further yield response will be obtained if the nitrogen is applied as a split dressing with 40–50% at growth stage 30 (usually early–mid April) and the balance at growth stage 32 (usually end April–early May). In crops that have low shoot numbers or are affected by take-all, 40 kg/ha N can be applied early in mid February–mid March to boost root growth and tillering. Winter wheat grown for breadmaking can also receive a further 40 kg/ha N during late stem extension to early grainfill (GS 39–59), often as a liquid urea spray to help boost grain protein content. Leaf scorching can occur if this application is made during hot sunny weather.

For winter barley, winter oats, rye and triticale where N applications are above 100 kg/ha, 40 kg/ha should be applied in mid February–mid March and the balance at the end March–mid April; if 100 kg/ha or lower, then apply as a single dressing at end March–mid April. When a malting winter barley is grown it is important to complete all N applications by the end of March in order to keep grain protein contents low. Owing to the crucial nature of these N timings, a revision of the Zadoks key has been published to explain in detail the accurate diagnoses of these growth stages (Fig. 3.4).

For spring-sown cereals, if the amount of N is greater than 70 kg/ha then 40 kg/ha of this can be incorporated in the seedbed and the balance applied at the 3- to 4-leaf stage (GS 13/14). When less than 70 kg/ha is used then apply all at GS 13/14.

Disease control

Detailed consideration of cereal diseases and their control is given in Chapter 6. Winter-sown cereals suffer more disease than spring-sown crops and this is especially true of barley. The first major consideration for disease control is varietal resistance. Choice of an appropriate variety for the region is of paramount importance since there is regional variation in the serious occurrence of several foliar diseases, e.g. yellow rust is not found very often in the south-west whereas *Rhynchosporium*, brown rust and *Septoria* are more prevalent in this region.

Unfortunately there are no varieties resistant to all diseases and any resistance is only temporary since fungal organisms can reproduce many times in a season and have the ability to eventually overcome resistance. Thus, varieties resistant to mildew frequently succumb to a new race of mildew within 3 years of widespread use. Cereal variety diversification schemes for mildew and yellow rust (NIAB) can help to reduce the effect of epidemic spread of these diseases on the farm. Until long-lasting disease resistance is achieved in varieties it will be necessary for growers to continue to use

Table 3.8 Nitrogen fertilizer recommended rates (kg/ha N) for cereals (adapted from MAFF, 2000). Note: these rates are for fertilizer + organic manure (N content of organic manure needs to be estimated)

	SNS index						
	0	*1*	*2*	*3*	*4*	*5*	*6*
Winter wheat							
Light sandy soils	160	130	100	70	40	0–40	0
Shallow soils over chalk		240	200	160	110	40–80	0–40
Medium and deep clay soils and shallow soil over rock		220	180	150	100	40–80	0–40
Deep fertile silty soils		180	150	120	80	40–80	0–40
Organic soils				120	80	40–80	0–40
Peaty soils						0–60	0–60
Winter barley (feed)							
Light sandy soils	160	130	100	70	40	0–40	0
Shallow soils over chalk		200	160	130	70	20–60	0–40
Medium and deep clay soils and shallow soil over rock		180	150	120	70	20–60	0–40
Deep fertile silty soils		150	120	80	40	0–40	0
Organic soils				120	70	20–60	0–40
Peaty soils						0–40	0–40
Winter barley (malting)							
Light sandy soils	120	80	40	0–40	0	0	0
Other mineral soils		130	100	70	0–40	0	0
Organic soils				70	0–40	0	0
Peaty soils						0	0
Winter oats, winter rye and winter triticale							
Light sandy soils	120	80	50	0–40	0	0	0
Other mineral soils		130	100	70	40	0–40	0
Organic soils				70	40	0–40	0
Peaty soils						0–40	0–40
Spring wheat							
Light sandy soils	160	130	100	70	40	0–40	0
Other mineral soils		180	150	120	70	40	0–40
Organic soils				120	70	40	0–40
Peaty soils						0–40	0–40
Spring barley (feed)							
Light sandy soils	125	90	60	40	0–40	0	0
Other mineral soils		150	120	80	40	0–40	0
Organic soils				80	40	0–40	0
Peaty soils						0–40	0–40
Spring barely (malting)							
Light sandy soils	100	70	40	0–40	0	0	0
Other mineral soils		120	80	50	0–40	0	0
Organic soils				50	0–40	0	0
Peaty soils						0	0
Spring oats, spring rye and spring triticale							
Light sandy soils	100	70	40	0–40	0	0	0
Other mineral soils		120	80	50	0–40	0	0
Organic soils				50	0–40	0	0
Peaty soils						0	0

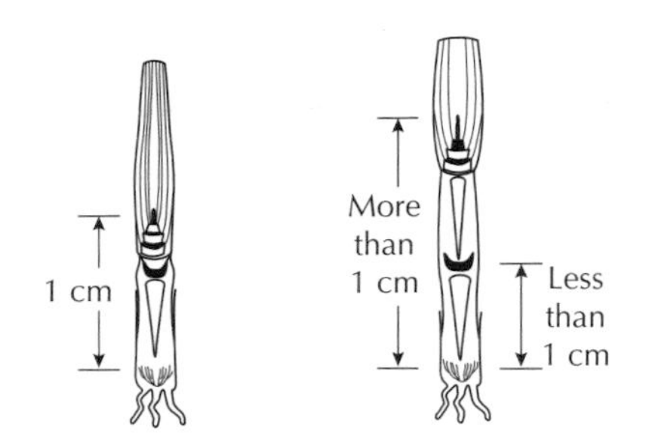

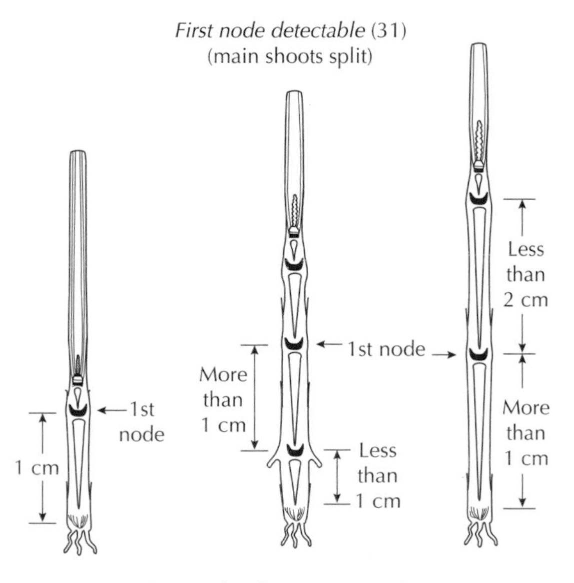

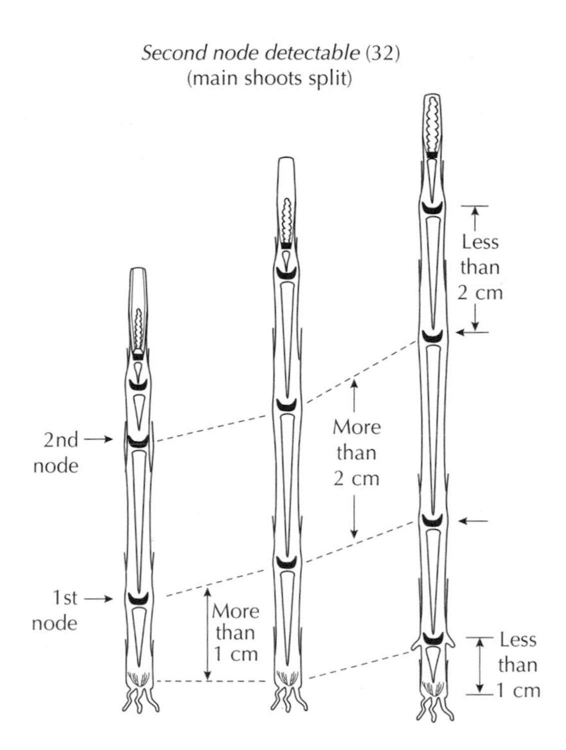

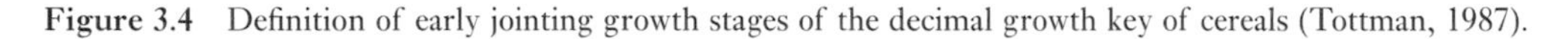

Figure 3.4 Definition of early jointing growth stages of the decimal growth key of cereals (Tottman, 1987).

fungicides. Problems with resistance also exist with the overuse of certain fungicides, e.g. eyespot resistance to carbendazim. Diversification of fungicides is also a recommended practice; thus when a series of fungicides is planned they should be chosen from different groups (Table 3.9). Timing of fungicide application is important and a knowledge of the disease and its epidemic spread pattern can help in deciding the timing of application. Some fungicides have a tendency to keep the plant greener for longer, particularly in the stem, e.g. strobilurins, which can lead to harvesting difficulties. For spring cereals usually only one fungicide is economically justified, usually to combat powdery mildew. A summary of the main diseases of winter cereals and the important time of their control is given in Table 3.10.

Plant growth regulators

Plant growth regulators are frequently used in winter cereals to protect against lodging. High risk situations are: weak or long strawed varieties, highly fertile conditions or high N fertilizer applied, poorly consolidated seedbed, and high incidence of eyespot likely. The growth regulators shorten and thicken the internodes on the stem making them resistant to bending and buckling. A summary of the timing of application is given in

Table 3.9 Cereal fungicide groups (after HGCA, 2000). Level of disease control: **** excellent; *** very good; ** moderate; * low; blank – none

	Activity ratings against specific diseases					
	Eyespot	*Mildew*	*Septoria nodorum*	*Septoria tritici*	*Yellow rust*	*Brown rust*
Triazoles[1]	*	**	***	***	***	***
Strobilurins[2]		***	**	**	***	***
Morpholines		***			**	**
Quinolene		****				
Spiroketalamine	***			***	**	
Anilinopyrimidine	****	***	***			
Imidazole	***	*	**	**	*	*
MBCs	*	*	*	*	*	*
Chloronitrile		*	**	***	*	*

[1] The precise activity of a specific triazole varies with the product – check the manufacturer's label recommendations.
[2] Formulations that are mixtures of strobilurins and other fungicides are available and improve the effectiveness and spectrum of diseases controlled.

Table 3.10 A summary of winter cereal diseases, their times of appearance and the target diseases for control. (Bold = target diseases)

Growth phase:	*Establishment and tillering*	*Early stem extension*	*Flag leaf emergence*	*Ear emergence Grain filling*
Decimal growth stage:	*11 13, 21 15, 23–25*	*30 31 32*	*37 39 45*	*51 59 69 71*
Approx. time				
W. wheat:	*Oct Feb*	*Mar Apr May*	*June*	*July Aug*
W. barley:	*Sept Oct Feb*	*Mar Apr*	*May June*	*July*
W. wheat Diseases present:	*Septoria tritici* Mildew Eyespot Sharp eyespot *Fusarium*	**Eyespot** *Fusarium* *Septoria tritici* *Septoria nodorum* Mildew	***Septoria* sp.** **Mildew** **Yellow rust** **Brown rust**	Sooty moulds
Control:	No control in autumn necessary	Target control against eyespot	Keep ear and top two leaves clean	Only control if milling variety
W. barley Diseases present:	Mildew Net blotch Brown rust *Rhynchosporium*	**Net blotch** ***Rhynchosporium*** **Eyespot** **Mildew** Sharp eyespot *Fusarium*	**Net blotch** ***Rhynchosporium*** **Mildew** **Brown rust**	
Control:	Only control in exceptional circumstances	Control foliar and stem-based diseases	Keep top 2–3 leaves clean	
Winter oats Diseases present:		Mildew	**Mildew** **Crown rust**	
Control:			Control if for human consumption	

Table 3.11. Plant growth regulators are not usually used in spring cereal production.

Plant growth regulators have also been used to manipulate tillering, although there are no manufacturers' recommendations for this type of use.

Pest control

The most economically important pests of cereals during the main crop growth phase are aphids. Populations can build up on the ear in summer and can lead to small grain size and sooty moulds growing in the

Table 3.11 Cereal plant growth regulators and their application timings

Active ingredient(s)	*Application timing*	*Crop*	*Notes*
Chlormequat Chlormequat + choline chloride	GS 30/31	Winter wheat Winter oats	Shortens and stiffens lower internodes; may be followed by second application at GS 31/32/33
Ethephon Ethephon + mepiquat chloride	GS 37	Winter barley	Shortens upper internodes and reduces necking and ear drop
Chlormequat followed by Ethephon	GS 30/31 + GS 33/34	Winter wheat	Combination of the two regulators gives excellent lodging control in high-value bread-wheat crops
Chlormequat Chlormequat + choline chloride	GS 13/15	Winter barley	For crops in Scotland, evens up tiller development going into winter

'honeydew' excreted by the aphids. Control at present is only by means of insecticides sprayed at a threshold average of 5 aphids per ear. Approved products include those that are selective against aphids and less damaging to beneficial insects. Aphid control is only usually economically justified on wheat crops. In an effort to reduce the use of insecticides in cereal production ecological research is defining more closely the natural predators of cereal aphids and strategies for maximizing biological control in the field.

Desiccation

Following a wet summer and/or lodging of a crop, desiccation may be necessary to kill green weeds in the crop prior to harvest. A rapid kill of green material can be obtained with the use of diquat. A slower but more thorough kill is obtained with glyphosate and this chemical kills couch grass rhizomes.

Minor cereals

There is always a specialist market in the UK for minor cereals and these can be very profitable but have very small market size.

Durum wheat

This wheat is used to produce semolina flour from which pasta is manufactured. It is a difficult crop to grow reliably in the UK because of quality considerations. Hagberg Falling Number is frequently low if the crop does not ripen under hot dry conditions and grain protein content can sometimes be low. It requires no vernalization and can be grown as a winter or spring crop. If sowing in the autumn then delay until the end of October. Otherwise treat as a winter wheat.

Triticale

This is a man-made hybrid between rye and durum wheat. It carries the environmental tolerance of rye and the quality characteristics of durum wheat. It can outyield wheat on poor soils, but its long straw can lead to problems of lodging. It is relatively disease resistant, but more disease is appearing as crop growth becomes more widespread. Only autumn-sown varieties are worth considering and husbandry is the same as for winter wheat. The high-protein grain makes it particularly suitable for pig rations, but as yet feed manufacturers are not giving large enough premiums on triticale to make it competitive with wheat.

Grain maize

Generally speaking the UK is at present too cold for the reliable widespread production of grain maize, but small areas of production are possible in southern Hampshire and low-altitude areas of Essex and Suffolk. The UK, however, imports large quantities of maize and maize products. The crop is grown in the same way as forage maize with a slightly lower plant population (10 plants/m^2). Specialized harvesting machinery is necessary and this is not widely available in the UK. As global warming increases, the opportunities for UK grain maize production could well increase.

Rye

A small but established market for rye grain exists for the manufacture of crispbread products and rye bread. The market is small and not expanding and is fulfilled by a small number of growers. Rye is also grown by dairy farmers as a forage crop to provide 'early bite' in March.

Oilseeds

Oilseed rape (*Brassica napus*)

Oilseed rape seed is grown for crushing to extract the oil which, depending on variety, can be used for human consumption (cooking oils and margarine), industrial use, or for production of biodiesel. At present the bulk of the UK crop is produced for human consumption. After crushing, a high-protein meal remains which can be valuable for inclusion in animal feed rations. To be suitable for human consumption varieties must be low in erucic acid (LEAR), and for the meal to be suitable for animal feed they must also be low in glucosinolates; such varieties are known as 'double lows'. Some industrial uses require higher levels of erucic acid (HEAR); non-food rape is currently the most popular set-aside crop.

Oilseed rape is the most popular UK cereal break crop; it fits well into cereal-based rotations, being broadly tolerant of soil conditions and using the same equipment but having different sowing and harvest dates, thus easing labour peaks, and providing an opportunity for cheaper grass weed control. As a brassica it does not share any soil-borne pests or diseases with cereals. At present oilseed rape is included in the Arable Area Payments Scheme (see Chapter 8). Yields average 3.5 t/ha for winter types and 2 t/ha for spring types.

Market

Crushers pay according to the following standards:

40% oil content
35 μmol/g max. glucosinolate
9% moisture
2% max. impurities

Variety selection

Selection from the varieties available should consider glucosinolate level, disease resistance, yield potential, height, lodging risk, and time to maturity. Each should be considered in relation to the characteristics of the site and/or the intended system of production, e.g. tall varieties will be more suitable for swathing. Glucosinolate level is of great importance in varietal selection. Research to date has given little promise of husbandry control over levels, and site and season have overriding but unpredictable effects. In order to be eligible for aid, the variety must be listed in the Common Catalogue. Hybrid rapes are available and require some modification to standard husbandry.

Soils and site

Well-drained soils (pH 6.0–7.0, 6.5–7.0 preferred), free from compaction, are necessary for good yields, medium to sandy loams being ideal for good tap root development. Soils and sites naturally low in sulphur may be important for low glucosinolate levels, hence light or chalky soils in areas of low atmospheric sulphur are favoured; the latter are found away from the heavy industrial areas of the London, Midland and northern conurbations. Rape should be grown on a 4-year rotation due to the risk of *Sclerotinia sclerotiorum*; other hosts are peas, spring beans and linseed, and those should not be grown in close rotation with other brassicas to avoid buildup of clubroot. Beet nematode can be encouraged in sequences including both rape and sugar beet, and thus this sequence should be avoided.

Winter types

Seedbed and establishment Rape seed is very small and requires a fine clod-free seedbed, and every effort should be made to conserve moisture. Optimum sowing date is from mid-August to early September (the latter suitable in southern England), when seed should be drilled to a depth of 2–3 cm on approx. 25 cm-row widths, followed by rolling. Late sowing is very detrimental to rape yields as growth made before the onset of low winter temperatures (growth ceases below 2°C) determines yield potential; conversely plants that have advanced too far into stem elongation due to early sowing can suffer badly from frosts; thus sowing date is critical. Ideally plants should have at least five leaves as they enter the coldest part of the year.

Spring plant population should be approx. 80 plants/m^2 and seed rate should be carefully calculated to achieve this because: small inaccuracies have big effects because the seed is small; varieties vary significantly in their TSW; and losses prior to spring can be significant as a result of failure to establish, losses to pests, and winter kill (50% losses for late-sown crops in poor conditions, 20% for optimal timing in good conditions). Rape responds to low plant density by branching more extensively, which compensates for yield but increases the extent of uneven ripening at harvest. Fifty plants per square metre is generally considered to be the minimum for a worthwhile crop; below this level it may be economic to redrill with a spring crop. Over-thick crops will underyield due to an inefficient canopy structure and increased disease, and may lodge, causing difficulties with ripening and harvest.

Establishment threats Weed competition during the establishment/autumn phase will reduce yield, with grass weeds and volunteer cereals having significant effects, particularly in later sown crops. Vigorous crops should compete well with broadleaved weeds, but chickweed and cleavers will cause yield reductions in many situations. A weed control programme which deals with grass weeds, volunteers, cleavers, and chickweed is recommended (see Chapter 6), to ensure yield losses are minimized; crop development stage must be taken into account when using herbicides as they vary in their 'safe' timings and few are safe once flower buds present (November in most regions). Additional benefits will be gained from sample purity (cleavers seed is similar size and shape to rape seed), including savings on subsequent cereal crop weed control, and easier harvest.

Pigeons can devastate rape crops, particularly 'gappy' ones; effective control is difficult and rape may have to be dismissed as an appropriate crop for high risk sites. Slugs, too, can decimate establishing rape and must be controlled in all but very low risk situations (see Chapter 6). Other pests associated with establishment are cabbage stem flea beetle (*Psylliodes chrysocephala*) and rape winter stem weevil (*Ceutorhynchus picitarsis*) which should be controlled on the basis of crop monitoring and threshold guidelines (see Chapter 6). Two diseases should be monitored in the autumn: light leaf spot (*Pyrenopeziza brassicae*) and stem canker (*Phoma lingam*); some varietal resistance is available in both cases but chemical control may be economic in some seasons (see Chapter 6).

Nutrients Nutrient requirements are given in Table 3.12. Timing of nitrogen addition is critical; rape is very responsive to spring nitrogen and the rapid growth encouraged by rising temperatures should not be checked by tardy application. If conditions permit and growth has begun nitrogen should be applied from mid February; in appropriate situations (e.g. light soils, applications >100 kg/ha) the dressing can be split, but applications after early April are wasted, except where flowering is likely to be delayed when slightly later applications may still be effective. Seedbed nitrogen should only be used on crops that will use it and will benefit from a boost to establishment to avoid losses.

Table 3.12 Nutrient requirements of oilseed rape (kg/ha); m indicates maintenance dressing (MAFF, 2000)

	SNS index						
	0	*1*	*2*	*3*	*4*	*5*	*6*
Nitrogen							
Winter OSR[1]							
Seedbed	30	30	30	0	0	0	0
Spring							
Mineral soils	220	190	160	120	80	40–80	0–40
Organic soils	—	—	—	120	80	40–80	0–40
Peaty soils	—	—	—	—	—	40–80	40–80
Spring OSR							
Light sandy soils	120	80	50	0–40	0	0	0
Other mineral soils	—	120	80	50	0–40	0	0
Organic soils	—	—	—	50	0–40	0	0
Peaty soils	—	—	—	—	—	0–40	0–40

	P or K index			
	0	*1*	*2*	*3*
Phosphate and potash				
Winter OSR[1]				
P_2O_5	100	75	50 m	0
K_2O	90	65[3]	40 m (2–) 20 (2+)	0
Spring OSR[2]				
P_2O_5	80	55	30 m	0
K_2O	75	50[3]	25 m (2–) 0 (2+)	0

[1] For crops with yield potential of 3.5 t/ha.
[2] For crops with yield potential of 2 t/ha.
[3] Reduce by 25 kg/ha on light sandy soils.

Rape has a high demand for trace elements during rapid spring growth and in some areas rape may suffer deficiencies of magnesium or sulphur. At soil index 0, 50–100 kg/ha magnesium should be applied every 3 or 4 years. When sulphur deficiency has been diagnosed (<4 mg/g of total sulphur, or N:S ratio >17, from leaf analysis), 50–75 kg/ha SO_3 should be applied in early spring.

Spring/summer threats Light leaf spot and stem canker may continue to be a problem, also *Botrytis*, *Alternaria*, and *Sclerotinia* may be encouraged by warm wet weather. Pollen beetle (*Meligethes* spp.), seed weevil (*Ceutorhynchus assimilis*), and pod midge (*Dasyneura brassicae*) are the main insect pest threats. Chemical control is available for these pests and diseases but should be applied according to threshold guidelines, and never when rape is in open flower due to the risk to bees; bees enjoy rape pollen and should be encouraged to forage in rape crops as they contribute to pollination.

Harvest Rape reaches maturity in late July/early August, but ripening is usually uneven due to the structure of the plant; seeds on the terminal (main, central) branch are older than those on the side branches, and maturity varies along each branch. Two methods are commonly used to even-up ripening: desiccation with diquat, or swathing into rows. In both cases combining is not possible until 7–14 days later, leaving the crop vulnerable to seed losses from pod shatter or poor weather. Choice of harvest method will depend on availability of a swather or high-clearance sprayer and suitability (e.g. swathing is preferable for exposed sites) and there are no consistent effects on yield or quality. Moisture content at harvest will normally be in the range 10–18%.

Spring types

Seedbed and establishment Good establishment is critical to the success of spring rape but can be difficult in dry springs. Seedbed requirements are as for winter types with no compaction or capping and moisture conserved. Seed can be sown from the beginning of March to end of April, end of March/early April is ideal in most years, at same depth, widths, etc. as winter types to achieve a plant population of 120 plants/m^2.

Nutrients Nutrient requirements are given in Table 3.11. All the nitrogen should be applied in the seedbed, except on light sandy soils at SNS index 0 where 50 kg/ha should be applied in the seedbed and the balance by early May.

Threats Spring rape is a popular crop in suitable areas due to its low input requirements; a quickly established crop will compete well with spring-germinating weeds, and though susceptible to the same diseases as winter rape it rarely suffers unless grown in close proximity to an infected winter variety. Downy mildew (*Peronospora parasitica*) may attack early-sown crops. Similarly spring rape is vulnerable to the same insect pests as the winter varieties but usually only pollen beetle requires control. (See Chapter 6 for further advice.)

Harvest The spring crop is later to maturity than the winter type, and is harvested in late August to late September. It suffers much less uneven ripening, however, due to its simpler, less branched structure, and can often be combined without any pretreatment. If uneven ripening is a problem then the procedures used for winter rape are equally applicable.

Linseed (*Linum usitatissimum*)

Market

Linseed is grown for oil extraction from the seeds; the oil has a range of industrial uses chiefly in the paint and varnish industries, and for the manufacture of linoleum and oilcloths. The cake remaining after extraction is included in ruminant feeds. Whole seed has a small market for horse and bird feed. Straw is high in fibre and can be used for, e.g., paper/board manufacture.

At present linseed is included in the Arable Area Payments Scheme (see Chapter 8), and may be grown on set-aside land as an industrial crop subject to certain conditions. Declining payment levels have reduced the popularity of the crop, however.

The advantages of linseed are its wide tolerance of conditions throughout the UK and its relative ease of production, fitting well into cereal-based systems; however, establishment success is critical and harvest can be problematic. Linseed can be grown throughout the UK, but harvesting problems will increase in later and/or wetter regions. Establishment can be difficult so sites with poor soil structure should be avoided. The main rotational threats are *Sclerotinia sclerotiorum* (other hosts are oilseed rape, peas, and spring beans) and wilt (*Fusarium oxysporum*), requiring a 4-year break, and fields infested with perennial grass weeds are a problem due to the very poor competitiveness of the linseed plant. Key characters on which to make varietal choice are regional suitability, time to maturity, and stem height and strength. Short-strawed varieties are less suitable for light or drought-prone soils. Winter varieties are available but generally show no advantages compared to spring types.

A wide range of soil types are suitable, but heavy clays should be avoided due to the difficulty of obtaining a fine seedbed for good establishment. Once established linseed is reasonably tolerant of drought, provided its tap root is able to penetrate down to available water; the flowering period is the most vulnerable to dry conditions. Optimum pH is 6.0–7.0.

Good seedbed conditions are essential for successful establishment. A fine tilth with no compaction and sufficient moisture should be provided by autumn ploughing, minimal spring cultivations and a roll prior to drilling. There is no advantage from early sowing into a cold, poor seedbed; appropriate seedbed conditions for quick establishment are of paramount importance. Optimum sowing period is mid-March to mid-April but acceptable yields can be achieved from sowings up to early May (late April in northern areas). Seed should be sown at 1.5–2.5 cm depth (3 cm may be necessary for dry seedbeds) on row spacings of 8–18 cm (narrower spacing gives better plant distribution). Even distribution of seed is difficult due to the 'flat' shape and the slipperiness of the seed. A plant population of 400–500 plants/m^2 is optimal for yield; a lower density offers less weed competition and encourages the plants to branch, leading to more uneven ripening; higher populations give weak individual plants with a lower yield potential, a propensity to lodge, and vulnerability to disease. Seed rate should take account of seed size, seedbed conditions, soil type, and row spacing; establishment ranges typically from 50 to 80%.

There is little response to P and K, which should be applied in the seedbed. Nitrogen should be applied in the seedbed or at time of unfolding of the first pair of true leaves, according to soil characteristics, sowing date and amount needed (Table 3.13).

The structure of the linseed plant renders it a poor competitor against weeds, especially during the establishment phase, so that good weed control is essential to prevent yield reduction and harvesting difficulties. Preparatory control, e.g. for perennial weeds, and post-emergence chemical control are favoured. (See Chapter 6 for further weed control advice.)

Some chemical control of diseases may be necessary, though the avoidance of thick, dense canopies and lodging will lessen the risk. The main threats are grey mould (*Botrytis*) and blight (*Alternaria linicola*), both at flowering, particularly in wet weather. (See Chapter 6 for further disease control advice.)

The crop matures in late August/September but can occur later in northern areas or in cool years. Desiccation with diquat is usually advantageous, primarily to dry out the green fibrous stems, but also to overcome uneven ripening, and achieve timely harvest. Desiccation should be carried out when the seeds rattle inside golden-brown capsules. There is little risk of seed loss from pod shatter following desiccation and the crop can stand until harvest is convenient; at least 10–20 days will be necessary to dry the stems, shrivel the leaves, and ripen the seed to dark brown with a moisture content of about 10–15%. The fibrous stems are the biggest problem at harvest as they are tough, need sharp knives, and have a tendency to wrap around the reel; the slippery thin seed flows easily and will be lost through any leaks in the combine and trailers.

The shape of the seed makes efficient, early drying vital as the seed packs down in store and has a high resistance to airflow. For storage the aim is to achieve

Table 3.13 Nutrient requirements for linseed (kg/ha); m indicates maintenance dressing (MAFF, 2000)

	SNS index						
	0	*1*	*2*	*3*	*4*	*5*	*6*
Nitrogen							
Light sandy soils	80	50	0–40	0	0	0	0
Other mineral soils	—	80	50	0–40	0	0	0
Organic soils	—	—	—	0–40	0	0	0
Peaty soils	—	—	—	—	—	0	0
	P or K index						
	0		*1*		*2*		*3*
Phosphate and potash[1]							
P_2O_5	80		55		30 m		0
K_2O	75		50		25 m (2–) 0 (2+)		0

[1]For crops with potential yield of 1.5 t/ha.

8% moisture from a maximum drying temperature of 65°C (higher temperatures can reduce oil quality); the recommended maximum storage depth is 80 cm (see Chapter 7 for further drying and storage advice).

Average yields are approx. 1.5–2 t/ha. Payment is based on a standard of 9% moisture and 38% oil content. Plant breeding effort continues to achieve higher yields and greater ease of production by means of a more efficient canopy structure and better partitioning from fibre to seed, though true dual-purpose varieties (seed and fibre) are also sought. Greatest effort is towards varieties with specialized oil profiles, including those suitable for human consumption (linola).

Arable ('field') legumes

Arable legumes are relatively large-seeded combinable dicotyledonous crops grown for their protein-rich seeds. They share the ability of all legumes to fix atmospheric nitrogen and are valuable contributors to sustainable crop rotations, adding 'no-cost' nitrogen to the soil (though the amount of N fixed can vary widely according to soil conditions and the health of the crop). Field beans and peas are the important UK arable legume crops currently, with lupins gaining in popularity; others such as navy beans, soya beans, chick peas, and lentils require further plant breeding effort to provide varieties consistently suitable and economically viable for the UK climate.

Field beans (*Vicia faba*)

Market

The majority of field beans are grown for harvesting 'dry' and sold as animal feed or used on farm through mill and mix units (though similar varieties are grown as broad beans for harvesting as a green vegetable). Their value as an animal feed lies chiefly in their crude protein content (approx. 24%) (see Chapter 18). Some smaller-seeded varieties enjoy a premium niche market as pigeon feed or as export for human consumption; a contract is required for these markets. Many varieties have high tannins in the seed coat which restrict their inclusion in feeds; some tannin-free spring varieties are available and are more attractive as a non-ruminant feedstuff though they tend to be lower yielding.

The Common Agricultural Policy has supported field bean production since 1978 in order to encourage the use of home-grown feed in place of imported soya. At present beans are included in the Arable Area Payments Scheme (see Chapter 8).

Advantages

Field beans are grown widely in the UK; they have a reasonable gross margin potential, are relatively tolerant and easy to grow, and fit well into cereal-based systems using the same equipment and spreading sowing and harvesting dates. They are an effective cereal break-crop as they provide an opportunity for grass weed control and share none of the cereal-associated soil-borne pests and diseases. As a legume they leave some residual nitrogen to benefit the following crop.

Disadvantages

The major disadvantages of field beans are their sensitivity to drought, their unstable yields, and their relatively late harvest. Winter beans can form extremely tall dense canopies in wet years leading to poor pollination, disease control problems, and harvesting difficulties.

Types

There are both winter and spring types of field bean available, with contrasting agronomic characteristics and which are suitable for different situations; yield and harvest date differences are not large. Winter beans are more frost hardy and form a much taller thicker leaf canopy than spring types; they can be ploughed in rather than drilled. Traditionally, beans were also categorized by seed size, with those having a TSW smaller than 530 g being tick (or tic) beans, and those having a TSW greater than 530 g being horse beans. Smaller beans are rounded and similar to a slightly elongated pea, larger beans are flattened and kidney shaped and are more difficult to handle, for example through augers. All winter beans are of the larger seeded horse type.

Varietal choice

Key characters on which to make varietal choice are end-use potential, maturity date, straw height, and disease resistance. Resistance to downy mildew is available in some spring varieties, but there is little difference in resistance to bean rust and chocolate spot between varieties at present. Shorter varieties should be chosen for soils not likely to suffer from moisture stress, and taller varieties for drier sites. It should be noted that shorter varieties tend to suffer more from bean rust (*Uromyces viciae-fabae*). Later-maturing varieties should be avoided in cooler, wetter areas of the country.

Suitable sites

Beans were traditionally grown on heavy soils, but a wide range of soil types are suitable where drought is not likely to be a problem. Generally spring beans are likely to be preferable to winter types in areas where spring sowing is appropriate and feasible as there are more varieties available with better marketing and husbandry characteristics. Winter beans are preferable in areas where spring sowing is problematic, summer droughts are more likely, and the yield differential will be pronounced, or where earlier harvest is important. Diseases of winter beans will easily cross infect to spring beans and close proximity of crop, trash, or volunteers should be avoided.

Sites should not have grown legumes in the previous 4 years to control footrot (*Fusarium* spp.), downy mildew (*Peronospora viciae*), and stem nematode (*Ditylenchus dipsaci*) though if the site is known to be infected with nematode an 8-year break should be practised (other hosts are legumes, oats, sugar beet, onions). Spring beans should be separated from oilseed rape and linseed by 4 years to control *Sclerotinia sclerotiorum*.

Waterlogged soils should be avoided as bean seeds are sensitive to poor aeration. Compacted soil is detrimental to beans but is less likely to be a problem for winter types. Light soils are unsuitable due to the risk of drought. Soil pH should be 6.5–7.5.

Establishment

Seed should be treated to help prevent damping off (*Pythium* spp.), and should be tested to avoid introducing stem nematode (*Ditylenchus dipsaci*).

Winter beans should be sown 4 weeks before frost is likely to occur to encourage good quick establishment without too much above-ground development; depending on the location this might be late October, November, or early December in mild areas. Ploughing-in winter bean seed to a depth of about 15 cm after broadcasting is a quick, cheap method of sowing and has often been shown to give a yield advantage over drilling as it allows the use of soil-residual herbicide, protects from pest damage (birds, rodents), and leaves a rough, free-draining surface which offers protection from cold winds. Winter beans are only hardy down to −12°C and the young shoots can be severely damaged by frost; this can lead to the production of many weak branches, which if excessive can lower yields. Optimum plant population is 25 plants/m^2, rising to 40 plants/m^2 for later sowings which are less likely to branch, thus reducing yield potential (optimum number of flowering stems is 40 stems/m^2). Accurate seed rates must be calculated to take account of the, often large, variations in seed size between varieties and seed lots (see this chapter). In addition, seed accounts for a high proportion of the variable costs of production, and the economic optimum plant population should be calculated for each season.

Spring varieties should be sown as soon as soil conditions permit (minimum temperature for germination approx. 5°C, and moist and aerated) in late February/early March; yield reductions are likely from sowings after mid-March, and the risk from drought exacerbated. Optimum plant population is 50–60 plants/m^2, but in many seasons the economic optimum is likely to be 40–45 plants/m^2. Spring varieties should be drilled on 15- to 35-cm row widths (the narrower widths should be used for shorter varieties) at a depth of 6–8 cm, especially if soil-residual herbicide is used. The seedbed should be well aerated.

Nutrients

Requirements are given in Table 3.14; no nitrogen is needed. Soil should be at pH 6.8–7.0.

Threats

Winter beans are uncompetitive to weeds in the establishment phase due to the low plant density and relatively slow development of a canopy. Some chemical control may be necessary, particularly to control grass weeds for the benefit of the rotation, though late-sown crops will encounter few weed problems for cultural reasons. The only major disease threat is from chocolate spot (*Botrytis fabae*) and possibly seed-borne leaf/pod spot (*Aschochyta fabae*). It is rare for winter beans to require insect pest control. (See Chapter 6 for detailed advice.)

Good early weed control is important for spring beans as they are uncompetitive in the establishment phase and tall/long weeds continue to compete and cause harvesting difficulties, e.g. volunteer rape, fat-hen, cleavers. The most serious disease threat is bean rust (*Uromyces viciae-fabae*), also chocolate spot and leaf/pod spot as for winter beans. Black bean aphid (*Aphis fabae*) is a common pest which damages flowers and pods as well as encouraging chocolate spot; stem nematode (*Ditylenchus dipsaci*) and bean weevil (*Sitona lineatus*) may also affect spring bean crops. (See Chapter 6 for detailed advice.)

Table 3.14 Nutrient requirements (kg/ha) of field beans (yield potential of 3.5 t/ha), and field and vining peas (yield potential of 4 t/ha); m indicates maintenance dressing (MAFF, 2000)

	P, K or Mg index			
	0	*1*	*2*	*3*
P_2O_5	85	60	35 m	0
K_2O	90	65	40 m (2–) 20 m (2+)	0
MgO	100	50	0	0

Harvest

Harvest is usually in September for both types, when pods are black and brittle and moisture content is below 20%, though up to 25% is combinable. Moist conditions should be chosen, if possible, to minimize pod shatter. Yield averages 3.5–4.0 t/ha.

Field peas (*Pisum sativum*)

Market

Also referred to as combinable peas, field peas are grown for harvesting 'dry' (and should be distinguished from vining peas grown for harvesting as a green vegetable to be frozen or canned). The majority of the UK crop is used for animal feed by the compounders, or directly on-farm through mill and mix units. A smaller but significant proportion of the national crop is used for human consumption as canned ('processed', 'marrowfat', and 'mushy') peas or dried peas, with potential for increase in the 'snack' pea market (comparable to peanuts). Their value lies chiefly in their crude protein content (approx. 24%) (see Chapter 18), though some current research effort is directed to their potential utilization as an oilseed crop. Peas destined for human consumption have additional quality requirements (colour and freedom from defects such as staining and splitting) compared to those for animal feed but can be sold at a premium of up to 50% of the feed price. Contracts are usually necessary for reliable sale for human consumption.

The Common Agricultural Policy has supported field pea production since 1978 in order to encourage the use of home-grown feed in place of imported soya. At present peas are included in the Arable Area Payments Scheme (see Chapter 8).

Advantages

Field peas are grown widely in the UK; they have a reasonable gross margin potential, are relatively tolerant and easy to grow, and fit well into cereal-based systems, using the same equipment and spreading sowing and harvesting dates. They are an effective cereal break crop as they provide an opportunity for grass weed control and share none of the cereal-associated soil-borne pests and diseases. As a legume they leave some residual nitrogen to benefit the following crop, and their early summer harvest date provides good entry opportunities.

Disadvantages

The major disadvantages of field peas are their sensitivity to poor soil/soil water conditions, their unstable yields, and their extreme susceptibility to lodging which frequently leads to difficult harvest operations. Pea yields can be highly variable from year to year and between sites. This yield instability is a result of extreme sensitivity to environmental conditions (for example, waterlogging at the pre-flowering and flowering stages), inter-plant competition (see Chapter 2) resulting in wide variation in individual plant size, and strong intra-plant competition arising from an indeterminate flowering habit leading to flower, pod, and seed abortion. 'Wild type' peas are climbers with long stems, many tendrils to anchor the plant, and with little need for a strong supporting stem, consequently peas rarely stand without falling over or lodging, even when short stemmed. Plant breeding effort, using the vast *Pisum* gene bank, continues to seek varieties with more efficient and productive plant and canopy structures, resistance to common pests and diseases, and seed components appropriate for a wider range of end uses.

Types

There are several categories of pea variety depending on their seed type (marrowfat, white, large blue, and small blue) and on their leaf type (normal, tare leaved, or semi-leafless; Fig. 3.5). All are suitable for animal feed

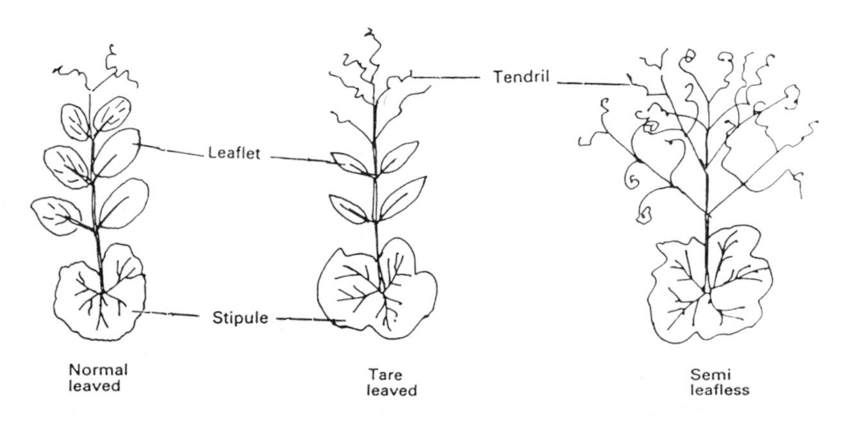

Figure 3.5 Field pea leaf types.

but only some varieties are suitable for human food use. Leaf type can be an indicator of potential for lodging, with normal leaved types having a heavier canopy and thus a greater propensity to lodge, particularly in fertile conditions. However, some normal leaved types have been selected to lodge early and thereafter to hold their pods clear of the ground, thus avoiding harvest difficulties. Tare leaved and semi-leafless types were bred to reduce the weight of the canopy and to increase the number of tendrils to create self-supporting canopies with less tendency to lodge.

Varieties

Key characters on which to make varietal choice are end-use potential, lodging and combining risk, resistance to pea wilt (*Fusarium opxysporum* f. sp. *pisi*) and downy mildew (*Peronospora viciae*), haulm length, and maturity date. Shorter varieties should be chosen for moist fertile soils, and taller varieties for dry poorer soils, to avoid combining problems. Late-maturing varieties should be avoided in cooler, wetter areas of the country. Varieties suitable for human consumption tend to be lower yielding and should therefore be avoided if the intention is to sell or use for feed. Autumn-sown varieties are available but are not widely grown.

Suitable sites

Peas are traditionally grown in lower rainfall areas as they are not compatible with wet conditions, particularly during the later stages of development when the canopy is vulnerable to disease and infection will reduce both yield and quality, and during the harvest period when rainfall adds further difficulty.

Peas should not be grown more frequently than 1 in 5 years on the same site. Main soil-borne threats which can be prevented by sound crop rotations are *Sclerotinia sclerotiorum* (other hosts are oilseed rape, linseed, spring beans, and most other brassicas and legumes), downy mildew *Peronospora viciae*, footrot complex mainly *Fusarium* spp. (other hosts are legumes), pea wilt *Fusarium opxysporum* f. sp. *pisi* (other hosts are beans), and pea cyst nematode *Heterodera goettingiana* (other hosts are legumes).

A wide range of soil types are suitable, but very heavy soils should be avoided as peas are very sensitive to waterlogging, particularly at emergence and flowering (for example, 5 days waterlogging at flowering can reduce yield by 75%), and compaction; also avoid very light drought-prone soils. Lighter free-draining soils are ideal. Optimum soil pH is 5.9–6.5; higher pH is acceptable but is likely to lead to manganese deficiency and a foliar manganese spray may be required (see Chapter 1). Stony soils can add to harvesting difficulties.

Establishment

A good seedbed is essential for successful pea establishment, compaction must be avoided, and peas should not be sown in wet conditions or 'forced' seedbeds. A good tilth should be obtained with the minimum of spring cultivations; ground should be ploughed in the previous autumn.

Seed should be treated to help prevent damping-off (*Pythium* spp.), downy mildew (*Peronospora viciae*), leaf and pod spot (*Aschocyta pisi*), and *Mycosphaerella pinoides*.

Peas should be sown as soon as soil conditions are favourable (minimum temperature for germination approx. 5°C, and moist and aerated) from mid-February; yield potential drops by approx. 125 kg/ha/week after the first week in March, with sowing after mid-April unlikely to be profitable. Seed is drilled on 10- to 20-cm rows at a depth of 4–5 cm, with care taken to achieve even spacing in order to reduce inter-plant competition. Guideline optimum plant populations are:

Marrowfats	65 plants/m^2
Whites and large blues	70 plants/m^2
Small blues	95 plants/m^2

However, as the seed cost is a high proportion of the variable costs of production (about 60%) the economic optimum plant population can vary from year to year, depending on seed costs, and from site to site, depending on likely output value. Accurate seed rates must be calculated to take account of variations in seed size (see this chapter) and likely field losses. Field losses can be high in peas due to the sensitivity of the seed and young seedling to soil conditions; seed lots should be tested for their vigour and only high-vigour seeds (up to 25 ϑ) used for early sowing or sowing in wet conditions. Expected losses for marrowfats are 10% when planted in March and 15% when planted in February, and for small blues are 15% in March and 20% in February. Other types show intermediate losses. An additional 5% expected loss should be included in the calculation for heavier wetter soils.

The seedbed should be rolled after drilling to conserve moisture and to provide a flat surface for effective pre-emergence weed control and ease of combining; rolling can also reduce loss of germinating seeds to bird pests.

Nutrients

No nitrogen is required. Rates for phosphate, potash and magnesium are shown in Table 3.14. On sandy or shallow soils that are low in organic matter where a sulphur deficiency is possible, 25 kg/ha SO_3 should also be applied. Germinating pea seed is sensitive to high concentrations of nutrient solution so fertilizer is best incorporated in the autumn to avoid seed scorch. Maintenance dressings are better applied at other courses in the rotation.

Weed control

The pea crop is not very competitive against weeds (particularly the semi-leafless types) and good weed control is therefore essential as infestation will reduce yields, increase disease risk, and add to harvesting difficulties. Pre-emergence herbicides are cheaper and effective against a broad range of weeds and are preferable unless soil type (organic or sandy) or conditions (dry or cloddy) prevent their use. Post-emergence application may be necessary for some troublesome weeds (oilseed rape volunteers, broad-leaved perennials, and well-established cleavers), or when a pre-emergence product is inappropriate or ineffective. It should be noted that varieties vary in their tolerance of some post-emergence herbicides, and that all varieties can be sensitive to post-emergence herbicides if their leaves are poorly waxed; this can result from harsh weather conditions, disease, or previous spray applications. If control has been unsuccessful, or if perennial grass weeds need controlling, a desiccant can be used pre-harvest. (See Chapter 6 for further weed control advice.)

Diseases

Most pea diseases are soil-borne or seed-borne, and are prevented by utilizing sound rotations (footrot, downy mildew, pea wilt), resistant varieties (downy mildew, pea wilt), healthy seed (leaf and pod spot, *Mycosphaerella pinoides*, bacterial blight), and seed treatment following appropriate testing (damping-off, downy mildew, leaf and pod spot, *M. pinoides*); most are encouraged by wet weather. Pea wilt and downy mildew are the greatest threats to yield. Botrytis or grey mould (*Botrytis* spp.) does not arise from soil or seed; it is encouraged by wet weather from flowering onwards, particularly in crops with a thick leaf canopy, and can be controlled with a protectant fungicide programme. (See Chapter 6 for further disease control advice.)

Pests

The pea aphid (*Acrythosiphon pisum*) is the most likely pest of the pea crop; it appears before flowering onwards and should be controlled if 1 plant in 5 is infested. Other potential pests are pea/bean weevil (*Sitona lineatus*), thrips (*Thrips angusticeps*) and pea moth (*Cydia nigricana*), all of which are controllable with insecticide, and pea cyst nematode (*Heteroda goettingiana*) which should not be a problem if sound rotations are practised. Pea moth affects quality rather than yield and control is only necessary in crops for human consumption. (See Chapter 6 for further pest control advice.)

Harvesting

Earliest varieties mature in July in warmer areas, though in the north harvesting can be in late August/early September. Ideally harvest should take place at 16–20% moisture content, but harvesting at up to 25% followed by drying may be necessary to avoid a reduction in quality during wet weather or to reduce splitting in crops for human consumption. If the crop is weedy or shows uneven ripening it may be desiccated with diquat when the lower pods are dry and brown and the seed dry, the middle pods are yellow and wrinkled and the seed rubbery, and the upper pods are green and wrinkled. The crop can be combined 7–10 days later.

Shedding losses can be high during harvest; to reduce losses harvest into the 'lay' of the lodging, use lifters, a slow drum speed with maximum fan, open up the concave, and use correct sieve sizes (see Chapter 24). For crops for human consumption avoid harvesting when the pods and haulm show surface moisture as this leads to peas coated with dust which lowers the quality.

Wet peas should be dried immediately to avoid spoilage from moulds. Splitting can result from incorrect drying and renders peas unfit for the human food market. Peas will store for 4 weeks at 17% moisture, for 6 months at 15% moisture, and long term at 14% or less (see Chapter 7).

Average yields are 4.0 t/ha. Payment for peas for feed is based on a standard of 14% moisture and 3% impurities, but price is variable and can be very low for poor samples. Quality samples for human consumption or seed attract a premium.

Root and vegetable crops

Sugar beet (*Beta vulgaris*)

Market

Sugar beet is grown exclusively for the sugar trade and is controlled in the UK by British Sugar PLC (BSC) by a system of contracts with growers. The closely related crop of fodder beet is not controlled and is grown mostly by dairy farmers in the south and south-west of the UK. Sugar beet needs to be processed to extract the sugar and growers are usually clustered around processing factories although some transportation supplements may be paid to assist growers far from the factories. A by-product of processing is sugar beet pulp which is sold to the feed supply industry, mostly for cattle feed.

Contracts with British Sugar guarantee a market but only for a fixed tonnage of beet at the top price (A quota). Excess production will be paid at a lower price (B quota) and excess to this may be rejected or paid at an even lower price. Payment is based on the weight of washed and correctly topped beet delivered to the factory at a sugar content of 16% (additions/deductions for deviations of 0.1%) modified by delivery and transport allowances. The payment is levied for research and development. The processing 'campaign' begins in September, with a higher price being paid for early delivery where yield is being sacrificed. Full details of the contract system can be obtained from British Sugar PLC, Peterborough, or the National Farmers' Union (NFU).

Refined beet sugar (Silverspoon) competes on the UK market with cane sugar (Tate & Lyle) which is imported under agreements with former UK colonies and developing countries. Thus the UK is only 50% self-sufficient in sugar whilst the EU is 135% self-sufficient. The EU sugar market is one of the last 'protected' markets and the world price is considerably lower than the EU price. At the time of writing, the removal of this market protection is being debated in the EU.

Growth cycle

Sugar beet originates from the Mediterranean coast; it has a salt requirement and shows sensitivity to cold temperatures. This is manifested by a slow germination rate at soil temperatures lower than about 8°C and a slow growth rate in early spring. Furthermore the crop is a biennial requiring exposure to cold temperatures (vernalization) to induce flowering. It is the first year's vegetative growth that is exploited as the sugar crop when the tap root swells and fills with sucrose. The vernalization requirement, however, can be fulfilled in the young seedling stage and flower production can occur in the first year of growth (bolting). For this reason the crop is sown in the spring and bolting risk is increased with early sowing or in cold springs. The crop canopy does not meet in the rows until July and this lowers the yield potential in the UK. The crop keeps growing well into October and can be harvested as late as November although to avoid undue soil damage many growers lift earlier and temporarily store the roots in a clamp.

Varieties differ in their bolting resistance, downy mildew resistance, and to some degree in virus yellows tolerance. Furthermore, small canopied varieties are available for fertile sites and on peat soils. Varieties are reviewed annually by NIAB and the BSC.

Soil types

Sugar beet must be grown on good, deep, well-drained, stone-free soils free of compaction and pans, which allow a good development of the tap root. Poor soil conditions lead to fangy roots which impede lifting and result in a crop that is difficult to clean, with a lower sugar content. Medium textured soils frequently give the best crops although crops are grown on a wide range of soil types. Soils must be of a good pH above 6.5.

Seedbed

The seedbed must be prepared with care for sugar beet to avoid compaction. Medium and heavier soils benefit from winter ploughing whilst light soils can be ploughed and furrow pressed in spring. Winter cultivations should not be worked in very wet conditions when smearing and pans can be formed, and double or cage wheels are recommended on cultivation tractors. Spring work should only be carried out on dry soils and the number of passes should be kept to a minimum to avoid wheelings. The final seedbed preparations should produce a fine and level seedbed which should be worked down on the day of drilling to facilitate moisture retention. Too fine a seedbed can result in capping of the soil following heavy rain and this can lead to poor crop emergence.

Sowing

In order to achieve high yields early sowing is recommended, but this carries with it a bolting risk. Early drilling and cold soils should be sown with bolt-resistant varieties. Normally drilling will commence as soon as possible after 20 March and be finished by 10 April.

Best crops are achieved with an evenly distributed plant population of about 75 000 plants/ha. Populations below 62 000 give serious yield decreases whilst above 100 000 impedes harvesting. Irregular stands interfere with harvesting equipment and can lead to a large wastage of unharvested roots and irregular topping. Row spacing is dictated by harvesting machinery, but yield is decreased if row widths exceed 500 mm.

Beet seed is a fused fruit containing several seeds and in the past had to be thinned after emergence or rubbed and graded before sowing, but all varieties used today are monogerm (single viable seed) and this enables the crop to be precision sown to a stand. An establishment rate of over 70% is expected and seed is generally sown at an interseed spacing of about 175 mm. For early sowings and on soils and fields known to give germination problems an interseed spacing down to 155 mm can be used.

Nutrient requirements

Sugar beet has a high requirement for potassium and a requirement for sodium (salt) because of its maritime origin, boron to counteract heartrot, and magnesium especially on light soils. It does not require much phosphorus and its nitrogen supply must be regulated to avoid an overstimulation of the leaf canopy which can reduce root yield and lead to high levels of nitrogenous compounds in the root. Recommended fertilizer rates are given in Table 3.15.

Soil pH should be 6.5–7.0 on mineral soils although a lower pH is tolerated on sandy soils (6.0–6.5) and peat soils (6.0). Use of magnesium limestone helps to provide adequate magnesium. Lime should ideally be applied to the previous crop in the rotation, but if this is impossible apply as two dressings and incorporate well to avoid acid layers which will lead to fanging.

Sugar beet germination can be impeded by fertilizers applied just prior to sowing. Most fertilizers (P, K, Mg, Na) are best applied in the autumn or winter prior to ploughing or on light soils just prior to seedbed preparation. Nitrogen is normally applied in spring to avoid winter leaching and 40 kg/ha is applied just post-drilling and the balance at the 2- to 3-leaf stage.

Weed control

Sugar beet is a slow growing crop in April and is very sensitive to weed competition. Many herbicides do not give adequate length of protection and it is common to spray three or more herbicides during establishment. Other methods of weed control include inter-row hoeing which is frequently used in conjunction with band spraying (spraying herbicide only on the crop rows). Inter-row weeds are sometimes left to grow quite large before being controlled, to hold the soil together and to prevent soil erosion (wind-blow) which can seriously damage the young beet. Another approach to this problem is to 'plant' chopped straw between the rows or even to sow cereals. Once the crop meets in the rows it suppresses further weed growth. Genetic modification technology has successfully produced sugar beet resistant to glyphosate (Roundup Ready Beet) which would enable a single herbicide to be used to give complete weed control in the crop. Consumer resistance to the introduction of genetically modified (GM) crops threatens the introduction of such varieties in Europe.

Of increasing problem in sugar beet crops has been the appearance of weed beet. This weed is a beet plant which flowers in its first year rather than producing a harvestable root. This should not be confused with bolters which flower because of a cold spring or early sowing. Weed beet if allowed to set seed can cause a problem in the field even when the rotation is long as a proportion of the seeds remain dormant in the soil. Weed beet cannot be controlled by spraying a herbicide but can be killed by using a weedwiper and a total translocated herbicide.

Pests and diseases

Sugar beet is susceptible to beet cyst nematode (*Heterodea schachtii*) which can only be controlled satisfactorily by using a sound rotation. Use of nematicides gives some suppression in infected soils.

Most pests of the crop are only damaging during establishment where damage leads to either plant death or subsequent plant distortion. Seedling pests include millipedes, symphylids, spring-tails, and pygmy mangel beetle. Wireworm can also be a problem if grass is in the same rotation. Blanket treatment for these seedling pests using an insecticide is not usually carried out unless a

Table 3.15 Nutrient requirements of sugar beet (kg/ha) (MAFF, 2000)

	SNS, P, K or Mg index					
	0	*1*	*2*	*3*	*4*	*5 & over*
Nitrogen						
All mineral soils (except deep silts)	120	100	60	30	30	0
Deep fertile silty soils	—	60	40	30	30	0
Organic soils	—	—	—	30	30	0
Peaty soils	—	—	—	—	—	0
Phosphate (P_2O_5)	100	75	50	0	0	0
Potash (K_2O)	150	125	100[1]	0	0	0
Magnesium (MgO)	150	75	0	0	0	0
Sodium (Na_2O)[2]	200	100	0	—	—	—

[1]Reduce to 75 if K index is high (i.e. 2+).
[2]Index 1 = <20 mg/l; index 2 = 20–40 mg/l; index 3 = >40 mg/l available Na.

problem is foreseen. Field mice also devour seed and are encouraged by spilled seed and shallow drilling. The most significant pests are peach potato and green aphids which carry beet virus yellows (BVY) and beet mild yellowing virus (BMYV) causing severe disease especially if infection occurs early. Crop losses can be as high as 40–50%. Control of aphid infestation begins with controlling the overwintering sites of the pest, i.e. in clamps, ground keepers, on spilt plants at the loading bays, and in gardens and neighbouring allotments. Severe winter frosts reduce aphid numbers in early spring. Early aphid attack can be prevented using soil-applied granular insecticides, but the expense of this form of control limits their use only to areas where severe attacks are likely (e.g. East Anglia). Foliar applications of insecticides are applied when one infected plant in four is recorded or when the BSC gives warnings. Usually one spray is sufficient but in bad years a second may be advisable. There is evidence that some aphids are building up resistance to organophosphate insecticides and the use of alternatives, e.g. carbamates, is recommended.

A serious root-infesting disease of sugar beet, rhizomania, is now established in the UK. Infected fields are quarantined and the growers banned from growing sugar beet or moving their equipment onto uninfected ground in an attempt to contain the spread. At present there is no known cure for the disease with the exception of ceasing to grow sugar beet. It is not known how long the organism can survive in the soil in the absence of the crop and therefore rotational control is difficult. The disease causes a proliferation of lateral roots and stops the correct development of the tap root thereby reducing sugar yields drastically. Because of the risk of rhizomania the BSC is seeking fresh ground for sugar beet and contracts for delivery to the Kidderminster factory have even been placed in Cornwall.

Irrigation

The crop is sensitive to drought during June when the tap root is extending and the leaf canopy is developing, and during July and August when the canopy is in full sucrose production for storage in the tap root and the drought leads to stomata closing which slows photosynthetic rate. The extension of the tap root (down to 1.8 m) provides adequate moisture for crop development during September. Irrigation water, up to 25 mm, should be applied in June when the soil moisture deficit approaches 50 mm. During July and August the soil moisture deficit should not exceed 50 mm and irrigation can be applied to bring the soil to field capacity if required.

Harvesting

There are several types of sugar beet harvester available and all perform three functions: they cut off the green leaf (topping), they dig the swollen part of the tap root out of the ground, and they elevate it into a hopper or trailer. The sugar beet 'campaign' begins in September with early lifted crops delivered direct to the factory. Later liftings can be temporarily stored on concrete mats where they can be washed. As the soil conditions deteriorate in October and November, later delivered crops are lifted and clamped under bales of straw to protect them from frost. Frosting severely damages the tap roots and leads to rejection at the factory. Clamps must be managed to keep the temperature below 10°C by ventilation or respiration will lead to sugar losses. The sugar beet crop yields about 40–50 t/ha of roots with a sugar content of 16%, giving 6–7 t/ha of sugar.

By-products

The leaves and crowns of sugar beet have feeding value to livestock but must be wilted for a few days to diminish high levels of oxalic acid in the leaves. If the tops are free of dirt they can be ensiled. The tops from a 50 t/ha root crop will have a feeding value equivalent of about 0.5 ha of marrowstem kale. Tops can be fed to both ewes (1 ha of tops lasting 250 ewes 1 week) and dairy cows (limit to a maximum of 19 kg/day per cow). The pulp of the processed beet roots also has feeding value to livestock. It is provided in several forms: dry powder, nuts or molasses. Feed value is equivalent to oats.

Potatoes (*Solanum tuberosum*)

Markets

Potato production now operates in a free market in the UK with the Potato Marketing Board having been abolished. The British Potato Council (BPC) now represents the interests of potato growers.

There are several types of market for potato tubers, requiring different varieties to be grown. Over 75% of fresh potatoes are now sold by the multiple retailers as 'prepacks', i.e. washed and packed in plastic bags. The market for processed potatoes (chips and crisps) continues to expand but processing capacity in the UK is limited and so product importation is necessary.

General market requirements are for smooth even tubers with shallow eyes for minimal wastage and reduced labour in peeling. The tubers need to be free of greening, growth cracks, disease (blight, scab, silver scurf, gangrene), pest damage (wireworm, cutworm, slugs), and free of mechanical damage (cuts and bruises). Cooking quality is related to variety, growth, and storage conditions. Early harvest in autumn of maincrop tends to give the best quality tubers and these are generally put into low-temperaure (2–4°C) storage because they have the best keeping quality.

Maincrop – ware potatoes

This is the largest volume market requiring heavy yields of saleable medium to large sized potatoes. It is divided into medium sized tubers for general domestic use and larger tubers for oven baking (bakers). More and more of the crop is now being retailed by supermarkets who require the crop to be washed, graded and bagged in small plastic bags. Bakers are sometimes sold in packs of two potatoes and must be very clean and blemish free. The traditional unwashed crop bagged in 25-kg paper sacks is still available especially from farm shops. The crop is harvested when fully mature, beginning in late September. Some crop is stored in ventilated cold stores and some is left in the field until required. There have been considerable advances in storage technology over the last 10 years in response to the market requirements for year-round supply of good blemish-free potatoes.

Earlies – new or salad potatoes

This is a high value sector of the market and is traditionally the provenence of the frost-free areas of the UK, particularly west Cornwall, south Pembrokeshire, east Kent and Jersey. The use of plastic film to advance the crop and give an early harvest, thus achieving a higher price, is now commonplace. The traditional crop is being undermined by produce from the Mediterranean (Cyprus, Greece and Eygpt). Prices are at their highest in April (up to £2000/t) and drop as more and more of the crop comes onto the market, reaching a base in July (about £200/t). The potatoes are lifted in an immature state and have a waxy texture after cooking. Later plantings and second early varieties give a continuity of fresh potatoes throughout summer until the maincrop arrives. Out of season early potatoes are produced by some growers in plastic tunnels where the potato is often fitted into a fallow period in lettuce production. A second crop of salad potatoes is sometimes produced in the south-west with the crop being replanted in July after harvest of the first earlies. The second crop is harvested in October. Sequential plantings of the salad potato Charlotte are becoming popular.

Second early potato varieties fit between the earlies and maincrop. These varieties achieve higher yields than earlies but harvest earlier than maincrops, in July and August. Depending on market price they may be harvested immature as salad potatoes or left to mature to give an early start to the maincrop season.

Processed potatoes

Potatoes sold for processing is an expanding sector of the market now taking over 20% of the total UK potato production, particularly for frozen chips and potato crisps. Only a very small amount is canned and this is usually early crop. Processing in the potato crisp industry requires specific named varieties with high dry matter content (over 20%), low reducing sugar content (below 0.25%), shallow eyes and regular shape. For frozen chips the potato must have high dry matter content (over 20%), low reducing sugar content (below 0.4%) and be regular with shallow eyes and a white creamy flesh which does not discolour upon cooking or suffer any taints.

Seed potatoes

This is a specialist market confined generally to regions where a cool summer means a lower incidence of aphids and consequently less chance of aphid-transmitted virus diseases, e.g. Scotland and Northern Ireland. The objective of seed tuber production is to produce a high yield of tubers in the range of 32–60 mm using clean certified seed so as to produce seed tubers which are true to type and as free of disease as possible. Some growers of early potatoes produce their own seed tubers and many maincrop growers keep the small tubers from a certified maincrop to provide 'once grown' seed tubers for the subsequent year.

A specialist method of seed production is now appearing in commercial practice with the availability of mini-tubers produced through a combination of micropropagation followed by intensive glasshouse raising at high plant populations.

Varieties

Growers should refer to the NIAB list for varieties of potatoes but need to determine their market before choosing a variety. Processors will often specify only one or two varieties. Early varieties are best chosen with experience of performance within a region and soil type.

Seed certification

Registered seed producers comply with DEFRA seed production regulations and can sell seed that is relatively virus free and is given a certificate of health with one or more of the following codes:

(1) Seed for multiplication purposes:
    VTSC – Virus tested stem cuttings: the highest grade of seed tubers that have been raised in aphid-proof glasshouses.
    SE – Super Elite seed: derived from VTSC of SE stocks.

E – Elite seed
1, 2 or 3 – Number of years of multiplication since VTSC.

(2) Seed for commercial crops:
CC – Certified Seed: not suitable for further seed multiplication.
NI – Indicates that the variety is not immune to wart disease.

In areas of intensive maincrop production it is advisable to purchase certified seed each year as disease spread through once grown tubers is likely. In less intense regions it may be possible to keep some once grown tubers for the following year's crop.

Seed size

The plant population of potatoes is a function of both seed size and tubers per unit area. As seed size increases, so does the number of potential shoots per tuber and logically these can be sown at lower seed rate. However, there is a complex competition between shoots per tuber and most growers prefer to plant smaller seed tubers. Seed of a size 30–60 mm is preferred for most crops with the lower range for earlies especially. Seed at the top end of the range or above tends to produce numerous small tubers in maincrop production and in Scotland these are retained for seed crop production.

Seed rate

Seed rate is determined by the average weight of the seed tubers and the number of tubers planted per hectare. The variety also influences the seed rate, with some varieties tending to produce either large or small tubers, and therefore seed rate has to be adjusted to force the variety to produce the required tuber size. Seed rate is also influenced by the target market. Seed tubers are very expensive and this can influence seed rate too. Lastly, the number of stems produced per tuber can be influenced by the pre-planting storage conditions, i.e. chitting, which can stimulate one or several shoots per tuber. Table 3.16 gives recommendations on seed rate for some common varieties. Seed rates are commonly 2.5–4.5 t/ha for varieties producing large tubers, e.g. Desiree and Pentland Crown, and 1.8–2.4 for varieties such as King Edward and Record which tend to produce undersized tubers. Seed rates for earlies, seed production and canning are given in Table 3.16.

Row width

Row width often varies depending on tractor tyre width and harvester width. On soils that tend to give green tubers then wider rows tend to allow more soil per row which counteracts greening. Maincrops tend to be grown in 710-mm rows although width can be 760, 800 or even 900 mm. Earlies tend to be grown on narrower rows because they are harvested before maturity and narrower rows tends to raise yields because of the better spatial distribution of the plants. Row widths of 540–600 mm are common. Seed and canning potatoes are likewise grown on slightly narrower rows of 660–760 mm. Spacing of tubers within the rows is determined by the row width and the seed rate and is given in Table 3.17.

Seed treatments (sprouting or chitting)

The process of sprouting leads to quicker emergence after planting and an earlier leaf canopy. This can be vital in obtaining an early harvesting crop of earlies and

Table 3.16 Guide to seed rates for potatoes. (Note: when growing new variety check on suitable seed rate – there are large differences)

Type of production	*Seed rate (kg/ha)*	*Remarks*
Earlies	3200–4500	Single sprouted seed required for earliest crops; multiple sprouted seed satisfactory for later lifting and second earlies. First early seed should be tightly graded and planted 200–360 mm apart according to size
Canning	Optimum 7500 Economic range 2500–5000	High stem density 160–220/m^2 required. Use multi-sprouted seed spaced 150–230 mm apart according to size. Size grade seed. Optimum seed rate depends on cost of seed and value of produce
Seed production Home-produced seed, e.g. Scotland	5000–7500	High seed rates are fully justified where cheap seed is available; home-produced ware size, which is of lower value than seed, is planted
Purchased seed	3800–5000	High seed rates are very expensive if seed is bought at high price, e.g. early growers producing once-grown seed. Use lower rate and multi-sprouted seed

Table 3.17 Spacing of potatoes according to seed rate

	Required seed rate (t/ha)												
	17.5	*20.0*	*22.5*	*25.0*	*27.5*	*30.0*	*32.5*	*35.0*	*37.5*	*40.0*	*45.0*	*50.0*	*75.0*
No. of sets/50 kg	*Spacing between sets (mm ×10) in 760 mm rows*												
400	94	84	74	66	61	56	51	48	43	41	37	33	21
500	76	66	58	53	48	43	41	38	35	33	29	26	17
600	63	56	48	43	41	38	33	30	30	28	24	21	15
700	53	48	43	38	35	30	28	28	25	23	21	19	12
800	48	41	38	33	30	28	25	23	23	20	19	16	11
900	43	38	33	30	28	25	23	20	20	18	16	15	10
1000	38	33	30	28	25	23	20	18	18	18	15	14	9
1100	35	30	28	25	23	20	18	18	15	15	14	12	7
1200	30	28	25	23	20	18	18	15	15	13	12	11	6

even for maincrops can lift yields by 3–5 t/ha. Well-sprouted seed tubers must be handled with care at planting and may even still be planted by hand in very early regions such as Jersey. For seedcrops sprouting is desirable since it leads to earlier bulking and maturity of the crop, allowing an early desiccation and avoidance of aphid risk, thus keeping the level of virus low in the seed tubers. Sprouting can advance development in the field by 10–14 days. Sprouting, is however, expensive and requires the availability of frost-free chitting houses or glasshouses. Large maincrop producers will only sprout a part of the crop, usually to spread out the harvest.

Sprouting is carried out in glasshouses or more usually in heated barns. Tubers are placed upright in special chitting trays and stacked up to 5 m high. Natural daylight or supplementary lighting (warm white fluorescent tubes) with a daylength of 8–12 hours is necessary for sprout development. Temperature must be controlled above a set minimum for a variety:

Below 4°C	Sprout development is inhibited
5.5°C	Fast sprouting varieties only grow
7°C	All varieties commence growth and fast sprouting varieties grow quickly
10°C	All varieties grow rapidly; fast sprouting varieties grow excessively.

For most varieties, a low temperature thermostat and additional heating are usually sufficient to keep the temperature above 7°C. Fast sprouting varieties can be more troublesome and may need to be cooled by ventilating at night to keep the temperature down.

Ideally, sprout length should be about 19–25 mm. If sprout length is advancing too quickly then the store should be cooled, and conversely if sprout length is too short then the temperature should be increased. Sprouts are very sensitive to frost and during freezing weather the store must be kept frost free. The store should also be kept well ventilated to avoid high humidities which form condensation at lower temperatures and can cause rots. A careful watch for aphids should also be kept and the store fumigated if necessary; this is a particular problem for earlies where the time in store can be very long.

Single-sprouted tubers are preferred by many growers and this can be achieved by commencing sprouting in autumn at 16°C when a single growth dominates and suppresses further sprout development (apical dominance). Multi-sprouted tubers can be achieved either by delaying the sprouting process until January or by rubbing off the main sprout and releasing the apical dominance. Single-sprouted seed is essential for earlies whilst for seed crops and canning multi-sprouts are preferred.

Many growers do not like to use fully sprouted seed because the care needed at planting does not fit in with the mechanization of large planting machines. An alternative method is to use mini-chitted seed with sprouts only 2–3 mm in length. With this method seed is first 'cured' at 13–16°C for about 10 days upon arrival at the farm, then cold stored at 3°C to suppress sprout development, and then 3 weeks before planting the temperature is raised to 7–10°C to begin the sprout development. With this method, seed does not need to be placed in special trays but can be kept in bags or in 500-kg bulk containers.

Sowing date

Sowing date varies with region, soil type and crop type. The potato foliage is very sensitive to frost and so should be sown at about the time of the last frosts so that it emerges in frost-free weather. Early crops in the frost-free growing regions are sown as soon as soil conditions allow. The earliest are sown in Jersey on the steep south-facing hillsides (cotees) in January and the remainder follow in February in Cornwall, Pembrokeshire and Kent. Second earlies follow earlies in March. Planting of maincrop varieties begins in late March in the south and early April in the north. Unsprouted seed may often be planted first as this will take the longest time to emerge. The aim is that planting be completed by the end of April.

Plastic film

Early potato production now routinely uses plastic film. Various types of film are available and are graded on the number of holes per square metre. Ventilation is essential or early blight may attack the crop; the holes also let rainfall and carbon dioxide into the developing crop. Some early growers also use spun fibre floating film (fleece) to good effect. The cost of the covers makes it economic to use them only on the most valuable crops, i.e. earlies and possibly second earlies. Covers can advance lifting of earlies by 10–14 days and their widespread use has undermined the natural advantage of the extremely early regions. Best responses to covers tend to occur in poor years when the greenhouse effect created by the cover is maximized.

Crops should be covered as soon as possible after sowing following the use of a soil-residual herbicide and should be covered within 14 days of planting. Timing of removal of the cover depends on weather factors and crop development. Optimum timing appears to be no longer than 4 weeks after 95% crop emergence.

Covers give some frost protection to the crop by virtue of raising the soil temperature by a couple of degrees, but well-developed crop touching the plastic will be easily scorched since the dew on the underside of the plastic frequently freezes during a radiation frost.

Soils

Early crops are grown on light soils which can be easily cultivated in early spring. Maincrops prefer deep moisture-retentive soils. Soils should preferably be well prepared and free from compaction and stones and clods which damage the crop when it is harvested. Most growers now use a stone and clod separator before planting a field as it has been shown that this results in a better quality crop. Various machines are available and some remove the stones and clods to a trailer for dumping but most bury the stones and clods under the furrow. Nearly all potatoes are grown in ridges arranged in bed systems to facilitate harvesting. The ridges may be drawn up twice in order to give some weed control by cultivation, but most growers rely on soil-residual herbicides for weed control. The residual herbicide may be mixed with a low rate of paraquat and applied just before emergence to kill germinated seedlings and prolong the effect of the residual chemical.

Potato fields are usually left to run slightly acid (pH 5.8–6.0) as this helps prevent potato scab. In a rotation, lime should always be applied to a field after a potato crop, never before.

Nutrient requirements

Potatoes respond well to applications of well-rotted manure provided they are applied in the previous autumn well in advance of planting. Maincrop potatoes yield best in fertile conditions and therefore fertilizer applications are frequently high. The crop does not remove all of this fertilizer and the residual amounts left in the soil benefit following crops such as winter wheat.

Early potatoes do not require as much nitrogen or potassium as maincrops because they are harvested before maturity; high nitrogen will delay tuberization and therefore harvest date.

Excess nitrogen reduces dry matter content of the tubers, thus reducing cooking quality; it delays tuber initiation and haulm maturity and thereby harvest date. It also predisposes the leaf to potato blight attack. Nitrogen should be reduced in fertile conditions and after heavy dressings with manure. Phosphorus encourages early tuber growth which is important for all crops especially earlies. It helps give good skin strength and counteracts blight susceptibility. Only soluble forms of phosphorus should be used. Early crops grown on acid soils in the west country give good responses to phosphorus. Potassium helps to produce a tuber that has a low dry matter content and is less likely to bruise, and also reduces the incidence of after-cooking blackening. Best responses to potassium are obtained with maincrops. Nutrient recommendations are given in Tables 3.18 and 3.19.

Different varieties of potato have different canopy structures or haulm longevities and nitrogen recommendations vary according to this character (Table 3.17).

Irrigation

Potatoes respond to irrigation, especially maincrops, and yields can be doubled in dry years. Irrigation water should not be applied until after marble-sized tubers are present in the crop or otherwise too many tubers will be stimulated and an undersized crop produced. The late application of irrigation to a droughted crop can result in growth cracks and deformities to the tubers, leading to rejections. This also occurs in unirrigated crops which experience heavy late-summer rainfall following a drought. Irrigation also helps reduce the incidence of scab but may lower dry matter percentage of the tubers.

Diseases and pests

The single most important disease of potatoes is potato blight (*Phytophora infestans*). This disease is so devastating to the crop that growers must routinely spray fungicides to keep the disease in check every 10 days during periods of risk. Fungicide resistance has made the job of complete protection more difficult in recent years. Most crops of earlies are harvested before conditions are favourable for blight spread and therefore escape the use of fungicides.

Table 3.18 Nitrogen requirements of potatoes (kg/ha) (MAFF, 2000)

Growing season		*SNS index*		
		0 and 1	*2, 3 and 4*	*5 and 6*
<60 days	Group 1	120–150	80–120	40–60
	Group 2	100–130	60–90	0–40
	Group 3	80–110	40–70	0–40
	Group 4	—	—	—
60–90 days	Group 1	180–220	140–170	100–120
	Group 2	140–190	100–130	60–80
	Group 3	120–160	80–110	40–60
	Group 4	60–90	20–40	0–40
90–120 days	Group 1	240–270	200–220	160–180
	Group 2	160–220	120–160	80–120
	Group 3	120–180	80–100	40–60
	Group 4	80–140	40–60	0–40
>120 days	Group 1	—	—	—
	Group 2	200–250	160–180	120–150
	Group 3	160–210	120–140	80–100
	Group 4	100–180	60–80	20–40

Group 1, short haulm longevity (e.g. Colmo, Estima, Maris Bard, Premier, Saxon, Shepordy, Wilja).
Group 2, medium haulm longevity (e.g. Atlantic, Hermes, Lady Roseta, Marfona, Maris Peer, Nadine).
Group 3, long haulm longevity (e.g. Desiree, Fianna, King Edward, Maris Piper, Russet Burbank, Pentland Dell, Pentland Squire, Saturna).
Group 4, very long haulm longevity (e.g. Cara).

The most important pest of potatoes is potato cyst nematode (PCN). This persistent soil-borne pest can only be controlled satisfactorily by crop rotation – 1 year in 6 cropping. However, many growers want to crop more frequently than this and use soil-applied granular nematicides to control the pest. Where insecticide granules are used for PCN some protection against aphids is given. Soil screening services are available. Aphids carrying virus are a problem, especially in seed crops, and aphicides must be used on all crops destined for seed.

Potato quality is severely affected by underground pests including slugs, wireworms and cutworms. Wireworms (larvae of the click beetle) are often worse 2 or 3 years after ploughing out grassland whilst slugs are a big problem during wet summers.

Harvest and storage

Earlies are mostly harvested without destroying the haulm whereas with maincrops routine desiccation is commonplace. Destruction of the haulm is important to prevent blight spreading into the tubers and it also starts the maturity phase of the tubers which commences with skin development. All potatoes need time to develop a skin which will protect them during subsequent handling. For early crops this occurs within a day or two after careful lifting. For other crops it will occur 10 days after destruction of the haulm or as much as 21 days if the halum was particularly vigourous.

Table 3.19 Phosphate, potash and magnesium requirements of potatoes (kg/ha) (MAFF, 2000)

	P, K or Mg index					
	0	*1*	*2*	*3*	*4*	*5 and over*
Maincrop (50 t/ha)						
Phosphate (P_2O_5)	270	230	180	130	50	0
Potash (K_2O)	350	325[1]	300[2]	150	0	0
Magnesium (MgO)	150	75	0	0	0	0
Early and seed (30 t/ha)						
Phosphate (P_2O_5)	270	230	180	130	50	0
Potash (K_2O)	220	195	170[2]	50	0	0
Magnesium (MgO)	150	75	0	0	0	0

[1] Reduce by 25 kg/ha on sandy soils.
[2] Reduce by 15–20 kg/ha if K index is high (i.e. 2+).

Many potatoes are now contract lifted or grown by large growers with sophisticated lifting machinery which is carefully set up to lift the crop without causing it damage. Early harvest of maincrops in mid-September to mid-October tends to give the least harvesting troubles, with a cleaner crop, less bruising, better healing, less dirt and better cooking quality.

Traditionally potatoes were stored in the field simply by delaying harvest. Nowadays the entire crop is lifted and some is stored for the later use. Two types of store are used: ambient temperature stores mostly in barns and farm buildings for short-term storage and low-temperature stores (2–4°C) for the long term. Stores must be regularly checked for hot spots or sites of rotting which can ruin a crop. Long-term storage beyond the point of natural dormancy places the crop at risk of sprouting. Chemical sprout suppressants are available, but many multiple retailer protocols now discourage their use and encourage storage at 2°C.

Crop yields

Earlies	7.5 –10 t/ha	Special attention is given to the price × yield equation!
Maincrops	30 –40 t/ha	Good crops can yield 50 t/ha or more especially with use of irrigation.

Vegetable crops

Quality is paramount for vegetable crop products. Quality requirements vary for each type of product and according to the market outlet – even between different supermarkets, and between different presentations for the same supermarket. General quality requirements include: size; uniformity; shape; colour; freedom from damage from handling, pathogens and pests; and freedom from chemical residues. Fresh vegetables are perishable and many types may only be stored for very limited periods, thus all-year-round production is often necessary using varieties and techniques to overcome climatic constraints. Most vegetables are grown under contract to processors, or to packhouses, who co-ordinate production and undertake grading and packing to supply the major supermarkets, catering outlets, or wholesalers (who supply the smaller retailers and caterers). Local niche opportunities exist for pick-your-own, farm shop or farmers' market sales, and community marketing, e.g. 'veg box' schemes, direct to the public. Prices can vary widely through the year according to supply and demand, with out-of-season produce generally attracting higher prices. Integrated crop management (ICM) protocols, or other quality assurance schemes, are an essential part of vegetable production for the major supermarkets.

Root vegetables

Carrots (*Daucus carota*)

Carrots are the most widely consumed vegetable in the UK and are grown all year round for the fresh market, and for canning, freezing and processing. Clean, straight, undamaged roots with good colour are required, of appropriate size depending on end-use. Varieties are grouped into several types which vary in their shape (e.g. Nantes type are cylindrical, Chantenay are conical), and maturity period (e.g. Amsterdam Forcing are early, Autumn King are late).

Best carrot soils are well-drained, deep, stone-free sands or light, loamy peats with good water holding capacity to allow easy root penetration and unrestricted root expansion; this ensures good shape and easy harvesting of clean roots. Irrigation is essential. Sites should be rotated with a 5-year break between carrot crops to prevent build-up of carrot cyst eelworm (*Heterodera carotae*) and violet root rot (*Helicobasidium purpurea*).

A fine, firm, level, clod-free seedbed should be achieved with minimum working; bed systems are preferable to eliminate compaction effects on root development. Dressed seed should be drilled as shallowly as possible into soil moisture at even depth. 'Out of season' production uses floating film covers for protection during establishment and/or for overwintering as appropriate; sowings are made between January and June, and for overwintering in October. An effective weed control programme is essential through the life of the crop as carrots are poor competitors (see Chapter 6). Chief pests are carrot fly (*Psila rosae*) and carrot willow aphid (*Cavariella aegopodii*), and both should be controlled according to timing of their life cycles and populations in relation to crop timing (see Chapter 6); pest monitoring systems are extremely useful for control within ICM principles.

Nutrient requirements are given in Table 3.20. On sandy soils 375 kg/ha agricultural salt should be ploughed in before drilling. Ideal pH is 5.8 on peats and 6.5 on sands; the crop is sensitive to overliming which can cause deficiencies of manganese and copper (and boron on sands).

Root size is influenced by plant population, sowing date, time to harvest, soil fertility, and water supply;

Table 3.20 Nutrient requirements of carrots, swedes, and parsnips (kg/ha); m indicates maintenance dressing (MAFF, 2000)

	SNS, P, K or Mg index				
	0	*1*	*2*	*3*	*4*
Maincrop carrots					
N	110	60	20	0	0
P_2O_5	200	150	100	50m	0
K_2O	275	225	175 (2–) 125m (2+)	35	0
Swedes and parsnips					
N	150[1]	100	50	0	0
P_2O_5	200	150	100	50m	0
K_2O	300	250	200 (2–) 150m (2+)	60	0
Carrots, swedes and parsnips					
MgO	150	100	0	0	0

[1] 100kg/ha only in the seedbed; remainder as topdressing at full establishment.

complete uniformity is unachievable, but, as each market requires different root size, grading of harvested crop can ensure a high proportion of saleable roots. Seed rate should be carefully calculated to take account of variety, situation, and required grade. Target populations will vary widely, for example a cylindrical type grown for small carrots (20–25mm shoulder diameter) will require about 170 plants/m^2, but the same yield (40t/ha) of large carrots (45–50mm shoulder) will be obtained from 16 plants/m^2. Field factors must be incorporated into calculations of seed rate: for cold soil/poor tilth 0.5; average conditions 0.6; good conditions 0.7. Spacing will depend on system, soil, and grade required; single, double, triple, and multirow systems are all used as appropriate.

Harvesting uses top lifting harvesters or elevator digger harvesters. Average yields are 40t/ha, exceptional crops can produce 100t/ha. Field storage can be achieved by earthing over or clamping.

Swedes (*Brassica napus*)

Most of the UK culinary swede production is from Devon, or from Scotland, supplying the market all year round from sowings made between March and July (or March transplants from January sowing). Soil should be well drained but moisture retentive, of pH 6.0–6.5, and on a rotation of 6 years to prevent clubroot and *Phoma*. Dressed seed is drilled into moisture 10–20mm deep to give approx. 10–15 plants/m^2 (a lower density will give an earlier harvest date), on 40- to 50-cm row widths. Nutrient requirements are given in Table 3.20. Main threats are clubroot, powdery mildew, *Phoma*, and cabbage root fly (see Chapter 6 for further advice on control). Yield is approx. 30–40t/ha.

Parsnips (*Pastinaca sativa*)

Parsnips are marketed all year round though demand is strongest in the winter months. The roots have good frost resistance and flavour and sweetness is improved by frosting. The market demands clean, 'white' straight roots without fangs, bruising or other damage. Size depends on outlet, from 35–65mm shoulder diameter for supermarket presentation packs to 150mm shoulders for processing. As for carrots, varieties show different shape characteristics from bulbous to long, tapering. Varieties also show variation in their susceptibility to bruising, and their resistance to cankers; orange brown canker (*Itersonilia* spp.) causes rot in the field, black canker (*Phoma* and *Mycocentrospora* spp.) causes rot at sites of damage.

For best quality crops soils should be deep, sandy, well drained and stone free. Heavy soils cause root fanging and make harvesting of clean roots very difficult; it is also difficult to obtain clean roots from peaty soils as the peat clings to the 'wrinkled' surface. A wide rotation with carrots (not less than 5 years) should be practised. A fine seedbed with no compaction should be prepared from minimum passes; bed systems are preferable. The main drilling period is from January to May, depending on site and target harvest date. Natural or pelleted seed is precision drilled up to 15mm deep to give a plant population of 40–60 plants/m^2 for smaller roots, or 20–30 plants/m^2 for larger roots, on 40- to 50-cm rows depending on system.

Carrot fly and canker are likely to be the most serious problems (see Chapter 6). Nutrient requirements are given in Table 3.20; organic soils may encourage manganese deficiency and a foliar spray of manganese sulphate will be economic if deficiency symptoms are noted. Soil pH should be at least 6.5 on mineral soils and 5.8 on peats.

Top lifter harvesters can be used if the tops are still present, but after they have died back digger elevator harvesters must be used. Parsnips are very susceptible to bruising and damage, and handling procedures should take account of this. Yields range from 15–30t/ha.

Alliums

Dry bulb onions (*Allium cepa*)

Grown for the fresh market, processing, and pickling, bulbs must be globe shaped with good appearance, no staining, appropriate colour, and thin dry necks. Many varieties are available with different colour, shape, maturity date, and keeping qualities. Varieties are either 'autumn sown' to be harvested in June/July, or 'spring sown' to be harvested in August/September.

Most onions are grown in the eastern counties as warm sunny conditions are necessary for quality and ease of harvesting. Soils should be well drained, deep, moisture retentive, stone free and free working, with a pH of 6.3–7.0 (5.5–7.0 on peats). Silts, brick earths, medium loams, sandy soils, and peats are most suitable. Irrigation is essential. Onions and leeks should not be grown on a rotation of less than 6 years to prevent build-up of white rot (*Sclerotium cepivorum*); crops are also susceptible to stem eelworm (*Ditylenchus dipsaci*) whose host range also includes legumes, oats, carrots, parsnips, and sugar beet.

Seedbeds must be fine, firm, level, and clod free with no compaction; bed systems are preferable. Overwintered crops should be sown in the second half of August (early August in the north of England); earlier sowing creates risk of bolting. Spring crops are sown in February/March. (Improved yield and quality can be obtained from module-raised transplants sown in January/February and planted out in March/April.) Seed is drilled into moisture 12–25 mm deep, on row widths to fit the system but usually 30–45 cm. Plant population will influence the size of bulbs: 65–85 plants/m^2 for ware production (over 25 mm diameter), 320 plants/m^2 for picklers (under 25 mm diameter). Field factor for seed rate calculation is 0.5 for cold soil, 0.7 for good conditions. See Table 3.21 for nutrient requirements. The injection of high-phosphate liquid fertilizer ('starter fertilizer') can improve growth and quality on soils at P index 3 or below. No more than about 20 kg/ha of nitrogen and 60 kg/ha of phosphate should be applied in this way, and should be deducted from recommendations given in Table 3.21.

Neck rot (*Botrytis allii*) must be controlled to prevent heavy losses in store; downy mildew (*Peronospora destructor*) and leaf spot (*Botrytis* spp.) should also be controlled. White rot and stem eelworm are controlled by rotation; onion fly (*Delia antiqua*) is controlled by seed dressing.

The crop is mature when tops go down. Drying procedure is crucial for quality; traditionally onions were undercut, windrowed and field dried for 7–10 days. Sophisticated drying/store rooms now allow the crop to be lifted immediately (after flailing tops), followed by a controlled-staged drying process in store. Quality is more secure by this method than by field drying. Yields range from 30–45 t/ha.

Table 3.21 Nutrient requirements of onions and leeks (kg/ha); m indicates maintenance dressing (MAFF, 2000)

	SNS, P, K or Mg index				
	0	*1*	*2*	*3*	*4*
Nitrogen[1]					
Bulb onions					
Spring sown	175	125	75	25	0
Autumn sown[2]	100	100	60	30	0
Leeks[2]	200	150	100	50	0
Phosphate	200	150	100	50 m	0
Potash	250	200	150 (2–) 100 m (2+)	0	0
Magnesium	150	100	0	0	0

[1] Nitrogen should be applied in the seedbed, except when more than 100 kg/ha is required (apply balance after full establishment). For autumn-sown onions seedbed N is only required on mineral soils.
[2] A further topdressing of 100 kg/ha may be required on soils other than organic or peaty soils.

Leeks (*Allium ameloprasum*)

A wide range of types and varieties and the use of transplanting techniques allow leeks to be harvested almost year round from July to May. The market requires clean, long, straight shafts with a good length of blanch; size depends on presentation required.

Leeks grow on a wide range of well-drained but water-retentive soils, with pH 6.5–7.0. Deep loams and peats are preferred; coarse sands make cleaning difficult as particles catch in the leaf sheaths giving a gritty unpleasant texture. Rotation is as for onions.

Direct sowing of seed into field position is possible for March/April sowing, when seed should be drilled 12 mm deep in beds on 25- to 30-cm rows or on 45- to 50-cm rows to allow earthing up. Transplanting of bed-raised or module-raised plants (sown January to April) is used for most leeks, especially early sown, as quality is improved (uniformity and long blanch). Planting out (April to July) uses row widths of 25–60 cm depending on bed system used. Final plant population is 20–50 plants/m^2 depending on size of leek required.

Main threats are white rot (*Sclerotium cepivorum*), rust (*Puccinia allii*), onion fly, stem eelworm, and onion thrip (*Thripstabaci*); see Chapter 6 for further advice. Leeks respond well to a heavy dressing of farmyard manure (60–70 t/ha) ploughed in. Nutrient requirements are given in Table 3.20. The injection of high-phosphate liquid fertilizer ('starter fertilizer') can improve growth and quality on soils at P index 3 or below. No more than about 20 kg/ha of nitrogen and 60 kg/ha of phosphate should be applied in this way, and should be deducted from recommendations given in Table 3.21.

Leeks are harvested by hand after undercutting, or by elevator diggers, or by complete harvesters. Yield is 20–30 t/ha.

Brassica vegetables

Cauliflowers (*Brassica oleracea* var. *botrytis*)

Cauliflowers are produced all the year round in the UK. The main types of cauliflower are:

Summer – July to September
Autumn – September to December
Winter-heading (known as 'broccoli' in Cornwall) – December to May
Overwintered – May to June.

The majority of the crop is produced on the Fen silt soils of the Wash in south Lincolnshire, but other significant areas of production are west Cornwall, south Pembrokeshire, east Kent (Isle of Thanet), Evesham and Lancashire. When in head the crop is sensitive to frost which causes blemishes on the curd and for this reason winter-heading types are mostly found in west Cornwall where the climate is milder. Winter-heading types are also produced in large numbers in north-west Brittany, France, and exported to the UK. Heading date is a function of vernalization requirement of the variety which is controlled genetically.

Virtually all crops are now F1 hybrid varieties with high uniformity and vigour leading to a high number of heads cut per field of class 1 curds. Growers are referred to the NIAB vegetable growers leaflet number 1 *List of Cauliflowers* and to seed merchants' catalogues for new varieties. The traditional varieties were open pollinated types which showed variable performance and a wide range of heading dates within a variety such that fields had to be trafficked and cut over a number of weeks.

The different groups of cauliflowers have differing husbandry requirements with regard to plant spacing, plant raising, fertilizer requirements and crop protection.

Rotational position

Cauliflowers should be rotated with non-brassica crops in a frequency of cropping of 1 year in 5 because of the risk of clubroot (*Plasmodiophora brassicae*), a soil-borne disease. On high pH soils cropping can be more frequent since this discourages clubroot build-up. Another problem with long-term cropping is stem canker (*Phoma lingam*). Cauliflowers, as all brassicas, can carry over beet cyst nematode and should not be mixed in rotation with sugar beet.

Cauliflowers are grown in rotation with early potatoes in Cornwall, Thanet and Pembrokeshire, but in Lincolnshire they often are the most important crop in the rotation. It is possible to transplant winter-heading cauliflower after an early harvested winter barley crop in the south-west.

Plant production

The entire cauliflower crop in the UK is produced from plantlets transplanted into the field and is never direct sown. Most plants are raised in modular trays (modules) by specialist growers under glass or in plastic tunnels and then delivered to growers for transplanting into the field. Traditionally plants were grown in seedbeds in open ground and lifted as bare-rooted plants (peg plants) for transplanting. Risks of seedbed failures and the increased price of F1 hybrid seed mean that this practice has all but disappeared on commercial farms but is still practised on small-holdings and on some organic farms.

The timing of sowing and transplanting is very important for all cauliflower types, with only narrow windows of opportunity. Module-raised plants need to be transplanted earlier than peg plants since they are generally smaller and slower to come away. Only very few varieties show tolerance of a wider window of sowing and transplanting and these are typically summer types with no or a very low vernalization requirement and which can be used as summer- or early autumn-heading varieties. Sowing and transplanting dates are given in Table 3.22.

Plant beds (for bare root transplants)

A well-cultivated plant bed must be prepared on clubroot-free land. Drill seeds in beds about 250 mm apart at a depth of 20 mm ensuring seed is placed into soil moisture. Sow about 30 g per metre run to give about 36 plants per metre run. Place seed 18 mm apart with a precision drill. Treat the seedbed against cabbage root fly if sown after April. If necessary protect seedlings against cabbage aphid and caterpillars using an insecticidal spray.

When selecting plants for transplanting reject small and weak plants and any with damage or aphids. Transplants should be protected from drying out too much during transplanting.

Module-grown plants are usually raised by specialists but large growers may raise their own. Early crops are often sown in modules with a large cell size, with later transplanted crops being grown in small cells (9–15 ml volume of compost per cell). Plants are routinely fed and watered and treated with insecticides and are usually

Table 3.22 Sowing and transplanting dates for the production of cauliflowers

Heading date	*Harvest months*	*Sowing dates*	*Transplant dates*
Early summer	July to August	February to early March	March to April
Mid summer	September	Mid March	Mid May
Early autumn	October to early December	Late April	Mid to late June
Late autumn	November to December	Mid May	Mid June to early July
Winter heading	December to May	Early May	Mid to late July
Overwintered (spring heading)	May to June	Late May to early June	July

dipped in an insecticide prior to planting to give cabbage root fly protection.

Soil types
Choose free-draining soils with a pH of 6.5–7.0. Cauliflowers thrive in good soils with good heart and high organic matter. The use of leys to achieve this is good practice, otherwise use farmyard manure. Alluvial soils with good water holding capacity are necessary to grow summer types successfully and crops should not receive any checks in the growing season or buttoning and poor curd quality can occur.

Cultivations
Cultivate land well and deeply plough for summer types and work to a plant bed quickly, retaining moisture. Land should be prepared well in advance of planting; a final pass just before planting gives some weed control and ensures an aerated bed for successful establishment. Ridging up of rows is practised in windy areas and in overwintered crops to divert rain away from the stem. Ridging gives some weed control also; it should be carried out once the plants are well established but completed before the crop meets in the rows to avoid damaging roots.

Plant population
The grower of cauliflowers is concerned with achieving the maximum number of good quality heads per hectare. Plant population influences both curd size and its quality. If population is too high then only small curds will be produced and with winter-heading types the quality will be markedly affected. Some supermarkets are now marketing 'baby,' curds and these are produced at very high populations. Supermarkets are now demanding smaller curds even for maincrop cauliflowers. It may be possible to increase seed rate slightly when

Table 3.23 Plant populations and spacings for cauliflowers

Crop	*Plant population/ha*	*Spacing (mm)*	
		Rows	*Within rows*
Early summer	30 300–37 000	600	450–550
Late summer and early autumn	27 700	600	600
Late autumn	23 800	600	700
Winter heading	19 050	700	750
Overwintered	23 500	650	600

using F1 hybrid seed since the increase in vigour may allow a higher degree of plant competition without detriment to curd quality. Higher plant populations are possible with summer-heading types and the winter-heading types require the most space.

Plant populations vary depending on the type of cauliflowers being grown (Table 3.23) and the harvesting machines being employed. Whatever harvesting machinery is being used, row width must be an exact multiple of the distance between the wheels. The trend over the last 10 years has been towards specialist harvesting rigs either tractor mounted or purpose built.

Nutrient requirements
Cauliflowers require fertile land to grow to best quality and nutrient requirements are given in Table 3.24. High levels of potash are required but heavy dressings should be applied well before transplanting to avoid scorch. Likewise, nitrogen applied at over 150 kg/ha should be split and half applied 1 month after transplanting. Winter-heading and overwintered types are not given as much nitrogen at transplanting but are topdressed later to encourage growth before heading. Thus winter-

Table 3.24 Nutrient requirements of cauliflowers and broccoli (kg/ha) (MAFF, 2000)

	SNS, P, K or Mg index						
	0	*1*	*2*	*3*	*4*	*5*	*6*
Nitrogen							
Summer/autumn cauliflower and calabrese[1]	290	250	210	175	120	40	0[2]
Winter/overwintered cauliflower							
Seedbed	75	75	75	0[2]	0[2]	0[2]	0[2]
Topdressing	Apply 50–200 kg/ha 6–8 weeks before harvest						
Phosphate (P_2O_5)	200	150	100	50	0	0	0
Potash (K_2O)	275	225	175[3]	35	0	0	0
Magnesium (MgO)	150	100	0	0	0	0	0

[1] Apply maximum of 100 kg/ha in seed/plant bed and topdress rest once established.
[2] Increase to 20–30 kg/ha if topsoil (0–30 cm) N levels are low on soil N minimum testing.
[3] Reduce by 50 kg/ha if K index is high (i.e. 2+).

heading types are topdressed in late autumn and overwintered types in early spring (Table 3.24). Boron is sometimes applied to avoid dead heart or stem rot.

Irrigation

Summer and autumn crops need moist soil to ensure good quality and need irrigating on demand with about 10 mm water at each irrigation. Other crops do not need irrigation although some growers trickle-irrigate the plants at transplanting in July to aid establishment. Bare-root transplants with their larger stem size appear to resist transplanting drought stress more than module plants. Their greater height also allows for deeper transplanting than modules, thus putting the roots into moister soil.

Harvesting and quality

The UK market is now dominated by the multiple supermarkets with packhouses, cooperatives and individual large growers having specific contracts with supermarket buyers. The wholesale market also still exists, e.g. Covent Garden.

Typical market requirements include:

- Good shape – well rounded, undercurved base, not 'blown' or breaking up;
- Good colour – as white as possible, not yellow or discoloured;
- Specific size – by weight, e.g. 700 g minimum; or by diameter, e.g. 12 cm minimum;
- Free of blemishes – no rot, frost damage, bruises, caterpillar droppings;
- Free of bracts – small rudimentary leaves growing through the curd;
- Not ricey – small white rice-like protrusions from the curd;
- Not too many wrapper leaves – well trimmed.

Most crops are cut to Class 1 standard but sold as Class 2s to avoid supermarket rejection. Class 2 quality curds are often sold on the wholesale market (for hotels, schools and hospitals) and command a lower price. Crops are not selected on nutritional or taste criteria.

Packaging varies according to the market. Curds may be overwrapped with clingfilm, but the majority are presented in supermarket crates which can be displayed directly on the supermarket shelf. Trimming and packaging is normally carried out on the harvesting rig. Overwrapping is frequently carried out in the packhouse. Most cauliflowers are packed in crates into which a set number of heads are packed according to head size, e.g. 6, 8, 11 or 12 per box or crate.

Cauliflowers are supported by an EU Intervention Board which can help to put the bottom into the market but this is only really taken advantage of during peak production times in the summer and early autumn months. Price fluctuates widely from week to week depending on supply which is very weather dependent. Growers often cooperate with a packhouse to try to even out the supply problem particularly where contracts are made with supermarkets which demand year-round supply.

Crop protection

Cauliflowers are susceptible to all the brassica diseases and pests. Of particular importance is cabbage root fly and crop protection is necessary during establishment. In organic systems crops are covered with fleece during egg-laying periods to reduce infestation. A major pest is aphids especially woolly cabbage aphid which causes leaf distortion, and aphicides are used when infestation is detected. Aphicides also give protection from infestation with cabbage white butterfly and diamond-back moth caterpillars which can defoliate the crop. Organic growers need to use Bt insecticide to achieve control.

Foliar diseases of importance are mildew (both downy and powdery mildew) especially during establishment. Late autumn and winter varieties can also be infected by ringspot (*Mycosphaerella* sp.) which can be controlled by a fungicide applied in early November.

Most seed is dressed with fungicide and insecticide to give crop protection during the first 6–8 weeks following sowing, although downy mildew cannot be controlled in this way.

Broccoli (*Brassica oleracea* var. *italica*)

Broccoli is also known as calabrese. The edible portion is a bright green succulent stem terminating in a tight head of green flower buds known as a spear. The most valuable portion of the plant is the primary spear but secondary spears can sometimes be produced which if the price is right may be worth harvesting. Calabrese has risen in popularity in the last 5 years at the expense of cauliflower and as a consequence the area grown has increased dramatically. This rise in popularity has also meant that there has been an improvement in the quality and number of varieties available.

Varieties

All commercial varieties are now F1 hybrids. Varieties do not have much vernalization requirement and mature some 75–90 days after sowing. Thus they behave as summer cauliflowers and produce spears from late May to late November in the UK. Early crops are obtained using module-raised transplants sown under glass in February or March whilst later crops are direct drilled to a stand in the open field. Earliness can be achieved by covering early crop with fleece or plastic. The spears are not protected by wrapper leaves and are easily damaged by hail during late summer storms.

There is plant breeding interest in producing winter-heading varieties for UK production since year-round supply to the supermarkets can only be achieved with imports from Spain and Italy during the winter months.

Soil type

The crop must be grown without any growth checks, but a wide range of soil types are suitable. Early crops are transplanted onto sandy soils which are accessible early in the spring but these crops may require irrigation. Moisture-retentive soils in good heart are preferred for direct-sown crops and soils not prone to soil capping are preferred. Soil pH should be 6.5–7 minimum since the crop is highly susceptible to clubroot.

Establishment

Early crops raised in modules must be transplanted at an early development stage (two true leaves) and large transplants should be avoided since these suffer too much of a check at transplanting, reducing yield. Prior to transplanting, modules should be soaked in a solution containing high nitrogen which helps establishment and reduces the checking effect. If the soil is dry at transplantation then establishment will be improved if trickle irrigation is applied with the transplants. Transplants must be protected against cabbage root fly.

Direct-sown crops are usually drilled to a stand using a precision drill. A well-prepared level seedbed is essential and some growers prefer to grow in beds where compaction is avoided. A plant population of 8.5–10 plants/m^2 is desired with the low population giving larger spears. Normal spacing is 500-mm rows and 230 mm within the row or in beds 1.68–1.80 m wide containing four rows 400 mm apart and plants 200 mm apart within the row. For the lower populations and larger spears the within-row spacing is increased to 250–300 mm.

Continuity of production

Transplanted crops often take longer to mature than direct-sown crops. Choice of variety is important especially for the early crops, with varieties being designated early varieties. The following programme gives continuity of production:

Transplanted crop		
Sowing date under glass	*Transplant date*	*Harvest period*
Mid February	Early April	Late May
Mid March	Mid April	Early June
Late March	Mid April	Mid June
Direct-sown crop		
Sowing date in field		
Mid March	–	Late June
Late March	–	Early July
Mid April	–	Mid July
Late April	–	Late July
Mid May	–	Early August
w/b 28 May	–	Mid August
w/b 4 June	–	Late August
w/b 10 June	–	Early September
w/b 17 June	–	Mid September
Late June	–	Late September
Mid July	–	Early October
Late July	–	Mid October
Early August	–	Late October–Early November
Mid August	–	Mid–late November

Efficient production of late-maturing crops can only be carried out successfully in the milder south and south-west of the UK and on Jersey.

Horticulture Research International and the Agricultural Development and Advisory Service (ADAS) have produced a computer model of calabrese growth and maturity prediction which is available to growers and grower groups to help plan production and supply.

Nutrient requirements

Like most brassicas calabrese prefers fertile soils and requires adequate levels of nitrogen and potassium. Recommended fertilizer rates are given in Table 3.24. Only apply up to 100 kg/ha of N in the seedbed and topdress the balance after emergence. If applying more than 75 kg/ha of K then apply well before drilling and work into the soil or germination and establishment may be affected. Calabrese has a magnesium requirement and compound fertilizers with added magnesium can be beneficial.

Irrigation

Irrigation is necessary to ensure quality spear production. Throughout the growing season 25 mm of irrigation can be applied whenever the soil moisture deficit reaches 25 mm. An application of 50 mm of irrigation 20 days before expected harvest helps to ensure quality spear production. Very dry soils at drilling should be irrigated before drilling, allowed to drain and then sown in order to ensure high germination.

Crop protection

The crop is susceptible to most brassica diseases and pests and needs to be protected in a similar manner to cauliflower. Herbicide choices are limited.

Harvesting and marketing

Spears must be harvested with a sharp knife whilst the heads are tight with no yellow petal showing. The stem must be succulent with no trace of woodiness. The preferred size is a head diameter of 60–80 mm with a maximum of 100 mm. The spear should be 140–150 mm in length and the stem diameter 15–20 mm with a maximum of 30 mm. Most crops need cutting over two or three times since maturity within a crop is slightly variable.

Harvesting early in the morning is preferable, especially in summer, to preserve high water contents in the spear. Harvested crop should be cooled immediately using a wet air cooler or a vacuum cooler. Conventional cool stores do not always give a satisfactory cooling rate and can lead to loss of water and flaccid spears.

Marketing through a cool chain is necessary to preserve quality. Some supermarkets demand overwrapping with cling film. Presentation will be dictated by the buyer but is commonly in single or small groups of spears. It is essential to check the buyer's requirements for presentation before cutting and preparing.

The crop yields from 2–12 t/ha with an average of 6 t/ha.

Spring cabbage (*Brassica oleracea* var. *capitata*)

This is the unhearted or semi-hearted 'spring greens' type cabbage, now produced all year round from different areas (December–March: south-west England; February–April: Kent; summer/autumn: Lincolnshire). Steminess and bolting must be avoided for quality. Yield is very variable.

Seed should be sown 2 cm deep into a fine, firm seedbed on free-draining soil of light to medium texture with adequate moisture and pH of 6.5–7.0. A bed system is preferable to avoid compaction and achieve uniformity. Plant population will influence size of heads and time to maturity; aim for 30–40 plants/m^2 for spring greens and 10 plants/m^2 for fully hearted crop, on approx. 35- to 40-cm rows.

Nutrient requirements are given in Table 3.25. Topdressing depends on heaviness of crop and timing should ensure a steady supply of nitrogen throughout the growing period.

Summer and autumn cabbage (*Brassica oleracea* var. *capitata*)

A range of varieties provides hearted cabbage for marketing from May to October (earlies are pointed 'Hispi' types, later are round). February and March sowings for early harvest are made under protection for transplanting; April and May sowings can be made outdoors, either for transplanting or direct into field position. Yields are variable but can reach 25 t/ha.

Soil should be of pH 6.5 (6.0 on peats), permit good root development and have good drainage; loams and peats are preferable for early crops, water-retentive medium to heavy soils are better for later crops. Beds for transplanting should be fine and level with no compaction; very good seedbeds are required if sowing direct. Plant population will influence size of heads and time to maturity. For summer harvest aim for 9 plants/m^2 on 30- to 40-cm spacings; for autumn harvest 4 plants/m^2 on 40- to 60-cm spacings. Nutrient requirements are given in Table 3.25.

Winter cabbage (*Brassica oleracea* var. *capitata*)

A range of types is available including winter whites, January King, Savoy, and red cabbage for fresh sales, storage, or processing. Harvest is from September to

Table 3.25 Nutrient requirements for cabbage and Brussels sprouts (kg/ha) (MAFF, 2000)

	SNS, P, K or Mg index						
	0	*1*	*2*	*3*	*4*	*5*	*6*
Nitrogen							
Cabbage (summer, Chinese, autumn, Savoy)	340	300	260	220	170	60	0[1]
Cabbage (white storage)	300	250	210	180	120	40	0[1]
Overwintered spring cabbage							
Seedbed	75	75	75	0[1]	0[1]	0[1]	0[1]
Topdressing	Up to 250 kg/ha applied 6–8 weeks before harvest						
Brussels sprouts	330	300	270	230	180	80	0
Phosphate (P_2O_5)	200	150	100	50	0	0	0
Potash (K_2O)	300	250	200[2]	60	0	0	0
Magnesium (MgO)	150	100	0	0	0	0	0

[1] Apply 20–30 kg/ha if soil testing shows a low soil N minimum in the topsoil (0–30 cm).
[2] Reduce by 50 kg/ha if K index is high (i.e. 2+).

March, depending on varietal characteristics and frost hardiness.

Free-draining, deep soils are essential. Both transplanting (seed sown in April/May and transplanted into field in June/July) and direct sowing into the field (May/June) are suitable; in either case moisture stress must be avoided and irrigation is often necessary. Plant population will influence size of heads and time to maturity. For supermarket quality aim for 4–6 plants/m^2, for larger heads for processing reduce to 2.5–3 plants/m^2. Average yield is 20–30 t/ha. Nutrient requirements are given in Table 3.25. Excess nitrogen will impair frost hardiness and potential for storage.

Legume vegetables

The main legume vegetables are broad beans, runner beans, dwarf beans, and peas, all harvested as green, immature seeds in the pod. Precise timing of harvest is vital for the necessary quality characteristics of sweetness and tenderness. All are highly perishable due to high respiration rates and quickly deteriorate; careful handling and cool chain transport are essential. All are grown for freezing, canning, or selling fresh.

Beans and vining peas for processing are confined to the eastern counties, close to the processing plants, and are grown under contract only in order to programme production to provide an even continuous supply over the growing season. At the optimum stage of maturity mobile viners and harvesters work round the clock to harvest the product for rapid transport and processing. Varieties and husbandry are dictated by the processor.

Beans and peas to be sold fresh are generally hand harvested, picking over the crop several times to maximize yield of the correct quality; thus labour costs are high. Variety lists and technical advice are published by the Processors and Growers Research Organisation (PGRO), Peterborough. For nutrient requirements see Table 3.26 (beans and fresh peas) and Table 3.14 (vining peas), and for general agronomy see field pea and bean sections of this chapter. Precise technical details vary widely according to site, type, variety, system of establishment and target harvest date. Effective weed, disease and pest control is vital to achieve the required quality, and irrigation is often beneficial.

Fodder crops

Fodder crops are grown for feeding direct (i.e. not ensiled) to stock at the appropriate stage in the life cycle of the crop, either in situ or as a home-produced feed supplement to conserved forage or bought-in feed. Forage maize is also included here as, although it is grown for ensiling before feeding and thus is not strictly a fodder crop, it is grown as an arable crop and must be harvested at the appropriate stage in its annual life cycle, unlike grass and other forage. Feeding value of different fodder crops will contribute to choice (see Chapter 18).

Forage maize (*Zea mays*)

Forage maize is restricted in its distribution in the UK by temperature. It is not frost tolerant and requires temperatures above 6°C to germinate and grow. The crop

Table 3.26 Nutrient requirements of vegetable legumes (kg/ha); m indicates maintenance dressing (MAFF, 2000)

	SNS, P, K or Mg index					
	0	*1*	*2*	*3*	*4*	*5*
Beans						
Nitrogen						
Broad	0	0	0	0	0	0
Dwarf and runner[1]	180	150	120	80	30	0
Phosphate	200	150	100	50 m	0	0
Potash	200	150	100 (2–) 50 m (2+)	0	0	0
Magnesium	100	50	0	0	0	0
Fresh ('market pick') peas						
Nitrogen	0	0	0	0	0	0
Phosphate	185	135	85	35 m	0	0
Potash	190	140	90 (2–) 40 m (2+)	0	0	0
Magnesium	100	50	0	0	0	0

[1] Apply no more than 100 kg/ha in seedbed; apply balance when fully established. Runner beans may require a further topdressing of 75 kg/ha at early picking.

is therefore a summer crop with a 5-month growth cycle (May to September). In recent years plant breeders have made considerable advances with varieties which can mature in less time (early maturing varieties) although the yield of these is slightly less. These advances combined with global warming have improved the reliability of the crop in the UK and pushed the potential area for growing the crop further north. The crop is favoured by dairy farmers since it produces a high quality, high yield silage with the minimum of arable expertise, and with the availability of contractors for drilling and harvesting the crop only a low investment in machinery is needed. Maize thrives on fertile soils and can be grown without the use of fertilizers on land that has had large amounts of slurry or farmyard manure applied. The crop commonly produces a fresh weight yield of some 30–50 t/ha (7.5–12.5 t/ha dry matter yield) with a dry matter content of 25–30%.

Growth cycle

Maize is a monocotyledonous crop with a growth cycle similar to other cereals. It has a vegetative phase when only leaves are produced, a stem extension phase, pollen shed (anthesis) and grain filling. It differs from other cereal species in that the male and female parts are carried in separate structures. The pollen is borne on the terminal tassel of the stem and is the last structure to emerge (tasselling) whereas the ovules are carried in the cob half way down the stem and use thread-like silks to catch the pollen which is wind-borne.

Suitable areas

Suitable areas for growing forage maize in the UK are mapped using long-term meteorological records and accumulating the average daily temperature from 1 May to 30 September after first subtracting 6°C [mean accumulated heat units (AHU) above 6°C]. For growing extra early maturing varieties, at least 1250 AHUs are needed, for second earlies 1300 AHUs, for mid-season 1350 AHUs and for late varieties 1400 AHUs (see Fig 3.6). Growers also need to take account of altitude when considering which variety to choose with approximately 25 AHUs lost for every 100 ft (30 m) above sea level. Furthermore, 25 AHUs are lost if the seedbed is cold and wet and 75 lost if the slope of the field is north facing. Very early sowing can also stress the crop if low temperatures are experienced, losing up to 50 AHUs if the crop is set back.

Once the average daily temperature is above 10°C the crop grows well and at temperatures above 15°C the C4 photosynthetic pathway of maize becomes a distinct advantage and growth rate is very rapid.

The crop is quite tolerant of low soil moisture but will yield better if moisture is available. For this reason yields are frequently higher in the west of the UK where summer rainfall is higher. In drier climates, e.g. central USA and Africa, the crop is frequently irrigated, but in the UK it is not economically feasible to irrigate. Summer droughts lead to reduced seed set as the critical pollen shed becomes out of phase with the receptivity of the silks.

Figure 3.6 Areas of England and Wales suited to growing forage maize based on accumulated heat units (AHU) over 6°C from 1 May to 30 September. (Adapted and redrawn from *Forage Maize Growing Guide*, Somences Coop de Pau.)

	1200 AHU	Very early varieties
Key	1250 AHU	Extra early varieties
	1300 AHU	Second early varieties
	1350 AHU	Mid-season varieties
	1400 AHU	Late varieties
	Note:	subtract 25 AHU for each 100 m above sea level and 25 if north facing or cold wet soil.

Seedbed

Forage maize can be sown on any soil type on which a good reasonably fine tilth can be produced in late April/May. This excludes the heavier soils which are cold and slow to warm up, thus slowing down germination, and which can be difficult in a late harvest. Available summer moisture often excludes light and shallow soils.

Ploughed land should be quickly worked down in the spring to avoid 'cobbly' seedbeds which impair both germination and soil-residual herbicide action. Seedbeds need to be reasonably fine but firm, and ring rolling after drilling is often desirable.

Rotation

Maize fits well into any rotation and does not carry over any pests, diseases or serious weeds of arable crops. It can be grown continuously, but large amounts of nitrogen and phosphate can build up in the soil if organic manure is regularly used. It can suffer serious damage from frit fly and should not follow a late ploughing out of a grass sward without at least 4 weeks between the crops. Since forage maize can be late to be harvested (October or even November) it can be difficult to get a winter cereal in as a succeeding crop. The crop is frequently sown on to a sacrifice field which has been used to dispose of slurry stores. This has been traditionally carried out through the winter but it is becoming recognized that this is an environmentally unsound practice. Winter slurry spreading leads to losses of nitrogen through volatilization of ammonia and run-off of slurry into water courses. Slurry should be spread on to the field in spring to minimize these problems.

Sowing

Sowing date depends on prevailing temperatures in spring and a good guide is to wait until the soil temperature is above 10°C. This occurs in late April in the south but more usually early to mid-May further north. Late sowings result in crops with poorly filled cobs which are late harvesting and result in poor quality silage.

The crop must be drilled at a depth of 50 mm using a precision drill usually on 75-cm row widths. Plant populations are relatively low – 11 plants/m^2 – and a seed rate of 12 seeds/m^2 is used. Higher seed rates can improve overall yield but depress seed set and lower the quality of the silage. Since precision drilling is used, seed is graded and often sold in packs by seed number. Growers and contractors need to adjust drills to the seed size of the particular variety and seed lot.

Seed is normally dressed with a fungicide and insecticide and a bird repellant can be added. This is to deter rooks which can devastate a field during germination. Harrowing or ring rolling after sowing removes the drill lines which also helps to deter rooks.

Pests and diseases

The crop is particularly sensitive to frit fly which eat out the apex of the plant. Since forage maize varieties do not tiller freely, a damaged plant is a virtual write-off. If bad frit fly attacks are common or expected then insecticidal granules can be broadcast or trickle applied at drilling. Other pests include rooks and badgers. Badgers can be important in dry summers when they knock plants over and feed on the grain as a protein supplement when slugs and snails are in short supply.

Several foliar diseases attack maize including *Septoria* leaf blotch and maize eyespot (*Kabatiella zeae*). Normally it is not economic to apply fungicides to forage maize in the UK, but in the west country during wet years control of eyespot may be necessary.

Nutrient requirements

Most forage maize is grown on land that has been well fertilized with farm wastes, particularly slurry, and if applications are heavy then no potash or phosphate fertilizers are needed but crops should be given a further 40 kg/ha of nitrogen in the seedbed or just prior to emergence. In the absence of heavy slurry applications then fertilizers should be applied as given in Table 3.27. Whilst the levels of potash recommended are high this is because the crop removes large quantities when silaged; however, there is not a large yield response to the application of either P or K. There is a trend towards higher yields from forage maize and crops are now planted earlier in late April in line with developments made in northern France, and in these situations the crop can respond to a 'starter fertilizer' applied in the drill rows of about 10 kg/ha of N, P and K. With the high-yielding crops nitrogen may be increased to 120 kg/ha in the absence of slurry applications.

Harvesting

The first frosts of autumn mark the end of the growing season and accelerate maturity, but crops should be harvested within 10–14 days of frosting or quality begins to deteriorate. The crop is harvested when it has dropped below 75% moisture content (25% DM) and the grain is at least at the hard doughy stage; below 70% when the grain is hard and flinty can improve ensilaging and leads to less effluent. Generally the leaves of the crop are dry and papery at harvest. Cool summers delay harvesting. Harvesting begins in the earliest areas with early/mid-maturing varieties in late September and can last to the end of October in late areas and with late varieties. In exceptionally late years harvesting may not be complete until early November.

Many crops are harvested using multi-row harvesters and it is important that the crop is well chopped and the grain well cracked as this will aid digestion by livestock. Many crops are contract harvested. Ensiling is relatively easy because of the high level of soluble carbohydrates located in the stems at harvest and additives are rarely necessary. The crop has an average D value of about 70 at harvest with a relatively low crude protein content of 9%.

Yield

Good crops of forage maize yield 50–70 t/ha of fresh matter (12.5–18.75 t/ha DM) which gives approximately 45–60 t/ha of silage.

Table 3.27 Nutrient requirements of forage maize (MAFF, 2000)

	SNS, P or K index						
	0	*1*	*2*	*3*	*4*	*5*	*6*
Nitrogen[1]							
All mineral soils	120	80	40	20	0	0	0
Phosphate[1] (P_2O_5)	110	85	60	20	0	0	0
Potash[1] (K_2O)	230	205	180[2]	110	0	0	0

[1] Assumes no organic manures applied. If organic manures are applied then fertilizer rates must be reduced appropriately (see MAFF, 2000).
[2] If K index is a high 2 (i.e. 2+) then reduce to 155 kg/ha.

Roots

Fodder beet and mangels (*Beta vulgaris*)

Fodder beet is a hybrid of mangels and sugar beet, and the three crops share many characteristics. Both fodder beet and mangels are capable of giving high yields of digestible dry matter suitable for feeding (after lifting) to ruminants, and the roots store well thus providing winter feed. Mangels provide 10–15% dry matter and fodder beet 15–20% dry matter, depending on variety, from yields typically up to 120 t/ha. Thus up to 12 t DM/ha can be expected from these roots, with fodder beets also supplying up to 5 t DM/ha from the tops.

All types require the warm, sunny conditions more typically found in southern England, and well-drained, highly fertile soils. Heavy soils make establishment and harvesting difficult. The use of monogerm seed eliminates the need for hand thinning; all fodder beet varieties are available in this form. A fine firm seedbed should be prepared to drill in late April or as soon as conditions are suitable, but variety susceptibility to bolting should be taken into account with respect to sowing date. Seed should be sown at 12–18 mm into moisture, at a row spacing of 50–60 cm (adapted to suit harvesting machinery), to achieve a plant population of 7–10 plants/m^2. Precision drilling will give the even stand needed to reach yield potential and ease machine harvesting. Once established, plants are relatively drought tolerant. Flea beetle can cause serious damage to young seedlings. Beet cyst nematode can be avoided by the use of sound rotations. Aphids should be controlled, particularly where there is risk of transmission of virus yellows. Varietal resistance should be used to counter powdery mildew (*Erysiphe betae*) and rust (*Uromyces betae*). Nutrient requirements are given in Table 3.28.

Lifting is usually in October or November (before severe frosts), when the outer leaves begin to wither, using modified swede or sugar beet harvesters, though hand harvesting is possible for most types except those whose roots sit well down in the ground (usually high dry matter fodder beet varieties). Only fully mature roots will store well, and damage should be avoided during lifting and handling; low dry matter varieties are more susceptible to damage. Depending on conditions some cleaning may be necessary as dirt will cause feeding problems; some varieties are more prone to fanging and thus a higher potential dirt tare. Roots can be stored with protection from frost, e.g. in earthed-up clamps, for several months. Fodder beet types produce useful amounts of tops, which should be kept clean during lifting, and wilted or ensiled before feeding.

Swedes (*Brassica napus*)

Swedes are a useful frost-hardy fodder crop yielding storable roots of 8–12% dry matter (depending on type), from a fresh yield of up to 60 t/ha. Green skinned varieties have higher dry matter content and can be stored for longer than the purple and bronze skinned types; the former are more popular in the north and Scotland. Some varieties are dual-purpose, being suitable for fodder and the culinary market (see under 'Vegetables' in this chapter for swede production for human consumption).

Crops are tolerant of a wide range of soils but they should be well-drained and moisture retentive with a pH of 6.0–6.5. There is some varietal resistance to clubroot and sound rotations should complement this. They are

Table 3.28 Nutrient requirements of fodder crops (kg/ha) (MAFF, 2000)

	SNS, P or K index						
	0	*1*	*2*	*3*	*4*	*5*	*6*
Fodder beet and mangels[1] (65 t/ha roots)							
Nitrogen (N)	130	120	110	90	60	40	0
Phosphate[2] (P_2O_5)	100	75	50	0	0	0	0
Potash[2] (K_2O)	160	135	110[2]	40	0	0	0
Forage swedes and turnips (65 t/ha roots)							
Nitrogen (N)	100	80	60	40	40	0	0
Phosphate[2] (P_2O_5)	100	75	50	0	0	0	0
Potash[2] (K_2O)	200	175	150[1]	80	0	0	0
Forage rape and stubble turnips (grazed)							
Nitrogen (N)	100[3]	90[3]	80[3]	60	40	40	0
Phosphate[2] (P_2O_5)	75	50	25	0	0	0	0
Potash[2] (K_2O)	100	75	50[2]	0	0	0	0
Kale (40 t/ha cut)							
Nitrogen (N)	130	120	110	90	60	40	0
Phosphate[2] (P_2O_5)	100	75	50	0	0	0	0
Potash[2] (K_2O)	250	225	200[2]	130	0	0	0
Forage rye and forage triticale (20 t/ha cut)							
Nitrogen (N)	80	60	40	20	0	0	0
Phosphate[2] (P_2O_5)	85	60	35	0	0	0	0
Potash[2] (K_2O)	170	145	120[2]	50	0	0	0

[1] For fodder beet and mangels sodium is required and 400 kg/ha agricultural salt should be applied and incorporated before sowing except on Fen peats and silts. If salt is not used, increase potash by 100 kg/ha.
[2] If K index is a high 2 (i.e. 2+) then reduce potash by 25 kg/ha.
[3] If being grown as a catch crop after cereals then reduce N applied to 75 kg/ha maximum.

best suited to cooler, moister parts of the UK. Seed is drilled from April to June for utilization from November until March, either in situ or after lifting and storing. Sowing date and utilization will be strongly influenced by the particular farm system, but swedes tend to be sown earlier in the north as a break crop, and later in the south as a catch crop (to avoid powdery mildew and cabbage root fly attack). Dressed seed should be drilled 10–20 mm deep into a fine firm seedbed at a row spacing of approx. 50 cm to give a population of 8–15 plants/m^2. A higher density gives smaller roots which are easier to feed and store. A residual herbicide is recommended, and flea beetles may need control. Powdery mildew (*Erysiphe cruciferarum*) can be a problem in dry conditions, and control may be economic if young plants are infected. It is not generally economic to control pests after establishment in swedes for fodder; seed dressings should be used to prevent problems as far as possible. Swedes are responsive to P and K; excess N reduces root yield by encouraging top growth, and roots are more tender and store less well. Nutrient requirements are given in Table 3.28.

Leafy brassicas

Leafy brassica forage crops should not be included in rotations with brassica cash crops such as oilseed rape or vegetables due to the potential for build-up of soil-borne diseases, notably clubroot. Powdery mildew (*Erysiphe cruciferarum*) and *Alternaria brassicae* are common to them all, the former particularly in dry weather.

Kale (*Brassica oleracea* var. *botrytis*)

Kale is a suitable crop for all but exposed sites, being tolerant of soil type, and typically yielding up to 7 t DM/ha at 15% dry matter content. It can be sown from April to June depending on type and target date for utilization (in situ or zero grazed): marrowstem types are sown early for utilization before Christmas (not frost hardy) when the thick fleshy stems are palatable and provide the bulk of the yield; thousand head types (tall and dwarf types available) are sown late for utilization of the leaves and sideshoots from Christmas to March (frost hardy); hybrid types have the advantages of both with thick high-yielding stems that remain palatable, and most varieties are frost hardy, allowing utilization from successional sowings from August to March (however, crops become less hardy with maturity and heavy N use). The crop is tolerant of soil type but heavy or poorly drained land may create utilization problems. It may be direct drilled or sown into a fine moist seedbed at 2.5–5.0 kg/ha broadcast (1.5 kg/ha precision drilled). Nutrient requirements are given in Table 3.28.

Forage rape (*Brassica napus*)

Forage rape is a useful green forage catch crop suited to western and northern moister areas, providing feed until heavy frosts are experienced. It may be sown from mid-April to August, into a fine firm seedbed with an optimum soil pH of 6.0–6.5 at 3.5–6.5 kg/ha (10 kg/ha broadcast). It is usually fed in situ to sheep from about 10 weeks after sowing, during the period from July until December, and is excellent for fattening lambs (care should be taken to introduce it gradually with grass or with other roughage available to avoid gorging). Forage rape can yield up to 4.5 t DM/ha at about 12% dry matter. Nutrient requirements are given in Table 3.28. (Hungry Gap kale and rape kale are also rapes but with much greater frost hardiness than forage rape varieties; thus they are sown into cereal stubbles for utilization in March/April.)

Quick growing 'continental/stubble' turnips (*Brassica campestris*)

These are quick-growing (ready for grazing 8–12 weeks after sowing), white-fleshed turnips grown primarily for the leaf as a forage crop, and should be distinguished from the longer-season white- and yellow-fleshed turnips grown for their roots. Seed can be sown between April and August, for summer or autumn grazing, at 3–6 kg/ha (6–8 kg/ha broadcast). Yield will be up to 3.5 t DM/ha at 7–10% dry matter. Nutrient requirements are given in Table 3.28.

Forage cereals (rye and triticale)

Early-bite can be produced using autumn-sown rye or triticale. These crops can produce early grazing for dairy cows. Crops can also be zero grazed (i.e. cut and brought in to housed stock). Crops should be completely grazed by mid-April before stem extension rapidly increases and palatability declines. Yields can be up to 20 t/ha fresh weight. Nutrient requirements are given in Table 3.28.

Fibre, fuel and other non-food crops

Non-food crops offer arable diversification possibilities with good potential for expanding markets in contrast to the food crops where, generally, demand is inelastic and market size is unlikely to grow within the EU. Many can also take advantage of being grown on set-aside land as industrial crops. The future is likely to see an expansion of non-food crop area, and varieties will be 'designed' to give unique composition profiles for a wide range of non-food uses from fuel to medicines. Wide-scale viability awaits the appropriate infrastructure and plant breeding advances.

Fibre crops

The fibre-producing crops hemp and flax are included within the Arable Area Payments Scheme. The fibre extracted from these crops is used as the raw material for a range of paper and textile products. Contracts are essential for the successful marketing of both crops.

Hemp (*Cannabis sativa*)

Hemp was grown widely in Britain for centuries but production ceased in the 1950s when its use as a drug became illegal. Low tetrahydrocannabinol (THC) varieties are now available and their production is permitted under Home Office licence. Outlets are limited at present but are likely to expand. The crop is easy to grow and requires very few inputs, but areas with low rainfall at harvest (late summer/autumn) are favoured due to the importance of drying to prevent deterioration during storage.

Approved-variety seed is sown at a rate of 80–90 kg/ha and a depth of 3 cm in spring as soon as soil conditions allow and there is no risk of frost. A firm fine seedbed is preferable for good establishment. Appropriate weed control may be necessary until a full canopy is produced. The crop can reach a height of 2–3 m and should be swathed, dried and baled in warm dry conditions. Deterioration of the fibre will occur if the bales are not dry. Yields of approx. 25 t/ha can be expected, but may be lower in dull years or for late sown crops. Good drying, bulk handling and storage facilities are vital.

Flax (*Linum usitatissimum*)

Flax is the same species as linseed and shares many of its characteristics, though each crop has varieties with appropriate features, for example flax varieties are taller with fewer capsules and seeds than linseed varieties. Seeds from flax plants grown for fibre are crushed for oil and contribute significantly to the EU crush, and 'dual purpose' varieties also exist (fibre can be obtained from linseed stems but is of poor quality particularly with respect to fibre length).

Flax is more demanding than linseed due largely to the higher plant density needed (up to 1800 plants/m^2) to produce the long fine stems for good fibre quality. The higher density leads to a greater sensitivity to low soil moisture (due to reduced root development) and a greater vulnerability to disease (particularly stem diseases). Very low rates of nitrogen are used to avoid production of coarse fibre, typically 20 kg/ha is applied in the seedbed.

Energy crops

The production of crops for energy has the advantage of creating a renewable fuel source which could use land surplus to meet requirements for food production, as well as providing a greater diversity of crop enterprises. The chief disadvantage of crops as a renewable energy source is their relative bulkiness per energy unit, making it necessary for processing to be carried out near the site of production if the energy yield is not to be used up for transport purposes. Such regional localized processing facilities are not currently in place across the UK.

In theory any crop can be a source of energy for heating or electricity since all crop dry matter is stored energy. However, an economically viable energy crop is likely to be an efficient converter of solar radiation into dry matter and yield close to the theoretical dry matter maximum for the UK, be harvested at low moisture content, be hazard free, and have a low relative cost of production. Perennial crops capable of efficient radiation interception year round and easily harvested could theoretically provide 50–60 t/ha dry matter (equivalent to about 30 kW/ha/year). In practice, short-rotation coppice (willow, poplar) and *Miscanthus* are the only currently viable candidates, and establishment grants are available for these. Annual arable crops cannot yield so highly but are more flexible for the arable producer; at present only oilseed rape is grown in the UK in this way, for the production of biodiesel (rape methyl ester), by a small number of producers under contract. Bioethanol can be produced from cereals, sugar beet, etc. and trials are underway in other EU countries. The expansion of the market for energy crops in the UK awaits further political commitment and industrial investment.

Medicine, flavouring, and other minor crops

The production area needed to supply the market from such crops as evening primrose (*Oenothera biennis*) and borage (*Borago officinalis*) (both grown for the extraction of gamma-linolenic acid), and condiment mustard (*Brassica juncea, Sinapsis alba*) is small, and the associated lucrative contracts are filled by a few favoured growers. However, there are potential niche markets for a wide range of crops in this general category for the entrepreneurial producer to develop. In the future many producers may become niche producers as genetic transformation technology enables plant breeders to develop new varieties with characteristics tailored to very specific end uses, for example antibodies. Exploitation of this technology will only be possible with the cessation of consumer resistance to the production of genetically modified crops.

References

*Published annually

HGCA (2000) The Wheat Disease Management Guide. Home Grown Cereals Authority, London.

MAFF (1984) *The Agricultural Climate of England and Wales*. HMSO, London.

MAFF (2000) *Fertiliser Recommendations*. Reference Book 209. The Stationary Office, London.

*DEFRA *Agriculture in the United Kingdom*. The Stationary Office, London.

*DEFRA *IACS Guide*. The Stationary Office, London.

*Nix, J. *Farm Management Pocketbook*. Wye College Press, Ashford.

Tottman, D.R. (1987) The decimal code for the growth stage of cereals, with illustrations. *Annals of Applied Biology*, **110**, 441–54.

Useful websites

www.defra.gov.uk
www.hgca.com
www.metoffice.com
www.niab.com
www.pgro.co.uk
www.potato.org.uk
www.silsoe.cranfield.ac.uk/cpf

4

Grassland

J.S. Brockman & R.J. Wilkins

Introduction

Grassland occupies a major part of the agricultural land of the UK, as shown in Table 4.1. The major function of grassland on farms in Britain has traditionally been to supply feed for livestock either through grazing or after conservation as hay or, more recently, silage. This still represents the predominant use of grassland. It is the most important source of nutrients for ruminants in Britain, supplying some 70% of the total metabolizable energy (ME). Increased attention is though now being given to other products and services that can be supplied by grassland. In the future grassland is likely to be managed for a wide range of objectives.

There are possibilities for using grassland as a source of biomass for the supply of energy, for the fractionation of grass to supply protein-rich products for human consumption and for the extraction of fibre and fine chemicals from grassland plants (Mackie & Wightman, 1997).

Grassland makes a major contribution to biodiversity and landscape. This is recognized in Agri-Environmental Schemes introduced in the last 15 years, with grassland forming a major part of many of the Environmentally Sensitive Areas (ESA). Over 1 million hectares of grassland are located within ESAs, with much of this subject to management agreements in which farmers are paid for adopting practices intended to maintain or further improve the environmental benefits from grassland, in many cases through increases in biodiversity.

Grassland also has a major role to play in the provision of access and enjoyment in the countryside for the public at large, as recognized in the Countryside Stewardship scheme. The management of grassland has a large effect on the quality of air and water. This is particularly important in relation to the predominance of grassland in overall land use. Grassland and the associated ruminants make major contributions to the emissions of the greenhouse gases, methane and nitrous oxide, and to the emission of the environmentally harmful gas ammonia. Benefits may, however, result from the sequestration of carbon in soils under long-term grassland, which may occur in quantities similar to those in forests. There are large variations between grasslands in the magnitude of losses to water of nitrate and phosphorus. Depending on location and management, grassland may either be an important source of pollution or alternatively (generally with extensive management) represent a preferred land use for the supply of water of high quality and with minimal content of pollutants.

It will be increasingly important to consider grassland in relation to these many aspects of production and environmental impact. This chapter is written principally from the standpoint of efficiency of agricultural production, but attention is also given to environmental effects. The first part of the chapter deals with plant production from grassland and includes consideration of grassland types, the characteristics of the main agricultural species and mixtures and fertilizer management. This is followed by discussion of herbage quality and utilization through grazing and conservation and a final section discussing management for environmental objectives.

Rough grazing

Rough grazing can be defined as uncultivated grassland found as unenclosed or relatively large enclosures on hills, uplands, moorland, heaths and downlands. Thirty-

Table 4.1 Grassland in the UK, 1999 (10^6 ha)

	England	*Wales*	*Scotland*	*Northern Ireland*	*Total UK*	*%*		
Grass								
Rough grazing[1]	1.1	0.6	3.8	0.3	5.8	33		71
Over 5 years	2.9	1.0	0.9	0.7	5.5	31	38	
Under 5 years	0.7	0.1	0.3	0.1	1.2	7		
Cereals	2.6	0.05	0.4	0.04	3.1	18		
All other crops	1.4	0.02	0.1	0.01	1.5	8		
Set-aside	0.5	0.01	0.1	0.01	0.6	3		
Total	9.2	1.8	5.6	1.1	17.7			

[1] Rough grazing includes both common and sole right areas.

Table 4.2 Estimated contribution of grass and forage to total ME requirements of UK ruminants

	Overall %	*Estimated for*		
		Dairying	*Beef*	*Sheep*
From grassland	60			
From rough grazing	5	65	80	90
From forage crops	5			
From other foods	30	35	20	10

three percent of all agricultural land is in rough grazing, and nearly two-thirds of this area is in Scotland. As shown in Table 4.2, the contribution of the rough grazing area to national ruminant nutrition is only 5% and this low figure indicates the low level of production associated with rough grazing. Rough grazing is, however, extremely important in relation to landscape and environmental impact. The natural vegetation has low yield and nutritional characteristics: the period of summer growth and the total yield are restricted by a combination of altitude, poor soil fertility, high rainfall, difficult access and the consequent poor control of defoliation management. The main types of rough grazing are as follows.

Mountain grazing

This is land over 615 m in altitude, where high rainfall, low soil temperatures and extreme acidity restrict the plant species that will grow and limit their growth. Winters are very cold in these areas, farms are isolated and the stock-carrying capacity of the land is very low.

Moors and heaths

Moors and heaths are found at lower altitudes and are often dominated by heathers (*Calluna* spp.) in relatively dry areas and purple moor grass (*Molinia caerulea*) or moor mat grass (*Nardus stricta*) in wet areas. Heather can be utilized by sheep and these areas can be improved by liming, selective burning and application of phosphates; grass will be encouraged, often the indigenous sheep's fescue (*Festuca ovina*).

Stock will eat moor mat grass in spring, but as the season progresses it becomes unpalatable, especially to sheep. Most natural species occur in patches and clumps, their occurrence reflecting the environment and the selective grazing of animals. Normally cattle will graze moor mat grass and purple moor grass more readily than sheep, and some improvement in overall production can be encouraged by regular grazing together with the use of lime and phosphate. Under these circumstances encouragement will be given to bents (*Agrostis* spp.), sheep's fescue and red fescue (*Festuca rubra*). Although these three grass species have a low production potential compared with grass species that are sown in lowlands, they have a greater value than *Molinia* and *Nardus*.

Some rough grazings occur on lime-rich soils, such as the downs in the south of England where the soil is too shallow or too steep for cultivations: here sheep's fescue and red fescue dominate the sward.

Permanent grassland

Permanent grassland can be defined as grassland in fields or relatively small enclosures and not in an arable rotation; 30% of all agricultural land is in permanent grassland and four-fifths of this area is in England and Wales. The 5.5 million hectares of permanent grassland represent many different types of grassland, and, as the above definition suggests, much of the area is on land not suited to arable cropping. It has been estimated that 30% of permanent grass in the UK suffers from serious physical limitations that impede use of machines: the main factors are rough and steep terrain, stones and boulders on the surface and very poor drainage (Green, 1982).

Several surveys have classified permanent grass on the basis of its botanical composition, an account of the classification being given by Davies (1960). In simple terms, swards can be placed in order of probable agricultural value as follows, ranking from poor to good quality (bearing in mind that grassland is rarely homogeneous in composition, and most natural species occur in irregular patches in the field).

Rough grazing

- heather (*Molinia, Nardus*);
- areas with red fescue;
- *Agrostis* with red fescue (with rushes and sedges on wet soils).

Permanent grass

- mainly *Agrostis*;
- *Agrostis* with a little perennial ryegrass (PRG) (less than 15% PRG, often with rough-stalked meadow grass, meadow foxtail, cocksfoot and timothy);
- second-grade PRG (15–30% PRG with *Agrostis*) (often with crested dogs tail, sheep's fescue, cocksfoot and tufted hair grass);
- first-grade PRG (over 30% PRG) (often with white clover, rough-stalked meadow grass, *Agrostis*, cocksfoot, sheep's fescue and timothy).

Table 4.3 gives a summary of surveys conducted by Baker (1962) and Green (1974). It shows that *Agrostis* dominated the permanent grasslands of the UK. There has, however, been a progressive increase in the content of perennial ryegrass, with Hopkins and Davies (1994) indicating that by the early 1990s about 70% of swards above 20 years in age had more than 15% of PRG. For swards of that age, the mean content of PRG was 24%, compared with an *Agrostis* content of 27%.

Table 4.3 Classification of permanent grass on botanical composition

	Percentage of fields surveyed containing	
	(Baker, 1962)	*(Green, 1974)*
15% or more of PRG	20	26
Mainly *Agrostis*	59	56
Fescue dominant	7	10
Others	14	8

Although permanent grassland with substantial physical limitations is unlikely to be ploughed and reseeded, periodic sward renewal may occur on the remaining area defined above as permanent grassland. Green (1982) noted that half of grass between 1 and 4 years of age was intended to last more than 4 years and some of it indefinitely, whilst most of the medium- and long-term swards were on land intended to grow grass continuously. Since that time mean sward age has increased, with further reduction in the frequency of reseeding.

Rotational grass

Rotational grass (temporary grass) can be defined as grass within an arable rotation. Only some 8% of all agricultural land is in rotational grass and it has one or more definite functions, quite apart from its role in supporting an animal enterprise. These functions include:

- being an essential part of rotation to
  - break disease and/or pest cycle
  - improve fertility
  - break weed cycle
  - improve soil structure.
- enabling the growth of grass species best suited for animal production, giving
  - better fertilizer response than indigenous grass
  - high yields of grass and animal product
  - good seasonal distribution of grass production.

It is estimated that about 250 000 ha of new swards is sown annually, with about half in rotational systems and half being reseeding of existing grassland. Rotational grass has a particularly important role in organic systems in which the use of inorganic fertilizers and technical chemicals is severely restricted.

Distribution and purpose of grassland in the UK

Grassland is not distributed evenly throughout the UK. For example, Cambridgeshire has only 9% of the agricultural area in grass, whereas grass represents 91% in Wales. Most grassland occurs in the western half of Britain and in Northern Ireland, where rainfall is high and the soils are heavier, and so less suited to arable cropping, than in the east. In some mainly arable counties grass is grown either as a break from regular arable cropping or on land that is unsuited for other crops. In the truly grassland areas, grass is grown as a highly specialized crop, often on farms growing little but grass.

Most of the grass grown in the UK is eaten by the ruminant animals, dairy cows, beef cattle and sheep. There is a great diversity in the way in which grass is farmed. This chapter outlines general principles that underlie economic grassland production, irrespective of the type of animal enterprise.

Grass as such is rarely a cash crop, unless sold as hay, dried grass or big-bale silage. On nearly every farm, grass is processed by ruminant animals which are adapted to digesting the cellulose and hemicellulose found in grass and also synthesizing proteins from simple nitrogen (N) compounds (see Chapter 16). Whilst it is possible to measure grassland production as weight of grass produced (and research results are often expressed in this way), it is important to realize that growing grass is only a part of the production process. Apart from producing grass, the farmer is concerned about the proportion of the grass that is grown which is eaten by stock, i.e. its degree of utilization. Also, ruminants are fed material of non-grass origin, e.g. cereals. Figure 4.1 illustrates the essential steps.

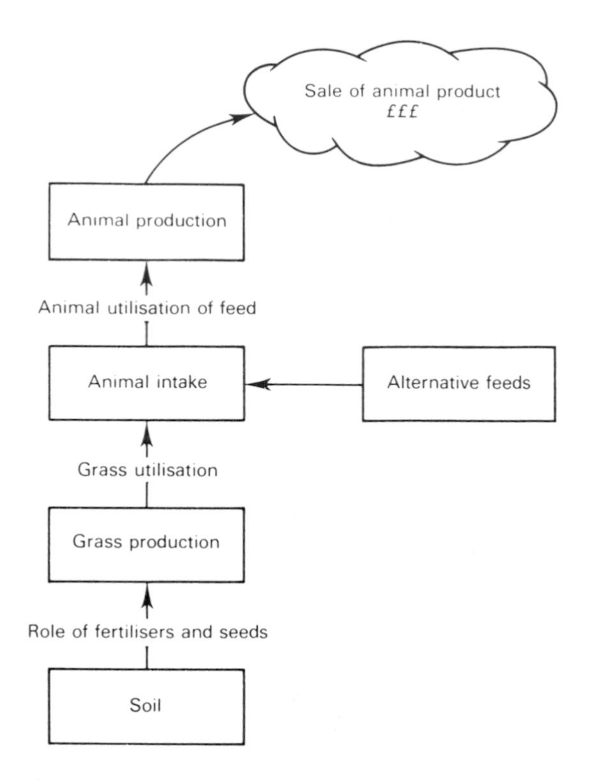

Figure 4.1 Factors affecting cash value derived from grassland used for animal production.

Although grass can supply most of the energy and protein required by ruminants, it has been estimated that in the UK some 30% of the energy used by ruminants comes from cereal feeds, as shown in Table 4.2. In pastoral countries, such as New Zealand, where animal products have a lower cash value, virtually all ruminant production is derived from grass; this illustrates that the relative proportion of grass and other feeds in the ruminant diet is related to the economic value of the animal product. It is possible that grass could make an even greater contribution to ruminant nutrition in the UK in the future; such a move would depend on improvements in grass production and utilization, or other feeds becoming relatively more expensive. On the other hand, it could be possible for grass to become less significant than now if alternative feeds became more attractive – either in monetary terms or from the point of view of convenience to grow and feed. At present, grass suffers in competition with some other feeds (e.g. cereals, maize) because it is not eligible for Arable Area Subsidy Payments.

Grassland improvement for agricultural production

The botanical composition of permanent grass and older rotational grass is related to the nature of the environmental factors operating there. The main constraints are as follows:

Within farmers' control
- Drainage
- Compaction
- Acidity
- Low fertility
- Weed problems
- Low stocking rate
- Poor grazing control

Outside farmers' control
- Steepness
- Altitude
- Climate
- Rough or stony terrain
- Aspect

If grassland improvement is being considered, the above factors can be used as a checklist. In fields where drainage, acidity and low fertility are not major problems, grassland production can often be improved by better defoliation management. The following section gives a brief account of basic grass physiology and shows how the farmer's management of grass can have a marked influence on the type of sward produced.

Nature of grass growth

Figure 4.2 illustrates a plant of perennial ryegrass (PRG) in a vegetative state. All the while the plant is not producing seed heads, the region of active plant growth (or stem apex) remains close to the ground and below cutting or grazing height. Growth of the plant continues in two possible ways:

- increase in size of existing tillers. As each leaf grows (Fig. 4.3), cells close to the stem apex divide and elongate so that the youngest part of the leaf is at its base and the oldest at the tip. As each leaf develops, its base surrounds the apex, so the stem apex is continually sheathed by developing leaves. This method of vegetative growth maintains the stem apex at ground level and protects it from damage caused when the oldest part of the leaves may be removed during cutting or grazing.
- tiller production. Each tiller bud contains a replica stem apex and further tillers will develop as the plant grows, giving rise to vegetative reproduction. These young tillers are also well protected from damage and rapidly develop a root system and an identity of their own – to further continue the process of vegetative growth and development.

When grasses have the ability to tiller freely and are encouraged to do so, they can spread laterally over the ground surface and form a thick or 'dense' sward, characterized by a large number of grass tillers per unit area.

During inflorescence development (sexual reproduction) the pattern of grass growth is quite different in two important ways. When inflorescence development

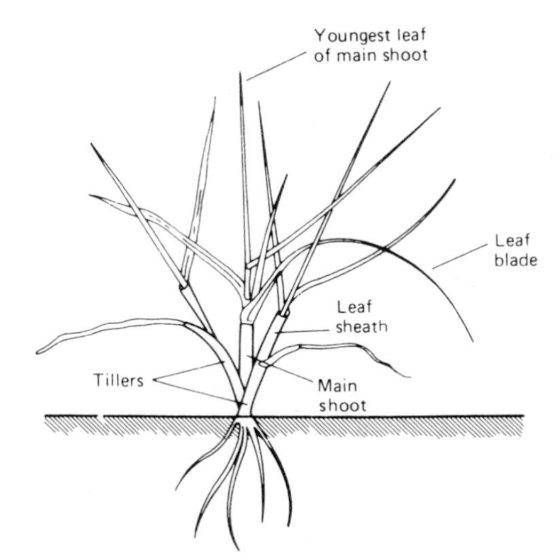

Figure 4.2 Vegetative development in grasses.

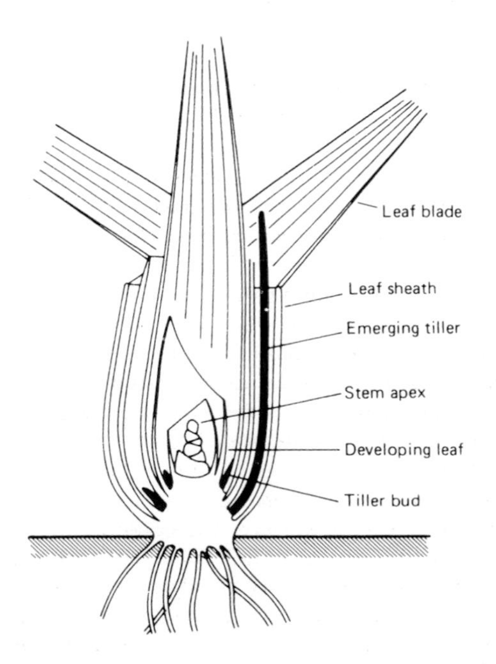

Figure 4.3 Position of stem apex and development of leaves and tillers in grass.

is triggered, by a mechanism based on day length (and slightly modified by temperature), two changes alter the normal vegetative growth:

- New tiller development is suppressed and existing tiller buds start to form the inflorescence.
- The stems of these inflorescence buds elongate rapidly, to carry the developing ear well above the ground.

When ear-bearing stems are defoliated, they are incapable of regrowth as the inflorescence was their apex, and plant regrowth can only occur from further developing tiller buds at the base of the plant.

When a grass plant bears visible ears, much of its impetus for vegetative growth is lost, and the longer the plant is left bearing ears, the slower will be the subsequent development of vegetative organs. Thus removal of a heavy, tall grass crop with ears emerged will leave a thin or 'lax' sward, with few tillers per unit area and a slow propensity to regrow. However, if regular frequent defoliation is then practised, the sward will become more dense, particularly if there is a high population of PRG, as this species tillers freely in the vegetative state. The effect of cutting date is well illustrated in Table 4.4, where a delay in taking the first silage cut gave a high silage yield but seriously delayed speed of regrowth.

Tillers and plants are in a state of continual change, the life of an undefoliated leaf of PRG being about 35 days in summer and up to 75 days in winter. If grass is left undefoliated, there is a rapid burst of vegetative (tiller) development in spring which is suppressed in May/June by tall inflorescence-bearing stems that restrict further vegetative development until autumn, followed by drastic tiller restriction during winter. Frequent defoliation in spring will restrict ear development, maintain tiller formation and lead to a dense sward. On soils of low to medium fertility, application of fertilizer N will further encourage overall plant growth and tiller production.

The timing and frequency of defoliations will have a marked influence on the seasonal distribution of grass growth, as shown in Fig. 4.4. It is important to realize that there is always an uneven pattern of grass growth during the growing season, but this pattern is exaggerated when spring defoliations are infrequent.

Most varieties of PRG can tiller freely compared with less desirable 'weed' species of grass, so that regular frequent defoliation enables PRG to compete favourably for space in the sward and become dominant. Also most varieties of PRG are comparatively prostrate in growth habit compared with weed grasses, and so defoliation close to the ground will further encourage development of a PRG-dominant sward.

The practical importance of the basic physiology described above was demonstrated over 60 years ago in classic experiments by Jones (1933). He showed on a relatively fertile lowland site that, provided there was a least 15% PRG in the sward, an *Agrostis*-dominant sward could become PRG dominant in 2 years, if subjected to regular, tight grazing. Similarly, he showed that a newly sown PRG sward would rapidly degenerate in composition if subjected to irregular high defoliation.

Some other plants, such as white clover and creeping bent (*Agrostis stolonifera*), can propagate vegetatively by stolons: under some circumstances close defoliation will favour these plants as well.

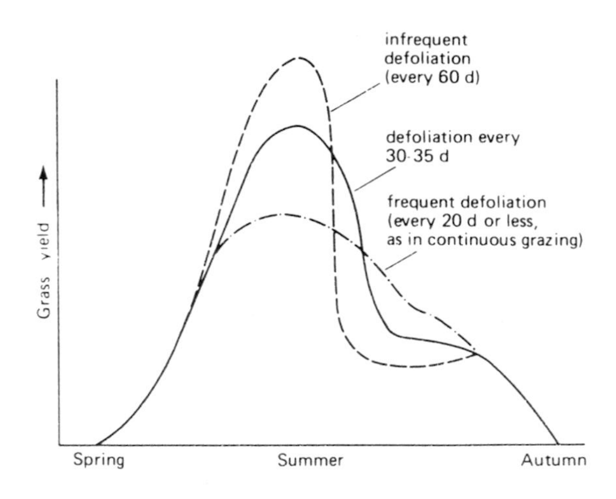

Figure 4.4 Effect of defoliation frequency on seasonal distribution.

Table 4.4 Effect of first cut date on regrowth (S23/Melle). (Based on Wolton 1980)

Date cut	*DM (t/ha)*	*Delay in regrowth (days)*	*Days to grazing stage (2.5 t/ha DM)*	*Total yield up to 20 July (t/ha)*
20 May	4.2	8	32	9.7
30 May	5.4	11	35	9.5
10 June	6.6	15	39	9.2
20 June	7.8	20	44	8.9

Why improve grassland?

It is important to analyse the factors limiting output from existing swards, before making an investment in reseeding or renovation. The production potential of most old grassland in Britain is high. Hopkins *et al.* (1990), in an experiment carried out on 16 sites in England and Wales, found that reseeding old swards with PRG led to a marked improvement in herbage production only in the first year following reseeding and when applications of N fertilizer were very high. The response directly following reseeding must be balanced against lost production in the year reseeding took place. Experiments at North Wyke with grazing beef cattle gave a similar response in terms of animal production (Tyson *et al.*, 1992). In both experiments, deficiencies in pH, phosphorous and potassium status were rectified with both series of swards, and good practice was followed for cutting and grazing management. Unless appropriate management can be imposed, the composition and productivity of the sward may fall rapidly after reseeding. Reseeding or renovation is likely, however, to be an attractive option when it results in:

- increased output by introduction of highly productive species, such as Italian or hybrid ryegrass;
- swards with improved ensiling characteristics, particularly through using grasses with high sugar content;
- increases in clover content and ability to produce well with low inputs of N fertilizer;
- improvement in soil conditions generally or in areas damaged by poaching or trafficking.

Methods of improvement

The farmer has three improvement strategies available:

(1) reseed – plough or chemically destroy the previous crop and reseed;
(2) renovate – introduce new species into the old sward, with or without partial chemical destruction;
(3) retair – improved defoliation technique, often with attention to drainage, pH and fertilizers.

Which improvement method?

	Reseed	*Renovate*	*Retain*
Speed of action	Very rapid	Quick	Slow
Cost	Expensive	Moderate	Very cheap
Stability	Can revert easily	Adjustable to acceptable level	Improvement based on farmer's ability
Special considerations	Other limitations must be corrected	Topography may limit	No real physical limits

Methods of grassland establishment

Grassland is established by sowing seeds either as part of a replacement/renovation procedure where grass follows grass or as part of a rotation involving other crops. Time of sowing will vary, but there are two 6-week periods in the year when sowing should take place. These are mid-March to April and August to mid-September. Sowing outside these periods is risky and can be unsuccessful due to summer drought or winter cold. The timing and method of sowing the seeds will be influenced by the purpose of the crop, soil texture and structure, climate and farm type. Where good white clover establishment is desired, then sowing should be either in spring or in early August, as sowing later will seriously limit the vigour of the clover plants and their ability to compete with grass.

Whatever method of sowing is used, it must be remembered that the grass seed has a relatively small food reserve and so must have an adequate supply of available nutrients near the surface of the soil. The biggest single demand is for phosphate, and except on high phosphate soils a fertilizer supplying 80 kg/ha of P_2O_5 should be applied at seeding time. Also grass seeds should be sown no deeper than 25 mm below the soil surface; where sowing is preceded by cultivations, care should be taken to ensure that the seedbed is level and fine. After sowing, grass establishment is enhanced greatly if the seedbed is well consolidated; the firmer the better, provided soil structure is not impaired or surface percolation rate reduced.

Undersowing

This can be a very successful technique provided a few simple rules are observed. The most important rule is to prevent the 'nurse' crop, usually spring cereals, from overshadowing the establishing grasses and thus depleting the density of the sward before the nurse crop is removed. Where cereals are undersown, an early-maturing, stiff-strawed variety should be used. Whilst farmers successfully undersow wheat, barley and oats, and the choice often depends on rotational considerations, barley is the most suitable, having short straw and an early harvest. Seed rate should not exceed 125 kg/ha, and, to avoid lodging, fertilizer N rate should be about two-thirds that which would have been used had the crop not been undersown. The seeds mixture may be broadcast or drilled, with drilling always being preferred if conditions are likely to be dry. The grasses may be sown immediately after the spring cereals are sown and the ground then harrowed and rolled; alternatively, some farmers prefer to wait until the cereal is established before sowing the grass seeds. If the cereals are sown in February, it is a good idea to delay sowing the grasses until March. When the cereals are taken through to harvest, the straw must be removed as soon as possible and often the grasses will benefit from a dressing of complete fertilizer, say 400 kg/ha of a 15 : 15 : 15 compound, followed by grazing 3–4 weeks later to encourage tillering.

Some farmers do not take the cereal to harvest, but cut the crop for silage when the grain is still soft; others grow cereal/forage pea mixtures for use in this manner. Removal of the crop at this earlier stage benefits grass establishment and is good practice, provided there is a use for the arable cereal silage that is made.

Whenever crops are undersown, it must be remembered that herbicide use in the cover crop may be restricted, particularly if the seeds mixture contains clover.

Seeding without a cover crop

This is now the most common method of establishment. Here the seed mixture is sown as a crop in its own right, the nature and timing of operations being adjusted to the needs of the grass crop.

Establishment must be rapid both to minimize weed competition and to reduce the period that land is unproductive. Thus adequate fertilizer must be applied to the seedbed: 80 kg/ha of both P_2O_5 and K_2O and 60–80 kg/ha of N if no clover is sown. If clover is in the mixture, it is advisable to use no fertilizer N in the seedbed otherwise clover establishment will be seriously impaired. An early grazing will benefit the establishment by encouraging tillering, but a September sowing should not be grazed before winter as it will drastically reduce the sward's ability to build up root reserves during the winter.

Cocksfoot and meadow fescue establish more slowly than ryegrass, with tall fescue the slowest of all. Seeding without a cover crop is best suited to ryegrass mixtures.

Direct drilling

Grass and clover seeds can be direct-drilled provided:

- the drilling date is within the periods given in the preceding section;
- the previous crop has been thoroughly killed off using a herbicide such as paraquat (if annual weeds are present) or glyphosate (if perennial weeds exist). If there was a dense mat of previous vegetation and paraquat is used, then two half-rate sprays at 3- to 4-week intervals should be made. After spraying with either chemical, at least 2 weeks should elapse before sowing;
- the seeds are placed 10–20 mm deep in mineral soil;
- slug control measures are taken;
- for autumn sowings, an insecticide is used to protect against frit fly.

The popularity of this technique has, however, declined as it has proved difficult to achieve even and reliable establishment, because of problems in achieving an even depth of sowing in the mineral soil and adverse effects from toxic compounds released from the decaying sward.

Slit-seeding

This technique is used for sward renovation, the principle being that at least some of the original sward is allowed to survive. Three basic rules are:

(1) Ensure aggressive weeds in the old sward are killed (particularly stoloniferous species), otherwise they will 'strangle' the newly emerging young plants.
(2) Use the narrowest width of drill feasible and cross-drill if possible, to get the most rapid knitting together of the sown species.
(3) Sow at a time of year to give seedlings ample time to grow.

If the old sward contains a reasonable proportion of perennial ryegrass but is beginning to fill with annual meadow grass (*Poa annua*), then slit-seeding may be the quickest and cheapest way of renovating the sward and boosting production. Also some farmers use this technique to introduce Italian ryegrass into old swards in an attempt to extend the grazing season and increase production. Furthermore, this method can be used to introduce white clover into an existing grass sward. Where there are no real problems from either grassy or broad-leaved weeds, then no herbicide treatment is needed. Otherwise, paraquat or glyphosate will have to be band-sprayed ahead of the slit-seeding of the grass.

There are several types of slit-seeder available, but the basic technique is to drill rows of grass seed into an existing sward with no previous cultivation. Some machines can apply a band of herbicide ahead of the drill; some cultivate the narrow band of the drill before the seed is sown; some can apply fertilizer and slug bait at the time of sowing.

The fertilizer rates and sowing periods are similar to those described in the preceding sections. In some cases, slit-seeding may be used on very old swards and on fields not suited to normal cultivations. It is essential to check soil pH and also ensure that bad drainage will not cancel out any good that may arise from sowing new grasses.

Scatter and tread

On an opportunist basis it is possible to renovate grassland by simply scattering the seeds on the soil surface under moist conditions. This is likely to be most successful if the sward has recently been cut or tightly grazed and has a low tiller number. Seeding should be followed immediately by turning in some grazing stock (preferably sheep) for 2–3 days to tread in the seeds. This technique is a gamble as its success depends on a thorough, quick tread and then some 2 weeks of continuing moist, but not wet, weather to allow the seeds to establish. The technique is cheap, and can be used to patch up worn areas in grazed paddocks. Also, some farmers report successful grass establishment following mixing of grass seeds in a slurry tanker and application with the slurry.

Steps to increase the white clover content of existing swards

Many swards contain little clover, and the following procedure, developed by research workers at the Institute of Grassland and Environmental Research (IGER), North Wyke, has been found to be an inexpensive way of increasing white clover content on suitable soils.

(1) Check soil suitability, particularly drainage and compaction; have soil analysed.
(2) Take silage (or hay) cut in June/July, having used a moderate N rate of up to 80 kg/ha.
(3) Unless sward is very open, spray paraquat at 1.5 litres/ha to kill 'bottom' grasses.
(4) If little clover is present, oversow or slot-seed white clover seed at 2 kg/ha; also apply slug pellets and Dursban as a precaution against frit-fly damage.
(5) If PK indices are under 3, apply 60 kg/ha of nutrient(s) in short supply. If pH is below 6, apply lime.
(6) Rotationally graze with cattle at approximately monthly intervals (if no cattle, use sheep). *Do not:*
    - use any fertilizer N;
    - apply any herbicides (including any 'clover-safe' ones);
    - stock continuously.
(7) Next spring *apply no fertilizer N*. Ideally, take a low-yielding silage (or hay) cut. If this cannot be done, continue management (6).
(8) By August of year 2, the clover should occupy at least 30% of the surface area when viewed from above. If below 30%, check out what is wrong. If 30% or over, start to manage it as a *grass/clover sward*.

Characteristics of agricultural grasses

In this section, the important characteristics of the commonly used grass species are reviewed, as a precursor to information on the formulation of seeds mixtures. An essential publication is the NIAB *Grasses and Herbage Legumes Variety Leaflet*, which is revised annually.

Perennial ryegrass (PRG)

Perennial ryegrass is very successful under British conditions; not only do PRG varieties account for well over half the grass seeds sown, but also the best permanent pastures have an appreciable PRG content. Provided they are defoliated regularly and are grown under conditions of medium-to-high fertility, they will give high yields of digestible herbage over a long season. PRG has the following valuable characteristics:

- tillers well and so forms dense swards under good management;
- fairly prostrate and good for grazing (but must be protected from prolonged overshadowing by taller species on cut swards);
- recovers well from defoliation, because of both high tiller numbers and good rate of replacement of root reserves;
- persists well and is suited to long-term grassland;
- good feeding value, with high digestibility and water-soluble carbohydrate (WSC) content.

PRG is classified into three main groups based on date of ear emergence in spring. The following heading dates are for central-southern England:

Very early and early group	Up to 20 May
Intermediate group	21 May–4 June
Late and very late group	5 June onwards

These heading dates are approximately 12 days later in northern Scotland: if a ruler is placed from Swindon to Aberdeen, then each twelfth of the distance represents 1 day later in heading.

In general, when early heading varieties of PRG are compared with late heading varieties, the early heading varieties:

- grow earlier in spring, but not as much earlier as the difference in heading date: for example, a 3-week earlier heading date would give 7–10 days' earlier spring growth;
- are more erect;
- tiller less freely;
- are easier to cut for conservation, because of both erect growth and good stem elongation at heading;
- do not grow well in mid-season and may be of lower quality because of continued heading.

Compared with early heading varieties, late heading varieties:

- are more prostrate;
- tiller well and are more persistent;
- give good mid-season growth (June–August);
- are of higher quality in mid-season.

PRG has a limited temperature tolerance compared with other grasses, being unable to withstand very hot summers where temperatures exceed 35°C (not found in the UK but can occur in some temperate continental areas) or very cold winters where temperatures persist below about −10°C. Lack of winter hardiness may be a problem with PRG in the north of Scotland, but varieties now available have markedly better winter hardiness than those used 30 years ago. Winter survival in PRG is enhanced if little or no autumn nitrogen is applied and the plants are defoliated to below 100 mm before the onset of winter.

Both diploid and tetraploid varieties of PRG are available. Compared with diploid varieties of the *same heading date*, tetraploid varieties:

- appear higher yielding and may have a greater yield above, say, 80 mm, but total DM yields are generally the same;
- have a higher WSC content;
- are more erect, with leaf of higher moisture content;
- tiller less freely and are less persistent;
- are more winter hardy and more drought resistant;
- have larger seeds and so require a higher seed rate.

Recent research by IGER indicates considerable differences between varieties of PRG in animal performance potential. Varieties with high contents of WSC have given increased milk yield by dairy cows, associated with increased intake and improved protein utilization (Moorby, 2001). Substantial and repeatable differences between varieties of intermediate PRG have been found for herbage growth and animal intake in swards grazed by sheep (Orr *et al.*, 2000). Reasons for these differences are under investigation.

Italian ryegrass (IRG)

These are grasses with an expected duration of 2–3 years and heading dates in the period 10–30 May (in southern England). They are very erect, and although capable of tillering they rarely tiller enough to prevent a steady decline in sward density from establishment onwards. IRG will grow for a long season, starting growth in spring before PRG and continuing well into the autumn under conditions of high fertility. However, the bulk of growth from IRG is in the April–June period, making it particularly useful for early spring grazing and heavy conservation cut. During the summer IRG tends to run into seedhead every 35–40 days and this can lower its feeding value at this time. Most varieties of IRG are less winter hardy than PRG. The spring heading date of IRG coincides with the early group of PRG. IRG will generally outyield PRG in the first harvest yield. IRG is especially useful as a rotation grass, where its 2- to 3-year duration is not a disadvantage. *Tetraploid* varieties are important, as tetraploidy emphasizes the natural advantages of IRG in terms of erectness, leaf size and WSC content. They tend though to be less persistent and to have rather lower yields than diploid varieties in the second year.

Hybrid ryegrass (HRG)

During recent years plant breeders have combined the vigorous pattern of growth found in PRG. A range of hybrids is now on the market, nearly all of which are tetraploid. There is considerable variation in persistence and in resemblance to either the IRG or PRG type. During the growing season hybrids generally behave in a similar way to IRG in early season, but then tend to switch to PRG traits, to produce high tillering leafy regrowth (Hides and Humphreys, 2000).

Progress is being made in breeding varieties that incorporate genetic material from both ryegrasses and fescues; these have the potential of having greater stress resistance than the ryegrasses, particularly in relation to drought and cold.

Westerwolds ryegrass

These are annuals. When sown in the spring or summer they flower in the year of sowing and do not persist over winter. They can be sown in autumn in mild areas and can provide very early spring growth and reduce the extent of nitrate leaching through taking up soil N during this period. Their main value is where high yields of grass are required for a short period, and they are best regarded as catch crops, otherwise IRG should be used, as it will give equivalent yields and longer duration for a similar or lower seed cost.

Timothy

Timothy is a persistent and winter-hardy grass that will grow well in cooler, wetter parts of Britain. Also, it has survived well in drought years in other areas. Varieties of timothy head 1–25 June – as late as and later than the latest heading of the PRG varieties. Timothy grows well in summer and autumn, but when grown alone it does not give a dense sward or very high yields. It is a good companion grass in seed mixtures and historically was included with meadow fescue and clover. Today it is used with PRG, particularly in Scotland where its winter hardiness is a valuable characteristic. Timothy seed is some three times lighter than ryegrass seed, so its weight in a mixture is an underestimate of its real importance.

Cocksfoot

Cocksfoot is a native grass that is very responsive to nitrogen and, with a very well-developed rooting system, is useful on light soils in low rainfall areas, where it shows drought resistance. It grows and regrows rapidly, is erect in habit and has relatively broad leaves for a grass; the leaves are also coarse and hairy. Cocksfoot heads in early to mid May and has an aggressive growth habit that makes it competitive in swards that are not closely defoliated during the main growing season. Over recent years cocksfoot has declined in popularity as it was found that its digestibility was below that of PRG and IRG.

Meadow fescue

This is a very adaptable grass that tolerates a wider range of climatic conditions than ryegrass. It heads mid to late May. It does not tiller as freely as PRG and so forms an open sward that is more prone to weed invasion, particularly as it is not aggressive. Meadow fescue makes a good companion grass in mixtures, for example with timothy and/or cocksfoot; also, it remains leafy during summer under conditions of fairly low fertility, unlike PRG which demands higher fertility if it is to remain leafy. As fertilizer use on grass has increased over the least 30 years, the importance of PRG has risen and the popularity of meadow fescue has declined, except where the winters are too severe for PRG – for example in Scandinavia. Historically, meadow fescue and timothy were the favoured grass companions to white cover. Meadow fescue could become more important again if there is a resurgence in the use of white clover or less aggressive legumes, such as birdsfoot trefoil.

Tall fescue

This is one of the earliest grasses to grow in spring. It is winter hardy, drought resistant and more persistent than PRG under regular cutting. It is very slow to establish, taking up to one season to become fully productive. Also it has a very rough, stiff leaf that is unpalatable to grazing stock, although high intakes may be achieved when the species is ensiled or dried. Tall fescue is used widely in France, but in the UK its main role is on grass-drying farms in eastern England, although a few farmers on light land have used tall fescue in areas they cut regularly for silage.

Herbage legumes

Legumes have root nodules that can be inhabited by rhizobial bacteria which can fix atmospheric nitrogen (N) and make it available to the host plant. Herbage legumes are grown either on their own ('straight') or mixed with grasses, where their ability to acquire fixed N can also benefit associated grasses.

This chapter provides information on white and red clover, alsike and the trefoils, whilst lucerne and sainfoin are described in Chapter 3.

White clover (WC)

White clover is scarcely ever grown straight in the UK: even white clover seed is produced from suitable defoliation management on a grass–clover sward. Unlike red clover, white clover is persistent under grazing, but is also commonly sown in swards used for both grazing and cutting. White clovers are classified according to their leaf size:

- *Small leaved*. These are very prostrate and have been selected from cultivars existing in old pastures (e.g. Kent Wild White). They are very hardy, but always remain small even when under favourable growing conditions; thus they do not stand up to grass competition when fertilizer N is used or when a conservation cut is taken. Their real value is on sheep-grazed swards, conditions of low fertility and at high altitudes.
- *Medium leaved*. Leaf size is not a consistent characteristic in the field: if a medium–large clover is used in a sward that is regularly and tightly grazed, its leaf size will be quite small. Also, there is a general tendency for persistence to decrease as leaf size increases. There has though been much recent progress in white clover breeding and medium-leaved varieties are now available which can be used very flexibly, with tolerance to cutting and grazing and have good persistence even when substantial quantities of N fertilizer are applied (Rhodes, 2001).
- *Large leaved*. These varieties do not persist when subjected to severe grazing by sheep, but may make a major contribution in swards grown with some N fertilizer for silage or for rotational grazing by cattle.

It is important to select the most appropriate type of white clover for the expected management conditions; unless new varieties with high flexibility are used, it is often advisable to sow a range of types in the mixture.

Red clover (RC)

Red clover was a crucial component in the Norfolk four-course rotation which was widely practised in earlier centuries. Its contribution was two-fold: to supply feed for ruminants and, because of N fixation, to increase fertility and the productivity of subsequent crops. Use of red clover fell dramatically during the twentieth century with the ready availability of N fertilizer. However, there has been a recent increase in interest in the species. Improved ensiling techniques have been developed and it has been discovered that the protein in red clover has some protection from breakdown in the silo, leading to improved animal performance. The development of organic farming has also recreated the need for legumes that give high yields over a 2- to 3-year period and increase subsequent crop yields.

There are two main types of red clover:

(1) *Early red* (also double-cut or broad red). These give early growth and high yields over two cuts and the possibility of autumn grazing. The varieties are not very persistent, particularly when grazed regularly, and are best grown for 2 years only, either straight or in mixtures with erect high-yielding grasses such as IRG or HRG for cutting.
(2) *Late red*. These flower 2 or 3 weeks later than early red and are more persistent, being suited to medium-term leys where a red clover constituent is required. Such leys should be cut periodically as red clover does not persist well if regularly defoliated more frequently than every 35 days.

NIAB publish a descriptive list of red clover varieties with both diploids and tetraploids in both groups. Tetraploid red clovers do not necessarily give a higher yield than diploid counterparts, and choice of variety can depend on selection for disease resistance. Red clover can suffer severely from two diseases:

- clover rot (*Sclerotinia trifoliorum*), and
- clover stem eelworm (*Ditylenchus dipsaci*).

Both diseases can be problems in areas where red clover has been commonly grown, e.g. in the arable areas of East Anglia. Where either disease is present, resistant varieties must be grown. Tetraploids give low seed yields and have larger seeds than diploids, so tetraploid seed is expensive.

Alsike

This is rather more persistent than red clover under grazing and is also more tolerant of wet acid conditions, more winter hardy and more disease tolerant. It may be sown with late red clover in mixtures, but is now seldom used in Britain.

Trefoil

There are many species in the genus *Lotus*, with two of agricultural significance. These are birdsfoot trefoil

(*Lotus corniculatus*) and marsh trefoil (*Lotus uliginosus*). The species are not currently widely sown in Britain, but may be of increased significance in the future because of important agronomic and nutritional properties.

Both species will thrive in stress conditions with low pH and low soil fertility. Birdsfoot trefoil will withstand drought and severe winter conditions, whilst marsh trefoil, as the name suggests, is more suited to wet conditions. In relation to feeding value, trefoils contain condensed tannins. After ingestion by ruminants, the condensed tannins will react with protein and reduce the rate of protein breakdown in the rumen. This appears to prevent the occurrence of bloat and may improve the overall efficiency of protein utilization. Tannin levels are higher in marsh trefoil than in birdsfoot trefoil, but there is some risk of adverse effects if the content of tannin is very high. However, in recent experiments in Britain and in Germany, levels of intake and animal performance have been higher with birdsfoot trefoil than with other legumes (Fraser *et al.*, 2000; Paul *et al.*, 2002).

The species though have agronomic weaknesses. They are particularly slow to establish and do not persistent well with frequent cutting or grazing, as discussed by Frame *et al.* (1998). Further work is required to clarify the contribution that these species can make in Britain, but movement to more extensive systems and to systems with a high premium on nutritional quality will increase their use.

Herbage seed production

The successful production of herbage seeds is very specialized and the majority of herbage seeds produced in the UK are grown on large arable farms, on land that is relatively free from 'weed' grass species and with adequate seed harvesting and handling equipment. Grass and clover seeds are small, and yields low (see Table 4.5), so that attention to detail is important. Also, the management of the crop prior to harvest should be adjusted so that the number and size of fertile seed heads is maximized and the presence of lush, leafy growth is minimized.

Depending on the species and variety being grown, the date of harvest can vary from the end of June (for early cocksfoot) to mid-September (for late timothy and some white clover crops). Thus the defoliation management in spring and early summer will depend on the date of expected harvest, the principal objective being to minimize leaf and maximize stem elongation and seed heads in the growth leading up to the cut.

Most seed crops are established specifically in the year prior to harvest. A firm, fine seedbed is essential and seeds are usually drilled. Weed control is very important and it is essential that no grasses develop other than the variety being grown. As there is a premium on a clean, vigorously growing crop, regular use of herbicides and application of high rates of seedbed P and K are advised. Also, spring-applied N rates of 100–150 kg/ha are often used, to encourage tillering and leaf growth, prior to defoliation and subsequent development of the seed-bearing stems.

Provided they remain free from other species and the yield potential is still high, seed crops may be harvested for at least 2 years (except IRG, HRG and red clover). In the years of seed production, there will be some grazing available from the field, and the threshed straw has a feed value equivalent to poor hay. Once seed pro-

Table 4.5 Some basic facts on herbage seed production

Species	*Type*	*Harvest period*	*Probable yield range (kg/ha)*
IRG		July–Aug	800–2000
HRG		July–Aug	800–2000
PRG	Early	Early July	750–1000
	Intermediate	Late July	800–1250
	Late	Up to mid Aug	600–1000
Timothy	Early	Mid Aug	375–650
	Late	Late Aug–Mid Sept	375–650
Cocksfoot	Early	Late June–Mid July	450–1000
	Late	Mid-July	450–1000
White clover	Large	Late July–Mid Sept	350–450
	Small	Late July–Mid Sept	150–350
Red clover	Early	Aug	400–500
	Late	Sept	350–500

duction has ended, there is the possibility that the sward can continue as a normal grassland field.

It must be emphasized that herbage seed production is a very specialized business. Existing growers have built up their expertise for specific varieties and usually continue to grow these as long as there is a market for them. Normally, selection of new varieties by existing seed producers and the introduction of new seed growers are done through seed houses and merchants, who have the necessary seed stock, the market outlets and the necessary inspection procedures.

Seeds mixtures

In contrast to arable crops, grassland is usually sown with a mixture of different varieties and species. Mixtures may give higher yields and more stable performance than pure stands and show greater tolerance to differing levels of inputs and stresses caused by pests, diseases and drought, as discussed by Turner (1997). However, the composition of the sward may be difficult to predict and the desirable characteristics of the best varieties will be diluted when they are used in mixtures. There are very many seeds mixtures available to farmers in the UK. Golden rules are to keep the mixture as simple as possible commensurate with its function and to use only recommended varieties.

The decision process in mixture construction given below is based on Turner (1997):

- Identify farm system – type of animal; site class; level of management;
- Identify ley required – cutting, grazing or dual purpose; short term, long term or permanent;
- Identify special needs – e.g. early spring production; silage quality; drought tolerance; disease resistance; winter kill; organic production;
- Select species and varieties – key characteristics have already been outlined;
- Determine best blend – ratio of diploids to tetraploids; balance between maturity groups (better to have similar maturity groups in different components of mixture for cutting, but a range of different maturity groups for mixtures used for grazing) and species;
- Check seed availability and cost.

Examples of mixtures for the three predominant methods of sward utilization are given in Table 4.6.

Fertilizers for grassland

Nitrogen

In grassland, the supply of N is so important that it has the effect of controlling yield potential and thus the requirements for other nutrients. The principles for efficient use of fertilizers and manures are outlined in Chapter 1. This section discusses features specific to grassland, highlighting important contrasts to arable crops:

- Much of grassland is grazed, with the grazing animal returning to the sward some 80–95% of the quantity of N it consumes.
- Grassland soils typically have high contents of organic matter, resulting in substantial supply of N through mineralization.

Table 4.6 Examples of species and variety types in mixtures for swards of differing intended duration and method of utilization (seed rates in kg/ka). Several varieties may be used in a species group, but no variety should be at less than 4 kg/ha for ryegrass, 2 kg/ha for timothy and 1 kg/ha for white clover. The shorter-term mixtures would probably include about 50% of tetraploid varieties and the long-term mixture about 25%

	Short-term cutting	*Medium-term dual purpose*	*Long-term grazing*
Duration (years)	2	5	10
IRG	20	—	—
Hybrid ryegrass	20	8	—
Early PRG	—	8	—
Intermediate PRG	—	8	10
Late PRG	—	8	10
Timothy	—	—	4
White clover	—	3	3

- Legumes in grassland may make major contributions to N supply through fixation and subsequent recycling.
- Swards are normally perennial and harvested several times during the year, with the previous treatments influencing current requirements and responsiveness to additional nutrients.
- Much grassland is farmed relatively extensively: in many situations the level of production required to support the stocking rate may be below the potential for the site.
- There is potential for high environmental losses of N from intensive grassland and its associated ruminant animals through nitrate leaching, ammonia volatilization and denitrification.

The grass plant is extraordinarily efficient in taking up N from the soil, provided there is sufficient moisture present and soil temperature is above the threshold for growth. However, the low N-use efficiency of the animal and the poor distribution pattern of returned N, particularly at grazing, means that there are generally large quantities of N cycling in grassland systems and care must be taken to reduce the risks of losses to the environment. However, increased knowledge of the various processes involved means that fertilizers and manures can now be used more precisely in grassland systems. This will result both in more economic use of applied nutrients and in reduced leakage of excess nutrients to the environment. Much of this recent information is incorporated in the fertilizer recommendations made by MAFF (2000). The key steps in determining the appropriate fertilizer N application, based on MAFF (2000), are listed below and then discussed further.

- Identify the soil nitrogen status (SNS) of the field.
- Identify the grass growth potential.
- Decide on the intended grass management and the required level of production.
- Determine the nutrients that will be applied in slurry and manures.
- Use tables to indicate the quantity of N fertilizer required for your situation.
- Adjust fertilizer N use according to weather and actual grass growth performance.

Soil nitrogen status

The SNS cannot yet be accurately predicted from simple analyses, but can be estimated from knowledge of previous grass management and N use, as indicated in Table 4.7.

Grass growth potential

This defines the maximum requirement of the crop for N (from all sources) and is related to summer rainfall and soil characteristics which determine the quantity of soil-available water. Table 4.8 is taken from MAFF (2000), but also indicates the analogous site classes as

Table 4.7 Soil nitrogen supply (SNS) status in grassland systems according to previous management (MAFF, 2000). Note: increase the SNS status by one class if more than 150 kg/ha of total N has been applied regularly as organic manure for several years. Reduce the SNS status by one class if grass was cut for silage and less than 150 kg/ha of total N as organic manure has been applied on average in previous years

SNS	*Previous grass management*	*Previous N use (kg/ha)*[1]
High	Long-term grass, high input includes: • Grass reseeded after grass or after 1 year arable • Grass ley in second or later year	Over 250
Moderate	First year ley after 2 or more years arable with last crop potatoes, oilseed rape, peas or beans, but not on light sandy soil	All
	Long-term grass, moderate input includes: • Grass reseeded after grass or after 1 year arable • Grass ley in second or later year	100–250 or substantial clover content
Low	First year ley after 2 or more years arable with last crop cereal, sugar beet, linseed or any crop on a light sandy soil	All
	Long-term grass, low input includes • Grass reseeded after grass or after 1 year arable • Grass ley in second or later year	Up to 100

[1] Typical fertilizer and available manure N used per year in the last 2–3 years.

Table 4.8 Grass growth classes (very good to very poor) (MAFF, 2000) and site classes (1 to 5) (based on Thomas *et al.*, 1991). Note: for altitudes above 300 m, reduce the growth class by one (increase site class by one)

Soil-available water	*Soil types*	*Summer rainfall*		
		Up to 300 mm	*300–400 mm*	*Over 400 mm*
Low	Light sandy soils and shallow soils (not over chalk)	Very poor/5	Poor/4	Average/3
Medium	Medium soils, deep clay soils and shallow soils over chalk	Poor/4	Average/3	Good/2
High	Deep silty soils, peaty soils and soils with groundwater	Average/3	Good/2–3	Very good/1–2

used by Thomas *et al.* (1991). Both of these relationships derive from multi-site experiments reported by Morrison *et al.* (1980).

Intended management

The value of grass to the farmer will be influenced by the efficiency with which the grass is utilized in the system and the value of the animal products. If grass is produced in quantities greater than the requirements of the animals carried on the farm, there will clearly be inefficient utilization; fertilizer strategy must be designed to avoid this situation. Assuming that utilization efficiency is high, rather higher application rates may be justified if the grass is used for milk production than meat production, because of the higher product prices.

Slurry and manures

Grassland systems normally involve production of considerable quantities of slurries and manures and these may make a major contribution to nutrient supply. Factors determining the efficiency of use of nutrients from these sources are discussed in Chapter 1, in MAFF (2000) and by Pain (2000). It is important to reduce the losses of nutrients from slurries and manures during storage and application and to check on their contents of nutrients. Equipment is now available for on-farm use which provides good assessment of the dry matter content of slurry and the content of readily available N (Chambers *et al.*, undated).

Quantity of N fertilizer required

Tables are given in MAFF (2000) to indicate the maximum N rate required for fields with contrasting SNS at high stocking densities for grazing by dairy cows (reproduced as Table 4.9) or by beef/sheep, for utilization as silage or hay and at grass establishment. It is important to note that nutrients applied as organic manures must be deducted from these figures. Research is in progress to produce a computerized decision support system enabling N application to be optimized for particular fields in relation to agricultural and environmental objectives. This system is not yet generally available.

Table 4.9 Maximum N rates for high stocking rates with grazing by dairy cows on a 28-day cycle (MAFF, 2000). Note: deduct available N applied as organic manures to calculate quantity of fertilizer N required

	SNS status (kg/ha)		
	Low	*Moderate*	*High*
Late February–early March	60[1]	60[1]	60[1]
April	75	60	60
May	75	60	50
June	60	60	50
July	60[2]	60[2]	40[2]
August	50[2]	40[2]	40[2]
Total	380[3]	340[3]	300[3]

[1] In mild areas or where early grazing is possible, N may be applied from early–mid February. In upland areas apply N from mid–late March.
[2] In dry summers, if growth has been severely restricted by drought, reduce or omit this application.
[3] These rates apply to average, poor or very poor grass growth classes. Up to an extra 40 kg/ha may be justified with good growth class and 80 kg/ha for very good growth class.

Adjustment during season

Fertilizer application should be reduced if grass yields (and N uptake) have been reduced by drought. Also, fertilizer application should be reduced if the quantity of grass grown has been above that required to satisfy the animal requirements for conserved forage and for grazing. A meter has been developed for the rapid in-field determination of mineral N in the soil under grass (Titchen & Scholefield, 1992). This can be used to adjust fertilizer N applications in mid and late season.

Clover nitrogen

Rhizobial bacteria in clover root nodules will fix atmospheric N and this N is used by the clover plant. As the clover plants grow, old roots die and some fixed N in these roots becomes available to grass roots. Also clover foliage contains fixed N and if the clover is grazed, then some of this N is returned to the soil in dung and urine.

The value of clover N depends on the quantity of clover in the sward. Where there is a very large amount of clover, likely to average 30% or more of the weight of herbage grown in the year, then clover N can be equivalent to 150 kg/ha of fertilizer N (and even 200 kg/ha under some favourable conditions). Figure 4.5 shows the differing seasonal growth patterns of grass and white clover. The best time to assess clover content is in July/August, when clover content must exceed 60% if the sward is to average 30% over the year. Scoring clover content in summer can give a guide to the value of clover in the sward. If 60% clover, assume 150 kg/ha N, if 30% clover assume 75 kg/ha, and so on.

Even with skilful management the application of fertilizer N will decrease the clover content, so that fertilizer N and clover N are not additive. As a rule of thumb, application of every 2 kg/ha of fertilizer N will decrease clover N by 1 kg/ha. This means that if 300 kg/ha of fertilizer N is applied over the season to a grass–clover sward, then clover N contribution will be nil (and pro rata for lower fertilizer N rates).

There is evidence though of very efficient cycling of N in grass–clover swards continuously stocked with sheep. Stocking rates, without fertilizer N, approached those for grass swards with 200 kg/ha of fertilizer, despite the mixed swards containing only approx. 10% of white clover (Orr *et al.*, 1990; Parsons *et al.*, 1991).

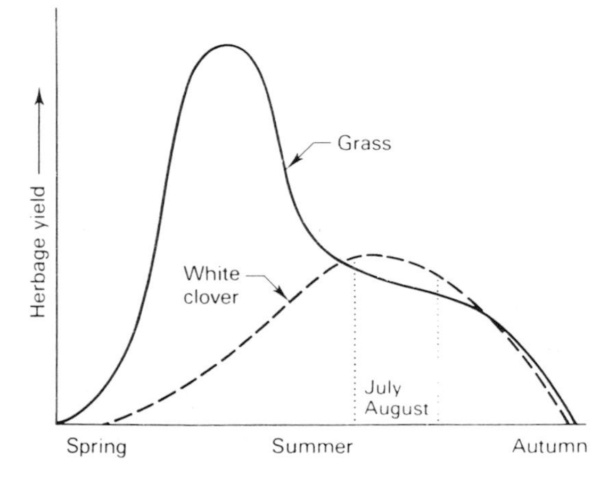

Figure 4.5 Relative seasonal production of grass and white clover in grass–clover swards.

Phosphate and potash

The soil can supply some or all of the phosphate and potash needed by grass, fertilizer being required to bridge the gap between supply and demand. As a guide, total phosphate demand is 60–80 kg/ha of P_2O_5 and that for potash is 150–300 kg/ha of K_2O, depending on the level of grass production (and hence available N supply). For both phosphate and potash, soil analysis can be a useful guide to the quantity of fertilizer needed (see Table 4.10), whilst herbage analysis can also provide a good indication of potash requirements.

Grazing animals return a substantial proportion of both the phosphate and potash eaten, virtually all the phosphate being in dung and most of the potash in urine. Dung is very unevenly distributed over the field and broken down very slowly to release the phosphate;

Table 4.10 Phosphate and potash recommendations for grassland

Annual stocking rate (LSU/ha)	*Amount of nutrient to apply (kg/ha)*								
	P_2O_5 soil index				*K_2O soil index*				
	0	*1*	*2*	*3*[1]	*Utilization*	*0*	*1*	*2*	*3*
<1.7	60	40	20	0[1]	Cut	100	80	40	0
					Grazed	80	40	20	0
>1.7	80	60	30	0[1]	Cut	200	120	90	45
					Grazed	80	60	30	0

[1] Some may be needed every 3–4 years to maintain level.

as a result, animal recirculation of phosphate is not assumed. Urine covers a much greater area of the sward than dung, and all the potash in urine is in a readily available state: thus allowance is made for potash recirculation on grazed swards.

Plant uptake of phosphate is steady and is well related to plant needs; it follows that the annual fertilizer requirement for phosphate can be applied in one annual application if desired, without causing nutrient imbalance. However, uptake of potash by the plant is related to the quantity of potash available to the roots and *not* to the plant's needs; thus grass can take up in a single growth period as much as twice the potash it needs for adequate growth. This excessive uptake is called 'luxury uptake' and can have two harmful consequences:

(1) Mg content is inversely related to K content in grass, so the unnecessarily high level of K leads to very low Mg content and a greater risk of hypomagnesaemia in stock eating the herbage, particularly in spring and autumn when herbage Mg contents in grass are naturally low.
(2) Growth following defoliation of this herbage may be restricted by lack of available potash.

Where high annual rates of potash are needed, some potash should be applied for each growth period, except none should be applied in spring to swards that will be grazed unless the soil K index is zero.

Sulphur

Application of fertilizer N to grass increases the synthesis of S-containing proteins, so the S requirement of grass is related to N use. At low N levels S requirements are modest, but high N grass needs the relatively large amount of about 40 kg/ha of S over a growing season. Few fertilizers contain much S and in recent years most available S has come from atmospheric pollution. As the pollution is reduced, more S is needed from other sources to meet the requirements of high N grass, as discussed by MAFF (2000).

Table 4.11 gives results from 23 sites of an ADAS experiment studying grass grown for silage. The results show two important aspects.

(1) S response interacted with N, with the largest yield response to added S at the highest N rate.
(2) Major S responses occur after utilization of the first growth in the season, as S deposition over winter is often adequate to support spring growth.

Table 4.11 Grass response to sulphur under regular cutting (t DM/ha)

	N rate (kg/ha)			
	0	*200*	*400*	*600*
Nil sulphur	5.0	10.3	12.2	12.5
50 kg S/ha	5.2	10.6	13.2	13.6
	% increase in DM yield from each cut due to S (mean of N rates 400 and 600 only)			
	Cut 1	0		
	Cut 2	10		
	Cut 3	20		
	Cut 4	18		

Recent research has shown that responses to S are greater on sandy soils and that rectification of S deficiency will improve N utilization and reduce N losses to the environment (Brown *et al.*, 2000).

Calcium

Calcium is applied in lime, and maintenance of a pH of 5.8 or above will ensure that acidity does not limit the growth of grasses and clovers *and* that adequate calcium is present in the soil.

Magnesium

Magnesium deficiency can occur in grazing animals as hypomagnesaemia. This is a problem that cannot be cured reliably by applying magnesium to the soil; the best prevention of the disease is to place magnesium compounds in the animals' food or water. However, where soil magnesium is low, it is sound practice to use a magnesium-containing form of lime when lime is applied.

Other nutrients

Grassland in the UK suffers rarely if ever from other nutrient deficiencies, although occasionally grazing stock may suffer from lack of minerals such as sodium or copper. The safest rule is to supply the deficient element directly to the animal rather than through the herbage: this ensures not only that the animals concerned receive a correct dosage but also that areas of grassland with adequate levels of minerals are not enhanced to reach toxic levels.

Patterns of grassland production

The basic pattern of grassland production is shown in Fig. 4.6 and shows that grass can be available for grazing over a long period of the year, with conservation as hay or silage removing surplus grass and making this grass available for winter feeding. As conserved grass has a lower feed-value than fresh grass, other feeds have to be used with hay and silage, normally a cereal-based concentrate.

Generally over 50% of total annual grass production has occurred by the end of May and some management practices can emphasize the spring peak even more than this. For example, heavy use of N during spring with much lower rates later in the year can result in 70% of annual production by the end of May; where one very heavy conservation cut is taken in early June, this cut can account for as much as 85% of annual production.

Spring peak in production will be emphasized by:

- very high N use in April–May, particularly as this can limit July–September production below that which would otherwise be obtained;
- taking one very heavy cut in spring, particularly as this will deplete tiller numbers and subsequent regrowth;
- use of a single grass variety, particularly an early-heading one;
- low summer rainfall reducing production from June onwards.

Spring peak will be minimized by:

- restricting N use in April–May, although this will give some reduction in annual grass yield, as grass is most responsive to N during this period;
- very frequent defoliation by cutting or grazing in the May–June period (Fig. 4.4);
- using a mixture of grasses with different spring growth patterns.

Effect of site class

Figure 4.7 shows the effect of site class (see Table 4.8) on the seasonal growth pattern of grass receiving optimum N supply throughout the season, and indicates that there are two parts to the growing season:

(1) up to the end of May, when site class has little effect on yield;
(2) from June onwards, when differences in site class can give a 100% variation in expected grass yields.

Farmers have to take into account such differences in grass growth when planning their grassland management strategy.

For example, if it is assumed that grazing animals require the same amount of grass every day throughout the season and the accumulated 'surplus' can be cut periodically for silage, then it might be possible to take

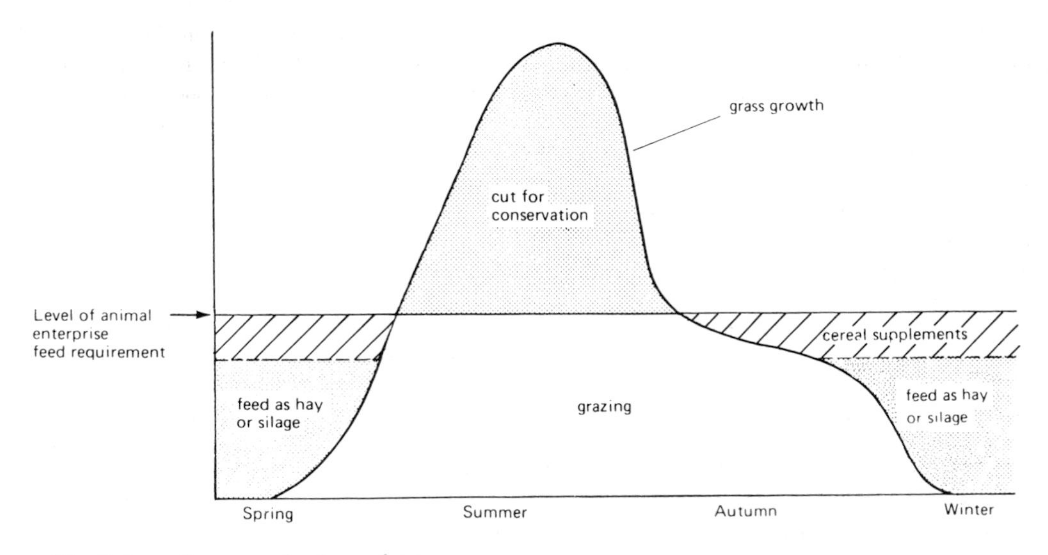

Figure 4.6 Pattern of grass production and use.

two to three cuts on site class 1 or 2 farms. On site class 5, only one spring cut will be possible and often the overall stocking rate will be lower as well.

Estimation of site class is based in part on average rainfall. If, in a particular year, rainfall is *above* average, the grass growth will be better than expected and surplus grass can be ensiled as an effective way of maintaining good grass utilization. In years when rainfall is *below* average, less silage will be made and on site class 4 or 5 the growth of grass may be below even that required for grazing; here other forage may be needed, e.g. 'buffer feeding' of silage.

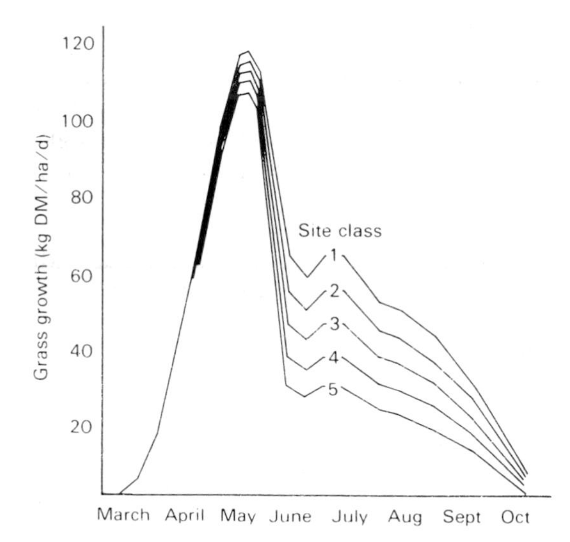

Figure 4.7 Seasonal pattern of DM production from a perennial ryegrass sward at five site classes.

Expression of grassland output on the farm

Farmers seldom weigh grass and so have little idea of grassland output as 't/ha'. Also, it is not the yield of grass *grown* that is important, but the quantity of grass *utilized* by the livestock system. There are two general ways in which the output of grassland can be assessed. There is a 'quick' way based on types of stock and overall stocking rate, and a more detailed method based on utilized metabolizable energy (UME).

Stocking rate method

The forage intake of ruminant stock is closely related to their liveweight and it is convenient to take a standard 'livestock unit' (LSU) of 550 kg and express all other stock on the basis of this LSU (sometimes called 'cow equivalent'; see Table 4.12). For example, at a particular time in the season, a farm has 250 ewes of average weight 60 kg plus 360 lambs of average weight 20 kg plus

Table 4.12 Cow equivalents of other ruminant stock (Forbes *et al.* 1980)

Liveweight (kg)	*Proportion of LSU*			*Liveweight (kg)*	*Proportion of LSU*	
10	0.04					
20	0.06					
30	0.08	Growing				
40	0.10	sheep		40	0.08	
60	0.13			60	0.10	Ewes
80	0.17		Growing	80	0.13	
100	0.20		cattle			
150	0.29					
200	0.38		Heifers			
250	0.47					
300	0.56					
350	0.64		Channel			
400	0.73		Island	400	0.56	
450	0.82		cows	450	0.61	Beef
500	0.91			500	0.68	cows
550	1.00	Friesian		550	0.75	
600	1.09	dairy				
650	1.18	cows				

40 heifers of average weight 200 kg. The total LSU would be

$$\left[250 \times \frac{60}{550}\right] + \left[360 \times \frac{20}{550}\right] + \left[40 \times \frac{200}{550}\right] = 54.9\,\text{LSU}$$

Growing animals change weight over the year, and some animals are sold and others bought, so that the annual expression of stocking rate as LSU depends on:

- conversion of all stock to LSU;
- allowance for changing stock weight and/or numbers.

This latter point is best assessed on a monthly basis.

If, in the previous example, the ewes were kept on the total grass area of 35 ha for the whole year, the lambs grew from 10–40 kg liveweight over 7 months and the heifers increased in weight over the year from 160 to 350 kg, then the annual LSU carried on the area is:

Ewes 250, weighing 60 kg, for whole year $= 250 \times \frac{60}{550}$

$= 27.3$

Lambs 360, average weight 25 kg, for 7 months

$= 360 \times \frac{7}{12} \times \frac{25}{550} = 9.5$

Heifers 40, average weight 255 kg, for whole year

$= 40 \times \frac{255}{550} = 18.5$

Total $= 55.3$

So stocking rate $= \frac{55.3}{35} = 1.58\,\text{LSU/ha}$.

An estimate of forage intake can be obtained from the equation:

$$y = 0.025x + 0.1z$$

where y = DM intake (kg/d), x = animal liveweight (kg) and, for dairy cows only, z = milk yield (kg/d).

For example, a 550-kg non-dairy LSU would have an appetite of 13.75 kg/d of DM. A dairy cow of this weight, and yielding 20 kg/d of milk, would have an appetite of 15.75 kg/d of DM.

Utilized metabolizable energy (UME)

The total energy value of the food eaten by the animal is its *gross energy* (GE). Some of this energy is passed out in faeces and the remainder is *digestible energy* (DE). Some of the DE is lost as methane from the rumen and some is passed out in urine; that remaining is termed *metabolizable energy* (ME). Some ME is lost as heat and the remainder, *net energy* (NE), is used for maintenance and production. Whilst NE is the nearest assessment to the production potential of a food, it has been found that ME is well correlated with production and is much easier to assess on a routine basis than NE. Feeding systems in the UK are based on ME. The ME required by animals to maintain their body functions, increase body weight, lactate and provide nutrition for any fetus carried are known and so it is possible to estimate the total quantity of ME utilized by animals. The ME values of all commonly used supplementary feeds are also known, so that, where grass is used along with other feeds, it is possible to produce a balance sheet that leads to an estimate of UME from grass, as shown in the example below. Note, this example illustrates the UME method of estimating output from grass; the exact UME standards for size and type of animal, its weight change and ME value of milk of specific quality can be obtained from published tables.

Example

Friesian herd of 100 cows, averaging 5500 litres of milk and using 1.3 t/cow of concentrates per year in addition to grazing and silage made from 50 ha of grassland. On average over a year each cow utilizes the following ME:

Maintenance	365 days, at 60 MJ/d	= 21 900 MJ
Weight change over year, none		
1 calf/cow		= 2400 MJ
5500 litres milk × 5.3 MJ/litre		= 29 150 MJ
Energy utilized per cow over year		= 53 450 MJ
Less energy from concentrates @85% DM		
Say 13 MJ/kg of DM		
1300 × 0.85 × 13		= 14 365 MJ
Difference is		
ME/cow utilized from grassland		= 39 085 MJ
	or	39.08 GJ
Stocking rate is 2 cows/ha, so		
UME/ha	2 × 39.08	= 78.2 GJ

Efficiency of grassland use

The UME figure gives a value for energy used; if the total quantity of ME available were known, then the percentage utilization of the energy could be calculated.

It is very difficult to measure grass yield on a farm, but some organizations predict ME output from site class, N supply and standard figures for the ME contents in concentrates and grassland feeds. The efficiency of grassland use is then calculated by expressing UME as a percentage of ME output.

Grass digestibility

In the 1950s workers at the Grassland Research Institute at Hurley found that the *digestibility* of grass (i.e. the percentage of grass eaten that was digested by the animal) was well correlated with the animal intake of grass and with the subsequent animal production from it. Since that time much advice and literature refers to digestibility. By convention, the term 'D-value' (i.e. the use of the capital letter D) should refer specifically to *the digestible organic matter as a percentage of the dry matter*. There is a close relationship between D-value and ME in grass, such that between D-values of 60 and 70, ME (MJ/kg of DM) = 0.16 D.

Figure 4.8 shows the effect of increasing grass growth period (maturity) on yield, N uptake, crude protein (CP) content and ME value. As grass matures it increases in yield but decreases in percentage CP and ME value; it is important to define the quality of grass required and defoliate it at the appropriate stage of growth. Generally, high grass quality is associated with low yields and vice versa (see Table 4.13). Not only does a feed low in ME provide a low feed value, but also the animals eat less of such a feed. Table 4.14 shows the double action of low ME value on intake and production.

Clearly the target ME value of grass must depend on the productive potential of the animals concerned and on whether the grass is grazed or conserved, the latter point being influenced by the loss of feed value which generally occurs when grass is conserved and the need for at least a moderate yield to justify the costs of conservation. Guidelines for ME levels for good production from grass are:

Utilization	*Minimum ME (MJ/kg DM)*	*Optimum ME* (MJ/kg DM)
Grazing dairy cow	10.5	11.5
Grazing other growing stock	10.0	11.0
Grazing dry cows	10.0	11.0
Grazing dry ewes	9.5	9.5
Silage and barn-dried hay	10.0	10.8
Field hay	9.0	10.0

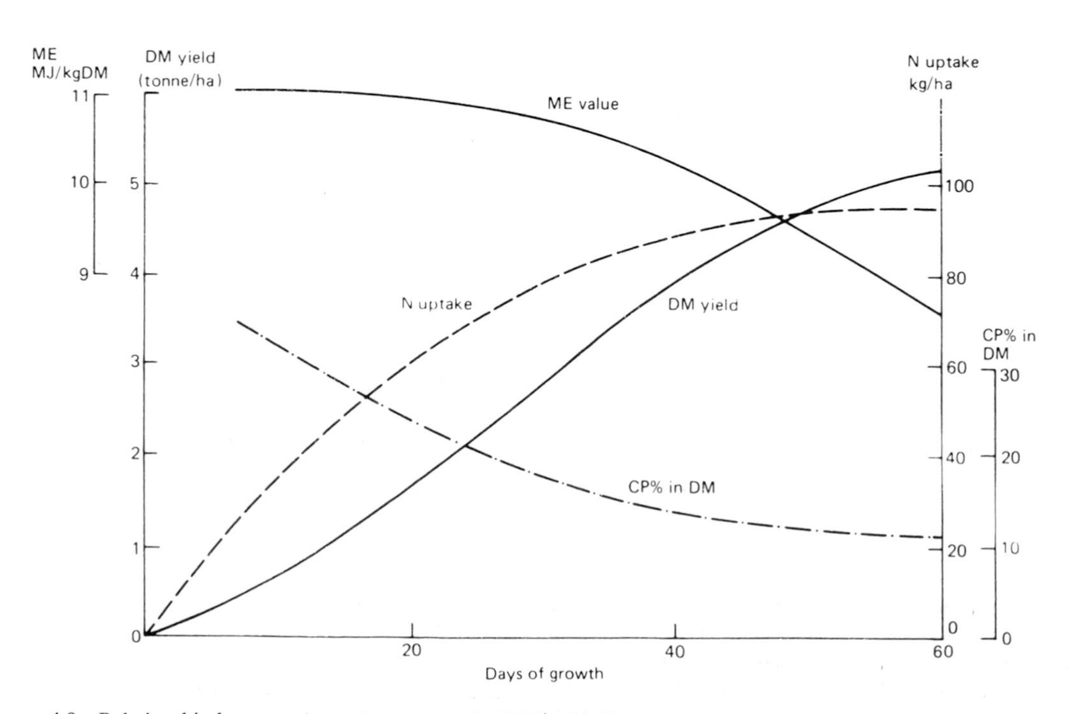

Figure 4.8 Relationship between days of grass growth, DM yield, N uptake, crude protein content and ME value.

Table 4.13 Effect of cutting frequency on grass yield and quality. (Source: Leaver & Moisey, 1980)

Three cut system			*Two cut system*		
Date	*DM yield (t/ha)*	*ME (MJ/kg DM)*	*Date*	*DM yield (t/ha)*	*ME (MJ/kg DM)*
(1) Field data					
Late May	4.6	10.6	Early June	7.8	9.6
Early July	3.2	9.8	Mid-August	3.7	9.0
Mid-August	1.8	9.6			
Total	9.6	Mean 10.0	Total	11.5	Mean 9.2
(2) Feeding data					
Concentrate feeding	Low	High		Low	High
Concentrate intake (kg/d)	5.0	9.3		5.0	9.3
Silage intake (kg/d DM)	11.2	9.0		10.3	8.4
Milk yield (litres/d)	19.9	20.8		17.3	10.7
Land required for 180-day winter (ha/cow)	0.25	0.21		0.20	0.16

Table 4.14 Effect of grass ME value on intake and production in dairy cows

Grass ME value (MJ/kg DM)	*Grass DM intake (kg/d)*	*ME intake (MJ/d)*	*Milk yield (litres/d)*
11.3	12.8	145	14.6
9.5	12.0	114	11.8
8.4	10.6	89	10.0

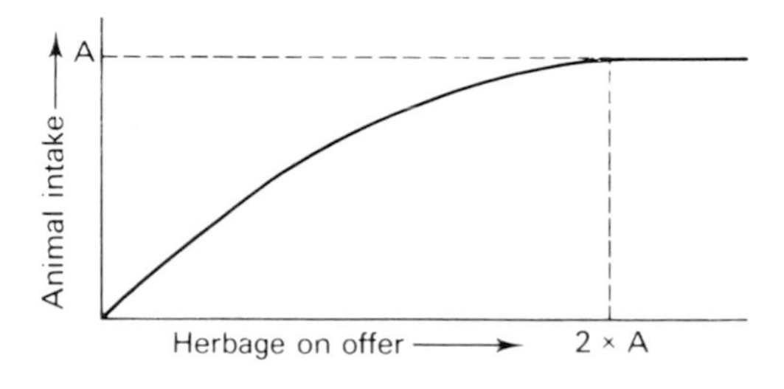

Figure 4.9 Relationship between herbage on offer at grazing and animal intake.

Output from grazing animals

Efficient grazing management requires resolution of the often conflicting requirements to achieve efficient utilization of the herbage grown, whilst at the same time achieving high rates of performance by individual animals.

Severe grazing will achieve high herbage utilization, but may depress the intake by individual animals and thus their performance, as discussed by Parsons and Chapman (2000). The impact of the quantity of herbage on offer on animal intake is illustrated in Fig. 4.9 and the effects of sward height on milk production by cows continuously grazed are shown in Fig 4.10. However, if animals are offered much more herbage than they are able to consume, much of this will decay and be wasted, with reduction in UME output. This is well illustrated by the results of an experiment at North Wyke in which permanent grass swards were grazed by beef cattle, with stocking rates adjusted to maintain sward heights (assessed with a rising plate meter) of 40, 60 and 80 mm. Individual animal performance was highest with lax grazing and the tallest swards, but stocking rates and UME outputs were highest with the shortest swards. Liveweight gain per hectare was maximized with the intermediate treatment (Table 4.15).

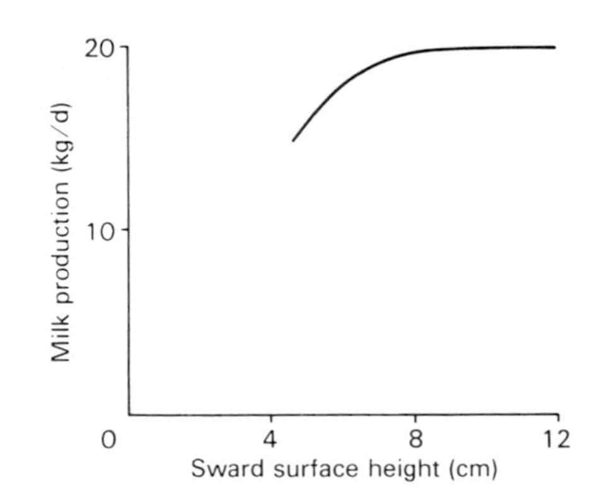

Figure 4.10 Influence of sward surface height on milk yield: average grazing season yields for spring-calving cows.

Table 4.15 Animal production and herbage utilization for steers grazing a permanent sward to different levels of severity (Wilkins *et al.*, 1983). The best treatment for individual variables is indicated in bold

	Sward height (mm)		
	40	*60*	*80*
Grazing days (per ha)	**1220**	740	530
Liveweight gain (g/head/d)	410	820	**900**
Liveweight gain (kg/ha)	500	**600**	480
UME (GJ/ha)	**57**	46	36

Table 4.16 Suggested target ranges of sward surface heights for continuous stocking (Hodgson *et al.*, 1986)

	Sward height (mm)
Sheep	
Ewes and lambs	40–60
Dry ewes	30–40
Cattle	
Dairy cows	70–100
Dry cows	60–80
Finishing cattle	70–90

Sward height

Sward height has become established as a valuable tool to aid grazing management on the farm. Experiments such as those discussed above have established the optimum height for maintaining swards for different animal systems. Target sward surface heights with continuous stocking are given in Table 4.16. In rotational systems, grazing should be continued until the sward height falls to these levels.

Sward height can be measured by placing a ruler (or similar marker) vertically in the sward, with the lower edge just touching the ground. Find the height at which a finger descending the ruler touches a green leaf. Points to watch are:

- Ensure the ruler is vertical and is not pressed into the soil or standing in a depression.
- Thirty to forty readings are taken in the field, avoiding bias and including a fair proportion of grazed and ungrazed areas.
- Contact with stems or seed heads is ignored.

It is also possible to obtain a good and rapid estimate of sward surface height by walking through the sward wearing wellington boots with marks at intervals of 20 mm.

An alternative approach is to measure compressed sward height with a rising-plate meter. The meter has a central spindle with a disk of fixed diameter and weight which slides up the spindle when the meter is placed in the sward. The height of the disk above ground level measures compressed sward height. The measurement will be influenced by both sward height and density, but is normally about 70% of the value for sward surface height. Rising-plate meters are available commercially and many measurements can be made rapidly. There is a correlation between compressed sward height and herbage mass (quantity of herbage DM above ground level), facilitating the use of these meters in grass budgeting (see later).

Various approaches can be used to help keep to the target sward height, such as adjustments to stock numbers and the areas available for grazing and cutting, the provision of supplementary feeds and changes in fertilizer treatment.

Grazing systems

Many areas of grass, particularly those for beef and sheep, are grazed on a very extensive basis which is not systematic in nature. On the other hand, most dairy farmers have a very definite basis to their grazing policy, and an increasing number of the more successful beef and sheep farmers are adopting grazing strategies similar in principle to dairy systems. This section will deal with the major methods used to graze dairy cows; some of the methods outlined are appropriate for other stock.

Whatever the grazing method or stock used, the principles outlined in the previous section apply. The achievement of the right grazing severity, as indicated by target sward height, is of much greater importance than the particular system used.

Two sward system

Here one area is regularly cut for conservation and the other regularly grazed. It is difficult to accommodate the seasonal growth pattern in the grazed area, but the system has advantages where part of the grass area is inaccessible to grazing (e.g. on a split farm) or where the cutting grass is in an arable rotation and does not justify fencing and a water supply.

Otherwise the integration of cutting with grazing is a powerful tool in smoothing out the availability of grass for grazing, because whereas only one-third of the total area may be needed for grazing in May, two-thirds may be needed in June–July and the whole area from August onwards.

For	*Against*
Useful if part of area cannot be grazed.	Risk of surplus grass in grazed area in spring and insufficient later.
	Poor feed quality in mid and late season.

Set-stocking

This implies a given number of stock on a fixed area for a long period, often the whole season. The term *set-stocking* is sometimes used erroneously for *continuous stocking* (see below). Set-stocking is advised under very extensive conditions only at low stocking rates. The problems are similar to those for the grazed area in the two sward system. There is a tendency for undergrazing in spring, leading to poor-quality mature herbage, followed by overgrazing in late summer and autumn resulting in low animal performance. One advantage is that provided the perimeter of the area is stockproof, fencing costs and water supply problems are minimal.

For	*Against*
Simple and cheap.	Does not match grass growth.
	Possible disease buildup.
	Poor feed quality in mid and late season.

Flexible continuous stocking

This refers to systems where the stock are allowed to graze over a large area for a fairly long period, say 2–3 months. Many farmers use this system for dairy cows and associate it with a high stocking rate, high and regular fertilizer N usage and a willingness to bring in other feeds if grass supplies become inadequate. As rate of grass growth slackens during the season, farmers can both reduce stocking numbers as cows dry off and increase the grazing area by adding some silage aftermaths. Thus part of the area may be grazed for the whole season, but neither the total area nor the stock numbers are fixed, and other feed is used as a buffer against periods of low grass availability. This system works particularly well in good grass growing conditions. It is well suited for management to achieve target sward heights.

Some farmers have a 'rapid rotation', where the area is divided into three or four blocks and each is grazed for about 1 day at a time. In other situations there is one area for day grazing and another for night grazing. Because of the frequent defoliation that occurs, these are best regarded as variants of flexible continuous stocking and are not types of rotational or block grazing.

Overall there should be 1 ha of available grazing for every three cows; at turnout in spring the whole area may be needed, but as growth picks up the area for spring grazing will be about 1 ha per five cows; following spring silage cuts the area may need extension to 1 ha per four cows, with the whole area in use from August onwards, depending on the season.

For	*Against*
Suited for management to achieve target sward heights.	No visible assessment of grass growth.
Reduces poaching as stock dispersed.	Regular check on performance essential.
Saves costs in fencing and water.	Cow collection can take time.
Increases sward density and clover.	Willingness to supplement feed at grass essential.
	Need to integrate cutting with grazing.

Three field system

The three field system is one that has much to commend it if the whole grass area can be grazed conveniently (this is often difficult with cows that have to walk to and from milking, and for this reason this method is more often used in beef and sheep systems). Basically this method formalizes a continuous stocking system into three periods in the season and in a very simple way adjusts to the expected pattern of grass production. The grassland area should be split by stockproof fencing into two areas, one area being approximately double the size of the other. In spring the stock graze the smaller area and the larger area is cut, preferably for silage as this will provide more reliable aftermaths than where hay is taken. As soon as the aftermaths are available, the stock are switched and the smaller area is taken for a conservation cut. Once the smaller area has available aftermaths, the whole area is grazed.

For	*Against*
Simple, self-adjusting to seasonal growth.	Inflexible.

Suitable for large animal groups.	Based on *expected* not *actual* growth.
Good parasite control; 'clean' grass.	Often unsuited to dairy grazing.

Block grazing systems

These are systems in which the grazing area is divided into a number of fairly large blocks with the aim of grazing on a rotational basis. Thus one block of grass is grazed with a large number of animals for a short period, usually 1–7 days, and then the animals are moved to other blocks whilst the grazed block regrows. After about 20–28 days, the block can be grazed again. Blocks that are surplus to grazing are cut for conservation. This basic principle is used for all classes of grazing stock, with the following common amendments.

For dairy cows. Occasionally farmers split their herd and do a leader–follower with the cows, putting the high yielders in the leader group, or putting dry and nearly dry cows as followers behind the main herd. Also, farmers often like cows to have a fresh allocation of herbage each day, so an electric fence can be used to ration the grass. This was formalized in the *Wye College* system where cows were given one-seventh of a block each day and one block lasted 1 week.

For ewes and lambs. The fence separating adjacent blocks can have spaces wide enough for the lambs to pass but not the ewes. Thus lambs can obtain the pick of the next block to be grazed by the ewes (*forward creep grazing*).

For cattle rearing. Young stock may be grazed one paddock ahead of older cattle so that the younger ones obtain the pick of the grass and the older cattle clean up, ensuring good utilization. This system is called *leader follower grazing* and has been used particularly in heifer rearing. This is one of the most effective means of combining high levels of animal performance (at least by the leaders) with efficient overall sward utilization.

For	*Against*
Efficient grass utilization.	Good fencing round each paddock essential.
Good parasite control for young stock.	Regular stock-moving decisions needed.
Areas large enough for conservation.	Many water points needed.

Paddock grazing

This represents a very formal method of rotational grazing, where the grazing area is divided into some 21–28 permanently fenced and equal-sized paddocks. The aim is that one paddock is grazed each day and the rotation around the area is completed as soon as the first paddock is ready for grazing again. Surplus paddocks can be taken out for conservation, but usually the operation is restricted by the small size of the paddocks. Paddock size must be related to the number of grazing animals, as the stock density must be sufficient to ensure that the grass is efficiently utilized during the short grazing period. A guide is to allow 100–125 cows/ha daily, depending on stocking rate.

Thus, for a highly stocked herd of 100 cows and where 25 paddocks are selected, each paddock should be about 1 ha in size. Every paddock must have a water point and independent access onto a track to avoid poaching.

Recent research has shown that milk production is increased if cows are moved into a new strip or paddock in the afternoon rather than the morning. This maximizes grazing intake late in the day, when feeding value is increased through accumulation of WSC (Orr *et al.*, 1998).

For	*Against*
Easiest rotational system to manage.	High fencing and water costs.
Gives objectivity to grazing plan.	Small areas for conservation.
Facilitates grazing to target sward heights and grass budgeting.	Some wasted land in trackways.
Intakes by high-yielding cows high, particularly if a large quantity of herbage is available.	

Strip grazing

This involves the use of an electric fence to give a fresh strip of herbage once or twice daily. Ideally, stock should be confined to the daily strip only by use of a regularly moved back fence to prevent the regrazing of young regrowth, but this is done rarely because of problems with animal access and water supply. Although this is the most sensitive system to allow adjustment for fluctuations in grass growth, it is not common because:

- It requires daily labour to decide on area allocation and to move the fence.
- There is a tendency for grass to become overmature ahead of grazing in large fields.
- There is a risk of serious poaching along the fence line in wet weather.

Grass budgeting

Grass budgeting can be used to improve the planning and management of grazing systems, as discussed by Mayne *et al.* (2000). Knowledge of probable grass yields and seasonal production patterns (as outlined earlier in this chapter), stocking rates and daily grass intakes can be used to calculate the areas that should be required for grazing through the year (or alternatively the quantity of N required to produce sufficient herbage on a fixed grazing area).

Grass budgeting is also valuable for fine-tuning of grazing management to take account of deviations in grass growth from normal growth patterns, particularly in rotational systems. Information on the quantity of grass currently available within the rotation (commonly referred to as 'grass cover'), grass growth rate and the quantity of grass consumed daily can be used to calculate the quantity of grass that will become available over the next weeks. This will provide early notification of the possible need to adjust the area available for grazing, the rotation length or the rate of use of additional feeds. The quantity of grass available and grass growth rates can be estimated by regular monitoring of compressed sward heights as paddocks regrow and the application of standard relationships between compressed sward height and mass. Information on current growth rates from plots cut under standard management is also becoming available on a regional basis through the internet (e.g. for Northern Ireland on www.greenmount.ac.uk).

Extending the grazing season

The relative importance of grass utilization by conservation rather than grazing increased markedly in the period from about 1970 to 1990, with the successful development and rapid adoption of improved methods of silage making. There was a tendency for the grazing season to be reduced and for silage also to be used as a buffer feed during the summer. This occurred despite many studies having shown that the cost of supplying nutrients in silage is considerably higher than that in grazed grass. Reduction in the real price of animal products in the 1990s led to re-examination of approaches to reduce the costs of milk and meat production, and there is now increased focus on approaches to extend the grazing season and reduce the quantities of silage that need to be made. Both sward type and grazing management have parts to play in extending grazing (Mayne *et al.*, 2000). Particularly important points are as follows:

- Duration of grass growth can be increased by using Italian ryegrass and early hybrid and perennial ryegrass varieties.
- The quantity of herbage available in late season can be increased by increasing the grazing area and not taking late silage cuts.
- Areas may be grazed in early season before closing for silage.
- In wet conditions, the restriction of grazing to 2–6 hours daily will minimize treading and poaching, whilst achieving high intake rates.
- The herbage growth period in the autumn should not exceed 8 weeks, because both yield and quality will fall.
- Avoid late application of N fertilizer in nitrate sensitive areas.
- Provide good cow tracks.

Experiments in Ireland have shown clear benefits in terms of increased milk yield per cow and reduced costs when grazing for a few hours per day has been introduced in either spring or autumn. For example, grazing for 3 hours daily over the period 29 October to 26 November reduced silage intake by 4.2 kg DM/day and increased milk yield by 2.1 kg/day (Mayne & Laidlaw, 1995).

Zero grazing

This implies that grass at grazing stage is cut and transported to stock that are either housed or kept in a 'sacrifice' area. Zero grazing is practised widely in some countries where grass is inaccessible for grazing (e.g. in 'strip field' systems) or where a range of crops needs to be grown to ensure continuity of supplies (e.g. in semi-arid areas). In temperate climates stocking rates can be higher under zero grazing because:

- The sward does not suffer the deleterious physical effects of treading and selective grazing.
- Herbage can be harvested at the ideal stage for the stock concerned.

- Utilization can be high as there is no field refusal of grass (although this implies that the stock eat most of the grass carted to them).

Both milk and meat production per hectare have been shown to be greater from zero grazing than from other grazing systems. However, zero grazing is not used commonly in the UK except for specific opportunist reasons such as to:

- reduce poaching in early spring and autumn;
- provide 'grazing' from distant or inaccessible fields;
- use other crops during crisis periods, as in drought.

The reasons for the lack of popularity of zero grazing despite its technical excellence for high production per hectare are:

- high labour and machinery costs for feeding *and* for removing slurry;
- complete dependence on machinery, with risk of breakdowns;
- output per animal tends to be depressed because opportunities for selection by the animal are reduced (but high stocking rate gives high output per hectare);
- problems of refused grass at feeding face, particularly as wet grass heats rapidly when heaped.

Buffer grazing and feeding

This term does not refer to a grazing system but to the provision of additional feed if and when that available from the grazing area is insufficient. The buffer (or additional) material can come from:

- another (often adjacent) grass area;
- a specifically grown forage crop;
- a bulk forage such as silage, hay or straw;
- a dry feed such as cereals, concentrates or dried grass.

Care is required in the use of buffer feeds when adequate quantities of grass for grazing are available, because the provision of additional forage is likely to reduce the intake of grazed grass and reduce grazing efficiency. The provision of the buffer may increase costs with no improvement in animal performance.

There is increased interest in the use of feeds with high energy value but low protein content as supplements for grazed grass. The CP content of grazed grass is often well in excess of animal requirements, resulting in excessive loss of N in excreta. A better nutritional balance and reduced N losses may result from the supplementation of grazed grass with maize silage. The practice of grazing grass during the day and feeding maize silage at night has become well established in the Netherlands. mainly to reduce N losses to the environment.

Conservation of grass

Conserved grass is used as the basis of ruminant feeding when grass for grazing is not available. Grass for conservation is either grown specifically for this purpose or taken as a surplus in a grazed area; often both situations apply in any one year on an individual farm. Where grass arises as a grazing surplus, an overriding factor is the rapid removal of the crop in order to allow regrowth for a further grazing (see Table 4.4).

When green crops are cut, biochemical changes occur and, if these are allowed to continue unchecked, degradation of the material will take place, releasing heat and effluent (water that contains much of the soluble cell contents). Grass may be conserved either by ensiling or by drying. There has been a substantial increase over the last 40 years in the total quantity of grass that is conserved and a major change in the predominant method of conservation. Silage represented less than 10% of the total DM conserved in 1960, but nearly 80% in 1990 (Fig. 4.11).

Silage may be made from fresh or wilted grass and from other forage crops. When long or chopped material is placed in a heap or a silo and access of oxygen is prevented, an anaerobic fermentation ensues. The acids produced by fermentation reduce pH and, provided the reduction in pH is sufficient and anaerobic conditions are maintained, the material is stable and no further biochemical changes take place.

Enzymic degradation of plant material ceases when material in aerobic conditions is dried to 85% DM. Drying in natural conditions in the field is called haymaking. Grass in the field usually has a DM content of 15–20%. This means that some 3.2–4.6 t of water needs to be evaporated in curing 1 t of hay – a difficult task in humid conditions. Sometimes hay that is almost dry is placed in a building and subjected to forced-draught ventilation to complete dehydration; this is called barn hay drying. On some specialized units grass at or near field moisture is taken to a high-temperature drier and dehydrated very rapidly; this is called green crop drying or grass dehydration.

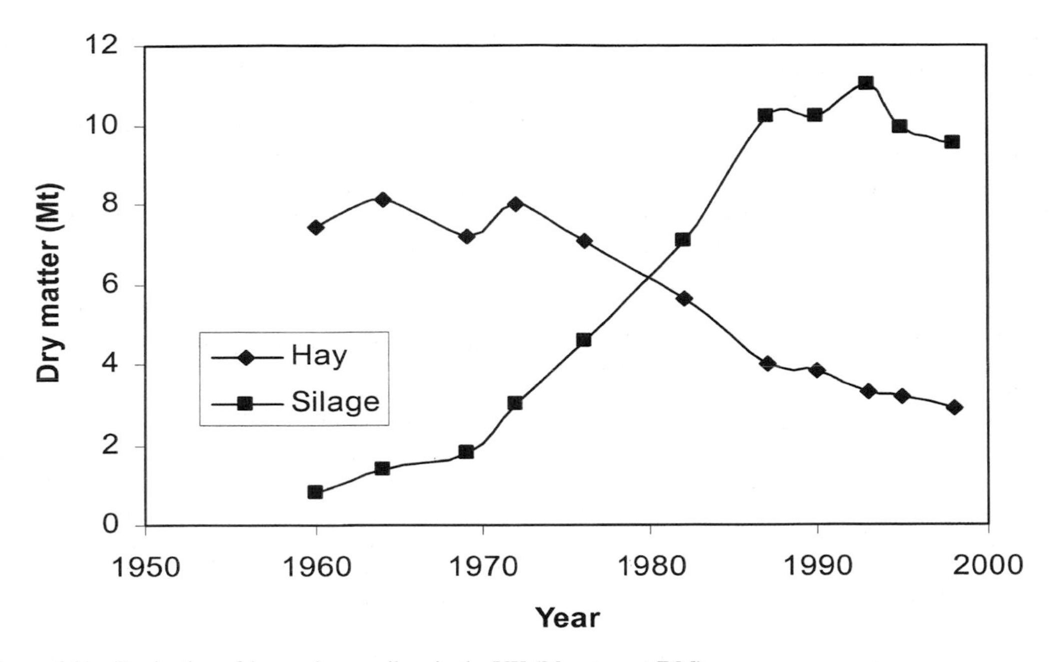

Figure 4.11 Production of hay and grass silage in the UK (Megatonnes DM).

Silage making

The process of ensilage consists of preserving green forage crops under acid conditions in a succulent state. When such green material is heaped, it respires until all the oxygen in the matrix is exhausted; during respiration carbohydrates are oxidized to carbon dioxide with evolution of heat. Continued availability of oxygen, as in a small outside heap of grass, will lead to enhanced oxidation with resultant decomposition and overheating. With restricted oxygen supply, bacterial fermentation will take place. Micro-organisms present on the crop, the machinery and in soil contamination fall into two categories, desirable and undesirable, as described by McDonald *et al.* (1991). The desirable bacteria can convert carbohydrate into lactic acid; these are mainly *Lactobacillus* species. Their even distribution and activity throughout the grass in the silo is encouraged by mechanical chopping of the grass, rapid consolidation and exclusion of air. Lactic acid is a relatively strong organic acid and its rapid production within the ensiled grass leads to a low pH and conditions that inhibit the lactic acid production bacteria and *all* other bacteria as well. The pH at which this 'pickling' occurs depends on the moisture content of the grass: the wetter the grass the lower the pH needed and the greater the quantity of lactic acid that has to be produced.

Silage DM%	*pH for stable silage*
18	3.8
20	4.0
22	4.2
26	4.4
30	4.6
35	4.8

Silage with a good lactic acid content is light brown in colour, has a sharp taste and little smell; it is very stable and can be kept for years if necessary provided nothing is done to permit oxygen to enter the material.

The undesirable organisms are:

- enterobacteria, which can ferment carbohydrates, but do not produce lactic acid – they may predominate in the early stages of fermentation;
- obligate anaerobes of the *Clostridium* species that can ferment carbohydrate and lactic acid to produce butyric acid and/or break down amino acids to produce ammonia and amines, causing reduction in the intake and protein value of silage;
- *Listeria monocytogenes*, a pathogen that proliferates as a consequence of air ingress, particularly in big bale silage;
- yeasts, which may develop when the silo is opened and cause heating and aerobic deterioration.

An indicator of clostridial activity is the ammonium-N content of silage, as the ammonia produced by these bacteria is retained in the silage.

	Fermentation quality		
	Good	*Moderate*	*Bad*
NH_4-N as % of total N	0–8	8–12	Over 12

Butyric silage is olive-green in colour, has a rancid smell and is unpalatable to stock. Also it has a higher pH than lactic silage and is unstable, continuing to deteriorate during storage. Clostridial activity in silage can be inhibited by:

- reducing moisture content of the grass, as this will lessen the quantity of acid needed to prevent decomposition;
- ensuring adequate WSC is present for lactic acid bacteria, or applying an acid to assist in lowering pH;
- adding inoculants of lactic acid bacteria;
- adding chemicals with specific anti-clostridial effects;
- avoiding contamination from soil or animal manure, both of which contain large numbers of clostridia.

Recent research in Germany has shown that the presence of nitrate in the ensiled material provides some protection against clostridial fermentation (Weissbach *et al.*, 1993). The risk of clostridial fermentation is increased in materials containing no nitrate, as may occur with low levels of N supply to the plant or a long growth cycle prior to harvest.

Intrusion of air during fermentation will delay or even prevent the achievement of a stable pH and will lead to excessive amounts of WSC being lost through respiration, thus lowering the nutritional value of the silage. Intrusion of air after the silage has reached a stable condition or during the feedout period will lead to further losses through aerobic deterioration. Any remaining WSC will be metabolized and lactic acid will be broken down by the action of yeasts and bacteria in the silage. The degree of heating gives a good indication of the extent of aerobic deterioration. Silages with high DM content are particularly prone to aerobic deterioration, whilst the volatile fatty acids (acetic, propionic and butyric) restrict aerobic deterioration. If a silage is prone to aerobic deterioration, then it is important to ensure rapid removal of material from the silo and to restrict the ingress of air. Some additives containing acetic or propionic acid will also reduce the problem.

Dry matter and nutrients may be lost at several stages in the ensiling process, as indicated in Table 4.17. Many of the losses arising from the effects of air – respiration, surface waste and aerobic deterioration – may be much reduced by good silage management. However, even in the best of conditions some 15–20% of the DM, and a rather higher proportion of the nutrients in the crop, are likely to be lost in the whole process. In a series of experiments carried out in different European countries, DM losses averaged 19% for unwilted silages made with additive and 17% for wilted silages with 30–40% DM and no additive treatment. The higher field losses with the wilted silages were compensated for by the absence of effluent and lower in-silo losses (Zimmer & Wilkins, 1984).

For the best fermentation, the crop should have a high WSC content and a low moisture content. Thus empha-

Table 4.17 Sources of dry matter loss during ensiling (%)

	Range of loss	*Comment*
Field		
Respiration	0–8	About 2% per day in field
Physical	0–8	Increases at high DM content
In silo		
Plant respiration	0–8	Increases with slow filling, poor consolidation and heating
Fermentation	0–10	No loss if completely homolactic; increased with clostridial fermentation
Surface waste	0–30	The surface waste present when silo is opened represents only part of this loss, as up to three times as much may have been lost through oxidation
Aerobic deterioration after opening	0–20	Generally higher in high DM silages; heating indicates high losses
Effluent	0–10	Closely related to DM content and silage density

sis is given to cutting crops when the WSC content is high and when weather conditions are suitable for rapid wilting. A WSC content of 3% of the fresh matter should ensure a lactic fermentation and the production of a stable silage. Wilting in good conditions will concentrate the sugars in the fresh matter. In order to reach the critical WSC content in the fresh matter, WSC in the DM must be above 15% in wet crops with 20% DM, but a value of only 10% is required if the crop has been wilted to 30% DM. In poor weather conditions, however, respiration after cutting will reduce WSC by at least 2% per day on a DM basis. This will lead to increased difficulty in achieving good fermentation and will reduce nutritional value. The maximum period of field wilting for silage should be 2 days.

In situations where the critical WSC content cannot be reached, farmers must consider the use of additives. These can be used to:

- *Increase water-soluble carbohydrate*. By adding extra WSC (e.g. molasses) the lactobacilli are better able to produce lactic acid. Also *enzymes* such as cellulase and hemicellulase can be used to convert some of the non-soluble material in the plant into WSC. Note there is no starch in grass and so amylase is not effective.
- *Increase effective bacteria*. Inoculants containing lactic acid bacteria can be used to improve the efficiency of conversion of WSC to lactic acid and ensure that preservation is not limited by shortage of suitable bacteria. It is important that bacteria are added at the rate of at least 10^6 per gram in order to dominate the indigenous microflora. Most inoculants are sold as freeze-dried preparations, but good results have also been obtained in a system in which freshly cultured bacteria are produced on the farm using a process similar to home brewing.
- *Add acid*. A range of organic and inorganic acids have been used as silage additives, but formic acid has been used most commonly and has produced consistent results. In the UK, formic acid is usually applied at 3 litres/t of grass. This produces an immediate reduction of pH to around 4.4 and reduces the quantity of lactic acid that must be produced to reach a stable pH. Consequently, the requirement for WSC in the crop is reduced. In some countries organic acids are applied at higher rates, with preservation being achieved by the direct effect of the added acid with minimal fermentation of WSC. Strict precautions are needed when handling acid additives as they may be dangerous chemicals. These hazards have led to reduction in the use of acids in Britain in recent years.
- *Add anti-clostridial agents*. Some organisms and chemicals have direct effects against clostridia. These include the virus clostridiophages, nitrites, nitrates and hexamine. These preparations, however, do not have such a proven track record as organic acids and some of the bacterial inocula.

A large number of silage additives are marketed in the UK. A voluntary registration scheme is operated by the United Kingdom Agricultural Supplies and Trade Association (UKASTA). In this scheme evidence supplied by the manufacturers is reviewed and the product may be approved in different categories in relation to evidence of effects on animal performance, aspects of nutritional value and effects on fermentation, aerobic stability and effluent production. This scheme is an important source of advice for the farmer.

Table 4.18 summarizes the Liscombe star system which can be used to warn of possible fermentation problems and the need for additive use. Whilst stress has been laid here on preventing clostridial fermentation and the Liscombe system provides guidance for the selective use of additives, many experiments have demonstrated animal performance benefits from the use of additives, even when untreated silages are apparently well preserved. Where high rates of animal performance are required, there is a strong case for routine use of an effective additive.

Storage

Storage of silage is in bulk silos (clamp or tower) or big bales (film wrapped in polythene).

Clamps

Clamps are found in a variety of forms, e.g. walled or unwalled, roofed or open, on the surface or in pits.

Towers

Towers are made of concrete or galvanized–vitreous enamelled steel. Material for ensiling is always added to the top, but, depending on type, silage is removed from either the top or the bottom.

Big bales

Big bales can be stored on level sites around the farm. They must be protected from wind which can damage the polythene, and from rodents which can eat holes in the film.

Making silage in clamps

Assuming the material to be ensiled is either high in sugars and low in moisture or having an additive

Table 4.18 The Liscombe star system. A total of five stars is needed for a good fermentation. Consider each factor, such as variety, growth stage, etc., and add up the stars (*)

Grass variety	Timothy/meadow fescue	*
	Perennial ryegrass	**
	Italian ryegrass	***
Growth stage	Leafy stage	
	Stemmy mature	*
Fertilizer nitrogen	Heavy (125 kg/ha +)	– *
	Average (40–125 kg/ha)	
	Light (below 40 kg/ha)	*
Weather conditions (over several days)	Dull, wet (less than 2% sugar)	– *
	Dry, clear (2.5 sugar)	
	Brilliant, sunny (3% sugar or more)	*
Wilting	None (15% DM)	– *
	Light (20% DM)	
	Good (25% DM)	*
	Heavy (30% DM)	**
Chopping and/or bruising	Flail harvester or forage wagon	*
	Double chop	**
	Meter/twin chop	***

How the star system works
For example a perennial ryegrass sward (**) in leafy silage growth stage ( ) and heavily manured (–*), being ensiled in dry weather ( ) and only lightly wilted ( ) with pick-up double chop (**) gives a total score of *** and will show a benefit from additive use.
In comparison an Italian ryegrass sward (***), in a leafy stage ( ), which is heavily manured (–*), in average weather ( ) but well wilted, 25% DM (*) and meter chopped (***) will give a total score of ****** and not require additive.
5 Stars – no additive needed.
4 Stars – use additive at recommended rate.
3 Stars – use additive at recommended rate.
2 Stars – use additive at higher recommended rate.
1 Star – use additive at higher recommended rate.
0 Stars – unsuitable conditions for making silage.

applied, the main principle during the filling process is to eliminate as much air from the matrix of herbage as possible and keep the material airtight. Polythene sheeting is an essential feature of silage making and is available for this purpose in 300 or 500 gauge and in widths up to 10 m. Walled clamps (bunkers) are often filled using a *wedge-filling* principle (sometimes called *Dorset wedge*).

On the first day of filling, the cut crop is stacked at one end of the silo, against an end wall that is either solid or has a polythene sheet lining. If the side walls are not solid, they too must have a polythene lining. The material is normally put into the silo using a push-off buckrake, and the buckraking tractor maintains the slope at the steepest reasonable angle. When the material has reached the maximum intended height, the slope is progressed forwards, leaving a fixed height of material.

One principal objective of the wedge system is to prevent warm air rising out of the silo, as this will encourage oxygen-rich cold air to come in at the bottom and sides. A polythene sheet is therefore placed over the grass each night, and when one section of filling is complete, the sheet is left in place so that the silo is gradually wrapped in polythene sheets. If the technique is carried out correctly, the oxygen in the air in the silo is soon used up, lactic acid fermentation proceeds, and the crop consolidates under its own weight, often resulting in a drop in crop height to two-thirds to three-quarters of the original. Because of this shrinkage the polythene sheet should not be fixed rigidly, but rather covered with a flexible and convenient material such as old tyres, sand bags, straw bales, or even a net, aiming to completely cover the silo. If the sheet is not held down tightly, it will flap in the wind, allowing in more air and eventually tearing the sheet.

The principles of this approach are extremely good, but with the increasing use of high-capacity equipment and of contractors, silos are now often filled within 2 days. In such circumstances overnight covering is of less importance and attention should be given to rapid and effective final sealing of the silo.

Fairly wet grass can be placed in a clamp, but it will produce effluent. Grass ensiled at 20% DM will release an average of 200 litres of effluent per tonne of grass ensiled, and the quantity of effluent decreases progressively until material ensiled at 28% DM should give no effluent.

Silage effluent

This is a real problem if it enters a watercourse, as it has a high biological oxygen demand (BOD), and will kill many oxygen-demanding organisms in the water, including fish. Even if it is planned to make high DM silage, it is essential to construct an effluent tank adjacent to the silo, taking care to ensure that *only* silage effluent can enter it (i.e. no surface or rainwater) (see Chapter 25). Silage effluent is very acidic, and all effluent conducting channels and ducts must be coated with acid-resistant material. Silage effluent contains some plant nutrients (say 2, 1 and 1.5 kg/1000 litres of N, P_2O_5 and K_2O respectively), but as it is very acidic it is very phytotoxic. It can be applied to arable land by tanker or a slurry irrigation system, but care is needed to ensure it does not get into land drains and hence into a watercourse.

Silage effluent can be given to stock, normally either cows or pigs. Care must be taken in collecting and storing the material to ensure there is no seepage. Also effluent deteriorates on storage and becomes unacceptable to stock, so a preservative such as fomalin should be added.

Generally, however, material should be wilted to a sufficiently high DM to restrict or prevent the production of effluent.

Making silage in towers

The crop has to be blown into the top of the tower and at feeding time the silage is removed by mechanical means; thus it is vital that the ensiled material is well chopped and sufficiently dry to remain friable after compaction in the tower during storage. For this reason, grass going into a tower must be above 35% DM and many tower operators prefer 45–55% DM (sometimes called *haylage*). Because the material going into a tower must be well chopped, and because a tower is almost airtight, compaction in a tower is good, and excellent fermentation is assured. In-silo losses are often only 5–10%, but field losses may exceed 10%. Tower silos are normally the starting point for mechanized feeding systems.

This method of ensiling is, though, now of little importance in the UK, although still widely used in North America. It suffers from high capital investment, low flexibility and the requirement for good weather and heavily wilted material.

Making big-bale silage

Big-bale techniques were first used in the late 1970s, but have become widely adopted with about 20% of the total silage made in the UK now being in this form. Grass is normally wilted to 25–40% DM and picked up with a big baler in bales weighing 600–800 kg. Originally the bales were put into individual polythene bags and then sealed, but practically all big bales are now wrapped with at least six layers of polythene film, using a special machine. This technique can be used with both round and rectangular bales. There have been recent developments with the introduction of chopper balers, which produce bales of higher density and of somewhat higher feeding value, and electro-mechanical applicators to allow even distribution of additives, as discussed by Merry *et al.* (2000). During storage of big bales it is important to avoid damage by wind, birds and vermin. Provided damage is prevented, DM losses in the silo are likely to be around 10%, somewhat lower than with clamp silos, but the high DM content that is required may lead to increased field losses.

Compared with clamp silage, big bags have the following features:

Advantages	*Disadvantages*
Effective way to deal with small quantities of grass.	High cost of polythene and requirement for disposal.
When baled by contractor, very low capital expenditure.	Risk of damage to film, with complete loss of bale.
Bales can be stored in field and moved when needed for feeding.	Some risk of *Listeria monocytogenes*, a pathogen, developing in surface layers.
Can have very efficient fermentation (low losses).	Mechanical feeding may be more complex.
Excellent when small amounts of silage needed (e.g. for buffer feeding).	
Very low losses from aerobic deterioration during feed out.	

Crops for ensilage

Many green crops can be ensiled and also arable by-products such as sugar-beet tops and pea haulms. By far the most common crop is grass (including grass–clover herbage), which should be ensiled when its digestibility is at least 65, and for high-quality silage, the D-value at cutting must be 67–70.

The crop will lose about 2 units of D during the ensilage period even where there is a good fermentation, and as much as 5 units of D can be lost if fermentation is poor.

Whole-crop cereals

Wheat, barley and oats can give yields of over 10 t DM/ha when cut for silage from 2 weeks after full ear emergence to 2 weeks before combine-ripeness. These are attractive crops with high yields coming in a single cut. With DM content at harvest from 30–60% there is no need for wilting, and direct cutting techniques can be used. There is little risk of clostridial fermentation, but the silages are prone to aerobic deterioration, with this problem being greatest with the more mature materials with high DM content. It has been found, however, that this problem can be solved by the addition of urea at ensiling. The urea breaks down to ammonia and gives storage in alkaline conditions with pH of above 8. Wetter materials can be ensiled without urea treatment and undergo a normal fermentation.

These crops have a low content of crude protein and modest digestibility of 62–63 D. However, animal performance may be reasonable, because the intakes of whole-crop cereal silages and mixtures with grass silages are higher than would be expected on the basis of digestibility alone (Browne *et al.*, 1995).

Special arable mixtures
These are sometimes grown for arable silage, usually based on the traditional oats–legume combination; this has the advantage of combining the high protein value of the legume with the good carbohydrate content of the cereal. Also this type of mixture requires less fertilizer N than straight cereals or grasses. Examples of such mixtures are:

125 kg/ha oats	125 kg/ha oats
35 kg/ha vetches	45 kg/ha forage peas
or 35 kg/ha beans	

Legumes
Lucerne and red clover may be grown as special crops for silage, either alone or in mixture with grass. Silage is commonly made from grass–white clover mixtures, also used for grazing. A recent European project (LEGSIL) provides comprehensive information on the production, ensiling and feeding of legume silages (Wilkins & Paul, 2002). These crops are characterized by low WSC content and high buffering capacity (this measures the quantity of acid required to lower pH). Consequently, ensiling at low DM contents without additive gives a very high risk of clostridial fermentation and poor silage quality. However, silages of good fermentation quality were consistently obtained in LEGSIL, provided (1) the crops were wilted to approx. 25% DM and ensiled with high rates of formic acid or (2) crops were wilted to 30–35% DM and ensiled either with a lower rate of formic acid or with a bacterial inoculum (Ecosyl). With these provisos, preservation was good in either bunkers or big bales. The resulting silages were stable in aerobic conditions and when fed to dairy cows gave higher intakes and higher milk yields than with grass silages.

Maize
Maize can make excellent silage provided it is well chopped before ensiling. A crop of maize with grain at the 'pasty' stage has a high carbohydrate and low protein content, a D-value of 63–65 and a DM content of about 30% as it stands in the field. In the silo, this crop ferments well without additives and it is very well suited to mechanical handling. Provided its low protein value is recognized at feeding time, it is excellent material for silage and a good contributor to stock nutrition. Mixtures of maize silage with grass silage generally give higher intakes and performance than that obtained with grass silage alone. Nitrogen-use efficiency by the animal is also improved by the use of maize.

Progress in plant breeding means that there are now varieties available that will consistently give high yields and reach adequate maturity in much of England. There may, however, be some problems with harvests in late October on heavy soils in wet conditions. Maize is, though, now established as an important silage crop in England, with over 100 000 ha grown annually.

Silage facts and figures

High yields of crop are essential for silage to justify the machinery, labour and fuel costs involved. Table 4.19 is a guide to the fertilizer rates and expected yields of grass silage cut at three different times in the season. High harvesting costs have led to a recent reduction in the quantity of silage made from light crops in late season. This herbage may be more economically used by grazing.

Density of silage

Clamp silo	Unchopped	20–25% DM = 720–800 kg/m^3
	Chopped	20–25% DM = 800–850 kg/m^3
Tower	Chopped	35–45% DM = 500 kg/m^3

Feeding values
See Table 4.20.

Haymaking

Haymaking is still a common method of conservation, involving the reduction of moisture from fresh grass at 80% to about 20% when the product can be stored. Successful haymaking should follow the following principles:

- Grass should be at the correct stage of growth when cut. As grass is allowed to mature its total yield increases and its moisture content falls, so it may be tempting to allow a very heavy, mature crop to develop before cutting. However, *digestibility* falls at a rate of one-third to one-half of a D-value unit/day once the seed-heads have formed; mature hay has low feed value and low intake characteristics, even though it may be well made.
- Losses should be kept to a minimum. Losses can arise in the following ways:
  - Respiration losses will occur in the field as the

Table 4.19 Good yields from grass cut for silage (t/ha)

Cutting period	*Fertilizer to apply (kg/ha)*	*Site class 1*			*Site class 3*			*Site class 5*		
		DM cut	*Grass cut @ 25% DM*	*Silage @ 25% DM*	*DM cut*	*Grass cut @ 25% DM*	*Silage @ 25% DM*	*DM cut*	*Grass Cut @ 25% DM*	*Silage @ 25% DM*
May	125 N (+30 P_2O_5 and 50 K_2O on index 0 and 1 soils)	6	24	20	6	24	20	6	24	20
July	100 N (+ at least 50 K_2O on soils below index 3)	4	16	12	3	12	10	Second cut likely in wet years only		
Aug/Sept	80 N (+ at least 40 K_2O on soils below index 3)	2.5	10	8	Third cut likely in wet years only					
Total over season		12.5	50	40	9	36	30	6	24	20
Expected number of cuts			3			2			1	

Table 4.20 Typical values for a range of silages

	DM (%)	*ME (MJ/kg DM)*	*D-value*	*Crude protein (% in DM)*	*DCP (g/kg)*
Grass silage					
Excellent	25	11.2	70	18	120
Good	25	10.7	67	16	105
Average	25	10.2	64	14	95
Poor	25	9.8	61	12	75
Whole-crop cereals					
Barley	30	9.3	62	10	50
Oats	30	8.6	57	9	60
Wheat	30	8.4	55	8	35
Oats and vetches	27	9.6	60	16	95
Lucerne	27	9.4	62	20	160
Forage peas	22	9.4	65	20	150
Maize	24	10.6	65	10	70

herbage continues to respire after cutting. Rapid drying will minimize these losses, which can amount to 2–10% of the DM yield.

- Mechanical losses occur if herbage is fragmented by machinery into pieces too small to be picked when the hay is collected for storage. Mechanical losses tend to become greater as the herbage becomes drier and when the action of the machines is abrasive. Mechanical losses are commonly above 10%.
- Leaching of nutrients can take place when the cut material is exposed to rain. Ideally hay is made and removed from the field without rainfall, when this loss is zero, but long periods of heavy rain can lead to a DM loss of up to 15% and a soluble nutrient loss far in excess of this.
- Some of the hay may itself be inedible due to dust and mould. Both these factors arise when hay is made under adverse conditions; the loss can be up to 15% of the DM and it emphasizes the importance of making hay under good climatic conditions.

When grass is cut and a wide area of swath is exposed to wind and sun, there is a rapid initial loss of moisture, as external moisture is lost and water from the outer cells of the leaves and stems. At this stage drying will take place without much sun provided the atmosphere has a low relative humidity. Also at this stage of drying the material can be treated quite roughly by machines. As drying proceeds it becomes progressively more difficult to remove water and the rate of moisture loss declines, together with an increasing risk of mechanical damage. For continued drying sun is necessary to provide heat, and wind is valuable for removing water vapour from within the swath. The rate of drying can be speeded up if the grass is cut with a mower conditioner.

Field drying rate is maximized if the cutting machine leaves the largest possible leaf area exposed to sun and wind, and not tight swaths. If bad weather threatens during the drying process, the material should be windrowed to present the smallest area to the rain. Also it is sound practice to swath-up the grass at night to minimize the effect of dew, ensuring the ground and surface of the swath are dry before the material is again spread next morning.

The rate of drying depends on the weather, proper use of machinery, rate of N used, varieties in the sward, stage of maturity at cutting, bulk of grass present, time of year and desired moisture content at transporting from the field. Most hay is baled. It should not be baled until it is fit for storage, i.e. with a moisture content of about 20%.

Barn hay drying

This is a very useful technique for reducing the risk of bad weather, minimizing losses and making hay from younger material (at a higher feed value) or where higher N rates have been used. The material is cured as far as possible in the field, certainly down to at least 30% moisture. It can then be baled, taking care not to over-compress the bale, and the bales carefully placed over a grid or ducts through which air can be blown. There must be no gaps between the bales, otherwise air will take the line of least resistance and fail to pass through the bales. If the outside humidity is low and only some 5–8% of moisture needs to be removed, then cold air blown for up to a week will suffice, often ceasing to blow at night when humidity might be high. During periods of sustained high humidity or when considerable moisture must be removed, the air must be heated. The air can be heated by either a flame or electric heaters, but in any event the heating of air is *very expensive*. Also, barn-drying installations cannot deal with a sudden large volume of material to cure. Barn drying is best regarded as a means of producing a limited quantity of quality hay.

Success in barn drying depends on:

- justifying the extra costs involved, rather by the higher quality of the hay than by a salvage operation on a mediocre crop;
- allowing moisture content to fall to below 30% (and certainly 35%) before baling;
- using bales at a low to moderate packed density;
- stacking the bales carefully to avoid cracks through which the majority of the air can pass;
- having sufficient fan capacity to obtain a good flow of air, with heating available if humidity is high or bales are wet.

Little hay is barn dried in the UK, despite the high-quality product that can be obtained. The technique is still widely used in areas of Europe where silage making is prohibited because of concerns of possible contamination of milk for cheese making with lactate-fermenting bacteria in silages.

Hay additives

Chemical and biological additives are sometimes used in haymaking, as discussed by Muck and Shinners (2001). The potential for such additives is extremely high as a successful additive would allow hay to be removed from the field at a similar DM content as that used for barn drying and reduce weather risks when compared with field curing. Additives based on propionic acid and on anhydrous ammonia are marketed, but additives have not had the impact in haymaking that they have had with silage. There are risks of high losses through volatilization at application, difficulties in achieving even additive distribution and possible problems of additive metabolism and loss occurring during storage.

Hay facts and figures

Quality in hay can be judged by its colour, which should be bright green/yellow, by its sweet smell and an absence of dust. Also the feeding value of hay can be determined by its analysis (Table 4.21).

Yields

Yields of hay vary considerably, and a high N rate should not be used to grow a hay crop because this aggravates the curing problem; a maximum N rate of 80 kg/ha for the growth period is recommended.

Light crop	=	2–3 t/ha of made hay
Medium crop	=	4.5 t/ha of made hay
Heavy crop	=	5 t/ha and over

Table 4.21 Typical values for a range of types of hay

	DM (%)	*ME (MJ/kg DM)*	*D-value*	*Crude protein (% in DM)*	*DCP (g/kg)*
Grass hay					
Excellent[1]	85	10.7	67	13	100
Good	85	9.6	64	11	60
Average	85	9.0	60	10	50
Poor	85	8.4	56	9	40
Grass–clover hay	85	10.0	64	15	110
Lucerne	85	8.8	55	20	140
Sainfoin	85	9.3	58	16	120

[1] Barn-dried samples can have higher values for ME, D and protein.

Table 4.22 Feed analysis figures for dried grass and lucerne

	DM (%)	*ME (MJ/kg DM)*	*D-value*	*Crude protein (% in DM)*	*DCP (g/kg)*
Dried grass					
Leafy grass	90	11.2	70	16	110
Average	90	9.6	64	15	95
Lucerne	90	9.6	60	24	170

A standard bale of hay measures approximately 0.9 × 0.45 × 0.35 m and weighs 20–30 kg depending on type of material and density (33–50 bales/t). A big round bale of hay measures about 1.5 m high and 1.8 m in diameter and contains some 25 times more material than a standard bale, weighing 500–600 kg. A big rectangular bale is approximately 1.5 × 1.5 × 2.4 m with the same weight as the round bale.

Approximate storage volumes

	m^3/t
Loose medium-length hay	12–15
Standard bales	7–10
Big rectangular bales	10

Green crop drying

When green crops are passed through a high-temperature drier, they are rapidly reduced to a stable moisture content with little or no loss of nutrient value. In the UK the majority of green crops dried are either grass or lucerne; both are traded under the general term 'dried grass'. Less than 1% of the grassland area of the UK is used for dried grass production, and most of the drying is done in a few specialized large units where the size of operation justifies the use of the large, very efficient high-temperature triple-pass drier, the smallest of which can produce some 3000 t of product per year, needing about 400 ha of land to provide material. Most green crop driers are members of the British Association of Green Crop Driers, who supply specialist information to members on a variety of topics related to the industry. Dried grass has a very high reputation as a supplement to silage in cattle and cow rations, and specialist markets have been developed, e.g. for horses and goats. It is unlikely, however, that green crop drying will expand in the foreseeable future, because it demands a high input of fossil fuel – up to about 200 litres of oil/t of material produced – quite apart from the large equipment needed to cut and transport the grass to the drier and the power needed by the mill and cuber to package the material after drying. The present driers in the UK have been in operation for several years and there has been little investment in new drying plant.

There is a subsidy payment from the EU for dried grass, because of the high output of protein that is obtained, the marketability of the product and the ability to use it in compound feeds.

In an efficient grass-drying unit, dry-matter losses are only about 3–5%, and in this respect the technique is far superior to hay and silage making (Table 4.22).

Grassland farming and the environment

As grassland is the most widely occurring crop in the UK it is inevitable that grassland farming has a broad interface with many aspects of the environment. This section will consider firstly grassland in relation to the pollution of water and the atmosphere and secondly grassland biodiversity.

Water and atmosphere

Problems from both point source and diffuse pollution may arise from grassland farming. The approach required to minimize risks is outlined in the *Code of Good Agricultural Practice for the Protection of Air* and the companion *Code of Good Agricultural Practice for the Protection of Water* (MAFF, 1998a and b).

Point source pollution of watercourses is covered by legislation to prevent contamination of water with silage effluent, slurry, dirty water and fuel oil. New facilities for storage of these materials must satisfy regulations and watercourses must not be polluted. The BOD (biological oxygen demand) of farm wastes is extremely high compared with domestic sewage, meaning that rigorous attention to the prevention of escape of water is needed. Clear guidance is given in the *Code of Good Agricultural Practice* (MAFF, 1998b). Substantial numbers of incidents of point source pollution still occur, particularly with dairying, leading to prosecutions of farmers, as well as reductions in water quality.

Figures for BOD are given below:

	BOD (mg/litre)
Raw domestic sewage	400
Dairy washings	2000
Cow slurry	35000
Silage effluent	67000
Milk	140000

Whilst it may cost a farmer money and time to ensure the prevention of point source pollution, techniques are available to achieve that objective. With diffuse pollution the effects are less obvious; they generally arise from the actions of a large number of farmers in an area and appropriate technical solutions are not yet available in all cases.

The following areas of concern are relevant to grassland systems:

- nitrates in water, giving adverse effects on drinking water quality and contributing to eutrophication of rivers and lakes and associated algal growth and effects on aquatic ecosystems;
- ammonia to the atmosphere, contributing to acid rain and, through N deposition, causing the deterioration of fragile ecosystems, particularly heathlands;
- nitrous oxide to the atmosphere – a powerful greenhouse gas;
- phosphorus compounds in water, contributing to eutrophication;
- methane to the atmosphere – a powerful greenhouse gas.

Agriculture, and particularly grassland farming, makes a major contribution to the total national emissions of all these pollutants. The EU in 1991 adopted the Nitrate Directive which required member states to introduce restrictions on agriculture in catchment areas within which water already exceeds 50 mg nitrate/litre (the maximum permitted in drinking water) or is at risk of so doing. Substantial areas of the country have been identified as Nitrate Vulnerable Zones and restrictions are imposed on farming methods in those areas in order to reduce nitrate leaching. Whilst legislation to limit diffuse pollution from P and the emissions of ammonia and greenhouse gases has not yet been implemented in Britain, it is probable that restrictions will be imposed within the next few years.

Nitrogen compounds

Three of the above five pollutants are N compounds, so that the efficiency of use of N in grassland systems is critical. Transformations and losses of N are discussed in Chapter 1 and in relation to grassland in Jarvis and Pain (1997). The inputs and outputs of N on a typical dairy farm in south-west England are shown in Fig. 4.12. The animal product represents only about 20% of the N coming on to the farm in fertilizers and feeds. The remaining N will either accumulate in soil organic matter or be lost through nitrate leaching, as ammonia gas or, after denitrification, as nitrogen gas or nitrous oxide.

Nitrate leaching occurs during the winter when the nitrate which has accumulated in the soil profile is eluted by excess rainfall. The quantity of mineral-N in the soil in autumn provides a reasonable guide to potential leaching. Some of this N will be lost by denitrification rather than leaching, but the quantity of potentially leachable N will be augmented by mineralization of soil organic matter in mild conditions. In grazing systems, urine patches are N 'hotspots' and a substantial proportion of N leaching is derived from urine patch areas. N leaching may be reduced by balancing N supply with crop uptake potential and by adopting cutting rather than grazing management.

The major sources of loss of ammonia in grassland systems are associated with housed rather than grazing animals. Large quantities of N are lost as ammonia from animal buildings, from slurry stores and after field application of slurry and manures. There are also substantial losses from excreta returned during grazing. Direct losses from N fertilizer are small, with the exception of urea. System losses may be reduced by covering slurry stores, by rapid incorporation of slurry and manure in the soil and by the use of shallow injectors or trailing shoe applicators for slurry application to grassland. Losses tend to be lower in systems based on farmyard manure rather than slurry.

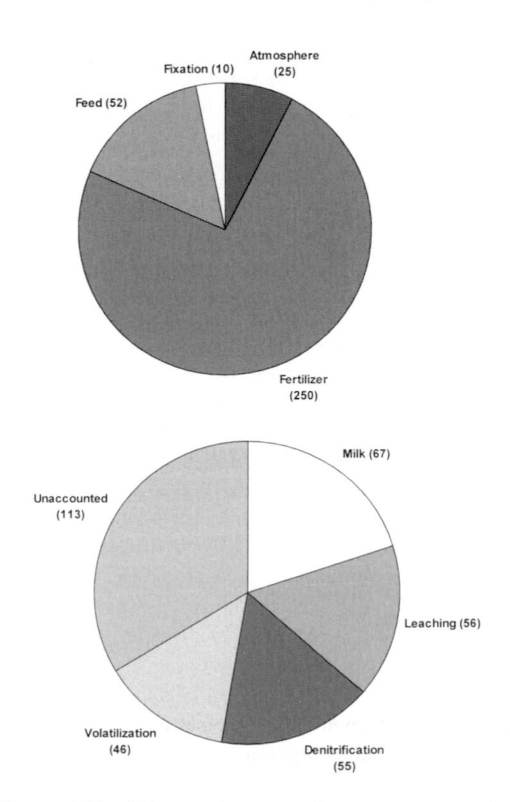

Figure 4.12 Nitrogen inputs and outputs on a typical dairy farm (kg/ha) (Jarvis, 1993).

Denitrification yields both nitrogen gas and nitrous oxide. Losses are particularly large in wet and waterlogged conditions when substantial quantities of mineral N are present in the soil, for instance after application of N fertilizer.

There is a risk that focusing on the reduction of one source of loss may increase loss of N in another direction. Drainage of grassland will, for example, reduce losses through denitrificaiton, but increase nitrate leaching. Reducing the grazing season will reduce leaching losses, but probably increase losses of ammonia, from the increased quantity of slurry that will be produced.

It is probably best to focus on the efficiency of N use in the system as a whole and seek approaches to reduce N inputs whilst maintaining plant and animal output. Table 4.23 indicates the scope for improvement in the various stages of conversion of N from soil to animal product, whilst Table 4.24 indicates the effects on output and N losses of various strategies for modifying the N use efficiency of a dairy farm.

Nutrient budgets are a valuable tool for focusing on overall N use and for analysing the possibilities for increasing efficiency at the farm level. Farm gate budgets are particularly valuable and calculated quite simply from the imports into the farm of N in fertilizer, feed, rainfall and any purchased slurry or manure in relation to the outputs from the farm in milk and meat and any crops or manures sold. The difference between inputs and outputs, expressed per hectare, is the N surplus of the system. If there are substantial quantities of legumes present, then an estimate of biological N fixation should be added to the input figures. The 'surplus'

Table 4.23 Efficiencies (%) of component parts of specialized intensive dairy farming systems in The Netherlands on sandy soil: (1) technically attainable and (2) realized in practice by skilled farmers (Jarvis & Aarts, 2000)

Component	*Technically attainable*		*Average realized in practice*	
	N	*P*	*N*	*P*
Soil: transfer from soil to harvestable crop	77	100	53	60
Crop: transfer from harvested crop to feed intake	86	92	71	75
Animal: conversion from feed to milk+meat	25	32	18	22
Slurry/dung+urine: transfer from excreta to soil	93	100	80	100
Whole farm: from inputs to outputs	36	100	16	27

Table 4.24 Approaches to increase N efficiency on UK dairy farms (model calculations in kg/ha) (Jarvis *et al.*, 1996; Jarvis & Aarts, 2000)

	Typical management	*Injected slurry; tactical fertilizer*	*Grass–clover with no N fertilizer*
N inputs			
Fertilizers	250	155	0
Fixation	10	10	144
Total[1]	337	242	210
N in milk	67	67	54
N surplus	270	175	156
N losses	160	86	79

[1] Includes also imported feeds and atmospheric deposition.

Table 4.25 P budgets for contrasting farm types in the UK expressed as kg P/ha (adapted from Haygarth *et al.*, 1998)

	Dairy	*Hill sheep*
Inputs		
Atmosphere	0.23	0.11
Fertilizer	15.96	0.44
Feed and straw	27.23	0.19
Total	43.42	0.74
Outputs		
Milk, meat and wool	16.16	0.15
Lost to water	1.00	0.26
Heather burning	—	0.09
Total	17.16	0.50
Inputs – Outputs	26.26	0.24

may be stored in soil organic matter or lost through the processes described above. Targets may be fixed for N surplus for different types of system. In The Netherlands nutrient budgets are widely used and farming systems have to operate within legally defined maximum levels of surplus (Van der Meer, 2001).

Phosphorus

The nutrient budget approach is also applicable to phosphorus. The major inputs of P are in fertilizers and in animal feeds. Much of the P in the soil is tightly bound and released gradually with time. Although only a small quantity of P is leached, it may have adverse effects on water quality. Larger quantities of P may be lost by particulate flow in soil after cultivation or poaching by animals and in circumstances when heavy rainfall follows manure or slurry application. Examples of P budgets are given in Table 4.25. There will often be opportunities to reduce P inputs in fertilizers and in feeds with no impact on productivity, at least in the short term, and opportunities to change management to reduce particulate losses of P.

Methane

Although some methane is lost from manures and slurries, this is small compared with the losses of methane from rumen fermentation in cattle and sheep. There is some opportunity to use feeds that give somewhat less methane per unit of ME. Rate of animal production, however, has a big effect on methane loss per unit of animal output. This arises because the quantity of feed used (and methane produced) for animal maintenance (no animal production) is 'diluted' when rate of animal output is high.

Grassland biodiversity

Many species of plants and animals may be found in grassland, but intensification in grassland use and management has generally led to a reduction in the number and diversity of species present in and associated with grassland. Decline in the area of grassland rich in plant species illustrates this trend. A report in 1987 concluded that 95% of neutral grasslands lacked significant wildlife interest and only 3% had not been damaged by intensification (Fuller, 1987). There is wide concern by society about this trend. A series of measures which provide financial incentives to manage grassland in ways that will maintain or enhance biodiversity and the environment have now been introduced. In 1985 the EU authorized member states to introduce special national schemes in environmentally sensitive areas to subsidize farming practices favourable to the environment. Subsequently it was agreed that such agri-environmental measures would receive financial support from Brussels and in 1992 countries were formally obliged to introduce such schemes.

The Environmentally Sensitive Area (ESA) scheme and Countryside Stewardship (CS) are particularly important in England, with similar voluntary schemes operating in the other countries of the UK. Essentially, within these schemes farmers are offered payments in return for complying with particular management practices. These practices have been defined in order to maintain or enhance particular features of the environment and, in the case of CS, also for provision of access to the countryside. Some 15% of the agricultural land in the UK is included in ESA areas, covering over 1 million ha of grassland with around one-third of this land now committed to the scheme. CS is available

Table 4.26 Generalized effects of agricultural management options on utilized output and on wildlife (adapted from Wilkins & Harvey, 1994). The direction of change is indicated by + and –; V indicates variable effects

	Utilized output	*No. of higher plants*	*Insects*	*Birds*	*Small mammals*
Increase fertility	++	– –	– –	–	0
Increase level of utilization[1]	++	– –	– –	–	– –
Increase reseeding	+	– –	– –	– –	–
Improve drainage	+	– –	– –	– –	?
Use herbicides	+	– –	– –	–	–
Use pesticides	+	0	– –	V	–
Earlier cutting	+	– –	– –	– –	–
Later cutting	–	++	++	++	++
Graze cattle rather than sheep	–	+	++	+	+
Graze intermittently rather than continuously	0	V	++	++	V

[1] Particularly through increase in grazing severity

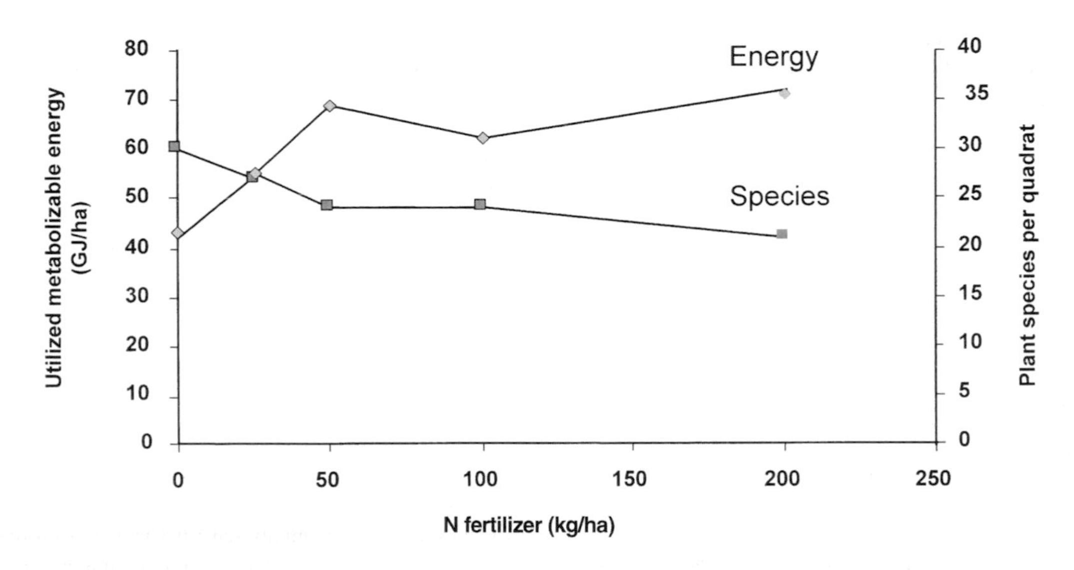

Figure 4.13 Output and number of plant species with different fertilizer N inputs.

throughout the country. A further important initiative has arisen from the Biodiversity Convention signed in Rio de Janeiro in 1992. Biodiversity Action Plans are being developed to protect and enhance specific important habitats, including grassland habitats.

The effects of agricultural management options on utilized output and on wildlife are summarized in Table 4.26. This highlights the conflicts between agricultural efficiency and biodiversity and the difficulties in combining these two objectives on the same area of land. Research carried out on species-rich grassland in Somerset illustrates this problem (Tallowin, 1996; Fig. 4.13). As N fertilizer inputs increased from 0–200 kg N/ha, there was progressive increase in UME, but a decline in the number of plant species present. Significant reductions in plant species were found within 3 years, even with the lowest N rate of only 25 kg N/ha.

Many of the ESA regulations will limit agricultural output and are intended to increase biodiversity. There are often several tiers in a particular scheme, with the higher tiers having more severe restrictions, but higher payments. Commonly the regulations will:

- limit the use of both fertilizers and manures;
- prohibit the use of herbicides, fungicides and insecticides;
- prohibit cutting for hay until after a particular date, such as 15 July (this will help ground-nesting birds and aid seed return).

The ESA scheme has generally been successful, with high voluntary participation rates and limiting intensification on the participating farms. However, there is a risk that, particularly with the lower tiers, the scheme may encourage management at a moderate level of intensity, with reduced agricultural efficiency, but delivering little environmental benefit. Simply reducing inputs to grassland that has previously been managed intensively may not lead to increases in the number of plant species present, as discussed by Wilkins and Harvey (1994) and Peeters and Janssens (1998). High levels of fertility from previous management, particularly in relation to soil P, will limit the development of new species. Also, there is a need for seeds or propagules of new species to be present in the seed bank or in surrounding vegetation. Seed of new species may need to be introduced to achieve a more diverse sward.

There is scope on most farms for identifying areas that may be best suited for management for increased biodiversity. These are often wet, steep or remote from farm buildings and will have had less intensive management in the past. Targeting actions to further increase biodiversity in these areas whilst managing for efficient agricultural production on the remainder of the farm may give greater benefits overall than managing the whole area at an intermediate level of intensity (Wilkins & Harvey, 1994). Areas targeted for biodiversity will provide some agricultural output, with grazing being an important management tool. The quality of the feed produced will often be low and research is now being directed to integrating the use of areas managed for biodiversity together with agriculturally improved grassland in efficient animal production systems.

Glossary of grassland terms

Based on Hodgson (1979) and Thomas (1980), with additions.

Anthesis: flower opening.

Biomass: weight of living plant and/or animal material.

Browsing: the defoliation by animals of the aboveground parts of shrubs and trees.

Buffer feeding (or grazing): provision of additional forage when quantity in grazing area is insufficient.

Canopy: the sward canopy as it intercepts or absorbs light.

Canopy structure: the distribution and arrangement of the components of the canopy.

Closed canopy: a canopy that either has achieved complete cover or intercepts 95% of visible light.

Cow equivalent (*CE*): an aggregated liveweight of animals equivalent to a standard cow of 550 kg (synonymous with LSU).

Crop growth rate (*CGR*): the rate of increase in dry weight per unit area of all or part of a sward.

Crop (*standing*): the herbage growing in the field before it is harvested.

Crown: the top of the tap-root bearing buds from which the basal leaf rosette and shoots arise (appropriate to clovers but *not* grasses).

Culm: the extended stem of a grass tiller bearing the inflorescence.

D-value: digestibility of organic matter in the DM eaten (as %).

Date of heading [*or date of ear* (*inflorescence*) *emergence*]: for a sward, the date on which 50% of the ears in fertile tillers have emerged.

Defoliation: the severing and removal of part or all of the herbage by grazing animals or cutting machines.

Density: the number of items (e.g. plants or tillers) per unit area.

Flag leaf: in grass this is the final leaf produced on the flowering stem.

Foliage: a collective term for the leaves of a plant or community.

Forage: any plant material, except in concentrated feeds, used as a food for domestic herbivores.

Grassland: the type of plant community, natural or sown, dominated by herbaceous species such as grasses and clovers.

Grazing: defoliation by animals.

Grazing cycle: the length of time between the beginning of one grazing period and the beginning of the next.

Grazing period: the length of time for which a particular area of land is grazed.

Grazing pressure: the number of animals of a specified class per unit weight of herbage at a point of time.

Grazing systems:

- *Continuous stocking*: the practice of allowing animals unrestricted access to an area of land for the whole or a substantial part of a grazing season.
- *Creep grazing*: the practice of allowing young animals (lambs or calves) to graze an area which their dams cannot reach.
- *Mixed grazing*: the use of cattle and sheep in a common grazing system whether or not the two species graze the same area of land at the same time.
- *Rotational grazing*: the practice of imposing a regular sequence of grazing and rest from grazing upon a series of grazing areas.
- *Set stocking*: the practice of allowing a fixed number of animals unrestricted access to a fixed area of land for a substantial part of a grazing season.

Harvest year: first full harvest year is the calendar year following the seeding year.

Herbage: the above-ground parts of a sward viewed as an accumulation of plant material with characteristics of mass and nutritive value.

Herbage allowance: the weight of herbage per unit of liveweight at a point in time.

Herbage consumed: the herbage mass once it has been removed by grazing animals.

Herbage cut: the stratum of material above cutting height.

Herbage growth: the increase in weight of herbage per unit area over a given time interval due to the production of new material.

Herbage mass: the weight per unit area of the standing crop of herbage above the defoliation height.

Herbage residual: the herbage remaining after defoliation.

Inflorescence emergence: the first appearance of the tip of a grass inflorescence at the mouth of the sheath of the flag-leaf.

Leaf: in grasses = lamina + ligule + sheath; in clovers = lamina + petiole + stipule. Note leaf is *not* lamina.

Leaf area index (LAI): the area of green leaf (one side) per unit area of ground. *Critical LAI* is LAI at which 95% visible light is intercepted; *maximum LAI* is the greatest LAI produced by a sward during a growth period; *optimum LAI* is the LAI at which maximum crop growth rate is achieved.

Leaf area ratio: total lamina area divided by total plant weight: *specific leaf area* is the lamina area divided by lamina weight.

Leaf burn or scorch: damage to leaves caused by severe weather conditions, herbicides, etc. This contrasts with *leaf senescence* which is a genetically predetermined process where leaves age and die, usually involving degradation of the chlorophyll in the leaves.

Leaf emergence: a leaf in grass is fully emerged when its ligule is visible or when the lamina adopts an angle to the sheath. In clovers the leaf is emerged when the leaflets have unfolded along the midrib and are almost flat.

Leaf length to weight ratio: lamina weight divided by lamina length.

Livestock Unit (LSU): see cow equivalent.

Net assimilation rate (NAR): defined as $1/A.\mathrm{d}W/\mathrm{d}t$ where A = total leaf area, W = total plant weight, t = time, units = $(\mathrm{g\,m}^2)/\mathrm{d}$.

Palatable: pleasant to taste.

Plastochron: the interval between the initiation of successive leaf primordia on a stem axis.

Preference: the discrimination exerted by animals between areas of a sward or components of a sward canopy, or between species in cut herbage. *Preference ranking* is based on the relative intake of herbage samples when the animals have a complete free choice of the materials.

Pseudostem: the concentric leaf sheaths of a grass tiller which perform the supporting function of a stem.

Regrowth: the production of new material above the height of defoliation after defoliation, often with initial regrowth at the expense of reserves stored in the stubble.

Rest period: the length of time between the end of one grazing and the start of the next on a particular area.

Seed ripeness: the stage at which the seed can be harvested successfully.

Seeding year: the calendar year in which the seed is sown.

Seedling emergence: a seedling has emerged when the shoot first appears above the ground.

Selection: (by animals) the removal of some components of a sward rather than others.

Shoot bases: the part of the sward below the anticipated height of defoliation. This becomes *stubble* after defoliation.

Simulated sward: an assemblage of plants intended to represent in convenient form a 'normal' sward.

Sod: a piece of turf lifted from the sward, either by machine or grazing animals.

Spaced plant: a plant grown in a row so that its canopy does not touch or overlap that of any other plant.

Standing crop: the herbage growing in the field before it is harvested.

Stem: the main axis of a shoot, bearing leaves.

Stocking density: the number of animals of a specified class per unit area of land actually being grazed at a point in time.

Stocking rate: number or weight of animals kept on a unit area of land for a long period (preferably for 12 months, so including grazed area *and* conserved area).

Sward: an area of grassland with a short continuous foliage cover, including both above- and below-ground parts, but not any woody plants.

Sward establishment: the growth and development of a sward in the seeding year. A *primary sward* is one that has never been defoliated. A *mixed sward* is one that contains more than one variety or species. A *pure sward* is one that contains a single stated variety or species.

Sward height: height above ground at which a descending finger first touches vegetative material (ignoring stems and seed heads).

Tiller: an aerial shoot of a grass plant, arising from a leaf axil, normally at the base of an older tiller. An *aerial tiller* is one that develops from a node of an extended stem.

Tiller appearance rate (TAR): the rate at which tillers become apparent to the eye without dissection of the plant.

Tiller base: the part of a growing tiller below the height of defoliation.

Tiller stub: the part of a tiller left after defoliation.

Turf: the part of the sward that comprises the shoot system plus the uppermost layer of roots and soil.

Utilized metabolizable energy (UME): the quantity of ME accounted for in animal maintenance and production. *If UME of grass*, then UME after deduction of ME supplied from non-grassland sources.

Winter burn: leaf burn in winter.

Winter hardiness: the general ability of a variety or species to withstand the winter.

Winter kill: death of plants in winter.

Acknowledgement (by R.J. Wilkins)

Dr. John Brockman, the sole author of this chapter in the previous edition of *The Agricultural Notebook*, died in 1998. I was asked at a late stage to undertake the updating and revision of this chapter. This I have done, whilst retaining the previous basic structure and much of John's material and ideas. I had close contact with John for many years in association with Seale-Hayne and with the British Grassland Society. I admired his work and much regret that he is not able to approve these revisions to his chapter.

References

Baker, H.K. (1962) A Survey of English Grasslands. *Proceedings of the 6th Weed Control Conference*, Brighton, **1**, 23–30.

Brown, L., Scholefield, D., Jewkes, E.C., Preedy, N., Wadge, K. & Butler, M. (2000) The effect of sulphur application on the efficiency of nitrogen use in two contrasting grassland soils. *Journal of Agricultural Science*, **135**, 131–8.

Browne, I., Allen, D., Phipps, R. & Sutton, J. (1995) *Mixed Forage Diets for Dairy Cows*. Milk Development Council, London.

Chambers, B., Nicholson, N., Smith, K., Pain, B., Cumby, T. & Scotford, I. (undated) *Managing Livestock Wastes. Making Better Use of Livestock Manures on Grassland*. Booklet no. 2. IGER, ADAS and SRI.

Davies, W. (1960) *The Grass Crop*. Spon, London.

Forbes, T.J., Dibb, C., Green, J.O., Hopkins, A. & Peel, S. (1980) *Factors Affecting the Production of Permanent Grass*. Grassland Research Institute, Hurley.

Frame, J., Charlton, J.F.L. & Laidlaw, A.S. (1998) *Temperate Forage Legumes*. CAB International, Wallingford.

Fraser, M.D., Fychan, R. & Jones, R. (2000) Voluntary intake, digestibility and nitrogen utilization by sheep fed ensiled forage legumes. *Grass and Forage Science*, **55**, 271–9.

Fuller, R.M. (1987) The changing extent and conservation interest of lowland grasslands in England and Wales: a review of grassland surveys. *Biological Conservation*, **40**, 281–300.

Green, J.O. (1974) *Preliminary Report on a Sample Survey of Grassland in England and Wales*. Grassland Research Institute, Hurley.

Green, J.O. (1982) *A Sample Survey of Grassland in England and Wales 1970–1972*. Grassland Research Institute, Hurley.

Haygarth, P., Chapman, P.J., Jarvis, S.C. & Smith, R.V. (1998) Phosphorus budgets for two contrasting grassland farming systems in the UK. *Soil Use and Management*, **14**, 160–167.

Hides, D. & Humphreys, M. (2000) Plant breeding from WPBS to IGER. *IGER Innovations*, **4**, 6–13.

Hodgson, J. (1979) Nomenclature and definitions in grazing studies. *Grass and Forage Science*, **34**, 11–18.

Hodgson, J., Mackie, C.K. & Parker, J.W.G. (1986) Sward surface heights for efficient grazing. *Grass Farmer*, **24**, 5–10.

Hopkins, A. & Davies, R.R. (1994) Changing grassland utilization in the United Kingdom and its implications for pollen production and hay fever. *Grana*, **33**, 71–5.

Hopkins, A., Gilbey, J., Dibb, C., Bowling, P.J. & Murray, P.J. (1990) Response of permanent and reseeded grassland to fertilizer nitrogen. 1. Herbage production and herbage quality. *Grass and Forage Science*, **45**, 43–55.

Jarvis, S.C. (1993) Nutrient cycling and losses from dairy farms. *Soil Use and Management*, **9**, 99–105.

Jarvis, S.C. & Aarts, H.F.M. (2000) Nutrient management from a farming systems perspective. *Grassland Science in Europe*, **5**, 363–73.

Jarvis, S.C. & Pain, B.F. (eds) (1997) *Gaseous Nitrogen Emissions from Grasslands*. CAB International, Wallingford.

Jarvis, S.C., Wilkins, R.J. & Pain, B.F. (1996) Opportunities for reducing the environmental impact of dairy farming managements: a systems approach. *Grass and Forage Science*, **51**, 21–31.

Jones, M.G. (1933) Grassland management and its influence on the sward. IV. The management of poor pastures. V. Edaphie and biotic influences on pastures. *Empire Journal of Experimental Agriculture,* **1**, 361–7.

Leaver, J.D. & Moisey, F.R. (1980) The silage makers' dilemma. Quality or quantity. *Grass Farmer*, **7**, 9–11.

Mackie, C.K. & Wightman, P.S. (1997) Set-aside and alternative uses of grass. In: *Seeds of Progress* (ed. J.R. Weddell). British Grassland Society Occasional Symposium, no. 31, pp. 46–55.

MAFF (1998a) *Code of Good Agricultural Practice for the Protection of Air*. MAFF Publications, London.

MAFF (1998b) *Code of Good Agricultural Practice for the Protection of Water*. MAFF Publications, London.

MAFF (2000) *Fertiliser Recommendations for Agricultural and Horticultural Crops (RB209)*. The Stationery Office, London.

Mayne, C.S. & Laidlaw, A.S. (1995) Extending the grazing season – a research review. In: *Extending the Grazing Season*. Proceeding British Grassland Society Discussion Meeting, Cheshire, April 1995. British Grassland Society, Reading.

Mayne, C.S., Wright, I.A. & Fisher, G.E.J. (2000) Grassland management under grazing and animal response. In: *Grass: Its Production and Utilization* (ed. A. Hopkins), pp. 247–91. Blackwell Science, Oxford.

McDonald, P., Henderson, A.R. & Heron, S.J.E. (1991) *The Biochemistry of Silage*. Chalcombe Publications, Marlow.

Merry, R.J., Jones, R. & Theodorou, M.K. (2000) The conservation of grass. In: *Grass: Its Production and Utilization* (ed. A. Hopkins), pp. 196–228. Blackwell Science, Oxford.

Moorby, J. (2001) Grass sugars make milk production sweeter. *IGER Innovations*, **5**, 36–9.

Morrison, J., Jackson, M.V. & Sparrow, P.E. (1980) *Report of Joint GRI/ADAS Grassland Manuring Trial*. Technical Report no. 27. Grassland Research Institute, Hurley.

Muck, R.E. & Shinners, K.J. (2001) Conserved forage (silage and hay): progress and priorities. *Proceedings of the 19th International Grassland Congress, Sao Pedro, Brazil*, pp. 753–62.

National Institute of Agricultural Botany (annual) *Grasses and Herbage Legumes Variety Leaflet*. NIAB, Cambridge.

Orr, R.J., Parsons, A.J., Penning, P.D. & Treacher, T.T. (1990) Sward composition, animal performance and the potential production of grass/white clover swards continuously stocked by sheep. *Grass and Forage Science*, **45**, 325–36.

Orr, R.J., Rutter, S.M., Penning, P.D., Yarrow, N.H., Atkinson, L.D. & Champion, R.A. (1998) Matching grass supply to grazing patterns for dairy cows under strip-grazing management. *Proceedings of the British Society of Animal Science 1998*, p. 49.

Orr, R.J., Cook, J.E., Atkinson, L.D., Clements, R.O. & Martyn, T.M. (2000) Evaluation of herbage varieties under continuous stocking. In: *Grazing Management* (ed. A.J. Rook & P.D. Penning). British Grassland Society Occasional Symposium, no. 34, pp. 39–44.

Pain, B.F. (2000) Control and utilization of livestock manures. In: *Grass: Its Production and Utilization* (ed. A. Hopkins), pp. 343–64. Blackwell Science, Oxford.

Parsons, A.J. & Chapman, D.F. (2000) The principles of pasture growth and utilization. In: *Grass: Its Production and Utilization* (ed. A. Hopkins), pp. 31–89. Blackwell Science, Oxford.

Parsons, A.J., Orr, R.J., Penning, P.D. & Lockyer, D.R. (1991) Uptake, cycling and fate of nitrogen in grass–clover swards continuously grazed by sheep. *Journal of Agricultural Science*, **116**, 47–61.

Paul, C., Auerbach, H. & Schild, G.-J. (2002) Intake of legume silages by sheep. In: *Legume Silages for Animal Production – LEGSIL* (eds R.J. Wilkins & C. Paul). *Landbauforschung Voelkenrode*, Sonderheft 234, 33–8.

Peeters, A. & Janssens, F. (1998) Species-rich grasslands; diagnostic, restoration and use in intensive livestock production systems. *Grassland Science in Europe*, **3**, 375–93.

Rhodes, I. (2001) A new era for white clover. *IGER Innovations*, **5**, 24–7.

Sheldrick, R.D. (ed) (1997) *Grassland Management in Environmentally Sensitive Areas*. British Grassland Society Occasional Symposium, no. 32.

Tallowin, J.R.B. (1996) Effects of inorganic fertilizers on flower-rich hay meadows: a review using a case study on the Somerset Levels, UK. *Grasslands and Forage Abstracts*, **66**, 147–52.

Thomas, H. (1980) Terminology and definitions in studies of grassland plants. *Grass and Forage Science*, **35**, 13–24.

Thomas, C., Reeve, A. & Fisher, G.E.J. (1991) *Milk from Grass*, 2nd edn. Scottish Agricultural College, Perth.

Titchen, N.M. & Scholefield, D. (1992) The potential of a rapid test for soil mineral N to determine tactical application of fertiliser nitrogen to grassland. *Aspects of Applied Biology*, **30**, 223–9.

Turner, R.W. (1997) Grass mixtures – an industry perspective. In: *Seeds of Progress* (ed. J.R. Weddell). British Grassland Society Occasional Symposium, no. 31, pp. 189–99.

Tyson, K.C., Garwood, E.A., Armstrong, A.C. & Scholefield, D. (1992) Effects of field drainage on the growth of herbage and the liveweight gain of grazing beef cattle. *Grass and Forage Science*, **47**, 290–301.

Van der Meer, H.G. (2001) Grassland and the environment. In: *Progress in Grassland Science: Achievements and Opportunities* (ed. S.C. Jarvis), pp. 53–67. British Grassland Society, Reading/IGER, Aberystwyth.

Weissbach, F., Honig, H. & Kaiser, E. (1993) The effect of nitrate on the silage fermentation. *Proceedings of the 10th International Conference on Silage Research, Dublin*, pp. 122–3.

Wilkins, R.J. & Harvey, H.J. (1994) Management options to achieve agricultural and nature conservation objectives. In: *Grassland Management and Nature Conservation* (eds R.J. Haggar & S. Peel). British Grassland Society Occasional Symposium, no. 28, pp. 86–94.

Wilkins, R.J. & Paul, C. (eds) (2002) *Legume Silages for Animal Production – LEGSIL. Landbauforschung Voelkenrode*, Sonderheft 234, 95 pp.

Wilkins, R.J., Newberry, R.D. & Titchen, N.M. (1983) The effects of sward height on the performance of beef cattle grazing permanent pasture. *Annual Report 1982*, pp. 82–3. Grassland Research Institute, Hurley.

Wolton, K.M. (1980) *AgTec*, Summer 1980. Fisons Fertilizer Division, Felixstowe.

Zimmer, E. & Wilkins, R.J. (eds) (1984) Efficiency of silage systems: a comparison between wilted and unwilted silages. *Landbauforschung Voelkenrode*, **69**.

Further reading

Useful books in addition to the references listed above:

Frame, J. (1992) *Improved Grassland Management*. Farming Press, Ipswich.

Hopkins, A. (ed.) (2000) *Grass: Its Production and Utilization*. Blackwell Science, Oxford.

Jarvis, S.C. (ed) (2001) *Progress in Grassland Science: Achievements and Opportunities*. British Grassland Society, Reading/IGER, Aberystwyth.

Lemaire, G., Hodgson, J., de Moraes, A. de F., Carvalho, P.C. & Nabinger C. (eds) (2000) *Grassland Ecophysiology and Grazing Ecology*. CAB International, Wallingford.

Sheldrick, R.D., Newman, G. & Roberts, D.J. (1995) *Legumes for Milk and Meat*. Chalcombe Publications, Canterbury.

See also:

British Grassland Society Symposium Proceedings published regularly on specific grassland and forage topics, published by British Grassland Society, Reading.

IGER Innovations, reporting research highlights, published annually by Institute of Grassland and Environmental Research, Aberystwyth.

Useful websites

www.britishgrassland.com – British Grassland Society, includes details of meetings and publications.

www.defra.gov.uk – Department for Environment, Food and Rural Affairs, includes statistics and publication details of Codes of Good Agricultural Practice.

www.greenmount.ac.uk/ – Greenmount College, Northern Ireland. Comprehensive grassland information including current grass growth rates in Northern Ireland.

www.iger.bbsrc.ac.uk/igerweb/ – Institute of Grassland and Environmental Research, includes the Institute's Annual Report, details of research projects and publications.

www.mluri.sari.ac.uk/ – Macaulay Land Use Research Institute, Aberdeen, includes details of research projects and publications.

5

Farm woodland management

A.D. Carter & H. Palmer

Introduction

Woodland covers 2 400 000 ha or about 10% of the total land area of Britain, of which farm woodlands make up about 350 000 ha. Less productive than many larger commercial forests, farm woodlands typically occur in small fragmented parcels and comprise a significant proportion of mixed broadleaved woods, especially in the lowlands of Britain, where they form important features in the landscape and may have considerable value for wildlife and game. Unfortunately, timber quality and productivity may be significantly reduced by a combination of factors, including difficult access, poor condition of the growing stock, small size and neglected or inappropriate management. Fortunately, both site conditions and the quality of the trees can be improved by the application of appropriate silvicultural techniques to create a quality timber resource with multiple benefits for the landscape, amenity, wildlife and game. New farm woodland planting is also expanding, as surplus agricultural land becomes available and grant incentives encourage farmers to diversify into woodland enterprises.

Despite this clear potential, a number of perceived obstacles deter landowners from either managing existing woodlands or planting new ones, including a lack of knowledge of silviculture and how to market small volumes of timber. The most enduring financial obstacle is the long gestation period involved in forest investments, requiring a wait of at least 20 years for most crops before thinnings even begin to offset the early establishment costs. This 'income gap' can be partially offset by some of the following options:

- Revenue from harvesting an existing crop can be used to offset the cost of establishment.
- Grants are available to compensate landowners for planting trees and other forms of woodland management.
- Development of shorter rotation crops (e.g. coppice).
- Integration of forestry with other revenue-generating activities in the early life of the crop, such as game management, recreation or growing Christmas trees.
- Agroforestry systems offer some potential to use the spaces between the trees in the early years for low-intensity grazing or other agricultural use.

Setting management objectives

While woodlands can be successfully managed for multiple benefits, competing objectives must be prioritized so that an appropriate silvicultural management plan can be formulated. For the majority of farm woodlands, maximum timber production may have to be compromised to a certain extent to incorporate secondary objectives, such as game and wildlife habitat or to meet the requirements of landscape designations. Once the overall objectives are set, the specific silvicultural operations most appropriate will depend on an assessment of the site conditions and the growing crop, together with a knowledge of the external environment, such as the market for timber produce, the wider policy framework and grant arrangements.

Forest products and markets

Timber is normally sold on a price per unit volume basis, which is set according to the volume available, the state and location of the timber (standing, stacked in the forest or at the roadside), ease of access, distance to markets and the method of sale (auction, tender, or private sale). The use of electronic marketing and auction systems is becoming increasingly popular. Each tree species will have recognized characteristics which must be met for particular markets, such as length, diameter (Fig. 5.1), straightness and freedom from defects. Some markets are much more exacting than others (e.g. veneers), demanding higher prices, so that the harvested timber may be graded and separated into different price bands according to quality. The difference between the market price for low and high quality is enormous. For example, oak veneers may fetch up to £300/m^3, while poor quality oak may only be worth £10/m^3 as firewood.

Small farm woodlands cannot compete with the large upland forests, which supply vast quantities of timber to bulk markets, such as pulp processors and particle board manufacturers. The farm woodland owner is best advised to produce high quality timber, to take advantage of local specialized markets (e.g. craft, furniture, domestic fencing) and to add value wherever possible by processing on the farm. For further information on timber properties of farm woodland trees, see Brazier (1990).

Marketing timber from farm woodlands may require significant effort – seeking out and negotiating with potential buyers, completing the statutory paperwork required by the Forestry Commission, selecting and marking timber to be felled, drawing up contracts and organizing and supervising contractors are all time-consuming activities. The more of this work the woodland owner can carry out themself, the greater the eventual returns will be.

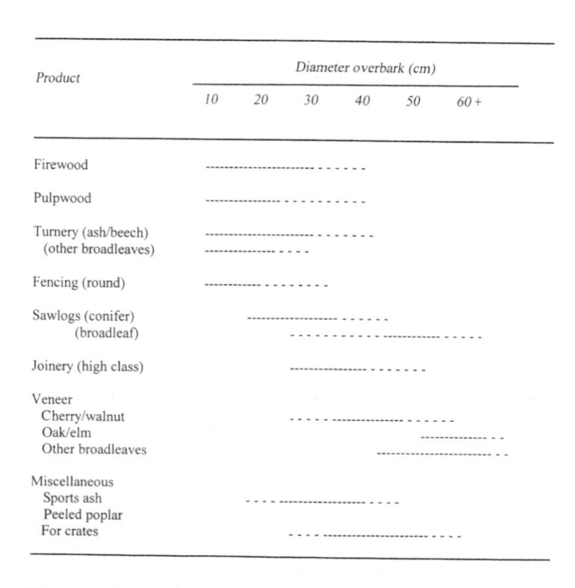

Figure 5.1 Diameter specifications for various end-uses (after Hibberd, 1988, by permission of the Forestry Commission, 2001).

Silvicultural options for existing woodlands

Clear felling

Clear felling involves the harvesting and removal of the whole forest crop at once. As the system of choice for maximum timber production, it enables the creation of a uniform and even-aged restocking, offering simplified forest treatment and associated economies of scale. The use of transplants enables the forest manager to take advantage of superior species, provenances or cultivars better adapted to the site.

The open nature of the clear fell creates substantially different forest conditions to those under the woodland canopy, which has a number of important environmental implications. The increased levels of light, rapid breakdown of organic matter and the exposure of mineral soil encourages the vigourous growth of weed species. The exposed soil may be prone to erosion and increased run-off, especially on sloping ground. The brash left from felling may provide a home for weevils and beetles which threaten future crops. The edges of neighbouring stands will be exposed to increased levels of light and risk of wind damage. Thin-barked species, such as beech, may suffer from sun scorch and oak trees may develop epicormic growth. On some sites, the large-scale clear felling of mature trees may be inadvisable due to amenity or landscape considerations.

Natural regeneration

The majority of existing woodlands have the potential to regenerate from seed that has fallen on the forest floor, given enough time and the right environmental conditions. For the forest manager, this 'free gift' can provide a low-cost means of restocking woodland. Natural regeneration can be used to enhance the amenity value of clear-felled forests by allowing native broadleaved trees to gradually restock pockets of open ground, such as along watercourses and the edges of forest tracks and rides, planting gaps, inaccessible slopes, rocky outcrops or on the forest margins.

Successful natural regeneration relies on interventions to enhance the natural process of regeneration in the forest. Where timber production is an objective of management, the parent crop must be of good quality, well suited to the site and capable of producing large quantities of viable seed. The most prolific seed bearers are ash, sycamore, birch, alder, Norway maple, Scots pine, sweet chestnut, cherry and rowan. Some species produce more seed in mast years (cycles of 3–5 years), such as oak and beech, although some seed may be produced every year. Common practice is to carry out a series of heavy late thinnings to encourage seed production from the best trees, by removing 40–60% of the crop, or overstorey.

Ground preparation may be required to break up weed growth, incorporate humus layers and expose the mineral soil, with the aim of producing a moist, friable soil surface free from competing vegetation. Rotovation, disk ploughing or the wallowing of pigs have proved successful in this respect. An assessment of the restocking of seedlings should be undertaken in the year following seeding, with gaps larger than 7 × 7 m planted up with transplants, if full stocking is required.

The young seedlings will require protection from grazing by herbivores, such as rabbits and deer, by either fencing (which can prove expensive for small areas) or individual protection with tree shelters. The retention of the overstorey allows rank weed growth to be controlled, as well as giving protection from frost and sun scorch. As the young trees become established, the overstorey is gradually removed, increasing the amount of available light to the growing crop. Subsequent operations include the respacing of the saplings to a minimum of 1500 trees/ha.

The creation of a fully stocked woodland through natural regeneration can prove a complex and consequently costly exercise. The key drawbacks for timber production include the reliance on seeding years rather than markets, the need for skilled management, overwhelming local populations of rabbits and deer, the possibility of adverse ground conditions (including brambles and rhododendron), and parent trees of inferior quality.

A number of silvicultural systems take advantage of the potential for natural regeneration:

- *Shelterwood* – a partial overstorey is retained for 2–30 years, depending on the species, creating an uneven-aged woodland structure for at least part of the silvicultural cycle. This can be advantageous where the complete removal of the forest crop may be too drastic for landscape or amenity reasons.
- *Continuous cover* – forest management involves the selection and harvesting of individual trees within any given compartment. The aim is to maintain a spread of trees of all ages throughout the forest at all times, including seed-bearing trees.
- *Group systems* – the size of the forest unit is reduced to a small felling coupe (30–50 m diameter), rather than the whole stand, normally centred on existing gaps or regeneration. Larger gaps (0.1–0.5 ha) are suitable for light-demanding species and are easier to manage, but have more weed growth, a higher risk of windthrow, and natural regeneration may not be able to penetrate the centre of the gap. Areas that fail to regenerate successfully can be enriched by planting. The smaller scale of operations is often very suited to farm woodlands.

Coppice

Coppicing is a silvicultural system suitable for mainly broadleaved species, which are capable of regenerating from the stumps, or stools, of a previous crop cut near to ground level. The coppice shoots arise from dormant or adventitious buds on the stool, at or below the cut surface. This vegetative regeneration has the advantage over seedlings in that ample supplies of carbohydrates are available from the parent stool and its root system. The vigourous early growth of coppice shoots means that weeding is not normally required, although the stools may need some protection from grazing damage in the first few years. Stools are normally cut in the dormant season, although the best time is early spring as the buds begin to swell (March to early April) and the roots have their maximum carbohydrate stores. Autumn and winter cutting leads to some loss of stools from bark separation and late spring frost damage. Regularly coppiced stools are known to survive many rotations, with some lime stools living to many hundreds of years old, although the ability to coppice generally decreases with the age of the coppice regrowth to 40 years or about 40 cm diameter.

The most vigourous coppice species are willow, sweet chestnut, hornbeam, lime, pedunculate and sessile oak, ash and hazel. Some species readily produce suckers, such as wild cherry and some poplars. Beech and birch coppice only very weakly. The vast majority of conifers do not coppice, with the exception of Coast redwood (*Sequioa sempervirens*).

Traditional coppicing arose by the regular cutting of natural mixed woodland. Stools that died were replaced by cuttings, seedlings or by a process of *layering* involving bending over a remaining coppice shoot from an existing, healthy stool, pegging it to the ground and covering with earth. Coppice is one the oldest forms of forest management, supplying fuelwood and roundwood for traditional rural crafts producing hurdles, thatching spars and tool handles. The length of rotation is set to the type of product required: short rotations of 5–7 years for small-diameter pliable material and longer rotations of 20–40 years for material to be turned, cleaved and squared. During the industrial revolution, coppice expanded to supply charcoal for the iron and glass industries and oak bark for tanning. Since the middle of the eighteenth century, however, the substitution of coal for fuelwood and the decline in demand for coppice products has led to many former coppices falling into disuse. The only significant market remains for sweet chestnut coppice for split fence palings and stakes, although local markets exist for fuelwood, pulp and specialized rural crafts.

For suitable species, coppice offers rapid early growth and assured establishment. Unfortunately, volume production may be less than half that achieved for the same species grown in high forest systems and overall timber production may be commonly 2–4 m^3/ha/year.

Coppice with standards

Coppice with standards is a two-storey system comprising an understorey (simple even-aged coppice) and the overstorey of uneven-aged standards, usually of seedling origin. Standards characteristically have short boles and a greater proportion of branches than high forest trees, especially when the coppice rotations are short. The proportion of coppice to standards is a compromise: ideally there are between 75 and 100 standards of all sizes per hectare. Oak is traditionally the most common standard, but may not be the most suitable, due to its branching habit. Trees with a more marked apical dominance, lighter canopy and less tendency to produce epicormics (e.g. ash, lime and cherry) may be more suitable. Thin-barked trees such as beech and sycamore tend to suffer from sun scorch when released and are unsuitable.

Common understorey species include oak, ash, hornbeam, beech, chestnut, sycamore, field maple, lime, hazel, alder, birch, cherry, sallow, elm and aspen. Where the density of standards is high, shade-tolerant species are favoured (e.g. lime, sycamore).

Stored coppice

Coppice can be converted to high forest by reducing the number of coppice shoots to one for every stool. The faster growth in the first few years does not continue through the rotation, so that volume out-turn equals that of maiden trees. Stored coppice typically produces poor quality timber, as a result of basal sweep, shake (internal splitting) and a tendency to decay.

Short rotation coppice for firewood and biomass

Firewood is traditionally a secondary product from the branchwood and low quality timber of broadleaved woodland, especially coppice. As a fuel, only air-dried wood should be burnt, since wood with a high moisture content gives off little heat and causes tar deposits in stoves and flues. The best timbers for firewood include ash, beech, birch, hazel, holly, oak and old fruit trees, which may be cut and split or chipped by machine.

High volume production can be secured over short rotations of coppice (5–10 years) from selected broadleaved hybrids planted at high densities on agricultural land for energy generation or industrial cellulose. Recent experimental trials of willow and poplar hybrids promise yields of up to 30 m^3/ha/year, equivalent to 13–15 tonnes of dry wood/ha/year or the energy produced from 7 tonnes of coal. Field-scale production is underway in some parts of Britain where a regional market exists, as a result of government initiatives to increase the production of energy from renewable resources.

High-yielding disease-resistant varieties of willow or poplar are planted as cuttings in a multi-clonal mix into prepared agricultural land. Regular herbicide and fertilizer applications are required to maintain rapid production of multi-stemmed stools. The above-ground biomass is mechanically harvested every 3–4 years in winter, followed by vigourous regrowth from the cut stools. This cycle can be repeated 6–10 times before the crop begins to lose productivity and has to be grubbed up and replanted. Fungal infection, such as *Melampsora* rust, is a serious threat to some crops which may be offset by the use of carefully selected resistant varieties. Further expansion of the biomass market will depend on the relative cost of fossil fuels and advances in energy generation technology. For further information see Armstrong (1999).

Christmas trees

The production of Christmas trees can be successfully incorporated into woodland management by utilizing 'dead ground' under electricity pylons or other temporary spaces in newly established mixtures and along forest rides. Traditionally, the most commonly grown species was Norway spruce, although more attractive and needle-retentive species, such as Noble fir (*Abies procera*) and Caucasian fir (*Abies nordmanniana*), are rapidly gaining ground. The objective is to produce healthy, well-shaped and bushy trees by planting at a close spacing of 1 × 1 m (10 000/ha) with careful attention to weeding and inspection for the presence of insect damage. Shearing may be required in the fourth year to restrict the length of the leading shoot and to create a better shaped tree. The trees are normally harvested in a range of sizes to suit local markets, with the main crop being harvested between 4 and 10 years.

Improvement of neglected woodland

The poor quality of many farm woodlands generally owes more to inadequate or inconsistent past forest management than any inherent poor site fertility. For a woodland owner keen to improve the timber quality of neglected woodland, management will generally be based on the following options:

- improvement
- enrichment
- replacement.

Improvement

Improvement assumes that there are sufficient trees of 'final crop' potential to make the production of a timber crop worthwhile. Between 150 and 300 straight, defect-free trees of marketable species which are evenly spaced throughout the stand should be identified. From the pole stage onwards (10 m height), all competing trees and climbers are removed from around the final crop trees to allow complete crown freedom. Pruning is advisable to produce clean stems and to remove epicormic growth. The felling of the poorer quality matrix may help offset the cost of these silvicultural treatments.

Enrichment

Enrichment involves the planting of additional trees in a stand to increase the overall stocking. The degree of enrichment required depends on the quality of the existing crop. Where some utilizable trees remain, the focus of enrichment is usually existing gaps or bare ground. The width of the planting gaps should be at least 1.5 times the height of the tallest neighbouring trees. If the existing crop does not contain sufficient utilizable stems, the overstorey can be partially cleared in swathes or bands at least 4 m wide. Enrichment species are chosen for rapid early growth to beat weed competition and coppice regrowth, and include Norway maple, cherry, Southern beech, sycamore, Western hemlock and Western red cedar. The individual establishment cost per tree is quite high and the use of treeshelters at 2- to 3-m spacing has proved most successful. Attention is needed to ensure that the growing trees are not shaded out by crown expansion of the remaining overstorey.

Replacement

Replacement involves the partial or full removal of the existing crop and its replacement by a new superior species, where the present stand is unlikely to produce a potential final crop. This is the most drastic option, and is rarely advisable, due to the loss of many other woodland benefits such as landscape and wildlife habitat. The system is effectively a clear fell, although most of the trees will not have reached conventional harvestable size. Regrowth from cut stumps can be killed by the use of a foliar herbicide (e.g. glyphosate) or ammonium sulphate on larger stumps. The incidence of honey fungus (*Armillaria* spp.) may preclude the use of some species for replanting. Many former broadleaved woodlands have been converted to more productive coniferous plantations by this method since the 1940s, with mixed success and enormous losses to wildlife and landscape value. A compromise may be to clear and replant a woodland in several stages by clear felling only part of the woodland at any one time (see group systems discussed in 'Natural regeneration' above).

Forest measurement

Forest measurement is usually required to provide an estimate of the quantity of timber for sale. In addition, a forest inventory can provide the woodland owner with useful information for planning and forecasting future production. A periodic inventory will show the current state of the stand, its performance over time and any consequent deviations from the expected which may require treatment (e.g. restricted drainage). The amount of time invested in forest measurement is a balance between the degree of precision required and the costs involved. Generally more precise measures are used for more valuable timber in larger quantities, and sophisticated electronic equipment is now available for woodland surveys and volumetric analysis.

Forest measurement and inventory planning are fully described in two Forestry Commission publications, Hamilton (1985) and Edwards (1983). Timber is normally measured as solid volume, although other measures exist for specific purposes (green weight, dry weight, number of pieces). A number of conventions exist.

Diameter

Diameter (cm) is normally measured using a girthing tape or electronic calipers around the circumference of the tree or log, rounded down to the nearest whole centimetre. *Diameter at breast height* (dbh) is measured at 1.3 m above ground level.

Length

Length (m) is measured by tape following the curvature of the log, rounded down to 0.1 m for lengths up to 10 m and 1 m for lengths over 10 m. *Height* (m) can be measured by an observer standing a known distance (at least 1.5 times the height of the tree) away from the base of the tree by either trigonometry or from direct readings from specifically designed instruments, such as the clinometer, altimeter or hypsometer. *Total height* is the straight-line distance from the base to the uppermost tip of the standing tree. *Timber height* is the height to the lowest point on the main stem where the diameter is 7 cm overbark. In broadleaves, timber height is the lowest point at which no main stem is distinguishable. The *top height* of the stand is the average total height of the 100 trees of largest diameter (dbh) per hectare and is used to estimate the yield class. A more practical method of estimating top height is given below for single-species even-aged plantations or for the dominant species in mixtures:

(1) Lay down random plots through the stand, the number of samples required depending on the size and variability of the crop.

Area of stand (ha)	*Uniform crop*	*Variable crop*
0.5–2.0	6	8
2.0–10.0	8	12
Over 10.0	10	16

(2) Record the total height of the single tree of largest dbh within 5.6 m of the plot centre.
(3) Calculate the average height of the samples taken to give an estimate of the top height of the stand.

Basal area

Basal area is the cross-sectional area of a tree at 1.3 m above ground level (dbh). Basal area (m^2) for an indivi-dual tree =

$$\frac{\pi \times dbh^2}{40\,000}$$

where *dbh* is diameter at breast height in cm.

The basal area of a stand (m^2/ha) is the sum of the basal areas of all the trees in the stand. The basal area of a stand can be measured by laying down sample plots or using a relascope (or angle gauge) to take direct readings (see Edwards, 1983). Basal area is a useful value as it gives a better measure of the level of stocking than tree density alone.

Volume

Volume (m^3) is measured overbark to 7 cm diameter, or where no main stem is distinguishable, not including the volume of branchwood. The measurement of timber volume is always an estimate, due to the variable nature of trees, and logs; in particular, the volume must be corrected to account for the taper of different species of tree. A number of useful tables have been produced in the aforementioned Forestry Commission booklets, which allow volume to be estimated by measuring the height and diameter of standing or felled timber. The following procedures are drawn from those publications.

Estimating the volume of a standing tree

The volume of a standing tree can be estimated by correcting the total height (or timber height for broadleaves) by the use of a 'form factor' to account for the taper. The degree of taper increases in widely spaced stands and among open-grown broadleaves. Appropriate form factors range from 0.5 for mature plantation conifers, 0.4 for open-grown conifers or close-spaced broadleaves to 0.35 for younger trees. To calculate the volume using this method, first find the basal area at breast height (in m^2) and multiply this by the total height (m) and the appropriate form factor.

Example

A mature plantation-grown Douglas fir with a total height of 22 m and dbh of 46 cm:

$$\text{Basal area} = \frac{\pi \times 46^2}{40\,000}$$

$$= 0.166\,\text{m}^2$$

$$\text{Volume} = 0.166 \times 22 \times 0.5$$

$$= 1.83\,\text{m}^3$$

An alternative method in Hamilton (1985) is based on the use of single tree tariff charts, by measuring dbh and total height.

Estimating the volume of a stand

The method of estimating stand volume will depend on the value of the timber and the method of sale. For moderate to high value crops the *tariff system* is appropriate. This system is based on the relationship between tree diameter and volume, which will vary depending on the tree species, age and site conditions. Tables have been produced for a wide range of conditions in British forestry, where this relationship is classified by tariff numbers from 1–60. By calculating the appropriate tariff number for any particular stand of trees, the volume of the stand is found by consulting the tariff tables in Hamilton (1985) or Edwards (1983).

Measuring the volume of felled timber

The volume of felled timber can be reasonably accurately calculated by measuring the diameter of the log at the mid point along its length (*mid point diameter*) and using Huber's formulae below:

$$\text{Volume}\left(\text{m}^3\right) = \frac{\pi \times \left(\text{mid point diameter}\right)^2}{40\,000} \times \text{length of log}$$

where diameter is measured in cm and the length is measured in m.

Logs over 15 m should be measured in two or more sections. Volumes can also be calculated using the mid point diameter tables in Edwards (1983) and Hamilton (1985). In addition, tables have been produced for sawlog and roundwood volumes on the basis of top diameter (normally underbark) and length. This assumes a standard taper of 1 : 120 and gives a quicker but less accurate measure of log volumes.

Other measures

An estimation of timber in stacks is sometimes useful, although the volume of the stack must be converted to a solid volume by the use of conversion factors, which take account of log diameter, length, taper, straightness and method of stacking. Typical conversion factors range from 0.55–0.65 for broadleaves and 0.65–0.75 for conifers, although for short, straight billets the figure may be as high as 0.85. Some timber merchants dealing in quality broadleaved timber may still refer to the traditional measurements of Hoppus and cubic feet.

The yield class system

The annual increase in timber volume for even-aged stands of trees is termed the *current annual increment* (CAI, m^3/ha/year), which rises rapidly in the early life of the plantation and falls gradually as growth rates decline in old age. The *mean annual increment* (MAI, m^3/ha/year), is the average annual yield up to any particular age, which reaches a maximum at the ideal rotation age for maximum volume production. The maximum MAI also defines the *general yield class* (GYC, m^3/ha/year) of the stand, which is a measure of the productivity of the trees under those site conditions (Fig. 5.2).

The GYC of an even-aged plantation can be estimated without detailed volume measurements, since a good correlation exists between top height, species, age and yield class. By estimating the top height of the stand, the yield class for that species can be read from the appropriate table in Edwards and Christie (1981) or Rollinson (1999). It should be remembered that these tables give estimated future volume production on the basis of normal silvicultural treatments and full stocking. A reduction of 15% from the total area is normally made for forest tracks, rides and stacking areas, although this should be increased where other unstocked ground exists.

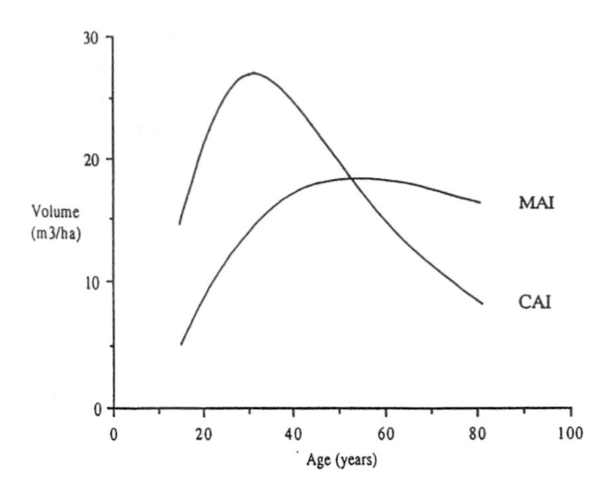

Figure 5.2 Volume increment in an even-aged stand (Sitka spruce GYC 18) (after Edwards and Christie, 1981, by permission of the Forestry Commission 2001).

Thinning

The spaces between trees in a young plantation quickly close, as the crowns of individual plants coalesce. Trees compete for light, moisture and nutrients and eventually some trees become suppressed or overtopped by their more dominant neighbours. The growth of these suppressed trees is arrested and they become susceptible to disease and decay. This natural process of self-thinning reduces the numbers of stems in a woodland by some 50–70% of the original planting density at harvest.

The removal of some trees artificially does not increase the total yield from woodland, although it does allow the forest manager to remove trees of poor quality and to selectively release the better trees. The effective utilizable yield and stand quality are therefore increased. Where the products of thinning are marketable, thinning yields a useful financial return throughout the life of a plantation.

The major determinants of timber quality are:

- *Sufficient length of clean stem or bole.* This is achieved by a high initial planting density, with a consequent rapid canopy closure. The competition from neighbouring trees suppresses the development of side branches and promotes apical growth (especially in broadleaves). In more open-grown conditions, high pruning of selected final crop trees (to 5 m) may be necessary.
- *Large girth.* This is achieved by reducing competition from neighbouring trees to allow for the full development of the canopy of the selected trees.
- *Evenness of growth.* For most conifers, the faster the rate of growth, the lower the timber density and strength, while for broadleaves, evenness of growth may be more important for some markets, such as veneers. Oak, by contrast, is stronger the faster the rate of growth (greater proportion of denser latewood).

The date of first thinning depends on timber species, site productivity and spacing and is normally within the range 18–30 years old or about 8 m in height. Faster growing trees on good sites are thinned earliest.

While the overall objective may be to maximize the financial return from the woodland, normally the first thinning is likely to yield very little utilizable timber, although its effect on the subsequent crop value may be considerable. Over the life of the crop, 80–90% of the trees could be removed as thinnings, without affecting the productive capacity of the stand. The amount removed at each thinning and the method of selection are determined by the thinning intensity, thinning cycle and thinning regime.

Thinning intensity

This is the rate at which volume is removed from the stand, often termed the *annual thinning yield*. The higher the intensity, the greater the amount of timber removed and consequently the larger the amount of growing space is left around the remaining trees. Higher intensities also produce larger-diameter trees, and consequently a greater financial return. The maximum intensity that can be maintained without a loss in overall volume production is termed the *marginal thinning intensity*. For a wide range of species and conditions, this point approximates to an annual rate of removal of 70% of the maximum mean annual increment (MAI). For a stand of yield class 20, the annual thinning yield at this point would be 14 m^3/ha/year, although this rate of removal only holds for stand within the normal thinning period, which varies according to species. See Rollinson (1999) for further information.

Thinning cycle

This is the number of years between each thinning, which is determined by the silvicultural requirements of the timber species, the thinning volume generated and the prevailing economic climate. Longer cycles allow for a greater volume out-turn, but may reduce volume production and increase the risk from windblow. Where small amounts of thinnings are used on the farm, it may be preferable to carry out light thinnings at more frequent intervals, with distinct silvicultural advantages, since the trees are released more gradually. Typical thinning cycles range between 3 and 6 years for conifers and up to 10 years for some slow-growing broadleaves.

Thinning yield

This is the actual volume removed at any one thinning, that is, the thinning intensity multiplied by the thinning cycle. For most commercial operations about 50–60 m^3/ha is the minimum amount of timber that can profitably be extracted. For example, a stand of Douglas fir (yield class 24) might have a thinning cycle of 3 years, which enables about 50 m^3/ha to be removed at each thinning (annual thinning yield 16.8 m^3/ha).

For even-aged stands of trees that are fully stocked it is possible to determine when thinning is required according to published thinning tables. Where stocking or management differs from normal (i.e. many farm woodlands), the point of thinning may have to be judged by visual inspection. A more accurate method is to estimate the basal area of the stand and to compare this with threshold basal area tables for fully stocked stands (see Table 5.1). If the basal area is equal to or more than the figure in the table, the stand is ready for thinning. The amount of timber removed is also best controlled by the volume or basal area of the thinnings, on the basis of marked plots (see Rollinson, 1999).

Thinning type (selective thinning)

This is the type of tree in the stand that is removed by thinning (see Fig. 5.3). It is normal practice to identify the final crop trees at an early stage, although it may be prudent to choose a larger number than required to allow for further selection at a later date. Thinning should aim to remove trees that are forked, spiral barked, coarse branched or leaning and any that look unhealthy or spindly. While marking for thinning, canopy and crowns must be continually observed so that excessive gaps are not produced after thinning. A *low thinning* removes trees that are currently in the lower levels of the canopy (sub-dominants and suppressed trees), while a *crown thinning* is directed to releasing the crowns of selected final crop trees, which may mean felling competing dominants. An *intermediate thinning* is the most common form of thinning, and comprises a low thinning to remove suppressed and subdominant trees, together with opening up of the canopy to release better dominant trees and create a more uniform stand.

Table 5.1 Before-thinning basal areas for fully stocked stands (basal areas in m^2/ha) (after Rollinson, 1999, by permission of the Forestry Commission, 2001)

Species	*Top height (m)*										
	10	*12*	*14*	*16*	*18*	*20*	*22*	*24*	*26*	*28*	*30*
Scots pine	26	26	27	30	32	35	38	40	43	46	—
Corsican pine	34	34	33	33	33	34	35	36	37	39	—
Lodgepole pine	33	31	31	30	30	31	31	32	33	34	—
Sitka spruce	33	34	34	35	35	36	37	38	39	40	42
Norway spruce	33	33	34	35	36	38	40	42	44	46	49
European larch	23	22	22	22	23	24	25	27	28	30	—
Japanese/hybrid larch	22	22	23	23	24	24	25	27	28	29	—
Douglas fir	28	28	28	29	30	31	32	34	35	37	40
Western hemlock	32	34	35	36	36	36	37	38	38	39	40
Western red cedar	—	49	50	51	53	55	57	60	63	66	70
Grand fir	—	39	39	39	39	39	39	40	41	43	45
Noble fir	—	45	46	46	47	48	49	51	52	54	—
Oak	24	26	25	24	24	25	25	25	25	26	—
Beech and sweet chestnut	22	23	25	25	27	29	30	31	33	35	37
Sycamore, ash, alder and birch	—	17	18	20	22	25	28	33	—	—	—

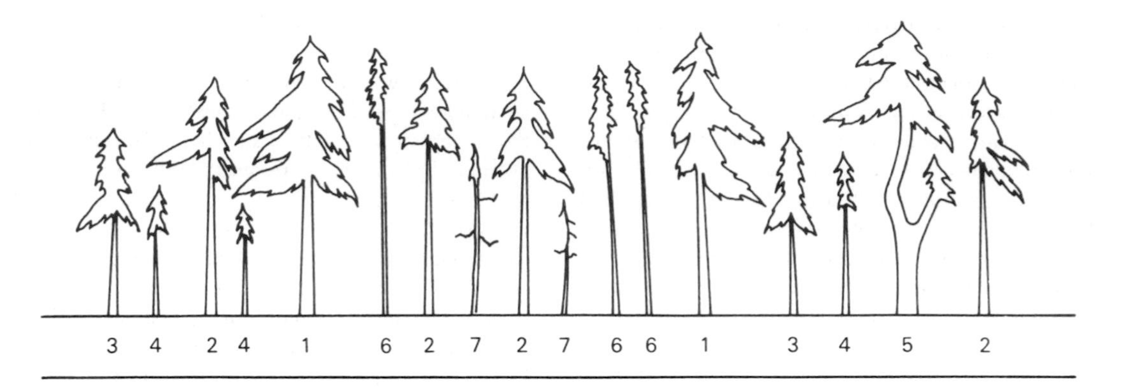

Figure 5.3 Classification of types of tree likely to be found in the crop. **1** Dominants; **2** co-dominants; **3** sub-dominants; **4** suppressed trees; **5** wolf trees; **6** whips; **7** leaning dead or dying trees.

Systematic thinning

The first thinning differs from most other thinnings in that very little, if any, financial return is likely, due to the small size of trees, and low volumes produced. In order to reduce the cost of this operation a number of systematic thinning systems can be used. All involve the removal of trees according to a predetermined pattern, irrespective of the merits of individual trees, and are often carried out by mechanical harvesters. Extraction and marking costs may be reduced greatly by the adoption of systematic methods of thinning, particularly in early thinnings as trees felled in selectively thinned stands are easily caught up in the dense thicket. *Line thinning* is a systematic form of thinning where trees are removed in lines along the planting rows or in more sophisticated systems as a series of interconnected lines (e.g. chevron thinning). The most common systems in practice are based on removing one planting row in four or five, although for close spacings (less than 2 m), two adjacent rows may be removed to allow for the passage of extraction machinery. The benefit of line thinning is restricted to those trees neighbouring the line removed; for 1 : 5 or 1 : 6 row thinning it may be advisable to carry out an additional selective thinning in between the rows. The removal of rows wider than 5 m should not be contemplated, due to the loss in volume production and especially on sites prone to windthrow.

No thin systems and delayed thinning

Where the danger of windthrow is particularly high, especially in the extreme north and west of Britain, a no-thinning system, with a reduced rotation age, may be preferable. Although total volume production may be reduced, the volume of sawlogs is not significantly lower than for thinned stands. Other recent developments include the use of chemical line thinning and self-thinning mixtures, which appear to be more cost effective than harvesting and have a lower risk of windthrow. Elsewhere, the decline in the real value of small roundwood in recent years has lead to a reappraisal of the economic advantages of thinning on large commercial forest estates. On some poorer, windfirm sites (GYC < 10), the time of first thinning may be delayed by up to 10 years, allowing for a greater out-turn of volume at a larger average diameter. For small farm woodlands, however, thinning may always be advisable on silvicultural grounds, in addition to the advantages of a more open canopy for wildlife, game and amenity.

Where previous thinning has been neglected, this may lead to a large proportion of weak or spindly trees. It is important at this stage not to correct the past neglect by a single heavy thinning, as this may open up the canopy to a greater risk of wind or snow damage, epicormic growth or sun scorch. Some species will recover better than others: for example, crown expansion is most rapid in oak, although ash will recover only weakly. In some older stands a decision has to be taken whether or not sufficient trees will respond from thinning, especially if there are already signs of windthrow or crown dieback. Any thinning that may be carried out should be restricted to 5% of the total basal area.

Harvesting

In the life of the woodland, a time comes for the harvesting of the mature final crop trees. The age of the trees at this point or *rotation length* is set according to the management objectives of the owner. For some markets, the rotation length is relatively fixed, according to specific size and diameter requirements, such as for sawlogs or veneers. The point of maximum volume production (maximum MAI) can be a useful indicator of the ideal rotation length, although it takes no account of the different prices obtained for different sizes of timber. Many commercial growers prefer the concept of *financial rotation*, whereby the maximum income from timber is the overriding objective for the site. In this respect the length of rotation can be drastically affected by the discount rate chosen, which in practice forecasts a harvesting date some 5–10 years before the point of maximum volume production. These concepts are useful in forest planning, although the exact date of felling will often be determined by external factors such as the local market conditions.

In recent years, the efficiency of forest-scale harvesting has been greatly advanced by the use of specialized felling, processing and extraction machinery for use in large commercial forests. The high capital expense of this machinery largely precludes its use in smaller farm woodlands, so that harvesting is likely to involve the use of chainsaws, tractor-mounted harvesters and smaller extraction machinery. A range of farm-scale machinery is becoming more widely available, including equipment that can be mounted on all-terrain motorbikes. The harvesting operation itself requires careful planning, to determine the course and direction of felling and the location of extraction routes and stacking areas. Chainsaws and other forest machinery are extremely dangerous, and woodland owners contemplating a DIY approach to woodland operations should undertake appropriate training in all relevant aspects before embarking on any harvesting themselves. The choice of harvesting system is a choice for the forest manager, and is determined by the nature of the terrain, the silvicultural system, the product specification, the ease of access, degree of mechanization available and landscape considerations. Where simple product mixes are required (e.g. sawlogs and pulp), the trees are normally cut to size in the wood and extracted separately (the *shortwood system*). For a more complex range of sizes and lengths, trees are extracted whole from the forest to a cleared area, where the timber is cut to size (the *tree length system*). The costs of felling and clearing by contractor will depend on the terrain and size of the timber cut, although it should be indicated whether the price includes the treatment of stumps with urea (against *Heterobasidion annosum*) or the burning of lop and top.

Control of tree felling

Woodland owners *must* apply to the Forestry Commission for permission to fell trees. Exceptions to this requirement include a small entitlement of up to 5 m^3 per quarter for own use, small-diameter trees (under 8 cm dbh) and for dead or dangerous trees. Felling is controlled by a Felling Licence from the Forestry Commission through the Woodland Grant Scheme and will be issued only if certain conditions are fulfilled, such as a commitment to replanting. Special permission may also be required for sites within areas designated as Conservation Areas, Sites of Special Scientific Interest (SSSI) or where the trees are protected by a Tree Preservation Order (TPO) imposed by the local planning authority. Owners failing to obtain all necessary permission risk prosecution and fines.

New farm woodlands

The establishment of new farm woodlands gives the landowner a greater degree of control than for existing woodlands over the size, location and most appropriate type of silvicultural system. New planting tends to be concentrated on areas of lower productivity or agriculturally marginal sites, such as steep slopes or awkward corners of fields. Other sites include extensions to existing woodland, shelterbelts or for amenity around farm buildings. Many farmers are now incorporating woodland planting with general conservation or landscape work around the farm.

Site and species selection

For new farm woodlands, the most important decisions are the correct choice of tree species and woodland design to match the site conditions and objectives of the landowner. Subsequent forest management will not be able to make up the lost ground as a result of a poorly adapted species growing in the wrong conditions. The location of new farm woodlands and species choice should be dictated with reference to the following site factors.

Climate and topography

While the mild climate of the UK is generally favourable for the growth of a wide range of trees, the choice of species for any particular area will be influenced by many factors. The most important to consider are:

- *Altitude* – increasing altitude results in reduced yields due to lower ambient temperature and shorter growing season (450 m above sea level is considered to be the 'economic' tree line).
- *Exposure* – exposed sites have lower overall productivity, and trees on the edge of woodlands will be shorter and more bushy with increased risk of windthrow.
- *Rainfall* – species with higher water requirements such as Douglas fir, larch, spruce, Western hemlock and Western red cedar thrive in the wetter north and west of Britain (rainfall 1000–1500 mm/year), while the drier eastern counties are most suited to less thirsty species such as Corsican pine and Scots pine.
- *Length of growing season* – summer warmth is required for some species such as sweet chestnut and walnut, which restricts economic growth of these species to the southern counties.
- *Incidence of drought or frost* – the incidence of seasonal frosts, often exacerbated by local topography, may cause injury to young trees, especially at the beginning or end of the growing season. The risk of frost damage is greatest in hollows or valleys where cold air from higher ground is able to collect. Only frost-hardy species should be planted in these locations.

Soils and ground vegetation

While healthy forest growth is attainable on a wide range of soil types, tree planting tends to be concentrated on those soils less suitable for agriculture, including less fertile mineral soils, gleys, podzols and deep peats. Adequate root development is essential for sustained growth through anchorage and supply of water and nutrients. Where root growth is constricted by impeded drainage, an iron pan, shallow or compacted soils, there is a greater risk of subsequent windthrow or moisture stress in drought periods, requiring further cultivations or drainage work to be carried out. A series of soil inspection pits should be dug pre-planting to check for soil depth, fertility, soil texture, drainage and the existence of hard pans. Further background information can be obtained through soil maps and land capability maps for forestry, available from the Soil Survey and Land Research Centre (SSLRC).

Ground vegetation can be used as an indication of soil and site conditions as a guide to tree species selection (Table 5.2), although care is needed since the type and pattern of vegetation present can be greatly modified by previous land management (most agricultural sites will have been improved at some time in the past). The systems most commonly in use are based on the presence of particular indicator species which denote site fertility, soil pH and drainage conditions.

Table 5.2 Assessment of site quality from vegetation present

Site quality	*Indicator species*
Good	Ash, beech, hazel, hornbeam, field maple, oak, bluebell (*Hyacinthoides non scriptus*), dog rose (*Rosa* spp.), primrose (*Primula vulgaris*), wild garlic (*Allium ursinum*), wild raspberry (*Rubus idaeus*), dog's mercury (*Mercurialis perennis*)
Moderate	Alder, bracken (*Pteridium aquilinum*), honeysuckle (*Lonicera periclymenum*), horsetails (*Equisetum* spp.), rhododendron (*Rhododendron ponticum*), soft rush (*Juncus effusus*)
Poor	Bilberry (*Vaccinium* spp.), cotton grass (*Eriophorum* spp.), deer grass (*Trichophorum caespitosum*), cross leaved heath (*Erica tetralix*), heather (*Calluna vulgaris*), moor mat grass (*Nardus stricta*), purple moor grass (*Molinia caerulea*), sphagnum moss (*Sphagnum* spp.)

The Forestry Commission has developed a useful Ecological Site Classification system as a guide to site assessment using a range of environmental indicators. This helps the woodland manager to select those trees most suited to the site conditions, including those species 'native' to the local area. See Pyatt (1997) for further information.

Species selection

The nature of the woodland site will impose certain restrictions on the choice of tree species for planting (Table 5.3). A fertile, sheltered site will enjoy a far longer list of possible tree species than an inhospitable one. Even in the most hostile locations, however, acceptable forest growth can be achieved by selecting species best adapted to the conditions. Following the site assessment, a list of suitable trees should be drawn up by relating the site conditions to their silvicultural characteristics. The final selection of species should be made on the basis of desired products, yield and other management objectives.

Tree species differ in their silvicultural characteristics according to their natural position within the woodland ecosystem. An important distinction relates to how tree species react to different light conditions. *Light demanders* tend to have thin crowns, light foliage and less dense timber and will grow rapidly in reasonably open

Table 5.3 Species characteristics and site requirements

Species	*Recommended sites*	*Unsuitable conditions*	*Silvicultural characteristics*	*Timber quality*
Scots pine (*Pinus sylvestris*)	Low rainfall areas on heather, poor gravel or sandy soils. Frost hardy	Wet or soft ground, chalk, limestone, high rainfall moorlands	Light demander, although slow growth rate and volume production. Useful as a nurse for hardwoods	General purpose softwood timber, good strength, takes preservative well
Corsican pine (*Pinus nigra* var. *maritima*)	Low rainfall areas and elevations on sand and clay soils especially near the sea. Plant only in southern and eastern England	High elevations, wet moorlands (increased risk of dieback from *Gremmeniella abietina*)	Higher volume production and better form than Scots pine, although more difficult to establish. Seedlings in Japanese paper pots give good results	Timber similar to Scots pine but coarser and a little weaker
Lodgepole pine (*Pinus contorta*)	Pioneer species on poor heaths, deep peats and sand dunes	All but the poorest sites where no other tree will grow	Coastal provenances give higher volume production but poorer form. Vulnerable to pine beauty moth in north Scotland	Timber similar to Scots pine
European larch (*Larix decidua*)	High rainfall areas on moist free-draining loams	Dry, poorly drained and very exposed sites or frost hollows. Avoid areas with rainfall under 750 mm	Deciduous conifer. Good nurse for hardwoods, although runs out of top growth quickly so thin from age 15–18. Susceptible to butt rot (*Heterobasidion annosum*)	Heavy and generally strong timber, best quality used for boat building. Heartwood is naturally durable, and makes a good farm timber for fencing, gates and other estate uses
Japanese larch (*Larix kaempferi*)	Mild and wet regions, less exacting than European larch. Pioneer in uplands on heather and grass	Dry, poorly drained and very exposed sites or frost hollows. Avoid areas with rainfall under 750 mm	Corkscrews on very fertile sites	
Hybrid larch (*Larix × eurolepis*)	Hybrid larch is slightly hardier than European larch	As for Japanese larch	Higher yielding than Japanese larch. Only use first generation seedlings	
Douglas fir (*Pseudotsuga menziesii*)	Sheltered valley slopes on well-drained and moderately fertile soils. Grows well in the wetter regions	Exposed or frosty sites, heather ground, wet, soft or shallow soils	A high yielding species although susceptible to windblow on shallow soils	Strong construction timber with a high strength : weight ratio
Norway spruce (*Picea abies*)	Moist grassy or rushy sites, most reasonably fertile soils and fairly heavy clays	All dry or exposed sites. More frost tolerant than Sitka spruce	Often grown on old woodland sites	Good general purpose timber, works and nails well. Stable in changing humidity conditions so suitable for building. Not suitable for preservative treatment or outdoor use

Table 5.3 *Continued*

Species	*Recommended sites*	*Unsuitable conditions*	*Silvicultural characteristics*	*Timber quality*
Sitka spruce (*Picea sitchensis*)	Wet exposed uplands in the north and west of Britain. Thrives on peats and grasslands in high rainfall areas	All sites liable to dry out and rainfall areas under 1000 mm	Will withstand very exposed conditions, although avoid previous scrub land where there is risk of honey fungus (*Armillaria* spp.)	Timber superior to Norway spruce, but too coarse for joinery. Good pulpwood
Western hemlock (*Tsuga heterophylla*)	Tolerant of high and low rainfall on acid mineral soils and better peats	Slow to establish on heather and open ground without shelter	Strong shade bearer, often grows best in mixture. Susceptible to butt rot (*Heterobasidion annosum*) and honey fungus on previous conifer sites	General purpose building timber and pulpwood
Western red cedar (*Thuja plicata*)	High rainfall areas on moderately fertile soils in sheltered sites	Exposed, poor and dry sites	Shade bearing and narrow crown, useful in mixtures or nurse species	Light coloured heartwood which is very durable. Cladding used outside for greenhouses and sheds
Lawson's cypress (*Chamaecyparis lawsoniana*)	Requirements not exacting but best on deep fertile soils with moderate to high rainfall	Dry infertile sites and heather ground. Avoid frosty, exposed and waterlogged sites	Slow growing shade bearer for use in underplanting or in mixtures (e.g. with oak)	Heartwood reasonably durable for general purpose uses. Small sizes used for fencing
Grand fir (*Abies grandis*)	Well-drained moist deep soils	All poor soils, dry, frosty or exposed sites	High volume producer, useful for underplanting, as moderately shade bearing	Soft white timber of only moderate quality
Noble fir (*Abies procera*)	Well-drained deep moist soils, tolerates acidity	Poor dry soils	Withstands exposure well. Tolerates drier sites than sitka spruce. A useful shelterbelt tree in Scotland	Poor timber quality exacerbated by high taper
Pedunculate oak (*Quercus robur*)	Deep clay loams with ample moisture, although will tolerate heavy clays	Shallow, infertile or poorly drained soils on exposed sites	Oak is a strong light demander and will not grow under shade. For top quality timber, close planting (at least 2500/ha) is required for suppression of sidebranches. Bole shading by underplanting or pruning will prevent problems from epicormics. Oak grown on lighter soils may suffer from 'shake'. Windfirm	Strong, naturally durable heartwood, although the sapwood does require preservative treatment if used outside. Timber quality (and price) varies tremendously. Prime quality oak is scarce and is used for veneers, planking and furniture. Lower grades for beams, fencing, gates and other estate purposes. The branchwood and lowest grades may only be fit for firewood or pulp

Sessile oak (*Quercus petraea*)	Best on deep porous brown earths, although will tolerate clay soils	As above and wet clays or fluctuating water tables (risk of shake)		
Beech (*Fagus sylvatica*)	Light well drained soils on chalk or limestone, deep sands and acid brown earths	Cold, wet poorly drained soils, dry infertile sands	Shade bearing, although may require a nurse (e.g. Scots pine) on exposed sites. An excellent amenity tree and suitable for underplanting. May suffer from severe squirrel damage, at the pole stage, therefore plant at close spacing to ensure sufficient trees for selection of the final crop	Timber strong fine even texture which takes a stain well. Used for wide range of interior uses: furniture, kitchenware, turnery and flooring. Lower grades used for firewood or pulpwood. Not suitable for exterior uses
Ash (*Fraxinus excelsior*)	Only thrives on moist, well drained fertile soils, ideally deep calcareous loams in valley bottoms. Wild garlic or dog's mercury are good indicators	Dry, shallow, heavy clay or badly drained soils, heaths and moorlands. Avoid frost hollows and exposed sites	Usually grown in mixture with beech, oak, cherry or larch. The stand needs to be thinned regularly to encourage large crowns	Spring timber with high shock resistance. Top grade ash (annual rings 1.5–6 mm) is used for sports goods, tool handles and furniture. Lower grade timber makes excellent firewood
Sweet chestnut (*Castanea sativa*)	Deep fertile soil in warm climate, ideally in southern England. Warm sunny acid sandy loam banks are ideal (pH 4.0–5.0)	Cold, wet, badly drained, exposed, frosty or infertile sites. Chalk, limestone or alkaline soils	Often grown productively as coppice for estate products. Timber trees over 80 years are prone to shake on free-draining soils	Sawn timber makes a strong substitute for oak. Coppice material used for cleft fencing, poles and stakes
Sycamore (*Acer pseudoplatanus*)	Moderately fertile free-draining soil. Grows on exposed upland sites and resistant to frost	Dry, shallow ill-drained or heavy clay soils	Moderate shade bearer and regenerates freely. Performs well on poor upland sites as windfirm. Prone to heavy squirrel damage	White timber used in turnery and where in contact with food. Figured sycamore valuable for veneer and furniture
Wild cherry (gean) (*Prunus avium*)	Deep fertile well drained soil, especially over chalk	Heavy soils, depressions and all infertile sites	Grows best in small groups in mixture with oak, ash or beech. Not attacked by squirrels, but may be severely affected by bacterial canker and deer browsing. Regular thinning and pruning necessary in summer for quality timber	Decorative timber with rich reddish-brown heartwood used for veneers, furniture and paneling
Common walnut (*Juglans regia*)	Sheltered south-facing slopes on deep fertile well-drained medium texture soils. Optimum pH 6.0–7.0	Only planted on carefully selected sites which meet requirements, avoiding frost hollows	Best grown in open situation or in small groups, although pruning will be needed to improve form in July/August	Valuable decorative dark brown timber, often dug out and taken to the sawmill with main roots as 'figuring' occurs in the base. Branches for firewood

Table 5.3 *Continued*

Species	*Recommended sites*	*Unsuitable conditions*	*Silvicultural characteristics*	*Timber quality*
Black walnut (*Juglans nigra*)	Chalk or limestone with more than 600 mm of overlying soil. Requires warm summers (ideally south and central England)			
Common alder (*Alnus glutinosa*)	River and stream banks and wet marshy places in lowlands and uplands. Frost hardy	All dry sites. For very acid sites prefer grey alder. Widespread incidence of *Phytophera* disease	Good shelterbelt tree (windfirm) for use on nutrient-poor sites or in coastal locations, as resistant to salt spray. Mixes with birch and ash and is a good nurse for oak on wet clay soils. Fixes atmospheric nitrogen in root nodules. Grows rapidly for the first 25 years and coppices vigorously	Timber dries red-brown, good for turnery, staining and polishing
Grey alder (*Alnus incana*)	Requires drier situation than common alder, but tolerates poorer sites	Dry infertile situations, heathlands, shallow and thin chalky soils	Widely planted for wind breaks, shelter and as a pioneer species for reclaiming derelict land. Extremely hardy	Timber worthless
Italian alder (*Alnus cordata*)	Drier chalk and limestone soils of southern England	Avoid dry thin acid or infertile soils	Pioneer species used for single row windbreaks	Timber worthless
Black poplar (*Populus robusta*) Black hybrids (*Populus* × *euramericana* var. *Eugenii*, *Gelrica* and *Heidemiz* T-78)	Base rich loamy soils in sheltered position with water table 1–1.5 m below the surface in summer	Exposed acid dry or infertile sites	Poplars often used for perimeter horticultural windbreaks	Light white timber used for pallets, chip baskets and pulpwood. The demand for match timber is much reduced

Balsam poplar (*Populus trichocarpa* var. *Fritzi Pauley* and *Scott Pauley*, *Populus tacamahaca* × *trichocarpa* var. 32)	Tolerates more acid soils than black hybrids and more suited to the cooler wetter parts of Britain	As for black poplar.	Balsam poplars subject to bacterial canker; plant only resistant clones. Preferred to black hybrids as they are fastigiate and trim more easily. Hybrid *Populus t×t* Clone 32 most commonly used. Other poplars include white and grey spp.	
Cricket bat willow (*Salix alba* var. *coerulea*)	Only margins of flowing streams or rivers with highly fertile soils	Useless elsewhere	Planted as sets 10–12 m apart, with all side shoots removed on the bottom 3 m of the stem for knot-free timber	Cricket bats
Silver birch (*Betula pendula*) Downy birch (*Betula pubescens*)	Brown earths, podzols, sands and gravels. Withstands frost and exposure. *B. pubescens* on poorly drained heathlands and waterlogged conditions at higher elevations		Pioneer species with light crown. Regenerates easily on mineral soils, although not widely planted. Used as soil improver and nurse for oak and beech	Strong, fine textured timber not naturally durable. Grown in Scandinavia with improved cultivars for veneer. Low grade timber makes good firewood
Southern beech (*Nothofagus procera*; *Nothofagus obliqua*)	Wide range of soils from heavy clay to deep sand	Badly drained, exposed or frosty sites	Both species are fast growing. Prefer *N. procera* in the wetter west country, *N. obliqua* in the drier east. Light-demanding species, but start under a thin canopy, thin early in life. Frost damage leads to stem cankers	Timber of both species similar to native beech, but with 20% less bending strength
Norway maple (*Acer platanoides*)	Moist deep free-rooting soils with high base status	Will not thrive on infertile soils, although will survive on thin chalk soils. Avoid frost hollows and exposed sites	Good amenity tree, suited for screens and mixed shelterbelts. Heavily damaged by squirrels	Timber as for sycamore; flooring, furniture, turnery and veneer. Good firewood
Hornbeam (*Carpinus betulus*)	Moist damp clays, chalk, limestone and acid brown earths. Frost hardy	Thin, infertile, dry and very acid sites	A substitute for beech on clay soils and where high frost resistance is required. Shade bearing and slow growing, therefore valuable as coppice understorey for shelterbelts or under oak to control epicormics.	Hard, heavy and tough timber giving a very smooth finish ('white beech' on the continent) used for carving and turnery. Makes very good firewood.

situations, although in shade they are quickly suppressed and die. *Shade bearers* tend to have more layered crowns and heavier timber and will only grow slowly at first, although they are able to survive moderate shade. Where maximum volume production is a major objective, light-demanding species are clear winners. Other important silvicultural characteristics include climatic and soil requirements, timber quality, response to silvicultural treatment and sensitivity to drought, frost, wind, exposure, insect attack, browsing and disease.

In addition to appropriate species choice, consideration should be given to the origin or *provenance* of the planting stock, since, for the same species, the silvicultural characteristics of the trees may be slightly different for seed drawn from different geographical sub-populations. Many commercially grown exotic conifers, for example, are native to the western seaboard of North America and seed from the northern populations (e.g. Alaska) will be more resistant to winter cold, although slower growing than seed drawn from more southerly populations (e.g. California). Since the first introductions of these exotic conifers, seed may now be obtained from seed orchards (or provenances) in Britain, even though the origin remains unchanged. Both seed origin and provenance are recorded in tree planting catalogues which must be matched with the environmental conditions of the planting site for optimum growth. Further information is available from Lines (1987).

Pure crops and mixtures

Pure crops are easier and simpler to manage than mixtures, although mixtures do offer other benefits if managed skillfully. Broadleaves are often grown in mixture with conifers to provide an early economic return from the coniferous thinnings to offset the delayed returns from the longer rotation of slower growing broadleaves. Appropriate mixtures may also be able to more fully utilize the site through differences in rooting depth and are far preferable on landscape, game and conservation grounds. Mixtures are often used where one species or *nurse* offers benefit to the main tree species, in the form of shelter against frost or cold, smothering weed growth, nutrition, support or suppression of side branches. Timely thinning is often required to prevent the nurse from outgrowing and suppressing the main tree species, so that care is needed in the choice of appropriate mixture. As a general rule, for broadleaf/conifer mixtures, the expected growth rate of the conifer should not be more than double that for the broadleaved species. Broadleaves should be established as blocks (9 or 25 to each block) within a conifer matrix, at an appropriate spacing to provide a final broadleaved crop.

Alternatively line mixtures can be used, although these should be lines of at least three of each species to reduce the risk of suppression. Normal practice is to remove the adjacent lines of conifers to the broadleaves at the first thinning followed by selective thinning thereafter. Greater care is required on hillside plantings to avoid regimented or geometric patterns of mixtures which are highly visible in the landscape.

Examples of suitably robust mixtures include:

- Oak, ash and cherry on moist, deep, fertile soils or clay over chalk.
- Oak and European larch on free-draining and lighter soils.
- Oak and Norway spruce on heavy acid clays.
- Oak and Western red cedar on free-draining soils.
- Beech with Western red cedar, Scots pine or Corsican pine.
- Sweet chestnut and European larch.
- Sitka spruce and European larch.

Establishment

The effective establishment of trees is an important management operation in the life of the woodland. The cost of establishing a woodland crop can seem very high; however, neglected or poor establishment practice, in an effort to reduce costs, will lead to all sorts of problems which will be difficult (and even more costly) to remedy later. On the other hand, careful planning and organization of site preparation, planting, weeding and tending may prove to be a good investment if it results in high quality woodland.

Site preparation

The aim of ground preparation is to create suitable planting conditions for rapid establishment. Where planting is to take place on the site of recently harvested woodland, the brash from the former crop must be cleared as part of the harvesting operation. Larger material can be cut up for firewood and the lighter lop and top either burnt, chipped or pushed into heaps between the planting lines. The whole clearance operation can often prove expensive. Additional problems with restocking include the re-wetting of the ground as soil moisture increases after felling, and compaction caused by heavy harvesting machinery.

On upland sites, the traditional method of ground preparation was by ploughing, using either a single furrow or double mouldboard, set at the appropriate planting distance. Ploughing is now rarely recommended because of its deleterious effect on soil erosion and water quality.

Mounding or '*dolloping*' is now the preferred technique, especially on heavy, wet, compacted or organic soils and restock sites, using a machine to dig out dollops of soil and deposit them on the planting position. This causes far less site disturbance than ploughing and creates a similar raised mound of cultivated mineral soil, through which the trees are planted. This can almost eliminate the need to weed during the first growing season, as the vegetation is completely buried under the inverted soil. The main advantage, however, is through improved root development, as a result of increased soil aeration, drainage, soil temperature and the release of nutrients from the breakdown of organic matter.

An alternative to dolloping, where drainage is not required, is *scarifying*, which involves the removal of surface vegetation by mechanical scraper. Scarifying is the preferred technique where sites are to be established at least partly by natural regeneration.

Site preparation on former agricultural land

Former agricultural land released for woodland planting is likely to be more fertile than land available for commercial upland forestry and different establishment practices are required. For further information consult Williamson (1992).

The initial preparation of the planting site will depend on the previous land use and soil conditions. Ploughing up former grassland or arable stubble can be counterproductive, as the disturbance of the soil both exposes viable weed seeds and creates a bare seedbed for the rapid invasion of arable weeds which are difficult and expensive to control. Conversely, undisturbed stubble may harbour herbicide residues which will reduce the quantity of weed seed in the surface layers of the soil and delay invasion by broadleaved annual weeds. Any straw can be chopped and left as a mulch to retain moisture and control the growth of volunteers and other grass weeds. A grass sward is the most controllable of ground conditions, and can be achieved on arable sites by underplanting a previous crop with a suitable grass mixture. Heavy soils present particular problems due to shrinkage and waterlogging. In a dry summer, heavy land will shrink, leading to a network of tiny cracks. On recently planted land, this shrinkage is often concentrated in the bare strips along planting lines or around the circumference of trees which are spot weeded, leading to excessive cracking and loss of plants due to moisture loss from exposed roots. The poor structure and texture of heavy soils can pose problems at planting too, which should be delayed until the soil has a suitable moisture content in the autumn. If the soil is too hard and dry, it will be difficult to firm in the roots, while if the soil is too wet, the sides of the planting hole will be smeared, which may crack and expose the roots in a dry summer. Some sites may suffer from compaction as a result of poaching around cattle feed troughs or heavy machinery travelling along the headlands. The presence of a soil pan at some depth could impede the growth of trees or increase the risk from windthrow. Ripping or subsoiling these areas will improve soil structure, although trees should be planted to the side of the ripped lines due to the risk of soil cracking.

Drainage

The prime function of forest drains is to remove surface water and to prevent waterlogging, which over prolonged periods will severely impede the growth of trees and may increase the risk of windthrow as root development is restricted. In the uplands, substantial areas of planting land suffer from excessive soil water, particularly on gleyed clays, peaty gleys and deep peat where the lateral movement of water is severely limited, requiring extensive open drainage systems. Open drains (60–90 cm deep) are cut to intercept surface water from above the plantation at a slight angle to the contour (no more than 2°), together with smaller collecting drains at 30- to 40-m spacing through the planting site. The environmental effects of increased levels of sediments in upland streams leaving afforested land have caused major concern in recent years, due to the impact on aquatic life. The Forestry Commission has produced a useful guide to silvicultural practices which can mitigate some of the worst of these effects (Forestry Commission, 2000a). In particular, drains should end 5–10 m from smaller streams and 15–30 m from main watercourses to reduce sediment loading.

Once fully established, the trees themselves draw water from the soil through transpiration which adds to the efficiency of the drainage system. Open drainage systems should be monitored regularly, although maintenance operations should be restricted to removing serious blockages and overflowing ditches. At harvesting, substantial damage to the drainage system can be avoided by working during drier periods and installing temporary culverts to remove surface water. Inevitably, some reinstatement work will be necessary, although the root channels formed by the previous crop will improve water movement, especially where these have broken through any surface pan or compacted layers.

Drainage work can be an expensive and time-consuming operation, normally involving the use of contractors with specialized tracked excavators. The majority of farm woodland sites are unlikely to merit anything other than localized drainage and the owner may be better advised to plant more tolerant species such as alder and willow or to incorporate those 'wet hollows' into areas of open ground within the woodland design.

Where planting is on former agricultural land with an existing below-ground field drainage system, owners should be aware that tree roots will rapidly block and break up these drains, rendering them useless. Where the existing system also serves surrounding agricultural fields, the below-ground drains in the new woodland may need to be replaced with open drains. For further information see Forestry Commission (2000a).

Fencing

New plantations will require protection from grazing animals (rabbits, hares, deer and farm stock) while the young trees become established. The choice of protection will depend primarily on the size and shape of the plantation and the density of planting. For sites smaller than 2–3 ha, individual protection by tree shelters is normally the cheapest method, while for larger areas, fencing becomes increasingly cost effective. Fencing may also be required on woodland sites bordering stock fields, roads or public places. Irregular-shaped and narrow belts of trees are the most expensive to fence, having a greater perimeter than square plantations.

The specification of the fence will depend on the type of animal present and the durability required. Rabbit fences should be at least 0.75 m high with 31-mm mesh wire netting, while for deer and stock a stronger fence is required to a height of 1.8–2.0 m. Spring steel wire is commonly chosen as a support for various grades of wire netting, since it will return to normal if accidentally stretched by an animal or fallen trees. If the fence is required for longer than 5 years, only treated wooden posts and struts should be used. The line of the fence should be chosen to avoid wet or shallow soils and snow hollows, which may allow entry of animals in winter. Where the fence cuts across badger runs, two-way gates should be provided. Further details of fence specifications are available in Pepper (1992).

Plant origin

The production of high quality timber depends on the selection of healthy planting stock grown from selected parents of superior quality from the correct choice of provenance. Under the Forest Reproductive Material Regulations, cuttings and plants of listed species (oak, beech, poplar and most conifers) intended for the production of timber may not be sold unless obtained from sources approved and registered by the Forestry Commission in Great Britain or other approved authorities elsewhere in the European Union. For unlisted species, seed from recommended sources is preferable to 'unknown' sources.

Planting stock

Although trees are available in a variety of sizes (BS 3936), smaller plants (up to a height of around 90 cm) offer the advantage of cheapness, ease of handling, faster early growth and less chance of dieback or drought stress than larger whips or standards. Plants should be ordered well in advance from a forest tree nursery to ensure delivery and quantity required (names and addresses can be obtained from the Royal Forestry Society of England, Wales and Northern Ireland). Only healthy plants should be accepted, which are stout and well balanced with plenty of fibrous moist roots. Check any suspect trees by nicking the bark; live plants will have a greenish-white appearance under the bark. The following types of planting stock are commonly available:

- *Bare-rooted transplants* are widely available as forest trees, being reasonably priced and less bulky than containerized trees, although greater care is needed in planting and handling. Transplants are raised in a seedbed for one or two seasons and then lifted and planted out in transplant beds for a further year or two, which encourages the growth of a vigourous root system. Nursery catalogues may refer to these plants in a coded system (e.g. a 1 + 1 plant has spent 1 year in the seedbed and 1 year in the transplant bed). In some nursery systems, plants remain in the same seedbed but are sown at a lower initial density and are undercut by a reciprocating bar passed through the seedbed to promote bushy root growth, e.g. 1/2 u 1/2 (this indicates a one-year-old plant which has been undercut midway through the growing season).

- *Container-grown plants* retain a growing medium around the roots of the tree at planting, allowing planting outside the growing season and a lower likelihood of mortality through poor handling. Plants are available in a range of containers and sizes, from reusable stiff plastic pots to degradable paper cells (Japanese paper pots) or flexible root trainers which produce small plants as 'plugs'. The plants are normally raised in polytunnels and hardened off outside prior to planting. Always be careful not to accept pot-bound trees, or very small, spindly specimens.
- *Cuttings and rooted sets* are used for poplars and willows which will grow from unrooted cuttings (20–25 cm by 1–2 cm diameter).

Time of planting

The planting season for bare-rooted transplants runs from late October until the end of March, the exact date determined by the onset of dormancy, which is later in a mild season, together with the ground conditions. Planting is most successful when coupled with higher soil temperatures, although delaying planting until after budburst renders trees susceptible to decay. Budburst can be delayed, while extending the planting season, by protecting the trees in cold storage before delivery. Broadleaves are best planted by late November to allow the roots to become settled-in over the winter and better able to withstand drought in the following season. Conifers are normally planted during early spring to avoid winter frost damage.

Plant handling

Although disease, frost or drought may account for some losses of trees in the first year of planting, a major cause of death is the result of poor handling between the nursery and the planting site. Particular problems include the following:

- *Root desiccation* is a significant cause of tree mortality from the exposure of bare roots to drying winds, particularly while the planting hole is being prepared. Bare-rooted transplants should ideally be kept in co-extruded opaque plastic bags, which are white on the outside to reflect heat and black on the inside to keep the roots cool. Fertilizer bags are not suitable due to the presence of chemical residues which may scorch the plants. Trees are traditionally stored prior to planting by 'heeling in', whereby the trees are lined out in a shallow, slanted trench (30 cm deep) to cover the roots. This practice should be avoided where storage is only required for a short period, as disturbance of the roots is likely. Where there is a danger of frost, straw or bracken should be placed over the trees.
- *Overheating* of trees in black or translucent plastic bags can lead to excessive moisture loss and eventually death. Trees should always be stored out of the bright sunlight, preferably in a cool dark shed.
- *Physical damage* can be caused by poor handling between the nursery and the planting site, leading to broken, bruised or damaged shoots, roots and buds. Plants should never be dropped off the back of a trailer, stacked in piles or treated roughly.

Planting

Before the planting operation, all site preparations should be completed and the plants and equipment safely delivered to the site. Ideal planting weather is mild and wet, and trees should not be planted in severe frost or while the ground is waterlogged. Sites can be marked out with coloured pegs or canes, and the first planting line laid out with sticks or twine, from which subsequent lines are measured. Experienced planters will judge the distance between plants by stepping or by the length of the planting tool.

Bare-rooted trees are normally planted by the *notch method*, which involves the cutting of a T- or L-shaped slit in the soil. Special planting spades, with a strengthened handle and straight blade, are preferable to garden spades. The first slit is levered open with the spade, allowing the tree to be inserted into the notch, until the old ground level mark (*soil collar*) of the tree coincides with the surface of the disturbed ground. The roots should be evenly distributed and the soil firmed down around the tree, which should remain vertical. For larger trees or standards, *pit planting* is the preferred technique. This involves the digging of a hole large enough to take the spread-out roots of the tree, the bottom of which may be broken up to aid drainage. Well-rotted organic matter or a sprinkle of bone meal may be added and, for large specimens, a stake should be driven in before inserting the tree.

On level and workable sites (free from large stones, stumps and ditches) mechanized planting by a tractor-mounted machine may be possible. This technique offers the advantages of quicker planting and improved survival from the reduced plant handling time, and may be more economical than hand planting on large sites with a simple planting design.

Spacing

The choice of spacing between the trees will reflect the management objectives for the woodland (Tables 5.4 and 5.5). If good quality timber is required, it is important to obtain a high initial stocking density, which will lead to straighter and less branchy trees, better returns from thinning and a greater number from which to select the final crop trees. Where the primary objective is for amenity or game, wider spacings will be more cost effective, although pruning of selected final crop trees can be undertaken if timber is a secondary objective. Where a grant is being paid to offset the costs of planting, a minimum average tree spacing or tree density per hectare will often be specified as part of the conditions of payment.

Treeshelters

These translucent open-ended tubes, placed over each tree, have revolutionized the establishment of broadleaves on sheltered lowland sites. They are also valuable for protecting naturally regenerated seedlings or small plantings (e.g. in hedgerows or enrichment schemes). The shelters act as 'tree greenhouses' providing the tree inside with a warm, moist microclimate. The results are very rapid initial growth and better root establishment, which increase tree survival, especially in dry summers. In addition, the shelters give protection against browsing animals and allow for the easier application of foliar herbicides. Height growth can be doubled for some species (notably oak), although this effect is reduced once the tree emerges from the top of the shelter. The advantage of treeshelters is therefore to speed up the expensive process of tree establishment by faster growth and increased survival over the first 3–4 years. Following this, the shelter continues to give support to the tree, until it disintegrates at between 5 and 10 years.

Treeshelters should not be used on exposed or waterlogged sites and any vegetation on the planting position should be screefed away. The base of the shelter will need to be pushed into the ground at least 5 cm to prevent air circulating up through the shelter and secured using a stout and preferably treated stake together with a special nylon tie. Different trees vary in their response to treeshelters (Table 5.6), which should only be used with appropriate species and using good transplants with a root collar of at least 6 mm. It is important to note that the use of treeshelters does not mean that weed control can be avoided (see section 'Weeding').

Table 5.4 Examples of recommended spacing for different tree species

Species and situation	*Approximate spacing (m)*
Christmas trees	0.75–1.00
Oak/beech (for timber production)	1.50–1.75
Slower growing conifers (pine and Norway spruce on poor sites)	1.50–1.75
Conifers/mixed broadleaves	2.0
Amenity mixed broadleaves in tree shelters	3.0
Poplar (pruned for timber)	8.0

Table 5.5 Tree requirements (number of trees required/ha planted on the square)

Spacing (m)	*No. trees/ha*	*Spacing (m)*	*No. trees/ha*
0.75	17 778	2.75	1322
1.00	10 000	3.00	1111
1.25	6 400	4.00	625
1.50	4 444	4.50	494
1.75	3 625	5.00	400
2.00	2 500	6.00	278
2.25	1 975	8.00	156
2.50	1 600	10.00	100

Table 5.6 Height increment improvement with treeshelters compared with mesh guards 3 years after planting (after Potter, 1991, by permission of the Forestry Commission, 2001)

>100% increase	Oak, beech[1], walnut, lime, sycamore, field maple, birch, hawthorn
50–100% increase	Douglas fir[1], grand fir[1], Sitka spruce[1], Norway spruce[1], Corsican pine[1], Japanese larch[1], Norway maple, alder, sweet chestnut, ash, holly, southern beech
<50% increase	Western hemlock[1], Western red cedar[1], wild cherry, black walnut, hornbeam, horse chestnut, rowan, whitebeam

[1] Best results in shelters no larger than 0.6 m. If deer pressure is high, fencing will be required. Beech is prone to severe fungal infection if grown in 1.2-m shelters without additional holes for ventilation.

Treeshelters have been the cause of great debate among foresters over the last 15 years, and they are by no means the best form of protection in all circumstances. Concerns have been raised over their stability in exposed conditions, the growth of trees after the shelter is removed, weeds in the shelter, the amount of maintenance they require and not least the unattractive appearance of lines of treeshelters in the landscape. Some of these problems have been overcome, such as the use of different coloured treeshelters to blend in with the ground vegetation and better advice on the maintenance and removal of shelters.

The final choice of protection method should be based on a very careful pre-planting assessment of the site, ensuring that any signs of potentially damaging browsing animals are taken seriously (Table 5.7). Where protection is required from domestic livestock (horses, cattle and sheep), treeshelters are only recommended where there is a very high management input, and in practice these animals should be fenced out of young plantations. Where the local population of browsing animals is particularly high (e.g. rabbits, roe deer), control of their numbers may be required to prevent overwhelming damage to the growing trees.

Beating up

Beating up involves the replacement of dead trees that are lost during the first year or two after planting. Losses of up to 10% are often considered acceptable and do not require beating up. Beating up is advisable where failures occur in groups and at wider spacings where losses are more noticeable. By using good quality plants and effective weed control for several years after the first planting, beat-ups should eventually be indistinguishable from the original crop.

Fertilizer application

The use of fertilizer on most farm woodlands on fertile mineral soils is unlikely to be economic as trees respond only weakly to nutrient supplements compared to most agricultural crops. In commercial upland plantations, fertilizer use is generally restricted to the establishment phase to improve initial growth rates of conifers until canopy closure, the decision to fertilize being based on soil or foliar analysis. After canopy closure, further applications are not normally necessary due to the shading of competing vegetation, improved nutrient cycling and the capture of atmospheric nutrients by the crowns of the trees. On restocked sites, the increased availability of nutrients from the breakdown of brash and litter normally precludes the use of fertilizers except for very poor heathland or moorland soils. Phosphate, potassium and nitrogen are normally applied as top dressings, using ground rock phosphate (50–75 kg P/ha), muriate of potash (75–100 kg K/ha) and prilled urea or ammonium nitrate (80–120 kg N/ha).

Examples of farm woodland sites where the use of fertilizer should be considered are:

- *Heathland mineral soils* (ironpans, podzols, gley soils) may suffer from phosphate deficiency. Sitka spruce planted on heather-dominated heathland or moorland suffers from a well-documented 'heather check' which restricts growth through the limited availability of nitrogen and may delay canopy closure by up to 10 years. Although heather

Table 5.7 Examples of suitable methods of protecting young trees from different types of damaging wildlife

Suitable protection		*Grazing animal*						
		Vole	*Rabbit*	*Hare*	*Roe deer Muntjac*	*Sika Fallow Red deer*	*Sheep*	*Cattle*
Voleguard		✓						
Shelter	0.6 m	✓	✓					
	0.75 m	✓	✓	✓				
	1.2 m	✓	✓	✓	✓			
	1.8 m	✓	✓	✓	✓	✓		
Spiral guard 0.6 m (sheltered sites only)		✓	✓					
Shrub shelter 0.6 m (wider diameter)		✓	✓					
Fencing recommended						✓	✓	✓

competition may be reduced by herbicide applications, additional inputs of nitrogen may be required. Alternatively, Sitka spruce may be planted in mixture with a suitable provenance of Scots pine, Lodgepole pine or Japanese larch, which appears to suppress the competing heather and increases the availability of nitrogen.

- *Peaty gleys and deep peats* are likely to suffer from phosphate and potassium deficiency where the peat depth exceeds 30 cm.
- *Chalk downland* may suffer from nitrogen and potassium deficiency on shallow soils.
- *Derelict and restored soils* often suffer from limited nitrogen availability, in addition to specific nutrient deficiencies or toxicity. Further soil analysis will normally be necessary.

If a crop appears to be showing signs of nutrient deficiency once established such as discoloration of leaves or needles, poor growth rate or early senescence, then samples of foliage can be sent to the Forestry Commission for analysis and rapid diagnosis of the problem. Remedial applications of fertilizer may subsequently eliminate the deficiency.

Aftercare

Weeding

The most significant threat to the successful establishment of woodland is competition from weed growth, especially on drought-prone sites. Grasses and broadleaved weeds compete effectively with the growing trees for water, light and nutrients, reducing height growth and potentially increasing mortality. On fertile sites, the growth of rank weed vegetation may also cause physical damage by smothering or collapsing over the trees in the autumn. The cutting back of these weeds and grasses will reduce light competition and physical damage, although root competition will not be affected. In fact, mown grass exerts an even greater competition for soil moisture than unmown grass. The standard recommendation is that a *weed-free zone of at least 1 m* should be maintained around young trees until they are well established and above the height of competing vegetation.

The weed control treatment most appropriate should be determined by regular inspection of the site prior to planting to reveal the type and abundance of weed species present, together with an assessment of the soil conditions and potential for weed invasion. The total removal of weeds from the site is rarely justified either by cost or on silvicultural grounds; in fact, there is some evidence that the retention of some weed growth may benefit the trees by providing shelter from frost or sun scorch. A totally weed-free plantation may also be prone to excessive soil erosion and has a substantially reduced wildlife value. Weed control therefore normally involves the selective spot treatment of competing vegetation around the trees or in strips along the planting lines, retaining weed growth in the central aisle. The timing of treatments will depend on the site, although the time of greatest potential moisture competition is early in the growing season (March–May) when tree roots are actively growing. Control should therefore aim to anticipate future weed growth before this becomes obvious on the ground. Current best practice involves pre-planting treatments to establish a weed-free planting position, together with additional treatment during the growing season if necessary. Weeding is usually carried out once or twice a year, for 3–6 years after planting, depending on the site.

The following forms of treatment are available.

Herbicides

Herbicides provide the most cost-effective form of weed control available. When used in accordance with the manufacturer's recommendations, the environmental risks are minimized. A wide range of chemicals are available for use on sites classified as 'farm forestry' sites (i.e. sites that have recently been either arable or improved grassland), with a much narrower range available for unimproved 'forestry' sites. Most herbicides will only affect a limited number of species or may only be active at certain times of the year in particular weather conditions. It is therefore important to match the weed control spectrum of the herbicide with the weeds found on the site. Herbicides may be applied by tractor- or ATV-mounted sprayers, although knapsack sprayers may be preferable in small or more irregular plantations. Herbicides can be divided into two main groups: foliar acting and residual.

- *Foliar-acting herbicides* are applied to the aerial parts of actively growing weeds, from where they are taken up through the stem and leaves. While some can be applied over the trees without harming them, few can be used during the growing period (May–July) without scorching the foliage or killing the tree completely.
- *Residual herbicides* are applied around the base of the tree and are most effective on moist, friable soils. The active chemical remains in the soil and is

absorbed by the roots of weeds as growth commences. These herbicides are particularly useful for broadleaves susceptible to foliar herbicides, often being applied in the dormant winter season to control weeds during the spring and early summer.

The need to reduce the cost of herbicide applications has led to the development of low-volume sprayers (weed wipers, spot guns or controlled drop application sprayers) which also reduce the environmental impact of these chemicals.

Current farm woodland practice involves the use of residual herbicides, such as propyzamide granules, in the winter (slowly volatilizes in cold soil) followed by the use of foliar sprays in the summer (e.g. glyphosate) to cope with localized weed problems. Where larger woody weeds are present (e.g. gorse, laurel, rhododendron), a more concerted approach is often required. Alternatives include the cutting back of shrubs by mechanical means followed by herbicide treatment of the regrowth from the stumps, or uprooting of shrubs by machines followed by chemical control of the early regrowth.

The storage and use of all herbicides is now subject to strict environmental standards under COSHH (the Control of Substances Hazardous to Health Act), and only fully approved contractors should be used for weed control operations. For further instruction on appropriate herbicides, storage, application and relevant legislation see Willoughby and Dewar (1995) and Willoughby and Clay (1996).

Mulching

Mulching is a technique normally only used in amenity or small-scale plantings since it tends to be expensive and time-consuming. More recently interest in mulching has grown since it offers an alternative to the use of herbicides. Mulching involves the prevention of weed germination and growth by covering the ground around the tree with either organic material or artificial sheeting. Mulches also retain soil moisture close to the surface by reducing evaporation losses, which promotes root growth and nutrient cycling. The use of mulches is more commonplace in climates where soil moisture is in very short supply. Artificial mulches include specially formed black polythene sheets or mats (125 μm thickness) with ultra-violet inhibitors to slow degradation, although squares of old carpet may be equally as effective. The mulching material should be laid securely with pins or weighed down with soil or stones to prevent displacment by the wind and to deter voles from burowing underneath. Successful organic mulches include straw, wood chips and crushed bark, although they may reduce tree growth by locking up available nitrogen as the mulch breaks down and are prone to being blown away by the wind. The mulch must be maintained for 3–4 years to prevent weed growth within 1 m of the tree to ensure good early growth.

Inter-row vegetation management

The inter-row vegetation between the weed-free spots or strips will grow unchecked if no weed control is carried out. This may lead to invasion by persistent perennial species or tall growing weeds which fall over and smother the trees in the autumn. The development of rampant inter-row vegetation may also impede access by farm machinery.

Occasional inter-row mowing (e.g. once or twice a year) can be undertaken in the first few years after planting. This will enable access throughout the crop and prevent weed species seeding into neighbouring fields. Machinery adapted to work within narrow rows may be required (e.g. mini-tractor or ATV-mounted equipment). As an alternative, the ground between the weed-free areas may be sown with a cover crop which is much more controllable or has uses other than weed suppression (e.g. game cover or conservation). Inter-row crops should be chosen for low productivity, minimizing competition for moisture with the trees and be easily controlled by herbicide in the weed-free zone around each tree (e.g. some fescues). Kale has been successfully used as a game cover crop (the instant spinney); it provides a sheltered habitat for game birds for about 3 years, in addition to the physical support it lends to treeshelters. A hardy variety should be chosen, sown in rows 50–60 cm apart, in June–July following tree planting. Fencing to exclude farm stock is essential, although game birds do not combine well with rabbit-netted plantations unless an area inside the rabbit fence can be mown to provide the birds with a place to dry off and warm themselves in the sun. Where the new woodland is part of a project to increase the value of wildlife habitats on the farm, the inter-row area may be sown with a low-productivity grass and wild flower mixture, which should be chosen to match the underlying soil type. This type of mixture produces a most attractive meadow habitat which is most appropriate for the edges of the woodland and areas adjacent to footpaths and other public spaces.

Cleaning

This refers to the removal of woody plant growth which is either in competition with the tree crop or physically damaging. The operation normally involves the cutting back of unwanted broadleaved trees or regrowth from hardwood stumps and the severing of harmful climbers such as honeysuckle (*Lonicera periclymenum*) and old man's beard (*Clematis vitalba*). Methods of cleaning depend on the nature of the site, although the work generally involves the use of clearing saws to cut or girdle shoots. Chemical treatment is a cost-effective operation, using glyphosate, applied either to foliage or cut surfaces. Where individuals or groups of crop trees have failed, naturally regenerated broadleaves or regrowth from stumps should be retained in these areas to suppress the development of side branches on neighbouring crop trees.

Inspection racks

As the new plantation reaches canopy closure, it becomes necessary to cut rackways through the thicket of trees in order to allow assessment of its silvicultural condition. These are normally made by cutting off the lower branches of two adjacent rows with a pruning saw, cutting flush to the trunk.

Brashing

This involves the removal of dead and dying branches from crop trees up to a height of about 2 m, normally undertaken during the thicket or pole stage prior to first thinning, to improve access for marking and extraction. Brashing may also be carried out where access for game beaters and amenity is required or to release broadleaves when grown in mixture with conifers. The persistence and difficulty of removal of the lower branches vary with species and crop spacing. Shade-bearing trees tend to have the most long-lived branches (e.g. Western hemlock), while larch tends to self prune well. Spruce tends to have quite tough branches, which are difficult to remove; larch branches break off easily with a stick and pine is intermediate between the two. The use of a special curved brashing or pruning saw is preferable to a billhook, especially with larger branches where extra care is need to reduce the risk of damage. Complete brashing of plantations is now rarely economic, so that only sufficient trees are now brashed to allow for inspection and marking, concentrating on the final crop trees. Where plantations are line thinned, the rows to be removed are left unbrashed.

Pruning

Pruning involves the removal of side branches above the height of normal brashing to produce longer lengths of clean knot-free timber. The operation is restricted to selected final crop trees, in order to achieve the standards of quality required for high-grade timber (e.g. planking, joinery, beams and veneers). Pruning should be carried out before the side branches become larger than 5 cm, for ease of working and to reduce the incidence of disease entry. A sharp pruning saw or chisel is used with an extendable handle for high pruning. The branches are sawn not quite flush with the trunk, but slightly proud to avoid bark damage. High pruning will inevitably remove some 'live' branches, although there are no great harmful effects if the amount of crown removed ranges between 25 and 45%.

Pruning regimes differ for conifers and broadleaves:

Conifers

Initial pruning to 4 m is carried out once the trees reach about 10 cm diameter or 20 years old. Further pruning to 5–6 m height is carried out once the trunk at this point reaches 10 cm diameter. The operation may be carried out at any time of year, although preferably during March to May for quicker healing. Pruning of conifers is not widely practised on larger commercial estates, although for smaller farm woodlands, if labour is available, it may be justified to increase the capital value of the woodland. An expanding market exists for pruned pine (Scots and Corsican), Douglas fir and European larch for 'boatskin' quality.

Broadleaves

Pruning begins early for broadleaves (5–10 years) to create a single main stem through the removal of competing shoots (*formative pruning*) and later by the removal of lower side branches from the trunk to about 5 m. The operation is particularly suitable for selected final crop trees grown for decorative veneers (e.g. oak, walnut, cherry and maple). Pruning should never be undertaken during flushing (March–May), since resistance to infection is lower at this time. For most species, winter is the optimum time for pruning, although cherry should be pruned between June and August to reduce the risk from canker and silver leaf disease (*Chondrostereum purpureum*). See Evans (1984) for further details on pruning broadleaved species.

Epicormic branches

Epicormic branches arise from adventitious buds around pruning scars or from dormant buds on the stem, which are triggered into sprouting by increased levels of light (e.g. following thinning). These epicormic shoots remain semi-moribund in normal woodland conditions, although a further increase in light levels will enable the shoots to develop into larger branches, which will severely affect the quality of the timber. Only small knots will be tolerated for high quality sawnwood and may have to be completely absent for decorative veneers. Pedunculate oak, sweet chestnut, poplar and cricket-bat willow are the worst affected. The growth of epicormics can be restricted by avoiding sudden changes in the environment of the stand (e.g. through heavy thinning) or by pruning. If carried out annually, epicormics can be removed by rubbing. Classical silvicultural treatment of oak stands to reduce the incidence of epicormics involved growing an understorey of beech, hornbeam, Norway spruce or Western red cedar to keep the boles of the oak trees shaded.

Forest protection

Fire

The chances of fire damage depend on the nature, size and location of the woodland. A key factor is the amount of inflammable material in the forest, either dead wood and brashings or dry undergrowth in young plantations before canopy closure. March to May is the most dangerous period, when a large amount of dead vegetation remains from the previous year, especially when combined with low humidity, high temperatures and windy conditions. Woodlands close to main roads or residential areas also tend to have a higher risk of fire outbreak. The risk of fires spreading can be reduced by cutting fire breaks, strips of at least 10 m width, which are kept free of inflammable material by mowing or cultivation. Japanese larch and alder do not easily catch fire and can be used as fire belts 10–20 m wide, between compartments. Forest tracks and roads also act as internal fire barriers. Other precautions that should be taken include:

- Inform the local fire brigade of the location of woodland access points and fire-fighting equipment.
- Ensure adequate access to woodlands by removing fallen trees and repairing broken culverts.
- Maintain fire-fighting equipment at strategic points, to include spades, buckets, axes, and provide birch brooms and beaters at entrance points to woodlands.
- Assess the proximity to water and if necessary dig additional ponds.
- For high-risk sites (along footpaths and roads) clear inflammable material and brash young conifers early.
- Ensure that all staff and contractors observe the fire precautions and are trained in the fire drill.
- Insure against fire.

Wind

Prolonged exposure to high winds significantly reduces tree growth rates and in extreme conditions will lead to stem damage from abrasion by neighbouring trees. Where root development is restricted by waterlogging or an ironpan, woodlands are more susceptible to windthrow during autumn and winter gales since the shallow root plate gives less anchorage. In persistently windy regions, the risk of windthrow can be reduced by shortening crop rotation lengths and either avoiding line thinning or not thinning at all. Site drainage and cultivation can improve tree anchorage and root development. Some species are more susceptible than others; Douglas fir and larches are particularly prone to windthrow on shallow clay soils. Windthrown trees can be economically harvested, although the cost of harvesting fallen trees, and the danger involved, is much greater than if they were standing. Opening up of the woodland can lead to further windthrow along exposed unstable edges, so windthrown pockets should always be cleared back to windfirm edges or existing compartment boundaries to reduce this risk.

In order to predict the likely effect of wind on the woodland, the Forestry Commission devised a Windthrow Hazard Classification (WHC) based on windiness, elevation, degree of exposure and soil conditions. The classification gave an indication of the height to which a tree crop can be grown before windthrow will become a limiting factor. The WHC system has now been superseded by the Forestry Commission's GALES computer software, which is a comprehensive management tool to assist with the planning of silvicultural operations on exposed sites. Further information on GALES can be obtained from the Forestry Commission Research Agency or see Quine (1995).

Insects

A large number of insects are dependent on trees for some part of their life cycle, either for food, by siphoning off plant sap, eating leaves, needles, shoots, bark or sapwood, or for reproduction, using the trunk, stump or shoots for laying eggs or building brood chambers. The vast majority of these associations cause insufficient damage to be of concern, while some insect–tree relationships are mutually beneficial. The most damaging insects are associated with particular stages in the life of the tree crop, together with site and climatic conditions favouring a rapid increase in insect population size. There is some evidence in this respect that trees already under some environmental stress (e.g. from incorrect choice of species for site) are more prone to serious damage. The most susceptible stage in the life of the woodland is at establishment, particularly where this is adjacent to neighbouring mature trees of the same species or where stumps from the previous crop are still present, which may provide the initial source for an insect outbreak. While some insect pests are fairly general in their tastes, others are specific to particular trees or groups, conifers tending to be more susceptible than broadleaves. The most important insects associated with particular stages in the life of the forest are as follows:

- *Establishment of all species on restock sites* may be prone to attack from large pine weevils (*Hylobius abietis*) and black pine beetles (*Hylastes* spp.), which will breed and multiply in the stumps and brash of the previous crop, to emerge and feed on the newly planted trees, causing severe losses. The long-established practice of dipping trees before planting with pyrmethrin insecticide has been recently banned because of environmental concerns (June 2000). Alternatives include delaying planting for several years to allow brash to break down and 'hot planting' which involves rapid planting very soon after harvesting before insect populations have had a chance to build up and cause trouble.
- From the *thicket stage* onwards, many tree species suffer varying degrees of defoliation from leaf and needle feeders. Most trees will recover from even quite severe defoliation, and control measures are normally only economic in large commercial woodlands. Specific examples of damage include pine looper moth (*Bupalus piniaria*) on Scots pine, pine beauty moth *(Panolis flammea)* on Lodgepole pine and the oak leaf roller moth (*Tortrix viridana*).
- On *windblown sites* or where *felled timber* is available, a number of secondary pests (weevils and bark beetles) will breed and multiply under the bark of damaged, dead or felled timber. Where subsequent populations are high, the surrounding healthy crop may be damaged, spruce, larch and pine being the most severely affected. The pine shoot beetle (*Tomicus piniperda*) is the most important pest, requiring good forest hygiene to control outbreaks in susceptible crops. Felled timber should be either removed or debarked within 6 weeks, during the period March to August, to prevent broods of insects from being raised.

The identification and control of insect pests is a specialized activity, so if insect damage is suspected, further advice should be sought from the Forestry Commission. Further information is available in Bevan (1987).

Disease

There are a number of fungal and bacterial diseases of trees, which can lead to deterioration of the main stem, destruction of roots, shoots or cambium and ultimately death or windblow. As with insect damage, the severity of the disease infection is often exacerbated by both climatic conditions and environmental stress, wounding of trees from extraction damage or pruning being examples. Control measures are rarely justified, so correct planting choice, using more resistant tree species, is to be advised in susceptible conditions. The most important diseases to be aware of are as follows:

Honey fungus (*Armillaria* spp.) can be a problem in old broadleaved woodlands, affecting both broadleaves and conifers. It spreads through the soil from infested wood, especially old broadleaved stumps, via a network of black strands (*rhizomorphs*) to infect young trees. Losses are sporadic and rarely affect a whole wood, although it may be wise to restock with a more resistant tree, such as Douglas fir or a broadleaved species.

The *stem and root rot* (*Heterobasidion annosum*) is the most serious disease of coniferous woodland, leading to deterioration of the lower trunk and roots. Larch, spruce, Western hemlock, Western red cedar and pines (on former agricultural land or alkaline soils) are most susceptible, especially on the sites of previous coniferous crops. The risk of disease can be reduced by treating the freshly cut stump surfaces of felled trees with a concentrated coloured solution of urea within 15 minutes of felling. More resistant species include Douglas fir, grand fir and Corsican pine.

Larch canker is caused by the fungus *Lachnellula willkommii* on European larch, favouring its replacement by hybrid larch. *Group dying of conifers* caused by the fungus *Rhizina undulata* spreads through litter from fire sites; therefore fires should be excluded from the forest interior. *Beech bark disease* develops from the association of the scale insect *Cryptococcus fagisuga* and the fungus *Nectria coccinea*, forming a matt of white greasy wool on the main stem and branches. Infected trees should be removed during thinning operations. Further information can be found in Gregory (1998).

Mammal damage

Mammals can cause damage to trees and woodlands by browsing on foliage, shoots and buds or stripping and fraying bark. Heavy browsing may devastate unguarded transplants and saplings, natural regeneration and coppice regrowth. While light woodland grazing may benefit some wildlife, very high local populations of rabbits or deer will reduce the structural diversity of the woodland ecosystem with fewer shrubs and less abundant ground vegetation. Management should be based on identifying signs of damage (Table 5.8), assessing local population density and protecting vulnerable areas by fencing or other control methods. General measures to protect young trees are included in previous sections on fencing and treeshelters. More specific measures include the following.

Grey squirrels damage pole-size trees by stripping bark from susceptible species (beech, sycamore, ash and pine) between May and July, causing deformities in the main stem and increased forking. The extent of damage can be reduced by controlling squirrels from April to July, by live cage trapping or using hoppers containing warfarin-poisoned bait. Warfarin is potentially hazardous to both people and wildlife, requiring training in its safe application. The use of warfarin is permitted only in approved areas, which effectively excludes much of Scotland, the Lake District and other areas where red squirrels may still exist. The designs of warfarin hoppers and guidelines for their use have become increasingly sophisticated over time in an attempt to reduce the risk of other species being poisoned. For maximum control the hoppers should be placed under large trees clear of vegetation at 200 m centres or 3–5 hoppers/ha. Research is continuing into alternatives to warfarin for the control of grey squirrels, with the introduction of immuno-contraceptives looking increasingly promising. Control by shooting is labour intensive and requires skilled staff to be effective.

Deer will browse the tender growth of most tree species, especially broadleaves such as ash, cherry, willow, hazel and rowan, leaving a ragged end to the nibbled shoots and clearly demarcated browse line throughout the woodland. Deer may strip bark from smooth-barked trees such as Norway spruce, larch, ash, willow and beech, leaving behind telltale broad teeth marks running up the peeled stem. Male deer mark out their territories and clean off velvet by fraying young saplings. Successful deer management depends on co-operation amongst neighbouring landowners,

Table 5.8 Identification of browsing and bark damage to trees (after Hodge and Pepper, 1998, by permission of the Forestry Commission, 2001)

Mammal	*Time of year*	*Type of browsing damage*	*Type of bark damage*
Bank vole	Winter and early spring	Bud removal, especially pine on restock sites	Short, irregular strips 5–10 mm wide, with incisor marks 1 mm wide in pairs. Can climb to 4 m
Field vole	All year, but greatest risk in winter		Bark signs as bank vole but restricted to height of surrounding vegetation
Rabbit	Winter and spring	Sharp-angled clean cut to shoots 0.6 m above ground. Removed portion eaten	Incisor marks 3–4 mm wide in pairs, running diagonally across the stem to 60 cm above ground
Hare	Winter and spring	As rabbits to 0.7 m above ground, but shoots often not eaten	
Squirrel	April–July		Incisor marks 1.5 mm wide in pairs, running parallel with stem on sycamore, beech, oak and pine trees 10–40 years old
Deer	All year	Ragged edge to damaged shoots (due to lack of teeth in front upper jaw) to 1.1 m above ground (roe deer and muntjac) or 1.8 m (fallow, red and sika deer). Fraying in March–May	Vertical incisor marks (red, fallow and sika deer)
Sheep and goats	All year	Coarse browsing of foliage to 1.5 m with uprooting of transplants	Diagonal incisor marks to 1.5 m above ground
Cattle and horses	All year	Coarse browsing of foliage to 2.0 m (cattle) and 2.5 m (horses) with uprooting of transplants	Diagonal incisor marks to 2.0 m (cattle rubbing) and 2.5 m (horses)

fencing around vulnerable areas and humane culling to ensure a healthy population. Design features to ensure safe and effective culling include the creation of sunny deer glades and wide grassy rides with strategically placed high seats. Shooting is strictly controlled by law (1995 Deer Act) although, properly managed, it may generate substantial income from stalking and venison. Further information is available from the Deer Initiative (or Deer Commission for Scotland) or see Mayle (1999).

Landscape design for farm woodlands

A woodland that is designed and managed using the principles of good landscape practice will enhance the surrounding landscape, and provide benefits for wildlife and public enjoyment. The Forestry Commission and other agencies (e.g. National Parks and local authorities) are increasingly enforcing the adoption of a professional approach to forest landscape design through control by felling licences, forest strategies and grant schemes. Landscape design skills have become an essential part of the forester's toolkit. There is strong evidence that more attractively landscaped forests command greater prices in the market.

On level ground, the landscape sensitivity of the forest is less than for highly visible forests on steep slopes at higher elevations, or in designated areas of great landscape value. Good landscape practice therefore begins by assessing the nature of the surrounding landscape, which will dictate the most appropriate design solutions.

For *hedgerow landscapes* in the lowlands, the design of woodlands should mirror the often geometric shape of the surrounding hedgerow pattern, the shape of the woodland interlocking harmoniously with the surrounding fields. For *open and upland landscapes*, the design of woodlands needs more careful consideration. In these areas the topography or landform will be the dominant force in the landscape so that any new woodland planting must reflect the scale and shape of the surroundings. Small woodlands may look out of scale and should be located close to existing woodlands or on lower slopes where a hedgerow pattern is more evident. The shape of the forest should follow natural boundaries and vegetation, rather than the contour or the edge of the fence. As a general rule the edges of the forest should rise up the valleys and hollows and fall down the shoulders of hills and ridges.

The landscape of all forests can be improved by attention to the following measures:

- *Diversity* creates visual interest and enhances the landscape value of the forest. Different textures and colours can be created by using different species, such as broadleaves along watercourses or compartments of larch, which change from fresh green in summer to brown in winter. A change of species can reduce the scale of the forest and allow more light into recreational areas, although the main species should make up two-thirds of the composition for aesthetic reasons. Diversity can also be increased by the provision of open ground, exposing outcrops of rock and creating a mosaic of different ages of trees.
- *Forest boundaries* are particularly important at the skyline, where the forest should either completely cover the skyline, or cross it at the lowest point, cutting diagonally across the main view or else curving gently over the skyline. The edges to the forest should be as natural as possible, varied in scale with the landscape by the use of irregular groups, different species or plant spacing and detailed shaping.
- *Forest operations* should be carried out in sympathy with the landscape. Planting should be organized to create open areas which follow the landform, and forest tracks should cut diagonally across the contour, rather than along it. The harvesting of the forest provides an excellent opportunity to improve the appearance of a previously poorly designed forest by introducing different species, a greater diversity of age structure and by reshaping the edges of compartments. Any such work will need to be carefully phased to avoid immediate drastic changes.

For larger woodlands, especially in sensitive locations, it will be essential to undertake a full landscape appraisal, possibly using professionals who will have access to specialist forest design software. For small-scale woodland design, photographs, acetates and overlays will suffice to show the effect of different forest operations, the phasing of harvesting and sites for new planting. For further information see Forestry Commission (1995).

Forest management for non-timber uses

Shelter woods

The provision of correctly designed and located shelter woods can benefit both crop and livestock enterprises on the farm. Shelter woods also have the potential to contribute to secondary objectives, including the provision of game cover, facilities for recreation and occasional timber production. The most important design features are the height and porosity of the shelter wood, which determine the extent and degree of wind protection (Fig. 5.4). The requirements of arable crops and livestock differ in this respect as follows:

- *Arable and grassland*. Shelterbelts within arable field crops or grassland will have the effect of reducing wind speed and moisture loss from the crop, leading to earlier harvests and greater productivity. Solid or impermeable windbreaks should be avoided, as they cause an upward deflection of the airstream, producing an area of low pressure to leeward of the barrier which results in intense turbulence more damaging than the original wind. More porous windbreaks (ideally 50% permeability) not only reduce turbulence but also reduce windspeed by up to 70% for a distance up to 30 times the height of the barrier. On sloping ground, the location of the windbreak should be carefully considered, as a trap is created for frosty air flowing down from higher up the hill. Single or double row windbreaks are normally used on horticultural and productive agricultural land, as wider windbreaks give no additional reduction in windspeed and may harbour pests such as pigeons or bullfinches and prevent normal drying, favouring fungal diseases such as grey mould (*Botrytis cinerea*). Shelterbelts may be inappropriate in some situations where they may shade neighbouring crops, harbour pests and vermin, and interfere with machine access, or on heavy wet soils where the drying effect of the wind is reduced. Suitable species for field crop protection are Balsam poplar hybrids (*Populus tacamahaca* × *trichocarpa*), grey alder, or Italian alder on drier soils, all at 1- to 2-m spacings.
- *Upland livestock*. Shelter woods afford protection to stock from the cold driving winds, reducing heat loss and hence improving survival and cutting food consumption. A dense impermeable windbreak is most suitable, as animals tend to pack in tightly to the lee of the wood during winter gales. Dense shelter woods create a narrow belt of calmer condi-

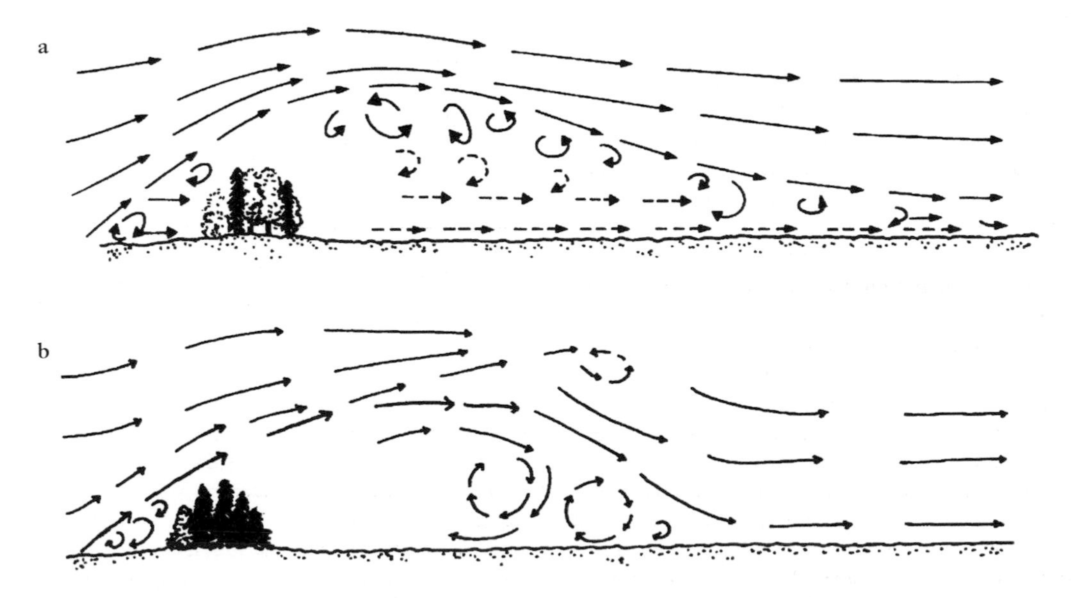

Figure 5.4 Pattern of windflow through shelter woods: **a** through a narrow and semi-permeable shelter wood: small rolling eddies create a long zone of shelter (15–30 times the height of the wood); **b** over an impermeable shelterbelt: large standing eddies and a short shelter zone (5–10 times the height of the wood).

tions in their lee up to 10 times the height of the trees and are also suitable for small areas such as lambing paddocks or farm buildings. Narrow belts of trees become draughty around the lower trunk with age and pose problems at maturity, as restocking will generally mean clear felling leading to a temporary loss of shelter. Neglected belts of conifers pose particular difficulties and require very gradual thinning to avoid catastrophic windthrow. Wider shelter woods should therefore be used (at least 20 m and 45 m if possible), which allow for the use of a mixture of shrub and tree species to create a more densely graded profile. Wider shelter woods allow restocking to take place progressively without loss of protection and have a greater proportion of utilizable timber together with other benefits for wildlife and game. Livestock can be allowed access to the woodland in extreme weather for shelter and feeding on ground vegetation, although this should only be a temporary measure as long-term entry by livestock will lead to root damage, poaching and browsing of natural regeneration. A mixture of species is advisable, especially oak, ash, beech, pine and sycamore with cherry, whitebeam, rowan and alder along the woodland margins. Sitka spruce and lodgepole pines are windfirm and are useful on exposed ground in the west.

The height and porosity of the shelter wood is controlled by the width of the wood, species choice and silvicultural management. Draughts can be reduced by planting shrubs and coppicing trees on the woodland edge. Lighter canopy trees such as birch and oak encourage the growth of dense ground vegetation and the wood should be thinned regularly.

Other factors to consider in the design of shelter woods include the local climate, topography and surrounding landscape. Narrow lines of trees will not thrive in very exposed conditions and a landscape appraisal may be a condition of grant aid, especially for long straight shelterbelts in a predominantly open landscape.

Pheasants

Farm woodland management for pheasants need not be in conflict with sound silvicultural practice. The value of woodlands can be substantially increased by incorporating some of the following specific measures to increase its potential for pheasants.

- Pheasants do not thrive in cold draughty conditions, so woodlands should be designed to give maximum protection to ground level using a mixture of taller trees and shrubs. The perimeter of the woodland may be planted with hedging species or Christmas trees, although temporary protection can be provided by strategically placed big-bales. Roosting trees should be retained on restocked sites and at the intersection of rides.
- The holding capacity of the wood can be increased by establishing ground-cover shrubs, which provide additional shelter and food for the birds. Suitable species include Butcher's broom (*Ruscus aculeatus*), Rose of Sharon (*Hypericum calycinum*), dogwood (*Cornus alba*), elder (*Sambucus nigra*), flowering nutmeg (*Leycesteria formosa*), privet (*Lingustrum vulgare*), raspberry (*Rubus idaeus*) and snowberry (*Symphoricarpus albus*).
- A rich ground layer should be maintained by avoiding planting trees that cast a dense shade (beech, sycamore and close-spaced conifers), favouring trees with a lighter canopy (oak, ash, birch, cherry, rowan, larch and pine) and thinning at the appropriate time to increase the amount of light reaching the forest floor. Coppice is particularly suitable as it maintains a higher proportion of ground vegetation and a diversity of ages in the stand.
- Pheasants are woodland-edge birds, using the forest for winter cover and roosting. By planting a series of smaller and longer woodlands, the proportion of edge can be increased.
- In larger woodlands, open spaces and rides (30–50 m wide) should be provided to allow birds room to fly up and over the guns. Grassy glades maintained by annual mowing provide favourable conditions for young birds.

Further advice regarding woodland design for game is available from the Game Conservancy Trust.

Conservation

The following section gives general guidance on the improvement of woodland for wildlife, although some woodlands will already have important conservation status or designation such as Sites of Special Scientific Interest (SSSI), National Nature Reserves (NNR) or ancient semi-natural woodland. These important sites will require a more specific management prescription, normally involving expert advice from either local or national wildlife organizations.

- Native trees support large numbers of species (especially invertebrates) and by planting trees native to the area the local genetic stock is preserved. Natural regeneration of native trees should be encouraged wherever possible.

- In commercial coniferous woodlands, native broadleaves should be planted along watercourses and woodland edges.
- Open space is an important woodland habitat with characteristic ground vegetation and associated wildlife. These areas can often be incorporated at planting by leaving bare ground around rock outcrops, stream sides and wetlands without reducing the overall productivity of the plantation (up to 20% open space is now accepted under the Woodland Grant Scheme). Woodland glades and rides (at least 5–10 m wide) can be managed by cutting back vegetation and mowing to create a graded profile.
- The ground vegetation that builds up following tree establishment supports a thriving population of small mammals and their predators. Following canopy closure, much of this vegetation is lost under the dense shade of the thicket of trees, especially conifers. By thinning earlier (up to 5 years) and slightly more heavily (up to 10% more volume removed), ground vegetation is more likely to survive.
- Structural diversity can be improved by planting an understorey of native shrubs, such as hazel, hawthorn, holly or juniper, particularly along woodland rides and edges.
- Harvesting creates the opportunity to increase the age diversity of uniform plantations by clearing smaller areas (0.2–0.5 ha) phased in over a longer period.
- Some mature trees should be retained beyond the normal rotation until physical maturity (at least 1%), to provide old-growth habitats and habitat continuity.
- Dead wood habitats are scarce in most commercial woodlands. Stacks of dead branchwood should be left to rot down (where forest hygiene permits) and standing dead trees retained for hole-nesting birds, specialized invertebrates and epiphytes. In addition, nest boxes and bat boxes can be provided in a variety of designs.
- The timing of forest operations should be planned to avoid breeding seasons and the passage of felling and extraction machinery organized to avoid disturbance to sensitive areas such as streams and wetlands. Herbicide use should be minimized and extreme care taken to avoid contamination of watercourses and other habitats.

The landowner may find it useful to incorporate some of these improvements into a conservation management plan for the woodland. Following an appraisal of the wildlife value of the site, appropriate management prescriptions should be proposed to reflect the overall aims for the woodland.

Forest investment

In financial terms, the forest plantation represents a long-term investment, where the owner faces a patient wait before revenue from the later thinnings and final harvest offset the early establishment expenditure. In order to compare forest investment with other alternatives, a discounted cashflow may be calculated, using an appropriate interest rate (net of inflation), to give a *net present value* (NPV) for the proposed woodland. The return from different rotations and species can be compared easily, since all future revenues and expenditure are discounted back to year zero. The *internal rate of return* (IRR) is calculated as the interest rate at which NPV = 0. The choice of discount rate is critical to the effect on NPV, since the higher the discount rate chosen, the lower the value of future returns, especially for longer rotation crops, such as oak. A figure of 3–5% is normally appropriate for most commercial woodlands, although lower rates may be acceptable where other less tangible benefits, such as landscape and wildlife, are considered important.

For a more immediate assessment of the financial impact of woodland planting, a *partial budget* or a *cash flow forecast* are useful tools. A partial budget involves a review of all the potential costs of planting set against any income over the first 5–8 years in the life of the new woodland. The costs of planting and aftercare should include proceeds foregone from land taken out of agricultural production and income should include grant aid, the sale of machinery or livestock and other cost savings to the enterprise. Fixed and variable costs should be included in the calculations. The woodland owner will then at least be aware of the short-term financial implications of a proposed woodland venture.

Integration of agriculture and forestry

An increasing interest in genuinely integrated farm forestry systems is apparent in some parts of the UK, where tight agricultural margins are forcing farmers to look at all the resources at their disposal. Integrated farming/forestry systems have the potential to provide many benefits. For example, labour can be deployed in the woodlands at less busy times of the year, marginal land may be more productive growing trees than being farmed and home-grown timber can be used on the farm or sold to local markets. A farmer or farm staff trained in woodland operations may be able to generate income from contracting work on neighbouring farms. Woodlands planted with a shelter objective can be highly beneficial to crops and livestock and income may be generated from letting out a game shoot.

The potential for intimate integration of agriculture and tree production – *agroforestry* – continues to be assessed by researchers in the UK. Agroforestry involves the establishment of widely spaced trees with grazing or arable crops in between the rows. The productivity of the agricultural component will depend on the tree spacing and resulting point of canopy closure. The timber objective is to produce a main stem of at least 40–50 cm dbh over a short rotation of 40–50 years, combined with pruning to remove side branches. Suitable tree species include ash, cherry, walnut, poplar, Douglas fir and hybrid larch. Economic models have shown that such systems may prove financially viable, particularly in marginal areas, although further field-testing is required. See Hislop and Claridge (2000) for further information.

Grants

The Woodland Grant Scheme (Forestry Commission)

The Woodland Grant Scheme (WGS) aims to encourage the management of existing woodland together with the expansion of private forestry in a way that achieves a reasonable balance between the needs of the environment, increasing timber production, providing rural employment and enhancing the landscape, amenity and wildlife conservation. The scheme can provide a substantial contribution towards the costs of woodland establishment and management, and a professional woodland adviser will be able to assist owners to make the most of the grants on offer.

National and local variations in the WGS now exist with different Forest Conservancies beginning to target funds towards the type of woodland considered most valuable in their region. This more discretionary approach involves the assessment of proposed woodland plans against defined criteria, with grants only being awarded to applications that score above a minimum standard. For specially designated areas, additional 'Challenge Funds' have been procured for woodland planting and are distributed by a competitive bidding process (e.g. the New National Forest in the Midlands and the Grampian Forest in northeast Scotland).

A number of general provisions must normally be met to obtain WGS funding. These include the maintenance of a broadleaved component of the woodland, protection of statutory designations (e.g. ancient monuments, SSSIs, rights of way), appropriate management of ancient woodland and compliance with various Forestry Commission guidelines such as landscape design, watercourse management, fire protection and nature conservation.

Grants are available for restocking, new planting or natural regeneration on areas larger than 0.25 ha or 15 m wide, although smaller areas may be accepted by agreement with the Forestry Commission. Payments are based on a stocking of at least 1100 trees/ha for broadleaves and 2250/ha for conifers. Plantations at wider spacings, coppice establishment and agroforestry projects are paid on a pro rata basis. Up to 20% of the total area may be left unplanted for woodland glades, forest margins or along watercourses, and up to 10% non-timber trees and shrubs may be planted for conservation or game.

Additional supplements may be available for planting on arable land or improved grassland (*Better Land Supplement*) or creating woodlands incorporating public access within 5 miles of selected urban areas (*Community Woodland Supplement*). Grants may also be available to contribute towards the cost of forest management where there is well-defined provision of public benefits such as access or high-quality wildlife habitats. The use of long-term management plans is currently being piloted with a 20-year time scale and all associated permissions and grants agreed in advance.

The Farm Woodland Premium Scheme (DEFRA)

The Farm Woodland Premium Scheme (FWPS) aims to encourage the planting of woods on farmland by providing the farmer with annual payments to compensate for the loss of agricultural revenue for up to 15 years after planting. The scheme is weighted towards better quality land in lowland areas and runs in conjunction with the Woodland Grant Scheme:

- Farmland that has been under arable cultivation for at least 3 years prior to the WGS application.

- Grassland that has been improved for at least 3 years prior to the WGS application, the sward comprising at least 50% ryegrass, cocksfoot, timothy or white clover.

Information packs about the Woodland Grant Scheme and Farm Woodland Premium Scheme are available from the Forestry Commission.

Other sources of grant assistance

Grants are available for approved amenity planting from a range of sources, such as the Countryside Commission, Countryside Council for Wales and Scottish Natural Heritage, operating through the relevant local authority. For woodlands of high wildlife conservation value, assistance may be available from English Nature (or equivalent organization), National Park Authorities or through charitable organizations such as the Woodland Trust. Additional help may be available for sites within other schemes such as Countryside Stewardship or Environmentally Sensitive Areas.

Further information on these and other local schemes can be obtained from the regional office of the Forestry Commission, professional woodland advisers, your local Farming and Wildlife Advisory Group (FWAG) officer or the Woodlands Advisor of the local authority.

Useful addresses and sources of advice

Arboricultural Association
Ampfield House, Ampfield, Nr Romsey, Hampshire SO51 9PA. Tel.: 01794 368717; www.trees.org.uk; email: treehouse@dial.pipex.com
Association of Professional Foresters
7–9 West Belford Street, Belford, Northumberland NE70 7QA. Tel.: 01668 213937; www.apf.org.uk
British Christmas Tree Growers' Association
18 Cluny Place, Edinburgh EH10 4RL. Tel.: 0131 447 0499; www.bctga.org.uk
Coed Cymru
Ladywell House, Newtown, Powys SY16 1RD. Tel.: 01686 26799
Deer Commission
Knowsley, 82 Fairfield Road, Inverness IV3 5LH. Tel.: 01463 231751
Deer Initiative (England)
c/o Great Eastern House, Tenison Road, Cambridge CB1 2DU. Tel.: 01223 314546
Deer Initiative (Wales)
c/o Victoria Terrace, Aberystwyth SY23 2DQ. Tel.: 01970 625866
Forest Research (publications and research)
Forest Research Station, Alice Holt Lodge, Wrecclesham, Farnham, Surrey GU10 4LH. Tel.: 01420 22255; email: ahl@forestry.gov.uk
Forest Research (Scotland)
Northern Research Station, Roslin, Midlothian EH25 9SY. Tel.: 0131 445 2176; email: nrs@forestry.gov.uk
Forestry Commission (local conservancy office – consult your phone directory or the FC website) www.forestry.gov.uk
Forestry Commission (main headquarters)
231 Corstorphine Road, Edinburgh EH12 7AT. Tel.: 0131 334 0303; www.forestry.gov.uk
Game Conservancy Trust
Fordingbridge, Hampshire SP6 1EF. Tel.: 01425 652381; www.game-conservancy.org.uk
Institute of Chartered Foresters
7a St Colme Street, Edinburgh EH3 6AA. Tel.: 0131 225 2705; www.charteredforesters.org/
National Small Woods Association
The Cabins, Malehurst Estate, Minsterley, Shropshire SY5 0EQ. Tel.: 01743 792644; www.woodnet.org.uk/nswa/
Royal Forestry Society of England, Wales and Northern Ireland
102 High Street, Tring, Hertfordshire HP23 4AH. Tel.: 01442 822028; www.rfs.org.uk
Royal Scottish Forestry Society
Hagg-on-Esk, Canonbie, Dumfriesshire DG14 0XE. Tel.: 01387 371518; www.foresters.org/rsfs/
Timber Growers Association Ltd
5 Dublin Street Lane South, Edinburgh EH1 3PX. Tel.: 0131 53871111; www.timber-growers.co.uk
Tree Council
51 Catherine Place, London SW1E 6DY. Tel.: 020 7828 9928; www.treecouncil.org.uk
Woodland Trust
Autumn Park, Dysart Road, Grantham, Lincolnshire NG31 6LL. Tel.: 01476 581111; www.woodland-trust.org.uk

References and further reading

Armstrong, A. (1999) *Establishment of Short Rotation Coppice*. FC Practice Note 7.
Bevan, D. (1987) *Forest Insects*. FC Handbook 1.
Blyth, J. *et al.* (1991) *Farm Woodland Management*, 2nd edn. Farming Press, Ipswich.
Brazier, J.D. (1990) *The Timbers of Farm Woodland Trees*. FC Bulletin 90.

Edwards, P.N. (1983) *Timber Measurement*. FC Booklet 49.
Edwards, P.N. & Christie, J.M. (1981) *Yield Models for Forest Management*. FC Booklet 48.
Evans, J. (1984) *Silviculture of Broadleaved Woodland*. FC Bulletin 62.
Ferris-Kaan, R. (1995) *Managing Forests For Biodiversity*. FC Technical Paper 8.
Forestry Commission (1990) *Forest Nature Conservation Guidelines*.
Forestry Commission (1995) *Forest Landscape Design Guidelines*, 2nd edn.
Forestry Commission (1998) *Forests and Soil Conservation Guidelines*.
Forestry Commission (2000a) *Forests and Water Guidelines*, 3rd edn.
Forestry Commission (2000b) *Forestry Practice Guide*.
Gregory, S.C. (1998) *Diseases and Disorders of Forest Trees.* FC Field Book 16.
Hamilton, G.J. (1985) *Forest Mensuration Handbook*. FC Booklet 39.
Hart, C.E. (1991) *Practical Forestry for the Agent and Surveyor*. Alan Sutton, Stroud.
Hibberd, B.G. (1988) *Farm Woodland Practice*. FC Handbook 3.
Hislop, A.M. & Claridge, J.N. (2000) *Agroforestry in the UK*. FC Bulletin 122.
Hodge, S. & Pepper, H. (1998) *The Prevention of Mammal Damage to Trees in Woodland*. FC Practice Note 3.
James, N.D.G. (1989) *The Forester's Companion*, 4th edn. Blackwell, Oxford.
Kerr, G. (1993) *Growing Broadleaves for Timber*. FC Handbook 9.
Lines, A. (1987) *Choice of Seed Origins for the Main Forest Species in Britain*. FC Bulletin 66.
Lucas, O.W.R. (1991) *The Design of Forest Landscapes*. Oxford University Press, Oxford.
MAFF (1993) *Farm Woodlands – Practical Guide*. HMSO, London.
Mayle, B. (1999) *Managing Deer in the Countryside*. FC Practice Note 6.
McCall, I. (1988) *Woodlands for Pheasants*. Game Conservancy, Fordingbridge.
Pepper, H.W. (1992) *Forest Fencing*. FC Bulletin 102.
Peterken, G. (1993) *Woodland Conservation and Management*, 2nd edn. Chapman and Hall, London.
Potter, M.J. (1991) *Treeshelters*. FC Handbook 7.
Pyatt, D.G. (1997) *An Ecological Site Classification for Forestry in Great Britain*. FC Technical Paper 20.
Quine, C. (1995) *Forests and Wind*. FC Bulletin 114.
Robertson, P.A. (1992) *Woodland Management for Pheasants.* FC Bulletin 106.
Rodwell, J.S. (1994) *Creating New Native Woodlands*. FC Bulletin 101.
Rollinson, T.J.D. (1999) *Thinning Control*. FC Fieldbook 2.
Savill, P.S. (1991) *The Silviculture of Trees Used in British Forestry*. CAB International, Wallingford.
Savill, P.S. (1997) *Plantation Silviculture in Europe.* Oxford University Press, Oxford.
Williamson, D.R. (1992) *Establishing Farm Woodlands*. FC Handbook 8.
Willoughby, I. & Clay, D.V. (1996) *Herbicides for Farm Woodlands and Short Rotation Coppice*. FC Field Book 14.
Willoughby, I. & Dewar, J. (1995) *The Use of Herbicides in the Forest*, 4th edn. FC Fieldbook 8.

Useful websites

The following sites are good starting places for UK forestry and woodland information:

- www.british-trees.com – general information and useful links.
- www.forestry.gov.uk – Forestry Commission website with details of grant schemes, market reports and research information.
- www.woodnet.org.uk – information for woodland owners on timber sales and contractors.

6

Crop health – responding to weeds, diseases and pests

R.E.L. Naylor

Introduction

Healthy crops grow better and produce greater quantity and higher quality yields than crops in 'ill-health'. Part of the task of ensuring crops are healthy involves *management* decisions (e.g. when to sow) as well as ensuring that the *physical* growing conditions which we can influence are as optimal as we can make them. Thus we cultivate soil and prepare 'a good seedbed'. We may add lime to raise the pH of acid soils and we may add fertilizer to ensure adequate crop growth. However, a major part of crop health is concerned with the *biotic* environment of crops, i.e. the other organisms present. These may be other plants (weeds), fungi, bacteria or viruses (disease organisms), or vertebrate or invertebrate animals (pests).

This chapter is mainly concerned with these other organisms, particularly what they do to crop stands in the UK and northern Europe, how we can minimize their occurrence and how we can minimize their effects when they do occur. The approach will be to identify the ways in which organisms can invade crops (how they get there, how they develop), to examine what they do to crops (how yield and/or quality is reduced), how we can ameliorate the effects on crops (management or control by various methods) and how we might minimize the recurrence in future seasons. This last point emphasizes that the problems in one specific crop should not be dealt with as an individual occurrence. Rather, thought needs to be given to establishing why the problem occurred, before defining appropriate management responses for the current crop as well as for future ones. This chapter will develop the general principles of maintaining crop health. The focus will be on conventional agriculture but with some highlighting of those problems in and management options suitable for low input and organic systems.

Abiotic challenges to crop growth

Crop under-performance is conveniently grouped according to the cause. Non-infectious, or abiotic, causes are features of the environment some of which we may be able to influence. Such environmental control may be achievable in enclosed or protected cropping. However, this is expensive and only high-value crops can provide return on the capital investment and operating costs. In field crops we have less control over crop growing environment, particularly the aerial environment, but we can modify the soil environment in a number of ways. Water and nutrient content are routinely modified as part of standard agricultural practice. Abiotic factors may cause plants to grow poorly or, in extreme circumstances, to die. Features of the environment that affect crop growth and the ways in which we can modify them are shown in Table 6.1.

Biotic challenges to crop growth

Biotic causes of plant underperformance are other organisms that interfere with crop growth and development and which in some cases may cause plant death. These organisms may be of many types, other plants, animals or micro-organisms. Some examples of the range of types are given in Table 6.2.

Organisms that interfere with plant growth usually achieve their yield-reducing effect by imposing gross damage, or above ground by reducing photosynthesis, or

Table 6.1 Abiotic factors and crop growth

Abiotic factor	*Plant challenge*	*Amelioration in field crops*	*Amelioration in protected crops*
Temperature	Too high		Provide ventilation and cooling
	Too low		Provide heating
Water	Too much	Install drainage	Install drainage
	Too little	Irrigate	Irrigate
Light			
Intensity	Too much	Plant a cover crop	Shade
	Too little		Install lighting
Photoperiod	Too much		Install light exclosures
	Too little		Provide supplementary lighting as night break
Soil pH	Too high	Acidify	Acidify
	Too low	Add lime	Add lime
Nutients	Too much (toxic)	Alter pH	Change soil
	Too little (deficiency)	Add nutrient	Add nutrient
Soil or air pollution		Grow resistant plants	

Table 6.2 Types of organisms that can cause crop 'ill-health'

Organism	*Propagules*	*Examples*
Annual plants (weeds)	Seeds	Blackgrass (*Alopecurus myosuroides*), cleavers (*Galium aparine*)
Perennial plants	Rhizomes, stolons	Couch (*Elytrigia repens*)
Parasitic plants	Seeds	Dodder (*Cuscuta* spp.), witchweed (*Striga hermonthica*)
Arthropods (e.g. insects, spiders, mites)	Eggs, larvae, pupae, adults	Peach–potato aphids (*Myzus persicae*), wheat bulb fly (*Delia coarctata*), grain mites (*Acarus siro*), large white butterfly (*Pieris brassicae*)
Nematodes	Free-living worms, cysts	Potato cyst eelworm (*Globodera rostochiensis*)
Molluscs		Slugs (e.g. *Deroceras reticulatum*)
Birds	Adults	Pigeons (*Columba palumbus*)
Mammals	Young and adults	Rabbits (*Oryctolagus cuniculus*)
Fungi	Spores	Potato late blight (*Phytophthora infestans*)
Bacteria		Soft rot of potatoes and vegetables (*Erwinia carotovera*)
Viruses	Many vectors	Barley yellow dwarf virus

below ground by impairing root functioning, or by interfering with translocation or physiologically by modifying growth and particularly development. These mechanisms will be detailed later in the chapter. The most convenient classification of biotic causes of crop ill-health is by the type of organism. The control or management of these biotic factors in crop production is termed crop protection and is of great practical and economic importance (Table 6.3).

Detailed knowledge of the biology and ecology of the specific organisms involved is essential for the development of effective, rational management and control systems. Nevertheless, certain principles are common and are given below. All organisms respond to their environment and abiotic factors influence both crops and the less desirable organisms within them. Thus, interactions between organisms are important, not least because one may be the vector that leads to the spread of another.

Table 6.3 Pests and diseases that attack a broad range of crop species

Common name	*Proper name*	*Possible losses (%)*[1]
Rabbit	*Oryctolagus cuniculus*	
Slug	*Deroceras reticulata*, *Arion hortensis* and other spp.	25
Leatherjackets	*Tipula* spp. (cranefly larvae)	
Wireworm	*Agriotes* spp.	
Cutworm	*Agrotis* and other spp.	
Aphids	Various species	15
Grey mould	*Botrytis cinerea* and other spp.	10
Damping-off	*Pythium*, *Alternaria*, *Fusarium* and other spp.	10

[1]Where no value is given it may be below 10% except when the pest infects a large proportion of the stand early in crop growth.

Features of crop production that contribute to control of biotic factors can be termed good husbandry and include crop rotation, crop hygiene and countermeasures. The latter include mechanical, physical, chemical and biological interventions.

The rest of this chapter will focus on the biotic causes of crop yield loss. This will be achieved by examining a set of general issues: these are

- Invasion (or how the organism gets to the site)
- Infestation (or how the organism develops within the crop)
- Ill-health (or the mechanisms whereby crop losses are caused)
- Impact (or the extent of crop under-performance)
- Interactions (or how other conditions and organisms influence the effect)
- Intervention (or what we can do about it)

Appreciation of the general issues will lead to suggestions for crop managers on how to minimize the occurrence or how to alleviate the symptoms of crop ill-health. The challenges of specific organisms will be summarized in tables. Where possible the impact of the biotic challenge to crop health will be quantified, but reliable information is not always available.

Invasion

It is obvious that new causes of ill-health in crops should not be introduced to the farm. This emphasizes the importance of using healthy planting material. The planting material is often seeds but may also be vegetative material such as tubers, bulbs or transplants. For brevity, the term seeds will be used subsequently, but the ideas are relevant to all types of planting material.

It is important to start the crop with high quality seed. This means seeds that are pure, healthy and germinable. The term 'seed purity' refers to samples that are free from weed seeds and seeds of other cultivars of the same species. In other words, the information on the label is accurate. Healthy seeds are free from the propagules of diseases or pests. These may be fungal spores, insect eggs, larvae or pupae or adults, nematodes, bacteria or viruses. These other organisms may be conveyed to the field with poor quality seeds loose in the seed container (and may be easy to see), on the surface of the seeds (not always easy to see) or inside the seeds (usually difficult to see). The seeds also have to be 'fit for purpose' which means they have the potential to germinate and produce normal seedlings. This is tested in an internationally standardized germination test with specific protocols agreed for each species. Seed lots that are traded in the UK have to bear a certificate from the International Seed Testing Association (ISTA) certifying purity, health and germination. Buyers can ask the seller to see the information on the ISTA certificate. To certify seed lots which have been produced in the UK the seed testing rules are applied by the Official Seed Testing Stations for England and Wales (at Cambridge), Scotland (at Edinburgh) and Northern Ireland (at Belfast) or by accredited testing stations. There are internationally agreed *minimum* standards for purity, health and viability that have to be achieved by each crop species, but many buyers or governments may set their own *higher* standards, particularly for imported planting material.

Although bought seeds have to conform to minimum standards of purity, health and germination, this does not apply to farm-saved seeds. Here there is the risk of sowing weed seeds into a prepared seedbed along with the crop, as well as sowing seeds that bear other organisms on their surface or within them. The viability of farm-saved seed and their potential for producing healthy seedlings is unknown. If farm-saved seeds are to be used it is crucial to ensure the seeds are (1) harvested from a weed-free area, (2) free from pests and diseases and (3) harvested and stored in good conditions. If possible, a germination test should be carried out before planting. The repeated growing of farm-saved cereal seed in order to preserve rare cultivars or types (e.g. bere barley) can lead to the buildup of seed-borne diseases, such as covered smut in barley (*Ustilago hordeii*).

Planting material is not the only way in which propagules can be introduced to fields or farms. They can also be introduced in soil attached to machinery. Farmers who have had pipelines installed under their fields have specified that the contractors have to prevent contamination by soil from other farms. For example, a major concern of Scottish farmers was to prevent the spread of the soil-borne potato cyst nematode which would render the tubers unfit for sale as high value seed potatoes. This was achieved through good management of topsoil stores and by thorough washing of machinery before it moved onto the farm.

Hay, straw and other crop residues are also a source of the spread of organisms from one farm to another. Straw is an important source of weed seed. Weed seeds in hay may survive the passage through an animal's gut and be germinable after spreading in faeces.

Infestation

Some weed seeds are comparatively large and visible: the seeds of wild oats (*Avena fatua* or *A. ludoviciana*) are up to about 15 mm long and those of cleavers (*Galium aparine*) about 2–3 mm diameter. Seeds of rushes are much smaller, about 0.5 mm long. The cysts of the potato yellow cyst nematode (*Globodera rostochiensis*) are less than 1 mm in diameter. This is large compared to fungal spores which are often about 0.001 mm diameter.

Although a single weed seed, fungal spore or bacterium may seem unimportant, each can become a 'focus of infection'. There will be a time lag before the organism can build up to a population level at which there is a high risk of plant infections, crop infestation and crop losses. The inoculum has the potential to develop into a serious problem for the grower.

Many of the sources of inoculum have been considered above under 'invasion'. Now we need to consider the buildup of a population in the field as a weed seed bank in the soil, populations of spores in soil and on crop residues, and pests in soil. This emphasizes the importance of considering soil management, crop residue management and crop rotation as contributions to crop protection. Crop rotation is important because if the biotic factor is surrounded by a crop in which it cannot grow well, then its population may decline. For pests and diseases that have narrow host preferences a sound crop rotation may lead to significant declines in population levels, i.e. in inoculum potential. Thus, for example, it is recommended that potatoes be grown for only 1 year in 7 to prevent the buildup of potato cyst nematode.

A small initial inoculum has the potential to develop to a large population. In some cases, specific environmental conditions may lead to rapid population growth. Knowledge of this process may be incorporated into a model or a prediction scheme. For example, predictions of the risk of barley mildew are derived from standard weather data and such forecasts can be delivered to farmers so that they can institute precautionary or preventive measures. Knowledge of crop residue management and the previous population size can help to predict the occurrence of specific weeds, e.g. blackgrass. Sometimes it is the vector of a disease that is predicted, for example, weather data can be used to predict the spread of aphids which can carry potato blight (*Phytophthora infestans*) or the insect vectors of sugar beet yellows virus.

The implications for crop management are to emphasize again the importance of information on what the biotic challenges have been in previous crops and to combine this with knowledge of the life history and ecology of the specific organism to develop a broad management strategy.

Causes of crop losses

Whatever the nature of the biotic factor in crop ill-health, there are five mechanisms whereby crop performance is altered.

(1) *Gross damage* generally leads to immediate or eventual death of the plants. For example, in damp conditions, slugs often eat the bases of seedlings and this causes them to die. This can be a problem in autumn-sown cereals and oilseed rape. A reduced crop stand can lead to reduced yield, particularly in crops where the seedling gives rise to a single 'product' and compensatory growth is poor (e.g. cabbages). The impact may be less in crops that can branch and in which compensatory growth is possible (e.g. cereals). However, if, for example, wireworms or slugs move along a crop row and kill a patch of plants then full compensation may not

be possible. Excessive rabbit grazing can completely defoliate large patches of young plants and kill them and pigeon grazing can remove the growing point of some crops.

(2) *Interference with photosynthesis* gives a direct reduction in the accumulation of crop biomass. There is a strong relationship between crop biomass and the amount and duration of green leaf material. Thus, any reduction of green leaf area is likely to lead to a reduction in biomass accumulation and of yield. Many pests eat sections of the leaf and directly reduce effective leaf area. Examples are caterpillars (insect larvae with biting mouthparts) and rabbits. Some diseases such as net blotch on barley (*Rhyncosporium secalis*) produce blotches on part of the leaf surface and thereby reduce the leaf area available for photosynthesis. Weeds, particularly where they establish early, can shade the crop and reduce the amount of photosynthesis taking place in crop leaves.

(3) *Interference with roots* can reduce the absorption of water and nutrients. There may be a reduction in the size of the root system through underground grazing by soil-dwelling animals such as wireworms (the larvae of click-beetles, *Agriotes* spp.). Root growth and development may be altered by some diseases, e.g. club root of brassicas (*Plasmodiophora brassicae*), so that water uptake is reduced. Weeds in crops can reduce the availability of water for the crop plants through their own water uptake. Many weeds have roots that spread wider and deeper than crop roots. In addition, some weeds may have a more efficient root system, enabling them to take up more water than the crop plants at times when water is of low availability. Also, they may have a greater water use efficiency within the plant. Because all plant nutrients are taken up in solution, reduction of water uptake is inevitably associated with reduced nutrient uptake.

(4) *Interference with translocation* leads to reduced growth of the desirable part of the crop plant. This is caused either by interference with water moving up the plants in the xylem, or by interference with the products of photosynthesis not moving in the phloem, or both. Insects with piercing or sucking mouthparts (e.g. aphids) divert plant assimilates from crop vascular bundles to their own bodies. Some plant diseases block the phloem or the xylem of the crop and thus interfere with water uptake or with assimilate distribution. In addition, fungi are reliant on assimilates produced by the host plant and thus divert part of the products to their own use.

(5) *Modification of plant growth and development* leads to alteration of the plant form. This may be an increase in the amount of leaf which can then not be supported by water uptake from the roots. Alternatively, root growth may be boosted at the expense of green leaf area. Increases in height of the plant may lead to weaker stems which lodge more easily and earlier. Delay in the onset of reproduction may reduce the opportunity for seed growth later in the season. Some fungi release hormones within the plant which modify plant growth. The most famous example is the fungus *Gibberella fujikuroi* which causes a serious disease of rice whose deforming symptoms are produced by the release of hormones now named gibberellins after it.

In many instances one of the above modes may be the main cause of crop ill-health but other effects may also occur. Interference with photosynthesis will alter the rate of growth of crops and can then reduce size of the root system leading to low nutrient uptake. Interference with root development reduces translocation so plants are more prone to drought. All these interactions make it very important to identify correctly the real cause of the crop ill-health before a management action is taken. Otherwise, the intervention made, which has a cost, may neither alleviate the symptoms nor eliminate the biotic factor causing the ill-health.

Impact

Yield losses due to biotic factors vary widely from field to field, from locality to locality and from year to year. This is to be expected given the patchy distribution of biotic factors and the range of measures we may choose to employ in order to try to limit the occurrence and extent of losses and effect and given the variability in the effectiveness of the measures.

It has been estimated that the elimination of all pesticide use in crops in the UK would lead to a drop in yield of potatoes and cereals of 40–50% and in sugar beet of over 60%. Seven years' worth of trial data from the Morley Research Centre in Norfolk show that the yield loss of oilseed rape due to not controlling broad-leaved weeds was equivalent to £70/ha. In some years, powdery mildew (*Erysiphe graminis*) can reduce yields of barley by 50%. Wheat bulb fly (*Delia coarctata*) can reduce yields by 40% where attacks are severe.

The amount of yield reduction from biotic influences depends very much on the time of the attack. Competition from small weeds in a crop is relatively minor and if these are controlled then the crop loss is small. In contrast, large weed seedlings in a slow-emerging crop can lead to significant crop losses due to the early onset of competition. Similarly, early infection by some diseases may be unimportant if the crop can 'grow away' later. In other cases, early infection causes severe losses but late infection may have little impact on crop yield. The same effects can be seen with pests. Thus, to plan rational intervention, it is important to obtain early recognition of causes of ill-health and to have the ability to predict the likely effect on yield. This enables a judgement to be made on whether the cost of intervention is likely to be recouped by the extra yield obtained (Fig. 6.1).

We can derive some general principles for rational management of crops from these considerations. Any threat to crop health can be magnified by other factors and intervention deserves to be considered. Early intervention reduces the risk of multiple factors combining. Detailed knowledge of the precise organism involved will help in the making of rational decisions. The timing of an intervention is likely to be crucial to its effectiveness. Excessive or ill-timed interventions could lead to high costs without the increase in yield to compensate.

Interactions

The interaction of biotic factors with each other and with environmental factors has already been mentioned. A major abiotic factor is the weather, which in the UK is often (even usually) variable from day to day. This makes it difficult to implement management strategies, particularly in relation to the most effective timing. This applies to cultivation and sowing time as well as to specific interventions such as spraying.

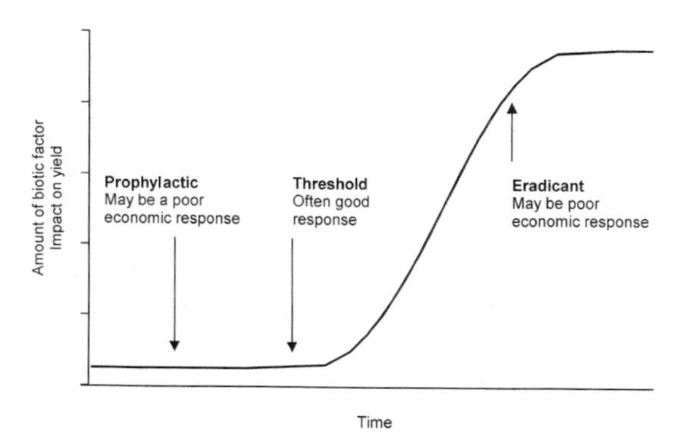

Figure 6.1 Importance of early detection of a biotic factor and the type of intervention possible.

The relationships between biotic factors are of considerable concern. Crop plants that have been defoliated can carry out less photosynthesis, can therefore grow less fast and are thus at greater risk from shading by tall weeds. Many organisms are vectors for others. Weeds of the Cruciferae family can harbour the disease club root which infects commercial brassicas such as oilseed rape, sprouts and cabbages. The effectiveness of a rotation is diminished if cruciferous weeds provide a host bridge across the rotation. Many wild grasses can develop ergots (the fruiting body of the fungus *Claviceps purpurea*) in which the seed is replaced by a sclerotium containing poisonous alkaloids which act as a vasoconstrictor and a plain muscle stimulant. When cattle graze these they can be poisoned. In addition, the fungus can infect cereals, particularly rye, and this has been a common disease of rye in the past. When the ergots have been ground up with the rye grains, serious human disease (St. Anthony's Fire) and deaths have resulted. There was a serious outbreak of ergotism in Manchester in 1927. The first European record of a legal requirement for grain quality was passed in Germany in the early sixteenth century AD in order to penalize millers who sold flour containing ground ergots.

Some pests complete their life cycle on two different hosts. The black bean aphid (*Aphis fabae*) feeds on bean plants in the summer but overwinters as an egg on the wild spindle tree (*Euonymus* sp.). This can make long-term control difficult if the alternate host is not also controlled. Many aphids also act as vectors of viruses.

All these interactions between organisms and with the environment tell us that the task of maintaining a healthy crop is not likely to be easy. It is crucial to recognize that we are dealing with a highly complex and interacting system and so we need to be aware that actions at one stage or on one component may have multiple effects or little effect. The 'shotgun' approach of applying chemicals prophylactically (as an insurance) to control every likely weed, disease or pest in case it might appear is not economically or environmentally viable (Fig. 6.1). It may be tempting to be lazy and to wait for problems to present themselves and then react with a spray in a 'fire-brigade' approach. This, however, is only a short-term cure of the symptoms, not a rational approach to combating the cause of the problem. As, rightly, restrictions on the use of pesticides increase because we understand more about the effects on the wider environment, so we will have to rely on a greater understanding of the biology and ecology of biotic factors. This means we will have to design crop protection systems that utilize that knowledge and integrate it into all the crop management decisions.

Intervention

The main ways of managing biotic challenges to crop growth are exclusion, eradication, reduction of the 'inoculum', protection of the host, modifying the environment and using competitive or resistant plants. Excluding the pathogen is achieved by legislation imposing plant quarantine.

Four aspects of intervention are important: identification, knowledge and information, timing and options. *Correct identification* of the nature of the cause of ill-health is crucial, because if the diagnosis is incorrect it is unlikely that a management plan can be designed with appropriate intervention. The intervention is then ineffective and the cost wasted. It is surprising how often poor diagnosis takes place, but understandable when you consider that many of the general symptoms of ill-health are common. For example, yellowing of cereal plants could be caused by not enough nitrogen, manganese deficiency, a wireworm attack on the seedlings, an infestation of frit fly (*Ocinella frit*) or snow rot (caused by the fungus *Typhula incarnata*). If in doubt seek professional advice rather than assume you have the answer. It is likely to be money well spent.

Knowledge of the biology of the biotic organism is essential if a sound strategy for management is to be developed. Fortunately, most of this knowledge is available and can be accessed either in written form or on the Internet or from a crop consultant. Previous research work has laid strong foundations for the management strategies we have developed in crop protection and it is wasteful not to exploit these opportunities.

The British weather is rarely settled for long periods and certainly not predictable or consistent from year to year. Not all farms have the same climate or soils or operate in the same way. This makes it difficult and inappropriate to advise a standard management strategy or one based on a single intervention. Instead, we have to develop comprehensive plans, which incorporate sets of options suitable for different circumstances. As considered below these options may include soil management, varietal choice and crop rotation as well as chemical and biological control options.

Timing is crucial to the success of any intervention (Fig. 6.2). Time of sowing influences potential yield but also influences which weeds might appear. In some crops a single, well-timed weed control operation is all that is required, but if the timing is too early then a further weeding may be necessary, while if the operation is delayed, e.g. by poor weather, then some yield reduction may have to be accepted. Similar considerations apply to pests and diseases. Disease and pest monitoring or prediction services provide advance warning of the likelihood of biotic challenges and enable precautionary measures to be taken.

There are many *options* for intervention. For all, we need to think about how they influence the invasion, infestation and impacts considered above. The prevention of invasion by using high quality planting material has been dealt with. It is difficult to prevent invasion of airborne propagules such as spores or flying insects. Some weed seeds can also be spread by wind. There is little that can be done to reduce the chance of their invasion from outside the farm. Attention to sound vegetation management should prevent, or at least severely restrict, on-farm spread. The invasion of soil-borne propagules can be restricted through taking appropriate hygiene methods when moving material and equipment from one farm or field to another. Other features of crop management can help to prevent problems arising. Damage to cereals from frit fly (*Oscinella frit*) can be reduced by ensuring there is at least a 6-week gap between ploughing down an old grass ley and drilling the cereal. Similarly, delaying drilling of winter cereals may incur a yield penalty but will reduce the risk of aphids transmitting barley yellow dwarf virus.

There are many ways of restricting the development of initial invasions in order to prevent the growth of populations to the point at which they cause significant crop losses. Obviously it is essential to prevent reproduction. A sound crop rotation and the use of resistant varieties can be effective in restricting the population growth of those organisms that are specifically adapted to particular crops. For the more general pests (e.g. slugs) or pathogens (e.g. many storage pathogens such as *Pythium ultimum*) this may not be effective. With weeds it is possible to hand-rogue low-density infestations. Although this can be effective at preventing the growth of populations of wild oats (*Avena fatua*) it is slow and expensive. Changes to soil and crop residue management may also reduce the opportunity for buildup of disease and pest populations. For example, ploughing (as opposed to surface cultivation) can help to reduce infestations of blackgrass (*Alopecurus myosuroides*). Other measures that encourage populations of natural predators can also assist in containing pest populations. Avoiding spraying insecticides in hedgerows or on field margins can allow populations of beneficial insects to build up. An example is the preda-

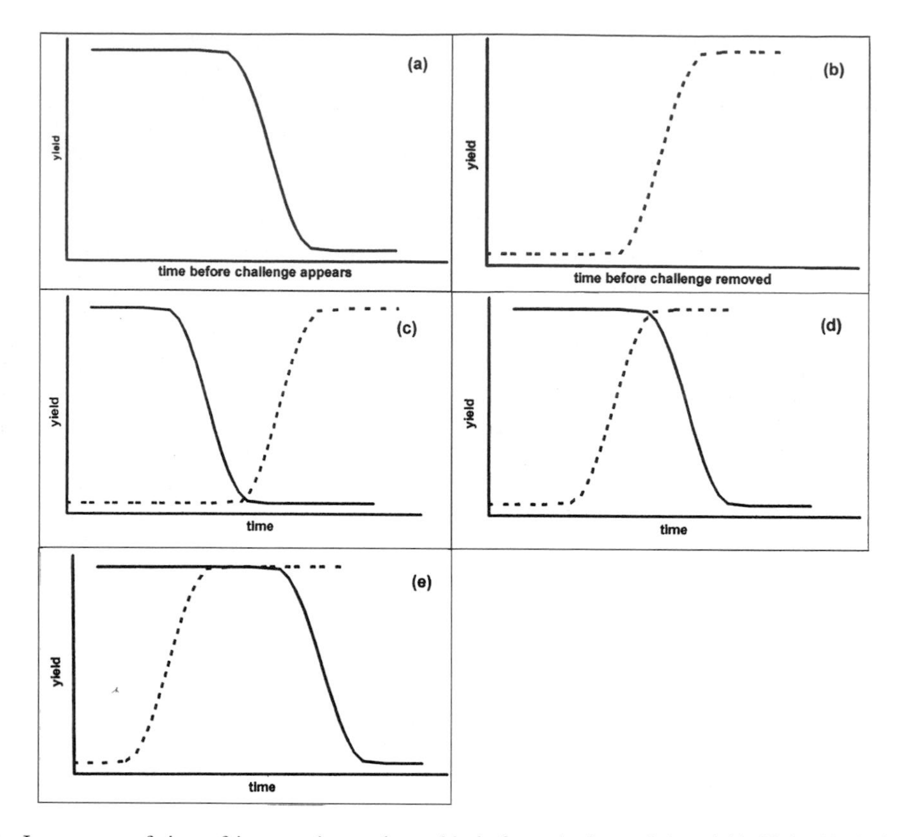

Figure 6.2 Importance of time of intervention against a biotic factor in determining yield. If the biotic factor appears late in crop growth or the crop can be protected during early stages then the response is as shown in **a**. If the biotic factor cannot be removed safely until some way through crop growth then yield responds as **b**. Depending on the degree of overlap of these responses, the crop manager may need to use more than one intervention **c**, may need to time a single intervention very precisely **d**, or may have a degree of flexibility in timing of an intervention **e**.

tion of aphids by spiders. Aphids are pests of many crop species and in addition to diverting plant assimilates to their own bodies are also vectors for a number of important crop diseases such as potato blight (*Phytophthora infestans*) and sugar beet yellows virus. Spiders feed on aphids but they are killed by the use of insecticides. The installation of beetle-banks is usually done to encourage wildlife, but may have similar benefits for crop protection.

Types of intervention

If a biotic population has built up to the level where significant yield losses are likely to occur then it is necessary to make interventions to protect the costs already incurred. The interventions may be of various types.

Mechanical interventions may be able to reduce the impact of weeds on crops and so harrowing and scarifying the soil surface may kill many of the weed plants. Mechanical methods are generally indiscriminate and injure or kill all the plants contacted. In grassland, mowing makes a contribution to weed management. *Non-mechanical interventions* use other methods of damaging weeds. Thermal weeders use heat to kill the weeds and are relatively indiscriminate except through control of direction. Applying mulches helps to smother weeds. Solarization is the practice of covering soil with sheets of white polythene to raise the temperature of the soil and kill the challenge organisms.

Chemical interventions exploit the advances of the past 50 years in applying compounds which selectively influence the growth of the biotic organism and have much less or no effect on the crop. Weeds can be controlled with herbicides which may act on contact with the foliage either by influencing photosynthesis or by being translocated to other parts of the plant where the compounds disrupt growth. Prophylactic control can be achieved using a residue of herbicide on the surface of the soil which kills seedlings as they contact it during emergence. Residual herbicides pose many problems of contamination of soils and waters and should be a 'weapon of last resort'. Diseases can be controlled with fungicides which are applied to crop plants and may either have a contact action on the fungus, or enter the plant and have a systemic action which protects the crop plants against any future infection. Similarly insecticides may have a contact or residual action. A wide variety of compounds is available for the management and control of weeds, diseases and pests, and this chapter will not provide details of these. Advice should be sought from reputable sources. Recommendations for agrochemical options for a wide range of field and vegetable crops and for grassland are available from the Agricultural Development Advisory Service (ADAS).

The *application* of agrochemicals is a highly technical subject. The chemical is formulated in an easily deliverable way as a liquid, solid or emulsion and combined with wetters (to make it spread over the surface of the plant), stickers (to prevent it being washed off by rain) and other compounds which may enhance its activity. Packaging is now sophisticated to minimize contact with humans and thus improve operator safety. Low volume applications (less than about 150 litres water/ha) conserve water. Ultra-low volumes avoid the use of water through specialized formulation. The machinery for delivery is now highly developed to ensure accurate placement and metering of quantities as well as to minimize wastage and contamination. The use of agrochemicals is widely regulated. There is a requirement for training of spray operators. There are codes of conduct for farm use of chemicals and for disposal of surplus chemicals and of containers. There are requirements for safe intervals between application and harvesting of the crop to ensure the produce is wholesome and safe to be consumed by humans or animals. Weather conditions, soil type and conditions and crop growth stage all strongly influence the safety and efficacy of chemical applications.

Recent years have seen the advent of *biological control* of biotic factors in crop production. As yet, few are used in field agriculture. The advantage of biological control is that it is generally selective with no side-effects, relatively cheap, self-perpetuating and persistent. In addition it is unlikely that pest resistance will develop. Disadvantages are that biological control does not offer extermination of the biotic factor, it may limit the subsequent use of chemical means of control and it is sometimes unpredictable. Biological control makes the system more complex by adding an extra species to the agricultural ecosystem and although the direct interactions with the other species are likely to be small, the other interactions and the responses to the physical environment may be less well understood. Microbial pesticides can also be considered a form of biological control. Such pesticides may be based on fungi, bacteria, viruses, nematodes or protozoa. A different form of biological control of insects is the use of pheromone traps. In these a sexual attractant lures insects to a trap where they die.

An important feature contributing to the management of biotic factors is the use of *competitive or resistant cultivars*. Plant breeders continue to put considerable effort into breeding new cultivars (or varieties) which perform well in the field. This often involves a large and persistent leaf area in order to maximize the opportunity for photosynthesis. Such cultivars are also likely to be effective in shading those weeds that germinate later than the crop. Crops of cereal cultivars that are leafier in spring have a lower leaf area of weeds. Such information on leafiness is available in leaflets from the National Institute of Agricultural Botany (NIAB) which performs and reports on the statutory comparative trials conducted prior to the registration of a new cultivar. Crop management that increases leafiness (e.g. adequate nitrogen) is also likely to aid weed suppression. Considerable information is available on the relative resistance of cultivars to a wide range of important diseases. Many disease organisms can eventually overcome the genetic resistance of new cultivars, usually by evolving a new genetic race. This is particularly so where the resistance is based on a single gene. In some new cultivars, several genes are used to give a broad-based resistance and this may be more durable.

Resistance to pests is based on a number of features of cultivars. Colour can be important. Lighter-green cultivars of onions seem to deter thrips. Red cabbage varieties suffer less from cabbage aphids and cabbage caterpillars, possibly due to the anthocyanin content making the plants less palatable. Spraying crop plants with a bitter compound can deter feeding by caterpillars (but of course it has to be removed or degraded before human consumption!). The surface characters of the plant are important to pests, and leaf waxiness and hairiness are components of resistance. Tissue hardness and content of toxins may deter pests. This can limit the usefulness of the product: for example the development of sorghum varieties with high tannin levels has deterred wild birds feeding on the grain but of course means the grain cannot be used in poultry diets. Phenology (the timing of plant development) can provide a way to avoid pests: early flowering pea cultivars may escape infestation with pea moth.

Weeds

Invasion and infestation

Weeds spread by seeds or by vegetative parts such as stolons, rhizomes or corms. Breaking up vegetative parts of perennials can serve to spread them rather than destroy them. The importance of clean planting material has already been stressed.

Ill-health and impact

Weeds interfere with crop growth and reduce yield (quantity) and the quality of yield. Various authorities have estimated an overall yield loss of about 10% due to weeds. In a survey of organic farmers in the USA, weeds were identified as more difficult to manage than pests and diseases. Weeds reduce yield mainly by shading crop plants and reducing the amount of photosynthesis that the crop can undertake. Some weeds are also very efficient at taking up water and can accentuate a water shortage in drought-susceptible crop species.

There is great variation in yield loss due to the specific crop, the particular season and the timing of the infestation. Early weed infestations can smother crops which are slow to establish and give yield reductions of 90%. In contrast late weed establishment in fast-growing aggressive crops may have little effect on yield. This emphasizes the value of detailed observations of young crops.

The extent of yield reductions due to weeds depends on the weed population. Assessing weed populations in the field can be problematic. The correct identification of weed species is important and has to be done at the seedling stage if management is to be at an early stage and effective. The density of weed plants needs to be considered. This is not straightforward because the distribution of weeds tends to be patchy. This also means that the whole field needs to be inspected, not just the area near the gate! A dense population (250 plants/m^2) of blackgrass (*Alopecurus myosuroides*) in a cereal crop may reduce yield by 25% (Table 6.4).

The timing of interventions is important in determining their effectiveness. In particular, the time of onset of weed competition and the length of the period during which weeds compete together determine the number of interventions necessary and indicate the appropriate time.

The effects of weeds may not just be on yield (quantity). There may also be difficulties in certification of seed contaminated by weed seeds (quality). Climbing weeds such as cleavers (*Galium aparine*) and bindweeds (*Polygonum* spp.) can slow down harvesting and retard crop maturation, thereby increasing harvesting and drying costs. In addition lodging may be increased, which both interferes with harvesting and reduces crop yield and quality.

Intervention

Many crop management decisions contribute to restricting the occurrence or promoting the long-term decline of weed infestations. Rotation of crop types, particularly between autumn-sown and spring-sown is useful because the unstable environment disadvantages some weeds and provides an opportunity to use a wider range of herbicides. However, rotation alone is not usually an effective means of weed management and needs to be combined with other measures. Selection of competitive crop varieties can also make a useful contribution to overall weed management. The management of crop residues is important, particularly in crops harvested for seed. Weed seeds returned to the crop surface with straw or those shed to the soil surface prior to the crop being harvested are important in the buildup of a seed bank. Measures to limit this in the past have included stubble burning, which was effective in killing up to 70% of blackgrass (*Alopecurus myosuroides*) and wild oat (*Avena fatua*) seeds. Shallow cultivation favours weed species such as blackgrass and barren brome (*Anisantha sterilis*). Changing to ploughing can restrict population growth of such species. Early crop establishment helps to shade out the weed plants and so methods that lead to earlier crop canopy development are useful. Chitted seed potatoes, the use of pregerminated seeds or transplanting of seedlings are examples of techniques that achieve this. Plant spacing also determines the penetration of light through the crop canopy and so an increased seed rate can make a contribution, as can narrower row spacing. A stale or false seedbed is used to provide an opportunity for weed seeds to germinate and be killed before the crop is sown. This technique helps to deplete the population of weed seeds near the surface of the soil which are the ones that develop into weed plants in the crop.

Stubble burning is now banned but was effective at reducing the weed seed rain – the weed seed population on the soil surface. A modern alternative is thermal weeding in which an energy source is used to damage weed seedlings. The heat may come from steam (usually in fixed installations) or gas (usually propane). Solarization is the process of covering the soil with clear plastic sheeting to heat the soil. In cool temperate regions this

Table 6.4 Important weeds in the UK

Common name	*Proper name*	*Main crops*	*Possible losses (%)*[1]
Annual grass weeds			
Blackgrass	*Alopecurus myosuroides*	Winter cereals	25
Wild oats	*Avena fatua*	Spring cereals	25
Winter wild oats	*Avena ludoviciana*	Winter cereals	25
Barren brome	*Anisantha sterilis*	Cereals	25
Annual broad-leaved weeds			
Corn marigold	*Chrysanthemum segetum*	Cereals	
Mouse-eared chickweed	*Cerastium fontanum*	Grassland	
Fat hen	*Chenopodium album*	Broad-leaved crops	
Hemp-nettle	*Galeopsis* spp.	Broad-leaved crops	
Cleavers	*Galium aparine*	Broad-leaved crops	40
Dove's-foot cranesbill	*Geranium molle*	Broad-leaved crops	
Mayweeds	*Matricaria* spp.; *Anthemis* spp.	Broad-leaved crops, cereals	
Black bindweed	*Polygonum convulvulus*	Broad-leaved crops	
Redshank	*Polygonum persicaria*	Broad-leaved crops	
Knotgrass	*Polygonum aviculare*	Broad-leaved crops	
Charlock	*Sinapis arvensis*	Broad-leaved crops	
Spurrey	*Spergula arvensis*	Broad-leaved crops	
Chickweed	*Stellaria media*	Broad-leaved crops	
Speedwell	*Veronica persica*	Broad-leaved crops	15
Perennial weeds			
Bracken	*Pteridium aquilinum*	Upland and hill grassland	
Buttercups	*Ranunculus* spp.	Grassland	
Couch	*Elytrigia repens*	Grassland	15
Docks	*Rumex* spp.	Grassland	
Japanese knotweed	*Reynoutria japonica*	Grassland	
Rushes	*Juncus* spp.	Wet grassland	
Thistles	*Cirsium* spp.	Grassland	

[1] Where no value is given it is likely to be below 10% except where the weed emerges early into a slow-emerging crop. Frequently broad-leaved weeds occur in mixed weed stands.

may increase weed seedling emergence and therefore contribute to a stale seedbed.

The main methods of weed management in crops are mechanical and chemical. Mechanical control uses machinery to remove weeds. The most effective way to kill the weeds is to completely bury them or to cut them below the surface. Harrows or tines can bury small annual weeds but are relatively ineffective against deep-rooted and perennial species. Rain soon after the operation can assist weeds in recovering. Many implements are available but all have the same design constraints of how to kill the weed without damaging the crop. The next 5 years should see the availability of high-technology machines which can be guided by machine vision to selectively remove those plants that the computer recognizes as 'not-crop'.

Chemical control of weeds makes an important contribution to weed management. For most major field crops there is a wide range of herbicides available. The mode of action of herbicides may be to inhibit photosynthesis, pigment production (especially carotenoids) lipid biosynthesis, cell division or to act as an auxin mimic.

Pesticide registration legislation is reducing the compounds available for use in other crops (e.g. peas, beans, linseeds, vegetables). The criteria for choice of herbicide include soil type and crop growth stage. Professional advice should be sought. When weed control is required within the growing crop, a selective herbicide must be used. If weed control can be done before planting, a non-selective compound can be used so long as it does not persist and damage the crop seedlings.

Biological control of weeds has not proved very successful in the UK.

Diseases

Invasion and infestation

There are many sources of inoculum of diseases. Seeds may be contaminated on the surface (e.g. wheat seeds carrying spores of bunt, *Tilletia caries*), but seed surface disinfection usually can eradicate this. More problematic are internal infections of seeds (e.g. wheat seeds infected with the mycelium of loose smut, *Ustilago nuda*). Crop residues and older parts of plants are common sources of disease inoculum. Wheat or barley leaves infected with glume blotch (*Septoria nodorum*), take-all (*Gaeumannomyces graminis*) or powdery mildew (*Erisyphe graminis*) are important inoculum sources. Volunteer plants, surviving from the previous crop, are important in supplying inoculum and provide a 'disease bridge' from one season to another. This can be important for the start of many rust outbreaks (e.g. *Puccinia striiformis*) and for potato light (*Phytophthora infestans*). Alternate or alternative hosts provide disease organisms with the option of surviving in the vicinity of the main host species. Club root of brassicas (caused by *Plasmodiophora brassicae*) can survive on related species in the same plant family such as the weeds charlock and shepherd's purse. Red clover plants can provide alternative sources for the pea mosaic virus to be spread by aphids. Clearly, management of crop residues, including volunteer plants, is an important part of disease management.

The means whereby spores are spread provides important information that may be of use in developing management strategies. The main agents of dispersal are wind, water, insects and other animals (including man). Wind is the most important and is the agent for spread of many important diseases worldwide including rusts, mildews and potato late blight. Trapping spores from the air may allow monitoring of the spread of the disease organism into a particular area so that warnings may be given and appropriate preventive actions taken. Raindrops splashing on the soil or on leaves are important in the spread of some diseases. Glume blotch of wheat (caused by *Septoria nodorum*) spreads up the plant in this way. Many bacteria are also spread by rain. Insect vectors of diseases include aphids, leafhoppers and whiteflies. Bark beetles have been the main vector involved in the spread of the Dutch elm disease (*Ceratosystis ulmi*). Management of these vector organisms can restrict the spread of fungal inoculum. Many other types of animals such as birds, mammals and nematodes are implicated in the spread of some diseases. Human activities may also spread fungi, particularly on the farm if careful attention to hygiene is not paid.

Infection of crop plants usually starts with the germination of a fungal spore. For soil-borne diseases this is often promoted by root exudates (substances such as sugars diffusing out of roots) from crop seedlings. Airborne spores landing on the leaf may germinate in response to water. Wounds on the surface of plants also provide an entry route into the plant. Wounds may be caused by machinery or damage to the plant surface by other organisms or by wind or hail.

Ill-health and impact

The most common effect of fungal infection of crop canopies is a reduction of photosynthesis. There is often a loss of chlorophyll which causes the leaf to look yellow (chlorotic). A similar effect occurs with infections of bacteria. In addition to reducing the amount of chlorophyll, some virus infections reduce the efficiency of chlorophyll in carrying out photosynthesis. A reduction in net photosynthesis may not only be caused by less photosynthesis occurring but also by a greater consumption of assimilates because of increased respiration. In many plants infected with rusts (e.g. yellow rust of wheat caused by *Puccinia striiformis* or beet rust caused by *Uromyces betae*) or powdery mildews (e.g. *Erisyphe graminis* on barley) there may be a doubling of the respiration rate by the time the fungus sporulates.

The disease of club root in crucifers is caused by *Plasmodiophora brassicae*. Infection of roots by the soil-borne pathogen leads to cell division and enlargement of root cells which inhibit nutrient and water movement from the root. The infected plants generally wilt and die. Interference with translocation is the main effect of tomato wilt, which is caused by the fungus *Fusarium oxysporum*. Wilting occurs because the fungal hyphae block the water-conducting xylem vessels. In addition, the fungus produces pectin plugs which block the vessels.

Estimates of crop losses due to diseases are about 20% around the world. Infection of potato with late blight (*Phytophthora infestans*) can cause losses of tuber yield of about 50%. Severe infections of wheat by glume blotch (*Septoria nodorum*) can cause yield reductions of 50%. Leaf blotch of barley (caused by *Rhyncosporium secalis*) can cause reductions in green leaf area leading to yield reductions of about 25%. Table 6.5 shows some diseases of specific crops and possible crop losses.

Intervention

The main ways of managing crop diseases are to exclude the pathogen, eradicate the pathogen, reduce the inoculum, protect the host, modify the environment and use resistant plants. Excluding the pathogen is achieved by legislation imposing plant quarantine. Seed certification is one form of legislation. Eradication is the aim of many management practices. If diseased plants can be recognized early then hand roguing may be an option. Eradicating alternate or alternative hosts may reduce the inoculum. Many routine 'tidying' activities can remove plant debris which may be a source of inoculum. For example, stubble burning was effective at reducing the overwintering stages of powdery mildew (*Erisyphe graminis*).

Where fungicides are used, the time of spraying is very important (Fig. 6.3). Spraying of healthy crops (prophylactic treatments) is one option. However, such preventive foliar spraying, before the disease has developed may give a poor economic response unless it is carried out after a warning that the disease is likely to invade the area. Curative applications are made later, when it is possible to detect the disease on the crop. Using a threshold value to decide when to apply chemicals, e.g. by using the percentage of the leaf area affected, can lead to effective control and high economic returns. However, missing the threshold and spraying later as an eradicant may be too late to prevent yield loss but may offer a cosmetic effect which is of value for quality.

An example of biological control is the wide use in orchards of a formulation of spores of the fungus *Trichoderma* to control silver-leaf disease (caused by *Chondrostereum purpureum*) which attacks apple, pear, plum and cherry trees.

Pests

Invasion and infestation

Some of the sources of inoculum of pests are similar to those of diseases. On perennial crops (e.g. fruit bushes and trees) many arthropod pests may overwinter on or in the plant. Treatment of the outside of the plant will not always influence those biotic factors overwintering internally. Crop residues and older parts of plants are common sources of pest infection and so routine pruning of older parts followed by burning the prunings is sound crop hygiene.

Ill-health and impact

Some pests are general feeders and attack many crops. They achieve their effects by gross damage, by removal of green leaf area or by severing roots. Rabbits eat the aerial parts of plants and can cause severe defoliation. They hide in field margins and the damage is worse near field edges. Defoliation of cereals may have only a small effect on yield because the plants can recover. However, severe defoliation may result in yield loss, particularly if the growing point has been damaged. Slugs are also general feeders at or just under ground level. They often move down a row of seedlings, producing a typical symptom of a group of adjacent seedlings which lean over and die.

Insect pests cause damage to crops by either eating the plants with biting mouthparts or diverting assimilates through their piercing or sucking mouthparts into their own bodies.

The larvae of butterflies and beetles both have strong biting mouthparts and may cause economic damage. Cabbage caterpillars (larvae of a number of butterflies including the large white, the diamond back moth and the cabbage moth) feed on foliage of many brassicas (cabbage, kale, swede, oilseed rape) and can severely reduce the leaf area, thereby reducing yield. In addition, in horticultural brassicas (cabbages, sprouts, cauliflowers) the excreta of the larvae can greatly reduce the value of the produce. Cutworms are the soil-inhabiting larvae of a number of related species of noctuid moths and they feed on seedlings and roots. The eggs of the turnip moth (*Agrotis segetum*) are laid on the foliage and stems of both weed and crop plants. Controlling weeds may help by reducing the number of sites for egg-laying. The larvae of a number of species of click beetles (*Agriotes* spp.) are known as wireworms and feed on the underground parts of plants and young seedlings, producing similar symptoms to slugs. They can be particularly problematic after a rotational grass ley has been ploughed and sown back to annual crops.

Aphids are typical of insects with piercing and sucking mouthparts that feed on cell sap. This causes leaf rolling and stunted growth. The black bean aphid (*Aphis fabae*) feeds on *Vicia* beans in the summer and can seriously reduce both the yield and the quality of crops. It overwinters on the spindle tree as an alternate

Table 6.5 Some diseases of specific crops

Crop	*Common name*	*Proper name*	*Possible losses (%)*[1]
Cereals			
Leaf diseases	Powdery mildew	*Erysiphe graminis*	10 (wheat), 30 (barley)
	Brown rust	*Puccinia recondita*	20
	Yellow rust	*Puccinia striiformis*	40 (wheat), less in barley
	Net blotch	*Pyrenophora teres*	(barley only)
	Leaf blotch	*Rhyncosporium secalis*	(barley only)
	Septoria	*Septoria tritici*	20 (wheat only)
Root and stem-base diseases	Take-all	*Gaeumannomyces graminis*	15 (less in barley)
	Eyespot	*Pseudocercosporella herpotrichoides*	15 (less in barley)
	Sharp eyespot	*Rhizoctonia cerealis*	
Ear diseases	Glume blotch	*Septoria nodorum*	25 (wheat only)
	Bunt	*Tilletia caries*	
	Loose smut	*Ustilago nuda*	
Potato			
Leaf and stem diseases	Blight	*Phytophthora infestans*	
	Leaf roll	Virus spread by aphid *(Myzus persicae)*	Up to 50
Tuber and storage diseases	Silver scurf	*Helminthosporium solani*	
	Gangrene	*Phoma exigua*	
	Blight	*Phytophthora infestans*	
	Powdery scab	*Spongospora subterranea*	
	Common scab	*Streptomyces scabies*	
Brassicas			
	Powdery mildew	*Erysiphe cruciferarum*	
	Canker	*Phoma lingam Leptosphaeria maculans*	
	Clubroot	*Plasmodiophora brassicae*	
	Light leaf spot	*Pyrenopeziza brassicae*	
Sugar beet			
	Downy mildew	*Peronospora farinosa*	
	Beet rust	*Uromyces betae*	
	Rhizomania	Virus spread by the fungus *Polymyxa betae*	
	Virus yellows	Virus spread by aphids	40–50
Peas and beans	Leaf, stem and pod spots	*Ascochyta pisi, A. fabae, Colleotrichum* spp.	
	Pea wilt	*Fusarium oxysporum*	
	Foot rot	*Mycosphaerella pinodes*	
	Halo blight	*Pseudomonas phaseolicola* (bacterium)	
Carrots	Violet root rot	*Helicobasidium brebissonii*	
	Black rot	*Stemphylium radicinum*	
Onions	Neck rot	*Botrytis allii*	
	Downy mildew	*Peronospora destructor*	
	Smut	*Urocystis cepulae*	
Forage grasses			
	Crown rust	*Puccinia coronata*	

[1] Where no value is given it may be below 10% except when the disease infects a large proportion of the stand early in crop growth.

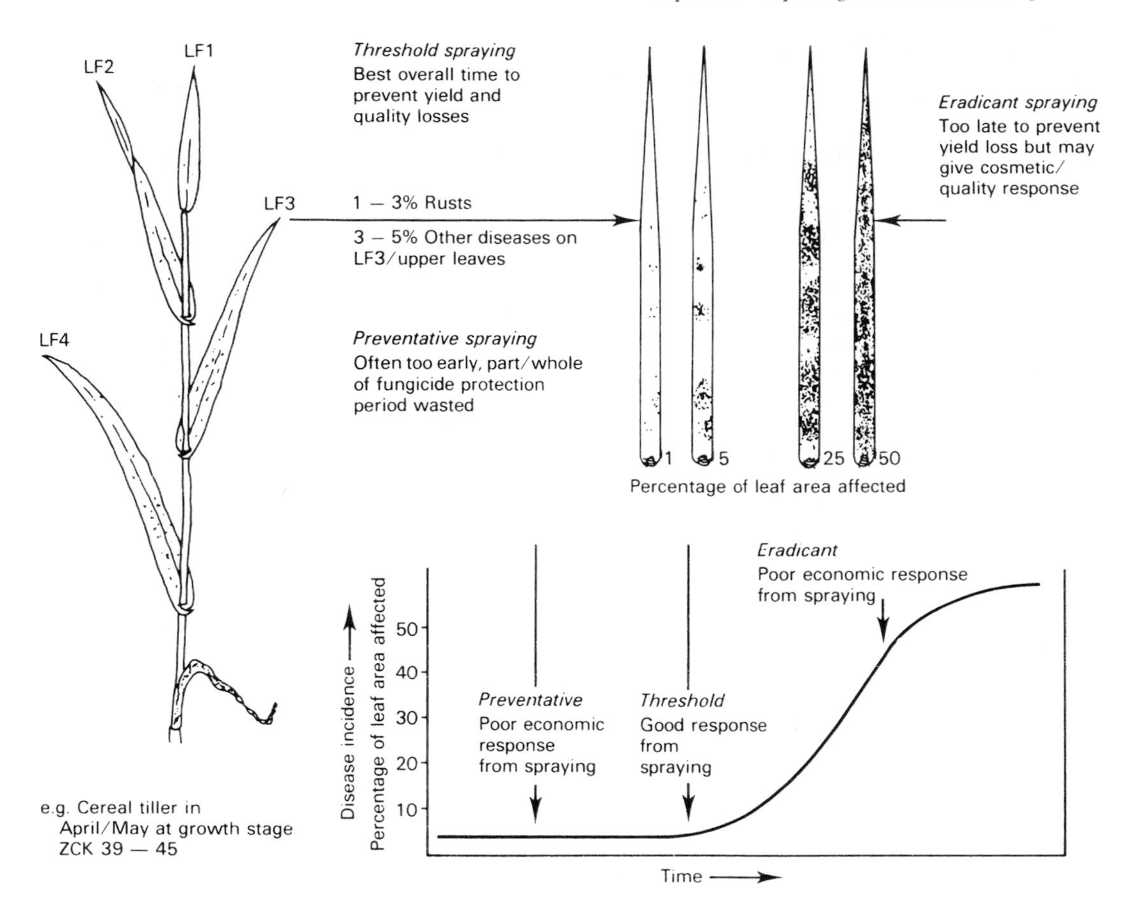

Figure 6.3 Effect of time of spraying on the likely yield response.

host in the winter. The peach–potato aphid (*Myzus persicae*) feeds on potatoes, sugar beet and swedes. It overwinters on peach trees and a wide range of other species, particularly in mild winters, and also in glasshouses and potato chitting houses. Attention to hygiene is therefore very important. In addition to direct losses, aphids are important as vectors for leaf-roll virus and severe mosaic virus (virus Y) of potatoes and of beet yellows virus (BYV) and beet mild yellowing virus (BMYV). Producers of 'seed' potatoes need to keep their crops free of contamination. The bird-cherry aphid (*Rhopalosiphum padi*) is an important vector which spreads barley yellow dwarf virus (BYDV) from other cereals and grasses. However, spread within barley crops is usually by the grain aphid (*Sitobion avenae*) and the rose-grain aphid (*Metapolophium dirhodum*).

Estimates of worldwide crop losses due to pests are about 25% (Table 6.6).

Intervention

In addition to attention to farm hygiene, many chemicals are available, mainly for controlling insects. However, there is considerable concern about the use of such chemicals.

There are examples of biological control of pests. The rabbit disease myxomatosis was first introduced as a way of controlling the numbers of rabbits. *Bacillus thuringiensis* can be sprayed on crops to control the caterpillars of cabbage white butterfly. In protected cropping a number of biological control agents are used, e.g. red spider mites can be controlled by predatory mites and a parasitic wasp is used to control whitefly.

Modern methods of pest management aim to integrate features from cultural and biological control and reduce the reliance on chemical control.

Table 6.6 Some pests of specific crops

Crop	*Common name*	*Proper name*	*Possible losses (%)*
Cereals	Frit fly	*Oscinella frit*	10
	Wheat bulb fly	*Delia coarctata*	25
	Aphids	*Sitobion avenae*, *Rhopalosiphum padi*,	10
	Cereal cyst nematode	*Heterodera avenae*	
Potatoes	Peach–potato aphid	*Myzus persicae*	
	Potato cyst nematodes	*Globodera rostochiensis* and *G. pallida*	
Brassicas	Slugs	*Deroceras reticulatum*	
	Pigeon	*Columba palumbus*	
	Flea beetles	*Phyllotreta* spp.	
	Cabbage stem flea beetle	*Psylliodes chrysocephala*	
	Pollen beetles	*Meligethes* spp.	
	Cabbage caterpillars	Various spp.	
Sugar beet	Millepedes, springtails and symphylids	Various spp.; *Onychiurus* spp.; and *Scutigerella immaculata*	
	Beet flea beetle	*Chaetocnema concinna*	
	Peach–potato aphid	*Myzus persicae*	
	Black bean aphid	*Aphis fabae*	
	Beet cyst nematode	*Heterodera schachtii*	
Peas and beans	Pea cyst nematode	*Heterodera goettingiana*	
	Pea and bean weevil	*Sitona lineatus*	
	Pea aphid	*Acyrthosiphum pisum*	
	Pea moth	*Cydia nigricana*	
	Pea midge	*Contarinia pisi*	
	Black bean aphid	*Aphis fabae*	
	Bean seed fly	*Delia platura*	
Carrots	Carrot fly	*Psila rosea*	
	Willow-carrot aphid	*Cavariella aegopodii*	
Onions	Onion fly	*Delia antiqua*	
	Stem and bulb nematode	*Ditylenchus dipsaci*	
Forage grasses	Frit fly	*Oscinella frit*	
	Weevils	*Sitona* spp.	
	Clover stem nematode	*Ditylenchus dipsaci* in clover plants	
	Ryegrass mosaic virus	Spread by the wind-dispersed mite *Abacarus hystrix*	

Integrated crop management

This chapter has emphasized that biotic challenges to crop plants provide multiple challenges to crop health. It is clear that no one approach to management of biotic factors will provide the cure for all factors, nor that it will always provide the cure for any one organism. What is needed is an approach in which knowledge and information are combined to develop a broad strategy for overall crop protection, which incorporates a range of measures. This has to be a broad strategy rather than a rigid plan because many other factors influence our ability to react – e.g. the soil may be too wet for cultivation or the weather may not be right for spraying.

The development of this broad strategic approach to the management of crops, especially their protection, is termed integrated crop management (ICM). It aims to combine a set of options, all of which contribute towards the maintenance of crop health. The greater attention to detail demanded of ICM serves to replace costly inputs by planned, timely interventions which increase profitability and benefit the environment.

One important feature of many integrated crop management strategies is the concept of a threshold which triggers a management response. A 'physiological' threshold level can be defined as the lowest density that causes economic damage. The level at which the benefit of intervention just exceeds the cost is termed an 'economic' threshold. However, this is less easy to define when management strategies other than chemical or biological are used because it is difficult to apportion costs to many of the components of a crop management system. Recently the view has been to apportion the cost over the full rotation. The application of threshold values is not simple because the relationship between pest incidence and damage depends on the environment and crop stage of growth and the relationship between intervention and yield benefit is also linked to environment and timing.

The features incorporated into ICM are:

- Site and soil management
- Crop rotation
- Variety choice
- Crop husbandry
- Crop nutrition
- Crop protection
- Pollution control
- Energy efficiency
- Conserving habitat and wildlife features on the farm

This list illustrates that there are many options that may be considered before the use of agrochemicals. No crop protection products can be applied to organic crops; nevertheless, serious weed, disease and pest problems do not always occur. Weeds are usually cited as the most problematic and the causes of the greatest yield loss in organic systems. Organic farmers rely on the features of good husbandry listed above including crop rotation, varietal selection and biological control. The success of organic farmers shows what can be achieved by the integration of a suite of crop management options which excludes agrochemicals.

Decision support systems

One feature emphasized in this chapter is the need to identify accurately the biotic challenge and then to select an appropriate set of responses for the crop management system on the farm. Unfortunately it is difficult for the grower to be an expert in everything. There is thus scope to provide growers with a tool that helps them make good decisions. Developments in computing are leading to the development of decision support systems (DSSs). These would first evaluate field histories and crop management systems to determine the biotic risks to the crop. They could guide the user to a correct identification of a biotic challenge when it was noticed. Reference to 'encyclopaedias' of weeds, pests and diseases and of agrochemicals and of other management factors together with current information on the crop, recent and forecast weather, would lead to a proposal for action. Ideally this would include a cost/benefit analysis. Some DSSs are available for weeds and in the UK a major research programme is underway to build such a DSS for arable crops (DESSAC). This would put an expert on every grower's computer desk and enable more informed decisions to be made. Various modules for this are being tested and are near to release. 'Watch this space!'

Conclusions

This chapter has described what causes crop ill-health and the mechanisms that lead to this. The range of management options available to crop managers is summarized. The individual crop manager has to decide the most appropriate set of options for the crops being grown in that season. There are great advantages in having a broad crop protection strategy integrated within the farm management system. To achieve this a great deal of informed planning is required to minimize the occurrence of anticipated biotic challenges. In addition the system needs to be flexible and the manager needs to be adaptable to deal with the unexpected biotic challenges. Access to knowledge and information is a key feature of the development of such an approach.

Ten tips

(1) Know what challenges are present and are likely to occur.
(2) Pay attention to farm hygiene to prevent spread of organisms.
(3) Choose appropriate crops and varieties that are competitive or resistant.
(4) Use good quality planting material.
(5) Fit the crops into a sound rotation.
(6) Manage the soil, water and crop nutrition appropriately.

(7) Incorporate mechanical control of weeds.
(8) Consider biological control.
(9) Incorporate chemical control where necessary.
(10) Think about the wildlife and environmental implications of what you do.

References

Carlile, W.R. (1995) *Control of Crop Diseases*, 2nd edn. Cambridge University Press, Cambridge.

Johnston, A. & Booth, C. (1983) *Plant Pathologists Pocketbook*, 2nd edn. Commonwealth Agricultural Bureau, Slough.

Jones, D.G. (1987) *Plant Pathology: Principles and Practice*. Prentice-Hall, Englewood Cliffs, New Jersey.

Radosevich, S., Holt, J. & Ghersa, C. (1997) *Weed Ecology: Implications for Management*, 2nd edn. John Wiley & Sons, New York.

van Emden, H.F. (1989) *Pest Control*, 2nd edn. Cambridge University Press, Cambridge.

Useful addresses

ADAS Headquarters, Woodthorne, Wergs Road, Wolverhampton, WV6 8TQ.

Arable Research Centres, Manor Farm, Lower End, Daglingworth, Near Cirencester, Glos, GL7 7AH.

Association of Independent Crop Consultants, Agriculture House, Station Road, Liss, Hampshire, GU33 7AR.

FWAG (Farming and Wildlife Advisory Group), National Agricultural Centre, Stoneleigh, Warwickshire, CV8 2RX.

LEAF (Linking Environment and Farming), National Agricultural Centre, Stoneleigh, Warwickshire CV8 2LZ.

Morley Research Centre, Morley, Wymondham, Norfolk, NR18 9DB.

SAC (The Scottish Agricultural College), Central Office, West Mains Road, Edinburgh, EH9 3JG.

Useful websites

www.abdn.ac.uk/agriculture/infoagric.html – one of many entries to agricultural web pages.

www.eppo.org/index.html – European Plant Protection Organization.

www.ifgb.uni-hannover.de/extern/ppigb/ppigb.html – a general introduction 40 pests, diseases and disorders.

www.ipmnet.org – newsletter on integrated pest management.

www.leafuk.org – the LEAF organization (Linking Environment and Farming).

7

Grain preservation and storage

P.H. Bomford & A. Langley

Introduction

Grain needs to be stored in good condition after harvest so that it can be marketed off the farm, or used on the farm, in the intervening period before the next harvest. Grain prices are generally lower at harvest time and can be expected to rise as the next harvest approaches. If livestock are to be fed with home-produced grain, careful storage can ensure a year-round supply of high quality feed.

A successful handling and storage system will preserve those grain qualities that are important for its proposed end use. Depending on specific market, some or all of the following qualities may be important:

- germination (for seed and malting);
- good baking or malting properties;
- good livestock feeding properties;
- freedom from moulds, mites and insects;
- freedom from impurities (other grains, small grains, weed seeds, etc.);
- freedom from discoloration or taint;
- moisture content and temperature.

The aim of any handling and storage system is to maintain the quality of the incoming material. Drying, cleaning and grading processes can contribute to maintaining and/or improving some of the quality aspects of the grain.

The majority of cereals crops are preserved in good condition by drying the crop and storing it under conditions that prevent deterioration. Grain stored in this way can be sold into any appropriate market, or fed to livestock on the farm. It is also possible to store grain safely for specific uses without drying by modifying the atmospheric conditions and by the use of chemicals.

The membership of grain quality assurance schemes requires that dry grain is stored under conditions that prevent the deterioration of the product. Grain moisture content and storage temperature, and the prevention of ingress of water, vermin (mice, rats, birds) and domestic animals are the main factors which must be addressed.

Conditions for the safe storage of grain

Stored grain deteriorates due to the action of moulds, insects and mites. Some of these organisms occur naturally on cereal crops and so are likely to be brought into the store with the crop. Others, such as residual populations of insects and mites, may be present in floor and wall crevices or crop residues in the 'empty' grain store.

The grain is protected from attack by a tough outer skin or pericarp and impermeable seed coat or testa. Unfortunately, these layers can be damaged during harvesting, allowing easy access by mould organisms to the starchy endosperm or the embryo. In one study, between one-fifth and one-third of all wheat grains were found to have threshing injuries to the skin of the embryo area. The embryo contains sugars, fats and proteins and is thus an excellent source of nutrients for the moulds. Damage to the embryo will impair the viability of the seed.

Factors influencing grain deterioration during storage (over a 35-week period) are illustrated clearly in the classic diagram (Fig. 7.1) which has been produced by research workers at the MAFF Slough Laboratory. In the figure, line A defines the conditions under which

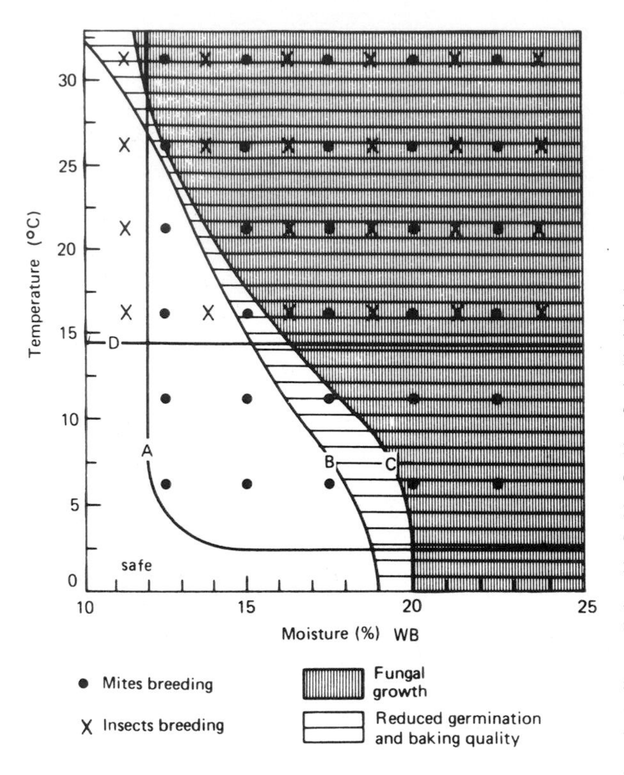

Figure 7.1 Combined effects of temperature and moisture content on grain condition and grain pests over a 35-week storage period. WB = Wet Basis. (crown copyright.) After Burges & Burrell, 1964.

mites will breed, and therefore infestations can build up. Breeding activity can be reduced markedly by keeping the temperature below 2.5°C, or the grain moisture content (MC) below 12%. These are difficult conditions to satisfy, both physically and economically, within the store, and protection from most major mite species is achieved if grain is dried below 14% MC (Wilkin, 1975). Careful cleaning and acaricide treatment of the store when empty will minimize the risk of mite populations persisting from one season to the next. Line B covers the risk of loss of germination due to embryo damage, and also the deterioration of baking (wheat) and malting (barley) qualities. Grain stored under conditions to the left of the line may be expected to retain all these qualities and thus be sold into all premium markets. Line C defines the conditions that favour mould growth which can lead to:

- crop weight loss as the mould organisms feed on the grain;
- crop discoloration due to visible mould growth;
- the production of mycotoxins (poisons produced by the mould organism which can kill or debilitate livestock and humans who eat contaminated foodstuffs);
- the production of mould spores (a very fine dust which can cause farmers' lung, a pneumonia-like affliction, when inhaled). It is essential to wear adequate organic dust respiration equipment if working with grain (or hay) which is suspected of being mouldy.

Moulds, insects and mites use oxygen, generate heat and liberate water and carbon dioxide as they breakdown food. Thus, any large concentration of these organisms will make the grain around it warmer and more moist. This will have the effect of making conditions even more favourable for their own development, and may also encourage the development of other pests or grain germination, which will happen if the local moisture content increases to 25–30%. Severe mould infestation results in grain being caked together, particularly at or just below the store surface and against the walls, and sprouted at the surface.

Line D shows that the effects (reduction in germination and feeding value, contamination with insect remains and faeces) of insects such as grain weevils and beetles can generally be minimized if the temperature of the grain is kept below 14.5°C. Note that grain moisture content has no effect on controlling insect activity. As is the case with mites, the best defence against insect infestation is 'good store housekeeping'. Control measures include:

- thoroughly cleaning the store prior to harvest and burning all old grain, crop residues and dust. If insect problems are anticipated, empty stores can be either sprayed or fumigated with an insecticide;
- applying an insecticide as the grain is conveyed into store;
- assessing insect populations by trapping;
- ensuring the grain is kept cool in store by ventilating it with ambient cold air;
- fumigating the crop with insecticides such as methyl bromide or magnesium phosphide if an insect infestation occurs. These substances are toxic and their successful use involves sealing the grain store and the deployment of specialized control and safety equipment. Some insecticides can only be applied by professional spray operators.

Figure 7.1 indicates that (with the exception of mites) the maintenance of viable grain in store for a 35-week storage period is achieved by keeping conditions within the area below line D and to the left of line B. If grain is intended for use or sale as livestock feed, then the right-hand boundary moves out to line C, allowing a slightly higher moisture content to be accepted.

The time factor in grain storage

Insect, mould or mite infestations build up over a period of time, the rate of population increase depending on the prevailing conditions. Moist and/or warm grain can be stored without treatment for a limited period with little or no harm occurring. The length of time grain can be held without loss of quality decreases with increasing storage temperature and/or moisture content. Table 7.1 shows the maximum recommended storage periods for bulk grain stored at various temperatures and moisture contents (remember that storage for any extended period at temperatures over 14.5°C entails the risk of insect buildup, however dry the grain).

Grain moisture content and its measurement

Grain moisture content is usually expressed as the ratio of the weight of water to the total weight of the sample. Standard laboratory methods of determining grain moisture content define the sample weight, the drying temperature and the drying period. For example, ISO 712 requires a 5 g sample of grain to be oven-dried for 2 hours at 130–133°C. After drying, the sample is reweighed to determine the weight of the water which has been driven off. The moisture content is found from the expression:

$$\text{MC\%}\left(\text{Wet basis}\right) = \frac{\text{Loss of weight on drying}}{\text{Original weight of sample}} \times 100$$

ISO 712 is the standard now used by all aspects of the grain trade.

For on-farm use, using accurate weighing equipment and ovens is both impractical and time-consuming. Equipment has been developed to satisfy the need for an 'immediate' evaluation of grain moisture content – Should harvesting take place? Should the drier be adjusted? Grain moisture meters measure an electrical property of the grain such as its resistance or capacitance, which varies with moisture content. A good quality instrument, well maintained, can be expected to give a reading that is correct to within ±0.5% of the true moisture content of the sample, and this reading can be available within minutes of the sample being taken. Grain temperature can affect the electrical properties of grain. Some meters require manual compensation; others automatically compensate for grain temperature.

Meters use either ground grain samples or whole kernels. Using a 'whole grain' meter encourages more samples to be taken, which in turn can lead to a more representative appraisal of moisture content. However, a whole-kernel instrument may give false results with recently wetted grain, with grain that has just passed through a drier, or with very cold grain since in these cases the external condition of the grain (in terms of moisture content or temperature) may not reflect the internal situation.

The inherent accuracy of the meter can be undermined by poor sampling techniques. Comprehensive sampling is necessary to ensure that the sample is representative of the bulk of grain. This can be best achieved by systematically extracting several small samples from different points and depths in the load or store, mixing them together and then taking the final sample from this mixture.

Table 7.1 Estimated maximum number of weeks of mould-free storage of barley at a range of moisture contents and temperatures (ADAS, 1985)

Moisture content (%)	*Storage temperature (°C)*				
	5	*10*	*15*	*20*	*25*
16	120	50	30	12	5
17	80	28	10	4	2
18	22	9	4	2	1
19	14	6	3	1.5	0.5
20	10	4.5	2	1	0.5
22	7	3	2	1	0.5
24	4	2	1	0.5	0.2
26	2.5	1	0.5	0.2	0.2

Drying grain

Drying grain, as a precursor to safe storage, has the benefit over other conditioning systems in that the grain, if dried carefully, can be sold into all potential markets (e.g. seed, malting, milling, animal feed). In the drying process, air is passed through the grain mass. Moisture within the grain kernels moves to the grain surface and evaporates, to be carried off in the air stream. The relative humidity (RH) of the drying air is an important parameter in the drying process. Air having a low RH

Table 7.2 Equilibrium moisture contents (%) of a range of seeds at 25°C. (After McLean, 1989)

Species	*Air relative humidity (%)*						
	40	*50*	*60*	*70*	*75*	*80*	*90*
Wheat	10.7	12.0	13.7	15.6	16.6	17.6	23.0
Barley	9.8	11.5	13.2	15.1	16.1	17.3	23.1
Oats	9.5	11.0	12.5	14.3	15.4	16.7	22.0
Oilseed rape	5.7	6.6	7.4	8.8	10.2	11.9	17.4
Linseed	6.3	7.3	8.5	10.0	11.0	12.5	18.2
Field peas	9.5	11.2	13.0	15.8	18.0	20.9	28.6
Perennial ryegrass	9.2	10.6	11.9	13.9	15.3	17.4	24.4

has a greater water-holding capacity compared with air with a correspondingly higher RH and faster drying will occur. The RH of the drying air can be lowered by increasing its temperature. An equilibrium relationship exists between the moisture content of grain and the RH of the air that surrounds it (Table 7.2).

For barley to be dried to a safe storage moisture content of 15%, the RH of the drying air must be less than 65%. Oilseed rape will reach a safe storage moisture content of about 8% if ventilated with air at 65% RH.

In addition to defining the final grain moisture content, the rate of moisture removal from grain is affected by the drying air's RH. Low air RHs dry grain quicker. However, to prevent over-drying, the drying process in driers that utilize a low air RH is stopped before the equilibrium condition is reached.

There are relatively few periods, especially in northern areas of the UK, from harvest-time onwards when the RH of the ambient air is suitably low for drying grain. In fact, air RH varies cyclically over a 24-hour period, being generally lowest between late morning and early evening. Control of RH is achieved by adjusting air temperature. Fuel oil and gas are the usual energy sources, due to their convenience and cost characteristics.

By removing water during the drying process, the dried grain will weigh less. This weight loss incurred is important in terms of estimating the final saleable quantity and can be calculated from the following formula;

$$X = \frac{W_1(M_2 - M_1)}{100 - M_2}$$

where X is the weight loss on drying; M_1 and M_2 are initial and final moisture contents (% WB); W_1 is the weight of grain before drying.

The above formula only considers moisture weight loss. Grain handling and cleaning also contribute to the total weight loss in terms of creating and removing dust, chaff and small and broken grains.

In addition to the losses incurred during drying and handling, a further weight loss of at least 1% can be expected over the storage period, due to respiration of the grain itself and to the activities of storage pest organisms.

Grain driers can be divided into high-temperature driers capable of drying grain rapidly within hours of its arrival from the field and low-temperature driers which complete the drying process over a number of weeks.

High-temperature driers

High-temperature driers extract moisture rapidly by using very low humidity air as the drying medium. Moisture evaporates faster from the grain surface compared to its movement through the grain. This results in the outer layer of the kernel being overdried. During storage, the moisture within the kernel redistributes to maintain uniform conditions.

As the hot air passes through the grain mass, it gives up some of its energy to evaporate moisture to water vapour and is thus cooled. The distance that the air travels through the undried, moist grain must be limited so that there is no risk of the air cooling to the point where it becomes saturated (i.e. can no longer carry all its water), and water condenses on the grain. In high-temperature driers, the thickness of the grain layer rarely exceeds 460 mm, while in low-temperature storage driers it may be as much as 3 m.

Increasing the drying temperature has the effects of increasing grain throughput and, in many cases, lowering the energy consumption per unit of water removed. However, if the drying temperature is too high, there is a risk of causing damage to germination, baking or malting properties, affecting the extraction of oils or starch and reducing the livestock feeding value of the grain. Table 7.3 shows recommended maximum air temperatures when drying grain for various markets.

The table stipulates air temperature as opposed to grain temperature. Unless the grain is completely dry, grain temperature will always be lower than air temperature due to the cooling effect of evaporating water. The greatest cooling effect will be where the grain is moist and the evaporation rate is high. Even at the end of the drying process, average grain temperatures will still be 10–20°C lower than drying air temperature.

These recommendations are based on research carried out in the 1930s. New recommendations are likely to be based on a grain temperature/exposure time factor rather than the temperature of the drying air. If in doubt, drier manufacturers' recommendations should be followed.

High-temperature driers can be classified as either continuous flow – a near-constant flow of grain moves through the machine and into store – or batch – a quantity of grain is loaded into the drier where it remains while it is dried, and is then unloaded into store.

Continuous flow driers

A continuous flow drier is part of a fixed flow process which can incorporate handling equipment, wet and dry grain storage facilities and perhaps some form of cleaning and/or grading equipment. A typical example of a drying facility incorporating a continuous flow drier (and some types of batch drier) would have the following components:

- a wet pit or inlet conveyor to receive grain from the field;
- temporary wet grain bins with self-emptying capability for ease of cleaning;
- a pre-cleaner to remove unwanted material prior to drying (can be used as a cleaner/grader prior to despatch);
- the drier;
- dry grain storage, either on-floor or in-bin;
- vertical and horizontal conveying equipment.

Table 7.3 Recommended maximum air temperatures for high-temperature grain driers (Blakeman, 1982)

	Moisture content (%)	*Maximum drying air temperature (°C)*
Malting barley	Up to 24	49
and seed grain	>24	43
Milling wheat	Up to 25	66
	>25	60
Grain for stockfeed		82–104

The main components of a continuous flow drier are a wet grain holding section, a grain drying section, a grain cooling section, a furnace, a fan and a grain flow control system.

Types of drier can be classified in terms of the relative direction of the grain and drying air flows. The most popular types are the 'crossflow' and the 'mixed flow' types. 'Counterflow' and 'concurrent flow' driers have also been developed and each type has characteristic features.

In crossflow driers, the grain moves in a layer 200–300 mm thick and air is directed through the grain at approximately right angles to the grain's travel direction. The grain requires to be turned or mixed as it passes through the drier to achieve uniform drying.

In a counterflow drier, drying air flows in the opposite direction to the grain so that the hottest air meets the driest grain. In practice, this is achieved by blowing drying air through a ventilated floor in the base of a bin. The grain is removed by a circular auger running on the ventilated floor. This results in fast and efficient drying, but the dry grain leaves the drier warm. This grain is usually moved into a cooling bin, containing a second counterflow system in which cooling air moves against the grain flow. A gentle temperature reduction is achieved and maximum heat transfer can be fully utilized.

In a concurrent flow drier, grain and drying air move in the same direction although the air moves at a much faster rate. The hottest air meets the wettest grain, where the cooling effect of evaporation will be greatest. By the time the air reaches the driest grain, its temperature will have been much reduced, minimizing the temperature stress on the grain. This may permit air temperatures above the recommended maximum, without causing grain damage. There is a faster throughput from a given drier size, and more efficient use of energy (Brooker *et al.*, 1974).

The mixed-flow drier combines features of all the above three types. Grain flows downward in a rectangular column in which are located alternate horizontal rows of inlet (drying) and outlet (cooling) ducts, spaced about 0.3 m apart. Heat transfer is efficient, and relatively high air temperatures can be used without the risk of overheating the grain. At a plenum (air inlet) temperature of 120°C and when drying from 20 to 15% MC, grain temperature only reached 50°C (Bruce and Nellist, 1986). Figure 7.2 shows a drying installation incorporating a mixed-flow drier.

Figure 7.2 A continuous flow drying system comprising of two self-emptying wet bins, a mixed-flow drier and associated conveying equipment. (Courtesy of Law-Denis Ltd.)

Drying does not cease once the grain has left the drying section. A further reduction of 1–1.5% MC can be removed while the grain is being cooled, utilizing the heat energy remaining in the grain from its passage through the drying section. The cooling section occupies a third to a quarter of the drier's volume and is generally capable of cooling grain to near ambient temperature before it is discharged from the drier. Problems arise when ambient conditions are warm or when only a small amount of moisture needs to be removed from the grain, and drier throughput is higher than rated. Under these circumstances, the cooling section can be the bottleneck of the drier in terms of grain throughput. If grain cooling facilities are available in the store, it may be possible to use the whole of the drier (including the cooling section) for drying and thereby increase drier throughput rate.

American research has demonstrated the value of 'tempering' grain. After leaving the drier, grain is held, still hot, in a sealed bin for a 12-hour period before being cooled. 'Tempering' will improve thermal efficiency as better use is made of the residual heat in the grain (Bakker-Arkema, 1984).

The drying capacity (in t/h) or 'throughput' of a continuous flow drier is usually defined when drying grain from 20 to 15% MC at a specified drying-air temperature (often 65°C). Drier output will be increased if a higher air temperature can be (safely) used, or if the grain requires drying by less than 5% points, and vice versa.

The amount of drying done to each grain depends on the length of time the grain is exposed to the stream of hot air in the drying section of the drier, and this is controlled by the rate at which grain moves through the machine. As incoming moisture content varies throughout the day, the throughput of the drier must be altered to maintain a constant final moisture content. Automatic drier control units which monitor grain moisture content or grain temperature near the end of the drying section are used to regulate the flow of grain through the drier.

As grain movement through a continuous flow drier is not instantaneous, any adjustments made to the machine will take time to come into full effect.

Energy consumption by continuous driers

In theory, 2.45 MJ is required to convert 1 kg of water into vapour at 20°C. Practical tests have shown an average energy use of 6.7 MJ/kg of water evaporated for a range of traditional crossflow continuous driers. The discrepancy between these figures is due to:

- energy required to operate fans and conveyors;
- energy lost in the form of latent heat locked up in water vapour;
- energy lost in the form of heated, unsaturated air either from the lower part of the drying section or from the cooling section of the drier.

Improvements in drier design have addressed some of these points. For example, airflow paths through the grain can be shortened to reduce the fan requirements, and cooling exhaust air can be recycled to exploit the energy it still carries. This latter improvement can give energy savings of up to 40% (Bakker-Arkema, 1984). Several designs of mixed flow drier can achieve specific energy figures of less than 4.0 MJ/kg under test conditions, although these are not often attained under farm conditions, due to the stop/start nature of their operation.

The furnace sizes required for different drier throughput ratings are shown in Table 7.4.

Due to this high energy demand, driers are fired by diesel (gas oil), LP gas (propane) or natural gas (if a piped supply is available). Over-drying of grain is wasteful of energy, increases the drying cost, reduces throughput and decreases the final grain weight. When drying grain on contract, the aim should be to over-dry by 0.5% MC. Reducing the drying air temperature will not save energy, and will certainly increase the cost of drying. Automatic control of moisture content can reduce costs in terms of both energy use and labour.

On farms, the major component in drying costs is the fixed cost of the equipment. Any cost-saving exercise should address this aspect first, before moving on to matters of energy consumption.

Batch driers

In batch drying, the drying cycle is split into a number of steps (load, dry, cool, unload) instead of a flow process. The sequence of events may require human intervention, but in some designs and systems, each step can be carried out automatically, so that the drier can be left to run unattended. The throughput achieved by a batch drier is dependent on the drying temperature, initial and final grain moisture contents and the amount of 'down-time'. It is important to empty and to refill the drier quickly, and, if possible, to avoid cooling within the drier.

The simplest batch drier comprises a ventilated tray which can support tractors, trailers and fork-lifts and is surrounded by three walls. The tray is loaded either directly from trailers or by a bucket to a depth of 1 m. A drying temperature of 50°C is used and fan and heater equipment is sized to provide a drying rate of 1% MC removal per hour. Tray capacity ranges from 10 to 30 t and a pair of trays allows drying to proceed in one tray while the other is being emptied and reloaded with wet grain. The emptying process mixes the grain sufficiently to produce a sample with a uniform moisture content.

Portable recirculating batch driers have the advantage of having integral loading, drying, cooling and emptying elements. The drying energy source can be fuel oil or gas and the ventilating fan is driven by either the tractor's power take-off (pto) or an electric motor. Grain is continually circulated by a central vertical auger during the drying (and cooling) process. This action polishes the grain and also allows light material to escape as the grain is ejected from the top of the auger. A wheeled chassis allows the unit to be transported between drying sites. Figure 7.3 shows a cross-section of a recirculating batch drier installation.

Another version of the portable drier is the trailer drier which has a fan and oil burner unit incorporated. This unit has a trailer body with a ventilated floor. Loading is achieved by a bucket and during the drying process, which takes place either outside or in an open shed (e.g. Dutch barn), the grain is stirred by auger units to promote uniform drying. Emptying is carried out by tipping the trailer in the dry grain store.

Some batch driers can be set to run 'continuously'. During the drying phase, sensors monitor a parameter (e.g. exhaust air temperature) which can be related to moisture content. When the appropriate exhaust temperature is reached, the burner unit is shut-down and the fan continues for a pre-set period to cool the grain. Automatic operation of the conveying system allows the dry grain to be unloaded. The drier is refilled automatically with wet grain and drying recommences. Provided a source of wet grain is available, the drier and its associated conveying equipment can be set to load, dry, cool

Table 7.4 Furnace capacity requirement of continuous flow driers used for drying grain from 20 to 15% MC at 65°C air temperature

Drying capacity (t/h of wet grain)	2.5	5.0	10.0	20	30	60
Evaporation rate (kg water/h)	150	300	600	1200	1800	3600
Furnace capacity (kW)	220	450	900	1860	2700	5300

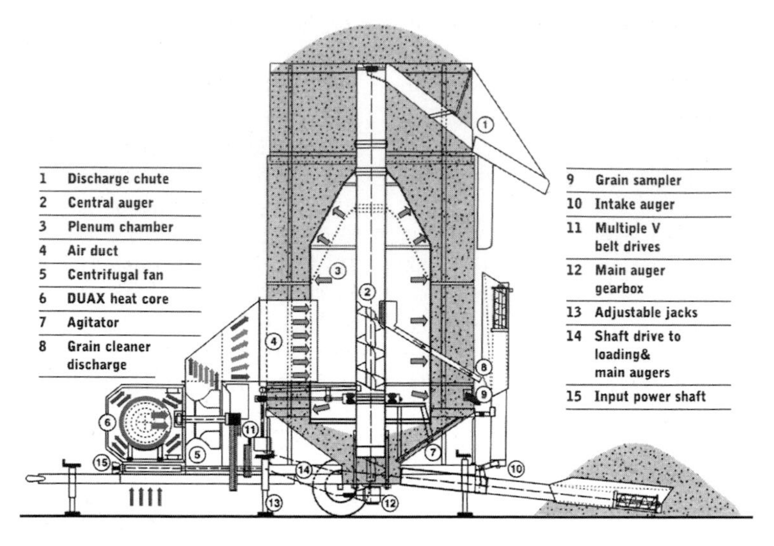

Figure 7.3 Cross-section of a recirculating batch drier. (Courtesy of Marksman International Ltd and Opico Ltd.)

and discharge grain 'continuously'. These units are installed on a permanent site.

Low-temperature driers

Low-temperature driers are characterized by the relatively deep layers of grain which are dried over a long time period. The drying area usually doubles up as the long-term storage compartment for the grain.

Drying can take place in bins or on a drying floor. An on-floor drying system comprises a ventilated grain floor, a fan and air heating source and an air distribution system (main air duct and lateral ducts under the ventilated floor). In-bin drying has the advantage of segregating batches of grain and each bin requires a level drying floor. Filling and levelling the bins can be relatively cheap and simple, but emptying the bins can be labour intensive or expensive if sweep augers are used.

The operation of any low-temperature drier relies on the fact that there is an equilibrium relationship between the moisture content of grain and the relative humidity (RH) of the air which surrounds it. Grain is brought to a safe moisture content by ventilating it for 2 weeks or more with air at the corresponding equilibrium RH. To avoid over-drying or re-wetting of the lower grain layers, careful control of drying air humidity is necessary. Although in the initial stages of drying, higher drying air RHs can be used, the aim in the final stages of drying is to use air with an RH of around 65% to achieve 15% MC.

Southern parts of Britain have more scope to use inexpensive ambient air for drying. However, in more northern regions, heat (provided by a gas or diesel fuel source) is required to achieve appropriate drying air conditions.

The design drying rate for this system is 0.5% MC reduction per day. It is therefore a slow process and is suited to farms either where only a small amount of drying has to be carried out at harvest or where relatively small quantities of grain are dried and retained for livestock use and the drier can double as a grain store.

Grain stirring units, consisting of vertical augers, which move through the grain mass and disturb and mix the grain are recent developments. Their use has resulted in improved airflow rates and has allowed higher drying temperatures to be used to improve the drying rate of the system.

Measurement and control of air relative humidity

Relative humidity can be measured manually with a wet and dry bulb hygrometer, which consists of a pair of identical thermometers. The bulb of one thermometer is covered with a wet cloth sleeve. After exposure to the air stream until readings stabilize, the two thermometers are read and the relative humidity is found from a slide rule or chart supplied with the instrument. All other instruments for measuring or controlling relative humidity must be calibrated regularly against the basic, and inexpensive, wet and dry bulb instrument.

If the relative humidity of the air is too high, it can be reduced by heating the air. Heating equipment which will raise the air temperature by 6°C is adequate to deal with most weather situations. Electrical heater banks (with their ease of installation, operation and control) are expensive options, and oil- or gas-fired heaters are available as cheaper options. A heat input of 1.1 kW is needed to raise the temperature of 1 m^3/s of air by 1°C, and thus reduce the relative humidity by 4.5 units.

Heater output must be adjusted to accommodate changes in ambient condition if the maximum drying rate is to be maintained. Automatic humidity controllers can sense the air relative humidity in the main air duct, and adjust, for example, the gas flow rate to maintain constant conditions.

Self-contained diesel engine-driven fans are used for grain drying, the waste heat of the engine being used to raise the air temperature by up to 4.5°C.

The low-temperature drying process

As the drying air moves upwards through the grain mass it picks up water and is cooled. Grain at the bottom of the store dries first, and a 'drying front' then moves slowly upwards through the grain until all the grain is dry after 2 weeks or longer. During drying, the grain above the drying front remains as wet (or wetter) as when it was harvested. This portion of the grain is preserved before drying by the cooling effect of the air which has dried the lower layers of grain.

If grain greater than 22% MC has to be dried, it should not be loaded to a depth of more than 1 m. This reduces the drying time, increases drying efficiency and also prevents the soft grains being squashed by the weight of a full depth of grain. If grains at the bottom were crushed, this would reduce the air spaces between them and restrict the flow of drying air.

Drying rate and efficiency are improved for grain of any moisture content if the grain store is filled in layers as drying proceeds, rather than piling the grain in one portion of the store first (Sharp, 1984). However, this can only be achieved if conveyors and throwers are employed. Heat should not be used for drying grain above 18% MC. Initial drying should be carried out with ambient air, with heat being applied only in the final stages of drying. Use of heat at an earlier stage will result in re-condensation of moisture in the upper layers of grain.

Progress of the drying front towards the grain surface can be followed by extracting samples from different depths with a sampling spear. The depth of the transition between dry and wet grain indicates the position of the drying front.

Air supply and distribution

An adequate quantity of air must be provided to make the drying period as short as possible. An airflow rate of 0.025–0.05 m^3/s per tonne of grain, is recommended (ADAS, 1982). Adequate fan capacity is a cornerstone of successful crop drying.

After being pressurized by the fan and heated if necessary, air reaches the grain via a main duct and a series of lateral ducts spaced at 1000–1200 mm under the grain. To limit the buildup of back pressure, ducts should have a minimum cross-sectional area of 1 m^2 for every 10 m^3/s of drying air. This will give a maximum air speed of 10 m/s in the ducts. Similarly, the size of the openings through which the air passes from the ducts into the grain should ensure that air speed at this point does not exceed 0.15 m/s. Lateral ducts should not exceed 10 m in length because the output of air along the duct becomes uneven beyond this length. A better distribution of air to the grain lying between the ducts is achieved when air escapes from the sides rather than the top of the duct.

Above-floor ducts convert any sound, level concrete floor into a drying floor, and can be removed if the building is needed for some other use. However, the operations of filling and emptying the store are made more difficult by the presence of above-floor ducts. Improved options are permanent below-floor ducts or wooden or concrete ventilated floors which are strong enough to support the weight of trailers or bulk lorries.

In grain bins, the entire floor is perforated and air passes up from a plenum chamber below the floor, through the perforations and into the grain. A further refinement to this type of floor is an arrangement of louvres which direct jets of air to 'sweep' the last few tonnes of grain out through an unloading chute using the power of the drying fan. This eliminates the manual task of sweeping or shovelling the last grain out of a flat-bottomed bin but results in a very dusty operation.

Achieving a uniform air flow throughout the grain mass requires that the grain density is uniform. Any differences in depth or compaction, or any local concentrations of wet grain or rubbish, will affect uniformity of air flow. Pre-cleaning the grain before it enters the store will remove most of the rubbish. If air flow to any part of the store is restricted, the grain in that part of the store will dry slowly or not at all. Air flow through the grain can be measured at many points on the surface using an airflow meter. Points having low readings can be investigated and corrected.

To ensure that moisture-laden air can escape easily from the roof space above the grain store, exhaust openings of at least 0.25 m^2 per m^3/s of air flow must be provided at eaves, ridge or gable ends.

Fans for on-floor drying

A fan moves air by creating a pressure behind it. The air output of the fan decreases as the back pressure or static pressure against which it is working increases. Back pressure builds up as the air is forced along ducts, round bends and through the grain. The greater the air speed, the higher the back pressure (Cory, 1991).

The pressure generated by a fan is measured with a water manometer, and fan air output (in m^3/s) can be determined from this reading by reference to manufacturers' performance curves. Working pressures for crop drying fans are 60–150 mm water gauge (WG). A fan is 40–85% efficient. The balance of its power consumption is lost as heat, which warms the air by up to 2.5°C (Sharp, 1982). This may be enough heat to dry a crop, but it is a disadvantage when the fan is used for cooling. Cooling systems sometimes operate under suction with the fan at the outlet; this way, the fan's heat does not enter the crop.

Some fans absorb more power as back pressure drops and air output increases, and the motor may become overloaded and overheat. Only non-overloading fans should be used in crop drying systems.

Centrifugal fans

Centrifugal fans develop pressure by accelerating the air in the rotor and then converting its kinetic energy into pressure energy in the diverging housing or volute. They are quiet in operation and bulky. A form of centrifugal fan with backward-curved (b-c) blades on the rotor is non-overloading, and is frequently used in grain stores. Pressures up to 300 mm WG are possible, according to rotor diameter and speed.

Axial-flow fans

Pressure is created by the rotation of an airscrew-like impeller in a close-fitting circular housing. Air passes in a straight line through the fan, giving high peak efficiencies. The fan alone can only produce a pressure of 80–120 mm WG, but stationary intake guide-vanes can increase this by 20–60%. Two fans contra-rotating in the same tubular housing can increase the pressure up to three times. The fan is non-overloading, and more compact than a centrifugal fan of the same output. It is noisy, particularly when guide-vanes are fitted. Silencers are available, or the fan can be shielded with bales to absorb the noise.

Table 7.5 shows the output of various crop-drying fans against a range of static pressures.

Grain cleaners and graders

Grain cleaning is carried out for two reasons:

(1) Pre-cleaning – the removal of short straws, chaff, dust, weed seeds and impurities prior to drying:
- avoids the expense of drying valueless material;
- reduces the risk of blockages and fire in high-temperature driers;
- prevents poor air flow conditions from occurring in bulk and bin drying systems.

(2) Cleaning and grading after drying or pre-despatch:
- removes shrivelled grain and impurities;
- improves the sample appearance;
- increases the specific weight (bulk density).

Cleaning and grading are achieved by sieving and aspiration. Sieves can have either round holes (separation

Table 7.5 Output of various crop-drying fans (air output, m^3/s) against a range of static pressures (manufacturers' figures)

Type and power of fan	*Static pressure (mm WG)*					
	50	*75*	*100*	*125*	*150*	*175*
3.8 kW b-c[1]	4.6	3.9	3.0	—	—	—
7.5 kW b-c	8.2	7.7	7.0	5.9	4.4	1.8
18.7 kW b-c	13.7	13.1	12.4	11.5	10.5	9.4
30 kW axial	10.0	8.8	7.1	4.7	1.8	—
52 kW axial	18.8	17.7	15.9	14.1	11.8	9.4

[1] b-c, Backward curved.

depending on diameter) or slots (separation depends on material width or height, whichever is the shortest). Sieves can have either a reciprocating action (flat sieve are mounted one on top of the other, largest at the top) or a rotary motion (cylindrical sieves are positioned in series, the smallest sieve nearest the grain inlet). Both types are available as two, three and four sieve units. Reciprocating sieve cleaners should be mounted on a rigid base so that no damage will result from their constant vibration.

If an air stream is passed up through a grain sample, material (light grains, awns, dust, short straws) having a lower terminal velocity than that of the air speed will be removed. This principle (aspiration) is used for pre-cleaning grain and is also combined with sieving units in cleaner/graders.

Figure 7.4 shows the material flow through a rotary cleaner fitted with an inlet aspiration unit.

Cleaning and grading machines have outputs to suit most drying systems and are typically within the range of 10–60 t/h. A dual-purpose machine will operate at twice the output when pre-cleaning compared to final cleaning and grading.

Handling grain

Grain can be moved by a range of equipment, and choice depends on the drying and storage systems in place. The characteristics of the common types of handling equipment are summarized below.

Augers

The auger is a versatile conveyor which will operate in many positions without adjustment. The grain is moved by a screw, or 'flight', which rotates inside a steel tube. Grain is drawn in at one end, and is normally discharged at the other end although intermediate discharge points are possible. There is considerable churning and abrasion of the conveyed material, particularly if the auger is only partially full. The churning effect may be used to advantage in mixing a preservative with grain, but augers should not be used for conveying fragile materials such as rolled barley or pelleted feeds, because of the risk of damage. Units can be mounted on wheeled chassis for easy movement.

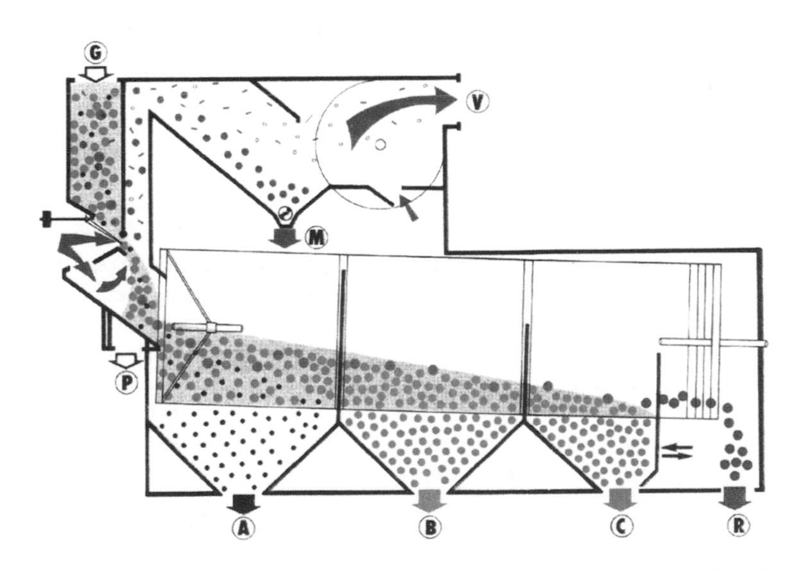

Figure 7.4 Cross section of a rotary cleaner fitted with an inlet aspiration unit. (Courtesy of Law-Denis Ltd.)
G – raw material in
P – by-pass
A – small material (e.g. weed seeds)
B – small grain (e.g. lights)
C – good grain
R – oversize material (e.g. straw, chaff)
M – heavy cleanings
V – light cleanings

The throughput of an auger is dependent on the transport material, its moisture content and the auger diameter and angle. Throughputs for dry wheat and an auger angle of 45° are shown in Table 7.6. The flight of the auger projects beyond the tube at the intake end. This portion must be guarded at all times with a welded-mesh screen to avoid the risk of serious personal injury.

Chain and flight conveyors

Grain is moved along a smooth wooden or steel trough by an endless chain fitted with horizontal scrapers or 'flights'. The grain can be fed into the trough at any point, and is discharged at the end of the conveyor or at intermediate discharge points if these are fitted. Some models are reversible. 'Double-flow' models are constructed with a transport trough under both top and bottom lengths of the chain and flight. This unit can be used to provide a continuous supply of grain to a drier (using the 'top conveyer'), with the excess grain being returned to the wet pit or bin by the 'bottom conveyor'.

Chain and flight conveyors are normally used for moving grain horizontally, although some can work at small slopes. They are quiet in operation, long-lasting, economical in power, and do little damage to the grain. Lengths up to 60 m are available with outputs ranging from 10 t/h upwards.

A development of the chain and flight conveyor is the 'angleveyor' which allows the grain to be moved in two dimensions (laterally and vertically). This is achieved by having a shaped trough for the grain and a guide track for the chain. These units are often used as an alternative to a wet pit and outputs in excess of 40 t/h are normal.

Belt conveyors

The grain is conveyed on an endless fabric-reinforced rubber belt which is supported on rollers. Grain can be loaded onto the conveyor at any point through a hopper which concentrates the grain in the centre of the belt. Discharge is over the end of the belt, or at a movable discharge point where the belt passes in an S-shape over a pair of rollers. The grain shoots over the upper roller and is deflected into a side discharge trough to right or left. A motorized discharge can move slowly back and forth along a conveyor unloading the grain evenly along the whole length. In conjunction with a grain thrower to project the grain across the store, a level layer of grain can be built up over the complete area.

Belt conveyors are quiet and handle grain gently. Conveyors with cleated belts can operate at steep angles, but the typical grain version has a smooth or textured surface without cleats, and operates horizontally. Belt conveyors are available in lengths up to 45 m and with capacities from 20 t/h. Power requirement is greater than for the chain and flight machine.

Table 7.6 Dry wheat throughputs for a range of auger diameters

Diameter (mm)	*Throughput (t/h)*
114	25
150	35
200	58
300	120

Bucket elevators

Bucket elevators are used to raise grain vertically. They consist of a casing in which a series of steel or plastic scoops or 'buckets' are bolted to a flat endless belt which moves vertically between a pair of pulleys. The motor and drive pulley are at the top, while the belt passes round a tensioning idler pulley at the bottom. Grain is normally fed into the 'buckets' near the bottom of the upward moving side of the elevator 'leg' and is discharged from a spout (by centrifugal force) on the downward moving side of the 'leg' near the top. Historically, elevators were located in a hole (pit) at least 1 m deeper than the bottom of the wet pit. This often causes flooding problems due to high water tables. The development of 'angleveyors' (see above) has allowed elevators to be installed at floor level which avoids the flooding problem and facilitates maintenance. The top of the conveyor is generally accommodated in a small extension ('penthouse') of the grain store roof where access for maintenance may be restricted.

Grain damage is not normally a problem. Wet crops, especially peas, can be squashed between the lower pulley and the belt. Cage pulleys are available to reduce this problem.

Capacities range from 5 t/h upwards and elevators can be 25 m high. 'Double-leg' models combine two separate elevators with a single drive unit. The two legs may be used separately; for example, to simultaneously load wet grain into, and remove dry grain from, a continuous flow drier. When high throughputs are required, for example when loading lorries, both legs can be used to raise the dry grain to the lorry loading discharge point.

Multiple valving devices, with controls at floor level, can be used to direct the grain from the elevator's discharge to alternative destinations.

Paddle elevators

Paddle elevators have the same function as bucket elevators but their construction allows their installation at an angle to provide both horizontal and vertical grain movement simultaneously. Instead of buckets, the conveying element consists of sections of flexible material which are fixed to a drive chain and fit tightly against the casing.

Pneumatic conveyors

These systems are versatile and layouts can range from simple to complex. A series of connectable pipe sections is used to convey grain both horizontally and vertically. Pipework can be arranged to accommodate any storage arrangement and the system is most useful in situations where either space above bins is limiting or there are numerous bins to fill. The systems have a high initial cost, and the power consumption of 1–2.5 kW/t per h is high, particularly for those models with a suction intake. The airstream is produced by a narrow, large-diameter fan driven by an electric motor or a tractor pto. The grain is carried along the closed pipe of 130–210 mm diameter by a stream of air moving at 70–100 km/h.

Intake may be by a flexible suction spout, which is very convenient for emptying flat-bottomed bins or clearing floor stores. Alternatively, the grain may be injected into the airstream through an air-sealed metering hopper. Discharge is from the far end of the system, which may be up to 200 m from the intake if sufficient power is applied, at outputs of 5–40 t/h. The pipe network system can be branched and valves used to allow grain to be moved to various locations.

The grain does not pass through the fan but considerable abrasion can occur as it moves along the pipes and round bends, resulting in the generation of dust at the discharge end. Fragile materials such as rolled grain or pelleted feeds should not be handled in this way.

Grain throwers

A floor store can be filled in a series of flat layers if a grain thrower is available. The thrower consists of a fast-moving conveyor belt or a rotating paddle-wheel which throws the grain up to 12 m horizontally, and to a height of up to 5 m. Larger models stand on the floor; smaller units may be mounted at the discharge point of a conveyor. Outputs range from 5 to 120 t/h, with a power consumption of 4 kW at a flow of 20 t/h.

An advantage of loading a floor store in this way is that any rubbish present in the grain is distributed evenly throughout the store and will not have any local effect on the drying process. Also, there is little damage to the grain, but a dusty atmosphere is created which can be a health hazard.

Intermediate storage hoppers

Moving grain off the farm can involve lorries of 38 t capacity or more. If grain-conveying equipment of 20 t/h is used, the lorry will be tied up for almost 2 hours and this may make the product less attractive to a potential customer. High-capacity conveyors are available, but would only be fully utilized for a few hours per year.

One way of speeding-up the loading rate is to use an intermediate storage hopper. These units of 40 t capacity are equipped with a V-bottomed hopper and are mounted high enough for a bulk lorry to be driven underneath. The hopper can be filled slowly with the store's existing conveyors, allowing plenty of time for the grain to be passed through cleaning equipment if required, and yet the lorry can be filled in 10 minutes or so with the minimum of effort or dust. In the likely event of there being a buyer's market for grain, it is this type of detail, plus a quality product, that will ensure a sale in a competitive situation.

Loading buckets

For bulk stores, a bucket on a manual handler should be capable of filling lorries at a rate of 1 t/h.

Management of grain in store

Storage requirements

Table 7.7 gives bulk density details for a range of crops. Table 7.8 shows capacity details for a range of round bins of different diameters. If on-floor drying is practised, a level grain surface is required. When storing dry grain in bulk on a floor, the grain can be loaded above the level of the walls and peaked in the centre. The angle adopted by the grain is known as the angle of repose and

Table 7.7 Crop bulk densities

	Bulk density (kg/m^3)	*Specific volume (m^3/t)*
Wheat	785	1.28
Barley	689	1.42
Oats	513	1.95
Beans	817	1.20

Table 7.8 Round bin capacities (wheat)

Bin diameter (m)	*Volume (m^3/m depth)*	*Capacity (t/m depth)*
2.4	4.5	3.5
3	7.1	5.5
3.7	10.8	8.4
4.3	14.5	11.3
4.9	18.9	14.8

varies for different species of grains and moisture contents. For dry grain, 30° is an approximate figure.

Monitoring grain temperature

Successful management of grain in store is largely dependent on monitoring grain temperature, since heating is a symptom of most of the problems that affect stored grain. The cause of any local increase in grain temperature must be identified and remedial action taken to prevent an escalation of the problem.

For large-scale storage, a permanently installed system comprising grain temperature sensors and a central monitoring station can be used. Manual checking should be carried out at least weekly. Automatic monitoring allows checking on a more regular basis and some systems can be linked either to alarms which sound if a pre-set temperature is exceeded or to ventilation fans. For smaller stores, a hand-held temperature-sensing spear can be pushed into the grain and the temperature recorded for a number of positions and depths in the grain store to allow a composite picture of grain temperature to be developed.

Aeration

The ability to ventilate a grain store with the grain in situ allows:

- general grain cooling;
- elimination of hot spots;
- maintenance of a uniform temperature within the grain bulk.

The basic principle is to ventilate the grain with relatively small volumes of cold air. The airflows involved will not dry grain. The required airflow is 0.5–0.7 m^3/min per m^2 of floor area, or 10 m^3/h per t of grain (ADAS, 1985). A 0.375-kW fan can aerate up to 50 t of grain at a time. A cooling system similar to that for drying grain can be used. However, as the airflows involved are much lower than those employed for drying, the air tends to percolate through the grain and fewer ducts are necessary. For large stores, where the grain is peaked, it is possible to use above-ground cylindrical ducts spaced sufficiently apart to allow tractor access between the ducts during unloading. To promote even air distribution in level-loaded stores, ducts should not be spaced more widely apart than twice the depth of the grain.

General cooling

Grain stored at a low temperature is less likely to deteriorate (see Fig. 7.1). Grain being loaded into store will be either at harvest temperature or at the exit temperature of the drier. In both cases, grain temperature is likely to be higher than that required for safe storage conditions. Without aeration facilities, reliance must be placed on ambient conditions and surface cooling to reduce grain temperature.

The cooling section of some continuous flow driers is the bottleneck to throughput. The output of a drier may be uprated by converting its cooling section into additional drying volume, and cooling the dry grain in store.

Moist grain, up to 21% MC, has been stored successfully using aeration equipment. This technique can be used in a wet harvest to maintain grain quality prior to drying, and has also been used on farms without drying facilities, to safely store grain for on-farm consumption.

Elimination of hot spots

When localized heating of grain is detected, a standard solution is to remove the grain from the store and during this process ('turning'), the grain will cool by exposure to ambient air. When aeration equipment is fitted, cool air can be blown through the grain to reduce spot temperatures without the need to move the grain.

Maintaining an even temperature throughout the grain

During the storage period, the layers of grain in contact with the walls of the store will cool, while the grain in the centre of the store will remain at its original temperature. Sunshine will heat the grain on the south side of a store. Any temperature inequalities of this sort will set up convection currents in the air spaces between the grains. This will result in moisture migration from the warmer grains and condensation on the cooler grains, particularly just below the surface at the centre of the store. This process can result in mould development, which may be followed by heating, insect activity and even sprouting. With regular aeration, grain temperature will be uniform throughout the grain mass and moisture movement will not occur.

Alternatives to grain drying

For grain that is to be used for livestock feeding, advantages of storing it without drying are as follows:

- Capital and running costs associated with drying are avoided.
- Grain can be handled into store at a fast rate, thereby avoiding harvest bottlenecks.
- Grain can be rolled easily, with less dust produced.

Storage of grain in sealed containers

If grain is stored under airtight conditions, grain respiration quickly reduces the oxygen concentration to less than 1% with a consequent increase in carbon dioxide. The grain is killed and fungal and insect activity is stopped. Some fermentation may take place if the moisture content is above 24% and this will taint the grain. As long as the container remains sealed, the condition of the stored grain is stable. When grain is removed from the store, care must be taken to allow only the minimum of air to enter, and to reseal as soon as possible. Once exposed to the air, the moist dead grain will deteriorate rapidly especially if the weather is warm. Normally, only enough for 1 or 2 days' use should be withdrawn at a time.

Although plastic and butyl bags can be used, the standard unit for sealed storage is an enamelled-steel tower silo. Capacities range from 15 to 800 t and the silo is filled by blowing grain up an external pipe to an opening near the top of the tower. Once filled, the tower is sealed. A breather bag in the silo is connected to the atmosphere and is responsible for adjusting the internal pressure as the silo heats up and cools down throughout the year. Unloading is by means of augers, the most sophisticated having a sweep arm which traverses the circular bottom of the silo, and a fixed arm which brings the grain from the centre of the silo to discharge outside through a resealable spout.

If grain above 22% MC is to be stored in a tower silo, the lower layers are likely to be compressed by the weight above and may bridge and be difficult to unload. A few tonnes of dry grain at the bottom of the silo will facilitate this initial stage of unloading, as will the use of a gentle method of filling such as a bucket elevator rather than the more convenient but damaging blower.

Extreme caution is required when entering a sealed silo due to the presence of carbon dioxide. It is important to ensure that all carbon dioxide has been removed before attempting to clear blockages or bridging problems.

Use of chemical preservatives to store grain

Treating undried grain with a chemical preservative such as propionic acid kills the grain and inhibits the activity of moulds and pests. The treatment rate is proportional to grain moisture content. After treatment, no special storage structure is required. The grain must be kept dry to avoid dilution or leaching of the preservative, and this requires the floor, walls and roof to be waterproof. Since propionic acid is highly corrosive, even to galvanized steel, the materials of the store must resist corrosion or be coated with plastic or a chlorinated rubber paint for their protection. Any equipment that handles either the preservative or grain that has been freshly treated must be thoroughly washed out with water after use to minimize corrosion. The preservative liquid or its fumes can cause irritation or injury to the skin, eyes, mouth or nose. Protective garments should be worn, and a supply of water kept close by in order to wash off splashes. The operator should stand upwind of working application machines or piles of recently treated grain.

The acid applicator consists of a pump, a flow regulator and a nozzle which draws the chemical from a container and sprays it on to the crop at the intake of a short (1.5 m) length of auger. The tumbling effect of the auger is effective in distributing the chemical over each grain, an important requirement of the system. Charts are provided so that the chemical flow rate can be adjusted to match the grain moisture content (up to 30% and more) and the (measured) throughput of the auger. The cost of the chemical is relatively high, so in order to be viable the handling and storage parts of the system must be inexpensive.

Since the grain is killed and tainted by the preservative, there is no market for the grain other than for stockfeed. Weed seeds are also killed. Because the chemical is absorbed into the grain, protection continues after the grain is removed from store. This allows large batches of feed to be prepared, even in warm weather, without risk of deterioration.

Urea, which, in addition to acting as a preservative, imparts a nutritive value to the crop, has also been used successfully.

References and further reading

ADAS (1982) Booklet 2416. *Bulk Grain Driers*. MAFF Publications, Alnwick.

ADAS (1985) Booklet 2497. *Low-Rate Aeration of Grain*. MAFF Publications, Alnwick.

Bakker-Arkema, F.W. (1984) Selected aspects of crop processing and storage: a review. *J. Agric. Eng. Res.* **30**, 1–22.

Blakeman, R. (1982) Fuel saving in grain drying. *Agr. Eng.* **37**, 60–67.

Brooker, D.B., Bakker-Arkema, F.W. & Hall, C.W. (1974) *Drying Cereal Grains*. Avi Publishing Co., Westport, CT.

Bruce, D.M. & Nellist, M.E. (1986) Simulation of continuous-flow drying. *Proc. BSRAE Association Members' Day*, 26–27 February.

Burges, H.D. & Burrell, N.J. (1964) Cooling bulk grain in the British climate to control storage insects and to improve keeping quality. *J. Sci. Fd. Agric.* **15**, 32–50.

Cory, W.T. (1991) Fans for today's agriculture. *Agr. Engr.* **46**, 2–8, 34–38, 66–70.

McLean, K.A. (1989) *Drying and Storing Combinable Crops*, 2nd edn. Farming Press, Ipswich.

Sharp, J.R. (1982) A review of low-temperature drying simulation models. *J. Agric. Eng. Res.* **27**, 169–90.

Sharp, J.R. (1984) The design and management of low-temperature grain driers in England – a simulation study. *J. Agric. Eng. Res.* **29**, 123–31.

Wilkin, D.R. (1975) The effects of mechanical handling and the admixture of acaricides on mites in farm-stored barley. *J. Stored Prod. Res.* **11**, 87–95.

Useful websites

http://www.alvanblanch.co.uk
http://www.law-denis.com
http://www.opico.co.uk

Part 2

Management

8

The Common Agricultural Policy of the European Union

P.W. Brassley & M. Lobley

Background, institutions and the legislative process

The need for a Common Agricultural Policy

When the European Community was created, it needed a Common Agricultural Policy for two reasons: first, the agricultural industries in each of the member states were subject to government intervention, and, second, if this intervention were to continue, it had to be compatible with the other provisions of the Community.

Left to themselves, in the absence of government intervention, the markets for agricultural products in developed economies will produce:

- fluctuating prices, and, as a result, fluctuating incomes for producers; and
- gradually declining prices, in terms of purchasing power, and, as a result, declining real incomes for producers in the long run.

The price of agricultural products, like that of any other product in a free market, depends upon the balance of demand and supply. If demand increases less than supply, the price will fall.

The demand for agricultural products in a developed country does not usually increase rapidly, unless those products can be sold cheaply on the world market. Technical change produces increased output per hectare and per unit of labour, so supplies tend to increase more rapidly than demand. Thus, in the long run, prices tend to fall. Demand is also relatively unresponsive to price changes, so short-run supply variations, caused by changes in weather and disease, produce short-run price fluctuations.

Short-run price fluctuations in a free market may result in the demise of businesses which would be viable at average levels of input and output prices, and the possibility of this event creates a disincentive to investment on such farms. In the long run, falling prices produce lower incomes for those farms that are unable either to reduce costs or to increase output. If such low incomes are unacceptable, they may cease trading altogether. It is usually found that farm incomes have to fall to very low levels before farmers leave the agricultural industry, and in any case the land that they no longer farm is often taken over by another farmer. The total output of the industry is therefore maintained, so there is no tendency for prices to rise. While farmers with low incomes may be found in all areas, some regions may be particularly disadvantaged by physiographical or structural factors (i.e. infertile land and small farms), and if agriculture is the major industry the whole region may be affected by income and outmigration problems.

In a free-trading economy, farm income problems may be further intensified by the availability of imported agricultural produces at prices below the domestic supply price. Imports will not only restrict domestic price rises but also increase the total import bill, which may be considered a problem in countries with balance of payments deficit problems. However, the government of a country that has no difficulty in exporting enough to pay for food imports may still consider it unwise to be totally reliant on imported food supplies: unforeseen circumstances may give rise to difficulties in obtaining supplies, or something may reduce the price advantage of imports in the long run. A capacity for rapid expansion of domestic agricultural output may therefore be desirable even though considerable reliance is placed on imports. In short, free markets for agricultural products in developed free market economies may produce low farm income, balance of payments and potential supply security problems.

Whether or not anything should be done about these problems is a political issue which different member states have approached in various ways. From the middle of the nineteenth century until the mid-1930s the United Kingdom essentially operated a free trade policy which resulted in a relatively small agricultural population and a high level of food imports. From the 1870s onwards, most continental European countries, including all of the larger ones, took the view that their interests were best served by maintaining a large rural population and a high level of food self sufficiency. It was not simply an economic issue: the small family farm was perceived to have a valuable social role, and latterly it was also seen as a guardian of the rural environment. Thus for many years the European model has stressed the desirability of addressing several rural policy issues through the medium of agricultural policy, a principle recently recognized under the term 'multifunctionality'. In the continental European countries, agricultural policies usually involved a system of import controls and price support, which kept price levels higher than they would have been in a free market and so maintained agricultural incomes and self-sufficiency levels. When the discussions which led to the formation of the Community were held in 1955–1956, it was agreed that the exclusion of agriculture from the general common market was impossible: without free trade in farm products, national price levels could differ and those countries with the lowest food prices would have the lowest industrial costs, thus undermining the common policies which would be introduced for other industries. Although there was a general similarity between the existing agricultural policies, the differences in detail were so numerous that the simple continuation of existing policies would have been impossible. Not only did the Community countries require an agricultural policy, they required a *common* agricultural policy.

The origins and development of the European Union

The treaty establishing the European Community, as it was then called, was signed in Rome on 25 March 1957, and the Community came into being at the beginning of 1958. The main objective of the original six member states was to promote economic development by removing the barriers to trade between them. As they gradually succeeded in bringing this about they were joined by other countries:

Date of accession	*Member states*
1958	Belgium, France, West Germany, Italy, Luxembourg, The Netherlands
1973	Denmark, Ireland, UK
1981	Greece
1986	Portugal, Spain
1995	Austria, Finland, Sweden

The EU also has association agreements with ten central and eastern European countries (CEECs) of which Estonia, the Czech Republic, Hungary, Poland and Slovenia are thought likely to join the EU before the remaining five (the Slovak Republic, Romania, Bulgaria, Lithuania and Latvia). Since the accession of the first five countries alone would increase the agricultural area of the EU by more than 20%, and the number of agricultural workers by more than 50%, enlargement clearly has major implications for the future of the CAP. Malta, Cyprus and Turkey have also applied to join the EU. The SAPARD (Special Accession Programme for Agricultural and Rural Development) has been established to assist the accession process.

The process of decision making in the European Union

The EU must make decisions about many areas of policy without, on the one hand, being bogged down in the process of consultation, or, on the other, neglecting the needs and views of 15 member states, over 300 million individual citizens, and numerous lobbies and interest groups. Decisions must be made about what policy shall be (primary legislation) and how it should work in detail (secondary legislation). Several institutions are involved in this process, of which the two most important are the Commission and the Council.

The *European Commission* is presided over by 20 Commissioners. They are appointed by agreement between the member governments, but are required to act in the interests of the EU and not individual member states. The Commission employs roughly 16000 officials, of whom about one-quarter are translators, and they work in special services such as the legal department and the statistical office, or in Directorates General which are concerned with policy areas. Thus there is a Directorate General for Economic and Monetary Affairs, another for the EU Budget, another for the Environment, and so on, and each is the responsibility of one of the Commissioners. The Agriculture Directorate General, with a staff of about 1000, is among the larger ones. In some ways, therefore, the Commission is like a national civil service, but it also has extra functions: it is responsible for proposing primary legislation; it handles the day-to-day administration of EU laws and policies resulting from this primary legislation, and may enact secondary legislation in order to do so; and it represents the EU in its relations with non-member states and other international bodies such as the World Trade Organization (WTO).

The *Council of the European Union* is the major legislative body of the Community, and consists of a minister from each member state, depending on the subject under discussion: agriculture ministers for agricultural matters, finance ministers for financial matters, and so on. The chair is held by an individual country for a 6-month period and rotates from country to country in alphabetical order, according to the name of each country in its national language: Greece (E for Ellas) therefore follows Denmark. It is normally the final decision-making body for all primary legislation, although major constitutional issues or especially insoluble problems may be passed on to the *European Council*, a meeting of the heads of government of the member states which has no legal basis in any treaty and which developed from earlier summit meetings. Its decisions have to be passed back to the Council of the EU to be given legal validity.

Since the Community is a partnership of the member states, it makes its decisions by negotiations in a series of committees. The annual review of agricultural prices is a good illustration of this process. The initiation of primary legislation is normally the responsibility of the Commission. The first moves are made by the appropriate department in the Agriculture directorate, usually in consultation with national civil servants, members of trade associations, and independent experts, all of whom may serve on expert working groups chaired by a member of the Commission staff. Other Commission departments with a legitimate interest in the proposals are consulted at an early stage, since agricultural policy measures might affect negotiations on, for example, external trade or monetary affairs. By the time the draft proposals are nearing completion the pressure groups make their views known to the Commission. There is a wide variety of these groups, from BEUC, the European Consumer organization, to the European Environmental Bureau (EEB), and various European trade associations such as the EDA (representing the dairy industry), CIAA (the food industry), and, perhaps most influential, COPA, the Committee of Professional Agricultural Organizations, which acts for farmers' unions in the Community. When the Agriculture directorate consider that the draft is complete, they submit it to a full meeting of the 20 Commissioners for their approval.

Once this approval is given, the draft becomes a Commission proposal and is submitted to the Council of the European Union and consultative bodes such as the European Parliament, the Economic and Social Committee and the relevant Management Committees (which are mostly concerned with secondary legislation – see below).

For most proposals, the Council, before it makes a decision, is required to receive the opinion of the *European Parliament*. The Parliament carries out this part of its work by nominating a committee which produces a report on the proposal which may then be debated by a plenary session of Parliament before becoming the Opinion which is passed on to the Council. In practice this opinion carries little weight and may often be ignored. The Parliament also has considerable indirect impact on the formulation of policy through its questioning of members of the Commission, both formally and informally.

Meetings of the Council are often complicated and lengthy, and so require detailed preparation. This preparation is the task of the *Committee of Permanent Representatives* (COREPER) for non-agricultural matters, and of the *Special Committee for Agriculture* (SCA) for agricultural legislation. Both of these committees consist of senior civil servants from the member states, meeting with members of the Commission. Detailed consideration of the proposals is carried out in working parties of national civil servants. The purpose of this procedure is to identify and, if possible, to resolve points of conflict. If it is possible to produce a draft proposal which is acceptable to all member states, it is returned to the Council of Ministers on the 'A' list, which can be passed by the Council with no further discussion. Otherwise it returns to the Council as a 'B' point, for further negotiation.

While this process of sorting out the agenda for the Council meeting is going on, the business of lobbying continues. Pressure groups, national government ministers and other politicians express their views on the Commission proposals, both at a European and at a national level. The President of the Council of Ministers is required to hold meetings with both COPA and, since 1980, BEUC, and to report on them to the rest of the Council. Nevertheless, most of the political pressure on individual ministers in the Council comes through national lobbying channels, so that national political problems, such as a forthcoming national or even local election, can have significant effects on the decisions made.

Thus any proposal, before a decision is taken on it in the Council, will have been the subject of comment by a wide variety of formal and informal, corporate and individual, Community and national, expert and lay sources. The issues that remain to be resolved by the Council should be the basic political ones. After debating a proposal, a decision must be made. Some legislation requires the unanimous approval of the Council, and a quasi-formal agreement, known as the 'Luxemburg Compromise', provides for a member state to exercise a veto if it believes that its 'vital national interests' are at stake. However, increasingly, and especially since the passage of the Single European Act, most decisions are taken on the basis of a *qualified majority*, which requires 62 or more votes in favour. Each country has the number of votes shown in Table 8.1 (although it should be noted that these numbers, and the number required for a qualified majority, will change after each member state has ratified the agreement made in Nice in December 2000).

Table 8.1 Votes in the Council of Ministers

Country	*Number of votes*
France, Germany, Italy, UK	10 each
Spain	8
Belgium, Greece, the Netherlands, Portugal	5 each
Austria, Sweden	4 each
Denmark, Ireland, Finland	3 each
Luxembourg	2

By this process the Council is said to 'adopt a common position'. Since the adoption of the Single European Act this is not the end of the decision-making process. The common position is then sent to the Parliament, which has 3 months to carry out a second reading and make its opinion known. It may:

(1) approve or take no decision, in which case the Council adopts the measure in question and it is effectively passed;
(2) reject the common position by an absolute majority, in which case the Council may maintain it and adopt it, but only if it can do so unanimously; or
(3) amend the common position by an absolute majority, in which case the Commission revises its proposals within 1 month and re-submits them to the Council, which may then:

- adopt the Commission proposal without change by a qualified majority;
- adopt the Commission proposal after amendment, but only if it can do so unanimously; or
- reject the proposal, in which case it lapses.

The proposals thus adopted fall into three categories. A *Regulation* is directly applicable to all people and governments in all member states. A *Directive* is binding as to its intention on governments, which must then pass legislation to give it effect, so that it is more flexible than a Regulation. A *Decision* is as binding and immediately applicable as a Regulation, but only on the people or governments to whom it is addressed.

Secondary legislation, which is concerned with the day-to-day running of the CAP (e.g. the administration of tenders to intervention stores, or setting the level of export refunds) is largely the responsibility of the Commission. In formulating its proposals, it may or may not take the advice of an advisory committee of representatives from all sides of the appropriate industry. The proposal is then considered by a *Management Committee* made up of national civil servants from the relevant division of the national ministries and Commission officials from the relevant Directorate General. There is a Management Committee for each of the main commodities, and others for agri-monetary affairs, plant health, structural policy and so on. The frequency of the meetings reflects the amount of work to be done: the beef committee may meet every fortnight, whereas the research committee may only meet a few times a year. The Committee gives its opinion on the Commission proposal using the same qualified majority system as the Council. After a proposal has passed through the Management Committee and been adopted by the Commission it becomes, in effect, law.

Community law is thus the primary and secondary legislation produced by the Council and the Commission, together with the law embodied in the Treaty of Rome and the various treaties of accession to the Community. It takes precedence over national law. It is the responsibility of the *Court of Justice* to decide whether or not Community legislation has been correctly applied, is being applied or is flawed. Some of the decisions of the Court have had a substantial effect on the way in which the Community is run, from deciding upon the powers of Parliament to defining what means may legitimately be used to prevent trade in food products.

The *Court of Auditors* exists to audit the expenditure of the Community and to ensure that its finances are properly managed. In this context it has published a number of reports critical of the operation of the CAP and has highlighted the problems of CAP fraud.

Formation and development of the CAP

The problems of the CAP

The CAP has proved a remarkably resilient policy. Its original objectives set out in the 1957 Treaty of Rome still remain (although see discussion of Agenda 2000 below) and the main mechanism of internal market support persisted largely unchanged until the 1990s. The main problems of the CAP also have a long history, notably the financially and politically costly surpluses and, later, widespread environmental problems. More recently, moves toward the liberalization of agricultural trade and the need to realign the CAP to more easily facilitate expansion of the Union can be added to the list of 'challenges' for the CAP in the twenty-first century. In order to understand these problems and the imperative for reform created by trade liberalization and EU expansion, it is necessary to understand the original purpose of the CAP and the main policy instruments employed.

The original objectives of the CAP as laid down in Article 39 of the Treaty of Rome are:

(1) to increase agricultural productivity by promoting technical progress and by ensuring the rational development of agricultural production and the optimum utilization of the factors of production, in particular labour;
(2) thus to ensure a fair standard of living for the agricultural community, in particular by increasing the individual earnings of persons engaged in agriculture;
(3) to stabilize markets;
(4) to ensure the availability of supplies;
(5) to ensure that supplies reach customers at reasonable prices.

Article 40 of the treaty lays down the broad guidelines for the various policy instruments by which these objectives are to be met. Detailed policy instruments developed later and it was not until 1968 that common prices were applied.

Outline of original price mechanisms

The underlying principle of the CAP price mechanism was that a producer should receive a price determined by market forces, but that these market forces were subject to control so that the market price was only allowed to fluctuate between predetermined upper and lower limits. Thus, the farmer was protected against excessively low prices and the consumer against excessively high prices.

This price support system recognized that farm products might be produced both within and outside the EU. As the demand for farm products is relatively constant, if markets were supplied only from within the EU the major reason for low prices would be oversupply. The CAP's *intervention* mechanism was designed to overcome this problem by purchasing farm products for storage when prices were low. If prices remained low, this produce could be sold on the world market with the aid of *export refunds*. The export refund acts to lower the cost of a product in order to make it competitive with other products outside the single market. The existence and use of this mechanism has caused most criticism of the CAP beyond the borders of the EU and it came as no surprise that export refunds were a key target in the GATT negotiations of the early 1990s (Fennell, 1997).

Many farm products could also be supplied from countries outside the EU (known as third countries). If these imports were available at less than the EU market price, that market price would be reduced. This effect could be mitigated by artificially increasing the price of products from third countries by applying variable levies. Conversely, when EU prices were high, perhaps as a result of shortages, the entry of produce from third countries would serve to reduce prices by increasing supply.

For most of its history, market control and internal price support have been relied upon to achieve the diverse objectives of the CAP although it is generally agreed that they have proved incapable of simultaneously satisfying all objectives (Fennell, 1997). In addition, the intervention mechanism has effectively insulated producers from the market, encouraging the production of products for which there is no genuine demand. Open-ended price support stimulated the long-term expansion of supply. Self sufficiency in wheat, for example, rose from 89% in the late 1950s to 101% in the mid-1960s (Fearne, 1991), while by the end of the 1970s milk production had increased by 1.7% annually and beef by 2% (Fennell, 1997). The result was persistent surpluses across a growing range of commodities. Although the precise meaning of 'surplus' can be debated, the European Commission is largely concerned with 'structural surpluses', i.e. long-term surpluses which arise from over-supply compared to demand rather than from accidental annual fluctuations due to weather conditions, etc. Even though surpluses had been in existence since the very inception of the CAP, a general unwillingness to take radical measures to deal with them (other than the ill-fated 'Mansholt Plan' – see below) resulted in major problems by the 1980s.

The problems and costs of surpluses stimulated through the operation of the CAP have received considerable public attention. In the UK surplus storage and handling costs increased nearly four-fold between 1980 and 1994 and, when surpluses peaked, in 1986 7 billion ECUs of the CAP budget was spent on handling and storage costs (Winter *et al.*, 1998). Some of the surpluses that could not be stored or converted into other products (such as industrial alcohol) were simply destroyed. Much of the surplus, however, was disposed of on world markets through the use of the EU export subsidy programme to deflate prices. Rather than dealing with the surplus problem internally, the heavy subsidization of exports essentially shifts the problems to 'third countries' (i.e. outside the EU), with impacts on food security within importing countries and on the markets of traditional food exporters. At the same time, the impact is also felt in the domestic budget as spiralling protectionism results in large parts of agricultural support throughout the world being mutually neutralizing. For example, it was estimated that in the early 1990s 25% of world spending on agricultural support merely offset the effects of other countries' policies, while in the US 40% of agricultural support in 1986/1987 offset the loss in profits deriving from low prices as a result of surplus dumping (Potter, 1998).

In addition to budgetary and surplus problems, during the 1980s, an environmental critique of the CAP developed, pointing to a direct link between the operation of the CAP which encouraged intensification and specialization and widespread environmental change in the countryside. It was argued that the operation of the CAP created conditions of confidence (through guaranteed prices and investment aid) in which farmers were encouraged to specialize, increase output and intensify production through new technology embodied in capital which was increasingly substituted for labour. Farmers were operating within 'a system which systematically establishes financial inducements to erode the countryside, offers no rewards to prevent market failure and increases the penalties imposed . . . on farmers who may want to farm in a way which enhances and enriches the rural environment' (Cheshire, 1985). High guaranteed prices were necessary if small and marginal family farms were to be maintained, but, since the level of support received by an individual farmer was largely determined by output, there was an incentive to bring uncultivated land into production (while at the same time larger farmers benefited more). High prices combined with reduced capital costs 'led to the logical choice of a specialised enterprise. . . . This specialisation, in turn, meant that hedges served little purpose, and the larger machinery that capital substitution and larger throughputs had induced, demanded larger fields' (Bowers and Cheshire, 1983).

The catalogue of post-war environmental change is now well rehearsed. Between 1949 and 1984 80% of chalk and limestone grassland, 40% of heaths, 50% of lowland wetlands and 30% of upland moors were either lost or significantly damaged (NCC, 1984), while by the mid-1980s, the area of species-rich, unimproved lowland grass was estimated to have declined by 97% over a 50-year period. Similar changes have been recorded throughout Europe.

Reforming the CAP

Early attempts to reform the CAP were typically piecemeal resulting from budgetary crises. During the 1980s a number of attempts were made to deal with the escalating surplus problem and associated budgetary issues. A ceiling on agricultural spending was introduced in 1984, whereby the increase in the farm budget was limited to 74% of the rate of the economic growth of the then EC. At the same time a stabilizer system was introduced, establishing a production limit beyond which support prices would fall. The crucial threshold, however, was set too high to have a significant impact on surpluses. Another important development in 1984 was the introduction of milk quotas. Milk quotas effectively introduced a super levy at the farm level on all milk delivered to dairies over the quota level. Although originally met with shock and dismay from the farming community, quotas soon became popular as production within quota continued to benefit from high prices and because quota was tradable and soon became a valuable capital asset. Towards the end of the decade, in 1988, a new and experimental approach was adopted to control production in the arable sector. A voluntary 5-year set-aside programme was established which provided compensatory payments to farmers for not cultivating arable land. In practice, uptake rates were low and the impact on total production consequently limited. Importantly though the voluntary set-aside experiment demonstrated that such an approach could have an impact on production levels and, significantly, if managed appropriately, set-aside land offered important environmental benefits (Potter, 1998; Winter *et al.*, 1998).

Despite these efforts surpluses were at a peak in 1986–1987, the budget was rising and EU agricultural support polices were coming under the spotlight of the GATT negotiations. At the same time it was clear that not only did much of the farm budget fail to reach farmers, but also that which did failed to satisfy concerns regarding the support of smaller and more marginal farms. In 1991 the European Commission published proposals for the reform of the CAP including the now famous 80 : 20 formulation: that is, that 80% of farm spending was going to only 20% of farmers and that these tended to be the larger, more efficient producers. Despite budgetary pressures and moves towards greater liberalization, the Commission also clearly outlined a major unwritten objective of the CAP by stating that:

> 'sufficient numbers of farmers must be kept on the land. There is no other way to preserve the natural environment, traditional landscapes and a model of agriculture based on the family farm as favoured by society generally' (CEC, 1991).

The original reform proposals were put forward in a document entitled *The Development and Future of the CAP* (CEC, 1991), also known as the MacSharry proposals after Ray MacSharry, the agriculture commissioner at the time. The reforms proposed by MacSharry explicitly stemmed from the failure of earlier measures to tackle the problem of surplus production, the encouragement of intensification and environmental damage through the coupling of farm support to output and, importantly, the perceived need to bring about a more equitable distribution of funds.

The distributional issue was particularly important and led to the promotion of 'modulation' (positive discrimination in favour of smaller farmers who would receive a greater degree of compensation). The reforms also had wider implications; in addition to redressing the distribution of agricultural support, the proposals reaffirmed the role of farmers as joint producers of food and managers of the environment. This role was to be enhanced explicitly through a series of agri-environmental schemes and was implicitly strengthened by a continued commitment to keeping farmers on the land – particularly the smaller family farmers who had been disadvantaged by the operation of earlier price support policies.

To summarize, the main objectives of the proposals were to:

- develop a competitive and efficient agriculture as part of a broader international economy;
- maintain a 'sufficient' number of farmers on the land;
- avoid/reduce surpluses and so prevent budgetary 'excesses';
- recognize, enhance and reward the dual role of farmers as food producers and countryside managers;
- redirect support in favour of those in greatest need.

Almost inevitably the original proposals were watered down in the final agreement, notably in terms of the removal of the redistributive modulation proposal. Nevertheless, for the first time sharp price reductions were achieved combined with the partial decoupling of support from output. The individual measures contained within the 1992 reforms are particularly complex and a detailed description and analysis is beyond the scope of this chapter.

Briefly, the final 1992 MacSharry reform package was based on:

- substantial cuts in support prices for certain commodities (for example, a 35% cut in the intervention price for grain and 15% for beef);
- the introduction of direct producer payments to partially compensate for the cut in institutional prices (through arable aid payments and headage payments in the livestock sector);
- the introduction of an annual set-aside scheme, participation in which was a condition of receipt of arable aid payments;
- a number of accompanying measures designed to facilitate early retirement from agriculture, the afforestation of agricultural land and, importantly, the improved environmental management of farmed land.

Despite the radical change in policy encompassed in the 1992 reforms it soon became apparent that this was simply a stage (albeit significant) in a much larger and longer reform process. The 1994 Uruguay Round Agreement (URA) of the GATT negotiations committed signatories to further reform in the future, particularly related to subsidies linked to production. Although the 1992 reforms had gone some way to making the CAP GATT-compatible, in 1995 the European Commission itself acknowledged that the CAP was still overly protectionist, bureaucratic and that support was insufficiently decoupled from production to meet the likely future demands of the WTO (the successor to GATT). In addition, if countries from central and eastern Europe were to be accommodated within the EU, further reform of the CAP was necessary. Following this analysis contained in a 1995 Agricultural Strategy paper (CEC, 1995) and the 1996 European Conference on Rural Development (popularly referred to as the Cork Conference), the Commission began to outline a 'third way' which emphasized the 'European model of agriculture' and its attendant support needs.

In his opening address to the Cork conference, agriculture commissioner Franz Fischler argued that 'rural society is a socio-economic model in its own right which must be preserved in the interests of European society as a whole' (Fischler, 1996). The Declaration that emerged from the Conference signalled broad policy principles, recognizing that the justification of the 1992 reform payments was increasingly being thrown into question. In their place

> 'the concept of public financial support for rural development, harmonised with the appropriate management of natural resources and the maintenance and enhancement of biodiversity and cultural landscapes, is increasingly gaining acceptance' (CEC, 1996).

The principles set out for the EU's future rural development policies in the Conference declaration reflect a continuing concern with the management and development of rural economy and society, albeit encompassing a shift away from an agriculturally centred notion of rural land management based on a sectoral policy towards a more territorially based rural policy. Thus, it was argued that a growing share of resources should be devoted to rural development and environmental objectives, and that an integrated policy framework be established encompassing 'agricultural adjustment and development, economic diversification, the enhancement of environmental functions and the promotion of culture, tourism and recreation' (CEC, 1996). Despite the clear intention of retaining rural-based EU spending, the declaration provoked criticism from those who saw a Cork-style policy as a threat to traditional agricultural policy entitlements (Lowe *et al.*, 1996).

Nevertheless, when, in 1997, the European Commission published its Agenda 2000 reform proposals, rural development was introduced as a 'second pillar of the CAP' while further price cuts were designed to preempt the next round of trade negotiations. In Agenda 2000 the Commission set out new objectives for the reformed CAP and in doing so began to outline the essential characteristics of 'the European model of agriculture' and its associated support policy:

- a *competitive* agricultural sector capable of exploiting the opportunities existing in world markets without excessive subsidy;
- *safe* production methods capable of supplying *quality products* that meet consumer demand;
- a *diverse agriculture*, reflecting the rich tradition of European food production;
- maintenance of *vibrant rural communities*;
- an agricultural sector *sustainable* in environmental terms which contributes to the preservation of natural resources and the natural heritage and maintains the visual amenity of the countryside;
- a *simpler*, *more comprehensible* policy which establishes clear dividing lines between the decisions that have to be taken jointly at Community level and those that should remain in the hands of the member states;
- an agricultural policy that establishes a clear connection between public support and the range of services that society as a whole receives from the farming community.

The new policy is based on a different model to the EU's major competitors and according to the European Commission a key difference can be found in the concept of

> 'the multifunctional nature of Europe's agriculture and the part it plays in the economy and the environment, in society and in preserving the landscape, whence the need to maintain farming throughout Europe and to safeguard farmers' incomes' (CEC, 1998).

Agenda 2000 displays a considerable degree of continuity with the approach adopted under the MacSharry reforms. Price cuts of between 15 and 20% for cereals, beef and milk producers are designed to improve competitiveness and are being phased in alongside increased direct aid payments designed to contribute towards a fair standard of living for farmers. Direct area-based payments for cereals, set-aside and oilseeds are all paid at the same rate, although protein crops receive higher payments. The economic impacts of these changes still remain unclear although MAFF's own forecasts suggest aggregate losses across the UK of £150 million a year (equivalent to 1% of gross output). This masks great variability between sectors. For example, the arable sector faces a 7% cut in producer returns, compared to 2% in the dairy sector. Beef producers on the other hand are, if anything, over-compensated and MAFF forecasts suggest a 5% rise in producer returns between 2000 and 2008 (MAFF, 1999).

The market reforms are also intended to contribute towards environmental enhancement which has potentially been strengthened through the option of member states to make direct compensation payments conditional on the observance of environmental requirements (cross-compliance). Non-compliance would result in a proportionate reduction or cancellation of payments which could then be redirected to finance agri-environmental or rural development measures. National discretion over a proportion of direct payments to the beef sector (the so-called national envelope) could be used to support extensive grazing, while the ability to modulate direct payments has been employed to boost agri-environmental spending. In England, modulation means that cuts in direct payments (of initially 2.5%, rising to 4.5%) will release funds which can be diverted into agri-environmental and rural development spending. In addition, under the Rural Development Regulation (RDR) all member states are required to implement agri-environment measures (see below). Despite these developments, however, spending on agri-environment and rural development measures remains a minor element of the CAP budget with most spending still devoted to market management and apparently open-ended compensation payments. In terms of the EU share of the budget, spending on rural development (which encompasses agri-environment schemes) is effectively frozen at approximately 10% of the budget until 2006 (see Fig. 8.1). The implications of these reforms in practice are explained in the following section.

The CAP in practice

Price mechanisms

The underlying principle established when the CAP came into being in the 1960s was that producers should receive a price determined by market forces, but that those market forces should be controlled so that prices fluctuated only between predetermined upper and lower limits. Thus the farmer was protected against excessively low and the consumer against excessively high prices. This principle remains in force, but now operates in association with the concept that producers should receive part of their income from compensation payments.

The most basic of all CAP support schemes (or regimes as they are often called) is that for cereals. The regimes for other commodities may be seen as variations of the cereals regime, the legal basis of which is set out in Regulation 1251/99.

An intervention price for cereals, proposed by the Commission and decided upon by the Council of Ministers, is established for a period of years. For example, the Berlin agreement of 1999 decided on the cereal intervention price until 2006. In principle, therefore, once the market price falls below the intervention price it will pay producers or traders to sell to intervention stores. In practice the system is more complex, because minimum quantities and quality standards must be observed, intervention stores are only open from November to May, and there are monthly increments in the prices paid. The prices are fixed in euros, so they can also change from day to day in countries with non-EMU currencies.

The second major component of the cereals regime is import control. Since world prices are often lower than EU prices it would be worthwhile, in the absence of any control, to import simply for sale into intervention. In practice, imports are controlled by the issue of import licences, and to obtain such a licence import duty

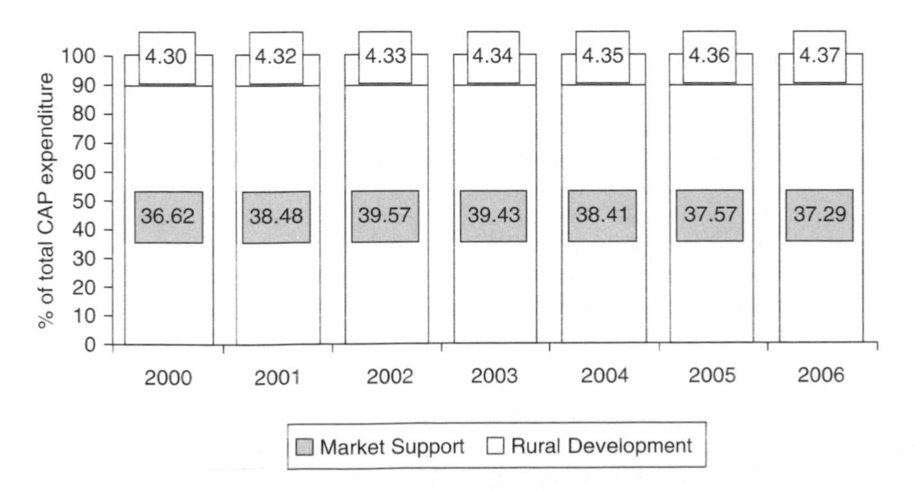

Figure 8.1 CAP spending 2000–2006 (billion euros, 1999 prices). (Adapted from CEC, 2000)

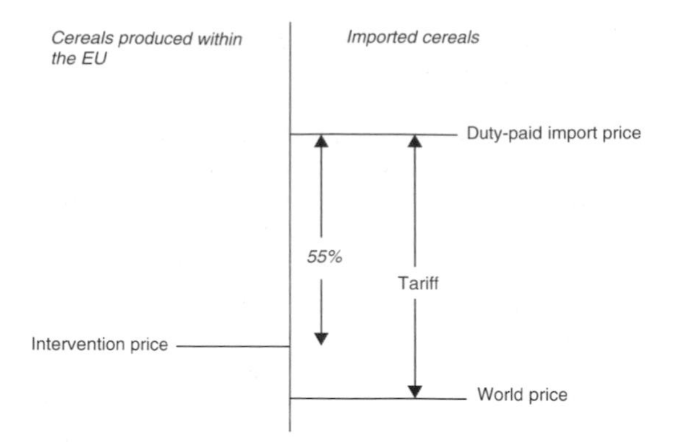

Figure 8.2 The cereal price mechanism.

must be paid (see Fig. 8.2). The level of duty is determined by the difference between the duty-paid import price and the world price. Again, complexities arise at this point. The duty-paid import price is fixed by international negotiation: at the time of writing it was the intervention price multiplied by 1.55, but future WTO negotiations could result in its variation. 'World price' is an imprecise term, and in practice the EU commission has identified several reference cereals which are used to specify prices for import duty calculations. Variations in insurance and freight costs and currency exchange rates must also be taken into account [for details see the CAP Monitor: Agra Europe (2001), pp. 10A/13–14].

Licences are also required for cereal exports, and the Commission uses export licences, and the refunds to which they entitle traders, to stabilize EU markets. Since EU prices are normally higher than world prices, EU traders will only find it worthwhile to export to non-EU countries with the aid of these export refunds. The Commission also has to take into account the commitment, under the Uruguay Round Agreement, to reduce both the volume of subsidized exports and the budgetary expenditure on export subsidies. In the unusual event of world prices being higher than EU prices the Commission may impose export levies.

Compensatory payments are the final component of the support regime. The logic of these is that they are paid in compensation for the lower intervention payment applying after the CAP reform packages of 1992 and 1999. The level of payment is determined in euros per tonne, and this figure is then multiplied by the historical average yield (for the period 1986–1990) for each region to give an arable area payment per hectare. Each member state decides upon the number of regions for compensation payments. In the UK there are seven regions (England, Scotland, Wales and Northern Ireland each divided into LFA and non-LFA areas) whereas in France there are 90 regions. Regional average yields are also decided by the member state, subject to the approval of the Commission.

Farmers who produce more than 92 tonnes (equivalent to about 15.5 ha at English regional average yields) of arable crops are only eligible for compensation payments if they *set aside* a proportion of their arable land. The base level of set aside is 10%, but this can be varied annually by the Council of Ministers. A simplified scheme operates for farmers growing less than 92 tonnes which does not require them to set any land aside. Some non-food crops, such as those intended for biomass or biofuels, are permitted on set-aside land.

The support regimes for other arable crops, such as oilseeds (oilseed rape, sunflowers and soya beans), protein crops (peas, broad beans, horse beans and lupins), linseed and sugar beet also include a set-aside requirement. Compensatory payments for oilseeds and flax are set at the same level as those for cereals, although the rate for protein crops is slightly higher. Fruit and vegetable prices are largely supported by aids to producers' organizations.

The basic support principles applying to arable crops are used for livestock and livestock products too: the farmer receives a return from the market, in which prices are controlled to some extent, and also receives compensatory payments, for some products at least. The most complex of these regimes applies to dairy products, prices of which are supported by intervention arrangements for butter and skimmed milk powder. Individual producers are affected by milk quotas, and from the 2005/2006 marketing year there will also be direct payments to producers in the form of dairy premiums, and in some cases, additional payments. Beef and sheepmeat producers are mainly supported by headage payments, although there are provisions for intervention buying (in the case of beef) or payments for the storage of sheepmeat in private cold stores. Pigmeat prices are occasionally supported by aids to private storage, but in practice pigmeat, like poultrymeat and eggs, is an unsupported product. Some idea of the relative importance of the various regimes may be obtained from Table 8.2, which shows the allocation of guarantee section expenditure in 1999.

Monetary arrangements in the CAP

Institutional prices in the CAP, such as intervention prices, together with headage payments, arable area payments, and so on, are denominated in euros. With the introduction of Economic and Monetary Union

Table 8.2 Allocation of guarantee section expenditure to various regimes in 1999. (Source: CEC, 2000, p. 24)

	Per cent
Arable crops	45.7
Fruit and vegetables	3.7
Sugar	5.4
Dairy products	6.4
Beef	11.7
Sheepmeat	3.6
Other products and measures	23.5
Total (€39.2 bn)	100.0

(EMU), 11 member states adopted the euro as their currency on 1 January 1999, and Greece joined the euro zone at the beginning of 2001. For these countries, therefore, there is no need for any special monetary measures for CAP purposes. For the non-participating countries – the UK, Denmark and Sweden – a system is needed to convert monetary amounts denominated in euros into national currencies.

For non-euro currencies, the value of prices, direct aids, export refunds and so on changes according to (1) the exchange rate published each day by the European Central Bank (ECB), and (2) the date of the relevant 'operative event'. The nature of the operative event depends upon the payment concerned. For intervention payments, it is defined as the beginning of the delivery or the acceptance of the tender; for export refunds or import tariffs it is the acceptance of the customs declaration; for arable area payments it is the beginning of the marketing year (1 July in the case of cereals); and for livestock headage payments it is also the beginning of the marketing year, which starts on various days at the beginning of January. The exchange rate used for any of these payments is the most recently published ECB rate prior to the operative event: i.e. for an event taking place on Friday, Thursday's rate will be used, but events falling on Saturday, Sunday or Monday will use the previous Friday's exchange rate.

Exchange rate fluctuations will therefore affect intervention prices and, perhaps most importantly for most farmers, arable area payments and livestock headage payments. If a national currency strengthens against the euro, such payments would decrease: suppose that €1 is worth 65p; if sterling then increases in value so that €1 = 60p, clearly the sterling value of a headage payment would decrease. Equally, weakening sterling would increase the value of such payments. There are provisions for compensating producers for currency revaluations in some circumstances, but the EU only contributes to half of the amount, the rest being paid by the national treasury.

Environmental and structural policy under the CAP

When the architects of the CAP were drawing up its objectives and policy instruments there was no explicit mention of the environment, although, if pressed, most would have argued that maintaining family farming would itself maintain the 'physical fabric' of the countryside. The earliest explicit mention can be found in the 1975 Less favoured Areas (LFA) Directive. The directive was designed to ensure the continuation of farming in order to maintain a minimum population level or conserve the countryside. Despite these objectives, the main emphasis of LFA policy has been to support farm incomes rather than the farmed environment (indeed, the operation of LFA policy became a major target of the environmental critique of the CAP).

It was not until 1985 that a UK proposal designed to allow support to be given to environmentally sensitive farming practices in certain areas was incorporated into European legislation (Article 19 of Regulation 797/85 on Agricultural Structures), although co-financing was not forthcoming until 1987. Specifically, Article 19 provided support 'in order to contribute towards the introduction or continued use of agricultural production practices compatible with the requirements of conserving the natural habitat and ensuring an adequate income for farmers' (CEC, 1985). In the UK following the 1986 Agriculture Act a series of Environmentally Sensitive Areas (ESAs) were designated in areas of national environmental importance where the continuation of certain farming practices was necessary for landscape and environmental reasons.

With the exception of the Netherlands and Denmark (and later France) other member states were slow to take up the provisions of Article 19 which was seen to reflect a narrow British preoccupation with landscape (Baldock and Lowe, 1996) and, despite the provision of co-financing, by 1990 few member states had made provisions under Article 19 although some 60 000 farmers were receiving EU premiums. This limited uptake contributed toward pressure for further agri-environmental reforms and in 1988 the European Commission produced two important discussion papers, *The Future of Rural Society* and *Environment and Agriculture*. The latter argued that 'society has to accept the fact that the farmer, as manager of the environment, is rendering a public service which merits an adequate remuneration' (CEC, 1988). The mould was set for more wide-ranging, multifaceted agri-environmental polices to assist in the maintenance of agricultural incomes, improve market balance (i.e. contribute to the reduction of surpluses) and to contribute to the maintenance and conservation of the farmed countryside. Environmental concern had become an important justification for agricultural support.

The 1992 agri-environment regulation

The 1992 CAP reforms may have heralded the most wide-ranging changes to the system of market support in the history of the CAP, but they also introduced an ambitious programme of agri-environmental reforms. The main environmental reforms were contained in one of three Accompanying Measures. Regulation 2078 on 'agricultural production methods compatible with the requirements of the protection of the environment and maintenance of the countryside' was designed to complement the extensification expected to follow from the market reforms and 'encourage farmers to serve society as a whole' by providing incentives to introduce or maintain farming practices that contributed to the maintenance of natural resources and the landscape. Significantly, aid was also to be provided to encourage farmers to reduce agricultural pollution, although this was only supposed to be available for schemes that went beyond good agricultural practice (Scheele, 1996). Regulation 2078 was compulsory to all member states and was designed to overcome some of the shortcomings of previous legalization which was seen as limited in its scope and had received only limited application, being confined mostly to the northern member states where the idea had developed. The regulation aimed to contribute to the Community's policy on agriculture and the environment and to 'contribute to providing an adequate income for farmers'. Overall, the emphasis of the Regulation was on stimulating and maintaining extensification, activities covered by Article 2 (a), (b) and (c) and often forming part of schemes established under 2(d) (see Table 8.3).

Table 8.3 Article 2 of EU Regulation 2078/92

Article 2 provides funding for schemes which encourage farmers to:

a.	reduce substantially their use of fertilisers and/or plant protection products, or keep to reductions already made, or introduce or continue with organic farming methods.
b.	change, by means other than those referred to in (a) to more extensive forms of crop, including forage production, or maintain extensive production methods introduced in the past, or convert arable land to extensive grassland.
c.	reduce the proportion of livestock per forage area.
d.	use other farming practices compatible with the requirements of protection of the environment and natural resources, as well as maintenance of the countryside and landscape, or to rear local breeds in danger of extinction.
e.	ensure the upkeep of abandoned farmland or woodland.
f.	set aside farmland for at least 20 years for purposes connected with the environment, in particular the establishment of biotope reserves or natural parks for the protection of hydrological systems.
g.	manage land for public access and leisure.

The regulation can be seen as an exercise in subsidiarity and has consequently been implemented in strikingly different ways in different member states (see Potter, 1998; Buller *et al.*, 2000). The acceptability of this approach to farmers can be seen in the rapid expansion in the area enrolled. By 1999 an estimated 20% of the agricultural area of the EU was enrolled in a scheme operated under Regulation 2078.

In England MAFF operated a range of schemes under the regulation, involving an expansion of existing initiatives (such as ESAs and the later Countryside Stewardship Scheme – CSS) alongside the introduction of new schemes such as the Habitat scheme, the Moorland scheme and the Organic Aid scheme. Different combinations of schemes operated across the UK with some restricted to particular areas such as Tir Cymen (and later Tir Gofal) in Wales and the Countryside Premium scheme in Scotland. Despite the expansion in the number of schemes, overall spending on the agri-environment measures remained low in comparison to mainstream CAP spending and in the UK the bulk of spending was devoted to just two schemes – ESAs and CSS. Nevertheless, ESAs have contributed to the maintenance of landscapes in designated areas, while the CSS has been more successful in bringing about positive environmental change and stimulating employment.

Structural policy

Although market support has dominated and continues to dominate CAP spending (albeit now in a largely different form) a parallel range of instruments known collectively as structural policy have been designed to speed up agricultural restructuring, not just in terms of the number of farms but also in terms of investment, marketing, etc. An early, and infamous, attempt to develop a radical approach to structural policy was presented in a Memorandum on the Reform of Agriculture in the EEC (CEC, 1969), more popularly referred to as the Mansholt Plan. In its essentials, the Plan envisaged a leaner European agriculture with a smaller number of larger production units created through the removal of 5 million of the Community's small and marginal farmers. The process of restructuring would be based on a series of 'carrots' and 'sticks' – the former in the shape of grants, pensions for farmers aged over 55 and assistance for younger farmers in finding alternative employment, while the latter took the form of a proposal for price reductions sufficient to force small marginal producers from the land (CEC, 1969). It was envisaged that around half of those leaving would retire and the other half take up employment in industry, the land released being used to create larger 'modern agricultural enterprises' and to reduce the agricultural area of the Community by some 15 million ha through afforestation and the creation of reserves and recreational parks.

There was sympathy with the desire to improve the structure of agriculture, although the proposal to 'force' small and marginal farmers from the land was greeted with hostility. In the event Mansholt's specific proposals were not acted on and a series of structural measures implemented in 1972 represented a watered down version of the 'Mansholtian' vision. Rather than a 'new agriculture', a policy was now presented which put the emphasis on the modernization of the existing community of farmers, reinforcing the political and symbolic importance of maintaining a large number of farmers on the land. Under the emergent farm structural policy, Directive 72/159 provided assistance for investment on farms that could be shown to be capable of reaching viability in a short period of time, while Directive 72/160 introduced a system of payments to those wishing to leave the industry, the idea being that land released could be used to achieve restructuring and viability in conjunction with Directive 72/159. There were few applicants for the outgoers scheme and little land released went to farmers receiving aid under the modernization plan (due to differences in the spatial pattern of uptake of the two schemes).

Early structural policies were criticized on the grounds that they concentrated too much on agricultural investment in northern member states. Subsequently, therefore, attempts were made to divert more of the funds available to Mediterranean countries, and to recognize that development in rural areas might involve other industries as well as agriculture. In the 1980s, for example, there was an 'integrated programme' for the Western Isles of Scotland, EU structural funds supported an integrated rural development programme in England's Peak District and there were also integrated Mediterranean programmes. Following the 1988 reform of structural funds the programming approach was developed further by combining spending from the European Regional Development Fund and the European Social Fund with FEOGA guidance section funding targeted at five key objectives, only some of which were relevant to rural areas.

Objective 1 regions were those in which income per capita was less than 75% of the EU average and included the whole of Greece, Portugal and Eire, and parts of other countries. Within these regions, EU funds paid up to 75% of the cost of 'Operational Programmes', involving, amongst other things, rural infrastructure and irrigation schemes.

Objective 5 was concerned with promoting rural development and was subdivided into: *5a* for speeding up the adjustment of agricultural structures in the framework of the reform of the CAP; and *5b* for facilitating the development and structural adjustment of rural areas. The significance of Objective 5 is that it meant that CAP reform was no longer just considered an end in itself but was also a means of promoting the development of rural areas as a whole (Fennell, 1997). Objective 5a included a range of measures to promote the marketing and processing of agricultural produce, improve efficiency, support co-operatives, farm diversification and support the income of hill farmers. Aid was provided through the Farm and Conservation Grant Scheme and the Processing and Marketing Grant Scheme although these were phased out to release domestic funds for 5b. The result was that the bulk of support under 5a was channelled through LFAs (Lowe *et al.*, 1998).

Objective 5b areas in the UK were areas of low population density, sensitive to developments in agriculture, where the environment was also under pressure. A number also included parts with LFA status. The primary objective was job creation and economic regeneration pursued through agricultural and rural diversification. New rules in operation from 1994 also introduced an environmental component.

The Rural Development Regulation

Under Agenda 2000 the European Commission declared its aim to establish rural development as the 'second pillar of the CAP'. This 'second pillar' is mainly linked to the new Rural Development Regulation (RDR) and also the reorganization of structural policy to concentrate on three new objectives (see below).

The new policy for rural development aims to establish a coherent and sustainable framework for the future of Europe's rural areas and to complement the reforms introduced into the market sectors by promoting a competitive, multi-functional agricultural sector in the context of an integrated strategy for rural development. The regulation is highly discretionary, strengthening subsidiarity and promoting flexibility through a 'menu' of actions to be targeted and implemented according to member states' specific needs. The RDR has three main objectives:

(1) to create a stronger agricultural and forestry sector, the latter recognized for the first time as an integral part of the rural development policy;
(2) to improve the competitiveness of rural areas;
(3) to maintain the environment and preserve Europe's rural heritage.

Agri-environmental measures are the only compulsory element of the RDR. In principle this represents a decisive step towards the recognition of the role of agriculture in preserving and improving Europe's natural heritage. That said, the RDR is largely based on a repackaging of previous measures under the old Objective 5 and the Accompanying Measures of the 1992 CAP reforms and still only attracts a small proportion of overall CAP spending (see Fig. 8.1).

In Britain, the RDR is implemented through a series of Rural Development Plans (one each for England, Scotland and Wales). Each plan contains an overview of the measures to be adopted plus an analysis of needs and policy responses at a more local level. Under the England Rural Development Plan (ERDP) the regional element is based upon the English Government Office regions. The ERDP is based on a significant expansion (and consolidation) of existing agri-environmental schemes, a reorientation of LFA support and the introduction of new schemes supporting training, diversification and rural development. The agri-environmental schemes expanding or continuing under the ERDR are the:

- Countryside Stewardship Scheme
- Organic Farming Scheme
- Woodland Grant Scheme
- Farm Woodland Premium Scheme
- Environmentally Sensitive Areas Scheme

Together these schemes will absorb over 70% of spending under the ERDP between 2000 and 2007 (see Table 8.4).

The system of support for farmers in LFAs, long a target of environmentalists, is significantly realigned under the new Plan. The new Hill Farm Allowance (HFA) scheme is designed to:

- contribute to the maintenance of the social fabric in upland communities through support for the continued agricultural use of land;
- help to preserve the farmed upland environment by ensuring that land in LFAs is managed sustainably.

Under the HFA, support for hill farmers is explicitly justified as a social and environmental measure. Unlike the old HLCA scheme which offered livestock headage payments, aid under the HFA is paid on an area basis,

Table 8.4 England Rural Development Plan indicative expenditure. (Source: MAFF data)

Scheme	*Seven-year total*[1] *(£ million)*	*Percentage of total budget*	*Percentage increase in annual spending by 2006/2007 compared with 1999/2000*
Agri-environment	1052	64.2	126
Of which:			
Environmentally Sensitive Areas Scheme	335	20.5	19
Countryside Stewardship Scheme	576	35.2	260
Organic Farming Scheme	141	8.6	99
Woodland Grant Scheme[2]	22	1.3	—
Farm Woodland Premium Scheme	77	4.7	87
Hill Farm Allowances[3]	239	14.6	—
Processing and Marketing Grant	44	2.7	New
Energy Crops Scheme	31	1.9	New
Rural Enterprise Scheme	152	9.3	New
Vocational Training Scheme	22	1.3	New
Total	1638	100.0	105

[1] Expenditure shown is on a financial year basis (2000/2001–2006/2007) and therefore differs from that in the ERDP which is on a FEOGA year basis (16 October 1999–15 October 2006).
[2] Excludes baseline expenditure of £117 million.
[3] Excludes expenditure in 2000 on HLCAs.

differentiated according to the type of land and 'modulated' according to the area claimed for. Claims for the first 350 ha attract the full payment rate, 50% payments are made on land between 350 and 700 ha and no payments are made above 700 ha. Top-up payments of 10 or 20% are available to farmers meeting additional environmental criteria.

In addition to incentives for environmentally friendly land management, new schemes have been initiated aimed at enabling the farming and forestry sector (and to some extent other rural businesses) to adapt to the changing rural economy. The *Rural Enterprise Scheme* (RES) is the most substantial in terms of its budget share and is designed to support the development of sustainable and diversified rural economies and communities. The scheme is regionally managed and although a wide range of different categories of project can be funded, each application is assessed on a competitive basis against regional priorities. A sliding scale of grant aid is offered depending on the commercial return expected. Projects with little or no commercial return but which offer community or environmental benefits attract the highest rate of grant aid. The development of processing facilities is assisted through the *Processing and Marketing Grant* (PMG) aimed at projects with eligible costs in excess of £70 000. As part of the rural development programme, DEFRA and the Forestry Commission jointly run an *Energy Crops Scheme* (ECS) providing establishment grants for short-rotation coppice and miscanthus (elephant grass). Investment in human capital is provided through a *Vocational Training Scheme* (VTS) which is again managed on a regional basis and offers grants of up to 75% for training in areas such as information technology, processing and marketing, and conservation land management. (Further details on all these schemes are available from DEFRA.)

New arrangements for structural funds

Although the 'second pillar' of the CAP is mostly pursued through the RDR, rural development in a broader sense is facilitated through structural policy which has been reorganized following Agenda 2000 to concentrate on just three 'Objectives':

- *Objective I* is essentially the same as previously (areas with less than 75% of the average EU GDP), but now also includes special measures targeted at the EU's sparsely populated arctic regions.
- *Objective II* combines former Objective 5b areas with former Objective 2 areas (which were declining industrial areas) which meet new Objective II criteria. Support is provided for economic diversification, training, environmental protection and improvement, etc.
- *Objective III* is a 'horizontal' measure, that is, it applies to all non-Objective I and II areas and supports training and education.

Problems and possibilities of the CAP

It would be a mistake to see the Agenda 2000 package as the end of the process of CAP reform; rather the constant flux of external influences and internal circumstances produces continuing pressures to change.

The internal circumstances may be divided into three main groups: those affecting the economic environment; those affecting the political environment; and the evolution of existing policy. Changes in the economic environment most obviously include short term market developments producing shortages or, more usually, surpluses. As there appears to be little evidence for the cessation of technical change, the underlying trend to produce more persists, even in the face of the tendency for farm size to increase as the older generation of farmers is only partially replaced by the younger. The people, and sometimes their houses, leave the industry, but the land remains, incorporated into neighbouring farms. At the same time, the rest of the food chain continues to evolve. A food processing and retailing industry increasingly dominated by large firms may be expected to exercise its market power whenever possible; equally, society as a whole may require the whole food chain to respond to food safety concerns. All these factors therefore affect the economic environment within which the agricultural industry works. Similarly, changing public attitudes to the environment and rural society impact upon the political environment, which is also likely to be affected by differing views of monetary union and EU enlargement in the first decade of the century. These factors would be likely to modify even a CAP that was perfectly in tune with policy requirements at the beginning of the decade, and few would claim that it was so, since there are still unreformed policy sectors (most notably sugar), market measures and direct compensation still account for the bulk of CAP expenditure, and the policy as a whole remains largely agricultural when there is an increasingly shared perception that the problems it addresses should be seen as more broadly rural.

The external influences on the CAP are partly economic and partly a result of international politics. The economic issues arise from world market fluctuations: high world prices decrease the costs of export refunds and reduce pressures for third-country access to EU markets. Low prices, which have historically been more common, have the opposite effect. More predictable is the impact of the Millennium Round negotiations of the World Trade Organization. Insofar as its payments create incentives to produce, and EU views of the desirability of multifunctional agricultural policies or new biotechnology products are controversial, there will be pressure for changes in the CAP. Thus Agenda 2000 may be seen as only one stage in a more or less continuous process of reacting to long- and short-term, internal and external pressures for change in the CAP.

References

Baldock, D. & Lowe, P. (1996) The development of european agri-environmental policy. In: *The European Environment and CAP Reform: Policies and Prospects for Conservation* (ed M. Whitby). CAB International, Wallingford, pp. 8–25.

Bowers, J. & Cheshire, P. (1983) *Agriculture, the Countryside and Land Use*. Methuen, London.

Buller, H. *et al.* (2000) *Agri-environmental Policy in the European Union*. Ashgate, Aldershot.

CEC (1969) *Memorandum on the Reform of Agriculture in the EEC*. Supplement to Bulletin No. 1. Office for Official Publications of the European Communities, Luxembourg.

CEC (1985) *Council Regulation (No. 797/85) on Improving the Efficiency of Agricultural Structures*. Official Journal of the European Communities, Office for Official Publications of the European Communities, Luxembourg.

CEC (1988) *Environment and Agriculture*. Office for Official Publications of the European Communities, Luxembourg.

CEC (1991) *The Development and Future of the CAP*. COM(91)100. Office for Official Publications of the European Communities, Luxembourg.

CEC (1995) *Study on Alternative Strategies for the Development of Relations in the Field of Agriculture Between the EU and the Associated Countries with a View to the Future Accession of these Countries*. COM (95) 607. Office for Official Publications of the European Communities, Luxembourg.

CEC (1996) *The Cork Declaration: A Living Countryside*. Office for Official Publications of the European Communities, Luxembourg.

CEC (1998) *Explantory Memorandum on the Future for European Agriculture*. Office for Official Publications of the European Communities, Luxembourg.

CEC (2000) *The CAP: 1999 Review*. Office for Official Publications of the European Communities, Luxembourg.

Cheshire, P. (1985) The environmental implications of European agricultural support policies. In: *Can the CAP Fit the Environment?* (eds D. Baldock & D. Conder). Institute for European Environmental Policy/Council for the Protection of Rural England/World Wide Fund for Nature, London, pp. 9–18.

Fearne, A. (1991) The history and development of the CAP 1945–85. In: *The CAP and the World Economy: Essays in Honour of John Ashtan* (eds C. Ritson & R.D. Harvey). CAB, Wallingford.

Fennell, R. (1997) *The Common Agricultural Policy: Continuity and Change*. Clarendon Press, Oxford.

Fischler, F. (1996) Europe and its rural areas in the year 2000: integrated rural development as a challenge for policy making. Opening speech to *European Conference on Rural Development*, 7 November, Cork.

Lowe, P., Rutherford, A. & Baldock, D. (1996) Implications of the Cork Declaration. *Ecos*, **17** (3/4), 42–5.

Lowe, P. *et al.* (1998) *CAP and the Environment in the UK*. Centre for Rural Economy, Newcastle.

MAFF (1999) *Restructuring the Agricultural Industry*. Working Paper 1, Economics and Statistics Group, Ministry of Agriculture, Fisheries and Food, London.

NCC (1984) *A Nature Conservation Strategy for Great Britain*. NCC, Peterborough.

Potter, C. (1998) *Against the Grain: The Environmental Reform of Agricultural Policy in the United States and European Union*. CAB International, Wallingford.

Scheele, M. (1996) The Agri-Environmental Measure in the context of CAP reform. In: *The European Environment and CAP Reform: Policies and Prospects for Conservation* (ed M. Whitby). CAB International, Wallingford, pp. 3–7.

Winter, M., Gaskell, P., Gasson, R. & Short, C. (1998) *The Effects of the 1992 Reform of the Common Agricultural Policy on the Countryside of Great Britain*. Rural Research Monograph Series No. 4, Cheltenham and Gloucester College of Higher Education.

Further reading

Agra Europe, *The CAP Monitor* (continuously updated).

Agra Europe (weekly).

Brassley, P. (1997) *Agricultural Economics and the CAP: An Introduction*. Blackwell Science, Oxford.

Ingersent, K.A. & Rayner. A.J. (1999) *Agricultural Policy in Western Europe*. Edward Elgar, Northampton, MA.

Ritson, C. & Harvey, D.R. (1998) *The Common Agricultural Policy*, 2nd edn. CAB International, Wallingford.

Useful websites

europa.eu.int/comm/dg06/index_en.htm – DG VI at the European Commission.

www.defra.gov.uk/

9

A marketing perspective on diversification

R.W. Slee[1]

Introduction

In the summer of 2000, a group of farmers in south-west France decided that too great a proportion of the consumers' franc/pound was captured by supermarkets and that it was possible to find alternative ways to the market. They established www.paysans.org to provide a short-chain marketing link with consumers in various parts of France. They deliver a range of produce in boxes/hampers ranging in price from £30–60 to most parts of France on a weekly basis and have seen their customer base expand rapidly over the first year-and-a-half of their existence.

In the UK, since the first market in Bath in 1997, farmers' markets have grown from humble beginnings to become a significant example of short-chain marketing of UK farm produce. They now sell a range of produce throughout the UK at nearly 300 markets. Their growth has been founded on recognition that there are discerning consumers who are seeking variety in what they purchase. These same consumers are often concerned about food quality and feel happier purchasing their food direct from the grower.

Neither of these initiatives is unique. Neither represents a significant share of consumer spending. However, both illustrate how it is possible to market farm produce in different ways in order to add value. Both are expanding components of a food market which attracts a declining portion of the consumers' pound. Both are marketing approaches that give farmers a bigger share of the final price of the product sold. They represent concrete examples of farmers taking the initiative and their rejection of passive acceptance of declining prices. In both cases, many growers are adding value by processing to offer butter, cheese, fruit juice or wine, as well as using these new marketing channels. This is not the only response to a continued cost–price squeeze, but similar responses have been adopted by an increasing number of farmers in various parts of the world over the last decade.

In addition to these new enterprises based on marketing products differently, there is a large range of types of diversification, many of which are based on the provision of services to a more mobile and affluent public. There has also been a tendency to think about alternative forms of production and to realize the capital values of old buildings.

Alongside these initiatives in new enterprises, marketing and adding value, there is growing evidence of a much more entrepreneurial spirit in the farm sector. This is partly a product of tightening margins in conventional production systems, but it is also driven by recognition that there have been profound changes in the countryside in recent decades, which create both obstacles and opportunities. These changes have led many farmers to alter their businesses to include unconventional enterprises on their holdings, probably to an unprecedented degree in the history of UK agriculture. Recent survey work in the UK by the National Farmers' Union indicates that the majority of UK farmers now have diversified their income base beyond conventional farming.

Alongside the increasingly entrepreneurial spirit evident in UK agriculture has come a need to look again at the ways in which we examine farm businesses. Farm management has become rather disconnected from general management thinking, in that it was heavily dominated by production management, and was much

[1] The author of this chapter would like to record his thanks to the editior, Richard Soffe, for background information on markets and marketing, which made his task immeasurably easier.

concerned in the recent past with gross margins. In addition, since most farmers can be seen as price takers rather than price makers, the end price received is taken as given. Further, the floor in prices offered by intervention buying for many products meant that the real rigours of the market place were never fully felt in much of the farm sector. However, with the major changes in farming and the emergence of more and more non-farm on-farm enterprises, there is a need to look again at the management toolkit used by those appraising farm businesses. Brunåker (1993) has argued the need to take a small business perspective on the examination of new enterprises. Others have argued for a stronger marketing perspective in the examination of farm diversification and a focus on (some would say a return to) mainstream business management approach such as the use of Return on Capital Employed (ROCE) as a performance indicator.

This shift towards a much more diversified business base in farming was at first viewed with suspicion and mistrust by many in the farming community, but is now endorsed by most people in all of the key institutions associated with agriculture. Before looking in more detail at the implications of this shift, we first need to understand it, and in particular to understand why it is likely to represent a sea change that cannot be ignored.

What factors have driven the change?

- Profound social changes have taken place in the way contemporary society views the countryside and in people's attitudes to conventional food production. The occupational community of rural areas – those that live and work the land – is now a minority in most regions and new social groups have moved into the countryside. These new groups have different values and demands. They have moved into rural areas for lifestyle reasons. Although often dismissed as incomers, they constitute an important market opportunity.
- Technical change in modern agriculture has dramatically affected the numbers that work on the land and the appearance and ecological values of rural areas. The number of farmers and farm workers has declined. Further, many of the elements of the new technologies are not well understood by typical new rural residents. Technical change has thus reduced farmers' representation at the same time as it has antagonized the new rural populace.
- Economic factors are a principal driver of change. Key milestones in EU policy change have often arisen because of attempts to keep down the growing costs of support to the farm sector. Problems of an imbalance between supply and demand have surfaced from the 1970s onwards. Given the tendency for many farm products to be characterized by static demand, the increasing output of most of the last 30 years has increased the costs of commodity-based support and led to 'dumping' of surpluses on world markets.
- Political change at both a national and global level is shifting the way that policy makers support the farm sector. The inclusion of agricultural trade in the Uruguay Round of GATT in the late 1980s and early 1990s and the continuing rounds of discussion at the World Trade Organization (WTO), with agriculture under the microscope, have led to a major reappraisal of the way in which governments around the world support their farm sectors. Countries or blocs with relatively high levels of support for their farm sectors, such as the EU, have had to make changes to their policies, which have reduced the scope for product price support and emphasized the need for new forms of support which cause less distortion to global markets.

These changes are profound in their significance to the UK farm sector. The adulation and respect in which UK farming was held in the early post-war period has been replaced by a more cynical and mistrustful set of values by much of the general public. These have been nurtured by a critical press that has plenty of events, such as BSE, the foot and mouth disease outbreak in 2001 and a growing incidence of *E. coli* 0157 to fuel its appetite for critical coverage. There has been a breakdown in trust between the consumer and producer, that increasing distance between the farmer and the consumer, a critical press and increased supermarket power have tended to exaggerate.

For almost 50 years after the Second World War, farmers in the UK received unparalleled support from the state. This post-war settlement between the farm sector and the government was replicated, albeit with some differences in the way support was given to farmers, in other parts of Europe. On entry to the EU, the UK's policy for farming converged with that of the EU. Now, with further eastward expansion of the EU imminent, and growing pressure from the WTO for further reductions in trade-distorting price support, the bases for that post-war settlement for farmers look fragile. Reforms in the CAP since 1992 have been deepened further in the Agenda 2000 reforms and these will place new demands on farmers to adopt a more entrepreneurial and market-oriented approach to their businesses.

Diversification of economic activities on farms and

increased attention to markets and marketing in both conventional and unconventional enterprises are both part of a necessary process of reappraisal of businesses. This need for profound changes in how farmers engage in business activity is forced on them by these inexorable policy shifts. Whilst under product-based support systems the impact of markets has been dampened to a considerable degree, under the new policy regimes a market orientation is becoming a necessity for those businesses that wish to thrive.

What do we mean by diversification?

Farm diversification has always been a problematic concept, in that in contemporary parlance it is associated with doing something different and unconventional with farm resources. Early discussion on diversification saw it purely in conventional agricultural terms, as a way of avoiding the risk of having 'all the eggs in one basket', in normal farm enterprises, or as a means of devising synergy between enterprises, such as improving the fertility of cropping land by the application of animal manure.

The problem is that what is 'different' can be a matter of tradition, common practice or policy support. When a leading US farmers' journal organized a conference on diversification in the late 1980s, the only common characteristic of the hundreds of ideas explored was that they were not mainstream enterprises in the area where they were being considered. So there can be no absolute definition of diversified versus conventional enterprises.

Yesterday's unconventional enterprise can become tomorrow's mainstream. Thus, oilseed rape is now a mainstream crop, but 30 years ago could be regarded as a diversified enterprise. Few other crops or types of livestock have been mainstreamed in the same way, though there is undoubtedly a far greater range of enterprises than in the relatively recent past.

Some authorities have attempted to distinguish between *structural* diversification where the farmer develops an enterprise completely different to agriculture, such as tourism, and *agricultural* diversification, where the enterprise is based on modified forms of food and fibre production. However, this distinction misses the fundamental point that most forms of diversification carry with them the necessity of moving into the real world of markets and consumers, outside the safety net of conventional product-based policy support.

Commentators have pointed out that diversification of economic activity by farm households often takes the form of household members working off the farm. In these cases, the diversification is of the labour input of the farm household, not the land, buildings or existing enterprises. The importance of this as a form of diversification is evident in the increased tendency for members of farm households to work off the farm and, in areas where small farms predominate, it is likely to be the most common response to falling household incomes.

Consequently, there can be no unambiguous definition of diversification. It is usually taken to be farm-based, benefits from no end-price support, and may involve either developing a new and different enterprise or adding value by marketing or processing to the output of a conventional enterprise. The extent to which a product is or is not 'alternative' can be debated endlessly. What is more important is that engaging in most of these enterprises requires a shift in business orientation to a much closer understanding of the market.

The background to diversification

Diversification of farming activity to include either novel and different enterprises or value-adding enterprises is by no means new. With hindsight, we may look back on the last 50 years of the twentieth century as one where the mainstreaming of narrowly defined production of agricultural and forest products led to an alienation of consumers from farming and other forms of land management. In many ways, they found the earlier countryside infinitely preferable to the so-called productivist countryside that resulted from the policy-distorted production practices of the late-twentieth century.

One antecedent to contemporary diversification is on-farm processing, which was much more common in the past. Before the Second World War, farming in the UK was characterized by higher levels of on-farm processing, often only for domestic consumption. The industrialization of food production in the post-war period destroyed most of the artisanal, small-scale food production, which is still relatively common in parts of southern Europe.

A second identifiable root of contemporary diversification is the demand for leisure in the countryside. Since the sixteenth century, urban wealth has been 'invested' in rural land, more in pursuit of leisure and amenity than for pursuit of profit. This movement peaked in Victorian times, when large areas of northern Scotland were purchased by rich industrialists and turned from sheep farms into sporting estates. It has continued since, with the rise of hobby farming and the purchase of second homes. It has long since become a commercial as well as a private activity. In the ninteenth century, farmhouses were let on Dartmoor for summer lodgings to tourists. In the late twentieth century, the conversion and/or sale of redundant barns/steadings has been an important way of creating new homes or tourist accommodation in areas of high planning constraint, thus realizing substantial sums for the seller.

The identifiable societal demands for rural living and leisure space have been expressed over a wider and wider area as a result of the flexibility of travel created by the private motor car. New residents can be treated by farmers with disdain as 'townees' who do not understand rural ways or alternatively as potential customers with substantial resources at their disposal. Whilst these new residents might often be critics of intensive farming, they may well be seeking livery for horses and ponies for their children, and are likely to be very supportive of local speciality food producers.

Types of diversification

Diversification can, at its simplest, be seen as the addition of a new income-generating enterprise to a (predominantly) farm business. However, in practice, diversification may also entail the realization of capital assets and the restructuring of a business, or the redeployment of labour or capital assets off the farm. The selling off of barns/steadings or of land for residential development can release cash and enable business adjustment. These different forms of business adjustment are indicated in Table 9.1.

It is difficult to come up with an all-embracing classification of types of diversification. However, it is immediately apparent that there is a huge range of types of diversification, including both services and products. The service–product distinction, whilst clear at the extremes, is not wholly meaningful in a practical context. By definition, a service is intangible, and cannot be taken ownership of by the consumer, whereas a product is something that the purchaser takes ownership of at purchase. Historically, most farms have dealt in products, rather than services, but in the twenty-first-century countryside, it is evident that the service functions of farming have become more and more important. Many services associated with farm diversification, such as farm tourism, include both service elements (the provision of hospitality) and product elements (the meal).

Table 9.1 Forms of diversification

	Income stream	*Capital asset realiszation*
On-farm	e.g. Farm tourism	e.g. Selling building plot, barn, etc.
Off-farm	e.g. Agricultural contracting; off-farm work	e.g. Adjusting share portfolio

Table 9.2 represents an attempt to classify the range of types of diversification. The classification includes three product categories and two service categories. The list of examples is meant to be illustrative rather than comprehensive.

Products

We can group products, for convenience, into three main groups:

(1) those that involve the primary production of different (alternative) crops and animals;
(2) those that involve adding value to farm output by on-farm processing and unconventional marketing;
(3) those that involve using the farm and its resources as a venue for some other kind of production such as game, forestry, rented buildings, etc.

In the *alternative primary production group*, there are many alternative animal and crop enterprises. In the case of many alternative production systems, there is no standard 'recipe-book' system that can be used either for production or for marketing. Knowledge is often held by relatively few people and transferring that knowledge into a different bio-physical environment may prove problematic and rely to a greater extent on trial and error than is the case with normal production systems where the production parameters are better known. Alternatively, it may rely on proprietary knowledge, which is sold along with the product to those starting up new enterprises with 'professional' support. At various times, diversification has been given a bad name by dubious selling techniques which promise large rewards if the buyer of the production package purchases the information from someone who has apparent understanding of both production processes and future markets. The selling of diversification packages for ostriches and snails represents, from different ends of the faunal speed spectrum, the dangers of seeking fast rewards and the high likelihood of slow profit.

There are two components of the *added value group*: an added-value-by-processing component and an added-value-by-marketing component. Those who have added value by processing might well seek to add further value by marketing unconventionally.

Table 9.2 The main groups of alternative enterprises on farmland. (Derived from Slee, 1989)

Group	Type	Enterprises
Unconventional agricultural enterprises	Animal products	Sheep milk Rare breeds Fish Deer and goats
	Crop products	Linseed Evening primrose Teasels
Adding value to conventional farm products	Animal products	Meat (direct sales, etc.) Skins/hides/wool Dairy products (direct sales/processing)
	Crop products	Milled cereals PYO and direct sales of vegetables
Use of ancillary buildings and resources	Woodland products	Fuel wood Craft timber products
	Redundant buildings	Industrial premises Accommodation
	Wetland	Fish Game
Tourism and recreation	Tourism	Bed and Breakfast Cottages/chalets Caravans/camping Activity holidays
	Recreation	Farm museums Visitor Centres Riding Game shooting Other shooting Fishing Farmhouse catering
Public goods	Wildlife	ESA/stewardship payments
	Access	Access agreements
	Historic sites	Heritage relief
	Landscape	Management agreements

Increasing product component (upward arrow); Increasing service component (downward arrow)

A farmer who decides to add value by pasteurizing milk on the farm as a producer–retailer, processing apples into cider, milling oatmeal or making sausages and other pork products from his home-produced pigs is seeking to add value by processing. The temptation of attempting to capture a greater proportion of the consumers' pound is considerable, but it is important to realize that adding value carries with it both costs and risks. The parameters of these production systems may be rather less well known than is the case for conventional farm production where raw, unprocessed material leaves the farm gate. For example, many people who have engaged in farmhouse cheese production have struggled to get the cheese production process right and have often had to adapt production processes or delicately adapt the nutritional regime for their cows to ensure milk quality that ensures good quality cheese.

Adding value by marketing differently is perhaps less risky. It is always possible to default to conventional markets if the alternative route does not pay. The principal problem that the producer faces is the ability to set up and sustain a marketing channel at reasonable cost in relation to the volume of produce available. Direct selling of vegetables such as potatoes is almost costless, beyond the initial investment in weighing apparatus and bags, with modest needs for additional labour to fill bags and carry out transactions. Direct selling of meats is more costly with a requirement to comply with Health and Hygiene regulations. One of the great virtues of direct selling is the ability of the customer to meet the grower. The close contact between the two in farmers' markets or farm shops is considered to be an important factor in building trust between consumer and producer. In direct marketing, there is clearly an important service component to the product. In selling an exotic vegetable, the grower can be directly asked 'What is the best way to cook this?' or 'What will it go with best for a special meal?' Someone who can deal effectively with consumers will be using such approaches to create additional custom and sustain trade.

The third group of products is those that are not conventional farm products and constitute a step away from conventional farming, but use what were formerly farm resources in different ways. They are not encapsulated by any simple term, but all entail the *alternative use of farm resources.* They include planting or regenerating existing woodland and converting farm buildings to alternative use. For convenience, recreational and tourist adaptations of farmland are considered separately.

Whereas farm woodland in the past was often neglected or used only for firewood and game, there has been considerable public expenditure in developing new farm woodland, much of it under the various farm woodland grant schemes, which have recently entered their third phase. In addition, there are many regionally specific schemes that give farmers in certain places significant supplements to plant trees. The rationale for much of this tree planting is not to provide timber but to enhance the service functions of the countryside, such as the New National Forest in the English Midlands. However, in some places, such as in north-east Scotland, the Grampian Challenge Forest places a strong emphasis on productive forestry.

The second type of alternative use of farm resources is the conversion of farm buildings to residential or tourist use. This single category of diversification in the use of farm resources has probably contributed more to the net assets of farmers in the last decade-and-a-half than any other form of diversification. However, it is not entirely unproblematic. Where it entails a change of ownership, farmers should be sensitive to the potential for objections if they seek planing permission for certain types of livestock development in the adjacent area.

Services

There are two main types of service: those that rely exclusively on the market place and are predominantly related to tourism and recreation; and those that are the product of government grants and subsidies under various environmental payment schemes which have become more important over the years.

Tourist and recreational enterprises constitute the single biggest grouping of diversified enterprise. There are naturally major regional variations in the opportunities for leisure-related diversification, with urban fringe areas often favoured more for recreation than tourist enterprises. However, even in urban fringe areas, there may be scope for farmhouse accommodation that serves a business-oriented market.

Public goods (so-called because the market will not deliver them without public sector assistance) have become increasingly important. Whereas initially there were only payments for farming development opportunities forgone as a result of nature conservation requirements under the 1981 Wildlife and Countryside Act, there are now many more payments offered under Environmentally Sensitive Areas (ESAs), Stewardship and other schemes. Many farmers openly acknowledge the importance of these environmental payments in sustaining their businesses during the recent financial difficulties. Often the features that are protected and/or enhanced under these public goods support schemes can contribute beneficially to tourist or recreational products.

Ways of analysing diversification

There is no 'right' way of examining the potential for diversification or the existing performance of a diversified enterprise. However, it is immediately evident that in most cases, the standard farm management pocketbook approach will not suffice. There is usually a need to individualize the analysis and not to accept standard figures.

Central to any consideration of alternative enterprises is the nature of small business management and an understanding of the market. This will be examined further below. This market understanding should not comprise simply an aggregate understanding of the market for whatever enterprise the farmer is contemplating or already engaged in. It should instead provide the enterprise manager with crucial information that subsequently informs a range of decisions on product design, market segment choice, marketing channels, etc. In the case of many diversified enterprises that attract the customer to the farm, location can be an absolutely crucial component of understanding the market. One key facet of marketing is to calibrate the impact of the location on the market. For a pick-your-own enterprise, a recreation or tourist enterprise, location is likely to be a critical determinant of market size, for which no amount of advertising can compensate.

Where the alternative enterprise takes up farmland, it is necessary to consider the opportunity cost of the land and other resources involved. What existing enterprise is displaced and at what cost? This analysis can readily be developed into a partial budget, which explores the costs saved and the revenue loss on the existing enterprise and contrasts this with the additional costs and revenues of the new enterprise (see Farm Business Management, Chapter 12).

Where more substantial investments in diversified enterprises are contemplated, a rather more detailed investment appraisal is recommended. Given the greater uncertainty associated with the development of diversified enterprises, sensitivity analysis which looks at all the key variables is highly recommended (see Farm Business Management, Chapter 12).

Small business management

Small firms have increasingly been seen as core contributors to rural economic development. Some of these firms will be independent of farming, but many have grown within farm businesses and some still constitute enterprises rather than separate businesses. Recent DEFRA evidence points to a process of enterprise growth in diversified farm enterprises (Hodge, 2001) with diversified sources of income growing at a time of great stress and reduction in farming incomes.

Brunåker (1993) argues that within the small business management literature, two areas provide a key to understanding the challenge of diversification. First, strategic management of the enterprise, and second, the examination of growth factors in small enterprises or businesses. Strategic management concerns in small business are broken down by Brunåker into two issues:

(1) Strategy as a position: by position, he refers to the need for the small firm to have a strategy both in relation to its competitors (by, for example, price leadership or quality differences) and in relation to its customers (by ensuring that the product or service offered satisfies a customer demand).
(2) Strategy as a business perspective: by perspective, Brunåker refers to the ability of the firm to develop strategic thinking in an uncertain environment: what is the vision of the key decision maker(s) and what actions must be taken to realize that vision?

It is widely recognized that there are huge variations in the performance of small firms. Some are 'flyers', others are conditioned by lifestyle considerations (or less than dynamic management) and tend to stay relatively small but may still contribute to aggregate economic activity, and yet others fail. Working on the assumption that failure (an all-too-familiar outcome for many enterprises) is to be avoided, what does the literature tell us about success factors?

First, we can consider a bundle of factors relating to the person starting up the enterprise or business. Does he have appropriate small business management skills and motivation and prior experience in similar types of business activity? Successful diversified enterprises on farms are often established by a spouse. A spouse who has a background of experience in another area of economic activity and has developed transferable skills may deploy these in a diversified enterprise. We are thus concerned to identify whether the person developing the enterprise can carry into it a bundle of personal qualities and skills that will help the enterprise to thrive.

Second, we can consider the strategy of the firm or enterprise:

- How does it skill its managers and its workers?
- How does it appraise the options and position its product(s) in the market?
- How does it adjust its product to changing market conditions induced by demand change or competition?

It is this bundle of strategic management skills that Brunåker reduces down to the twin decisions of (1) strategy against competitors; and (2) strategy in relation to customers.

However, behind these positioning exercises lie some basic appraisal skills that should be considered in relation to any enterprise. Strengths, weaknesses, opportunities and threats (SWOT) analysis can be defined as "A distillation of the findings of the internal and external audit, which draws attention to the critical organisational strengths and weaknesses and the opportunities and threats facing the company" (Kotler and Armstrong, 2001). In SWOT analysis, we are looking for factors that are critical to success or failure of an enterprise, not an endless list of factors that just might be important. We are also concerned with distinguishing between threats that can be managed and wider threats that cannot be countered by a strategic response.

SWOT analysis is normally conducted on an existing business or enterprise in a way that exposes its 'room for manoeuvre'. Ideally, we would like to see any diversified enterprise with few threats and a number of new opportunities. In practice, we may see enterprises with many threats and modest potential. SWOT analysis has a place in any business appraisal, both prior to the establishment of a new enterprise as a core part of any ex-ante appraisal and as part of a review procedure for an established enterprise.

What do we mean by marketing?

Marketing is often thought of by primary producers (and non-marketers) as a necessary way of getting rid of something that has already been produced. This approach to marketing is widely held by those firms that are product- or production-oriented, whose management concerns focus on getting production systems right and using technology to its maximum potential. However, such an approach to marketing horrifies those who see marketing as a much more all-embracing customer-centred way of approaching business activity.

Some formal definitions

The following three definitions sum up some key features of marketing: its customer centredness; its holistic, all-embracing nature in informing all business decisions; and the extent to which it enables the firm to manage and take at least partial control over what in the past it had passively accepted.

> "The management process responsible for identifying, anticipating and satisfying customer requirements profitably." [Chartered Institute of Marketing UK (CIM); Smith & Chaffy, 2002]

> "Marketing is the process of planning and executing the conception, pricing, promotion, and distribution of ideas, goods and services to create exchanges that satisfy individuals and organisational objectives." [American Marketing Association (AMA); Kotler, 1999]

> "Marketing Management is concerned not only with finding and increasing demand, but also with changing or even reducing it. This marketing management seeks to affect the level, timing and nature of demand in a way that helps the organisation achieve its objectives. Simply put, marketing management is demand management." (Kotler, 1999)

Macro- and micro-marketing environment

The typical response of the bulk commodity-producing farmer to marketing is that these marketing concepts can offer little to him as a producer of a largely undifferentiated product. Such a producer may be relatively constrained in terms of micro-marketing management, in that there may be limited opportunities to improve his business performance through the manipulation of marketing mix variables. However, the same producer can and should consider the wider external conditions under which his business operates. These contextual variables can be thought of as the macro-market environment. Figure 9.1 illustrates the broad grouping of factors that should be considered.

The macro-market environment represents the external environment in which all firms operate, but over which they have no control. The individual firm (or enterprise manager) needs to consider closely the implications of these macro-market determinants. The examination of this external environment is often termed PEST analysis after the initial letters of the four factors: *p*olitical; *e*conomic; *s*ocio-cultural; and *t*echnical.

Economic factors

The principal economic factors that lie behind the push for diversification can be considered in the context of supply and demand. Increased self-sufficiency resulting from the application of output-increasing technology, more open trading conditions and competition from cheaper producers in other countries has created enormous problems for the conventional farm sector in recent years. At the same time, the demand for most bulk agricultural commodities is stagnant or in decline. The message is simple, if unpalatable: in more open trading conditions, farm product prices will show a tendency to drop, and marginal producers will be forced out of business. It is to escape from this inexorable cost–price squeeze that farmers seek to diversify.

The economic situation facing farmers in the UK has been exacerbated in recent years by macro-economic factors. The UK government's refusal, to date, to join the Euro and the strength of the pound have meant that, even without markets closed by BSE, food exports to

Figure 9.1 The marketing environment – macro and micro.

other EU member states would have been negligible. The combined effects have resulted in the recent farm income crisis being particularly severe in the UK.

However, in relation to diversified enterprises, the situation may not be so negative. Given that most diversified products are not commodities but occupy niches in the market, there is some scope for product differentiation. Rising incomes and 'demassifying' markets have meant that there is an increasing tendency for individualistic products to be sought out by more discerning consumers. However, on the negative side, the strong pound has made it much cheaper for UK tourists to travel abroad and for overseas tourists to come to Britain, resulting in difficulties in the tourism sector.

Socio-cultural factors

A number of important changes are discernible in the socio-cultural environment as it affects consumer perceptions of farming and diversification. The contemporary consumer is much more critical of modern agricultural practice, and often looks back to a (maybe) golden age when things were apparently happier and food was healthier. The contemporary loss of trust amongst consumers over food quality has profound consequences and has driven an increasing proportion of consumers into vegetarianism and organic consumption.

A second major axis of socio-cultural change has been growing concern for the environment, which is strongly developed among northern European countries. The membership of the RSPB is now over a million people. Its membership has risen dramatically in the last two decades. Many other environmental organizations have also experienced rapid growth. The effect of these changes might be seen as largely negative for farmers, but these same social changes have triggered substantial payments to farmers in a range of stewardship-type payments (see also Chapter 10 on Farming and Wildlife).

Political factors

Political influences have been enormously important in determining rural economic well-being. It is almost inconceivable to imagine what state the farm sector and the rural economy would be in if free markets had operated in the post-war period, as they did for most of the century after the repeal of the Corn Laws in 1846. The post-war settlement in the 1947 Agriculture Act and the Treaty of Rome, which was 'inherited' on UK entry to the EU, underpinned farm prosperity for much of the past 50 years. Under both, substantial support was given to farmers. Although the cost of farm support is by no means insignificant, policy change has not come about because we cannot afford to pay farmers, but because political masters choose not to do so. These political masters have responded to both intra-EU pressures to reduce budgets and WTO pressures to reduce trade distortions.

Whereas in the past, farming organizations had major influence in the debate over rural policy, their power is now limited by the much greater strength of environmental pressure groups. There has been a distinct shift from agricultural to rural policy. The often fuzzy concept of sustainability has risen up the political agenda and increasingly influences policy.

Technical factors

Technical change impacts on rural businesses in all kinds of ways. Farms now require fewer workers because of the substitution of people by large, more sophisticated machines. New technology means there is potential for increased output, as is likely to be the case with genetically modified organisms, but such technical change can also engender public disquiet. Equally, new technologies can affect marketing opportunities. Individual farms can market their wares on the web. Consumers can obtain detailed information on one farmer's Aberdeen Angus production systems and go on to order their meat direct from him. The earlier difficulties of marketing farm holidays at individual farm level can now be largely circumvented through web-based advertising.

Any appraisal of markets should go beyond the marketing mix variables, which can be controlled to a degree, and consider those external macro-market variables that cannot be controlled. It is important to understand the nature of change in these factors and consider how they will impact on the future prosperity of different types of enterprise. The changes will not favour all types of diversification, but they will in general tend to favour diversified enterprises over traditional commodities. They should set the context for consideration of alternative forms of business activity.

Market research

Market research comprises the collection of information to inform product and marketing decisions (Lancaster and Massingham, 1988). Market research is normally divided into two types: primary market research and secondary market research. Whilst primary market research involves the collection of new information, secondary market research involves trawling existing sources of information. Both types of market research can be conducted in-house or by outside consultants. There is always a danger that consultants reorganize information gleaned from those for whom they are apparently working, and present it in glossy format without offering any new insights. Those using outside consultants are advised to save their own views as a template against which to compare the findings of market research consultants.

Market research can be used for a range of inquiries and should be able to deliver answers to the following questions:

- How big is the actual or potential market and who are your actual and potential customers?
- What are the trends in the market in general and the sub-segment occupied by you?
- What are the characteristics of the market in terms of other competitors, market segments, etc?
- How distinctive is your product and what are the likely reactions of consumers to any price changes?
- How effective is your promotion?
- How appropriate are the sales outlets you have chosen?

Where primary market research is conducted, it should be guided by principles of good questionnaire design and effective sampling procedures. Students in colleges and universities are often keen to learn how to conduct such studies and may provide a cheap, if not wholly reliable, source of labour. The case for a disinterested outsider conducting market research is strong, as the promoter of the enterprise may be unwittingly positive, even when the evidence does not allow such conclusions to be drawn.

As a general rule, primary market research should be avoided until the gaps exposed by secondary market research have been revealed. Primary market research can be relatively expensive, but may be necessary where markets are rather local and where useful secondary sources of information are lacking.

Secondary market research consists of a number of types of information:

- published reports (which are produced by major market research companies like MINTEL) and can be very expensive, and official reports from The Stationery Office such as the Family Expenditure Survey or other socio-economic data;
- the so-called grey literature, including reports for enterprise agencies and tourist boards which are often unpublished but may be available for consultation; and
- internal sources such as suggestions and comments books in accommodation enterprises and sales records in different types of market outlet.

Segmentation, targeting and positioning

Marketing textbooks talk about 'STP marketing' (Kotler, 1999). Segmentation of markets, targeting particular segments, then positioning a product or service within it comprise the core strategic marketing concerns of most firms.

Segmentation consists of trying to break down the market into its component parts. Segmentation can be on a number of criteria, such as regularity of purchase, income, or demographic characteristics. Segmentation is always purposive, exploring the way in which markets can be divided up to help make a decision as to whether particular segments should be targeted. Segmentation is product-specific. What is an appropriate segmentation for one product or service may not be appropriate for another product.

Targeting involves making a decision on whom to target. For example, in a farm cider enterprise, should the target market be farm-gate sales, pub sales or off-licence sales? Further, should there be a strong local regional bias or should the London market be targeted? Given that our cider farm may have several different bottle sizes, we may need to produce a matrix of products and outlets and develop an idea of which segments absorb which product. A farm shop may decide to target particular demographic segments such as affluent females over 30 years of age. Various 'direct marketing' options exist to target these, through promotions, specific advertising, (e.g. glossy magazines) and purchasing a list of addresses from a 'list broker'.

Positioning involves taking a strategic position in the market. Is a farm tourism enterprise going to be the best because of its décor or its food, or because it's the cheapest or what? Depending on the asset bundle, the location, the existing resources of labour, etc., a decision must be taken to maximize advantage to the enterprise.

Marketing mix management

The marketing mix is the bundle of variables over which the enterprise manager may be able to exercise some control. In the classic marketing texts there are four marketing mix elements – the 'Four Ps': product, price, place and promotion, to which a fifth P is often added where there are service enterprises – people.

Product

Marketers regard the product (or service) as comprising a relatively narrowly defined core product and various additional components. Some of these additional elements may come as an expectation; others are recognized as distinct augmentations of the core product. The challenge is to provide a level of augmentation that the customer desires and is willing to pay for, rather than provide additional levels of augmentation that are not sought. Modest augmentation of a product such as fresh-cut flowers in a room or a piece of home-made cake with a cup of tea on arrival at farm accommodation may effectively differentiate a farmhouse holiday enterprise from the hotel alternative.

In practice, the consumption of goods and services is wrapped up with issues of personal identity. Consumption behaviour is guided by social values and any cues to the consumer that his/her values are being nurtured by the consumption experience are worth flagging up by the seller. Many urban visitors' interests in conservation and environment and pursuit of rural authenticity may be indicated by appropriate wording in brochures such as 'home-grown food', 'local produce', 'farm nature trail', etc.

Price

When bulk products are sold in mass markets the farmer is a price taker. In more differentiated markets, farmers have more room for manoeuvre. Whilst the general economic laws of supply and demand are normally operative, marketers need to consider the positioning of the product, the nature of competition and the value perceived by the visitor (who must be persuaded to make the purchase on the basis of the offer price). Occasionally, a low price sends out alarm signals. Those involved in, for example, direct sales will have to balance cheapness against perceptions of quality ('are these second-grade produce?'). The level of uniqueness of a product determines the extent to which the supplier can adjust price; consequently, the speciality food producer or the tourism enterprise with a special view or next door to an important historic site have to make pricing decisions rather than accept a going-rate price.

Place – location and distribution

With many products, such as farm tourist enterprises, the place of sale is fixed. However, before entering into such an enterprise the manager should consider whether or not the location is appropriate for the type of enterprise. Will people find it easily? Will they be discomforted in accessing the product down a bumpy road? Location is such a crucial variable in so many visitor enterprises that its impact on visitor numbers needs to be estimated and the consequence of someone setting up in competition in a more optimal location carefully considered.

Place can also be considered in relation to place of sale. Is a farmer's market more appropriate a place to sell speciality vegetables than a wholesaler, restaurants or the catering trade? When place of sale is considered, there are also decisions to make about which marketing channels to use.

With more specialized diversified enterprises, especially those dealing with relatively small volumes of produce, the organization of efficient and low-cost supply chains can be a considerable challenge. In the case of speciality foods, there are potential problems with quality loss as a result of breaking cool chains or inappropriate storage. Such problems are particularly likely with dairy products, especially speciality cheeses.

With speciality foods, the final market may be rather dispersed and there may be few easy ways of accessing it. If, for example, speciality foods are sought by a range of country restaurants with relatively small individual demands, it may be essential to piggy-back your products on others delivered to the same destinations. Although this is by no means costless, it may be far more cost effective than purchasing a delivery van with cool storage facilities.

Promotion

There are so many facets of promotion that some marketers talk about a 'promotional mix'. As a general rule, word of mouth is seen as a highly desirable form of promotion, especially in the form of recommendations from satisfied customers. So, for a similar reason, is any free promotion such as positive press coverage. Rather than pay for an advertisement for a farm park, a press release with 'New animals arrive at XYZ Farm Park' may have a similar impact in raising visitor awareness.

When promotional decisions that require expenditure must be made, it is important to think about the clientele:

- Who are they?
- How can they be most effectively reached?

Rarely will there be a single-stranded approach to promotion:

- Leaflets and brochures should provide all the vital information.
- Advertisements may be placed in different magazines or newspapers. In addition to print, consider radio, TV and increasingly the internet.
- Websites have become increasingly important.
- Public Relations management can help. If any awards have been obtained such as food awards, or

tourism awards, they should be flagged up in the promotional material as this offers the potential consumer an external validation and testimony on the product. A positive story on a TV programme can be worth a fortune in advertising.

However, a degree of monitoring is essential. Marketers need to monitor effectiveness of different promotional elements. If a particular promotional strategy is not working, it needs to be dropped quickly.

People

In service-related diversification, the nature and quality of interaction between the provider (host) and the consumer (visitor) are likely to be central to the overall valuation of the 'product' by the consumer. An informative and friendly host who carefully elicits the visitor's interest and then adapts the product to maximize the value for the visitor is effectively changing and customizing the product (or at least the extended or augmented product) for each customer. Of course, such a response from the provider-host requires a strong information base of personal knowledge and a sensitive approach to ascertaining consumer-visitor interest.

One issue of importance in diversified enterprises is that at times of peak visitor usage, there is a tendency to use casual staff to meet peak needs and those staff may be amongst the least well trained and least knowledgeable about the consumers' interests. Thus, at visitor attractions, farm tourist accommodation, farm shops or any other diversified enterprise with a strong service component, there is a potential problem if the front-line workers are relatively unskilled.

It is impossible to underestimate the importance of personal qualities in service-related diversified enterprises. Some people have an open welcoming approach, which at times can constitute a core part of the visitor experience. This uncontrived friendliness and welcome and a reaching out and genuine interest in the visitor's wellbeing are features of the best farm holiday accommodation. It may seem self-evident, but it is nonetheless enormously important. It is likely to generate positive recommendations and is worth more than considerable expenditure on advertising.

SMART planning

Once the basic principles of business activity and marketing are understood, what is the next action. First, the manager of the enterprise needs clear objectives. These should be SMART:

- *S*pecific
- *M*easurable
- *A*chievable
- *R*ealistic
- *T*ime-bound

The objectives should be precise. What exactly does the manager want to achieve? They should be measurable. Can an occupancy rate of 40% be set for the first year of operation? They should be achievable. If other established local operators have occupancy rates of 35% how can a new entrant expect to achieve 40%? Fourth, the objectives should be realistic. In the light of the above comments, we may wish to set the occupancy objective at 25% (and go on to consider the knock-on effects on profitability, etc.). Finally, we need to know the time period in which the objectives will be achieved. Is it a year, 3 years or when?

The application of marketing to diversified enterprises

The centrality of marketing in diversified enterprises is highlighted by the fact that there is usually no market information or product prices support on these enterprises. There may be some support with set-up costs, but such support is normally conditional on there being a development proposal and marketing plan for the diversified activity. A marketing perspective is very much a precondition for the success of diversified enterprises, as well as obtaining any public support for their establishment or improvement.

This section looks at some actual examples of diversification to examine how a number of practitioners have thought through and addressed the challenges posed by the development of alternative enterprises.

Products

Example 1

John has operated a mixed farm in southern England for the last 40 years. For a number of years he has had a significant pig enterprise. He has stuck with traditional breeds and, as a result of a downturn in prices and pressure from his main buyer to move to more modern breeds, he decided to try his hand in farmers' markets and adopt a shorter marketing chain to the final consumer. He used a local butcher to assist in the preparation of the meat and, dressed in a straw hat and white coat, was amazed that he was completely sold out of meat in a couple of hours. He passed business cards to people who asked about repeat custom and rapidly built up a market of loyal customers. He is a highly effective communicator and a natural salesman which helps enormously in direct selling.

Following this initial success, he created a brand, 'Lashbrook Unique Country Pork', and established a farm shop and set up a mail order dimension to extend his market and keep the repeat customers happy. The next step has been to set up a website and move towards e-business. The farmers' markets, shop, mail order and the e-business are now an important part of his overall business activity.

What can we learn from this example?

- Enthusiasm and commitment are crucial ingredients in success.
- Once into his first diversified enterprise, the other components dropped into place almost 'naturally' to create real synergy in a number of enterprises.
- Personality and charisma are important in a customer-oriented business.
- He must now make decisions about how to sustain his original farming enterprise and how to sustain the retailing operation. If his own skills are vital for both, at what point does he seek to bring in additional labour?

Example 2

Isabella and Alasdair Massie live on an arable farm in north-east Scotland. With a combined background in domestic science, health and safety consultancy and retailing, but their hearts and home in rural Aberdeenshire, they have converted farm buildings to establish a small food business producing an expanding range of relishes and jams. Isabella's Preserves was established 5 years ago and, since that time, has won a series of food awards for quality, including a Food From Scotland Excellence Award in 1998 and a Gold Award in the Great Taste Award 2000. The business has developed incrementally, making maximum use of second-hand equipment such as freezers to store the highest quality ingredients. When established, they moved out of rented premises to converted buildings on their own farm.

The product range has expanded considerably from modest beginnings and now includes ingredients from all over the world, but great attention is paid to the quality of the ingredients, the distinctiveness of the product and to labelling and packaging. Where possible, local ingredients are used, including local raspberries and strawberries and Scottish tomatoes. The whole operation is highly market-oriented, but in a way that does not compromise product quality. The Massies attend many shows and fairs, which help to enhance brand awareness and create a promotional platform for their products.

What lessons can be learned from this example?

- A broad range of skills is essential to make such a small business work. The key skills of market awareness and technical catering skills are rarely all available. The combination of skills in this business has underpinned its success.
- Brand awareness is essential and has been assiduously cultivated, both with the public through shows and fairs and through applying for and being successful in obtaining food awards.
- It is possible to break into an area (relishes and preserves) already strongly represented in the region through Baxter's Foods, with new niche products, but rather than just targeting at a quality market, they aim to sell into the food gifts market.

Services

Example 3

Glenlivet is a large tenanted estate of approximately 23 000 ha to the north-east of the Cairngorms Mountains, with about 30 tenanted farms, incorporating improved pasture, rough grazings, open moorland, rivers and forestry. Apart from whisky production, which employs very few people directly, the area has been highly dependent on farming and forestry. Both of these sectors have experienced adverse economic conditions in the last decade. The Crown Estate, the owners of the land, commissioned a study in 1991 to look at future development opportunities. Their long-term interest meant that they had to look beyond agriculture and the resultant plan proposed various forms of diversification. At about the same time, the area was designated as an Environmentally Sensitive Area.

With the support and encouragement of the estate, a number of farms have diversified into tourism, with both self-catering and bed and breakfast accommodation. Initially, they produced a leaflet to promote their activity and established a range of joint marketing initiatives with the Crown Estate. They have now developed a website (www.tghh.ukgateway.net) to promote the available accommodation, which is mainly but not exclusively farm-based. The website stresses that it is developed by an association not an agency and offers a warm and personal welcome to those that log onto it and go on to visit the area. The result of joining the forces of a holiday enterprises with a marketing group has led to beneficial effects on the businesses, helping to maintain and increase their occupancy rates at a time when the general trend in Scotland has been downwards. The strength of this group is the strong voluntary association of the members and their commitment to providing a really warm welcome to their visitors.

However, the tourism activity does not take place in a vacuum. Although there are some existing visitor attractions, such as distilleries, the Crown Estate recognized that more had to be done to put Glenlivet on the tourists' map. It has developed a ranger, service, interpretive facilities, trails and cycle routes, and actively promoted farmer adoption of environmental improvement schemes under the Cairngorm Straths Environmentally Sensitive Area.

What are the key lessons of the Glenlivet experience?

- If you are a tenant farmer you need the estate owner on-side.
- You often cannot go it alone: a group of active farmers (and others) involved in the same activity can have a bigger effect.
- You cannot look at tourist enterprises in isolation: you have to think what else the visitor might do and offer a range of services.
- Good promotion and group-based activities with attention to quality can pay significant dividends.

Example 4

Willie is a crofter in the north West Highlands who decided 10 years ago that he wanted to generate a full-time income from a croft, which he took on from his father. At the time he took over the croft, there were over 300 sheep and about 20 hill cattle and their respective offspring. There were diversification grants available under an EU co-financed Rural Diversification Grant Scheme, which he used to build a new house for holiday letting. He lets it through an agency, which takes quite a large cut, but it generates very good occupancy rates, exceeding the average for the region by nearly 10 weeks a year.

He also signed up to the Scottish Countryside Premium Scheme (an agri-environmental scheme) to maintain low grazing intensity and has dramatically reduced the sheep numbers. This action was taken before the prices of store livestock plummeted in recent years. He has also taken on a substantial area of birch woodland under a farm woodland grant scheme. The delivery of the environmental public goods is of considerable interest to his visitors and is remunerative to the croft. In 2000, he converted a small ruined barn into a bunkhouse, which he lets predominantly to hill walkers. He markets the bunkhouse over the internet and in national brochures.

He recognizes the problems of his location (up a single-track road for 7 miles) but also exploits its picturesqueness and proximity to attractive mountains and a famous waterfall. He will pick up and transport people at no cost, holds stocks of food, which he sells on at cost or a small profit, and hires bikes to visitors to enable them to visit a wider hinterland. He clearly augments his core product with many additional services and is highly attentive to his guests and much respected for his efforts. He is not a typical profit-maximizing businessman, but he has taken advantage of environmental land management schemes and other grants to build a secure and diverse range of enterprises, which generate a living wage for him and his family. He is what we might think of as a natural marketer, conscious of a changing market place, and constantly adapting his product range to suit it.

What lessons can we draw from this case?

- It is vital to understand the strengths and weaknesses of a particular location.
- Different types of promotion may be necessary for different products/services.
- The ability to customize the product and to offer a personalized and attentive service to guests pays dividends.
- Different types of diversified enterprise synergize to strengthen the business, reduce risk and offer a more broad-based range of products.

Products and services combined

Example 5

John used to manage a large farm in the Midlands, which was sold off by the owner when property prices were high. He felt a strong attachment to the area and decided to take as redundancy 'payment' a small 10-ha holding and the associated buildings which had been an outlying farm on the complex he previously managed.

He realized immediately that if he wanted to make a living from the farm he would have to diversify. His diversification strategy has been driven partly by the farm assets, partly by his skills and partly by his ability to work out the market possibilities. With 10 million people within a 50-mile radius, there is a real opportunity to sell produce from the farm gate. Whilst some neighbours had proved that there was a market for pick-your-own soft fruit, he decided to grow speciality fruit and vegetables and to sell it from the end of the farm road. He specializes in growing asparagus, over 80% of which is sold direct to the final consumer.

After the asparagus season is over, he sells picked strawberries and specializes in the good-flavoured varieties. He is slowly adding additional speciality fruit and vegetables to broaden his product range, although he is acutely aware of a need for reliable casual labour at peak season. He has adjusted his lambing time for his small sheep flock with a view to direct marketing his lambs to the same client base that he has built up on his fruit and vegetable enterprise.

The converted barn in which he lives with his family is being developed additionally for bed and breakfast. When the conversion was carried out, all rooms were made en-suite. He made the decision to go into bed and breakfast after a colleague running a high-quality fishery nearby complained of his inability to find suitable accommodation for the fishermen who visited the fishery from all over the UK. With no advertising whatsoever, John has already developed a useful supplementary enterprise.

The success of the bed and breakfast enterprise has led him to take advantage of a farm diversification grant to create two self-catering units to a high standard in another old barn. This work remains incomplete, but he is confident that there will be a market from fishermen and others who have come on repeat visits and personal recommendation.

What are the ingredients in success of this bundle of enterprises?

- the location, with a huge nearby market for the picked fruit and vegetables (and the prospect for a growing number of farmers' markets nearby as his product range expands);
- the relatively affluent hinterland of people who lack the time or energy to go to PYO enterprises;
- the ability to work with another nearby enterprise which effectively guarantees him visitors at times of the year when other bed and breakfast facilities are relatively quiet;
- because of its central location in the Midlands, he also has an opportunity to tap into business rather than holiday tourism.

What do we learn overall from these examples?

- Often the success of the diversified activity depends on pulling a number of elements together, perhaps on a single unit, often though in association with others in ways that ensure synergistic benefit.
- All of the enterprises are acutely aware of the need to satisfy the demands of their customer and to go beyond the basic levels of service. They differentiate their product/service both by their personalities and by augmenting the product in a variety of ways.
- Their particular skills add value to the bundle of raw materials that they have. They have consciously looked at the resources they have and built on their strengths and developed appropriate opportunities around their particular skills.
- They have kept costs as low as possible, but have maintained quality rigorously.
- They have a strategic vision of what they want from their enterprise/bundle of enterprises.
- They have not been hidebound by traditions and have been prepared to innovate.
- They have all added substantially to the capital value of their properties, increasing their value should they want to sell out at any time.

Conclusion

To conclude, diversified enterprises can develop on the back of personal interests and intuition. They may succeed. However, the risk of failure of such enterprises is high and to enhance the prospects for survival of such enterprises, a stronger business and market orientation can be seen as a vital ingredient in success. In the absence of a strategic approach to management, an awareness of the market and marketing and a shift from old ways of thinking, many diversified enterprises will struggle. Others will succeed, and an understanding of these success-inducing factors is needed to guide the path forward into an uncertain future.

References

Brunåker, S. (1993) *Farm Diversification – New Enterprises on Competitive Markets*. Dissertations 7. Department of Economics, SLU, Sweden.

Hodge, I. (2001) Rural economy and sustainability. Paper presented to the *Agricultural Economics Society Conference on Agriculture and the Rural Economy*, November 2001.

Kotler, P. (1999) *Marketing Management*, Millennium edn. Prentice-Hall, London.

Kotler, P. & Armstrong, G. (2001) *The Principles of Marketing*, 9th edn. Prentice-Hall, Englewod Cliffs.

Lancaster, G. & Massingham, L. (1988) *Essentials of Marketing*. McGraw Hill, Maidenhead.

Slee, B. (1989) *Alternative Farm Enterprises*, 2nd edn. Farming Press, Ipswich.

Smith, P.R. & Chaffey, D. (2002) eMarketing eXellence: The Heart of eBusiness. Butterworth Heinemann, London.

10

Farming and wildlife

D. Fuller

Introduction

Since the Second World War the main pressure on farmers has been to increase production. Concerns about self-sufficiency, the policies of successive governments, the impact of the Common Agricultural Policy and demand from consumers all pushed farmers to increase production targets. Responses to these pressures were highly successful; too successful in fact, as the huge changes of intensified agriculture brought about overproduction and subsequent production control mechanisms such as quotas and set-aside. It has also become evident that during this period of huge change the wildlife of the countryside suffered drastically and there is now an increasing concern about the environmental impact of intensive farming.

Over three-quarters of the UK countryside is still put to agricultural use with less than 5% of the land directly devoted to conservation purposes. If there is to be a legacy of biodiversity for future generations it will require the proactive support of the agricultural community. Farmers have always acted as custodians of the countryside, but now, with a greater public awareness of the consequences of an intensified agriculture and other emerging forces, they have a truly key role in caring for the countryside and its wildlife. The skill required will be to balance the primary essential of food production with earning a living and with stemming and reversing the declines in wildlife in the countryside.

The emerging forces

There are five main forces pushing farmers to balance production with maintaining biodiversity in the countryside:

(1) Market forces – increased marketing opportunities through rapidly increasing consumer demands for food produced safely and with reduced inputs.
(2) Economic considerations – increased output is no longer a guarantee of increased income and thus inputs must be examined more closely.
(3) Changing policy – government and EC policy is shifting away from commodity support to rewarding environmental gains.
(4) Public pressure – increased public awareness and concern about the environment.
(5) Quality of life – dramatic declines in wildlife populations indicate a decline in the general well-being of the countryside.

Market forces

The pace of change is largely being set by consumer demand. The main trends in demand are:

- Organic food – concerns about the impact of pesticides on the food chain and parents wishing to provide for their children what they perceive as the best food available are bringing about a growing demand for organic food. Supermarkets report a significant increase in the purchase of organic food and in some instances are unable to meet consumer demand. There is assistance to enable farmers to convert to organic production and certified organic produce can realize attractive premiums for the grower. Organic farms are also perceived by consumers to be more wildlife friendly and there is some evidence of this from research by the British Trust for Ornithology.
- Food safety – consumers are increasingly concerned about how safe their food is to eat and the effect it

could have on their long-term health. These concerns have been heightened by various outbreaks of *Salmonella* and BSE in particular. There are parallel concerns in the arable sector over the levels and intensity of pesticides used on crops and the possible effects these might have on the food chain and on wildlife. Consumers are looking for food grown in an assured environmentally safe and healthy way as well as being cheap. Supermarket chains are responding to this by introducing their own schemes which encourage growers to produce crops in the way in which consumers expect and which are environmentally friendly.

- Leisure activities – the opportunities for consumers to get to the heart of the countryside through access schemes and farm diversification into tourist facilities bring the consumer far closer to farm life and activity. They have certain expectations to see wild birds, animals and other wildlife within the working environment and the richer the biodiversity, the more content they will be that the environment is in safe hands.

Economic considerations

Intensive farming is not cheap – pesticides are much more expensive than natural pest controls, and while blanket spraying may encourage high levels of production, it does not guarantee that this production will be profitable. Sotherton and Page (1998) argue that farming for biodiversity through even simple techniques can lead to savings of $15/ha or more on arable land. Current management thinking stresses the need to reduce waste – it is often more profitable to cut costs than increase production.

Changing policy

Alongside this, an increasing number of grants are designed to encourage environmentally friendly farming. The schemes offer payments to farmers who agree to manage their land for the positive benefit of the environment. The schemes in England are usually operated by MAFF and are partly funded by the European Union and include:

- *Environmentally Sensitive Areas (ESAs)*. These are designated parts of the country of particular landscape or wildlife value which are threatened by changes in farming practice. Incentives are offered to farmers to adopt agricultural practices that will safeguard and enhance the rural environment. Over 10% of the agricultural land in England is covered by 22 ESAs.
- Outside the ESAs the *Countryside Stewardship Scheme* is the government's main incentive scheme and operates throughout England. It offers assistance to farmers who are prepared to protect, enhance, restore and recreate certain wildlife habitats and is targeting arable land in two pilot areas, both of which have been oversubscribed.
- *Set-aside* offers important opportunities to increase biodiversity as it can offer space for plant species together with food and cover for many birds, animals and insects that have been endangered by intensive farming.
- Other schemes are available which encourage conversion to organics, planting new woodlands and recreating habitats alongside watercourses and lakes. Similar schemes also operate in Wales and Scotland.

Public pressure

Public opinion is moving increasingly behind measures that support biodiversity and conservation. The result of a government survey of public opinion shows that the environment is one of the top five concerns, just behind law and order and ahead of the economy. Almost half of those surveyed were 'very worried' about the use of insecticides and fertilizers and about the loss of plants and animals.

Membership of conservation organizations continues to grow, with the RSPB, for example, having over one million members. Such organizations are proving to be powerful pressure groups, lobbying government and seeking to work with farmers to change farming practices.

People are looking to spend more time in the countryside and have an expectancy to see a varied landscape and a range of wildlife.

Quality of life

While farming for biodiversity makes good management sense, it also has vital longer-term implications. Changes in farming practice – in particular the switch from pasture to arable, the decrease in spring-sown crops with resultant loss of winter stubbles, the switch from unimproved to improved pasture, drainage of wetlands, the increasing use of pesticides and inorganic fertilizers and the switch from hay to silage – have led to dramatic declines in much of our wildlife.

- A number of important habitats have disappeared. Over 250 000 miles of hedgerow have been lost since 1945, along with 95% of herb-rich meadows, 80% of downland grassland, 60% of lowland heath and 50% of ancient woodland.
- There have been dramatic declines in a number of bird species that are dependent on farmland. The Common Bird Census completed by the British Trust for Ornithology in 1993 showed that in the last 25 years we have lost more than half the population of many farmland birds including the song thrush, skylark, linnet, bullfinch and lapwing. Grey partridge, tree sparrow and corn bunting, once abundant species, have declined by 80%, disappearing from haunts where they were once common and becoming confined to limited areas.
- Many other species of wildlife are also under threat. Mammals such as harvest mice and brown hares are declining, while numbers of the pipistrelle bat have fallen by 70% over 25 years. The water vole is no longer found in two-thirds of its previous haunts and the dormouse has disappeared from half its former range. Insects such as butterflies and grasshoppers, amphibians including frogs, toads and newts, reptiles including adders and lizards, and a range of plants such as bluebells, cornflower and pennyroyal are all in decline.

It is not just environmentalists who are concerned about these changes. The rural community stands to lose most from any deterioration of the environment. Farmers have always had a deep understanding of the countryside and the wildlife that lives in it, and many are committed to halting the declines and encouraging the return of key species.

The process of farming for wildlife

Farming for wildlife and biodiversity should be an integral part of effective management practice. Having understood that there are problems over declines of wildlife in the countryside the process of farming for biodiversity can be seen as a four stage process:

(1) Find out what you have now.
(2) Look after what you have got.
(3) Improve/restore what you have.
(4) Create new habitats.

Find out what you have now

The first practical step that a farm should take is to carry out a biodiversity audit to determine what the farm has now. This should be an inventory of all the characteristics of the farm including area and types of grassland, field margins and boundaries by type, non-cropped areas such as set-aside and rough areas, trees and woods, watercourses, buildings, and of course arable land. Out of this audit a plan for further actions should be created. This might involve maintaining existing, restoring or creating new habitats and should balance the financial costs and benefits of different actions. In many cases simply preserving what is there might be as beneficial to wildlife as the more demanding and costly option of creating something new.

Look after what you have got

In many cases the most important action a farm can take is to maintain its current level of biodiversity, and to avoid this deteriorating further. This is particularly important for wildlife habitats like:

- unimproved or semi-improved pasture
- spring-sown arable fields
- well-established hedges
- woodland, scrub or old trees in fields
- unpolluted ponds, streams or ditches

Ways to maintain these habitats in good wildlife-friendly condition include:

- timing activities such as mowing or spraying to minimize damage to wildlife;
- maintaining a rotation system that includes some spring-sown crops;
- not increasing inputs beyond the current level;
- rotating hedge trimming on a 3-year cycle and trimming just one side of a hedge at a time to promote hedge growth;
- avoiding pesticide drift onto hedges, into streams, ditches or ponds.

Improve/restore what you have

Not every option has to be an expensive one. It may be possible to take quite small steps to improve existing habitats, or restore those that have suffered, for the benefit of wildlife. Little or no-cost examples could include:

- filling in gaps in a hedge and planting tussocky grasses to smother weeds;
- creating grass margins around the edges of fields to provide winter shelter for beetles and other natural predators;
- managing set-aside as wild bird cover;
- converting some silage to hay production to give more time for ground-nesting birds and for wild flowers to set seed;
- linking wildlife-friendly habitats by planting hedgerows or grassy margins.

Create new habitats

While very valuable, the creation of new habitats is not necessarily the best option. It invariably takes longer to realize any benefits for wildlife and is likely to be the most costly option. Nonetheless, this may be worth considering if a small area of the habitat exists already and can be expanded; some species that would naturally colonize it have survived nearby; and financial support is available. Examples can include:

- creating a new pond
- planting a new broad-leaved woodland
- re-establishing grasslands

In all cases it is essential to get expert advice before embarking on the establishment of any new habitat.

Managing for wildlife

Within this four-stage process of farming for wildlife there is a whole range of benefits to be gained and a whole variety of management actions that can bring about those benefits. Field boundaries and margins can be among the most fruitful areas to develop for wildlife, but the importance of everyday crop management and grassland management should not be ignored; neither should the benefits that can be derived from wildlife-friendly set-aside management. Virtually every farming management decision can have an effect on wildlife.

Managing hedgerows

Hedgerows are a very valuable resource, providing food, shelter, breeding areas and corridors for wildlife. Wildlife-friendly management techniques include:

- Rotate hedge cutting on a 2- to 3-year basis. Cut different parts of hedges each year to create a variety of growths; a dense, 2-m-high hedge holds the greatest variety of wildlife.
- Manage hedges between November and February after breeding is over and the berries have been eaten. Never cut between April and July.
- Maintain structure and fill gaps by planting, coppicing or laying.
- Avoid ploughing right up to the base of the hedge – leave a margin of at least 1 m which could be planted with tussocky grasses.
- Leave old trees, but consider pollarding some hedgerow trees.
- Avoid drift of pesticides and fertilizers into the hedgerow.
- Avoid grazing by stock.

Managing field margins and game strips

Just as with the hedgerow boundaries, field margins and game strips can provide food, shelter, breeding areas and corridors for wildlife and can at the same time be beneficial to the arable operations. Suitable management techniques include:

- Create a 2- to 6-m uncropped strip round the outside of the field (or a 20-m strip under set-aside).
- Allow natural colonization or seed tussocky grasses such as cock's-foot and Yorkshire fog that smother weeds such as cleavers and barren brome and provide shelter for natural predators like spiders and beetles.
- Plant a mix of brassicas and cereals and leave for 2 years to allow broad-leaved annuals to colonize.
- Do not use fertilizers or spray with insecticides or broad-leaved herbicides.
- If neccesary use selective herbicides to control cleavers or grass weeds and apply only in November or December.
- Mow in the first year to control weeks and help tussocky grasses to form.
- If the weed burden is heavy, use a 1-m sterile strip between the crop and the margin.

As an alternative to an uncropped strip it is sometimes worth considering having conservation headlands whereby the 6-m headland is still cropped but is not sprayed with insecticides in summer or fertilized. For larger fields it is also worth considering the creation of a beetle bank across the field to encourage natural predators.

Crop management and protection

Arable land is host to many hundreds of species, some of which are now rare. Nearly 300 varieties of wild plants, including some of our rarest, grow on tilled land and the vast majority of invertebrates present do no harm at all and are part of the food chain. Few mammals and birds live in standing crops but many make use of them. The height and variety of crops grown will dictate the variety of wildlife.

- Aim for diversity of crops and plan long-term rotations, extending the intervals between repeat cropping.
- Choose varieties resistant to pests and diseases, avoiding those needing high inputs to achieve full yields.
- Leave winter stubbles and other crop residues untouched as long as possible, preferably until the following spring.
- Harvest from the centre of a field outwards.

A better understanding of the role of natural predators and the development of selective pesticides and more accurate delivery mechanisms have made it much more practical to integrate efficient crop protection with wildlife conservation. The key is advance planning:

- Create a long-term strategy, identifying the main threats using past knowledge.
- Calculate economic pest thresholds, aiming to control, not cure.
- Consider all available alternatives, choosing only the right pesticides.
- Keep in touch with developments.

Set-aside

Set-aside rules are variable, but both rotational set-aside and 'permanent' set-aside can be valuable feeding and breeding habitats for wildlife on the farm at no expense to arable cropping. For non-rotational set-aside it is well worth considering a 20-m-wide grassy field margin or conservation game strip. Natural regeneration of non-rotational set-aside produces particularly good wildlife habitat.

Grassland management

Both pastures and meadows can support a wide range of plants and animals when managed appropriately. Where silage is taken this is less beneficial than when hay is made, when ground-nesting birds in particular can benefit. However, if cutting of either hay or silage is done too early then the effect can be detrimental to wildlife. There also needs to be a minimum of 6 weeks between silage cuts if birds are to have time for a second breeding attempt. Beneficial management includes the following:

- Continue to graze unimproved pasture without under- or overgrazing, keeping summer grazing to a low intensity.
- Cut hay late 1 year in 5 and normally not before early July in the south and August in the north.
- Do not reseed or use artificials or slurry.
- Maintain or restore winter flooding regimes and high spring water tables where appropriate.

Advice and assistance

There are many other more detailed aspects of farming for wildlife than given here, and it is essential that for the more specialized farming expert advice is sought. However, many of the actions that will encourage wildlife as an integral part of farming on the majority of farms are simple and can often be done with little effort and at little or no cost. Others are more demanding and, in some cases of restoration, for example, can be costly. In most instances, however, there is now assistance available through agri-environment schemes and the like and in some cases advice can be obtained free of charge. Even for some of the simple actions it is sensible to take advice before commencing on the actions, and for the more complicated, such as digging new ponds, it is essential.

A way forward

Modern farming can be balanced with the need to prevent further declines in, and to enhance the future of, wildlife in the countryside. It does, however, need to be planned in advance and well managed. Advice and assistance are generally available to aid this process, which will enable the public's expectations to be fulfilled.

References and further reading

Andrews, J. & Rebane, M. (1994) *Farming and Wildlife: A Practical Management Handbook*. RSPB, Sandy.

British Agrochemical Association (1997) *Arable*

Wildlife: Protecting Non-Target Species. British Agrochemicals Association, Peterborough.
Dodds, D., Appleby, M. & Evans, A. (1995) *A Management Guide to Birds of Lowland Farmland*. RSPB, Sandy.
Sotherton, N. & Page, R. (1998) *A Farmer's Guide to Hedgerow and Field Margin Management*. Game Conservancy Trust. Fordingbridge.

Useful websites

www.allertontrust.org.uk – The Allerton Trust.
www.bto.org/ – British Trust for Ornithology.
www.fwag.org.uk/ – Farming & Wildlife Advisory Group.
www.gct.org.uk – Game Conservancy Trust.
www.lbcnc.org.uk – Land Based Colleges National Consortium.
www.defra.gov.uk/ – DEFRA.
www.rspb.org.uk – RSPB.

11

Organic farming

N. Lampkin

Definitions and historical perspectives

Since the early 1990s, organic farming has expanded rapidly in the United Kingdom, other parts of Europe and around the world. The expansion has been fuelled by strong interest from consumers and policy-makers, reflecting the perceived potential of organic farming to contribute to environmental, animal welfare, social and nutritional goals. More recently, increasing attention has also been paid to the rural development potential of organic farming in the face of declining incomes from many conventional[1] farming systems. However, organic farming as a concept is not new, dating back to the beginning of the twentieth century, and the concept is much more than the common perception of no use of artificial fertilizers and pesticides.

What is organic farming?

Organic farming can be defined as an approach to agriculture where the aim is to create integrated, humane, environmentally and economically sustainable production systems. This encompasses key objectives related to achieving high levels of environmental protection, resource use sustainability, animal welfare, food security, safety and quality, social justice and financial viability.

Maximum reliance is placed on locally or farm-derived, renewable resources (working within closed cycles) and the management of self-regulating ecological and biological processes and interactions (agro-ecosystem management; see Altieri, 1995), in order to provide acceptable levels of crop, livestock and human nutrition, protection from pests and diseases, and an appropriate return to the human and other resources employed. Reliance on external inputs, whether chemical or organic, is reduced as far as possible. In many European countries, organic agriculture is known as ecological agriculture, reflecting this reliance on ecosystem management rather than external inputs.

The term 'organic' refers not to the type of inputs used, but to the concept of the farm as an organism, in which all the component parts – the soil minerals, organic matter, micro-organisms, insects, plants, animals and humans – interact to create a coherent and stable whole.

In recent years, the market for organic food has developed strongly and is now often seen as a main feature of organic farming, However, the market initially developed as a means to support the broader goals of organic farming, rather than an end in itself, at a time when official support was non-existent. This allowed organic producers to be compensated financially for restricting their production practices, effectively internalizing costs that could be considered as externalities of conventional agriculture (Pretty *et al.*, 2000).

Detailed descriptions of the principles and practices of organic farming can be found in various publications (e.g. Lampkin, 1990; Blake, 1994; Newton, 1995), as well as the detailed codes of practice contained in the standards documents of the various certification bodies operating in each country (see 'Role of the market and regulations').

[1] The term conventional is used to refer to a wide range of non-organic systems that reflect the majority or normal practice in a particular region. Both high- and low-intensity holdings can be included in this usage of the term.

Why organic farming?

The development of organic farming can be traced back to the 1920s, although many of the underlying ideas of self-reliance and sustainability feature also in earlier writings (for reviews, see Boeringa, 1980; Merrill, 1983; Conford, 2001; Reed, 2001). Steiner (1924) laid the foundation for biodynamic agriculture (Sattler and Wistinghausen, 1992), grounded in his spiritual philosophy of anthroposophy, which later was to have a significant influence on the development of organic farming. At about the same time, Dr Hans Müller founded a movement for agricultural reform in Switzerland and Germany, centred on Christian concepts of land stewardship and preservation of family farms. Later, Dr Hans-Peter Rusch contributed important ideas relating to soil fertility and soil microbiology, which led to the further development of organic-biological agriculture in central Europe (Rusch, 1968).

In the English-speaking parts of the world, King (1911) in *Farmers of Forty Centuries* used the long history of Chinese agriculture with its emphasis on recycling of organic manures as a model of sustainability, while McCarrison (1936) focused on nutritional issues and the influence that methods of food production might have on food quality and human health. Stapledon's work on alternative husbandry systems in the 1930s and 1940s (see Conford, 2001) and Sir Albert Howard's work on the role of organic matter in soils and composting (Howard, 1940) were also of key importance. These writers provided a powerful stimulus to Lady Eve Balfour (Balfour, 1976), who founded the Soil Association in 1946. The key emphasis at that stage, as the name suggests, was on soil fertility and soil conservation, with the dust bowls of the 1930s a recent event. The links between a healthy, fertile soil and crop and livestock health, food quality and human health were central to the mission of the organization.

Since the 1960s, wider concerns have influenced the development of organic agriculture, including food safety concerns due to pesticide residues, BSE and other issues, social concerns over working conditions in agriculture, the loss of jobs and rural population decline, as well as animal welfare concerns and environmental concerns over the loss of wildlife species and habitats, pollution, and the use of non-renewable resources.

These issues have come to be reflected in the broad concept of sustainable agriculture (Pretty and Hine, 2001), which emphasizes the use of systems and practices that maintain and enhance food supplies, safety and quality; financial viability of farm businesses; resource use sustainability; ecological impacts (including impacts on other ecosystems, pollution, biodiversity); and social and cultural wellbeing of rural communities.

The objective of sustainability lies at the heart of organic farming and is one of the major factors determining the acceptability or otherwise of specific production practices and technologies. However, organic farming does not represent the only approach to achieving agricultural sustainability, with integrated crop production in Europe (LEAF, 2001) and low-input sustainable agriculture (LISA) in the US (Francis *et al.*, 1990) providing less-restrictive options, while permaculture (Mollison, 1990) and agro-forestry (Young, 1997) place more emphasis on the potential of integrating perennial crops (trees and shrubs) to achieve sustainability.

A perfectly sustainable agriculture is not achievable because sustainability is a multi-objective concept and there are inevitably conflicts and trade-offs between the different objectives. A key issue, therefore, is the relative contribution that the different approaches can make. This is subject to considerable debate [see Stockdale *et al.* (2000), Stolze *et al.* (2000) and Tinker (2000) for contrasting views] – not just because of the lack of evidence in some areas, but also because of the need to take account of the weightings placed on individual sustainability objectives by different parts of society, with clear differences observable, for example, between the food industry and environmentalists.

Growth of organic farming

Recent years have seen very rapid growth in organic farming (for detailed statistics, see Foster and Lampkin, 2000). In 1985, certified and policy-supported organic production accounted for just 100 000 ha in the EU, or less than 0.1% of the total agricultural area. By the end of 2000, 4 million ha, 3% of the total EU agricultural area, was managed organically, representing a 40-fold increase in 15 years (Lampkin, 2001a). Over the same period, the number of certified holdings increased from 6000 to 140 000 (Fig. 11.1).

These figures hide great variability within and between countries. Several countries have now achieved 6–12% of their agricultural area managed organically, and in some cases more than 30% on a regional basis. Countries like Austria, Italy, Spain, Sweden and Switzerland, and more recently the UK, have seen the fastest rates of growth. In the UK, the organic land area grew from 6000 to 50 000 ha between 1985 and 1996, but increased dramatically to 525 000 ha in 2000. The number of farms has increased from 865 in 1996 to 3500 in 2000.

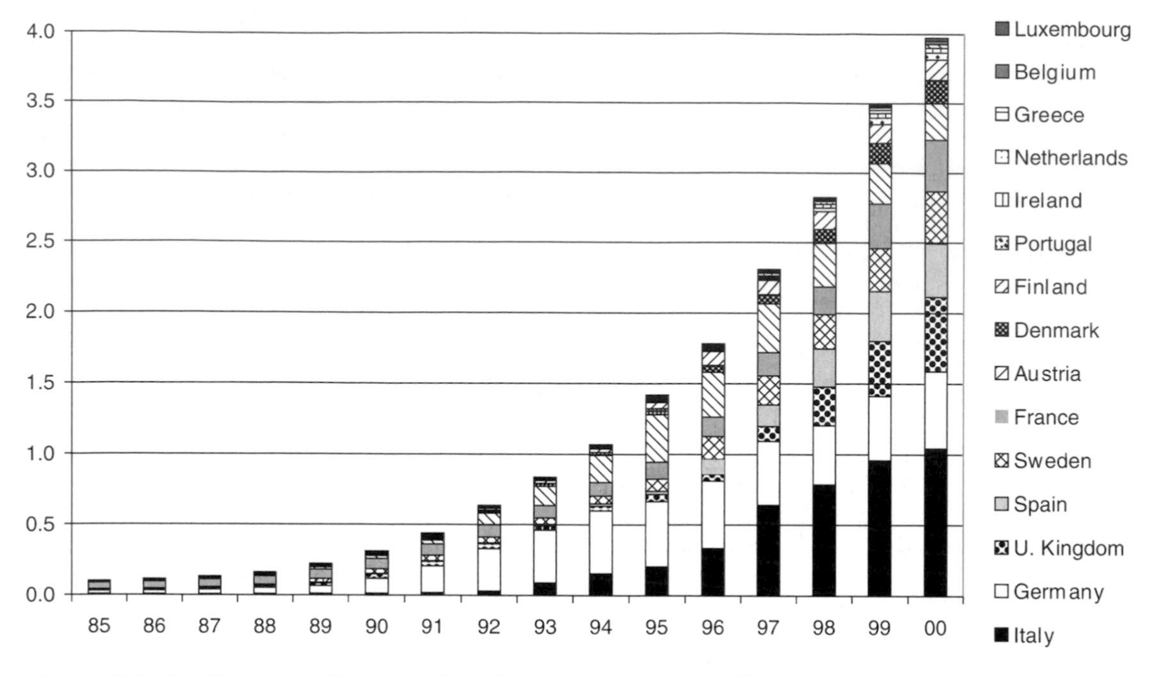

Figure 11.1 Certified organically managed and in-conversion land area (Mha) in the European Union, 1985–2000 (year end).

Alongside the increase in the supply base, the market for organic produce has also grown, but statistics on the overall size of the market for organic produce in Europe are still very limited. Some estimates have suggested that the retail sales value of the European market for organic food was of the order of 7–9 billion Euro in 2000/2001 (Soil Association, 2001a). The UK share of this is approximately 15%, with the UK organic market valued at £260 million retail sales value in 1997/1998, rising to £390 million in 1998/1999, £600 million in 1999/2000 and £800 million in 2001 (Soil Association, 2001a). However, the rate of land conversion in the UK has not kept pace with demand until recently, leading to increased reliance on imports, estimated at 70% of retail sales in 2001 compared with 75% in 2000 (Soil Association, 2001a).

Organic farming in practice

Soil fertility and soil management

The soil is one of the most important resources in agriculture and the maintenance of the long-term fertility of soils is a key objective of organic farming. Here a distinction needs to be made between fertility, in the sense of the inherent capacity of the soil to sustain production, and productivity, which could entirely be due to the input of external resources. Soil biological activity (microbes, fungi, nematodes, earthworms, arthropods, etc.) plays an essential role in the maintenance of soil fertility. Organic matter plays a critical role in maintaining a stable soil structure and soil biological activity, by providing soil micro-organisms, fungi and earthworms with the energy to make nutrients available to crop plants. Organic farming favours supportive practices such as green manuring, straw incorporation and the utilization of other crop and livestock residues. There is now a substantial body of research evidence to indicate that these management practices, as well as the restrictions on fertilizer and pesticide inputs, have a significant beneficial effect on soil biological activity (e.g. Reganold *et al.*, 1993; Mader *et al.*, 1996).

Soil cultivations in organic farming should aim to provide a soil structure that will allow for deep rooting by crops, providing adequate aeration and drainage so that plant roots can exploit available nutrients in the full soil profile. This is particularly important as highly soluble nutrient sources applied to the soil surface are avoided. At the same time, cultivations should aim to maintain the biologically active surface layers in the top 15–20 cm; therefore shallow ploughing or minimal cultivations in combination with subsoiling-type operations may be preferred, but the chosen method of cultivation will also need to fulfil other objectives, in particular weed control requirements.

Manure management

Farmyard manures and slurries, often referred to inappropriately as livestock 'wastes', represent a valuable resource in organic farming. If mismanaged, they represent not only a significant environmental problem, but also a productive loss with financial consequences for the farm business.

Great care has to be taken to avoid pollution by manures or slurries (MAFF, 1998; ADAS, 2001). Storing manure outside in a field is unsatisfactory due to the risk of nutrient leaching and pollution, but this may be minimized by thoughtful siting and covering with a plastic sheet during winter. Careful composting and slurry aeration can enhance the quality of farmyard manures and slurries by improving the smell, nutrient stability and weed control.

Manure and slurry applications should be planned into the rotation with applications directed at points of greatest nutrient off-take, for example where green matter is harvested and sold off the farm or conserved as forage.

Crop nutrition

The principle of reliance on farm- or locally derived renewable resources in organic farming means that, where possible, fertility should be obtained from within the farm system. In the case of nitrogen, which is freely available from the atmosphere, biological nitrogen fixation provides the basis for nitrogen self-sufficiency. This requires an appropriate proportion of legumes in the rotation and good management of crop residues and livestock manures.

In principle, energy self-sufficiency through solar energy capture by photosynthesis and other forms of renewable energy should also be possible, but in practice some reliance continues to be placed on fossil energy for mechanization and other purposes. The development of bio-fuel crops, in particular for bio-diesel, might in future provide a means of achieving the energy self-sufficiency objective in practice.

Despite the aim to improve the recycling of other crop nutrients, there will always be some net export from the holding as crops and livestock are sold. The first step is therefore to minimize unnecessary losses from the system, by avoiding leaching and erosion. Attention should also be paid to the sales of nutrients off the farm, for example to avoid high potassium losses from the system through selling of straw and conserved forage crops. Secondly, some reliance can be placed on the release of nutrients from soil minerals through normal soil formation processes. Purchases of livestock feeds and straw can also add a substantial amount of nutrients to the cycle.

If sales exceed the purchases and rate of natural regeneration, then it will be necessary to supplement nutrients such as potassium, phosphate, calcium, magnesium and trace elements from external sources. These should preferably be in an organic form, or in a low-solubility mineral form, so that the nutrients are released and made available to plants indirectly through the action of soil micro-organisms. In this way, luxury uptake by crop plants can be avoided. In some cases, such as potassium, this may not always be possible. Production standards therefore allow restricted use of potassium sulphate in cases of demonstrated need and provide a list of inputs for soil improvement and crop nutrition that organic farmers can use. Nutrient budgeting can be used to check whether the returns of manures and/or slurry are adequate to compensate for nutrient removals in a balanced rotation (Lampkin, 1990).

Crop production

Rotations

The rotation is the core of most organic farms, based on the principle that diversity and complexity provide stability in agricultural as in other ecosystems. Beneficial interactions between enterprises, and between the farm and the external environment, should be exploited fully. The role of the rotation is therefore to:

- ensure sufficient crop nutrients and minimize their loss;
- provide a self-sustaining supply of nitrogen through legumes;
- minimize and help control weed, pest and disease problems;
- maintain soil organic matter and soil structure;
- provide sufficient livestock feed where necessary;
- maintain a profitable output of cash crops and/or livestock.

There is no 'blueprint' organic rotation. There are as many different types of organic system as there are conventional systems, and each uses rotation in different ways to satisfy the requirements listed above. In some cases, such as all-grass farms or perennial cropping, there may be no rotation at all. Instead, diversity is achieved through species mixtures in space rather than over time.

Polycultures

Polycultures provide an important alternative to rotations in many situtations, but may also be included within a rotational context. The term includes a variety of approaches including strip cropping, intercropping, under-sowing and crop and variety mixtures. In many contexts, these are more productive than growing single crops, due to improved exploitation of available space, light and growth factors, complementarity between nitrogen-fixing and nitrogen-demanding species, and structural/shade support offered by specific species. Polycultures form the core of many tropical organic farming systems, and polycultures including perennial tree and shrub species are a key element of permaculture (Mollison, 1990) and agro-forestry (Young, 1997) systems.

The main perceived disadvantage of polycultures in the context of commercial UK agriculture is that they tend to be less easy to mechanize and more labour intensive. However, many crop mixtures can be mechanized, for example cereals and grain legumes or variety mixtures of the same species.

Weed control

Herbicides are not permitted under organic production standards. Weed control is achieved primarily through preventive measures, including rotation design, soil cultivations, crop variety selection (e.g. tall cereal varieties to shade out weeds), under-sowing, use of transplants, timing of operations and mulching with crop residues or other materials. Direct intervention to correct weed problems when they occur should be seen as secondary. Options include mechanical, thermal and biological controls of different types. In practice, the need for significant additional weed control (in addition to techniques of cultivation around crop establishment) in arable crops and grassland is relatively low, although in some cases perennial weeds such as some grass weeds, docks and thistles can become problematic. For horticultural crops, mechanical control and hand-weeding may be needed, the extent of which depends on the extent of mechanization of the holding and on weather conditions and timeliness.

Pest and disease control

Pest and disease control should also be achieved primarily through preventive measures, including balanced crop nutrition, rotation design, variety selection for resistance, habitat management to encourage pest predators, and organic manuring to stimulate antagonists to soil-borne pathogens. Where direct intervention is required, organic production standards permit a number of biological controls, such as *Bacillus thuringiensis* for cabbage white caterpillar control, as well as certain non-synthetic pesticides and fungicides, several of which require prior approval by the certification body. These products should be used as a last resort as they still have the capacity to disrupt beneficial insects and the ecosystem interactions on which the stability of the organic farming system depends. The general requirement for undressed seed may increase pest and disease risks at germination, particularly for forage crops, so seed quality and hygiene become very important.

Livestock production

Role of livestock on organic farms

Except in the case of stockless horticultural and arable farms, livestock form an integral part of organic farming systems. Ruminant livestock are able to utilize and provide a financial return to the legume and other herbage species that contribute to nitrogen fixation and the fertility-building phase of the rotation. They can digest cellulose and hence obtain energy from herbage that would not otherwise be available for human consumption and the return of livestock manures provides a nutrient source for subsequent crops. Livestock thus play an important role in nutrient and energy cycling. They can also contribute to weed, pest and disease control through grazing and forage conservation. As with crop production, an emphasis on single species can lead to problems with internal parasites or weeds (e.g. bracken in grassland). The mixing of species, such as sheep and cattle or sheep and poultry, can contribute to the control of parasites and improve grassland management.

As far as possible, livestock enterprises should be land-based and supported from the farm's own resources. This effectively excludes intensive, permanently housed pig, poultry and feedlot cattle production, which depend heavily on bought-in feeds and lead to livestock 'waste' disposal problems. The manures produced should be capable of being absorbed by the agricultural ecosystem without leading to disposal or pollution problems. Stocking rates should reflect the inherent carrying capacity of the farm and not be inflated by reliance on 'purchased' hectares. The EU regulation defining organic livestock production (1804/1999) stipulates a maximum stocking rate of ca.

2.0 LU/ha, although farms may stock more intensively if they are able to form an agreement with another organic farm to take the surplus manure and the combined stocking rate on the two holdings does not exceed the specified limit.

Livestock nutrition

The aim in organic livestock feeding is to rely primarily on home-produced feeds which are suited to the animal's evolutionary adaptation and which, as far as possible, avoid the use of feedstuffs that are suitable for direct human consumption. In the case of ruminants, this means the ration must be predominantly forage (>60% DM). This is normally comprised of grass/clover grazing and conserved forage, although there is clearly potential for other forage crops such as arable silage, fodder beet and maize. The use of cereals and pulses in ruminant diets is aimed at balancing the diet rather than stimulating additional production. Pig and poultry rations will inevitably rely on cereals and pulses although there is also a need to provide some green material and indeed considerable reliance can be placed on forage crops in pig diets.

New organic livestock production standards aim to achieve fully organic diets for all animals by August 2005. In the interim, annual limits on the quantities of dry matter fed from conventional sources have been specified as 10% in the case of herbivores (i.e. ruminants and horses) and 20% for all other species (i.e. pigs and poultry). All other feedstuffs must be organically produced, although up to 30% may come from land still in conversion. This proportion can increase to 60% if the feed is produced on the own holding. The standards contain a positive list of permitted feed materials and only components listed can be used. Pure amino acids and coccidiostats are not permitted.

Trace elements and mineral supplements are not fed as a routine but are supplemented where specific deficiencies occur, usually in the form of rock salt or seaweed meal. Problems such as hypomagnesaemia are less common on organic farms because of reduced fertilizer use, but pastures may be dusted with calcined magnesite or magnesium supplements fed if necessary.

Livestock health

Animal health in organic farming is based on preventive management and good husbandry. The aim, through good stockmanship and attention to detail, is to optimize breeding, rearing, feeding, housing and general management in order to achieve stability and balance in the farming system, maximize the natural health of the animal and minimize disease pressure and stress. All organic livestock farms are required to have an animal health plan as part of the livestock management, which should be drawn up with the help of a specialist organic advisor and the veterinary surgeon.

Preventive treatment is restricted to limited use of vaccination and homoeopathic nosodes for known farm problems or strategic use in the context of a health plan. Growth promoters, hormones and the routine (prophylactic) use of antibiotics (e.g. dry cow therapy) are not allowed. Where possible, treatment of ailments is approached by aiding the animal's own resistance and the use of complementary therapies such as homoeopathy. Conventional treatment should be used in all cases where it is necessary to prevent prolonged illness or suffering, but a maximum limit of three courses of treatment within a year and longer withdrawal periods are imposed under organic production standards. A medicine book recording all cases and treatments (including complementary medicines) must be kept.

Hovi *et al.* (2000) have produced a comprehensive guide to preventive management strategies and treatments for livestock diseases in organic farming. Further information on animal health and welfare issues in organic farming can also be found on the NAHWOA website.

Livestock behaviour and welfare

Animal welfare is a key objective of organic farming and organic production standards contain a number of provisions designed to achieve this, including, as a minimum, that the MAFF/DEFRA welfare codes for different livestock species should be applied, and that animals should be free to exhibit their normal behaviour patterns. However, the animal welfare case is often disputed (e.g. House of Commons, 2001), in part because welfare is often seen in the narrow context of animal health and the restrictions placed on the use of prophylactic treatments. An understanding of basic animal ethology and welfare considerations (e.g. Foelsch, 1978) is therefore fundamental to the interpretation and implementation of production standard requirements.

The Farm Animal Welfare Council's Five Freedoms (covering hunger, thirst, pain, disease, stress and freedom to exhibit natural behaviour patterns) provides a basis for assessing the welfare provisions and impacts of organic farming – on many the organic standards provide for more than the legal minimum, in particular with respect to housing, outside access and freedom to exhibit normal behaviour patterns, but some areas are open to debate, for example:

- Do the restrictions on feedstuffs permitted in organic farming cause a welfare problem in high genetic-merit animals due to insufficient nutrient concentrations leading to hunger? Many organic producers are reluctant to use high productivity/ high genetic-merit breeds for this reason.

- Do the restrictions on the use of prophylactic medications, and on the number of treatments before an animal loses its organic status, result in a failure to protect animals from pain and disease? Organic standards require farmers to implement health management plans and to treat to avoid prolonged suffering, but is this sufficient?
- Do restrictions on mutilations such as beak clipping represent a greater welfare threat because of the risk of feather-pecking and cannibalism? Are the alternative management procedures proposed for organic free-range poultry production sufficient to minimize the risks of a problem developing?

These questions are not simple to answer and there are many different views on the subject. Further research is underway to assess the animal health and welfare implications of organic farming (NAHWOA website), but there is also a need to keep organic standards under review in order to ensure that the animal welfare goals are achieved in practice.

Livestock housing

Outside access and housing appropriate to animal welfare and behavioural needs are required under organic production standards. Battery cages, tethering, fully slatted floors, etc. are prohibited, with the emphasis instead on free-range systems, particularly for pig and poultry production. Where livestock have to be housed, group housing is required (for calves after 1 week), natural bedding materials should be used and the standards specify minimum space allowances for different livestock species. Housing enrichment, for example scratching posts, and overhead shelter on poultry ranges are recommended.

Livestock breeding and rearing

Emphasis is placed on maintaining closed herds and flocks, i.e. breeding replacements on the farm, so as to minimize the risk of importing diseases from elsewhere. Other breeding objectives may include high lifetime yields from forage for dairy cows, suitability of outdoor systems for pigs, and internal parasite resistance in sheep. Breed choice will also be related to the quality requirements of target markets. Breeding males may be brought from non-organic sources as part of the 10% non-organic livestock replacement allowance and the use of A.I. is permitted. Calves have to be fed on organic whole milk or organic milk replacer up to the age of 12 weeks, and, therefore, multiple suckling by a nurse cow is often preferred in dairy herds, but rearing at foot may also be an option in some cases. In other cases, natural weaning may be practised. Although these practices add to production costs, it is often argued that the benefits in terms of subsequent health, production and longevity more than compensate.

Organic farming as a business

Financial viability is as important for an organic farmer as for any other farmer. A business that is not viable is not sustainable. In order to maintain financial viability while restricting the use of practices and technologies that could enhance yields, organic producers have had to focus more on quality, adding value, and marketing of their products. In recent years, the strong market for organic products has resulted in substantial differences in the prices for organic and conventional products, although in fact organic prices have remained relatively stable, while conventional prices have declined significantly. Conventional and organic markets have effectively become decoupled, so that the notion of a fixed percentage premium for organic products is outdated.

The financial and physical performance of organic farming in Europe has been reviewed by Lampkin and Padel (1994) and Offermann and Nieberg (2000). Lampkin and Measures (2001) provide detailed information on typical gross margins for organic enterprises in the UK, while Fowler *et al.* (2000 and 2001) provide survey results for different organic farm types in England and Wales.

Enterprise gross margins

In a northern European context, organic crop yields can be substantially lower (40–50%) than conventional, particularly where intensive production methods are used. However, this does not necessarily apply elsewhere. In the United States and Australia, where conventional methods are typically less intensive than in Europe, the relative differences are less significant (10–20%), and Pretty and Hine (2001) have shown clearly that in many resource-poor, developing countries, sustainable agricultural approaches (including organic farming) have the potential to actually increase yields through more effective utilization of internal resources.

However, the yield reductions experienced in the UK are such that higher prices are essential to compensate, despite the savings on input costs. Current organic prices for crops more than compensate for these losses, resulting in substantially higher gross margins on organic farms (Table 11.1). For milk, higher gross margins per cow (and per litre) can be achieved, but the price premium would need to be restored from the end 2001 price of 25 ppl (pence per litre) to the early 2001 price of 30 ppl to achieve similar margins per hectare.

Whole farm performance

Despite the relatively good performance at gross margin level, the rotational and other constraints on enterprise mix on organic farms means that high gross margins for individual enterprises do not necessarily translate into high whole farm gross margins. In particular, the financial returns to the fertility-building phase of the rotation may be significantly lower than to the main cash crops, but account needs to be taken of the hidden benefits of interactions between the enterprises (non-financial internal transfers), for example with respect to weed control and parasite control. It would therefore be a mistake to use gross margin performance as a main criterion for selecting production enterprises in organic farming.

Recent surveys of organic farms in England and Wales (Fowler *et al.*, 2000, 2001) and other European studies (Offermann and Nieberg, 2000) have confirmed that organic farms can make incomes that are comparable to similar conventional farms (Table 11.2), although much of this is due to the declining profitability of conventional farms in recent years rather than an increase in the profitability of organic farms.

Various studies indicate that labour use is higher on organic farms, typically in the range 10–25% (Lampkin and Padel, 1994; Jansen, 2000; Offermann and Nieberg, 2000). These increases are associated with the introduction of more labour-intensive (but high-value) crops and/or production techniques, on-farm cleaning, grading, processing and marketing of produce, and the diseconomies of scale associated with a greater diversity of enterprises. There is little evidence that labour requirements for *existing* enterprises increase substantially, although labour use per animal may increase as a result of reduced stocking, preventive health management, increased observation and where intensive livestock enterprises are converted to free-range systems.

Conversion to organic farming

The conversion (or transition) from conventional to organic farming systems is a complex process subject to several physical, financial and social influences which differ from those associated with established organic farming systems, involving a significant number of innovations and restructuring of the farm system as well as changes in production methods. The time required and the difficulties associated with the necessary changes depend on the intensity of conventional management and the condition of the farm before conversion, the extent to which new enterprises and marketing activities are introduced, and any yield and financial penalties related specifically to the conversion process. As a consequence, the conversion period may well be longer than the statutory 2-year conversion period prescribed by regulation before full organic certification and premium prices are obtained.

Prior to and during conversion, farmers face a range of challenges, including personal (family), social (peer pressure) and institutional resistance to the decision to convert (Padel, 2001), in addition to the direct financial and technical impacts on the farm business. Intensive

Table 11.1 Comparative gross margins for selected organic and conventional (*Conv.*) enterprises, 2001 prices

		Wheat		*Potatoes*		*Milk*		
	Unit	*Organic*	*Conv.*	*Organic*	*Conv.*	*Unit*	*Organic*	*Conv.*
Yield	t/ha	4	8	28	42	l/cow	5800	6400
Price	£/t	180	80	200	100	ppl	25	20
Output	£/ha	950[1]	870[1]	5600	4200	£/cow	1450	1280
Seeds	£/ha	90	50	1200	550	Feeds	280	230
Fertilizer and sprays	£/ha	35	200	100	600	Forage	60	90
Input	£/ha	140	260	2400	2550	£/cow	490	480
Margin	£/ha	810	610	3200	1650	£/cow	960[2]	800[3]

[1] Incl. £230/ha arable area payment.
[2] £1440/ha @ 1.5 LU/ha (£1730/ha @ 30 ppl).
[3] £1760/ha @ 2.2 LU/ha.

Table 11.2 Model farm estimates of financial changes during and after conversion (*Conv.*), by farm type, early 2001 prices (Lampkin, 2001b)

Farm type	*Size (eff. ha[1]) (acres)*	*Annual whole farm gross margins less conversion-related fixed cost charges (£/eff. ha[1])*			
		Conv.	*Staged transition[2]*	*Crash transition[2]*	*Organic*
Specialist dairy	60 (150)	1109	1153	1391	1754
Mainly dairy	145 (363)	772	886	1097	1337
Stockless arable	210 (525)	592	767	734	897
Mainly arable	260 (650)	473	489	511	562
Lowland livestock	180 (450)	365	411	484	555
Upland livestock	83 (206)	247	380	425	429
Hill livestock	130 (325)	190	258	276	235

[1] Effective area (rough grazing adjusted).
[2] Five-year average.

preparation, including visits to and contacts with other organic farmers, as well as detailed planning can help address many of these potential problems (Padel and Lampkin, 1994; Lampkin and Measures, 2001). Once the decision to convert has been taken, two main options exist within the context of organic production standards:

(1) *Staged* conversion involves the conversion of parts of the farm, typically 10–20%, in successive years, using a fertility-building legume crop as an optimal entry point into organic management. The learning costs, capital investment and risks can be spread over a longer period, sometimes up to 10 years or the full length of a rotational cycle, and are more easily carried by the remainder of the farm business.
(2) *Single-step* or crash conversion involves converting all the land on the farm at one time. This enables the farm to gain access to premium prices sooner, but means that all the risks, learning costs and financial impacts of conversion are concentrated into a short period of time. Rotational disadvantages can arise because not all of the farm can be put down to fertility-building crops at the same time. If mistakes are made, the impacts are likely to be more severe and the approach may turn out to be more costly, despite the earlier access to premiums.

There is some evidence of a decline in yield during the conversion period greater than that which would be expected in an established organic system (Dabbert, 1994). This is because biological processes such as nitrogen fixation and rotational effects on weeds, pests and diseases take time to become established. In many cases, avoidable conversion-specific yield reductions may be due to mistakes or inappropriate practices, such as the removal of nitrogen fertilizer without taking action at the same time to stimulate biological nitrogen fixation using legumes. Forage crops are the one area where conversion-specific yield decline may be inevitable, with production lost as a result of reseeding grassland with new mixtures containing legumes, or waiting for clover to establish naturally following the withdrawal of nitrogen fertilizer. The loss of output can place significant pressure on stocking rates for livestock on predominantly grassland farms, particularly where permanent grassland is involved.

The costs of conversion vary widely according to individual circumstances, arising from a combination of output reductions, new investments, information and experience gathering and changes in production costs. However, lack of access to organic prices during conversion, and changes in eligibility for subsidies, may be more important factors influencing the cost of conversion. Conversely, the Organic Farming Scheme in the UK (see DEFRA website) with regional variations in Scotland, Wales and Northern Ireland, provides direct financial support to farms in conversion, but unlike other EU countries, does not provide support for continuing organic production.

Model farm calculations prepared for DEFRA's Organic Conversion Information Service (Table 11.2) indicate that, depending on farm type, costs of conversion may be relatively low or non-existent for farms of average intensity (Lampkin, 2001b). These models take account of prices for organic and conventional products in early 2001 and include the Organic Farming Scheme payments, as well as the flexibility built into the main arable and livestock support schemes for organic farmers. However, they are sensitive to the organic price assumptions, as the difference between organic and conventional prices can account for up to £800/ha in the specialist dairy and stockless arable organic results.

Role of the market and regulations

The profitability of organic farming is highly dependent on the prices received, and these are only possible as a result of the development of a distinct market for organic products. The first efforts to develop such a market began in the late 1960s and early 1970s, as a means to support the broader goals of organic farmers, and effectively to allow consumers willing to pay a means to compensate farmers for internalizing external costs by restricting the use of certain inputs and production technologies. Since then the organic market has become a well-established, global phenomenon worth more than 15 billion Euro in retail sales value, with Europe, the United States and Japan the key consuming markets.

The very success of the organic market brings with it certain risks. Firstly, that the market becomes an end in itself, not a means to an end, and that the original goals of organic farming are devalued, thereby leading to the possible loss of consumer confidence in the integrity of organic farming. Secondly, a market of this size clearly invites fraud, and there is a need for legislation to avoid this and to protect consumers and *bona fide* producers. As a consequence, organic farming is the only approach to sustainable agriculture (see above) that is enshrined in legislation in Europe (EU, 2001), in the United States (USDA, 2000) and at international level through the *Codex Alimentarius* agreement (FAO, 2001).

Under EC Regulation 2092/91 there has been a legal requirement since 1993 for all crop products sold as organic in the European Union to be certified. In August 2000, the EC Regulation 1804/1999 came into full effect, extending the legal requirement for certification to organic livestock and livestock products. This legislation, and the relevant production standards based on it, emphasize control of the production process, not the end product. The UK Register of Organic Food Standards (UKROFS, 2001) is the national authority responsible for implementing the EU legislation regarding organic production in the UK. UKROFS is responsible for licensing organic sector bodies to carry out farm inspections and certification and for commissioning surveillance inspections to ensure procedures are followed correctly. The bodies licensed by UKROFS in 2000 to carry out producer inspections were Organic Farmers and Growers Ltd. (OFG), the Scottish Organic Producers Association Ltd. (SOPA), Organic Food Federation (OFF), Soil Association Certification Ltd. (SACert), Bio-dynamic Agricultural Association (BDAA/Demeter), Irish Organic Farmers and Growers Association Ltd. (IOFGA) and Organic Trust Ltd. (see Lampkin and Measures, 2001, for further details and contact addresses).

Alternative business organizations

The emphasis on social justice goals and close links with consumers has led to a range of initiatives to develop new models of business activity as well as alternative production methods. There are close links between organic production and many Fair Trade initiatives designed to improve the returns to producers in developing countries, particularly with respect to tea, coffee, cocoa and banana producers (Maxted-Frost, 1997; Lockeretz and Geier, 2000; Fairtrade Foundation website). Community supported agriculture (CSA) is another increasingly popular approach, particularly with small producers in North America, Europe and Japan (CSA website). The simplest models are based around box scheme subscriptions, whereby consumers commit to a fixed weekly subscription, sometimes paid in advance, and receive a box of vegetables in season in return. Variants on this theme include subscription gardening, where a fixed fee is paid to the grower to look after the consumer's plot, but the consumer can determine what is to be grown and can participate in the growing and harvesting of the crop. Other models include the consumer investing a capital stake in the holding and participating in the farm decision-making process. Various alternative models with respect to worker participation can also be found, ranging from worker co-operatives to the large Sekem company in Egypt, which has developed a school, university, community hospital and other facilities for its workforce and their families on the back of a thriving organic food, fibre (cotton) and clothing business (Maxted-Frost, 1997; Sekem website).

Organic farming and society

The key goals of organic farming with respect to environmental protection, animal welfare, food security, safety and quality, human health and nutrition, resource use sustainability and social justice are ones for which the market mechanism does not normally provide an adequate financial return and are normally seen as public goods and services, of benefit to society as a whole rather than the individual. These goals are increasingly important to policy-makers too, leading to increasing interest in the potential of organic farming as a policy option. Organic farming has established a complex set of principles and practices that are believed to contribute to achieving these objectives, but that does not guarantee that the objectives are achieved, and much debate centres around the extent to which these objectives are achieved in practice (Tinker, 2000; House of Commons, 2001). It is not possible in a chapter such as this to provide a detailed and comprehensive assessment, so some 'heroic' generalizations will be needed and readers should consult sources identified in the following sections for further information.

Agricultural sustainability can be seen as a measure of the performance of different systems with respect to all these goals, as well as the financial viability and hence sustainability of individual farm businesses. However, the key factor determining the relative sustainability of different systems in such a multi-objective context is the weighting placed on the individual objectives by different parts of society. A high weighting on environmental factors may favour one approach, while a high weighting on yield will favour another. Thus it may prove impossible to come to a conclusive view on the relative merits of the different approaches to sustainable agriculture, such as organic farming, integrated crop management, agroforestry and permaculture.

Environmental impacts

The impact of organic farming on the environment has been reviewed most recently by Greenwood (2000), Soil Association (2000) and Stolze *et al.* (2000). There is now a significant body of research indicating the beneficial effects of organic practices on soil structure, organic matter levels and biological activities, as well as plant, insect, bird and wild animal biodiversity. However, differences can vary depending on farm type, the relative intensity of the conventional and organic systems compared, and the management ability and interest of the individual farmer, so that better performance is not necessarily guaranteed in all cases. As Stolze *et al.* (2000) point out, these benefits are clearer on a per unit land area basis, but the reduced yields from organic farming may mean that the benefits per unit food produced are not as great. There is an ongoing debate as to whether the reduced yields might require additional land currently not in production to be brought into production, with potential negative environmental consequences. However, this assumes current production structures will be retained, including the current level of feeding crop suitable for human consumption to livestock.

In terms of non-renewable resource use (and the related pollution risks/greenhouse gas emissions), several studies indicate that organic farming has the potential to reduce resource use and pollution, not only on a per unit land area basis, but also per unit food produced (ENOF, 1998), with significant implications for future global food security in the context of diminishing resources.

Food quality, nutrition and health

The impact of organic farming on food quality, nutrition and human health has been a core concern of organic farming since the research of McCarrison (1936). The issue has been subject to recent reviews (Williams *et al.* in Tinker, 2000; Soil Association, 2001b). The evidence on food quality is less conclusive that that for environmental benefits, with some studies showing benefits with respect to increased valuable nutritional components (vitamins, minerals, trace elements, secondary metabolites) and reduced harmful components (nitrates, pesticide residues), while other studies show little or no differences and some authors have raised the theoretical risk of increased levels of potentially harmful components such as *E. coli* 0157 and mycotoxins in organic foods, but with little evidence to substantiate this (FAO, 2000; FSA, 2000; Soil Association, 2001b). Some animal studies have shown beneficial impacts on fertility and morbidity from organic diets.

Such differences as have been identified tend to be specific to particular crops or farming situations, so that it is difficult to generalize an overall benefit from organic food. However, it is clear from the focus of many agricultural research programmes that the way food is produced does affect its quality. Therefore it is reasonable to expect that quality differences, for better or for worse, could exist between organically and conventionally produced foods. There is a clear need for further research on this topic, which is now more likely to take place than in the past, as the resistance of governments to funding such research is waning.

Food security and developing countries

The relatively large crop yield reductions observed in the northern European context have led many to question whether organic farming is capable of meeting the food needs of a growing global population. This is a complex question, which has to take account of both distribution and production issues, as well as the increasing demand for meat as incomes increase and the role of livestock production as a direct competitor with human food needs. In addition, the large yield reductions experienced in northern Europe are not reflected in other studies from countries where conventional production is less intensive (Lampkin and Padel, 1994) and Pretty and Hine (2001) have demonstrated the potential for yields to be increased in resource-poor countries (where the ability to pay for external inputs, in particular agro-chemicals, is severely limited) through the adoption of ecological management principles. The experience of Cuba in pursuing organic farming as a key part of its food security strategy in the face of US economic sanctions is particularly relevant in this context (Pretty and Hine, 2001; Food First website).

Social impacts

Social impacts are perhaps the least considered aspect of organic farming, although social goals have long been part of the organic farming concept. The International Federation of Organic Agriculture Movements' standards (IFOAM website) include a section on social justice which covers workers' rights and expectations of appropriate working conditions, rewards for labour, and educational opportunities. In a European context, basic rights are covered by national legislation, and have therefore not been a focus of organic farming standards and legislation, but there is a need to look critically at working conditions, employment and income levels on organic farms (Jansen, 2000; Offermann and Nieberg, 2000). In general terms, the case can be made that employment and incomes can be maintained or increased on organic farms, but even securing current farming businesses and existing employment and income levels might be beneficial for rural communities in the context of the dramatic structural changes currently taking place in conventional agriculture. There is clearly a question whether organic farms will not in the longer term be exposed to the same economic pressures for specialization and rationalization as conventional farms, and it may be that local marketing and processing initiatives are more important than production in terms of the rural development potential of organic farming (OMIARD website).

Organic farming and agricultural policy

More than 80% of the expansion in the organic land area in Europe up to 2000 took place since the implementation in 1993 of EC Regulation 2092/91, defining organic crop production, and the widespread application of policies to support conversion to and continued organic farming as part of the agri-environment programme (EC Regulation 2078/92). The former provided a secure basis for the agri-food sector to respond to the rapidly increasing demand for organic food across Europe. The latter provided the financial basis to overcome perceived and real barriers to conversion.

The increasing role of EU policy support of this type during the 1990s has arisen because of a gradual convergence of policy goals with the underlying objectives of organic farming, including environmental protection, animal welfare, resource use sustainability, food quality and safety, financial viability and social justice. Organic farming is also perceived to contribute to reducing problems of overproduction and to rural development. In addition, organic farming offers three potential advantages over other, more targeted policy measures: it addresses many of these goals simultaneously, it utilizes the market mechanism to support these goals, and it has achieved global recognition.

Agri-environment support

The agri-environment measures came into effect in 1993, although the majority of organic aid schemes under EC Regulation 2078/92 were not fully implemented by EU member states until 1996, and significant differences between the schemes implemented exist (Lampkin *et al.*, 1999). By October 1997, more than 65 000 holdings and nearly 1.3 million ha were supported by organic farming support measures at an annual cost of more than 260 million ECU. Organic farming's share of the total agri-environment programme amounted to 3.9% of agreements, 5.0% of land area and nearly 11% of expenditure, the differing shares reflecting in part the widespread uptake of baseline programmes in France, Austria, Germany and Finland. There were wide variations between countries in terms of the significance of organic farming support, both relatively and absolutely, within the agri-environment programme. Financial support for organic farming has continued to be provided, in many cases at increased levels, under the agri-environmental measures in the Rural Development regulation (EC Regulation 1257/1999).

Mainstream commodity support

Like their conventional counterparts, organic farmers also qualify for the mainstream commodity support measures, including arable area payments, livestock headage payments, and support for capital investment and in less favoured areas where available. In most EU countries, the mainstream commodity support measures are seen as beneficial, at least for organic arable producers (Lampkin *et al.*, 1999). Set-aside in particular is seen to have potential to support the fertility-building phase of organic rotations during conversion and on arable farms with little or no livestock.

Only in a few cases have significant adverse impacts of the mainstream measures on organic farmers been identified. In some cases, special provisions have been made to reduce these, for example flexibility with respect to use of clover in set-aside in the UK. There is a case that, since organic producers are producing significantly less output, and the organic market is under-supplied, then organic producers should not face compulsory set-aside. In 2001, the European Commission moved in this direction, permitting organic producers to use set-aside land for forage production.

The loss of eligibility for livestock headage premiums as a result of reduced stocking rates following conversion is seen as potentially more problematic, but this can be mitigated by extensification payments and quota sales or leasing where applicable.

Several countries have made use of investment aids and national/regional measures to provide additional assistance, including special derogations for organic producers.

Rural development and structural measures

Organic farming is seen in many countries in Europe as having significant potential for rural development, in terms of its capacity to supply premium markets and thereby to support rural incomes and employment. Organic farming projects were favoured under the marketing and processing support and structural measures in the 1990s and this has continued under the new Rural Development Programme and structural measures under Agenda 2000. Some countries, for example Denmark, France, the Netherlands, Sweden and Wales, have developed integrated action plans for organic farming (Lampkin *et al.*, 1999), which fully utilize the support available under these measures and aim to ensure a better balance between support for supply growth through the agri-environment programme and demand growth through market-focused measures.

Information programmes, including support for research, advice, training and demonstration farms as well as consumer information, are also seen in many countries as essential counterparts to the other programmes (Lampkin *et al.*, 1999) and have been supported at EU level through the Framework research programmes and national funding, as well as through the provision of specific training and advice under the Rural Development Programme.

Future potential for organic farming

There is currently renewed debate about the potential for organic farming in Europe. The spread of BSE to other European countries and the outbreak of foot-and-mouth disease in early 2001 have led to many calls for a fundamental review of the future direction of agriculture, including an increased role for organic farming. Moves to develop a European Action Plan for organic farming have been supported by several agriculture ministers at an international conference held in Copenhagen in 2001 (MFAF, 2001) and at the June 2001 Council of Ministers meeting in Gothenburg.

Several countries have set targets for organic farming to grow to 10 or 20% of total agriculture by 2005/2010. Although growth trends in individual countries have varied considerably, with periods of rapid expansion followed by periods of consolidation and occasionally decline, overall growth in Europe has been consistently around 25% per year for the last 10 years, i.e. exponential growth. Continued 25% growth each year would imply a 10% share of EU agriculture managed organically by 2005 and nearly 30% by 2010. Clearly this rate of growth cannot be sustained indefinitely; a slower rate of growth of 15% each year would still result in 5% of EU agriculture by 2005 and 10% by 2010.

A target of 20% by 2010 at the European level, as currently under discussion, would imply a seven-fold increase in the size of the sector, resulting in ca. 1 million holdings and 30 million ha managed organically, and a retail market potentially worth more than 50 billion Euro. This level of growth has significant implications for the provision of training, advice and other information to farmers, as well as for inspection and certification procedures. It also has implications for the resourcing of existing organic support schemes under the Rural Development Programme, as the cost could increase to more than 6 billion Euro annually. It is an open question whether policy-makers, farmers and consumers will respond to the challenge to make this sort of expansion possible.

References

ADAS (2001) *Managing Livestock Manures.* Booklet series including organic farming. ADAS, Gleadthorpe.

Altieri, M. (1995) *Agroecology – The Scientific Basis of Alternative Agriculture*, 2nd edn. Intermediate Technology, London.

Balfour, E.B. (1943) *The Living Soil.* Faber and Faber, London. Reprinted 1976 as *The Living Soil and the Haughley Experiment.* Universe Books, New York.

Blake, F. (1994) *Organic Farming and Growing.* Crowood Press, Swindon.

Boeringa, R. (ed) (1980) Alternative methods of agriculture. *Agriculture and Environment.* Special Issue 5. Elsevier, Amsterdam.

Conford, P. (2001) *The Origins of the Organic Movement.* Floris Books, Edinburgh.

Dabbert, S. (1994) Economics of conversion to organic farming – cross sectional analysis of survey data in Germany. In: *The Economics of Organic Farming – An International Perspective* (eds N.H. Lampkin & S. Padel), pp. 285–93. CAB International, Wallingford.

ENOF (1998) *Resource Use in Organic Farming.* Conference proceedings. European Network of Organic Farming, Barcelona.

EU (2001) *Organic Farming – Guide to Community rules.* Directorate-General for Agriculture, European Commission, Brussels, http://europa.eu.int/comm/agriculture/qual/organic/brochure/abio_en.pdf

FAO (2000) *Food Safety and Quality as Affected by Organic Farming.* Agenda Item 10.1, 22nd FAO Regional Conference for Europe. Food and Agriculture Organisation, Rome, http://www.fao.org/docrep/meeting/X4983e.htm

FAO (2001) *Guidelines for the Production, Processing, Labelling and Marketing of Organic/Bio-dynamic Foods.* Codex Alimentarius Commission, Food and Agriculture Organisation, Rome, http://www.fao.org/organicag/doc/glorganicfinal.doc

Foelsch, D. (ed) (1978) *The Ethology and Ethics of Animal Production.* Birckhaeuser, Basel.

Foster, C. & Lampkin, N.H. (2000) *European Organic Production Statistics 1993–1998.* University of Wales, Aberystywth, http://www.organic.aber.ac.uk/library/European%20organic%20farming.pdf

Fowler, S., Lampkin, N.H. & Midmore, P. (2000) *Organic Farm Incomes in England and Wales 1995/96–1997/98.* Report to MAFF. University of Wales, Aberystwyth, http://www.organic.aber.ac.uk/library/Organic%20Farm%20Incomes.pdf

Fowler, S., Wynne-Jones, I. & Lampkin, N. (2001) *Organic Farm Incomes in England and Wales 1998/99.* Report to MAFF. University of Wales, Aberystwyth, http://www.organic.aber.ac.uk/library/Organic%20Farm%20Incomes%201998–99.pdf

Francis, C.A., Flora C.B. & King, L.D. (eds) (1990) *Sustainable Agriculture in Temperate Zones.* John Wiley, New York.

FSA (2000) *Food Standards Agency View on Organic Foods.* Position paper. Food Standards Agency, London.

Greenwood, J.J.D. (2000) Biodiversity and environment. In: Tinker, P.B. (ed) *Shades of Green – A Review of UK Farming Systems.* RASE, Stoneleigh, pp. 59–72.

House of Commons (2001) *Organic Farming.* Second report of the House of Commons Agriculture Committee. The Stationery Office, London.

Hovi, M., Roderick, S., Wassink, G. & Oakeley, R. (2000) *Compendium of Animal Health and Welfare in Organic Farming.* Veterinary Epidemiology and Economics Research Unit, University of Reading.

Howard, A. (1940) *An Agricultural Testament.* Oxford University Press, London.

IFOAM (2001) *Basic Standards of Organic Agriculture.* International Federation of Organic Agriculture Movements, Tholey-Theley.

Jansen, K. (2000) Labour, livelihoods, and the quality of life in organic agriculture. *Biological Agriculture and Horticulture,* **17**, 247–78.

Jordan, V.W.L. (1990) Low input farming and extensification. *Agricultural Progress,* **65**, 101–103.

King, F.H. (1911) *Farmers of Forty Centuries – Permanent agriculture in China, Korea and Japan.* King, Madison.

Lampkin, N.H. (1990) *Organic Farming.* Farming Press, Ipswich.

Lampkin, N.H. (2001a) *Organic Farming Statistics.* University of Wales, Aberystwyth, http://www.organic.aber.ac.uk/stats.shtml

Lampkin, N.H. (2001b) *Organic Farming Conversion Models.* Unpublished report. University of Wales, Aberystwyth.

Lampkin, N.H. & Measures, M. (eds) (2001) *2001 Organic Farm Management Handbook.* University of Wales, Aberystwyth.

Lampkin, N.H. & Padel, S. (eds) (1994) *The Economics*

of Organic Farming – An International Perspective. CAB International, Wallingford.

Lampkin, N.H., Foster, C., Padel, S. & Midmore, P. (1999) *The Policy and Regulatory Environment for Organic Farming in Europe. Organic Farming in Europe: Economics and Policy*, Vol 1. Universität Hohenheim, Stuttgart–Hohenheim.

LEAF (2001) *Handbook on Integrated Farm Management.* Linking Environment and Farming, Stoneleigh.

Lockeretz, W. & Geier, B. (eds) (2000) *Quality and Communication for the Organic Market.* Conference proceedings. International Federation of Organic Agriculture Movements, Tholey-Theley.

Mäder, P., Pfiffner, L. Lützow, M.v., Fliessbach, A. & Munch, J.C. (1996). Soil ecology – the impact of organic and conventional agriculture on soil biota and its significance for soil fertility. In: *Fundamentals of Organic Agriculture* (ed T. Ostergaard), Conference proceedings, pp. 24–46. International Federation of Organic Agriculture Movements, Tholey-Theley.

MAFF (1998) *Codes of Good Agricultural Practice for the Protection of Water, Soil and Air.* Ministry of Agriculture, Fisheries and Food (now DEFRA), London.

Maxted-Frost, T. (1997) *Future Agenda for Organic Trade.* Conference proceedings. Soil Association, Bristol.

McCarrison, R. (1936) *Nutrition and Health.* McCarrison Society, London.

Merrill, M.C. (1983) Eco-agriculture – a review of its history and philosophy. *Biological Agriculture and Horticulture,* **1**, 181–210.

MFAF (2001) *Organic Food and Farming – Towards Partnership and Action in Europe.* Conference proceedings. Danish Ministry of Food, Agriculture and Fisheries, Copenhagen, www.fvm.dk/konferencer/organic_food_farming

Mollison, B. (1990) *Permaculture – A Practical Guide for a Sustainable Future.* Island Press, Washington, D.C.

Newton, J. (1995) *Profitable Organic Farming.* Blackwell Science, Oxford.

Offermann, F. & Nieberg, H. (2000) *Economic Performance of Organic Farms in Europe.* Organic Farming in Europe – Economics and Policy, Vol 5. University of Hohenheim, Stuttgart.

Padel, S. (2001) Conversion to organic farming – a typical example of the diffusion of an innovation? *Sociologia Ruralis*, **41**, 40–61.

Padel, S. & Lampkin, N.H. (1994) Conversion to organic farming – an overview. In: *The Economics of Organic Farming – An International Perspective* (eds N.H. Lampkin & S. Padel), pp. 295–313. CAB International, Wallingford.

Pretty, J. & Hine, R. (2001) *Reducing Food Poverty with Sustainable Agriculture – A Summary of New Evidence.* Final Report from the SAFE-World Research Project. University of Essex, Colchester.

Pretty, J.N., Brett, C., Gee, D. *et al.* (2000) An assessment of the total external costs of UK agriculture. *Agricultural Systems,* **65**, 113–136.

Reed, M. (2001) Fight the future! How the contemporary campaigns of the UK organic movement have arisen from their composting of the past. *Sociologia Ruralis,* **41**, 131–45.

Reganold, J.P., Palmer, A.S., Lockhart, J.C. & Macgregor, A.N. (1993) Soil quality and financial performance of biodynamic and conventional farms in New Zealand. *Science,* **260**, 344–49.

Rusch, H.-P. (1968) *Bodenfruchtbarkeit – eine Studie biologischen Denkens.* Verlag K F Haug, Heidelberg.

Sattler, F. & Wistinghausen, E.v. (1992) *Bio-Dynamic Farming Practice.* Bio-Dynamic Agricultural Association, Stourbridge.

Soil Association (2000) *The Biodiversity Benefits of Organic Farming.* Soil Association, Bristol.

Soil Association (2001a) *Organic Food and Farming Report 2001.* Soil Association, Bristol.

Soil Association (2001b) *Organic Farming, Food and Human Health – A Review of the Evidence.* Soil Association, Bristol.

Steiner, R. (1924) *Agriculture – A Course of Eight Lectures.* Rudolf Steiner Press/Bio-Dynamic Agricultural Association, London.

Stockdale, E.A., Lampkin, N.H., Hovi, M. *et al.* (2000) Agronomic and environmental implications of organic farming systems. *Advances in Agronomy,* **70**, 261–327.

Stolze, M., Piorr, A., Haering, A. & Dabbert, S. (2000) *The Environmental Impacts of Organic Farming in Europe.* Organic Farming in Europe: Economics and Policy, Vol. 6. University of Hohenheim, Stuttgart.

Tinker, P.B. (ed) (2000) *Shades of Green – A Review of UK Farming Systems.* Royal Agricultural Society of England, Stoneleigh.

UKROFS (2001) *UKROFS Standards for Organic Food Production.* UK Register of Organic Food Standards, DEFRA, London.

USDA (2000) *National Organic Program.* United States Department of Agriculture, Washington, DC, www.ams.usda.gov/nop/facts/index.htm

Young, A. (1997) *Agro-Forestry for Soil Management.* CAB International, Wallingford.

Useful Websites

www.defra.gov.uk/farm/organic – Department of Environment, Food and Rural Affairs (DEFRA).
www.efrc.com – Elm Farm Research Centre.
http://europa.eu.int/comm/agriculture/qual/organic/index_en.htm – European Union.
www.fairtrade.org.uk – Fairtrade Foundation.
www.fao.org/organicag – FAO Organic Agriculture.
www.foodfirst.org/cuba/index.html – Food First.
www.ifoam.org – International Federation of Organic Agriculture Movements (IFOAM).
www.irs.aber.ac.uk/omiard – Organic Marketing Initiatives and Rural Development (OMIARD).
www.nal.usda.gov/afsic/csa – Community Supported Agriculture (CSA).
www.organic.aber.ac.uk – Organic Centre Wales.
www.organic-europe.net – Organic Europe.
www.organic-research.com – CABI Organic Research.
www.organicts.com – Organic Trade Services.
www.sekem.com – Sekem.
www.soilassociation.org – Soil Association.
www.veeru.reading.ac.uk/organic – Network for Animal Health and Welfare in Organic Agriculture (NAHWOA).

12

Farm business management

M.F. Warren

What is business management?

Business management is difficult to define precisely, and is best described in terms of the activities performed by the manager of a business – whether employee or owner. A manager's job is to make decisions – not just the day-to-day ones that everyone has to make, but decisions concerning the next month, the next year, or even the next 10 years. The manager is deciding how best to use the resources at his or her disposal in trying to achieve the objectives of the business. A manager's job is also concerned with ensuring that his decisions are having the desired effect. The outcome of each decision must be measured and checked to see that it matches up to expectations. If actual progress does not compare well with the expected performance, the manager should take the appropriate action to bring the two into line. Viewed in this way, the manager's job can be seen as *planning* (making decisions which affect the future operation of the business) and *control* (monitoring progress of the decision and taking necessary corrective action), in order to achieve the *objectives* of the business.

In a one-person business, where the owner is also the manager and the workforce, the business has only to satisfy the owner's objectives. In most farm businesses, however, there are other people involved – family, partners, employees, maybe even a paid manager. These people have their own objectives, and this adds another dimension to the manager's job – making sure that everyone is working towards the same end. A major objective of many farm workers, for instance, is to do a job to be proud of. The task of the manager is to use that fact to help the business achieve its goals rather than allow it to cause a nuisance.

This dimension of the manager's job – marshalling resources, co-ordinating and harmonizing objectives and efforts – can be summed up as *organization.* Management thus becomes a process of planning, organizing and controlling (Fig. 12.1). This chapter is concerned largely with the planning and control functions: the following chapter includes some of the skills needed in organization.

Setting primary objectives

Setting objectives is the key to the management process. Unless the manager is sure what he or she is trying to achieve, the business can have no clear direction. Objectives can be roughly grouped into primary and secondary objectives. Primary objectives are the overall aims of the business – the long-term goals for which the business is striving. Secondary objectives are the shorter-term goals that the business needs to meet in order to achieve the primary objectives.

A primary objective might be, for instance, to increase the wealth of the business by 10% in the next 5 years; one of the associated secondary objectives might be to achieve an average wheat yield of 10 t/ha over this period. Primary objectives are reflections of the long-term expectations of the manager. These expectations will range far and wide, but for convenience can be classified into personal, physical, responsibility and financial:

(1) *Personal expectations* are those that are concerned with the lifestyle and personal preferences of the manager. They often include time for family, social and leisure activity, and acquisition of status or influence through, for instance, being a magistrate or county chairman of the NFU.
(2) *Physical expectations* relate to the productivity of the farm, to innovation, and to general technical prowess.

Figure 12.1 The management process.

(3) *Responsibility expectations* involve satisfaction of the responsibility the manager feels towards other groups in society. These can include employees, customers, neighbours and the local community.
(4) *Financial expectations* relate to the generation of income and wealth. These are the crucial expectations, not because they are particularly important in themselves (although some people are highly motivated by the acquisition of money for its own sake), but rather because they govern the ability of the business to satisfy the personal, physical and responsibility expectations.

It is because of the importance of financial expectations that the following sections relate particularly to financial management. A further reason is that, to plan and control effectively, a manager needs a widely accepted and flexible measure by which to judge performance. Money provides that measure.

Financial measures

Where the finances of a farm are concerned, three factors are crucial: survival, efficiency and stability. These will dictate the financial objectives of the business, and the way in which the manager plans and monitors progress.

(1) Survival, in financial terms, is a matter of having sufficient cash available at any time to pay for essential inputs, such as labour, fertilizers, diesel, finance charges and so on. Cash availability is measured through net cash flow – the difference between cash coming into the business and cash going out.
(2) Operational efficiency is also reflected in the money coming in and going out of the business. There are other factors to take into account, however, such as changes in the levels of stocks and of bills outstanding. The financial measure that attempts to measure efficiency by taking these and other factors into account is known as net *profit*.
(3) Financial stability is imparted by wealth. Wealth implies ownership of assets – money and things of value – but it must also take account of outstanding debts, or liabilities. A business that has £1 000 000 assets and £999 000 debts is likely to be less stable than one with £10 000 assets and £50 debts. The appropriate measure of financial stability is thus net capital – the difference between assets and liabilities.

Planning to achieve primary objectives

Once the primary objectives have been established, the planning process can get under way. The first, and usually most protracted, stage is to formulate an overall or long-term plan. The time period covered by such a plan will be one or more years, and it will indicate the decisions and actions needed for the business to have the best possible chance of achieving the primary objectives. This process relies much on subjective judgement and can never be made foolproof. A structured approach can, however, minimize the risks of making grossly erratic judgements. One such approach is as follows:

(1) Assess the farm and its resources.
Take a hard look at the *strengths* and *weaknesses* of the business. Assess the quantity and quality of the resources available such as the land, buildings, expertise of the employees, ease of access to markets, and so on. Take account of factors that could limit freedom of action; not only physical and financial factors such as the slope of the ground and the amount of capital available, but also legal and administrative constraints such as town and country planning laws, restrictive clauses in a lease, or the likelihood of obtaining a quota or contract.
(2) Appraise the present and future environment of the business.
Identify the *opportunities* that are open to the business, and the *threats* to its welfare. There is more to this than keeping abreast of technological change. It involves being aware of the economic

and political climate affecting the business, and making judgements about how that climate is likely to change during the period planned for (see Chapter 8). It also requires an awareness of changes in the habits and attitudes of society in general (e.g. food consumption habits and attitudes towards farming). The likely actions of competitors must be anticipated. The search for opportunities should not be confined to farming, but should include all possibilities for using the business resources effectively.

(3) Formulate alternative plans.
List all the possible ways of using the resources of the business to achieve its primary objectives. 'Possible', in this context, must be judged by reference to the strengths and weaknesses, opportunities and threats listed earlier. It is essential to look further than the 'obvious' solution, and consider as wide a range of alternatives as possible. This is, in many ways, the most difficult part of the planning process: it is hard to be objective and imaginative about a business that one knows intimately. One way of overcoming this problem is to enlist the help of family and/or friends in a 'talking shop' aimed at stimulating new ideas.

(4) Appraise the alternative plans and select the one most likely to succeed.
The likely outcomes of the alternative plans must be measured and compared. Given the importance of financial objectives, and the fact that money provides a common unit of measurement, it is critical at this stage to prepare budgets. A budget is a calculation of the financial effects of a course of action, and given that the financial objectives of a business can be measured in terms of cash flow, profit and capital, budgets should be expressed in these terms.

The plan whose budgets indicate the nearest result to the financial objectives will then be chosen, unless another plan more nearly achieves the non-financial objectives specified, and the manager is content to forgo the extra financial benefit for the sake of achieving non-financial expectations. He or she may happily put up with a low-key, low-profit system for the sake of minimizing mental and physical strain.

The resulting overall plan can then be used as a focus for short-term planning. The manager will need to:

- set secondary objectives. These are the targets that must be achieved for the overall plan to work. They may be expressed purely in financial terms, as the margin that a particular enterprise should achieve in a given year. They may also include key physical targets, such as the yield produced by a major enterprise.
- formulate short-term plans. This is a continual process aimed at enabling these targets to be achieved. It ranges from formal budgeting of enterprise margins for a season, or cash flow for a month, down to day-to-day decisions about organization of the workforce.

Once plans have been formulated and acted upon, the control process can start. On a week-by-week, month-by-month basis, progress can be monitored by comparing key results with the short-term targets (most valuable in dairy and intensive livestock enterprises). On an annual basis, control will involve monitoring the progress of overall plans. Given the importance of financial objectives at this level, this implies comparing financial expectations with what has actually happened. This *budgetary control* can only take place if the manager has available two key sets of information – the *budgets* which show the detail of financial expectations, and the *accounts* which record the actual performance. Both budgets and accounts use the same measures – cash flow, profit and capital.

Information for financial management

Budgets

Budgeting for profit

A profit budget in conventional form

A budget shows the expected financial performance of a business over a future time period. This period is normally a year and, for convenience, is usually the financial year used by the farm's accountant. Many farm financial years start from Michaelmas (end of September) or Lady Day (end of March); other popular dates are the beginning of January and the beginning of April.

The easiest way of beginning the budgeting process is to start from the practical basics of the farming system – how many hectares of crops are to be grown during the year in question? how many head of livestock are to be kept? On this framework can be built a detailed picture, by estimating the quantities of inputs that are likely to be consumed and their prices, and the quantities of products that are likely to be produced and their prices.

To reflect fully the business performance of the farm, though, some adjustments need to be made to these figures. The first is that allowance should be made for any goods or livestock likely to be on the farm at the end of year. Although these 'stocks' will not yet have been sold, they will have been produced or purchased by the business during the year, and thus ought to be included. Similarly, any bills yet to be paid at the end of the year (creditors) or income due from goods that have been sold and not yet paid for (debtors) must be included as 'belonging' to the budget year. Conversely, any stocks, creditors or debtors likely to be outstanding at the *beginning* of the budget year are taken out of the calculation – they 'belong' to last year's production.

The second type of adjustment is to remove any payment or receipt concerned with personal, tax or capital matters. The amount the owner of a business takes for his or her own use is not determined by the efficiency of the business, but by lifestyle: the same applies to tax payments, which also vary according to how good the accountant is and the policy of the prevailing government. Capital items relate to lump sums, such as the receipt and repayment of loans, and purchase of land, buildings and machinery. Since these do not arise out of the normal trading of the business, they are left out of the calculation with the exception of 'wasting assets' such as buildings and machinery. The cost of these is spread over their useful life in the business: the resulting annual cost is known as 'depreciation' (see 'Estimating outputs and costs' below).

Finally, any farm products consumed without payment by the farmer, family and employees should be added in. These 'benefits in kind' or 'household consumption' are just as much part of the farm production as those sold. Similarly, goods and services paid for on the farm account but consumed without payment should be credited back to the farm.

The result (shown in Fig. 12.2) is a budget showing the expected net profit for the year concerned.

Enterprise budgeting

A profit budget of the type just illustrated is useful in a limited way. One can tell whether the operation of the farm is likely to make a positive or negative contribution

Trading and profit and loss account, year ended 31 March 200x

	This year		*Last year*	
Farm income	£	£	£	£
Milk	98 159		88 521	
Calves	7 700		11 569	
Cull cows	8 750		17 580	
Grain	22 400		512	
Straw	2 500		335	
Area payments	7 600		5 642	
Miscellaneous	1 500		500	
		148 609		124 659
Less: Purchases				
Feed	7 200		26 513	
Livestock	14 400		725	
Vet. and med., A.I., etc.	4 800		4 958	
Fertilizer	14 740		2 160	
Seed	3 590		12 589	
Sprays	4 120		154	
Regular labour	20 000		2 561	
Machinery running costs	15 000		1 583	
Rent and rates	14 000		13 521	
Miscellaneous costs	10 650		9 852	
Bank interest	9 243		10 256	
Depreciatio of machinery	9 000		8 250	
		126 743		93 122
Benefits in kind		1 151		958
Increase in valuation of stocks		50		1 361
Net profit		23 067		33 856

Figure 12.2 Example of a profit/loss statement.

to the owner's financial welfare (i.e. profit or loss). It is also possible to see whether the result represents an improvement on previous years. It is difficult, however, to see how particular costs and returns relate to each other, and to see how the financial quantities reflect physical performance. The *value* of milk produced is easily apparent, for instance, but not the *amount* of milk produced in total and, more important, per cow.

To overcome these problems, it is possible to rearrange the information in the budget on an enterprise basis. Conceptually, the easiest way of doing this is by calculating a net profit for each enterprise (an enterprise being an output-producing part of the business, such as a particular crop or livestock type, or a service such as holiday accommodation or contracting).

The form of enterprise accounting most commonly found in UK farming is the gross margin system. This involves identifying for each enterprise the value produced (the *output*) and the *variable costs.* The latter are costs that both are easily attributable to the enterprise concerned and vary in direct proportion to small changes in the scale of the enterprise. Thus, for a livestock enterprise, livestock purchase, feed, veterinary and medical costs, transport and so on are regarded as variable costs. Each one is readily identifiable with the enterprise, and a 5% increase in the size of the flock or herd, for example, would tend to increase each of these by a similar proportion. Typical variable costs for a crop enterprise include seed, fertilizer and spray costs.

Subtracting the variable costs from the output gives, for each enterprise, a gross margin. When the gross margins from the various enterprises are added together the farm overhead or fixed costs can be deducted to give the net profit for the business (Fig. 12.3). This net profit should be the same as that shown by the budget in conventional form described above – it is only the layout of the information that has changed, not the information itself.

The main advantage of this method is that it avoids the allocation of the fixed costs between enterprises. Such allocation can involve difficult and arbitrary decisions, particularly when general farm costs are concerned, such as the cost of the farm Land Rover, office expenses, finance charges, etc. Gross margins are also useful when considering small changes to the farm system. The financial effects of switching one field from spring barley to winter wheat can be seen by substituting the gross margin of one crop for that of the other in the budget (Fig. 12.4).

There are situations when the gross margin method is insufficient. If it is important to know exactly how much a particular enterprise contributes to the business (when pricing a contract, for instance), it will be necessary to calculate a net margin. In its extreme form, this involves allocating all the costs of the business (including the general overheads mentioned earlier) between the enterprises. A useful compromise is to allocate to enterprises all those fixed costs that can easily be attributed, leaving only such costs as general overheads and finance charges to be deducted on a whole-business basis (Giles, 1986).

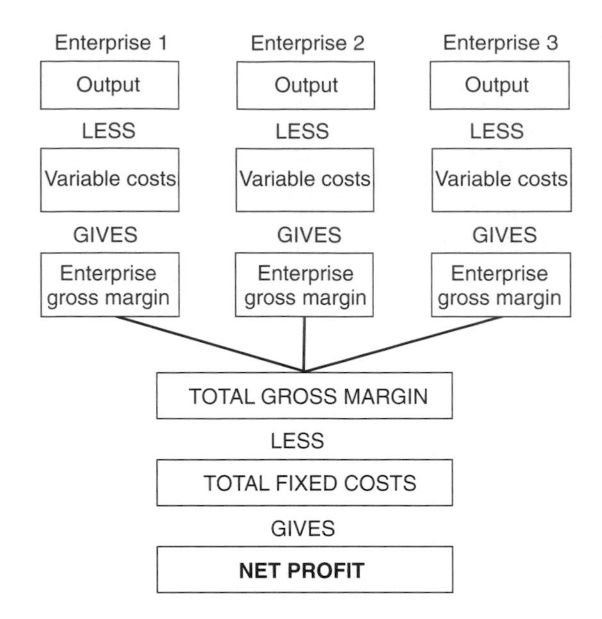

Figure 12.3 The gross margin system.

		£
Gross margin gained:	10 ha wheat at £645.50/ha	6 455
Gross margin lost:	10 ha barley at £527.50/ha	5 275
	Net gain	1 180

Figure 12.4 Net gain from substituting 10 ha winter wheat for spring barley.

Where the farm is large enough to support its own secretarial services and/or its enterprises are fairly self-contained, the benefits of using this system can outweigh the effort involved in allocating labour and machinery costs (including analysis of timesheets when recording). For most general-purpose management accounting, though, the gross margin system remains a valuable compromise, provided that fixed cost implications are not overlooked.

Preparing an enterprise gross margin budget

Estimating an enterprise gross margin involves two tasks: identifying the information needed, and organizing the information so that it can easily be used. Calculating the likely output of an enterprise needs information about any value produced by that enterprise. As well as sales (adjusted for changes in debtors), therefore, the calculation must include transfers of products between enterprises within the business. Thus home-grown barley fed to livestock enterprises must be valued at market value at time of transfer, and credited to the barley enterprise as part of its output. In addition, any produce consumed as benefits in kind (milk consumed in the farmhouse, for instance) should be counted as output. Finally, allowance should be made for any increase or decrease in the valuation of stocks of output items (crops in store or in ground, livestock and livestock products).

Similar conditions apply to the variable costs. Allowance must be made for transfers of home-grown inputs from other enterprises, for household consumption, and for changes in stock valuation of each input. The manner of organizing this information is partly a matter of taste, but the important criteria are that the result should be easy to read, and show readily the type of information that is important to the manager. One way of laying out a crop gross margin is shown in Fig. 12.5. Note that the five columns give information not just on financial amounts in total and per unit of production, but also physical information such as yields and usage per head or hectare. This degree of detail is necessary for the budget to be used effectively in budgetary control, as it enables the manager to identify discrepancies between actual and planned results, and also to explain how they have arisen.

Figure 12.6 shows a similar calculation for a breeding livestock enterprise. Note that the cost of buying and transferring in livestock, though strictly a variable cost, is deducted in calculating the enterprise output. There is no particular reason why you should follow this practice, but it has become an established convention in agricultural management and most published gross margin information is expressed in this way.

When budgeted gross margins have been prepared for each of the planned enterprises, the net profit can be calculated by totalling the gross margins and deducting the fixed costs. The major fixed costs are likely to be those of regular labour, machinery running costs and depreciation, rent and/or property depreciation and maintenance costs, and finance costs such as interest and leasing charges. Overdraft interest charges, being dependent on fluctuating overdraft levels, can be left on one side for now. This results in a budgeted net profit before overdraft interest; the final net profit can be calculated after using the cash flow budget to estimate interest costs.

Ha 40	*Total quantity*	*Tonnes/ hectare*	*Price/ tonne*	*Total* £	*£/Ha*
Grain					
Sales	220	5.5	102	22400	560.00
Transfers out	25	0.6	90	2250	56.25
Valuation change	−10	−0.3	90	-900	−22.50
Private consumption	0				
Total grain	235	5.9		23750	593.75
Straw					
Sales	80	2.0	31	2500	62.50
Transfers out	25	0.6	25	625	15.63
Valuation change	+5	0.1	25	125	3.13
Private consumption	0				
Total straw	110	2.8		3250	81.25
Area payments				7600	190.00
ENTERPRISE OUTPUT				34600	865.00
	(tonnes)	*(kg)*	*(£)*	*(£)*	*(£)*
Fertilizer (nitrogen)	11	275	100	1100	27.50
Fertilizer (compound)	20	500	120	2400	60.00
Seed	9	225	250	2250	56.25
Sprays				2800	70.00
Other costs				230	5.75
VARIABLE COSTS				8780	219.50
GROSS MARGIN				25820	645.50

Figure 12.5 Example of gross margin calculation for winter wheat.

Average numbers 70	*Total quantity*	*Quantity/ head*	*Price/ unit*	*Total £*	*£/Head*
Milk					
Sales	455000	6500	0.215	97803	1397.19
Private consumption	1655	24	0.215	356	5.08
Total milk	456655	6524	0.215	98159	1402.27
Calves					
Sales	65	0.9	118	7700	110.00
Transfers out	0				
Valuation change	0				
Total calves	65	0.9	118	7700	110.00
Cull sales					
Sales	15	0.2	583	8750	125.00
Valuation change	1			800	11.43
Less l/stock purchase	-16	-0.2	900	-14400	-205.71
Net replacement cost	0			-4850	-69.29
ENTERPRISE OUTPUT				101009	1442.98
Concentrate feed: homegrown	25	0.36	90	2250	32.15
Concentrate feed: bought	48	0.69	150	7200	102.86
Total concentrates	73	1.04	129	9450	135.01
Vet & med., A.I.				4800	68.57
Bedding: home-grown	25	0.36	25	625	8.93
Other				2160	30.86
Forage costs				5890	84.14
VARIABLE COSTS				22925	327.51
GROSS MARGIN				78083	1115.48
Stocking rate (head/ha)		2.0	**Gross margin/ha**		2230.95

Figure 12.6 Example of gross margin calculation for a dairy enterprise.

Estimating outputs and costs
The value of a budget depends on the quality of the information used in its compilation. Estimates of outputs and variable costs can be built up from forecasts of prices and quantities. The best guide for yields and quantities of inputs is usually the performance of previous years, taking bad years with the good. A good farm management handbook will also help, particularly where the manager has little previous experience of a particular enterprise (see, for instance, textbooks mentioned under 'Farm management data' at the end of this chapter). The biggest danger in estimating is being too optimistic – the most realistic estimate is usually produced by incorporating a healthy pessimism into budget figures.

Prices are affected by a great many factors, including the state of world trade, the political climate, changes in consumption habits, and quality of the product. The individual manager is unlikely to have the time or the expertise to weigh up all these factors for every commodity. The best procedure is to make use of the opinions of those who do have the time and expertise. Economic researchers are employed by a number of organizations: their conclusions are frequently reported in the farming and other media. The job of the manager then becomes one of locating the various price forecasts, reconciling any conflicts between them, and relating them to the business in question.

In estimating fixed costs, previous years can again be a useful guide, allowing for intervening wage awards and increases in prices. The handbooks mentioned above contain guidance where previous years' figures are not available or not relevant (where the system has been changed, for instance). Tables show the likely number of man- and tractor-hours needed for particular enterprises, from which can be calculated likely running costs.

Two methods are in common use for estimating depreciation. The straight-line method takes the original cost of the asset (machine, building or property improvement) divided by the likely life of the asset. Thus, an asset costing £10 000 and expected to last 20 years would have an annual depreciation of £500 per year. This method is useful for budgeting several years in advance, the constant annual payments making estimation quick and easy. It does not, however, result in a realistic estimate of the depreciation of machines, which tend to lose value more quickly in the first years of their lives than in the later. To allow for the latter effect, the diminishing balance method can be used. Here a constant annual percentage depreciation rate is applied to the depreciated value from the end of the previous year. Figure 12.7 shows the depreciation pattern of an asset costing £10 000, depreciated at a rate of 20% per year.

Ideally, the depreciation rate should be assessed individually for each machine, taking account of its likely life and resale value. Some farm management handbooks contain tables of appropriate rates. In practice, this is too complicated a procedure for most management purposes, and in both budgets and accounts, machines and buildings tend to be grouped or 'pooled' with other similar assets. A depreciation rate is chosen for the type of assets in the 'pool' (e.g. tractors), and is applied to the written-down value from the last year, adjusted for sales and additions during the current year. Figure 12.8 illustrates the process.

Assessing the profit budget

Before moving on to the cash flow budget, it is important to make an initial assessment of the implications of the profit budget. The first obvious check is whether the budget shows a net profit or a loss, and whether that figure matches up to the profit objective. Another useful check is whether the profit is sufficient to cover basic family drawings and known tax commitments. A comparison between the result of the budget and known results for previous years will help to give a feel as to whether the budget is realistic (it is very easy to be overoptimistic in budgeting).

It is likely that such an assessment of the budget will indicate deficiencies, and encourage adjustments in the projected gross margins as a result. Indeed, several such adjustment stages may be necessary before the manager is sufficiently satisfied (or convinced that no more profitable plan is possible) and moves on to test the effects on cash flow.

Budgeting for cash flow

Once a profit budget in gross margin form has been compiled, preparation of a cash flow budget is relatively easy. The aim is to present a picture of the flow of cash in and out of the business over the budget year.

An outline cash flow budget is illustrated in Fig. 12.9. For each month, the likely payments and receipts of cash are estimated. By taking the total cash payments from the total cash receipts for the month, an estimate is formed of the likely net cash flow for that month. When the net cash flow is added to the bank balance at the beginning of the month, the estimated bank balance at the end of the month is achieved. If this is done for each month of the budget year, the result is a picture of the way in which the bank balance of the business is likely to vary during that year.

The first stage in compiling a cash flow budget is to make a list of cash receipts and cash payments likely to affect the business bank balance during the budget year. A profit budget, especially one in gross margin form, makes an excellent basis for this list. It is important to remember, however, that one is now concerned only with cash items. Changes in debtors, creditors and stock valuations must be excluded, as must depreciation and benefits in kind. All personal, tax or capital receipts or payments must be included.

Year	*Opening value*	*Depreciation*	*Closing value*
1	10 000	2 000	8 000
2	8 000	1 600	6 400
3	6 400	1 280	5 120
4	5 120	1 024	4 096
5	4 096	819	3 277
6	3 277	655	2 621
7	2 621	524	2 097
8	2 097	419	1 678
9	1 678	336	1 342
10	1 342	268	1 074
11	1 074		

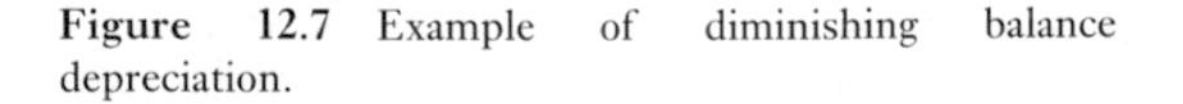

Figure 12.7 Example of diminishing balance depreciation.

	£
Value of machinery at beginning of year	30 000
plus purchases of machinery	20 000
less sales of machinery	−5 000
Value before depreciation	45 000
Depreciation at 20%	9 000
Value of machinery at end of year	36 000

Figure 12.8 Example of 'pool' calculation.

	TOTAL	Oct	Nov	Dec	Jan	Feb	Mar	Apr	May	June	July	Aug	Sep
					12500								
Income													
Milk	97803	8910	8894	8823	8751	8680	8609	8446	7705	6530	6417	7430	8608
Calves	7700	2100	1400	1050	700	350						700	1400
Cull cows	8750			350		350		700	350	700	3850	2450	
Wheat	22400				22400								
Wheat straw	2500	2500											
Area payments	12500				12500								
Sundries	1500		300			200		500		200		300	
VAT charged on outputs	0												
VAT refunds	0												
Capital – grants	0												
machinery sales	5000				5000								
Personal receipts	0												
SUB-TOTAL	158153	13510	10594	10223	49351	9580	8609	9646	8055	7430	10267	10880	10008
Payments													
Feed: cows	7200	3000		2500		1000		700					
Vet & med.	4800	500	700	600	600	400	300	200	200	200	200	400	500
Livestock purchases: cows	14400										6750	3600	4050
Misc. dairy costs	2160	180	180	180	180	180	180	180	180	180	180	180	180
Seed	3590	2420					1170						
Fertilizer	14590		14590										
Spray	4120		4120										
Misc. crop variable costs	490								490				
Wages – permanent	20000	2000	1500	1500	1500	1500	2000	1500	1500	1500	1500	2000	2000
Power and machinery	15000	2000	1000	1000	1000	1000	1000	2000	1000	1000	1000	1000	2000
Rent and rates	14000						7000						7000
Misc. fixed costs	8000	650	750	650	650	650	650	750	650	650	650	650	650
Overdraft interest	4967	417	408	541	554	378	351	395	373	353	380	415	403
Loan interest	4276		2138							2138			
Capital – buildings	0												
machinery	20000				20000								
capital repayments	7000			3500						3500			
	0												
Personal – drawings	15000	1250	1250	1250	1250	1250	1250	1250	1250	1250	1250	1250	1250
tax	5000				2500						2500		
SUB-TOTAL	164593	12417	26636	11721	28234	6358	13901	6975	5643	10771	14410	9495	18033
NET CASH FLOW	−6440	1093	−16042	−1498	21117	3222	−5292	2671	2412	−3341	−4143	1385	−8025
OPENING BANK BALANCE	−50000	−50000	−48907	−64948	−66446	−45329	−42107	−47399	−44728	−42316	−45656	−49800	−48415
CLOSING BANK BALANCE	−56440	−48907	−64948	−66446	−45329	−42107	−47399	−44728	−42316	−45656	−49800	−48415	−56440

Figure 12.9 Example of a cash flow budget.

Once the list is complete, the pattern of timing of receipts and payments can be worked out. Each type of receipt is examined in turn, and a judgement made concerning the way in which that receipt will be split between the months of the year. With many items, this will require careful estimation of the pattern of physical production – the way in which milk receipts will occur during the year, for instance, will be governed by the combined lactation curves of the cows in the herd.

It is a normal trade practice to allow between 2 and 4 weeks' credit on sales, so many cash receipts will be paid into the bank the month after the actual sale has taken place. This must be allowed for in deciding in which month the cash is likely to enter the bank account. Payments by opening debtors should be assumed to be received in the first month of the year. The process is repeated with each payment item, again ensuring that allowance is made for credit taken. It is sometimes possible to negotiate extended credit (3 months, for instance) with suppliers of items such as fertilizer and machinery. This must be allowed for in deciding timings.

The budget can now be completed by entering receipts and payments in the body of the table according to the timings decided, and calculating monthly bank balances. Beginning with the column for the first month of the budget year, total payments for the month are subtracted from total receipts to give the month's net cash flow. When added to the bank balance expected for the beginning of the year, this gives the expected bank balance for the end of the first month. A positive figure denotes money in the bank; a negative figure indicates an overdraft.

This closing bank balance for the first month becomes the opening balance for the second month, and the above calculation is repeated for this and each of the remaining 10 months of the year. If the closing balance of any month is negative (i.e. shows an overdraft), the interest due for that month can be estimated by multiplying the balance by one-twelfth of the likely overdraft rate. The result is entered under 'overdraft interest' in the following month's column.

A vital part of the budget is the 'total' column, used to summarize the cash flow for the whole year. This is compiled by adding the contents of each line across the sheet, and then finding the annual net cash flow and closing balance as for a monthly column (the opening balance being that used in the first month's column). As well as providing a useful summary of the year's expected cash flow, and a means of cross-checking with the contents of the profit budget, it provides an important internal check. If the closing balance calculated in this manner matches that shown in the final month's column, one can be confident that the arithmetic in the budget is correct. If it does not, the calculations must be checked and rechecked until the error is located.

A valuable aid to the interpretation of the cash flow budget is a graph of the monthly closing bank balances, as shown in Fig. 12.10.

Assessing the cash flow budget

Before moving on to the budget for capital, it is important to weigh up the implications of the cash flow budget. As before, the main test is whether the budget indicates that the cash flow objective (probably expressed as the maximum tolerable overdraft) will be achieved. A parallel test is whether the overdraft at any time is likely to exceed the existing limit agreed with the bank manager. The next concern should be to check for significant variations in bank balance between months, and for any unusual movements in key items.

If these checks indicate potential problems, there are four possible strategies open to the manager. Ways can be sought of *increasing cash flow*, and thus lifting the overdraft away from problems. Possibilities include cutting or postponing private drawings and capital expenditure, and a general reduction of overhead costs. Another possibility is that enterprise gross margins could be examined for possibilities of increasing output and/or reducing variable costs, although most avenues

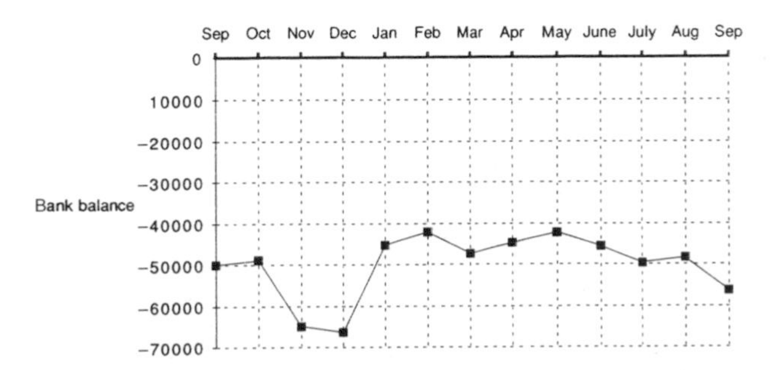

Figure 12.10 Cash flow chart.

should have been explored when assessing and adjusting the profit budget.

The second opportunity is that of *delaying payments* to reduce overdraft peaks. Buying fertilizer as it is needed, for instance, or negotiating longer trade credit from a machinery supplier could avoid the overdraft limit being exceeded and help to cut interest charges. Another way of reducing overdraft peaks is by *advancing receipts*, such as selling the wheat off the combine rather than storing, or pressing debtors for earlier payment.

These possibilities should be tested by reworking the cash flow budget and seeing the effect on the cash flow. Each change will have a cost – lost discounts on fertilizer, lower prices on corn sold early – and the costs and benefits of the change should be examined (a partial budget is an ideal tool for this; see 'Planning for change' later in this chapter). Some changes antagonistic to the manager's values and objectives (for instance pressing debtors for prompt payment, or cutting personal expenditure) may be unacceptable in the first instance.

The final possibility, when all other options have been considered and the overdraft is still over the bank's existing limit, is to seek to negotiate a new overdraft limit. The existence of a carefully worked and reworked cash flow budget will be of great help in this process. If this fails, and/or the new limit goes against personal objectives with respect to the maximum borrowing, a complete rethink of the business objectives and plan is necessary. Otherwise, the next stage is to prepare a budgeted balance sheet.

Budgeting for capital

The budgets for profit and cash flow provide the raw material for a projected balance sheet. This is a statement of the wealth of the business at a particular time. Wealth denotes ownership of assets – money and other things of value, such as stocks, machinery and land. Against these assets must be set liabilities, however – debts such as outstanding trade creditors, overdraft and loans.

A typical farm balance sheet is shown in Fig. 12.11. The difference between assets and liabilities is shown as *net* capital (also known as 'net worth' or 'owner equity'). This figure represents the net value of the business to its owner(s). In a company business net capital will be

Projected balance sheet at 31 March 200x	*This year* £	£	*Last year* £	£
Fixed assets				
Bungalow (at cost 198x)	70 000		70 000	
Plant, machinery and vehicles	36 000		30 000	
Breeding livestock (dairy herd)	50 800	156 800	50 000	150 000
Current assets				
Trading livestock (calves)	280		280	
Crops in store	41 950		42 700	
Crops in ground	3 000		3 000	
Miscellaneous stores	9 000		9 000	
Trade debtors	9 265		8 910	
Cash	0		0	
	63 495		63 890	
Current liabilities (due within one year)				
Trade creditors	1 349		1 200	
Bank overdraft	56 440		50 000	
	57 789		51 200	
Net current assets (liabilities)		5 706		12 690
Total assets less current liabilities		162 506		162 690
Deferred liabilities (due after one year)				
Bank loan		25 000		27 100
Capital account				
Opening net capital	135 590			
Add: net profit	23 067			
	158 657			
Less: Private drawings	15 000			
Income tax	5 000			
Goods for own use	1 151			
Net capital		137 506		135 590
Capital employed		162 506		162 690

Figure 12.11 Example of a balance sheet (tenant).

expressed in a slightly different form, as a combination of share capital and reserves.

The figure used for bank balance or overdraft is the closing balance from the last column of the cash flow budget. The debtors and stock valuations are those used previously in compiling the profit budget. The values of machinery and buildings are those after depreciation (used in the profit budget). The value of land can be taken as that at the beginning of the year, plus or minus any sales or additions during the year (which will be shown in the cash flow budget).

Apart from the overdraft, the liabilities include creditors, which have already been determined in compiling the profit budget. The value of loans can be taken as the amounts outstanding at the beginning of the year, plus or minus any additions or repayments during the year (shown in the cash flow budget). Figure 12.11 also shows the previous year's balance sheet figures in a separate column, thus enabling an easy assessment of likely changes over the year.

A further refinement is the inclusion of a capital account. This traces the growth or otherwise of net capital over the year, by reference to profit and injections of money from outside the business (both adding to net capital), and private drawings, tax payments and benefits in kind (all reducing net capital). As well as showing the relationship between profit and drawings, the capital account acts as a useful arithmetical check. The result of the calculation should be identical to the closing net capital shown by the balance sheet: if it is not, the budget calculations should be checked and rechecked until the error is found.

Assessing the projected balance sheet

Much of the message of the projected balance sheet can be seen at a glance. A positive net capital (i.e. assets more than liabilities) confirms that the business is still likely to be solvent (in business) at the end of the year. Comparison of the opening and closing balance sheets shows how the financial position of the business is likely to change over the budget year – whether net capital is likely to increase, how much total assets and total liabilities are likely to change, and to what degree major changes in individual items are responsible.

Additional clues can be derived from the application of various ratios. Space precludes mention of all such, and indeed it is doubtful whether it is useful to know more than a few easily remembered calculations [but see Warren (1998) for more detailed discussion]. To check overall stability of the finances, the *percentage owned* can be calculated (net capital as a percentage of total assets). This both shows the owner's stake in the business and gives an impression of the 'safety margin' enjoyed by the business. A business with 40% owned has rather less buffer against changes in financial fortunes than one with 80% (typical of many long-established owner-occupied farms). The degree to which this percentage is likely to change during the course of the budget year is a vital indication of the financial health of the business. The example farm (Fig. 12.11) has a low and declining percentage owned of 62%.

An additional overall check is provided by the *long-term debt to equity ratio*. This expresses long-term borrowing (loans and any other finance with a fixed repayment pattern) as a percentage of the net capital. Such borrowing imposes demands for both interest and repayment, irrespective of the general cash flow of the business. Thus the higher the ratio, the higher the risk faced by the business, and the greater the need for a high return on capital (see below) to cope with these demands. In an occupation such as farming, which commonly gives low returns on capital, a ratio of above about 25% would give rise to concern – but this is only a very general guide. The example farm has a ratio of 18%.

Short-term stability can be measured by the ratio between the liquid assets of the business (very short-term assets such as cash and trade debtors) and current liabilities (those at short call such as overdraft and trade creditors). This is the *acid-test ratio*. A ratio of 100% indicates that even if the current liabilities were suddenly called in, the business would have enough short-term reserves to cope. The lower this ratio, the more susceptible is the business to running out of cash. As before, changes over the year are more important than static measures – a declining ratio is a warning signal. The example farm has a ratio of 110% – down on the opening figure of 125%.

Finally, budgeted net profit and net capital may be linked by calculating a return on capital, such as *return on owner equity*. This is net profit expressed as a percentage of net capital, and indicates how efficiently the owner's capital is used within the business. The farmer who finds that his return on net capital will be less than 3% while he could get an assured 10% by investing in government securities (say), needs to ask himself some searching questions – such as whether satisfying personal objectives (for instance pleasure in running a farm) is worth the financial cost involved. Profits can be highly variable between years, and before making a decision based on return on capital, outline budgets should be used to check likely trends over the following 2 or 3 years. The example farm has a return on net capital of 16.8%, which will go some way to compensating for the low-stability measures.

As with the other budgets, if the picture indicated by

the assessment of the projected balance sheet does not meet the objectives of the owner, alternative plans must be investigated and budgeted. First, though, the valuations of the assets in the balance sheet should be checked. If these are based on past balance sheets, asset values may be out of date. This should be checked, assets conservatively revalued, and ratios recalculated, before any decision to rebudget is made.

Accounts

While budgets estimate the likely future effect of decisions, accounts provide the historic information to show the actual effects of those decisions.

Recording cash flow

The primary source of historic information about a business over a year is the record of the business transactions made during that year. Various methods exist of making this record, ranging from merely collecting invoices in a cardboard box, to using a computer accounting system. That which is most commonly found on farms is the *cash analysis* system of book-keeping. As its name suggests, this is a method of recording net cash flow. It is generally based on monthly intervals, recording expenditure and receipts as the payment is made or received. The main features of a cash analysis book are shown in the example of a monthly payments and receipts analysis shown in Fig. 12.12. The broad pages accommodate a large number of columns. Those at the left are used to record the essential information about each transaction: date, details, cheque number (paying-in slip number for receipts) and the amount of money involved.

The remaining columns allow the amount paid or received to be categorized, ideally by reference to eventual use in gross margin accounts. Thus payment analysis columns could be divided into variable and fixed cost groups, with the former including columns for feed, livestock purchase, veterinary and medical, seeds, sprays, fertilizers, contract hire and so on, and the fixed cost columns including those for regular labour, machinery running costs, rent and rates, finance charges and miscellaneous overheads. In addition to these 'trading' items, some of the transactions will relate to personal, capital or tax items (VAT, income and capital taxes): hence the columns at the right of the sheet.

The receipts sheet is laid out in a similar manner, with the analysis columns categorizing receipts by type of output. The extent of the division is limited partly by the number of columns on the sheet and partly by the need for clarity and simplicity.

At the end of each month, a line should be drawn beneath the last entry on each page, and the total found for each column. Taking the total of the payments page 'bank' column from the total of that on the 'receipts' page gives the bank account at the end of the month – a negative balance denoting an overdrawn account. The accuracy of this balance (and thus the recording for the whole month) can be checked by comparing the balance with that shown on the bank statement (which should be requested to show the cleared transactions up to the end of each month).

The two figures will rarely be identical, due to delays between cheques being written and their reaching the bank, and to non-cheque items such as standing orders and direct debits and credits, which may not have been recorded in the cash analysis book. If, though, after making adjustment for these items, the balances in cash analysis book and bank statement still disagree, a mistake is indicated (either at the bank or in the book-keeping) and must be located before proceeding.

Monitoring cash flow

The totals of the analysis columns can now be transferred to the 'actual' column for that month in the cash flow budget (see 'Budgeting for cash flow' above). Ideally the headings used in both cash analysis book and cash flow budget will be the same, making the transfer of information between the two as easy as possible. Once this is done, and the actual closing balance entered at the bottom of the column, the two sets of information can be compared in order to detect discrepancies. The aim in this comparison should be to establish the main causes of discrepancy, to anticipate the likely effect over the remainder of the year, and to identify appropriate remedial action (see Fig. 12.13).

The cash flow budget is an essential tool for the latter two tasks. If there is little that can be done to bring actual performance into line with the budget, it may be time to revise the budget for the rest of the year. Should this revision imply a higher overdraft facility than has been agreed with the bank, it can be used to good effect in renegotiating that facility well before it is actually needed.

For a more complete description of the mechanics of cash recording and control, a specialist text is recommended such as Hosken and Brown (1991) or Warren (1998).

Recording profit and capital

Actual profits or losses, and actual net capital, are measured by exactly the same methods as used in the budgets. Thus a profit and loss account, in conventional

PAYMENT ANALYSIS

Date	Details	Stub no.	Bank	Dairy feed	Vet and med.	L/stock purchase	Misc. dairy	Crop costs	Wages	Power	Rent, rates, misc. o/hds	Interest and charges		Capital and	Personal and tax	VAT paid
1/10	Opening bank balance b/fwd (o/d)		50000.00													
5/10	Wages	120	438.40						438.40							
5/10	A. Dealer: tractor repair	121	575.31							489.63						85.68
5/10	Farmco Ltd.: 25t dairy feed @ £142.50/t	123	2850.00	2850.00												
9/10	Adams: vet bill	124	405.38		345.00											60.38
13/10	Newgro Ltd.: corn and grass seeds	125	2560.32					2560.32								
13/10	Fastfuel Ltd.: tractor diesel	126	532.58							532.58						
13/10	Telephone bill	127	162.85								138.59					24.26
13/10	Wages	128	484.56						484.56							
20/10	Wages	129	523.87						523.87							
20/10	High St. Garage: petrol and car service	130	300.66							255.88						44.78
20/10	Browns: dairy supplies	131	550.83				468.79									82.04
28/10	Wages	131	427.35						427.35							
28/10	Fencing wire	132	447.06								380.48					66.58
29/10	A. Dealer: tractor	133	21737.50											18500.00		3237.50
29/10	Transfer to personal a/c	134	1250.00												1250.00	
	TOTAL PAYMENTS		83246.66	2850.00	345.00		468.79	2560.32	1874.18	1278.09	519.07			18500.00	1250.00	3601.21

INCOME ANALYSIS

Date	Details	Stub no.	Bank	Milk	Calves and culls	Crop sales	Sundry income		Capital	Personal	VAT charged	VAT refunds
5/10	J. Smith: 15 calves	927	1568.00		1568.00							
20/10	M.M.B.: milk sales (direct credit)		9234.32	9234.32								
20/10	F. Bloggs: contract ploughing	928	327.83				279.00				48.83	
29/10	A. Dealer: sale of tractor	929	7637.50						6500.00		1137.50	
	TOTAL INCOME		18767.65	9234.32	1568.00		279.00		6500.00		1186.33	
	BALANCE CARRIED FORWARD (O/D)		64479.01									
			83246.66									

Figure 12.12 Example of cash analysis for one month.

	OCTOBER	
	Budget	Actual
Income		
Milk	8910	9234
Calves	2100	1568
Cull cows		
Wheat		
Wheat straw	2500	
Sundries		279
VAT charged on outputs		1186
VAT refunds		
Capital – grants		
machinery sales		6500
Personal receipts		
SUB-TOTAL	13510	18767
Payments		
Feed: cows	3000	2850
Vet. & med.	500	345
Livestock purchases: cows		
Misc. dairy costs	180	469
Seed	2420	2560
Fertilizer		0
Spray		0
Misc. crop variable costs		
Wages	2000	1874
Power and machinery	2000	1278
Rent and rates		
Misc. fixed costs	650	520
Overdraft interest		
Loan interest		0
VAT payments		3601
Capital – buildings		
machinery		18500
capital repayments		
Personal – drawings	1250	1250
tax		
SUB-TOTAL	12000	33247
NET CASH FLOW	1510	-14480
OPENING BANK BALANCE	-50000	-50000
CLOSING BANK BALANCE	-48490	-64480

Figure 12.13 Example of cash flow monitoring.

or gross margin form, is used to measure net profit or loss, and a balance sheet to measure the net capital.

It was suggested above that the selection of headings in the cash analysis book should be influenced by the needs of gross margin accounts, with payments divided into fixed and variable categories. The result will be that, at the end of a year, the monthly totals of each of the 'trading' analysis columns (i.e. excluding capital, personal and tax columns) can be accumulated to give the raw material for the gross margin enterprise accounts. Given the limitations on the number of analysis columns in the cash analysis book, it may be necessary to use a simple coding system during the year to distinguish between items within the same column. A letter 'P' could be written alongside appropriate items in the 'feed' payments column, for instance, to indicate pig feed as opposed to dairy or sheep feed.

From this point, it is a relatively simple matter to compile gross margin accounts for the year, using the procedure described for the preparation of budgeted gross margins. Fixed costs are likewise derived from the relevant analysis column totals, and deducted from total gross margin to arrive at an actual net profit.

Analysis and interpretation of profit and loss accounts

Interpretation of a profit statement depends on comparison: with previous years' figures, with other businesses, and with the relevant budget. All of these methods can be used with a profit and loss account in conventional form, but provide much more useful management information when used with a profit statement in enterprise account form, such as the gross margin accounts discussed above.

Comparison with previous years gives a valuable picture of trends within the business, and management 'alarm bells' will be rung by major differences between the performance of the immediate past year and those preceding it. As always, the comparison should start with the 'bottom line' – the net profit – and work back through total revenues, expenses and stock valuations to build up a picture of cause and effect concerning the variations. Given the physical and financial detail shown in each enterprise gross margin account, such comparisons can be used in building up a list of clues to help management explain how discrepancies arose, and how to correct them in the future (or capitalize on them, where performance is better than before).

Previous-year comparisons are limited by variations in the general environment (e.g. weather, prices) between years, and are also insular, relating as they do to just one business. To avoid the latter problem in particular, comparisons can be made with other businesses. Various comparison media are available in UK agriculture. Some are based on surveys, such as the annual Farm Business Survey (FBS) conducted by Regional Agricultural Economics Centres. The results of this survey are published in regional reports, most of which present their results in enterprise account as well as whole-farm form. Comparing farm gross margin results with these survey results thus gives rise to a number of new clues as to where farm profitability could be improved.

The value of survey results is often limited by the time taken to collect and process the information, so that the latest FBS results available in a region are likely to be at least 1 year out of date at publication. Survey regions are large, and the number of farms of a particular type in each sample usually relatively small. A further drawback is that such surveys are often not designed to collect detailed physical enterprise information: this inevitably limits the diagnostic possibilities of comparisons.

An alternative source of comparison information is a costing service. Such services are offered to farmers by a number of organizations, including quasi-government agencies (such as the MLC Pigplan service), commercial companies and other agricultural institutions. These services, particularly those relating to dairy and other intensive livestock, usually depend on monthly updating of information from the farmer, and use computers for the collation and analysis of that information.

As a result they can process information rapidly and provide reasonably up-to-date comparisons. Moreover, being designed specifically as management aids for the farmer, they usually contain a thorough breakdown of the most important physical information concerning the enterprises concerned. Thus a farmer using MLC's Pigplan can trace a low gross margin back to such details as low gradings, high feed used, low liveweight gain, poor market prices, and so on. In these respects, the use of comparisons via a costing scheme has advantages over the use of survey data. On the other hand, it must be remembered that the results of a costing scheme relate to a self-selected sample of farmers, and that the accuracy of the information fed into the costing scheme by farmers is not subject to the same rigorous control over accuracy as would be the data collected for survey purposes. As long as the user remembers that any form of comparison is merely a process of looking for clues rather than answers, this should not pose too great a problem.

The third, and potentially most useful, method of comparison is that with budgets. This is part of the budgetary control described earlier as an integral part of the management process, and is analogous to the budget/actual comparisons explained in the section on monitoring cash flow. The process of comparison is as before, but this time using the budgets prepared at the beginning of the year in question, rather than previous years' figures, or those derived from other farms. This form of comparison has the advantage of being specific to the farm in question, and relating to the most recent year for which results are available. Most importantly, it measures the farm's performance against the objectives set at the beginning of the year, as part of the planning process.

Similar strictures apply to this as to the other forms of comparison – that it is concerned with collecting clues rather than answers, and that to follow up the clues needs the application of down-to-earth farming knowledge, such as that provided by the other chapters in this volume.

Recording and analysis of net capital

Once cash analysis book and profit statement have been completed for a particular year, the compilation of a closing balance sheet is a simple matter, using information contained in those accounts and the methods described in the earlier section on budgets. Similarly, analysis of the balance sheet is exactly as described for the projected balance sheet, with the sole additional facility of comparison between budgeted and actual balance sheets.

In summary, effective control of a business depends on *budgets* anticipating the outcome of decisions in terms of cash, profit and capital. It also requires *accounts* to show actual performance in the same terms, allowing the manager to monitor progress and take corrective action.

Physical information

Calculation of reliable gross margin accounts, particularly in the detail suggested above, depends on good physical records. Examples include records of output, such as milk yields, pig liveweights and grain yields; and records of inputs such as amounts of feed consumed and fertilizer applied.

Physical records are important for other reasons. Breeding records and charts, for instance, are vital in maintaining and improving the genetic potential of a breeding livestock enterprise. Short-term control often depends on the use of physical records. While the cost of water may not be sufficiently large to justify recording solely for enterprise accounts; for example, water records can give valuable advance warning of changes in the health of an intensive poultry flock or pig herd. Other physical records, such as those of labour use and management time, may be useful in making the use of time more efficient.

There is no 'right' way of keeping physical records, but certain guidelines should be borne in mind:

- Never record for the sake of recording – be sure that the benefit gained from using each record will justify the trouble taken. It is easy to find oneself keeping records that are rarely, if ever, used, and which could be safely discontinued.
- Make it as easy as possible for the records to be entered. This implies careful design of recording sheets, but also careful positioning of the sheets,

near to the place where measurement is made, well lit and with a pencil handy.

- Make it as easy as possible for the records to be interpreted. This is particularly important where the records are used in day-to-day control: key information should be highlighted and use made of graphs so that problems are clearly and rapidly shown up. It also applies where records are used as the basis for financial accounts. A little thought at the design stage about the way in which the information is to be used could save a great deal of trouble later.

For more information on the keeping of physical records, Hosken and Brown (1991) is recommended.

Planning for change

There are times when to use a full-blown set of budgets would be inappropriate. An example is where a number of partial changes to an established system are being considered. To avoid unnecessary work and potential confusion, a device is needed that will allow consideration of only the *net* financial effects of the change to be calculated. Such a device is the partial budget. Partial budget principles can be applied to any of the three financial measures – profit, cash flow and capital – but are most commonly used to test the effects on profitability of a given change. These effects are grouped under four headings: revenue gained; costs saved; revenue lost; and extra costs.

The result is a calculation such as that shown in Fig. 12.14. This is a budget for a 'normal' year – in other words, when the change has been fully established. It is possible to compile a partial budget for a specific year, such as the first year the change is implemented, if profitability is likely to vary significantly between years. The normal year budget is the most important, though: if this does not show a benefit, there is no point in looking more closely at particular years.

Note that all costs and revenues are expressed on an annual basis. Thus expenditure made (or saved) on buildings and machinery is represented by depreciation (initial cost divided by the useful life in years), and the cost of building up (or reducing) a breeding herd or flock of livestock is shown by the extra annual costs of replacement incurred by those animals. Note also that the budget is kept clear and uncluttered by including only those items that are likely to change. If a change is likely to reduce workloads, but it is unlikely that a worker can be laid off or overtime reduced, wages costs can safely be omitted.

The best use of a partial budget of this sort is as an 'initial screening' device, allowing a quick and simple check of the relative merits of a number of alternative changes to the farm system. Those changes that partial budgets show to be likely to lose money, or make insufficient profit to justify the effort and inconvenience caused, can be scrapped without further ado, and the

COSTS OF CHANGE	£	BENEFITS OF CHANGE	£
Revenue lost		***Extra revenue***	
Cows: milk: 10 @ 6524 litres @ £0.215/litre	14023	Milk: 60 ewes @ 350 litres @70 p/l	14700
Culls: 2/year @ £583	1166	Lambs: £50/ewe	3000
Calves: 0.95/cow @ £118	112	Cull ewes:18/year @ £10	180
		Cull rams: 1/year @ £20	20
		Wool: 2 kg/ewe @ £0.30/kg	36
		Ewe premium: £15/ewe	900
Sub-total	15301	Sub-total	18806
Extra costs		***Cost saved***	
Ewe purch.: 18/year @ £60	1080	Cows: concs. (1.04 t/cow)	1350
Ram purch.: 1/year @ 500	500	Vet & med., A.I., etc.	686
Concs.: 200 kg/ewe @ £160/t	1680	Other	398
Vet & med.	500	Forage: £198/ha × 5 ha	841
Other	450		
Forage	600		
Extra labour	500		
Capital: 5000 over 5 years	1000		
Sub-total	6310	Sub-total	32715
TOTAL COSTS	21611	TOTAL BENEFITS	22111
BALANCE – EXTRA PROFIT	**500**	BALANCE – LOSS OF PROFIT	0

Figure 12.14 Partial budget for reducing the dairy herd by ten cows, and releasing 5 ha for 50 milking evens and four rams.

remainder built in to revised whole-farm budgets to test the effects on cash flow and balance sheet, as well as profit. Where the change requires investment, it is important to relate the extra profit to the capital used. There are, broadly, two ways of doing this – by including an interest charge in the partial budget, and by calculating a marginal return on capital.

The ideal way of achieving the former is to compile a cash flow budget. For this, however, the cash flow budget would have to relate to a particular year, rather than a 'normal' year, and its complexity of calculation militates against its convenient use in initial screening. The alternative is to make an estimate of the amount borrowed, and apply to this an interest rate.

Rather than estimate the amount of borrowing outstanding year by year (which again would lose the simplicity that makes the partial budget so effective), a common practice is to take the initial amount borrowed and assume it will be paid off in equal instalments over the life of the project. The average borrowing outstanding in a 'normal' year is then taken as the initial borrowing divided by two. At worst it is possible to assume that all the net initial capital is to be borrowed, but only if no firm information is available concerning the amount that is to be borrowed.

Net initial capital is the amount of money to be invested in the first instance. It includes both fixed capital (items such as property, machines, breeding stock, etc.) and working capital (the cash needed to finance the running costs of the change until it begins to bring cash in to defray those costs). It also takes account of any fixed and working capital set free by the introduction of the change. An example, related to the earlier partial budget, is shown in Fig. 12.15.

Using this example, and assuming all the capital needs to be borrowed, the average borrowing would be £5078/2 = £2539. At an interest rate of, say, 10%, the interest charge in a 'normal' year would be £254, and

FIXED CAPITAL		
Building and plant	5000	
Breeding stock – ewes	3600	
rams	2000	
LESS sale of cows: 10 @ £600	–6000	
		4600
WORKING CAPITAL		
Half running costs of sheep	2115	
LESS half running costs of cows	–1638	
		478
TOTAL NET INITIAL CAPITAL		5078

Figure 12.15 Calculation of net initial capital.

the extra profit after interest would be 750 – 254 = £496. There are various problems with this type of calculation, not least because of the many assumptions and generalizations which have to be made in the process. Moreover, it only measures the *cash* cost of borrowing, and puts no value on any non-borrowed capital used.

An alternative is to relate profit to capital through calculating return on capital. The appropriate measure here is *marginal return on capital*, calculated by dividing the extra profit before interest (from the partial budget) by the net initial capital required, and multiplying by 100. In the example above, for instance, the rate of return would be $750/5078 \times 100 = 15\%$.

The resulting percentage rate can firstly be compared with interest rates payable on borrowed money. If the rate of return is significantly more than the interest rate, it looks financially worthwhile; if less, the change is a non-starter from a financial point of view – the finance costs will outweigh the return. If the rate of return is only marginally more than interest rates, the farmer must decide whether it is worth the risk that profit might be less than forecast (or capital needs greater). If the change passes this test, its marginal return on capital can be compared with those of other possible changes as a basis for selection between projects.

Among the drawbacks of this use of marginal return on capital are that it rests on large assumptions, takes no account of the pattern of costs and returns either within or between years and can give pessimistic results. The latter can be countered by using average rather than initial capital in calculating the return, but this tends to give overoptimistic results – it is perhaps better to tolerate a pessimistic result than to risk making decisions based on a result that may never be achieved.

More sophistication is available in the shape of 'discounted cash flow' techniques. These are essentially partial budgets which take account of differences in costs and returns between years of a project, and allow flows of money in the near future to be weighted more heavily than those in the far future. Tax payments can also be incorporated. The result is a 'net present value' showing the consolidated margin of future cash flows over the amount invested, after allowing for the time factor. This can in turn be used in calculating an 'internal rate of return', which can be used in the same way as the marginal return on capital described above.

The very sophistication of the techniques gives rise to difficulties in application, calculation and interpretation. Without training, and preferably a computer, it is likely to be better to use the crude but simple devices described earlier in this section. If these are treated purely as initial screening measures, to be followed by full budgets where these appear to be warranted, the

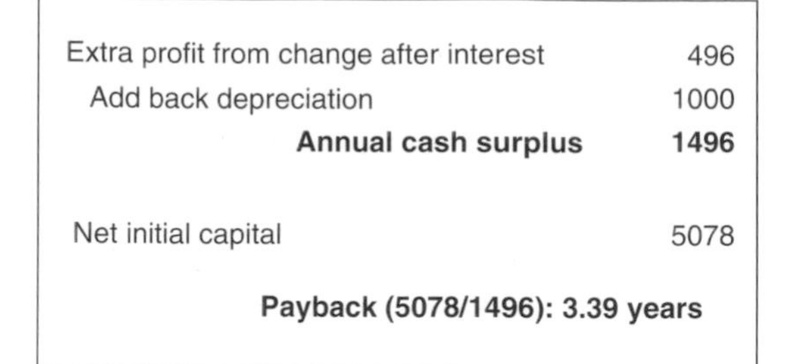

Extra profit from change after interest	496
Add back depreciation	1000
Annual cash surplus	**1496**
Net initial capital	5078
Payback (5078/1496): 3.39 years	

Figure 12.16 Example of simple payback calculation.

crudity is acceptable. Further reading is suggested at the end of the chapter.

One additional device can be useful in planning for change. This is *payback* – the number of years needed for a change to repay the net initial capital. For this it is necessary to estimate the net cash inflow for each year of the project's life. One way of doing this is merely to take the extra profit from the partial budget, exclude depreciation (the main non-cash expense) and include interest. It is possible to make more precise estimates using simple cash flow budgets. The net cash flows are then progressively totalled, starting from year one (allowing for negative flows), until they accumulate to the net initial capital invested. The payback of the project is the number of years it takes to get to this point. A quicker and cruder method is shown in Fig. 12.16, where a 'normal year' cash flow is calculated and divided by the net initial capital.

Payback is of use in its own right, in giving a 'feel' of the worthwhileness of the project. Compared with that of other possible projects, it gives a measure, though imprecise, of the relative riskiness of the projects. If size, marginal returns and personal preferences were equal between three projects with paybacks of 3, 6 and 10 years, the rational decision-maker would go for that with the lowest payback.

Whatever partial budgeting devices are used for the testing of the effects of a change from an established system or a whole-farm budget, it will be necessary, for all but the smallest of changes, to eventually test the effects of the change using whole-farm budgets for profit, cash flow and capital.

Sources of capital

To put a business plan into operation (and to maintain it) usually needs investment of lump sums of money – capital. That capital may be wholly or partly on hand, in the form of savings, retained profits or the proceeds of the sale of assets. The farmer may be lucky enough to be able to claim grants to defray some of the costs of investments, or even to have relatives or other well-wishers willing and able to make outright gifts of money. Though such capital may appear to be free of charge, there are costs attached to the use of any capital sum. If nothing else, there is the opportunity cost arising from the lost benefits from using the money in some other way. Other, non-financial costs arise from legal obligations, in the case of a grant, or 'moral' obligations, in the case of a gift, for instance. Nevertheless, the costs are usually relatively low, and these sources should be exploited to the full before turning to raising money from outside the business through borrowing or shared ownership.

Borrowing

The most common alternative is to borrow money. Borrowed money incurs opportunity costs and obligation costs as described above. In addition, it incurs a direct cost – that of interest (the exception being some private loans from relatives). This is a fee charged by the lender of the money, expressed as an annual percentage of the amount borrowed. The rate of interest depends on both external factors, such as the state of the economy, and internal factors, such as the security offered and the apparent creditworthiness of the business.

Interest rates are often expressed in relation to the clearing banks' base rate, which reflects the level of interest in the economy as a whole and is published in the financial sections of quality newspapers. An interest rate may be variable (i.e. it fluctuates over time with changes in the base rate) or fixed (i.e. it stays the same over the length of the borrowing term).

Rates can be applied in various ways. For instance they can be related to the daily balance outstanding (as with an overdraft), to the amount outstanding at the beginning of a particular year, or applied to the initial amount borrowed ('flat rate'). To allow the cost of different forms of borrowing to be compared, each rate should be expressed as an Annual Percentage Rate (APR). The APR on an overdraft is normally close to its stated rate, while that of a flat rate can be nearly double the expressed rate.

A lender will usually require some form of security against default on payments by the borrower. This security can be in the form of title to specific fixed assets, enabling the bank to sell those assets to recover the debt if necessary. A 'floating charge', commonly used for overdrafts, is a generalized version including more movable assets of the business. Alternatively, security can be provided in the form of a guarantee from an individual or an institution that, if the borrower defaults, the guarantor will repay the money outstanding.

Sources of borrowed capital

It is difficult to avoid generalizing when referring to the characteristics of various sources of borrowing. The following notes should be judged in this context.

The high-street banks are the main source of borrowed capital for farming, and the largest part of this is in the form of *overdrafts*. An overdraft is created when more money is drawn out of an ordinary bank current account than was deposited there in the first instance. Interest rates are low (typically 2–3% above base rate), and are applied to the daily balance outstanding. Unauthorized overdrafts are charged a penal rate which can be more than 15% above the base rate, so it is important to agree an overdraft facility with the bank manager in advance of needing it. When an overdraft facility is negotiated or reviewed, an arrangement fee may be charged. This fee can be as much as 1% of the overdraft facility. Overdrafts are intended for short-term finance, and the bank will normally expect the account to return to credit at some stage during the year. It is normally secured by a 'floating charge' on all the assets of the business, and is technically repayable on demand.

Bank loans, on the other hand, are subject to a contract binding each party. When a loan is arranged, the borrower is given the amount of the loan as a lump sum. This amount is repaid over time in regular instalments, with interest payments (usually at a fixed rate) reflecting the decline in the loan outstanding. As long as these instalments are paid, and the other terms of the contract are met by the farmer, a loan cannot be called in before it is due (or the business becomes insolvent). To protect itself in the event of insolvency, the bank will normally expect the loan to be 'secured' on a specific asset – in other words, if the business defaults on the loan, the bank has a legally valid claim to the title of the asset, and can sell it in order to recover the money owed.

A loan is normally intended for longer-term finance than an overdraft. Interest rates are higher (e.g. 3–6% above base rate) and arrangement fees are again likely to be payable. Against the higher costs must be set the greater security of the loan and the regular and predictable interest and capital payments, making budgeting easier.

The high-street banks are not the only institutions offering loans to farmers. Loans secured on land and property ('mortgages') are available from a variety of sources, including building societies (for house purchase and building) and the Agricultural Mortgage Corporation (for land purchase, buildings and other longer-term needs).

Trade credit is a popular form of short-term finance, being quick and easy. It is obtained by not paying bills immediately. Most suppliers will allow a period of grace for payment of invoices, usually 2–4 weeks, though sometimes as short as 7 days. Using this period to the full can help keep down the business overdraft. Extending credit beyond this limit will, however, incur penalties – at the least, loss of goodwill from the supplier, and at the worst a credit charge or 'loss of cash discount'. These charges are invariably much more expensive than overdraft finance.

Two forms of borrowing are particularly relevant to machinery purchase. *Hire purchase* or *lease purchase* uses the asset purchased as the security for the borrowing: repayments are arranged so that the recoverable value of the asset is always greater than the debt outstanding. It is convenient, as purchase and finance can be arranged in one operation, but it can be considerably more expensive than bank finance.

Finance leasing involves a machine (or other asset) being bought by a finance company (often a subsidiary of a major bank) and rented to the farmer. After the initial rental period of 2–5 years (depending on the life of the asset) the farmer may be able to continue the lease into a 'secondary' rental at a nominal rent, or even to buy it. If the machine is sold on behalf of the finance company, he is usually able to retain a large proportion of the proceeds. Maintenance, servicing and insurance are all the farmer's responsibility. Thus the effect is of purchase, even though the arrangement is technically a lease. Rentals are quoted in terms of interest rates, and the finance companies' ability to exploit tax concessions helps to keep rates low even compared with overdraft finance. An attractive characteristic is that spreading the cost over several years enables machines to be acquired without immediate effects on the overdraft. This can be dangerous if done to deceive the bank manager as to the true state of affairs (at the time of writing it is not obligatory for sole traders and partnership to show leased assets on their balance sheets).

Calculating the cost of borrowing

It is crucial, if considering borrowing capital, to be able to estimate the likely costs to the business. Apart from administration costs (such as arrangement fees), the annual costs are composed of two elements – the repayment of the amount borrowed (needed for the cash flow calculations), and interest charges (needed for both the profit and the cash flow budgets).

In calculating the financial commitments arising from

an overdraft, there is no substitute for a cash flow budget as described earlier in this chapter.

Where interest on loans or other forms of finance is quoted at a 'flat' rate, the annual interest charge is given by the initial amount borrowed multiplied by the flat interest rate. The annual capital repayment is given by the initial amount borrowed divided by the number of years of the term. Thus £10 000 borrowed over a term of 5 years at a flat rate of 8% will give rise to interest payments of £800 per year (10 000 × 8%) and capital repayments of £2000 per year (10 000/5 years).

A very common method of repayment is that known as the 'annuity' or 'normal' method. This adjusts the proportions of interest and repayment throughout the term of borrowing, to keep the combined annual payment constant while allowing interest payments to reflect the gradual repayment of the capital. Tables are provided in most farm management handbooks and financial textbooks to enable estimation of interest and repayment charges for a particular year in the life of a loan. Part of such a table is shown in Fig. 12.17. From this it can be calculated that in year 3 of a 5-year loan of £10 000 at 12% a business will have to pay £800 interest (80 × 10) and £1970 capital repayment (197 × 10) (see Fig. 12.17).

Shared ownership

An alternative to borrowing is to share all or part of the business with others, in return for an injection of capital. The simplest form of sharing is a partnership, where two or more people agree to run the business in common, sharing profits and control in return for investment of capital. Each of the partners is jointly and severally liable for debts incurred by the other partners as well as by him or herself. In other words, if the business fails, the partners may have to sell all their personal assets if that is the only way that the creditors can be repaid – even if the cause of the failure was the action of only one of the partners. This is the motivation behind the formation of many private *limited companies*. A company can be formed with as few as two people, and has the advantage of liability being limited to the capital invested in the company – at the expense of higher formation and administration costs, and a certain lack of privacy.

Formation of a *co-operative* can allow shared investment in just part of a business, ranging from a single machine to a complete marketing and distribution service for produce. Liability is limited, and formation costs are similar to those of a company.

Any form of shared ownership raises complex and difficult questions, particularly where taxation is concerned. Generalization is impossible, and a reputable solicitor and accountant should be consulted before entering into any such arrangement.

Coping with risk and uncertainty

Implicit in all the above discussion is the difficulty of planning and budgeting for the future. This arises from the fact that nothing about the future is certain. The sources of this uncertainty, and the consequent risk to the business owner on making decisions about future events, are many and varied. In farming, physical factors are prominent, such as the possibility of disease, pest damage, drought, flood, fire, health of farmer and employees, and so on. Economic fluctuations also have effects – the state of world trade, the strength of the pound, and the growth of incomes of the population as a whole are some of the influences on product and input prices at the farm gate.

Just as important can be political factors, affecting

	Interest rate					
	8%		10%		12%	
Year	Interest	Repayment	Interest	Repayment	Interest	Repayment
1	80	170	100	164	120	157
2	166	184	84	180	101	176
3	52	199	66	198	80	197
4	36	215	46	218	56	222
5	19	232	24	240	30	248
	£ Payment per £1000 borrowed					

Figure 12.17 Extract from a loan repayment and interest table.

such things as tax allowances, agricultural support prices and quotas on production. Social and personal aspects also have an effect – such as changes in general food consumption habits, hardening attitudes of non-farming neighbours to noisy or smelly activities, marriage of the farmer's son or daughter, or the divorce of the farmer.

There are ways of organizing the business to minimize some of the effects of uncertainty. Some of the factors mentioned can be insured against (fire, for instance). Prices of some products and inputs can be protected by selling and buying forward, or by using the futures markets. The farm system can be designed with a wide spread of different enterprises, so that if one enterprise performs badly, the others can compensate. The farmer can avoid trying any new products or production methods to avoid the risk of their failure.

Not all risks can be avoided by such methods, and reducing risk usually has a cost. A farmer may prefer to take the risk rather than incur that cost. Insurance incurs the cost of the premium. Forward buying and futures trading incur the cost of losing potential revenue if spot market prices improve and the farm is tied to a contract price. Diversification and conservatism involve lost opportunities to increase profits – either by losing the chance of specializing in products that the farm and farmer are good at producing, or by forgoing the uptake of new, potentially profitable enterprises and production methods.

If risk cannot be limited beyond a certain point, it can at least be allowed for in the planning and control process. Reference has earlier been made to the importance of erring on the pessimistic side when budgeting, and to the need for adequate research to ensure that the forecasts on which the budgets are based are as accurate as possible. Most important of all is the process of budgetary control, whereby the effects of a deviation from the expected performance can be readily identified and appropriate action taken.

None of these methods gives the farmer much advance warning of how much risk he or she faces – and this can be crucial in making a decision. The key question concerns the 'downside risk': 'How much do 1 stand to lose if things go wrong?'

The simplest solution in concept is to prepare two sets of budgets, one based on the inputs and outputs that the farmer thinks are most likely, and the other based on the worst possible performance. The result is effective, with the farmer being able to see what he or she stands to lose if things go wrong and to decide whether or not to take this risk. Preparing two sets of budgets is time-consuming, though, unless one has available a suitably programmed computer. Even with a computer, interpreting the results can be difficult: so many variables may have been changed that it is impossible to pick out the factors most responsible for the variations.

The latter problem may be overcome by the calculation *of breakeven points.* Those factors that seem most likely to affect the outcome of the profit budget are identified. Two such factors on a dairy/arable farm might be the amount of concentrate fed to the cows, and the likely price for winter wheat. The budgetary margin (net profit or loss in a whole-farm budget, extra profit or loss in a partial budget) is then expressed in terms of that factor. For instance, in the partial budget example used earlier, the extra profit estimated was £496 after interest. At a price of 65 p/litre of ewe's milk, this represents 496/0.65 = 2712 litres (45 litres/ewe). This is the amount by which the milk yield would have to fall before the extra profit started to become a loss of profit. The breakeven milk yield is thus 350 – 45 = 305 litres/ewe.

Working out breakeven points for various components of a budget does two things. By looking at each breakeven point, the farmer can make subjective judgements as to which of the breakeven points is most likely to be reached: this gives an indication of those items that most contribute to risk, and which thus should be monitored most closely. It also helps the farmer to judge whether or not the risk is worth taking. In the example above, for instance, the farmer can ask himself, 'How likely is it that I will achieve only 341 litres/ewe, and am I prepared to take that risk?'.

The strength of the breakeven point technique – that it allows the effect of particular factors to be singled out – is also its weakness. It is not possible to identify the combined effect of more than one factor varying at the same time. While this may not be a huge problem in a simple partial budget, it is a severe limitation in budgets with many component variables.

An alternative is a form of *sensitivity analysis.* This uses tables to show the way in which the budgetary margin is likely to vary with changes in the value of one component of the budget. The example in Fig. 12.18 is based on the partial budget from earlier in the chapter. It shows how the extra profit is likely to change from the value in the budget (£496) if ewe milk yield were to vary from the 'most likely' assumption of 350 litres used in that budget. The lower figure is, in the farmer's view, the worst likely outcome, and the higher figure is the best that he or she feels is possible.

This gives the farmer a feel for the 'sensitivity' of the budgetary margin to variations of a particular component of the budget, and thus the degree to which the successful outcome of the plan depends on the farmer's ability to achieve this level of performance. Better still, this table can be combined with others to show the effect

MILK YIELD PER EWE		
300 litres	350 litres	400 litres
−1454	496	2446
Extra profit from change, after interest		

Figure 12.18 Sensitivity analysis: one-way table.

	MILK YIELD PER EWE		
MILK PRICE	300 litres	350 litres	400 litres
55p/litre	−3554	−1604	346
65p/litre	−1454	496	2446
70p/litre	−404	1546	3496
	Extra profit from change, after interest		

Figure 12.19 Sensitivity analysis: two-way table.

of interaction between components. In Fig. 12.19, for instance, the milk yield table has been extended to incorporate variations in milk price. Here the decision-maker has the opportunity to see what the consequences would be of poor performance on both fronts – and to decide whether or not to take the risk. At the cost of simplicity and clarity, such a table can be extended to show the combined effects of variation in three or even four budget components.

The methods described merely present the farmer with information about the likely consequences of deviation from the forecast performance assumed in the budgets. They leave the farmer to use subjective judgement as to how likely these deviations are to occur. Other techniques exist which allow probabilities of different outcomes to be incorporated in the calculation, and these are described in texts such as Barnard and Nix (1979) and McCrimmon and Wehrung (1986). For most practical purposes in farm management the simpler devices described above will be sufficient, particularly if combined with a rigorous system of budgetary control.

References and further reading

Barnard, C.S. & Nix, J.S. (1979) *Farm Planning and Control*, 2nd edn. Cambridge University Press, London.

Giles, A.K. (1986) *Net Margins and All That*. Study No. 9. The Farm Management Unit, University of Reading.

Giles, A.K. & Stansfield, J.M. (1990) *The Farmer as Manager*, 2nd edn. CAB International, Oxford.

Hosken, M. & Brown, D. (1991) *The Farm Office*, 4th edn. Farming Press, Ipswich.

Lumby, S. & Jones, C. (1999) *Investment Appraisal and Financing Decisions*, 6th edn. International Thomson Business, London.

McCrimmon, K.R. & Wehrung, D.A. (1986) *Taking Risks*. Collier Macmillan, London.

Mott, G. (1993) *Investment Appraisal*, 2nd edn. Pitman, London.

Warren, M.F. (1998) *Financial Management for Farmers and Rural Managers*, 4th edn. Blackwell Science, Oxford.

Farm management data

Agricultural Business Consultants Ltd. (published twice-yearly) *The Agricultural Budgeting and Costing Book*. ABC, Atherstone.

Nix, J.S. (published annually) *Farm Management Pocketbook*. Imperial College at Wye, Ashford, Kent.

Scottish Agricultural College (published annually) *The Farm Management Handbook*. SAC, Edinburgh.

University of Wales *Organic Farm Management Handbook*. University of Wales and Elm Farm Research Centre, Aberystwyth.

Each Regional Agricultural Economics Centre (usually based at a prominent university or college) also publishes annual reports and handbooks.

13

Farm staff management

M.A.H. Stone

Introduction

Between 1990 and 1999 the total labour force engaged in farming in Britain declined by over 30 000, continuing a long-established trend. However, farming still makes significant use of labour, providing work for 592 000 men and women (including spouses) in 1999 (MAFF, 2001). The total labour force is spread between 240 000 farm holdings, the majority of which are less than 100 ha in size (MAFF, 1999).

While it is important for all enterprises to make the best use of their staff, effective use of labour is essential for the large proportion of farm enterprises using very small amounts of labour. This chapter is written with these farms very much in mind. It is likely that on those farms that employ only a small amount of labour, farmers and managers will not have the benefit of frequent practice in a wide range of people management skills. Therefore, this chapter aims to break down each process into stages, to define the terms used and to detail options.

There are three Ps which support any business: *people*, *product* and *profit*. All the three Ps need to be adequately addressed for the business to be successful. People are highly flexible but also highly complex. Dealing with this resource will be discussed under the following headings:

- Job analysis and job design
- Recruitment and pay
- Interviewing and discrimination
- Control, guidance and negotiation
- Training and development
- Maintaining and increasing performance
- The employment environment
- References and further reading
- Glossary of key terms used

Job analysis and job design

Systems and methods of work need to match both the goals of the business and the capacity of the workforce available. However, the use of labour is not based on rational economic decisions alone; there are many other influences. Social, cultural and legislative influences provide a frame of reference within which organizational decisions about labour use are made.

The demand for labour is rarely constant, and some of the many causes of change are:

- growth or decline in the business
- the seasons
- cycles of production
- change in demand
- change over the product or service life cycle
- change in the product
- change in the manpower itself
- change in methods and equipment

The purpose of a job is most commonly defined in functional terms:

- to do
- to provide
- to support
- to be responsible for.

This is the first step, for until you have decided exactly what the job is designed to achieve, you will be unable to determine:

- how much labour you need
- who will be best to do the job
- how the job can be done best or even the best format for the labour provision, e.g. full/part time.

This process is essential to ensure you are getting value for money from your labour. Each time the business requires additional labour or the workforce changes, the work to be done must be analysed and then the job/s designed or redesigned. This process helps to clearly determine what the organization wants the job holder to do.

The majority of job vacancies are those replacing leavers or for additional staff doing similar jobs. For this reason, the best source of information about what a job consists of is the person/s currently doing the job. The manager must carry out a *job study*, or, in its more scientific form, *work study*, to determine working methods and required outputs.

Actively seeking the involvement of employees in the design, evolution and redesign of jobs is to be encouraged. The positive outcomes of worker involvement in job analysis and job design arise because the job holder:

- is close to the work
- has evolved the process by learning from experience
- can gain recognition for his/her work and skill
- involvement is fostered
- health and safety risks can be assessed

However, there can be some problems because the job holder:

- may want to hide the ease of some of his/her work
- may fear management motives regarding pay, grading or bonus payments

The best results will be achieved in a climate of real dialogue between employee and employer.

Job information can also be obtained from leavers at an exit interview (an interview with an employee before they leave, conducted for the benefit of the firm). The employer may discover frustrations, suggestions and details concerning employee relations that will aid future job planning. This is best done after you have written the leaver's reference.

Job analysis and work-study allow jobs to be broken down into the technical/non-technical functions required and for worker skills to be compared to the job. This allows managers to set standards, which can be built into a Job Description and Person Specification. A *Job Description* is a working document, which details jobs, tasks, duties and responsibilities in a measurable way. A Job Description can be used to:

- supply information to candidates
- as a reminder at the interview
- as a guide for induction, training and appraisal

A *Person Specification* can be thought of as a profile of the ideal candidate for the job defined in the Job Description. It should:

- detail the knowledge, skills, and attitudes required
- prioritize the criteria you will use to both select for interview and recruit

Labour turnover allows the leavers in a set period to be compared to the total workforce, normally expressed as a percentage:

$$\frac{\text{Leavers in the period}}{\text{Staff employed}} \times 100$$

Monitoring labour turnover allows the manager to check for trends, which might highlight problems. Some turnover cannot be stopped, and it is not desirable to try, for new staff bring new ideas and approaches to a business. Turnover in jobs with no promotion or development prospects should be planned for, maybe even advertised as a stepping-stone job.

If labour turnover proves a problem, you will need to know if this is a problem of your business alone, or a local/typical problem. You can find answers or help with this from such places as:

- local employers/employer groups
- department of employment/job centres
- chambers of commerce
- professional or industry groups

Labour profiles can be calculated so that seasonal manpower requirements are known and may be plotted as a bar chart. While not a precise tool, these charts are also extremely helpful when considering a change in the size of an enterprise or the introduction of a new one. A labour profile can also show casual labour requirements. When compiling these profiles or gang-work day chart, a useful source of data on average/premium worker performance is the *Farm Management Pocket Book* by Professor John Nix (published annually).

Gang-work day charts

Gang-work day charts (Fig. 13.1) aim to ensure that enough labour and machines are available to carry out operations at the correct time and within the time available for the operation. It is seldom necessary to chart the whole year; usually only peak periods need be chartered. These charts may also be useful when

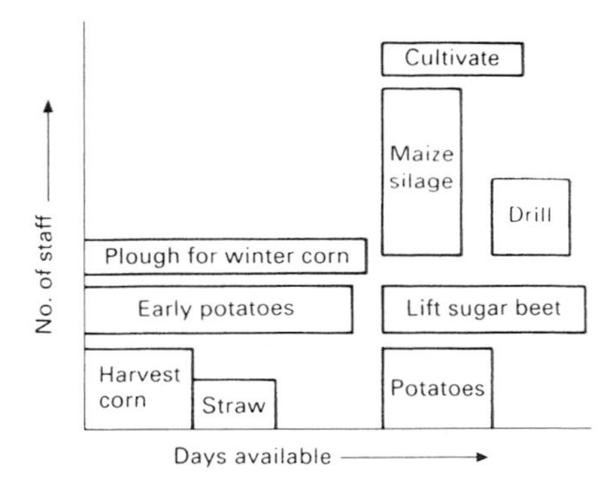

Figure 13.1 A simplified example of a gang-work day chart.

considering the viability of *annual hours* contracts for farm staff.

Construction of a gang-work day chart

Stage one – collect data on:

- cropping and stocking
- regular labour employed
- machinery and implements available
- gang size and machines required for each operation
- rate of work of each operation
- earliest start and latest finish date for each operation
- number of days on which field work is possible in each month

Stage two – tabulate the information for each operation, under the headings:

- area
- time period
- working days available
- work rate and gang size

Then calculate the gang-work days required for each operation.

Stage three – draw the chart.
Operations using large gangs are the least flexible and should be put in first. Initially assume operations can start at the earliest start date. Charting is a matter of trial and error, like a jigsaw. A chart with no spaces has no flexibility for poorer-than-average seasons or other delays. Casual labour may be written in or 'ballooned' on top.

One element in *workforce planning* often not recognized by farmers in time is that they and their workforce grow old together. The position then facing the heir or, equally, an appointed manager is that all the staff may have to be replaced fairly quickly with the consequent loss of knowledge and skill.

Recruitment and pay

Recruitment

Employers need to make quick, accurate, relevant and legally defensible employment decisions. Managers must plan and organize to get and keep labour. Rerecruitment most commonly occurs because of poor preparation for, or poor follow-up to, recruitment.

Constant recruitment when the business is not growing can quickly have a negative impact on the credibility and efficiency of the organization. For example:

- Too much concentrating on recruitment can become a vicious circle, by lowering the morale of experienced staff who may then leave.
- Output can fall as initial training increases at the expense of output and developmental training.
- Short-term coverage can become the norm, leading to tiredness, increased accidents, errors of skill and judgement, withdrawal of good-will and strained working relationships.
- Management credibility falls and with it falls the ability to motivate and control.

We all frequently make initial judgements (first impressions) on the basis of little evidence and at great speed. To be objective in recruitment we must be aware of our bias and suspend judgement until we have sufficient relevant evidence to make a choice. The employer should concentrate on skills and behaviours because they are observable, describable and measurable. Past behaviour is the best guide to future behaviour that we have.

When measuring an applicant against a job description/person specification, attention should also be paid to the future. What skills and behaviours will aid the development of the business? The relationship of behaviour to a job is illustrated in Fig. 13.2.

Recruitment is a two-way selection process; potential employees also need information and time to make a rational decision. The process can be viewed as attempting to deter the unsuitable as well as attracting the suitable. The advantage of supplying negative as well as positive information to potential applicants is that selection can be made on the basis of employee/employer choice without the sales pitch element of recruitment.

As an employer, try to build into jobs those elements

Figure 13.2 Behavioural aspects of a technical job.

that will appeal to your target audience as well as serving the needs of the business; clear responsibilities in jobs for recent graduates, for example.

Your own time is a commonly overlooked cost when running a business. If you do most jobs for yourself, you can get very caught up duplicating the work best done by others. Use the skill and experience of job centres, newspapers and agencies, much of which is free; e.g. use job centre application forms rather than design your own.

Advertising

The options include:

- job centres (national coverage)
- local papers
- national papers
- specialist journals (group or interest)
- agencies (recruitment and advertising)
- school and colleges

Payment and reward systems

The options include:

- payment for employee's time
- payment for employee's output
- payment for employee's service or loyalty

The rules covering eligibility and the period or date of review should be included in both the offer of employment and the written terms and conditions of employment.

Time rates

This means payment for time worked, at a monetary rate per hour, including provision for extra or differing working hours. Some of the options for the length of the working period are:

- day, week, month
- season
- year
- salaried positions
- fixed hour contracts
- flexi-time

Production/performance rates

This refers to pay related wholly or in part to the achievement of production, sales, profit or other identified targets. For example:

- *piece work* is pay strictly limited to the quantity of output
- *commission*, a method of payment that aims to give the employee motivation to sell, for each sale directly increases pay

Payment, based on performance or results, relies on clear and effective measurement criteria. The manager must ensure work is available in sufficient and measurable quantity to allow workers to earn a reasonable and consistent wage. The rules must be clear from the start and not require quick or frequent adjustment. The system must have sufficient flexibility to deal with changes in production and sales that may occur. Most organizations generate a mixture of productive and non-productive work. It can be hard to measure accurately and consistently the contribution to profit or success of all employees.

Service and loyalty rewards

Many businesses offer benefits in kind in addition to money, to reward, or try to encourage loyalty and long service. For example:

- subsidized housing, child care and food
- uniforms/clothing
- loans/mortgages, non-contributory pensions
- training and development funding
- cars/transport
- professional membership fees

The manager will need to measure the returns to the business of such rewards. These rewards are not of equal value to all employees or over time; one option may be to offer your staff the choice from a range of benefits.

Establishing rates of pay and pay scales involves more than just establishing the going rate for a job; issues of power, personality, tradition and organizational culture are also involved. Pay levels are a way of establishing or reinforcing a hierarchy within an organization. As well as rewarding skill and experience, pay levels can signify levels of power and authority. Whatever pay system or method is chosen, compliance with the national minimum wage level and the working time directive must be ensured.

The opinion as to how fair a pay system is will depend upon the position of the individual within the system. For example, staff of differing ages offer different things to an organization, but they may not fully understand the contribution of the other.

The effort–reward bargain

The effort–reward bargain is the personal perception of the exchange of effort, knowledge and time for the benefits of monetary and non-monetary reward. This perception will change over time and in comparison to the rewards of others.

Teams

Although many business goals are set and achieved through teams, managers must ensure that rewards to the individual have some relevance to individual performance.

Administration

Administration of pay and remuneration must be workable. Simplicity of measurement has meant that hourly pay is the most popular basis for reward. With this method, the control and measurement of output volume and quality will be a reflection of motivation and discipline. Before introducing a more complex form of payment, you must be sure the potential benefits of the new systems outweigh the potential costs.

Interviewing and discrimination

Positive indicators to look for on application forms/CVs are that the applicant has:

- transferable skills
- relevant experience – not just employment
- learnt and developed
- non-work interests
- motivation to do this particular job
- provided a full and clear account of his/her career

The telephone screen

When screening potential applicants by telephone, the employer should:

- describe the job requirements
- solicit candidate reaction
- check if experience is comparable
- check wage/salary requirements
- describe the selection process
- listen carefully

The selection interview

The selection interview is a conversation with a purpose, the purpose being to gain evidence of each applicant's suitability for the job and to enable an objective assessment to be made against the person specification. Good interviews:

- allow the participants to get acquainted
- help the organization make the best decision by collecting information in order to predict how successfully the individual will perform in the job
- help the applicant make the right decision by providing full details of the job and organization
- allow candidates to feel they have been given a fair hearing

Preparing for the interview

Stress

A selection interview is not a natural situation and can put a lot of strain on the applicant; a potentially first-rate employee could be so nervous that you never discover how good he or she is. A good rule is that you should aim to stretch the candidate, but not to stress him/her.

Time

Give yourself enough time and tell interviewees how long you plan to keep them, before they come.

Interruptions
Try to minimize them; they destroy concentration and do not allow for relaxation of interviewer or candidate.

The environment
If there is more than one applicant, make arrangements for their arrival, a place to wait, toilets and refreshments.

Documentation
You will need:

- person specification
- job description
- application form or CV
- interview plan/prepared questions/notes
- payment if travel expenses are to be paid
- notes from any references taken up

The WASP interview method and acronym

- Welcome
- Acquire
- Supply
- Plan and Part

Welcome

- Establish rapport by adopting a friendly approach and showing interest in the candidate.
- Greet the applicant, giving your name and position.
- Explain purposes of the interview.
- Outline the interview structure.
- Tell the candidate you will be taking notes.

Acquire

- Establish applicant's experience, knowledge, skills and attitude.
- Ask questions using the 'funnel technique':
  - opening questions
  - probing questions
  - summarizing questions
- use a separate funnel of questions to deal with each new topic (See 'Questions' below and Fig. 13.3).

Supply
Tell applicant about:

- the job, using the job description
- the conditions of employment
- the business/firm

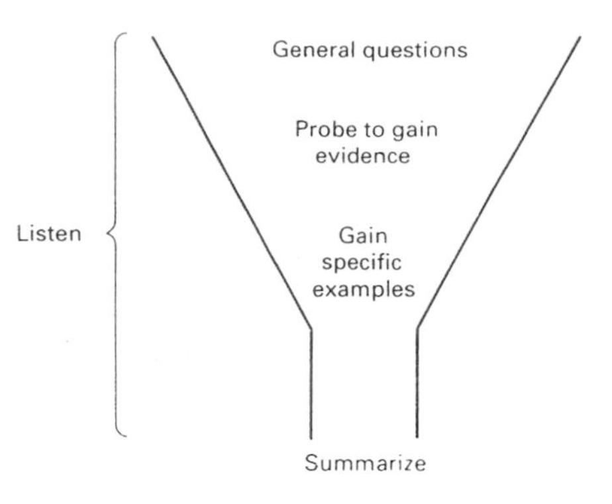

Figure 13.3 Behavioural questions and 'the funnel'.

- the company benefits
- career prospects
- the downside of the job as well as the highlights

Plan and part

- Tell the applicant when a decision will be made.
- Tell them how they will be informed.
- Pay travelling expenses.
- Thank applicant for attending the interview.
- Pass to second interview/tour of the farm/arrange medical/testing.

Even if an applicant seems unsuitable, do not cut the interview abruptly short but consider all applicants fully for the job.

Even if you are impressed with an applicant, do not say anything that could be construed as a job offer. You may see a better applicant later, or a colleague may pick up an adverse point that you have missed. Employers should make it very clear to applicants at what stage a job offer is made as acceptance of the offer forms part of a binding contract between applicant and employer.

Questions

Avoid long or multiple questions; encourage the applicant to speak by keeping your questions short and simple. You should always be encouraging in your manner, give the applicant's answers your full attention and make use of verbal and non-verbal prompts. Try not to involve your own opinions on specific subjects mentioned.

A behavioural approach to interviewing is based on gaining evidence of past performance behaviour as an indicator of future job performance. Investigating any of the areas circled in Fig. 13.4 could provide evidence of relevant behaviours or skills.

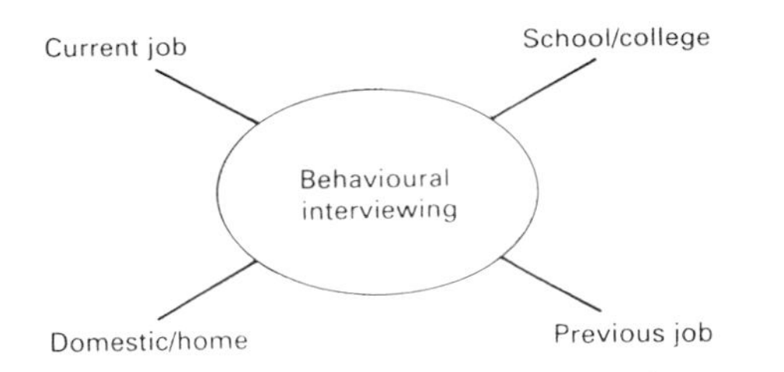

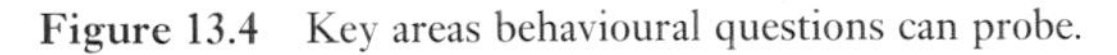

Figure 13.4 Key areas behavioural questions can probe.

Example general opening questions

- What opportunities have you . . . ?
- Give me a recent example . . .
- Describe a recent activity . . .
- How did you go about . . . ?
- Tell me about the last occasion . . .
- When have you been involved . . . ?
- What experience have you had . . . ?
- Describe a situation in which you . . .
- To what extent . . . ?

Example follow-up probes

- Describe specifically how . . .
- What exactly did you do . . . ?
- Tell me more specifically . . .
- How did you go about . . . ?
- What was your role . . . ?
- What were your reactions . . . ?
- Walk me through the process step by step . . .
- How many times . . . ?
- Describe exactly what you did in detail . . .
- How were you personally involved . . . ?
- Would you choose to do it this way again?
- What went wrong/What went well?
- How did you assess your success?
- How would you approach a similar situation again?

Summarize

- Am I right in saying . . . ?
- So you are saying . . .
- It is fair to say . . .
- What you are saying is . . .

Body language
Listen to the candidate and probe for more information where the body language signals do not seem consistent with the spoken message. This questioning process is equally applicable to other types of interview situation:

- counselling
- disciplinary
- grievance
- appraisal
- dismissal

How to improve listening skills
Following the acronym LISTEN:

- Look interested, do not ignore the candidate, yawn or look out the window.
- Inquire with questions.
- Stay on target.
- Test your understanding.
- Evaluate the message.
- Neutralize your feelings, do not discriminate or allow personal opinions to influence your decision.

Listening is not talking; resist the temptation to launch into a long discussion about your ideas or the company.

Areas of questioning you should NOT engage in during an interview

- age
- race
- marital status (or living with someone)
- dependants
- child care situation
- housing situation
- willingness to work different hours, unless this is a bona fide job requirement

Minority/ethnic groups or females should not be asked questions not routinely put to white or male applicants.

Discrimination

In an employment context, discrimination is the application of unsound or irrelevant criteria to the process of selection, access and decision making. Equal opportunities form part of good management practice. It is a reality that half the workforce is female, that significant numbers of people are disabled and that Britain is a multi-cultural country. These people can and want to make a full contribution. Skills and behaviour qualities are not found only in specific geographical areas, among one sex or specific age or racial group; exclusion on these grounds is therefore not rational or legal.

Everyone experiences discrimination at some time, but most experiences do not last for life; those of race, colour and sex often do. These experiences will affect those discriminated in their approach to getting a job. Discrimination in employment can persist even when members of these groups reach the interview stage due to the absence of cultural or social reference points, the lack of common experience between interviewer and interviewee.

Fair, legal and efficient recruitment processes are those that focus on the behaviour that will make an individual successful in your organization. This approach will enable you to deal positively with discrimination issues and assist the manager in making sound employment decisions about other unfamiliar groups, such as school leavers, the long-term unemployed or women returning to work.

Keep things simple; the more complex a recruitment system the more chance that it will not be carried out in full or to the same standard every time.

Psychometric tests are an increasingly common form of assessment. These tests are devised by professionals and therefore involve costs. The values of the test must be compatible with those of the organization and its staff. Do not be tempted to use tests designed for other organizations.

Remember all applicants (including their relatives and friends), their contacts, official and commercial, along with current employees form attitudes and expectations based upon the publicity and promises delivered in the recruitment process. Those you deal with will have an opinion as to how they have been treated and so act as either good or bad ambassadors for you and your business.

Internal applicants should receive the same consideration and follow the same procedures as external candidates.

The Disability Discrimination Act

The Act establishes a framework of rights for disabled people covering employment and access to goods and services. The Act provides a statutory right of non-discrimination for disabled people. The Act applies to businesses employing 15 or more staff and covers both physical and mental impairments, which have substantial and adverse long-term effects on a person's ability to carry out normal day-to-day activities. This also applies to individuals suffering from HIV or AIDS.

The Act requires employers to make reasonable adjustments to the workplace to overcome the practical effects of disability. A new body, the National Disability Council, will monitor the effects of the legislation. Individuals who believe that they have been unlawfully discriminated against within their jobs or when applying for a job may take a claim to an employment tribunal (Crushway, 2000).

Control, guidance and negotiation

Identifiable responsibilities in conjunction with house rules and disciplinary procedures provide a clear framework for the management and guidance of staff behaviour throughout the employment relationship.

House rules

House rules are guidelines that should be written and displayed, that state the actions or behaviour demanded by the organization in set circumstances, covering, for example, such things as:

- the use of telephones, equipment or vehicles for private use
- arrangements for visitors
- uniform and safety equipment policy

House rules also signal the organization's attitude towards such things as:

- smoking
- the protection of property and stock
- drinking, on or prior to duty, especially for those who work with machinery

When small farms expand or develop they can become vulnerable to dishonesty, from within and externally; *audits* can help. In an employment context an audit can be seen as the scrutiny or review of records and systems, in the same way that a financial audit checks the books. Periodic audits of all systems and records can reduce the temptation of staff to break the rules through the knowledge that systems and records are checked. It is likely that as an organization grows, more financial and access controls will be required.

Discipline can seem very legalistic, but there is a clear distinction between what constitutes a breach workplace discipline and what constitutes a breach of law. For example, to discipline a member of staff for removing equipment, the manager does not need to prove a worker intended to steal from the firm but simply that equipment was taken off site without permission.

Employers should set out the responsibility they expect each of their staff to take. Some firms will ask staff to sign these statements to signify their understanding and agreement. This approach is commonly found within large established firms. These firms are not by nature less trusting, but have implemented these policies in the light of experience or incidents.

The purpose of a system of workplace discipline is not punishment but behaviour change. The legalistic way in which many of these situations are described and the emotions that can be generated often make the employee feel that punishment is at the root of the process, in the same way that breach of school rules can lead to detention.

Staff are not always willing to see the person disciplining them as someone reasonable, to whom they may have caused problems. Individuals who are the subject of workplace discipline commonly feel threatened and many react accordingly!

Procedures

Establishing a system of discipline allows the manager or employer to introduce order into what can be situations of conflict and high emotion. Procedures help to set the agenda and provide the manager with a tool to:

- calm
- control
- gain consensus
- gain acceptance
- reduce uncertainty

Steps to ensure disciplinary procedures are seen as fair include:

- publicity – everyone knows the rules/system
- detail the procedures in the contract or terms and conditions of employment
- agree procedures with workers, their representatives or organizations
- following the 'best practice' advice of professional and advisory bodies such as ACAS
- always follow the procedures
- be seen to be consistent and impartial

Remember, staff not involved in a disciplinary situation will be interested in the handling and outcome, due to curiosity and self interest.

Investigation

This must be your first action. You do not have a disciplinary situation on the basis of one person's statement. By starting the disciplinary process by the investigation of incidents or accusations, you are immediately giving those concerned the chance to have their say (and let off steam). By not jumping to conclusions and taking the time to check the situation you are also allowing time for calming.

During an investigation the manager will need to:

- explain the process
- talk to the relevant people in private
- try to ignore pure opinion
- make notes
- be aware of bias
- probe for the whole picture
- *remember people do things for a reason*

If the matter under consideration is serious and/or may take some time to investigate, you may wish to suspend the employee, on *full pay*.

Threats are often associated with disciplinary situations; they can come in the form of judgmental phrases, made on the spur of the moment. These do not help the orderly conduct of procedure, or the outcome. Do not endorse the threats or judgmental remarks of others and, if possible, follow these up and prevent repetition.

Do not commit yourself to anything at the outset except to investigate as a preliminary to starting the disciplinary process. Try to ensure the person being disciplined understands why he or she is being disciplined. It is also worth stressing that time will be spent investigating the incident because you are interested in a positive outcome.

Time scales

Disciplinary issues should be resolved quickly. The uncertainty and stress of a long wait are unlikely to contribute positively to the business. The only legitimate forms of delay are:

For the manager

- investigation
- to allow cooling off
- time to consider your actions

For the subject of the disciplinary process

- to prepare evidence
- to find a supportive colleague willing to attend
- to consult with unions or other bodies

Once a decision has been reached a line should be drawn under events. The only circumstances in which the event can be relevantly revisited are those of repetition and only then in private. As with criminal matters, once a sentence or fine has been paid the matter should be seen as closed.

Retrospective action

If a matter deserves investigation and/or implementation of disciplinary procedures, it must be fully dealt with at the time. If a blind eye is turned to a problem it is not legitimate to use the incident as evidence at a later disciplinary situation. If you ignore something, you can be perceived as condoning it. This will express to staff how seriously the issue is taken; ignored more than once and it becomes custom and practice.

Fundamental steps in handling disciplinary situations

- investigation
- reference to personnel/responsible level
- representation
- notice of meeting
- presentation of evidence
- fair hearing
- meeting/notes
- pause for consideration
- mitigation
- make decision
- right of appeal

Steps in a performance warning

- Explain what is being discussed.
- Re-state the standard required.
- Outline the shortfall.
- Discuss the target standard.
- Give a time span for improvement.
- Provide assistance.
- Impose warning or offer guidance.

Steps in a conduct warning

- Explain what is being discussed.
- Detail the incident of poor conduct.
- State the standard expected.
- Outline the consequences of repetition.
- Give a time span for improvement.
- Provide assistance.
- Impose warning or offer guidance.

Dismissal

This is the involuntary termination of the employment relationship by the employer. Dismissal can be the culmination of the disciplinary process, or due to gross misconduct.

You can start the disciplinary process at any stage if the level of seriousness warrants this, but the common disciplinary actions short of dismissal are:

- an initial verbal warning
- a follow-up written warning
- the final written warning
- dismissal

Staff need to be aware of the actions that will result in disciplinary action and at what level. Failure to make this clear could result in complaints of constructive dismissal.

Employment tribunals

When the disciplinary system fails the result may be a claim for unfair dismissal at an Employment Tribunal. The main concern of the tribunal is to establish if:

- there has been any breach of employment law
- appropriate procedures were followed

Some possible outcomes of tribunals are:

- out of tribunal settlements
- compensation
- re-engagement
- re-instatement
- legal precedent

Ill health

Termination on health grounds can be a delicate issue, but failure to address the issue may prove costly, especially for small organizations. The common steps required prior to dismissal are:

- ask for medical checks
- seek independent advice
- keep the employee informed of the likely outcomes

Retirement

It is sensible to plan for the changes in the level and type of input to the organization during working life. Some options close to retirement are:

- a stepped decline in effort/responsibility
- gradual handover
- consultancy

The grievance interview

It is important to deal with grievances promptly to prevent them from escalating. The objectives are:

- to allow the employee to air the problem
- to discover the causes of dissatisfaction/worry
- to remove the problem/improve the situation

When conducting the interview:

- let the employee state the case in his/her own way
- distinguish the facts and the feelings
- listen, do not argue
- get the employee to explore his/her own motives
- ask the employee what he/she thinks the solution is
- do not commit yourself to anything at this stage
- fix a date to come back to him/her

Following up

Investigate the facts, obtain information and communicate by the agreed date. Check again later to see if the situation has improved.

Negotiation and bargaining

Industrial relations issues are closely tied up with power, control, conflict and personal interest. What helps to cloud industrial relations issues further is their linkage to politics, ideology and propaganda. You as a manager must stick closely to the issues for negotiation. Any help you can give your employees and their representative groups to do the same will be rewarded.

Once one party deviates from the substantive issues both sides can quickly take refuge behind the posturing that has characterized many famous disputes. Once this posturing takes place it allows those who have no part in, or knowledge of, the issues to stick their oar in, looking for proof that the 'other side' is abusing their position or being irresponsible (the very stuff of gossip and newspaper circulation). Once you have lost sight of the substantive issues it is very hard to refocus the participants!

Types of negotiation and bargaining include:

- distributive bargaining: this is a fixed sum game, which can be fine for simple issues. There are winners and losers and the situation can be characterized by suspicion and defensiveness.
- integrative bargaining: this suits common problems that require a mutual solution. It is a positive sum game: the outcome can be more for all, because co-operation can increase the size of the cake. It also takes longer and requires greater honesty.

Stages of negotiation and bargaining

Negotiation is not a one-off incident, there will always be some history to consider. Trust levels grow slowly and are best built up in low-stress situations.

(1) *Opening moves*
These are commonly a statement of your position and a rejection of the other side's position. This is one extreme of your bargaining position, the other being the resistance point, beyond which you cannot go. For successful negotiation there needs to be some overlap in the bargaining range of each side.

(2) *Exploratory stage*
Here you try to determine where your opponent stands. This process may consist of questions, threats and abuse. Frustration and time pressures are brought into play; your aim is to move your opponent's resistance point.

(3) *Consolidation*
Here a change of concentration occurs, moving from divergent issues to convergent issues. Contentious issues are avoided and conflict is minimized.

(4) *Decision-making*
This starts when a compromise seems possible. Only then are concessions made to move from discussion to agreement. The goal changes from an ideal settlement to an acceptable one. It is sensible to help the other side accept the decision.

Negotiation has similarities to a game: it has rules, conventions, form, accepted moves and there is usually an audience who are passionate about the result. For example, offers once made are normally not withdrawn and moves are normally towards the other side's position.

High initial demands may help to:

- widen the bargaining range in your favour
- change the other side's expectation of what is reasonable
- lead your opponent to raise his/her estimate of your resistance point
- give you scope to make big concessions

However, you may lose credibility if your opponents refuse to negotiate or the final settlement is small compared to the demand.

If you play to win in the short run you may not achieve this in the long run.

Training and development

Spending time and money on the training and development of staff is easy. Managing the investment to get value for money is not. Before resources are allocated to the development of people you need to know the level from which you are starting. This is sometimes called a skills audit.

Skills audit

When looking for and measuring skill, the starting point should be goals and objectives to which the skills will be applied. In other words, what do you want your staff to achieve? The next step is breaking down the defined goals and activities of the organization into measurable standards and objectives for each member of the team.

Once you have defined what the staff should be aiming for you can start to study the skills they will need. A skills audit could also look for unused or under-used skills among the workforce, foreign language skills for example.

The skills audit will produce a comparison of the skills applied and the outcomes achieved. However, a shortfall in a standard of performance is not automatically a training need.

Questions for a skills audit

- Are there defined standards of performance?
- Are the standards set at a correct level?
- Are these standards being met?
- Are our competitors achieving the same results by more simple/cheaper methods?
- Are our competitors achieving better results by the same/different means?

Training

A training need is an identifiable skill or body of knowledge, the learning or enhancement of which would generate positive, measurable benefits to the individual and/or the organization. Employers should only pay to eliminate training needs that are achievable, relevant and when the need for the training is clear to the trainer and trainee. For example:

- to improve standards
- to increase performance
- to generate higher profit
- in preparation for growth
- to facilitate change
- to motivate

A clear objective for any expenditure should have a quantifiable measure of success, a standard. Training to a standard:

- helps to achieve continuity
- lets staff know what is expected
- allows trainers to plan and be effective
- sets a measurable level of satisfactory performance
- lets the customer know what to expect, including customers within your own organization
- you can measure the standard and can then know when to stop training
- helps to focus competitive activity on the standard (not other staff)

Standards of performance

Ideally these should be written; if so it is a good idea to make these available to the people that need them, the staff as well as the manager.

Setting standards and the recording of training

For many organizations standards of performance, training programmes and training records are all separate documents and systems. This need not be the case. The standards for the job can be both your training programme and record. A more detailed hard copy of standards may help trainers, but for the day-to-day monitoring and improvement of standards these are not required.

By establishing such systems you are providing the climate in which those that work for you can express their individual motivation to do a good job and improve their own and colleagues' skill. The manager will be facilitating, by providing the tools for the job.

Some sources of information on training needs

- job analysis
- appraisal
- workplace assessment
- changes in the nature of a job, equipment or markets
- general communication

Workplace assessment

A true measure of job performance can often only be gauged on the job. The person best suited to assess job performance is someone who:

- knows what the standards are
- knows what the obstacles to achievement are
- knows, or can see, ways of improvement

Workplace assessment is undertaken in order to:

- give feedback on job performance
- help measure and build consistency
- assess the impact of pressure and environment on standard delivery
- identify true long-term strengths and weaknesses; of individuals and their environment

An environment of trust and a collaborative approach to organizational development are required for best results. If standards of performance are outlined from the start, the regular assessment of the performance of starters will form a natural part of the training and induction process. After establishing such an environment, workplace assessment can focus on the introduction or change of standards and the upkeep of existing standards.

Workplace assessment is one way in which managers can show they are committed to getting it right. However, this approach will only gain acceptance if shortfalls in standards that are the responsibility of the firm or management are pursued with the same vigour as shortfalls by staff. For example:

- lack of equipment
- lack of maintenance
- lack of staff
- poor planning
- poor communication from above

For workplace assessment to work best, it must be seen as a collaborative process. Remember, you are not there to conduct tests, but to facilitate staff achievement. During workplace assessment, new or amended duties can be agreed, progress can be monitored and praise given. Problems or barriers to progress can be approached jointly.

Training and learning

Training should be carried out prior to the need for the skill. The work method can then be assessed in action to determine the need for:

- additional training
- further clarification of the standard
- amendment of the standard

This is a cyclical process.

The trainer also benefits from training because it:

- improves his or her handling of and confidence with people
- makes the job easier
- supports staff while not doing their job for them
- develops trust and respect
- is part of managing instead of doing

When offering training or development, be systematic; give the trainee a structure to their learning, such as the Kolb model in Fig. 13.5.

There are limits to learning and the acquisition of skill, imposed by:

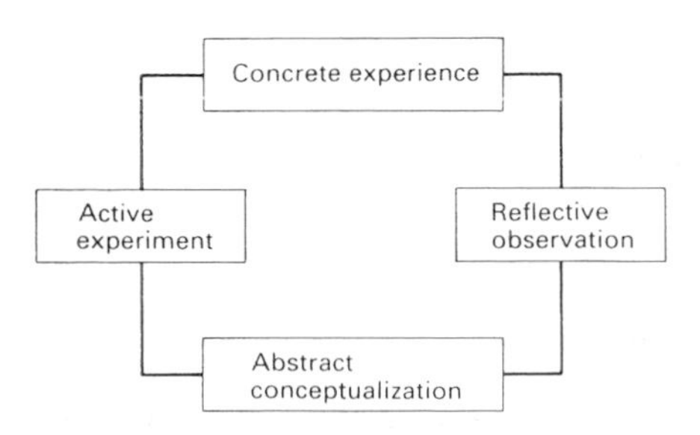

Figure 13.5 The Kolb model of learning and development.

- the situation and personal make-up of the person to be trained (learning styles)
- the training techniques and the environment they are to be practised in
- current skill limitations
- intelligence limitations
- lack of confidence
- fear of the unknown
- personal inhibitions
- lack of practice
- attitudes that are brought to the training
- effect of peer pressure or ridicule
- dynamics of the training group
- people only learn what they want to learn
- loss of concentration (after 20 mins)

Key aids to learning are as follows:

- Visual aids
- Mnemonics
- Use of more than one sense
- Build from known to unknown
- Show relevance to work situation
- Provide constructive feedback
- Use clear objectives
- Build on success

As well as technical competence, a trainer needs to develop a working relationship with those to be trained and establish a climate in which learning can take place. Learning is the gradual extension of knowledge, skill and attitudes. Good trainers can be summed up by the acronym TRAINER:

- **T**actful
- **R**espectful
- **A**ccurate
- **I**maginative
- **N**eat and tidy
- **E**nthusiastic
- **R**eliable

The benefits of increased individual performance feed into company and personal career progression. These effects can be cumulative; development can progress faster if the individual is familiar with the methods and objectives of the organization.

Investing in people can be thought of as ensuring that the organization has a maintenance and development programme for staff. The call to treat training and development as an investment is not new; at an 'Investors in People' conference Sir John Harvey Jones chided British companies for investing more in plant and machinery than their staff and thereby risking failure in competitive markets (Harvey Jones, 1993).

Induction

Induction is a programme aiming to deliver the smooth and effective introduction of people into an established working environment. Break these programmes down according to what the employee needs to know by when. Use a variety of methods to allow a new employee to take on board potentially vast amounts of new information.

Special consideration will need to be given to different groups, such as:

- school leavers
- long-term unemployed
- graduates
- women returning to work after a career break

Maintaining and increasing performance

Appraisal

An appraisal is a meeting between an individual and his/her immediate manager to discuss the job. The purpose is to maintain, guide and develop job performance. Best practice for appraisal suggests that past performance should not be looked at in a judgmental way, but reviewed to aid the process of work planning and goal setting for the period ahead. More sophisticated systems make provision for appraisees to carry out some self-assessment and provide comment and guidance to those that manage them (Self and Upward Appraisal).

Appraisal benefits the manager by:

- allowing time to plan the work load and its allocation with the team for the period ahead
- giving the manager feedback about his/her own performance
- helping foster a joint approach to work
- providing a recorded plan

Appraisal benefits staff because:

- career planning and development can be discussed
- feedback about performance can be gained
- future work and priorities can be clarified
- it provides a recorded plan

Job chats are an increasingly common alternative or supplement to the appraisal – shorter than the appraisal and less formal, but still a forum for the discussion of issues arising from the job, past, present and future.

The major threats to positive appraisal are:

- lack of trust
- lack of understanding/knowledge of the job
- lack of time
- being too closely linked to pay

When appraisal is too closely linked to pay, individuals will concentrate on the assessment of past performance, negating the appraisal as a forward-looking collaborative exercise.

Activity plans

Typically, activity plans are written plans for activities covering a fixed period, including details of success criteria, support required and an interim review date. The plan should be reviewed at the conclusion of the period and a new one agreed. Plans can cover months or weeks and cover normal areas of work and development as well as specific projects/objectives.

Management by objective

This management tool aims to break down the organization's goals into sub-goals and, in turn, individual goals. The starting point and concentration is the goal and not the method. This opens up the possibility of more individuality of method, so long as the goals are common/compatible.

Feedback

Feedback is a powerful aid in learning about ourselves and the impact we have on others. To make this process constructive requires skill and practice on the part of the giver and recipient of feedback. The recipient needs something to build upon, positive or negative, and this information must be both understandable and useable.

We often fail to give positive feedback because we forget, we are embarrassed or we are worried that someone may become too conceited. On the other hand, negative feedback is rare because we fear the recipient may get upset and we are not ready or willing to deal with this. We may be worried that it will affect the relationship we have with that person in a lasting way; that he/she will not really understand what we are saying, or will distort what we say. Finally, we may say nothing because we do not believe the feedback will have any effect.

Although not always a good idea, giving feedback does provide an opportunity for change. The skills of giving feedback are as follows:

- Be clear about what you want to say in advance.
- Start with the positive.
- Be specific.
- Select priority areas.
- Focus on the behaviour rather than the person.
- Refer to behaviour that can be changed.
- Offer alternatives.
- Be descriptive rather than evaluative.
- Own the feedback (this is what I think).
- Leave the recipient with a choice.
- Consider what it says about you.
- Give the feedback as soon as you can after the event.

When receiving feedback:

- listen to the feedback rather than immediately rejecting or arguing with it
- be clear about what is being said
- check it out with others rather than relying on only one source
- ask for the feedback you want but do not get
- decide what you will do as a result of the feedback

Coaching

Coaching is a method of developing and/or correcting skills, attitudes, behaviours and techniques. With this approach the trainer, manager or mentor must have a basic agenda/outline of the subjects to be covered by the trainee. The key point is that situations that occur in the normal working day are used as triggers for training. One way to think of coaching is the use of mini case studies. In this way principles and applications can be looked at together and in a context that the trainee is more likely to understand.

It is important that the trainee understands the technique of coaching and is aware of when it is being practised, especially early in the coaching relationship, so as take advantage of the learning situations and not feel patronized.

Mentoring

Mentoring can work very much in the same way as coaching, but with mentoring the initiator of training or the coverage of issues is more likely to be by the trainee. For best results the choice of mentor must at least partially lay with the mentee. The mentor needs to be able to put him- or herself in the place of the trainee. The biggest expert is not always the best trainer. The mentor must have an understanding of the mentee's personal goals and the need to make developmental activity personally relevant.

When looking at a scheme of mentoring you must decide what the basis of the relationship will be. Below are some popular models:

Description	*Application*
• Patron–Protégé	Commercial
• Expert–Trainee	Training
• Confidant–Colleague	Partnership
• Educator–Disciple	Developmental

Poor standards

Poor standards or the failure to fulfil potential may be the result of:

- poor initial training
- forgetfulness
- cutting corners
- faulty/inadequate equipment
- lack of interest
- personal problems
- personality clashes

An acronym for the steps to corrective action is OSCAR:

- **O**bserve
- **S**tep in
- **C**alm use of interpersonal skills
- **A**sk questions
- **R**epeat training

Incentives

Considerations for the manager include the following:

- What are the aims of the incentives?
- Why are incentives needed over and above existing payment/remuneration policies?
- What happens when the scheme ends?
- What happens if the incentive does not work?

There are numerous circumstances that may cause incentive schemes to fail, many of them outside the control of an organization.

Incentives are introduced through:

- planning
- consultation
- explaining to staff why an incentive is needed and how the organization and the individual are going to benefit
- trial
- time to test the theory and your assumptions
- review

The goals of an incentive need to be precise. Vague goals, such as to increase production or productivity, are the first step to failure. You need to know:

- how much
- to what quality
- for how long
- for what reward

Fairness

An incentive scheme is a deal between you and your staff. The required effort must deliver the promised benefits, for both individuals and organization. It is rarely possible to apply the same scheme accurately to staff on different jobs. This is where life gets complicated. There are two main choices:

(1) development of additional criteria to allow more groups of staff to participate, while maintaining a benefit for the organization
(2) exclusion from a scheme

Teams

Incentives work best, administratively and practically, when applied to the individual because the motivation of a team is not the same as individual motivation. The evolution of team motivation is subject to group dynamics. The existing working relationships will need to be considered when designing incentives.

Quality and quantity

Incentives, which concentrate on increasing output quantity, require safeguards for quality. These can be contentious issues, especially if quality problems result in scheme changes which lower income. The desire to be flexible and to take time out for training and peripheral work can be reduced by incentives. What gets measured and paid for gets done. Short-term income generation can become the opportunity for the worker, while longer-term concerns become problems for the manager. All rewards should be received as close to the effort that earned it as possible.

The employment environment

Data Protection Act

The Act refers to data subjects; these are individuals about whom an organization holds or processes data. This includes everyone, not just employees. Interpretation of the 1998 Act is still evolving. For example, although the Act deals mostly with data held electronically, it also allows for employee access to some structured manual systems maintained for a specific purpose, for example sickness records.

The 1998 Act extended the coverage of data protection to include any information stored about individuals in a 'relevant filing system', a relevant filing system being one that is structured, either by reference to individuals or by reference to criteria relating to individuals, and thereby readily accessible. These include:

- databases
- computer systems, email and CCTV should be considered a computer system
- some manual record systems

The data stored must be relevant, accurate and up to date, so out-of-date information should be identified and destroyed, unless there are other reasons for keeping it, which can be justified. Justifiable reasons include:

- the data subject giving his/her consent
- processing the data is necessary to fulfil a contract to which the data subject is party, i.e. a contract of employment
- processing the data is necessary for legal obligations, taxation records for example

If the employer uses data about employees to provide statistical information, employees should be told how the information held will be used. Employees' consent must be sought regarding sensitive personal data such as:

- racial or ethnic origins
- political opinions
- religious or other beliefs
- membership of trade unions
- physical or mental health or condition
- sex life
- actual or alleged offences and legal proceedings or sentences

The onus is on the employer to tell employees what is kept, where, for what purpose and who may have access. It is good practice to:

- inform all employees what data are kept, the purposes for which the data are used, and who may have access to the data
- ask all employees to give consent to those data uses
- help employees to keep their personal details up to date by providing them with a copy annually to check and amend

Similar systems should be considered for external data subjects.

Individuals can access their data by making a request (usually in writing) and paying a fee if applicable of not more than £10. The business has 40 days to respond. Exceptions to the right of access include where the data are held/produced for management forecasting or planning and where release of the data would be likely to prejudice the conduct of the business (Crushway, 2000). Examples may include:

- manpower planning
- succession or career planning

Guidance on the provisions, exemptions and applicability of the Data Protection Act can be obtained from The Office of the Data Protection Commissioner, Wycliffe House, Water Lane, Wilmslow, Cheshire, SK9 5AF or www.dataprotection.gov.uk. This should be the first step before setting up a system.

The internet

It is increasingly common for businesses to have internet access within the workplace. Like any other business tool/practice it is sensible for the business to make it clear to staff how the business expects staff to use this technology. Key issues include:

- the image of the business portrayed to outsiders
- copyright and disclaimers
- harassment and defamation

- privacy, confidentiality and monitoring
- private usage
- the status of email compared to paper communication
- the Data Protection Act
- viruses and system security/protection

Fire regulations

Fire regulations require:

- an established means of raising the alarm
- an evacuation procedure
- a safe and protected means of escape
- an assembly area
- records of training, equipment maintenance, fires and false alarms

First aid

Legal and effective coverage requires more than the correct ratio of first aiders to staff, supported by the correct level and maintenance of equipment. Employers need to consider the geographic distribution of staff and that a first aider is available throughout the working day and week.

Health and safety at work

This is a vital area of farm staff management and should be a central consideration of job design, work planning, recruitment, training and the development of your staff. This complex area and its effects upon efficiency and legal implications are covered elsewhere in this book.

Glossary of key terms used

Detailed below are working definitions for many of the terms used in the text, defined with reference to the context within which they are used in the text.

Activity plans: written plans for activities covering a fixed period, including details of success criteria, support required and review dates.

Appraisal: a meeting between an individual and his/her immediate manager to discuss the job, the purpose being to maintain, guide and develop job performance.

Behavioural interviewing: an approach to interviewing based on gaining evidence of past performance behaviour as an indicator of future job performance, regardless of background, the behaviours required being derived from the job description and person specification.

Coaching: a method the manager can use to develop and/or correct the skills, attitudes, behaviours and techniques of subordinates. The manager uses situations that occur in the normal working day as mini case studies to carry out training and development. Principles and applications are looked at together and in a context that the trainee is likely to understand.

Constructive dismissal: the situation when an employee leaves an organization, not because the employee wanted to, but because he/she felt it was the only course open to him/her; e.g. as the result of fear or frustration.

Discrimination: the application of unsound or irrelevant criteria to the process of selection, access and decision making.

Distributive bargaining: a negotiating situation in which the parties/sides seek to determine the allocation of a fixed quantity of resources, a fixed sum game.

Equal opportunities: a situation in which there are no discriminatory barriers to prevent employees or applicants from making a full contribution to the organization.

Exit interviews: where the manager interviews an employee before the employee leaves so as to gain feedback on the performance of the job and organization and to ensure positive suggestions based on experience are not lost to the organization.

Feedback: a method by which we learn about ourselves and the impact we have on others. Feedback provides individuals with an opportunity for change.

Flexitime and fixed hour contracts: systems designed to allow the manager to better balance the work to be done and the labour available; be that in the short term with flexi-time or over the longer term with fixed hour contracts.

Gang-work day charts: a method of graphically representing plans for both labour and machine usage and their interaction over time.

Induction: a programme to deliver the smooth and effective introduction of people into an established working environment.

Integrative bargaining: a negotiating situation in which both parties to the negotiation can gain through mutual co-operation, a positive sum game.

Job analysis: where the manager assesses the work to be done, breaks this work down into the technical/non-technical duties required and sets standards which can be built into a job description and person specification.

Job chats: a discussion of issues arising from a job, past, present and future. Generally, job chats are shorter and less formal than an appraisal. They can be a useful alternative or supplement to the appraisal.

Job descriptions: a working document, which details jobs tasks, duties and responsibilities in a measurable way. This will aid recruitment, training and planning.

Job design: the process by which systems and methods of work are matched to both the goals of the business and the capacity of the workforce available.

Labour profiles: a graphical representation of the labour required. They allow the manager to build up a picture of the labour needed in each time period by adding together the labour input needed for each task in that period.

Labour turnover: a statistic based on the number of staff leaving compared to the total workforce.

Management by objective: a management practice which works by breaking down organizational goals into team and, in turn, individual goals. The starting point and concentration is the goal and not the method.

Mentoring: a system in which the mentee or trainee can trigger training relevant to his/her work needs through access to a mentor (more experienced colleague). The mentor provides relevant development by being able to put him- or herself in the place of the trainee and understanding the mentee's personal goals.

Person specifications: a written record of the ideal candidate and basis for any advertising, against which the manager can compare applicants.

Psychometric testing: a system by which job or promotion candidates are compared with a psychological profile of the ideal applicant, as defined by the organization and the test designer.

Workplace assessment: the method by which an assessment of an employee's competence can be made in the workplace in relation to the standards defined for the job.

Work study: a formalized way of measuring the work that makes up a job; this can involve defining approved methods and allocating times for the completion of tasks.

References and further reading

Armstrong, M. (1999) *Handbook of Human Resource Management Practice*, 7th edn. Kogan Page, London.

Armstrong, M. & Barow, A. (1998) *Performance Management*. IPD, London.

Chartered Institute of Personnel Management (published periodically) *Codes of Practice*. CIPM, Winbledon, London.

Crushway, B. (ed) (2000) *Essential Facts: Employment*. Gee-Professional Publishing, London.

Farm Management Journal.

Fowler, A. (1990) *Negotiation Skills and Strategies*, 2nd edn. IPD, London.

Handy, C. (1991) *Age of Unreason*. Arrow, London.

Harvey Jones, Sir J. (1993) *Personnel Today*, 9 Feb.

HSE (1997) *Managing Health and Safety: An Open Learning Workbook*. HSE, London.

Honey, P. (1997) *Improve Your People Skills*, 2nd edn. IPD, London.

MAFF (1999) *Agriculture in the UK: 1999*. The Stationery Office, London.

MAFF (2001) www.maff.gov.uk

Mill, C. (2000) *Managing for the First Time*. IPD, London.

Morris, S. (1993) *Discipline, Grievance and Dismissal: A Manager's Pocket Guide*. Industrial Society, London.

Nix, J. (2000) *Farm Management Pocket Book*. Wye College, London University, Ashford (published annually).

Osbourne, D. (1996) *Staff Training and Assessment*. Cassell, London.

Parker, S. (1998) *Job and Work Design*. Sage, London.

Parslow, E. (1999) *The Manager as a Coach and Mentor*, 2nd edn. IPD, London.

Pugh, D., Hickson, D. & Hinnings, C. (1996) *Writers on Organisations*, 5th edn. Penguin, Harmondsworth.

Reid, M., Barrington, H. & Kenney, J. (1992) *Training Interventions: Managing Employee Development*, 3rd edn. IPM, London.

Salaman, M. (1998) *Industrial Relations: Theory and Practice*, 3rd edn. Prentice-Hall, London.

Smith, D. (1998) *Developing People and Organisations*. Kogan Page, London.

Vroom, V.H. & Deci, E.L. (1992) *Management and Motivation*, 2nd edn. Penguin, Harmondsworth.

Wood, S. (ed) (1988) *Continuous Development*. IPM, London.

14

Agricultural law

T.J.F. Felton

Introduction

This chapter provides an outline of the law as it relates to the ownership, possession and use of agricultural land. The role of agriculture and agriculturalists in relation to methods, quality of production and ownership rights over land is a matter of genuine public concern. Issues such as conservation, pollution of land, air, water and access to the countryside remain high on the public agenda. The result has been the confirmation of agriculture's role in the area of environmental law. Europe in particular is a rich source of new law as it attempts to both protect the environment and restructure the CAP, with its knock-on effects as regards the ownership and transfer of quota.

It is an imperative of modern farming that those involved in the industry take a positive attitude towards the legal framework that increasingly surrounds their everyday working lives so that they may operate most effectively within it. This chapter seeks to provide a starting point for the creation of such an attitude. It does not give, nor does it seek to provide, a comprehensive guide to all the rules and regulations that apply to the industry.

An understanding of the basic principles of the law as it affects the agricultural industry should be used in the following ways:

- to plan and manage effectively within the law;
- to understand how the law is developing, to monitor change and to react appropriately;
- to voice an opinion through the industry's pressure groups, for example the NFU, the Country Landowners' Association (CLA), the Rural, Agricultural and Allied Workers National Trade Group of the Transport and General Workers' Union (TGWU) and the Tenant Farmers Association.
- to be alert to public attitudes and aspirations;
- to seek professional advice at the most appropriate and cost-effective time.

Advice should be sought to prevent problems arising, not as a damage limitation exercise. In selecting a solicitor to provide advice, care should be taken to make certain that your choice of individual or practice has the skills required to advise on rural business matters. The role of the solicitor may be explained as the general practitioner of the legal profession, but through him/her it is possible to obtain the services and advice of specialists in advocacy, drafting and specific areas of law. Members of this second branch of the profession are known as barristers. Although the Bar is essentially a referral profession, opportunities are increasing for identified professionals such as rural practice chartered surveyors to access legal advice from the Bar using such schemes as 'Direct Professional Access' and 'Bar Direct'. In company with sound legal advice, farmers should also secure the services of competent insurance brokers to obtain cover against the risks involved.

The English legal system

Any greater understanding of the substantive law which controls an industry must commence with a review of the system that creates and administers that law. English law shares with our agriculture a long history. Many of its features retain traces of medieval origin, but in recent times large accretions and alterations have shown it to be as lively and innovative as modern farming. Our law remains different in substance and procedure from continental European practice – although this is not to say that ideas of justice differ – and it should be noted that in the ruling of the Court of Justice of the European Union there now exists a unifying factor of particular importance in relation to agricultural law. Agricultural law is not a separate part of English law; the same general principles apply in agriculture as in other areas.

Law consists of those rules of conduct that the courts will enforce. In this connection we include in the term 'court' all those bodies recognized by the judges as having an obligation to act judicially. So in addition to the civil and criminal courts we recognize various tribunals and other bodies set up by Act of Parliament.

Civil courts decide disputes between fellow citizens where one, the claimant, alleges that he/she has suffered injury or loss by the unlawful act or omission of another, the defendant. If the defendant is adjudged legally responsible for the injury, he/she will be required to compensate the claimant, normally by a money payment called damages. When a state agency causes such an injury, an action brought against it will also be a matter of civil law. The criminal courts decide cases where the state is involved as prosecutor against a citizen accused of committing a criminal offence. Accused persons, if found guilty, are punished by fine or imprisonment or both.

The two sets of courts are kept separate; Magistrates' Courts and the Crown Court hold criminal trials while County Courts and the High Court hear ordinary civil cases. Certain matters are, however, reserved for specialized tribunals of which the Agricultural Land Tribunal, the Lands Tribunal and Employment Tribunals are particularly relevant to the agricultural industry. At the top of the hierarchy of courts the Court of Appeal and the House of Lords, sitting as a court, provide an appeal structure with the possibility of reference to the European Court on questions of European Union law.

Most hearings in court are concerned with matters of fact: in criminal trials the prosecution will have to prove 'beyond reasonable doubt' that the defendant has committed the crime of which he/she is accused; in civil trials the court will decide on the balance of probabilities, whether the events concerned took place as alleged by the claimant or not. However, in some cases the decision depends on a question of law – for example the Dairy Produce Quotas Regulations 1984 refer to 'land used for milk production'. In *Puncknowle Farms Ltd* v. *Kane* (1985)3 All ER 790, the High Court had to decide whether this phrase included only the farm area used for current milk production or whether it also included land used for dry cows and for young heifers, i.e. land used to support the dairy herd as a whole. The court decided that, in the context of apportionment of quota on partial transfer of a holding, the phrase included land used for the support of the dairy herd as a whole, and thus included the land used by followers, as well as by cows in-milk.

It is to be hoped that when an agriculturalist wishes to know the law on a certain topic, a source will already exist in clear terms and be readily available without the need for litigation. There are two principal sources of law: decided cases and statute. Case law has been built up into a system termed the common law because when a superior court makes a decision turning on a point of law, that decision becomes a precedent binding on judges dealing with subsequent cases involving the principle. Statute consists of Acts of Parliament and Statutory Instruments, i.e. orders and regulations made under the authority of Acts of Parliament by duly authorized ministers. EC Treaty provisions are incorporated into English Law by virtue of the statutory force to Regulations passed by the council of the European Union. European Union Directives normally require legislative action by the UK Parliament before implementation, but they are capable in certain circumstances of having direct effect. Decisions of the European court also create precedents which UK courts must follow. For example, in *Von Menges (Klaws)* v. *Land Nordrhein–Westfalen* No. 109/84 1986 CMLR 309, the claimant had kept a herd of dairy cows until 1980 when he obtained a premium for going out of milk production. He undertook, as required by Regulation 1078/77 EEC, not to market milk or milk products for 5 years. In 1981 he let his farm to another farmer who planned to use it for milk-sheep. He asked in the German courts for a declaration that the marketing of sheep's milk would not be contrary to the undertaking given. The local administrative court referred to the European Court, under Article 177 EEC, the question of the meaning of 'milk and milk production' in the Regulation. The court held that the words concerned ewe's milk and ewe's milk products as well as cow's milk and cow's milk products, since otherwise the premium payments would only encourage the replacement of dairy cows by milk-sheep, leading to new surpluses. This decision is binding upon courts throughout the European Union which of course includes UK courts, who will in future follow this interpretation of 'milk and milk products' in this context.

A statutory provision overrules any common law precedents that directly conflict with it. Nowadays the definition of criminal offences and the powers of courts to penalize offenders depend almost exclusively upon statute, and a vast range of legislation covers fiscal and commercial matters together with the social activities of government, e.g. housing, employment, public health and social security. Even in the field of property law, contract and tort (see below), formerly the domain of common law, parliament has codified or amended many of the rules developed in the courts.

Thus, to find the law relevant to a particular topic it is necessary to know if statutory rules apply. For instance, in respect of security of full-time farm workers occupying service cottages, the position is governed by one of two statutory codes: the Rent (Agriculture) Act 1976 for occupancies entered into prior to 15 January 1989, and the Housing Act 1988 Part 1 Chapter 3, which relates to occupancies that commenced on or after that date. Questions concerning the rights of the farmer and worker when employment comes to an end can only be resolved by reference to these acts.

When the topic is one that is covered by common law, the decisions of relevant cases can be found in the Law Reports, where decisions of significance are recorded. In practice, a legal practitioner will depend upon books of reference and texts dealing with specific areas of the law to enable him or her to find the statutes and cases that are relevant. The agriculturalist must know about changes in the law relating to his or her business. Details of many new statutory rules are publicized by government agencies, e.g. on employment law by the Department of Employment and on safety regulations by the Health and Safety Commission. However, to keep up to date a farmer should read the professional journals.

Criminal offenders are liable to punishment; those committing civil wrongs are liable to pay damages or to have their activities stopped by an order termed an injuction made by a court. These are the means by which the law is enforced. By the standards of other legal systems, enforcement of civil judgments in England is reasonably effective. Debtors may have their property sold up; bankrupt persons are subject to severe business disabilities. Flouters of injunctions may be imprisoned for contempt. Although legal procedures can be protracted, and nothing is to be gained by suing an impecunious defendant, the farmer should be prepared, when it is business-like to do so, to assert legal rights just as much as he or she should be careful to fulfil legal duties. However, before embarking on expensive litigation a farmer would do well to seek advice on available methods of alternative dispute resolution, such as mediation or expert determination which may lead to a satisfactory conclusion far more expeditiously and economically.

Human Rights Act 1998

No commentary on the English legal system at the start of the twenty-first century would be complete without mention of the Human Rights Act 1998 (HRA) which came into force on 2 October 2000. It was heralded as bringing about the most important constitutional changes for over 300 years. It is legislation that by its very nature will affect every aspect of law in this country and the way in which public authorities carry out their duties. As such it will affect farmers and landowners.

The HRA incorporates the European Convention on Human Rights (ECHR) into our domestic legal system. This means that the ECHR is enforceable in our own courts. The ECHR to which the UK has been a signatory since 1950 guarantees certain fundamental freedoms. These include:

- the right to life
- the right to respect for private and family life
- freedom of expression
- the right to a fair and public trial by an impartial independent tribunal
- the right to peaceful enjoyment of possessions

The HRA operates in three key ways. First, it requires all legislation to be interpreted and given effect as far as possible compatibly with Convention rights. Second, it makes it unlawful for a public authority to act in a manner incompatible with Convention rights. Third, UK courts and tribunals must take account of Convention rights in any case they hear, thus providing the opportunity for the common law to be developed. In addition, a minister in charge of a bill must confirm that the bill is compatible with human rights legislation. If he/she is unable to provide such a confirmation, he/she must request the permission of the house to proceed with the bill.

Remedies under the HRA

Essentially a court or tribunal can provide any remedy that is within its powers and that is just and appropriate; for example, award damages, quash unlawful decisions, make an order preventing a public authority from making a decision that would otherwise be unlawful.

After a year on the statute books it is generally considered that the more dramatic effects predicted for the Human Rights Act in the area of property law have not emerged. However, only time will reveal its full impact.

Legal aspects of ownership, possession and occupation of agricultural land

Most agricultural enterprises are run by owner-occupiers or tenants of agricultural holdings. In many cases these will be single individuals or partnerships, and the persons involved will be legally responsible for any obligations that arise. Where the enterprise is run by a limited company, the company is recognized as a legal person with rights and duties separate from those of its members. The directors of an enterprise organized in this way must, however, recognize their obligations under the Companies Acts which contain provisions designed to prevent them from using the advantages of corporate personality to defraud creditors, members and employees of the company.

English law recognizes only two ways by which land may be held: freehold and leasehold. If it is desired to tie up freehold land in the ownership of succeeding generations within a family, this can only be done by the creation of a trust set up in accordance with certain rules – known as Equity – developed in the Court of Chancery, now a part of the High Court. This definitely requires professional expertise.

A freeholder has the largest possible freedom to decide how to use his or her land that the law recognizes. It has never been possible for freeholders to do whatever they pleased to the detriment of neighbours, but in addition to the limits already imposed by the law of nuisance have been added the constraints of town and country planning and the compulsory acquisition of land. Public and private rights of way may also diminish a freeholder's privacy, and land can be lost by adverse possession. A freeholder's rights may be reduced by restrictive covenants and his/her obligations increased by a mortgage.

Although ownership of land normally includes rights over what lies in and beneath the soil, mineral rights may be held separately from the freehold, and coal and oil deposits belong to the State. Water may be taken from surface streams by riparian owners for normal agricultural purposes such as watering stock (not for aerial spraying), but it is generally the case that for other purposes it may only be taken under a licence from the Environment Agency and the same applies to most water taken from underground sources.

Farm business tenancies

By the final decade of the twentieth century the freedom of landlords and tenants to negotiate the terms under which property was let had been greatly circumscribed by statute – in no area more so than agriculture. As a result there was a fall-off in let land from some 90% in 1910 to 36% of the total agricultural land available in 1991. In 1991 the government produced a consultation paper entitled *Agricultural Tenancy Law Proposals for Reform*, the objectives of the reform being:

- to deregulate and simplify;
- to encourage the letting of land;
- to provide an enduring framework which can accommodate change.

Having undertaken a full consultation process, the government introduced the Agricultural Tenancies Bill to parliament in November 1994; the new legislation took effect on 1 September 1995. From this date all new tenancies created are entitled 'farm business tenancies'. The terms of such tenancies are freely negotiable between the landlord and tenant within a much simplified legal framework designed primarily to prevent disputes arising. The legislation:

- defines farm business tenancies;
- requires the service of notices to quit;
- prescribes arrangements for compensation;
- prescribes fall-back procedures for disputes;
- provides fall-back procedures on rent reviews.

To fall within the definition of a farm business tenancy the Act stipulates a number of conditions.

Business condition

The first is the business condition. S.1(2) requires that during the life of the tenancy all or part of the land comprised in the tenancy is 'farmed for the purposes of a trade or business'. Thus it is not possible to diversify completely out of agriculture. The condition also precludes hobby or recreational farming from falling under the 1995 legislation.

Agriculture condition

Once the business condition has been satisfied it is also necessary to demonstrate that the character of the tenancy is primarily or wholly agricultural [s.1(3)]. In determining whether the use of the land is agricultural, a definition of agriculture is provided at s.38(1) and the court is required to consider the terms of the tenancy, any commercial activities undertaken on the land and other relevant circumstances.

Notice condition

S.1(4) creates the notice condition. This allows the landlord and tenant to identify that the tenancy being created is to be a farm business tenancy by exchanging appropriate notices prior to the commencement of the tenancy. For a successful exchange of notices the written notices must:

- be exchanged before 'the earlier of the beginning of the tenancy and the date of any written agreement creating the tenancy';
- identify the land to be let in the proposed tenancy; and
- state that the person giving the notice intends the proposed tenancy to be and to remain a farm business tenancy.

Notice to quit

Where a tenancy is for a term of less than 2 years the agreement will come to an end on the term date. If a periodic tenancy has been created the common law principles of notice will apply. For example, a monthly periodic tenancy will require 1 month's notice.

Any farm business tenancy for a term of 2 years or more can only be terminated by a notice to quit of at least 12 months and less than 24 months. Should the landlord fail to serve notice the tenancy will continue as a tenancy from year to year until it is properly terminated.

Compensation

Compensation is available for any tenant's improvement which is a physical improvement made wholly or partly at the tenant's expense. However, to be eligible the tenant must have obtained the landlord's consent to the improvement. A tenant may also seek compensation for 'intangible advantages', for example milk quota.

Resolution of disputes

The Act contains provision for the resolution of disputes by arbitration or alternative dispute resolution.

Rent reviews

The Act provides freedom for the parties to agree rent review dates and periods. If the agreement is silent as to rent review the Act provides for rent to be reviewed every 3 years to an open market rent. If they so wish the parties may agree a rent formula outside the statutory framework.

All tenancies created under legislation existing pre-1995 remain intact. As a result an account of the old regime is given below; both the owner-occupier and the tenant farmer have duties as occupiers towards neighbours, visitors and trespassers. These duties are in general the result of principles developed from case law, but they also arise as a result of the Environmental Pollution Act 1990 and from Health and Safety legislation. As mentioned above, recent legislation has continued to impose new constraints on agricultural activity and the use of land in the interests of production limitation, conservation and public access.

Farm tenancies under the 1986 Act regime

The law relating to the letting of agricultural property under the old regime differs from the general law on leaseholds. It is therefore appropriate to consider it in more detail. In the first half of the twentieth century the normal agricultural tenancy developed under the common law was a tenancy from year to year terminable at 6 months' notice by landlord or tenant. Statutory provisions have now changed this law radically.

Security of tenure

Security for tenant farmers was first introduced as a permanent measure in the Agricultural Holdings Act 1948 by placing restrictions on the landlord's right to give notice to quit to a tenant. This Act together with all others dealing with agricultural tenancies has now been consolidated in the Agricultural Holdings Act 1986. If the landlord serves a 12 months' notice to quit and the tenant does nothing the notice will be legally effective. However, the tenant may, within 1 month, serve a counter-notice if he/she does not wish to go. In this case the landlord's notice will not have effect unless the Agricultural Land Tribunal consents to its operation.

The tenant will be deprived of his/her security of tenure only on a limited number of grounds, i.e. if the purpose for which the landlord requires possession is:

- in the interests of good husbandry;
- in the interests of sound estate management;
- in the interests of agricultural research or education, or for the provision of smallholdings and allotments; or
- that greater hardship will be caused by the withholding than by the granting of consent.

If the Tribunal finds one or more of these grounds proven, it may nevertheless refuse consent 'if in all the circumstances it appears to them that a fair and reasonable landlord would not insist on possession'.

The tenant's right to serve a counter-notice is excluded by a strictly limited list of cases covering consent by the Tribunal for the reasons given above; failure to remedy a breach of the contract of tenancy; bankruptcy of the tenant; bad husbandry; failure to pay rent due by the tenant; also the intended use of the land for non-agricultural purposes for which planning permission has been granted or is not needed. Under the Act there are special rules regarding notices to remedy breaches of the tenant's obligations to keep fixed equipment including hedges, ditches, roads and ponds in good order, and the tenant may go to arbitration if he/she disputes the landlord's claim. An arbitrator can be chosen by agreement of the parties or one of them can ask the President of the Royal Institution of Chartered Surveyors to appoint one from an approved panel.

Until the Agriculture (Miscellaneous Provisions) Act 1976 the death of a tenant enabled the landlord to serve an incontestable notice to quit. This was then changed to enable security of tenure to be claimed for up to three generations including the first occupier, i.e. there can be two succession tenancies. In 1984 the law was changed to enable new tenancies to be created without such security, and the rules relating to succession are now included in the Agricultural Holdings Act 1986. This means that security of tenure for tenants' successors now applies in effect only to tenancies that were in existence during the period 1976 to 1984. Under these rules eligible persons can apply to the Tribunal for a tenancy of a holding whose tenant has died. To be eligible a person must be the widow, widower, brother, sister or child, natural or adopted, of the deceased; have derived his or her livelihood from agricultural work on the holding for 5 years out of the previous 7 (up to 3 years spent at college or university will count) and not be the occupier of another viable commercial unit. An eligible applicant must also be found by the Tribunal to be suitable to take over the holding; suitability is judged by the agricultural experience, age, health and financial standing of the applicant.

If a landlord serves a notice to quit within 3 months of the death of a tenant and if no application is made by a suitable successor, this will end the tenancy. If an application is made, the landlord can dispute the application on grounds of unsuitability, ineligibility of the applicant, or for reasons of good estate management or hardship, but the 'fair and reasonable landlord test' still applies. Since 1984 succession is also possible subject to the above conditions on the retirement of a tenant aged 65 or over, either by agreement between landlord and tenant or on the tenant's application to the Agricultural Land Tribunal.

The law on security of tenure and succession is complicated, and there have been many rulings given by the courts on matters of detail. Moreover, the issuing of the relevant notices must be made according to strict time limits and in prescribed form. Both landlord and tenant should insist upon skilled advice from qualified lawyers or surveyors.

Obligations of landlord and tenant

The landlord and tenant of an agricultural holding will be bound by the terms of their agreement, but there has been considerable intervention by statute so as to modify and extend their contractual obligations. Either party can insist on a written agreement, and terms may be fixed by arbitration. The parties can fix the rent at the start of the agreement and change it at any time by agreement. The 1986 Act provides for the rent reviews at 3-year intervals and either party can demand an arbitration to fix it. The rules are to be found in Schedule 2: the rent has to be fixed by reference to a number of factors including the productive capacity and related earning capacity of the holding. Scarcity of lettings has to be disregarded when taking the rents of similar holdings into consideration as comparables.

Although the parties may agree their respective maintenance and insurance obligations, 'model repair clauses' are set out in the Agriculture (Maintenance, Repair and Insurance of Fixed Equipment) Regulations.

Improvements undertaken by the landlord with the tenant's agreement may lead to an increase in rent. If the tenant carries out improvements he/she will be entitled under certain circumstances to compensation at the end of the tenancy in accordance with the Act and relevant regulations. The Act also deals with the tenant's right to remove fixtures and the landlord's right to purchase them. When a tenancy comes to an end there will normally be a settlement of claims as between landlord and tenant, by arbitration if necessary, for disturbance and delapidations.

When milk quotas were introduced in 1984 it became evident that the value of a landlord's property could be greatly diminished if a tenant gave up a tenanted farm's quota. In general, however, for a tenant to do so now would be a breach of the contract of tenancy. Tenants meanwhile felt that, where the milk output of a farm had been increased during their term of occupation of a holding, the consequent size of the quota allocated should be reflected in compensation payable to them by the landlord when a tenancy was terminated. In the Agriculture Act 1986, S13 and Schedule 1, provision is made for such compensation for existing tenancies.

In general, subject to the comments made above, it may be said that a tenant farmer has freedom to crop a holding as he/she sees fit despite contrary indications in the tenancy agreement. There are, however, limits on a tenant's freedom in the last year of a tenancy when items of manurial value may not be sold or removed from the holding and the tenant may not establish a cropping scheme at any time that is not in accordance with the practice of good husbandry.

Occupier's liability

Any person recognized by the law as the occupier of land owes duties towards persons who enter upon it and toward persons who are in its vicinity. These duties are imposed by the law of tort which covers wrongs caused by a person's failure to carry out duties imposed by law as contrasted with duties imposed by a contract. The common law has evolved a number of such duties which have been recognized in judicial decisions as falling within the categories of trespass, nuisance, negligence and strict liability, to which further rules have been added by statute.

A farmer is more likely to suffer from trespassers against whom an action for damages or an injunction may be brought, than to be a trespasser him-/herself. However, if a farmer's animals trespass on to neighbouring land, the farmer will be liable for the damage they cause unless it was the fault of the neighbour or another person and could not have been reasonably anticipated. In this connection it should be remembered that a farmer is responsible for keeping his/her own stock in, and cannot complain if his/her animals escape through a neighbour's fence. Much of the law on liability for animals is covered by the Animals Act 1971 which also changed the rules about animals straying on to or off the highway. A person who negligently allows this to happen is now liable for damage done. However, there is no duty to fence in animals grazed on common land by those who have the right to do so if it is customarily unfenced. The Act also permits a farmer to shoot dogs worrying livestock. However, the farmer must notify the police within 48 hours in order to have a defence if he/she is sued by the dog's owner.

Persons entering land lawfully, known technically as 'visitors', as guests or for payment or because they have a statutory right of entry (e.g. Health and Safety Inspectors) are owed a duty by the occupier defined in the Occupiers Liability Act 1957 as 'a duty to take such care as in all the circumstances of the case is reasonable to see that the visitor shall be reasonably safe in using the premises for the purposes for which he is invited or permitted by the occupier to be there'. The Act permits the occupier to 'restrict, modify or exclude' the common duty of care by contract or adequate notice, but since the passing of the Unfair Contract Terms Act 1977 it is normally impossible for an occupier to exclude liability for personal injury caused by his/her own negligence.

Trespassers were regarded by the common law as entering upon other people's land at their peril until, in the 1970s [see *British Rail Board* v. *Herrington* (1972) AC 877] the courts began to make rulings that implied that a farmer would be expected to take care to prevent child trespassers from encountering hazards. This idea has been confirmed and widened by the Occupiers Liability Act 1984 which imposes a duty of care on occupiers towards 'persons other than visitors' such as trespassers. The duty is to take reasonable care to give such persons protection against known dangers, for example by putting up warning notices, although this would probably not be sufficient where child trespassers are known to be at risk. Farmers allowing persons to enter their land for recreational or educational purposes are allowed to exclude their duty of care towards such persons so long as they are not allowed entry as part of the farmer's business enterprise.

'Mass trespass' has caused considerable loss to farmers in certain areas since the formation of 'hippy convoys'. Trespass is in itself not a crime, but under the Public Order Act 1986, 39, police have powers to order trespassers to leave land where two or more have entered as such and done damage or used threatening behaviour or brought 12 vehicles onto the land. Failure to leave in response to such an order is an offence. The above legislation has proved inadequate to meet the needs of farmers who have suffered from the incursions of 'travellers', a phenomenon that has continued to persist throughout the 1980s and 1990s.

Legislation to deal with this problem and the difficulties caused by those who unlawfully enter on to private land to take direct action against legitimate field sports is contained in the Criminal Justice and Public Order Act 1994. The offence of 'aggravated trespass' by 'disruptive trespassers' was created to deal with the latter situation. To bring a successful prosecution the following three elements have to be established: an act of trespass in the open air; the presence of a 'lawful activity' on that or adjoining land; and an intention to interfere with that activity. The Act also contains provision for local authorities and the Home Secretary to ban open-air assemblies of 20 or more persons which may lead to a 'serious disruption to the life of the community'. The purpose of this section was to control the 'rave' culture.

As a result of the leading case of *Rylands* v. *Fletcher* (1868) LR 3 HL 330, an occupier is strictly liable for the escape of any potentially dangerous thing kept on his land if it escapes and injures neighbouring property or people. Defences to such a claim are very limited and the occupier is liable even if the escape was caused by a contractor. The liability is, however, limited to non-natural uses of land. Both the doctrine of *Rylands* v. *Fletcher* and the tort of nuisance were extensively reviewed by the House of Lords in the 1993 case of the *Cambridge Water Company* v. *Eastern Counties Leather plc*, 'the Cambridge Water Case'. Two important points to note from the judgement are:

- Foreseeability of harm of the type suffered is a requirement of liability under the rule in *Rylands* v. *Fletcher*.

- The fact alone that usage of a particular item is common to an industry does not bring that use within the definition of 'natural' for the purposes of the rule in *Rylands* v. *Fletcher*.

A farmer will cover his/her liabilities in tort (including common law nuisance) as occupier by insurance, but it should be remembered that insurance covers only for legally enforceable claims. Agriculture as an industry has a bad record so far as accidents are concerned. The utmost vigilance is required, for the law does not attempt to compensate for the incompensatable such as the death of a child drowned in a slurry pit, and damages cannot make good injury caused to an active adult crushed under a runaway machine.

Access to the countryside and public rights of way

The latter half of the nineteenth century and the entire twentieth century witnessed increasing pressure for public rights of access to the countryside over and above those provided by public rights of way. This pressure was ultimately successful when the Countryside and Rights of Way (CROW) Bill received the Royal Assent in November 2000. CROW provides a further illustration of the development of the countryside from agricultural production to public resource. From a legal standpoint it is now necessary to consider public rights of access to the countryside first, under those rights and obligations created under Part 1 of CROW and popularly but inaccurately known as the 'right to roam' and, second, under those statutory and common law rights relating to public rights of way as amended by CROW.

The Countryside and Rights of Way Act 2000

It is important to note that the rights and obligations created by CROW will be introduced over a number of years. The consequences of this are two-fold. First, it allows farmers and landowners the opportunity to take part in and influence the manner in which the Act is implemented; second, it requires farmers to remain vigilant and responsive to the gradual introduction of the legislation. Influence over the way in which the new rights are managed may be gained by lobbying either directly or through pressure groups the National and Local Access Forums which CROW requires to be created to oversee implementation of the legislation.

The Act itself is made up of five parts each subdivided into chapters. These deal with:

(1) Access to the Countryside
(2) Public Rights of Way and Road Traffic
(3) Nature Conservation and Wildlife Protection
(4) Areas of Outstanding Natural Beauty
(5) Miscellaneous and Supplementary

Access to the countryside

Part I introduces a new statutory right of access for open-air recreation to 'access land' defined as 'open country' which is 'land which is wholly or predominantly mountain, moor, heath, down' and registered common land. The Act also includes a power to extend this definition to coastal land and for landowners voluntarily to dedicate irrevocably any land to public access. Land will be identified as 'access land' on a map to be issued by the countryside bodies. Following extensive consultation both locally and nationally the countryside bodies will be responsible for determining the extent of any mountain, moor, heath and down. Any land over 600 m and registered common land immediately qualifies as access land. It is imperative that those who own or occupy land that is or may be subject to the new regime become actively involved in the consultation process so that maps accurately reflect the extent of 'access land'. As indicated above, one opportunity for this will be through representation on a Local Access Forum. A right of appeal may be made to the Secretary of State (or National Assembly for Wales) against the inclusion of land on provisional maps as access land. Schedule 1 of the Act lists land excepted from access; this includes 'land on which the soil is being, or has at any time within the previous 12 months, been disturbed by any ploughing or drilling undertaken for the purposes of sowing crops or trees'.

Rights and liabilities of landowners

During the debate leading to the passing of the CROW bill much time was given to the issue of the liabilities of persons with an interest in land towards those seeking access. Clause 12 of the Act provides that the right of access does not increase the liability of a person interested in the land in respect of the state of the land or things done on it. Clause 13 amends the Occupier's Liability Act of 1957 to reduce the liability of occupiers to those exercising their statutory right of access to that owed to a trespasser under the Occupiers' Liability Act of 1984 (see 'Occupier's liability' above). However, it should be noted that those who seek to deter access by the production of misleading information will be penalized under clause 14 which makes it an offence to display a notice containing false or misleading information. The use of any path or area of land in exercise of the statutory right of access will not provide the basis for a claim for the existence of a right of way or of a town or village green.

Exclusion or restriction of access

Part 1, Chapter 2 of the CROW Act defines the situations under which access may be excluded or restricted. Land may be closed as of right for a period of 28 days. However, the days on when this right is available are in themselves limited. For example, it must not include more than four Saturdays or Sundays and these Saturdays or Sundays are limited to certain periods of the year to avoid restrictions at the height of the summer. Above the 28 days, further closures may be applied for on a number of grounds including land management; nature conservation and heritage preservation; defence or national security; or exclusion or restriction of access in the case of emergency. Landowners and occupiers should confirm, prior to an application to restrict access, that the days they are seeking to achieve this on are permitted days. The ability to exercise dogs is also restricted (Schedule 2) in that they must be on a lead both in the vicinity of livestock or during the period beginning with the 1 March and ending with the 31 July each year. These restrictions may be relaxed by direction of the relevant authority.

CROW 2000 – conclusion

There can be little doubt that the introduction of CROW 2000 will come to represent a significant development in the public's perception of their relationship with the countryside. However, despite the inevitable teething problems and individual difficulties that many will have to face, the likely reality of CROW is that a highly managed right of access will emerge in areas generally well-defined by map. It is important that landowners and farmers play their part in that management process. The foot-and-mouth outbreak in February 2001 and the response of government to provide powers for local authorities to close public rights of way and impose fines of up to £5000, the immediate response of many such authorities and organizations representing ramblers, suggest that there is a growing awareness of the necessity for a truly responsible approach to access in the countryside.

Public rights of way

The legal provisions that relate to public rights of way occur as the result of both common law and statute, are numerous and can be complicated. However, if farmers are to maintain any credibility as 'stewards of the land' with those who have access to it, they must demonstrably be aware of their legal duties. Below is an outline of the main responsibilities of farmers and landowners with regard to public rights of way over their land as amended by CROW 2000. Note: the provisions of CROW will be introduced over a period of time. Where reference is made to this legislation, those seeking to rely on it should confirm that the amendment has taken effect.

Redesignation of roads used as public paths

Most members of the public are aware of the designations footpath and bridleway. To these has been added the restricted byway (CROW 2000). Section 43 repeals section 54 of the Wildlife and Countryside Act 1981. It redesignates all roads used as public paths (RUPPS) as restricted byways; these are defined as including a right of way on foot; on horseback or leading a horse; and for vehicles other than mechanically propelled vehicles.

Obstructions

No person may wilfully obstruct the free passage along a highway (highway includes both footpaths and bridleways and restricted byways) (S137 Highways Act 1980). This means that if a person 'without lawful authority or excuse, intentionally as opposed to accidentally, that is, by an exercise of their own free will, does something or omits to do something which will cause an obstruction, he or she is guilty of an offence' [*Parker LCJ Arrowsmith* V. *Jenkins* (1963)]. Any person may serve the Highway Authority with notice requesting it to secure removal of an obstruction (CROW 2000). The Highway Authority must, within 1 month of the date of service, inform the complainant of the intended action. If the complainant is not satisfied that the obstruction has been removed an application may be made to the Magistrates' court.

Gates and stiles

The maintenance of a gate or stile that crosses a footpath or bridleway is the responsibility of the landowner (S146 Highways Act 1980). It must be maintained in a safe condition so that there is no unreasonable interference with the public's rights of access.

Overhanging vegetation

Where any overgrowth interferes with a public right of access, the Highway Authority or District Council may serve a notice under S154 of the Highways Act 1980 requiring the owner or occupier to cut back the growth. S154 has been amended by CROW to include overhanging vegetation that endangers or obstructs horseriders.

Bulls: Wildlife and Countryside Act 1981 S59

Bulls may not be kept in fields crossed by rights of way unless the following conditions are met:

- the bull is less than 10 months old, and
- it is not a recognized dairy breed and it is running with cows or heifers. Dairy breeds include Ayrshire, British Friesian, British Holstein, Dairy Shorthorn, Guernsey, Jersey and Kerry.

Misleading notices

S57 National Parks and Access to the Countryside Act 1949 makes it an offence for any person to place or maintain, on or near any way shown on the definitive map, a notice that contains false or misleading statements which are calculated to deter the public from using the way.

Barbed wire: Highways Act 1980 S164

Where barbed wire is placed on land adjoining a highway in a position in which it is likely to be injurious to persons or animals using the highway, the Highway Authority may serve a notice on the occupier of the land requiring him/her to remove the nuisance.

Farmers and landowners should also be aware that other activities that interfere with the public rights of access may be a nuisance at common law or under statute. In recent years the Environmental Protection Act 1990 Part 3 S79 has added to the list of statutory nuisances, which a farmer might breach when hindering the public's rights of access.

Chief obligations imposed by the 1990 Rights of Way Act

Under the Act farmers may plough or disturb the surface of a crossfield footpath or way as long as it is not convenient to avoid it. Farmers may not disturb field edge paths. When the surface of a public right of way has been 'disturbed' (disturbance includes all necessary operations for cultivation):

- The period allowed for restoration of the surface is 24 hours; however, where there is a sequence of operations leading up to sowing, a period of up to 14 days is allowed.
- When restoring the path or way the farmer is required to indicate the line of the route on the ground.
- A specific duty is imposed on farmers to prevent crops from encroaching on to paths by:
  - growing through the surface;
  - overgrowing on to them from the sides.
- Paths and ways that have been disturbed by cultivations must be restored to the following minimum widths:
  - crossfield path 1 m
  - crossfield bridleways 2.5 m
  - field edge path 1.5 m.
- Offences committed under the Act are punishable by fine which may be increased when an offence is repeated.
- If a farmer fails in his/her obligations under the Act, the local authority may act in default and seek reimbursement from the farmer.
- Excavations and engineering activities that disturb the surface of a public right of way may only be carried out if you first get written permission from the Highway Authority.

As indicated, the above is merely an outline of the obligations placed upon those who have public rights of way over their land. However, likely future pressures suggest that it is the wise farmer who has a complete knowledge of the law as it affects his or her particular circumstances.

Extinguishment of rights of way

A further important addition to public rights of way law is the extinguishments of unrecorded rights of way. A cut-off date of 1 January 2026 has been introduced. This in effect creates a 25-year deadline for the investigation into alleged rights of way, the production of evidence to prove their existence and orders seeking to make appropriate amendments to the definitive map, on which all public rights of way are recorded.

Legal constraints on the development of land

The early 1990s saw a radical reshaping and the introduction of major changes to the statutory planning legislation in this country. This included the closer integration of agriculture into the planning system. In 1990 the Town and Country Planning Act (TCPA 1990) came into force consolidating previous legislation. On 25 July 1991 the Planning and Compensation Act (PCA 1991) received the royal assent. This Act is extensive in the number of areas it affects and fundamental in a number of the changes it has brought about. The 1991 Act is an enabling act, i.e. it gives the power to the Secretary of State to introduce secondary or delegated legislation which is used to complete, amend and update the primary legislation. It is therefore extremely important when considering planning law to make certain you are working with up-to-date information. The planning legislation is complemented by government circulars and Planning Policy Guidance Notes (PPGs) which detail national policies. These do not have the force of law but are material factors to be taken into account when, for instance, planning applications are considered. Particularly relevant to farmers are PPG 7, 'The Countryside: Environmental Quality and Economic and Social Development' and PPG 9 'Nature Conservation'.

Use of land for agriculture is permitted because this activity does not constitute a development under the TCPA 1990. Other activities, i.e. building, mining, quarrying or the carrying out of engineering works, are development, as is a material change of use, e.g. the conversion of farm buildings to dwellings or for light industry. In such cases planning permission must be obtained from the local planning authority – normally the district council.

By virtue of s.59 of the TCPA 1990 and Article 3 of the Town and Country Planning General Permitted Development Order (GPDO) 1995 certain developments on agricultural land are deemed not to require planning permission. However, farmers should ensure that the development complies with all the requirements of the GPDO, for example that the building does not exceed 465 m^2 subject to conditions as to its siting in relation to other buildings, other recent constructions, roads and its height. The further following requirements should also be noted.

Agricultural holdings of less than 5 ha no longer benefit from previous permitted development rights to construct farm buildings. More limited development rights have been introduced for such holdings. Farmers should now check before making use of agricultural permitted development rights with their local planning authority as to whether the authority require to give prior approval for certain details of the development. In the case of an agricultural building this would include siting, design and external appearance. Under this 'determination' procedure the planning authority has 28 days to decide whether prior approval will be required for:

- the siting, design and external appearance of agricultural and forestry buildings;
- the siting and means of construction of a private way;
- the siting of excavations or waste deposits with an area exceeding 0.5 ha;
- the siting and appearance of fish tanks.

If no response is forthcoming from the planning authority within 28 days, the farmer is entitled to proceed.

It is suggested that before starting any new 'development' project farmers and managers should obtain the answers to the following questions:

- Does the project involve 'development' within the definition of the 1990 Act, i.e. building, engineering, mining or other operations in or over land or a material change of use of the land, e.g. conversion of farm buildings to dwellings?
- If yes, are there permitted development rights available under the GPDO? If yes, contact the local planning authority to determine whether approval is required for some details. If no, planning permission should be sought in the normal way.

The value of a careful preparation to the application cannot be overstressed. This should include the canvassing of neighbours and local opinion and an understanding of the factors, e.g. development plans, or PPGs which the planning authority will take into account when reaching its decision. If the application is successful, it may well have conditions attached which substantially restrict the way in which the development may proceed. In order to be valid these conditions:

- must relate to the permitted development;
- must serve some useful planning purpose;
- may not be manifestly unreasonable;
- may be declared invalid for uncertainty.

Agriculture no longer holds the privileged position it once did in the planning world, and although the Rural White Paper (2001) *A Future for Our Countryside* indicates a desire to assist farm diversification through a positive use of the planning system, further restrictions on agricultural development, as has happened, for example, with the requirement for environmental impact assessment on identified developments, become an increased possibility.

Nature conservation and production methods

The Wildlife and Countryside Acts 1981 and 1985 provide the cornerstone for the protection of birds and wild animals. Apart from the destruction of common pest birds it is an offence to kill wild birds, destroy their nests or disturb them near their nests. The Secretary of State for the Environment, working with English Nature and the Countryside Commission for Wales, has powers to make orders to protect Sites of Special Scientific Interest, and either the Minister or a local authority may enter into a management agreement with owners and occupiers of land to preserve and enhance the natural beauty of the countryside. The land involved will be subject to restrictions as to the use for which compensation is payable, that will be binding not only upon the owner or occupier who makes the management agreement, but also upon successors in title.

The EC Council Directive on the Conservation of Natural Habitats and of Wild Fauna and Flora (Directive 92/43/EEC) – the Habitats Directive – came into force in June 1994. The following designations are provided for:

- Special Areas of Conservation (SACs)
- Special Protection Areas (SPAs)

Sites designated as such will be subject to restrictions when a proposed development is likely to have a significant effect on the area. The local planning authority will be required to assess the impact of a development and respond accordingly. PPG 9 has been issued to advise on the planning implications of regulations which implement the EC Habitats Directive.

The Countryside and Rights of Way Act 2000

Part III of the Countryside and Rights of Way Act 2000 (CROW) seeks to enhance nature conservation and wildlife protection. The following issues have been addressed.

Biological diversity

A new duty has been placed on government departments and the Welsh National Assembly to have regard to biodiversity conservation and maintain lists of species and habitats for which conservation steps should be taken or promoted.

Sites of Special Scientific Interest (SSSIs)

CROW creates new procedures designed to enhance the protection of SSSIs including:

- the provision of powers to conservation agencies to refuse consent for damaging activities and to encourage positive land management;
- a statutory duty for public bodies to further the conservation and enhancement of SSSIs;
- increased penalties for damage to SSSIs by owners, occupiers and other parties.

Wildlife protection

CROW amends and updates the Wildlife and Countryside Act 1981 to increase legal protection for threatened species by:

- providing increased search and seizure powers to the police and making certain offences 'arrestable';
- creating a new offence of reckless disturbance;
- allowing courts to impose heavier fines and prison sentences for most wildlife offences.

Areas of Outstanding Natural Beauty(AONBs)

The management of AONBs will be improved by:

- the production of management plans by local authorities for AONBs in their area;
- creating the mechanism for the creation of conservation boards where there is local support for such a scheme. The role of a conservation board would be to take over the role of the local authority in the production of a management plan and other management matters concerning an AONB.

Limitation of production and agri-environmental schemes

Limitation of production is intimately related to the Common Agricultural Policy. The introduction of milk quotas by EEC Council Regulations in 1984 introduced a new and complicated area of law to the agricultural scene. This scenario has had further complexities added in the form of the Suckler Cow Premium Scheme and Sheep Annual Premium Scheme. It should be noted that they differ from milk quota in two basic respects. First, they are owned by the producer and, second, they are a right to receive premium each year and not to produce. With the continuously shifting sands and perilous nature of this area of law, those involved in land and/or quota transfers are respectfully referred to take expert advice.

Another approach to product limitation coupled with

conservation is to be found in the agri-environmental schemes, for example, the *Environmentally Sensitive Areas Scheme* (introduced by the Agriculture Act 1986) of which there are now 22 in England and Wales. Payments are made to farmers in such areas for farming in accordance with traditional methods, thus maintaining established landscape patterns and improving the environmentally beneficial aspects of farmland. This 'carrot' approach to the protection and enhancement of the countryside has also manifested itself in schemes such as that for 'Countryside Stewardship'. Such initiatives are complemented by further schemes that are part of the Rural Development Programme, the response to the European Rural Development Regulation; they include:

- Energy Crops Scheme
- Farm Woodland Premium Scheme
- Hill Farm Allowance Scheme
- Organic Farming Scheme
- Processing and Marketing Scheme
- Rural Enterprise Scheme
- Vocational Training
- Woodland Grant Scheme

Farmers would do well to contrast the approach that has been adopted with the more prescriptive alternatives.

Pesticides and pollution

The law relating to the use, supply and storage of pesticides is controlled by the Control of Pesticides Regulations 1986 (SI 1986 No. 1510) made under the Food and Environment Protection Act 1985, and the Control of Substances Hazardous to Health Regulations 1988 (1988 No. 1657), more familiarly known as COSHH, made under the Health and Safety at Work Act 1974 (see Chapter 15 of this book). The term pesticide includes herbicides, insecticides and fungicides but does not include those substances applied directly to livestock (e.g. sheep dips, which are regulated by separate legislation, namely the Medicines Act 1968). The legislation seeks to protect both human health and the environment by requiring Ministry approval for all pesticides and prescribing the training necessary for the use of such products. Those farmers or workers seeking to use pesticides who were born after 31 December 1964 must obtain a certificate of competence before using an approved pesticide.

All farmers should be aware of the two statutory approved Codes of Practice that accompany the legislation. These are the Code of Practice for suppliers of pesticides to agriculture, horticulture and forestry, and the Code of Practice for the Safe Use of Pesticides on Farms and Holdings (the 'COSHH Combined Code').

Those farmers whose actions cause a deterioration in water or air quality or who cause other statutory nuisances may find themselves held legally responsible under the Water Resources Act 1991 or Environmental Protection Act 1990. Under the former it is a criminal offence to cause or knowingly permit any poisonous, noxious or polluting matter, or any solid waste matter, to enter any controlled waters. In the Control of Pollution (Silage, Slurry, and Agricultural Fuel Oil) Regulations 1991 (SI 1991 No. 324), minimum standards are set for the construction of silage, slurry and fuel oil installations. These regulations apply to constructions built after 1 March 1991. However, it should be appreciated that the Environment Agency may give notice to a farmer to achieve the statutory standards on an installation built before 1 March 1991 if its condition presents a 'significant' risk of pollution.

Farmers should always seek to follow the Codes of Good Agricultural Practice for the Protection of Water, Soil and Air. These codes have statutory approval but do not in themselves create any criminal liability. The Water Code does not furnish a defence for charges brought under the 1991 Act. This is a major change from the position under the previous legislation (Control of Pollution Act 1974) where a farmer could state by way of defence that he/she had acted in accordance with the Code.

As a result of regulations made under the Environmental Protection Act, air quality has sought to be protected by the prohibition of straw and stubble burning as from 1993. Further emissions of odours or smoke from farms may create statutory nuisances under the Act as defined by S79(1). If a local authority is satisfied of the existence of such a nuisance, it must serve an abatement notice. Failure to comply with such a notice, without reasonable excuse, constitutes a criminal offence.

Conclusions

A knowledge of the law as it affects agriculture and the business environment, the mechanisms by which law is created and the direction in which it is progressing is now a necessity for the agricultural land manager. Those working on the land may feel isolated, but their actions and the results thereof, be it with regard to quantity and quality of production, the effect on the land, landscape or labour force, are matters of general public concern.

The last decade has shown that, if agriculture will not respond of its own volition to the wishes of the society it serves, then laws will be introduced to seek to achieve those ends. However, changes in the law provide new opportunities too; the let land market has begun to expand again and financial incentives, albeit limited, are being put in place to encourage different approaches to farming and the marketing of products. Knowledge of the law, acquired by education and timely professional advice, should be used in a positive fashion to manage and plan effectively.

Further reading

As a general introduction to the English legal system and the general principles of English law:

Keenan, D. (1998) *Smith and Keenan's English Law*, 11th edn. Pitman, London.

On most aspects of farm tenancies, land ownership and occupation:

Lennon, A.A. & Mackay, R.E. (2002) *Agricultural Law, Tax and Finance*. Longman, London.

For occupier's liability, trespass, nuisance and other civil matters related to land:

Anon. (2000) *Clerk and Lindsell on Torts*, 18th edn. Sweet and Maxwell, London.

For detailed planning information:

Anon. (2002) *Encyclopedia of Planning Law*. Sweet and Maxwell, London.

Useful websites

Gateways and portals to legal information on the web

http://library.ukc.ac.uk/library/lawlinks/
http://www.venables.co.uk/
http://www.bailii.org/
http://www.justask.org.uk/

UK government and the administration of justice

http://www.open.gov.uk/lcd/lcdhome.htm
http://www.courtservice.gov.uk/
http://www.detr.gov.uk/
http://www.defra.gov.uk/

Europe

http://europa.eu.int/

Access to the countryside

http://www.pill.plym.ac.uk/footpath/

15

Health and safety in agriculture

D.J. Mattey

Introduction

Health and safety legislation has existed and been developed in most industries for over 150 years. However, it was not until the mid 1950s that any comparative requirements were applied to the health, safety and welfare of people at work in agriculture, with the introduction of The Agriculture (Poisonous Substances) Act 1952 and The Agriculture (Safety, Health and Welfare Provisions) Act 1956. These Acts required general welfare provisions, but more significantly allowed the introduction of specific regulations covering basic areas of injury concern. For example: First Aid; Ladders; Children; Workplaces; Stationary Machinery; Circular Saws; Field Machinery; and Tractor Cabs. The regulations imposed an absolute duty of compliance – mainly, but not exclusively, on employers to protect workers employed in agriculture.

The concept behind that type of legislation was that the individual regulation acknowledged the generic risk, and prescribed the remedial action required. Whether the person on whom the duty was placed had met a particular requirement in practice became a matter of fact, in that the action taken could be directly compared with the regulatory requirement. For example: a guardrail was either in position at a height of 3 ft 6 inches, or it was not.

By the early 1970s the legislative approach generally was such that each industry had separate specific acts and regulations – and a corresponding range of enforcing inspectorates under different government departments. For example: MAFF responsible for agriculture; the Department of Employment for factories.

The Robens Report in 1972 reviewed the national health and safety legislative framework. It observed that there was 'a mass of ill-assorted and intricate detail in existing legislation' and that there should be 'a comprehensive orderly set of provisions under a new enabling Act . . . should contain a clear statement of basic principles of safety responsibility . . . supported by regulations and non-statutory Codes of Practice. The scope of new legislation should extend to all employers, employees, and self-employed'. An important objective also expressed by Robens was the need for all industries to move towards 'self-regulation'.

The Health and Safety at Work etc. (HSW) Act 1974 was the result, and transformed the legislative approach by moving from the 'absolute duty' or prescriptive requirement to a concept of 'goal setting' (self-regulation). It is interesting and significant that the concepts and principles behind the HSW Act were reflected 20 years later in the more modern EU Directive-based legislation (discussed later).

The agricultural industry itself had been faced with a major transformation, developing from an intensive-labour, low-productivity industry, to one with a reducing workforce, greater mechanical and technical development and achieving far greater output. Increases in efficiency, new techniques and practices also introduce new hazards and corresponding risk of injury. An analysis of fatal injuries for the period 1986/1987–1998/1999 provides an indication of the extent and source of those hazards. Table 15.1 shows the status of those injured, while Table 15.2 shows the kind of accident.

Expressing the injury cases as 'incidence rates' (the rate per 100 000 workers) provides a comparison with other industries. The fatal injury numbers in agriculture over the period 1995/1996–1999/2000 average 52/year and show no downward trend, with the incidence rates remaining high – particularly for the self-employed. It is also a feature of agriculture that injuries occur across a far wider age range than any other industry – the hazards existing primarily because many families live in the

Table 15.1 Fatal injuries to employees and self-employed people in agriculture, 1986/1987–1998/1999

		86/87	*87/88*	*88/89*	*89/90*	*90/91*	*91/92*	*92/93*	*93/94*	*94/95*	*95/96*	*96/97*	*97/98*	*98/99*
Employees	No.	27	21	21	23	25	18	21	16	14	20	20	20	16
	Rate[1]	8.6	6.8	7.0	8.1	9.0	6.7	8.2	6.4	5.7	7.8	7.6	6.7	5.4
Self-employed	No.	17	31	25	30	27	32	19	22	32	20	35	20	30
	Rate[1]	6.9	12.7	10.3	12.3	10.9	13.0	6.9	9.1	11.4	8.3	14.3	8.7	14.8
Employees and self-employed	Rate[1]	7.8	9.4	8.5	10.1	9.9	9.7	7.5	7.7	8.7	8.0	10.8	7.5	9.3

[1] Rate per 100 000 workers.

Table 15.2 Fatal injuries to employees and self-employed people in agriculture, 1986/1987–1998/1999

Kind of accident	*Employees*	*Self-employed*	*Total numbers*	*Percentage*
Struck by moving vehicle	49	62	111	18.4
Trapped by something collapsing or overturning	29	52	81	13.5
Fall from a height	39	56	95	15.8
Contact with machinery or material being machined	37	42	79	13.1
Struck by moving, including flying or falling, object	28	52	80	13.3
Contact with electricity or an electrical discharge	37	16	53	8.8
Asphyxiation	19	19	38	6.3
Injury by an animal	12	23	35	5.8
Other	12	18	30	5
Total	262	340	602	100

workplace. People well over the conventional retirement age and very young children are often victims and need particular protection from risk in workplace activities.

The responsibility for ensuring adequate health and safety conditions rests primarily with those who create the risks, and the legislation has been framed accordingly. The Health and Safety at Work etc. Act 1974 provides a legislative framework to promote, stimulate and encourage high standards of health and safety at work.

The Health and Safety at Work etc. Act 1974 (HSW Act)

The Act is in four parts:

- Part 1 – dealing with health, safety and welfare in relation to work.
- Part 2 – relating to the Employment Medical Advisory Service.
- Part 3 – relating to Building Regulations.
- Part 4 – miscellaneous and general provisions.

Part 1 is the most relevant for the purposes of this chapter. The objectives of this Part are:

(1) securing the health, safety and welfare of people at work;
(2) protecting people other than those at work against risk to their health and safety arising from work activities;
(3) controlling the keeping and use of explosive or highly flammable or otherwise dangerous substances, and generally preventing people from unlawfully having and using such substances;
(4) controlling the release into the atmosphere of noxious or offensive substances from premises to be prescribed by Regulation.

(It should be noted that the storage and use of pesticides is also subject to the Food and Environment Protection Act 1985.)

The HSW Act applies to employment generally. Thus duties are placed on all people at work – employers, employees, self-employed; manufacturers, suppliers, designers and importers of equipment and materials used at work; and people in control of premises (for example landlords).

The Act does not distinquish between different industries, and was superimposed on earlier related legislation, which became 'relevant statutory provisions'. However, the longer-term intention to progressively replace that earlier legislation has resulted in the repeal of virtually all of the two original agriculture-related acts and their regulations, mentioned earlier.

The HSW Act established two bodies, the Health and Safety Commission (HSC) and the Health and Safety Executive (HSE). The Commission comprises a Chair, and up to nine part-time members all appointed by the Secretary of State (DETR), after consulting employer organizations, trade unions, Local Authorities and other organizations for nominations. The Commission is responsible for promotion of the objectives of the HSW Act, encouraging research and training, provision of an Information and Advisory Service, and making proposals to government for Regulations. In practice HSC has developed arrangements for wide consultation on any proposals to change or introduce Regulations.

The Commission has also established a number of Advisory Committees, one of which is the Agriculture Industry Advisory Committee (AIAC). The Chair is the Chief Inspector of Agriculture, its 12 members are drawn from the industry itself, others with particular expertise and knowledge are appointed as assessors, and there are members of two Working Groups, covering Machinery and Health.

The Health and Safety Executive consists of three full-time members, appointed by the Commission and with the approval of the Secretary of State. The Executive's duties include making arrangements for the enforcement of the legislation and carrying out any of the Commission's functions as requested. It also advises the Commission on new legislative proposals. In practice the Executive and its staff form the operating arm of the Commission, and are known collectively as HSE. It should be noted that responsibility for health and safety at work as such lies with HSC/HSE and not with DEFRA, and similarly no responsibility was transferred to the Scottish Parliament or Welsh Assembly – although all are 'stakeholders' and maintain an active interest.

HSE's organizational structure differs from that of other specialized public bodies in that its business units are highly integrated. That integration with a high degree of collaboration between units allows HSE to take a holistic approach to the diverse, complex and difficult task of regulating health and safety at work. In practice the units cover four areas of activity – Operations, Policy, Science and Technology, and Central Services.

Policy staff are involved in advising HSC on the need for changes in legislation or standards, and in negotiations over EU proposals and development of associated legislation. Science and Technology staff provide specialist advice, information and medical support to other parts of HSE, and central government, on industrial health and safety matters, including extent and nature of risks, the effects of hazards on the individual and the environment, and on appropriate standards.

Operational staff form the largest group, with the majority of Inspectors working in the Field Operations Directorate (FOD) and covering most industries – including agriculture – within a regionally based structure. This is the main enforcement arm of HSE, and it secures compliance with legal requirements and accepted standards, interpreting the latter so far as is possible on a national basis. They operate through a mixture of preventive inspection, investigation of accidents, advice, publicity and enforcement.

An important aspect of their work is in contact and liaison with the industry and internationally in developing strategies and industry-specific advice on standards and 'best practice'. Staff engaged in this area of work in the Agriculture and Wood Sector are based in Nottingham and Stoneleigh (NAC) and are responsible to the Chief Inspector of Agriculture.

Basic statutory requirements of the HSW Act

The Act imposes duties on everyone involved with work activities. Those duties are expressed in general terms in the Act itself, with more specific requirements covered by Regulations and supported by Approved Codes of Practice (ACoP). An ACoP approved by the Commission has a particular standing in law. If a Code appears to a Court of Law to be relevant, then it is admissible in evidence. If the guidance within the Code has not been followed in order to meet the requirements of a Regulation, it is for the defendant to show that he/she has complied by equally effective means. An example in agriculture is the Prevention of Accidents to Children in Agriculture Regulations 1998 (which replaced the original Avoidance of Accidents to Children Regulations 1958) and the introduction of an associated ACoP 'Preventing accidents to children in agriculture'.

Some of the duties are qualified by the term 'so far as is reasonably practicable'. This is an important qualification which is not defined in the Act but has been interpreted by the Courts. It implies an assessment of risk balanced against the cost, in terms of money, time or trouble, involved in the measures to avert the risk. Thus if the risks to health and safety are very low and the cost or technical difficulties of the action to avoid the risks are very high, then it might not be reasonably practicable to take that action. The balance, however, does not include consideration of the financial standing of those who have the duty of compliance.

Employers have a general duty to ensure, 'so far as is reasonably practicable', the health, safety, and welfare at work of their employees by:

(1) provision and maintenance of safe equipment and systems of work;
(2) ensuring the safe use, handling, storage and transport of articles and substances;
(3) provision of information, instruction, training and supervision;
(4) maintaining safe premises, and other places of work;
(5) provision of a safe healthy working environment, and with adequate welfare facilities.

The concept of planning work operations by identifying hazards, evaluating risk and taking action to eliminate or reduce those risks is embodied in a requirement for employers with five or more employees to prepare a written safety policy statement, and bring it to their notice.

Self-employed persons (*as well as employers*) are required to conduct their business activities 'so far as is reasonably practicable' so that they do not put persons at risk who are not their employees. This applies both on and off their premises, and includes, for example, members of the public and children. The self-employed are also required not to put themselves at risk.

Employees are required to take reasonable care of their own health and safety, and of others who might be affected by what they do. They must also cooperate with their employers and others in meeting statutory requirements, and must not interfere with or misuse anything provided to protect their health and safety.

Persons who have control to any extent of non-domestic premises are required to ensure 'so far as is reasonably practicable' a safe place of work. This does not apply to any employee who may have been put in control of premises at any particular time by his/her employer.

Designers, manufacturers, importers and suppliers of articles and substances for use at work must ensure 'so far as is reasonably practicable' that products are safe and without risk when used in prescribed circumstances. They are required to carry out tests as necessary and make information available about the uses for which the products have been designed and tested (HSW Act Section 6 duties were amended by the Consumer Protection Act 1987).

People in general (*i.e. the public*) have a duty not to intentionally interfere with or misuse anything provided in the interests of health and safety. This could apply, for example, to fencing, warning signs, machinery and guards.

Enforcement

Inspectors are appointed under the HSW Act and derive their powers from the Act. They are issued with a warrant which specifies those powers. Essentially this authorizes the Inspector to carry out statutory provisions: in effect powers of entry; powers to make examinations and investigations, and to require that premises or equipment remain undisturbed for those purposes; to take samples; to require persons to answer questions. Inspectors also have powers to institute legal proceedings in England and Wales, put cases to the Procurator Fiscal in Scotland, and also issue *Improvement Notices and Prohibition Notices*. The circumstances in which Notices are issued is circumscribed by detailed requirements in the HSW Act (Sections 21–23), but in summary:

(1) An *Improvement Notice* may be issued if in the Inspector's opinion there is, or will continue to be, a legal contravention; and will require the matter to be remedied within a specified time.
(2) A *Prohibition Notice* may be issued, with either immediate or deferred effect, where in the Inspector's opinion there is a risk of serious personal injury; and will require that the activity shall cease until the deficiencies have been remedied.

There is an Appeals procedure against the service of Notices, details of which appear on the Notice. Such cases are dealt with by Industrial Tribunals.

Prosecutions can be tried summarily (Magistrates Court in England and Wales) or on indictment (Crown Court in England and Wales). Fines in the lower Courts are up to £5000 for some offences, but breaches of

General Duties of the HSW Act (Sections 2–6) and failure to comply with Notices can attract £20 000 or 6 months' imprisonment, or both. The higher Courts can impose unlimited fines or 2 years' imprisonment, or both.

The effect of European Directives

In meeting obligations under European Community law, Member States need to ensure that their law is consistent with European Directives. In Great Britain, the Regulations made to implement Directives with a health and safety context have mostly been made under HSW Act s.15, with s.80 allowing repeal or modification of existing 'statutory provisions' as necessary. (It should be noted that European Directives should not down-grade existing standards.)

Regulations specific to agriculture have reduced significantly since 1993 when most 'old' regulations were repealed, and six sets of new Directive-initiated regulations, applying to all sectors of industry, were introduced. These Regulations, known colloquially as 'the six pack', made more explicit some of the duties of the earlier legislation. More significantly they emphasize the importance of health and safety management, and of assessing risk and selecting appropriate preventative and protective measures. The six sets of regulations are as follows.

The Management of Health and Safety at Work Regulations 1999

These regulations can be regarded as key direction in the management of risk concept, and set out broad duties applying to most work activities. Principal provisions relate to risk assessment, arrangements for health and safety, health surveillance, etc. They include duties of employees to use equipment, etc. in accordance with any training or instruction given (these requirements reflect those already in force under s.7 of the HSW Act). There is also an Approved Code of Practice, which may be regarded as a cornerstone of the modern approach to health and safety legislation.

Provision and Use at Work Regulations 1998 (PUWER) (replaced original 1992 regulations)

Work equipment is defined to include everything from hand tools to complete machinery installations, but not substances or livestock. The Regulations deal with the suitability of the work equipment, maintenance, provision of suitable training and information, etc.

In an agricultural context these regulations require:

(1) equipment to be inspected after installation if significant risk could result from incorrect installation;
(2) operators of chain saws to have a certificate of competence *unless* work is part of an agricultural operation *and* the person has used the chainsaw before 5 December 1998;
(3) a prohibition on carrying passengers on mobile equipment unless it is suitably equipped (this applies from 5 December 1998 for new equipment and from 5 December 2002 for existing equipment);
(4) a requirement (subject to some exemptions) for roll-over protection on mobile equipment, if there is a risk of overturn in use (new equipment from 5 December 1998 and existing equipment from 5 December 2002).

There is also an Approved Code of Practice.

Manual Handling Operations Regulations 1992

These Regulations establish a hierarchy of measures, such as avoidance of dangerous practices so far as is reasonably practicable, and assessing those that cannot be avoided and taking action to reduce risk of injury – for example by provision of mechanical assistance.

There is also an Approved Code of Practice.

Workplace (Health, Safety and Welfare) Regulations 1992

These Regulations cover matters such as maintenance, ventilation, lighting, temperature, etc. They include provisions for preventing falls from roofs and falls into dangerous substances. However, workplaces which are fields, woods or other land forming part of an agricultural or forestry activity are exempt from most of the requirements, except those relating to sanitary conveniences, washing facilities and drinking water (Regulations 20–22).

There is also an Approved Code of Practice.

Personal Protective Equipment at Work Regulations 1992

These are short Regulations dealing with provision, maintenance and accommodation for equipment (PPE), and assessment of its suitability, etc. Also included is provision of information, instruction and training. It is important to note that the provision of PPE is regarded as the last resort in the hierarchy of measures to protect against risk. Engineering controls and safe systems of work should always be considered first. Employers' duties in this respect are framed in the Management of Health and Safety at Work Regulations 1992 (mentioned earlier).

Health and Safety (Display Screen Equipment) Regulations 1992

These Regulations cover the use and working positions of those involved with display screens. They are not a major feature of agricultural operations but the Regulations apply, with some exemptions for workplaces on agricultural and forestry land away from the main buildings.

Lifting Operations and Lifting Equipment Regulations 1998 (LOLER)

These Regulations replaced existing legal requirements relating to the use of lifting equipment, for example in construction and docks. Although not one of the original 'six', they supplement requirements contained in the Provision and Use at Work Regulations 1998, and created duties for the first time in agriculture on this specific type of equipment. Provisions include adequate strength and stability; safe load markings; positioning and installation to reduce risk of loads falling onto people; examination (by a competent person) where equipment could create a risk in event of failure. Evidence of examination must be provided by persons who sell, loan or hire out lifting equipment. Not all equipment which 'lift' is defined as lifting equipment, and further advice on agricultural application is available from the HSE.

Other significant Regulations

Control of Substances Hazardous to Health Regulations 1999 (COSHH)

These Regulations require employers to control exposure to hazardous substances to prevent ill health. They have to protect employees and others who may be exposed. Hazardous substances include:

- substances used directly in work activities (e.g. disinfectants and dairy cleaning agents);
- substances generated during work activities (e.g. fumes from welding);
- naturally occurring substances (e.g. grain dust).

COSHH sets out seven basic measures that employers, and sometimes employees, must take. It follows the concept of Hazard identification; Risk assessment; Decide on precautions; Prevent or Control exposure, but also requires that control measures are used and maintained; exposure is monitored; health surveillance is carried out where necessary; and employees are properly informed, trained and supervised.

Note: pesticides are also subject to the Food and Environment Act 1985 (FEPA), and the Control of Pesticides Regulations 1986.

Health and Safety (Consultation with Employees) Regulations 1996

Employers must consult their employees on health and safety matters. Where employees are members of a 'recognised' trade union, the Safety Representatives and Safety Committees Regulations 1977 also apply.

The Health and Safety (First Aid) Regulations 1981

These place general duties on employers and self-employed persons so that first-aid facilities are provided for employees, and self-employed can render first aid to themselves.

The Reporting of Injuries, Diseases and Dangerous Occurrences Regulations 1985 (RIDDOR)

These Regulations place responsibilities on employers, the self-employed and others in control of premises and sites. They set out reporting arrangements for fatalities and specified major injuries; certain diseases; specified dangerous occurrences and gas incidents.

In the case of death or any of the specified major injuries the requirement is to notify the responsible authority (HSE for agricultural activities) by the quickest practicable means, and to forward a written report within 10 days. In the case of an injury involving more than 3 days off work, a written report must be sent (to HSE) within 10 days. If a doctor notifies an employer that an employee suffers from a reportable work-related disease a written report must be sent (to HSE). Should something happen which did not result in a reportable injury, but clearly could have done so, it may be necessary to report it immediately (to HSE).

A record of any reportable injury, disease or dangerous occurrence must be kept (for 3 years). Definitions of reportable events and appropriate forms of written report appear in the Regulations.

Noise at Work Regulations 1989

These Regulations aim to reduce hearing damage caused by loud noise, and identify noise levels at which action is required. Employers must assess the exposure risks, and take action when the levels are reached.

The Electricity at Work Regulations 1989, as amended 1996 and 1997

These Regulations require precautions to be taken against risk of electrocution or injury from electricity. Equipment must be safe, and properly maintained, and in only exceptional circumstances should work be carried out on live systems.

The Health and Safety (Young Persons) Regulations 1997

These Regulations require that employers ensure that risk assessments for those under 18 years old take particular account of inexperience, immaturity and lack of awareness of relevant risk.

Prevention of Accidents to Children in Agriculture Regulations 1998 (PACAR)

These Regulations replaced the original regulations brought in in 1958 for the protection of children aged under 13 years. They remain unaltered in intent, and prohibit the riding on or driving of vehicles in agricultural operations. However, the key feature associated with these Regulations is the introduction of a comprehensive Approved Code of Practice.

Construction Health Safety and Welfare Regulations 1996

The Construction (Design and Management) Regulations 1994 (CDM)

Many building and structural activities on farms are subject to these Regulations, which require that health and safety is managed throughout all stages of the work, from design and planning to work on site, subsequent maintenance and repair.

Supply of Machinery (Safety) Regulations 1992

These Regulations implement the European Machinery Directive and require all UK manufacturers and suppliers of new machinery to ensure that the products that they supply are safe. Essential health and safety requirements (ESRs) are identified, (including provision of adequate instructions for safe use), and the circumstances for affixing a 'CE' mark as indication of conformity are outlined.

Food and Environment Protection Act 1985 (FEPA)

This is an enabling Act, in three parts. Parts 1 and 2 deal with contamination of food and deposits in the sea. Part 3 deals with pesticides and takes effect by means of Regulations, as follows:

The Control of Pesticides Regulations 1986 (as amended) (COPR)

These Regulations apply to pesticides (which include different types such as herbicides, fungicides, etc.) used in agriculture, horticulture and forestry; and to other uses outside agriculture. They apply to animal husbandry – but not to those treatments applied as veterinary medicines, for example sheep dip and warble fly dressing, which are separately covered by the Medicines Act 1968.

The Regulations prohibit the advertisement, sale, importation, supply, storage and use of a pesticide unless it has been approved. All manufacturers, importers and suppliers must obtain approval for each product. The approval system for agricultural use is the responsibility of MAFF (Pesticides Safety Directorate).

Anyone who advertises, sells, supplies, stores or uses a pesticide is affected by the Regulations. Certificates of Competence are required for anyone who stores approved pesticides for the purpose of sale, and by those selling or supplying approved pesticides. Similarly contractors and persons born after 31 December 1964 who apply approved pesticides must hold a Certificate of Competence. The Regulations also require any person who uses pesticides to confine the application to the land, crop, structure, material or other area intended to be treated.

Conclusions

The legal framework within which people have to work and which requires them to accept responsibilities for health safety and welfare is well developed. However, injuries still occur at unacceptable rates. The financial cost to the agricultural industry is immense yet it is believed that few are motivated to improve health and safety performance as a route to increased profit. It is interesting that times of business hardship are often blamed for lack of investment in health and safety provision, but as the statistics show, injury rates have changed little from times when business was relatively better. Similarly the totality of the cost of pain, grief and suffering to the injured, and those who either suffer the loss or have to support an injured person, is probably incalculable. It is also likely that much of the hidden costs of injury are uninsured, thus falling directly on the industry and individual.

In some circumstances the law has catered for particularly dangerous situations through the imposition of detailed requirements which effectively prevent personal judgement. The compulsory introduction of tractor safety cabs is an example which has reduced injuries from overturn, and noise-induced deafness, whilst at the same time creating a better ergonomic and general working environment. Similarly the detailed requirements for the formal approval and use of pesticides have led to protection of individual health and the environment.

However, it should be remembered that the law in itself does not produce safe and healthy working conditions. It is people who create risks, whether they be manufacturers or users of equipment or substances, or otherwise engaged in the diverse activities which make up agriculture. It follows that those who actually create risk should be responsible for eliminating or reducing that risk. Advances in technology and farming practices have radically changed the industry over the period that health and safety legislation has been developed in agriculture and it could have been assumed that the agricultural industry had taken on management of health and safety as an integral part of business. Unfortunately the injury picture does not seem to support that assumption. Many of the injury situations involve human error (or human factors) and highlight deficiences in appreciation of risk. It does seem that the risk management concept has not been applied to health and safety in the same way that it has been applied to other aspects such as finance, crop and animal husbandry, and machinery management. Training and educational fields are probably the areas where additional attention could be focused, and greater advantage taken of the numerous information, guidance sources and potential initiatives which could be developed with industry stakeholders. Long-term improvement in health and safety performance does depend on individuals and organizations within the industry itself having the motivation to improve, and to recognize and address the challenges – with the Regulator (HSE) acting as a catalyst to prompt action.

This brief description of health and safety in agriculture in Great Britain is by way of guidance only. It is not intended to be a comprehensive or an official authorative interpretation of Acts and Regulations, copies of which can be obtained through the Stationary Office or good booksellers.

Further reading

HMSO (1972) *Safety and Health at Work – Report of the Committee 1970–72 (Lord Robens)*. Ref. SBN 10150340. HMSO, London.

HSE (1998) *Successful Health and Safefy Management*. Ref. HSG65. HSE Books, Sudbury.

HSE (1999) *Five Steps to Risk Assessment*. Ref. HSG183. HSE Books, Sudbury.

HSE (1999) *Your Essential Guide to Health and Safety in Agriculture*. Ref. MISC. 165. HSE Libary, Sheffield.

HSE (2000) *Fatal Injuries in Farming, Forestry and Horticulture 1999–2000*. Ref.C12. HSE Books, Sudbury.

Further information

Information and advice on many health and safety aspects can be obtained from trade, professional and industry bodies, training and educational establishments, but the most comprehensive source is probably the Health and Safety Executive. Contact can be made at any of the regional or local offices listed in the telephone directories.

HSE produces priced and free publications and a catalogue, available by mail order from HSE Books, P.O. Box 1999, Sudbury, Suffolk, CO10 6FS. Many are also available from good booksellers. It also produces a range of videos for managers, workers, students and trainers in farming, forestry and horticulture. The current catalogue is available from HSE Videos, P.O. Box 35, Wetherby, Yorkshire, LS23 7EX. HSE also provide a helpline – 'Infoline', Tel.: 0541 545500 or write to HSE's Information Centre, Broad St., Sheffield, S3 7 HQ.

The widest potential information source is the HSE home page on the World Wide Web: http://www.open.gov.gov.uk/hse/hsehome.htm.

Part 3

Animal Production

16

Animal physiology and nutrition

R.M. Orr & J.A. Kirk

Introduction

Animal production involves the conversion of feedstuffs, mostly of plant origin, to animal products in the form of meat, milk, eggs and wool. In this chapter those physiological and metabolic processes which are especially relevant to this conversion are considered. The chapter is divided into nine sections:

(1) Regulation of body function
(2) Chemical composition of feedstuffs and animals
(3) Digestion
(4) Metabolism
(5) Voluntary food intake
(6) Reproduction
(7) Lactation
(8) Growth
(9) Environmental physiology and behaviour

Since animal production involves various species of farm animal, an attempt is made throughout to include reference to the comparative aspects of the subject. In addition, since animal productionists are concerned with the efficiency of the conversion of feedstuffs into products, reference is made where appropriate to this important aspect of livestock science.

Regulation of body function

In order that animals can survive their environment and changes in that environment, they must possess the necessary coordinated mechanisms to maintain their own internal environment in a steady state. These physiological reactions that maintain the steady states of the body have been designated 'homeostatic' and are achieved by the combined action of the various organ systems of the body. This coordinated action is carried out by the nervous and endocrine systems. These systems must initiate the necessary adjustments that enable animals to respond to external environmental changes such as the availability of food, temperature and light.

The nervous system

The nervous system in mammals is an extremely complex collection of nerve cells which plays an essential part in the functioning and behaviour of the animal. The basic units of the nervous system are the nerve cells or neurons and the associated receptors.

The ability of animals to maintain homeostasis is dependent on the nervous system receiving and responding to information from specialized receptor cells. To this end mammals have evolved receptors that are sensitive to a wide range of physical (e.g. temperature), chemical (e.g. specific components of the blood) and mechanical (e.g. muscle stretch) stimuli. When any response to information received at receptors is necessary, the information is carried from receptors to the cell or tissue (effector) where appropriate adjustments are made. This is the function of the neurons.

There are three different types of neurons:

(1) sensory or afferent neurons which carry information from receptors to the spinal cord and brain;
(2) motor or efferent neurons which carry information from the spinal cord and brain to the effector organ; and
(3) interneurons or association neurons which make connections between sensory and motor neurons within the spinal cord and brain.

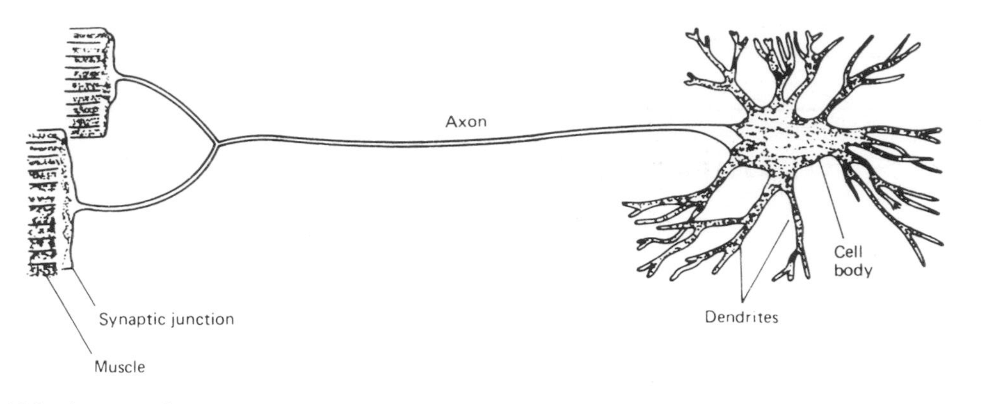

Figure 16.1 A nerve cell.

Neurons consist essentially of a cell body with an elongated projection, the axon, and numerous shorter projections, the dendrites (Fig. 16.1). For our purposes the cell body can be thought of as performing two functions. First, it receives information from the axons of other cells both directly onto its surface membrane and also via the dendrites and, secondly, it acts to sum all the effects of inputs, which may be either excitatory or inhibitory, that it receives. Should this sum of inputs exceed a critical level, then an impulse is triggered which is then carried along the axon to the synaptic region which in turn passes the neuronal signal to the next stage in the nervous system, either another neuron or, alternatively, a muscle or gland. Propagation of signals along an axon is dependent on changes in the electrical potential between the inside and the outside of the axon membrane which results from changes in the permeability of the membrane to calcium, sodium, potassium and chlorine ions. Passage of neuronal signals at synaptic regions involves the release of a chemical transmitter substance into the small gap between the synaptic region and the membrane of the next neuron. The nature of the transmitter substance determines whether the influence on the next neuron is excitatory (e.g. acetylcholine and noradrenaline) or inhibitory (e.g. γ-amino butyric acid). These transmitter substances released by the synapse are received by specialized sites on the dendrites of the next neuron which has the effect of changing the electrical potential of the neuron.

It is the nerve cells – the basic building blocks – from which the various components of the nervous system of mammals are built (Fig. 16.2). Division of the nervous system between central and peripheral is made simply on the basis of whether the fibres or cell bodies lie outside the spinal cord. The peripheral nervous system can be further subdivided into the somatic nervous system and the autonomic nervous system. The former is involved in the control of the skeletal muscles in the body whereas the latter controls specific target organs in the body such as the heart, lungs, blood vessels and intestines. The autonomic nervous system has two subdivisions, the sympathetic and the parasympathetic. In many cases these two divisions have opposite effects on the target organs and glands. The sympathetic nervous system acts to increase the level of activity or secretion in an organ whereas stimulation of the parasympathetic will have the opposite effect.

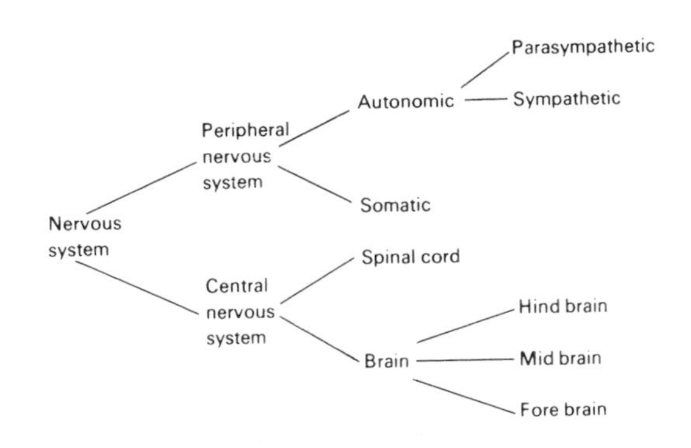

Figure 16.2 The nervous system of mammals.

The central nervous system consists of the spinal cord and brain. The spinal cord is contained in a continuous channel within the vertebrae, which form the backbone. At each vertebra there are two openings in the base through which nerves can pass in and out of the spinal cord. Sensory nerves enter through openings on the dorsal surface, motor nerves leave through openings on the ventral surface. The brain itself is a vastly complex organ consisting of billions of nerve cells, interconnected by neurons. However, certain areas of the brain (Fig. 16.3) have been found to have specific functions.

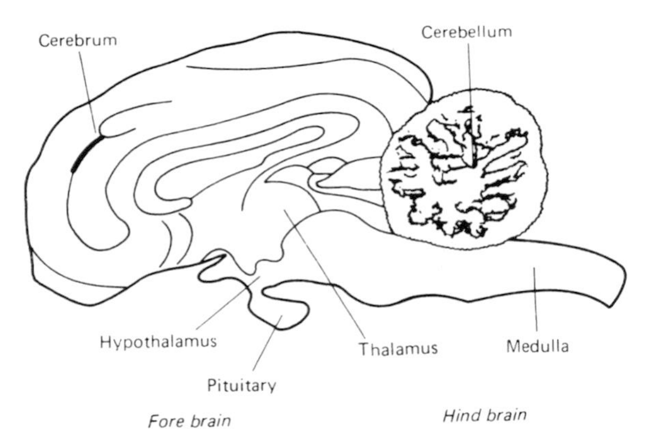

Figure 16.3 Midline section of the brain showing the location of major structures.

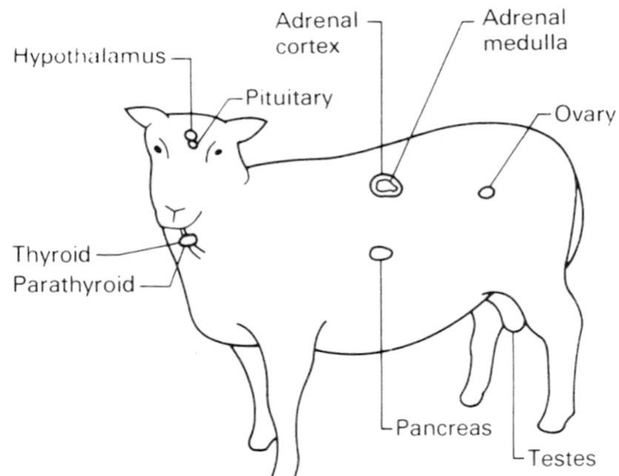

Figure 16.4 Location of the major endocrine glands.

(1) The hind brain: this consists of the medulla and the cerebellum. The medulla is involved in the regulation of the heart, breathing, blood flow and posture. The cerebellum is involved in the coordination of muscle activity.

(2) The fore brain: this consists of the thalamus, the hypothalamus and the cerebrum. The thalamus plays an important role in the analysis and transmission of sensory information between the spinal cord and the cerebral cortex. Lying directly beneath the thalamus is the hypothalamus which, despite its small size, has been shown to be involved in the control of basic behaviours such as hunger, thirst and sexual behaviour. The hypothalamus also has close connections with the pituitary – a hormone-secreting gland – which, under the control of the hypothalamus, secretes hormones into the bloodstream. In mammals, the cerebrum constitutes by far the major part of the brain. Definite areas of the cerebrum have been shown to control specific motor and sensory function.

The endocrine system

The endocrine system, like the nervous system, acts as a means of communication in the regulation of the physiological and biochemical functions of the body. It differs from the nervous system in that its actions are slower and more generalized. The endocrine system employs chemical messengers known as hormones. These substances act in an integrated fashion to regulate activities such as growth, reproduction, milk production, intermediary metabolism and adaptation to external environmental factors including light, temperature and stressors (stimuli that evoke a stress response).

The endocrine system consists of various endocrine glands, the locations of which are shown in Fig. 16.4. These glands produce two types of hormone: steroids (which are fat-soluble substances formed from cholesterol) and protein hormones (which are water-soluble). Hormones are secreted directly into the blood to be carried to target tissues or organs where they exert their effects, which may be either excitatory or inhibitory. The pituitary gland is said to be the 'master' endocrine gland because of the control it exerts over many of the other endocrine glands through the action of its trophic hormones, such as thyroid-stimulating hormone (TSH), adrenocorticotrophic hormone (ACTH), and the gonadotrophins, follicle-stimulating hormone (FSH) and luteinizing hormone (LH). The activities of the pituitary are in turn controlled by the region of the mid-brain known as the hypothalamus. Hypothalamic-releasing factors are responsible for stimulating or inhibiting the release of specific anterior pituitary hormones into the blood circulation. A complete list of the hormones produced by the various endocrine organs, their chemical nature, target tissues and main physiological effects are shown in Table 16.1.

The blood concentration of a given hormone will depend on its rate of secretion and rate of removal from the blood. Normally, hormones are continuously being secreted in small amounts and inactivated by the liver and kidney. The rate of inactivation tends to remain the same; therefore changes in blood concentrations of hormones are mainly due to changes in secretion rate, which are controlled either by the nervous system or by

Table 16.1 Endocrine structures and hormones

Endocrine structure	*Hormone*	*Chemical nature*	*Major activity*
Hypothalamus	Growth hormone releasing hormone (GHRH)	Polypeptide	
	Prolactin inhibiting hormone (PIH)	Polypeptide	
	Luteinizing hormone releasing hormone (LHRH)	Polypeptide	All are neurohormones regulating the release of the pituitary hormone indicated
	Follicle stimulating hormone releasing hormone (FSHRH)	Polypeptide	
	Thyrotropin hormone releasing hormone (TRH)	Peptide	
Anterior pituitary	Growth hormone (GH)	Protein	Stimulates growth
	Adrenocorticotrophic hormone (ACTH)	Protein	Stimulates release of hormones of the adrenal cortex
	Thyroid stimulating hormone (TSH)	Protein	Stimulates release of thyroxine
	Follicle stimulating hormone (FSH)	Protein	Regulates development of ovarian follicles in females and spermatogenesis in males
	Luteinizing hormone (LH)	Protein	Triggers ovulation and stimulates progesterone and testosterone release
	Prolactin	Protein	Stimulates production of milk by mammary gland
Posterior pituitary	Oxytocin	Peptide	Stimulates uterine contraction and involved in milk ejection
Adrenal gland			
Cortex	Vasopressin	Peptide	Reabsorption of water by the kidney tubules
	Aldosterone	Steroid	Regulates water and electrolyte balance
	Cortisone and corticosterone	Steroid	Regulator of carbohydrate metabolism
Medulla	Adrenaline and noradrenaline	Amino acids	Regulator of carbohydrate metabolism in muscle and liver, constriction of peripheral vessels and contraction of smooth muscles
Thyroid	Thyroxine	Amino acid	Stimulates oxidative metabolism
	Calcitonin	Polypeptide	Regulates calcium levels in body fluids (decreases)
Parathyroid	Parathyroid hormone	Polypeptide	Regulates calcium levels in body fluids (increases)
Pancreas	Glucagon	Polypeptide	Raises blood glucose levels
	Insulin	Polypeptide	Lowers blood glucose levels
Digestive tract			
Duodenum	Secretin	Polypeptide	Stimulates release of pancreatic enzymes
	Cholecystokinin	Polypeptide	Stimulates release of pancreatic enzymes and regulates the gall bladder
Stomach	Gastrin	Peptide	Stimulates secretion of HC1 and pepsin by gastric mucosa
Testes	Androgens (e.g. testosterone)	Steroids	Stimulates development of secondary sex characteristics in male; maintains accessory sex organs
Ovary	Oestrogens	Steroids	Stimulates development and maintenance of female secondary sexual characteristics
	Progesterone	Steroid	Stimulates uterus in preparation for ovum implantation; development of mammary gland for lactation
	Relaxin	Polypeptide	Relaxation of pelvic tissues at parturition

the composition of hormones and other chemicals in the blood. Hormones control their own secretion rates or those of other hormones by means of a feedback mechanism which is either positive or, more commonly, negative feedback.

The action of hormones on their target tissues is not well understood. The initial interaction of a hormone with its target cell is for it to become attached to a specific receptor either at the cell membrane or within the cytoplasm. The number of receptor sites and the proportion of those that are occupied are thought to determine the degree of responsiveness of the target tissue. Hormones are thought to act on cells in one of two ways: (1) by changing the activity of critical enzymes in the cells' metabolic pathways; and (2) by changing the transportation of substances across the cell membrane (for example insulin increases the uptake of glucose into the cell).

Since the endocrine system plays such a fundamental role in the control of animal functions, there has been considerable attention in recent years on the use of natural and synthetic hormone preparations in animal science, particularly in the areas of promoting increased growth and the regulation of reproductive processes.

Chemical composition of the animal and its food

The animal and its food contain the same types of chemical substances. These constituents may be conveniently illustrated by Fig. 16.5. Comparisons of the chemical composition of the animal and its food (Table 16.2) show that the predominant fraction in the dry matter of most foods is the carbohydrate fraction, although in some foods protein predominates. In contrast, the animal body contains very little carbohydrate and is largely composed of protein and fat. Thus, in simplistic terms, animal production involves the conversion of inputs that are mainly in the form of carbohydrates with lesser quantities of protein into outputs that are mainly composed of fat and proteins. The minerals (ash) and vitamins are present in relatively small proportions and are collectively termed micronutrients.

Proteins

One of the most characteristic chemical features of all organisms is their content of proteins. In both plants and animals, proteins perform the important function of enzymes which are responsible for metabolic processes. In the animal they act as the structural proteins of the skin (collagen and keratin), connective tissues, tendons and bones (collagen) and muscle (actin and myosin), as hormones (e.g. insulin), as antibodies and as substances performing transport and osmotic functions in the blood (e.g. haemoglobin, albumin).

The fundamental units from which all proteins are constructed are amino acids. All amino acids contain the elements carbon, hydrogen, oxygen and nitrogen. The nitrogen content of most proteins varies between 15 and 17% with an average of 16%. Measurement of protein content of foods is based on estimation of nitrogen content. Traditionally protein content is expressed as crude protein (CP) where $CP = N \times (100/16)$. Three amino acids, cystine, cysteine and methionine, also

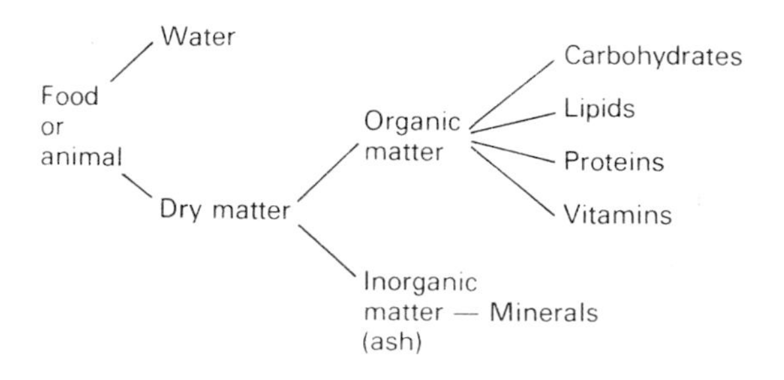

Figure 16.5 Composition of the animal and its food.

Table 16.2 Composition of the animal and feedstuffs (%)

	Water	*Carbohydrate*	*Protein*	*Fat*	*Ash*
Pig, 30 kg	60	<1	13	24	2.5
Adult cow	58	<1	16	20	4
Hen	56	<1	21	19	3
Grass – leafy	80	12	5	1	2
Wheat straw	10	75	3	2	9
Wheat grain	13	72	12	2	2
Turnips	90	7	<1	1	1
Soyabean meal	12	35	45	2	6

Table 16.3 Major amino acids occurring in proteins

Essential	*Non-essential*
Arginine	Alanine
Histidine	Aspartic acid
Isoleucine	Citrulline
Leucine	Cystine
Lysine	Glutamic acid[1]
Methionine	Glycine
Phenylalanine	Hydroxyproline
Threonine	Proline[1]
Tryptophan	Serine
Valine	Tyrosine

[1] Amino acids also essential to the chick.

contain the element sulphur. Thus the sulphur content of proteins is a function of the proportion of sulphur-containing amino acids present, e.g. wool has a high sulphur content reflecting its high cystine content. Proteins are synthesized from a pool of 20 different amino acids (Table 16.3). These conform to the general formula

$$H_2N - \overset{\displaystyle R}{\underset{\displaystyle H}{\overset{|}{\underset{|}{C}}}} - COOH$$

in which a central carbon atom has attached to it an amino group ($—NH_2$), a carboxylic acid group (—COOH), a hydrogen atom and a variable side chain designated here by the letter R. The condensation of the amino group of one amino acid with the carboxyl group of the next, forming peptide bonds, is the mechanism by which proteins are produced. Proteins generally contain several hundred amino acid residues and are thus large molecules with molecular weights ranging from 35 000 up to several hundred thousand. Proteins are synthesized in plant and animal cells where the cell nucleus contains genetic material which determines the nature of the newly synthesized protein. The physical and chemical characteristics of proteins are altered by the different proportions of amino acids, the sequence in which they occur in the protein, the degree of cross-linking that occurs between different parts of the molecule and by the presence of other compounds in their structure. For example, some proteins contain lipids (lipoproteins), carbohydrates (glycoproteins) or mineral elements (haemoglobin, casein). From the viewpoint of the nutritionist the proportion of the different amino acids present in a protein and those characteristics of the protein which may influence their availability to the animal are of primary importance.

Whereas plants and micro-organisms are capable of synthesizing all of the different amino acids found in proteins provided they have an adequate supply of inorganic nitrogen and organic compounds capable of supplying the other elemental components, higher animals do not have this capability. Thus, some amino acids are required in the diet of most animals and are referred to as essential amino acids. Those not specifically required in the diet are called non-essential amino acids (see Table 16.3).

The ability of a feed protein to furnish the animal with essential amino acids is thus a parameter that particularly influences its value to the animal. In practical animal nutrition the amino acids most likely to be limiting in their supply are lysine, methionine and tryptophan. Ruminants do not require dietary amino acids to the same extent as monogastric species such as pigs and poultry. This is because the microflora in their digestive tracts are capable of synthesizing amino acids (both essential and non-essential) from the simple organic compounds found in the rumen. This supply of amino acids is, however, inadequate to meet the demand of high-producing animals.

The ease with which food proteins may be hydrolysed during digestion may influence their nutritional value. This is influenced by the solubility characteristics of the protein. Feed processing techniques (e.g. heat or formaldehyde treatment) which bring about denaturation of the physical structure of a protein may influence its solubility characteristics.

In ruminant nutrition the proportion of dietary protein hydrolysed by the rumen microbial population is referred to as its degradability. Protein hydrolysed by rumen microorganisms is termed rumen degradable protein (RDP) and that which is resistant to hydrolysis is undegradable protein (UDP).

A comparison of the crude protein, rumen degradability parameters and essential amino acid content of animal feeds is given in Table 18.3.

Carbohydrates

Carbohydrates are the main product of photosynthetic activity in plants and contribute up to 80% of the dry matter in many feedstuffs. The nutritive value of these carbohydrates to the animal is variable and dependent on both the chemical structure of the carbohydrate and the digestive capacity of the animal. Generally speaking, the variety of carbohydrate types in plants is far wider

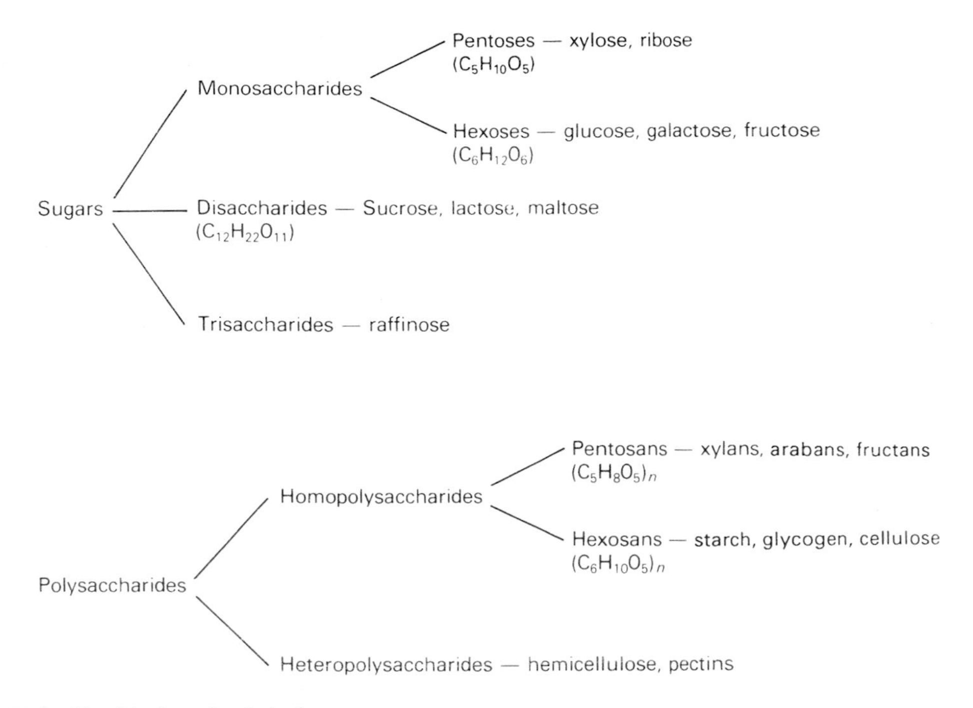

Figure 16.6 Classification of carbohydrates.

Table 16.4 Carbohydrate content of plant tissues (% of DM). From Van Soest, P.J. (1982) *Nutritional Ecology of the Ruminant*, O & B Books Inc., Corvallis, with permission.

Component	*Tropical grasses*	*Temperate grasses*	*Cereal seeds*	*Alfalfa*	*Green vegetables*
Sugars	5	10	Negligible	5–15	20
Fructans	0	1–25	0	0	—
Starch	1–5	0	80	1–7	Low
Pectin	1–2	1–2	Negligible	5–10	10–20
Cellulose	30–40	20–40	2–5	20–35	20
Hemicellulose	30–40	15–25	7–15	8–10	Low

than in animals where it accounts for only a very small proportion of the body composition.

Carbohydrates are composed of the elements carbon, hydrogen and oxygen. The general formula $C_n(H_2O)_n$ may be used to represent any carbohydrate and indicates that the ratio of hydrogen to oxygen atoms is always 2 : 1, as in water. A classification of commonly occurring carbohydrates is given in Fig. 16.6 and the carbohydrate content of a range of plant tissues is shown in Table 16.4.

From the point of view of the nutritionist a distinction may be made between those carbohydrates that occur within plant cells, either as simple sugars or storage reserve compounds (e.g. starch, sucrose and fructans), and those that occur in the cell wall performing a structural role (e.g. pectins, cellulose, hemicellulose). Such a distinction is based on the digestibility characteristics of carbohydrates, the former being readily hydrolysed whereas the latter are relatively resistant, their degradation being largely a function of microbial activity in the digestive tract.

The non-structural carbohydrates of plant material can be further categorized in terms of their cold-water solubility. The term 'water-soluble carbohydrates' refers to the monosaccharides, oligosaccharides and some polysaccharides, principally fructans, and distinguishes them from the starches that are the principal component of most seeds. Cereal grains are the major source of starch to farm animals.

Although small amounts of various free monosaccharides may be detected in feeds (e.g. glucose, fructose), most are of sufficiently low concentrations to be of little importance in nutrition. Sucrose is the main sugar in the sap of plants and in the case of root crops serves as the primary form of energy storage. Many plants convert sucrose into other forms for storage. Temperate grasses store fructans in leaves and stems and starch in the seed. There are two general types of fructans, the levans that occur in grasses, and inulins that are characteristic of the Compositae, e.g. the Jerusalem artichoke. Fructan content may account for up to 25% of dry matter in perennial ryegrasses, especially in cool growing seasons.

The water-soluble carbohydrate of forages is especially important in the ensilage process (see Chapter 4) and may also influence their palatability. The water-soluble carbohydrate content is chiefly influenced by the physiological conditions of growth; high light intensity and photosynthetic rate increase the content. Hence marked diurnal variation in the water-soluble carbohydrate content occurs in the living plant. Respiration of cut and drying forage may markedly reduce sugar content.

Starch is the most important storage carbohydrate in plants. Two types of starch exist, amylose and amylopectin. Both have crystalline structures which are disrupted by heating. The temperature at which this occurs is called the gelatinization temperature and these changes in the starch structures upon moist heating are partly responsible for the improvement in utilization resulting from steaming, flaking, micronizing and pelleting. Physical processing of the gelatinized starch is often required to prevent recrystallization (retrogradation). Excessive heat treatment may cause caramelization of the carbohydrates which has a detrimental effect on utilization. The conflicting results that have been obtained in feeding studies of processed grains are most likely due to these interacting effects of gelatinization, retrogradation and caramelization.

Of the structural carbohydrates the most abundant are the celluloses along with lesser quantities of hemicelluloses, principally xylans. The former are polymers of glucose, the latter polymers of xylose. The proportion of these carbohydrates in the plant increases as it matures. Their nutritional availability to the animal varies from total indigestibility to complete digestibility and depends on the animal to which it is fed and on the degree of lignification. Differences between animals in their ability to utilize these carbohydrates are largely a function of microbial activity in their digestive tracts – ruminants and to a lesser extent other herbivores (e.g. horses, rabbits) having a greater capacity than monogastric species.

Lignin is a complex substance, a phenylpropanoid polymer of high molecular weight which associates with the structural carbohydrates in the cell wall to form an amorphous matrix. It is particularly resistant to degradation and is the main factor limiting digestibility of forages. The lignin content of plant cell walls increases with maturity and accounts for the reduction in digestibility of herbage as it matures. Cereal straws are examples of highly lignified feeds. Chemical treatment of cereal straws with alkalis such as sodium and ammonium hydroxide breaks the mainly ester bonding between lignin and the structural carbohydrates in the cell wall and increases the susceptibility of the cell wall to degradation in the digestive tract of animals. The efficiency of the treatment is dependent upon the proportion of lignin—carbohydrate bonds that are broken.

In animal tissues the main carbohydrates represented are glucose, which features in the energy metabolism of animals, and glycogen which is synthesized in muscle and liver from glucose and acts as a readily available form of energy storage. The amount of glycogen and its rate of metabolism are of particular relevance to changes which occur in the muscles of meat animals after slaughter. Lactose is a disaccharide synthesized in the cells of the mammary gland, representing over 95% of the total carbohydrate present in milk.

Lipids

The lipids are a diverse group of substances which share the common property of being relatively insoluble in water and readily soluble in organic solvents such as ether or chloroform. Most animal feeds contain up to 5% of lipid in the dry matter whilst the animal body may contain up to 40% of lipids. A variety of different types of lipids are found in plant and animal tissues, performing a range of important biochemical or physiological functions. A chemical classification of lipids and some of their functions are given in Fig. 16.7.

Fats and oils are constituents of both plants and animals and account for about 98% of all naturally occurring lipids. They have the same chemical structure and properties, differing in that oils occur in liquid form in plants, and fats chiefly in the solid form in animal tissues. Structurally, they are mainly triglycerides (neutral fats) in which three fatty acids are joined by ester linkages to glycerol. The fatty acids have the general formula $CH_3(CH_2)_nCOOH$. As well as differing in chain length – naturally occurring fatty acids are mainly of chain length 4–18 carbons — they also differ in their degree of unsaturation (Table 16.5). Most naturally occurring unsaturated fatty acids are of the *cis* configuration at the double bond although milk and

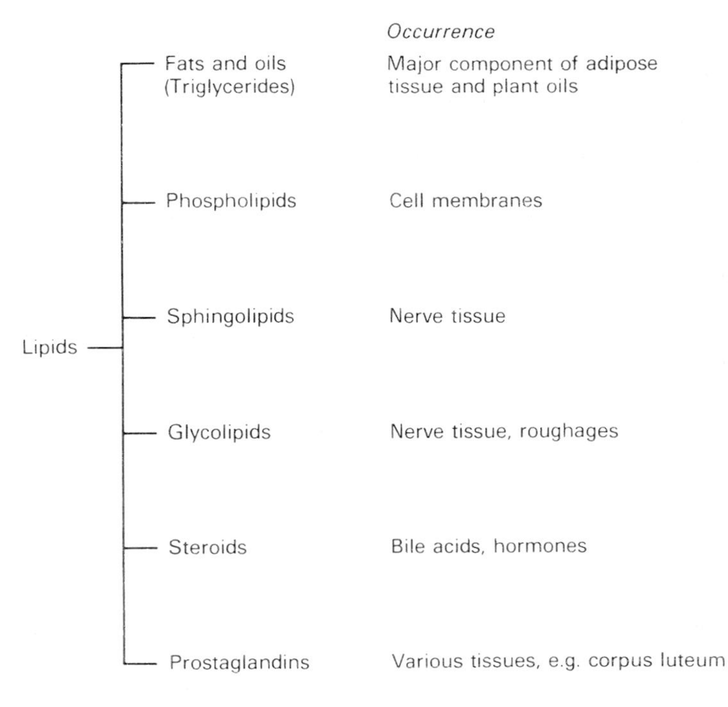

Figure 16.7 Chemical classification of lipids.

body fat from ruminant animals contain significant amounts with the *trans* configuration.

In most triglycerides there is a mixture of fatty acids, and in both fats and oils there is a mixture of triglyceride types. Animal body fats are characterized by a high proportion of saturated and mono-unsaturated fatty acids – particularly stearic and oleic – plant and fish oils by a high proportion of polyunsaturated fatty acids (Table 16.6). Three of the fatty acids, linoleic, linolenic and arachidonic acids, are known as the essential fatty acids since they cannot be synthesized by the animal body and must be provided in the diet from vegetable sources. However, arachidonic acid can be formed from dietary linoleic acid. The essential fatty acids are precursors of the prostaglandins.

The energy content of triglycerides is considerably greater than that of other nutrients – whereas 1 g of carbohydrate has a gross energy content of about 16 kJ, 1 g of triglyceride has a gross energy content of about 40 kJ. Thus fat may be included in animal diets as a means of increasing their energy density. Triglyceride, unlike the glycogen of liver and muscles, can be laid down in virtually unlimited amounts in adipose tissue (white fat) and serves as a more economical means of energy storage. Much of the white adipose tissue of the body is deposited under the skin (subcutaneous) with smaller proportions of the body's fat to be found around the kidneys, between muscles (intermuscular fat), and

Table 16.5 Commonly occurring fatty acids. The first number after C indicates the number of carbon atoms and the second number indicates the number of double bonds present. Δ Indicates position of double bond from—COOH end of chain

	Abbreviated designation
Saturated fatty acids	
Acetic	C2:0
Propionic	C3:0
Butyric	C4:0
Caproic	C6:0
Caprylic	C8:0
Capric	C10:0
Lauric	C12:0
Myristic	C14:0
Palmitic	C16:0
Stearic	C18:0
Arachidic	C20:0
Unsaturated fatty acids	
Palmitoleic	C16:1 Δ^{9}
Oleic	C18:1 Δ^{9}
α-Linoleic	C18:2 $\Delta^{9,12}$
Linolenic	C18:3 $\Delta^{9,12,15}$
Arachidonic	C20:4 $\Delta^{5,8,11,14}$
Ecosopentaenoic (EPA)	C20:5 $\Delta^{5,8,11,14,17}$
Docosohexaenoic (DHA)	C22:6 $\Delta^{4,7,10,13,16,19}$

Table 16.6 Fatty acid composition (%) of some common fats and oils

Fatty acid	*Palm*	*Soyabean oil*	*Rapeseed oil*	*Butter*	*Beef tallow*	*Lard*	*Fish oil*
C14 and less	1	1	—	15	2	1	5
C16:0	45	12	5	23	35	32	13
C18:0	4	4	2	9	16	8	4
C18:1	40	27	56	35	44	48	22
C18:2	10	50	25	3	2	10	6
C18:3	—	6	9	—	0.4	1	4
C20 and above	—	—	3	—	—	—	37

between and within muscle fibres. Species and strain of animals are important factors determining the distribution of fat depots. For example, 'beef' breeds of cattle have lower proportions of internal fat depots than 'dairy' breeds of cattle. The composition of the body fat may be influenced by the type and amount of dietary fat consumed.

The membranes of cells and intracellular organelles in animals are largely comprised of a group of lipids termed phospholipids characterized by their phosphoric acid component. The most commonly occurring are the lecithins which also contain the water-soluble vitamin choline. The phospholipids are higher in unsaturated fatty acids than the triglycerides of adipose tissue. Arachidonic acid is especially prominent and acts as a depot of this metabolically important fatty acid. The phospholipids are, along with another group of compound lipids, the sphingolipids, major components of nervous tissue. As substances with emulsifying properties phospholipids fulfil important functions in lipid transport in the blood in which they combine with simple fats, cholesterol and proteins to form lipoproteins. Blood lipoproteins are classified on the basis of their density which is a reflection of the proportions of lipid and protein in their make-up.

Whereas triglycerides are the major lipid components of concentrate feeds and seeds, the lipids in roughages are mainly in the form of glycolipids which in addition to containing fatty acids, chiefly linoleic and linolenic, have a carbohydrate component in the form of galactose.

The steroids are a large group of physiologically important compounds in plants and animals. All are derivatives of cyclic alcohols, the parent compound being the sterol nucleus.

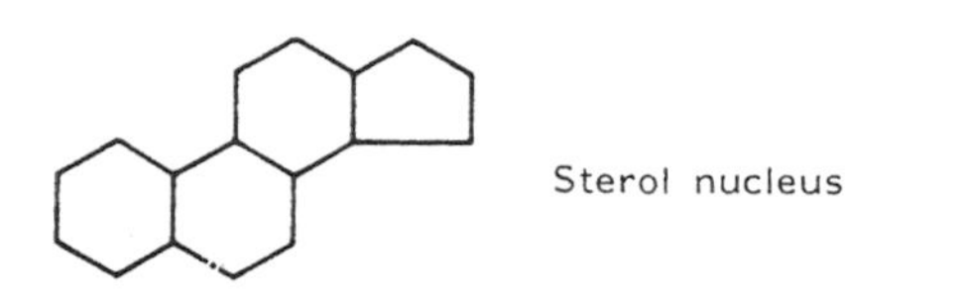

In animal tissues the sterol cholesterol is the most common, occurring in cell membranes, nervous tissue and the blood. It is important as a precursor of such substances as the bile acids, sex hormones (oestrogens, androgens and progestins), hormones of the adrenal cortex (cortisol, corticosterone, aldosterone) and vitamin D. In recent years cholesterol has received much attention in the context of levels in the blood being associated with the condition of coronary heart disease in humans.

Micronutrients

Micronutrients are dietary components that need to be present in only relatively small quantities compared with carbohydrates, lipids and proteins. There are two broad groups of micronutrients: the minerals, which are elements other than carbon, hydrogen, oxygen and nitrogen required by the animal, and the vitamins, which are organic compounds required for particular body functions. Details of the occurrence of individual micronutrients in feeds and their role in the animal are given later (see 'Mineral nutrition and metabolism').

Water

Water is contained in all plant and animal cells. In foods the water content varies widely from over 90% in certain roots to 10–15% in dried feeds such as cereal grains and hay. Such variation in the water content of feeds makes it essential that comparisons of the nutritive value are made on a dry matter basis. The water content of the animal body is also highly variable. Since muscle tissue contains some 75% of its weight as water whereas fat contains only 12–15%, the water content of the whole animal varies with the proportions of these two tissues, being highest in young, lean animals and least in mature, fat animals.

Water is obtained by drinking and from food. Additionally, water is produced by metabolic processes during the oxidation of nutrients within the body. This is termed metabolic water. The oxidation of 1 kg carbohydrate yields 0.6 kg water whilst for every kilogram of fat oxidized, 1.1 kg water is produced. Metabolic water production is usually of the order of about 5–10% of the total water consumed. Animals lose water through faeces and urine, water vapour in expired air and through skin. Milk represents a major loss in lactating animals, whilst the pregnant animal deposits large quantities of water within the uterus particularly in late pregnancy.

Within the animal, body water is the major component of all body fluids and internal secretions. In the ruminant especially, large amounts of water are secreted into the digestive tract through saliva and digestive juices. Although the bulk of this water is reabsorbed from the digestive tract, faecal water losses are considerably higher in ruminants than in other species. The high specific heat and latent heat of evaporation of water aids in the regulation of body temperature, whilst chemically water takes part in hydrolytic reactions and in the absorption of digested nutrients.

Water intake and requirements are influenced by dietary and environmental factors. Thus water intake is related to dry matter intake and to the composition of the dry matter, especially in relation to the mineral salt content. The levels of sodium chloride and to a lesser extent the protein content of the diet influence urinary excretion and therefore requirements for water. Increases in the environmental temperature which increase water losses through respiration and sweating result in greater water consumption. All stock need adequate supplies of water at all times; insufficient water means reduced DM intake with resultant depression in productivity.

Digestion

The process of digestion breaks down complex food materials in the diet to simple products which can be readily absorbed from the digestive tract into the blood and lymph.

Anatomy of the digestive tract

Farm mammals may be divided into either simple-stomached (monogastric) animals such as the pig, or ruminants which possess a compound stomach of four regions, cattle and sheep being the main examples.

The monogastric digestive tract

The arrangement of the monogastric digestive tract is shown in Fig. 16.8. Food is digested by physical and chemical means while in the digestive system. The process begins in the mouth with the physical breakdown of food by the teeth. Saliva from the salivary glands helps the mastication and swallowing of the food. Movement of food through the digestive tract is brought about by waves of alternate contraction and relaxation of the muscular layers of the digestive tract wall, an action that is termed peristalsis.

Chemical digestion in the stomach begins with the secretion of acidic gastric juices. The digesta passes from here into the duodenum and the rest of the small intestine, where it is digested further by the actions of bile, pancreatic juice and intestinal juice. A more detailed description of the chemical processes involved is given later in this section. Digestive products are absorbed mainly in the small intestine. The large intestine is chiefly concerned with absorption of large quantities of water from the digestive waste, or faeces, before it is voided from the body (defaecation).

The ruminant digestive tract

The ruminant stomach, unlike that of the monogastric animal, is divided into four compartments: rumen, reticulum, omasum and abomasum. The first three of these

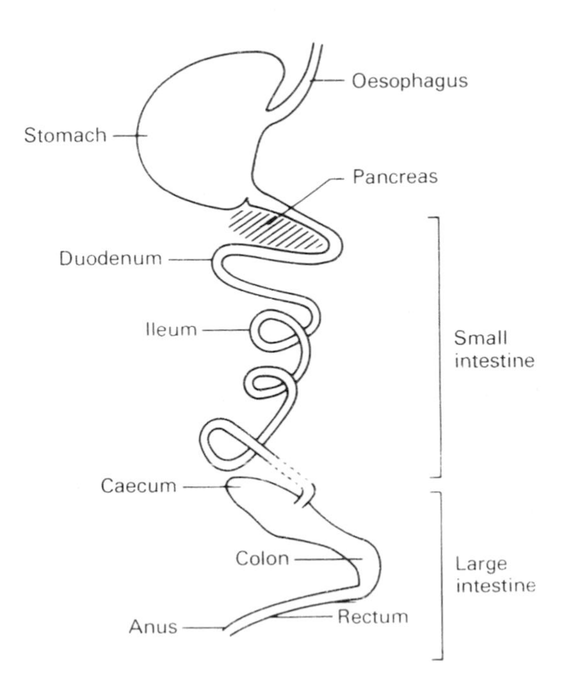

Figure 16.8 Diagrammatic representation of a monogastric digestive tract.

are known as the forestomachs and as such have no digestive glands. The rumen is the largest of these compartments, comprising more than 60% of total stomach capacity. The fourth stomach, the abomasum, closely resembles the structure and function of the stomach of the monogastric animal. The passage of food through the ruminant stomachs is somewhat complicated (see Fig. 16.9). After swallowing, food enters the rumen to be mechanically mixed and subjected to fermentation by the host population of microbes. Food may spend 48 hours or longer in the rumen. From time to time, coarse material is regurgitated into the mouth to be rechewed (chewing the cud) and then returned to the rumen for further digestion. The more fibrous the diet, the greater the rumination time, which may be as much as 8 hours a day. In total, a cow may regurgitate 50 kg of food per day. Eventually the finer material in the rumen passes into the omasum, where water is extracted, and then on to the abomasum. Here, any bacterial activity within the digesta is inhibited by the acidity of the gastric secretions. Microbial protoplasm is digested by the proteolytic enzymes present. From this stage onward the ruminant digestion resembles that of the pig.

The pre-ruminant digestive tract

At birth the rumen and reticulum are very much underdeveloped compared with the adult; thus the animal is said to be at a pre-ruminant stage. Consequently, little or no rumen fermentation takes place in newborn animals. Instead the liquid milk diet bypasses the rumen-reticulum and goes directly to the abomasum, where milk proteins are clotted and partially digested. The rumen bypass is made possible by the presence of an oesophageal groove running from the oesophagus to the omasum. Suckling or the drinking of a liquid diet stimulates the groove to close over into a muscular tube so that the contents are prevented from entering the rumen or reticulum. As the young animal is slowly introduced to a more solid diet, the reflex closure mechanism gradually diminishes, thus allowing solid material to enter the rumen so that fermentation can commence (see description of calf rearing in Chapter 19).

Biochemistry of digestion

Few of the dietary nutrients present in an animal's food can be directly absorbed by the animal and must therefore be modified to make them available. This is accomplished in the digestive tract by enzymes either produced and secreted into the digestive tract by the host animal or of microbial origin, the micro-organisms, living in symbiotic association with the animal. The former is most characteristic of monogastric species whereas the latter is characteristic of the ruminant. Neither monogastrics nor ruminants are, however, solely reliant on either host-produced or microbial enzymes for nutrient breakdown.

Carbohydrate digestion

Monogastrics

Only non-structural plant carbohydrate such as starch and the various sugars are subject to hydrolysis by

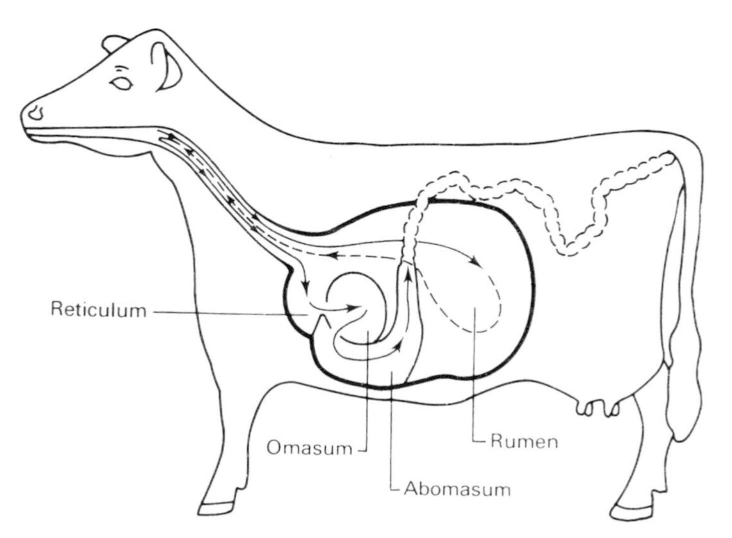

Figure 16.9 Diagrammatic representation of the pathway of food through the ruminant digestive system of the cow.

digestive secretions. There are no enzymes present in digestive secretions capable of hydrolysing cellulose and hemicellulose.

Saliva contains significant quantities of only one enzyme, α-amylase, and like pancreatic amylase this enzyme initiates the breakdown of starch into a mixture of maltose and dextrins. With an optimum pH of 7, salivary amylase activity is short-lived since as the food reaches the stomach the pH falls to a value of around 2. In the duodenum the food mixes with pancreatic and duodenal secretions and bile, the pH rises and conditions become suitable for the action of pancreatic amylase which continues the breakdown of starch principally to maltose.

The mucosal cells of the small intestine produce a number of carbohydrases. These are located on the brush border of the mature cells and include maltases, lactase and sucrase. Lactase activity is maximal in the suckling animal and declines as milk makes a declining contribution to the diet. These carbohydrases can hydrolyse the appropriate simple sugars to their monosaccharide constituents which are then readily absorbed from the small intestine, mainly by 'active transport' mechanisms. Thus the main end product of carbohydrate digestion is glucose.

Cell wall carbohydrates pass through the small intestine of pigs and poultry and reach the caecum where they are substrates for the microbial population that exists in this region. This microbial population can ferment cellulose and hemicellulose and any starch evading breakdown in the small intestine to volatile fatty acids and gases in a similar fashion to ruminal fermentation (see next section), but it is quantitatively limited. These volatile fatty acids are absorbed from the rear gut and have in the case of the pig been estimated to supply between 10 and 20% of the absorbed energy. In monogastric herbivores such as the horse the hindgut is enlarged to allow greater contribution from microbial fermentation.

Ruminants

All the dietary carbohydrates are subject to some fermentative degradation in the rumen. The major end-products of this fermentation are volatile fatty acids (VFA), mainly in the form of acetic, propionic and butyric acids, along with the gases carbon dioxide and methane. VFA production represents up to three-quarters of the effective energy value of the diet. The rate at which dietary carbohydrates are fermented to these end-products is dependent on type, soluble carbohydrate being fermented more rapidly than starches which in turn are fermented more rapidly than cell wall carbohydrates.

The concentration of VFAs in the rumen is a function of their rates of production and rates of absorption. Absorption of VFA is in the free form without active transport. The acids produced during fermentation are partially neutralized by the buffers present in the saliva. Where diets containing high quantities of readily fermentable carbohydrates are fed, acid conditions may result. Such lowered pH conditions can interfere with rumen fermentation and may lead to acidosis in the host (see 'Rumen-related metabolic disorders'). Where such diets are being fed, violent fluctuations in rumen pH may be prevented by more frequent feeding, the inclusion of certain minimum amounts of long roughage to induce greater production of saliva, and the inclusion of agents such as sodium bicarbonate in the diet.

The proportion of the various types of VFA produced by the fermentation of carbohydrate is dependent on diet. Typical values for a range of diets are given in Table 16.7. The acetic + butyric/propionic (non-glucogenic VFA/glucogenic VFA) ratio is of particular relevance to the efficiency of dietary energy utilization, and efforts to manipulate this ratio are a feature of ruminant production. For example, the inclusion of ionophore-type and other antibiotics such as monensin to feeds can increase the proportion of propionic and

Table 16.7 Effect of diet on the molar proportions of VFA in rumen liquor

Diet	*Molar % of VFA*			
	Acetic	*Propionic*	*Butyric*	*Higher*[1]
Grass silage	76	13	7	4
Fresh grass	63	20	13	4
60% hay: 40% concentrates	60	25	12	3
40% hay: 60% concentrates	55	30	10	5
20% hay: 80% concentrates	51	36	10	3
10% hay: 90% concentrates	45	39	11	5

[1]Higher VFAs include valeric, isovaleric and caproic acids.

thus narrow the ratio, whereas increasing the frequency of feeding promotes an increase in the proportion of acetic and thus a wider ratio. The former change is of benefit to the efficiency of growth in animals. Differences in the proportions of VFA end-products are a function of the type of microbial population present in the rumen in that different microbial species use different nutrients as their principal substrates. Thus, for example, whereas *Bacillus amylophilus*, utilizes starch, *Bacillus succinogenes* is a principal utilizer of cellulose.

Under some circumstances a portion of the fermentable carbohydrates may escape rumen fermentation. This is mainly in the form of structural carbohydrate which is sufficiently lignified to prevent microbial degradation. Further fermentation of this fraction may take place in the caecum, leading to a further source of VFA to the animal, but in practice the extent of this is very limited. Under some circumstances dietary starch may also escape rumen fermentation and reach the lower gut. This is a possibility when high-cereal diets processed in particular ways are fed (e.g. finely ground maize). This starch can be digested in the small intestine in the same way as occurs in monogastric species with glucose as the principal end-product. Although this starch is used more efficiently, the small intestine of the ruminant is more limited in its carbohydrase activity than the monogastric. Gas production resulting from the fermentation of carbohydrates is lost by eructation and in energy terms accounts for about 7% of the food energy going into the rumen. This is in the form of methane and results from the reduction of carbon dioxide by hydrogen. A wide range of compounds have been shown to be capable of depressing methane production with the aim of increasing the efficiency of the fermentation. Although many of these are unsuitable to be used commercially it is one of the modes of action of ionophore-type antibiotics.

Protein digestion

Monogastrics

The enzymes concerned with protein digestion may be considered either as exopeptidases (enzymes that hydrolyse peptide bonds at the ends of polypeptides) or endopeptidases (those that hydrolyse peptide bonds within a polypeptide). The exopeptidases are either carboxypeptidases or aminopeptidases; the endopeptidases include pepsin, trypsin, chymotrypsin and dipeptidases. With the exception of pepsin, which operates in the acid environment of the stomach, the protein-digesting enzymes are to be found in the small intestine. Trypsin and chymotrypsin are secreted in the pancreatic juices; the other enzymes are secreted from cells within the mucosa of the small intestine. A feature of these proteolytic enzymes is their specificity for peptide bonds involving particular amino acid types. For example, chymotrypsin has a preference for peptide bonds involving the carboxyl group of aromatic amino acids.

The end result of such proteolytic activity is the hydrolysis of dietary protein to free amino acids. These amino acids are absorbed by active transport into cells lining the small intestine and subsequently enter the portal blood and are transported to the liver. Small quantities of short peptides may also enter the absorptive cells of the small intestine where they are hydrolysed to allow free amino acids to enter the blood.

During the passage of food through the gut, considerable quantities of endogenous protein are added in the form of digestive secretions and desquamated mucosal cells. This protein is itself digested and absorbed like dietary protein and complicates the assessment of dietary protein requirements.

Ruminants

In the ruminant animal consideration must include not only the digestive utilization of dietary protein but also the utilization of other sources of dietary nitrogen, termed non-protein nitrogen (NPN) (Fig. 16.10). Although it is technically more correct to refer to both the true protein and NPN in terms of 'nitrogen', feed manufacturers and farmers tend to use the concept of crude protein (N in feed × 6.25).

Within the reticulo-rumen, microbial activity results in the degradation of all the NPN and large and variable amounts of true protein. Ammonia is produced from these degradative processes. Collectively the NPN and true protein degraded in the rumen are termed the *rumen degradable crude protein* (RDP) fraction of the feed. Rumen degradable crude protein can be further partitioned into a *quickly degraded crude protein* (QDP) fraction and a *slowly degraded crude protein* (SDP) fraction. Quickly degraded crude protein includes the NPN and all the true protein that is quickly and completely degraded to ammonia. This is essentially the feed nitrogen that is readily soluble in cold water. Slowly degraded crude protein describes the fraction that releases ammonia at a slower rate and is essentially the total RDP in the feed minus the QDP fraction. It should be noted that this SDP fraction has a variable rate of degradation and is quantitatively influenced by the effect of level of feeding on its rate of outflow from the rumen. A further component of dietary crude protein, that fraction which is resistant to degradation by microbial activity in the rumen, is termed *undegradable crude protein* (UDP). The ratios of RDP/UDP and of QDP/SDP in rations are variable and dependent on the sources of

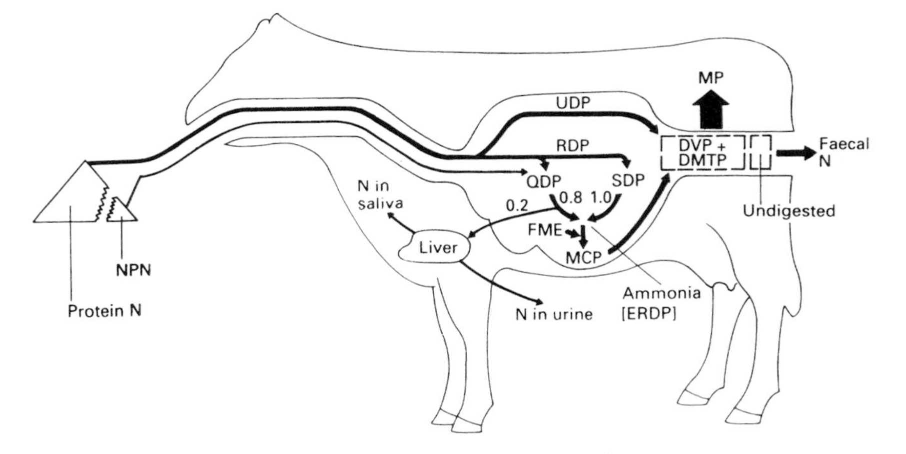

Figure 16.10 Summary of nitrogen utilization by the ruminant.

crude protein and the effect of any processing procedures such as ensilage, heat or formaldehyde treatment.

The true protein within the RDP fraction is initially hydrolysed by bacterial proteases to yield free amino acids. These may be subsequently utilized by microorganisms to synthesize microbial crude protein (MCP), but the bulk of these free amino acids are further degraded to produce ammonia, organic acids and carbon dioxide. This ammonia is the major source of nitrogen for the synthesis of MCP. Non-protein nitrogen within the RDP fraction is likewise degraded to release ammonia; in the case of the most commonly used commercial source, urea, it is rapidly hydrolysed by bacterial ureases. The rate of MCP synthesis and the efficiency with which the RDP is converted to MCP is dependent on the rates of ammonia release and energy availability in the rumen. The availability of energy to microbes in the rumen is dependent on the dietary intake of *fermentable metabolizable energy* (FME). Fermentable metabolizable energy consists of the metabolizable energy (ME) which rumen microbes can utilize, that is the ME in the diet minus the ME in dietary fat and dietary volatile organic acids such as those found in silages (see Chapter 18, section on metabolizable energy). The maximal efficiencies of conversion of QDP and SDP to MCP are 0.8 and 1.0 respectively. This gives us the expression *effective rumen degradable crude protein* (ERDP = 0.8 QDP + 1.0 SDP). The rumen microbes' ability to synthesize MCP in relation to FME intake is recognized to be affected by level of feeding and to vary between 9 g MCP/MJ of FME (maintenance) and 11 g MCP/MJ of FME (3 × maintenance).

In situations where there is an excess of ERDP, especially QDP, in relation to FME, ammonia concentrations build up in the rumen and significant absorption of ammonia into the blood may occur. On reaching the liver it is converted to urea, an energy-demanding process, and is mainly excreted in the urine, although a small quantity (approx. 20%) is recycled to the rumen via the saliva. In cases of extensive ammonia absorption from the rumen, ammonia toxicity may occur (see 'Rumen-related metabolic disorders'). However, protein rationing systems aim to minimize ammonia losses by providing an optimal ERDP/FME ratio in the diet.

In addition to ERDP and FME, the rumen microorganisms require sulphur and phosphorus for protein synthesis. Sulphur is required for the synthesis of the sulphur-containing amino acids. A ratio of N : S of 10 : 1 is considered optimal, and phosphorus is a component of the nucleic acids which are involved in protein biosynthesis.

The MCP resulting from ruminal biosynthesis reaches the abomasum and small intestine in the form of microorganisms carried by fluid out of the rumen once they lose their attachments to food particles. They pass into the abomasum and small intestine together with any dietary UDP. The processes of digestion of these proteins in the ruminant abomasum and small intestine are little different from those occurring in the monogastric animal. Therefore the mixture of amino acids available for absorption from the small intestine is comprised of those that make up MCP and those present in UDP. It is thus markedly different in composition from dietary protein. The sum of these two forms of protein absorbed as amino acids from the small intestine make up the metabolizable protein (MP) supply of the animal. In the metabolizable protein rationing system (see Chapter 18 under 'Ruminant – metabolizable protein') it is assumed that MCP contains 75% of its total nitrogen as amino acids (the remaining 25% being in nucleic acids) and has a true digestibility of 0.85 to give the *digestible microbial true protein* (DMTP) fraction. The true digestibility of UDP, the *digestible undegraded protein* (DUP), is related to the crude protein contained in the acid detergent fibre fraction, the *acid detergent insoluble protein* (ADIP) fraction according to the equation DUP = 0.9 (UDP – ADIP).

The metabolizable protein system allows nutritionists to formulate rations that will provide animals with the appropriate amounts and types of dietary protein to meet their requirements. It does not, however, take into account the ability of that dietary supply and its subsequent digestive utilization to furnish the animal with the correct amounts of individual essential amino acids. Although this is less important than in the monogastric because a major advantage of the rumen microbes is their ability to synthesize essential amino acids in MCP, the future development of even more precise systems will necessitate this information. For example, it is currently known that in silage-based diets methionine is the limiting amino acid for milk protein synthesis. By protecting supplementary pure methionine from rumen decomposition inside a coat of hydrogenated lipid, it is possible to supply this and perhaps other limiting amino acids.

A feature of the nitrogen content of microorganisms reaching the small intestine is that a proportion (25%) of it is in the form of nucleic acids. The pancreatic juices of ruminants contain high activity of nucleases which break down the nucleic acids. The phosphorus released in this process is absorbed and recycled to the rumen via the saliva, but the other end-products are of no value to the animal and represent a further source of nitrogen loss in the digestive utilization of dietary nitrogen by the ruminant.

Lipid digestion

The digestion of lipids requires that they become miscible in water before they can be absorbed through the villi of the intestine. In this respect fat digestion and absorption differ from those of carbohydrate and protein.

Monogastrics

The small intestine is the site of fat digestion and absorption. As it enters the small intestine, fat becomes mixed with bile salts from the gall bladder in the form of taurocholic and glycocholic acids which have emulsifying properties reducing the size of fat particles and giving an increased surface area for digestive hydrolysis.

The major enzymes with lipolytic activity are to be found in the duodenum and are secreted in the pancreatic fluids. Pancreatic lipase hydrolyses triglycerides to a mixture of mono-, di- and triglycerides. In the presence of bile, these contribute to the production of dietary fat forming micelles which after disruption on contact with the microvilli can enter the mucosal cells. Other lipid components in the form of fat-soluble vitamins and sterols are likewise components of these micelles. Within the mucosal cells the various lipid fragments absorbed are resynthesized into triglycerides and phospholipids and along with sterols and protein combine to form particles called chylomicrons. These pass into the lymph and then into the general circulation.

Differences in the efficiency with which different fats are digested and absorbed exist. Unsaturated fatty acids are digested better than saturated ones and digestibility decreases with chain length. Synergistic effects between fatty acids exist such that absorption of saturated fatty acids is greater in the presence of unsaturated fatty acids.

Ruminants

Ruminant diets normally contain only some 3–5% lipid either as triglyceride in concentrate feeds or as galactolipids in forages but may contain up to 10%. The fatty acid composition can vary widely, but in forages and cereals high proportions of linoleic and linolenic acid are present. In the rumen, microbial lipases hydrolyse a high proportion of these fats releasing fatty acids. Of the polyunsaturated fatty acids some 80–90% are rapidly hydrogenated to saturated and monounsaturated fatty acids. In addition this may result in the production of fatty acids in the *trans* configuration. Ruminal microorganisms also synthesize long-chain fatty acids, many of which are of the odd-numbered and branched chain type. These processes account for the presence of these unusual fatty acids in both ruminant body and milk fat.

Unlike VFA, the long-chain fatty acids resulting from dietary and bacterial fat are not absorbed until the digesta reaches the small intestine where mechanisms of absorption are similar to those of the monogastric.

Although the modification of fat in the rumen is restricted to hydrolysis and hydrogenation, dietary fats themselves may inhibit fermentation with consequent effects on the extent of digestion in the rumen. The digestion of cellulose is particularly affected and may result in a 'high propionate type' fermentation. This occurs when the fat content of the diet is around 7–10% but is dependent on the type of fat and manner of incorporation in the diet. For example, unsaturated fat types have a greater effect on the fermentation than saturated fats, and free fat a more deleterious effect than fat present in 'whole' oilseed meals.

In recent years systems of 'protecting' fat by coating the fat with undegradable protein have been developed. The inclusion of additional fat is of greatest relevance to dairy cows where there is often the need for more energy-dense diets, although it may also be used to manipulate the amount and composition of milk fat. Appropriate processing of whole oilseeds may produce a similar 'protected-fat' system.

Metabolism

Metabolism is the name given to the sequence of chemical processes that take place in the tissues and organs of the animal. Some of these processes involve the breakdown of compounds (catabolism), others involve the synthesis of substances (anabolism). Catabolic processes are frequently oxidative in character and are primarily concerned with generating energy for mechanical work and for the chemical work of synthetic processes. Thus the anabolic processes of carbohydrate, protein and fat synthesis in the body are inextricably linked to the catabolic processes. The common currency of energy production and energy utilization in catabolic and anabolic processes is the substance adenosine triphosphate (ATP).

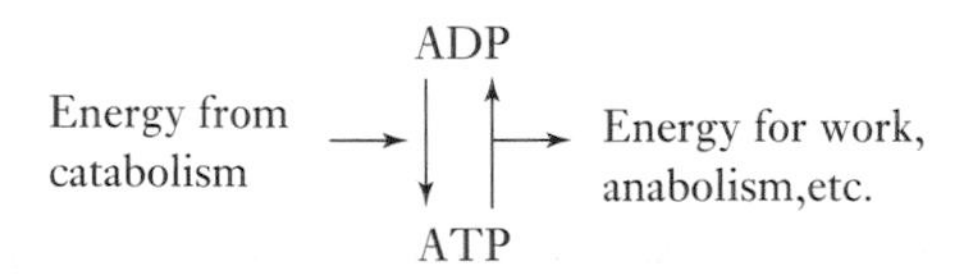

The starting point of metabolism may be looked upon as the substances absorbed from the digestive tract.

Dietary nutrient	*Major end-product of digestion*	
	Ruminant	*Monogastric*
Carbohydrate	VFA – acetic, propionic, butyric acids	Glucose
Proteins	Amino acids	Amino acids
Fats	Mono-, di- and triglycerides, fatty acids	Mono-, di- and triglycerides, fatty acids

As can be seen, these end-products are different in ruminant and non-ruminant species and give rise to differences in metabolism between these animal groupings.

In the coverage of metabolism given here only a general outline of the processes involved is given. Further detail of particular pathways may be obtained by referring to appropriate textbooks.

Glucose metabolism

The major pathways by which glucose is catabolized to yield energy as ATP involve firstly the glycolytic pathway which results in the production of pyruvate, and secondly the citric acid cycle which results in the complete oxidation of pyruvate to carbon dioxide and water (Fig. 16.11).

Glucose for these pathways is mainly obtained in the monogastric through absorption from the digestive tract. Glucose that is absorbed but not immediately catabolized may be stored as glycogen in muscle and liver cells. The amount of glucose that can be stored as glycogen is relatively limited and the bulk of glucose that is not catabolized is stored in the animal's adipose tissue. The concentration of glucose in the general circulation is kept within fairly narrow limits through the action of a variety of hormones but especially insulin and glucagon. Thus, although the rates of absorption of glucose from the digestive tract fluctuate with meal patterns, the levels of glucose in the blood are kept relatively constant. The storage of surplus glucose as glycogen and the ability to reconvert glycogen, especially that in the liver, back to glucose is one of the means by which blood glucose levels are regulated. Liver cells are also able to synthesize glucose from certain metabolites (e.g. amino acids, oxaloacetic acid) in order to maintain blood glucose levels. This is termed gluconeogenesis. Body fat reserves may also be called upon as an energy source in order to spare glucose catabolism when rates of absorption are lower than rates of utilization.

In addition to acting as a principal energy source for the organs and tissues of the animal, glucose is used as the precursor for lactose synthesis in the mammary gland of lactating animals and is the principal energy-supplying substrate to cross the placenta and be used by the growing fetus.

For the complete oxidation of glucose the presence of oxygen in cells is necessary. Under anaerobic conditions the pyruvate produced by glycolysis is converted to lactic acid. This is of relevance in the muscle cells of the animal post-slaughter where the catabolism of cell glycogen to lactate produces a fall in the pH of the muscle. The amount of lactic acid, produced and the resulting pH can thus be influenced by the reserves of glycogen at slaughter. When animals have low muscle

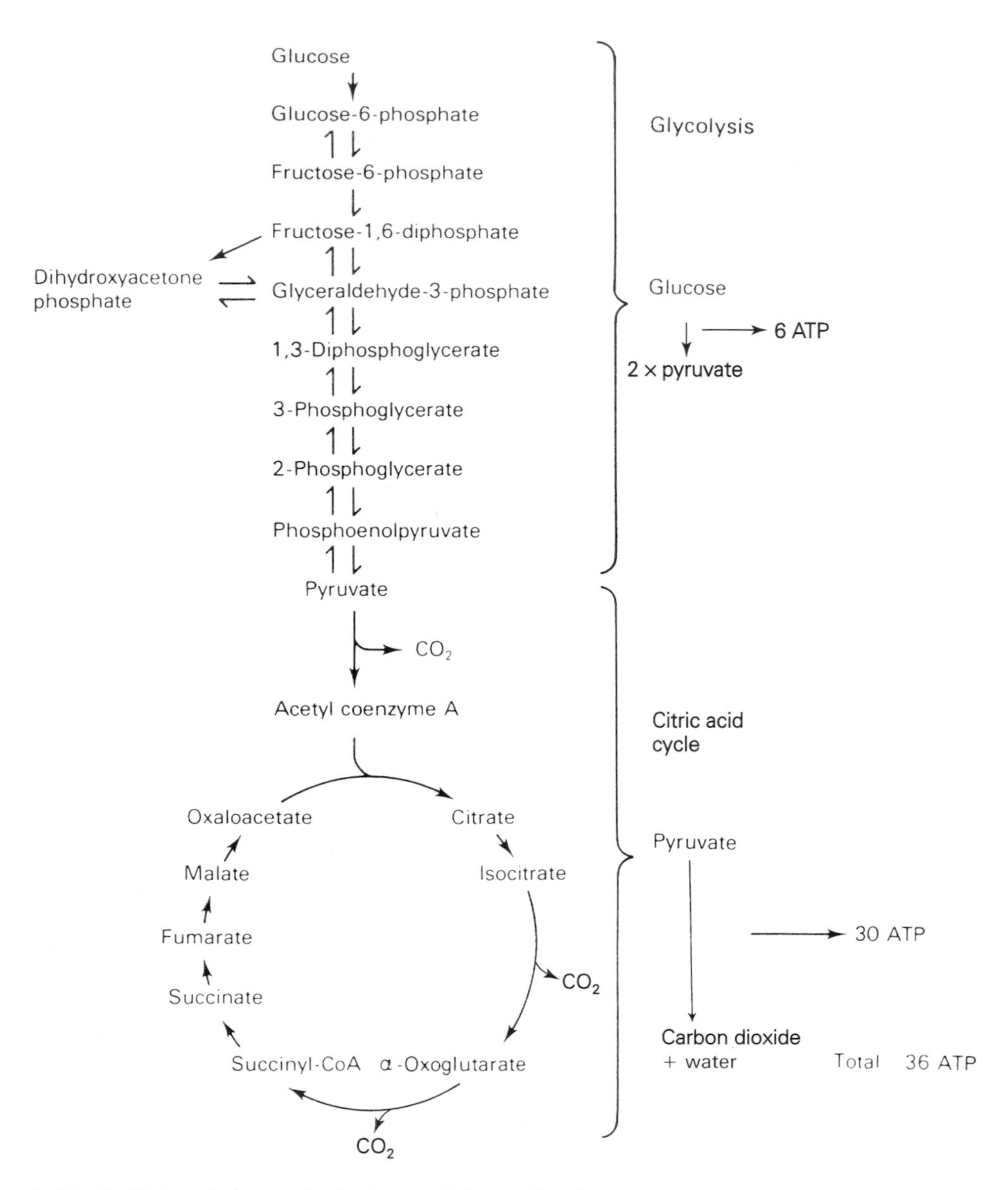

Figure 16.11 Oxidation of glucose via glycolytic and citric acid pathways.

glycogen reserves at slaughter, only a relatively small fall in pH results. The meat from such animals (most commonly bulls) is usually characterized as being dark in colour and firm and dry in texture (DFD meat), both of which are undesirable characteristics from the viewpoint of the consumer. Under some circumstances in animals that show excessive adrenalin release at slaughter, most commonly pigs with particular genetic characteristics, glycogen is quickly metabolized to lactic acid, giving a rapid fall in muscle pH while the carcass is still warm. This tends to give rise to meat that is of pale appearance and soft and exudative in texture (PSE). Again these are undesirable characteristics to the consumer.

VFA metabolism

The volatile fatty acids are absorbed across the rumen wall down a concentration gradient. Rates of absorption are thus mainly dependent on rates of production in the rumen although other factors such as rumen pH may have an influence. Although fluctuations in the rates of absorption exist and are dependent on feeding regime and ease with which dietary carbohydrates are fermented, the diurnal variation in absorption of energy-supplying metabolites is considerably less in the ruminant than the fluctuations in glucose absorption in the monogastric.

Of the three main VFAs produced in the rumen, only acetic appears in the peripheral circulation. This is because propionic is rapidly converted to glucose by the liver and butyric to 3-hydroxybutyric acid during passage across the rumen wall. Acetic acid is the major VFA absorbed and as such is catabolized via the citric acid cycle by a variety of tissues to provide energy. When acetic acid is absorbed in amounts surplus to immediate energy needs, the surplus is synthesized into long-chain fatty acids and stored in the adipose tissue. Although the acetic acid acts as a precursor for fat synthesis, the energy required for this anabolic process is supplied by glucose. Likewise in the lactating ruminant acetic acid is the precursor for milk fat synthesis within the mammary gland.

Although little glucose is absorbed from the ruminant digestive tract, levels of glucose in ruminant peripheral circulation are only slightly lower than those found in the monogastric. A major source of this glucose is propionic acid although gluconeogenesis from amino acids also makes a significant contribution. Whilst glucose is not used as a principal energy source in the ruminant, it is utilized in many ways as in the monogastric. Thus, for example, it is the main source of energy for the nervous tissue and the fetus, for glycogen and lactose synthesis and as an energy supplier for fat and protein biosynthesis.

Although usually absorbed in somewhat lesser quantities than acetic acid and propionic acid, butyric acid can be used as an energy source by a variety of tissues after being converted to 3-hydroxybutyrate in its passage through the rumen wall. Like acetic it may also serve as a precursor for both body and milk fat synthesis.

The proportions of the different VFAs absorbed into the peripheral circulation can have an influence on the endocrine balance within the animal such that the partition of nutrients into different body processes may be affected. Thus high-concentrate, low-forage diets which favour the production of propionate in the rumen depress milk fat secretion in the lactating ruminant and promote lipogenesis in adipose tissues. It is thought that VFA ratios indirectly affect partition through their influence on insulin and growth hormone production.

Fat metabolism

Body fat metabolism is used as a means of regulating the energy metabolism of the animal. Thus dietary energy consumed in excess of immediate requirements is deposited as fat in adipose tissue and is released from adipose tissue as free fatty acids when the dietary supply of energy is inadequate. Such fluxes in fat reserves are characteristic of lactating and pregnant animals which have to call on fat as an energy source when food intake is inadequate to meet requirements and replace this fat at other times in their production cycle. Even on a diurnal basis, particularly in monogastric animals being fed discrete meals, a cycle of deposition and mobilization of fat occurs. Fat is a considerably better energy store than either glycogen or glucose because not only does it have more than twice the energy content per unit weight than these nutrients, it also has very little water associated with it (approx. 12% in fat) whereas glycogen and protein are heavily hydrated (approx. 75%). The energy stored per kilogram of tissue is about 32 MJ for fat and 4 MJ for glycogen and protein. Thus fat can supply a starved animal with energy for several months whereas the energy reserves in glycogen are sufficient for only about a day.

A principal site of fat metabolism is the liver. Both catabolism and anabolism of fatty acids are located in this organ. Fatty acids may be completely oxidized by the processes of β-oxidation and the citric acid cycle to carbon dioxide and water and release energy. Such catabolism also takes place in muscle. Glucose and acetate are the major precursors for fatty acid synthesis. The relative importance of these two substances depends on species, the former being more important in monogastrics, the latter more important in ruminants. Adipose tissue can also synthesize fatty acids and is the major site of fatty acid synthesis in some species, including pigs and ruminants.

Fat circulates in the blood in a variety of forms. Fat resulting from the digestion and absorption of dietary fat is in the form of particles called chylomicrons. These particles consist mainly of triglyceride with small amounts of protein. Mostly they are processed in the liver, but they may release their fatty acids to adipose tissue or mammary tissue in the lactating animal, and thus contribute to body fat and milk fat synthesis. It is in this way that the fatty acid composition of dietary fat may influence that of the body and milk fat. Fatty acids that have been synthesized in the liver are transferred for storage in adipose tissue in the form of lipoproteins. It is also in this form that liver-synthesized fatty acids are transferred to the mammary gland for direct inclusion in milk fat. The other form in which fat circulates in the plasma is as free fatty acids (FFA), also termed non-esterified fatty acids (NEFA). These are mostly transported in the plasma bound to albumin. Although the amount of NEFA in the plasma is very small, this fraction is important metabolically because lipid is released in this form from adipose tissue and transported to the liver or other tissue to supply energy in times of need. The levels of NEFA are typically elevated in the undernourished animals. These NEFA may also be taken up by the lactating cell and be incorporated into milk fat.

Amino acid metabolism

Amino acids resulting from the digestion of dietary protein are absorbed from the small intestine into the portal blood. These amino acids contribute to the metabolic 'pool' of amino acids in the blood and tissues. In the body, amino acids are also constantly being liberated into the blood from the breakdown of tissue proteins. This pool serves as a source of amino acids, some of which are used to build nitrogenous substances such as muscle or milk proteins or certain hormones whilst the vast majority are catabolized, their nitrogen being excreted as urea (Fig. 16.12). The process of tissue protein synthesis and degradation is referred to as the 'turnover' of body proteins. Clearly in the growing animal synthesis exceeds degradation.

Muscle accounts for up to 60% of body protein. Thus the growth of animals involves a substantial uptake by muscle tissue of amino acids from the pool for protein synthesis. The efficiency with which amino acids in the pool are utilized is dependent on the proportions of the different amino acids that make up the pool. The proportions of amino acids in the pool are mainly influenced by the profile of amino acids absorbed from the intestine. The idealized profile is where the amount of all essential and non-essential amino acids absorbed from the intestine is the same as that required for tissue protein synthesis. Only the total of non-essential amino acids supplied need be considered since there exists in the liver a capacity to synthesize non-essential amino acids that are in deficit from others that are in surplus by the process of transamination. Such an idealized profile does not exist in practice and some imbalance in the supply of amino acids exists (Fig. 16.13). In this example lysine is the amino acid in greatest shortage, termed the 'limiting' amino acid, and is responsible for the restriction in protein synthesis.

In monogastric species the profile of amino acids absorbed from the intestine is a reflection of the amino acid composition of dietary protein. In pig diets the amino acid most likely to be limiting is lysine, levels of which are especially low in cereals. Hence considerable attention is given to ensuring adequate lysine levels in practical diets through the inclusion of protein feeds high in lysine or of synthetic lysine. Diet formulation

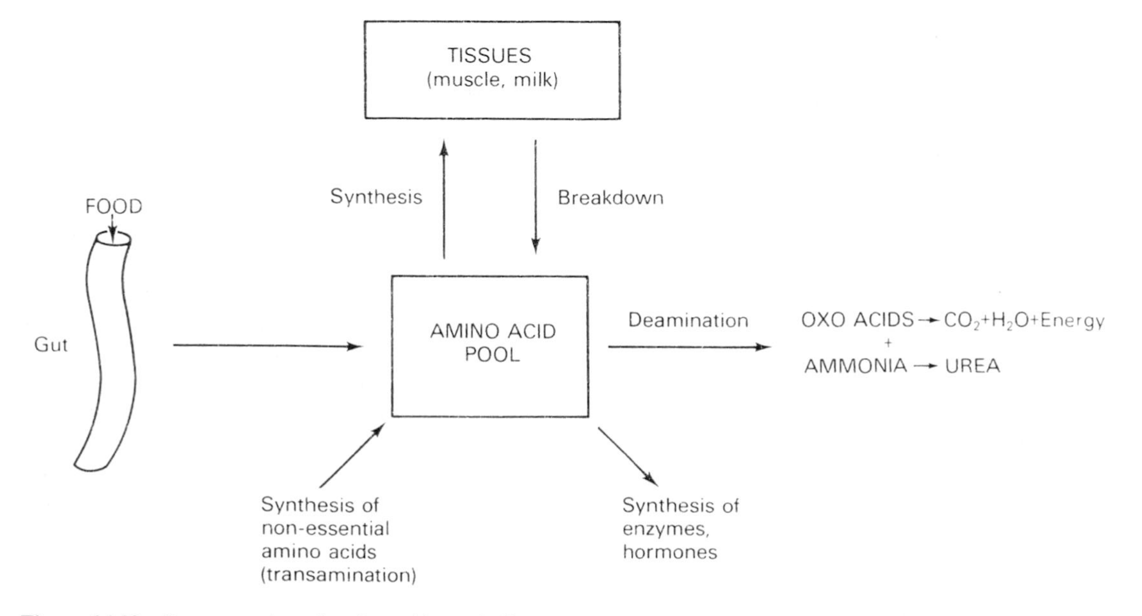

Figure 16.12 Representation of amino acid metabolism.

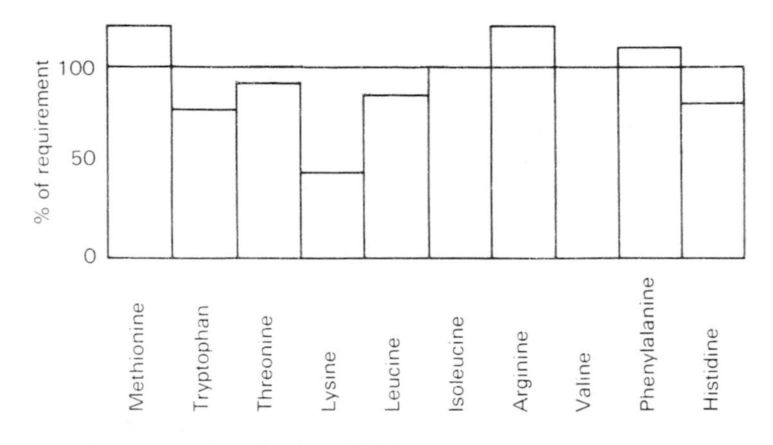

Figure 16.13 Diagrammatic representation of amino acid balance.

programmes for monogastric species aim to optimize the essential amino acid profile of the diet.

In ruminant species the profile of absorbed amino acids making up the metabolizable protein is influenced by the composition of the microbial protein produced from dietary RDP and any DUP present in the diet. In many production situations the supply of amino acids from microbial protein is sufficient to meet the animal's requirements. In certain situations of high productivity, however, in particular the high-yielding cow, the supply of particular essential amino acids may be limiting. In grass silage based diets, methionine is most likely to be limiting, whereas in corn silage based diets, lysine may be limiting. Through the inclusion of DUP of appropriate amino acid composition or synthetic amino acids protected from rumen degradation, an appropriate balance may be obtained.

Amino acids that are surplus to requirements for protein synthesis are converted by the liver in a process known as deamination to oxo acids and ammonia. The ammonia is rapidly synthesized to urea which is subsequently excreted in the urine. The oxo acids may be used as energy sources mostly by way of the citric acid cycle. Surpluses of amino acids occur when an excess of protein is fed and/or where the profile of absorbed amino acids differs significantly from requirements. A further situation where the extent of deamination may be elevated is in starvation where tissue protein may be catabolized to supply energy and amino acids used to maintain blood glucose levels.

Mineral nutrition and metabolism

The essential mineral elements are designated as either macro- (or major) or micro- (or trace) elements depending on their concentration within the animal body (Table 16.8). Although animal tissues and feeds contain about 45 mineral elements, only about half have been shown to have an essential function. Three general functions may be identified for minerals:

Table 16.8 Essential mineral elements

Major	*Trace*
Calcium	Iron
Phosphorus	Zinc
Magnesium	Copper
Sodium	Cobalt
Potassium	Manganese
Chlorine	Molybdenum
Sulphur	Selenium
	Iodine

(1) structural function in bones, e.g. Ca, P, Mg and fluorine;
(2) as electrolytes in body fluids maintaining acid/base balance, osmotic pressure and inducing excitation of nerves and muscles, e.g. Na, K, Cl and Ca;
(3) as integral components of enzymes and other biologically important compounds. Trace elements especially function in this capacity.

Animals obtain most of their mineral requirements from the feed they consume. Animal feeds vary widely in their mineral composition, and individual feeds may vary in their mineral content according to the soil, fertilizer and other environmental influences on its production. The trace mineral composition of feeds tends to vary more widely than that of the major elements. As well as the absolute amount of individual mineral present in feed, the availability of the element in terms of the proportion absorbed and utilized is also important. The availability of individual mineral elements may be influenced by other dietary components. Thus, for example, a surplus of phosphate may reduce the availability of calcium through the formation of insoluble salts. An excess of molybdate may likewise reduce the availability of copper. In some instances, availability of certain ions may be increased through the formation of soluble chelates. Such interactions present difficulties in the determination of requirements and in some instances the identification of particular deficiencies and imbalances.

In situations where feedstuffs fail to provide the animal with sufficient amounts of particular minerals to meet requirements mineral supplementation may be given in concentrated forms such as finely divided powders that may be efficiently distributed through feeds, suitable licks containing the deficient elements, or 'bullets' and 'needles' that lodge in the reticulum and slowly release the mineral throughout the animal's lifetime. The latter are particularly useful for the supply of trace elements and as a means of giving supplementary minerals to animals kept under extensive systems of production.

Calcium, phosphorus and magnesium

These three elements are found together in the animal's skeletal tissue. The calcium and phosphorus exist mainly in the form of crystalline hydroxyapatite [$Ca_{10}(PO_4)_6(OH)_2$] which mineralizes the organic matrix. About one-third of the magnesium in bones is bound to phosphate, the remainder being absorbed on the surface of the mineral structures. The main component of the organic matrix is protein in the form of collagen. Some 98% of the body calcium, 80% of the body phosphorus and 65% of the body magnesium are present in bone tissue. The remaining amounts of these elements are present in the blood either as free ions or in complexed forms and in the soft tissues. In these tissues they have diverse functions. Thus, calcium participates in neuromuscular activity and has a role in blood clotting; phosphorus functions in energy metabolism as a component of ATP and is present in the phospholipids of cell membranes and certain proteins; whilst magnesium, like calcium, has an involvement in the excitation of nerves and muscles, and acts as an enzyme cofactor in metabolism.

The overall metabolism of these elements is illustrated in Fig. 16.14. An important element of this metabolism is the maintenance of constant levels of calcium

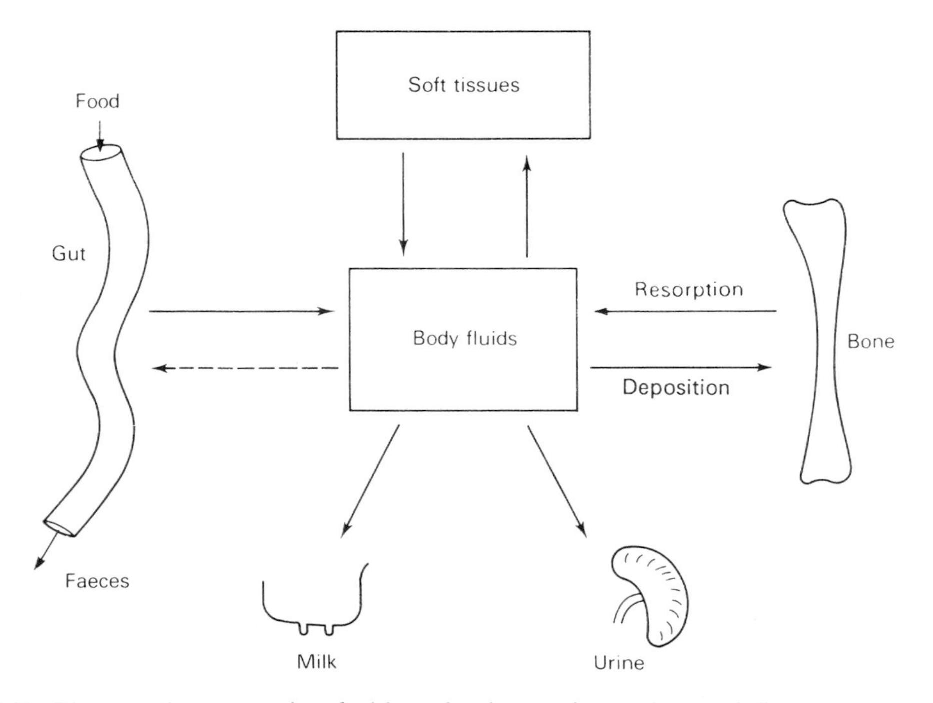

Figure 16.14 Diagrammatic representation of calcium, phosphorus and magnesium metabolism.

and phosphorus in the blood. This is achieved through the interaction of the two hormones, parathyroid hormone and calcitonin, and the active metabolite of vitamin D_3, 1,25-dihydroxycholecalciferol [$1,25(OH)_2 D_3$] which control the absorption from the digestive tract, influence the balance of resorption and deposition in bone and influence the extent of excretion in faeces and urine. Parathyroid hormone (PTH) is secreted in response to a fall in blood calcium, and its action in raising blood calcium through increasing absorption efficiency from the gut, increasing resorption relative to deposition in bone and depressing urinary excretion, is mediated through [$1,25(OH)_2D_3$]. The effects of calcitonin are antagonistic to PTH but do not involve $1,25(OH)_2D_3$. PTH and $1,25(OH)_2D_3$ are likewise involved in the regulation of plasma phosphate concentration. The amounts of ionized magnesium in blood serum are normally in the range of 20–40 mg/litre. Although there is constant exchange of magnesium ions between serum and bone surfaces, there do not appear to be the same type of homeostatic regulatory mechanisms of the type that exist for calcium.

Simple deficiencies of calcium, phosphorus or vitamin D result in bone abnormalities such as rickets in the young and osteomalacia in the mature animal. The metabolic disease milk fever, most commonly found in dairy cows just after parturition, is the main problem associated with a breakdown in the regulation of calcium metabolism (see later section on 'Metabolic disorders in livestock'). Because of the animal's lesser ability to mobilize bone phosphorus, low levels of blood phosphate may readily arise when intakes of phosphorus are inadequate. Low levels of blood phosphate have been associated with poor fertility in animals. Low levels of serum magnesium (hypomagnesaemia) may arise through dietary insufficiency, but especially in ruminants low coefficients of absorption may be a contributory factor. Absorption which occurs from the reticulo-rumen may be adversely affected by high levels of ammonia, potassium and phosphates. Outbreaks of either chronic or acute hypomagnesaemia (grass staggers) occur in ruminant livestock as discussed later.

The amounts of calcium and phosphorus found in different feeds vary widely whereas most of the commonly fed diets for farm animals contain sufficient magnesium to meet body needs. As a generalization, green forages are relatively high in calcium but low in phosphorus whilst animals on diets high in concentrates (cereals and oilseeds) are likely to be receiving insufficient calcium but sufficient phosphorus. A number of supplementary sources of both minerals are available to correct problems of either insufficiency or imbalance in the dietary supply of calcium and phosphorus, e.g. ground limestone, mono-, di- and tricalcium phosphates, and sodium phosphate. These supplementary sources supply different amounts of the two elements and thus particular supplements may be used in given situations. Phosphates used as supplements must be defluorinated before feeding since fluorine is toxic.

Sodium, potassium and chloride

These three minerals are found mainly in the soft tissues and body fluids and together are largely responsible for the maintenance of osmotic pressure, fluid balance and acid-base balance and also have, along with calcium and magnesium, an important function in neuromuscular activity. Sodium is the major cation in body fluids, almost exclusively in the extracellular fluid and blood, whereas most of the potassium is found intracellularly. Chlorine is the major anionic constituent of body fluids. It is a constituent of hydrochloric acid secreted in the gastric juice.

Deficiency of these elements is unlikely since the potassium content of most foods is high and although sodium and, to a lesser extent, chlorine are not always present in sufficient amounts, the practice of including common salt or sodium bicarbonate in the diet means that these elements are supplied in sufficient amounts. The intake of potassium is frequently much higher than that of sodium, but a balance between the two is maintained by the hormone aldosterone which ensures the excretion of excess potassium and reabsorption of sodium by the kidney. Excessive intake of potassium may, however, limit the absorption of magnesium in the ruminant.

In situations of inadequate water intake or where either an excess or imbalance of sodium and chloride in the diet exists, animals may be unable to regulate osmotic and acid-base balance, allowing the development of either an alkalosis or acidosis. This is more likely in monogastric species and in young animals. There is now evidence that, since these elements are all involved in the maintenance of acid-base and water balance, an optimum balance must be achieved between them and that imbalances, even in the absence of deficiency or toxicity, can adversely affect production. For example, in chickens it has been suggested that the sum of $Na + K - Cl$ should be between 250 and 300 mmol/kg of feed.

Sulphur

Sulphur in the body and food is largely present in the form of protein since sulphur is a constituent of the amino acids cysteine, cystine and methionine. Wool is particularly rich in cystine and contains about 4% of sulphur. Some sulphur is contained in the vitamins biotin and thiamine. In monogastrics sulphur deficiency is reflected in a shortage of these essential sulphur-containing amino acids which are often limiting in tissue protein synthesis. In the ruminant the micro-organisms utilize elemental sulphur or sulphate to synthesize these sulphur-containing amino acids. Deficiency in the ruminant is unlikely in most situations, but there is evidence that animals receiving a high proportion of their nitrogen from urea, or animals grazing pastures growing on soils low in sulphur, may benefit from supplementation.

Iron

More than half the iron (60–70%) of the animal body is present as a constituent of the protein haemoglobin found in red blood cells and functioning in the transport of oxygen from the lungs to the tissues. Small amounts are to be found in myoglobin which serves as an oxygen store in muscle and in certain enzymes, but the remainder of body iron is present in the form of the storage compounds ferritin and haemosiderin. These are to be found in the liver, spleen, kidneys and bone marrow.

Although iron is poorly absorbed from the digestive tract – a coefficient of absorption of about 5–10% pertains in the adult – that which is absorbed is efficiently retained. Thus iron liberated from the breakdown of red blood cells and haemoglobin is efficiently recycled in the resynthesis of haemoglobin in the bone marrow and only small amounts are lost through excretion in the faeces (as components of bile) and in urine.

Iron requirements in the adult are generally low and easily met by dietary sources, but young suckling animals are at risk due to the low level of iron in milk and the poor reserves of the newborn. The pale colour of the meat in veal calves is due to the low levels of iron-containing pigments. Suckling piglets are especially vulnerable to deficiency and should be routinely given supplementary sources either orally or by intramuscular injection. Piglets reared out of doors acquire sufficient iron by rooting in the soil. Supplementing pregnant females with iron compounds may have the effect of increasing the iron reserves of the newborn, but dietary intake of iron has no influence on the level of iron in the milk of lactating animals.

The main consequence of iron deficiency is anaemia, characterized by low blood haemoglobin, subnormal growth and diarrhoea. Low levels of haemoglobin may arise for reasons other than iron deficiency. For example, parasitic infection, insufficient dietary protein and the presence of the unusual amino acid S-methyl-cysteine sulphoxide, which is present in brassicas, may contribute to the development of anaemias.

Cobalt

The chief role of cobalt is as a constituent of the vitamin B_{12} molecule. In the ruminant, micro-organisms synthesize this vitamin provided they have a source of dietary cobalt. Vitamin B_{12} is an essential cofactor in the metabolism of propionic acid in the liver.

Cobalt deficiency occurs in grazing areas where the levels and uptake of soil cobalt are low. The condition is given various names, for example pine, and is characterized by a general unthriftiness, poor appetite and anaemia. Prevention of deficiency depends on oral administration of cobalt either as salts in the diet or as 'bullets'. The latter are small pellets that contain cobalt oxide which remain in the reticulum and release cobalt slowly.

Copper

Copper is widely distributed in the body and functions as a cofactor in several enzyme systems. It is involved in the formation of haemoglobin, is present in certain pigments found in the hair and has an involvement in the production of the characteristic physical properties (crimp) of wool. The liver has an ability to concentrate copper and thus acts as a storage organ.

Various conditions arise, especially in ruminant stock, when there is a shortage of copper. These range from anaemia resulting from copper being necessary for the utilization of iron in haemoglobin synthesis, bone disorders resulting from copper being required to produce the structural integrity of the organic matrix, pigmentation failure, diarrhoea and impaired fertility. A condition in lambs known as swayback is associated with copper metabolism. Affected animals are unable to walk properly, suffering incoordination of the muscles and rear limbs. Postmortem examination reveals damage to the spinal cord which arises through inadequate synthesis of the myelin sheath in which copper has a role.

Such conditions may arise from a simple deficiency due to grazing areas with low levels of soil copper but frequently are due to poor utilization of ingested copper. The absorbability of copper is influenced by a number of other dietary components. The presence of molybdenum and sulphate may, for example, make copper unavailable through the formation of insoluble copper thiomolybdate in the rumen. Such interactions influence the requirement of the animal for copper.

Whilst copper is an essential element it may also be highly toxic when consumed in quantity. In this context there is considerable species variation. Pigs can tolerate high levels of copper (it is routinely included at up to 175 ppm to stimulate growth), whereas sheep and to a lesser extent cattle are particularly susceptible to copper toxicity. This toxicity is due to the accumulation of copper in the liver and its subsequent liberation into the blood where it causes haemolysis.

Molybdenum

This is an element required in traces for certain enzyme systems in both the animal and ruminal micro-organisms but which is toxic at higher levels. In practice, deficiency is rarely encountered. Toxicity is related to its interaction with copper (see above) and may occur when animals graze herbage with a high molybdenum content. Molybdenum uptake by herbage is influenced by soil pH, being greater at higher pH, such that molybdeosis can follow overliming of pastures. Control is through oral administration of copper sulphate.

Zinc

Zinc is distributed widely in the animal body with higher concentrations in bones, skin, hair and wool and the testes. Its major role within the animal is as an integral constituent of certain enzyme systems. Most diets contain sufficient zinc for farm animals, but insufficiency may arise through inadequate absorption from the small intestine. Excesses of calcium and copper in the diet may depress zinc absorption as may also the phytic acid present in cereals and oilseeds. Deficiency initially results in skin lesions, termed parakeratosis in growing pigs, a condition readily alleviated through the addition of zinc to the diet.

Manganese

This element occurs in most tissues but is present in higher concentrations in bone, liver, kidney and the pancreas. It is a component of enzymes involved in the synthesis of the mucopolysaccharides that occur in bone. In pyruvate carboxylase, an enzyme central to fat and carbohydrate metabolism, it has a role along with the vitamin biotin, and it is a cofactor for enzymes involved in the synthesis of cholesterol and urea. Deficiency is rare in farm livestock, but when it occurs it can cause retarded growth, skeletal deformity and reduced fertility in breeding animals. The element is widely distributed in feeds but only some 5–10% is absorbed. The presence of excessive calcium and phosphorus may further depress this coefficient of absorption. Manganese tends to be concentrated in the exterior layers of cereal grains so that bran is a good dietary source.

Iodine

The sole role of iodine in the body is as a constituent of thyroxine, a hormone secreted by the thyroid gland and unique in having an inorganic constituent. Thyroxine is primarily involved in the regulation of energy metabolism, but may also influence development and fertility. The secretion of thyroxine is regulated by hypothalamic (thyroid releasing factor, TRF) and pituitary (thyroid stimulating hormone, TSH) factors which are subject themselves to negative feedback. A deficiency of iodine leads to the production of insufficient thyroxine to prevent TSH from continually stimulating the thyroid, leading to its enlargement, a condition known as goitre.

Iodine deficiency is generally a problem in regions with soils containing low levels of iodine. Where such situations exist, supplementation with iodized salts is necessary, taking care to avoid excessive levels. Some iodine deficiencies can arise in situations of apparently adequate intake due to the presence of dietary constituents termed goitrogens. These occur most commonly in brassicas in the form of thiocyanates and thiocarbamides, and interfere with thyroxine synthesis. Whereas the goitrogenic effects of the thiocyanates can be overcome through the addition of extra iodine to the diet, the effects of the thiocarbamides can only be partially overcome by this measure.

Selenium

The role of selenium as an essential nutrient is closely related to that of vitamin E in that both micronutrients are involved in the prevention of oxidation of tissue lipids. As a component of the enzyme glutathione peroxidase, selenium is involved in the destruction of peroxides produced by the oxidation of unsaturated fatty acids before they can have an adverse effect on cell membranes. It is also necessary for the production of pancreatic lipase.

A deficiency of selenium causes a variety of syndromes in farm animals. These range from nutritional muscular dystrophy (white muscle disease) found most commonly in calves and lambs and characterized by degeneration of skeletal muscle; mulberry heart disease found in piglets in which the heart muscle is affected and sudden death may occur; and the exudative diathesis found in chicks in which damage to the cell walls leads to leakage of fluids from cells and oedema of the breast. As well as being cured by inclusion of trace amounts of selenium to the diet these conditions also respond to vitamin E supplementation.

The margin between sufficiency and toxicity levels of selenium in the diet of farm animals is a relatively fine one and considerable care is required in providing supplementary sources. Selenium in foods exists either as the inorganic form of selenites or in organic forms bound to sulphur-containing amino acids, e.g. selenomethionine. The organic forms are more available than the inorganic forms. The amounts of these different forms in feeds can vary widely and feeds originating from areas with low soil selenium may be deficient.

Vitamins in metabolism and nutrition

Vitamins are organic compounds required in extremely small quantities for the normal function, health and productivity of animals. As a generalization, animal cells are unable to synthesize these substances and must therefore obtain them from exogenous sources, either the diet or in some instances through microbial synthesis within the digestive tract. Individual vitamins are required for specific metabolic roles, frequently as integral parts of various enzyme systems. Thus deficiencies reveal a variety of disorders and symptoms that are frequently non-specific especially in the marginal deficiencies that are more likely to pertain in farm livestock.

Vitamins encompass a variety of chemical structures but may be readily classified according to their solubility characteristics into fat-soluble and water-soluble vitamins (Table 16.9). All the vitamins have chemical names but many continue to be known by letters of the alphabet by which they were designated prior to knowledge of their chemical identity.

The solubility characteristics of vitamins have implications for their absorption and storage. Fat-soluble vitamins are absorbed along with fats in micelles and may be stored in fat-containing tissues whereas water-soluble vitamins are absorbed mainly by passive diffusion and there is little or no capacity for body storage, necessitating frequent intake in the diet.

Most foodstuffs contain some vitamins or in some cases the precursors from which the animal derives vitamins (provitamins), but the amount of individual vitamins in feeds varies widely. Monogastric animals and young ruminants must obtain most of their vitamin requirements from feedstuffs, but in the case of the ruminant the microbial population in the rumen synthesizes the B vitamins and also vitamin K which subsequently become available to the animal. In most instances the supply of these vitamins is sufficient to meet the ruminant's requirement for them, but in situations of high productivity there may also be a need for dietary supplementation. Microbial synthesis of these vitamins also occurs in the rear gut of animals, both ruminants and monogastrics, but very little is absorbed and is only of real value to animals that practise coprophagy.

Table 16.9 Vitamins of importance in animal nutrition

Vitamin	*Chemical name*
Fat-soluble vitamins	
A	Retinol
D_2	Ergocalciferol
D_3	Cholecalciferol
E	Tocopherol
K	Phylloquinone
Water-soluble vitamins	
B_1	Thiamine
B_2	Riboflavin
—	Nicotinamide
B_6	Pyridoxine
—	Pantothenic acid
—	Biotin
—	Folic acid
—	Choline
B_{12}	Cyanocobalamin
C	Ascorbic acid

Vitamin A

All animals have a dietary requirement for vitamin A. In considering the supply of this vitamin two groups of compounds are of interest. One group are the carotenoids, the most important of which is β-carotene. Carotenoids are principally found in the leaf tissue of plants and to a lesser extent in seeds, and are precursors of vitamin A. The other group of compounds are forms of vitamin A itself which are only found in animal products and thus, from the viewpoint of farm animals, are present as supplementary rather than naturally occurring components of the feed.

The utilization of dietary carotenes differs between species. In the pig, sheep and goat they are largely converted to vitamin A in the intestinal mucosa prior to absorption whereas in cattle and poultry some carotene escapes conversion and appears in the blood. Such absorbed carotenoids can be converted to the active vitamin in the liver and kidney but are also evident in the pigmentation of egg yolks, poultry carcasses and the milk and fat of cattle. The efficiency of conversion of carotene to vitamin A varies from almost 100% in poultry to less than 30% in ruminants. In all animals the conversion efficiency declines with increasing intake of either carotenes or vitamin A. Surpluses of vitamin A in the body may be stored in the liver and protect the animal during periods of vitamin A insufficiency.

Vitamin A performs a variety of functions in the body, but many of these relate to the maintenance of the integrity of epithelial tissues. Thus, in vitamin A deficiency, keratinization of epithelia in the respiratory tract, genitourinary tract, alimentary tract and cornea occurs. Such damage to the membranes in these tissues not only leads to poor absorption from the digestive tract and also respiratory and reproductive problems but also allows ready entrance of bacteria such that secondary infections may arise.

Vitamin A has an important role in the formation of bones, especially in the formative stages. Deficiency causes retardation of growth, bones being shorter and thickened. The synthesis of certain glycoproteins, a major constituent of the organic matrix, is depressed.

The animal normally receives its vitamin A from carotene in plant materials. The carotene content is high in green crops, especially in young leafy material, whilst roots and cereals are generally poor sources. The carotene content of feeds can be influenced by harvesting and storage processes.

Both enzymic and non-enzymic oxidation may lead in some instances, e.g. haymaking and storage, to up to 80% destruction of carotene in feeds. When vitamin A supplements are added to animal feeds it is usually protected in a gelatin-carbohydrate coating containing antioxidant to prevent loss in activity during feed storage.

Vitamin D

Two forms of vitamin D are of practical importance, namely D_2 and D_3. These active forms of the vitamin, chemically known as ergocalciferol and cholecalciferol, are formed from the effects of ultraviolet irradiation on the steroids ergosterol and 7-dehydrocholesterol which are the respective provitamins. Ergosterol occurs commonly in plants and is transformed during the sun-curing of forages to the active form of the vitamin. The provitamin of D_3 is synthesized in animal tissues, especially skin. Vitamins D_2 and D_3 have similar effectiveness for mammals but D_2 is virtually inactive in poultry.

The role of vitamin D is chiefly concerned with calcium and phosphorus metabolism and the mineralization of bone (see the earlier section headed 'Calcium, phosphorus and magnesium'). It is now clear that vitamin D undergoes metabolic change in the liver (to 25-hydroxyvitamin D_3) and the kidney (to 1,25-dihydroxyvitamin D_3) before it can perform its role of influencing the intestinal absorption, bone mobilization and urinary excretion of calcium and phosphorus. The mode of action of $1{,}25(OH)_2\ D_3$ is similar to that of steroid hormones through inducing the production of messenger RNA.

Intensively kept livestock that do not receive direct access to sunlight and which receive little of the vitamin from cereal-based diets require supplementation whereas grazing animals receive adequate amounts from irradiation. Synthetic metabolites of vitamin D have been used in the prevention of milk fever.

Vitamin E

Vitamin E refers to a group of substances called tocopherols, the most important of which is α-tocopherol. Although first identified as having a role in reproduction, it is now known to function primarily as an antioxidant preventing peroxide damage to cell membranes resulting from the oxidation of polyunsaturated fatty acids. In this role it is supported by the selenium-containing enzyme glutathione peroxidase. The conditions that result from selenium deficiency (see 'Selenium') may also arise if vitamin E is deficient.

Green foods and cereal grains are good sources of the vitamin although some of the activity may be lost during storage, e.g. high-moisture storage of cereal grains. Where supplementation is required, it is usually provided as atocopherol acetate. The need for supplementation is increased if the diet has increased levels of polyunsaturated fatty acids as, for example, when calves are turned out to grass in spring.

Vitamin K

Vitamin K is required for normal blood clotting through its involvement in the synthesis of prothrombin. Deficiency of vitamin K is rare in mammals since it is synthesized by the micro-organisms of the rumen in ruminants. In pigs, hind-gut synthesis combined with a degree of coprophagy is usually sufficient to meet requirements, although young piglets housed on slatted floors require supplementation. Likewise, poultry are routinely supplemented with vitamin K concentrates or with feeds such as lucerne meal that are high in vitamin K. The dietary requirement is increased during periods of treatment with antibiotics, which reduce intestinal synthesis.

A relative vitamin K deficiency can occur when its action is blocked through antagonists such as dicoumarol which is found in mouldy clover hay. It can be prevented by giving extra vitamin K. Agents that similarly block the action of vitamin K, e.g. warfarin, have found use as rat poisons and in the treatment of thrombosis in humans.

B vitamins

The group of vitamins referred to as B vitamins function primarily as components of enzyme systems involved in carbohydrate, fat and protein metabolism. Ruminant animals usually obtain their requirements from microbial synthesis in the digestive tract, and since the B vitamins are present in significant quantities in a wide variety of feeds it was formerly thought that only in particular situations did pigs and poultry require supplementation. The need for supplementation, however, is constantly being increased through the improved performances demanded of animals such that not only monogastrics but even high-producing ruminants may now benefit.

Table 16.10 Metabolic roles and deficiency symptoms associated with the B vitamins

Vitamin	*Metabolic role*	*Deficiency symptoms*
Thiamine (B_1)	Decarboxylation in carbohydrate metabolism	Nervous disorders – cerebrocortical necrosis (CCN), polyneuritis
Riboflavin (B_2) } Niacin }	Hydrogen transfer in energy metabolism	Non-specific; curled toe paralysis (chick)
Pantothenic acid	Part of coenzyme A	Non-specific; 'goose-stepping' (pigs)
Pyridoxine (B_6)	Amino acid metabolism	Non-specific; skin lesions, anaemia
Biotin	Fatty acid metabolism	Skin and hoof lesions, reduced fertility
Choline	Phospholipids, methionine metabolism	Fatty liver, perosis (birds)
Folic acid	Nucleic acid synthesis	Non-specific; anaemia, reproductive problems
Cobalamin (B_{12})	Nucleic acid synthesis, propionate metabolism	Non-specific; dermatitis (pigs), poor feathering (poultry)

Some of the metabolic roles and symptoms arising in deficiency are given in Table 16.10. Identification of problems resulting from B vitamin insufficiency is frequently difficult because they are often associated with non-specific symptoms such as reduced appetite, poor performance, diarrhoea and general poor condition.

In meeting the requirements for individual vitamins not only is the vitamin content of feedstuffs relevant but also the availability and stability of the vitamin. Thus, for example, thiamine deficiency in ruminants results not from a lack of thiamine in the digestive tract but from a combination of thiamine destruction by thiaminases and reduction in availability through the action of antagonists. Antivitamin factors have likewise been associated with biotin (streptavidine), folic acid (sulphonamide), niacin and pyridoxine. As far as stability of vitamins in feeds is concerned, the effect of heat and light during processing and storage can lead to loss of thiamine, pyridoxine and riboflavin activity. In the case of biotin, levels in feed do not relate to activity of the vitamin. Thus, for example, in wheat and barley, although biotin is present in significant amounts, it is almost totally unavailable, whereas in feeds such as soya and fish meal all of the biotin is available.

Dietary supplementation using commercial forms of vitamins is primarily dependent on the ability of dietary raw materials to supply the vitamin but also on the cost of the vitamin. Thus, since excess of water-soluble vitamins is not harmful, supplementation may in some instances (e.g. riboflavin) be carried out merely as a precautionary measure in cases where the vitamin is inexpensive, but in the case of a highly expensive vitamin such as biotin there is greater need to justify and be precise in supplementation. A further important parameter in deciding whether supplementation is necessary is the degree of confidence that can be placed in tables of composition and requirements.

Vitamin C

It is often considered that since farm animals can synthesize vitamin C there is no need for its consideration in the diet. Some evidence exists, however, that extra dietary vitamin C may help animals deal with stress resulting from adverse environmental conditions such as high temperature and ill health. Such an effect of vitamin C may result from its role in the synthesis of corticosteroids. Beneficial effects of supplementation have been reported for poultry kept at high temperatures and for early-weaned piglets.

Vitamin C is also known to function in collagen synthesis, as an antioxidant and as an enhancer of iron absorption from the gut.

Metabolic disorders in livestock

A number of disorders occur in farm animals which appear to be related to an imbalance in the dietary supply of and demand for nutrients. These disorders tend to occur in animals with a high demand for particular nutrients and are sometimes referred to as 'production diseases'. They occur where the classic physiological mechanisms which normally deal with fluctuations in supply, demand and partition of nutrients are unable to cope or adapt quickly enough, thus leading to metabolic imbalances.

Ketosis

This condition occurs in both the cow and the ewe. Typically it occurs in cows (bovine ketosis/acetonaemia) in early lactation when milk yield is at a peak, and in ewes (pregnancy toxaemia/twin lamb disease) in late pregnancy when carrying two or more lambs. The metabolic origins of the condition are an imbalance in the animal's energy supply and demand. The energy intake in these conditions is limited in the case of the pregnant ewe by the growing fetus restricting rumen volume whilst in the dairy cow maximum voluntary food intake is not achieved until some time after peak yield. In both ewes and cows, animals that are overfat tend to have more restricted food intake.

Glucose is the energy supplying metabolite particularly in demand in the pregnant animal for the metabolism of the fetus and in the lactating cow for milk synthesis. Thus a characteristic of the condition is low blood glucose levels. Further to this the imbalance in energy supply and demand leads to the metabolism of body fat reserves. The metabolic consequence of this is that in its attempts to maintain glucose supplies through gluconeogenesis the mobilized body fat is incompletely oxidized and leads to the formation of the ketone bodies that are characteristic of the condition. The accumulation of ketones can be recognized through blood analysis, the use of the Rothera test on milk samples, or the presence of the sweet odour of acetone on the animal's breath. As well as ketone bodies being produced, fatty acids mobilized from adipose tissue may lead to fat deposition in the liver ('fatty liver' syndrome). This restricts gluconeogenesis and further exacerbates the problem. The low availability of glucose eventually leads to nervousness, inappetance and in the cow a drop in milk yield. The drop in milk yield in the cow can lead to spontaneous recovery as the supply and demand for energy gets back in balance, but in sheep the condition is often fatal unless the fetuses are aborted.

Prevention of the condition involves taking measures that allow the animal to keep energy metabolism in balance. Avoidance of overfatness, maintenance of an appropriate energy intake through ration formulation and avoidance of sudden changes in diet that may affect the rumen function are essential. The inclusion of appropriate amounts of protein, particularly UDP, has also been shown to be an important preventative measure.

Milk fever (parturient paresis)

Milk fever is a metabolic disturbance affecting high-producing dairy cows normally within 2–4 days of calving. It is less frequently found in sheep in late pregnancy. The condition arises from an imbalance in the animal's calcium metabolism which leads to a dramatic fall in blood calcium from the normal of 2.5 mmol/litre to less than 1.5 mmol/litre. The clinical signs include initial excitement and change in muscle tone developing into muscle tremor, stiffness of gait and eventual recumbency and coma. The condition arises through the inability of the mechanisms which normally regulate calcium metabolism to adjust to the sudden increase in demand for calcium for milk synthesis at the onset of lactation. Under normal circumstances any imbalance between the supply of calcium in the diet and demand for calcium is taken care of through the mobilization and resorption of calcium from and into skeletal reserves. Such regulation is mainly brought about through the actions of parathyroid hormone, calcitonin and vitamin D. Higher yielding and older cows are more susceptible, the latter being due to the lesser ability to mobilize calcium from the skeleton.

Preventative measures suggested include the feeding of low-calcium diets in late pregnancy to stimulate the regulatory mechanisms into action. In practice, it is difficult to feed diets with sufficiently low levels of calcium. Alternatively, treatment with vitamin D_3 and its analogues at appropriate times pre-calving can lead to the mobilization of bone calcium, but the timing of such treatment is critical. Therapy involves the intravenous administration of calcium borogluconate solution. Other compounds such as calcium hypophosphate and magnesium sulphate are necessary in certain cases. Recovery from therapeutic treatment is usually rapid although relapses may occur.

Grass staggers (hypomagnesaemia tetany)

This is a disorder of cattle and sheep and is characterized by low levels of blood magnesium resulting from an inadequate absorption of magnesium from the digestive tract. In its acute form, which most commonly occurs in lactating cows shortly after 'turn-out' in spring and less commonly during periods of rapid growth in autumn, animals show progressively hyperirritability leading to staggering gait, tetany, violent convulsions and death. A chronic form is found particularly in beef cattle being fed poor-quality winter feeds and in out-wintered animals.

The inadequate absorption of magnesium from the gut may be due to a restricted food intake, low magnesium content in the diet or low absorption efficiency. Very often a combination of these factors is involved. Unlike the situation with calcium metabolism the animal has little ability to buffer fluctuations in supply and demand for magnesium since there is little skeletal reserve and this is largely immobile. A steady supply of magnesium in the diet is therefore essential. Amongst a variety of factors that have been implicated in influencing the efficiency of magnesium absorption are rates of nitrogen and potassium fertilizer application to grass, rumen pH and the stress that sudden adverse weather conditions can cause.

It has been well demonstrated that daily administration of magnesium oxide (calcined magnesite) in the concentrate part of the feed will prevent grass tetany during seasons when it is prevalent. Dosage levels of 50 g/day for cattle and 7 g/day for sheep are adequate. The use of magnesium bullets which lodge in the rumen and slowly release magnesium are a useful alternative in more extensively managed stock.

Rumen-related metabolic disorders

The health of ruminant animals may be adversely affected by imbalances in rumen metabolism. Such imbalance may occur through the excessive or sudden ingestion of concentrates, leading to *acidosis*, and when excessive non-protein nitrogen and quickly degradable protein is fed, leading to *ammonia toxicity*. The problem of acidosis results from the production of very high levels of lactic acid and low rumen pH. This may influence gut function, causing rumen stasis and epithelial damage, or alternatively the rapid entry of lactic acid into the blood may so upset the animal's acid–base balance that hypotension and respiratory failure ensue.

Sudden ingestion of concentrates brings about the acute form of the condition, but chronic forms may be seen in cattle given very high concentrate diets with little or no forage. Such diets are associated with characteristic changes in the microbial population of the rumen, Gram-negative organisms being replaced by Gram-positive organisms such as lactobacilli.

Ammonia concentration in the rumen is a function of its rate of production from dietary RDP and NPN and its rate of utilization for microbial protein synthesis. The latter is mainly dependent on energy availability in the rumen. An imbalance in the production and utilization of ammonia in the rumen leads to accumulation of ammonia in rumen liquor. This eventually spills over into the blood, upsetting the animal's acid–base balance and producing signs of hypertension and in extreme cases respiratory failure.

Prevention of both these conditions is dependent on care in ration formulation and feeding regimes. Observation of appropriate ratios of forage to concentrate and of the amount and type of degradable protein to fermentable metabolizable energy is essential. The mixing of ingredients and frequency of feeding need also to be considered.

Blood chemistry and nutritional status

Under normal circumstances the concentration of blood metabolites is kept within well-defined limits and it is only in circumstances of metabolic imbalance that deviations from the 'normal' occur. Identifying deviations from normal in the composition of blood allows the diagnosis and correction of subclinical conditions which may adversely affect production. The analytical details of levels of metabolites present in a blood sample taken from an animal are referred to as its 'metabolic profile'.

Metabolic profile testing has to date been mainly used to monitor the nutritional status of dairy herds. Animals considered to be representative of high-yielding, medium-yielding and non-lactating groups in the herd are sampled and the blood analysed for such metabolites as glucose, 3-hydroxybutyrate, free fatty acids (indicators of energy status); albumin, globulin, urea and haemoglobin (all indicators of protein status); packed cell volume (an indicator of blood dilution) and various minerals (e.g. Ca, P, Mg). After analysis of samples the results are processed by computer and presented to allow comparison of levels of individual metabolites with 'normal' values. 'Mini metabolic profile tests' which restrict blood analysis to glucose, urea and albumen have also been developed.

Full interpretation of the data requires information on both the feeding regimes and performance of animals. Only then may recommendations on changing the nutritional inputs be made.

Since blood has several practical disadvantages as a test medium, recent studies have investigated the potential of milk to evaluate the health and nutrition of dairy cows. Analytical developments may lead to routine monitoring of milk components through the use of probes in milk jars. This is already possible in the case of acetone, an early indicator of energy imbalance.

Voluntary food intake

Voluntary food intake may be defined as the weight of food eaten by an animal per unit of time when given free access to feed. Its importance as a parameter influencing both the biological and economic efficiency of production has been increasingly realized in recent years. The level of intake dictates the rate of production, the proportion of the intake going towards production and thus the efficiency of feed conversion. The level of intake may also influence the products of production. For example, when intake is excessive, then excessive fat deposition may occur in growing and lactating animals. Thus an aim in production is to match intake to the required level and type of production.

An ability to predict and manipulate the voluntary food intake of farm animals requires a knowledge of those factors that influence it. The control of intake is undoubtedly complex and is the integration of a combination of both negative feedback and positive feedforward mechanisms. It is considered to involve information relating to characteristics of the feed,

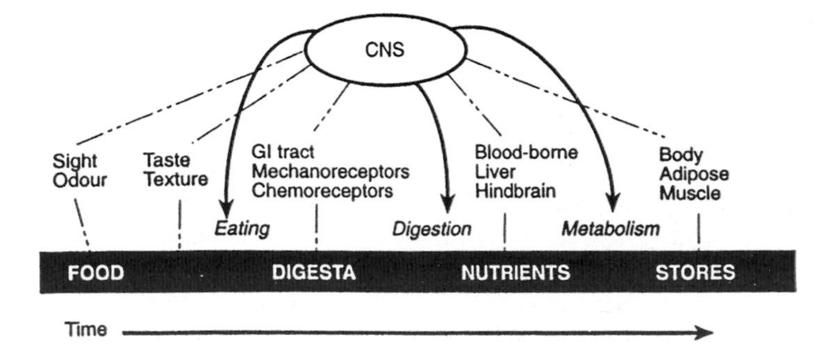

Figure 16.15 Satiety cascade. (From D'Mello, 2000.)

responses to the presence of the feed in the digestive tract, nutrients derived from the digesta and the responses of the liver, adipose tissue and other organs and tissues to the availability and metabolism of absorbed nutrients. This cascade of signals is co-ordinated by the central nervous system (CNS) into the animal's feeding behaviour (Fig. 16.15). In the central nervous system it centres in the hypothalamus which receives and integrates the neural and endocrine information that regulates the 'drive' to eat. Within this complexity, however, a fundamental parameter being regulated appears to be energy balance (energostasis). Thus if animals are fed a range of diets of differing energy concentrations they will alter their dry matter consumption through altering either meal frequency and/or meal size in order to achieve energy balance.

Mechanisms of energostasis

The mechanisms by which animals are capable of regulating their energy balance are imprecisely understood. Many of the early theories on the control of food intake have proposed single factors. These include such classic ideas as the *chemostatic theory* in which it is proposed that levels of primary blood metabolites, such as glucose in the monogastric and volatile fatty acids in the ruminant, provide the feedback information; the *thermostatic theory*, which proposes appetite to be linked to the animal's thermoregulatory mechanisms; and the *lipostatic control theory*, which proposes that the feeding drive is in some way controlled by the body's fat reserves. It is unlikely, however, that energy balance is regulated by any single mechanism and that whilst some aspects of these theories are involved, the centres in the brain concerned with the control of feeding are more likely to receive a variety of feedback signals which they integrate to determine feeding behaviour. As well as the involvement of primary blood metabolites it is likely that metabolic hormones, such as insulin, and gut peptides, such as cholecystokinin, have a part to play in energostasis. The discovery of the role of leptin, a hormone secreted by adipose cells in proportion to their size, in the control of food intake in rats may also prove central to the understanding of energostasis in farm animals and help explain why intake declines with increasing levels of fatness.

Limits to energostasis

Although the underlying mechanism by which animals regulate their intake is related to energy balance, a number of factors may limit the capacity of animals to achieve 'energostasis'. The ultimate limiting factor to the intake of food must clearly be the physical capacity of the digestive tract, the rates of digestion and absorption and the rate of passage of food through the digestive tract. Thus, although animals may seek to control their energy consumption through physiological mechanisms, they may fail to achieve this when fed bulky and/or indigestible feedstuffs. In practice the extent of dietary dilution necessary to produce limitations to gastric distension is less likely to pertain in the monogastric and is more likely to arise in the ruminant animal.

Evidence for the physical limitation of intake comes from observations relating intake to the available energy concentration of the diet. This is illustrated in Fig. 16.16. The range of energy concentrations over which energostasis may be accomplished or physical limitations to intake prevail is dependent upon the physiological status of the animal and characteristics of the feed other than simply energy concentration. Thus, for example, in pregnant or overfat animals where the developing uterus or internal fat deposits restrict rumen volume, physical limitation occurs at a lower dietary energy concentration. As far as feed characteristics are concerned these features which influence the rate of

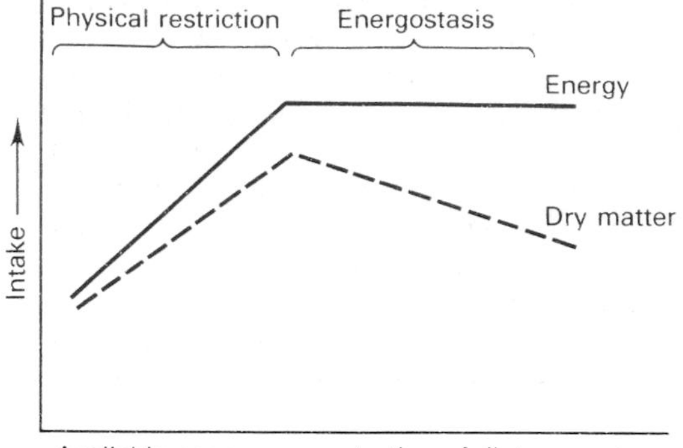

Figure 16.16 Representation of relationship between energy concentration of the diet and energy and dry matter intake.

passage of food through the gut are relevant. Thus digestibility and rate of digestion and those features of a feed such as chemical composition, level of processing and water content which may influence these parameters are influential. The extent of gut fill is monitored by stretch receptors in the gut wall and this information is relayed to the brain via the nervous system where it influences feeding behaviour.

Prediction of intake

Since a knowledge of the dry matter intake which an animal will eat is essential for the precise and economic formulation of diets, the ability to predict intake is of the utmost importance. For ruminants, particularly dairy cows, many relationships have been proposed. The simplest take account of body weight and milk yield, the main determinants of energy requirements, as in, for example:

$$\text{Dry matter intake (kg/day)} = 0.025W \times 0.1Y$$

where W is body weight (kg) and Y is daily milk yield (kg).

Such a simplistic prediction equation has obvious drawbacks in that characteristics of the feed, such as physical form, palatability, acidity and chemical composition and characteristics of the animal other than body weight and milk yield (e.g. pregnancy, degree of fatness, liveweight-change and stage of lactation), all of which are known to influence intake, are not taken into account. An example of a more complex equation, predicting the dry matter intake of lactating dairy cows, is:

$$\text{DMI (kg/day)} = 0.076 + 0.404c + 0.013W - 0.129n + 4.121\log_{10}(n) + 0.14Y$$

where DMI is dry matter intake, c is kg of concentrate dry matter, n is the week of lactation, W is body weight (kg) and Y is milk yield (kg).

Prediction of intake in pigs has received less attention since, until recently, the practice in pig feeding has been to restrict feeding to prevent excessive fat deposition. The use of boars along with intensive selection for leanness now means that growing pigs may be fed *ad libitum*. As is also the case with poultry, age, weight and also environmental temperature are parameters that feature in prediction equations.

Manipulation of intake

The objectives in the manipulation of intake are dependent on the species involved. In pigs there has been some manipulation of appetite by genetic selection. In the past, selection for leanness has also led to a reduction in intake. For the future it would appear that, if genetic increases in potential lean tissue growth are to be realized, then there is a need to breed for increased appetite. Studies on how monogastrics interact with their feed have shown pigs to have preferences for 'wet-feeding' systems, leading to higher dry matter intakes. Although both pigs and poultry exhibit 'nutritional wisdom' in terms of diet selection to meet their requirements, there is little understanding of the mechanisms involved and as yet no commercial application of this characteristic.

With ruminant animals, production is frequently limited by physical restraints on the intake of forage and the objective is to increase total dry matter intake, especially the proportion of forage in the diet.

The greatest scope for manipulating intake at present lies in altering the characteristics associated with the feed that affect intake. Thus the processing of foodstuffs can have a major effect on the intake of feeds. The grinding and pelleting of forage, the chemical treatment of forages with alkali, and the processing of cereal grains either physically or through techniques such as micronization that gelatinize the starch may all influence intake through their effects on the rate of passage and/or digestibility of feeds. The method of feeding may likewise influence total daily intake. The mixing of the various dietary components as in 'total forage diets' can increase intake, as may the method of feeding employed in feeding systems involving the separate allocation of compound and roughage. Thus the more frequent feeding of compound through the use of 'out-of-parlour' feeders and the mechanized feeding as opposed to self-feeding of silage may increase intake.

In the grazing animal there is now a conceptual basis for understanding the influence of plant morphology and sward structure on herbage intake in terms of the effect of such parameters as bulk density, sward height and shearing strength. Quantification of such relationships so that sward variables may be objectively manipulated is now being undertaken.

The facility to manipulate the animal in ways that will influence intake is more limited. Since body fatness has a clear influence on intake, attaining appropriate body condition targets in the production cycle of pregnant and lactating animals is important. Social interaction can influence feeding behaviour such that group size, pecking order and social facilitation of feeding are characteristics with potential for manipulation. Feeding behaviour and intake are affected by photoperiod, and although the mechanisms involved differ in different species, it may be possible to manipulate this parameter to advantage. The observation that Brahman cattle consume considerably more forage dry matter in relation to body weight when compared with Friesians has led to the suggestion that selection for the characteristic of rumen volume and/or giving animals early experience of very bulky, low-quality diets may be worthwhile.

Reproduction

Reproduction is of crucial importance to all animal populations, both for herd or flock replacement and as a means of genetic improvement. Reproductive physiology is the science of reproductive functions within the animal. As such it holds the key to increased reproductive efficiency. It may also be able to provide explanations when things go wrong – such as subfertility or infertility in breeding stock. The manipulation of reproduction through the application of reproductive technology is now standard practice in most modern livestock enterprises. The following section seeks to provide a basic outline of the physiology of reproductive processes and to review the more recent advances in the technology of reproduction.

The female reproductive system

Anatomy

In all mammals the reproductive system of both sexes is bilaterally symmetrical, which means that organs consist of right and left paired structures. The female reproductive system is comprised of two ovaries, two oviducts or fallopian tubes, two uterine horns, a uterus, a cervix, a vagina and an external opening, the vulva (see Fig. 16.17). Although the female reproductive anatomy is essentially similar between farm species, there exists a distinction in the degree of fusion of the uterine horns and their relative length with respect to the body of the uterus. This relationship depends on the litter size of the species. Thus the pig has long uterine horns compared to the body of the uterus since the many embryos of the sow develop within these horns. In contrast, the cow normally has a single offspring which develops in the uterine body, therefore the uterine horns are relatively much shorter. The ewe reproductive tracts falls somewhere between these two extremes.

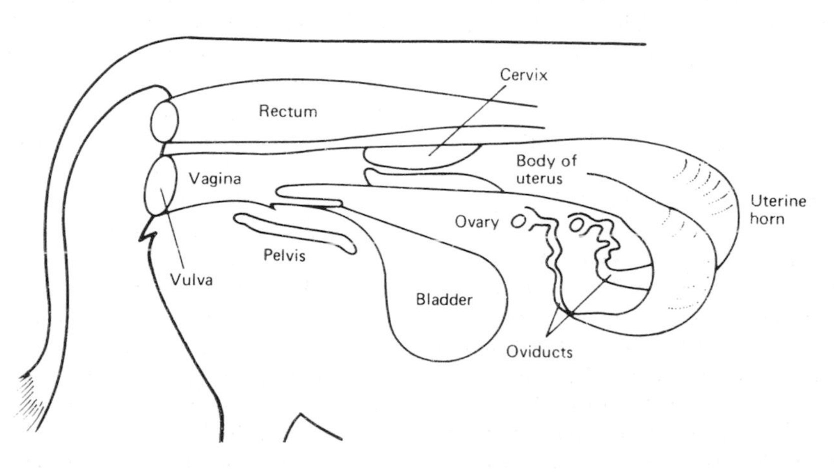

Figure 16.17 Reproductive tract of the cow.

Sex determination

The sex of an offspring is determined by its genetic makeup or chromosomes which are donated by its parents. A female mammal has a pair of identical sex chromosomes, designated XX, whilst those of the male are non-identical, XY. The primary reproductive organ, the ovary, produces female gametes or eggs which carry only half the chromosome complement of the nuclei of the parent cells. Therefore each egg cell can carry only one sex chromosome and this must be an X chromosome. Male gametes or sperm cells on the other hand can be either X- or Y-bearing. Thus, at fertilization, the sex of the resulting embryo is determined by the 'sex' of the fertilizing sperm cell. Within ejaculated semen there are approximately equal numbers of X- and Y-bearing sperm; thus the chances of a fertilized egg being male or female is 50 : 50. This is known as the primary sex ratio. The secondary sex ratio is the number of male:female offspring at birth.

The genetic sex of the female is determined at fertilization. The phenotypic expression of that sex, or, more simply, the appearance of 'femaleness', is largely determined during embryonic development. Of primary importance in this respect are the embryonic ovaries which secrete the female sex hormone, oestrogen. This hormone determines that the embryo reproductive tract shall develop into a female one rather than a male reproductive system.

Impairment of normal embryonic sexual development sometimes occurs, as for instance in the case of the freemartin heifer. In the small proportion of twins that occur in cattle, when a male and female embryo share the uterus, the production of male hormone testosterone, from the testes of the male embryo, causes the reproductive system of the female embryo to be effectively 'masculinized'. The result is a female calf, or freemartin, which is usually infertile.

Sexual development

The ovaries are the site of production of the female gametes or oocytes. Their production is known as oogenesis. At birth there are something like 200 000 primary oocytes already present in the ovaries, far in excess of those needed for the animal's entire reproductive life. During their development into mature oocytes, female gametes go through a stage of meiotic division which results in a halving of the chromosome number of each egg cell. Oocytes are surrounded by layers of ovarian cells and the whole structure is termed a follicle (see Fig. 16.18). Follicle development is controlled by the gonadotrophins from the anterior pituitary, chiefly FSH. Maturing follicles produce increasing quantities of the steroid hormone oestrogen, which influences both growth of the reproductive tract and the sexual behaviour of the animal.

Follicle development does not reach completion until puberty occurs.

Puberty

This signals the beginning of the female's active reproductive life when the animal commences breeding activity. Thus puberty is indicated by the first heat or oestrus during which the female is receptive to the male. The age at which puberty is reached is dependent on the animal's body maturity which is related to its body weight and liveweight gain. Onset of puberty coincides with the point of inflexion on the animal's growth curve (see Fig. 16.35). Breed size influences the onset of puberty; smaller, faster maturing breeds reach puberty before larger breeds. Other factors are nutrition, health, season of birth and the presence or absence of a mature male. The introduction of a mature boar to late prepubertal gilts advances their onset of puberty. Data on the age at puberty for farm species are given in Table 16.11. Puberty in the male is discussed later in this chapter.

Oestrous cycles

All females of farm species, once puberty has been attained, exhibit sequences of reproductive activity known as oestrous cycles. They may be continuous throughout the year as in cattle and pigs or may be restricted to a specific breeding season as in the case of sheep, goats and deer, all of which are autumn breeders. The non-breeding time of the year is known as anoestrus. In sheep seasonal anoestrus is much longer for temperate or northern breeds, such as Scottish Blackface, than it is for breeds of more equatorial origin such as Merinos, which tend to breed for most of the year. For seasonal breeders, daylength or photoperiod is the cue to onset of breeding activity. Photoperiodic changes influence the activity of the pineal gland, via the optic pathway. The pineal regulates seasonal changes in reproduction through its control of gonadotrophin release from the anterior pituitary. Melatonin, a hormone secreted by the pineal, is produced in increased amounts during darkness. This hormone has an important role as a mediator between environmental light patterns and seasonal differences in gonadotrophin secretion (see later section on melatonin and out-of-season breeding).

The sequence of events in the oestrous cycle of the sow, ewe and cow are essentially similar, although their precise timing and duration do differ. (For a more detailed comparison see Table 16.11). The oestrous cycle may be divided into four main stages: oestrus, metoestrus, di-oestrus and pro-oestrus.

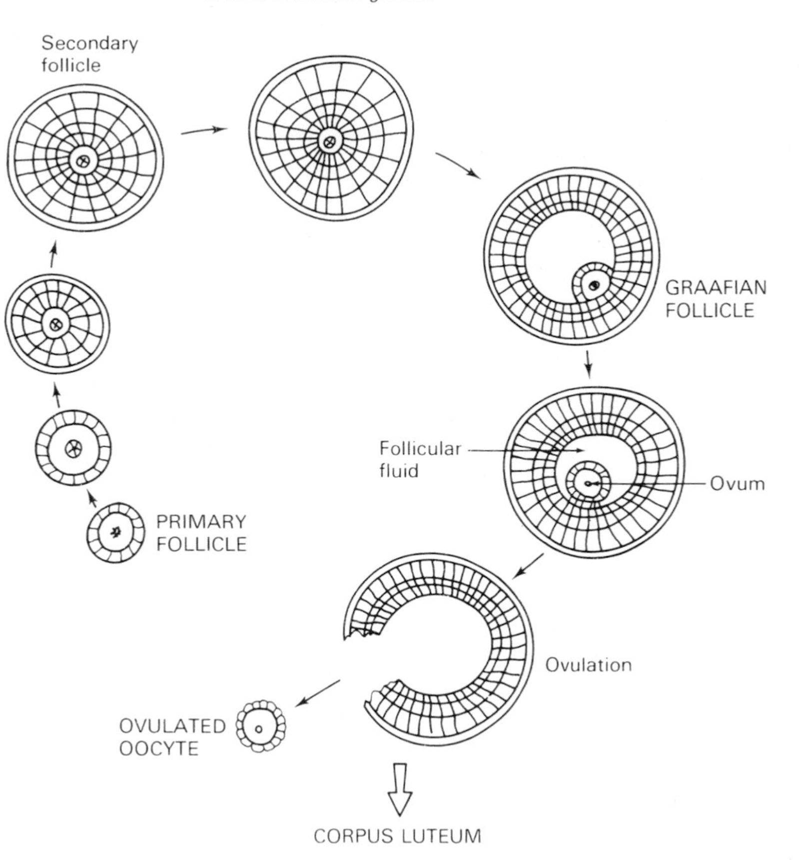

Figure 16.18 Diagrammatic representation of the sequence of follicle development, ovulation and corpus luteum formation during the ovarian cycle.

Table 16.11 Some data on normal female reproduction

Animal	*Onset of puberty*	*Age at first service*	*Length of cycle (days)*	*Duration of oestrus (hours)*
Cow	8–12 months	15–18 months	21 (18–24)	18 (14–26)
Ewe	4–12 months (first autumn)	First autumn or second autumn	17 (14–20)	36 (24–48)
Sow	4½–10 months	7–10 months	21 (18–24)	48 (24–72)

Animal	*Time of ovulation*	*No. of ova shed*	*Length of gestation (days)*	*Optimum time for service*
Cow	10–15 h after end of oestrus	1	282 (277–290)	Mid to end of oestrus
Ewe	Near end of oestrus	1–4	149 (144–152)	16–24 h after onset of oestrus
Sow	Middle of oestrus	10–20	114 (111–116)	15–30 h after onset of oestrus

Oestrus

This is the stage of the oestrous cycle when the female comes into heat and is characterized by physiological and behavioural changes, such as reddening of the vulva; increased mucus secretion of the cervix and vagina; increased irritability and vocalization; mutual grooming activity; and, most importantly, a willingness to stand for the male. In cattle in particular, in the absence of a male the oestrous female will allow mounting by other females, who themselves are often in oestrus or approaching it. For a more detailed coverage of behaviour and detection in oestrus, see chapters on cattle, sheep and pigs (19, 20 and 21).

Duration of oestrus and length of the oestrous cycle differ between individual animals and may be influenced by such factors as age of animal, its standing in the social hierarchy, stage of the breeding season, time of year, climate, nutritional status, health and stress.

In pigs and sheep, egg release (ovulation) occurs at the end of oestrus. In contrast, ovulation in the cow occurs some 10–12 h after the end of behavioural oestrus. There is evidence that the interval between onset of heat and time of ovulation is significantly shortened when females are naturally mated compared with those artificially inseminated.

Met-oestrus

This is the stage immediately post-oestrus when the corpus luteum forms in the ovary from the cells of the post-ovulatory Graafian follicle (see Fig. 16.18).

Di-oestrus

This is the longest stage of the oestrous cycle and coincides with the maturation of the corpus luteum. It is otherwise referred to as the luteal phase.

Pro-oestrus

This is of relatively short duration when one or more follicles undergo the final stage of maturation. Animals in pro-oestrus may show a reddening of the vulva and increased sexual activity, although they will not stand to be mounted.

Hormonal control

The oestrous cycle is regulated by pituitary and ovarian hormones. Generally speaking, the concentrations of these hormones in the blood reflect the physiological changes occurring in the animal (see Figs 16.19a and b). The endocrine events in the oestrous cycle of the cow have been assumed as typical of all farm mammals.

If the beginning of oestrus is day 0, then this is the stage at which Graafian follicles reach maturity under the influence of the gonadotrophins FSH and LH. In the cow, ovulation occurs shortly after the end of oestrus as a result of the increases in the levels of the luteinizing hormone (LH). Thus a mature egg is released into the oviduct and the follicle cells within the ovary form the corpus luteum. The corpus luteum gradually matures during di-oestrus, secreting progesterone, whose function is to prolong the luteal phase of the oestrous cycle until the fate of the ovulated egg has been decided. Progesterone, through its effects on the pituitary, inhibits the release of FSH and thus prevents maturation of further follicles. Progesterone, in conjunction with the low levels of circulating oestrogen at this time, helps prepare the uterus lining or endometrium for the possible arrival of a fertilized egg. Sometimes animals become acyclic due to the presence of a persistent corpus luteum, or luteal cyst as it is generally termed. Thus a return to oestrus is indefinitely delayed.

Normally, however, in the absence of fertilization, the corpus luteum naturally regresses at about day 16 or 17 of the cycle. Luteal regression is accompanied by a fall in progesterone. The cause of these events is the release by the uterus of a hormone known as prostaglandin.

With the end of the luteal phase, follicle development accelerates. Maturing follicles are the major source of oestrogen production, the chief oestrogen in the cow being oestradiol. During this pro-oestrus stage, blood concentrations of oestrogen are therefore greatly elevated. Oestrogen is the hormone responsible for the physiological and behavioural changes taking place during pro-oestrus and oestrus.

The number of follicles that mature and are eventually released by the ovaries is termed the ovulation rate. Although influenced by gonadotrophins, ultimately ovulation rate is genetically determined. Thus some species and breeds have a greater ovulatory capacity than others. Other factors influencing ovulation rate are age, stage of the breeding season and nutritional status.

The practice of flushing, where the level of nutrition is increased shortly prior to mating, usually results in an improved ovulation rate.

Synchronization of oestrous

Synchronization of oestrus, or controlled breeding, is possible in farm species through the manipulation of oestrous cycles by means of hormone administration. This can be achieved in two ways:

(1) by premature regression of the corpus luteum (luteolysis) using prostaglandins;
(2) by prolongation of corpus luteum activity beyond the normal length of the luteal phase, using progesterone; followed by sudden withdrawal of treatment.

In both cases animals return to heat shortly after treatment, in a synchronized fashion. See Fig. 16.20 for a summary of hormonal changes associated with oestrous synchronization.

Prostaglandins

Prostaglandin $F_{2\alpha}$ or its synthetic analogues, when administered to a breeding female with an active corpus

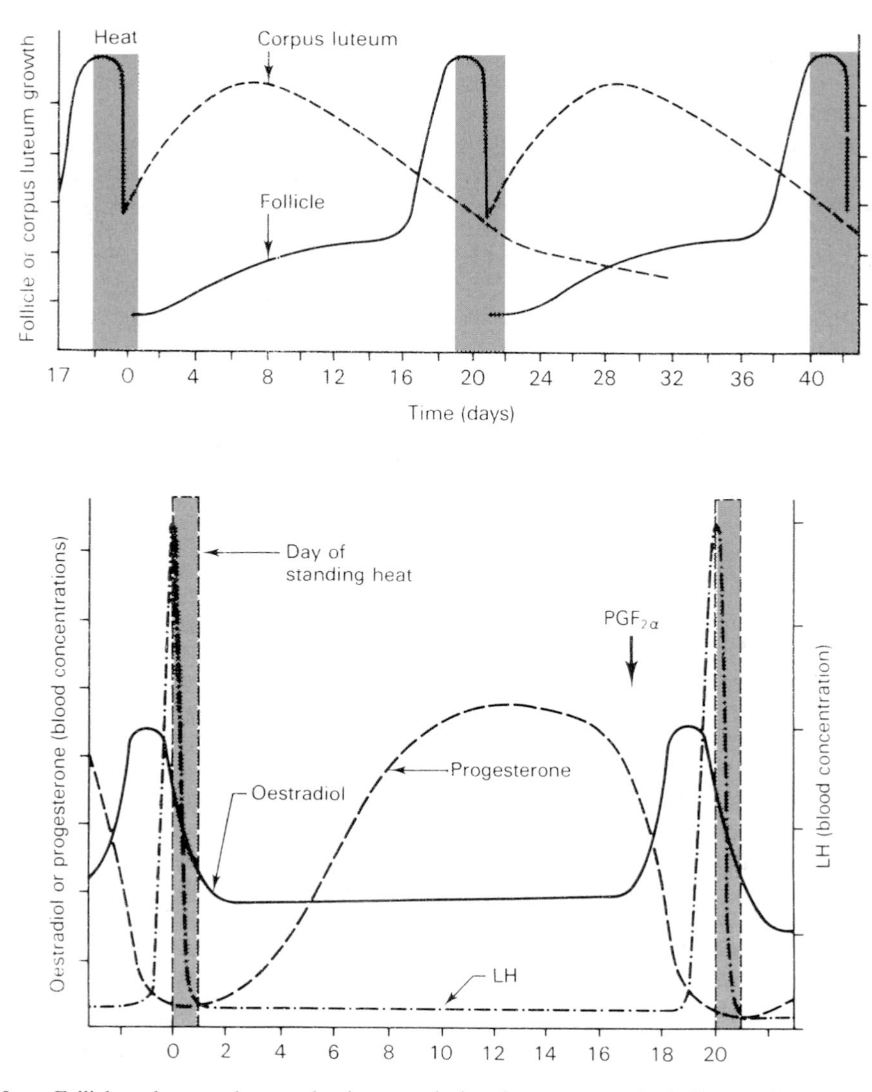

Figure 16.19 a Follicle and corpus luteum development during the oestrous cycle. **b** Changes in concentrations of the various hormones associated with the oestrous cycle.

luteum, causes an effective shortening of the oestrous cycle. For instance, in cattle, the injection of prostaglandin causes the corpus luteum to immediately regress and oestrus normally follows 2–3 days later. Insemination can then be carried out 3–4 days after prostaglandin treatment. However, not all cows respond to prostaglandin in the same manner. Animals between days 1 and 5 and days 17 and 21 of the oestrous cycle are unaffected by prostaglandin administration. Between days 1 and 5, the corpus luteum is immature and therefore unresponsive to prostaglandin and cows between days 17 and 21 have no corpus luteum present. The solution to this problem has been either to use two injections 11 days apart (when all animals should then be between days 5 and 17 of the cycle) or to use one injection and combine this with oestrous detection of unaffected animals. The use of prostaglandins to synchronize oestrous in pigs is largely ineffectual since the corpora lutea of the sow will only regress if treated between days 12 and 15 of the 21-day cycle.

Progesterone

Use of progesterone or its synthetic analogues (collectively referred to as progestagens) simulates the activity of a corpus luteum thereby artificially prolonging the luteal phase of the oestrous cycle. Thus once

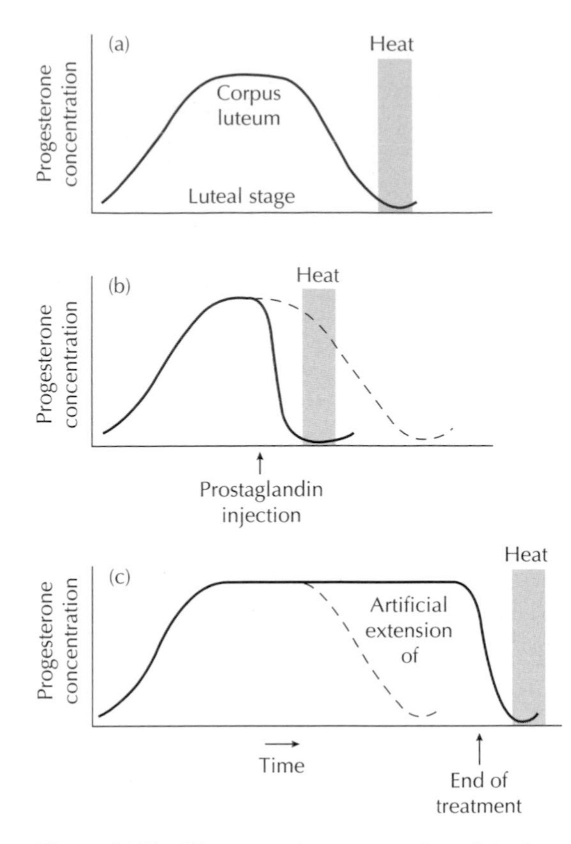

Figure 16.20 Diagrammatic representation of the hormonal events of the oestrous cycle in **a** a normal cycle, **b** after prostaglandin injection, and **c** after progestagen treatment.

progestagen treatment is ceased, the animal returns to oestrus shortly afterwards. Greater synchrony of oestrus is achieved by either combining the progestagen with oestrogen or by $PGF_{2\alpha}$ injection following progestagen withdrawal.

In cattle, progestagens are administered intravaginally with oestradiol, in the form of a coil known as a PRID (progesterone-releasing intravaginal device). PRIDs have an advantage over prostaglandins in that their effectiveness is not dependent on the presence of a corpus luteum and unlike $PGF_{2\alpha}$ inadvertent use on pregnant animals is not likely to cause abortion. In sheep, progestagen-impregnated vaginal sponges are used in a similar fashion to PRIDs. However, immediately following sponge removal, PMSG may be injected to stimulate ovulation rate when synchronizing sheep at the beginning of their breeding season. (For further details of oestrous synchronization procedures in sheep see Chapter 20.)

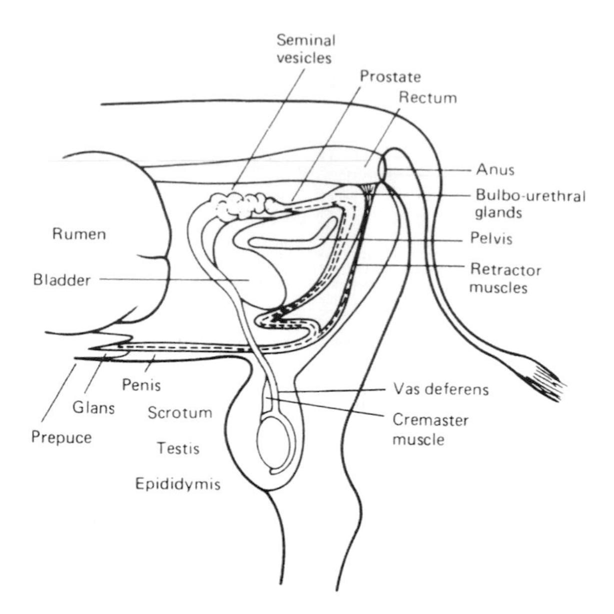

Figure 16.21 Reproductive tract of the bull.

The success of the above techniques depends on the degree of oestrous synchrony that can be achieved. Unfortunately, there is an inevitable variability in animals' responses to treatment, which means that in practice synchronization is never perfect. Pregnancy rates following synchronization are generally similar to those of untreated animals.

The male reproductive system

Anatomy

The male reproductive system is comprised of paired primary and secondary sex organs, together with accessory sex glands (see Fig. 16.21). The primary sex organ is the testis whose function is to produce sperm and male sex hormones. The testes are situated outside the abdominal cavity in the scrotal sac. The secondary sex organs consist of the reproductive tract, extending from the testis to the urethra. The chief role of the reproductive tract is to transport sperm. The urethra, in addition, carries urine from the bladder. Accessory sex organs (the seminal vesicles, the prostate gland and the bulbourethral glands) are situated at the base of the penis and are responsible for seminal fluid production (see later section).

Sexual development

The possession of a Y chromosome by the male fetus causes the undifferentiated gonad to develop into a testis rather than an ovary. The production of male hormone, testosterone, by the fetal testis regulates the subsequent development of the male reproductive tract. The entire reproductive system is thus fully differentiated prior to birth and by then the testes have normally descended into the scrotal sacs via an opening in the ventral abdominal wall known as the inguinal canal. In a small proportion of males, the testes may remain within the abdomen, a condition known as cryptorchidism. Since sperm production depends on the testes being maintained at 2–4°C lower than body temperature, such animals are invariably infertile.

After birth, significant further development is limited until puberty.

Puberty

Puberty in the male is characterized by both physical and behavioural changes. The marked increases in testosterone output at puberty stimulate sperm production and the development of secondary sex characteristics associated with the mature male. These characteristics include further growth of the reproductive tract, an increase in muscle formation, development of a masculine voice and body odour, together with an increase in sexual and aggressive behaviour. The exact onset of puberty, in contrast to the female, cannot easily be determined. Thus puberty in the male may be said to have occurred when enough sperm can be produced to successfully impregnate a female. For example, in the bull, the ejaculate should contain a minimum of 5×10^6 spermatozoa with at least 10% motility. Such a definition, however, takes no account of sexual behaviour of the animal. Sexual activity may reach full intensity considerably later than puberty. Thus puberty should not be confused with full sexual maturity. For bulls, puberty usually occurs at between 7 and 9 months of age, whilst sexual maturity may not be attained until 4–5 years of age.

Male sexual behaviour has two components: libido (or sex drive) and the ability to copulate (mating behaviour). Whilst libido is largely genetically determined, mating behaviour may depend on the social conditions in which the male is reared.

Castration, which is the removal or destruction of the testes, results in a loss of both sperm production and libido. Males castrated soon after birth fail to develop the characteristic male appearance. Growth is also affected by castration (see section on growth). It is important not to confuse castration with vasectomy which is the sectioning of the vas deferens to render the male infertile. Unlike castration, vasectomy has no apparent effect on sexual behaviour or the animal's ability to sexually arouse the female.

Sperm production

Sperm is produced by the seminiferous tubules of the testis (see Figs. 16.22a and b). The germinal epithelium only begins significant sperm output at puberty. Spermatogenesis is temperature-dependent, the temperature of the testes being lower than that of the body. This is achieved by countercurrent heat exchange between the blood of the spermatic artery and vein supplying the testes. Maintenance of testes temperature is also helped by the action of the cremaster muscle which in cold conditions draws the testes closer to the body and in hot conditions moves them further away.

Sperm production is seasonal in wild species. In farm species, seasonal breeding in the male is not so clearly defined. However, the ram shows marked changes in sperm production and testis size depending on the season of the year. There are, however, breed differences and this should be borne in mind when choosing the ram for out-of-season breeding of ewes. Sperm production, as well as being daylength dependent, is affected by environmental temperatures. For example, hot weather may seriously affect sperm production capacity in the boar.

Spermatogenesis

Sperm production by the seminiferous tubules is controlled by FSH and LH. The Leydig cells are stimulated by LH to produce testosterone, which in turn stimulates spermatogenesis. The process of spermatogenesis is outlined in Fig. 16.23.

Spermatogenesis is characterized by a massive cell proliferation accompanied by the eventual halving of the chromosome complement (meiosis) of each spermatozoon. By the time spermatozoa have collected in the lumen of the seminiferous tubule, they have attained their distinctive appearance, as shown in Fig. 16.24. They do not, however, become fully motile until they have been transported to, and resided in, the caudal epididymis. The complete sequence of spermatogenesis takes between 40 and 60 days in farm species.

Testis diameter is a useful criterion for sperm production. The testis of the ram noticeably increases in size as the breeding season approaches.

Seminal fluid is added to spermatozoa prior to ejaculation by the prostate, seminal vesicles and bulbo-urethral glands. A comparison of sperm and semen output in farm species is given in Table 16.12.

Artificial insemination (AI)

The technique of artificial insemination permits the semen from a genetically superior male to inseminate a large number of females. The number of inseminations that can be performed using the semen from a single

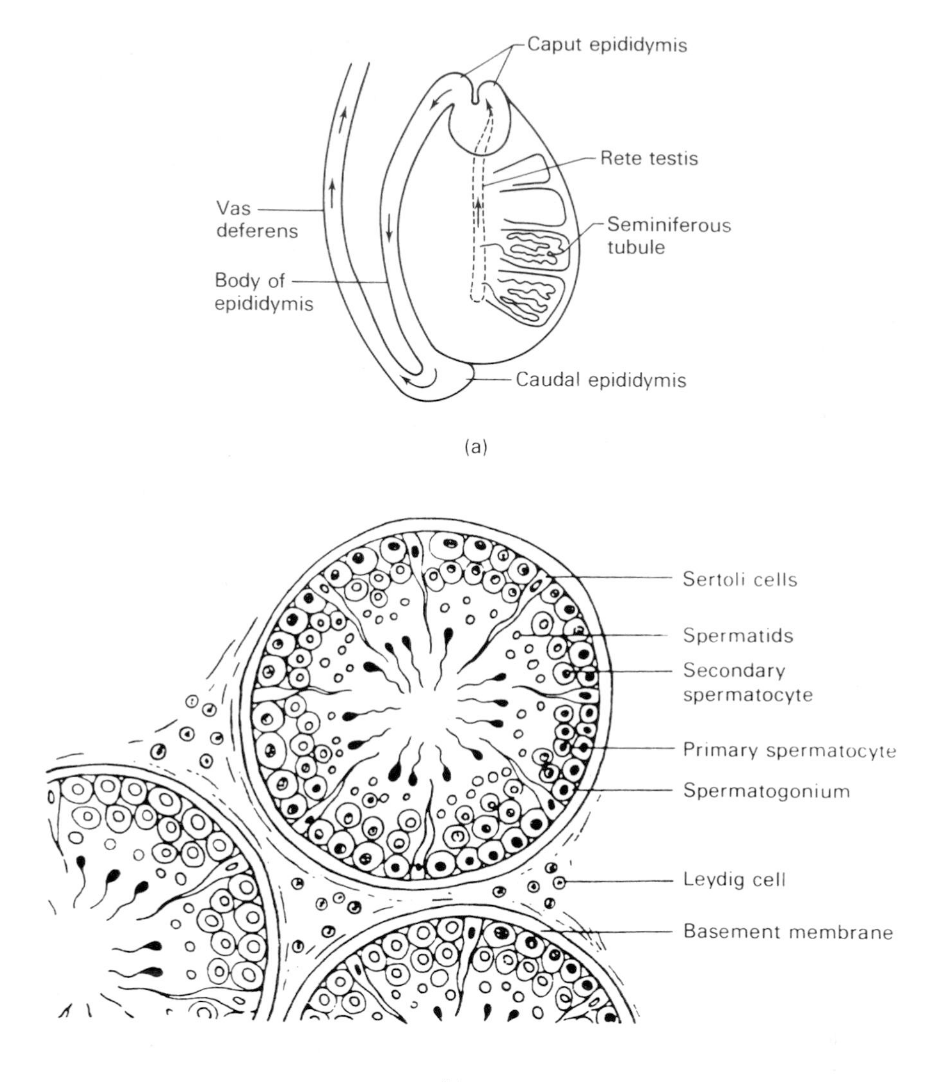

Fig. 16.22 **a** Diagrammatic representation of a longitudinal section of the testis of the bull (arrows represent passage of spermatozoa). **b** Cross-section of the seminiferous tubules of the testis.

ejaculate depends on the extent to which the semen can be diluted without impairing its fertilizing capacity. As many as 1000 cows can be inseminated with the diluted semen from one ejaculate of a bull. This figure is far less for the boar and the ram (see Table 16.12). Another factor limiting the widespread use of AI in pigs and sheep is the difficulty of long-term preservation of semen. Unlike cattle semen, which can be stored deep-frozen, boar and ram semen do not survive the process without some reduction in their fertilizing capacities. Thus AI in pigs and sheep has largely been confined to the use of fresh semen.

Artificial insemination involves a number of distinct stages: semen collection; examination and evaluation of semen; semen dilution and storage; and, lastly, techniques of insemination.

Semen collection

One commonly used method in cattle is the use of a teaser female or a dummy, with the ejaculate collected into an artificial vagina. With pigs, a dummy mounting stool is used and the penis is grasped manually. A technique which is especially successful in the ram is that of electro-ejaculation, where a rectal probe provides a low

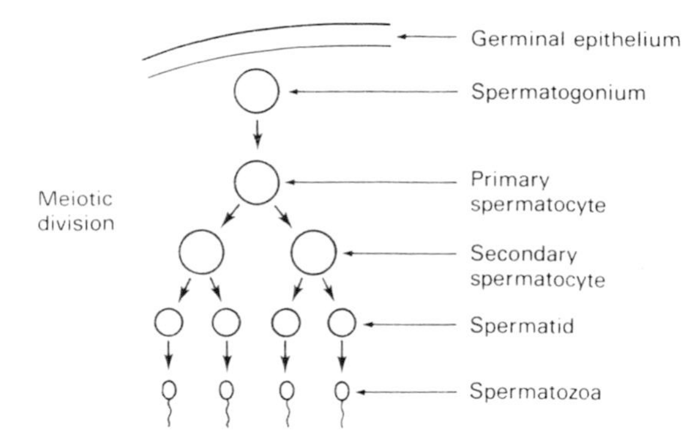

Figure 16.23 Spermatogenesis. (From Peters & Ball, 1996.)

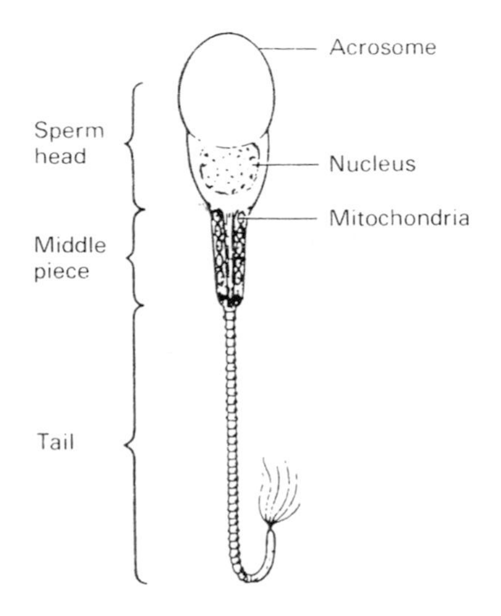

Figure 16.24 Structure of a spermatozoon (or sperm cell). (From Peters & Ball, 1996.)

voltage stimulation of the musculature of the reproductive tract.

Examination and evaluation of semen

During collection and examination every effort is made to avoid temperature shock to the semen, which might result in a drastic reduction in sperm viability. Semen examination involves an evaluation of a number of aspects of sperm quality: sperm numbers; morphological appearance of the spermatozoa; and sperm motility. The sperm count is important in that it determines the extent to which the semen can be diluted, which in turn dictates the number of inseminations possible from a single collection. Morphological examination assesses the quantities of abnormal or immature spermatozoa present. For instance, in the case of a young male or a sire that has been overused, a high proportion of immature spermatozoa may be present. In these instances the semen would not be used for commercial inseminations. Sperm with poor motility would also be rejected on the basis of its probable low fertilizing capacity.

Semen dilution and storage

Semen is diluted with chemical 'extenders', so chosen because of their ability to preserve the life of the sperm and at the same time have a negligible effect on the fertilizing ability of spermatozoa. Such diluents contain nutrients, a buffer to protect against pH changes, and, in the case of deep-frozen semen, a chemical such as glycerol to protect against the effects of chilling. Antibiotics are also included to prevent bacterial contamination. In pigs and sheep, where fresh semen is used, such chemical extenders may prolong its effectiveness for up to 3 days. Once diluted, bull semen is stored in liquid nitrogen at −196°C in plastic straws. Each straw contains sufficient diluted semen for a single insemination.

Insemination techniques

Insemination is performed using specially designed catheters or insemination 'guns'. Such devices in the case of cattle and sheep permit the semen to be deposited further up the female reproductive tract than would be achieved by natural insemination. This allows a more dilute semen to be used without adverse effects on conception rates. It is important to note that with any artificial insemination procedure, stressful circumstances surrounding the event may have detrimental effects on its successful outcome. The timing of insemination is also vital relative to the stage of oestrus of the female, and this is dealt with more fully in Chapter 19.

Factors affecting male fertility

In summary there are a number of factors that affect optimum fertility in the male:

- Genetics – there are considerable individual and breed differences in male fertility.
- Age of the animal – performance increases with age from puberty.
- Nutritional status – e.g. rams should be in good body condition at mating to sustain performance throughout the mating season.
- Environment – daylength and ambient temperature affect level of sperm production.
- Frequency of use – both overuse and underuse affect fertilizing capacity of the male (a mature boar may be used 2–3 times per week and a bull 3–4 times per week).

Table 16.12 Data on sperm and semen production in farm animals. Source: courtesy of Hunter, R.H.F. (1980) *Physiology and Technology of Reproduction in Female Domestic Animals.* Academic Press, London

Species	*Volume of ejaculate (ml)*	*Expected sperm concentration (× 10^6/ml)*	*Total no. of sperm per ejaculation (× 10^9)*	*No. of collections per week*	*Site of semen deposition at: Mating*	*AI*	*Potential no.of inseminations per ejaculate*	*Volume of inseminate after dilution (ml)*	*No. of motile sperm (×10^9)*
Bull	4–8	1200–1800	4–14	3–4	Anterior vagina	Cervix and/or uterus	400	0.25–1.0	5–15
Ram	0.8–1.2	2000–3000	2–4	12–20	Anterior vagina	External cervical os	40–60	0.05–0.2	50
Boar	150–500[1]	200–300	40–50	2–4	Uterus[2]	Uterus[2]	15–30	50–100	2000
Stallion	30–150	100–250	3–15	2–6	Uterus[2]	Uterus[2]	5	20–50	1500[3]

[1] Includes gelatinous secretion of the bulbo-urethral glands.
[2] The ejaculate makes passing contact with the cervical canal.
[3] Total number deposited in three inseminations during the prolonged period of oestrus.

- Health – e.g. physical damage to the penis may cause sufficient pain to affect libido. Damaged legs or feet may hinder mating ability.

Mating

Mating, the process which ensures that spermatozoa are deposited in the reproductive tract of the oestrous female, represents the culmination of a complex sequence of behavioural and physiological events. Prior to actual mating the male and female seek each other out by a combination of visual and olfactory signals. For example, the bull and the ram both display the characteristic Flehmen response which involves curling the upper lip and sniffing the air. This enables the male to detect the sexual pheromones given off by the oestrous female. In a similar way, the female may be sexually aroused by the odours of the mature male. Pheromones are believed to play a part in individual animal recognition. That a male may show a clear preference for some females over others is important with regard to mating management of herds or flocks. For example, running several rams with a flock or using males in rotation takes account of this fact.

Mating itself consists of mounting the female, thrusting, intromission and ejaculation. The exact site of deposition of sperm in the female reproductive tract depends on the species. In the cow and the ewe, sperm is deposited in the vagina, whilst in the sow, sperm is placed beyond the cervix into the uterus. Once in the female reproductive tract, sperm need to make their way to the site of fertilization, the oviduct. This may take as little as 2 h in the case of the pig, but a much longer time may be necessary in cattle and sheep. Only a few hundred of the billions of sperm in the ejaculate manage to reach the oviduct. Fertilization of the egg is still not possible until the spermatozoa have completed the process of sperm capacitation. This term refers to a minimum period of 'acclimatization' to the uterine conditions which is necessary before sperm have the capacity to successfully fertilize an egg.

Spermatozoa and eggs undergo the effects of ageing while in the female reproductive tract. For sperm the effective fertilizable lifespan is 1–2 days and for the egg it is 10–12 h from time of ovulation. Given the rapid ageing of an ovulated egg and the delay in time between insemination and when spermatozoa are able to effect fertilization, it is clear that mating or artificial insemination must precede ovulation by an appropriate amount of time if conception rates are to be optimized (see Chapter 19 for advice on timing of AI in cattle).

Fertilization

Once spermatozoa have become fully capacitated, the acrosomal portion of the sperm head releases enzymes which enable the sperm to penetrate the protective membrane, the zona pellucida, which envelops the egg. There then follows a fusion of nuclear contents of the egg and sperm cells. The zona pellucida undergoes a subsequent chemical change which prevents entry of further sperm. The fused egg and sperm are termed an embryo or zygote.

Pregnancy

Embryonic development

Whilst in the oviduct the embryo begins to divide mitotically thus doubling its cell mass at each successive division. The embryo, by the time it has reached the uterus after about 3–4 days, has developed into a hollow ball of cells.

Implantation

It is generally considered that pregnancy only becomes established once the embryo has successfully attached itself to the lining of the uterus wall or endometrium. This process is known as implantation.

Fetal development

An embryo may be regarded as having attained the status of a fetus once recognizable organ development is apparent; heart, limb buds, liver and spinal cord are all present at an early stage of pregnancy. The majority of growth taking place in the first half of pregnancy, however, is mainly that of the fetal membranes and the placenta. Most of the growth of the fetus itself is confined to the latter stages of pregnancy (see Fig. 20.7).

A summary of the sequence of the major events of pregnancy is shown in Fig. 16.25.

Maintenance of pregnancy

Hormones, especially progesterone, play an important part in the establishment and maintenance of pregnancy following mating and fertilization. In the cow, for example (see Fig. 16.26), following ovulation and service the blood and milk progesterone levels remain identical up to day 17 or 18 whether the egg has been fertilized or not. After this time, in the non-pregnant animal, progesterone levels fall, as described earlier (Fig. 16.26). In the pregnant animal, however, progesterone levels remain elevated due to the continued presence of a corpus luteum. It is thought that the embryo itself plays a major role in preventing the prostaglandin $F_{2\alpha}$-induced regression of the corpus luteum that occurs if fertilization has not taken place. Thus the continuing production of progesterone after day 18 in the pregnant animal both ensures the maintenance of a uterine endometrium to receive the fertilized egg and acts on the hypothalamus/anterior pituitary to inhibit gonadotrophin release. Progesterone therefore prevents further oestrous cycles. The high circulating levels of progesterone characteristic of pregnancy eventually fall at the end of pregnancy, shortly before parturition.

The source of progesterone through pregnancy differs in farm species. In the sow, the corpus luteum persists throughout the whole of pregnancy and is the main source of progesterone. In the cow and the ewe, in contrast, the responsibility for progesterone production switches from the corpus luteum to the placenta approximately halfway through pregnancy. There are obvious practical implications for prostaglandin treatment. Administration of prostaglandin to pregnant animals with an active corpus luteum would result in an abortion.

Embryonic and fetal mortality

It is estimated that 30–40% losses in potential offspring occur between fertilization and parturition in farm species. In cattle, where normally only one egg is ovulated, the death of an embryo or fetus means a failed pregnancy. In contrast, in pigs, where there is a much higher ovulation rate, success or failure may be calculated in terms of the proportion of ovulated eggs (as measured by the total number of corpora lutea on both ovaries) surviving to term.

Some two-thirds of all pre-natal deaths can be put down to embryonic losses. Much of this loss tends to occur either in the first few days following fertilization or at around the time of implantation. Genetic abnormalities, hormonal insufficiency or failure due to an aged egg are thought to be among the chief causes. Any nutritional problems or imbalance occurring at this stage will only serve to exacerbate such losses. If embryo mortality is largely attributed to failure of the fertilized egg, then fetal losses are much more likely to be the result of some deficiency on the maternal side, such as placental inadequacy or uterine infection. A dead fetus may result in an abortion or in some cases it may be retained within the uterus in a mummified condition, where the fetal fluids have become reabsorbed by the uterus lining.

Pregnancy diagnosis

An accurate diagnosis of pregnancy is vital to the reproductive efficiency of breeding livestock. Any pregnancy diagnosis service should also be early, fast and reliable so that animals not successfully inseminated may be rebred as soon as possible. In addition, as in the case of sheep, a determination of the number of fetuses present allows the farmer to feed the mother accordingly. Techniques that can offer early diagnosis suffer the disadvantage of the high rates of early embryo mortality and are therefore less able to give an accurate estimate of

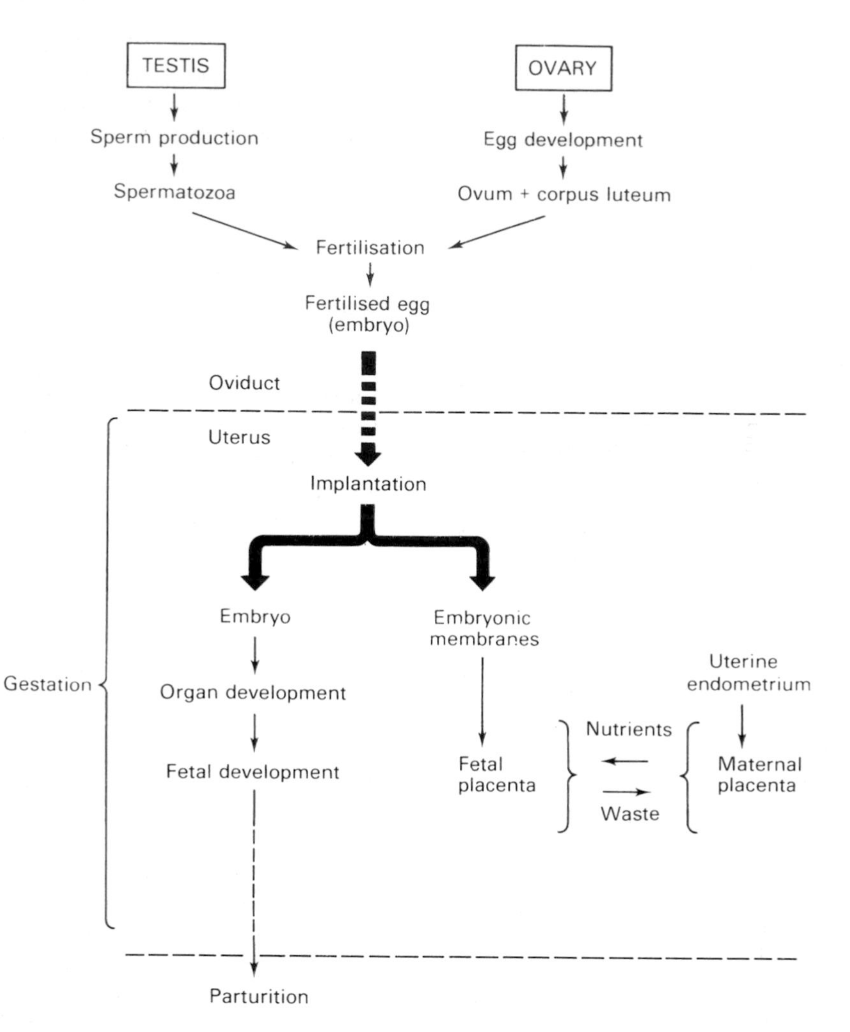

Figure 16.25 Schematic representation of the principal events of gestation.

viable offspring at the end of pregnancy. Some of the more commonly employed pregnancy diagnosis techniques are outlined below.

Hormone assay

This involves the measurement of hormones found in the blood or milk of pregnant animals, by means of a highly sensitive, laboratory-based procedure known as radioimmunoassay. Hormone assay techniques have been confined commercially to dairy cattle. There are two types of hormone diagnosis available: an early progesterone test and a later oestrone sulphate test.

Progesterone test

This method allows pregnancy diagnosis to be made as early as 21–24 days after ovulation. Samples of blood or, more conveniently, milk, are taken at this time and the levels of progesterone present are determined.

Pregnant animals will have high progesterone values, while low values will indicate that the animal has returned to oestrus (see Fig. 16.26). Although the laboratory procedure is highly accurate, the technique does produce a proportion of false positive diagnoses, which may be due either to cows that have abnormal-length oestrous cycles or to insemination at the wrong time of the cycle. There is a further drawback implicit in such an early dignosis, in that an animal correctly diagnosed as pregnant may subsequently lose the embryo.

Oestrone sulphate test

This is a much later test (at around 15 weeks) which can confirm pregnancy with almost 100% reliability. This is

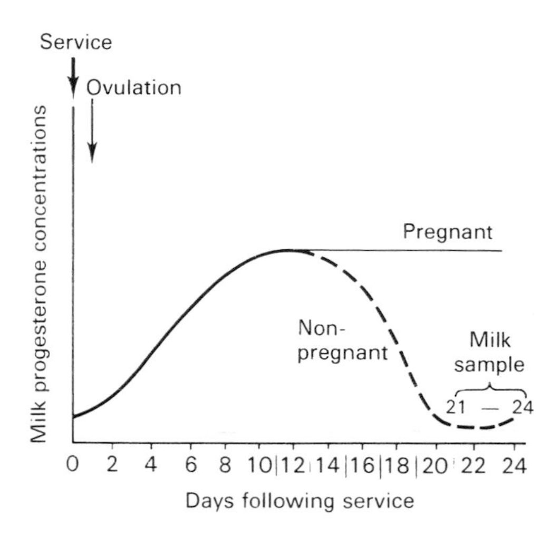

Figure 16.26 Milk progesterone profiles in a pregnant and non-pregnant cow following service and ovulation.

because the method, unlike the progesterone test, measures a hormone, oestrone sulphate, which is derived from the fetus itself. The test is of limited practical use because of the lateness of the diagnosis.

Rectal palpation

The technique involves the insertion of an arm into the rectum of an animal in order to manually detect (palpate) the presence of a fetus within the underlying uterine horns. This procedure is largely limited to cattle and needs to be performed by a trained operator. The fetus can be palpated from around day 40 up to the fifth month of pregnancy. After this time palpation becomes impossible until after month 7 because the uterine horns become greatly enlarged and 'disappear' below the pelvic brim.

Ultrasonic techniques

Such procedures consist of ultrasonic detection of fetal or placental structures. A variety of techniques have been developed, all of which involve the same basic principle. A probe is applied to the surface of the animal's abdomen which transmits and receives an ultrasonic beam. The reflected signal, which is able to detect the presence of the fetus and associated tissues, is then converted into an auditory or visual output. Recently, real-time ultrasound scanners, used routinely in human medicine, have been developed for farm use. This device allows an image of the uterine contents, including the fetus, to be displayed on a screen. This method has been commercially used in sheep to detect fetal lambs between 50 and 100 days of pregnancy (see Chapter 20 for a fuller description). With all the available ultrasonic methods, a skilled operator is needed to correctly interpret the results obtained. Commercially, ultrasonic pregnancy diagnosis has been restricted to pigs and sheep.

Parturition

Gestation length

The onset of parturition or birth is determined by the length of gestation. Gestation length is largely determined by the fetus. Whilst it is the mother that decides the hour of birth, it is the fetus that decides the month, week and day. Other factors that influence gestation length include species, breed, litter size, sex of the fetus, age of the dam and the genotype of the sire.

Initiation of parturition

As stated above, it is the fetus that is chiefly responsible for initiating birth. Recent evidence has shown that once the fetus achieves its maximum growth, hormonal changes cause progesterone production to be 'switched off'. Thus the withdrawal from circulation of the hormone responsible for maintaining pregnancy signals the beginning of parturition. Increases in oestrogen and prostaglandin $F_{2\alpha}$ follow, with a resultant release of oxytocin, which is responsible for causing uterine contractions (i.e. labour). A diagrammatic summary of events in the sheep is shown in Fig. 16.27.

Stages of labour

The process of parturition may be divided into three stages:

- *The first stage* – the preparatory stage involving dilatation of the cervix and relaxation of the pelvic ligaments.
- *The second stage* – the expulsion of the fetus through the pelvic canal.
- *The final stage* – the expulsion of fetal membranes and the initiation of uterine involution.

Any difficulty or prolongation of the birth process is termed dystocia.

Induction of parturition

Simulation of the events which initiate parturition (as outlined in Fig. 16.27) can be brought about by exogenous hormone treatment of late pregnant animals.

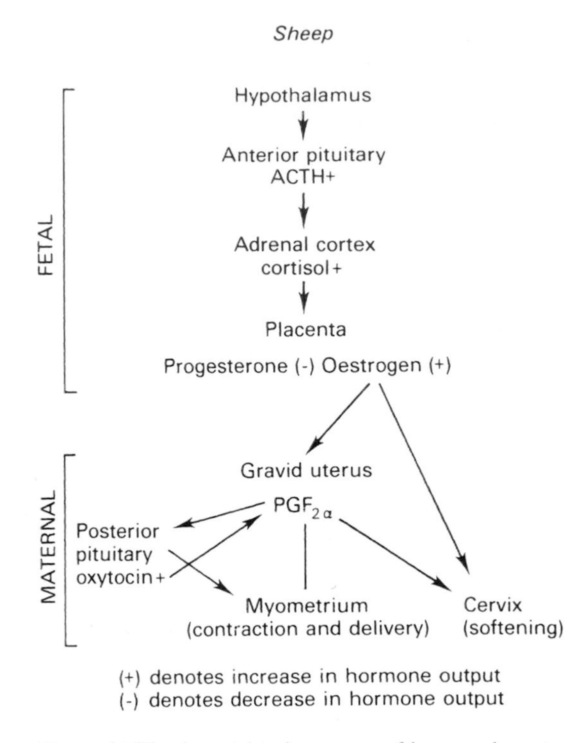

Figure 16.27 A postulated sequence of hormonal events controlling the onset of parturition in the sheep. [Courtesy of First, N.L. (1979) Mechanisms controlling parturition in farm animals. In: *Animal Production* (ed. H. Hawk), pp. 215–257. Allanheld Osmun, Monclair, New Jersey].

Corticosteroids, prostaglandin $F_{2\alpha}$ and their synthetic analogues, when administered to animals nearing the end of their natural gestation, will induce parturition within 2–3 days of treatment. The method is not without its problems, however. For example, in cattle such inductions of parturition have been associated with increased calf mortality and greater incidence of retained placentas. Used commercially the procedure allows a more exact timing of birth to fit in with management and labour requirements. (For a detailed description of induction in lambing see Chapter 20).

Lactational anoestrus

This describes the situation, following parturition, when active lactation is accompanied by minimum ovarian activity. Behavioural oestrus and complete resumption of breeding cycles are not observed for several weeks in cattle, sheep and pigs, depending on such factors as milk yield, nutritional status and suckling intensity.

Recent advances in reproductive technology

This section will attempt to review some of the more recent technological developments not already mentioned earlier in this chapter.

Superovulation

This process involves the injection of hormones such as FSH and pregnant mare serum gonadotrophin (PMSG) to increase the number of ovulatory follicles and hence the number of eggs released by the ovaries. Superovulation has been used in association with embryo transfer to allow the genetic potential of superior breeding females to be more fully exploited.

Immunization against ovarian steroids

Immunization against ovarian steroids is a newly developed technique which improves the ovulation rate without the risk of excessively large litters that sometimes results from superovulation. The female is injected with ovarian steroid and the immune system is stimulated to produce antibodies against the hormone. Normally, within the animal ovarian steroids regulate, by a process of negative feedback, the gonadotrophins that are responsible for follicle development and ovulation. Immunization is believed to interrupt this negative feedback and thereby produces an increase in ovulation rate. The ovarian steroid androstenedione, commercially available as Fecundin, has been successful in improving lambing percentages. Two injections are required, at 8 and 4 weeks prior to mating.

Melatonin and out-of-season breeding

The pineal gland and its secretion, melatonin, have a central role in the photoperiodic control of reproduction in seasonal breeders. Melatonin, when fed in the afternoon to out-of-season sheep, has the effect of creating a chemical 'darkness' which convinces the animal that it is experiencing a shortened daylength. This results in an advance of the breeding season. Melatonin is licensed for use in sheep as a subcutaneous implant marketed under the trade name Regulin.

Embryo transfer

The commercial application of the technique of embryo transplantation has been largely confined to cattle. A number of eggs are produced by superovulation in a donor cow and, after fertilization, non-surgically collected and transplanted into the uteri of recipient cows. In this way a greater number of calves can be produced from a genetically valuable female than would be achieved normally during its reproductive lifetime. It is possible to freeze embryos for long-term storage and still achieve acceptable pregnancy rates with the thawed embryos. This facility therefore offers a useful extension to the scope of embryo transfer.

Sex determination

This can be achieved either by sexing of embryos or by attempting to separate the X- and Y-bearing spermatozoa. Although laboratory techniques exist to do this, no commercial method is yet available for use in animal production. The Y chromosome of the male produces a protein known as the HY antigen, which it is possible to detect by immunological means. Alternatively, it is possible to karyotype the cells of the embryo by a procedure that examines and identifies the sex chromosomes. A successful method of sex determination would have enormous benefits for the AI industry, given that half the number of dairy calves produced consist of males that may be unwanted. Being able to select the sex of offspring would enable quick genetic improvement of breeding replacements.

A rapid milk progesterone test

A rapid, highly sensitive progesterone assay is now available commercially, employing ELISA procedures (enzyme-linked immunosorbent assay), which can be used 'on-farm' by untrained operators. A bulk milk sample is obtained between 21 and 24 days after service in the same way as for the laboratory-based radioimmunoassay. The ELISA method shares the same disadvantage as its more-established counterpart in the number of false positives that are produced (see earlier section on pregnancy diagnosis).

Lactation

The growth and development of the mammary gland and the subsequent manufacture and secretion of milk represent an important phase in the reproductive cycle of mammals. This section describes the structure and development of mammary tissue, the processes involved in the synthesis, secretion and release of milk, together with the underlying hormonal control of these events.

Anatomy of the mammary gland

The structure of the adult mammary gland of the cow is shown in Fig. 16.28. The majority of the gland consists of closely packed lobules, each made up of numerous clusters of alveoli. It is the epithelial cells lining the lumen of each alveolus, the secretory cells, that are the site of milk synthesis. The secreted milk is stored in the alveolar lumen prior to passing into the small ducts leading from each alveolus. The arrangement of the alveoli and ducts is rather similar to a large bunch of grapes, with the alveoli being the grapes and the ducts the stem branches. The ducts merge into ever larger ducts until the largest of these, the collecting ducts, empty into the gland cistern. In the cow prior to milking, 40–50% of the milk is held in the gland cistern and larger ducts, while the rest is retained within the small ducts and alveolar lumen. In contrast, in the sow approximately 90% of the milk is held in these latter structures.

Continuous with the gland cistern is the teat cistern and the streak canal through which the milk is eventually secreted. The streak canal is surrounded by a sphincter composed of circular smooth muscle fibres. In cows that tend to leak milk, this sphincter is not tight enough, whilst the opposite is the case with slow milkers. The mammary gland of the sow differs in that each teat is served by two separate streak canals, each leading to the teat cistern and gland cistern. The sow normally has seven pairs of glands distributed bilaterally on the ventral surface of the body from thorax to inguinal region, a single teat serving each gland. The cow has two pairs of glands, found in the inguinal region, the rear two accounting for approximately 60% of total milk production. The ewe has a single pair of glands, inguinally situated.

The udder of the cow may weigh as much as 15–30 kg and as much again when full of milk. Support for the udder is therefore very important. This is achieved by means of suspensory ligaments which are attached to the body wall and pelvic girdle by tendons (see Fig. 16.29). The lateral suspensory ligament is fibrous and inelastic while the median suspensory ligament is more elastic; thus, when the udder fills with milk, the median ligament stretches to allow the udder to expand.

Mammary development

At birth there is little mammary development present apart from teats, connective tissue and a rudimentary duct system. There is no secretory tissue, its place being taken by fat. At puberty there is an increase in ductal development, but it is only when the animal is pregnant that significant development of the duct and lobule-alveolar system commences.

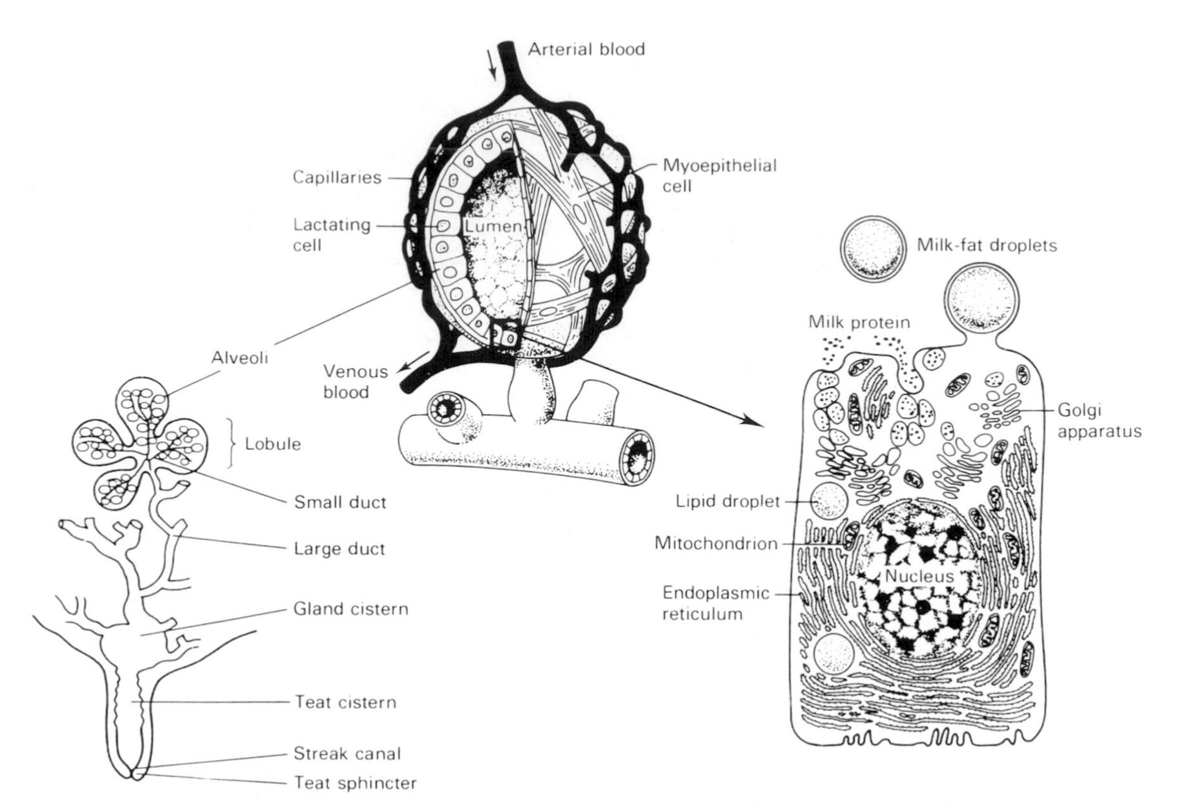

Figure 16.28 Structure of the mammary gland of the cow.

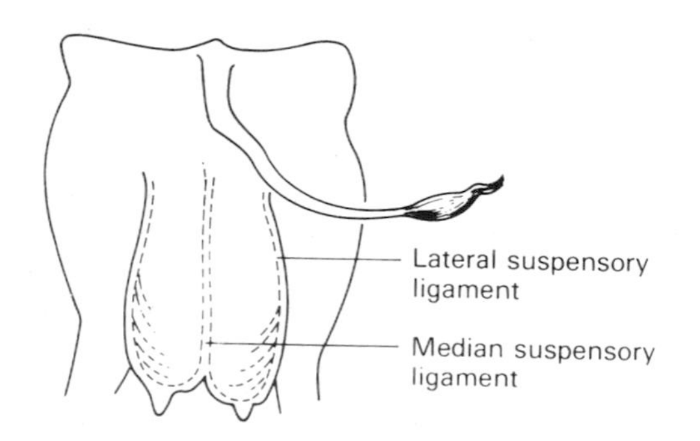

Figure 16.29 Support structures for the udder of the cow.

Mammary gland differentiation is under the control of a complex battery of hormones from the anterior pituitary, ovary and adrenal gland. A summary of their effects on mammary development is given in Fig. 16.30. Lactogenesis, which is the initiation of milk secretion, occurs towards the end of pregnancy and at parturition. It is thought that the trigger for this event may be the decline in progesterone concentrations at parturition, although prostaglandin $F_{2\alpha}$ may also be involved. Following parturition there is a proliferation in the secretory cell population which, together with an increased output of each secretory cell, is responsible for the dramatic mammary secretion at this time. Suckling activity, which is itself related to the number of offspring, encourages greater secretory response. Conversely, a decline in suckling or milking leads to a diminished milk output. At weaning or 'drying off', the mammary gland regresses or involutes and this is marked by a dramatic reduction in the secretory cell population.

Milk biosynthesis

Whilst the composition (Table 16.13) and quantity of milk secreted by the dams of the different domestic species vary considerably, the milk from all species has the same basic qualitative composition. Milk essentially consists of two phases: an aqueous phase, in which are partitioned the main proteins (casein) in colloidal suspension and lactose, the remaining proteins, minerals and water-soluble vitamins in solution; and a lipid phase

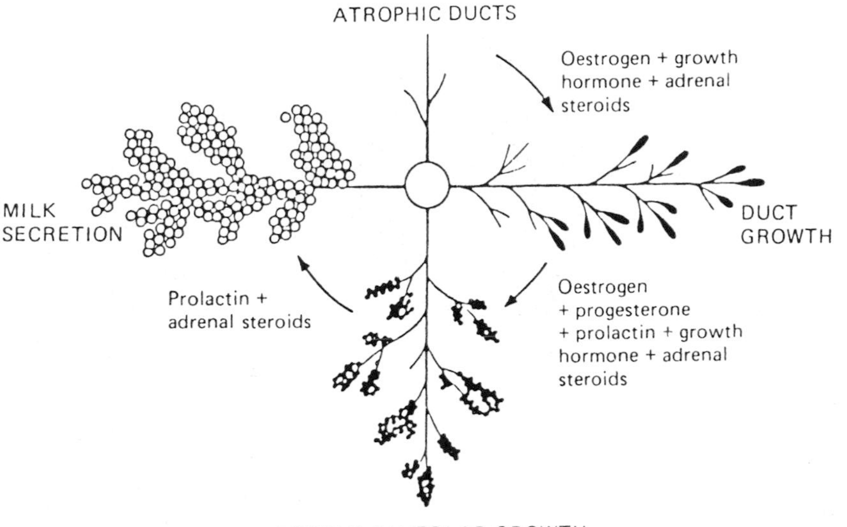

Figure 16.30 Summary of hormonal control of mammary developments. (From Peters & Ball, 1996).

Table 16.13 Approximate average composition of milk of different species (% by weight)

	Total solids	*Fat*	*Protein*	*Lactose*	*Minerals/ water-soluble vitamins*
Cow	12.5	3.7	3.4	4.7	0.7
Sow	20.0	8.0	6.0	5.2	0.8
Ewe	17.0	5.5	6.2	4.5	0.8
Rabbit	35.0	18.3	14.0	2.0	0.7
Human	12.4	3.8	1.0	7.0	0.6

containing triglycerides, phospholipids, sterols and fat-soluble vitamins. The lipid in milk is about 99% triglyceride and exists in the form of globules ranging from 3–5 μm in diameter surrounded by a complex membrane. This is comprised of the other lipid components consisting of phospholipids, cholesterol and fat-soluble vitamins and which is acquired at the time of secretion from the lactating cell.

The protein fraction in milk consists of two major protein groups, the caseins, which exist in the form of colloidal particles called micelles and of which there are four types – α_s, β, γ and κ-caseins – and the serum (whey) proteins consisting of principally β-lactoglobulin and α-lactalbumin.

Because of the commercial importance of cow's milk as a food, more is known about it than about the milk of other mammals and the mechanisms of synthesis of milk have been much investigated. In this section, therefore, synthesis will be described with reference to synthesis in the ruminant and attention drawn where appropriate to differences in the monogastric.

Site of biosynthesis

The site of biosynthesis is the vast number of epithelial cells which line the alveoli, each cell in an alveolus discharging its milk into the lumen or hollow part of the structure (see Fig. 16.28). A feature of the secretory cells is the large number of mitochondria present, being indicative of the high energy demand of biosynthesis and also the marked hypertrophy of the Golgi apparatus and endoplasmic reticulum which occurs at the onset of lactation, these being the sites of synthesis of milk constituents. The alveoli of the mammary gland are well supplied with blood capillaries and it is from the arterial supply that the metabolites used in milk synthesis are drawn. Arterio-venous (AV) difference studies have shown that acetate, 3-hydroxybutyrate, free fatty acids, glucose, amino acids and proteins are withdrawn from the blood as it passes through the mammary tissue and act as precursors in milk synthesis. The origins of these precursors are illustrated in Fig. 16.31.

Milk protein synthesis

Arterio-venous difference studies have shown that the total free amino acids withdrawn from blood are apparently sufficient for synthesis, but since the balance of

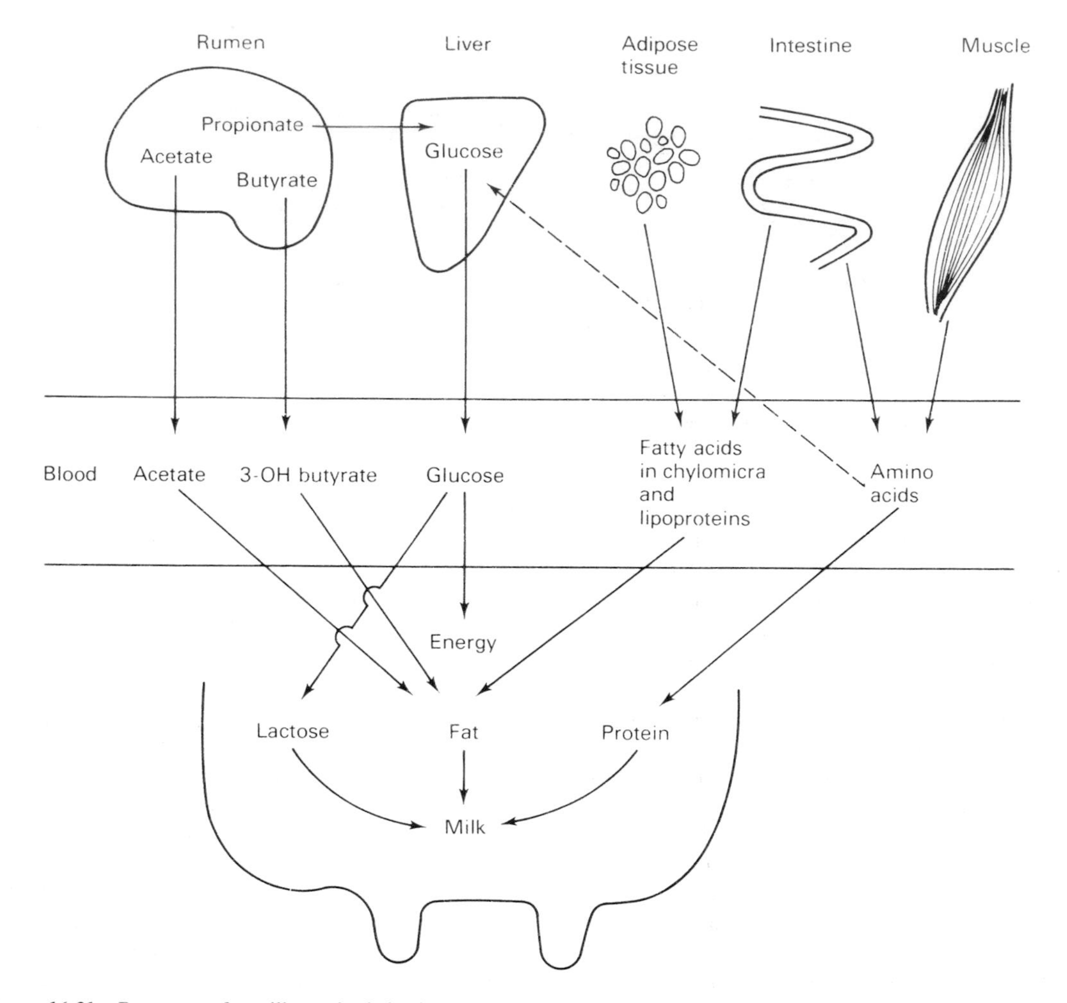

Figure 16.31 Precursors for milk synthesis in the ruminant.

absorbed amino acids is not the same as that present in the milk proteins, a certain amount of amino acid synthesis must occur. Of the amino acids incorporated into milk protein the provision of the amino acids essential for protein synthesis is of greatest concern since the provision of non-essential amino acids can be achieved by synthesis within the lactating cell. A number of studies have indicated that under the dietary circumstances encountered by dairy cows in the UK, methionine is likely to be the first limiting amino acid for milk production and that there may be value in giving 'protected' methionine supplements. Protein synthesis involves the assembly of amino acids into a polypeptide chain through peptide bonds. The sequence of the amino acids in the various milk proteins is dictated by the usual method of transcription by ribonucleic acid on the ribosome of the cell.

Following the release of the protein from the endoplasmic reticulum they are 'packaged' in secretory vesicles derived from the Golgi apparatus and transported to the apical membrane of the cell before being secreted by exocytosis into the lumen of the alveolus.

Milk fat synthesis

The fatty acids present in the triglycerides in milk fat range from C_4 to C_{18} in chain length and in comparison with other fat sources are characterized by having a relatively high proportion of short chain fatty acids. The origins of the different fatty acid groups vary (Fig. 16.32). Thus, short chain fatty acids and most of the

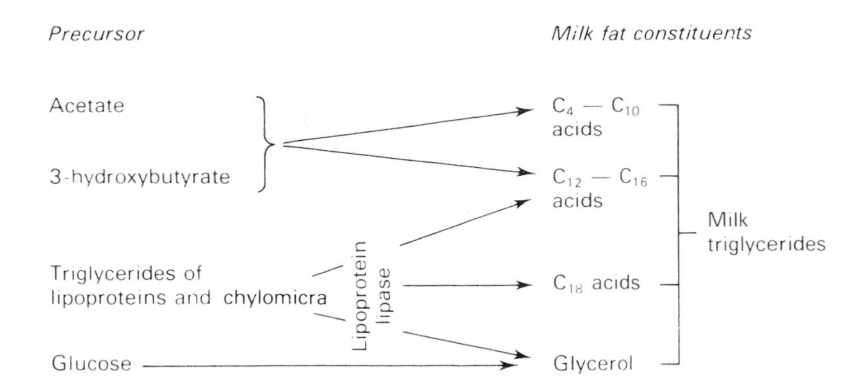

Figure 16.32 Origins of milk fat triglycerides.

medium chain fatty acids are synthesized within the lactating cell using acetic acid and to a lesser extent 3-hydroxybutyrate as precursors, whereas the long chain fatty acids and some of the medium chain fatty acids are derived directly from circulating lipoproteins in the plasma. Fatty acids are removed from circulating lipoproteins by the enzyme lipoprotein lipase present in the capillary membrane. Whereas acetic and 3-hydroxybutyric acids are mainly derived through absorption from the rumen, the fatty acids present in circulating lipoproteins may be derived from the mobilization of adipose tissue or absorption of dietary or microbial lipid from the small intestine. The presence of odd-numbered and branched chain fatty acids in milk fat is due to the latter source.

Non-ruminant animals appear to use glucose for the synthesis of milk fat. In all species the metabolism of glucose by the lactating cell is the precursor of the glycerol moiety for triglyceride synthesis and in addition is the main energy source for both fat and protein synthesis in the lactating cell.

In the secretion of fat from the lactating cell, droplets of fat coalesce and migrate towards the apical membrane of the cell where they are enveloped in membrane as they push their way into the lumen of the alveoli.

Lactose synthesis

Of the major milk constituents lactose is chemically the simplest. Synthesis takes place from glucose withdrawn from the blood as it passes through the mammary tissue. In ruminant animals this utilization of glucose for lactose synthesis represents a major drain on the animal's glucose reserves. Lactose secretion would appear to share the same secretory route, via vesicles, with the milk proteins, together with water, ions and other water-soluble materials.

Metabolic control and nutritional manipulation of milk biosynthesis

The rate of synthesis of milk components by the lactating cells is dependent on the availability of milk precursors and the ability of the animal to partition these precursors into milk production. The latter appears to be mainly a function of the genetic make-up of the animal but it is also influenced by diet.

The ability of animals to partition nutrients into milk production as opposed to body gain is controlled by a variety of hormones. It appears that the most important are growth hormone (GH), insulin, glucagon and the insulin-like growth factors (IGF-1 and IGF-2). Research has shown that GH, also termed somatotropin, has a major role in partitioning nutrients away from deposition in body tissues and towards milk production. Research using bovine somatotropin (BST) derived from recombinant-DNA (rDNA) technology has shown that the injection of BST can increase yields in cows by between 10 and 40%. The BST is thought to operate indirectly via IGF-1 by supplying the mammary gland with more nutrients through increasing the rate of body fat mobilization and hence the levels of plasma free fatty acids, acetoacetic acid, 3-hydroxybutyric acid and lipoproteins available to lactating cells. It may also increase uptake of glucose and amino acids by the mammary gland. Although widely used around the world, BST is not licensed for use in the EU.

The greatest effects of diet on the secretion rates of milk constituents are caused by the form of dietary carbohydrate and the fat content of the diet. Dietary protein, within normal limits, has lesser but economically important effects.

It has long been recognized that the ratio of acetic + butyric : propionic acid in the rumen, which is largely a function of type of dietary carbohydrate (see the earlier

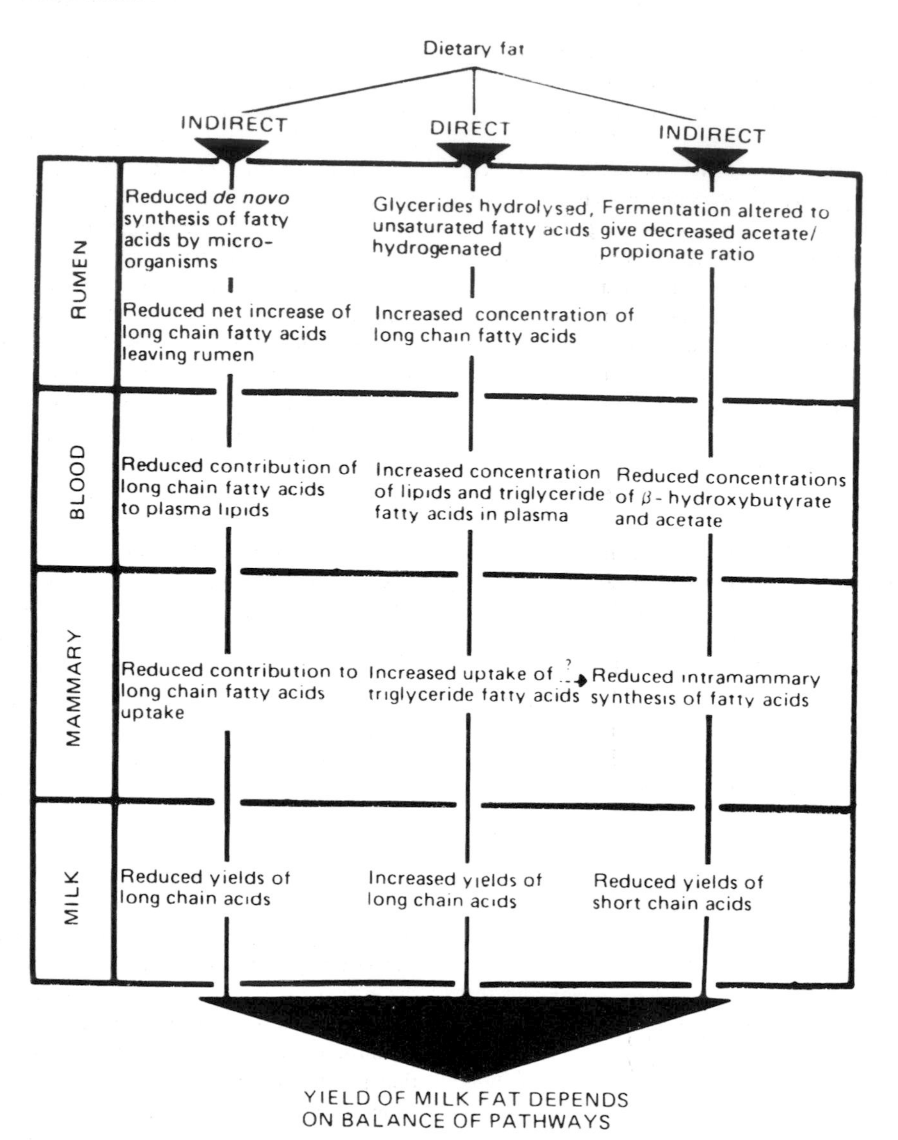

Figure 16.33 Effects of dietary lipid on aspects of metabolism related to milk fat synthesis. [Courtesy of Storry, J.E. (1981) *Recent Advances in Animal Nutrition*. Butterworths, London.]

section headed 'Carbohydrate digestion'), has an influence on milk fat content. Thus increasing the roughage : concentrate ratio increases the yield of milk fat and vice versa. This is due to the fact that propionic acid derived from the fermentation of concentrates in the rumen is gluconeogenic and elevates plasma insulin. As a consequence there is reduced mobilization of fatty acids from body fat, and milk fat secretion falls due to a reduced supply of precursors for milk fat synthesis.

The effects of dietary fat on milk composition are complex. They are mediated through direct effects on the availability of diet-derived fatty acids in the plasma and the indirect effects of dietary fat on rumen fermentation (see 'Carbohydrate digestion'), and on the *de novo* synthesis of short chain fatty acids in the rumen (Fig. 16.33). Thus the yield of milk fat depends on a balance of these effects. Not only total yield but also the composition of the milk fat can be influenced through the inclusion of dietary fat. The use of 'protected' fat systems can allow the incorporation of dietary fat without some of the indirect effects of dietary fat being incurred.

Milk protein synthesis is influenced by the amount and balance of amino acids delivered to the mammary gland in the blood. This is dependent on supplies of digestible microbial true protein (DMTP), mainly a function of dietary fermentable metabolizable energy (FME), and digestible undegradable protein (DUP) (see earlier section on 'Protein digestion'). Along with increasing FME and DUP in the diet, increasing the supplies of limiting amino acids such as lysine and methionine has been shown to benefit milk protein production.

The production of 'designer milks' as sources of functional nutrients, e.g. conjugated linoleic acid, that benefit human health is currently the subject of research.

The milk ejection mechanisms

As a result of secretion, milk accumulates in the alveoli and ducts of the mammary gland. This milk is ejected from the alveoli and ducts by contraction of the smooth muscle fibres (myoepithelial cells) which surround the alveoli. This process of milk 'let down' or, more accurately, milk ejection, is the result of a neurohormonal reflex (see Fig. 16.34). The reflex is made up of a neural sensory input and a hormonal output. Tactile stimulation of the teats (through the butting and suckling of the offspring) provides the stimulus which evokes the release of oxytocin from the posterior pituitary. Oxytocin then acts on the myoepithelial cells to cause their contraction, and hence milk ejection. The time from initial stimulation to the start of alveolar contraction is less than 1 min; in the cow the effective duration of milk ejection is 2–5 min, and in the sow it is about 10–30 s.

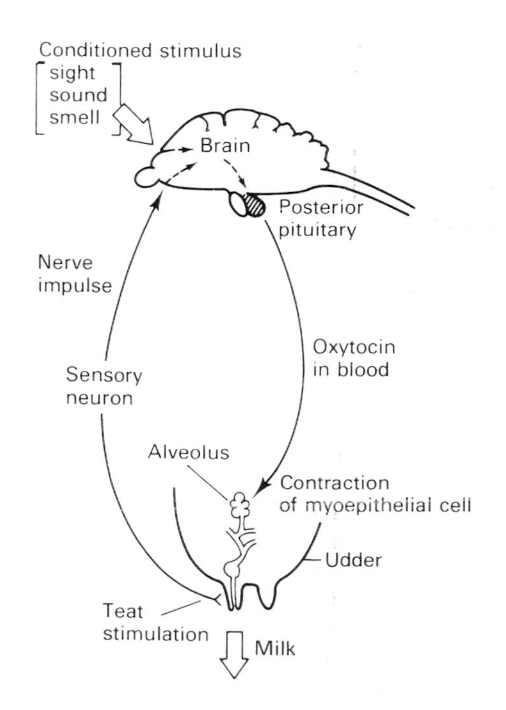

Figure 16.34 Diagrammatic representation of the milk ejection reflex.

Because a proportion of milk is stored in the gland cistern and larger ducts, initially milk is passively withdrawn from the gland, to be followed by the active milk-ejection phase. The completeness with which milk is removed from the gland depends on the amount of oxytocin released from the posterior pituitary, which itself depends on the strength of the initial stimulation.

Natural suckling is more efficient in this respect than either hand or machine milking. Once the pituitary store of oxytocin has been depleted, there is a substantial delay while more milk is synthesized and the store of oxytocin is replenished.

In machine-milked dairy animals the normal routine prior to milking is sufficient to trigger a milk ejection reflex. This is a good example of what is termed a conditioned reflex. The milk ejection reflex can be inhibited by emotional disturbance. It is likely that this is due to a release of adrenaline which, by its vasoconstrictive action on the blood supply to the udder, reduces oxytocin availability to the gland.

Growth

A knowledge of the processes involved in animal growth and an ability to manipulate these processes may be used to influence the efficiency of production. In the production of animals for meat a knowledge of mechanisms involved in the control of muscle, fat and bone deposition is relevant such that we may efficiently produce animals with a high proportion of muscle and an optimal amount of fat at market weight. Growth is also a characteristic of animals that is related to reproductive performance (see earlier section).

The nature of growth

Growth has two aspects. The first is the increase in mass (weight) per unit time of the whole animal or part of the animal. The second involves changes in form and composition resulting from differential growth of the component parts of the body. In studies of meat animals we are primarily concerned with the growth of the major tissues of the carcass which are muscle, fat and bone and with the proportions of these three major tissues in the carcass.

A growth curve for weight or size in the whole animal follows the pattern shown in Fig. 16.35. From birth the animal will typically grow along a sigmoidal curve, showing acceleration at about puberty and slowing down as maturity is approached. Animals are usually slaughtered around the pubertal phase of their growth curve at between one-half and two-thirds of mature body size.

The liveweight growth of farm animals is the gross expression of combined changes of the carcass tissues, organs and viscera and gut fill. The animal productionist is primarily interested in the growth of the carcass tissues comprising the muscle, bone and fat. The carcass weight of the animal is comprised of the liveweight less the viscera, gut fill, blood, hide, head and feet and may be influenced by a variety of biological factors, particularly level of fatness and diet type and by carcass handling procedures.

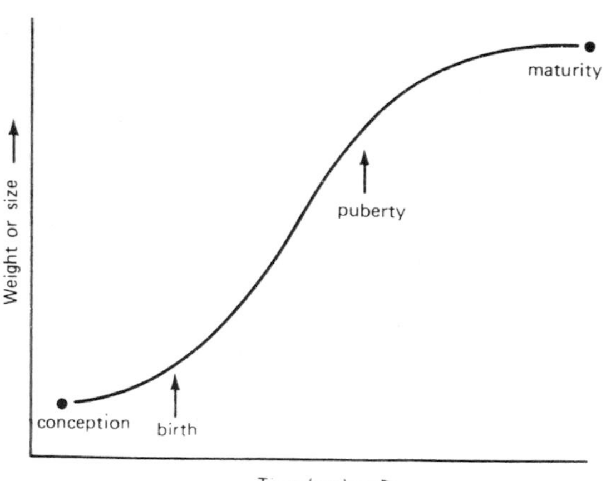

Figure 16.35 The sigmoid growth curve.

The carcass is an extremely variable commodity reflecting the temporal growth of its component tissues. Although each of the carcass components follows a sigmoidal growth curve similar to that of liveweight, the different growth curves are not in phase with one another. This is due mainly to the fact that the different tissues vary in priority for the available nutrients. The bone tissue reaches maximal growth rate prior to maximal muscle growth, with adipose tissue being the latest of the body tissues to attain peak growth intensity (Fig. 16.36). Thus, carcass composition in terms of the proportions of muscle, fat and bone changes as an animal grows (Fig. 16.37).

The chemical composition of the body changes with age in a way that clearly substantiates the phasic development of the tissues. The most marked changes with age are the decrease in the proportion of water in the body and the increase in the proportion of lipid. The almost inverse relationship between the water and fat content of the body reflects the lower water content of fat (10%) compared with that of muscle tissue (75%). The percentages of protein and mineral matter decline only slowly as growth proceeds, mainly as a result of changes in lipid content. If chemical composition is expressed on a fat-free basis the chemical composition of the body remains remarkably constant during growth, reflecting that the bones and muscles of the limbs and trunk remain in proportion to one another.

Tissue growth

The growth of individual tissues is a reflection of the increase in both the size and number of cells. An increase in number of cells is called *hyperplasia*, and an increase in cell size is termed *hypertrophy*. Hyperplasia

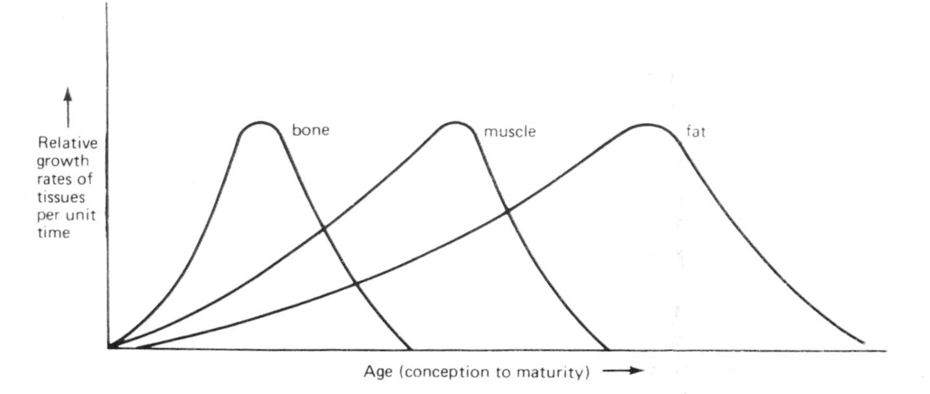

Figure 16.36 Relative growth rates of body tissues from conception to maturity.

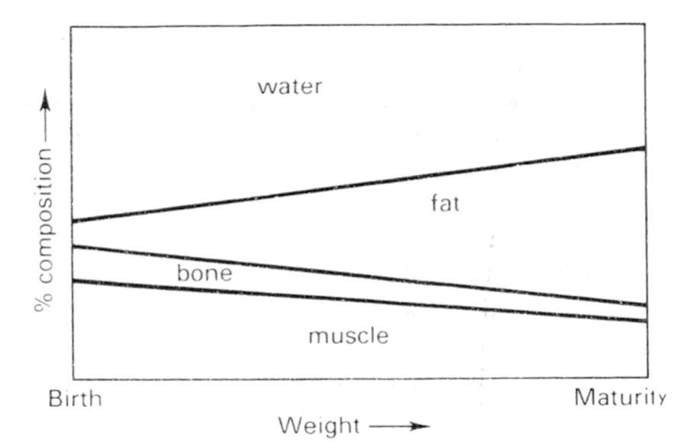

Figure 16.37 Diagrammatic representation of changes in carcass composition.

and hypertrophy of all body cells occur during embryonic life.

Bones grow during both prenatal and postnatal periods through the interaction of three different cell types: chondrocytes, osteoblasts and osteoclasts. *Chondrocytes* are cells that produce cartilage, osteoblasts produce bone collagen and other bone components, and osteoclasts break down bone during the process of resorption. Bone grows in length by ossification of the epiphysial cartilage. Once this is complete the bone will stop growing in length. In mature animals bone is composed of approximately 50% mineral [$Ca_{10}(PO_4)_6(OH)_2$] and 50% organic matter and water, on a weight basis.

In muscle tissue development the number of muscle cells is fixed early in life. Thus the growth of muscle that follows birth is increasingly due to hypertrophy of cells through protein deposition. Accretion of protein in muscle represents the net effect of two processes: protein synthesis and protein degradation. Control of protein deposition may be exerted through either or both of these processes although growth clearly involves synthesis exceeding degradation. The supply of nutrients, especially amino acids and energy, and the hormonal status in the animal, particularly the balance of insulin, growth hormone, the glucocorticoids and the sex steroids, are influential.

Fat consists of fat cells (adipocytes) and supporting connective tissue. The period during which hyperplasia of adipocytes occurs is unclear, but some postnatal as well as prenatal development does occur. Hypertrophy of adipose tissue occurs through the net deposition of fatty acids in adipocytes. Adipocyte size may vary considerably according to the animal's energy balance, reducing in size in times of negative energy balance and increasing in times of energy surplus. There are two types of adipose tissue, which are referred to as *white fat* and *brown fat*. Most body fat is white and it functions as a depot of stored energy. Brown fat is very active metabolically and can be used to maintain body temperature, which is especially important in neonatal animals (see the later section headed 'Environmental temperature and animal production').

Factors influencing growth

The patterns of tissue growth and consequent changes in chemical composition of the body are influenced by several interrelated environmental and genetic factors.

Genetic effects

Genetic differences occur in that animals of the same species vary in their mature size and weight, a feature that is reflected in differences in carcass composition. Generally, animals that reach the asymptote of the growth curve at an early age have lower mature body weights. The converse also applies. This is exemplified in the Angus and Friesian breeds of cattle and largely reflects differences in the timing of adipose tissue growth. Differences in mature weight are, however, probably not solely responsible for differences in composition between breeds. Thus, for example, Soay sheep have been shown consistently to contain less fat than Down breeds at the same stage of maturity. Although it is often implied that animals that have large mature body weights are more efficient because of their lower maintenance energy requirement per unit of body weight, there is little evidence that this is the case and the overriding factor governing efficiency within a species is the choice of slaughter weight.

In addition to these differences in total body size and body composition there are genetic influences on the proportions of muscle and bone tissue. These are most noticeable in the comparisons of dairy and beef breeds of cattle, the latter generally having more muscle relative to bone when comparisons are made at equal levels of fatness and stage of maturity. In cattle and sheep, although not pigs, there are clear breed differences in the distribution of body fat, especially between the carcass (subcutaneous and intramuscular) and intra-abdominal (perinephric fat, omental fat and mesenteric fat) depots. Thus 'dairy' breeds of cattle and sheep breeds such as the Finnish Landrace tend to deposit a higher proportion of fat intra-abdominally than 'beef' breeds of cattle and sheep such as Suffolks and Hampshires.

Differences in the growth and composition of body tissues occur between the sexes. In cattle and sheep, if comparisons of body composition are made at equal weights or ages, entire males have less fat than castrates and castrates less fat than females. The situation differs slightly in pigs where gilts have less fat than castrates at equal body weight. Differences in mature weight between males and females explain some of the differences in composition attributable to sex. The rest is most likely due to differences in the levels of testosterone found in males, females and castrates. The effect of androgens is specifically shown in the more pronounced development of the forequarter muscles of the bull and boar compared with the female or castrate. The influence of sex on muscle to bone ratio is unclear, but at the same level of fatness it is usual that entire males have higher muscle : bone ratios than females. This illustrates further that the impetus for fattening supersedes the impetus for muscle growth at lighter weights in females than in males.

Nutritional effects

Nutrition is generally the dominant factor influencing the expression of growth potential in farm animals. Both quantitative (plane of nutrition) and qualitative (diet composition) variation in nutrition influence growth. The general influence of plane of nutrition on growth develops from the priorities that exist for available nutrients as an animal grows (Fig. 16.36). On the high plane of nutrition the growth curves are telescoped together, whereas on a low plane the sequence is extended. The most profound effects are on the deposition of fat. This is illustrated in Fig. 16.38 and shows that not only does a plane of nutrition above maintenance influence the proportion of the different tissues synthesized but also that a plane of nutrition below maintenance can lead to 'negative' growth, especially of fat tissue in the first instance.

The main features of diet composition that can influence growth are the energy and protein content of the diet and the ratio of energy : protein in the diet. Fat growth relative to muscle and bone growth is dependent on the level of energy intake. Increasing the level of energy intake at a specified protein intake level increases the deposition of fat relative to muscle. Additionally with any specified level of energy intake, animals increase in fatness if their muscle growth is restricted by the amount, and in the case of nonruminants the quality, of protein in the diet. Thus a balance of energy and protein is necessary for the production of carcasses with the desirable level of fatness. The plane of nutrition can be used to adjust the growth rate in different stages of the growing/fattening period. Thus, for example, by restricting the energy intake of pigs the amount of fat in the carcass at slaughter may be limited and the time to slaughter in cattle may be manipulated through adjusting the plane of nutrition in order to make best use of price fluctuations.

A further nutritional influence that requires mention here is that of 'compensatory growth'. Animals whose liveweight growth has been retarded by restriction of feeding exhibit this phenomenon which is characterized by more rapid than normal growth when introduced to a high plane of nutrition. Compensatory growth is characterized by rapid deposition of cellular protein so that the maximum protein/DNA ratio in the cell is achieved. Although in biological terms the issues involved in compensatory growth are complex, in economic terms the phenomenon can be put to good use. Under UK conditions the growth rate of cattle may be restricted during the winter (stored period) with 'compensatory growth' being achieved on summer grass.

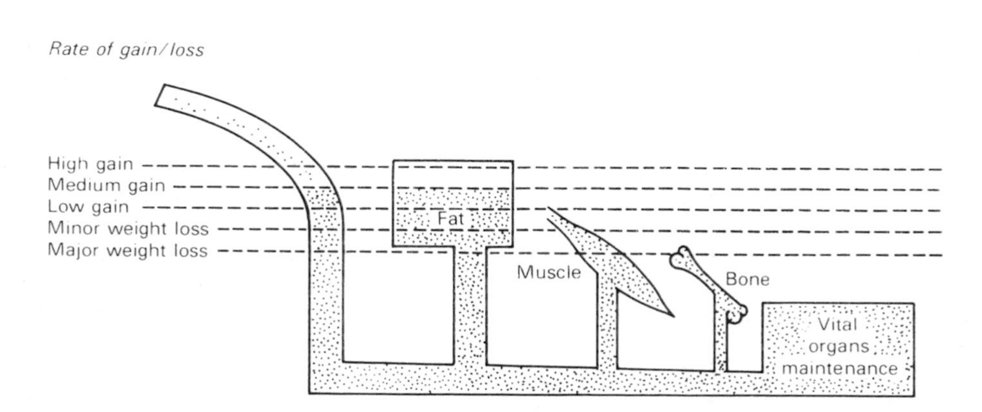

Figure 16.38 A model of priorities for nutrients during growth. [After Berg, R.T. & Butterfield, R.M. (1976) *New Concepts in Cattle Growth*. Sydney University Press, Sydney.]

The hormonal control and manipulation of growth

The endocrine system is a major regulator of animal metabolism and probably all hormones, either directly or indirectly, influence animal growth. Those considered to have specific effects on growth are the hormones of the anterior pituitary, the pancreas, thyroid, adrenals and gonads. These hormones affect body growth mostly through their effects on nitrogen retention and protein deposition. The precise mode of action of the individual hormones is only partly understood and is complicated by the fact that in some cases at least their effects depend on the sex, species of animal and balance of other hormones. When workers have attempted to correlate the concentration of hormones in the blood and growth, no clear relationships have emerged.

Although there is little evidence of an association between blood growth hormone (GH) levels and growth in livestock, some workers have shown there to be greater growth hormone secretory activity in genetically superior cattle and pigs. More recently it has been shown that exogenous growth hormone may produce an anabolic response in farm livestock. The anabolic response of GH appears to be mediated through a group of peptide hormones called insulin-like growth factors (IGFs) also termed somatomedins. Not only do these increase growth rate but also they produce leaner carcasses. The secretion of GH from the pituitary is controlled by the balance of an inhibitory hypothalamic peptide, termed somatostatin, and a stimulatory GH releasing factor (GHRF) which is also a hypothalamic peptide. Experimentally, removing or neutralizing the effects of somatostatin by active immunization techniques and supplementing the effect of GHRF have positive effects on growth.

It is now generally accepted that the natural steroid hormones – oestrogens, androgens, progestins and glucocorticoids – exert an effect on growth. The difference in growth rate, conformation and mature body size of male, female and castrate animals suggests that the sex hormones play an important role in the control of growth. There can be no doubt that these differences are in the main androgen-dependent. Deprived of the anabolic effect of androgens, castrates divert energy intake into the synthesis of fat rather than muscle and are thus less efficient in food conversion efficiency terms. Castration can only be justified on the basis of its effects on sexual and aggressive behaviour and in some situations the possible risk of carcass taint. Such characteristics only reveal themselves in the period close to slaughter and the technique of late castration using immunological techniques has been proposed to maximize the anabolic advantage of the testes. Immunological castration involves the active immunization of animals against gonadotrophin releasing hormone (GnRH). This has the effect of inhibiting the secretion of follicle-stimulating hormone (FSH) and luteinizing hormone (LH) such that spermatogenesis ceases and testosterone secretion declines sharply.

Several synthetic compounds have been used to promote growth and in many cases increase the protein content of the carcass. Classically, androgenic compounds would be expected to increase protein synthesis and hence protein deposition in muscle. However, the synthetic androgen trenbolone acetate has been shown to decrease protein synthesis and exerts its growth-promoting effect through decreasing protein degradation in muscle. This it appears to do through depressing the catabolic effects of the glucocorticoids. As with trenbolone acetate, the oestrogenic anabolic agent zeranol, which has been widely used commercially, would also appear to promote growth through reducing the rate of protein turnover rather than through stimulating muscle protein synthesis. It has also been suggested that oestrogens stimulate circulating GH concentration and change thyroid hormone status, and that this accounts in part for their growth stimulatory effects.

The administration of exogenous anabolic agents to animals to increase the rate of growth and alter the composition and conformation of animals through reducing fat and increasing muscle is widely practised in many parts of the world but is not permitted within the EU. In addition to the use of the growth hormone of the various species using rDNA technology the main substances used are agents like trenbolone acetate (a synthetic androgen) and oestrogenic substances such as hexoestrol and zeranol. The ability of these latter agents to influence growth is dependent on their producing an optimum balance of androgens and oestrogens in the animal. Thus response to implants is determined by the sex of the animal; best responses have been obtained by the exogenous administration of androgens to females, oestrogens to entire males and a combination of androgenic and oestrogenic agents to castrates.

One of the major concerns of the meat-producing industry in recent years has been the excessive amount of fat deposited in carcasses. The origins of this concern have been primarily due to public concern about the health risks associated with the consumption of animal fats although excessive fat production also represents a source of inefficiency in livestock production. Whilst leaner carcasses may be achieved in the short term by slaughtering at lighter body weights and in the long term by appropriate selection programmes, as exemplified by the reduction of backfat in pig carcasses, a group of substances termed β-agonists have been shown to have potential. Agents such as clenbutarol and cimeterol which are substituted catecholamines have been shown to reduce fat deposition and increase the protein content of the carcass. Their mode of action on fat would appear to be due to a reduction in lipogenesis and an increase in lipolysis, whilst the effects on protein accretion appear to be by a combination of reduced protein degradation in muscle and an increase in the rates of protein synthesis. As yet, however, β-agonists have not been approved for commercial use.

Environmental physiology

Throughout their evolution animals have been subjected to environmental influences, and farm animals, like their ancestors, have to employ a range of structural and functional strategies to adjust to changes in their environment. The term environmental physiology describes those physiological mechanisms involved in the process of homeostasis, and the external stimuli, such as temperature, daylength and stocking rate, that necessitate such responses.

The nature and importance of photoperiod in regulating seasonal activities such as reproduction have already been mentioned in this chapter and are also discussed in Chapter 20. Perhaps the most important environmental influence, particularly with regard to animal productivity, is that of temperature. This governs the amount of heat energy lost from the animal, which in turn determines the amount of dietary energy retained by the animal.

Maintaining thermal balance

Farm animals maintain a high (37.5–39°C) and relatively constant body temperature despite large fluctuations in environmental temperature. The combination of physiological and behavioural changes which regulate body temperature is controlled by the hypothalamus.

The maintenance of a constant body temperature requires that there be a balance of heat production within the body and heat loss from the body. Heat is produced as a byproduct of the metabolism of ingested nutrients used for maintenance and the various productive processes. Heat transfer between the body and the environment occurs by means of conduction, convection, radiation and evaporation. Since generally the animal body is at a higher temperature than the environment, heat may be lost by all these methods. The partition of losses between the various routes of heat exchange varies according to the air temperature, with evaporative losses being high at high air temperature, whereas radiation and convection are the main avenues of loss at low and intermediate temperatures. In addition to air temperature, other interacting environmental factors – wind, precipitation, humidity and sunshine – are determinants of rates of heat transfer. Wind, precipitation and humidity particularly affect evaporative losses. Wind also affects convection losses by increasing air movements around the animal's body, thus destroying the insulating layer of air trapped in the animal's hair or wool, and sunshine especially affects the net radiation exchange between the animal and its environment in animals kept outdoors.

The mechanisms by which animals can regulate body temperature fall into two categories: alteration in the rate of heat production and the rate of heat loss. The animal has a number of mechanisms by which the rate of heat loss can be altered: by varying the degree of vasodilation, sweating, piloerection (erection of body hair), and numerous behavioural responses such as change in posture, activity and food intake. It is these mechanisms that come into play first of all in the regulation of body temperature. The range of temperature within which body temperature can be regulated without changes in metabolic heat production is termed the thermoneutral zone. The air temperature at which these mechanisms are no longer successful in dissipating sufficient heat to prevent a rise in body temperature is termed the upper critical temperature. The temperature below which the animal must increase its rate of heat production is termed the lower critical temperature (see Fig. 16.39). These critical temperatures are not fixed and are especially dependent on the animal's metabolic body size and level of production.

Environmental temperature and animal production

To date much of the precise work on the effects of the environment on the animal, in temperate regions, has been concerned with intensively kept pigs and poultry and has primarily studied their interaction with the nutrition of the animal. Energy is the main nutrient affected, higher environmental temperatures releasing more food energy for productive purposes but having secondary effects on protein metabolism, food intake and carcass composition.

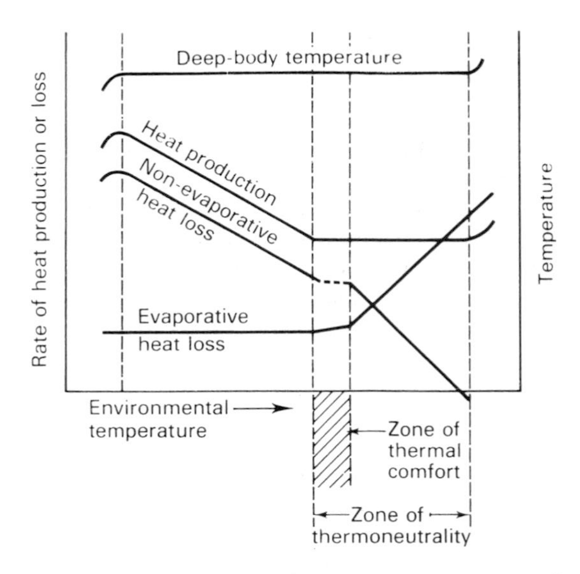

Figure 16.39 Thermoregulatory responses to a range of environmental temperatures.

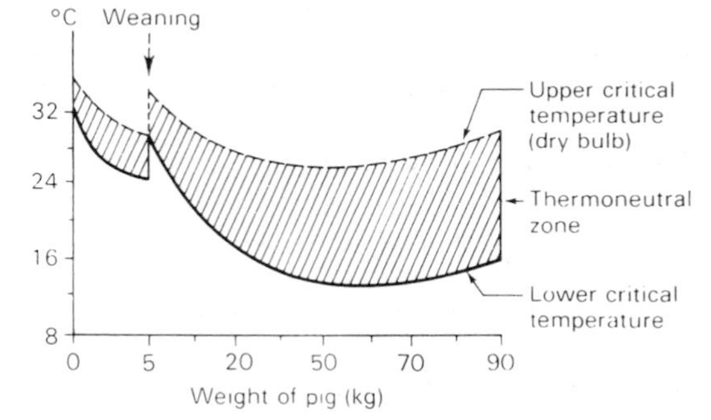

Figure 16.40 Relationship between upper and lower critical temperatures and body weight in a growing pig. [Courtesy of ADAS (1982) *Pig Environment*, Booklet 2410.]

Of all the farm species the pig is perhaps the most vulnerable to the effects of environmental temperature. It has little body hair for insulation and few sweat glands to help with evaporative cooling. However, pigs do make certain behavioural adaptations to temperature, such as huddling together in cold weather and wallowing in hot conditions. The lower critical temperature of pigs changes markedly with age and body size. Small pigs with a larger surface : body mass ratio will lose relatively more heat from their skin surfaces than will their older and larger counterparts. Thus a newborn pig is particularly vulnerable to a cold environment. Figure 16.40 shows how the lower critical temperature of a pig improves with increase in body weight. Apart from the level of dietary energy, lower critical temperature is also affected by factors such as ventilation rate, wetness and type of flooring, and provision of straw.

In the light of these effects it is clear that where quantitative data on environmental effects are available, they may be taken into account in the planning of feeding programmes, the design of buildings, and optimizing of investment in insulation and supplementary heating.

In the outdoor situation experienced in the UK, most interest in environmental effects centres on the combined effects of extreme cold, wind and rain. The effects on adult animals are primarily discomfort and loss of production, but in the young and newborn such weather conditions are a threat to life. Surveys have shown that perinatal lamb mortality can vary from 5–45% on hill farms and that, at the higher mortality rates, extreme cold is the main contributor. In extreme conditions, shivering thermogenesis is often insufficient to maintain body temperature in the neonate. These animals may increase heat production by non-shivering thermogenesis, a process mediated by the effects of cold on the sympathetic nervous system and particularly by the action of noradrenaline on brown adipose tissue. The breakdown of this special type of adipose tissue results in much higher levels of heat production than are produced from other catabolic processes, and serves to help prevent hypothermia in the neonates of many species. It appears likely that variations in the amounts of brown adipose tissue present in the body at birth may explain in part the genetic variation in the ability of the newborn lambs to survive adverse conditions.

Environmental physiology and animal welfare

In recent years the welfare of farm livestock has come under increasing scrutiny, particularly the more intensive methods of animal production. In such systems, an animal may be housed in a largely artificial environment. The extent to which animals are able to adapt to their surroundings will depend mainly on the design of the accommodation and the husbandry skills of the stockman.

Failure to interpret correctly the animal's physiological and behavioural needs may have serious consequences for both animal productivity and animal welfare. Animals that are physiologically or behaviourally deprived may suffer stress, a response that in the long term can result in an increased incidence of disease through a lowered resistance to infection and reduced growth rates. Even off the farm it has been shown that pre-slaughter stress results in a loss of carcass quality and weight, as in the case of dark-cutting meat of beef animals and the pale, soft, exudative meat of pigs stressed immediately prior to slaughter. Thus attention to the conditions during transport, auction and lairage and at slaughter may have benefits for animal productivity as well as animal welfare.

The Farm Animal Welfare Council (FAWC) in their codes of welfare recommendations offer guidance to the farmer on the physiological and behavioural needs of farm animals that will help safeguard their welfare.

References and further reading

Chesworth, J.M., Stuckbury, T. & Scaife, J.R. (1998) *An Introduction to Agricultural Biochemistry*. Chapman & Hall, London.

D'Mello, J.F.D. (ed) (2000) *Farm Animal Metabolism and Nutrition*. CABI, Wallingford.

Frandson, R.D. & Spurgeon, T.L. (1992) *Anatomy and Physiology of Farm Animals*. Lea & Febiger, Philadelphia.

Hafez, E.S.E. & Hafez, B. (2000) *Reproduction in Farm Animals*. Lippincot, Wilkins & Wilkins, Philadelphia.

Kay, I. (1998) *Introduction to Animal Physiology*. Bios Scientific Publishers, Oxford.

Larson, B.L. (ed) (1995) *Lactation*. Iowa State University Press, Ames.

Lawrence, T.L.J. & Fowler, V.R. (1997) *Growth of Farm Animals*. CABI, Wallingford.

McDonald, P., Edwards, R.A., Greenhalgh, J.F.D. & Morgan, C.A. (2002) *Animal Nutrition*. Pearson Education, Harlow.

Mepham, B. (1987) *Physiology of Lactation*. Open University Press, Milton Keynes.

Peters, A.R. & Ball, P.J.H. (1996) *Reproduction in Cattle*, 2nd edn. Blackwell Science, Oxford.

Rook, J.A.F. & Thomas, P. (1983) *Nutritional Physiology of Farm Animals*. Longmans, London.

17

Animal welfare

J.C. Eddison

Introduction

The welfare of animals under the care of humans has been of concern for many years. Most livestock producers both recognize the economic benefit of managing their animals in a humane manner and would be able to identify essential elements common to all high-welfare husbandry systems. These would include:

- housing designed both for the animals and the producer;
- stocking levels appropriate to production and welfare;
- consistency of management practices;
- high-quality stockmanship;
- an awareness of the needs of the animal.

The aim of this chapter is to provide an introduction to the key principles that underpin practical animal welfare.

Background

The momentum for the current debate regarding the welfare of agricultural livestock in particular originated in the 1960s. The concerns expressed at that time were, perhaps, most famously voiced in a book entitled *Animal Machines* (Harrison, 1964). In this book, Ruth Harrison described aspects of intensive livestock production that many people found abhorrent, and which she described by the term *factory farming*. As a result of the adverse publicity caused by the contents of both this book and other publications, the UK government established a Technical Committee (commonly referred to as the Brambell Committee after its chairman, Prof. Brambell) to review the welfare of animals kept under intensive livestock husbandry systems (HMSO, 1965). One of the many proposals of that committee was the establishment of a standing committee to advise government on farm animal welfare. This group is now known as the Farm Animal Welfare Council and its functions include both the review of scientific and other evidence and the provision of advice to government on legislative changes that may be necessary.

Public awareness and debate concerning welfare issues are not confined to the UK: many countries throughout the world have enacted welfare legislation and so has the European Union. Although differences do exist between nations with regard to their attitudes towards welfare legislation, we have to accept that the issue of animal welfare is now important in agriculture throughout much of Europe and many other countries throughout the world. Moreover, with the continual enlargement of the EU, new member states will have to conform to the growing EU welfare legislation. Not only is there a greater public awareness of welfare issues but also, as will be outlined later, the interests of welfare and productivity have a great deal in common. As a consequence, all those individuals with livestock in their charge need to have a basic understanding of animal welfare for two very important reasons. First, so that they can manage their stock in such a way that the animals are maintained in an environment that is as humane and stress-free as possible; and, second, so that the animal welfare debate can be conducted in an informed and reasoned manner with contributions from all those involved with, or with an interest in, livestock production.

This chapter is divided into three sections. The first introduces and defines, where possible, some important terminology and reviews some of the biological principles that underlie welfare. The second section outlines methods by which welfare can be assessed. Finally, the chapter concludes with a discussion of the factors that influence how we determine acceptable levels of welfare.

What is welfare?

Many emotive words are used during discussions of welfare and so, before any reasoned welfare debate can take place, we must have a clear understanding of what is meant not only by the term *welfare*, but also by terms such as *pain, suffering* and *stress*.

Welfare

Although there have been numerous attempts to define welfare scientifically, there is no firm consensus because, as Duncan and Fraser (1997) stated, welfare is not a purely scientific concept, but one that also involves value judgements by humans about the quality of life of the animals in their charge. For example, there is a commonly (but not universally) held view that animals such as outdoor pigs and free-range hens are necessarily in a high state of welfare because they are outside, unconstrained by artificial housing systems, and able to perform their full behavioural repertoire. However, is such a *naturalistic* view of animal welfare really valid? Others would argue that animals that are housed can be cared for more effectively than those that are outside, and are, therefore, kept in a better state of welfare. Although the different definitions have been derived from various perspectives, some general principles have arisen from the debate.

The Brambell Committee (HMSO, 1965) stated that the meaning of the word welfare embraced '. . . both the physical and mental well-being of the animal'. Equally important as defining the meaning of welfare, the members of that committee felt the need to state what welfare is not. They were quite adamant that an animal should not be judged to be in a state of good welfare merely because it is growing or producing some non-meat product such as milk, eggs or wool. In this context, they viewed production as merely indicative of an animal being fed adequately. Furthermore, by including both the mental and physical states of the individual in their definition of welfare, they clearly did not simply equate welfare with health (in the layman's sense of the term), although they recognized that good health is a necessary prerequisite of good welfare.

Hughes (1976), in his examination of how behaviour can be used to assess welfare, proposed the following definition: '. . . a state of complete mental and physical health, where the animal is in harmony with its environment'. As it stands, this definition is very similar to that of the Brambell Committee, but Hughes went further and discussed the difficulties of interpreting evidence concerning the welfare of animals. For example, an animal may modify its behaviour when it is introduced to a new environment. The difficulty for anyone who wishes to judge the welfare of an animal in a novel environment lies in distinguishing between behavioural changes that are part of an adaptive response to the novel surroundings, and those that are indicative of impaired welfare due to the inadequacies of the new environment.

Broom (1986) developed the themes explored by Hughes and defined the welfare of an individual as '. . . its state as regards its attempts to cope with its environment'. He argued that in some environments, an individual would have to exert very little effort to cope, while under other conditions much more energy might be required to survive and, in some circumstances, individuals might not be able cope at all; these latter individuals would be regarded as in a poor state of welfare.

A difficulty raised by Broom is that it is far easier to recognize individuals that are not coping (and whose welfare is poor) than to discriminate between those individuals who are in a state of average or good welfare. Moreover, one cannot regard the absence of indicators of poor welfare as positive evidence of good welfare. To be able to identify those animals in a state of good welfare, and thereby isolate those factors that promote a state of well-being, would enable us to take action to improve the conditions of other, less-advantaged individuals. This difficulty in recognizing good welfare has been echoed by the Farm Animal Welfare Council (1993) which identified as a research priority the development of new measures of welfare and also the search for measures of good welfare in particular. How welfare is actually assessed is discussed later in this chapter.

The ability of an animal to cope with its environment, and the very act of switching-on its coping mechanisms, depend to a great extent on the individual's perception of its situation. If an animal perceives its environment to be satisfactory, then it will not have any stimuli to which it has to respond: it does not perceive itself to have diminished welfare. Conversely, if an individual is provided with all its needs but, at the same time, it *wants* some additional commodity or change to its environment, e.g. a pen mate or more food so that it can spend more time feeding (see Appleby and Lawrence, 1987), then that animal will perceive itself to have decreased welfare and, in the context of the definitions of welfare by both the Brambell Committee and Hughes, it is not in a state of physical and mental harmony with its environment. This view of welfare, in which it is dependent upon the animal's perception of its environment, has been proposed most notably by Duncan (Duncan and Petherick, 1989; Duncan, 1993) and Dawkins (Dawkins, 1990). They argue that, although poor health, stress and low biological fitness (the ability to reproduce) may accompany poor welfare, it is really how the animal *feels* about its condition that determines the welfare of the individual.

A third argument, that high welfare is achievable only when animals can perform their full behavioural repertoire, has an evolutionary basis. Supporters of this viewpoint (e.g. Kiley-Worthington, 1989) argue that each behaviour has evolved under some advantageous selection pressure, and so we must provide the opportunity for animals to perform all their natural behaviours, otherwise they will suffer. Critics of this argument (e.g. Poole, 1996) point out that many behaviours performed by wild animals are useful to an individual in time of danger (e.g. predator avoidance) and, as such, are not necessary in animal husbandry systems. However, one cannot totally dismiss this naturalistic view of animal welfare. There are many examples of animals demonstrating a high motivation to perform specific behaviours. In fact, the concept of *behavioural needs* is central to the issue of animal welfare (Nicol, 1987; Jensen and Toates, 1993; Horrell *et al.*, 2001). A good example of this being put into practice was provided by the work of Stolba and Wood-Gush (1984) who studied the behaviour groups of pigs in an extensive park, noting the location and the way in which the pigs performed key behaviours. They then designed a housing system that incorporated features that enabled the pigs to perform those key behaviours in an appropriate context.

These three broad approaches to welfare (*coping*, *feeling* and *naturalistic*) have been reviewed in detail by Duncan and Fraser (1997) and, whilst there are certainly differences between them, there also is a great deal of commonality. All authors are in total agreement that welfare is judged at the level of the individual. This can be illustrated by a very simple example. In a cubicle house for dairy cattle, most individuals might be coping with their environment very well, but there will be some variation around the average. At the bottom of the scale there may be weaker or subordinate individuals that might not be able to lie in the best locations and may, perhaps, regularly have to lie in damp, dirty or draughty areas. Those individuals would have to expend more energy coping with their environment than the others and may be exposed to a greater disease risk. They may be able to cope with this less-than-ideal situation, but they would be said to be in a poorer state of welfare than the cows that do not have to lie in draughts or in dampness. They may also not feel so good about their situation, nor be able to perform important behaviours as frequently as they might wish. This simple example demonstrates that welfare is a continuum, with poor at one extreme and good welfare at the other, irrespective of the definition that we use.

A further point to make in this respect is that individuals can respond to a given stimulus in a diversity of ways, and therefore it is extremely important to assess the welfare of an individual using a variety of measures rather than relying upon a single index. For example, Duncan and Filshie (1979) compared two strains of hen: one flighty and another regarded as placid. They observed the behaviour and monitored heart rate of both strains when the hens were confronted with an approaching human. The flighty strain exhibited a stronger behavioural response to the stimulus than the placid strain. However, the heart rate of the placid strain was elevated for far longer than the flighty strain. The two measures, behavioural and physiological, provided a more detailed description of how the hens had responded to the approach of the human. Independently, neither measure accurately described the responses.

Taking into consideration the fact that many homeostatic mechanisms exist (e.g. physiological, behavioural, neurological and cognitive) and that individuals vary in the way they respond to similar stimuli, it is important to employ a combination of measures (including both physiological and behavioural indices) to assess welfare.

Pain

Although various definitions of *pain* have been proposed, there is a consensus about its meaning. Fraser and Broom (1990) offered the following definition: '. . . pain is a sensory stimulus which is itself aversive'. However, the real difficulty with pain as a concept is not what is meant by the term itself, but that it is a very personal experience and, therefore, there are great problems in assessing pain levels or in comparing the intensity of pain between individuals.

There are some points in Fraser and Broom's definition that are extremely important to the animal welfare debate. First, the definition correctly emphasizes the importance of the sensory or nervous system in pain detection. This is particularly important since much time has been devoted to assessing whether the experience of pain is similar across all vertebrates. The dangers of attributing human-like feelings to other groups of animals have hindered discussions of the effect of specific surgical practices such as beak trimming of poultry and dehorning of cattle. In fact, pain receptors (or *nociceptors*) are found throughout bird and mammal species (which include all traditional farm animals) and are also found in many other groups of animals. Not only are there structural similarities between species in this respect, but evidence also exists to show that similar physiological processes in different groups occur after traumatic surgery. For example, Gentle (1986) showed that neurological processes occurred in chickens after beak trimming which directly parallelled those that occurred in human amputees and which, in the latter case, are associated with extreme pain. Whilst such evidence is not conclusive proof, a cautious approach would be to assume that animals do experience pain and that beak trimming, for example, is painful to the individual.

The second point of importance to emerge from the definition of Fraser and Broom is that the sensation of pain itself is aversive; this is quite distinct from any effect on the individual that may result directly from the specific cause of the pain.

Clearly, there is an adaptive advantage to be gained from responding to pain. For example, removing a paw from a hot surface will minimize injury. However, different groups of animals vary in the way that they respond to pain or try to avoid it. Some, such as sheep, remain silent when subjected to a painful experience, whilst dogs, in contrast, scream loudly. Since the neural anatomy of mammals is fairly similar, it is unlikely that this variation in response is because one species feels pain and the other does not.

There are a number of reasons that could explain variations in response to pain. One possibility is that evolutionary pressures have led to the suppression of the visible expression of pain in prey species such as sheep. An injured sheep that announced itself as such would attract the attention of predators which could lead to rather dire consequences.

Mechanisms other than behavioural modification (which is the easiest to detect for a human observer) exist that enable animals to cope with pain. Natural analgesics (β-endorphin and enkephalins) have been found to be secreted within the brain. These neuropeptides, which are in fact natural opiates, will reduce the sensation of pain.

The variations in the way animals cope with pain make the task of assessing it very difficult. However, it is important to have some methods through which assessment can be made. A very useful checklist was proposed by Morton and Griffiths (1985) which provides a starting point (Table 17.1). However, irrespective of the tools that may be available, only attentive stockmanship will ensure that an animal that is in pain will be identified quickly. In fact, the key to the achievement of high levels of welfare is high-quality stockmanship.

Suffering

The relationship between pain and suffering was explained in the Brambell Report (HMSO, 1965), suffering being used as a collective noun that includes not only pain, but also fear, frustration and exhaustion. Dawkins (1980) defined suffering as '. . . a wide range of unpleasant emotional states' and added that the loss of a companion (bereavement), anxiety and conflict should also be included. While there may, of course, be other forms of suffering, Dawkins' definition seems to be perfectly adequate.

An important aspect of the concept of suffering to note is that, while pain may be extremely intense and perhaps even transient, other forms of suffering can be much less intense, but, because of their much longer duration, can be equally distressing to the individual animal. This is illustrated by the chronic frustration experienced by predators that are kept in confined enclosures in some zoos. This frustration is manifested in stereotypic pacing along well-worn paths within their enclosures (for review, see Lawrence & Rushen, 1993).

Stress

Of all the terms associated with animal welfare, stress has probably been the subject of the greatest number of interpretations. Its use in discussions of animal welfare has been made even more difficult because it is used very widely, and precisely, in many scientific disciplines, and it is also in common usage in non-scientific conversation.

In many situations, livestock have to cope with adverse conditions. Examples would include: vibrations during transport; the disruption of dominance hierarchies when new individuals are introduced into established groups; and the frustration experienced by farrowing sows and laying hens in the absence of suitable nesting material. Most people would agree that some or all of these circumstances can be stressful to the animals that are involved. However, if we are to engage in objective discussions of stress, we need to have a clear understanding of what is meant by the term.

Table 17.1 A simple checklist of assessing pain and distress. (After Morton & Griffiths, 1985.)

	Normal (0)	*Mild* (1)	*Moderate* (2)	*Severe* (3/4)
Appearance	Coat loses sheen, hair loss, starey – harsh			
	Failure to groom, soiled perineum			
	Discharge from eyes and nose			
	Eyelids partly closed			
		Eyes sunken and glazed		
	Hunched up look			
	Respiration laboured, abnormal panting			
		Grunting before expiration; grating teeth		
Food/water intake	Reduced		Zero (prolonged)	
	Faecal/urine output reduced		Zero	
Behaviour	Away from cage mates, isolated		Unaware of extraneous activities or bullying from mates	
	Self mutilation			
		Restlessness, reluctant to move, recumbent		
	Change in temperament			
		Squealing, howling, etc. especially when provoked		
Clinical signs	Strong pulse		Weak pulse	
Cardiovascular	Cardiac rate increased or decreased			
	Abnormal peripheral circulation			
		Pneumonia, pleurisy		
Digestive	Altered faecal volume, colour, consistency			
	Abnormal salivation			
	Vomiting (high frequency)			
		Boarded abdomen as in peritonitis		
Nervous		Lameness and arthritis		
(musculoskeletal)	Twitching		Convulsions	

Much of the way that biologists view the term stress results from the work of Selye, who identified a consistency of response to adverse conditions across several species (Selye, 1956). He formulated the notion of a General Adaptive Syndrome (GAS) to describe the way an individual adapts to a set of adverse conditions. With the benefit of techniques unavailable to Selye, this universal stress response is no longer viewed as correct. There is now evidence of considerable differences in the way different species, and also individuals within the same species, respond to stressful situations. One lasting element of Selye's investigations into stress, however, is his division of an animal's response to a stressful situation into three stages: *alarm*, *resistance* and *exhaustion*. This has proved to be a very useful structure within which to discuss stress and, following the example of many other authors (for reviews, see Broom and Johnson, 1993; Moberg and Mench, 2000), it will be used here.

The *alarm* stage is a relatively short period during which an animal becomes aware of the noxious stimulus and takes short-term compensatory action. An individual may be able to cope with a novel environment or adversity by physiological or behavioural modification. In such cases we would say that the individual has *adapted* to its conditions (Hughes, 1976). This may require the rapid provision of energy which can be made available in the short-term by the mobilization of glycogen stores in the liver. The glycogen is converted to energy through the action of the hormones adrenaline and noradrenaline which are produced in the adrenal medulla under stimulation of the sympathetic nervous system. These have been referred to as *flight* or *fight* hormones because their activity is often associated with agonistic interactions or extreme defensive action. However, their action is not solely confined to aggression; they provide short-term energy for any purpose (for a general discussion of hormonal action, see Campbell *et al.*, 2000).

The release of adrenaline will also cause the constriction of peripheral blood capillaries and increased blood supply to skeletal muscles. It has been suggested that this is the causal mechanism underlying the interruption of milk flow in frightened cows: the blood supply to the udder is reduced, leading to a reduction in oxytocin and a halt to milk release.

If the animal's response to adversity is longer term, other mechanisms have to be activated (Selye's *resistance* stage) in order to maintain homeostasis. The adrenal cortex, stimulated by *adrenocorticotrophic hormone* (ACTH) released by the hypothalamus, produces cortisol, hydrocortisone and corticosterone (glucocorticoids). These have the effect of releasing energy from non-carbohydrate sources, such as protein, over a period of several hours.

Neither the alarm nor resistance stages are physiologically damaging to an individual: the various response or adaptation mechanisms have evolved to assist the animal to cope with adverse conditions. For example, increased energy provided to increase body temperature in cold conditions would ensure survival. Moreover, it has been suggested that brief activation of adrenal cortex activity early in life can have positive benefits (Fraser and Broom, 1990).

The control mechanisms underlying the way in which an individual adapts to its environment can be divided into two distinct groups. The first of these is termed *negative feedback control* where the animal responds to change, and the second is *feed forward control* where action is taken in anticipation of change. These terms are borrowed from control engineering.

It is relatively easy to understand how an animal can respond to a stimulus through negative feedback control. Sensory input is continually being received and integrated in the brain which, in turn, initiates appropriate actions by, for example, muscle, endocrine and physiological systems. For example, in cold weather sows will lie closer together on straw in order to minimize heat loss. In addition to centrally processed actions, peripheral reflex actions act in order to counteract stimuli such as the vibration of lorries during transport. Under those circumstances, the tension of several muscles in an individual is adjusted in order to maintain its balance in the face of the vibration of the vehicle and the jostling of the other animals being transported.

Modification in physiological states or behaviour which occur in anticipation of environmental change may be brought about through a variety of routes. Environmental cues, such as diurnal changes in light and seasonal changes in daylength, are strongly correlated with endocrinological and behavioural changes in many species (e.g. the reproductive and migratory cycles of birds and several mammals). Predictable components of the management of a husbandry system are learned quickly by the animals. Pigs, for example, will soon learn the timing of feeding and modify their behaviour during the time immediately prior to the arrival of food. Anticipation of aversive experiences will stimulate avoidance or escape behaviour: sheep that have been poorly handled will be reluctant to move along a race. Whilst predictable events that are aversive will be stressful, it has been shown that the occurrence of similar stimuli that are unpredictable will be much more stressful (Weiss, 1971).

The final (*exhaustion*) stage of Selye's description of the stress response occurs when the reserves of the animal have been exhausted and it is no longer able to cope with the conditions and, unless rapid action is taken, it will die. Whilst the short-term secretion of glucocorticoids can be beneficial, chronic activation of the adrenal cortex results in many problems. These include reductions in: food consumption, protein synthesis, growth, gonadal activity (ACTH being secreted in preference to GH and other gonadotrophins in the pituitary), lymphocyte count and immune response. It also leads to changes in cardiovascular activity, gastric ulcers and increased disease susceptibility and has been proposed as the cause of sudden deaths, particularly in pigs and poultry (Fraser, 1975). This is the final stage of the stress response, when the body's defence mechanisms are exhausted, resulting in a reduction of the animal's biological fitness (i.e. its reproductive potential or chances of survival decline). Accordingly, Broom (1983b) suggested that it is only this stage that should be referred to as stress, defining it as: 'the process by which environmental factors over-tax control systems in an individual, thus activating responses whose effects are prolonged and ultimately detrimental to that individual'. Moberg (2000) expressed very similar ideas, but, in order to add clarity, he uses the term *distress* to describe the state when an individual's welfare is being adversely affected, and uses the term *stress response* to describe the whole process.

There is also a genetic component in stress: breeds and strains of both pigs and poultry have been shown to inherit a susceptibility to stress (Mills *et al.*, 1988). For example, the halothane gene in pigs is both beneficial for growth and body conformation and, when present as the homozygous genotype, makes individuals possessing it more susceptible to stress.

Stress terminology can be summarized as follows: *stress* is a detrimental process to which an individual may be subjected by a *stressor*, the result of which is that the individual exhibits a *stress response* and, if this response is not adequate to maintain homeostasis and the animal's welfare is diminished, the animal is said to be *distressed*.

How do we assess welfare?

Various authors (Fraser, 1975; Broom, 1983b) have emphasized the importance of using several indicators to assess welfare. However, what should we measure? There is no simple S.I. unit for welfare and we have also seen that there is not even a standard definition on which all scientists agree. Therefore, in order to perform such an assessment, we need to refer to the various definitions for guidance on what we should measure.

The view taken by the Brambell committee (HMSO, 1965) and Hughes (1976) was that welfare was concerned with the *well-being* of an individual. So, in order to gauge welfare using this definition, we need to ask questions like: what is good or bad for a particular animal? This logically leads on to the problem of *who* should decide what is good or bad for an individual? It is also important to note that this problem is quite independent from another important question of how bad conditions can become before we say that the welfare of an animal is deemed to be unacceptable: this is discussed in the final section.

There is a great danger of ascribing human feelings to animals without evidence to show that there is such a similarity of feeling. Therefore, wherever possible, the animal should be the appropriate decision-maker regarding its needs or wants. Duncan and Poole (1990) reviewed various methodologies used to assess the needs of animals. These needs might be environmental (e.g. preferred floor types for laying hens) or behavioural (e.g. the opportunity to build nests). They concluded that simple preference tests were flawed; for example, individuals may choose to satisfy short-term needs to the detriment of their longer-term welfare. The use of operant conditioning techniques, where an individual works for a reward, offers a better guide to an animal's motivation for a particular environmental feature or opportunity. The amount of work that an individual is prepared to perform in order to obtain a reward is regarded as a measure of its motivation for the reward. Some difficulties with this technique have been identified (Dawkins and Beardsley, 1986), but they are not insurmountable, and so they are useful as long as some care is taken when designing experiments. These methods offer scientists the opportunity to gain an insight into the animal's perception of its environment which, given the definitions of welfare discussed earlier, is extremely important.

In order to summarize the needs of farm livestock (i.e. what is good for them), the Farm Animal Welfare Council (1979, 1992) adapted ideas from Brambell's report (HMSO, 1965), and established a pragmatic framework within which welfare can be discussed and on which new developments can be based. This framework has become known as the *Five Freedoms* and they incorporate the basic prerequisites for good welfare:

(1) freedom from thirst, hunger or malnutrition;
(2) appropriate comfort and shelter;
(3) prevention, or rapid diagnosis and treatment, of injury and disease;
(4) freedom to display most normal patterns of behaviour;
(5) freedom from fear.

It is important to note, though, that satisfaction of these criteria does not automatically ensure that good welfare will necessarily follow: they provide a guide to the basic requirements of a husbandry system. These *Freedoms* have been incorporated into the Codes for Recommendations for the Welfare of Livestock (MAFF, 1983) and are implicit in all UK welfare legislation.

An assessment of welfare that incorporates use of the Five Freedoms measures the *provision* of the animal's environment with respect to welfare. Broom's (1986) definition, in contrast to those of Brambell (HMSO, 1965) and Hughes (1976), is couched in terms of the animal's ability to *cope* with its environment: by measuring how well an individual is coping we would be measuring welfare and, therefore, would measure the functional capability of the individual. The two approaches complement each other.

Mason and Mendl (1993) highlighted the difficulties of measuring welfare and identified points that had to be considered when designing experiments. For example, different individuals may cope with adversities in their environment in various ways (e.g. Duncan and Filshie, 1979) thereby making the choice of indicators quite difficult. Although Duncan and Poole (1990) correctly identified behavioural indicators as useful because no disturbance is caused when they are measured, they did recommend the use of a multitude of different types of measure for assessing welfare. For convenience, these have been grouped under the following headings: physiological and biochemical, production and behavioural.

Physiological and biochemical indicators

Several physiological and biochemical correlates of welfare have been identified. The example of increased heart rate in fearful poultry has already been described. Similar increases in cardiovascular activity have also been recorded when sheep were introduced to a flock, loaded into a vehicle and during transportation (Baldock and Sibly, 1990). However, care must be taken when interpreting such results since perfectly normal activities, such as play, running, courtship and mating, will also elicit similar responses.

Under chronically poor welfare conditions, the plasma concentration of glucocorticoid hormones will increase. However, simple assays of hormone levels can be misleading for several reasons: diurnal variation occurs in many species; there is considerable inter-individual variability in hormone activity; and the duration of the intervals between stimulus, response and the time when samples are taken can have a significant impact on the magnitude of observed response. All of these factors will have a profound effect upon the magnitude of the measurement. Furthermore, like heart rate, the secretion of glucocorticoids is not confined to times of adversity; elevated levels will also be observed, for example, during courtship.

One methodology that does seem to be useful is ACTH challenge, in which an injection of ACTH is administered and the adrenal response is measured. An individual that has been chronically subjected to adverse conditions will have enlarged adrenal glands which will produce more corticosteroids than smaller adrenals in an animal that has not been distressed. This method seems to overcome the problems associated with taking simple measures of glucocorticoid activity, and has been used to demonstrate the distress associated with mixing and space limits in dairy cattle as well as the effects of stocking density in pigs.

In addition to the enlargement of adrenal glands, the development of gastric ulcers is a symptom of chronic stress.

Biochemical changes in skeletal muscles *post-mortem* can occur in response to adverse conditions during the period prior to slaughter. Pale, soft and exudative muscle (PSE) and dark, firm and dry meat (DFD) are both indicators of earlier poor welfare.

Rapid glycolysis (the breakdown of glycogen stores) immediately prior to slaughter is a problem particularly associated with pigs which can result from long or arduous journeys. It causes the production of high levels of lactic acid and a dramatic decrease in muscle pH after slaughter which leads to a decrease in the water-binding properties of the muscle. Meat derived from pigs stressed in this way is pale grey, soft and exudes water: it is of low value.

Long-term depletion of glycogen stores leads to an unpopular and, therefore, low value meat for different reasons. With low levels of glycogen, there is little or no production of lactic acid in the muscles after slaughter, and the pH remains high giving rise to a dark-cutting meat that is firm and dry. This particular problem is found in cattle and pigs and can be caused by a variety of factors including fighting during the mixing of unfamiliar individuals during transport or in lairage at the slaughterhouse (Warriss *et al.*, 1990; Gregory, 1998).

Production indicators

It has already been stated that normal growth and reproduction are indicative of adequate food provision but not necessarily of high welfare (HMSO, 1965). It is important to remember that animal species utilized by man to provide meat or other products have been genetically selected for their muscle growth and reproductive performance. As a consequence, growth and reproductive performance will generally be affected only when individuals are subjected to extreme or chronic adversity. If individuals do not grow at the normal rate, fail to come into oestrus, or have reduced conception rates or litter sizes, we can probably conclude that they are not in a state of good welfare. They need to be examined swiftly in order to identify the underlying reasons and to take appropriate action.

The immediate (or proximate) causes of reduced growth, loss of weight, or diminished reproductive performance are generally either illness, nutritional deficiencies, digestive abnormalities or behavioural factors. Obviously, insufficient food allowance, rations that are deficient in particular nutrients (e.g. protein deficiencies) or trace element deficiencies in pasture will limit growth or cause metabolic diseases (e.g. swayback in sheep due to copper deficiency). Disease may have a direct effect on growth and it may also depress appetite. Observations of many farm species have demonstrated that subordinate individuals may not display normal reproductive behaviour in the presence of more dominant conspecifics (members of the same species), thus leading to reduced productivity.

Ultimately, however, such factors may really be symptoms of inadequate or inappropriate housing, or poor management. For example, if insufficient bedding material is available for group-housed cattle or pigs, then some individuals may be forced to lie on cold or wet flooring thereby diverting energy to thermoregulatory activity that would otherwise be used for reproduction or growth. Such conditions will also provide sites for disease organisms to multiply. Limited space will prevent subordinate individuals from avoiding more dominant animals. Lack of protection when feeding has been shown to result in reduced access to food by subordinate animals (Albright, 1969; Bouissou, 1970). Several solutions to such problems are available. Sufficient feeding stalls for all individuals in a group can be provided (e.g. Morris and Hurnik, 1990). Alternatively, transponder-based systems with a smaller number of feeders have been designed for cattle and pigs in which every individual has its own food allowance and is protected from the rest of the group while it feeds. Such systems prevent food stealing and therefore do not reinforce this behaviour in the more dominant individuals.

Advances in technology applied to animal production can be used to provide information to stockpersons concerning animals that are potentially at risk. Daily action lists generated by computerized feeding systems for cattle and pigs indicate which individuals have not consumed all of their food allowance and thereby give the stockperson an early warning of possible problems.

Production indicators are not all confined to growth or production deficiencies. Laying hens that are stressed may retain the egg in the oviduct longer than normal which can lead to the deposition of extraneous calcium carbonate in the eggshell (Mills *et al.*, 1987).

Welfare is a concept that is based upon the individual; therefore herd averages of production or food utilization do not provide an adequate index of welfare. However, if group averages do fall, they may well be indicative of diminished, or a potential decrease in, welfare that requires rapid action.

Behavioural indicators

The behaviour of an animal can provide an insight into many aspects of its welfare and, because animals are not disturbed while their behaviour is being observed, behavioural indicators can be very useful indeed. Abnormal behaviour can indicate illness or disease, inadequacies of the physical or social environment, and even nutritional problems. For example, if a sow has difficulty in walking or lying down, it may be suffering from a lameness disorder, it may be fatigued due to disease, the stocking density might be too high, or its pen may be too restrictive for easy movement.

One of the major reasons why behavioural indicators are so fundamental to welfare assessment is that animals are highly motivated to perform certain behaviours (e.g. nest-building by pre-parturient sows or pecking for food by hens). When frustrated, individuals may perform vacuum activities such as displaying nest-building behaviour without a suitable substrate such as straw. Alternatively, their behaviour may be redirected towards themselves or conspecifics such as is the case with feather-pecking in battery hens or belly-nosing penmates in newly weaned piglets who would still suckle their mothers if they had the opportunity. On occasions, frustration may lead to stereotypic behaviour (e.g. repeated pacing in battery hens or bar-biting and sham-chewing in tethered sows). The importance of such motivation was recognized by the Farm Animal Welfare Council and incorporated into the Five Freedoms: the freedom to display most normal patterns of behaviour.

The diversity of abnormal behaviour is clearly illustrated by the categorization used by Fraser and Broom (1990) in which they catalogued four main groups: stereotypic or repetitive, non-functional behaviour; behaviour directed towards self or inanimate objects; behaviour directed towards other individuals; and abnormal function. While space limitations preclude a detailed discussion here of the diversity of behavioural indicators of welfare, it is important to emphasize some key principles. Fortunately, there are several substantive reviews of this topic (Kiley-Worthington, 1977; Broom, 1983a; Fraser and Broom, 1990; Lawrence and Rushen, 1993).

A behaviour may be regarded as abnormal because its performance is observed only rarely. More frequently, however, it is either the inappropriate context in which the behaviour is performed or the excessive repetition without obvious function that signifies its abnormality. Moreover, the fact that a behaviour is performed frequently by a large number of individuals in a group does not necessarily mean that such behaviour is normal: outbreaks of tail-biting amongst piglets certainly cannot be viewed as normal!

The performance of such behaviour is indicative of a problem, and so the task of the stockperson is threefold:

(1) to be aware that the behaviour is abnormal;
(2) to identify the causal factors;
(3) to take appropriate action to effect a remedy.

The key to solving welfare problems is contextual, and it is important to remember that the source of a problem not only may be related to the current physical and social environment of the animal, but also may have an experiential or evolutionary component.

Evolutionary factors were used earlier to explain the differences between sheep and dogs in their expression of pain. The importance of early experience was illustrated by individuals that may be reluctant to move along a race or up a ramp into a vehicle because they have experienced aversive stimuli in the past (e.g. rough handling). Poor stockmanship in one context can have profound effects on later behaviour.

The social structure of a herd, particularly its stability, is extremely important. The introduction of new individuals to a group can be especially traumatic and lead to fighting and serious injury.

What is an acceptable level of welfare?

The preceding sections have reviewed both the various definitions of welfare and the diversity of scientific indices of welfare. The role of the welfare scientist is twofold: to determine *how* to assess welfare and also to carry out actual assessments. On the other hand, the task of determining what is an acceptable *level* of welfare, and to identify the cut-off point between acceptable and unacceptable welfare is not for the scientist to perform, but is one for which society in general must take responsibility.

The minimum level of welfare that society demands is a function, and inevitably a compromise, of many factors. In order to understand how acceptable levels of welfare are determined (and how they might change in the future), we need to examine more closely the relationship between production and welfare. Furthermore, we need to identify those other factors that contribute to the attitude of society towards welfare.

Ultimately, the need for an adequate supply of affordable food and other animal-based products (e.g. wool) drives animal production. So, in addition to deciding on levels of welfare, society determines the required level of animal production. In post-war Britain, the need for cheap food generated the pressure for livestock producers to turn towards more intensive methods. Frequently, debates concerning animal welfare in agriculture have been portrayed as a conflict between two quite distinct groups: producers satisfying the demand for food and their need to make a living, and the welfare lobby who are protecting the interests of livestock. However, this is a very simplistic view of the situation and is inevitably a gross misrepresentation for two simple reasons. First, in many ways the interests of food production are served by high welfare. And, second, there are many pressures other than production itself that determine husbandry methods.

The close correlation between the interests of production and welfare has been very clearly illustrated by earlier examples. Stress can reduce meat quality, disrupt egg production and depress milk yield. Moreover, responses to chronic stress also include the diversion of resources from growth to energy production.

Many factors influence the way livestock are managed. Individually, each may be quite laudable, but their interaction with production or with each other may lead to associated welfare problems. This point is well illustrated by an example relating to food hygiene and human health.

The very natural anxiety about food hygiene particularly at time of slaughter has led to legislation (EC, 1991, 1992) that lay down very stringent requirements for hygiene in slaughterhouses in the EU. The cost of upgrading abattoirs to the required standards has led to many closures. As a consequence, animals have to travel farther and for longer on their way to slaughter and their welfare suffers. In addition to the welfare consequences, meat quality will, on occasions, be affected detrimentally. In this situation, production and welfare are certainly not in opposition!

In recent years, many welfare assurance schemes have been established by both retailers and independent welfare organizations (e.g. the RSPCA's Freedom Foods) in response to consumer pressure. Producers who supply livestock to these schemes adhere to strict welfare guidelines and, in return, receive a price premium.

Welfare issues are clearly neither always simple nor straightforward: they can rarely be reduced to a choice between two simple options. Furthermore, in the case of farrowing crates, the welfare of one group of individuals (the piglets) is advantaged at the expense of another (the sow).

Figure 17.1 illustrates some of the factors that influence the way we manage our livestock. The minimum level of welfare that society will accept is a function of the relative importance of these and other factors. Their relative strengths will fluctuate over time and so will acceptable levels of minimum welfare. Given that welfare is measured along a continuum, the level of welfare that society deems to be acceptable is a moving point which, during the past 30 years, has been moving further up the scale, requiring from farmers a higher level of welfare for livestock. So, while it is impossible to fix permanent minimum standards, it is important to ensure that, when any legislation or recommendations concerning livestock are prepared, the full welfare implications are taken fully into consideration.

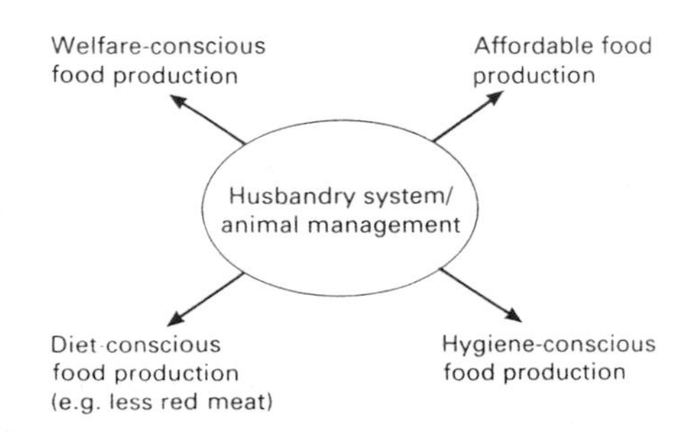

Figure 17.1 Some of the pressures that determine the husbandry methods by which livestock are reared.

References

Albright, J.L. (1969). Social environment and growth. In: Hafez, E.S.E. & Dyers, I.A. (eds) *Animal Growth and Nutrition*. Lea and Febiger, Philadelphia.

Appleby, M.C. & Lawrence, A.B. (1987) Food restriction as a cause of stereotypic behaviour in tethered gilts. *Animal Production* **45**, 103–110.

Baldock, N.M. & Sibly, R.M. (1990) Effects of handling and transportation on heart rate and behaviour in sheep. *Applied Animal Behaviour Science* **28**, 15–39.

Bouissou, M.-F. (1970) Rôle du contact physique dans la manifestation des relations hierarchiques chez les bovins: consequences pratiques. *Annales de Zootechnie* **19**, 279.

Broom, D.M. (1983a) Stereotypies as animal welfare indicators. In: Smidt, D. (ed) *Indicators Relevant to Farm Animal Welfare*. Martinus Nijhoff, The Hague. pp. 81–7.

Broom, D.M. (1983b) The stress concept and ways of assessing the effects of stress in farm animals. *Applied Animal Ethology* **11**, 79 (Abstract).

Broom, D.M. (1986) Indicators of poor welfare. *British Veterinary Journal* **142**, 524–6.

Broom, D.M. & Johnson, K.G. (1993) *Stress and Animal Welfare*. Chapman & Hall, London, 211 pp.

Campbell, N.A., Mitchell, L.G. & Reece, J.B. (2000) *Biology: Concepts and Connections*. Addison Wesley Longman, San Francisco, 860 pp.

Dawkins, M.S. (1980) *Animal Suffering: The Science of Animal Welfare*. Chapman and Hall, London, 149 pp.

Dawkins, M.S. (1990) From an animal's point of view: motivation, fitness, and animal welfare. *Behavioral and Brain Sciences* **13**, 1–61.

Dawkins, M.S. & Beardsley, T.M. (1986) Reinforcing properties of access to litter in hens. *Applied Animal Behaviour Science* **15**, 351–64.

Duncan, I.J.H. (1993) Welfare is to do with what animals feel. *Journal of Agricultural and Environmental Ethics* **6** (Suppl 2), 8–14.

Duncan, I.J.H. & Filshie, J.H. (1979) The use of radio telemetry devices to measure temperature and heart rate in domestic fowl. In: Amlaner, C.J. & MacDonald, D.W. (eds) *A Handbook on Biotelemetry and Radio Tracking*. Pergamon, Oxford, pp. 579–88.

Duncan, I.J.H. & Fraser, D. (1997) Understanding animal welfare In: Appleby, M.C. & Hughes, B.O. (eds) *Animal Welfare*. CAB International, Wallingford, pp. 19–31.

Duncan, I.J.H. & Petherick, J.C. (1989) Cognition: the implications for animal welfare. *Applied Animal Behaviour Science* **24**, 81.

Duncan, I.J.H. & Poole, T.B. (1990) Animal welfare. In: Monaghan, P. & Wood-Gush, D.G.M. (eds) *Managing the Behaviour of Animals*. Chapman and Hall, London, pp. 193–232.

EC (1991) *Council Directive amending and consolidating Directive 64/433/EEC on health problems affecting intra-Community trade in fresh meat to extend to the production and marketing of fresh meat*. OJ L268/69.

EC (1992) *Council Directive amending and updating Directive 77/99/EEC on health problems affecting intra-Community trade in meat products and amending Directive 64/433/EEC*. OJ L54/1.

Farm Animal Welfare Council (1979) *Farm Animal Welfare Council*. Farm Animal Welfare Council, Surbiton.

Farm Animal Welfare Council (1992) *The Five Freedoms*. Farm Animal Welfare Council, Surbiton.

Farm Animal Welfare Council (1993) *Report on Priorities for Animal Welfare Research and Development*. Farm Animal Welfare Council, Surbiton.

Fraser, A.F. & Broom, D.M. (1990) *Farm Animal Behaviour and Welfare*, 3rd edn. Baillière Tindall, London, 437 pp.

Fraser, D. (1975) The effect of straw on the behaviour of sows in tether stalls. *Animal Production* **21**, 59–68.

Gentle, M.J. (1986) Neuroma formation following partial beak amputation (beak trimming) in the chicken. *Research in Veterinary Science* **41**, 383–5.

Gregory, N.G. (1998) *Animal Welfare and Meat Science*. CABI publishing, Wallingford, 298 pp.

Harrison, R. (1964) *Animal Machines: The New Factory Farming Industry*. Vincent Stuart, London, 186 pp.

HMSO (1965) *Report of the technical committee to enquire into the welfare of animals kept under intensive livestock husbandry systems*. Command: 2386. HMSO, London.

Horrell, R.I., A'Ness, P.J., Edwards, S.A. & Eddison, J.C. (2001) The use of nose-rings in pigs: consequences for rooting, other functional activities, and welfare. *Animal Welfare* **10**, 3–22.

Hughes, B.O. (1976) Behaviour as an index of welfare. *Proceedings of the 5th European Conference on Poultry Welfare, Malta*, pp. 1005-18.

Jensen, P. & Toates, F.M. (1993) Who needs 'behavioural needs'? Motivational aspects of the needs of animals. *Applied Animal Behaviour Science* **37**, 161–81.

Kiley-Worthington, M. (1977) *Behavioural Problems of Farm Animals*. Oriel Press, Stocksfield.

Kiley-Worthington, M. (1989) Ecological, ethological, and ethically sound environments for animals: towards symbiosis. *Journal of Agricultural Ethics* **2**, 323–47.

Lawrence, A.B. & Rushen, J. (eds) (1993) *Stereotypic Animal Behaviour. Fundamentals and Applications to Welfare*. CAB International, Wallingford, 212 pp.

MAFF (1983) Codes of recommendations for the welfare of livestock. MAFF Publications. (Individual codes for each species of farm livestock.)

Mason, G. & Mendl, M. (1993) Why is there no simple way of measuring animal welfare? *Animal Welfare* **2**, 301–19.

Mills, A.D., Marche, M. & Faure, J.-M. (1987) Extraneous egg shell calcification as a measure of stress in poultry. *British Poultry Science* **28**, 177–81.

Mills, A.D., Faure, J.-M. & Lagadic, H. (1988) Génétique du stress. *Recueil de Médecine Vétérinaire* **164**, 793–800.

Moberg, G.P. (2000) Biological response to stress: implications for animal welfare. In: Moberg, G.P. & Mench, J.A. (eds) *The Biology of Animal Stress: Basic Principles and Implications for Animal Welfare*. CABI Publishing, Wallingford, pp. 1–21.

Moberg, G.P. & Mench, J.A. (eds) (2000) *The Biology of Animal Stress: Basic Principles and Implications for Animal Welfare*. CABI Publishing, Wallingford, 377 pp.

Morris, J.R. & Hurnik, J.F. (1990) An alternative housing system for sows. *Canadian Journal of Animal Science* **70**, 957–61.

Morton, D.B. & Griffiths, P.H.M. (1985) Guidelines on the recognition of pain, distress and discomfort in experimental animals and an hypothesis for assessment. *Veterinary Record* **116**, 431–6.

Nicol, C.J. (1987) Behavioural responses of laying hens following a period of spatial restriction. *Animal Behaviour* **35**, 1709–19.

Poole, T.B. (1996) Natural behaviour is simply a question of survival. *Animal Welfare* **5**, 218.

Selye, H. (1956) *The Stress of Life*. Longmans, Green & Co., London.

Stolba, A. & Wood-Gush, D.G.M. (1984) The identification of behavioural key features and their incorporation into a housing design for pigs. *Annales de Recherche Veterinaire* **15**, 287–98.

Warriss, P.D., Brown, S.N., Bevis, E.A. & Kestin, S.C. (1990) The influence of pre-slaughter transport and lairage on meat quality in pigs of two genotypes. *Animal Production* **50**, 165–72.

Weiss, J.M. (1971) Effects of coping behavior in different warning signal conditions on stress pathology in rats. *Journal of Comparative Physiology and Psychology* **77**, 1–13.

Useful web sites

http://www.applied-ethology.org – the International Society for Applied Ethology (ISAE) is the premier learned society for the study of the behaviour and welfare of animals managed by humans. It has a formal relationship with Elsevier scientific publishers of *Applied Animal Behaviour Science* (a major journal publishing research in animal welfare and behaviour).

http://europa.eu.int/comm/food/fs/aw/index_en.html – the European Union's internet entry point to EU welfare policies and legislation.

http://www.defra.gov.uk/animalh/animindx.htm – the Department of the Environment, Food and Rural Affairs pages on animal health and welfare.

http://www.fawc.org.uk – Farm Animal Welfare Council (FAWC).

http://www.rspca.org.uk – the Royal Society for the Prevention of Cruelty to Animals: a major animal charity which has developed the Freedom Foods welfare assurance scheme.

http://www.ufaw3.dircon.co.uk – the Universities Federation for Animal Welfare, a charity devoted to animal welfare which funds scientific research into animal welfare. It also publishes the journal *Animal Welfare*.

18

Applied animal nutrition

R.M. Orr

Introduction

Feed represents the major cost to animal production. Thus, the efficiency of its use can have a considerable impact on the performance of an enterprise.

In general terms efficient utilization of feeds is dependent on the following elements:

- a knowledge of the nutrient value of the range of feedstuffs available;
- a knowledge of the nutrient requirements of animals;
- an ability to formulate diets from a mixture of feedstuffs which will meet the nutritional requirements of animals as determined by their genetic potential and within their potential voluntary feed intake. Clearly, an important element of this formulation process is that of meeting nutrient requirements most economically, but, increasingly, formulation must also take account of the quality of the product, environmental impact and animal behaviour, welfare and health.

In this chapter the procedures used in determining the first two elements are given attention as are characteristics of the major feeds, in terms of feeding value and limitations to their use, that are used in diet formulation. The aspect of feeding particular classes of livestock is dealt with in the chapters on the individual species.

Nutrient evaluation of feeds

The value of a feed is dependent on how much of particular nutrients in the feed the animal is able to use to meet the requirements of the various body processes: In the utilization of feed there are inevitable losses of nutrients due to inefficiencies in digestion, absorption and metabolism. Losses through incomplete digestion are a major factor in determining feeding values, and digestibility measurements are in some instances used to give an indication of feeding value – this is particularly true in the evaluation of roughages. For the purpose of diet formulation, however, the nutritive value of foods is determined in terms of their ability to furnish the animal primarily with energy and protein and, to a lesser extent, with minerals and vitamins.

Digestibility measurements

The digestibility of a food is defined as the proportion of the food that does not appear in the faeces and is assumed to be absorbed. This is strictly termed 'apparent' digestibility because a very small fraction of faecal excretion is in the form of fragments of gut mucosa and therefore not of dietary origin, but in practice this is not taken into account. The apparent digestibility of a feedstuff may be determined in animals restrained in metabolism crates which allow the precise measurement of food intake and of faecal excretion over a specified period of time. From this and the analysis of feed and faeces, digestibility coefficients for the food dry matter and also for individual food nutrients may be calculated, i.e. organic matter, protein, energy.

The percentage of digestible organic matter (DOM) in the dry matter of a feed is termed the D-value. The D-value is widely used to describe the nutritive value (especially energy potential) of roughage feeds, particularly hay and silage.

Since digestibility assessments using live animals are tedious to conduct and unsuitable for routine estima-tions of large numbers of samples, alternative laboratory techniques (*in vitro* methods) have been developed. The earliest *in vitro* methods were developed for roughages and aim to simulate ruminant digestive processes. A two-stage process is carried out. The first stage involves incubation of a sample with rumen liquor to simulate rumen fermentation, the residue from this being further incubated with an acid pepsin solution to simulate the digestion of protein in the lower gut. More recently, methods based on the use of fungal cellulases have been used successfully to predict the digestibility of forages and other feeds. Following boiling with neutral detergent solution which hydrolyses the cell contents, the remaining cell wall or neutral detergent fibre (NDF) fraction is incubated with a cellulase solution, the residue from which can be ashed to determine the digestibility of the organic matter in the dry matter. This is referred to as *neutral detergent cellulase digestible organic matter* (NCD). For compounded feeds, additional steps involving a preliminary treatment of the feed with amylase to digest starch and the inclusion of gammanase to the cellulase digestion phase to break down any mannans that may be present are necessary. This is termed the *neutral detergent cellulase + gammanase digestibility* (NCDG).

The apparent digestibility of feedstuffs may vary between animals due to such factors as type of animal (e.g. ruminant versus non-ruminant; sheep versus cattle). The main characteristic of food that affects digestibility is its chemical composition, particularly the fibre, protein and lignin content. Thus, for foodstuffs such as cereal grains and protein concentrates which are fairly constant in chemical composition, digestibility values are also fairly constant, but for forages where composition varies widely according to growth stage, wide variation in digestibility values may be found.

The other main characteristic that can have an effect is the method of feed preparation and processing involved. The effects of processing on the digestibility of cereal grains are complicated by different responses in the different species of farm animal, but such techniques as pelleting, rolling, milling and cooking through micronization and steaming can be of use in particular situations. Likewise, chemical processing through alkali treatment can be used to effect improvements in digestibility of both grains and forages. Physical processing of forages, especially grinding, decreases digestibility through increasing rate of passage. The decreases in digestibility of foods observed when the level of intake is increased are likewise largely due to rate of passage effects. Because the level of intake affects digestibility, measurements are usually made at the maintenance level of feeding.

Measurement of feed energy

Animals use energy for a variety of body functions ranging from energy required for essential muscular activity and maintenance of body temperature to energy required for the synthesis of body tissues. It is primarily energy intake that determines the level of performance that an animal can sustain.

The energy content of feeds may be expressed in the following ways: gross energy (GE), digestible energy (DE), metabolizable energy (ME) and net energy (NE). To understand the meaning of the terms involves a recognition of the ways in which food energy is utilized by the animal.

The large number of chemical bonds that make up the organic compounds contain the energy present in the food. This chemical energy, termed gross energy or heat of combustion, is determined by the technique of bomb calorimetry in which a known weight of food is completely oxidized and the heat liberated measured. Of this gross energy in a food only a portion is eventually available to provide for the animal's energy requirements. In the utilization of the gross energy there are several routes of energy loss, namely:

- losses of indigestible energy in the faeces;
- energy losses in the urine, mainly in the form of nitrogen-containing substances such as urea;
- losses of energy as methane produced during the fermentation of dietary energy in the rumen and to a much lesser extent in the large intestine;
- heat produced from inefficiencies in metabolic processes of the animal and dissipated from the body surfaces, and to a lesser extent heat produced from inefficiencies in the metabolism of microbial cells in the digestive tract (heat of fermentation). The sum of these forms of heat is termed the *heat increment of metabolism*.

The energy remaining after these losses are accounted for represents the energy the animal utilizes for maintenance and production. This may be represented by Fig. 18.1.

The unit of measurement used in the assessment of food energy is the joule (J). More typically it is customary to use the kilojoule (kJ) or the megajoule (MJ).

$$1\,\mathrm{MJ} = 1000\,\mathrm{kJ} = 1\,000\,000\,\mathrm{J}$$

Gross energy value

This is measured by burning a known weight of the food and recording the total heat produced. The gross energy value of a food is dependent on the proportions of

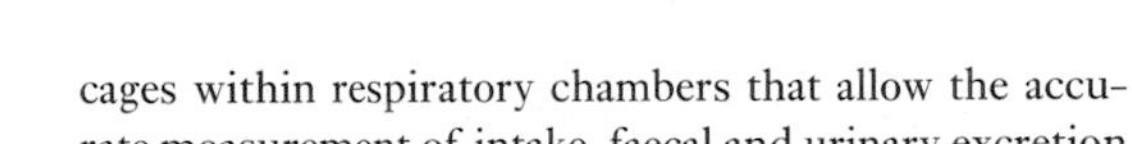

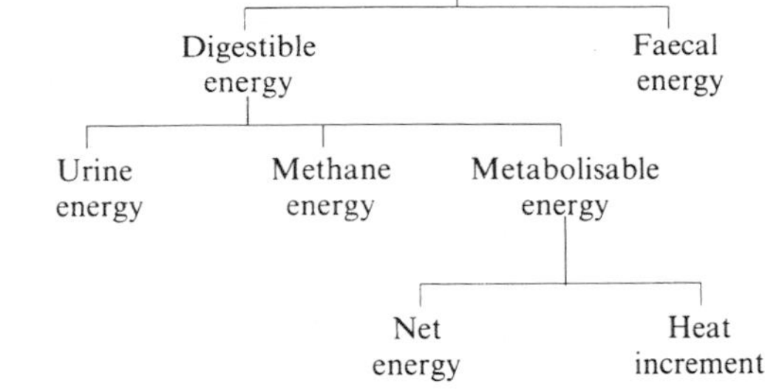

Figure 18.1 Routes of energy loss.

carbohydrate, protein, fat and ash that are present. The approximate energy values of these components are: carbohydrate 17.5 MJ/kg DM, protein 23.5 MJ/kg DM, fat 39 MJ/kg DM and ash 0 MJ/kg DM. Thus the fat and the ash content of a food have potentially the greatest influence on gross energy values. In practice, most animal foods have a low and fairly constant amount of these components, mainly consisting of carbohydrate with lesser amounts of protein, such that the gross energy of a wide range of foods is fairly consistently around 18 MJ/kg DM. Although gross energy tells us the potential energy in a feed, it is the least satisfactory assessment of energy value since no account is taken of the availability of that energy to the animal.

Digestible energy

Digestible energy (DE) is a measure of the gross energy of the food minus that part of the food energy lost in the faeces. Strictly speaking, this is the 'apparent' DE of a food and is determined in digestibility trials in which the intake and faecal excretion of gross energy is measured. The DE value of foods is dependent on the gross energy value and the factors that influence the digestibility of foods.

Metabolizable energy and fermentable metabolizable energy

The metabolizable energy (ME) of a food is the digestible energy less the energy lost in the urine and as methane. Typically about 8–10% of the digestible energy is lost in urine. About 7–12% of the digestible energy is lost as methane in ruminants whereas in the pig losses are negligible. Thus, for pigs metabolizable energy values are approximately 0.9 of DE values and for ruminants are approximately 0.81–0.86 of DE values.

Metabolizable energy values are determined by feeding trials in which animals are kept in metabolism cages within respiratory chambers that allow the accurate measurement of intake, faecal and urinary excretion and methane production. ME values are primarily influenced by the digestibility of the feed. Although the proportion of digestible energy lost in urine and methane is relatively consistent, urine losses are affected by the protein content of the diet.

In relation to ruminant feeding the measure *fermentable metabolizable energy* (FME) content of a feed or diet is required in order to use the metabolizable protein (MP) system. The FME is defined as the ME minus the ME present as oils or fat in the feed (ME_{fat}) and the ME contribution of fermentation acids (ME_{ferm}) present in silages.

Thus:

$$[\text{FME}](\text{MJ/kgDM}) = [\text{ME}] - [\text{ME}_{\text{fat}}] - [\text{ME}_{\text{ferm}}]$$

It describes the amount of ME in a feed that is available to the microbes in the rumen for use in the synthesis of microbial protein. Dietary fat (ME = 35 MJ/kg DM) does not supply energy to the rumen microbes such that with a fat content of only 4% the FME is reduced by 1.4 MJ/kg DM. Fermentation acids, mainly lactic and acetic in well-preserved silages, cannot be further fermented to provide energy in the rumen. For a silage with 100 g lactic acid/kg silage DM, a deduction of 1.51 MJ from the estimated ME must be made. Where lactic acid values are not known, ME_{ferm} may be taken as 0.1 ME.

Net energy value (NE)

The metabolizable energy supplied by the diet may be looked upon as the absorbed nutrients available for the various metabolic processes in the tissues. Thus, in the ruminant, volatile fatty acids, and, in the monogastric, glucose, can be looked upon as some of the main components of ME. The metabolic pathways into which these and other metabolites are directed so that forms of energy (e.g. ATP) that are useful to the animal may be created are not 100% efficient, some of the energy being converted into heat which is termed the heat increment (HI). Some heat is also produced from the work of digestion whilst in the ruminant heat arises from the inefficiencies in the metabolism of rumen microbes (heat of fermentation). The HI of a food is measured by the technique of animal calorimetry, and when subtracted from the ME gives the NE value. Typically the HI is greater in the ruminant (35–65% of ME for mixed diets) than in the monogastric (10–40% of ME for mixed diets).

In the monogastric considerably less HI results from the metabolism of fat (9%) than of carbohydrates

(≃17%) or proteins (≃25%), but in contrast to the ruminant the efficiency with which ME is used for the different body processes, such as maintenance and growth, is very similar. The NE value of a food to the monogastric is therefore a function of its ME value and the nutrients comprising the ME, especially the proportion of fat present.

In the ruminant the extent of heat production in the metabolism of ME and therefore the NE value of a feed are dependent on the purpose for which the ME is utilized. This is because the products of digestion (the constituents that make up the energy of ME) are used at different levels of efficiency for maintenance or conversion to the various productive functions of lactation, growth and reproduction. The situation is further complicated by the fact that the principal energy-containing end-products of digestion, the VFAs – acetic, propionic, butyric – are used with differing efficiencies. Thus, diet composition, which influences the molar proportions of VFA, as well as physiological function, have an effect on the efficiency of utilization of ME. The feature of feeds that correlates most closely to this effect of diet composition is its *metabolizability* (q_m) which is defined as the proportion of [ME] in the [GE] of the feed:

$$q_m = [ME]/[GE]$$

A mean value for the gross energy (GE) of ruminant diets that may be used in this calculation is 18.8 MJ/kg DM. This is illustrated in Fig. 18.2 and means that individual feeds cannot be assigned single NE values.

Using the linear equations given in Fig. 18.2, the NE values of feeds when used for maintenance (NE_m), lactation (NE_l) and growth (NE_g) can be calculated as follows:

$$NE_m = ME \times k_m$$
$$NE_l = ME \times k_l$$
$$NE_g = ME \times k_g$$

Thus, for example, for two diets (A and B) with ME values of 10 and 13 MJ/kg DM respectively and taking an average GE of 18.8 MJ/kg DM for both diets, the NE values will be as follows:

MJ/kg DM	*Diet A*	*Diet B*	*Ratio A/B*
ME	10.0	13.0	0.77
NE_m	6.89 (0.69)	9.68 (0.74)	0.71
NE_l	6.06 (0.61)	8.60 (0.66)	0.70
NE_g	4.21 (0.42)	7.08 (0.54)	0.59

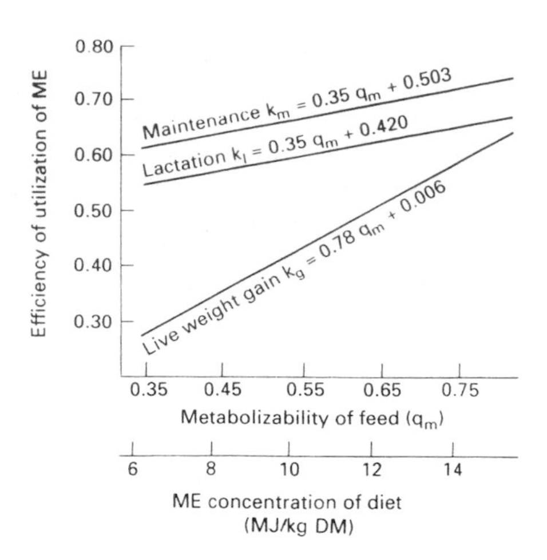

Figure 18.2 Effect of diet and production on the efficiency of utilization of ME.

This illustrates how the efficiency of utilization of ME (values in parentheses) and NE values differ according to the ME value of the diet and the purpose to which the ME is put.

Protein evaluation of feeds

Animals require protein to provide them with amino acids. These amino acids are used for the synthesis of body tissue proteins and proteins in products such as milk and wool which leave the body. Simple-stomached animals obtain their supply of amino acids from the digestion and absorption of dietary protein whereas in the ruminant the situation is complicated by the presence of the rumen which means that the supply of amino acids comes from a combination of microbial protein produced from dietary rumen degradable protein (RDP) or non-protein nitrogen (NPN) and dietary undegradable protein (UDP). Thus, different approaches to the evaluation of foods as protein sources are necessary for the monogastric and the ruminant.

An estimate of the crude protein in feeds is the starting point of all protein evaluation systems, subsequent measures of protein quality being dependent on its digestive and metabolic use by the animal.

Crude protein (CP)

Proteins in food are the main nitrogen-containing components. Thus, by measuring the nitrogen content of a food, it is possible to derive its protein content. Two assumptions are made in calculating crude protein content from nitrogen content. First, it is assumed that all nitrogen in a food is present in protein. In reality this is incorrect in that some nitrogen is present in such compounds as amides. For most feeds this nitrogen is less than 5% of the total, but for root crops it may be considerable. Secondly, it assumes that all food protein contains 16% nitrogen although, in practice, a range of 15–18% for different protein types exists. Because of the inaccuracy of these assumptions the protein content of feeds is expressed as crude protein thus:

$$\%CP = \%N \times (100/16)$$

or more commonly

$$\%CP = \%N \times 6.25$$

The crude protein content gives no indication of how efficiently the protein is utilized.

Measures of protein quality

Monogastrics

The usefulness of proteins in the diet of monogastrics is dependent on their digestibility and absorption and the subsequent retention of amino acids in the animal's tissues.

Most simply the *digestible crude protein* content of a diet is determined from the results of a digestibility trial where it is assumed that the difference between the nitrogen in the feed and faeces represents the quantity of nitrogen absorbed. As this does not take into account the fact that some of the faecal nitrogen arises from nitrogen secreted back into the digestive tract in the form of digestive secretions, it is more accurately known as the '*faecal apparent digestibility coefficient*'. More recently, for pigs it has been shown that as no useful amino acids are absorbed in the rear gut a better measure of protein digestibility is *ileal apparent digestibility* (D_{il}). This is determined in cannulated animals through the measurement of nitrogen disappearance prior to the food reaching the distal ileum and is usually about 8% lower than faecal digestibility values although for some feeds it may be as much as 14% lower. The technique can also be used to obtain apparent ileal digestibility coefficients for individual amino acids in a feed. This is useful because individual amino acids behave differently in the gut and frequently have digestibility coefficients different from that of the whole protein.

Although it might be thought that, once absorbed, amino acids are available for incorporation into tissue proteins, utilization efficiency is very much dependent on the balance of amino acids absorbed. There are a number of methods of assessing this aspect of protein quality (e.g. apparent biological value and growth assays), but in practice the efficiency with which digested protein is utilized is largely dependent on the supply and balance of certain essential amino acids, e.g. lysine, threonine, tryptophan and methionine; the amino acid that is available in least supply in relation to requirement determines the extent of protein synthesis (the limiting amino acid). For poultry, evaluation of protein sources is based upon the content of the three main limiting amino acids, lysine, methionine and tryptophan.

In mixed diets for pigs the term *protein value* (V) can be derived from a comparison of the ileal-digested and utilizable amino acid composition of the diet with the profile of utilizable ileal digested amino acids in an *ideal protein*. The ideal protein is broadly similar in terms of essential amino acid composition to that of pig tissue and milk protein. That amino acid in the comparison giving the lowest ratio determines the protein value of the diet. In cereal-based diets this is frequently lysine, although in some situations threonine and methionine may be limiting. The use of pure synthetic amino acids can do much to boost the protein value of pig diets.

Using the above parameters and knowing the efficiency of utilization of absorbed ideal protein, the term *ideal protein supply* (IP) from a diet may be calculated as

$$IP = \text{Feed intake} \times CP \times D_{il} \times V \times v$$

where CP is crude protein concentration, D_{il} is ileal digestibility of the CP, V is protein value and v is the efficiency of use of an ideal protein. The efficiency of use of ileally digested ideal protein (v) varies from 0.70 to 0.95.

Ruminants – metabolizable protein

Metabolizable protein (MP) is defined as the total digestible true protein (amino acids) available for metabolism after digestion and absorption of the feed in the animal's digestive tract. The background to this is covered in Chapter 16. Metabolizable protein has two components:

- *the digestible microbial true protein* (DMTP) which is a measure of the amino acids being made available for the animal's metabolism from microbial crude protein synthesis in the rumen;
- *the digestible undegraded feed protein* (DUP) which is that fraction of a feed that has not been degraded during passage through the rumen, the *undegradable protein* (UDP), but which is digested in the small intestine.

For the calculation of MP a number of parameters describing the extent and rates of degradability of feed proteins as they pass through the digestive tract are required. They include the following:

- *total dietary crude protein* (CP) which is the total N × 6.25 and includes nitrogen from true protein and NPN sources. This CP contains a fraction with a variable rate of degradability in the rumen, the rumen degradable crude protein (RDP), and a fraction resistant to degradation in the rumen, the undegradable crude protein (UDP).
- *quickly degradable crude protein* (QDP) which is determined by placing the dried, ground feed in a Dacron bag and washing it in cold water. The crude protein extracted is the *a* fraction. Thus:

$$[\text{QDP}](\text{g/kg DM}) = a \times [\text{CP}](\text{g/kg DM})$$

- *slowly degradable crude protein (SDP)* which is determined from knowledge of the following values:
  - the '*b value*' which is the crude protein fraction in a feedstuff that can potentially be degraded in the rumen but does not form part of the QDP fraction;
  - the '*c value*' which is a measure of the fractional rate of degradation of the '*b*' fraction. Fractional rates of degradability per hour vary for different feed protein sources, varying from 0.08 (ex soyabean) to 0.50 (wheat);
  - the '*r value*' which is a measure of the fractional outflow rates of protein from the rumen. These are highly correlated with level of feeding in the animal. The following fractional outflow rates as affected by level of feeding have been derived:

Maintenance	0.02/hour
2 × maintenance	0.05/hour
3 × maintenance	0.08/hour

The *b* and *c* values have been determined for individual feeds using the '*in sacco*' method which involves incubation of small samples of feed in the rumen in nylon bags. These bags have pores small enough to retain the feed but large enough to allow bacteria to enter the bag. Thus:

$$[\text{SDP}](\text{g/kg DM}) = \{(b \times c)/(c + r)\} \times [\text{CP}]\text{g/kg DM}$$

- *undegradable crude protein* (UDP) in a feed is given by

$$[\text{UDP}] = [\text{CP}] - \{[\text{QDP}] + [\text{SDP}]\}$$

- *digestible undegradable crude protein* (DUP) which is the UDP digested and absorbed from the small intestine. It is influenced by the amount of crude protein bound to the fibre fraction of the feed, the *acid detergent insoluble crude protein* (ADIP), and has an average digestibility of 0.9. DUP may be calculated according to the equation:

$$[\text{DUP}](\text{g/kg DM}) = 0.9\{[\text{UDP}] - [\text{ADIP}]\}$$

- *effective rumen degradable protein* (ERDP) describes the amount of crude protein synthesized by rumen micro-organisms from the QDP and SDP fractions. The efficiencies of nitrogen capture being 0.8 and 1.0 respectively, ERDP may be calculated according to the equation:

$$[\text{ERDP}](\text{g/kg DM}) = 0.8[\text{QDP}] + [\text{SDP}]$$

Example Calculation: For a feedstuff such as ground barley (CP = 129 g/kg DM) which has CP degradability factors of:

$a = 0.25$ $\quad r = 0.05$
$b = 0.70$ $\quad [\text{ADIP}](\text{g/kg DM}) = 2.5$
$c = 0.35$

$$\begin{aligned}
[\text{QDP}](\text{g/kg DM}) &= 129 \times 0.25 = 32.3 \\
[\text{SDP}](\text{g/kg DM}) &= [\text{CP}] \times \{(b \times c)/(c + r)\} \\
&= 129 \times \{(0.7 \times 0.35)/(0.35 + 0.05)\} \\
&= 79 \\
[\text{ERDP}](\text{g/kg DM}) &= 0.8[\text{QDP}] + [\text{SDP}] \\
&= 105 \\
[\text{UDP}](\text{g/kg DM}) &= [\text{CP}] - \{[\text{QDP}] + [\text{SDP}]\} \\
&= 18 \\
[\text{DUP}](\text{g/kg DM}) &= 0.9\{[\text{UDP}] - [\text{ADIP}]\} \\
&= 14
\end{aligned}$$

It is from these values that values for the synthesis of *microbial crude protein* (MCP), *microbial true protein* (MTP) and finally metabolizable protein (MP) supplied by a diet can be calculated. The synthesis of MCP in the rumen is dependent on the amount of fermentable metabolizable energy (FME) in the diet and the level of feeding. Thus the values for MCP synthesis per MJ FME (value y) at different levels of feeding (L) relative to the ME requirement for maintenance are described by the equation:

$$y(\text{gMCP/MJFME}) = 7.0 + 6.0\{1 - e^{(-0.35L)}\}$$

Thus when $(L = 1)$ $y = 9\,\text{gMCP/MJFME}$
$(L = 2)$ $y = 10\,\text{gMCP/MJFME}$
$(L = 3)$ $y = 11\,\text{gMCP/MJFME}$

and it follows that:

$$[\text{MCP}](\text{g/kg DM}) = y \times \text{FME}(\text{MJ/kg DM})$$

In diet formulation the objective is for ERDP supply to be equal to the potential MCP supply from the available FME. If ERDP supply exceeds its requirement for MCP synthesis, then FME is limiting and the excess ERDP will be wasted and result in elevated levels of blood ammonia and urea. If ERDP supply is less than ERDP requirement, then the diet is ERDP limited and

$$[\text{MCP}](\text{g/kg DM}) = [\text{ERDP}](\text{g/kg DM})$$

Of the MCP produced by the rumen microbes about 0.25 is present as nucleic acids, which cannot contribute to the synthesis of body tissue or milk protein. The MTP content of a diet is therefore 0.75 of the MCP. MTP is 0.85 digestible in the small intestine so that digestible microbial true protein (DMTP) in a diet is:

$$\begin{aligned}[\text{DMTP}](\text{g/kg DM}) &= 0.75 \times 0.85 \times [\text{MCP}] \\ &= 0.6375 \times [\text{MCP}]\end{aligned}$$

To this source of true protein absorbed from the small intestine we must add the DUP to obtain the MP supplied by a ration:

$$[\text{MP}](\text{g/kg DM}) = 0.6375[\text{MCP}] + [\text{DUP}]$$

Micronutrients

The mineral composition of feeds is usually described in terms of the total quantity (g or mg/kg) of the individual elements present. Since availability of minerals in food varies greatly and is influenced by a number of factors there is no attempt to include this parameter in evaluation.

Vitamin concentration of food is usually expressed as milligrams or micrograms per kilogram. However, the levels of fat-soluble vitamins are normally expressed as International Units (IU).

1 IU vitamin A = 0.3 mg crystalline vitamin A alcohol
1 IU vitamin D = 0.025 mg or crystalline vitamin D_3
1 IU vitamin E = 1 mg α-tocopherol acetate

Chemical analysis and the prediction of feed value

The most widely used routine methods of evaluating the nutrient content of feedstuffs are based on chemical procedures. Such analyses are only of value, however, if the feed analysed is representative of the entire batch. To this end it is important that appropriate sampling procedures be followed. Using appropriate sampling equipment a number of core samples should be taken from different areas of the feedstuff concerned, these samples thoroughly mixed together and a suitable size of subsample taken for analysis. It is important that samples for analysis are appropriately packaged, labelled and stored prior to analysis. For example, silage samples not analysed immediately should be frozen.

The longest-standing method of chemical analysis is the Weende or proximate analysis in which the feed is divided into six fractions (Table 18.1). The system has many criticisms, particularly with respect to adequately describing the carbohydrate fractions and the overestimation of protein which results from the measurement of total nitrogen rather than solely protein nitrogen. Whilst some elements of it are still in use to describe feed composition, a number of other analytical procedures are now used. Since the proximate analysis fails to differentiate the carbohydrates in a feed in terms of their nutritional value, several methods have been developed to measure the highly available carbohydrate as starch and sugar, whilst the less readily available carbohydrate or 'fibre' fraction may be characterized according to the Van Soest scheme of analysis (Fig. 18.3) Within this, the *neutral detergent solubles* (NDS) fraction represents a

Table 18.1 The fractions of proximate analysis

Fraction	*Components*	*Procedure*
(1) Moisture	Water and any volatile compounds	Dry sample to constant weight in oven
(2) Ash	Mineral elements	Burn at 500°C
(3) Crude protein (= N × 6.25)	Proteins, amino acids, non-protein nitrogen	Determine nitrogen by Kjeldahl or Dumas method
(4) Ether extract (oil)	Fats, oils, waxes	Extraction with petroleum spirit
(5) Crude fibre	Cellulose, hemicellulose, lignin	Residue after boiling with weak acid and weak alkali
(6) Nitrogen-free extractives	Starch, sugars and some cellulose, hemicellulose and lignin	Calculation (100 minus sum of other fractions)

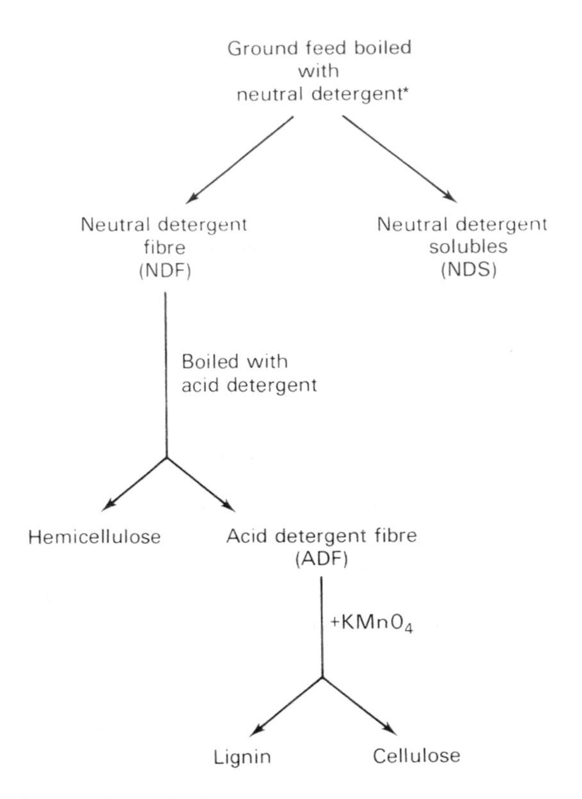

Figure 18.3 The Van Soest scheme of analysis. (*Starch-containing feeds are pre-treated with amylase.)

rapidly digested/fermented fraction consisting mainly of proteins, lipids, starches and sugars. The *neutral detergent fibre* (NDF) consists mainly of the cell wall materials such as cellulose, hemicellulose and lignin and represents a fraction that is relatively indigestible to the monogastric and slowly fermented in the ruminant. The determination of *acid detergent fibre* (ADF), consisting mainly of cellulose and lignin, is especially useful for forages as it is closely correlated to digestibility. The term *modified acid detergent fibre* (MADF), which has its origins in a variation of Van Soest's ADF method, has been used extensively in the UK to predict the ME value of forage feeds. It is, however, largely outmoded and like other conventional analytical methods has been replaced in routine analysis by the introduction of near infra-red (NIR) spectroscopy.

NIR spectroscopy represents a radical departure from conventional analytical methods, in that the entire feed sample is characterized in terms of its absorption properties in the NIR region (1100–2500 nm) of the electromagnetic spectrum, rather than separate subsamples being treated with various chemicals to isolate specific components. By selecting particular wavelengths and using computer interpretation of the spectra, rapid measurement of those chemical components already referred to can be made in straight concentrate, forages and compounded feeds. In addition, it can be used to predict energy values, digestibility, protein degradability values and even the ingestibility characteristics of feeds. Being a predictive technique, this method of analysis is, however, only as precise as the calibration of the instrument with reference samples of known chemical composition, and feeding value. Thus in the case of standard compositional measures it is only as reliable as the 'wet chemical' analysis of feeds and in the case of feeding values such as digestibility, energy, protein degradability and ingestibility is dependent on feeding and metabolism trials carried out on feedstuffs using animals.

The accuracy of any feeding system is only as precise as the measures of nutritional value of component feeds. This is especially true of forages whose feeding values can vary considerably, and since energy is the primary nutrient in diet formulation it is important to be able to assess energy values. To this end a number of prediction equations which relate the metabolizable energy values of forages to chemical composition and measures of digestibility carried out *in vitro* have been developed. A selection of these is shown below:

Grass silage

$$\begin{aligned}\text{ME}(\text{MJ/kg DM}) &= 15.0 - (0.0140 \times \text{MADF})\\ &= 5.45 + (0.0085 \times \text{NCD})\\ &= 2.91 + (0.0120 \times \text{IVD})\end{aligned}$$

Maize silage

$$\begin{aligned}\text{ME}(\text{MJ/kg DM}) &= 13.38 - (0.0113 \times \text{MADF})\\ &= 3.62 + (0.0100 \times \text{NCD})\end{aligned}$$

Grass hay

$$\begin{aligned}\text{ME}(\text{MJ/kg DM}) &= 15.86 - (0.0185 \times \text{MADF})\\ &= 4.22 + (0.0086 \times \text{NCD})\\ &= 2.63 + (0.0109 \times \text{IVD})\end{aligned}$$

where

MADF = modified acid detergent fibre (g/kg DM)
NCD = neutral detergent cellulase (DOMD) (g/kg DM)(see under 'Digestibility measurements')
IVD = *in vitro* digestibility (DOMD)(g/kg DM) (see under 'Digestibility measurements')

Fermentable metabolizable energy (FME) values for forages are mainly of relevance to silages where some of the energy is present as fermentation acids. As the extent of fermentation is related to the dry matter of the ensiled crop, a prediction of FME can be made from:

$$\text{FME}(\text{MJ/kg DM}) = \text{ME} \times \{(0.467 + (0.00136 \times \text{ODM}) - (0.00000115 \times \text{ODM}^2)\}$$

where ODM is oven dry matter (g/kg) of the silage.

Although the energy values of concentrate feeds is less variable than that of forages there are occasions where it is useful to be able to predict their energy value. This is especially the case with compounded feeds. The following equations have been developed in the UK for:

Ruminants:

$$\text{ME}(\text{MJ/kg DM}) = 0.0140\ \text{NCDG} + 0.025\ \text{oil}$$

Pigs:

$$\text{DE}(\text{MJ/kg DM}) = 17.49 + 0.0157\ \text{oil} + 0.0078\ \text{CP} - 0.0325\ \text{ash} - 0.0140\ \text{NDF}$$

Poultry:

$$\text{ME}(\text{MJ/kg feed}) = 0.01551\,\text{CP} + 0.03431\ \text{oil} + 0.01169\ \text{starch} + 0.01301\ \text{sugar}$$

where NCDG is the neutral detergent cellulase + gammanase in a feed (see section on 'Digestibility measurements'), and other measures are expressed as g/kg DM (pig and ruminant feeds) or g/kg as fed (poultry feeds).

Raw materials for diet and ration formulation

A great variety of feedstuffs is used for the formulation of animal diets and rations. Whilst pig and poultry diets consist almost entirely of concentrated feedstuffs, the requirements of ruminant livestock are met by a combination of roughages and concentrates, the contribution of concentrates to total requirements being greatest in dairy cow feeding and least in the feeding of sheep and suckler cows.

In considering feedstuffs it is useful to distinguish between the feedstuffs that are relatively high in fibre and/or water content (bulky feeds or roughages) and those that are low in fibre and water (concentrate feeds). Figure 18.4 indicates into which of these major groupings different feedstuffs fall.

The following terms that are frequently used in animal feedingstuffs terminology are also worth defining at this stage.

Compound feeds A number of different ingredients (including minerals, vitamins, additives) mixed and blended in appropriate proportions, to provide a properly balanced complete diet or, in the case of ruminants, a technically designed supplement to natural foods (e.g. grass or roughages).

Straights Single feedingstuffs which may or may not have undergone some form of processing before purchase, e.g. wheat, barley, flaked maize, soyabean meal, fish meal.

Protein concentrates Products specially designed for further mixing before feeding, at an inclusion rate of <5%, with planned proportions of cereals and other feedingstuffs either 'on farm' or by a feed compounder. These usually contain protein-rich ingredients such as fish and oilseed meals which may be supplemented with minerals and vitamins.

Additives Substances added to a compound or a protein concentrate for some specific purpose other than as a direct source of nutrient, e.g. coccidiostats, antibiotics, flavourings.

Supplements Products used at <5% of the total ration and designed to supply specific amounts of trace minerals, vitamins and non-nutrient additives.

In animal feeding a basic consideration is the nutritional value of available feeds. This information has been accumulated over many years and is being continually updated as more precise measures of nutritive values are obtained. Details on composition and nutritive values of the more widely used feedstuffs for the various classes of livestock are given in Table 18.2. More

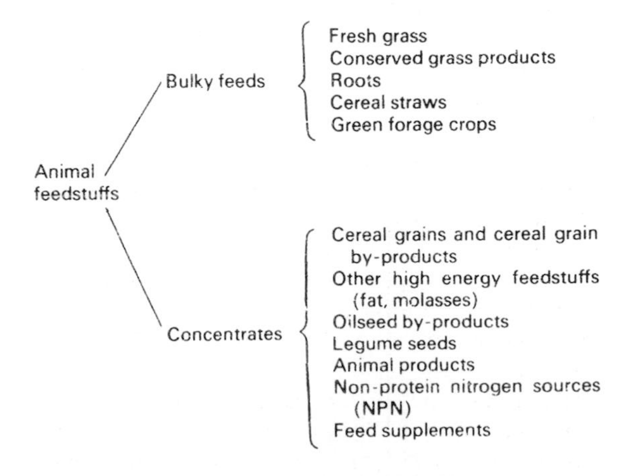

Figure 18.4 Classification of animal feeds.

Table 18.2 Guide to the chemical composition and nutritive values of some commonly used feedstuffs. A–F = as fed

Feedstuff name and number		Dry matter content (%)	Chemical composition								Nutritive value – energy		
			CP (% of feedstuff DM)	EE	CF	NFE	Ash	Starch + sugars	NDF	ADF	Metabolizable energy[1]		Digestible energy
											Ruminants (MJ/kg DM)	Poultry (MJ/kg A–F)	Pigs (MJ/kg A–F)
(1)	*Pasture grass* (average)[2]	20.6	13.9	2.2	24.2	52.1	7.6	10.0	62.0	30.0	11.2 (10.5)	—	—
(2)	*Conserved grass products*[2]												
(2.1)	Silage (average)	28.9	15.9	3.7	27.1	44.6	8.7	3.0	54.0	36.0	11.0 (8.1)	—	—
(2.2)	Hay (average)	85.8	12.2	1.8	30.1	48.1	7.8	11.3	68.8	38.0	9.2 (8.6)	—	—
(2.3)	Dried grass (average)	89.0	19.7	3.8	22.4	42.7	11.4	13.5	54.5	28.0	10.7 (9.4)	—	8.90
(2.4)	Maize silage	30.4	8.8	3.0	17.2	66.1	4.9	25.0	55.0	30.0	11.3 (9.0)	—	—
(3)	*Roots*												
(3.1)	Cassava (dehydr, ground)	88.0	2.8	0.3	8.7	84.8	3.4	74.0	15.4	6.4	13.2 (13.3)	14.43	15.3
(3.2)	Fodder beet	18.0	6.8	0.3	5.9	79.1	7.7	66.7	19.5	9.5	11.9 (11.8)	—	2.9
(3.3)	Mangels	10.0	9.2	0.8	6.2	76.9	6.9	—	—	—	12.4 (12.1)	—	1.97
(3.4)	Swedes	10.5	10.8	1.7	10.0	71.7	5.8	60.0	23.3	12.0	13.9 (13.3)	—	1.7
(3.5)	Turnips	10.0	12.2	2.2	11.1	66.6	7.8	55.1	25.4	7.0	13.0 (12.9)	—	1.4
(3.6)	Sugar beet pulp (dried)	90.0	9.9	0.7	20.3	65.7	3.4	8.2	37.2	21.3	12.8 (12.6)	—	12.9
(4)	*Cereal straws*												
(4.1)	Barley	86.0	3.7	1.6	48.8	39.2	6.6	3.2	84.5	51.0	6.4 (5.9)	—	—
(4.2)	Oat	86.0	3.4	2.2	39.4	49.3	5.7	2.0	74.9	52.3	7.2 (6.7)	—	—
(4.3)	Wheat	86.0	2.4	1.5	42.6	47.3	6.2	—	80.9	50.2	6.1 (5.7)	—	—
(5)	*Green forage crops*												
(5.1)	Kale	14.0	15.7	3.6	17.9	49.3	13.6	17.5	44.0	25.0	11.8 (11.1)	—	—
(5.2)	Rape	14.0	20.0	5.7	25.0	40.0	9.3	—	—	—	9.5 (—)	—	—

(6)	*Cereal grains and cereal by-products*												
(6.1)	Barley	86.0	10.8	1.7	5.3	79.5	2.6	59.5	23.1	6.4	13.3 (12.7)	11.13	13.2
(6.2)	Wheat	86.0	12.4	1.9	2.6	81.0	2.1	70.0	11.0	2.5	13.7 (13.1)	12.18	14.3
(6.3)	Oats	86.0	10.9	4.9	12.1	68.8	3.3	44.1	37.6	12.8	12.1 (10.7)	11.05	11.4
(6.4)	Maize	86.0	9.8	4.2	2.4	82.3	1.3	70.0	11.7	2.8	13.8 (12.4)	13.22	14.9
(6.5)	Maize (flaked)	90.0	11.0	4.9	1.7	81.4	1.0	—	—	—	15.0 (13.3)	—	15.3
(6.6)	Sorghum	86.0	10.8	4.3	2.1	80.1	2.7	70.3	13.1	5.4	13.4 (12.6)	12.98	14.3
(6.7)	Brewer's grains (dried) – barley	90.0	20.4	7.1	16.9	51.2	4.3	7.0	56.5	20.0	11.5 (9.3)	7.91	7.6
(6.8)	Distillers' grains (dried) – barley	90.0	30.1	12.6	11.0	44.3	2.0	6.3	50.3	16.0	12.4 (10.5)	9.37	—
(6.9)	Bran – wheat	88.0	17.0	4.5	11.4	60.3	6.7	25.9	47.5	13.7	10.1 (9.3)	8.45	8.9
(6.10)	Fine wheat middlings	88.0	17.6	4.1	8.6	65.0	4.7	34.0	40.0	10.2	11.9 (10.4)	11.80	11.8
(6.11)	Rice bran	90.0	14.5	2.0	15.0	55.0	13.5	27.0	44.0	25.0	7.1 (6.9)	6.5	8.0
(6.12)	Maize gluten feed	90.0	21.0	3.8	3.9	68.5	2.8	25.5	42.5	9.9	11.8 (10.3)	9.1	11.8
(7)	*Other high-energy feeds*												
(7.1)	Fat and oil (FGAF)	98.0	—	100	—	—	—	—	—	—	33.0 (0)	35.5	33.0
(7.2)	Molasses	75.0	4.5	0	0	84.1	6.9	65.0	—	—	13.1 (13.0)	11.5	14.1
(8)	*Oilseed by-products*												
(8.1)	Extracted soyabean	90.0	50.3	1.7	5.8	36.0	6.2	14.5	13.4	9.2	13.3 (12.7)	10.67	13.6
(8.2)	Extracted rape seed	90.0	41.3	3.4	10.4	36.6	8.2	14.5	36.5	19.1	13.3 (12.1)	7.36	12.2
(8.3)	Extracted groundnut (decort.)	90.0	50.4	2.1	27.3	31.6	4.7	14.0	19.5	14.0	13.7 (11.3)	11.4	13.7
(8.4)	Extracted sunflower	90.0	42.3	1.1	18.1	31.2	7.2	7.5	47.0	32.0	9.6 (8.8)	7.5	12.6
(9)	*Legume seeds*												
(9.1)	Field beans	86.0	26.5	1.5	9.0	59.1	4.0	45.5	21.1	12.5	13.1 (12.7)	10.00	12.2
(9.2)	Peas	86.9	26.1	1.4	6.0	66.5	3.2	49.5	19.0	7.5	13.5 (13.0)	10.86	14.34
(10)	*Animal products*												
(10.1)	Fish meal (white)	90.0	70.1	4.0	0	1.8	24.1	—	—	—	14.2 (11.0)	11.46	11.8
(10.2)	Herring meal	90.0	76.2	9.1	0	4.4	10.2	—	—	—	14.8 (11.8)	13.35	13.5
(10.3)	Meat and bone meal (medium protein)	90.0	52.7	4.4	0	1.7	41.2	—	—	—	9.7 (7.8)	7.1	10.5
(10.4)	Dried skim milk	95.0	37.2	1.1	0	53.2	8.5	7.9	—	—	14.1 (13.3)	12.26	15.6

[1] Fermentable metabolizable energy (FME) values of ruminant feeds in brackets.
[2] Values highly variable and samples require to be routinely analysed.

Table 18.2 Continued

Feedstuff number[3]	*Nutritive value, protein*							*Amino acid composition*						*Mineral composition*	
	Parameters for MP system, ruminants				*Digestibility coefficients, pigs*		*Digestible crude protein, poultry (% A–F)*	*Glycine + serine (% of total feed A–F)*	*Lysine*	*Methionine + cysteine*	*Isoleucine*	*Threonine*	*Tryptophan*	*Calcium*	*Phorphorus*
	a	*b*	*c*	*ADIP (g/kg DM)*	*Faecal*	*Ileal*								*(% of total feed DM)*	
	(decimal proportions)														
(1)	0.34	0.57	0.09	7.5	—	—	—	—	—	—	—	—	—	0.50	0.29
(2)															
(2.1)	0.63	0.26	0.14	8.1	—	—	—	—	—	—	—	—	—	0.62	0.30
(2.2)	0.22	0.60	0.08	7.5	—	—	—	—	—	—	—	—	—	0.55	0.28
(2.3)	0.37	0.63	0.04	14.4	0.36	0.30	—	—	0.75	0.44	0.68	0.68	—	0.92	0.33
(2.4)	0.66	0.19	0.20	16.9	—	—	—	—	—	—	—	—	—	0.39	0.18
(3)															
(3.1)	—	—	—	—	0.40	0.25	—	—	0.09	0.03	0.05	0.05	0.01	0.20	0.03
(3.2)	0.25	0.65	0.44	5.6	—	—	—	—	0.035	0.02	0.024	0.036	0.02	0.22	0.22
(3.3)	—	—	—	—	—	—	—	—	—	—	—	—	—	0.22	0.22
(3.4)	—	—	—	—	—	—	—	—	—	—	—	—	—	0.42	0.33
(3.5)	0.25	0.65	0.34	10.0	—	—	—	—	—	—	—	—	—	0.44	0.33
(3.6)	0.24	0.70	0.05	10.0	0.50	0.45	—	—	0.69	—	0.37	0.37	0.09	0.96	0.09
(4)															
(4.1)	0.30	0.50	0.12	6.3	—	—	—	—	—	—	—	—	—	0.39	0.08
(4.2)	0.30	0.51	0.11	3.8	—	—	—	—	—	—	—	—	—	0.24	0.09
(4.3)	0.30	0.50	0.12	5.0	—	—	—	—	—	—	—	—	—	—	—
(5)															
(5.1)	0.25	0.65	0.27	14.4	—	—	—	—	—	—	—	—	—	2.14	0.31
(5.2)	—	—	—	—	—	—	—	—	—	—	—	—	—	0.93	0.42

(6)															
(6.1)	0.25	0.70	0.35	2.5	0.76	0.72	9.0	0.81	0.35	0.34	0.35	0.30	0.15	0.05	0.38
(6.2)	0.39	0.57	0.13	3.8	0.82	0.75	8.8	1.00	0.31	0.39	0.40	0.31	0.12	0.03	0.40
(6.3)	0.72	0.23	0.40	2.5	0.75	0.70	8.5	1.00	0.38	0.34	0.36	0.34	0.12	0.09	0.37
(6.4)	0.26	0.69	0.01	8.1	0.78	0.73	6.7	0.78	0.26	0.29	0.32	0.32	0.08	0.02	0.27
(6.5)	—	—	—	—	—	—	—	—	0.29	0.33	0.36	0.36	0.10	—	0.29
(6.6)	—	—	—	—	0.75	0.70	8.4	0.65	0.22	0.31	0.38	0.30	0.08	0.03	0.68
(6.7)	0.19	0.67	0.07	24.4	0.70	0.60	14.1	1.80	0.52	0.55	0.94	0.55	0.21	0.32	0.78
(6.8)	0.84	0.12	0.11	81.3	0.60	0.55	—	—	—	—	—	—	—	0.31	0.33
(6.9)	—	—	—	—	0.60	0.50	11.1	1.50	0.60	0.50	0.55	0.45	0.25	0.16	0.84
(6.10)	0.34	0.57	0.11	2.5	0.70	0.60	15.0	1.43	0.64	0.40	0.49	0.49	0.22	0.13	0.91
(6.11)	0.04	0.78	0.06	8.8	0.50	0.40	—	—	0.50	0.34	0.45	0.34	0.44	0.19	1.10
(6.12)	0.38	0.61	0.12	5.0	0.65	0.55	22.3	5.51	0.55	0.80	0.55	0.70	0.13	0.28	0.80
(7)															
(7.1)	—	—	—	—	—	—	—	—	—	—	—	—	—	—	—
(7.2)	1.0	0.00	—	—	0.30	0.25	—	—	—	—	—	—	—	—	—
(8)															
(8.1)	0.08	0.92	0.08	14.4	0.87	0.80	42.8	3.94	2.94	1.35	2.52	1.81	0.60	0.23	1.02
(8.2)	0.32	0.61	0.16	22.5	0.70	0.60	26.1	3.32	2.04	1.26	1.36	1.58	0.44	0.59	0.94
(8.3)	0.22	0.77	0.09	12.5	—	—	40.8	—	1.67	1.25	1.67	1.36	0.52	0.12	0.51
(8.4)	0.24	0.71	0.35	16.3	0.75	0.65	24.8	2.91	1.60	1.49	1.71	1.56	0.50	0.41	1.33
(9)															
(9.1)	0.33	0.72	0.09	3.1	0.75	0.68	21.1	2.08	1.50	0.36	0.93	0.84	0.21	0.19	0.67
(9.2)	0.56	0.44	0.09	30.0	0.75	0.68	20.6	1.05	1.6	0.59	0.97	0.87	0.20	0.09	0.83
(10)															
(10.1)	0.30	0.63	0.02	—	0.92	0.80	59.0	7.12	4.40	1.26	2.46	2.46	0.61	7.93	4.37
(10.2)	—	—	—	—	0.92	0.80	66.6	—	5.62	1.62	3.10	2.91	0.58	3.26	2.22
(10.3)	—	—	—	—	0.65	0.58	41.2	8.00	2.61	1.48	1.26	1.55	0.29	11.44	6.00
(10.4)	—	—	—	—	0.95	0.90	27.5	2.82	2.63	1.26	2.42	1.58	0.47	1.17	0.96

[3] For feedstuff names refer to *Feedstuff name and number* column in **Table 18.2** on previous pages.

comprehensive information may be obtained from appropriate publications (see 'Further reading/information sources').

The value of a feedstuff is not solely described by its compositional analysis or nutritive value. Limits to the use of a feed may be imposed by the presence of antinutritional or toxic factors or by particular organoleptic or physical properties. In the consideration of individual feeds the possibility of such factors operating will be given attention.

Grazed herbage

Grazed herbage provides a major source of nutrients for ruminant animals. As a feedstuff, however, it has the disadvantage of being variable in nutritive value and in addition its efficiency of utilization is difficult to control. Aspects of this are covered in Chapter 4, the salient points being that whilst a number of factors such as climate, soil fertility and botanical composition of the sward may affect nutritive value, the most important factor is stage of growth (Fig. 18.5).

Early in the growing season grass has a high water, organic acid and protein content and a low content of structural carbohydrates and lignin, making it highly digestible to the animal. As the plant matures and the yield of forage increases, there is an increase in the proportion of stem tissue and a decrease in leaf tissue. This is reflected in increased levels of structural carbohydrates and lignin (ADF), decreased protein and thus declining digestibility and metabolizable energy values. A further consequence of these changes in nutritive value that occur is that the animal's voluntary intake of herbage declines as it matures. Thus, systems of utilization of grassland must take into account those variables together with animal production targets and other management constraints, e.g. amount of herbage required for winter use. A relatively high degree of control of these factors may be exerted on lowland pastures and has led to the development of systems which allow the measurement of the yield and efficiency of utilization of ME from grass (see Chapter 4). In the case of hill pastures the restricted growing season and difficulties in controlling herbage utilization lead to a very cyclical pattern of nutrition in grazing animals. Although during the growing season (May–October) grass in hill pastures will be in the range of 65–55 D, during the winter months D-values may be as low as 40 with little or no protein. Such low levels of protein restrict the rate of herbage breakdown in the rumen and consequently adversely affect the voluntary intake of the animal.

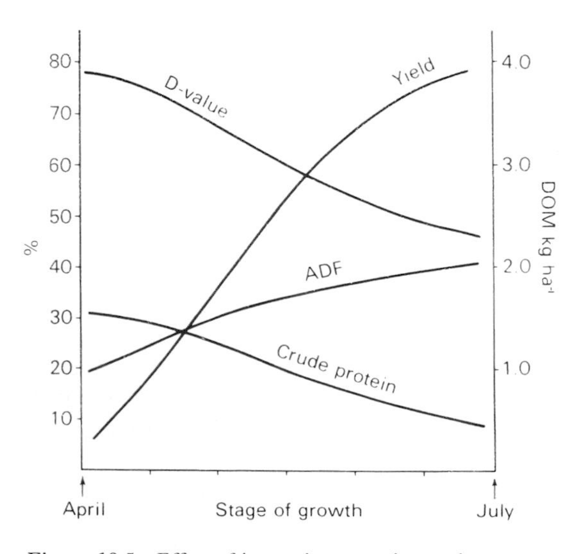

Figure 18.5 Effect of increasing maturity on the composition and yield of digestible organic matter (DOM) of a typical ryegrass sward.

The presence of clover in pasture increases the protein supply to the animal and since the effects of stage of growth on the feeding value of clover are less obvious, due to the lower degree of lignification of stem tissue, the inclusion of clover in a sward results in a reduced rate of decline of nutritive value with advancing maturity. In the grazing of pastures with a high clover component a degree of care must be taken to guard against 'bloat' which may occur due to the stimulation of foam production in the rumen by certain cytoplasmic proteins.

Conserved grass

A feature of all conserved grass is that the nutritive value is dependent on the quality of grass used and the efficiency of the conservation process. Thus, the aim of all conservation processes is to minimize the losses of nutrients from the grass utilized and to produce a feedstuff that fulfills the production objectives of the livestock to which it is to be fed.

The main methods of conserving grass are as hay, silage and artificially dried grass. Most grass in the UK is now conserved as silage. Because of the costs of production, artificially dried grass is produced on a relatively small scale for specialized purposes. Details of these processes and factors affecting the efficiency of the processes are given in Chapter 4, Grassland.

As the nutritive value of conserved hays and silages can vary considerably, it is advisable that appropriate samples be routinely analysed so that most prudent use of both conserved forage and supplementary concentrate feeds may be made. Since grass is generally cut at an earlier growth stage for silage than for hay, the nutritive value is generally higher. Thus when compared with the best quality hays good quality silages have approximately 10% more ME and up to 50% more CP. Retention of the vitamins and minerals in grass during hay and silage making processes is variable, but in general the mineral content of both materials reflects levels in the crop unless effluent losses or leaching losses are high; silages are good sources of carotene but contain negligible amounts of vitamin D whereas the opposite situation pertains with sun-cured hays.

During haymaking, losses of nutrients through oxidation of water-soluble carbohydrates, protein breakdown, leaf shatter and the possibility of leaching losses when the crop is subjected to continuous wetting and redrying combine to increase the proportion of cell wall material and thus reduce the digestibility when compared with the original grass. The calcium and phosphorus content of hay is frequently only half of values for grass. Absence of green colour is a good indication that carotene has been destroyed.

Some reductions in nutrient losses may be achieved through baling at lower dry matters (50–60%) and resorting to some form of barn drying. Failure to achieve dry matters of 85% in store may result in fungal spoilage which represents a hazard both to human health, since it causes farmers' lung, and to animals due to the fungal toxins being a cause of abortion in cattle.

The feeding value of silage whilst primarily dependent on the stage of growth of grass used may be further influenced by the extent of nutrient losses incurred during the ensilage process. The digestibility of silages is, however, usually very similar to those of the original grass except where there have been high losses through continued respiration either during wilting in the field or prior to anaerobic conditions being achieved in the clamp. Restricting the extent of the fermentation or increasing the efficiency of the fermentation in the clamp through the use of additives results in greater retention of water-soluble carbohydrates (WSC). These give the silage a higher energy value. Although nutrients such as WSC, proteins and minerals may be lost in the effluent of silages made from unwilted grass, experimental evidence indicates that greater milk production may be obtained from high moisture silages. This is probably the result of an influence on rumen fermentation altering the partition of nutrients in the animal.

A further factor that may influence the feeding value of silages is the extent of nutrient losses that are incurred due to the exposure of silage to air prior to its consumption. Oxidation of any residual WSC and organic acids by micro-organisms is especially likely and the management of the clamp face and the method of feeding should aim to restrict these losses.

The ensilage of whole crop cereals, particularly forage maize, gives a product of higher energy value than grass silage, but lower levels of CP. Such silages may be usefully supplemented with NPN compounds.

Whilst the nutritive properties of conserved forages may be defined to an extent by chemical analysis, one of the main problems in feeding roughages when they are a major component of the diet is to predict how much animals will consume. This is more of a problem with silages where factors such as digestibility, dry matter content, chop length and the fermentation pattern are operating. Intakes tend to be enhanced by high dry matter, high digestibility and reduced chop length. Whilst intakes of silages with high pH and high ammonia levels resulting from poor fermentation are low, at the other end of the scale high levels of acidity, especially in maize silages, may also restrict intake. NIR spectroscopy techniques are offering considerable potential for predicting the voluntary intake of forages.

Typical artificially dried grass is a product that, by virtue of its relatively low fibre content, occupies an intermediate position between the bulky feeds and the concentrates. Due to the rapid removal of water at high temperatures, dried grass has a similar feeding value to the original grass when expressed on a dry matter basis. Since artificial drying is an expensive process, only young leafy grass of high D-value and high crude protein content is used. An effect of high-temperature drying is to decrease the degradability of protein. Intakes of dried grass do not present a problem since not only is it highly digestible but also its cost dictates that its use is limited to a supplemental role.

Cereal straw

The gross energy content of cereal straw is similar to that of grains but because of its low digestibility even to ruminants much of this energy is lost in the faeces and makes no contribution to the animal. A further constraint to the use of straw is that a slow rate of digestion and therefore rate of passage through the gut limits the intake.

Since the levels and digestibility of the protein in straw are negligible and since there are very few vitamins and minerals present, the only nutrient straw can supply is energy. The availability of this energy is limited by the very high levels of lignin protecting the fibrous carbohydrates from microbial breakdown. In addition when straw is a major component of the diet the low levels of nitrogen available to the rumen micro-organisms may be a further factor limiting its digestibility. It has, however, been observed that considerable differences in nutritive values exist not only between the different cereal types, oat straw being superior to barley, which in turn is of higher quality than wheat, but also between varieties. Typically, spring varieties of barley straw give higher nutritive values than winter varieties with digestibility values ranging for 40–60. Likewise for wheat straws digestibilities ranging from 30–50 have been found for different varieties. It should be noted that growing conditions may also influence nutritive value although the influences of fertilizer treatment, season and daylight are largely unknown.

A great deal of research has been carried out with the objective of improving the nutritional value of straw. Chemical treatments with alkali in the form of sodium and ammonium hydroxide lead to improvements in nutritive value and dry matter intake. The extent of this improvement is dependent on the quality of the original straw – greater improvements are obtained with poor quality materials – and the method of treatment chosen. A number of 'on-farm' methods of treatment have been devised ranging from the low-cost stack methods to high capital investment oven-treatment methods. In the production of nutritionally improved straw (NIS) by feed compounders the straw is additionally ground and pelleted. As with the feeding of untreated straw, however, it is important that the remainder of the ration supplements the low levels of protein and nitrogen present. The use of ammonium hydroxide as the alkali helps overcome this problem. Due to their increased digestibility the intake by animals of treated straw is much higher than that of untreated straw, allowing much wider inclusion in the diets of ruminant animals.

Both treated and untreated straw may be used to effect in devising feeding strategies. Thus, in the feeding of dairy cows it may be used in the case of dry cows to save silage; in the case of freshly calved cows alkali-treated straw may be used as a chemical 'buffer' to increase appetite; as a buffer feed for cows at grass as a means of maintaining milk fat content; or as a means of ensuring adequate rumination so preventing displaced abomasum in animals being fed very high density diets. Traditionally straw has formed an important part of the ration for suckler cows and 'store' animals, but for situations where higher levels of performance are required from growing beef animals the level of inclusion of straw in the ration and therefore the potential benefit from treatment are severely limited. Some farmers have used both treated and untreated straw to effect as an alternative roughage source in the feeding of ewes.

The economic advantage from using treated straw is difficult to assess but is largely dependent on the cost of the original straw and the price of alternative energy sources.

Roots and tubers

The root crops, turnips, swedes, mangolds and fodder and sugar beet, are characterized by having a high water content (75–90%) and low crude fibre (5–10% of DM). The main component of the dry matter is sugar. Thus roots are highly digestible and the principal nutrient they supply is energy.

The crude protein content is in the range of 50–90 g/kg DM, a fairly high proportion of this being in the form of NPN. As suppliers of vitamins and minerals (potassium excepted) roots are poor and rations containing a high proportion of roots require appropriate supplementation.

Whilst roots can be fed and are palatable to pigs their high water content and bulkiness limit levels of incorporation for young growing pigs although diets for sows may contain up to 50% of their DM as roots. Roots are more commonly used in the diets of ruminant animals, either being fed *in situ* to sheep or carted and fed to housed animals. Although roots are capable of producing very high yields of energy per hectare, the costs and difficulty of harvesting represent a considerable problem as far as the feeding of housed animals is concerned. A further problem is that where large quantities of roots are fed to housed animals, recognition should be made of the large quantity of urine that is excreted.

Sugar beet is not primarily grown for animal consumption, but sugar beet pulp, the residue left after sugar has been extracted at the factory, is widely used in ruminant diets where in energy terms it can substitute for barley. The higher fibre content compared with barley makes it a useful component of the concentrate diet when cows go out to grass in spring, but account must be taken of its low protein value when substituting for cereals. This product is sold either directly to farmers as fresh 'pressed pulp' with about 28% DM and is therefore a perishable commodity or as dried pulp which may be readily stored for many months. Dried pulp is available in shredded form, or as pellets or nuts. Its feeding value is frequently modified prior to sale by the addition of molasses, magnesium, minerals and vitamins.

The feedstuff variously termed tapioca, manioc or cassava is a raw material which acts as one of the main cereal substitutes. It is extracted from the tropical root crop cassava and is principally comprised of starch with low levels of protein (2.5%) and oil (0.5%). The energy content is similar to that of wheat and barley but when used as a replacer for these cereals requires greater protein supplementation. Raw manioc contains cyanogenic glucosides which can be enzymically hydrolysed to hydrocyanic acid. Processing must eliminate these glucosides since the hydrocyanic acid has the effect of depleting the body reserves of the essential sulphur-containing amino acids methionine and cystine. A further quality control characteristic that must be assessed is the ash content since on occasion it may acquire excessive levels of silica from the soils on which it is grown. This reduces the energy value. As ground manioc is very dusty it is normally included in pelleted feeds rather than meals.

Green forage crops

The main green forage crops fed are kale, forage rape and stubble turnips. These brassica crops can provide a useful source of succulent feed during autumn and winter. Composition varies between the different brassica species and varieties, being particularly affected by the leafiness. As a generalization they compare favourably with the feeding value of good quality silage in energy terms and are rich in protein, minerals and vitamins.

These crops are generally fed *in situ*, kale principally to dairy cows, and rape and stubble turnips mainly to fattening lambs. For the latter they may be fed without cereal supplementation, but cereals can usefully complement the high protein supply of these crops to give increased liveweight gain.

In feeding these brassica crops account has to be taken of certain agents that may prove harmful when brassicas are fed in excess. The presence of goitrogens which affect iodine utilization (see Chapter 16) and of the chemical S-methyl-cysteine sulphoxide (SMCO) which can produce haemolysis and consequently anaemia mean that their use in diets should be limited. The latter problem is minimized if the crop is grazed before it is too mature and before secondary leaf growth occurs. For dairy cows it is recommended that intake of these brassica forage crops be limited to a maximum of 30% of the total dry matter intake and particular attention paid to the health of animals. Additionally, it should be remembered that the high levels of calcium present in brassicas may upset the Ca : P ratio in the diet, necessitating phosphorus supplementation.

A further problem is that rape can cause a skin condition termed yellows in white-faced lambs. In this condition the face and ears become sensitive to light.

Cereal grain and cereal grain by-products

Cereal grains and their by-products are the main ingredient of rations for pigs and poultry and provide the major source of energy in compound feeds fed to ruminant animals. The main cereal grains used in the UK are barley, wheat and oats which are mainly homegrown along with imported maize and sorghum.

All cereal grains are rich in carbohydrate which is mainly in the form of starch. Starch accounts for about 70% of the seed, varying between grain types. In raw material terms the crude protein of cereals is rather low and is the most variable item ranging from 6–14%. This crude protein is of relatively poor quality being low in the essential amino acids lysine, methionine and tryptophan and containing 10–15% of the nitrogenous compounds in the seed as NPN.

The crude fibre levels in cereals vary with species, being lowest in maize and highest in oats. Its level has a direct bearing on the digestibility and therefore energy value of the whole grain. The oil content also varies with species, oats and maize having higher levels than other cereals. Cereal oils are unsaturated, the main acids being linoleic and oleic, which leads them to become rancid fairly quickly after processing.

Cereals are relatively good sources of phosphorus but much of this is present as phytates which adversely affect its availability to livestock. In general, cereals are deficient in calcium and contain varying levels of other trace minerals. Of the vitamins most cereals are good sources of vitamin E, although under moist storage conditions much of this may be destroyed. Except for yellow maize, cereals are low in carotene and vitamin A and are deficient in vitamin D and most of the B vitamins, thiamine excepted.

The feeding value of cereal grains is relatively constant during prolonged storage, but occasionally poor harvest and storage conditions can lead to the presence of fungal toxins such as aflatoxin, zearalenone, vomitoxin and ochratoxin. *Aspergillus*, *Penicillium* and *Fusarium* moulds are particularly responsible for the production of these toxins. Affected grains are usually thin and shrivelled with a pinkish colour. In the case of vomitoxin maximum concentrations in feed should be <1 ppm for pigs and dairy cows and <5 ppm for poultry and other cattle. Cocktails of mycotoxin in contaminated feed are, however, potent at lower levels than are individual mycotoxins.

Maize

Of the commonly used cereals yellow maize has the highest metabolizable energy content. With a metabolizable energy value of 15 MJ/kg DM it is especially useful for broilers where yellow carcass pigmentation resulting from its high carotene content is desirable. For laying hens it supplies not only xanthophylls that enhance the colour of egg yolks but also linoleic acid which is a necessary dietary component for birds to produce eggs of a satisfactory size. The high proportion of unsaturated fatty acids means that its inclusion in pig diets must be restricted to 35% of the diet otherwise there is the likelihood of soft fat depots high in polyunsaturated fatty acids. The yellow pigmentation in body fat resulting from the feeding of yellow maize is considered undesirable in pig and ruminant carcasses. White maize has all the attributes of yellow maize without the problems of carcass fat pigmentation.

Wheat

Although not appropriate for the production of yellow broilers, wheat is suitable for all other feeding situations. It frequently forms up to 70% of diets for poultry, can be included at high levels for pigs provided that care is taken to control the fineness of grinding, whilst for ruminants inclusion levels of up to 30–40% in compound feeds may be used provided the remainder of the diet contains sufficient fibre. The gluten content of wheat can have a beneficial effect on the quality of heat-processed extruded cubes or pellets.

Barley

Traditionally barley is considered to be particularly suitable for pig feeding, having an appropriate amount of both fibre and oil which is associated with the production of saturated carcass fat. The higher level of fibre than is found in wheat or maize means that the upper inclusion rate for poultry must be limited to about 30% in many cases. High levels of barley in broiler diets have been associated with wet droppings.

Oats

Oats are normally only used for ruminants and horses. Their low energy and high fibre means that they are seldom used for pigs or poultry. In recent years varieties of 'naked' oats have received some attention as being especially suitable for young piglets, since without the husk the grain is of high energy value due to its comparatively high oil content and the protein quality is somewhat better than in other cereals.

Sorghum

From a nutritional viewpoint the grain sorghums (milo, kaffir and hybrids) resemble wheat. They have a low fibre content and in comparison with maize contain more protein and less oil. Certain varieties have a high content of phenolic compounds including tannins which not only influence palatability but also lower protein digestibility and reduce energy values for pigs and poultry.

With respect to other cereals, rye can only be tolerated at low inclusion rates because of its content of B-glucans and phenolic compounds which reduce performance and result in wet droppings in poultry and are toxic in quantity. Rice and millet may be used as an alternative to wheat once the husk has been removed. Triticale, a cross between durum wheat and rye, is similar to wheat but has a higher protein content.

Cereal preparation and processing

The aim in all cereal processing methods is primarily to increase the efficiency of utilization of the nutrients. Such improvement may result simply from an increased nutrient availability, but other factors such as changes in palatability and nutrient density may also contribute towards improvements in performance.

Most grain processing methods have as their main objective improvement in the availability of the starch present. Some of the techniques involve solely physical change, others chemical and some a combination of both physical and chemical; in addition, some processes are carried out 'wet', others 'dry'; some involving heat treatment, others under cold conditions. Mechanical alterations of the grain are the most widely used and involve physical disruption of the grain such that the starch is made more available. Grinding, rolling and crushing are the most common processes employed. It is worth noting, however, that such treatments have little or no effect on the nutritive value of barley or wheat offered to sheep.

Processing procedures which bring about chemical changes through the gelatinization of the starch include such techniques as steam flaking, micronizing, popping and pelleting. The 'flaking' process has long been applied to produce 'flaked maize' in a process involving steaming of the grain either at atmospheric pressure or in a pressure chamber followed by rolling. Micronizing and popping are both 'dry heat' processes. In the former grain is passed under gas-fired ceramic tiles, the radiant heat from which produces rapid heating within the grain causing it to soften and swell. It is then crushed in a roller mill which prevents reversal of the gelatinization process. Popping is the exploding of grain through the rapid application of dry heat. Popped grain is usually rolled or ground prior to feeding. Popping has been shown to be particularly effective in processing sorghum grain.

When feeding concentrate mixtures to certain livestock classes it is common practice to produce it in the form of pellets. From a management point of view it reduces wastage, prevents selection of ingredients and makes for easier storage and handling. Additionally, there are in some instances nutritional advantages to pelleting, this being due in the main to improvements in the available energy content.

Recent studies involving the treatment of grain for ruminant consumption with alkali appear to indicate potential as a means of increasing the availability of energy.

Cereal by-products

When cereal grains are processed for human consumption a number of by-products are produced which are used extensively as livestock feeds. The main sources of by-products are the flour milling industry, the brewing and distilling industries and, imported from the USA, by-products from maize processing.

Wheat by-products

When wheat is milled to produce flour for human consumption about 28% of the grain becomes available as by-products of the process. Wheat millfeeds are usually classified and named on the basis of decreasing fibre as bran, coarse middlings and fine middlings or shorts. These arise from the removal of the outer layers of the kernel. Wheat bran is comprised of the coarser fraction (about 50% of wheatfeed) and is usually fed to ruminant species. The finer fractions are lower in fibre and widely used in formulating diets for pigs and poultry. Since the protein in wheat grain is concentrated mainly in the outer layers, crude protein levels in these residues are higher than the whole grain. Likewise wheatfeeds are relatively good sources of most of the water-soluble vitamins, except niacin.

Brewing/distillery by-products

The main by-products arising from the brewing and distilling industries are brewers' grains and distillers' grains (draff), which are essentially the part of the grain that remains after the starch has been removed in the malting and mashing processes. They may be purchased without being dried and fed either fresh or after ensiling or alternatively may be purchased after being dried. Since both feeds have a relatively high fibre content their use is limited to ruminant rations. As well as the fibre fraction being more concentrated by the loss of the starch, so also are the crude protein and oil fractions. In the case of distillers' grains the relatively high lipid content (80–90 g/kg DM) is known to interfere with the cellulolytic action of the rumen microflora, but this can be overcome to an extent by the addition of suitable amounts of calcium which results in the formation of insoluble calcium soaps.

Other by-products of these industries, namely malt culms, dried brewers' yeast and dried distillers' solubles, may also be fed to livestock.

Rice bran

In recent years rice bran has become increasingly important in animal feeds and up to 10% may be included in compound feeds for ruminants. It is also used in sow diets. Rice bran consists mainly of the bran and outer part of the grain and is obtained as a by-product when rice is 'polished' for human consumption. It tends to be a variable product, but good quality rice bran contains very little of the less nutritious hull fragments and is rich in oil (14–15%) making it susceptible to rancidity. This oil is usually extracted either by solvent, leaving less than 1% oil in the product, or by expeller press in which case up to 1% oil may remain. Solvent extracted brans have ME values of the order 6.5–7.5 MJ/kg DM whereas those with higher oil content can have ME values of 11 MJ/kg DM. Such differences emphasize the need for users to ascertain the composition of rice bran before purchase or use in a formulation.

Maize by-product feeds

The wet milling of maize is used in the USA for the production of starch, sugar and syrup. A number of by-products result from this process including bran, germ meal and gluten. The latter is the most important, and large amounts of maize gluten are exported from the USA into the EU.

Maize gluten is a good source of energy and protein for both dairy and beef rations. Energy levels are similar to those of barley with crude protein of the order of 20%. Up to 20% of the total DM may be included in ruminant rations and it may also be included on a more limited scale in pig and poultry diets. The quality of maize gluten from different sources can vary considerably such that it is essential to monitor the quality closely.

Other high energy feeds

Whilst feed grains are the main energy-supplying concentrate feeds, other feeds are routinely used to supply energy to livestock.

Fats and oils

These are of particular value in increasing the energy density of the ration since their energy value is more than twice that of digestible carbohydrate. Fats may also improve rations by reducing dustiness and increasing palatability. The inclusion of fats in animal diets has found greatest application in milk replacers for suckling animals and in the diets of pigs and poultry where energy density is a factor controlling total energy intake. In recent years fats have also been included in dairy cow and sow diets. Energy costs can represent up to 75% of total formulation costs for animal feeds, and fat is frequently the cheapest source.

A major source of fat is feed grade animal fat (FGAF), often referred to erroneously as tallow. This is the fat rendered from meat and bone meal and in fatty acid compositional terms is typical of animal fats, being high in palmitic, stearic and oleic acids. In recent years FGAF has been increasingly blended with vegetable oil by-products to produce blended fats that have fatty acid profiles appropriate to the class of livestock to which they are fed. The vegetable oil by-products come mainly from edible oil refining and consist of a mixture of neutral oil and free fatty acid. A further source of vegetable oil is oil recovered from such processes as potato crisp manufacture. The blending of oils to produce an appropriate fatty acid profile is of relevance in that this parameter can influence the digestion, absorption and consequently metabolizable energy value. Synergistic effects between fats frequently exist as evidenced by mixtures of tallow and soyabean oil having higher metabolizable energy values than either of the individual sources. In nutritional terms the most important aspect is the ratio of saturated to unsaturated fat and the specific requirements for polyunsaturated fatty acids such as linoleic acid for pigs and poultry.

Current opinion is that hard fats based on palm or FGAF are most suited to ruminants, and high levels of free fatty acids which are unacceptable to monogastrics are thought to be beneficial to energy values. The use of 'protected fat' systems (Chapter 16), either through the mixing of fat with a carrier such as vermiculite or as calcium or magnesium soaps, allows the incorporation of extra fat into ruminant diets without the adverse effects on cellulose digestion normally associated with supplementary raw fat. It also presents an opportunity to manipulate milk fat output, and through the incorporation of polyunsaturated fatty acids the fatty acid composition of milk fat and carcass fat. These protected fats are usually free-flowing powders which are easily incorporated into the diet without special equipment.

For poultry diets a typical profile for supplementary fat is 35% saturated fatty acids, 20% linoleic acid and less than 50% of the total fatty acids as free fatty acids. Relatively soft fats are also recommended for pig rations provided it is within the limitation of producing soft carcass fat. It is frequently observed that the utilization of the non-fat components of the non-ruminant diet are improved by adding fat to the ration. This is probably due to an effect of additional fat on rate of passage of food through the gut.

A further source of dietary fat that has received attention in recent years is the inclusion of full fat oilseeds in the diet. Most attention has been given to full fat soyabeans. When used in the diet of non-ruminants there is the need for severe physical processing in order to make the oil fully available. Extrusion techniques and to a lesser extent micronization are most effective. Appropriate processing of whole oilseeds also provides a means of supplying additional fat to dairy cows within a 'protected fat' system. Extensive heat treatment, considered to be overheating in the preparation of whole oilseeds for monogastrics, is recommended in the processing of oilseeds for dairy cows.

Molasses

This is a by-product of sugar refining. It is very low in protein, the main constituent being sugar, giving it an energy value of about 85% of the value of cereal grain. It is mainly used in beef and dairy rations and also sow diets where it may be of particular value as a pellet binder or as a component of feeds which include NPN sources. The limiting factor to its inclusion, usually 5–10%, is the difficulty of mixing it into concentrated feeds. This requires specialized equipment, making it very difficult to use 'on-farm'.

Citrus pulp

Dried citrus pulp is prepared from the residue resulting from the manufacture of citrus juices and consists of a mixture of pulp, peel, seeds and cull fruits. Since a variety of fruits go into it, it has variable composition and requires routine analysis before use. It is similar in feed value to dried sugar beet pulp with a lower protein content (5–8%). It is mainly used for dairy and beef rations and quantities of up to 50% of the total DM in the diet may be used if desired. At such levels it may produce taint in milk. This product is not very palatable to monogastrics although up to 10% may be included in sow diets.

Oilseed residues

Several oil-bearing seeds are grown to produce vegetable oils for human consumption and industrial processes. The residues that remain after the extraction of the oil are rich in protein and are of great value as livestock feeds.

Among such high protein feeds are soyabean meal, rapeseed meal, sunflower seed meal, cottonseed meal, groundnut meal, palm kernel meal and sesame meal.

Oil is extracted from these seeds by hydraulic pressure (expelled) or solvent (extracted). Most oilseeds are now subjected to the latter treatment, the efficiency of oil extraction being much higher, leaving little oil (<1%) in the residue, whereas expeller methods leave up to 6% of the oil in the residue. The amount of oil left in the residue affects the energy value of the feed.

Fibre levels will also affect the energy value of the feed and in some cases, e.g. groundnuts, fibre levels will vary according to whether the seeds have been decorticated (removal of husk) prior to processing.

Of those oilseeds used widely, soya has the best quality protein followed closely by sunflower and rapeseed. Soya is slightly deficient in methionone, whereas sunflower and rapeseed, whilst being slightly higher in methionine, are deficient in lysine. Groundnut meal, although high in crude protein content, has very low levels of methionine and is low in both lysine and tryptophan.

In addition to nutritive values a number of anti-nutritional factors have to be taken into account when feeding certain oilseed residues.

In rapeseed meal the presence of glucosinolates, which under the action of myrosinase produce compounds (isothiocyanates and oxazolidinethione) that are goitrogenic (see Chapter 16), can limit its inclusion in animal feeds. In recent years newer varieties with lower levels of glucosides have been produced. These newer varieties contain less than 1% w/w glucosinolate and apart from that grown for industrial purposes now account for almost all the rapeseed grown. Further problems with rapeseed meal can occur when it is fed in excess of 5% to laying poultry in that it can increase the incidence of haemorrhagic fatty liver and through the presence of sinapine promotes the accumulation of trimethylamine which can cause a fishy taint in eggs. In some instances the presence of anti-nutritional factors can be nullified by heat during their processing. This is the case with the trypsin and urease inhibitors present in raw soya and with the yellow pigment gossypol which is present in cottonseed meal. Such heat processing reduces the solubility of the protein and in addition reduces the quality of the protein through involving amino acids such as lysine in browning reactions with carbohydrates. A particular problem associated with groundnut meals is that they are prone to infestation by the mould *Aspergillus flavus* which results in their subsequent contamination with the mycotoxin called aflatoxin. Current regulations prohibit the importation of groundnut meal containing >0.05 ppm aflatoxin B_1.

Legume seeds

The seeds of legumes such as peas, beans and lupins are useful sources of both energy and protein although it is for the latter nutrient that they are primarily included in rations.

The protein in field beans is particularly rich in lysine, but the low levels of methionine restrict the extent of its inclusion in the diets of pigs and poultry. Varietal differences in the digestibility of the protein present exist and are related to the amounts of tannins, vicin and convicin present. Similar varietal differences in the digestibility of peas exist. As a consequence of their only moderate protein content the extent of their use tends to be restricted, especially in the diets of young growing stock.

Lupins have a significantly higher protein content than peas and beans, but they tend to be deficient in lysine, the sulphur-containing amino acids and tryptophan. For pigs and poultry the poorer protein quality restricts the amount that may be included as an alternative to soyabean meal to about 10% of the diet. Further limiting factors to their use at present are the levels of the alkaloids, lupenine and sparteine, which by imparting a bitter taste depress feed intake, and also the presence of α-galactoside sugars, which are fermented in the rear gut of monogastrics, causing flatulence.

Animal protein supplements

Feedstuffs of animal origin other than rendered fats are principally included in diets as supplemental sources of high quality protein which can remedy deficiencies in the essential amino acid composition of the rest of the diet. Although this role has long been recognized in the feeding of pigs and poultry, recent thinking on the protein nutrition of ruminants indicates that high quality protein, which is resistant to degradation in the rumen, may be of value.

Protein supplements of animal origin are derived from slaughterhouse wastes as meat meal, meat and bone meal, blood meal and feather meal; from milk and processed milk chiefly in the form of skimmed milk powders and wheys; and from fish and processed fish as fish meals. Although the bulk of these by-products are simply available as dried ground meals of the raw materials, a recent innovation is the production of protein hydrolysate of these waste products. These are produced through the action of proteolytic enzymes and have the advantage of being virtually odour free and readily produced in a variety of formats to meet market demand, e.g. in extruded admixtures with grain or grain by-products. These protein hydrolysates are especially useful for inclusion in poultry and pig starter rations and with their high degree of solubility may also be included as an ingredient of low antigenicity in milk replacer formulae for calves. A further attraction of protein hydrolysis is that it allows less-biodegradable materials such as poultry feathers to become useful protein sources.

A problem with the major animal meals is that their composition tends to be variable according to the composition of the raw materials used. This is especially true of their oil and ash content. A further variable factor is that during processing overheating of these products can markedly reduce their digestibility to monogastrics and also reduce the available lysine content. Variations in processing temperatures can also cause variability in the rumen degradability characteristics of these products.

The use of animal protein supplements has undergone legislative review in many parts of the world in recent years. This is due to the risk of transmittable spongiform encephalopathies (TSEs) to animals and ultimately humans. Within the UK, the use of mammalian meat and bone meals has been banned from use in any animal feed since 1996, whilst the use of fish meals in ruminant diets has been prohibited since 2001.

Fish meals

These are excellent sources of protein, being especially high in lysine and methionine, and also of minerals. In the UK fish meals are limited in use to pig and poultry diets. The composition of fish meals is dependent on the type of fish used, demersal species, such as cod, producing meals with a low oil content (2–6%) whilst pelagic species, such as herring, produce meals with fairly high oil levels (7–13%). Such variation in oil contents influences the metabolizable energy values, but the use of high oil content meals is more limited because of the possibility of their imparting fishy taints or 'soft' fat in the product.

The oil is an especially good source of the very long chain omega-3 essential fatty acids (EPA and DHA) which are considered to reduce infection and improve immune status. Feeding fish meals or fish oils can also improve the omega-3 content of meat and eggs. Such products can contribute to increased consumption of these fatty acids by humans where they are likewise considered to have a number of health benefits.

As a component of the diets of early weaned pigs fish meal has certain advantages over a protein source such as soyabean meal in that it appears to have a relatively lower antigenicity.

Milk by-products

The main product that can be used in compounded animal feedstuffs is skimmed milk powder. The extent of its use is largely confined to inclusion in high quality milk replacers for young stock. Such powders have been extensively denatured such that protein digestibility to monogastrics is reduced. If included in pelleted feeds the level of inclusion needs to be limited to avoid pellets that are too hard. Liquid milk by-products, such as wheys and ultrafiltrates from cheese-making operations, are frequently used in liquid feeding systems for pigs, but levels of inclusion have to be monitored because of their high lactose and mineral contents.

Non-protein nitrogen sources (NPN)

Feedstuffs that contain nitrogen in a form other than protein are termed non-protein nitrogen sources. Such compounds can serve as useful components of ruminant diets in that they provide a source of nitrogen for rumen micro-organisms to synthesize microbial protein. Although a wide variety of compounds can be used as NPN sources the market is dominated by urea. When urea is fed it is initially broken down to ammonia and carbon dioxide by microbial urease. This ammonia may then be utilized along with appropriate oxo-acids in the synthesis of microbial protein. The efficiency of urea utilization is particularly dependent on the availability of oxo-acids and of energy to meet the needs of protein synthesis. These are affected by the amount and type of dietary carbohydrate. Starch appears to be the best source. Failure of micro-organisms to incorporate ammonia rapidly into microbial protein leads to a loss of nitrogen through urinary excretion; in extreme cases of ammonia production outstripping utilization, toxic levels of ammonia in the blood may result. A variety of factors must be considered in utilizing urea in feeds. These may be summarized as follows.

- The diet to which urea is being added must be suitable in terms of its energy, protein and mineral status to allow efficient use of NPN.
- Diets containing urea must be introduced gradually to allow rumen micro-organisms to adapt.
- Urea should not be used in pre-ruminant diets or where the level of NPN in the diet of adult ruminants is already fairly high, e.g. silage.
- Levels of inclusion should be appropriate to the class of stock being fed. For example, levels exceeding 1.25% in the concentrate ration will affect production in dairy cows, particularly in early lactation, whilst for beef cattle and suckler cows levels should

not exceed 2.5% with restricted feeding or 3% with *ad libitum* feeding.

Feed supplements (nutrients)

In formulating a ration the primary aim is to fulfil the animals' requirements for energy and protein and sometimes fibre. Should the ration prove to be lacking in micronutrients, then additions of the appropriate minerals, vitamins or amino acids may be made. Nutrient supplements are commercially available which allow the addition of small amounts of specific nutrients without changing the general make-up of the initial formulation.

Supplementation of rations with specific amino acids is mainly of concern to non-ruminants where cereal-based diets may be sub-optimal in such amino acids as lysine, methionine and tryptophan. In the feeding of high-yielding dairy cows there is evidence that supplementation of silage-based diets with 'protected' methionine may give economic responses.

Mineral supplementation may be provided in the form of licks or feeding blocks which allow the animal free access to a suitable combination of minerals or alternatively those specific minerals in deficit in a ration may be added to the ration in the form of a powder. In using mineral supplements the interrelationships among minerals must be recognized since excessive amounts of one mineral can cause a deficiency of another. Many proprietary supplements containing mixtures of macro- and trace minerals appropriate to particular production situations are available commercially.

As with minerals, any vitamins in deficit in a formulation must be made good by supplementation. For ruminants fat-soluble vitamins are the major consideration whilst for pigs and poultry both the water- and fat-soluble vitamin content of the diet may require supplementation. A variety of balanced vitamin premixes formulated for particular circumstances and usually containing the vitamins in a chemically pure form such that only very small amounts are required are available commercially. In assessing the need for vitamin supplementation it is important to recognize the variability in the vitamin content of feedstuffs and also that vitamins are easily destroyed by agents such as heat, light and oxidation.

Nutrient requirements

The requirements of animals for nutrients are initially derived in net terms and subsequently converted to and expressed in the same terms and units of measurement that are used to describe the nutrient content of foods in the rationing process. Thus, for example, the energy requirements for maintenance of a ruminant are determined in net energy terms, then converted to and expressed in metabolizable energy terms. Nutrient requirements are a measure of what the average animal requires for a particular function. Tabulated data on requirements generally include a safety margin over and above what the average animal requires in order that animals with requirements higher than the average are adequately fed. Such values are referred to as recommended nutrient allowances.

Recommended nutrient allowances may be expressed in two ways, either as a quantity or as a proportion of the diet. Thus the lysine allowances of a growing pig may be expressed either as 12 g/day or 8 g/kg of the diet, on the assumption that the pig is consuming 1.5 kg/day of diet. In practice, the allowances for ruminants are expressed in quantitative terms on a daily basis whereas those for pigs and poultry are expressed as a dietary concentration of the nutrient in the feed.

The remainder of this chapter is devoted to a survey of how requirements are derived and of the factors influencing the requirements of animals. The total requirement for particular nutrients of animals is arrived at factorially by adding together the requirements for maintenance and the various production functions of lactation, growth and reproduction. For ruminants the nutrient allowances for maintenance and the productive functions are usually given separately whilst for pigs and poultry nutrient allowances for maintenance are usually combined with the appropriate production function. Tabulated data on nutrient requirements and allowances are contained in the following publications: *The Nutrient Requirements of Farm Livestock. No. 1 Poultry* (1975) ARC, London; *Energy and Protein Requirements of Ruminants* (1993) AFRC, CAB International, Wallingford; *The Nutrient Requirements of Pigs* (1981) Commonwealth Agricultural Bureaux, Slough; *Nutrient Allowances for Pigs* (1982) MAFF/ADAS, Booklet 2089; *Advisory booklet Nutrient Requirements of Sows and Boars* (1990) AFRC, HGM Publications, Bakewell.

In this section energy and protein requirements will be considered for each of the body processes. Micronutrient requirements will be considered at the end of the section.

Energy and protein requirements for maintenance

The maintenance requirements of an animal refer to the amounts of nutrients an animal requires to keep its body composition and weight constant when it is in a non-productive state.

Energy

Since the energy used for maintenance must leave the body through heat, the *net energy requirement* for maintenance may be determined by measuring the heat produced by an animal when kept in the fasted state (fasting metabolism, FM). Such measurements may be made in animal calorimeters and show that energy requirements for maintenance are roughly proportional to the surface area of the animal or more closely, and of more practical value, related *to body weight* (W), or a function of body weight, namely $W^{0.73}$ or $W^{0.67}$. This is termed the metabolic weight of the animal.

A number of factors other than body weight may influence energy requirements for maintenance. Those with greatest influence are the activity of the animal and environmental conditions, although lesser effects related to breed, sex and age have also been identified. Clearly animals that have to forage for food expend more energy than stall-fed animals. Likewise animals kept under adverse environmental conditions expend additional energy. These additional energy expenditures are debited to the maintenance requirement.

Environmental effects are chiefly mediated through the interacting effects of temperature, wind speed and rain. Within certain limits (the thermoneutral zone) the animal can maintain thermoregulation by limiting heat loss/gain through altering peripheral blood circulation and behavioural mechanisms. Under more extreme conditions, however, the animal may have to use energy to maintain thermoregulation either to produce heat or to dissipate heat. The environmental temperature at which the animal is forced to produce heat to maintain thermoregulation is referred to as the *lower critical temperature*, and the environmental temperature at which it is required to use energy to dissipate heat is the *upper critical temperature*. Neither of these values is constant, being influenced by wind speed, rain and the effective insulation of the animal as influenced by coat or fleece and subcutaneous fat.

To convert net energy requirements for maintenance to the same units of measurements used in the energy evaluation of feeds, the efficiency with which food is used to meet net energy requirements, the addition of allowances for activity and a safety factor (+5%) must be taken into account.

Thus, for example, the maintenance ME requirements of cattle, sheep and goats, M_m, are given by:

$$M_m(MJ/d) = (F + A)/k_m$$

where F = fasting metabolism

A = activity allowance as defined below, and

k_m = efficiency of utilization of ME for maintenance (see section headed 'Net energy value')

For lactating cows fasting metabolism (F) is given by:

$$F(MJ/d) = 0.53(W/1.08)^{0.67}$$

where the factor 1.08 converts liveweight (W) to fasted body weight, and activity allowance (A) assuming 500 m walked, 14 hours standing and nine position changes to be:

$$A(MJ/d) = 0.0095\ W$$

Similar equations exist for other cattle, sheep and goats, also incorporating, where appropriate, the effects of housing and terrain on the maintenance requirement.

For pigs, energy for maintenance (E_m), expressed as ME, can be estimated from:

$$E_m(MJ/d) = 0.440\ W^{0.75}$$

and effects of ambient temperatures (T) less than the lower critical temperature (T_c) on additional energy requirements for maintenance (E_h) are calculated to be 0.012 MJ ME per kg $W^{0.75}$ for each °C below the lower critical temperature ($T_c - T$). Thus

$$E_h(MJ/d) = 0.012\ W^{0.75}(T_c - T)$$

It should be noted that T_c is not a fixed value and is dependent on such factors as overall metabolic activity, as influenced by body size, feed intake and rates of growth or milk production, and also the environment in which the animal is kept.

Protein

Proteins in body tissues are constantly being renewed. During this *turnover* of body tissues the amino acids resulting from breakdown are not re-utilized with 100% efficiency for synthesis into new protein. Thus the diet must provide the animal with sufficient protein to remain in nitrogen balance. This *net requirement* for protein for maintenance can be determined by measuring the faecal and urinary protein losses (N × 6.25) from the animal when it is fed a protein-free diet, termed the *basal net protein* requirement (NP_b), and the dermal losses of protein as scurf and hair (NP_d). These combined losses are more closely related to metabolic body weight ($W^{0.75}$) than body weight.

In ruminants these net protein requirements are converted into a *metabolizable protein* (MP) requirement taking into account the efficiency (k_n) with which MP is utilized. The efficiency of utilization of MP for maintenance (k_{nm}) is 1.00.

Thus in cattle, for example, the MP requirement for maintenance (MP_m) is:

$$MP_m(g/d) = (NP_b + NP_d)/1.00 = 2.30\ W^{0.75}$$

$$\text{where } NP_b\ (g/d) = 2.1875\ W^{0.75}$$

$$\text{and } NP_b(g/d) = 0.1125\ W^{0.75}$$

Similar equations are available for sheep and goats.

In expressing the protein requirements of pigs the concept of *ideal protein* (see section on 'Measures of protein quality – monogastrics') *requirement* is used. The ideal protein requirement for maintenance (IP_m) can be calculated from

$$IP_m(kg/d) = 0.0013\ W^{0.75}$$

Energy and protein requirements for lactation

The nutrient requirements of the animal for milk production are dependent on the yield of milk being produced and the composition of that milk.

Energy

The energy value of milk, which is a measure of the net energy required for milk production, is dependent on its compositional analysis. It may be determined by measurement of its gross energy using a bomb calorimeter or more usually by prediction equation using information on its compositional analysis. The energy yielding components of milk, i.e. the fat, protein and lactose, are considered to have energy concentrations of 38.5, 24.5 and 16.5 MJ/kg respectively. However, on a routine basis the constituents of cow's milk commonly determined are fat and protein. With a knowledge of these values, fairly accurate predictions of milk energy value can be made using the equation:

$$EV_l(MJ/kg) = (0.376 \times \%\,\text{fat}) + (0.209 \times \%\,\text{protein}) + 0.946$$

Taking into account the efficiency with which metabolizable energy is used for milk production (k_l) (see section headed 'Net energy value') the ME requirement for lactation (M_l) is calculated for dairy cows from

$$M_l(MJ/d) = Y \times EV_l/k_l$$

where Y is the milk yield in kg/d.

Provision of dietary metabolizable energy to lactating dairy cows is complicated by the relationships between milk yield, the partition of dietary energy and the voluntary food intake of the animal through the lactation. The situation is illustrated in Fig. 18.6.

In the early part of lactation it is usually the case that cows are unable to consume sufficient energy to meet energy requirements for milk production and call on body reserves, chiefly fat, resulting in a loss of body weight.

The amount of dietary ME for milk production to which this weight loss is equivalent may be calculated as follows:

NE value of liveweight change = 19 MJ/kg
Efficiency of use of mobilized body reserves for milk synthesis = 0.84

Thus the ME from liveweight loss available for milk synthesis in the lactating cow = 19 × 0.84 = 16 MJ/kg. The mobilized nutrients are then utilized with the same efficiency (k_l) as absorbed nutrients such that the dietary ME equivalent for milk production of liveweight loss is $16/k_l$ MJ/kg.

The tissue that is mobilized in early lactation must be replaced in later lactation through the inclusion of energy in the ration over and above that required for maintenance and production. The amount of dietary ME required for body gain may be calculated from the NE value of liveweight change and the efficiency of utilization of ME for body gain (k_g) to be $19/k_g$ MJ/kg.

Whilst these calculations are appropriate for rationing purposes it is worth pointing out that the partitioning of dietary energy between milk production and body tissue is an individual characteristic of the cow, related in physiological terms to the hormonal – especially the growth hormone/insulin – balance of the animal. In

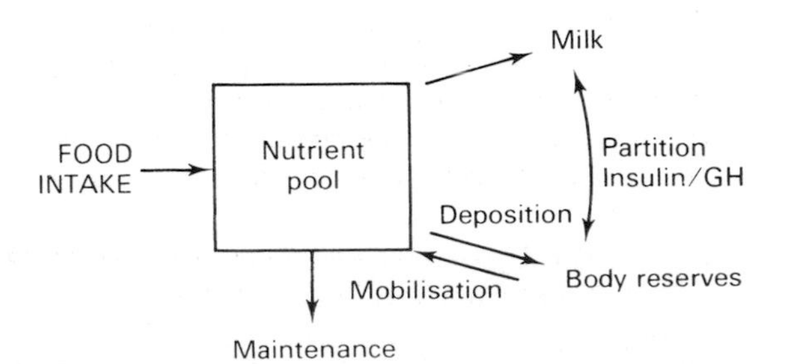

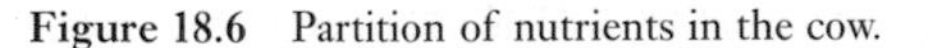

Figure 18.6 Partition of nutrients in the cow.

addition, the composition of the diet may affect partition, diets with a high energy concentration encouraging the deposition of energy in body tissue at the expense of milk production.

As with the cow the energy requirements for milk production in the sow may be obtained from food or from body fat. The amount of energy being obtained from the latter source depends on the feed intake of the sow, a characteristic that itself may be influenced by level of body fatness; sows fed well in pregnancy and carrying more fat at the start of lactation frequently eat less food and lose weight faster than sows fed less well during pregnancy. Given the energy value of average sow's milk to be 5.4 MJ/kg and the efficiency of use of dietary ME into milk production (k_l) to be 0.70, the ME required per kg milk produced is 5.4/0.7 = 7.7 MJ/kg. When the energy from mobilized body fat is used for milk synthesis, the efficiency factor is 0.85; for catabolized body protein it is much lower at 0.5. Thus assuming 1 kg body fat to contain 39.3 MJ, 1 kg of body fat can convert into 33.4 MJ of milk energy (39.3 × 0.85) or 6.2 kg of milk (33.4 ÷ 5.4) or the equivalent of 47.7 MJ of dietary ME (6.2 × 7.7). When food ME is converted to energy in body fat, the efficiency factor is 0.75, i.e. 52.4 MJ (39.3/0.75) is required per kg fat deposited.

Protein

The net protein requirements for milk production are dependent on the protein content of the milk. Protein levels in cow's milk are normally reported as CP (% per litre) of which 0.95 is *true protein*. Given that milk has a mean density of 1.03 the true protein content can be calculated as:

$$\begin{aligned}\text{True protein}(\text{g/kg}) &= \text{CP\%} \times 0.95 \times 10/1.03 \\ &= 9.22\,\text{CP\%}\end{aligned}$$

Using an efficiency of utilization of absorbed amino acids for milk protein synthesis (k_{nl}) of 0.68 the metabolizable protein requirement for milk production (MP_l) can be calculated from

$$\begin{aligned}\text{MP}_l(\text{g/kg milk}) &= (\text{True protein content of milk})/0.68 \\ &= 9.22\,\text{CP\%}/0.68 \\ &= 13.56\,\text{CP\%}\end{aligned}$$

Similar equations, taking into account the differences in the protein content of their milk, are available for sheep and goats.

In the sow, since the supply of protein is already expressed in terms of available absorbed *ideal protein* (see section headed 'Measures of protein quality – Monogastrics') which takes into account all the inefficiencies in the utilization of dietary protein, the protein requirement for milk production expressed as an *ideal protein requirement* is the same as the amount of protein the sow produces in her milk, i.e. protein content (g/kg) × yield (kg).

When animals change weight during a lactation, there are extra requirements to produce liveweight gain and contributions of protein to milk production when animals mobilize body tissue. In cows and ewes the MP requirements for liveweight gain, calculated on the basis of the net protein content of gain and an efficiency of use factor (k_{ng}) of 0.59, are 233 and 140 g/kg respectively. When ruminants mobilize tissue protein, it is assumed that the mobilized protein is used with an efficiency of 1.0 and thus adds an equivalent amount of protein to the MP supply of the animal. The supply of MP per kg weight loss is estimated to be 138 and 119 g/kg for cows and ewes respectively. The efficiency with which this source of MP can be used for milk synthesis is exactly the same as all other sources of MP (k_{nl} = 0.68). In the sow it is estimated that the conversion of body tissue protein into milk protein is around 0.8–0.9.

Energy and protein requirements for reproduction

The nutrient requirements for reproduction may be considered from the following viewpoints:

- how the nutrient intake of animals may influence reproductive potential in terms of its effect on such parameters as age at puberty, ovulation rate, conception and re-breeding; and
- the determination of the nutrient requirements of the pregnant female animal for fetal and extra-uterine growth in preparation for lactation.

In general, the requirements of pregnancy increase exponentially from insignificant levels in early pregnancy and it is especially in the last trimester of pregnancy that it becomes necessary to make special dietary provision (Fig. 18.7).

Energy

The plane of energy nutrition during the *rearing* period can influence the age at which puberty is achieved in cattle and sheep. Whilst earlier puberty may be achieved by increased levels of feeding, caution must be taken to ensure that excessive fat deposition does not occur. In practice the level of energy fed and its effect on age at

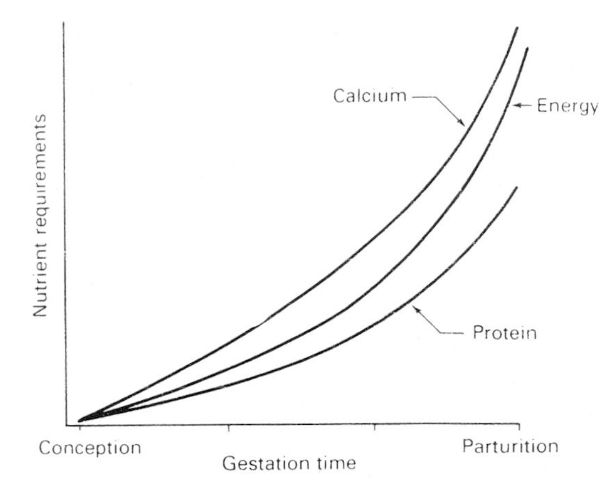

Figure 18.7 Effect of stage of gestation on the relative requirements for nutrients during pregnancy.

puberty are determined by the requirements of the production system chosen. In pigs, plane of nutrition would appear not to have such a marked effect on age at puberty.

The *period before mating* is considered important since ovulation rate may be influenced by the condition of the animal. In ewes, improving the body condition can have a particularly marked effect on the fertility of hill breeds of sheep, but more prolific breeds and crosses of these with hill breeds respond less dramatically. The more short-term practice of 'flushing' ewes prior to mating would also appear to improve fertility especially of ewes with lower condition scores. There is also reason to suppose that a similar relationship operates in the sow. From the viewpoint of feeding, the most practical advice is to ration energy according to body condition, taking account not only of those animals that require an improvement in body condition but also of overfat animals which tend to have reduced fertility. In the dairy cow mating must occur during early lactation at which time body weight loss is frequently occurring. Excessive weight loss is associated with infertility and the aim in feeding over this period is to restrict the extent of weight loss.

The period immediately *post-conception*, when the developing embryo is nourished by direct absorption from its fluid environment, is critical in that faulty or inadequate nutrition at this time can jeopardize the successful implantation of the fertilized ova in the uterine wall. Adequate nutrition must therefore be continued over this period although subsequently there is little extra requirement for energy until the last trimester of pregnancy when accelerating fetal growth increases energy demand. Although severe energy undernutrition in early pregnancy can affect placental growth, some loss in body condition can occur in ewes without adverse effects on fetal development.

Just as an awareness of body condition governs feeding in the premating period, so also it is an important parameter to be monitored as pregnancy progresses. Whilst over-thin sows and ewes will produce weak, underweight young with reduced survival rate, the development of over-fatness, especially in mid-pregnancy, can lead to pregnancy toxaemia and problems at parturition. Over-conditioning in late pregnancy may also lead to reduced voluntary food intake in the subsequent lactation. In the feeding of ewes during the last 6 weeks of pregnancy it is important to take into account the likely number of fetuses being carried. In this context the use of pregnancy scanning techniques has greatly improved the precision of rationing.

With dairy cows it is practice to terminate a lactation some 8 weeks before the birth of the next calf. Restoration of body tissue converted to milk in early lactation should take place prior to 'drying off' since the conversion of feed energy into liveweight gain is more efficient in the lactating than the non-lactating state. Provision of extra energy during the dry period is not generally recommended because not only is it less efficiently used but also there is the risk of excessive fattening and it has little influence on the birthweight of the calf.

Equations which may be used to calculate the net energy (NE) retention in the gravid uterus are:

For cattle:

$$\text{NE retention } (\text{MJ/d}) = 0.025\ W_c\left(E_t \times 0.0201\ e^{-0.0000576t}\right)$$

where E_t is total energy retention at time t (days from conception) and can be calculated from:

$$\log_{10}(E_t) = 151.665 - 151.64\ e^{-0.0000576t}$$

and W_c is the calf birthweight in kg and can be calculated from

$$W_c(\text{kg}) = \left(W_m^{\ 0.73} - 28.89\right)/2.064$$

where W_m is the mature body weight of the dam.

For sheep:

$$\text{NE retention } (\text{MJ/d}) = 0.25\ W_0\left(E_t \times 0.07372\ e^{-0.00643t}\right)$$

where W_0 is the total predicted weight of lambs at birth (kg), E_t is total energy retention at time t (days from conception) and can be calculated from

$$\log_{10}(E_t) = 3.322 - 4.979\ e^{0.00643t}$$

For sows:

Net energy deposition in uterus and mammary tissue $(MJ/d) = 0.107\ e^{0.027t} + 0.115\ e^{0.016t}$

where t is days of gestation.

Metabolizable energy requirements for concepta growth can be derived from these net energy values using an efficiency of ME use for growth of the concepta (k_c) of 0.133 for the ruminant and 0.5 for the sow.

Additional energy requirements need to be satisfied in pregnant animals that are themselves still growing. Sows, for example, do not reach their mature body weight until their fourth parity whilst cows' mature weight is not reached until their third lactation. There is still considerable uncertainty about the quantitative aspects of requirements for this aspect of pregnancy, because of the impact of pregnancy itself on energy utilization. There is some evidence that the maintenance requirements of the pregnant animal are reduced in early pregnancy whilst later in pregnancy the so-called heat of gestation increases energy needs for maintenance.

Protein

In the pregnant animal the net protein requirement is represented by not only the products of conception, fetal and uterine growth, mammary tissue growth and protein turnover losses, but also lean tissue growth in the maternal body. Clearly these various requirements are not easily separated out one from another, and, further to this, precise estimation of requirements is complicated by changing efficiency of utilization during pregnancy. Especially towards the end of pregnancy there appears to be more efficient utilization of protein, termed pregnancy anabolism, probably resulting from the animal's changed hormonal status. Requirements are clearly at their greatest during the latter part of pregnancy when uterine and fetal growth predominate.

Estimation of the metabolizable protein requirement for pregnancy (MP_c) in ruminants can be obtained from the tissue protein (TP) deposition in the products of conception – mainly influenced by stage of pregnancy and birthweight of the calf or lambs – and an efficiency of utilization of MP for pregnancy (k_{nc}) of 0.85. Equations developed to incorporate these parameters are:

$$\text{For cows } MP_c(g/d) = 1.01\ W_c\left(TP_t \times e^{-0.00262t}\right)$$

$$\text{For ewes } MP_c(g/d) = 0.25\ W_c\left(0.079\ TP_t \times e^{-0.00601t}\right)$$

where t is number of days from conception, W_c is predicted birthweight of the calf/lambs (kg), and TP_t (kg) can be derived from:

$$\log_{10}(TP_t) = 3.707 - 5.698\ e^{-0.00262t} \text{ in cattle}$$

$$\log_{10}(TP_t) = 4.928 - 4.873\ e^{-0.00601t} \text{ in sheep}$$

For sows, ideal protein requirement for uterine and mammary tissue (g/d)

$$= 0.0036\ e^{0.026t} + 0.000038\ e^{0.59t}$$

where t is number of days from conception.

Energy and protein requirements for growth

The commercial measure of growth in farm animals is liveweight gain (LWG). In determining the nutrient requirements for LWG the major problem to be faced is that the composition of LWG is not constant. The major influences on the composition of LWG are:

- the physiological age or weight of the animal, and
- the rate of LWG.

Detailed coverage of these influences on the composition of LWG is given in Chapter 16.

Energy

The net energy content of the LWG made by an animal is dependent on the proportions of protein, fat, water and ash present. Thus the energy content is represented by the protein and fat fractions. In general, the gain made by young animals has a low energy content since it has a high proportion of water (in lean) and ash (in bone) and little fat. As animals grow and mature the energy content of gain increases, this being due to the declining proportions of water present and the increasing proportions of fat.

The rate of LWG influences the net energy content of the gain since the greater the rate of gain the higher the proportion of fat produced.

In specifying energy requirements for growth it is necessary, therefore, to indicate the rate of growth anticipated. This is especially the case in beef production where it is desirable to achieve target LWG at each stage of the particular production system (see Chapter 19). In some situations the target LWG is the maximum growth potential for some or all of the production cycle. This is normally achieved in practice by allowing *ad libitum* intake of high-energy feeds. Some restriction of energy intake is frequently desirable as animals approach slaughter weight as a means of reducing the fat content of the carcass, a principle applied in the production of pigs for bacon where overfatness of carcasses is penalized.

In the rationing of growing ruminants a particular problem arises in that energy requirements are not affected simply by the energy content of the gain but also by the metabolizable energy concentration (total ME in diet/total DM in diet, M/D) of the diet which has a marked effect on the efficiency of utilization of metabolizable energy. What it means in practice is that a higher proportion of the metabolizable energy in a largely concentrate-based diet is laid down as body tissue than in a largely roughage-based diet. The origins of such differences lie in the types of rumen fermentation and particularly the VFA end-products that different diets produce. Thus increasing the proportion of propionate decreases the heat increment of metabolism. This extra dimension to assessing the energy requirements for growth has led to the development of a Net Energy System, sometimes referred to as the Variable Net Energy System, which takes account not only of the effects of the weight of the animal and rate of gain but also the M/D of the diet.

Whilst energy requirements for gain are mainly influenced by the factors considered above, the breed and sex of the animal also have an influence. Thus for cattle, later-maturing breeds require less energy per unit of gain at equal body weights than early maturing breeds; the energy content of gain at a particular weight is greater in heifers than in steers which in turn is greater than in bulls.

An equation that may be used to predict the energy value of weight gains (EV_g) in cattle which takes account of these factors is:

$$EV_g(MJ/kg) = \frac{C_2(4.1 + 0.0332W - 0.000009\,W^2)}{(1 - C_3 \times 0.1475\,\Delta W)}$$

where $C_3 = 1$ when plane of nutrition > maintenance and 0 when < maintenance; C_2 corrects for mature body size (±15%) and sex (±15%) of the animal, ranging from 0.70 for late-maturing bulls to 1.00 for medium-maturing castrates and 1.30 for early-maturing heifers; and ΔW is liveweight gain (kg/d), such that the energy retained in the animal's body per day (E_g) is given by:

$$E_g(MJ/d) = \Delta W \times EV_g$$

and the ME required to support this gain is given by:

$$\text{ME requirement}(MJ/d) = E_g/k_g$$

where k_g is the efficiency of utilization of ME for growth (see section headed 'Net energy value'). Similar equations are available for growing sheep and goats.

An approach used in pigs is to consider the stoichiometry of protein and lipid synthesis in liveweight gain.

The lack of information on the extent of variation in tissue turnover, especially of protein, with body size and rate of growth makes it difficult to predict with any certainty the energy costs of retention. Values of the order of 50–70 MJ/kg protein retention and 50–55 MJ/kg fat retention, at efficiencies of 0.44 for protein and 0.75 for fat, have been suggested. In terms of what this means as far as energy requirements for body tissue gain are concerned, the differences in water associated with protein and fat must be taken into account.

Protein

Growth rate in animals is primarily determined by the energy intake. The same principles that govern the net energy requirements for gain also affect the net protein requirements for gain. In reality the requirements are not so much for protein but more specifically for the various amino acids that contribute to growth.

In cattle the net protein requirement for gain (NP_f) is influenced by sex and breed, which affect mature body weight and also rate of gain. For castrates of medium mature body size,

$$NP_f(g/d) = \Delta W(168.07 - 0.16869\,W + 0.0001633\,W^2) \times (1.12 - 0.1223\,\Delta W)$$

where ΔW is liveweight gain in kg/d.

Taking into account the efficiency of use of MP for body gain (k_{ng}) of 0.59 allows calculation of MP requirement for gain (MP_f) as:

$$MP_f(g/d) = NP_f/0.59$$

Correction factors of ±10% are used to take into account each of the effects of sex and breed size, giving, for example, correction factors of 1.20 for late-maturing bulls, 1.00 for medium-maturing castrates and 0.80 for early-maturing heifers. Similar equations can be used for prediction of MP_f in sheep and goats.

For pigs, the protein requirement expressed in terms of ideal protein (see section headed 'Measures of protein quality – Monogastrics') is precisely equal to the amount of protein deposited in the tissue. Information on the precise rates of protein retention in relation to body weight and the influences of sex and genetic make-up on these rates of retention is required if the potential rates of retention are to be maximized but not exceeded.

Micronutrient requirements

Micronutrients have a vast number of roles in the animal body and failure to meet the animal's requirements for particular micronutrients can at least impair the performance and at most produce clinical conditions. The precision of tabulated data on allowances for micronutrients is limited by the methodology of determination and by the fact that numerous interactions occur between individual micronutrients. Such interaction can influence availability in terms of absorption and their functional activity in a metabolic pathway. In addition, a variety of mechanisms exist in the animal to provide a relatively constant micronutrient milieu for metabolic activity under circumstances of variable dietary intake, thus ensuring the normal physiological functioning of the animal. Nevertheless, such homeostatic mechanisms can break down under more extreme dietary supply situations, leading to impaired physiological function through either an inadequacy or an accumulation of micronutrients.

Allowances for individual mineral elements have been determined either through factorial methods or by feeding trial techniques. In the former method the net requirement for the element is determined by adding together the endogenous losses of the element that occur from the animal when fed a diet free of the element – equivalent to maintenance requirement – and the amount of element that is present in the product of the animal. For a product such as milk this is relatively easy, but for estimation of the mineral composition of liveweight gain it is a somewhat more laborious procedure. To convert the net requirement to a dietary requirement the availability of the particular element in terms of absorbability has to be taken into account. In the performance or feeding trial method of estimation, diets containing different amounts of the element are fed and their influence on performance monitored. The main difficulty is in setting the parameters that relate to the optimum intake of the element. For example, in the case of an element such as calcium, maximum growth rate may be obtained at a level of intake that is inadequate in terms of bone strength. The ability of the animal to store some mineral elements further complicates the situation.

Estimates of requirements for vitamins are obtained from feeding trials. The same criticisms made for the use of this method for estimating mineral requirements also apply to vitamins.

In practice, levels of supplementary inclusion in particular situations are arrived at empirically, taking into account any conditions such as feed composition and health status of the herd which may influence requirements.

Further reading/information sources

AFRC (1993) *Energy and Protein Requirements of Ruminants.* CAB International, Wallingford.

Bolton, W. & Blair, R. (1977) *Poultry Nutrition*, 4th edn. MAFF Bulletin No. 174. HMSO, London.

D'Mello, J.P.F. (ed) (2000) *Farm Animal Metabolism and Nutrition.* CAB International, Wallingford.

Ewing, W.N. (1997) *The Feeds Directory.* Context Publications, Heather, Leicestershire.

Fisher, C. & Boorman, K.N. (eds) (1986) *Nutritional Requirements of Poultry and Nutritional Research.* Butterworths, London.

Givens, D.I., Owen, E., Axford, R.F.E. & Omed, H.M. (2000) *Forage Evaluation and Ruminant Nutrition.* CAB International, Wallingford.

MAFF (1990) *UK Tables of Nutritive Value and Chemical Composition of Feedingstuffs* (eds D.I. Givens & A.R. Moss). Rowett Research Services, Aberdeen.

MAFF (1992) *Feed Composition – UK Tables of Feed Composition and Nutritive Value for Ruminants*, 2nd edn. Chalcombe Publications, Canterbury.

McDonald, P., Edwards, R.A., Greenhalgh, J.F.D. & Morgan, C.A. (2002) *Animal Nutrition.* Pearson Education, Harlow.

Mounsey, A.D. (ed) (2000) *Handbook of Feed Additives 2000/2001.* HMG Publications, Bakewell.

Theodorou, M.K. & France, J. (eds) (2000) *Feeding Systems and Feed Evaluation Models.* CAB International, Wallingford.

Whittemore, C. (1993) *The Science and Practice of Pig Production.* Longmans, London.

Wiseman, J. & Cole, C.J.A. (1990) *Feedstuff Evaluation.* Butterworths, London.

19

Cattle

J.A. Kirk

Introduction

Domesticated cattle are nearly all descended from two major species: *Bos taurus*, which includes the European types, and *Bos indicus*, to which the Zebu cattle belong. Selection from these has led to the development of a number of well-defined breeds. These breeds vary from types used primarily for milk production to those developed for beef production. In some breeds an attempt has been made to combine the desirable qualities of both dairy and beef cattle to provide dual-purpose cattle.

Since the 1950s a number of exotic breeds have been imported into the UK, mainly from Europe. The rationale behind this importation of breeding stock has been to improve beef production although some are dual-purpose breeds in their country of origin. Most of these immigrants have been large-bodied, fast-growing cattle producing lean carcasses (i.e. Charolais, Limousin, Simmental), characteristics particularly valuable for satisfying consumer demand for lean meat. Similarly, the dairy sector has benefited from the importation of Holstein cattle from Canada, the USA, Denmark and New Zealand. Fat cattle, calves and milk are an important sector of total UK agricultural output (Table 19.1) and Table 19.2 shows how the pattern of UK meat consumption has changed.

Health scares of cattle

The cattle industry has throughout the 1990s been dominated by a number of health scares, all of which have affected production, and the consumption of milk and beef.

Bovine spongiform encephalophathy (BSE)

BSE is a degenerative disease of the central nervous system of adult cattle, characterized by the development of sponge-like formations in the brain. BSE has the same causal agent as scrapie in sheep. In Britain it was made a notifiable disease with a compulsory slaughter policy in 1988. BSE has a long incubation period; symptoms may take years to develop and usually appear when the animal is 3–6 years old. Infected cattle can only be identified when clinical symptoms, nervous and unco-ordinated behaviour, are exhibited. Confirmation of the disease is through post-mortem examination of brain tissue. The disease is believed to have been caused by animals eating contaminated food from recycled ruminant by-products used as a source of protein. The practice of ruminant protein being used as a ruminant food has been banned since July 1988.

In March 1996 the Secretary of State announced a possible link between BSE, a neurological disorder in cattle, and a new human form of Creutzfeld-Jacob disease (CJD). After this announcement beef sales in Britain fell by approximately 40%. Many EU states declared a ban on British beef and by the end of the month, March 1996, the EU Commission placed a world-wide ban on the export of British beef, semen, embryos and any products containing bovine material. The value of exports resulting from the industry is extremely valuable. The total financial value is approximately 1 billion pounds Stirling per year.

The result of the export ban coupled with the decline in beef consumption resulted in the collapse in the markets for beef cattle, cull cows and calves. The crisis was not entirely confined to farmers but affected livestock hauliers, slaughterers, manufacturers, retailers and renderers. There was a similar reduction in the

Table 19.1 UK agricultural output, 1998 and 1992

	1992 £ *million*	*%*	*1998* £ *million*	*%*
Livestock				
Finished cattle and calves	2034	14.5	1990	13.4
Finished sheep and lambs	1034	7.4	1133	7.7
Finished pigs	1091	7.8	881	5.6
Poultry	919	6.6	1347	9.1
Other livestock	1058	7.5	153	1.0
Total livestock	5217	37.2	5503	37.2
Livestock products				
Milk	2930	20.9	2718	18.4
Eggs	437	3.1	400	2.7
Wool	41	0.3	24	0.2
Others	33	0.2	21	0.1
Total livestock products	3441	24.6	3162	21.4
Total crops	3534	25.2	4292	29.0
Total horticulture	1817	13.0	1839	12.4
Total agricultural output	14009		14796	

consumption of beef throughout the EU. Legislation was introduced to ban the sale of animals over 30 months of age (The Over Thirty Months Scheme, OTMS) which is still in existence at the time of writing (April 2002). Also introduced was the Calf Processing Aid Scheme (CPAS); this scheme ended in July 1999 and nearly 2 million calves had been taken out of the farming system.

Milk from BSE-suspected cattle cannot be sold even though there is no evidence of BSE being carried in milk from infected animals.

Escherichia coli

Again, this bacteria became headline news in 1996 when an outbreak occurred in central Scotland. All animals have *E. coli* in their intestines. However, *E. coli* 0157-H7 at extremely low levels of infection (fewer than ten bacterial organisms) may cause serious illness and even death in humans. Cattle are the main source of infection, which may be present not just in their intestines but also on the hide, having been transferred by faeces. After slaughter whilst the animals are being dressed, the hide being removed and eviscerated, bacteria may be transferred to the meat.

As a result of this outbreak the Meat and Livestock Commission introduced the Meat Hygiene Service Clean Livestock Policy to reduce the delivery of wet and dirty animals, especially cattle and sheep, to abattoirs.

Bovine tuberculosis (bovine TB)

Bovine TB is an infection of cattle primarily caused by the bacteria *Myobacterium bovis*. The number of cattle in the UK infected with bovine tuberculosis is increasing. This is giving cause for concern, as bovine TB has been a serious problem in humans through drinking milk from infected animals. Human tuberculosis or 'consumption' used to be a major health problem. Pasteurization of milk kills TB bacteria and cooking should destroy any infection in meat. The main risk to humans is to those working in close contact with cattle. Tuberculosis in cattle was virtually eradicated by subsidised tuberculin testing of all cattle and slaughtering any reactors to the test. Testing is still compulsory and minimum testing intervals are stipulated in EU legislation (Directive 64/432). Infected animals are still slaughtered and farmers paid 100% of market value.

The 'Kreb Report' was written by a group of independent scientists and outlines the strategy for government action. Much of the controversy associated with bovine TB is due to the suggested link between badgers and cattle. It has been claimed that badgers spread the disease and infect cattle. This led to a badger culling trial in specified areas; it is this that has caused much controversy.

Table 19.2 UK meat consumption per head (kg). [After MLC (1991 and 1999) *Beef Yearbook*.]

	1982	*1991*	*1994*	*1995*	*1996*	*1997*	*1998*
Beef and veal	18.3	17.6	16.2	15.4	12.6	14.4	15.1
Mutton and lamb	7.3	7.4	5.9	6.1	6.3	6.0	6.5
Pork	12.9	13.9	13.5	12.6	13.2	14.2	14.5
Bacon and ham	8.4	7.5	7.8	8.0	8.0	7.8	8.1
Poultry	14.4	20.2	24.8	25.4	26.7	26.3	27.7
Total	61.3	66.6	68.2	67.4	66.6	68.7	72.0

Foot-and-mouth disease

Foot-and-mouth disease is an extremely contagious viral disease that affects cloven-footed animals. There are several types of virus with strains of different types. It is another notifiable disease with a compulsory slaughter policy. The majority of infections come from imported meat as the virus can live in frozen meat. At the time of writing, the UK has suffered a serious outbreak of foot-and-mouth disease. The extent to how many animals have been slaughtered is at present 4 million animals. Without doubt there will be effects on beef, lamb and pig meat supplies in the short and long term. There will be a shortage of beef for slaughter and dairy cows including replacement heifers for milk production.

Grassland management

Grazed grass is the cheapest and one of the best quality feeds on the farm for both beef and dairy cattle. Unfortunately grazed grass is often underutilized. The most important factor in efficient grass utilization is the average grass height of grazed and ungrazed areas. Correct stocking is necessary to achieve the correct degree of sward height control. Refer to the chapter on Grassland for more detail.

Milk production

Milk product surpluses arose because of the increased amount of milk being produced and reduced consumption. In 1986 it was estimated that accumulated intervention stocks stood at: butter 1 500 000 t, milk powder 1 100 000 t and beef 620 000 t. Associated with this was the massive cost of the Common Agricultural Policy for the initial purchase of these, other products and their storage. It was for this reason that in March 1984 the EEC Agricultural Ministers agreed on a policy for curbing the increase in EEC milk production by the introduction of a system of quotas. The agreement allowed for a super-levy to be charged on all milk produced above a specific quantity or 'quota'. Since the introduction of milk quotas in 1984 the UK dairy industry has seen dramatic changes (Table 19.3). The number of dairy herds has fallen by 33%, whilst dairy cow numbers have fallen by 23%; the result being an increase in the average size of dairy herd. During the same period milk yield per cow has increased by approximately 2000 litres per lactation. These factors combined have resulted in an overall decline in milk production. Tables 19.4 and 19.5 compare production criteria between countries in the EU. The three countries Germany, France and the UK are the largest producers and supply over 50% of total EU production. Table 19.3 also illustrates the variability in cow numbers and average herd size between member countries.

Table 19.3 Dairy cattle numbers in the UK. [After MMB (1992) *UK Dairy Facts and Figures* and The Dairy Council (2000 edition) *Dairy Facts and Figures.*]

Year	*No. of herds (×10)*	*No. of dairy cows*	*Average herd size*	*Average milk yield (litres/cow/year)*
1985	48 827	3 147 000	64	4770
1990	41 248	2 884 000	66	5050
1995	36 000	2 601 000	71	5380
1997	34 125	2 585 000	73	5530
1999	32 700	2 439 000	78	5960

The decline in liquid milk consumption has resulted in a greater percentage of milk being used for manufacture rather than sold as liquid milk (Table 19.6). UK consumption patterns of milk have changed, with sales of whole milk significantly decreasing, whilst semi-skimmed milk sales increased by 50% (Table 19.7). As consumption of skimmed milk is fairly static the increased consumption of semi-skimmed milk is reflected in the decreased consumption of whole milk. In 1995 sales of semi-skimmed milk were, for the first time, greater than those of whole milk and semi-skimmed milk accounted for 50% of household purchases in 1997. A further change in the liquid milk market during the 1990s was the move to purchasing milk in retail outlets rather than doorstep deliveries; by the end of the decade this had declined to about one quarter (Table 19.8). Two major reasons for this are the opportunity for buying in large containers and the cheaper prices in large multiple retail stores.

Milk sold for processing commands a lower price than that in the liquid milk market (Fig. 19.1). Milk price as paid to the producers has to take into account the quality and lower price of milk sold for manufacture. Butter and cheese account for a large amount of the milk being processed by manufacturers. Thus the more milk being processed into these products results in a lower producer price. Unfortunately for milk producers the decline in liquid consumption is paralleled by the decline in consumption of these products, thus lowering overall demand for milk (Table 19.9).

A third major change to the industry during the 1990s was deregulation with the abolishing of the Milk Marketing Board (MMB) in 1994. Since that date there

Table 19.4 Dairy production in the EU (1999). [Adapted from The Dairy Council (2000 edition) *Dairy Facts and Figures*.]

Country	*Number of herds (×1000)*	*Number of cows (×1000)*	*Milk yield (litres/cow/year)*	*Average herd size*	*Production of milk (litres) (×1000)*[1]
Austria	86	720	4607	8	3 253
Belgium	20	633	4963	32	3 120
Denmark	13	670	6845	51	4 520
France	146	4 476	5635	31	24 170
Finland	29	383	6430	13	2 408
Germany	186	5 193	5774	28	27 512
Greece	24	184	4102	8	7 282
Irish Republic	39	1 268	3988	32	5 090
Italy	102	2 078	5350	21	11 300
Luxembourg	1	47	5657	37	259
The Netherlands	37	1 643	6838	44	10 850
Portugal	70	362	5526	5	1 950
Spain	106	1 261	4713	12	5 993
Sweden	16	468	7276	30	3 253
UK	36	2 496	5954	72	14 578
EU Fifteen	911	21 883	5632	24	119 000

[1] Excludes estimates of quantities suckled by young animals.

Table 19.5 Cattle population in the EU (1999). [After The Dairy Council (2000 edition) *Dairy Facts and Figures*.]

Country	*Total cattle and calves*	*Total cows*	*Dairy cows*	*Other cows*	*% Dairy cows in total cattle population*
Austria	2 153	875	698	177	79.8
Belgium	2 970	1 160	619	541	53.4
Denmark	1 976	817	681	136	83.4
France	20 196	8 487	4 419	4 068	52.1
Finland	1 068	403	374	29	92.7
Germany	14 657	5 495	4 709	786	85.7
Greece	590	305	168	137	55.1
Irish Republic	6 707	2 393	1 261	1 132	52.7
Italy	7 357	2 799	2 135	664	76.3
Luxembourg	207	76	45	31	59.3
The Netherlands	4 079	1 657	1 570	87	94.7
Portugal	1 403	676	351	325	51.9
Spain	6 203	3 048	1 236	1 812	40.6
Sweden	1 680	605	447	158	73.9
UK	11 281	4 344	2 438	1 906	56.1
EU Fifteen	82 545	33 142	21 152	11 990	63.8

Table 19.6 Utilization of milk (million litres) in the UK and average consumption (litres per head per week). [After MMB (1992) *UK Dairy Facts and Figures* and The Dairy Council (2000 edition) *Dairy Facts and Figures*.]

	1985	*1989–90*		*1991–92*		*1998–99*	
Liquid	48%	6582	48%	6643	49%	6906	49.7%
Manufacture	52%	7245	52%	6853	51%	6988	51.3%
Average consumption	2.36	2.27		2.24		2.20	

Table 19.7 Consumption of milk by varying fat content (% of household milk purchases). [After The Dairy Council (2000 edition) *Dairy Fact and Figures*.]

Year	*Whole milk*	*Semi-skimmed*	*Skimmed*
1990	60.3	25.7	13.4
1991	56.9	30.2	11.9
1992	51.0	35.2	12.6
1993	45.7	40.9	12.3
1994	43.4	42.9	12.7
1995	39.6	46.1	13.3
1996	37.7	48.3	13.1
1997	37.7	48.3	13.1
1998	35.4	50.6	12.9
1999	33.7	51.9	13.5

Table 19.8 Household purchases of liquid milk in Great Britain. [Adapted from The Dairy Council (2000 edition) *Dairy Facts and Figures*.]

	Total	*Doorstep (million litres)*	*Retail*	*Doorstep as a % of total*
1990	5503	3708	1795	67.4
1991	5454	3532	1923	65.0
1992	5395	3251	2144	60.3
1993	5332	2967	2365	55.6
1994	5253	2605	2648	50.0
1995	5134	2193	2941	42.7
1996	5084	1925	3159	37.9
1997	5045	1766	3279	35.0
1998	4983	1659	3323	33.3
1999	4924	1484	3440	30.0

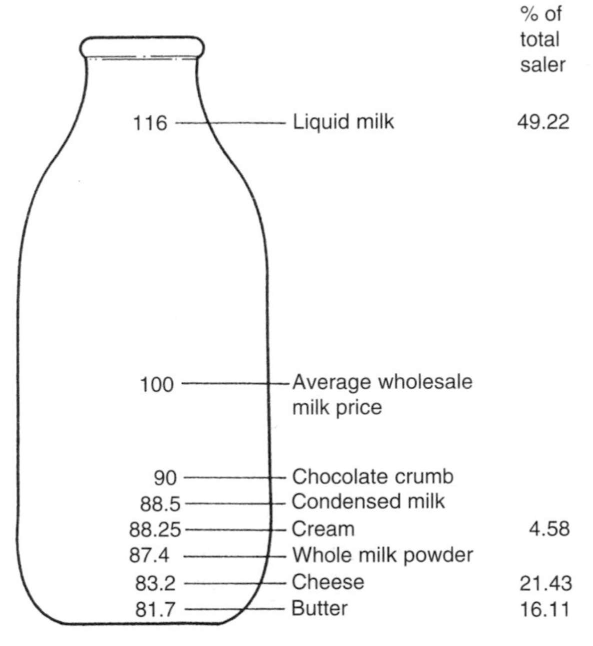

Figure 19.1 Relative values of milk products and the percentage of milk sales by product (1992). [Source: MMB (1992) *UK Dairy facts and figures*.]

Table 19.9 Household consumption of dairy products (litres/head/year). [After The Dairy Council (2000 edition) *Dairy Facts and Figures*.]

	1996	*1997*	*1998*	*1999*
Whole milk	40.6	37.1	36.1	33.1
Semi-skimmed	48.9	51.0	49.3	50.0
Skimmed	7.2	8.2	8.6	8.7
Total milk	96.7	95.0	94.0	91.8
Dried and other milk	3.6	3.3	2.9	2.8
Dairy desserts	1.2	1.2	1.4	1.6
Yogurt and cremo frais	6.7	6.7	6.6	6.8
Cream	0.9	0.8	0.9	0.8
Ice cream and ice-cream products	5.0	5.0	4.6	4.3
Butter (kg)	2.0	2.0	2.0	1.9
Cheese (kg)	5.2	5.1	4.9	5.0

have been changes in the structure of the raw milk market. Deregulation offered producers a variety of alternatives for selling their milk. These range from dairy companies to producer representative organizations such as Milk Marque which was the largest voluntary co-operative. After investigation by the Monopolies and Mergers Commission, Milk Marque was reorganized in April 2000 into three separate co-operatives. Since deregulation, comparison of producer prices is very difficult. Purchasers of milk have varying methods of calculating farm prices and not all purchasers publish the information of calculating of producer price.

After deregulation, the Milk Development Council (MDC) was established. This organization continues some of the responsibilities of the extinct MMB. These activities include the preparation of industry statistics, genetic evaluation and livestock improvement of dairy cattle and research.

In the UK there is a close relationship between milk production and beef production. Dairy herds commonly use beef bulls for crossing, and the crossbred calves, together with purebred bull calves from dairy herds, are fattened for beef. A number of heifers from the dairy herd are reared as replacements for suckler beef herds (Fig. 19.2).

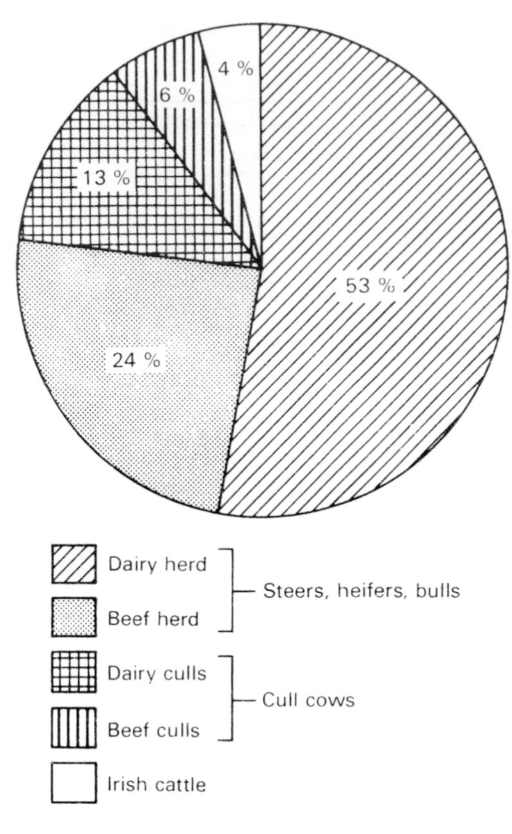

Figure 19.2 Sources of home-produced beef, 1985. [After MLC (1997) *Beef Yearbook*, reproduced with permission.]

Definitions of common cattle terminology

At birth
male: bull calf, bullock calf if castrated
female: heifer calf, cow calf
First year
male: yearling, year-old bull
female: yearling heifer
Second year
male: two-year-old bull, steer, ox, bullock
female: two-year-old heifer
Third year
male: three-year-old bull, steer, ox, bullock
female: three-year-old heifer, becomes cow on bearing calf

Heifer – usually applied to a female over 1 year old, which has not calved. An unmated animal is known as a maiden heifer and a pregnant one as an in-calf heifer. In some areas the term first-calf heifer is used until after the birth of a second calf. A barren cow is barren, cild or farrow and when a cow stops milking she is said to be yeld or dry.

Stirk – limited to males and females less than 2 years old in Scotland. It is usually applied to females only in England, the males being steers.

Store cattle – stores are animals kept usually on a low level of growth for fattening later.

Veal – the flesh from calves reared especially for this purpose and normally slaughtered at about 16 weeks of age.

Bobby veal, slink veal – flesh from calves slaughtered at an early age, often only a week or two old. These animals tend to be of extreme dairy type, thus making them undesirable for rearing as beef cattle.

Cow beef – beef from unwanted cows, often used in processing.

Bull beef – beef from entire male animals, the majority produced as an end-product but some as a by-product from redundant breeding stock.

Ageing

The development of the incisor teeth can be used as an indication of the age of cattle.

The dental formula for a full mouth is:

Permanent molars	*Temporary molars*	*Incisors*	*Temporary molars*	*Permanent molars*	
$\frac{3}{3}$	$\frac{3}{3}$	$\frac{0\ 0}{4\ 4}$	$\frac{3}{3}$	$\frac{3}{3}$	= 32

There are four pairs of incisor teeth. These are found in the lower jaw; the upper jaw has no incisors but is a hard mass of fibrous tissue known as the dental pad. Starting in the middle of the jaw the pairs are known as centrals, medials, laterals and corners. The times at which the temporary incisors are shed and replaced by the permanent incisors are important (Fig. 19.3). Individual animals will vary from these figures. Breed, management and feeding all have an influence.

Calf rearing

The foundations of the future health and well-being of the calves are laid by good feeding and management throughout the first 3 months of age. The rectal temperature of a healthy calf is 38.5–39.5°C.

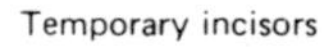

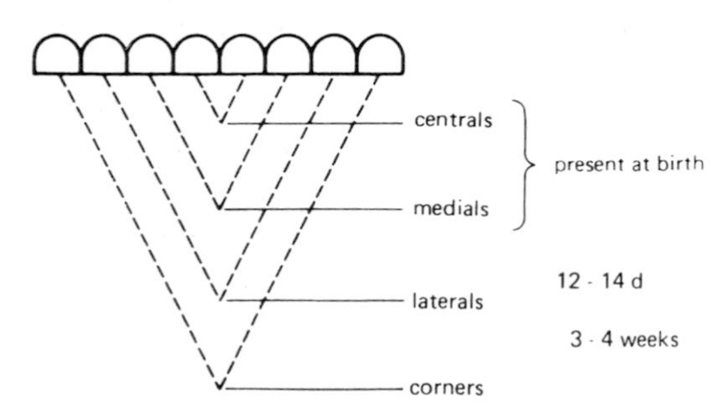

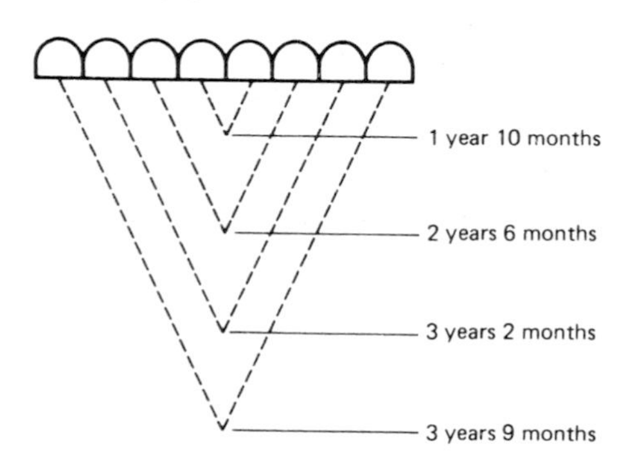

Figure 19.3 Development of the incisor teeth as an indication of the age of cattle.

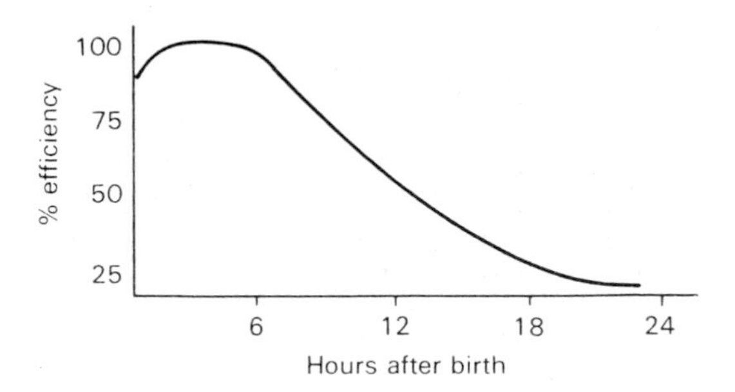

Figure 19.4 Effective uptake of antibodies from colostrum in the young calf.

Table 19.10 Composition of colostrum (first 24 h after calving) and of milk. (After Roy, 1980.)

	Colostrum	*Milk*
Fat (%)	3.6	3.5
Non-fatty solids (%)	18.5	8.6
Protein (%)	14.3	3.25
Casein (%)	5.2	2.6
Albumin (%)	1.5	0.47
Immunoglobulin (%)	6.0	0.09
Ash (%)	0.97	0.75
Calcium (%)	0.26	0.13
Magnesium (%)	0.04	0.01
Phosphorus (%)	0.24	0.11
Iron (%)	0.20	0.04
Carotenoids (μg/g fat)	25–45	7.0
Vitamin A (μg/g fat)	42–48	8.0
Vitamin D (μg/g fat)	23–45	15.0
Vitamin E (μg/g fat)	100–150	20.0
Vitamin B (μg/g fat)	10–50	5.0

Common to all the systems of rearing is the need for an adequate supply of colostrum (the secretion drawn from the udder at the time of parturition). At birth the calf is virtually free of bacteria but quickly becomes infected with organisms from its surroundings. The blood of the newborn calf contains no antibodies until the calf has received colostrum. As can be seen in Table 19.10, colostrum contains a high percentage of proteins especially immunoglobulins with their attendant antibodies, minerals, vitamins (especially the fat-soluble A, D and E) as well as carotene which is the cause of the yellow coloration. It is vitally important that the calf receives colostrum during the first 24 hours of life. The calf's ability to absorb the antibody protein is greatest during the first 6–12 hours of life. This is due to the calf's stomach wall changing and becoming impermeable to the immunoglobulins. Hence, to obtain the maximum protection, the calf should be fed colostrum during the first 6 hours of life (Fig. 19.4).

Colostrum feeding should continue for 4 days. Some farmers achieve this by leaving the calf to suckle; others remove the calf at birth. If the calf is removed at birth, it is easier to teach it to drink from a bucket. Frequently, the cow produces a greater quantity of colostrum than is needed by the newborn calf. Any excess colostrum can be diluted with water at the rate of two parts of colostrum to one of water and fed to older calves in place of milk or milk substitute. Alternatively colostrum may be frozen and kept in case of an emergency when it may be fed to newborn calves. The composition of colostrum changes to milk during the first 4 days' milking. If pre-calving milking is practised, this change may take place before the calf is born.

The zinc sulphate turbidity (ZST) test enables the concentration of circulating antibodies in the calf's blood to be determined. There is a high correlation between the results of this test and calf health, showing that it is essential that calves get adequate amounts of colostrum. This is undoubtedly one of the major factors involved in ensuring that disease and ill health are kept under control (Table 19.11).

Although the newborn calf has the same four stomach compartments, rumen, reticulum, omasum and

Table 19.11 Relationship of colostral status to calf performance at 5 weeks. (Source: Thickett *et al.*, 1979.)

	Colostrum rating		
	Low	*Medium*	*High*
No. of calves	87	158	182
% Treated for illness			
No treatment	17.2	37.3	48.9
Treated once	27.6	41.2	30.2
Treated twice	33.3	12.0	13.8
Treated more than twice	21.9	9.5	7.1

Table 19.12 Proportion of tissues by weight of a calf's stomach. (Source: Warner & Flatt, 1965.)

Stomach compartment	*Age (weeks)*			
	0	*4*	*12*	*20–26*
Rumen-reticulum	38.0	52.0	64.0	64.0
Omasum	13.0	12.0	14.0	22.0
Abomasum	49.0	36.0	22.0	14.0

abomasum, as that of the adult, they vary in size (Table 19.12) and development. The same relative proportions would also apply to the volume of the various compartments. The stomach of the newborn calf is similar to that of a monogastric animal. The oesophageal groove conveys liquids or semi-liquids directly to the omasum and from there they pass to the abomasum. It is in the abomasum where true gastric digestion takes place. The age at which transition to ruminant digestion takes place depends on the diet the young calf is fed. The longer a calf is fed milk, the less is the urge to consume other feeds and the later will be the development of the rumen.

Ideally calves should be kept for at least 7 days on the farm on which they were born. Newly purchased calves should be allowed to rest after a journey. If the calves have only had a short journey and were recently fed, additional carbohydrate, e.g. glucose, may predispose the animals to diarrhoea. Generally, calves arriving before noon should be left to rest in a dry, draught-free, well-bedded area and then fed in the evening with a drink of an electrolyte/glucose solution. Calves arriving mid to late afternoon should be left until the following morning before being fed the same solution. Proprietary electrolyte solutions are available, or a solution of glucose (22 g) with the addition of sodium chloride, common salt (4.5 g), per litre of water can be used.

First drink	1.5 litres electrolyte/glucose solution
Second drink	1.5 litres electrolyte/glucose solution
Third drink	50 : 50 milk powder/electrolyte solution
Fourth drink	Milk powder alone at standard concentration

Numerically the most common method of rearing calves is 'by hand'. Artificial milk replacers are used instead of cow's milk because the cost per unit of milk replacer is less than the equivalent amount of milk. Calves destined as replacements in the dairy herd are almost always reared by hand. This method of rearing requires specialist accommodation and is more exacting in terms of labour but enables economies in food costs to be achieved. Milk substitutes basically consist of skim milk powder and added fat, protein, vitamins and minerals; whey powder is sometimes included. Most artificially reared calves are 'early' weaned.

Early weaning

In this method the calf is weaned onto a diet of dry food by 5 weeks of age. The early introduction of concentrates, hay and water by day 7 after birth encourages the development of the rumen. To achieve the desired intake of concentrates it is essential that they be palatable and fresh. Milk replacer is restricted, again to encourage the consumption of hay and concentrates. Calves should be eating 0.75–1.00 kg concentrates daily with a liveweight of 65 kg by weaning at 5 weeks. The advantages of early weaning are that concentrates are much cheaper per kilogram of gain than milk replacer.

'Milk' feeding can be practised in several forms. Once- or twice-daily feeding of milk substitutes is the common practice on farms. *Ad libitum* feeding of milk substitute was first made possible with the advent of high-fat milk substitutes and automatic dispensing machines. The introduction of 'acid' milk replacers has enabled *ad libitum* feeding of cold milk to be practised without the need for sophisticated machines. Acid milk replacers are classified according to two types:

(1) Medium acid based on skimmed milk. These have a pH of 5.5–5.8 and a protein content of 24–26% with 17–18% fat.
(2) High acid usually based on whey from cheese manufacture. These have a pH of 4.4–5.8 with 19–20% protein and 12–15% fat.

The acidity has a positive effect in helping to reduce the incidence of digestive upsets. The pH of the abomasum prior to feeding is 2.0–2.8, but after feeding conventional milk replacer of pH 6.2–6.5 the abomasum pH rises to between 4.5 and 6.2. It then declines to pre-feeding levels after 3–5 hours. Satisfactory digestion depends on enzyme action and the optimum pH for this to occur is between 3 and 4. The conclusion therefore is that by feeding high acid milk replacer the abomasal pH is kept near optimum levels for enzyme activity and below the pH at which most organisms can survive. Milk replacers may be prepared for 3 days' feed supply at a time and the formulation aids the normal digestive process of the calf.

The suckling action by the calf ensures correct closure of the oesophageal groove, allowing milk to enter the abomasum without spilling into the developing rumen, where it can ferment and possibly cause digestive troubles. The argument that cold feeding is bad for calves probably grew out of the bucket feeding system where a calf may suffer a physiological shock when consuming large amounts of cold milk in a short period of time. With an *ad libitum* feeding system the 'little and often' effect of food consumption ensures that the small amount of milk taken in at any one time is rapidly warmed up to body temperature.

There are three important elements if the use of *ad libitum* feeding of acidified milk is to be applied correctly:

(1) Ensuring that calves are consuming enough milk. One of the easiest ways of achieving this is to feed the calves individually for the first 5–10 days by means of a bucket and teat. Any initial difficulty with calves not drinking may be overcome if the milk is fed warm; later the calves readily consume the milk replacer if it is fed at progressively cooler temperatures until cold feeding is practised.
(2) Calves should remain on cold acid milk *ad libitum* for 3 weeks. At no time should the supply of milk be allowed to run dry as excess consumption may take place when replenished and this may lead to scours.
(3) After 3 weeks, substituting cold water for milk replacer during the night may reduce the intake of milk. This encourages the consumption of concentrate food. Calves must have access to palatable concentrates at all times. These should preferably be sited near the teats where milk is drunk, thus encouraging consumption.

Criticism of *ad libitum* feeding is usually made on the grounds of the overconsumption of milk, linked to the cost of feeding greater quantities of milk replacer. Commercial farmers claim that although it may cost more to rear a calf, the animal experiences less stress and fewer health problems and as a result is a better calf. Added to this is a reduction in labour coupled with a more flexible work routine.

When reared in groups and machine-fed, care should be taken to ensure that individuals know how to suck and that there is no bullying so all have an opportunity to feed. Evenly matched batches and close observation for the first few days are essential.

A pen 1.8 m long and 1 m wide will accommodate an individual calf up to 8 weeks old. Less pen space is needed by calves reared in groups; an area of 1.1 m^2 per calf should be sufficient up to 8 weeks, this being increased to 1.5 m^2 per calf by 12 weeks. Whatever system of housing is used, warmth, a dry bed and, particularly, prevention of draughts are important to the well-being of young calves.

Systems of natural rearing

The three most commonly practised systems of natural rearing are single suckling, double suckling and multiple suckling.

Single suckling

This is by far the most popular and is carried out mainly on hill and marginal farms, by pedigree beef breeders, on lowland farms with inaccessible grass and on some arable farms utilizing grass as a break crop. In this method the cow rears her own calf with milk being the main source of nutrients. The calf remains with the dam until weaning at approximately 6 months of age. Cows are normally calved in either autumn or spring. The advantage of autumn calving is that calves are old enough to make full use of the grass in spring and are heavier at weaning in the following autumn. The disadvantage is that cows and calves often have to be housed during some of the winter period. Spring calving has the advantage that cows can be overwintered outdoors; the disadvantages are that the young calf cannot make as good use of spring grass, and spring grass may cause a flush of milk in the cows resulting in the calves scouring. Concentrates should be fed before weaning so eliminating any loss in condition later.

Double suckling

The objective is for each cow to rear two calves together. After calving, a second calf should be introduced, and allowed to suckle with the cow's own calf. Particular care should be practised at first until the cow willingly accepts the new calf. Success of the system depends on a supply of newborn calves as required, cows having an adequate supply of milk (often a cull dairy cow or a Friesian cross cow) and good management.

Multiple suckling

This necessitates nurse cows yielding a suitable quantity of milk, say 4000–5000 litres, and a supply of suitable calves. Each suckling period usually lasts about 10 weeks. In the first 10 weeks of lactation four calves are suckled, in the second 10 weeks three new calves are suckled, in the third period another three and for the last period two calves, giving a total of 12 calves reared during the lactation. Cows are often removed from the calves between feeds.

Veal production

For veal production calves capable of high rates of liveweight gain are required, Friesian bull calves normally being used. These were traditionally reared on an all-milk diet and slaughtered at between 140 and 160 kg liveweight at 14–18 weeks of age. It should be noted that the recently published welfare codes recommend that calves have access to roughage feed. Friesian bull calves will have a killing-out percentage of 55–60%.

Correct feeding is critical if an adequate return on capital is to be achieved. Milk replacers of the high-fat type with at least 15% and up to 25% fat are required for maximum gains. The aim is for a daily liveweight gain to slaughter of 1 kg/d or more. Good husbandry with particular attention being paid to hygiene and observation of animals for ill health is of paramount importance.

Calves were normally housed in buildings with a controlled environment and kept in individual pens on slatted floors. New systems of rearing veal calves involve loose housing in barns with Yorkshire boarding sides, floors bedded in straw and the animals fed *ad libitum* replacer from machines.

Management of breeding stock replacements

Rearing policy from weaning depends on two main factors: the season in which the replacement is born (autumn or spring) and the age at which it is to be calved. Age at calving is important as it is related to conception, dystokia and milk yield in first lactation, overall lifetime milk production and the herd calving pattern.

Conception is largely influenced by liveweight, oestrus being associated with weight. Table 19.13 gives the target liveweights for various breeds.

Dystokia problems require particular consideration and help to determine the appropriate weight and age at first calving. Calving problems are greater in younger heifers, especially if calved before 22 months of age. The size of both the calf and the heifer are important, and both are influenced by the feeding of the heifer. Condition scoring is a valuable management aid and heifers should score between 3 and 3.5 6–8 weeks prior to parturition. If the condition of heifers is correct at this time, restricted feeding during the last weeks of pregnancy will not affect heifer size but will help to reduce calving difficulties by minimizing the growth of the calf.

Choice of bull affects calving difficulties, but the choice is also influenced by the values of the calves as herd replacements. Sires from large beef breeds cause the greatest problems in Friesian heifers. As Friesian bulls tend to cause more problems than either of the two smaller beef breeds, Aberdeen Angus and Hereford, their use on heifers is not recommended unless there is a need for a large number of dairy herd replacements. Another advantage in favour of the Hereford and the Angus is that both colour-mark their calves thus adding value. Whatever the choice of breed, individual variation within breeds has a great effect on the incidence of difficult calvings.

First lactation yield is lower in heifers calved early, milk yield being closely related to liveweight at calving (Table 19.13). In subsequent lactations, differences in milk yield are minimal. Evidence suggests that because early-calved heifers (2 years) are kept in the herd to the same age as later calved (3 years) they average one lactation more. Thus their lifetime yield is increased, as well as providing the extra calf (Table 19.14).

Heifer rearing should be planned so that replacements enter the herd to fit the intended calving pattern. This is one of the main determinants in maintaining a system of block calving. Once the age and the month at which the heifer is to calve have been decided, growth rates to achieve the necessary target liveweights at

Table 19.13 Target liveweights (kg)

Breed	*Weight at service*	*Calving weight*
Jersey	230	340
Guernsey	260	390
Ayrshire	280	420
Friesian	330	500
Hereford × Friesian	320	500
Aberdeen Angus × Friesian	290	430

Table 19.14 Milk production according to age at first calving. (Source: Wood, 1972.)

	Age at first calving (months)				
	23–25	*26–28*	*29–31*	*32–34*	*35–37*
Herd life (lactations)	4.00	4.03	3.84	3.81	3.78
Lifetime yield (kg)	18747	18730	17964	17991	17657

Table 19.15 Target weights for 2-year calving. [Source: MLC/MMB Joint Publication *Rearing Replacements for Beef and Dairy Herds* (reproduced with permission).]

	Weight (kg)
Autumn-born heifer	
Birth	35
Turnout (6 months)	150–170
Yarding (6 months)	275–300
Service (15 months)	330
Turnout (18 months)	375–400
Calving (24 months)	500–520
Spring-born heifer	
Birth	35
Turnout (3 months)	80–100
Yarding (6 months)	140–160
Turnout (12 months)	250–280
Service (15 months)	330
Yarding (18 months)	400–420
Calving (24 months)	500–520

service and calving can be calculated. Differences between growth rates for autumn- and spring-born heifers are largely due to the higher weight gains achieved at grass (Table 19.15). Maximum use of grass means economical rearing and the advantage of compensating growth.

Autumn-born heifer calving at 2 years old

Autumn-born calves should be weaned at 5 weeks of age when consuming at least 0.75 kg of concentrates. The concentrate should be palatable and contain 17% crude protein. At between 5 and 12 weeks calves should be fed concentrates containing 15% crude protein *ad libitum* up to a maximum of 2 kg/d and hay to appetite. Silage as a partial substitute for hay can be fed from 6 weeks of age. From 3 months to turnout in the spring reducing the crude protein to 14% can cheapen the concentrates fed. Hay or silage should be fed *ad libitum*. By 6 months each calf will have consumed 50 kg of early weaning concentrates, about 300 kg of rearing concentrates and 330 kg of hay.

To achieve the desired gains during the first grazing season (6–12 months of age) a continuous supply of good quality grass and control of parasitic worms are essential. Calves should be vaccinated against husk (lungworm) before turnout unless clean pastures are available (clean pastures being those that have been free of cattle since the previous midsummer). Supplementary feeding of concentrates (1–2 kg/d) after turnout prevents a check in growth, which might otherwise occur. A change to clean silage aftermath after dosing against stomach worms in midsummer is generally recommended. At this time cereal feeding (9% protein plus vitamins and minerals) may be introduced when grass becomes scarce or very wet and lush.

The grazing system should be integrated with the conservation area; 0.25 ha/animal can be divided into three sections. One section is grazed until the end of June, whilst the other two are cut. After this time the two conserved areas are grazed and the other one cut; finally all three areas are grazed.

The second winter is best subdivided into two halves: from yarding until service at 15 months and from service until turnout at 18 months. Cattle benefit from dosing against internal parasites at yarding and from the use of an insecticide against warble fly. A liveweight gain of 0.6 kg/d is important to ensure a target service weight of 330 kg and a good conception rate. Conserved forages form the basis of the ration (25 kg silage or 6 kg hay/d) and this is supplemented by 2.5 kg of concentrates (12% protein). The concentrates can be based on barley and the protein content and quantity of concentrate adjusted according to the quality of roughage.

Identification of bulling heifers is often found to be a problem. Careful observation for oestrus, and heat detection devices may prove to be valuable aids. After service the liveweight gain may be reduced to 0.5 kg/d; this allows the concentrate level to be reduced. The overall concentrate use during the first 12–18 months should be about 250 kg of concentrates and 4–5 t of silage or 1.25 t of hay.

During the second grazing season target weight gain should be about 0.7 kg/d. With good grassland management no concentrates are necessary until steaming-up (generous feeding of cow pre-calving) takes place in late summer. Excess steaming-up should be avoided to prevent overstocking of the udder prior to calving. The best guide to the level of feeding during late summer is body condition score. This should be between 3 and 3.5 6 weeks before calving.

Spring-born heifer calving at 2 years old

The management and feeding of the newborn calf until weaning is the same as for autumn-born calves. As young calves are too small to make efficient use of grass, concentrate (16% crude protein) feeding should be continued after turnout. Hay should also be available during this time (Table 19.16). If growth rate from grass alone falls below 0.5 kg/d, concentrate feeding should be restarted. The target stocking rate should be about 10 calves/ha. This may be achieved by a similar system of grassland management as for autumn-born calves. It is important to ensure that grazing is clean, as very young calves are extremely susceptible to parasites. At yarding calves should weigh between 140 and 160 kg and have consumed 100 kg of concentrates and 125 kg of hay from weaning.

From yarding at 6 months to turnout in the following spring at 12 months, a growth rate of 0.6 kg/d should be maintained. At yarding animals should be dosed against internal parasites and may be dressed against warble fly during the winter. The basis of the ration will be either silage or hay, the amount fed being 18 kg or 5 kg/d, respectively. The level of concentrate feeding depends on the quality and quantity of the conserved forages, a guide being 2–3 kg/d of 16% crude protein. If the growth rates are not maintained, the target service weight of 330 kg at 15 months of age in the spring will not be achieved and conception rates will be poor.

The second grazing season growth rates from turnout to service should be 0.8 kg/d. After service, target growth rates can be reduced to 0.6 kg/d. Supplementation with mineralized cereals (barley) should commence in early autumn.

During the second winter the aim should be to produce a Friesian heifer calving down at about 500 kg. This can be achieved by feeding 2 kg/d of concentrates and up to 30 kg of silage. This will take approximately 4.5 t of silage or 1.25 t of hay and 360 kg of concentrates. Steaming-up prior to calving can be practised and the heifer's body condition score should be between 3 and 3.5 6 weeks prior to calving.

Heifer replacement rearing enterprises compete for resources with milk-producing animals. The two major economies that can be made in the resources required in the production of replacement heifers are the reduction in the number required and reduction of the age at which they calve. Both these enable considerable economies to be made in land, labour and capital invested in livestock and buildings. However, young heifers grown well enough to calve at 2 years of age need a high plane of nutrition. This necessitates the feeding of greater quantities of concentrates, thereby increasing the cost of concentrate feed for animals calving at 2 rather than 3 years of age. As far as total feed cost is concerned, heifers calving at around 2 years of age cost nearly as much as those calving a year older. Thus intensification of heifer rearing with a greater reliance on concentrates has important repercussions in that the land and labour saved can be made available for milk production or other more profitable enterprises.

In addition to these direct savings in resources the heifer calving at the younger age will have a longer herd life with a greater total lifetime milk production (Table 19.14). These higher lifetime yields make up for a slightly lower first lactation yield; 4300 kg for a 2-year-old heifer compared with 4500 kg for a 3-year-old.

As far as total feed is concerned it appears that heifers calving at 2 years of age cost nearly as much as those calving a year older. The financial saving in grazing and forages for the younger calving animals is substantially eroded by the need for a higher concentrate input to maintain growth rates. This is of especial importance during the winter periods unless the diet comprises ample good quality forage.

Even if there is no great saving in the cost of producing younger calving heifers there is a saving in capital investment in young stock because of the quicker turnover in animals and a reduction in the number of followers kept. Indirect benefits may result from the successful adoption of such a system of rearing, the intensification required having repercussions throughout the whole dairy enterprise, especially with respect to better grass production, conservation and utilization. The effects that the more rigorous discipline involved in rearing heifers to calve at 2 years imposes are likely to extend right across the rearing enterprise and benefit the whole farm economy.

Table 19.16 Approximate feed quantities used

	Autumn born	*Spring born*
Milk substitute (kg)	13	13
Concentrates (kg)	700	920
Silage (t)	5.5	7.25
and hay (kg)	100	
or hay (t)	1.9	2.0

In beef suckler herds calves out of heifers have a slightly slower growth rate than calves out of 5-year-old cows (Table 19.17). These slower calf growth rates are largely due to the early-calved beef heifers having a slightly lower milk yield. However, beef heifers calved first at 2 years of age produce more calves over their lifetime than those calving at 3 years.

Beef production

Recent developments in beef production systems, using imported breeds, better grassland management and more efficient use of concentrate feeds, have resulted in greater liveweight gains and a move from grass to yard finishing. Baker (1975) suggests four factors that need to be considered when choosing a system for beef production:

(1) financial resources. These include cash flow requirements and the capital availability;
(2) physical resources, e.g. the area and quality of grassland available, field structure and the availability of water;
(3) date of birth of calves – autumn or spring;
(4) type of cattle – pure dairy, dual purpose or beef, or crosses.

Growth is usually measured by the change in liveweight of the animal. As this includes changes in the weight of feet, head, hide and the internal organs, including the content of the gut, besides carcass tissue, it is not always a reliable indicator of the final amount of saleable beef. Liveweight gain follows a characteristic sigmoid curve (see Chapter 16), the rate of growth being influenced by nutrition, breed and sex.

An important phenomenon in some beef systems is compensatory growth. When animals have been fed on low quantities and/or low quality feed, growth will slow down or cease. This is known as a store period. When full levels of feeding are resumed, these animals will eventually catch up animals that have been kept on full feeding levels; this is compensatory growth. It is exploited in beef systems so winter feeding can be kept at as low cost as possible and when spring feeding is started compensatory growth takes place, taking advantage of relatively cheap grazed grass.

Table 19.17 Effect of age at calving on calf growth (percentage difference in 400-day weights compared with calf out of a 5-year-old cow). [Source: MLC/MMB Joint Publication *Rearing Replacements for Beef and Dairy Herds* (reproduced with permission).]

	Percentage difference
Heifer	
First calving at 2 years	–8
First calving at $2\frac{1}{2}$ years	–5
Cow	
Second calf	–3
Third calf	–2
Fourth calf	0

The development of the animal varies as different tissues mature at different rates; nervous tissue first, followed by bone, muscle and finally fats (for a fuller explanation see Chapter 16). If at any time the energy intake of the animal is in excess of that needed for the growth of earlier-maturing tissues, bone and muscle, then fat will be deposited. In time the fattening phase is reached when fat grows fastest. Fat is deposited in the body in a certain order: subperitoneal fat (KKCF, kidney knob and channel fat) first, then intermuscular fat, subcutaneous fat and, finally, intramuscular fat (marbling fat). In the later-maturing breeds, which have characteristically high growth rates, the fat deposition phase occurs at a greater age and weight than in the early-maturing breeds fed on similar diets.

Sex effects are also important: bulls grow faster than steers and these in turn grow faster than heifers. In parallel, bulls are later maturing than steers and steers later than heifers. As a result, heifers of early-maturing breeds quickly reach an age when they are depositing fat and are slaughtered at a younger age. Bulls of the late-maturing breeds need high levels of feeding before fattening commences (Fig. 19.5).

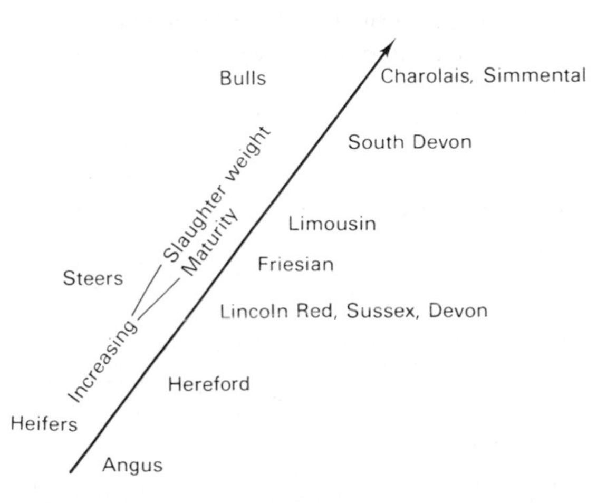

Figure 19.5 Ranking of cattle of different breeds and types in terms of slaughter weight and rate of maturity. [From MLC (1982), reproduced with permission.]

In practice, different types of cattle are managed differently and are slaughtered at different levels of carcass fat cover to suit the requirements of different sections of the meat trade. An understanding of the relationship between production system and cattle type is fundamental in planning production for a particular market.

The amount of fat in the animal influences killing-out percentage (sometimes called the dressing percentage), i.e. the yield of carcass from a given liveweight. As the animal gets heavier and fatter, the killing-out percentage increases. Systems of describing carcasses have been developed in many countries. The purpose of beef carcass classification is to describe carcasses by their commercially important characteristics. The development of the classification system sought to improve the efficiency of marketing throughout the whole industry. The description enables wholesalers and retailers to define their requirements. Producers should be able to obtain higher returns by producing animals to match these market requirements. The result is a flow of information from the consumer through the retailer to the farmer. Producers can then assess the economics of providing one type of carcass from others and plan their systems of management accordingly (Fig. 19.6).

The beef carcass classification scheme is based on a grid system, which describes fatness on the horizontal scale and conformation on the vertical scale (Table 19.18). Table 19.19 gives the average composition of carcasses in each of the EU fat classes. Fatness and conformation are determined by visual appraisal. Conformation relates to shape and takes into account carcass thickness, blockiness and the fullness of the round. Fatness is always referred to before conformation, e.g. a carcass falling in fat class 2 and conformation R would be described as 2R. The scheme also includes information on weight, sex type and sometimes age. The MLC undertakes many demonstrations of the system throughout the UK every year and anyone interested should contact the nearest MLC Regional Fatstock Offices.

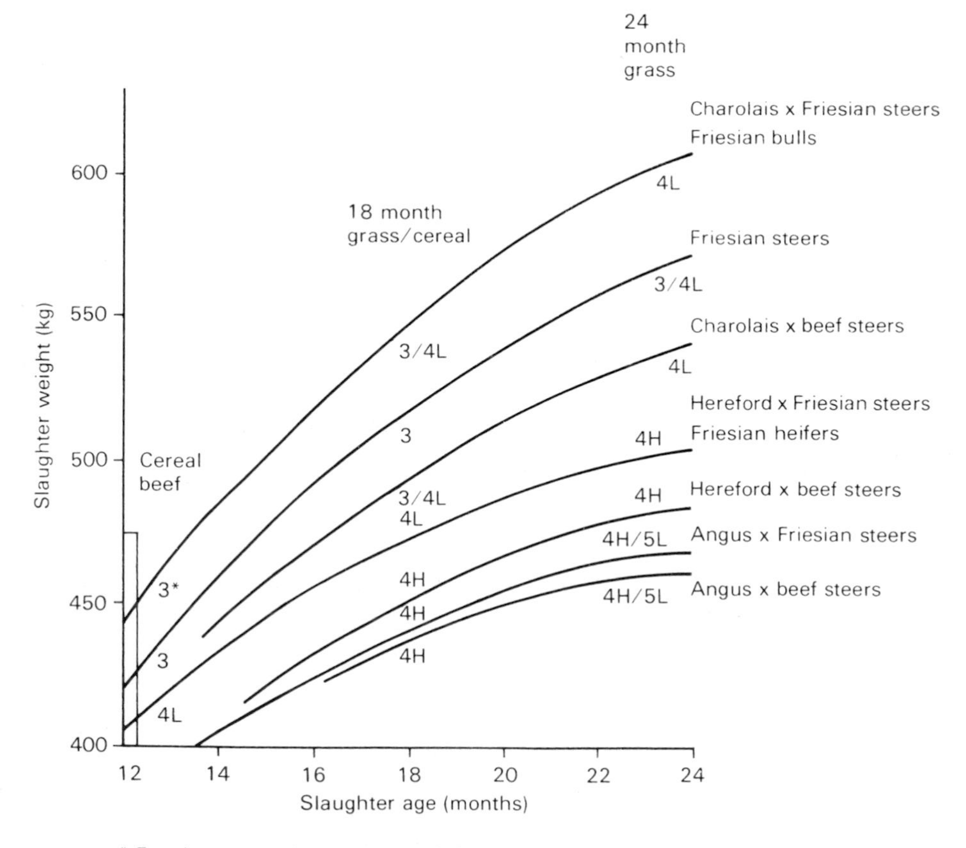

Figure 19.6 Average liveweight, slaughter age and carcass fat for different breeds and types of cattle. [After MLC (1982), reproduced with permission.]

Table 19.18 Percentage distribution of Great Britain clean beef carcasses in the EU classification grid (1998). [Source: MLC (1989) *Beef Yearbook* (reproduced with permission).]

		Fat class							
		Leanest *1*	*2*	*3*	*4L*	*4H*	*Fattest* *5L*	*5H*	*Overall*
Very good	E			0.1	0.1				0.2
C	U+		0.1	0.4	0.6	0.2			1.3
O									
N	–U		0.5	2.4	4.7	2.3	0.2		10.2
F									
O	R	0.1	1.3	7.4	21.7	11.6	0.9	0.1	43.1
R									
M	O+	0.1	1.2	7.1	15.1	6.4	0.7	0.1	31.5
A									
T	–O	0.1	0.9	3.5	5.4	1.5	0.2		11.7
I									
O	P+	0.1	0.3	0.7	0.6	0.1			1.8
N									
	–P								0.1
Very poor overall		0.5	4.5	21.7	48.9	22.1	2.0	0.3	

Table 19.19 Average composition and saleable meat yield of *R* conformation carcasses by fat class. (*R* conformation is the most common conformation class; it accounts for about 40% of classified carcasses) [Source: MLC (1982) *Selecting Cattle for Slaughter* (reproduced with permission).]

Carcass composition	*Fat class*				
	1 and 2	*3*	*4L*	*4H*	*5L*
	Percentage of carcass weight				
Lean	69	64	61	58	53
Fat	12	18	22	25	31
Bone	19	18	17	17	16
Saleable meat					
Meat yield	74	72	71	70	69
Fat and lean trim	7	10	12	13	15

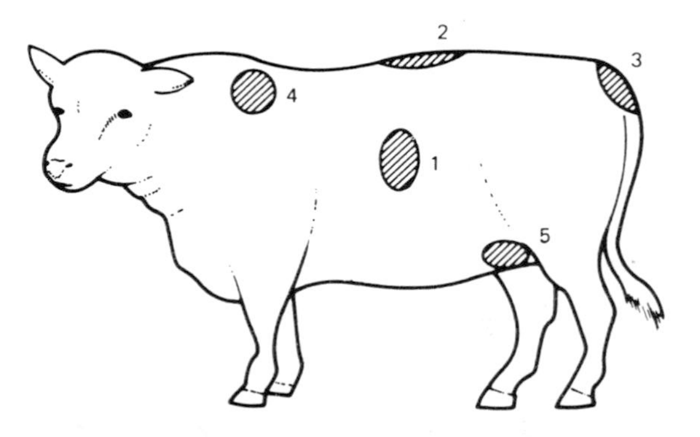

Figure 19.7 Key handling points.

Selecting cattle for slaughter

The two most important aids to the selection of cattle for slaughter are weight and fat cover (finish). Knowledge of the weight of animals and their previous performance means that decisions can be based on finish. Once animals reach their minimum weight, the decision to sell should be based on finish. Excess fat is wasteful and has to be trimmed off carcasses, besides which feed turned into unwanted fat is wasted and is very expensive.

The best way of assessing carcass fat cover is to handle the cattle. Figure 19.7 shows the five key points for handling:

(1) over the ribs nearest to the hindquarters;
(2) the transverse processes of the spine;
(3) over the pin bones on either side of the tailhead and the tailhead itself;
(4) the shoulder blade;
(5) the cod in a steer or the udder in a heifer.

If producers are doubtful about the fat levels in animals they have selected for slaughter, they should check their classification reports or follow carcasses through the abattoir.

An EEC directive banning the sale of carcasses from animals in which hormone growth promoters were used commenced 1 January 1988. The British Government implemented a ban on the use of hormone growth promoters from 1 December 1986. This ensured that Britain was in line with other member states of the European Community.

Beef production systems

Cereal beef

Cereal beef, more commonly known as barley beef, is so named because of the system of production. Dairy-bred animals are fed on an all-concentrate ration, usually based on barley. This encourages rapid liveweight gain and animals are slaughtered at 10–12 months of age weighing 430–470 kg. Killing-out percentage is usually 54–56%, thus yielding a carcass of 230–265 kg.

Calves are reared on an early weaning system (see earlier section headed 'Early weaning'). Weaning takes place at 5 weeks of age on to a concentrate ration containing 16% crude protein. At 10–12 weeks of age a diet usually based on rolled barley supplemented with protein, vitamins and minerals giving a mix of 14.5% crude protein is fed *ad libitum*. From 6–7 months the protein level of the mix can be reduced to 12% thus reducing costs (Table 19.20).

The system is best suited to late-maturing animals as early-maturing animals become overfat at light weights.

Table 19.20 Liveweight and protein levels (% CP) of diet

Period	*Liveweight (kg)*	*% CP in diet*
5–12 weeks	100	16
3–6 months	250	14
6 months to slaughter	430–470	12

Table 19.21 Cereal beef targets (Friesians). (Source: Allen & Kilkenny, 1980.)

Period	*Gain (kg/d)*		*Economy of gain (kg feed:kg gain)*	
	Bulls	*Steers*	*Bulls*	*Steers*
0–5 weeks	0.45	0.45		
6–12 weeks	1.0	0.9	2.7	2.8
3–6 months	1.3	1.2	4.0	4.3
6 months to slaughter	1.4	1.3	6.1	6.6
Overall	1.2	1.1	4.8	5.5
Slaughter age (days)	338	345		
Slaughter weight (kg)	445	404		
Carcase weight (kg)	231	211		

The Friesian is most commonly used because of its ready supply from the dairy herd. As bulls are later maturing than steers and have better food conversion efficiencies (FCE), there have been an increasing number of entires kept on this system. Animals being housed throughout facilitate keeping bulls. Their faster and leaner growth, compared with steers, means that bulls can be slaughtered at the same age (10–12 months) weighing 470 kg. For a comparison of bulls and steers see Table 19.21.

Practical problems that may be encountered include respiratory diseases and bloat. With animals housed throughout, building design is of particular importance both for the handling and physical management of stock and for reducing respiratory diseases. Good ventilation and draught-free buildings significantly reduce the incidence of pneumonia. Bloat or rumen tympany is a greater problem when animals are kept on slats rather than in bedded pens where they can consume roughage. If 1 kg/d of hay or barley straw is fed, this normally prevents bloat.

Cereal beef production is sensitive to the relative prices of calves, barley and beef. Its main advantages are that it makes no direct use of land, as all feeds can be bought-in, and it is not seasonal, hence there is an even cash flow once a regular throughput of animals is established. Table 19.22 gives average performance figures.

Maize silage beef

Continental producers have developed this system of fattening dairy-bred calves and it is now a well-established system for bull beef production (see Tables 19.23 and 19.24). British interest has been aroused as

Table 19.22 Average performance of cereal beef units – purchasing calves in spring 1998 and finishing in spring 1999; mainly continental crosses. [Source: MLC (1998) *Beef Yearbook*.]

Performance	
Feeding period (days)	388
Weight (kg)	
At start	48
At sale	544
Carcase weight	296
Killing out %	54.6
Daily liveweight gain (kg)	1.3
Feed	
Milk replacer (kg)	16
Calf concentrate (kg)	140
Finishing concentrate (kg)	2143
Feed conversion ratio (kg feed/kg gain)	4.7

Table 19.23 Maize silage beef production (Friesians). [Source: MLC (1978) *Beef Improvement Services* (reproduced with permission).]

	Bulls	*Steers*
Gain (kg/d)	1.0	0.9
Slaughter weight (kg)	490	445
Slaughter age (months)	14	14
Feed DM (kg/kg liveweight gain)	5.6	6.0
Protein concentrate (kg)	500	500
Maize silage DM (t)	1.75	1.65
Stocking rate (cattle/ha assuming 10 t DM/ha)	5.7	6.1

Table 19.24 Slaughter weights for maize silage beef. (Source: Allen & Kilkenny, 1980.)

	Slaughter/weight (kg)	
	Bulls	*Steers*
Friesian	480–500	430–450
Charolais, Simmental, Blonde d'Aquitaine, Limousin, South Devon × Friesian	510–540	460–480
Devon, Lincoln Red, Sussex × Friesian	480–500	430–450
Hereford × Friesian	450–480	400–430
Aberdeen Angus × Friesian	420–450	380–410

maize silage systems are fully mechanized and the crop provides a useful arable break crop.

Calves are reared to 12 weeks in the same way as for cereal beef. Maize silage is then introduced and fed *ad libitum*. A protein supplement must be fed to bring the overall crude protein of the diet to 16%, the main problem with maize silage being its low protein content. As with the cereal beef the system favours animals of late-maturing type.

Maize silage can also be used for finishing suckled calves. It is important to ensure that the silage has a high dry matter content (i.e. 25% DM). Silage with low dry matter levels will result in lower feed intake and lower liveweight gains.

Eighteen-month grass/cereal beef

Autumn-born calves are reared on a conventional early weaning system. Friesian and Hereford × Friesian calves are commonly used. The autumn-born steer should weigh approximately 200 kg when turned out in the spring. Calves born during the winter and early spring weigh less and have lower weight gains at grass. The daily gain of heifer calves is 20% poorer than steers and

Table 19.25 MLC Beefplan average results for 18-month beef units using calves born autumn 1997 finishing spring 1999 – mainly continental crosses. [After MLC (1999) *Beef Yearbook*.]

Performance	
Feeding period (days)	580
Weight (kg)	
At start	47
At turnout	201
At yarding	324
At slaughter	531
Carcase weight (kg)	285
Killing out %	54.0
Carcase conformation score	3.4
Carcase fat class score	3.7
Calf performance	
Daily gain (kg)	
First winter	0.8
At grass	0.8
Second winter	0.8
Stocking (cattle/ha)	
Overall stocking rate	3.7
Grazing stocking rate	5.5
Feed (kg)	
Second winter concentrates	436
Second winter forage	4448

they finish at lighter weights. Bulls grow more rapidly and produce carcasses about 10% heavier than steers (Baker, 1975).

During the grazing season the aim is to achieve a daily liveweight gain of 0.8 kg. To achieve this, grass must be managed to provide a continuous supply of high quality herbage. Supplementary feeding with rolled barley may be practised to maintain the target growth rate; this will usually be necessary for the first few weeks after turnout in the spring and after late August when herbage quality and quantity begin to decline.

The animals should weigh about 320–350 kg at yarding. The target rate of liveweight gain is 0.8–1.0 kg/d. Winter feeding is usually based on silage, which is supplemented with mineralized rolled barley. The amount of barley fed will depend on the quality of the silage and the desired weight gains. Cattle are slaughtered at 15–20 months of age weighing 450–550 kg. Table 19.25 gives MLC Beefplan average results.

Grass silage beef

Grass silage beef is also known as the Rosemaund Beef System or Storage Beef. The system normally uses dairy-bred calves – Hereford × Friesian or Friesian bull calves – fed on grass silage to appetite with rationed compound feeds and housing the animal throughout. Animals are slaughtered at 11–15 months of age. A comparison of results from units using Hereford × Friesian and Friesian calves is given in Table 19.26.

Calves are reared on an early weaning system, and are gradually changed from the early weaning ration to a rearing compound. High quality silage is introduced at this time. At 12 weeks the rearing compound is restricted to 2 kg/animal/d and this level of compound is fed until slaughter. If silage quality is less good the level of compound may be increased to 4 kg/animal/d (see Table 19.27 for further details).

The system depends on the ability to obtain high growth rates from young animals; it is therefore important that the quality and quantity of silage are adequate. Rosemaund results show that when grass is cut at a D-value of 70 or more, satisfactory liveweight gains can be achieved.

Stocking rates are dependent on silage yields and tonnes of silage used per animal. High stocking rates lead to a high working capital requirement and hence high interest charges.

Suckled calf production

Suckled calf production is practised under a variety of environmental conditions, from hill land to lowland. Beef suckler cows vary from the larger and milkier types to the smaller and hardier types. The number of purebred beef calves has declined, crossbreds becoming more prevalent, and this introduces the heterosis effects of greater fertility and calf viability.

The availability of buildings and winter grazing, the bulk of calvings taking place in either the autumn or late winter/early spring, affects the choice of season of calving. A short calving season enables the cows to be fed as one group without over- or underfeeding of individuals and also gives a more uniform batch of calves to be sold or fattened.

The use of condition scoring has added some precision to the management of suckler herds. The body condition of breeding cows at service and at calving is particularly important in reducing the incidence of barren cows. Herds that have a condition score of above 2, with the optimum being 2.5–3, have the better calving intervals and the greatest number of reared calves (Tables 19.28 and 19.29).

Suckler calf production usually produces calves for sale post-weaning as stores; this is especially true of hill and upland producers where supplies of winter feed are scarce. Lowland farmers will then fatten these. Lowland suckler herds are more likely to carry their calves through to slaughter (Table 19.30).

Table 19.26 Performance of Hereford × Friesian bulls, Friesian bulls and Friesian steers on grass silage beef system. [After MLC (1986) *Beef Yearbook* (reproduced with permission).]

	Hereford × Friesian bulls	*Friesian bulls*	*Friesian steers*
Performance			
Days to slaughter	393	364	395
Mortality (%)	3	3	2
Weight at start (kg)	105	110	111
Weight at slaughter (kg)	489	483	498
Daily gain from 12 weeks (kg)	0.98	1.05	0.98
Feeds			
Concentrates (kg)	824	954	976
Silage (t)	5.4	5.7	5.5
Stocking rate (head/ha)	7.87	7.44	6.88

Table 19.27 Feeding bulls from weaning to sale (storage beef). (Source: Hardy & Meadowcroft, 1986.)

Time from calf arrival (weeks)	*Liveweight (kg)*	*Feeding*
5–6	65–75	Weaned: *ad libitum* early weaning concentrate
6–8	75–85	Calf concentrate replaced by rearing compound fed to appetite
8–12	85–105	Silage first offered to appetite
12–16	105–130	Compound gradually reduced to 2 kg/day
16 to sale	130 to slaughter	Silage offered to appetite plus 2 kg compound/d rising to 4 kg if needed

Season of calving tends to determine the production system. Autumn calving is most popular, as calves sold in the autumn store sales are older and heavier. However, autumn calving involves more buildings and a greater requirement for winter feed.

Grass finishing of stores

Suckler-bred stores are purchased in the autumn sales and may be overwintered before being finished off on grass the following summer (Table 19.31). Alternatively the stores may be finished on a cereal-based diet over winter for sale the following spring (Table 19.32).

Table 19.28 Relationship between body condition and reproductive performance (beef cows). [Source: MLC (1978) *Beef Improvement Services* (reproduced with permission).]

Body condition score	*Calving interval (days)*	*Calves weaned per 100 cows served*
1–2	418	78
2	382	85
2–3	364	95
3+	358	93

Scores are on the scale 1 (very thin) to 5 (very fat).

Dairying

Milk production is the largest enterprise in UK agriculture with an annual net sum received in 1998 by producers of some £2718 million (Table 19.1). This accounts for about 18.5% of the total value of all UK agricultural output.

Lactation curves

If the milk yield of cows is plotted against time, a graph of a lactation curve is produced. The standard lactation is 305 days and the annual cycle can be conveniently split into four distinct parts of early, mid and late lactation and the dry period (Fig. 19.8). After calving, milk yield will rise for a period of 4–10 weeks when peak milk yield

Table 19.29 Body condition score targets

Stage of production	*Target score*	
	Autumn calving	*Spring calving*
Mating	2.5	2.5
Mid pregnancy	2.0	3.0
Calving	3.0	2.5

Table 19.30 Average performance of suckler herds (1992). [After MLC (1999) *Beef Yearbook* (reproduced with permission).]

	Lowland		*Upland*		*Hill*	
	Spring	*Autumn*	*Spring*	*Autumn*	*Spring*	*Autumn*
Cow performance/100 mated						
Calving spread (weeks)	13	14	11	12	12	11
Calves born live	91	92	91	91	89	89
Calves purchased	2	2	1	2	2	1
Calf mortality	3	2	3	3	3	3
Calves reared	91	91	90	90	89	89
Calves						
Age at sale/transfer (days)	224	317	213	331	215	315
Weight at sale/transfer (kg)	269	338	268	345	250	323
Daily gain (kg)	1.1	1.0	1.05	1.0	0.95	0.95
Feeds						
Cow concentrates (kg)	165	278	122	181	144	214
Calf concentrates (kg)	75	200	87	217	55	187
Silage (kg)	3021	5306	3961	5656	4285	5747
Stocking						
Stocking rate (cows/ha)	1.8	1.5	1.6	1.3	1.0	1.0
N fertilizer (kg/cow)	72	74	55	43	28	43

Table 19.31 Average results for units overwintering and grass-finishing stores, purchased autumn 1997–98. [After MLC (1999) *Beef Yearbook* (reproduced with permission).]

Performance	
Feeding period (days)	333
Weight (kg)	
At purchase	292
At slaughter	486
Carcase weight (kg)	271
Daily gain (kg)	
Over winter	0.6
Over grazing period	0.7
Feeds	
Concentrates (kg)	346
Silage (kg)	2086
Stocking (cattle/ha)	235
Overall stocking rate	3.6
Grazing stocking rate	4.5
Gross margin/head	135
Gross margin/ha	450

Table 19.32 Average results for winter cereal-finishing suckled calves, 1988–89. [After MLC (1999) *Beef Yearbook* (reproduced with permission).]

Performance	
Feeding period (days)	157
Weight (kg)	
At start	325
At slaughter	554
Carcase weight (kg)	307
Feeds	
Concentrates (kg)	1358
Silage (kg)	418
Prices	
Suckled calf cost (p/kg liveweight)	127
Gross margin/head	144

will be achieved. The time taken to reach peak yield varies with breed, individual, nutrition and yield.

Once a cow has reached peak yield the subsequent decline is approximately 2.5% per week or 10% per month. Many producers are now beating this performance with milk declining by 2% per week, post peak yield. The decline in heifers is about 7–8% per month. The daily peak yield of cows is approximately 1/200th of total 305-day yield. Thus a cow giving 30 kg at its peak will have a total yield of approximately (30 × 200) 6000 kg. The heifer's peak yield will be approximately 1/220th of total 305-day yield. Thus a heifer giving 18 kg at peak will yield approximately (18 × 220) 3960 kg. A useful 'rule of thumb' guide is that two-thirds of the total yield will be produced in the first half of lactation.

It is clear that the height of the curve at peak milk yield has a great influence on the total lactation yield. The factor that is most likely to limit the level at which cows reach their peak lactation is nutrition.

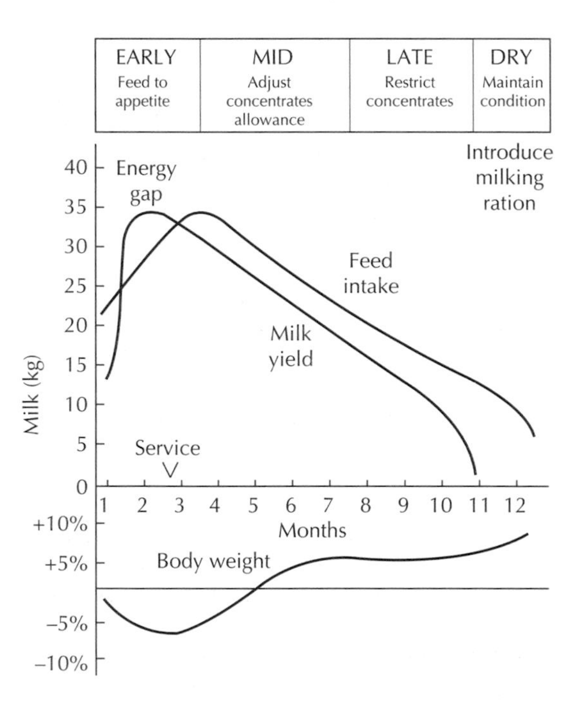

Figure 19.8 Relationships of feed intake, milk yield and body weight to lactation for a dairy cow producing 35 kg milk daily at peak. [From the Scottish Agricultural Colleges (1979), reproduced with permission.]

Feeding dairy cattle

The nutrition of cattle can be divided into two distinct requirements: first, energy and nutrients to provide maintenance (maintenance requirement); and second, to provide nutrients for growth, the development of the unborn calf and milk (production requirement).

Accurate feeding of dairy cows involves long-term planning as the requirements and allowances at any one time can be influenced by previous nutrition. The practice of feeding prior to calving is a good example of the nutrition in one period affecting the production in another. During the last 2 months of pregnancy – the latter part of the dry period – the unborn calf grows rapidly and there is extensive growth and renewal of mammary tissue. The practice of steaming-up (the generous feeding of a cow pre-calving) is designed to fulfil these two needs, as well as providing reserves of body tissue which the cow can catabolize during early lactation and accustoming the rumen to consuming increasing quantities of concentrates.

Steaming-up is usually started some 6–8 weeks prior to calving, beginning with a small quantity of concentrates, which is then increased each week. Traditionally, in the week before parturition the level of concentrates fed would be about one-half to three-quarters of the amount that it is expected will be needed at peak lactation. The amount of concentrates fed will depend on the body condition of the animals, the expected level of milk yield at peak lactation and the quality of bulk fodders. For efficient and economic feeding, bulk fodders should be analysed for their nutrient value. Heifers are not usually given more than 3 kg of concentrates/day unless their body condition is very poor.

As the date of calving becomes close the animal may have a temporary loss of appetite. This should be regained within 3–4 days after calving. As the appetite increases, the cow is commonly fed for the quantity of milk being produced and an extra 1 kg/d of concentrates in an attempt to increase future production: this is usually termed 'lead feeding'. However, too rapid an increase in concentrate feeding can lead to digestive upsets. It is important to realize that feeding dairy cows should not be viewed entirely in terms of concentrate nutrients as the bulk food part of the ration can contribute a substantial part of the total nutrients. This is increasingly the case since the introduction of milk quotas and at a time of falling returns to milk producers.

Correct feeding pre- and post-calving is essential to prevent excess weight loss in the early part of the lactation. Some weight loss is inevitable, as the cow's appetite will not reach its maximum until after peak lactation. This difference in time taken for an animal to reach peak lactation and peak DMI means that a nutrient gap occurs. The cow catabolizing fat makes up this deficiency in nutrients, and a daily liveweight loss of at least 0.5 kg can be expected. Too great a loss in weight can be conducive to a higher incidence of ketosis (acetonaemia). The manipulation of body weight and condition in dairy cows has become necessary to sustain high milk yields. The pattern in Table 19.33 is commonly suggested.

Frood and Croxton (1978) showed that the ability of a cow to reach a predetermined level of milk was closely related to its condition at calving. Cows whose condition score was below 2 at calving did not achieve their predicted milk yield, and those whose score was above 2.5 yielded more than their predicted yields. Animals with a high condition score at calving, i.e. having ample body reserves, gave a higher earlier peak milk yield than animals in poor condition whose body reserves could not furnish enough nutrients during the time of low appetite and high energy demands.

Table 19.33 Liveweight change during lactation of dairy cows. [Source: HMSO (1984) Crown copyright (reproduced with permission).]

Week no.	*Liveweight change (kg/d)*	*Change during 10 weeks (kg)*	*Net effect on liveweight (kg)*
0–10	–0.5	–35	–35
10–20	0.0	0	–35
20–30	+0.5	+35	0
30–40	+0.5	+35	+35
40–52	+0.75	+63	+98

The condition at calving and subsequent weight loss can have repercussions on conception rates at service. To ensure good conception rates cows should have a condition score of at least 2.5 at service. Below this score fertility is adversely affected. Cattle that are 'milking off their backs' must have their diet supplemented with protein and minerals additional to those normally included in the diet.

Once peak milk yield has been achieved and production starts to decline, concentrate use can be reduced and intake of bulk foods increased. The level of concentrates fed will be determined to some extent by cow condition. Animals that have lost much weight will need liberal feeding until after they are served and are in calf. Concentrate levels should still be fed slightly in excess of milk production in an attempt to prevent the decline in yield. Liveweight during the period of mid-lactation should be stable (Table 19.33).

During late lactation cows have a maximum DMI relative to milk yield. The aim is to maintain milk production and restore the body condition of the cow. Maximum use of good quality forage with little or no concentrates should achieve the aims. Suggested condition scores for dairy cows are as follows:

Calving	3.5
Service	2.5
Drying off	3

It should be stressed that the above deals with obtaining the maximum amount of milk from a cow. Farmers coping with quotas by reducing the yield per cow will usually rely on feeding fewer concentrates and rely on a greater proportion of milk from forage.

Flat-rate feeding

With the traditional method of feeding, differing amounts of concentrates are fed according to the stage of lactation and yield. On a flat-rate feeding system all lactating cows are fed the same daily amount of concentrates throughout the winter regardless of their individual yields. Bulk food is usually silage because of its better feed value; this must be of good quality (at least 10 MJ/kg DM) and must be fed *ad libitum*. If there is not enough silage available, other good quality bulk foods can be fed in its place. On most farms the optimum level of concentrates per cow will be between 6 and 10 kg/d; this depends on the yield of the cows and the quality of the forage fraction of the diet. The system works as the amount of concentrates fed rather than their pattern of feeding is the important factor.

Flat-rate feeding of concentrates is most effective and easier to manage if the herd has a tight calving pattern.

Complete diets

In this system all the dietary ingredients are mixed in such a manner that individual ingredients cannot be selected out from the rest and the mixture is offered *ad libitum*. The whole system can be mechanized and does away with the need for equipment designed to feed concentrates both in and out of the parlour. It is more common with larger dairy herds where cows can be split into at least three groups according to the stage of lactation and appropriate rations can be fed to each group.

Self-fed silage

This system became popular after the Second World War with the advent of loose housing. The silage depth should not exceed 2.25 m for Friesians or they will not be able to reach the highest part of the clamp. If the depth of silage is greater than 2.25 m, the upper part will have to be cut and thrown down, otherwise there is a tendency for cows to tunnel into the clamp and animals can be trapped when the overhang collapses. This is especially dangerous if tombstone barriers are used as the cows can have their necks trapped between the uprights of the barriers. The width of the feeding face should be 150 mm/cow. It is unwise to have a greater width than this or the silage will not be consumed quickly enough, resulting in spoilage and secondary fermentation.

The most common barrier to keep the cows from the silage face is either an electric wire or an electrified pipe, the latter being less likely to break than wire and providing a better electrical contact. The distance between the barrier and the feed face can control consumption. The layout of buildings is important as cows need ready access to the silage. The width of the face is relative to the number of animals that will feed from it and one of the problems encountered with such a feed system is that significant herd expansion is difficult because of the inability to increase the width of the silo face.

Milk quality

Milk is a major source of protein, fats, carbohydrates, vitamins and minerals in the human diet. All these are present in milk in an easily digested form. The greater part of milk is water. The remainder is solid, which can be divided into two main groups, the milk fat (butterfat) and the solids-not-fats (SNF) (see Fig. 19.9).

The total solids content of milk varies. Milk fat is a mixture of triglycerides and contains both saturated and unsaturated fatty acids. Casein is the main protein

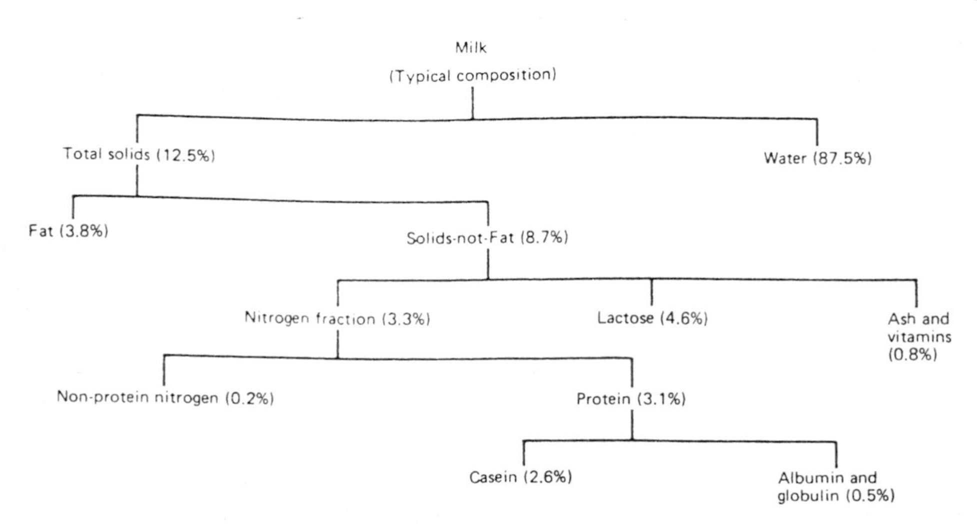

Figure 19.9 Typical composition of milk from a Friesian dairy cow.

contained in milk. There is generally an inverse relationship between milk yield and the milk fat and protein percentages. The higher the yield, the lower the percentage composition of these components.

The breed of cow is an important factor affecting milk composition. Friesians, one of the heaviest milk-yielding breeds, have some of the lowest fat and protein contents. Jerseys on the other hand have low yields but high fat and protein content. Individual variation within a breed can also cause a large effect on milk composition. Old cows tend to produce milk with a lower total solids content because yields tend to increase up to the fourth lactation and udder troubles increase with age. These factors have a depressing effect on fat and solids-not-fats. A regular intake of heifers into the herd will help to maintain milk quality, because of the age structure of the herd.

Incomplete stripping of milk from the udder can cause day-to-day variations in milk quality. The first milk to be drawn from the udder at milking contains low levels of fat when compared with the last milk to be drawn off; solids-not-fats change very little during the milking period. Fat percentage is usually lower after the long period between milkings. Part of this is due to the higher udder pressure which causes lower fat secretion.

The fat content of milk often drops when cows are turned out to lush grass in the spring. This is a function of the corresponding increase in yield and because of low fibre levels. The problem can be mitigated to some extent by feeding 2 kg of hay, straw or long roughage before the cows go out to grass and by restricting the grass so that the more fibrous stem fraction is eaten as well as the leaf. Long fibre is necessary for the rumen fermentation to produce acetic acid, acetate being the main precursor of milk fat.

Progressive underfeeding and poor body condition are responsible for cows producing milk with low protein levels. If energy supply is deficient either during the winter or as a result of grass shortage during the summer, then solids-not-fats will drop. Subclinical mastitis can be the cause of a reduction in protein levels of milk.

If the above factors are not the cause and there is still a milk quality problem, then the answer might be to change the breed or to use a progressive breeding programme with bulls selected with milk quality characteristics.

The milk pricing system is usually based on:

- compositional quality payments – based on the fat, protein and lactose content of the milk produced;
- contemporary payments – this means that compositional quality payments are based on test results obtained in the month being paid for;
- seasonal adjustments.

Different wholesale purchasers of milk will have their own scheme for valuing payment for milk. As milk is valued by the value of its component parts, any factors that affect the monthly quality of milk, such as low fat at turnout in spring, can have a marked effect on producers' returns. Producers attempting to alter the balance amongst milk constituents, fat, protein and lactose are restricted by seasonal, lactation and breed factors. Variation in milk constituents with stage of lactation and with season is shown in Figs 19.10 and 19.11 respectively.

Oestrus detection (see Chapter 16)

Oestrus (bulling or heat) is the time during which the cow will stand to be mounted by the bull. This period

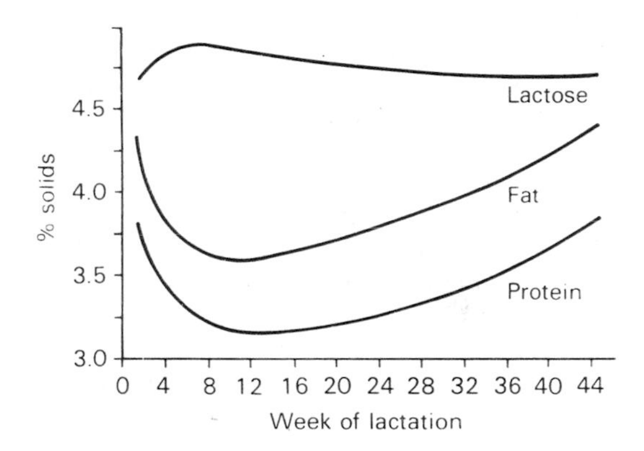

Figure 19.10 Variation in milk constituents with stage of lactation. [Source: MMB (1983) *The New Pricing Package*.]

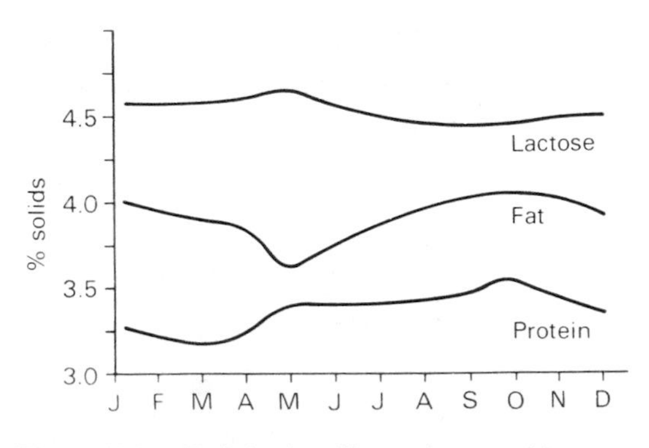

Figure 19.11 Variation in milk constituents with season. [Source: MMB (1983) *The New Pricing Package*.]

lasts on average 15 ± 4 hours. During the winter the period may be greatly reduced; a shorter period of 8 hours is also common with heifers. The average interval between oestrus is 20–21 days.

Failure to detect animals on heat can be a major problem, especially with the increasing size of herds. Poor conception rates are often blamed on poor artificial insemination techniques, bull fertility and disease, but the actual problem may be poor management.

The first management factor necessary to improve oestrus is good cow identification. There are a number of methods by which cows can by identified clearly, but freeze branding is probably the best. In conjunction with identification should be the keeping of good records. It is essential that simple, accurate and complete records are available whilst cows are being observed for signs of oestrus. Detecting cows in oestrus is part of good stockmanship as cow behaviour is the best indication of heat.

Signs indicating oestrus are:

- mounting other cows;
- sore, scuffed tail-head and hip bones;
- mud on flanks;
- mucus from vulva;
- restlessness; steaming animals.

Stock people must be allowed adequate time for observation of the herd. The most productive times are after the animals are settled after milking and feeding. Mid morning and late evening are particularly productive. Observations should ideally take place three or four times a day and the herd should be watched for at least half an hour at a time. Any sign of oestrus should be written down and not left to memory. The target should be to detect 80% of all cows in oestrus. Various aids may be used, including vasectomized bulls, pads or paint placed on the cows' rumps, all of which have proved useful.

The timing of service is important. If cows are served too early or too late in the heat period, poor conception rates may result. Ideally, cows should be served in the middle of the heat period between 9 and 20 hours after the start of heat to obtain the best results (Fig. 19.12).

The aim should be for a calving interval of 365 days. If the cow is to have a lactation of 305 days and a dry period of 60 days, with pregnancy lasting between 278 and 283 days, cows should conceive 81–86 days after calving. If service is delayed until the first heat after 60 days – a common practice – the result will be a slightly higher conception rate but a longer calving interval.

Oestrus can now be controlled and synchronized by the use of prostaglandins. The technique is useful where a group of heifers needs to be inseminated. The animals will come into heat at the same time and the result will be a tighter calving pattern. In older cows the technique can be used on individual animals rather than the herd as a whole. The technique is not a cure for infertility and will only bring an animal on heat that is cycling normally (see section on prostaglandins in Chapter 16).

The practice of calving the herd over a period of about 10 weeks is not as common as would be expected

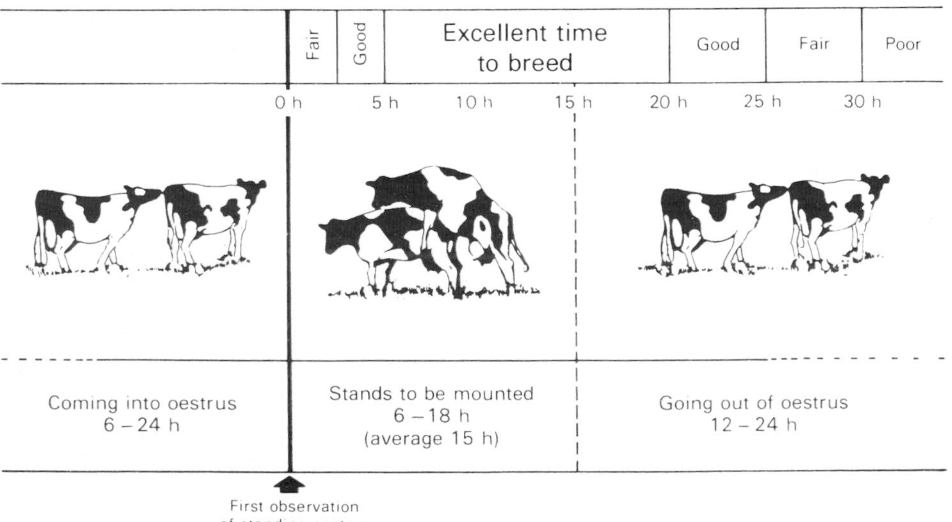

Figure 19.12 Optimum time for insemination. [From HMSO (1984), reproduced with permission.]

despite the potential simplification of management routine. All the main tasks, calving, oestrus detection, service and drying off, can occur during a specific time period for the entire herd. Coupled to this is the fact that the feeding management of the herd can be simplified by having uniform groups of animals at similar stages of lactation. Heifer rearing becomes simpler as the animals have a smaller age range. Not all farmers are attracted to such a routine, but the regime can create conditions more conducive to the achievement of better results. The move by wholesale purchasers to pay premiums for an even monthly supply of milk will necessitate a move to all-year-round calving to meet this requirement.

Milk production

Milk production and the premises under which milk is produced are controlled by the Milk and Dairy Regulations 1959, a requirement being that all dairy farms are registered by the Ministry of Agriculture. Premises, stock and methods of production are open for inspection at any time.

The milking process (see Chapter 16)

Efficient milking is largely dependent upon a good milking routine. This should be simple and consistent, providing a 'let-down' stimulus for the cow and a routine series of operations which the milker has to perform on each animal. Milk ejection or 'let-down' is controlled by the hormone oxytocin. Milk let-down is a conditioned reflex in response to a stimulus. Quiet handling of animals prior to milking is essential; if cows are nervous or frightened, milk let-down will be inhibited. The natural stimulus for cattle is the calf suckling; in dairy herds a substitute stimulus in the form of feeding and/or udder washing is used. Besides providing this stimulus, udder washing removes dirt which may contaminate the milk as it leaves the teats. Methods of udder washing include sprays or buckets of water; the warm water often includes an antibacterial agent and cloths or paper towels are used to clean and dry the udder and teats. Disposable paper towels are preferable to cloths as there is less chance of infection being passed from cow to cow. Before or after udder washing, fore-milk (the first milk from the udder) is removed by hand into a strip cup to reveal any signs of mastitis in the form of clots, flakes or watery milk. This is a useful indication of clinical mastitis.

Teat cups should be applied as soon after washing as possible. Maximum rate of milk flow is reached after about 1 min; later the flow rate declines quite rapidly. Once milking has started it should be accomplished as rapidly as possible. Overmilking should be avoided as this may damage the udder, thus predisposing mastitis. Stripping – the removal of the last milk – can be achieved by applying downwards pressure to the teat cup with one hand and the udder massaged downward with the other, the forequarters being done first. In parlours the stripping of milk and automatic removal of teat cups, when the milk flow slows to a predetermined rate, is becoming more common.

Teat disinfection by dipping teats in a cup containing an approved iodophor or hypochlorite solution helps prevent the spread of bacteria causing mastitis.

After milking the milk will be at an ideal temperature (37°C) for the growth and multiplication of most bacteria. As the milk is collected once a day by bulk tankers some must be kept overnight. It is important to cool the milk quickly to prevent its deterioration. Cooling takes place in a refrigerated bulk tank and milk needs to be below 4.5°C by 30 min after milking. Before milk is allowed into the bulk tank it is passed through a filter.

Milk is routinely tested for keeping quality by direct or indirect tests and penalties are imposed for failure to meet the required standards. Stringent penalties are also incurred if antibiotics are detected in milk; these usually result from the intra-mammary treatment of mastitis.

Inadequate cleaning of the milking equipment resulting in a buildup of residue can cause poor keeping quality. Hand washing is normally used for bucket milking plants. A cold water rinse of approximately 10 litres of water can be drawn through the clusters of each unit; this removes the film of residue left after milking. The equipment is then dismantled and washed once in either detergent or a detergent and sterilant solution at 50°C, then finally rinsed in clean water also containing a sterilizing agent.

With pipeline systems where it is impractical to dismantle the equipment, cleaning is done in situ by circulation cleaning. Here the success of the operation relies on the properties of the chemicals and heat for the disinfectant effect. Again the basic process starts with a rinse of cold water; this is followed by the circulation of a detergent solution. The initial temperature needs to be fairly high (80–85°C) as the solution will be cooled during the first cycle round the plant. A final rinse with cold water, which may contain sodium hypochlorite for sterilization, is circulated. The most common sterilizing and disinfecting agents are sodium hypochlorite, hypochlorite, bromates and iodophors. They should be used in accordance with the manufacturers' recommendations.

Another method of cleaning equipment in situ is by the use of acidified boiling water. Here hot water (96°C) is flushed through the plant for 5–6 min and allowed to run to waste; this pre-rinses and warms up the plant. Nitric or sulphamic acid is mixed with the water; this has no disinfecting effect but removes the milk deposits from the equipment. The aim is to heat the plant to 77°C for at least 2 min to achieve disinfection.

The circulation cleaning method involves the use of expensive chemicals and takes more time (15 min) whereas the acidified boiling water system needs more water (13–18 litres/unit) at a much higher temperature, thus using more energy, but it only takes 5–6 min.

To prevent a buildup of scale on equipment it may be necessary to use a milkstone remover once per month.

Bulk tanks may be cleaned by hand, using long-handled brushes, or by a mechanical spray. As tanks are cooling mechanisms and often contain an ice bank at the base, a cold system of cleaning is normally used. This uses an iodophor or bromate cleaning agent at mains water temperature. The tank is rinsed with cold water by hand, immediately after emptying, and the cleaning agents applied. In automatic systems the rinse is sprinkled into the tank; this is then followed by a solution containing the chemicals. The inside is then rinsed before milking. Particular attention should be paid to the outlet, paddle, dipstick and underneath the lid and bridge of the tank.

Cattle breeding

The aim of selecting breeding animals is to produce a future generation with improved performance. Improvement is brought about by increasing the frequency of desirable genes and by creating favourable gene combinations. The genotype of an animal is its genetic constitution. The phenotype of an animal is the sum of the characteristics of the animal as it exists and is the result of both genetic and environmental effects. It is possible for animals to have the same genotype but different phenotypes owing to environmentally produced variation, i.e. how it is fed, housed and managed. Environment does not affect all characters to the same extent. For example, the normal homozygous black coat colour of the Aberdeen Angus is always dominant to the recessive red of the Hereford; thus all Angus × Hereford cattle are black with a white face. Environment has no effect as these characters are controlled by the presence of one pair of genes. The expression of many characters of economic importance – body growth, milk yield, milk quality, fertility – is controlled by many genes and environment also exerts an effect. Characters in which there is a close resemblance between parent and offspring, whatever the environment, are characters of high heritability (Table 19.34).

Greater progress can be expected if the breeding programme is concentrating on characters of high heritability. As the sire often serves many cows, a great deal of attention needs to be placed on his selection. The two main methods of evaluating bulls are: (1) performance testing and (2) progeny testing.

Performance testing involves measuring an animal's individual performance – growth rate, feed conversion efficiency – and comparing them with animals from a comparable group, which have been subjected to similar conditions of feeding and management. Performance testing can only be used to measure characteristics in the live animal. Assessing carcass composition used to necessitate slaughtering the animal. However, new techniques such as ultrasonics have to a large extent overcome this particular difficulty.

Progeny testing involves the examination of an animal's offspring. The characters are measured and compared against the progeny of other sires. Again progeny should be kept under similar environmental conditions. It is particularly useful for testing dairy bulls as milk production is only measurable in the female and the characters in question are often of low heritability. The method was used by the Milk Marketing Board and is still used by other organizations for evaluating its dairy bulls for artificial insemination (AI) purposes. The bull's progeny can be compared with daughters sired by other bulls in the same herds. The record should reflect the differences attributable to the bull's genetic constitution and is referred to as a contemporary comparison.

Table 19.34 Heritability of breeding stock characters. [After Johansson & Rendel (1968) (reproduced with permission).]

Trait	*Heritability*
Birth weight	0.4
Weaning weight	0.2
Daily gain, weaning to slaughter	0.3–0.6
Food conversion	0.4
Mature weight	0.4
Wither height	0.5–0.8
Heart girth	0.4–0.6
Dressing percentage	0.6
Points of carcase quality	0.3
Cross-section of eye muscle	0.3
Bone percentage	0.5
Lactation yield	0.2–0.3
Fat content of milk	0.5–0.6

These contemporary comparisons can provide guides when selecting a bull for AI. The higher the contemporary comparison the greater the chances are that the bull will pass on genes resulting in improved daughters. The greater the number of daughters used to evaluate a bull in this way, the more reliable the comparison figure. The number of daughters used is referred to as a 'weighting' which appears with the contemporary comparison figure.

When selecting animals for multiple traits the identification of overall superior animals is very difficult. The method that is expected to give the most rapid improvement of economic value is to apply the selection simultaneously to all the component traits, appropriate weighting being given to the trait's relative economic importance, its heritability and the genetic and phenotypic correlations between the different traits. After this has been calculated a single index is the result. A selection index may contain any number of traits. MLC/Signet's Beefbreeder service has the stated objective: 'To improve the financial margin between the value of saleable meat and the cost of feed, taking into account the cost of difficult calvings'. The system collects data on individual animals in the herd, which are then analysed using Best Linear Unbiased Predictor (BLUP). There are many anachronism for estimating the breeding value (EBV) of animals, INDEX, PROFIT and BLUP are just a few of those commonly used.

References and further reading

Allen, D. (1990) *Planned Beef Production and Marketing*. BSP Professional Books, Oxford.

Allen, D. & Kilkenny, B. (1980) *Planned Beef Production*. Granada, London.

Baker, H.K. (1975) Grassland systems for beef production from dairy breed and beef calves. *Livestock Production Science* **2**, 121.

Frood, M.J. & Croxton, D. (1978) The use of condition scoring in dairy cows and its relationship with milk yield and liveweight. *Animal Production* **27**(3), 285.

Hardy, R. & Meadowcroft, S. (1986) *Indoor Beef Production*. Farming Press, Ipswich.

HMSO (1984) *Dairy Herd Fertility*. Reference Book 259. HMSO, London.

Johansson, I. & Rendel, J.L. (1968) *Genetics and Animal Breeding*. Oliver and Boyd, London.

MLC (1978) *Beef Improvement Services*. Data summaries on beef production and breeding. MLC, Milton Keynes.

MLC (1982) *Selecting Cattle for Slaughter*. MLC, Milton Keynes.

MLC (annual) *Beef Yearbook*. MLC, Milton Keynes.

MLC/MMB Joint Publication *Rearing Replacements for Beef and Dairy Herds*. MLC, Milton Keynes.

MMB (annual) *UK Dairy Facts and Figures*. MMB, Thames Ditton.

MMB (1983) *The New Pricing Package*. MMB, Thames Ditton.

MMB (1986b) *EEC Dairy Facts and Figures*. MMB, Thames Ditton.

Roy, J.H.B. (1980) *In the Calf*. Butterworths, London.

Scottish Agricultural Colleges (1979) *Feeding the Farm Animal – Dairy Cows*. Publication No. 42. SAC, Aberdeen.

The Dairy Council (2000 edition) *Dairy Facts and Figures*.

Thickett, W.S., Cutherbert, N.H., Brigstocke, T.D.A. & Wilson, P.N. (1979) The inter-relationship between liveweight gain, performance, water intake and colostrum status in calf early-weaning systems. BSAP *Winter Meeting Paper* No. 18.

Warner, R.G. & Flatt, W.P. (1965) Herbage digestion. In: *Physiology of Digestion in the Ruminant*, (ed. R.W. Dougherty) Butterworths, London.

Wood, P.D.P. (1972) MMB *Better Management* No. 7:1.

20

Sheep and goats

R.A. Cooper

Sheep

The international picture

There were 1061 million sheep in the world in 1999, and in most countries populations are static or decreasing. Details of their distribution are given in Table 20.1. Production of sheepmeat by the world's major producers is shown in Table 20.2, with EU output detailed in Table 20.3. In the EU, flock sizes are declining in most countries. Mean carcass weight in most situations is 15–16 kg, but in areas where sheep are kept as dual-purpose milk–meat animals (e.g. in Italy and Greece) carcasses may only weigh 7–9 kg as lambs are slaughtered at 6–8 weeks of age.

Sheep in the UK

The total ewe population of the UK is 20.05 million head (June 2000), a drop of 3% on 1999 levels. Data for breeding animals are given in Table 20.4, which also indicates the importance of sheep in hill and upland areas. Table 20.5 gives a breakdown of the sheep industry by flock size. Although there are still over 37 000 holdings with flocks of less than 100 ewes, the trend continues towards fewer, larger flocks, with over 50% of ewes in flocks of more than 500, and 24% in flocks of over 1000 ewes. The national sheep flock is made up of some 70 'pure' breeds and more than 300 crosses, but few of the pure breeds, and even fewer of the crosses, are of major significance except in localized circumstances. Details of individual breeds may be obtained from the National Sheep Association publication *British Sheep*. Flock categories, important characteristics and major breeds are given in Table 20.6.

Sheep production systems in the UK are many and diverse, because of the varied conditions under which sheep are kept and the multiplicity of breeds and crosses available. They are linked, however, by an interdependence based on a substantial cross-breeding programme. A generalized outline of this programme, which is referred to as stratification, is shown in Fig. 20.1. Under the very harsh conditions experienced on many hill farms, ewes cannot thrive for more than four seasons. As a result, the hill farm retains many of its ewe lambs as replacement breeding stock while the sale of draft ewes provides an important part of hill farm income. Drafted into the better conditions on upland and marginal farms these ewes have several years of productive life left, and given better climate and nutrition are capable of lambing levels well beyond the 80–100% they have achieved previously.

On the upland farms to which these ewes go there is often little scope for finishing lambs satisfactorily. Put to a longwool ram these ewes produce a crossbred lamb with the benefits of heterosis evident in terms of fertility, milkiness and general vigour. There is a strong demand for such animals for lowland breeding flocks, where these attributes may be complemented by those of suitable Down rams for the production of prime lamb. Sixty per cent of ewes tupped under lowland conditions are derived from longwool × hill matings. Table 20.7 gives details of the most important of these crosses.

Factors affecting flock performance

Given the diverse nature of sheep production in the UK it is difficult to discuss flock performance in other than general terms. The standards that flocks recorded by the Meat and Livestock Commission (MLC) are achieving

Table 20.1 World distribution of sheep (×10[6]) (1999). [Source: *FAO Production Yearbook* (1999), reproduced with permission.]

Africa 239.8		*N. America 11.3*		*S. America 80.8*		*Asia 411.0*		*Europe 1153.3*		*Oceania 162.9*	
South Africa	28.7	USA	4.7	Argentina	14.0	China	131.0	E. Europe	18.9	Australia	117.1
Ethiopia	22.1	Mexico	5.9	Uruguay	15.2	India	57.9	UK	21.46	N. Zealand	45.7
Sudan	42.8			Brazil	18.3	Turkey	30.2	Spain	23.7		
Algeria	18.2			Peru	13.7	Iran	53.9	France	10.1		
Nigeria	20.5			Bolivia	9.0	Pakistan	24.0	Italy	11.0		
Morocco	16.3					Afghanistan	14.3	Greece	9.04		
Somalia	12.0					Mongolia	14.0	Ireland	5.39		
								Russian Fed.	14.0		

Table 20.2 Sheepmeat production (×10[3] t) (1999)

EU	1043	N. Zealand	517
Former USSR	508	N. America	154
Australia	614	Near East	1414
S. America	267	Far East	1877
Africa	1140		

Table 20.3 Production and consumption of sheepmeat in EU (1998/99)

	Production (10[3] t)	*Consumption (10[3] t)(1998)*
EU total	1043	1423
UK	402	353
France	142	299
Spain	245	242
Greece	130	144
Italy	50	100
Germany	45	99
Ireland	85	29

Table 20.5 Distribution of breeding ewes by flock size (million head)

Flock size	*No. of ewes*	*Percent of Flocks*	*Percent of ewes*
1–99	1.573	44.8	8
100–499	7.997	41.1	40
500–999	5.596	10.0	28
>1000	4.884	4.1	24

Table 20.4 Numbers and distribution of breeding ewes in the UK (×10[6]) (1999)

	England	*Wales*	*Scotland*	*N. Ireland*	*Total*
Severely Disadvantaged Areas (SDAs)					
At higher subsidy rate	1.976	2.556	2.087	0.389	7.008
At lower subsidy rate	1.002	0.820	1.265	0.282	3.369
Disadvantaged Areas (DAs)	0.575	0.100	0.079	0.250	0.905
Lowland	6.279	2.469	0.377	0.734	9.863
Total	9.257	5.845	3.729	1.405	20.24

are shown in Figs 20.2 and 20.3. Under lowland conditions the top one-third producers achieved a gross margin 50% above average in 1999 (£589 vs. £392 per ha). The factors contributing most to this superiority were average return per lamb (48%) and stocking rate (35%). In hill flocks performance per ewe is more important. Top one-third flocks achieved higher returns from lambs sold per ewe (£30.2 vs. £24.7) and had lower variable costs (£12.7 vs. £14.4). It is apparent from these data that there is room for improvement in both hill and lowland flocks. Details on good grassland management can be found in Chapter 4. The improvement of lambing percentage, which should be defined as lambs reared/100 ewes to ram, is discussed below.

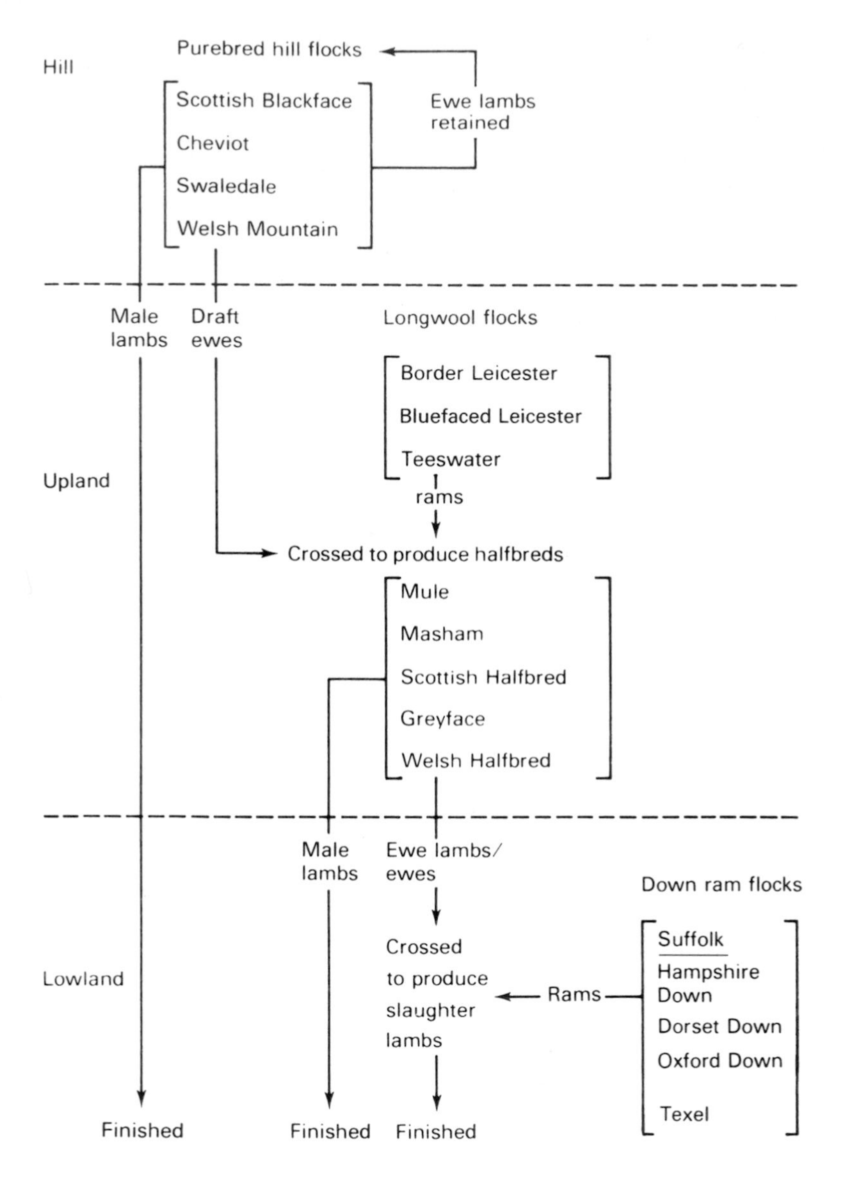

Figure 20.1 Stratification of the UK sheep industry.

Oestrus

The sheep is a species in which there is generally an annual rhythm of breeding activity, the onset of the breeding season corresponding with a period of decreasing daylength and cessation occurring as daylength increases in spring. There are, however, wide variations within as well as between breeds, and a much-reduced cyclicity is evident at lower latitudes. Some idea of the range of dates involved is given in Table 20.8. These values can be taken as no more than guides because of the interactions of many other factors as well as daylength. The response of the ewe to photoperiod is known to be mediated by the production of a hormone, melatonin, by the pineal gland, during the hours of darkness. It has recently been demonstrated that the administration of exogenous melatonin, in the form of a subcutaneous implant, may be used to bring mating forward by up to 6 weeks. Implantation from mid-May (Suffolks) or early June (Mules) should be followed by the introduction of teaser rams 35 days later and of entire rams 17 days after that. Peak matings will occur 52–62 days post-implantation. This technique has also been shown to increase lambing percentage by up to 20%.

Table 20.6 Breed types, important ewe characteristics and main breeds

Type	*Desirable characteristics*	*Main breeds*
Hill breeds	Regular lambing	Scotch Blackface
	Milking/mothering ability	Welsh Mountain
	Ability to rear 100% crop	Swaledale
	Hardiness	Cheviot
	Good wool weight and quality	Herdwick Gritstone
Upland breeds	Milking ability	Kerry Hill
	High fertility	Clun North Country Cheviot
Longwools	High fertility and prolificacy Milking ability	Blue-faced Leicester
	Growth rate	Border Leicester Teeswater
Down breeds	Growth rate	Suffolk Oxford Down
	Carcass quality	Hampshire Down Texel Dorset Down Charollais

Table 20.7 Important crossbreeds and their derivation

Cross	*Sire breed*	*Dam breed*
Mule	Blue-faced Leicester	Swaledale, Blackface, Speckledface, Clun
Greyface	Border Leicester	Scotch Blackface
Masham	Teeswater, Wensleydale	Dalesbred, Rough Fell, Swaledale
Scottish Halfbred	Border Leicester	Cheviot
Welsh Halfbred	Border Leicester	Welsh Mountain

The sudden introduction of rams during the period immediately before the onset of oestrus may also bring forward breeding activity by a few days. More importantly it can stimulate a considerable degree of synchrony. When associated with the use of vasectomized, or 'teaser', rams the technique can be used to obtain a tighter lambing pattern when 'entire' rams replace the teasers after 14 days.

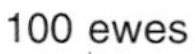

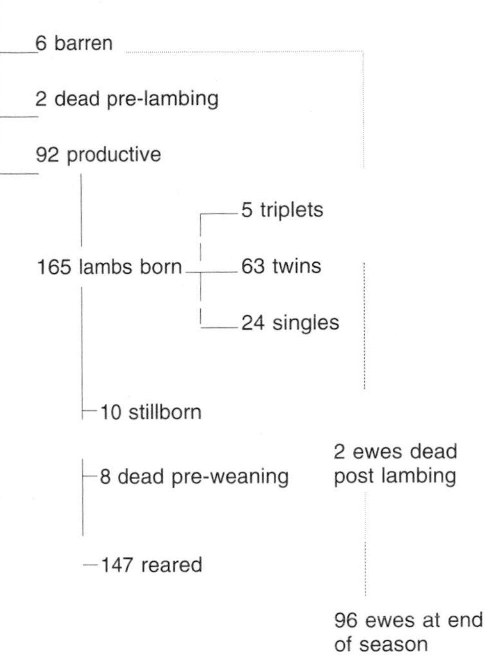

Figure 20.2 Average physical performance of lowland flocks (1999 data).

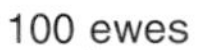

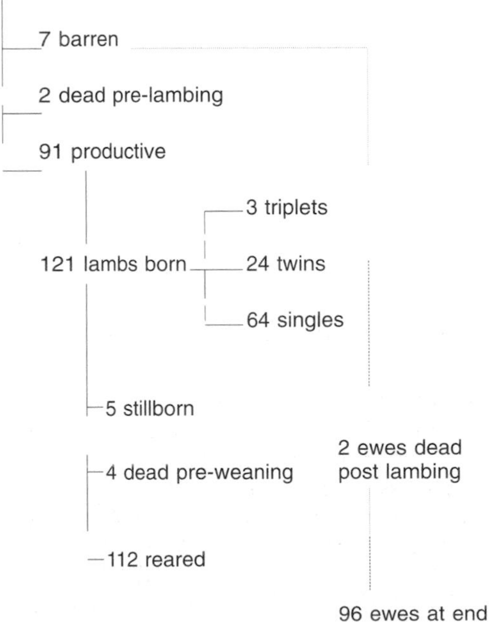

Figure 20.3 Average physical performance of hill flocks (1999 data).

Table 20.8 Breeding seasons for some British breeds. [Adapted from Hafez, E.S.E. (1952) *J. Agric. Sci.* **42**, 189, with permission.]

Breed	*Onset*		*Cessation*	
	Mean	*Range*	*Mean*	*Range*
Blackface	25 Oct	26 Sept–10 Nov	19 Feb	17 Jan–3 Apr
Welsh Mountain	25 Oct	11 Oct–11 Nov	17 Feb	1 Feb–25 Feb
Border Leicester	9 Oct	24 Sept–14 Oct	10 Feb	18 Dec–11 Mar
Romney	4 Oct	16 Sept–19 Oct	2 Mar	20 Jan–23 Mar
Suffolk	3 Oct	12 Sept–23 Oct	17 Mar	7 Feb–20 Apr
Dorset Horn	24 July	15 June–21 Aug	2 Mar	22 Jan–22 Apr

Ovulation and fertilization

One or more 'silent' ovulations often precede the onset of behavioural oestrus. Ovulations occur every 17 days (range 14–19 days) throughout the breeding season. Oestrus lasts for about 30 hours (range 24–28 hours) with ovulation occurring towards the end of this period. One of the prerequisites for a good lambing percentage is a high ovulation rate, and subsequently a high fertilization rate with minimal embryo mortality. Genotype will have an influence here, although nutritional factors may confuse the situation. Within a flock ovulation rate will be influenced by the following factors.

Nutrition

It has been known for many years that putting ewes onto a rising plane of nutrition for 2 or 3 weeks prior to tupping ('flushing') has a beneficial effect. Traditionally ewes were weaned onto hard conditions, in part to assist drying off, in part to put them into lean condition before 'flushing'. This increasing body condition, or dynamic effect, gives some 6–8% more twins compared with the performance of ewes mated whilst on a maintenance diet, and 12–16% more twins when compared with ewes mated while in negative energy balance. More recently a 'static' effect has also been recognized, this being dependent on the level of body reserves in the ewe, ewes with greater body reserves having a higher ovulation rate. For ewes on a level plane of nutrition an extra 3.5 kg bodyweight at mating represents a potential 6% extra twins (comparisons within flock).

Recognizing the impracticality of weighing ewes on commercial farms, the former Hill Farming Research Organization devised a simple, rapid and accurate method of judging the body condition of ewes by palpating the spinal column over the loin area. Scoring, on a scale from 0–5, is possible to the nearest 0.5 (see Appendix 1). Body condition scoring may be used on sheep of any breed, and at any time of the year, as a guide to the adequacy of nutrition of the flock. It is particularly valuable in the pre-tupping and pre-lambing periods. Pre-tupping it can be used to divide the flock into groups according to their needs in terms of the duration and degree of flushing required. Lowland ewes should have a condition score of at least 3.5 at mating and hill ewes 2.0–2.5. In a normal flushing period (approximately 3 weeks), lowland ewes having an initial score of 3 or above can achieve this target easily. The relationship between score at tupping and lambing percentage is shown in Table 20.9, but many breed differences exist and some of these are illustrated in Fig. 20.4. A scoring at least 6 weeks before tupping can identify those animals in need of preferential treatment and a second, 3 weeks later, can be used to monitor response. An improvement in condition from score 2.5 to score 3.5 is equivalent to a gain of some 6 kg liveweight. Body fat content (f) may be predicted using the equation $f = 8.69$ score $+2.69$.

Table 20.9 Relationship between condition at mating and lambing percentage. [After MLC (1980) *Sheep Improvement Services, Body Condition Score of Ewes*, reproduced with permission.]

Breed	*Average score at mating*	*Live lambs/100 ewes to ram*
Scotch Halfbred	3.8	184
	2.6	120
Welsh Halfbred	3.5	158
	2.7	133
Masham	3.4	178
	2.2	127
Clun	3.6	156
	2.0	131

Age

Ewe lambs may attain puberty in their first year. There is a nutrition/date of birth interaction, and with lightweight or late-born lambs oestrus may not be attained.

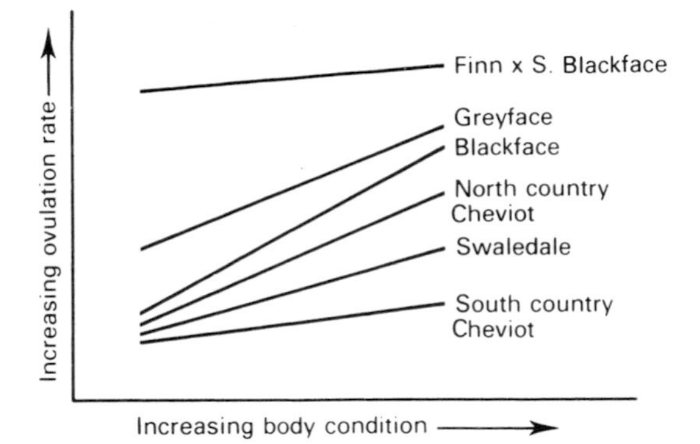

Figure 20.4 Breed effects on the relationship between condition at mating and ovulation rate. [From ARC (1986) *Science and Quality Lamb Production*, with permission.]

Although most animals that attain puberty will rear one lamb at best, there is some evidence that animals that are bred as lambs may be more prolific in later life. Subsequently, mean ovulation rate will continue to rise to 5 or 6 years before falling off again. Hence the draft hill ewe has only just achieved her potential when she is sold.

Season

Many ewes begin the season with a 'silent' oestrus. Mean ovulation rate then rises for three or four cycles and will likewise drop at the end of the season. On average, therefore, lambing percentages will be lower in early- or very late-lambing flocks and are very low in most out-of-season lambings (unless exogenous hormones are used). These problems are exacerbated by higher numbers of barren ewes at such times. The Finnish Landrace and its crosses do not appear to be affected in this manner and are thus widely used in 'out-of-season' production systems.

The relationship between tupping date (and hence lambing date) and lambing percentage in lowland flocks is shown in Table 20.10. Decisions on lambing date are likely to be influenced also by factors such as climate, altitude and custom. In hill situations the introduction of the ram into the flock will in part be dictated by tradition. Traditional dates in Scotland, for example, are 10 November for Cheviots and 25 November for Blackface. Under normal conditions all lowland sheep will ovulate and produce at least one ovum; the majority should produce two or more. Genotypically there is no reason why this should not be similar for hill breeds, the lower ovulation rates experienced in hill flocks being largely the result of environmental effects.

If all ova shed are to be fertilized, then rams must be both fertile and active. Low libido (sexual drive) may well be a problem in rams in early lambing or out-of-season situations. At the beginning of the breeding season all rams should be examined to ensure that they are healthy, sound in feet and legs and without any abnormality, especially of the testes. Microscopic examination of a semen sample is an additional check though results need careful interpretation, particularly early in the season and especially if electro-ejaculation has been used. Correct ewe to ram ratio will depend on circumstance. A strong ram should be able to cover 50–60 ewes adequately, but under extensive conditions the ratio may need to be lower and where an attempt has been made to synchronize oestrus it will need to be no more than 10 : 1. An inexperienced ram lamb may be used (carefully) in his first year, but only at a low ewe : ram ratio (20 : 1) and only with mature ewes. It is particularly important that ram lambs do not have to compete with experienced older rams at this time.

Table 20.10 Effect of mating date on mean lambing percentage in lowland flocks. [MLC (1978) *Sheep Facts*, reproduced with permission; West of Scotland College of Agriculture Research and Development Publication No. 7.]

Lambing date	*Lambing %*	
	MLC data	*WSCA data*
January	148	137
February	155	145
1–15 March	160	—
16–31 March	167	159
April	168	—

The practice of applying ochre to the breastbone of rams (keeling or raddling), or the use of suitable harness and crayon, is a helpful management tool. Rams treated in this way leave a mark on all ewes mated and this allows the monitoring of the progress of the breeding programme and, supplemented by more permanent markings, can form the basis of subdivision of the flock prior to lambing. In larger flocks this can lead to significant savings in concentrate costs. With such a system the colour of the raddle should be changed at least every 16 days, going progressively from lighter to darker colours.

Embryo mortality

A significant percentage of fertilized ova are lost in early pregnancy, especially in the first 30 days. Losses as high as 40% have been suggested, but the mean is probably about 25%. Some loss of embryos is to be expected, but steps can be taken to minimize it. In particular the avoidance of stress and the adequate nutrition of the ewe are critical; above-maintenance levels of feeding should be maintained for the first month of pregnancy. Embryo mortality also increases with increasing ovulation rate, so that ewes that have been flushed pre-mating are likely to be more sensitive to post-mating drops in feed intake. Embryo loss is also greater in early and out-of-season breeding and in ewe lambs. In some countries a further factor is likely to be heat stress, which can cause very high embryo losses if it occurs at the early cleavage stage.

During the middle part of pregnancy few losses appear to occur. The ewe is able to withstand periods of undernutrition – hill ewes often losing 5–10% of bodyweight during this period – without significant effect. Losses in excess of this, or in the later stages of pregnancy, may lead to fetal death, reduced birthweights and increased perinatal mortality.

Perinatal mortality

Perinatal mortality (including stillbirths and abortions) is a major source of loss in many sheep flocks. Surveys suggest that as many as four million lambs may be lost in the UK every year. Main causes of loss are: abortions and stillbirths 20–40%, starvation and exposure (hypothermia) 35–55%, infectious diseases 5–10% and misadventures 5–10%. Organisms of the *Pasteurella*, *Toxoplasma*, *Brucella* and *Rickettsia* groups may be implicated in so-called abortion storms in sheep. (It should be noted that several of the abortion-inducing organisms are zoonotic. It is important to stress that pregnant women should not be exposed to sheep close to lambing.) Vaccination can help avert the problem of abortion in sheep. Additionally, appropriate vaccination (e.g. against clostridia), given to the ewe prepartum, can help reduce postnatal lamb mortality by increasing antibody levels in colostrum.

The nutrition of the ewe in the 6 weeks prelambing is extremely important. During this period the fetus makes 70% of its growth. Undernutrition of the ewe at this time will lead to reduced birthweights and an increased susceptibility to hypothermia (see Fig. 20.5), or even to the death of some or all fetuses *in utero*. Additionally, correctly fed ewes are much less likely to suffer from pregnancy toxaemia (twin lamb disease), a metabolic disorder which may cause the death of both ewe and lambs. The accuracy with which pregnant ewes can be fed is increased if the number of fetuses carried is known. Pregnancy diagnosis using real-time ultrasonic scanning and carried out 50–100 days post-mating can detect pregnancy to an accuracy of 98% and fetus number to 95% or better. Such knowledge allows appropriate division of ewes, facilitating differential feeding and making lambing management easier. Savings of concentrates of up to 20 kg/barren ewe and 10 kg/single-bearing ewe are possible and will often cover the cost of scanning. Reduced ewe and lamb mortalities may further contribute to the cost-effectiveness of the technique.

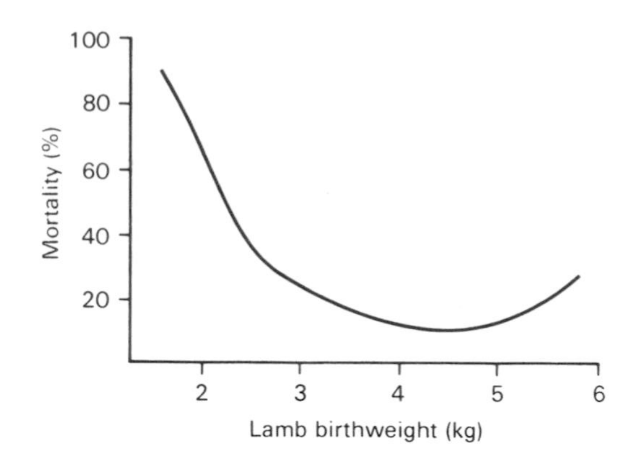

Figure 20.5 Relationship between lamb birthweight and mortality. [From Maund, B. (1974) *Drayton EHF Annual Review*.]

Table 20.11 Intensive care routine at lambing

(1)	Check lamb breathing: clear mucus from airways.
(2)	Treat navel with iodine, chloromycetin or suitable alternative.
(3)	Individually pen ewe and lamb(s).
(4)	Draw milk from both teats of ewe.
(5)	Put lamb to teat and initiate sucking.
(6)	Use stomach tube to administer colostrum if necessary.

Hypothermia

Once the lamb is born, careful shepherding can do much to ensure its survival. The lamb's ability to withstand the challenges of climate and infection is critical. A well-planned care routine (see Table 20.11) can do much to reduce these problems. Hypothermia is one of the major causes of postnatal mortality. The newborn lamb has a coat with little insulating value, a large surface area : weight ratio, and is wet. It must produce heat as rapidly as it is losing it or it will die. Lambs have only small fat reserves, especially those that are small, and rely heavily on feed energy. The small lamb may quickly burn up body reserves in a cold, wet environment and thence be too weak to suckle properly. Such a situation will lead to death within a few hours of birth. The larger lamb may be able to utilize reserves for up to 24 hours, but must then replace them if it is to survive.

Body temperature is a good guide to the lamb's status. Workers at the Moredun Institute have developed a programme based on lamb body temperature (see Fig. 20.6)

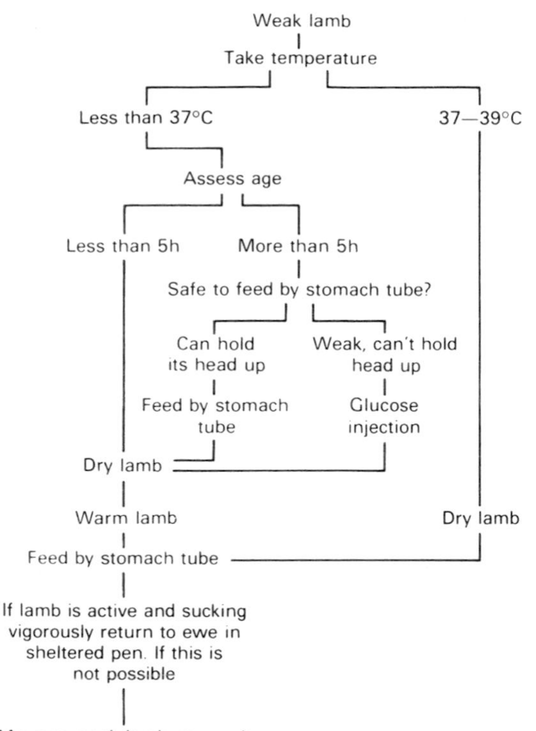

Figure 20.6 Course of action taken with lamb suspected of being hypothermic. [From Eales, F.A. & Small, J. (1986) *Practical Lambing*. Longman, London, reproduced with permission.]

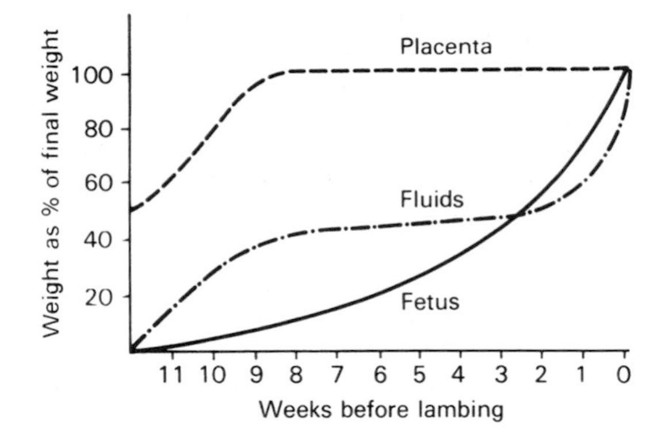

Figure 20.7 Differential development of components of conceptus. [From Robinson, J.J. *et al.* (1977) Studies on reproduction in prolific ewes. *Journal of Agricultural Science*, **88**: 539, reproduced by courtesy of the editor and publisher.]

Table 20.12 Growth of fetus during twin pregnancy. [After Wallace, L.R. (1948) Growth of lambs before and after birth in relation to level of nutrition. *Journal of Agricultural Science* 38: 93.]

Stage of pregnancy (days)	*Fetus weight (g)*
28	0.001
56	89
84	1000
112	4250
140	10000

which, if rigorously applied, should dramatically reduce the number of lambs lost to hypothermia.

Meat production

Growth and nutrition

Under UK conditions the main output of the sheep unit is in the form of lamb – either finished or store. The money obtained for each lamb is a function of weight, either live or carcass, quality and the date of sale.

As can be seen from Fig. 20.7 the development of the components of the conceptus is variable, with most placental development having finished by 6 weeks prior to lambing while most fetal development is yet to take place (see Table 20.12). The importance of ewe nutrition during this period has been discussed in relation to the survival of the newborn lamb. It is also important in terms of the subsequent growth of that lamb, especially in the case of multiple births.

From a daily requirement of 12–18 megajoules (MJ) of metabolizable energy in late pregnancy, the energy demands of the ewe will increase to 20–30 MJ at peak lactation. Especially in the flock lambing before grass becomes available, this energy demand may not be met from the diet and the ewe will begin to 'live off her back'. Ideally the ewe will lamb down at condition score 3.5 if she is going to be able to milk satisfactorily. Ewes suckling more than one lamb respond by producing more milk; some 40% more if suckling twins, 50–55% more for triplets, and there is a case for running ewes suckling triplets separately and feeding them accordingly. (Better still is the fostering of one triplet lamb onto a single-bearing ewe.) Notwithstanding this extra milk, lambs reared as multiples are unlikely to grow as fast as those reared as singles. Average figures for daily milk yield are given in Table 20.13. For the lowland ewe the average of 1.6 kg produced/day will support daily gains of 300 g in the single lamb. When producing 2.54 kg the ewe is supporting some 460–500 g of gain, but this is only 230–250 g/lamb. By the third month the relevant figures are 220 and 130 g/d for singles and twins respectively.

The effects of this on lamb liveweight are demon-

Table 20.13 Average milk yields during the first 12 weeks of lactation

Type of ewe	*No. of lambs*	*Yield (litres)*		
		Month 1	*Month 2*	*Month 3*
Hill	1	1.35	1.30	0.95
	2	2.25	1.85	1.25
Lowland	1	1.60	1.55	1.25
	2	2.55	2.10	1.45

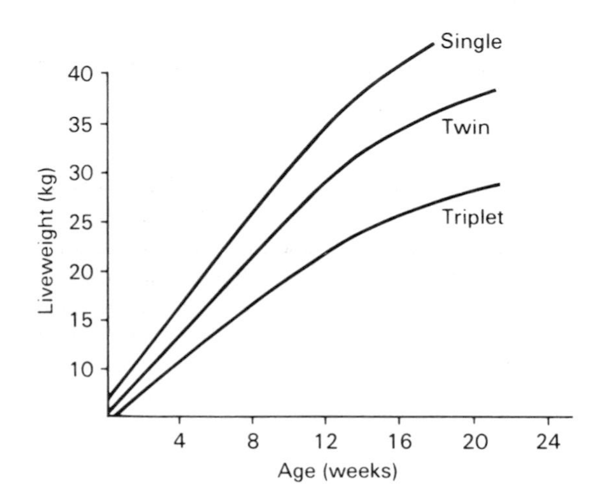

Figure 20.8 Typical growth curves for lambs. (From Spedding, C. (1970) *Sheep Production and Grazing Management*. Baillière, London, reproduced by courtesy of the publishers.)

strated in Fig. 20.8. For the single lamb, growth rate is almost linear, and with 'milky' ewes the lamb will need no supplementary feed. On the other hand triplet lambs start at lower weights, grow more slowly, become dependent on alternative feeds earlier and when grazing are thus much more prone to roundworm infestation. There are few advantages to be gained by leaving lambs with their dams beyond 12 weeks of age and, indeed, in situations where grass is in short supply, there may well be advantage in weaning at 10 weeks and allowing the lambs access to the best grazing.

A second factor affecting daily gain beyond the 10- to 12-week period is the shift, in terms of components of any gain, away from lean and towards fat. The characteristics of growth in the lamb are similar to those for other species, namely that initially bone growth predominates, followed by lean tissue and finally fat. Animals in which the waves of growth for bone, lean and fat are steep and close together are called early maturing. In sheep the tendency is for smaller breeds, such as the Southdown, to be early maturing and to begin to lay

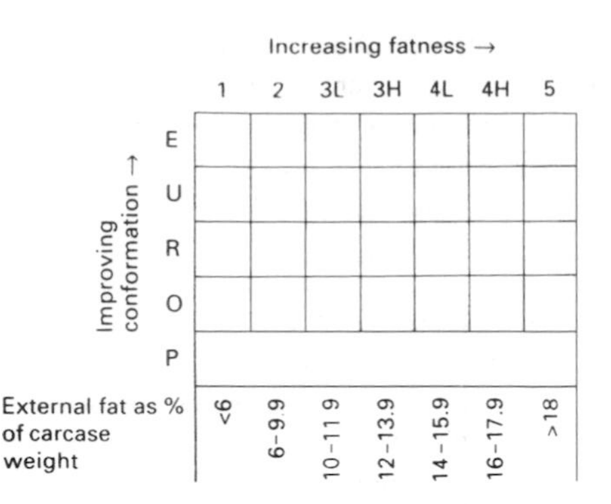

Figure 20.9 Sheep carcass classification grid.

Table 20.14 Relationship between carcass weight and carcass classification

Carcass weight range (kg)	*Percent carcasses in fat class* 2	3	4	5
16–17	29	60	7	—
18–19	18	63	15	1
20–21	11	60	24	3
22–23	7	52	33	6
24–25	4	40	41	14

down fat at relatively low carcass weights, and for heavier breeds to be later maturing and capable of producing, given time, a heavier carcass at any given level of fatness.

The current demand is for leaner carcasses. The Meat and Livestock Commission (MLC) has developed a classification scheme for lamb carcasses which can be used to assess and describe any carcass in terms of its conformation (muscling) and fatness (see Fig. 20.9). On this basis, current demand is for a carcass falling into conformation classes E, U or R and into fat classes 2 and 3 L. An indication of the relationship between carcass weight and classification is given in Table 20.14. It is possible to predict, with a reasonable degree of accuracy, the likely carcass weight of any breed or cross of lamb at a fat score of 2–3 L, using the formula $(WD/2 + WS/2)/2 \times$ estimated killing-out percentage (where WD is mature weight of dam breed and WS is mature weight of sire breed). Thus, for example, using a Suffolk ram (90 kg) on a Scotch Halfbred ewe (90 kg) will produce a 20-kg wether carcass at a killing-out percentage of 45. For ewe lambs this weight will be reduced by about 10%. Choice of sire breed is a matter of individual judgement. It is essential that the breed of ram chosen suits the lamb finishing system to be used, and it must be remembered that a heavier lamb will take longer to produce and that this may be inconvenient. For example, lambs aimed at the early market need rapid growth and early maturity, and will tend to be sold at lower carcass weights, while animals to be finished on roots as late lambs or hoggets (animals sold after 1 January) need to be capable of producing a heavy carcass without laying down too much fat.

Systems of lamb production

If lambs are categorized according to date of sale, then in the UK early lambs, sold before the end of June, account for only 16% of total output, lambs finished off grass 35%, lambs finished on forage crops 30% and hoggets 19%. There are many systems of lamb finishing. Important elements of the main ones are as follows.

Early lamb

Early lambs are aimed particularly at the Easter market, prices peaking between mid March and late April as hogget numbers decline. Dorset Horns and their crosses are able to lamb regularly in the period September to December, as are selected strains of other breeds. The system is characterized by high lamb prices and high stocking rates for weaned ewes (25 ewes/ha), but these points tend to be offset by lower lambing percentages and higher concentrate costs.

Traditionally Dorset Horn lambs have been run on root crops, when only modest rates of gain are necessary. For later-born lambs the need is for concentrate feeding, with high quality feed (12.5 MJ, 170 g DCP) being fed from 3 weeks. On such diets an FCE (feed conversion efficiency) of 3.5 : 1 is possible, with lambs gaining 350–400 g/d and going off at 34 kg liveweight in 12–14 weeks. This type of system fits well on arable units where grass is limiting, and where 'off-peak' buildings and labour are available.

Lambs off grass

The traditional system of production in many areas is lambs off grass. Such systems should aim for high lambing percentages (170+), high stocking rates (15–18 ewes/ha) and as many lambs sold 'off-the-ewe' as possible (75%). Lambs not sold by weaning may then be finished on silage aftermaths or forage crops, or sold as stores. Performance of lambs on such a system is improved by the adoption of forward-creep-grazing techniques, which should routinely produce increases in growth of up to 5 kg over a 10-week grazing period.

Lambs and hoggets off forage crops

Those lambs that have not been finished by 1 October in any year are generally termed store lambs. Many such animals change hands at this time of year, but some will be finished on their home farms. Animals available will be wethers of the hill breeds, weighing 23–30 kg, wethers of the hill crosses (Mules, etc.) at 25–35 kg and the remains of fat lamb flocks at 30–40 kg. Feeds for these animals may be grass, silage, catch-crop roots, main-crop roots and arable by-products. It is important that lambs are matched to the feeding system proposed and that stocking rates are carefully monitored. For example, main-crop roots are expensive to grow and the aim must be for high yields and high stocking rates. On the other hand, if stocking rates are too high, individual animal performance may suffer. Details of potential yields and stocking rates for the main crops are given in Table 20.15.

Grass finishing

Grass finishing of lambs is suited to Down crosses for finishing by December, and halfbred wethers for February–March sale. Where dairy cow paddocks are being grazed, sheep must be off by December to avoid subsequent drop in grass yield; potential silage ground may be grazed until March. Relatively low rates of gain (70–100 g/d) produce carcasses with acceptable fat levels, but Down-cross lambs will become fat if held too long.

Silage finishing

This depends on availability of surplus silage of good quality (DM 25–35%, D-value 65+, ME 10.5 MJ+, pH 4, NH_4N <10%). The system needs lambs of 30 kg liveweight with good frame (e.g. halfbred wethers). Growth rates are variable at around 75 g/d. Where good silage is available (100–120 kg/lamb) it is only necessary to supplement it with small quantities (30 g/d) of a low-degradability protein to improve silage intake. For silage of D-value 65, up to 250 g/d of a 15% protein barley/protein mix will be necessary. This amount will increase to 450 g at D-value 63 and 750 g at D-value 61. These quantities may need increasing further if silage intake is limited (e.g. for long-chop or poorly fermented material). At the same time, in some situations silage

Table 20.15 Yields and stocking rates on forage crops

Crop	*Season of use*	*Yield (t/ha)*	*Stocking (lamb days/ha)*
Stubble turnips	Sept-Jan	40–50	2500 (i.e. 60 lambs for 6 weeks)
Rape	Oct–Feb	30–35	2500
Swedes	Dec–Mar	65–100	8000
Beet tops	Oct–Dec	?	1800
Grass	Oct–Mar	?	600

intake may be suppressed by concentrate feeding. A 10-cm feed face is required for *ad libitum* silage feeding and a 35- to 40-cm trough for concentrates. One of the problems of this system is that where lambs are held to take advantage of the higher prices in late February/March they may be occupying buildings needed for the breeding flock.

Finishing on roots

The performance of lambs on roots is very variable (between 50 and 200 g/d). Short-term systems, aiming at merely 'finishing' well-grown but lean lambs, will utilize stubble turnips. Long-term systems, usually involving swedes or hardy turnips, aim to have animals putting on lean before finishing and depend on the rise in lamb prices in February/March for their profitability. One of the main problems with this system is that of balancing wastage and performance. High individual-animal performance is normally associated with high waste (up to 60% in a bad year). Where heavier stocking is practised, to reduce crop waste, lamb growth suffers. The feeding of up to 200 g barley/d can help maintain outputs and reduce variation between animals. Block grazing tends to produce lower wastage rates, without affecting lamb performance, and has a lower labour requirement than strip grazing.

Finishing on cereals

The high cost of cereal-based diets means that this system is only applicable to February/March sales. Lambs must be capable of going to 40 kg liveweight without going to fat, but as with root-based finishing the production of overfat lamb is a major hazard. Ideally lambs are introduced to cereals whilst still at grass, having been vaccinated against pulpy kidney (*C. welchii*), and preferably also against *Mannheimia* (*Pasteurella*), at least 2 weeks earlier. Lambs may be housed when eating 500 g/d and gradually brought to *ad libitum* feeding, but shy feeders should not be brought in. The ideal cereal for lamb is whole barley, but addition of protein and minerals is then difficult. A compromise 13% protein ration might be as follows: whole barley 350 kg, rolled barley 300 kg, beet pulp shreds 230 kg, soyabean meal 100 kg, sheep minerals 20 kg. On such a diet lambs will gain 1 kg/week, at an FCR of 5–5.5 : 1, and will tend to get fat. Regular handling of all animals is necessary if fat class 4 carcasses are to be avoided.

The use of concentrate feeding may also be considered in systems involving 'out-of-season' lambing [when concentrates (12.5 MJ, 17% CP) are introduced at 7 days and fed *ad libitum* to slaughter], or when 6-week weaning of one lamb from twins is practised (see later section headed 'Early weaning of hill lambs'). For such concentrate-fed lamb, milk substitute should be offered *ad libitum* for 3–4 weeks followed by 1 week of restricted access. A feed conversion of 1 : 1, on a dry matter basis, is possible for the liquid diet, with 4 : 1 being achieved on concentrates. Gains of 300 g/d are possible, with 10 kg milk substitute and 100 kg concentrates necessary per lamb.

Milk production

In recent years there has been an upsurge of interest in milking sheep in the UK. Elsewhere in the world the practice is widespread. Within Europe, Italy, Spain, Portugal, Greece and France have well-established industries (Table 20.16). In Greece sheep's milk accounts for over 35% of all milk produced.

As may be seen from Table 20.16, yields of sheep's milk are extremely variable. In most of Europe, ewes are used as dual-purpose animals, producing both meat and milk, and are suckled for varying lengths of time. Under UK conditions it is likely that milking ewes will be weaned as soon as the lamb has taken colostrum. Many breeds may be milked satisfactorily but yields will vary. 'True' milking breeds, such as the Friesland and British Milksheep, will produce up to 450 litres in a 180- to 220-day lactation. More 'dual-purpose' animals, such as the Clun or Mule, will yield upwards of 150 litres. The lactation curve of a sheep is similar to that of the cow. 'Let-down' occurs in two phases, with an initial flow of milk from the udder cistern followed some 20 s later by that from the alveoli. Udder washing may reduce this delay. Ewes' milk is very rich, having approximately 70 g fat, 50 g lactose and 50 g protein/kg. This creates a high energy demand (7.8 MJ/kg) as well as tending to create difficulties with milkstone deposits on equipment.

As with the dairy cow, the most difficult time in terms of feeding the ewe is in early lactation when appetite is limited. This is especially true of ewes yielding in excess of 3 litres. The nutrition of the ewe is discussed later. One of the major problems likely to be encountered when ewes are lambed in winter is that of low water intake. Ewes may be encouraged to drink by using a feed such as sugar beet pulp presented soaked and covered with water.

Wool

At one time wool was a major source of income for the sheep farmer. Today wool accounts for less than 2% of gross output. Yields of wool vary greatly between breeds, with longwools such as Teeswater and Devon

Table 20.16 Outline of sheep's milk production in Europe (1993)

Country	*No. of ewes* (×10^3)	*Average flock size*	*Main breeds*	*Average yield* (*kg*)	*Annual production* ('*000t*)
Italy	5000	10	Sarda	100–250	780
			Gentile di Puglia	30–35	
			Sopravissana	50–55	
			Churra	120	
Spain	2500	60	Manchega	100	305
			Castellana	90	
			Chios		
Greece	2500	80	Vlahico	90	670
			Karagunico		
			Merino		
Portugal	1000	60–100	Churra	120	97
			Bordaliera		
			Manech		
France	1000	100–350	Corse	110	243
			Lacaune	160	

Longwool producing 5–6 kg/head while hill breeds such as the Swaledale and Herdwick will only yield 2 kg.

Wool is produced by follicles in the skin. Two distinct types of follicle are identifiable. Primary follicles, normally associated with an erector muscle and a sweat gland, produce coarser, medullated fibres of hair or kemp. In improved breeds many of these coarser fibres are shed soon after birth. Secondary follicles, associated only with a sebaceous gland, produce wool fibres. The ratio of primary : secondary follicles determines the fineness of the fleece and varies between breeds. For example, the fine-woolled Merino may have 25 secondaries per primary, while most British breeds will have no more than eight.

Fineness is a major criterion in wool grading, the standard measure of fineness being the Bradford Count. Bradford Count is the number of hanks of yarn 510 m long that can be spun from 450 g of wool. Values will vary, from those of the Merino at 80+ to those of the Blackface at 27–30. Fibre diameters will vary from 23 μm at 60 to 38 μm at 45. There are also within-fleece differences, with breech wool being coarsest and shoulder wool finest. Breech wool also contains a higher percentage of kemp. A second quality factor in wool is its crimp; that is the number of corrugations per 25 mm. Crimp values vary from 8–28 and are closely correlated with fineness.

The growth of wool is photoperiodic and cyclic, 80% of growth taking place between July and November. Wool fibres are almost entirely keratin – a protein with a very high cystine content (12–14%). Notwithstanding this, the main relationship between nutrition and wool growth is in terms of energy input. Wool continues to grow even when the sheep is in negative energy balance. It does, however, grow more slowly and is finer under such conditions. While wool that is uniformly fine along its length is desirable, the development of a thinner area in an otherwise thicker fibre, a condition known as tenderness, is to be avoided since it reduces the value of the fleece and may even cause premature shedding. Tenderness develops in conditions of underfeeding, for example in ewes suckling triplets, or during bad attacks of parasitic gastroenteritis or liver fluke.

At shearing, care should be taken to preserve the quality of the fleece. Sheep must be dry when shorn; double cuts, which reduce effective staple length, should be avoided; organic matter such as straw and faeces should be kept out of the wool; and the fleece should be carefully packed and stored in the dry. Considerable penalties are incurred for badly presented or marked fleeces. Details of grades, prices and penalties are published annually by the British Wool Marketing Board, who have a monopoly on all wool sales.

No yield advantage accrues from shearing sheep more frequently than once yearly. Sheep are traditionally shorn in late spring or early summer, but recent years have seen the development of winter shearing of housed ewes. This technique allows increased stocking densities (up to 15% more) and tends to produce cleaner fleeces. It also reduces heat stress in the ewes, leading to increased dry matter intakes of the order of 10%. This extra intake is reflected in higher lamb birthweights and lower perinatal mortalities. In the prolific ewe flock at Drayton EHF, shearing increased birthweight by 600 g/lamb and reduced mortality from 16 to 7%.

Winter shearing does not fit all situations. To avoid

problems the house must be well bedded and draught-free. A minimum of 8 weeks must elapse before turnout, to allow some regrowth, and even then it is undesirable to turn shorn ewes out before early March in sheltered areas of the south or early April in the north. Ewes with condition score below 2.5 should not be shorn.

Flock replacements

Flock replacement costs are a major element in the profitability of a sheep enterprise. In 1999 the costs averaged £7.79/ewe in MLC recorded lowland flocks and £8.90/ewe in upland flocks. Keeping these costs down is a question of minimizing the number of ewes to be replaced each year and of keeping down the cost of each replacement.

On most hill farms some ewes will be lost and some will need to be culled. The majority of ewes leaving the farm will be drafted out at 4–5 years of age simply because they are no longer able to cope with conditions. Thus, on average some 25% of the flock will need to be replaced each year. Set against this is the fact that many of these ewes will be sold as draft breeding stock rather than culls, thus making a substantial contribution to the output of the flock. Traditionally, and of necessity, replacement ewes will be retained from within the lamb crop. This is essential if the ewe lamb is to become a productive member of the flock. Within a hill flock, groups of sheep become territorially organized or 'hefted'. This tendency is encouraged by careful shepherding, for it reduces the need for fences and assists the survival of the sheep since they know of, and will seek out, areas of shelter and sources of food during bad weather. From a husbandry viewpoint one of the major problems created by the need to keep ewe lambs as replacements is what to do with them over their first winter. If they stay on the hill, then they add to feed requirements at a difficult time, and some indeed may not survive the winter. On the other hand, the traditional 'tacking' or 'agistment' of ewe lambs onto lowland farms has been made more difficult by a reduction in the number of farmers willing to co-operate, and by escalating costs, both of agistment and of transport. One compromise that is becoming more common, despite the cost, is the provision of housing for these animals on the home farms.

Under lowland conditions most ewes being sold off will be culls. Criteria taken into account when deciding on which animals to cull may include condition of udder, feet and mouth, previous history, condition and temperament. Many of these factors are influenced, to a greater or lesser extent, by management. It could be argued, for example, that culling because of bad udder or bad feet should only occur as an extreme measure, and that if many sheep are involved it may be that flock management is suspect. Culling on teeth condition will depend on two main management factors: stocking rate and winter feed policy. The need for a ewe to be 'sound' in mouth relates to her ability to feed. As stocking rates increase, so competition for grass increases and the ewe needs to be able to graze closer to the ground, a facility made more difficult if the ewe is 'broken-mouthed'. Similarly, if winter feeding involves the use of 'hard' roots, such as swedes or turnips, the broken-mouthed ewe will be unable to compete and will tend to lose condition. Additionally, the feeding of hard roots tends to increase teeth loss. By $4\frac{1}{2}$ years of age up to 75% of ewes may still be full-mouthed if fed on hay and concentrates; in root-fed flocks the figure may be as few as 35%.

In addition to the above factors, opportunity plays a part in determining culling rate. In years when replacements are expensive, culling levels tend to be lower than in years when they are relatively cheap. The cost of each replacement can also be influenced by the age of the animal at purchase; ewe lambs being appreciably cheaper than shearlings. Having opted for the cheaper ewe lambs, the farmers must next decide whether to put them to the ram in their first year or not. Puberty in sheep is influenced by age, bodyweight and daylength, older heavier animals being more likely to attain puberty than younger or lighter ones. The advantages of breeding from ewe lambs are that it increases lifetime lamb production per ewe and decreases replacement cost, but against this must be set the likelihood of more mis-mothering problems and difficulties in getting their lambs away fat. In the end the decision may well depend on the source of the lambs. For homebred animals, whose management has been aimed at producing an animal suitable for tupping at 7–9 months, the system can work quite well. For the producer who is using long-wool crosses, and who of necessity has to purchase all his replacements, it may be much less viable.

Feeding the ewe

Theory

As has been emphasized, the nutrition of the ewe is a major factor in determining flock performance. Feed requirements vary with liveweight, physiological state and environment. Details of these requirements, and suggested systems for meeting them, are discussed in the MLC publication *Feeding the Ewe*. The main points are discussed below.

Voluntary feed intake

Dry matter intake will vary with physiological state and with diet, as well as with bodyweight. Intakes will normally be within the range 60–70 g/kg $W^{0.75}$ (where $W^{0.75}$ is metabolic bodyweight), but may be as high as 100 g/kg $W^{0.75}$ for concentrate diets or as low as 45 g/kg $W^{0.75}$ on silage-based diets, especially if the silage is acidic (pH below 4) or poorly fermented. Appetite is depressed by up to 20% in late pregnancy and very early lactation but recovers quickly to reach a peak some 2–3 weeks after peak milk yield is achieved. The increase in intake between late pregnancy and peak milk yield may be as high as 60%.

Energy requirements

Energy requirements vary according to bodyweight, stage of pregnancy and milk yield, and are reduced by 0.75–1.2 MJ/d in housed ewes. Maintenance requirements, based on $M_m = 1.8 + 0.1\,W$, are usually in the range 5–10 MJ/d. During pregnancy, requirements increase significantly as fetal growth accelerates in the last 6–8 weeks (see Fig. 20.7), being 20% higher 6 weeks before lambing and up to 150% higher immediately prelambing. During lactation, energy requirement varies with milk yield, which is in turn influenced by the number of lambs suckling. Average milk yields are given in Table 20.13. Requirements for energy, based on the above considerations, are given in Tables 20.17, 20.18 and 20.19.

Protein requirements

Protein requirements also vary according to the physiological status of the ewe. Maintenance levels are generally between 65 and 100 g CP/d. For weight gains (increases in body condition) before tupping, these values increase by up to 100% (Table 20.17), but during mid-pregnancy maintenance levels are again adequate. During the last 6 weeks of pregnancy, requirements increase to between 110 and 235 g CP/d, depending on ewe weight and number of lambs carried (see Table 20.18). Throughout this period the provision of 10 g CP/MJ ME in the diet should be adequate, but where ewes are expected to utilize backfat during late pregnancy there will be a need for supplementary protein,

Table 20.17 Mean daily allowances for ewes up to mid-pregnancy

Liveweight (kg) and condition change[1]	*Pre-mating*		*Post-mating Month 1*			*Months 2 and 3*		
	ME (MJ)	*CP (g)*	*ME (MJ)*	*CP (g)*	*MP (g)*	*ME (MJ)*	*CP (g)*	*MP (g)*
0.0	6	65	6	65	70	5.5	55	58
40 + 0.25	9	95						
+0.50	12	120						
0.0	8	85	9	90	91	7	70	71
60 + 0.25	12	125						
+0.50	17	165						
0.0	10	105	11	110	103	9	90	83
80 + 0.25	16	155						
+0.50	22	210						

[1] Level of body condition change in last 28 days pre-mating.

Table 20.18 Mean daily allowances for ewes in late pregnancy

Liveweight (kg)	*No. of fetuses*	*Weeks pre-lambing* 6			4			2		
		ME (MJ)	*CP (g)*	*MP (g)*	*ME (MJ)*	*CP (g)*	*MP (g)*	*ME (MJ)*	*CP (g)*	*MP (g)*
40	1	6.7	80	68	7.4	95	73	8.2	110	78
	2	7.1	95	72	8.2	120	78	9.5	150	87
60	1	8.8	110	81	9.8	125	86	10.8	140	92
	2	9.4	130	86	10.9	155	91	12.7	185	100
80	1	10.9	135	93	12.1	150	98	13.4	165	104
	2	11.8	155	97	13.7	190	103	15.8	235	112

Table 20.19 Mean daily allowances for lactating ewes (based on yield data from Table 20.13)

Liveweight (kg)	*No. of lambs*	*Month of lactation* 1			2			3		
		ME (MJ)	*CP (g)*	*MP (g)*	*ME (MJ)*	*CP (g)*	*MP (g)*	*ME (MJ)*	*CP (g)*	*MP (g)*
40	1	16.3	200	152	15.9	200	148	13.2	150	125
	2	23.3	300	215	20.3	240	188	15.6	190	146
60	1	20.3	210	183	19.9	205	180	16.7	170	151
	2	27.6	315	249	24.2	250	218	19.0	205	171
80	1	22.3	220	195	21.9	210	192	18.7	190	163
	2	29.6	335	261	26.2	265	230	21.0	220	183

preferably in the form of a material of low degradability such as soyabean meal. For the lactating ewe, protein requirements depend on stage of lactation and number of lambs suckling (see Table 20.19). Here, too, the use of an undegradable protein is advantageous, particularly in situations where ewes are utilizing back-fat reserves. In terms of metabolizable protein (MP), requirements for maintenance may be calculated as $2.57\,g\,W^{0.73}$, that for wool growth may be taken as 20.4 g/d and that for lactation as 72 g/kg milk produced (see Tables 20.17–20.19). Calculation of requirements is complicated. Acceptable approximations (for maintenance and pregnancy) are given in Table 20.18.

Feeding in practice

In practice it may be neither practical nor economically desirable to meet the ewe's requirements in full at all times. Most flocks contain ewes with a range of bodyweights and carrying or suckling varying numbers of lambs. The aim should be to feed accurately at critical times in the production cycle, using the body condition of the ewes as a guide. The need for ewes to be in good body condition, and on a rising plane of nutrition, at tupping has already been discussed. The ewe should not be allowed to drop in condition for the first month of pregnancy, but during the second and third months a moderate degree of underfeeding, leading to weight loss of up to 5%, is acceptable. Forage-only diets, with an energy concentration of 8–9 MJ/kgDM, are adequate at this stage. During late lactation it is essential that ewes be brought back to good body condition (condition score 3–3.5). Pregnancy scanning can allow grouping of ewes according to number of fetuses and differential feeding. Otherwise lowland ewes should be assumed to be carrying twins and hill ewes singles. Concentrate feeding will normally begin, some 6–8 weeks before lambing, at 100 g/head/d. This amount should gradually increase to a maximum of 600 g/d at lambing. With housed ewes and good-quality silage, concentrates may be fed for 4 weeks only, and to a maximum of 400 g/d. At all times the body condition of the ewe should dictate concentrate-feeding level. Forage will usually be offered *ad libitum*.

In many hill situations such feeding is economically more difficult to justify and practically often very difficult, not only because of the difficulty of reaching the ewes but also because it is undesirable to interrupt the normal grazing behaviour of the flock. In such conditions the judicious use of feed blocks may be considered. Such blocks should be sited at strategic points, allowing one block/30 ewes/week. These blocks should be offered from late January and where possible should be supplemented by small amounts of grain feeding in the last few weeks of pregnancy (up to 200 g/head/d). Additionally storm-feeding provision will often be necessary. Adequate quantities of hay, ideally 1 bale/ewe, should be stored on the hill to be fed when grazings are snow-covered.

For the lactating ewe the amount of hand feeding will depend on date of lambing. Where grazing is immediately available, then no supplementary feeds may be needed. In early lambing flocks, root crops may be utilized or forage feeding continued. In these conditions the use of an undegradable protein source will maintain milk yields and lamb growth rates. Under normal circumstances the aim should be to restrict concentrate feeding to 45 kg/ewe/year, with about 0.5 t silage or 215 kg hay/ewe being needed during winter housing. Suggested rations for a 75-kg lowland ewe in late pregnancy are given in Table 20.20.

Grazing management

Good grazing management is one of the most important factors influencing sheep flock profitability. Variable costs will be affected by the extent to which the flock needs supplementary feeding, and by forage variable costs, while gross margin per hectare is very dependent

Table 20.20 Suggested rations for a 75-kg ewe in late pregnancy (kg/d)

Ration	*Silage*	*Hay*	*Swedes*	*Kale*	*Concentrates*[1]	*Barley*
1		0.64	4		0.6	
2		0.7		4		0.5
3	4				0.6	
4		1.0			0.6	

[1] 13 MJ/kg DM; 160 CP/kg DM.

on stocking rate. In 1999, average stocking rate was 12.3 ewes/ha while the top one-third of producers kept 14.5 ewes/ha. These higher stocking rates were achieved at higher levels of nitrogen per hectare (136 versus 114 kg) but similar levels per ewe.

The use of ewes per hectare as a measure of stocking rate can be misleading, since it takes no account of ewe size. In the MLC publication *Prime Lamb from Grass* an attempt has been made to overcome this problem by interrelating ewe weight, prolificacy and fertilizer usage, rather than ewe numbers. Representative values for average-quality land are given in Table 20.21, but individual farm circumstances may markedly affect these values. Although there is a trend towards higher fertilizer N levels leading to higher stocking rates (+10 kg N = 0.3 ewes/ha), many farmers fail to fully exploit this potential. A second problem associated with the recording of grassland usage, especially on mixed farms, is that of allocating variable costs. The simplest way of standardizing figures is to use the livestock units system (see Chapter 4).

The times of the year when problems of high stocking rates are likely to be most apparent are in early spring and in the latter half of the grazing season. In early spring the provision of adequate grazing for the lactating ewe is important in terms of milk yield and lamb growth rates. It is easier to obtain this early grazing if the flock can be kept off the grass over winter. Where sheep are run at grass over winter, the effect on subsequent performance is variable and will depend on season and timing. Up to January there should be no effect on yield, but grazing in January and February will reduce early growth by 20%, or even more in a wet year. On the other hand, most swards have recovered from such an early grazing by silage or hay making. Wherever possible, therefore, sheep should be outwintered on fields destined for conservation.

In-wintering of sheep is becoming more common. More than two-thirds of lowland flocks are lambed indoors but only about 40% are housed throughout the winter. (The question of housing is examined thoroughly in Bryson, 1984.) In-wintering is expensive and unlikely to be justifiable in terms of increased ewe output alone. It can only be defended on grounds of easier shepherding and if grassland utilization is improved as a result.

Table 20.21 Relationship between ewe weight, nitrogen use and stocking rate on average land. [After MLC *Prime Lamb from Grass*, reproduced with permission.]

Targets	*Nitrogen (kg/ha)*		
	0–75	*75–150*	*150–225*
Weight of ewe carried (kg)	750	900	1050
Weight of lamb produced (kg)	600	750	850
Stocking rates to achieve above targets (ewes/ha)			
Welsh halfbred	13	16	18
Mule	11	13.5	15.5
Scotch halfbred	10	12.5	14.5

In the latter half of the grazing season the main problem is competition between ewes and lambs, which can lead to reduced rates and increased parasitic gastroenteritis in the lambs. This situation is most acute on all-grass farms where sheep are the only enterprise, since there is less scope for utilizing silage aftermaths or alternating grazing between species. In such conditions the forward creep grazing of lambs is worth considering and may give benefit of up to 500 g extra growth/lamb/week. Overall, the aim should be to keep the sward-surface height at between 6 and 8 cm, using conservation as a means of utilizing grass not needed for grazing.

Increasing lamb numbers per ewe

As was shown in Figs 20.2 and 20.3, average flock performance is well below what is theoretically possible. Under most hill farm conditions a 110% lambing is considered satisfactory. Where higher lambing percentages are required, manipulation of ewe nutrition, as discussed earlier, can lead to increased numbers of lambs being born. There is then a need to adopt an alternative management strategy.

Early weaning of hill lambs

The main problems associated with twins from hill ewes are loss in body condition in the ewe, which may be severe enough to cause barrenness, and reduced growth rates in the lambs. Removal of one lamb, leaving a female where possible, can overcome these problems.

Weaning at 24 h, once the lamb has taken sufficient colostrum, allows the ewe to return immediately to the hill. The 'weaned' lamb must be fed on milk substitute and concentrates. A higher labour and feed cost, coupled with generally poor lamb performance, make this a difficult system to operate successfully. Where adequate in-by grazing is available, it is more satisfactory to allow both lambs to suckle for 5–6 weeks before weaning one. The ewe may then be returned to the hill while the weaned lamb will either remain at grass, a technique demanding clean grazing and lamb-proof fencing, or be concentrate-fed indoors. This latter system is more likely to produce a finished lamb, with growth rates in excess of 200 g/d, and can improve the utilization of buildings used to house the flock through the winter.

Use of more prolific ewes

Prolificacy in sheep is a trait with low heritability (0.1). The development of a flock of highly prolific ewes will generally involve the use of animals with some Finnish Landrace blood (e.g. the Finn–Dorset) or of the relatively new Cambridge breed. Where such animals are used as breeding females, then lambing percentages in excess of 250% may be expected. As mean lambing percentage in a flock increases, so the number of triplet and quadruplet births increases (see Table 20.22). A high level of stockmanship is necessary in such flocks. Chief features of the management in these situations are as follows:

(1) careful organization of mating to facilitate organized lambing;
(2) generous feeding in late pregnancy;
(3) closely supervised lambing, normally indoors in March/early April;
(4) weaning by mid July to allow ewes to regain body condition prior to tupping.

Table 20.22 Relationship between lambing percentage and distribution of litter size

Mean no. born	*Litter size distribution (%)*				
	1	*2*	*3*	*4*	*5*
1.5	49	50	1		
2.0	12	76	12		
2.5	9	43	41	5	1
2.6	10	36	41	13	2

Additionally, the combination of Finnish Landrace blood and the smaller birthweights of triplet/quadruplet lambs means that careful selection of terminal sires is necessary if lambs are to be finished successfully.

Increased frequency of lambing

Theoretically it is possible for ewes to lamb every 6 months and some may do this. However, ewes may not recover from one lambing in time to be mated again and the problems of operating such a system are many. In particular, low conception rates following the mating of lactating ewes necessitate very early weaning, and failure to conceive greatly increases barrener percentage or spreads the period of lambing to an unacceptable degree. More commonly ewes will be lambed every 8 months and often two flocks will be run, 4 months out of synchronization, so that ewes failing to conceive in one flock may be 'slipped' into the other and so given a second chance (see Fig. 20.10). In a frequent-lambing flock a variety of breeds may be used, but although exogenous hormones or daylight manipulation will generally be used, highly prolific ewes having a long breeding season are to be preferred. In this respect ewes of the Finn–Dorset type are probably ideal.

Although daylength may be manipulated to increase breeding frequency (see Fig. 20.11), the method is seldom commercially viable and exogenous hormones will be used. An intravaginal sponge, impregnated with progesterone, is left in place for 12 days. On its withdrawal an intramuscular injection of 500–750 units PMSG is given and ewes will show signs of oestrus 36–72 h later. A detailed calendar describing such a system is given in Appendix 2.

Critical to such systems are the management of the rams, the close supervision of lambing, the abrupt weaning of the lambs at 1 month of age and the careful nutrition of the ewe. Close supervision of lambing may

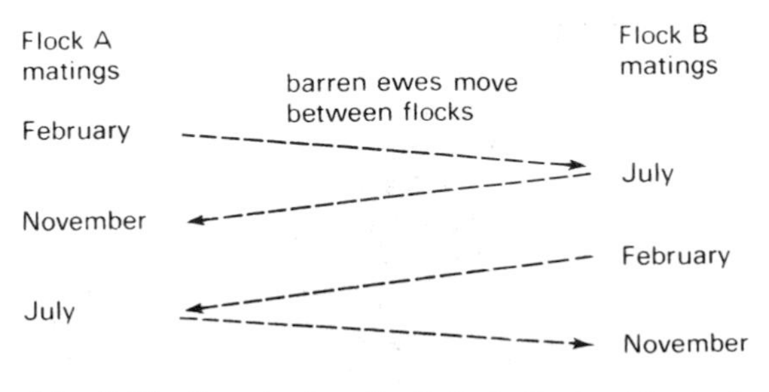

Fig. 20.10 Organization of split flock allowing 'slipping' of animals failing to conceive.

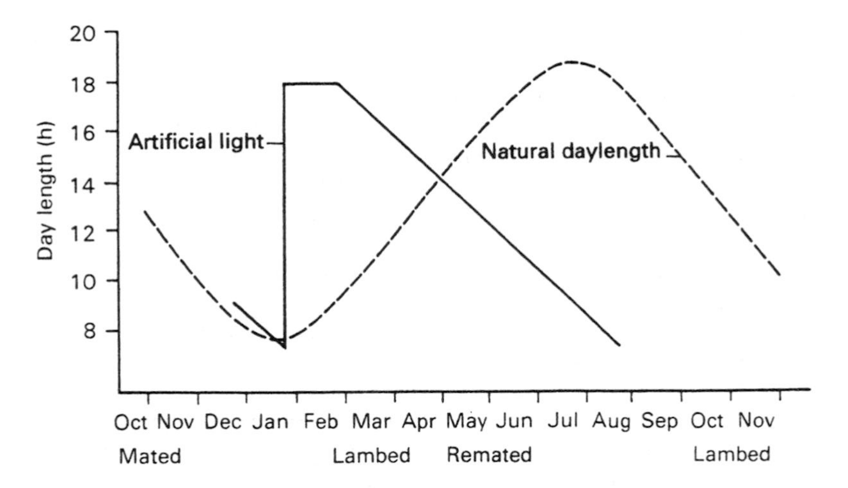

Fig. 20.11 Manipulation of daylength to control oestrus. [From Robinson, J.J. *et al.* (1975) *Annals Biol. Anim. Biochem. Biophys.* **15**, 345, reproduced by courtesy of the editor and publishers.]

be made easier if lambing is induced. The use of corticosteroid injection (e.g. 20 mg dexamethazone) on the evening of day 142 of pregnancy will result in ewes beginning to lamb some 36 h later. Lamb viability and growth rates, and ewe fertility, are not affected by such treatment.

Diagrammatic representation of the nutrition of a frequently lambing ewe is given in Fig. 20.12. Of particular importance is the protein input. In early lactation the inclusion of undegradable protein, such as soyabean meal, facilitates backfat utilization and improves milk yield, while the reduction in total protein at the end of week 3 of lactation speeds the drying-off process and encourages backfat deposition.

The frequent-lambing system is characterized by high variable costs. It has been suggested that an annual production in excess of 2.5 lambs/ewe is necessary for the system to be viable, but the high stocking rates possible with the system (25 ewes/ha) can help compensate for lower margins per ewe. A detailed discussion of this aspect of sheep production may be found in Littlejohn (1977).

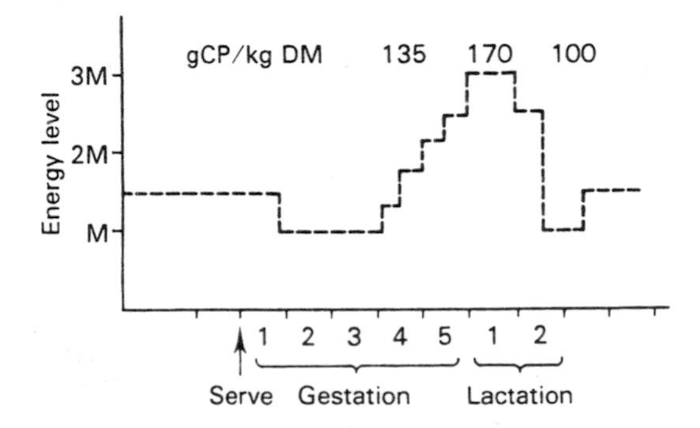

Fig. 20.12 Outline of nutrition of frequently lambing ewe.

Goats

There are more than 714 million goats in the world, largely in Asia and Africa (Table 20.23). Within Europe there are regional differences in the uses to which goats are put. In France and the UK the main output is milk and cheese. Here the most important breed is the Saanen (and its crosses), and average yields are 700 litres/year (France) and 800–1000 litres/year (UK). In Spain and Italy approximately 30% of goats are milk-type, with the remainder dual-purpose or meat-producing, whereas in Greece almost all animals are dual-purpose and milk yields are lower. An outline of main breed types and milk yields is given in Table 20.24.

Goat production in the UK is still in its infancy, with approximately 77 000 breeding does of which some 22 000 are milked. The main output is milk; interest in mohair production grew during the early 1990s but has now waned. There is a small but significant (and growing) interest in cashmere production. Estimated goat numbers, together with average milk production data, are given in Table 20.25.

Milk production

The lactation curve of the goat is similar to that of the cow. Potential lactation yield may be predicted by

Table 20.23 World distribution of goats (×10^6) (1993). [Source: *FAO Production Yearbook* (1998), reproduced with permission.]

Africa	*206*	*Asia*	*396*	*N. America*	*10.2*	*Europe*	*17.8*	*S. America*	*26.0*
Nigeria	24.3	India	117.5	Mexico	8.8	Greece	5.3	Brazil	12.6
Ethiopia	17.0	China	148.4	USA	1.4	Spain	2.6	Argentina	3.4
Somalia	11.0	Pakistan	47.4	Haiti	0.9	France	1.2	Bolivia	1.5
Sudan	37.8	Iran	25.8			Italy	1.4	Peru	2.1
Kenya	7.7	Indonesia	15.2			Russian Fed.	2.1	Venezuela	4.0
Tanzania	10.0	Bangladesh	33.5			Bulgaria	1.0		

Table 20.24 Types and production of goats in Europe

Country	*Main products*	*Breed(s)*	*Yield (litres/year)*	*Annual production (10^6 litres)*
France	Milk 50%	Alpine	770	496
	Meat 50%	Saanen	770	
Greece	Milk/meat	Local breeds	120	460
Italy	Milk 30%	Sarda	150–200	131
	Meat 70%	Maltese	450	
Spain	Milk 30%	Murciana	500–850	317
	Meat 70%	Malaguena	600	
		Delas Mesetas	300	
		Canaria	700	

Table 20.25 Goat breed distribution and lactation data for the UK

Breed	*%*	*Lactation yield (kg)*	*Fat (%)*
Crossbred	53	800+	3.5
Saanen	17	920	3.6
Anglo-Nubian	12	750	4.5
Toggenberg	12	880	3.3
Alpine	5	920	3.5

Table 20.26 Typical composition of milk of different species

	Lactose (%)	*Fat (%)*	*Protein (%)*
Cow	4.5–5.0	3.5–4.0	3.0–3.5
Ewe	5.2–5.5	5.5–11.0	4.5–7.5
Goat	4.0–5.0	3.0–4.5	2.8–3.6

multiplying daily yield at peak by 200; beyond peak, yield may be expected to decline at 2–2.5% per week. In the non-pregnant goat, lactation can continue for up to 2 years. Whilst this will increase lactation yield, annual production is unlikely to equal that of two lactations in the same period. The seasonality of breeding activity in the goat makes it difficult to sustain milk production throughout the year. Many producers overcome this problem by manipulating daylength so that does may be bred in April or May to kid in September and October (see Fig. 20.11).

Table 20.26 shows that the analysis of goat's milk is similar to that of the cow. The fat in goat's milk has a higher content of small globules (28% versus 10% < 1.5 μm) and a higher percentage of short-chain fatty acids (15% versus 9%). It is some of these short-chain acids – caproic, caprilic and capric – that give goat's milk its characteristic flavour. The protein in goat's milk is characterized by smaller casein micelles and by an increased β-casein content (67% versus 43%). The absence of carotene leaves the milk looking very white, although the level of vitamin A is higher than in cow's milk, as is that of nicotinic acid. On the other hand, goat's milk has only 10% of the vitamin B_6 level found in cow's milk and is deficient in folic acid.

There is a specific demand for goat's milk and goat's milk products from those allergic to cow's milk. Estimates suggest that 7.5% of babies and 2.5% of adults may be allergic to cow's milk and that 60% of these are probably not allergic to goat's milk. There is no quota on goat's milk production, but it is now covered by the same Dairy Products Regulations as milk from the dairy cow.

Meat production

In many parts of the developing world goats are an important source of meat. Total production exceeds 2.9 million tonnes. In the EU, production is estimated at 81.700 tonnes, with Greece accounting for 57% of the total and Spain 21.5%. Goat's meat may come from cull adults, from kids weaned and slaughtered at 6–12 weeks or from intensively finished animals – usually surplus males.

On intensive systems, goats tend to have higher dry matter intake (DMI) than lambs of similar weight, but poorer efficiencies. Daily gains of up to 250 g may be expected at FCRs of 4–6 : 1. Killing-out percentages are similar to lamb, but conformation is poorer and carcass composition is different, with goats having half as much subcutaneous fat but almost twice as much kidney fat. There is some evidence that Angora goats may be fatter than other types. Castration will tend to slow down growth rates and increase fat levels.

Fibre production

Most goats have a 'double-coat' fleece with long guard hairs covering finer under-wool. It is this under-wool that is combed out, in February/March, to produce cashmere. Cashmere is 'deciduous' and must be combed out before it falls out and is lost. Two combings, a month apart, may be necessary. Cashmere fibres should be less than 6 cm long and 13–16 μm diameter. Yields vary between breed types and range from 50–250 g. Sixty percent of world cashmere output comes from China (7300 tonnes from 52 million goats). In the Angora goat the under-wool is longer and coarser and is clipped twice yearly, in autumn and spring, to produce mohair. Mohair fleeces can weigh up to 5 kg, with fibres up to 15 cm long and with a diameter of 20–40 μm. Major mohair producers are Turkey, South Africa and Texas.

Table 20.27 Nutrient requirements of housed dairy goats

Yield (kg/d)	*Nutrient*	*Liveweight (kg)*		
		50	*60*	*70*
0	DM (kg)	1.5	1.8	2.1
	ME (MJ)	8.0	9.2	10.3
	DCP (g)	51	59	66
	MP (g)	43	50	56
1	DM (kg)	1.7	2.0	2.3
	ME (MJ)	13.1	14.3	15.4
	DCP (g)	106	114	121
	MP (g)	86	92	98
2	DM (kg)	1.9	2.2	2.5
	ME (MJ)	18.2	19.4	20.5
	DCP (g)	161	169	174
	MP (g)	129	1.35	141
3	DM (g)	2.1	2.4	2.7
	ME (MJ)	23.3	24.5	25.6
	DCP (g)	216	224	231
	MP (g)	171	178	184

Feeding goats

Dry matter intakes, at 80 g/kg $W^{0.75}$, are higher than for sheep, especially in lactating animals. Appropriate metabolizable energy inputs are of the order of 0.4 MJ/kg $W^{0.75}$ for maintenance and 0.7 MJ/kg $W^{0.75}$ during the last 8 weeks of pregnancy. For a 60-kg goat these values equate to 9- and 15 MJ/d respectively. During late pregnancy it is important to monitor body condition to avoid does becoming overfat and thus in danger of suffering from ketosis in early lactation. Mean energy requirement for milk production is 5.1 MJ/litre (6 MJ/litre for Anglo-Nubians). During early lactation it is often impossible to meet energy requirements from the diet. Backfat utilization can provide 2.8 MJ/100 g backfat loss.

The protein requirement of a goat for maintenance is approximately 7 g DCP/MJ ME (equal to 60 g for a 60-kg doe). This amount should be increased by 60 g/d for the last 8 weeks of pregnancy and by 55 g/litre of milk produced (see also Tables 20.27 and 20.28).

In lactating goats water intake is critical. In temper-

Table 20.28 Nutrient requirements of growing goats

Liveweight (kg)	*Nutrient*	*Liveweight gain (g)*			
		50	*100*	*150*	*200*
10	DM (kg)	0.45			
	ME (MJ)	4.5	6.0	7.5	9.0
	DCP (g)	45	55	65	75
	MP (g)	40	55	71	86
30	DM (kg)	1.30			12.3
	ME (MJ)	8.3	9.8	11.3	90
	DCP (g)	60	70	80	95
	MP (g)	53	67	81	
50	DM (kg)	1.50			
	ME (MJ)	11.5	13.0	14.5	16.0
	DCP (g)	71	81	91	101
	MP (g)	66	79	92	105

ate areas intakes of 140 g/kg $W^{0.75}$ plus 1.4 litres/litre of milk (i.e. 4–7 litres/d) may be expected. Goats do not like very cold water and intake may drop significantly in winter, adversely affecting yields. Warm water may need to be provided in such situations. In concentrate-fed goats, reduced water intake can lead to the development of urinary calculi (bladder stones), a condition to which Angoras seem especially prone. Addition of 1% salt or ammonium chloride to the diet (as fed) reduces the danger of this condition developing.

Goats are browsers, but will graze when necessary. In grazing trials comparing sheep and goats, and using grass/clover swards and natural vegetation, goats have been shown to preferentially eat indigenous species such as bent grass (*Molinea caerulea*) and rush (*Juncus* spp.), and to select against clover. This suggests that goats may have an important role in the improvement of hill grazing. On the other hand, problems with internal parasites (anthelmintics may not be used on lactating goats without withdrawing milk from sale) and difficulties with fencing have caused many goat's milk producers to house their animals year-round. The goat's tendency to produce tainted milk if fed on aromatic material is an additional hazard.

The most common feed for housed goats is hay, although good silage may also be used. Maize silage is particularly useful for raising milk yield and quality but needs careful balancing. Goats appear to be more susceptible to listeriosis than sheep and care must be taken when using big-bale silage. Sugar beet pulp is a useful additional feed, especially early in lactation, when its palatability stimulates appetite and its digestible fibre helps maintain butterfat levels. Most producers favour concentrate feeds presented as a coarse mix, but there is no evidence that these are any better than pelleted diets; they are certainly more expensive.

Compared with silage, hay is generally lower in energy and much lower in protein. A switch to silage feeding, where possible, will thus lead to a saving in concentrate requirements. Yearly requirements for a dairy goat are given in Table 20.29. In fibre production the most important consideration is avoidance of nutritional stress, particularly in terms of energy input. Feeding at levels equal to those required by a dairy goat of similar size is unlikely to produce problems in this respect.

Table 20.29 Annual feed requirements for a housed dairy goat (kg)

	Silage (10.5 ME)	*Hay (8.5 ME)*
Forage	3500	700
Beet pulp	110	110
Concentrate	200	450

Reproduction in the goat

Does are seasonally polyoestrous, but less so than ewes. Attainment of puberty depends on date of birth and level of nutrition and is generally at 4–6 months, with full sexual maturity reached at 6–8 months in does and 8–10 months in bucks. Oestrous cycle length is very variable, about at 21 days mean. Oestrus lasts 4–40 h, with ovulation 30–36 h after onset. Unmated does will have three to five cycles, with up to seven recorded for Angoras. Does in oestrus become vocal and restless, showing active male-seeking activity. Tail fanning tends to cause a wet area around the tail. Behaviour is more marked in the presence of an odorous male (the odour being produced by a secretion from a sebaceous gland at the base of the horns). Problems of oestrus detection in goats are less marked than in cattle because seasonality pre-ordains likely breeding periods and because breeding is unlikely to coincide with peak yield. Techniques for synchronizing oestrus, or for stimulating out-of-season breeding, are equally applicable to goats and sheep (though melatonin implants are not licensed for use in goats in the UK).

Gestation length varies between breeds within the range 148–153 days, with Anglo-Nubians tending towards shorter and Saanens towards longer gestations. Litter sizes vary between and within breeds, and are largely a reflection of environment [heritability (h^2) = 0.1]. Mean litter size is approximately 1.75 under UK conditions.

Intersexuality

A well-recognized problem in goats is that of reduced fertility resulting from intersexuality. The condition is related to the presence or absence of horns. Polledness is under the control of a single gene (designated P). Three genotypes and two phenotypes are identifiable:

PP = homozygous polled
Pp = heterozygous polled
pp = homozygous horned

In females, PP genotypes are generally infertile while Pp genotypes have a 50 : 50 chance of being subfertile: these genotypes are indistinguishable. Fewer males are infertile if they are PP genotypes (up to 50%) but almost all will be subfertile. Careful inspection of the vulva of polled kids can often indicate whether or not they are likely to be infertile, the presence of a pea-like protrusion indicating a problem.

Appendix 1
Body condition scores

(0) Extremely emaciated and on the point of death. Not possible to detect any tissue between skin and bones.
(1) Spinous processes prominent and sharp. Fingers pass easily under ends of transverse processes. Possible to feel between each process. Loin muscles shallow with no fat cover.
(2) Spinous processes prominent but smooth; individual processes felt as corrugations; transverse processes smooth and rounder, possible to pass fingers under the ends with little pressure. Loin muscle of moderate depth but little fat cover.
(3) Spinous processes have little elevation, are smooth and rounded, and individual bones can only be felt with pressure. Firm pressure required to feel under transverse processes. Loin muscles full with moderate fat cover.
(4) Spinous processes just detected with pressure, as a hard line; ends of transverse processes cannot be felt. Loin muscles are full with thick covering of fat.
(5) Spinous processes cannot be detected, even with firm pressure; depression between layers of fat where processes would normally be felt. Loin muscles are full with very thick fat cover.

Appendix 2
Calendar for frequent-lambing flock

July
14 Insert sponges.
27 Withdraw sponges. Inject 750 units PMS.
29 Mating. Allow adequate rams (1 : 10) and rotate between paddocks every 8 h or use AI.

August
15 Mate repeats.
20 Remove rams.

December
23 Begin lambing.

January
9 'Repeats' lamb.
27 Wean first lambs. Insert sponges into all ewes.

February
10 Withdraw sponges. Inject 750 units PMS. Wean 'repeats'.
12 Mate.
28 Mate 'repeats'.

March
3 Remove rams.

July
9 First ewes lamb.
25 'Repeats' lamb.

September
20 Wean all lambs.

November
5 Introduce rams (normal mating, no hormonal treatment).

April
1 First ewes lamb.

June
14 Wean all lambs.

July
14 Insert sponges.

Etc.

Further reading

AFRC (1998) *Nutrition of Goats*. CAB International, Wallingford.
Alderman, G. (1993) *Energy and Protein Requirements of Ruminants*. CAB, Wallingford.
Boden, E. (ed) (1991) *Sheep and Goat Practice*. Baillière Tindall, London.
Bryson, T. (1984) *Sheep Housing Handbook*. Farming Press, Ipswich.
Eales, F.A. & Small, J. (1986) *Practical Lambing*. Longman, London.
Gall, C.G. (1981) *Goat Production*. Academic Press, London.
Littlejohn, L. (1977) *A Study of High Lamb Output Production Systems*. Scottish Agricultural Colleges, Aberdeen.
MLC (1983) *Artificial Rearing of Lambs*. MLC, Milton Keynes.

21

Pig meat production

P. Brooks & M. Varley

Introduction

Man has had a special and close relationship with the pig ever since it was domesticated. The pig is one of the few true omnivores in nature and as a consequence it can prosper in many situations and live on many materials that would fail to support other animals. The pig was originally the cottager's animal, kept by rural families to convert transient surpluses and food wastes into high quality protein and much valued fat (lard) that could be used for cooking. The pig was able to live on a huge variety of materials that its human owners could not or did not want to eat. It had the added advantage that once it had been slaughtered, salting, drying, smoking, fermenting or a combination of these processes could be used to preserve the meat of the pig. Thus the food that derived from the family's pig could be rationed out over a long period of time. The pig was seen as a valuable asset by families and played an important part in their food economy. In less developed economies the pig still fulfils such a role.

In Europe, in the Middle Ages, individually owned pigs were often herded as groups in woodlands where they could take advantage of abundant acorns and beech mast to supplement a diet that consisted mainly of vegetable matter, fruits, carrion and whatever they could root up by way of insects, worms and small mammals. The industrial revolution radically changed man's relationship with the pig. Initially, country people migrating to the developing towns took their pigs and dairy cows with them, but public health problems in the sprawling slums rapidly made the keeping of animals untenable. Farming practices had to change to keep pace with differing demands. Aspects of food processing like the production of butter, cheese and bacon that had previously taken place in individual kitchens became the basis of new rural industries. Butter and cheese production in particular produced large quantities of skimmed milk and whey. The cottagers had always fed these by-products to their pigs. Consequently, it was logical that when butter and cheese were produced on a large scale the by-products should continue to be fed to pigs. This led to the development of pig farming in areas where milk by-products were abundant. It was also logical that much of the pig meat should be preserved in some way in order that it could be moved the long distances from point of production to the point of consumption (i.e. the towns). Companies developed to turn the pig meat into preserved products such as bacon, cooked ham and in continental Europe into preserved sausage.

Although the pig industry has changed radically in the last 200 years much of its success is still based on the same two factors that made the cottager's pig such a valuable asset; first, the ability of the pig to produce high quality human food from the low-value residues resulting from food processing, and secondly, the suitability of pig-meat for preservation and processing.

Trends in the global market for pig meat

When viewed on a world scale pig meat is still by far the most important meat commodity (Fig. 21.1). The growth in poultry meat production since the Second World War has been dramatic. Nevertheless, it is notable that pig and poultry meat production have been increasing in parallel in recent years. The popularity of pig meat comes from the fact that in addition to producing fresh meat products suitable for roasting and grilling, it has extremely good manufacturing properties. Pig meat

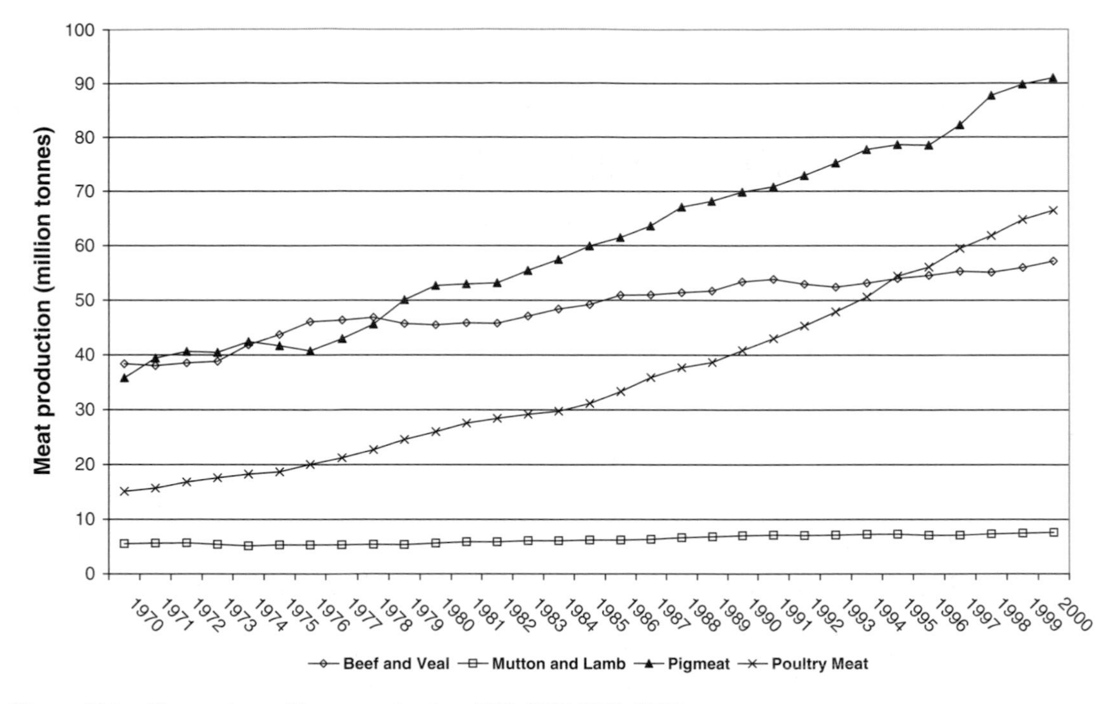

Figure 21.1 Changes in world meat production, 1970–2000 (FAO, 2001).

can be found in a huge range of manufactured products ranging from bacon to salami, Parma ham to pâté. It is this huge diversity of form in which pig meat can be presented to the consumer that results in a continuing and growing demand for the product. In contrast, sheep meat does not lend itself to manufacturing and consequently demand for the product is sluggish.

The growth in pig meat production/consumption has varied in different parts of the world and in different countries. Generally, growth in consumption has followed trends in both population and economic growth. There has been a repeating pattern that as countries industrialize there is a growth in the demand for pig and poultry meat. However, there are important ethnic differences. Traditionally, pig meat has had an important place in the cuisine of European and Asian countries (Table 21.1). In the past, countries with large grassland resources and relatively low population densities (e.g. the USA, Brazil, Argentina, South Africa, Australia) have tended to eat more beef. However, concerns about the impact of red meats on human health, and particularly the linkage of saturated fat consumption to coronary heart disease, has increased the demand for pig and poultry meat in these countries. Pork is an insignificant component of the diet of people in many African and Middle Eastern countries where Islam is the predominant religion, and pork consumption is not permitted. In these countries increased affluence has resulted in very rapid growth of their poultry industry.

The most significant growth in pork production and consumption has been in China and the developing economies in the Far East (Fig. 21.2). This has been a function both of population growth and increasing per capita consumption. Given the continuing development of these economies it can be anticipated that demand for pork will continue to grow.

As the data in Fig. 21.2 show, pig meat production has continued to increase in the EU despite a slowing in population growth. There have been considerable changes in the EU pig industry in the last decade. The most notable has been the increase in pork consumption in the Mediterranean countries of the EU (Fig. 21.3). Both production and consumption of pork have increased dramatically in Spain and the most recent data suggest that Spain now has the highest per capita consumption of pork in the EU and has become the second largest producer of pig meat (Table 21.2). In stark contrast, the UK now has the lowest per capita consumption of pig meat and produces less than 6% of the EU total. There is considerable intercommunity and extracommunity trade in pig meat. Traditional exporters such as Denmark, the Netherlands and

Table 21.1 Meat consumption (kg per capita) in selected countries (1996) (Food and Agriculture Organization, 2001)

	Pig	*Bovine*	*Poultry*	*Total meat*	*Kg beef consumed per kg pork*
Austria	66.2	21.7	15.1	103.0	0.33
Denmark	65.3	19.1	16.0	100.4	0.29
Germany	53.5	16.5	13.6	83.6	0.31
The Netherlands	50.9	18.3	20.6	89.8	0.36
Hungary	49.8	6.0	23.5	79.3	0.12
Czech Republic	49.7	15.4	14.8	79.9	0.31
Poland	49.4	9.9	10.2	69.5	0.20
Portugal	38.4	13.6	21.7	73.7	0.35
Belgium–Luxembourg	38.2	22.3	21.2	81.7	0.58
Italy	36.0	24.1	18.5	78.6	0.67
France	35.3	27.6	25.6	88.5	0.78
Finland	32.6	19.1	10.3	62.0	0.59
Ireland	32.6	13.1	26.0	71.7	0.40
USA	28.0	43.8	44.6	116.4	1.56
Canada	27.1	33.1	30.7	90.9	1.22
Romania	26.4	7.7	13.1	47.2	0.29
China	26.3	2.9	7.4	36.6	0.11
Greece	24.4	20.6	15.8	60.8	0.84
UK	24.1	15.1	26.0	65.2	0.63
Australia	18.4	42.2	26.8	87.4	2.29
Japan	17.9	10.3	14.7	42.9	0.58
New Zealand	16.9	38.9	24.3	80.1	2.30
Malaysia	13.9	5.4	32.0	51.3	0.39
Mexico	10.3	15.3	16.2	41.8	1.49
Brazil	9.3	37.5	22.1	68.9	4.03
Thailand	8.6	5.5	14.3	28.4	0.64
Uruguay	8.2	56.5	12.5	77.2	6.89
South Africa	3.3	12.7	12.5	28.5	3.85
Indonesia	3.0	2.1	4.7	9.8	0.70

Belgium still produce considerably more pig meat than they consume. Germany and the UK continue to be big importers.

The factors underlying the changes in production and consumption trends are interesting. The growth in consumption in the Mediterranean countries of the EU reflects the improved economic environment that has resulted from EU membership. The increased production in Spain has a number of causes. Spain is a country that has a considerable land mass and in rural areas a relatively low population density. Thus it has been easier for developers to build large new units without causing environmental problems or provoking an adverse reaction from local inhabitants. In addition, the low pig density has been attractive to investors as the risk of disease transfer between units has been less than in the traditional, pig-dense areas in Europe. An improved infrastructure, high cereal production capacity and a climate that minimizes heating costs in pig buildings have also been important factors.

Many traditional pig-dense areas on the western seaboard of Europe have come under increasing pressure in recent years. These include the Netherlands (which has the highest livestock density in Europe), parts of Denmark, Belgium, Brittany and Humberside in England. Increasingly draconian environmental legislation has made it less profitable to produce pigs in these areas. In addition, non-farming inhabitants in these areas have made it increasingly difficult for producers to obtain planning permission to build new or extend old units. The situation is most acute in the Netherlands where the government has determined that the environmental issues can only be resolved by a 20% reduction in the size of the pig herd. This move has resulted in outward investment by producers to areas of Europe that are less regulated and less densely

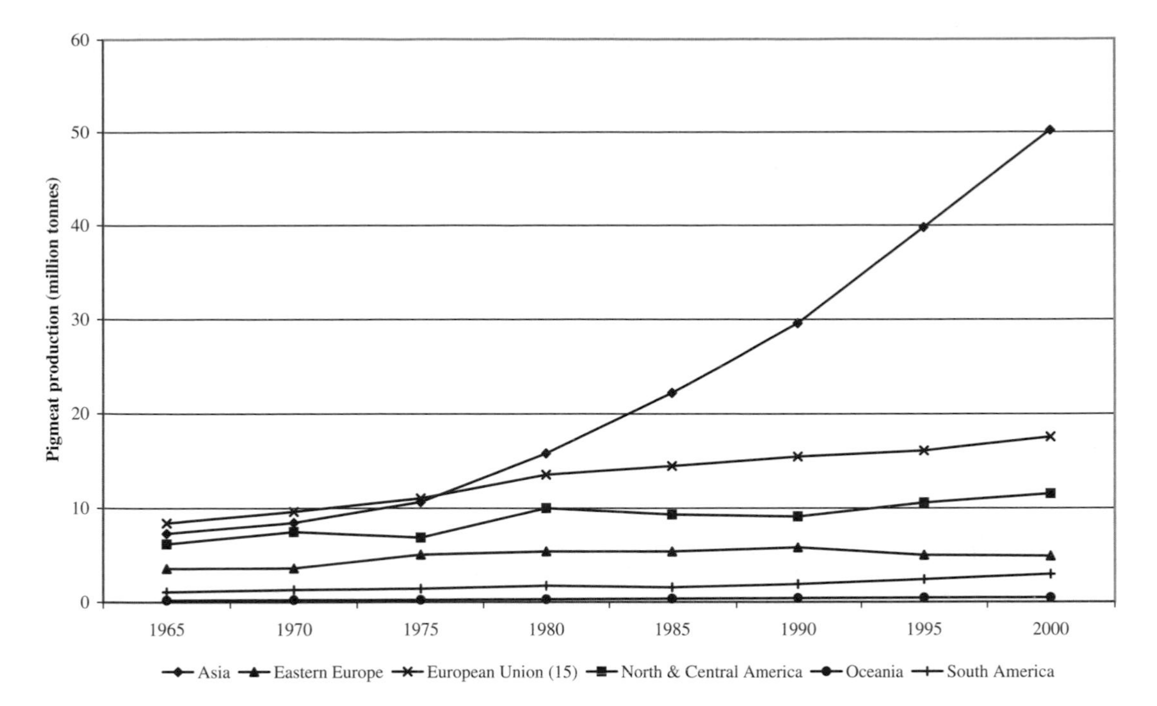

Figure 21.2 Regional changes in pork production, 1970–2000 (FAO, 2001).

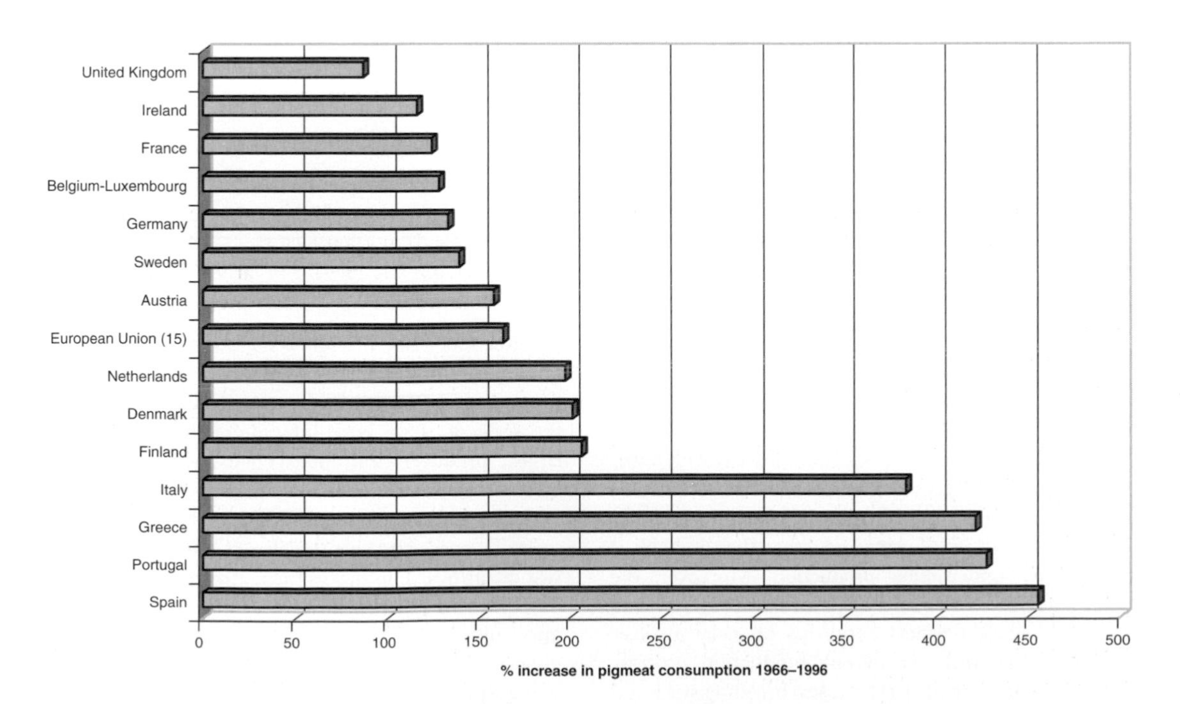

Figure 21.3 Percentage change in pork consumption in EU countries, 1966–1996 (FAO, 2001).

Table 21.2 Pig meat production, consumption and self-sufficiency in the EU (1999) (MLC, 2001)

Country	*Production ('000 tonnes)*	*Production as % of EU total*	*Pig meat consumption (kg per head)*	*Self-sufficiency (%)*
Spain	2892	16.0	66.1	112
Denmark	1642	9.1	65.8	490
Austria	520	2.9	57.8	107
Germany	4113	22.8	56.8	85
Belgium/Luxembourg	1005	5.6	44.6	222
Portugal	344	1.9	43.5	77
The Netherlands	1711	9.5	41.6	283
France	2378	13.2	37.2	106
Sweden	325	1.8	36.8	101
Irish Republic	251	1.4	36.3	190
Italy	1472	8.2	35.1	67
Finland	183	1.0	34.4	103
Greece	138	0.8	32.3	41
UK	1048	5.8	23.0	77
Total	18 022			

Table 21.3 The ten largest pork producers in the USA (Freese, 2001)

2001 Rank	*Name of operation*	*Sows (2001)*	*Sows (2000)*
1	Smithfield Foods	710 000	695 000
2	Premium Standard Farms	211 000	201 000
3	Seaboard Farms	185 000	175 000
4	Triumph Pork Group	140 000	—
5	Prestage Farms	122 000	122 000
6	SMS of Pipestone	120 000	—
7	Cargill	109 675	109 000
8	Tyson Foods	107 000	110 000
9	Iowa Select Farms	100 000	96 000
10	Christensen Farms	80 500	74 000
	Total	1 885 175	

populated like Poland, East Germany and Hungary. In the short term, it has also resulted in the export of weaned pigs from the Netherlands to Spain where they are grown to slaughter weight.

The movement of pig production to areas with lower human population density and less regulatory control is a global trend. Pig meat is a globally traded commodity and, increasingly, agribusinesses rather than individual farmers are undertaking production. Moreover, the trend is for the pork chain to be integrated. That is to say that a single business entity has financial involvement in and control over not only production, processing and wholesaling of pork, but also breeding stock and feed provision. In this respect the pig industry is becoming ever more like the poultry industry. This trend is most apparent in the USA where the 'top ten' pork producers account for 1.8 million sows (Table 21.3). To put the scale of these pork integrators into context, the largest producer Smithfield Foods/Murphy controls more sows than the entire UK pig industry.

Such operations can gain considerable economies of scale. They are also able to deliver uniform products in vast quantities. This makes them attractive suppliers for multiple retailers and in export markets where uniformity and consistency of supply are of paramount importance. Large processors also gain efficiency by selling specific parts of the pig carcass and/or particular pork products to different markets, which may be in various parts of the world. It is often the price that can be realized for the cheapest parts of the carcass, rather than for the most expensive cuts, that determines profitability.

Corporate farmers have no geographical allegiance and will operate where resources such as land, labour and feed inputs are cheapest and where regulations are most easily and cheaply met. It used to be necessary to produce pig meat close to markets; however, this is no longer such an important constraint. It is notable that Taiwanese investors have decided to invest heavily in pork production in Canada, where grain is abundant and there is plenty of land for the disposal of effluent, rather than continue importing grain to Taiwan and producing the pigs there where there are significant environmental and animal health problems due to the high pig density. The only significant exceptions to the growing domination of the agribusinesses are the countries, like Denmark, that have a long tradition of well-disciplined farmer cooperatives. Although still based on small farmer-operated units, pig production in Denmark is organized to produce different types of slaughter pig to satisfy the domestic market and export markets in Japan, the UK and Germany.

As noted above, the UK pig industry has had the lowest growth rate over the last 30 years and the per capita consumption of pork is the lowest in Europe. There are a number of reasons for this. Historically, the UK has never been self-sufficient in pig meat. The UK imported pig meat, in the form of processed product such as bacon and ham, mostly from Denmark, but also from the Irish Republic and the Netherlands. For many years health controls protected the UK industry from competition in the fresh meat market. However, accession to the EU changed this and opened up the UK market to foreign competition in the fresh pork sector as well. The low-cost producers in the Netherlands and Belgium took advantage of this situation. In the 1990s the UK pig industry was put at a further competitive disadvantage as a result of the UK government unilaterally taking the decision to ban sow stalls. This decision added considerably to production costs as producers were forced to change their housing systems to comply with the new regulations. This move will continue to disadvantage the UK industry until the EU welfare regulations bring the remainder of Europe in line with UK regulations (planned for 2012). The industry was further disadvantaged by the major UK retailers demanding Quality Assurance standards that imposed more expensive production practices on their UK suppliers; like insistence on the provision of straw bedding rather than the use of totally slatted floors. These requirements increased production costs in the UK compared with foreign competitors. Although UK retailers claimed that they required similar standards from overseas suppliers, there was evidence that this was not always the case. In particular, ensuring that pork included in processed products has been produced to specific standards is particularly difficult. The final factor that influenced the competitiveness of the UK industry was Britain staying out of the Euro zone. A strong pound meant that foreign competitors had an advantage in the UK market.

The impact of all these factors can be seen in the change in gross margins of producers (Table 21.4). The combination of declining profitability and uncertainty about the future regulatory framework led to a progressive decline in the UK sow herd (Fig. 21.4). At the time of writing in 2001, the UK industry lacked the confidence to invest in the restructuring that will be necessary to maintain a viable pig production industry. The structure of the UK industry militates against its viability in the future. Unlike the Danes or the Bretons, the UK producers have never had a tradition of strong and disciplined cooperatives to control and market their products. Conversely, environmental regulations and planning constraints in the UK make it extremely unlikely that pork integrators will take over UK pork production. There are real fears that the UK pig meat industry is declining to a size that will not support the necessary infrastructure to ensure its survival in the twenty-first century. It is possible that UK producers may be left supplying a relatively small market for niche products while imported products will supply the major commodity market for pork and pork products. As the WTO reduces the barriers to trade, these products may come not only from our EU partners but also from low-cost producers in Brazil, Thailand, the USA and Eastern Europe.

Table 21.4 Gross margins (£) for breeding and feeding herds (1992–1999) based on pigs with an average sale weight of >85 kg (MLC, 2000)

	1992	*1993*	*1994*	*1995*	*1996*	*1997*	*1998*	*1999*
Gross margin per pig	27.63	20.11	14.58	27.50	45.88	23.75	10.82	6.87
Gross margin per sow	450.50	367.88	224.77	473.40	853.22	448.24	191.41	117.09

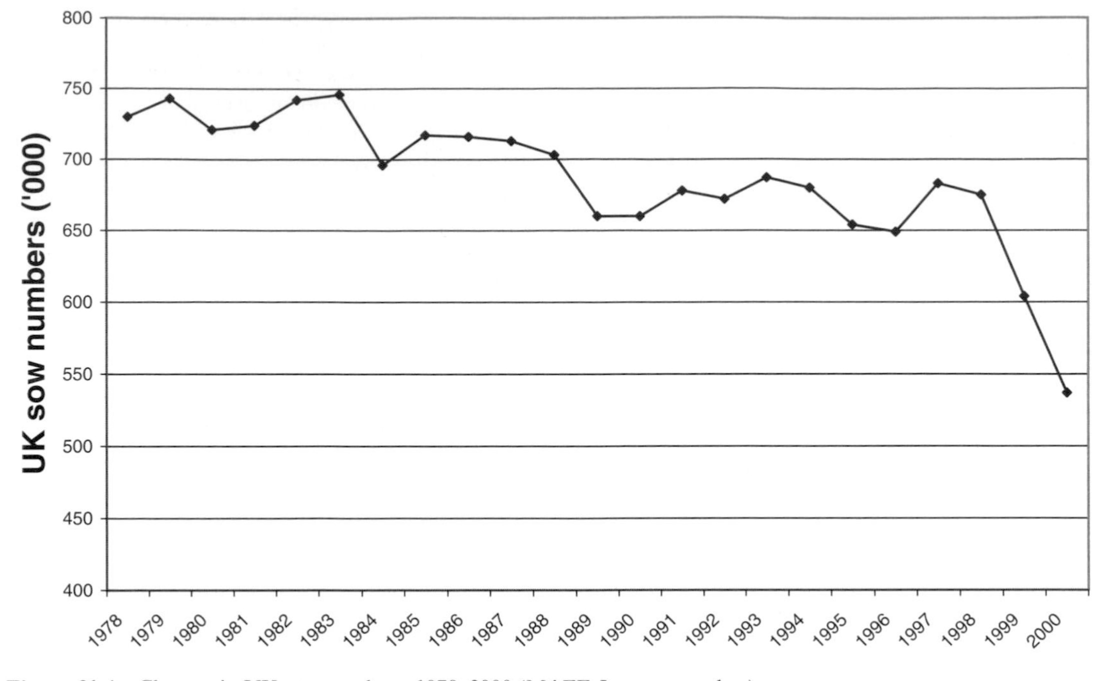

Figure 21.4 Changes in UK sow numbers, 1978–2000 (MAFF, June census data).

Table 21.5 Regional distribution of pigs in the UK (2000) (DEFRA 2002)

	Percent of national pig herd
North-east	1.8
North-west	5.3
Yorkshire and Humberside	29.5
East Midlands	9.8
West Midlands	6.5
East	25.6
South-east	8.5
South-west	13.0

Structure and performance of the UK pig industry

The structure of the UK pig industry has altered significantly since the end of the Second World War as a result of social and economic pressures. In the early 1950s, pigs were considered a secondary enterprise on mixed farms. Herds were small and only a small number of specialist producers existed. Arable farms growing their own cereals could process milled barley and wheat into pig meat, straw was freely available for bedding and manure disposal was not a problem using cereal stubbles in the autumn months.

Today, the pig industry is in relatively fewer hands and there is a continuing trend towards specialist production, although as pig units have increased in size the large-scale arable farmers of the eastern counties of England have found it easier to expand with minimal slurry disposal problems and readily available straw. There has been a gradual shift of production to the east of Britain partly because of the drier climate but mainly because of the logistics involved in the production, processing and transport of the raw materials involved such as corn and straw (Table 21.5). In 1964 around 75 000 holdings in the UK had sows on them. By the middle of the 1970s there were only about 22 000 holdings in the UK with pigs and by the year 2000 this had dropped to 12 416 holdings. More importantly there are less than 8500 units with sows on them and the industry is dominated by 14.1% of these units (1196) which have in excess of 200 sows. These units account for 75.2% of the nation's pigs (Table 21.6). In reality the 52% of farms with less than 10 sows must be regarded as hobby farms and of no commercial significance. In the current economic climate it is extremely doubtful whether a unit with less than 100 sows is financially viable.

Although the UK is a net importer of pig meat, it also exports pig meat to other European countries

Table 21.6 Holdings by total number of sows in herd (DEFRA, 2002)

Number of sows	*Number of holdings*	*%*	*Number of pigs*	*%*
1–10	4378	51.7	14 240	1.8
10–20	757	8.9	10 136	1.3
20–50	802	9.5	25 723	3.2
50–100	626	7.4	45 768	5.7
100–200	714	8.4	101 356	12.7
200+	1196	14.1	599 417	75.2
Total	8473	100.0	796 640	100.0

Table 21.7 UK imports and exports of pig meat ('000 tonnes) (MLC, 1998)

	Pork		*Bacon and ham*	
	Imports from	*Exports to*	*Imports from*	*Exports to*
Total	151.4	247.4	232.6	6.6
Denmark	56.0	9.5	106.4	
Irish Republic	32.0	10.5	6.9	4.4
France	24.7	24.1	4.5	
The Netherlands	21.0	41.1	107.2	
Belgium/ Luxembourg	5.9	13.9		
Germany	7.8	81.3	5.3	0.3
Other EU	1.3	31.6	2.3	1.5
Non-EU		32.7		

Table 21.8 Performance of UK breeding herds (1999) (MLC, 2000)

	Average	*Top 1/3*	*Top 10%*
Average sow herd size	354	364	369
Sow replacement rate (%)	39.6	43.3	35.6
Successful services (%)	87.1	90.5	91.2
Litters per sow per year	2.25	2.34	2.39
Non-productive days	37	26	18
Total pigs born per litter	11.96	12.38	12.36
Pigs reared per litter	9.79	10.26	10.50
Pigs reared per sow per year	22.0	24.0	25.1
Average weaning age (days)	25	24	24
Sow feed per pig reared (kg)	67	62	65
Feed cost per pig reared (£)	7.14	6.68	7.22

Table 21.9 Comparison of results for indoor and outdoor breeding herds (1999) (MLC, 2000)

	Indoor	*Outdoor*
Average sow herd size	254	626
Sow replacement rate (%)	40.5	38.4
Successful services (%)	82.7	93.7
Litters per sow per year	2.26	2.23
Non-productive days	35	42
Total pigs born per litter	12.20	11.67
Pigs reared per litter	11.90	9.50
Pigs reared per sow per year	22.10	21.80
Average weaning age (days)	25	24
Sow feed per pig reared (kg)	61	73
Feed cost per pig reared (£)	6.63	7.72

(Table 21.7). However, the tonnages do not reflect the relative value of imports and exports, as much of the pork exported consists of cheaper cuts and cull sow carcasses, which are sold to other EU countries for incorporation into continental sausage and pâtés.

Although technical performance is generally good, there remains a great variation in performance between individual herds. The top 10% of producers rear about three pigs more per year than average producers despite having a similar weaning age and similar feed cost per pig reared (Table 21.8). It is apparent that there is no one overriding factor that is affecting performance. Rather the better performing herds achieve their improved performance by a combination of small improvements to a number of components that contribute to overall productivity. In recent years there has been a growing trend for sows to be kept outdoors. Although there are no official figures it is estimated that approaching 30% of UK sows are now housed outdoors. Outdoor production used to be regarded as a low-cost low-output system. However, the management of outdoor units is now of a very high standard and results of outdoor units now compare favourably with those achieved in indoor units (Table 21.9). As capital costs are much lower on outdoor units it follows that these can be very profitable enterprises.

Typical performance of pigs in the rearing stage (i.e. the period immediately following weaning) is shown in Table 21.10. Although there are some differences in pig performance between the average and the best producers it is apparent that the price that producers pay for their feed is one of the biggest factors contributing to differences between units. This is also true in the case of feeding or finishing herds (Table 21.11).

Table 21.10 Rearing herd performance (1999) (MLC, 2000)

	Average	*Top 1/3*	*Top 10%*
Average number of pigs	1377	1899	913
Weight of pigs at start (kg)	6.8	6.8	6.8
Weight of pigs produced (kg)	35.3	39.4	36.2
Mortality (%)	2.4	2.0	2.3
Feed conversion ratio	1.75	1.66	1.54
Daily gain (g)	469	513	514
Feed cost per tonne (£)	181.19	166.99	163.45
Feed cost per pig reared (£)	9.05	9.01	7.38

Table 21.11 Feeding herd performance (1999) (MLC, 2000)

	Average	*Top 1/3*
Average number of pigs	1377	1671
Weight of pigs at start (kg)	21.8	23.6
Weight of pigs produced (kg)	91.3	93.4
Mortality (%)	3.5	2.8
Feed conversion ratio	2.63	2.60
Daily gain (g)	630	659
Feed cost per tonne (£)	121.04	105.30
Feed cost per pig reared (£)	22.10	19.08
Average sale value (£)	50.83	51.05

Pig housing and animal welfare

Pig production requires specialist buildings. Each class of pig has different environmental and space requirements and to meet these the pig unit needs a range of different buildings. From the 1960s until the end of the century producers developed 'confinement systems'. In such systems pigs are totally housed in controlled environment buildings, although it has to be said that the degree of environmental control actually achieved is very variable. These systems make management more predictable and reduce labour demands. They have enabled the construction of very large pig units with a very high pig density. This can be both a blessing and a curse in terms of manure management. Large units produce very large volumes of effluent and can also result in aerial pollution. Conversely, if well managed, keeping a large number of animals on one site can justify the expenditure in more sophisticated effluent management systems. For a number of years the trend was for units to both produce weaner pigs and grow them to slaughter weight on the one site (often referred to as farrow to finish units). Such units avoid many problems with transporting pigs but have the disadvantage that because they have pigs of all ages on the unit there can be a continuous recycling of disease.

In recent years, many larger units have reverted to 'multi-site' production. This means that different parts of the production process are conducted in different buildings on different sites. Multi-site production enables buildings to be operated on an 'all in all out' principle (Fig. 21.5). For example, a nursery may be completely filled with the pigs weaned in a single week and completely emptied 4 weeks later. Depopulation of the entire building enables thorough cleaning, disinfection and drying before the next batch of pigs enter. This process prevents the buildup of disease and prevents one batch of pigs contaminating the next. Although it is possible to practise an 'all in all out' policy for rooms within quite a small pig unit this approach is much easier to operate in large integrated units or in co-operatives. A further advantage is that it allows staff to become expert at managing a particular class and age of stock.

Gestating sows

Throughout most of the world, a majority of sows are still confined in individual stalls during gestation and in farrowing crates during lactation. In these the sow has sufficient room only to stand up and lay down. Initially, the main reason for adopting this form of housing was that it stopped sows from bullying each other and enabled them to be individually fed and watered. Such housing was cheaper to construct and also made it easier to manage the sows. This was particularly the case with lactating sows where it is necessary for the stockperson to have easy access to the piglets without the risk of being attacked by the sow. Confinement systems also made it easier to automate routine tasks like feeding and the removal of effluent, so each person on the unit could manage more sows. Although gestation stalls reduce fighting by sows this is at the expense of placing very severe restrictions on sow movement and behaviour and often causing other forms of damage by not allowing the sow to exercise. In the UK animal welfare organizations orchestrated a very effective campaign against sow stalls and gained sufficient public support that the practice has been outlawed since 1999. As noted earlier, this has placed the UK producer at a competitive disadvantage compared with other European producers. Within the EU only Sweden has adopted similar regulations.

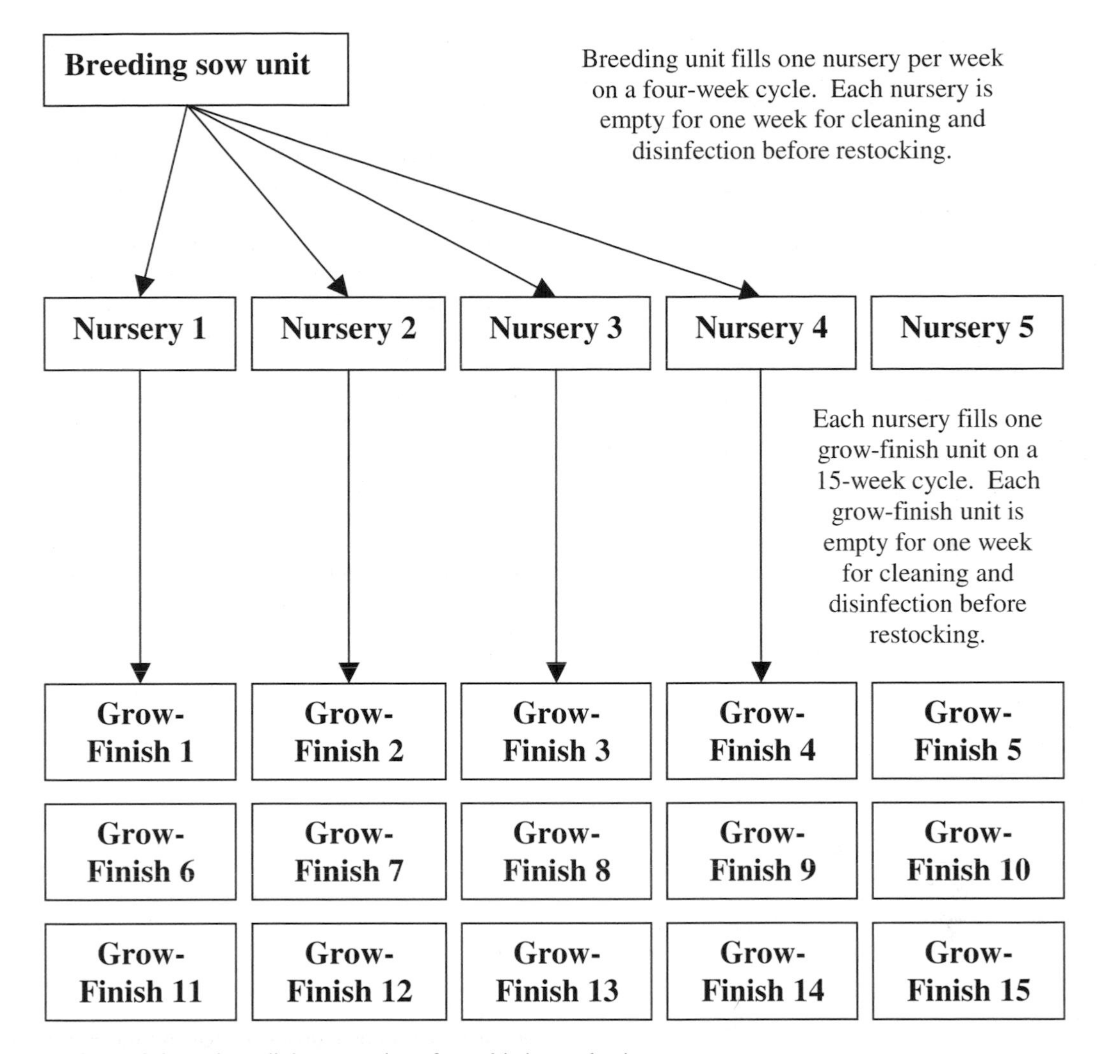

Figure 21.5 Schematic outlining operation of a multi-site production system.

However, it is pleasing to note that consumer pressure is forcing at least some producers in the rest of Europe to adopt similar standards to the UK in order to continue supplying the UK market. In the rest of the world animal welfare has not become a significant concern with consumers and as a result confinement housing is still the norm and vast new pig production facilities are still being built incorporating gestation stalls.

Effective, alternative sow housing systems have been developed in the UK. At one extreme producers have responded by adopting very low-technology outdoor housing systems for sows. At the other extreme they have used new technology to produce systems in which sows are housed in groups and fed using computer-operated feed stations (electronic sow feeders). These feeding stations recognize individual sows when they enter, deliver an individually determined ration and protect them while they consume it. Such systems work best where generous quantities of straw bedding are used to produce a deep litter lying area. If the unit is well designed it should not be necessary to remove all the bedding at regular intervals; merely cleaning soiled material from the edge of the bed should suffice. In loose housing systems sows should be allowed at least 1.5 m^2 of lying area per sow. Well designed and managed systems using group housing, straw bedding and electronic sow feeders probably represent the highest standards of welfare currently afforded to sows.

The average consumer believes that keeping sows outdoors allows them a more 'natural' life. However, it does subject the animals to the vagaries of the British climate and when sows are up to their bellies in mud during the winter or suffering from sunburn in summer this can hardly be said to be protecting their welfare. On ideal sites, that is sites that are flat and freely draining but have soils with sufficient holding capacity that they are not easily turned to mud, stocking rates of up to 25 sows per hectare can be achieved. Outdoor systems integrate well with large arable units where the sows can fit in well in the rotation. Outdoor pigs can be a very effective 'cleaning crop' saving on one or more applications of herbicide and/or cultivations. They also leave residual nutrient value in the ground that can be used by a subsequent crop. Conversely, there is a danger that outdoor sows may pollute watercourses with their excreta if they are kept on ground with little or no plant material capable of taking up the nutrients that they produce.

Lactating sows

On indoor units, sows are usually moved to farrowing crates about 3–7 days before parturition, to allow them to acclimatize to their new surroundings. Farrowing crates are designed to minimize the loss of newborn piglets due to overlying by the sow in the early days after farrowing. This is the major cause of death in the neonatal period. Farrowing crates commonly include a 'creep' area where piglets have access to a supplementary dry food and a heat lamp to keep them warm. The crate is necessary to restrain the sow and prevent her overlying her piglets. Welfare groups have now turned their attention to the use of farrowing crates, which they consider to be unnatural. Despite a considerable amount of research in recent years no viable alternative to the farrowing crate has yet been devised. Certainly for the first 3–4 days post farrowing a crate is necessary to protect the piglets from damage or overlying by the sow. However, some older farrowing crate systems are giving cause for concern. For many years geneticists have been breeding pigs with a higher lean growth potential. A consequence of this is that the mature size of the sow has been increasing and in some cases they are becoming too large for their crates. Suggested crate lengths for sows of different weights are given in Table 21.12.

In Europe designs are now emerging for farrowing houses that allow the sow to be confined for 3–4 days and then for the crate to be opened up, allowing the sow some freedom of movement within the pen. However, these have not yet gained wide acceptance and they generally increase housing costs and labour input.

Growing pigs

To perform at maximum efficiency the growing pig needs to be maintained within its thermo-neutral zone, that is at a temperature above which it is not having to divert additional food energy away from production for body temperature maintenance (lower critical temperature, LCT) and below the temperature at which it has to limit feed intake to avoid problems of having to dissipate heat. In general, the LCT increases as the pig increases in size. This is because pigs produce heat in proportion to their mass (weight) and lose heat in proportion to their surface area. As pigs grow their surface area to mass ratio declines so proportionally they lose less heat. A guide to the temperature requirements of pigs is given in Table 21.13. However, any such guidelines must be treated with caution as the LCT is influenced not only by the weight of the pig but also by air speed, humidity, the insulating value of the floor on which the pig rests and the ability of the pig to huddle together with other pigs and reduce its surface area. Growing-finishing pigs are normally kept in groups and as they grow they need proportionally more floor space

Table 21.12 Suggested length of farrowing crate in relation to sow weight

Sow weight (kg)	*Length of crate (mm)*
150	1550
200	1700
250	1850
300	1950
350+	2300

Table 21.13 Suggested minimum temperatures for different classes of pig. Temperature ranges are given as fed intake, air speed, floor type and group size all affect requirements. Perforated floors and low feed intakes increase the temperature requirement while the provision of solid floors, bedding and high feed intakes will reduce the temperature requirement

Category of pig	*Temperature*	
	°C	*°F*
Sows	15–20	59–68
Suckling pigs in creeps	25–30	77–84
Weaned pigs <6 weeks	27–32	81–90
Weaned pigs >6 weeks	21–24	70–75
Growing pigs 30–60 kg	15–21	70–75
Finishing pigs 60–90 kg	13–18	55–64
Finishing pigs >90 kg	10–15	50–59

(Table 21.14). In order to produce pigs of more uniform carcass composition and to limit the deposition of fat, pigs may be rationed and fed a number of discrete meals each day in a trough. This approach is more commonly adopted if pigs are fed liquid diets, which can be delivered using automated and computer-controlled systems. Throughout Europe as a whole about 30% of growing-finishing pigs are fed on liquid diets. If pigs are fed discrete meals, in troughs, all the pigs in the pen must be able to eat at the same time. As they increase in weight they need more trough space per pig (Table 21.15). Where pigs are of high genetic merit and/or are sold on a contract that does not have a demanding back fat thickness requirement, the pigs may be fed *ad libitum*. In this case pigs will be fed using either a multi-space or a single space *ad libitum* feeder. Guidelines for feeder space allowance are given in Table 21.16.

Water is the most important nutrient given to the pig as it enables the pig to maintain mineral homeostasis, excrete the waste products of protein digestion (as urea) and also helps the pig to control its body temperature. Whether pigs are fed on dry or liquid diets, rationed or fed *ad libitum*, they must be provided with a separate supply of clean water. Where pigs are restrictively fed it is recommended that they should be provided with one drinker point per 10 pigs. This can be increased to one drinker per 15 pigs if they are fed *ad libitum*. It is important that drinkers deliver a sufficient flow of water (Table 21.17) as pigs will only spend a limited amount of time drinking and if they do not obtain sufficient water during that period they will reduce their feed intake and hence their growth rate.

Ideally, the pen in which pigs were housed would grow with them in order to accommodate their changing needs. As this is not feasible growing pigs usually move through two or three different sizes of pen during their lifetime. Every few years some producers try to overcome the need to move pigs by trying to devise pen and building designs that will accommodate the needs of the pig from weaning to market weight. Unfortunately, until now all attempts to do this have foundered, as they inevitably require a compromise at some stage, which reduces productivity.

One of the most contentious issues in modern pig production is whether pigs should be provided with straw bedding or whether they can be kept on totally perforated floors without compromising their welfare. Straw provides a substrate for the pig to root in, investigate and chew. It also adds comfort to a lying area and if the pig is fed below its voluntary feed intake limit it can also provide gut fill. On the debit side, straw can carry disease (from bird and vermin droppings), produces dust and carries fungal spores that can exacerbate respiratory disease. Straw-based systems generally have higher operating costs. Supporters of systems using perforated floors point out that such floors separate pigs from their excreta, keeping them cleaner and reducing disease transfer. Straw and solid floored systems can be problematic under high ambient temperature conditions. Pigs do not perspire and can only take advantage

Table 21.14 Floor space allowances for growing-finishing pigs (Farm Assured British Pigs, 1998)

Average liveweight (kg)	*Total area (kg)*
10	0.15
10.1–20	0.20
20.1–30	0.30
30.1–50	0.40
50.1–85	0.55
85.1–110	0.65

Table 21.15 Feeder space allowances for growing-finishing pigs of different weights (Farm Assured British Pigs, 1998)

Weight of pig (kg)	5	10	15	20	30	40	60	80	100	120
Feed space/pig (mm)	100	130	150	160	185	200	230	255	275	290

Table 21.16 Feeder space allowances for growing-finishing pigs of different weights fed *ad libitum* (Farm Assured British Pigs, 1998)

Pig weight	*Single-space feeders (pigs/space)*		*Multispace ad libitum feeders*	*Liquid ad libitum feeders*
	Meal	*Pellets*	*(mm/pig)*	*(pigs/space)*
Weaners (5–25 kg)	12	12	33–47	10
Growers (25–45 kg)	12	15	38–55	16
Finishers (45–120 kg)	10–20	10–20	55–75	16

Table 21.17 Suggested water flow rates from nipple- and bite-type drinkers (Farm Assured British Pigs, 1998)

	Flow rate (litres/min)
Weaners (from 5 kg)	0.5
Growers (30–50 kg)	0.7
Finishers (50–110 kg)	0.7
Pregnant sows	1.0
Lactating sows	1.5

Table 21.18 Heritability estimates for pigs

	Heritability
Reproductive characteristics (low heritability)	
Ovulation rate	0.10–0.25
Embryo survival	0.10–0.25
Numbers born	0.10–0.20
Survivability	0.05–0.10
Readiness to rebreed	0.05–0.01
Milk yield	0.15–0.25
Milk quality	0.30–0.50
Longevity	0.10–0.20
Growth and carcass quality (moderate heritability)	
Daily gain	0.30–0.60
Lean tissue growth rate	0.40–0.60
Appetite	0.30–0.60
Back fat thickness	0.40–0.70
Carcass length	0.40–0.60
Eye muscle area	0.40–0.60
Ham shape	0.40–0.60
Meat quality	0.30–0.50
Flavour	0.10–0.30

of evaporative cooling if they are allowed to wallow. In solid floored systems they can only do this by rolling in their own excreta. This is unhygienic and increases disease transfer. If pigs are housed on perforated floors they can be provided with showers to cool them without compromising cleanliness and hygiene. A part-slatted floor in which the solid portion could be bedded would seem a sensible compromise. Unfortunately, it is difficult to operate perforated floor systems in combination with bedded material. Consequently, it is difficult to find a flooring compromise that will satisfy the thermal needs of the pig and maintain its welfare under all thermal conditions.

Genetics and pig improvement

The contribution of genetics to the improved performance of modern pigs cannot be overemphasized. Over the last 40 years, the performance characteristics of pigs have altered enormously. A major factor was the foresight of pig breeders in using established principles of quantitative genetics and abandoning some of the traditional methodology of pedigree breeding. Pig breeders have followed a similar course to the poultry industry and adopted the techniques of mass selection for multiple objectives and used large populations to ensure statistical validity in the selection of parents to breed the next generation. The large populations available in breeding companies together with the adoption of advanced software tools like Best Linear Unbiased Prediction (BLUP) have enabled geneticists to continue to improve the performance of the pig. Geneticists have been helped by the fact that there were very clearly defined breeding objectives. For many years pigs have been purchased on the basis of 'deadweight and grade', that is the weight of the pig and the amount of back fat that they had on the carcass. Fortunately for the pig industry it is possible to measure back fat thickness accurately on the live pig using ultrasonics. The ease of measurement of characteristics such as growth rate, food conversion ratio and back fat thickness, coupled with the fact that the carcass characteristics of the pig are moderately heritable (Table 21.18), enabled breeders to make significant improvements to pig genotypes. The success of breeders has been most clearly in the reduction in back fat thickness that has been achieved in the national herd (Table 21.19). Although some of the reduction is due to better nutrition and feed management, much of the reduction has been a result of improvements in the genotype. In comparison, litter size, which has a low heritability, has increased little over the last 30 years.

For three decades British pig breeding companies led the world and developed lucrative markets exporting British genetics to the world. During that period the major companies developed into multinational companies and as the UK pig industry has declined they have progressively moved their operations to the larger markets in the USA and continental Europe.

Breeds

The concept of breed is almost irrelevant in the modern pig industry as parent lines have been developed that have had infusions of genetic material from a variety of so-called breeds. The basis of most commercial pig production is a white-skinned hybrid (in strict genetic terms a cross bred). The hybrids themselves are created from purebred parent lines maintained within nucleus

Table 21.19 Changes in pig performance (1970–1999). (MLC, 2000)

	1970	*1975*	*1980*	*1985*	*1990*	*1995*	*1999*
Litters/sow/year	1.9	2.0	2.18	2.25	2.23	2.25	2.25
Piglets born alive/litter	10.3	10.4	10.3	10.4	10.7	10.8	11.0
Piglets reared/sow/year	16.3	17.5	19.8	20.9	21.1	21.6	22.0
Annual sow disposals	—	33.9	35.9	38.1	40.0	42.6	42.0
P_2 at 100 kg (mm)[1]	—	22	19	14.5	13.0	11.5	11.0
FCR in feeding herd (kg/kg)	3.8	3.4	2.9	2.8	2.70	2.58	2.61

[1] P_2 measurement is taken using an optical probe to measure the back fat and rind thickness at a point level with the head of the last rib and 6.5 cm from the dorsal mid-line.

populations. The bulk of UK slaughter pigs still derive most of their genetic material from the Large White and the Landrace. As purebreds the Large White and Landrace are extremely prolific and when crossed together show heterosis, producing 5–10% more pigs per sow per year than either of the parent breeds. When the F1 hybrid (50% Large White and 50% Landrace) is backcrossed to one of the parent breeds or to a terminal sire of a different breed (or synthetic boar line) they will produce 10–15% more pigs than the original pure breeds. As the data in Table 21.19 illustrate, such sows can average 11 pigs per litter and rear in excess of 22 pigs per year. Another important attribute of the Large White and the Landrace breeds is that they both have a white skin. This is required by the meat trade in the UK as bacon, pork chops and roasting joints are usually sold with the skin on, and consumers do not wish to see coloured skin on these products.

Pig breeding companies have also incorporated genetic material from Belgian Pietrain pigs. The Pietrain breed had a very large eye muscle area and particularly well-developed hams. They also had a very high lean content, which was due in large part to the presence of the halothane gene. Heterozygotes carrying this gene have increased lean growth and higher carcass lean percentage, but, unfortunately, homozygotes are 'stress susceptible'. This condition comprises two undesirable characteristics. First, stress-susceptible animals can die suddenly in stressful situations like transportation or mixing with other animals. Second, such animals can produce meat that suffers from a condition known as Pale Soft and Exudative (PSE). Although it has no effect on the eating quality, PSE meat has an unattractive appearance and poor manufacturing characteristics; hence the value of the meat is reduced. Initially breeders incorporated the halothane gene into terminal sires in order to gain the benefits of increased lean growth and carcass lean percentage. This presented no problems if producers were strict in their use of first-cross females. However, if they retained the F2 generation for breeding, up to 25% of the progeny could be stress susceptible and produce PSE meat. In recent years breeders have bred the halothane gene out of their populations while maintaining high lean growth potential in their genotypes.

Another pure breed that has an impact on modern hybrids is the American Duroc. The Duroc is a very hardy breed compared with the European white breeds and some strains can grow extremely quickly. For these reasons the Duroc has been included in cross-breeding programmes for outdoor production. Because sows are not confined in outdoor production it is important to have sows that are not only hardy but also docile and easy to handle. For this reason producers traditionally favoured crossbred females that contained 50% British Saddleback. The Saddleback is renowned for its docility, ease of handling and good mothering characteristics. Some crosses including the Duroc as an alternative to the Saddleback have not been as good-natured and tractable as the traditional crosses, which has led some producers to reject them.

Currently, there is more interest in using Durocs in confinement systems. The reason for this is an anomaly in the way in which the Duroc lays down fat. Normally, intramuscular (marbling) fat is not laid down until a very late stage of development and then only when the carcass has achieved a very high fat content. The Duroc is unusual in that it has around 4% intramuscular fat in its carcass at normal levels of total fat. The importance of this small intramuscular fat is that it can improve the cooking characteristics of the meat. For this reason some Quality Assurance schemes now specify that slaughter pigs should be 25% Duroc; however, controlled studies suggest that few consumers are able to detect any noticeable improvement in the meat that is produced from such crosses.

In 1987 a small number of Chinese pigs was imported into Britain. The stimulus for this was the high prolificacy of these pigs. Some anecdotal accounts suggested that breeds, such as the Meishan from central China, could regularly produce 20 piglets in one litter. The objective was to incorporate the genes for prolificacy in British pigs to produce hyper-prolific hybrids. However, the challenge that faced the breeders was to do this while maintaining an acceptable carcass. Although very prolific, the Chinese breeds are very early maturing and extremely fat. This has been achieved and although it has not produced white hybrids that produce 20 pigs per litter, small but significant improvements in litter size have been made.

Breeding plans

Systematic crossbreeding is the basis of commercial pig production (Fig. 21.6). The majority of producers purchase crossbred (hybrid) females and boars (or semen) from breeding companies. Most hybrid females are based on this same first cross animal. In the context of pig breeding, hybrid is thought of as an 'improved crossbred' where simultaneous genetic improvement is made in the parent lines as well as selection for their combining ability. Increasingly, breeding companies are producing specialized sire and dam lines. This enables particular traits to be more effectively selected. For example, in the dam line selection can focus on traits such as prolificacy and milking ability, while in the sire line the focus can be on growth, carcass and meat quality attributes. The boar breed used may be one of the parent breeds of the F1 female, in which case this is known as a 'back-cross'. More frequently the boar will be of another breed, a crossbreed or a 'synthetic' sire-line. It is not advisable to retain female progeny from the slaughter generation as potential breeding herd replacements as further crossing with crossbred boars or synthetic lines will produce offspring with extremely variable genetic composition. This can adversely affect both the uniformity and the productivity of the slaughter generation.

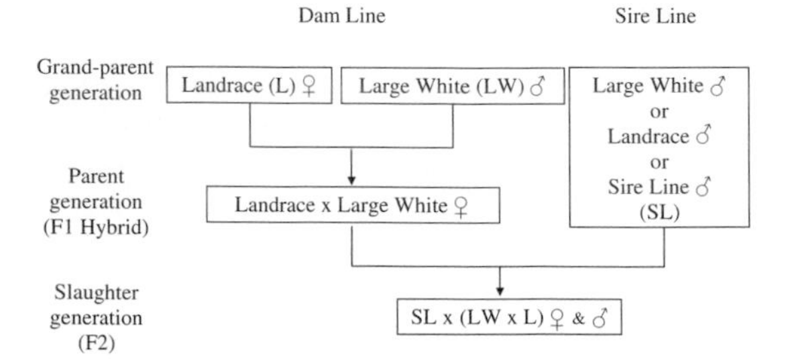

Figure 21.6 Typical breeding programme for the production of meat pigs.

Some producers do produce their own gilt replacements from purebred parent stock. To give farmers flexibility in their replacement policies, breeding companies sell grandparent females directly to commercial farmers who then carry out the cross-breeding programme to generate their own gilt replacements. This approach has the advantage that it reduces the need to import so many animals onto the unit and thereby reduces disease risk. However, it has the disadvantage that it is more difficult to ensure that replacement animals are introduced to the unit at the correct time to maximize building use. It also means that a proportion of the herd has to comprise purebred sows that are less prolific than the F1 females. It can also distract management from the primary task of producing slaughter pigs. Producers who purchase all their replacement gilts from a breeding company incur higher replacement costs and have a higher risk of introducing disease onto their units. On the other hand, if they purchase their replacements from the same breeding company herd health need not be compromised and they will capitalize on the continuous genetic improvement being made by that company.

Artificial insemination took a long time to become popular with pig producers. There were three main reasons for this. First, there was a ready supply of good quality boars available from breeding companies at attractive prices. Second, there was no inseminator system for the pig industry partly because of cost but also because of the health risks of having operators moving from one farm to another. Third, for a long time it was not possible to freeze boar semen so fresh semen had to be used for insemination. As a consequence of the last two factors reproductive performance was impaired by the use of AI and the additional gains in carcass value were insufficient to offset the loss in productivity. Natural service can produce conception rates of 80–90% and litters of 10–11 pigs. However, in the past many producers could achieve only 60–70% conception with AI and also found a 0.5–1.0 pigs reduction in litter size. This situation has now changed. Improved semen diluents have extended semen life to 4–6 days and the use of pooled rather than single-sire semen has improved conception rates by as much as 10%. In addition to these technical advances insemination techniques have improved. As herd sizes increase it is easier to ensure that at least one person on the unit is properly trained and becomes proficient in insemination. Some larger units maintain their own boar stud and collect fresh semen to use on their own sows. However, this is not essential and on large units it is possible to place a standing order for semen in the knowledge that the number of inseminations each week can be predicted. It is difficult to determine the extent to which AI is being used currently. In the UK between 15 and 20% of matings are by AI. In France and Germany the percentage is probably between 35 and 40%. It is interesting to note that AI usage is much higher in Spain (circa 75%) and in the USA (circa 70%). The increased usage in these two countries is consistent with the development of very large, integrated breeding units.

Initially, AI was promoted as a way of improving the genetic merit of breeding stock as it offered small producers access to the best genetic material and provided them with the opportunity to replace their existing herd with superior animals over a period of time. However, this is no longer the principal objective on large units as they generally import all their breeding stock. Units producing pigs for slaughter use AI in order to gain the benefits of superior growth rate and lean percentage in the slaughter pig.

Most of the leading pig breeding companies operate their own AI stations and dispatch orders of semen by post. The AI process itself is carried out by the stockperson using a disposable catheter and each insemination takes between 10 and 15 minutes. Timing of the insemination is crucial to success and a rigorous heat detection programme must be used to determine the onset of oestrus. The first insemination should be carried out about 24 h after the onset of oestrus (not the first time that oestrus is actually observed) and a second insemination given between 8 and 16 h later. In commercial herds producing pigs for meat, producers may use a combination of natural service for a first insemination and AI for the second. This approach has three advantages. First, the use of a boar ensures good oestrus detection. Second, most conceptions result from the second mating so the genetic merit of the offspring usually reflects the AI sire. Third, the combination treatment conserves 'boar power' and generally produces an improvement in fertility.

There have been many attempts to use deep-frozen semen for pig AI. To date the results are not good enough to justify its use in commercial units. However, it does have a place in the breeding programmes of multinational breeding companies who can use it as a convenient way of moving genetic material between countries.

Sow and gilt reproduction

The breeding sow exists for only one purpose: to produce and rear piglets to a stage at which they can be raised satisfactorily and economically without her. For the last three decades there has been an assumption that profitability was directly related to sow productivity. This is not strictly true. Profitability is not synonymous with biological output. Marginal additional increments of output may require additional expensive resources and significantly increase the demands on management. The preoccupation of producers with biological output should be replaced by a preoccupation with profit. This means that in most cases they should aim to optimize rather than maximize the biological output of the sow.

Culling and the replacement gilt

The management of replacement gilts is extremely important for the profitability of the whole unit. As indicated in Table 21.19 average annual culling rate exceeds 40%. As sows produce just over two litters per year on average this means that gilt farrowings will represent around 20% of the herd total (Table 21.20). Gilts generally produce smaller litters than sows, with maximum litter size not being achieved until the sow reaches fourth or fifth parity. Although gilts produce smaller litters than older sows there is little difference in numbers reared by first to seventh litter sows, with rearing ability declining thereafter (Table 21.20). A worrying feature is that around 15% of females only produce two litters, therefore never reaching the age at which they might be expected to be most productive (Table 21.21). Nearly 40% of disposals of first and second litter sows are because of reproductive problems. Many of these reproductive problems are due to failures in management of the gilt.

Whether gilt replacements are home produced or purchased from a breeding company they will have been

Table 21.20 Pigs born, weaned and weaning to last service interval by parity. (MLC, 1999)

Parity	*% of total*	*Alive*	*Weaned*	*Days weaning to last service*
Gilt	19	10.6	9.5	
2	17	10.9	9.9	11.2
3	16	11.5	9.7	9.7
4	13	11.7	9.5	9.0
5	11	11.9	9.7	8.4
6	9	11.5	9.4	9.1
7	7	11.0	9.4	8.4
8	5	11.1	9.3	8.7
9	3	11.2	9.1	10.7
10	1	9.4	8.2	6.9
>10	<1	9.3	8.0	16.8
Average 3.8	100	11.2	9.6	9.5

Table 21.21 Sow disposals (%) by parity groups. (MLC, 1999)

Reason for disposal	*Parities*			
	1–2	*3–7*	*8–15*	*All*
Failure to come on heat	9.2	7.4	3.9	6.4
Failure to conceive	25.5	24.5	7.3	18.3
Numbers born	2.1	8.3	4.9	6.1
Rearing ability	2.1	5.8	1.5	3.6
Died	18.4	13.9	4.9	11.2
Health	9.9	5.1	2.3	4.8
Aborted	2.8	2.8	2.6	2.7
Legs/feet	7.2	3.9	1.2	3.4
Physical damage	3.5	4.1	1.2	2.9
Old age	1.4	10.7	64.7	29.3
Miscellaneous	17.9	13.8	5.5	11.3
Percentage of total	15.3	47.2	37.5	100

reared away from the breeding herd. It is very important to ensure that they do not introduce disease into the herd and equally as important to ensure that they become acclimatized to the herd they are entering. This implies that incoming gilts have a period of quarantine (usually 3 weeks) from delivery before they enter the sow herd. Following this they need to have another 3 weeks of acclimatization at least before they are mated. It is possible to combine these two functions by housing the gilts away from the main herd and introducing sows to them that are going to be culled. The acclimatization period is important because if the gilt is responding to a disease challenge her reproductive performance is likely to be adversely affected.

Puberty

The sooner that replacement gilts are mated and produce their first litters the lower will be the overall replacement costs. Puberty or first heat in the female pig occurs when the animal is about 6 months of age at circa 90 kg liveweight. There is enormous variation around this age and some gilts attain puberty as early as 130 days and some as late as 280 days. There has been much research carried out to establish the factors controlling puberty. Genotype, nutrition and season of the year all influence the time of first heat, but the most important influence is the 'boar effect'. If prepubertal gilts are suddenly exposed to contact with mature boars, this stimulates the first heat and normal cyclicity in a high proportion of gilts. There are a number of components of this effect such as sight, sound and tactile contact with a boar. The biggest component is that of boar odour and the pheromones (airborne hormones) contained within the boar's characteristic smell.

Gilts at 160–180 days of age should be introduced to mature boars and given daily contact for at least 20 min/day in the same pen as the boar. From the time of first introduction to males gilts should be housed within the same air space as or in pens adjacent to mature boars. This maximizes the exposure of the gilts to boar pheromones. Stimulation with mature boars usually necessitates relocation of the gilts within the piggery. Gilts also respond to the stimulus provided by relocation and it helps to initiate the first oestrus. It is common for gilts delivered to a farm from a breeding company to exhibit oestrus within 7–10 days of delivery. However, they should not be mated at this time, as they will not have acclimatized to their new environment. A final stimulus to puberty attainment is the mixing of groups of gilts into new social groups. This alone will cause many gilts to reach puberty within a few days of mixing. The use of this practice in a management programme may need careful control to avoid physical damage to the animals as a result of fighting when forming a new social group.

Despite the abundant data and research on gilts, failure to exhibit oestrus is a perennial and serious problem on some farms. There are now a number of pharmaceutical products that can be used very effectively as 'cycle starters'. These products are usually based on combinations of the gonadotrophic hormones (FSH and LH) or their analogues.

Oestrous cycles

Following first heat, most gilts will exhibit regular oestrous cycles at approximately 21-day intervals. This varies from about 18 days to 23 days in individual animals. However, in some gilts the first heat may be followed by a number of so-called silent heats. In a silent heat ovulation takes place but the overt signs of oestrus, including receptivity to the boar, are absent. Some gilts have very short oestrous periods or show only a weak oestrous response in the presence of the boar. Both these situations make successful mating more difficult to achieve.

A major decision is the stage, relative to puberty, at which mating should be carried out. Ovulation rate (the number of ova shed) is very poor at pubertal heat and mating gilts at this time can lead to unacceptably low litter size. Most producers wait until the second or third heat before mating gilts. By this time ovulation rate will have increased and a larger litter is likely to be produced. For many producers, the differences in litter size that result from mating gilts at different heat periods should be a secondary consideration. A more important concern should be that gilts are mated so that they fit into the farrowing pattern of the unit and optimize the use of expensive farrowing accommodation.

The general recommendation would be that gilts should be 220–230 days of age, 130–140 kg liveweight, have 18–20 mm P_2 backfat and be mated at their second or third oestrus.

Oestrus and mating

The outward signs of oestrus are much more apparent in gilts than in sows and oestrous females are more readily identified when females are kept in groups. Oestrous females will tend to go off their food and in the pro-oestrus period will mount other females. Gilts will normally exhibit vulval reddening and swelling at this time under the influence of rising levels of the oestrogen hormones. These outward changes in the genitalia are frequently absent in older sows. At the onset of oestrus, gilts will often respond to the application of firm hand pressure on their backs and will stand rigidly, often holding their ears erect. This standing response (known as lordosis) tends to be shown better if there is a boar in the vicinity. Once this response has been elicited, it can be extremely difficult to move the animals around. The presence of this response is an unequivocal sign that the female is in oestrus, but the absence of the response cannot be taken as a clear indication that she is not. In the absence of a boar for heat detection it is possible to identify around 60% of animals that are on heat. With a boar in the same pen as pro-oestrus gilts or sows there is a much higher probability of detecting heats. Some farms use vasectomized boars for heat detection and for the stimulation of first heat. Using sterile boars removes the risk of mating at the wrong time or with the wrong boar.

Oestrus normally lasts for 2–3 days and ovulation takes place in the middle of this oestrous period. The aim with both natural service and AI is to have fresh semen in the reproductive tract to await the ova being shed. This is important because the sperm cells need to mature for a number of hours in the female's tract before they are able to fertilize the ova. If sperm are introduced into the female's tract either too early or too late, then the sperm and ova are out of phase in their physiological maturity. This can result either in a failure to conceive or in reduced litter size.

With natural service, it is best to carry out heat detection twice a day to ascertain the onset of oestrus. Ideally, service is carried out 30 hours after the beginning of oestrus and in practice this is on the morning after the first time that oestrus is detected. Mating is then repeated the next day. For sows detected in heat in the afternoon, it is better to wait until the next morning before the first mating is carried out. There is evidence that allowing three matings during the course of the 2-day oestrous period improves conception rates and litter size. Increasing the number of matings maximizes the chances of placing fresh semen in the female's tract at just the right time. On the debit side, this practice increases the number of boars or semen samples needed. If sows or gilts are mated twice each and the optimum number of services a week for each boar is about four times, then this means that for every 20–25 females in the herd a farmer will need one boar. If triple serving is practised, the ratio needs to be increased to around 15 females per boar to avoid the overuse of boars.

Feeding the gilt

Many of the problems of poor performance in gilts are a result of producers not appreciating the extent to which the gilt has been changed over the last 25 years (Table 21.22). Gilts of modern genotypes have a much higher lean growth potential and a reduced propensity to deposit fat than they did in the past. Feeding regimes used to be based on the premise that gilts started their breeding lives with large back fat reserves and that these could then be depleted over a number of parities. This is no longer the case; therefore, the nutritional management of the replacement gilt has become more critical. It has now become important to feed the gilt to ensure that she gains sufficient fat before mating. One recent study found that gilts with 13 mm fat or less at mating for the first time produced only 2.8 litters and 22 pigs in their reproductive life whereas gilts with 17 mm back fat or more produced 3.75 litters and 30 pigs in their lifetime. To deposit enough fat pre-mating gilts should be fed a good quality diet containing 13.0–13.5 MJ DE and 6–8 g lysine/kg. The quantity of food given should be between 2.5 and 4.0 kg/d, depending on the weight

Table 21.22 Changes in the characteristics of gilts. (After Boyd, 1999)

	1975	*2000*
% lean in carcass	45	55–60
% fat in carcass	27	15–18
P_2 at first service (mm)	30–35	18–20
Lean tissue growth rate (g/d)	200	340
Liveweight at third parity (kg)	195	250

of the gilts on arrival in the breeding unit, the housing environment, group size and feeding system.

From the time gilts are introduced to mature males, most farmers 'flush' their gilts. Flushing is the abrupt transition to a higher plane of feeding for 2 weeks prior to ovulation and this is best achieved by feeding to appetite during this period. This practice optimizes ovulation rate and in turn improves litter size. Although a high feed intake is beneficial before ovulation, a high feed level is detrimental after mating as it reduces embryo survival. Therefore, immediately after service the feed scale should be adjusted again to a low plane of about 1.8–2.0 kg/d for at least 3 weeks post mating. This will prevent the excessive loss of fertilized embryos that will occur if gilts are overfed in the first 3 weeks of pregnancy.

Pregnancy

Gestation in the pig lasts about 115 days and the variation around this mean value is generally small although some sows will naturally give birth as early as 110 days after mating or as late as 119 days. The initial days of pregnancy are the most critical in terms of whether the pregnancy will be maintained and it is in the initial weeks that litter size is determined. Initially the fertilized ova are free-living entities within the lumen of the uterus. By day 12, the initial attachment of embryonic membranes begins and this process of implantation continues until about day 20 of pregnancy. Because implantation is similar in some respects to the host-graft relationship seen in organ transplantation or in skin grafting, the process is a very delicate one. The developing embryos can be immunologically rejected by the uterus and development stops. Any stress on the sow or gilt at this time causes an imbalance in the hormone status of the sow and more embryos die. Similarly, any difficulty experienced by individual embryos in securing a supply of nutrients across the placental wall will also lead to embryonic death. Once implantation has taken place the sow is more able to buffer embryos from an adverse external environment.

The final litter size is set initially by ovulation rate. Throughout gestation the number of potential piglets is reduced by losses at different stages. Under average conditions 25–30% of all fertilized ova will be lost in the first 3 weeks of gestation. Another portion of loss occurs in the fetal stages when the developing fetuses compete for available space and nutrients within the uterus. This latter percentage of loss is less significant than early embryo losses.

The early detection of sows that are non-pregnant can be a valuable management tool and a variety of ultrasonic devices are available that can be used for the determination of pregnancy. Non-pregnant sows or gilts can then be moved back to the service area, or culled, without incurring further costs. Pregnancy diagnosis is possible as early as 28 days after service, but repeated testing through pregnancy is necessary to pick up sows that have resorbed their fetuses after initially conceiving a litter. It should be noted that conception rates are very high on most commercial units and many producers would consider that the time invested in detecting the occasional non-pregnant animal would be better spent on other management functions.

Feeding the pregnant sow

Pregnant sows need nutrients for maintenance, maternal growth and to sustain the developing conceptus (embryos, fetal membranes and placenta). A factorial approach has been used to calculate energy and protein requirements for sows (Table 21.23). The gilt at mating has generally only reached 30–40% of her mature body weight so she needs to continue to grow throughout her reproductive life, but at a diminishing rate. The calculations in Table 21.23 show that there is considerable variation in the needs of sows of different weights. In theory, it should be easy to meet these differing needs

Table 21.23a Energy requirements of pregnant sows. (After Close & Cole, 2000)

Body weight at mating (kg)	*Net weight gain (kg)*	*Maintenance (MJ DE/d)*	*Maternal gain (MJ DE/d)*	*Conceptus gain (MJ DE/d)*	*Total requirements (MJ DE/d)*	*Feed allowance[1] (kg/d)*
120	45	20.1	8.6	1.6	30.3	2.3
150	35	22.5	6.7	1.6	30.8	2.4
200	25	26.7	4.8	1.6	33.1	2.5
250	15	30.4	2.9	1.6	34.9	2.7
300	10	34.3	1.9	1.6	37.8	2.9

[1]Diet containing 13 MJ DE/kg.

Table 21.23b Energy requirements of lactating sows. (After Close & Cole, 2000)

Body weight at farrowing (kg)	*Maintenance (MJ DE/d)*	*Milk production (MJ DE/d)*		*Total (MJ DE/d)*		*Feed requirement (kg 14.5 MJ/kg diet/d)*	
		10 piglets	*12 piglets*	*10 piglets*	*12 piglets*	*10 piglets*	*12 piglets*
150	21.1	60.8	73.0	81.9	94.1	5.6	6.5
200	26.2	60.8	73.0	87.0	99.2	6.0	6.8
250	30.9	60.8	73.0	91.7	103.9	6.3	7.2
300	35.5	60.8	73.0	96.3	108.5	6.6	7.5

when sows are confined in sow stalls or fed in individual feeders. However, because of the labour involved in providing individual sows with different allowances producers have tended to feed all animals as if they were 'average' and made few concessions to their varying needs. Electronic sow feeders have provided one solution to the problem as individual allowances can be programmed into the computer that controls feed delivery.

Traditional feed scales allowed for high weight gain in pregnancy and much of the gain was catabolized in the ensuing long lactation. This process is inefficient and has the effect of reducing the appetite of lactating sows. The emphasis is now on the conservation of body fat reserves throughout the reproductive cycle and indeed the reproductive life of the sow. Sows are allowed to make modest weight gains in pregnancy and can then be encouraged to take large amounts of food in lactation, which can be used directly for milk production. Diets for pregnant sows are formulated to contain around 13 MJ DE/kg. Although factorial estimates show that the energy requirement of the sow increases by 17–20% during the course of pregnancy it is unusual for producers to adjust intakes (unless they use electronic sow feeders).

Protein requirements for sows have also been well researched and it has been demonstrated that a sow will perform satisfactorily on a diet containing as little as 12% crude protein. However, it is not very helpful to express the requirement in terms of crude protein because it is the daily supply of essential amino acids that is important. For first-pregnancy gilts, the lysine requirement is 14–15 g/d and for older sows 10–11 g/d. Assuming the feeding levels in Table 21.23 this would imply that a diet containing 13 MJ DE/kg should contain 5.5–6.0 g lysine/kg.

Parturition

Parturition and the time that sows deliver their offspring is a crucial phase in the reproductive cycle. If things go wrong, piglets will either die before they are expelled from the uterus or die in the early hours or days after birth. Sows should be transferred to their farrowing quarters at least 3 days before the expected time of farrowing. This allows them to acclimatize to the new environment before the litter is delivered. Ideally sows should be part of a group farrowing within a few days of each other and housed in a previously cleaned farrowing room. Farrowing sows as a group facilitates cross-fostering of piglets between sows to even up litter size. It also allows the sows to be weaned together and the farrowing room to be emptied and cleaned before restocking. This in turn helps to prevent the transfer of disease within the unit.

Labour is initiated 48 hours or so prior to any visible signs. The ovaries respond to a hormone of the prostaglandin series produced in the uterine wall, which initiates regression of the corpora lutea in the ovaries and as a result blood progesterone levels fall. Without the support of progesterone, uterine contractions begin and labour commences. A number of other hormones then control the frequency and the strength of the uterine contractions and amongst these are oxytocin, relaxin and oestrogen. Partly because of the complexity of the hormonal events the whole process is prone to dysfunction. Any stress on the sow prior to parturition can cause delayed farrowing and prolonged delivery. This in turn leads to oxygen starvation in those piglets born last and they may be stillborn.

When a sow is about to deliver she will exhibit 'nest building' activity or characteristic pawing with the front legs. Finally she will settle down and at the appearance of a bloody discharge she will begin to deliver piglets. The interval between the births of successive piglets may vary from a few minutes to an hour or two. From beginning to end, the whole process may take from 1–12 hours or even longer. If parturition exceeds 6 hours the percentage of piglets that are stillborn increases exponentially and an obstetric examination may be required.

It can aid management if sows farrow close together and during the normal working day. In order to achieve this sows can be treated with synthetic versions of naturally occurring prostaglandin hormones. These are extremely effective for inducing parturition. A single injection is given 26 hours before delivery is required and almost 100% of sows respond to this by going into labour at the prescribed time. This makes it possible to avoid nighttime and weekend farrowings when it is unlikely that a stockperson will be around to supervise the deliveries. Care must be taken when using prostaglandin analogues, for if farrowing is induced too early, the fetuses will not be mature enough to survive the external environment. Piglets induced 2 days or more before the expected date can lack vigour and will only survive if the postnatal care if very good.

Lactation and weaning

The piglet is born with very low fat reserves and no placentally transferred immunity. Therefore it is vital that it suckles immediately it is born, first to gain immunoglobulins from the sow's colostrum which will help protect it from infection, and second, in order to obtain energy from the sow's milk. Sow's colostrum and milk is more concentrated than that of other farm species (Table 21.24). Sows nurse their piglets at intervals of around 50 minutes and milk is only available to the piglets at this time. The number of piglets being suckled influences the quantity of milk produced. Daily milk production builds up to a peak at the end of the third week of lactation and then declines steadily. Modern hyperprolific sows can produce 10–12 kg milk/d, which is almost twice the amount produced by sows 30 years ago. Even such high milk yields are insufficient to support maximum growth rate in the piglets after the first week of lactation. In order to overcome this limitation to piglet growth pigs are introduced to solid (creep) food as early as possible. The creep feed must be supplied in small quantities and kept very fresh and palatable in order to attract the piglets and stimulate them to eat.

Table 21.24 Composition of sow's milk and colostrum (g/100 g milk). (After Darragh & Moughan, 1998)

Component	*Colostrum*	*Mature milk*
Total solids	24.8	18.7
Protein	15.1	5.5
Non-protein nitrogen	0.3	0.3
Lactose	3.4	5.3
Fat	5.9	7.6
Ash	0.7	0.9

As the sow is anoestrus during lactation, producers have tended to reduce weaning age in order to get the sow pregnant again and increase the number of pigs produced per sow per year. In Europe it became the norm to wean piglets at around 3 weeks of age. Weaning at less than 21 days' lactation tended to be counter productive, as sows took longer to return to oestrus, had poorer conception rates and tended to have slightly smaller subsequent litters. In the UK the Welfare Regulations now prohibit weaning at less than 3 weeks of age. In the USA there has been a move to wean much earlier at 10–12 days. This was not done to improve the productivity of the sow but as a means of improving piglet health. The theory was that in the first 2 weeks of life the piglet was protected from a number of diseases by the immunoglobulins that it acquired from its mother's colostrum. The passive immunity gained from the mother declines to a low point at around 3 weeks post weaning. Therefore, it was argued that removing the piglets from the sow before this time and transferring them to clean buildings on another site would prevent them from contracting the diseases present on their original unit. If this system is operated very carefully it can be effective, but experience has shown that few producers can manage the system well enough to realize the supposed advantages.

In Europe the trend is for weaning to occur later than 21 days. The reason for this is that a number of antibiotic growth promoters have been banned and without these it is proving difficult to wean piglets at young ages and prevent them from developing enteric disease. Weaning the piglet at a later date allows its immune and digestive system to mature and makes it less susceptible to contracting post-weaning diarrhoea. In Sweden, where the use of antibiotic growth promoters was banned first, producers generally wean piglets at 4 or 5 weeks of age, and with careful management of sows during lactation have seen little reduction in overall sow productivity. With continued consumer pressure for antibiotics and other antimicrobials to be removed from pig diets it seems likely that piglets will be weaned at 4–5 weeks of age.

Feeding the lactating sow

As noted earlier the modern sow may be producing 10–12 kg milk per day with a total solids content close to 20%. The production of over 2 kg dry matter per day by a sow weighing 200–250 kg puts a considerable strain on the sow. A sow suckling a litter of 12 pigs has a daily energy requirement of around 100 MJ DE. Unfortunately, while the geneticists have increased the milk production of the sow they have not increased her appetite proportionally; in some genotypes appetite has actually been reduced. The modern sow, like the high-yielding dairy cow, is generally in energy deficit as she approaches peak lactation and she will have to mobilize body reserves to maintain milk production. In theory, it is not possible to overfeed a lactating sow and on many units around the world sows are fed *ad libitum*. However, *ad libitum* feeding does not necessarily maximize feed intake. Allowing the sow to eat all she wants in early lactation can make the sow overproduce milk at a time when the piglets cannot utilize it. The sow will reduce intake to compensate and then fail to achieve maximum intake thereafter. Work by the Meat and Livestock Commission in the UK has shown improvements in intake and performance if sows are fed on an increasing scale of feeding that matches the demand for nutrients for milk production (Table 21.25). This scale has much higher maximum feed allowances than many other scales and has been found to stimulate feed intake in the sow. In a comparison of herds using the Stotfold sow feeding strategy with herds using other feeding strategies, the former were found to produce 0.9 pig more per sow per year while consuming 8% less food.

There is considerable variation in the specification of commercial diets for lactating sows. In selecting diets, account has to be taken of the appetite of the genotype used. For genotypes with a high appetite a diet containing 13.5–14.0 MJ DE and 8–9 g lysine/kg may be adequate. For genotypes with lower appetite the specification may need to be increased to 14.5–15.0 MJ DE and 11–12 g lysine/kg.

A major problem in managing lactating sows and their litters is providing a suitable thermal environment for both. Suckling pigs need a much higher ambient temperature than their mothers (see Table 21.13). If piglets are too cold they will huddle against the sow and are more likely to be crushed. Therefore, it is normal to provide heated areas for the piglets away from the sow. Traditionally, these areas have been heated using infra-red lamps, but in recent years there has been increasing use of pads heated either electrically or using hot water. Heated pads provide a comfortable, warm resting place for the piglets and keep them away from the sow except at feeding times.

If sows are kept at high ambient temperatures they will reduce their voluntary feed intake and hence their milk production. In countries where summer tempera-

Table 21.25 Example of the 'Stotfold' feeding system for lactating sows (Stotfold Pig Development Unit, 2001)

Day	*Feed allowance (kg/d)*				
	Gilt <10 pigs *Sow <9 pigs*	*Gilt 10 pigs* *Sow 9 pigs*	*Gilt 11 pigs* *Sow 10 pigs*	*Gilt 12 pigs* *Sow 11 pigs*	*Gilt 13 pigs* *Sow 12 pigs*
1	2.5	2.5	2.5	2.5	2.5
2	3.0	3.0	3.0	3.0	3.0
3	3.5	3.5	3.5	3.5	3.5
4	4.0	4.0	4.0	4.0	4.0
5	4.5	4.5	4.5	4.5	4.5
6	5.0	5.0	5.0	5.0	5.0
7	5.5	5.5	5.5	5.5	5.5
8	6.0	6.0	6.0	6.0	6.0
9	6.5	6.5	6.5	6.5	6.5
10	7.0	7.0	7.0	7.0	7.0
11	7.0	7.5	7.5	7.5	7.5
12	7.0	7.5	8.0	8.0	8.0
13	7.5	8.0	8.5	8.5	8.5
14	7.5	8.0	8.5	9.0	9.0
15	8.0	8.5	9.0	9.5	9.5
16	8.0	8.5	9.0	9.5	10.0
17	8.5	9.0	9.5	10.0	10.5
18	8.5	9.0	9.5	10.0	10.5
19	9.0	9.5	10.0	10.5	11.0
20	9.0	9.5	10.0	10.5	11.0
21	9.5	10.0	10.5	11.0	11.5
22–28	9.5	10.0	10.5	11.0	11.5

tures are invariably high strategies are employed to keep the sows cool in order to maintain feed intake. These include dietary manipulation such as feeding more nutrient-dense, high-fat diets, feeding at cooler times of the day and using liquid diets. Liquid feeding ensures that the sows take abundant water as well as increasing feed intake. Even in temperate climates the use of liquid feeding may increase intake by 15–25%. Techniques are also employed to cool the sow. Water sprays or drippers can provide evaporative cooling, but these are not very helpful when humidity is high. Jetting air over the sow's head can be effective in reducing her temperature, as around 20% of heat loss takes place from the head and ears.

For mature sows the daily allowance offered in a diet containing 13 MJ DE and 15% CP might be 1.8 kg plus 0.45 kg for each piglet suckling. A sow suckling a litter of ten piglets would therefore be offered 6.3 kg each day or around 82 MJ DE. This allowance would not be offered at the beginning of lactation and from the gestational scale of say 1.8 kg/d an extra increment of 0.45 kg would be added each day until the set amount is reached.

The weaning to remating interval

At weaning suckling stimulus is removed and oestrous cyclicity resumes. Under normal circumstances sows weaned at 21–28 days' lactation will exhibit oestrus and ovulate between 3 and 10 days following weaning. A majority of sows show heat at either day 4 or day 5 after weaning, but a sow can return to oestrus on any day post-weaning and this means that oestrous checks must be made on all weaned sows on a daily basis to ensure that all sows are mated again as quickly as possible. Missed heat periods represent lost profit and the post-weaning period is one of the critical points in the reproductive cycle. Anoestrus, or absence of any signs of heat after weaning, is one of the major reasons for culling sows from the herd and even the best herds will have 2–4% of these every cycle. At worst this figure can rise to 10–20% and leads to the complete disruption of the farrowing programme. The cause of anoestrus is multifactorial, but poor sow body condition, season and health status are often implicated. A rigorous heat detection policy is essential to ensure that there are no sows recorded as anoestrus that were simply missed by the stockperson.

There is no advantage in the withdrawal of either food or water in the first 24 hours after weaning. It was once thought that this practice might help to dry the sow off. It has been shown repeatedly that this gives no benefit in terms of the time to return to oestrus or the percentage of sows showing oestrus within 10 days of weaning. On the contrary, this is a period during which it is important for the sow to begin to recover from the negative energy balance and loss of body fat in the preceding lactation. Many producers feed sows to appetite during the period from weaning to remating. The aim should be an intake of around 4–6 kg feed/d as this will hasten the return to oestrus and maximize the ovulation rate. It is also important that the sow restores depleted protein reserves so the diet used in lactation (which has a higher protein content than a pregnancy diet) should be fed during this period.

The suckling pig

Piglets are born without any protective antibodies in their bloodstream and as a consequence can easily become infected with diseases. Piglets gain passive immunity by ingesting the sow's colostrum. Colostrum contains a very high concentration of immunoglobulins and immediately after birth these can pass straight across the piglet's gut wall into the bloodstream. Within hours of birth the piglet's gut loses the capacity to transfer large immunoglobulin molecules straight into the bloodstream. However, the immunoglobulins in colostrum and subsequently in milk still exert a protective function in the lumen of the gut. The amount of colostrum ingested in the first 12 hours of life determines the piglet's survival prospects for many weeks to come because after birth the blood concentration of antibodies falls at a regular rate. Piglets do not start to produce significant amounts of their own antibodies until 2–3 weeks of age. This means that the pig weaned between 2 and 4 weeks of age has a comparatively poor immune status. Its passive immunity, gained from its mother's colostrum, has fallen to a low level and its own active immunity is not yet providing it with much protection. From 3 weeks of age onwards the piglet's blood level of immunoglobulins rises and its ability to withstand disease challenge increases.

In order to give the piglets the best chance of survival, as much supervision as possible should be given in the postnatal period. The stockperson should assist small and weakly piglets to find a teat and suckle so that they gain colostrum and the immunity that it provides. The use of prostaglandin analogues to induce farrowing during normal working hours may help to facilitate the supervision of suckling. Cross-fostering can be used to even-up litter size and to make sure that every piglet in the litter has a functional teat available to it. Piglets should only be cross-fostered between sows that have farrowed within a few hours of each other. Preferably, piglets should be fostered onto a sow that has farrowed more recently than their own mother to ensure that they gain a good intake of colostrum.

Modern hyperprolific sows are not able to produce enough milk to maximize the growth rate of their piglets. Therefore, piglets are normally offered additional food and water away from the sow. When sows are housed in farrowing crates this can be placed anywhere outside the crate that the sow cannot reach. If sows are housed in pens that do not confine them, a railed off area known as a 'creep' is provided and the food for the piglet is placed in this area; hence the name 'creep feed'. Creep feed should be extremely palatable and easily digested as the piglet does not have a full complement of enzymes in its gut at this stage and cannot digest more complex diets. Creep diets usually contain a high proportion of milk products such as skim milk or whey powder and in addition cooked cereals and high quality fish meal. Consuming small quantities of solid food helps stimulate the development of the piglet's enzyme systems and teaches it to look for food and water to meet its nutritional needs.

The creep should be kept fresh and attractive by offering only small amounts at any one time. A clean water supply should also be provided adjacent to the food. Good management of creep feed and water for piglets can make a useful contribution to weaning weights. Even if it does not, the importance of teaching piglets to find food and water cannot be overemphasized. Piglets that have become familiar with finding food and water while still suckling the sow will be much better equipped to make the transition to solid food after weaning.

The newly weaned pig

On commercial units weaning is usually an abrupt event rather than the gradual process that occurs in nature. The piglet is usually moved to different accommodation where it must locate new sources of food and water and mixes with new individuals from other litters. The combination of these stressors can reduce immune competence and result in post-weaning diarrhoea. At best, post-weaning diarrhoea causes chronic loss in growth and at worst can result in high piglet mortality. For the last three decades many of these problems have been overcome by the addition of antibiotic growth promoters and copper sulphate to the diet. The use of these antimicrobial feed additives can reduce the proliferation of the organisms responsible for post-weaning diarrhoea. More recently zinc oxide, another powerful antimicrobial, has been added to weaner diets with the same aim. These products have undoubtedly been effective in reducing enteric disease in the weaned pig, but they do so by treating the symptoms rather than correcting the cause of the problem.

There is now growing concern about the use of these antimicrobial feed additives in animals. The antibiotics used for this purpose were carefully selected because they had no role in human medicine and were considered safe to use in animals. Unfortunately, in recent years evidence has accumulated that the use of some antibiotic growth promoters could produce resistance not only to themselves but also to other antibiotics used in human medicine that had somewhat similar chemical structures. To overcome this problem the use of some antibiotic growth promoters has been banned in the EU. Consumer concerns are such that it is likely that there will be a complete ban on the use of antibiotic growth promoters in the future. Unfortunately, there has not been international acceptance of the need to control the use of these products. Therefore, they will continue to be used in some countries and these countries will gain a competitive advantage by so doing. More importantly, bacteria can continue to become antibiotic resistant in those countries and with increasing trade in meat products and international travel, consumers in countries that have banned antibiotic growth promoters may still be exposed to resistant organisms.

There is no evidence of bacteria becoming resistant to copper sulphate or zinc oxide. The pressure to remove these from diets is due to environmental concerns. High levels of copper and zinc in the effluent produced by pigs could have deleterious effects on the soil and watercourses. However, this concern is probably exaggerated as calculations show that with sensible policies for the spreading of effluent environmental impact should be minimal.

The removal of antimicrobial feed additives from weaner pig diets presents pig producers with major problems. Diets for newly weaned pigs are already very carefully formulated to reduce the challenge to the piglet's digestive system. The emphasis is now being placed on developing diets that actively promote gut health. A factor known to be responsible for a high proportion of the outbreaks of enteric disease in the post-weaning phase is the type of protein source used in the feed. Even though a diet may meet the nutritional requirements for maintenance and growth, some types of protein may cause hypersensitivity reactions in the piglet's gut. Plant proteins are worse than animal proteins in this respect and soyabean proteins may be amongst the worst. It is easy to formulate a high-energy diet with maize meal, wheat and corn oil and to balance this with skimmed milk powder as a protein source. These feed ingredients are highly nutritious and very palatable to young piglets and will cause few problems. The problem is that skimmed milk powder is a very expensive ingredient. The price of diets can be kept down by the inclusion of heat-treated soyabean meal. The use of these soyabean-based diets needs some caution, although it is not impossible to use them. 'Tolerising' piglets to the protein source can avoid many of the problems. This entails giving piglets the potentially harmful ingredients in small but regular amounts for as long as possible before weaning. When piglets are subsequently put onto a high level of the same feed ingredient after weaning, their immune system does not overreact with an allergic response because the animal is familiar with the source of protein.

A large number of different nutritional strategies are being tried in an attempt to produce a gut environment that will favour the beneficial lactic acid bacteria and minimize the enteropathogenic bacteria. Probiotics, which are dried cultures of lactic acid bacteria, can be added to the diets. Lactic acid bacteria are normal inhabitants of the gut and a healthy flora of these organisms tends to inhibit the further colonization of the gut wall by potentially pathogenic bacteria such as *E. coli*. Feeding liquid rather than dry diets can help stimulate feed intake in the weaned pig. The problem with liquid feeding has always been ensuring that the food stays fresh and palatable. Without good management of liquid feed it can become a source of enteropathogens. A recent approach has been to ferment liquid feed for weaner pigs using lactic acid bacteria. This approach uses the production of lactic acid to preserve the feed, in the same way that lactic acid bacteria preserve grass silage or yoghurt. If the fermentation is properly controlled the feed remains palatable and the piglet benefits not only from consuming the lactic acid but also from consuming large numbers of lactic acid bacteria.

The newly weaned pig has a much less acid environment in its stomach than the older pig. The acid conditions in the stomach provide a first line of defence against pathogens and also have an important role in protein digestion. To overcome the deficiency in stomach acidity in the weaned pig it is becoming more common to acidify their diets with mixtures of organic acids. Herbs and spices are also being added to diets. Many of these enhance the flavour of the feed and increase palatability. More importantly, many herbs and spices also have antimicrobial properties and help promote a favourable gut environment. A further approach is to include in the diet materials that stimulate the growth of the beneficial organisms in the gut. Some complex carbohydrates like fructo oligosaccharide and mannan oligosaccharide have been found to have a positive influence on the gut microflora and to promote in a natural way the same changes that have previously been achieved with antimicrobials.

Water delivery systems for weaned piglets need some care in their selection and operation. Nipple drinking systems have been used for some time and in general they are maintenance-free and almost impossible for piglets to foul. Unfortunately, some types may not be easy for small piglets to operate particularly in the immediate post-weaning period. If the piglet does not learn to drink quickly after it is weaned it rapidly becomes dehydrated and ill. On many units one of the first limiting factors to post-weaning growth is water intake. Low water intake limits dry matter intake. Providing water in a more accessible form can help overcome this problem. Many producers have found improved performance by using turkey drinkers in weaner pens for the first week or so after weaning. Producers have accepted that the additional labour involved in maintaining these devices is repaid by improved performance.

It is likely that a combination of a number of these strategies will be needed if producers are to develop management strategies that will enable them to maintain performance in the absence of feed additive antimicrobials. Experience in Sweden, which was the first country to ban antimicrobial growth promoters, has been that radical changes to management were needed. The most notable trend there has been the increase in weaning age. Increasing the age at which the pig is weaned overcomes a number of problems. The weaned pigs are more immune-competent and can cope better with a disease challenge, they have developed independent eating and drinking and so adapt better to removal of their mother as a source of food, their digestive systems are more mature so they can cope with less perfect diets and they are not so vulnerable to low environmental temperatures. It seems likely that for some producers the way to overcome the problems associated with weaning pigs at very young ages will quite simply be not to do it!

Growing-finishing pigs

Nutrition

The nutrient requirements of growing and finishing pigs have been the subject of vast amounts of research around the world as nutritionists have striven to keep pace with the improvement in the pig's genetic capability for lean growth. The differences in appetite (voluntary feed intake) and protein deposition rate (lean growth potential) vary considerably between different genotypes and make it impossible to give generalized recommendations for the nutrition of growing-finishing pigs. In order to optimize the nutrition of an individual production unit it is necessary to have information not only about the genotype, but also about the health status, the building environment in which the pig is to be kept, the stocking density and the market for which the pigs are being produced. All of these factors will affect feed intake and the way in which the pig uses its food to deposit lean and fatty tissue. Computer growth simulations are frequently used to model pig performance on individual units and to examine the potential impact of different feeding strategies on both biological and economic performance. Many producers purchase complete diets from a compound feed manufacturer and these companies normally provide computer growth modelling and market analysis as a service to their customers. Producers who wish to produce their own diets and use a proportion of home-produced cereals in their diets normally retain a nutrition consultant who can undertake similar exercises. The sensitivity of pigs to changes in nutrition and the price differentials paid for carcasses of different fat content make it imperative that producers avail themselves of this technology. In a harsh economic climate the ability of the nutritionist to fine tune the performance of a unit can make the difference between profit and loss.

It is beyond the scope of this chapter to provide an exhaustive account of current understanding of growing pig nutrition, but a brief outline of some of the underlying principles will be given.

The nutrient requirement of the pig can be determined using a factorial approach, that is to say by determining the nutrients required for different functions and adding these together to produce an overall requirement. A significant component of the energy requirement of the pig is the food needed for body maintenance. The pig needs to maintain body temperature. As a generalization, the pig produces heat (from metabolizing food) in proportion to its body weight and loses heat in proportion to its body surface area. As the pig grows its surface area to weight ratio changes and this has an impact on heat loss which in turn affects its lower critical temperature (LCT) and its maintenance requirement. Recent estimates of maintenance requirement suggest that modern high-lean genotypes have a higher maintenance requirement than less highly selected genotypes (Fig. 21.7). The maintenance requirement of modern genotypes can be calculated using the equation:

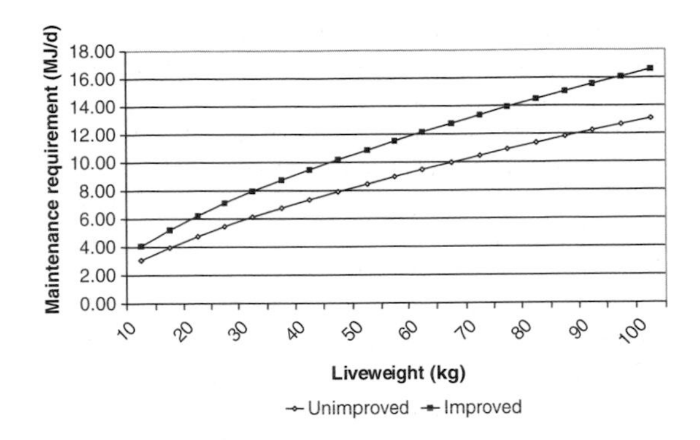

Figure 21.7 Effect of genotype on maintenance requirement of the pig.

$$M = 1.00\,W^{0.61}$$

where M is maintenance requirement in MJ ME per day and W is weight in kg.

The pig also needs energy in order to grow lean and fat tissue. The pig requires 44.1 MJ ME to deposit a kilogram of protein and 53.6 MJ ME to deposit a kilogram of fat. As each kilogram of protein represents approximately 4.3 kg lean tissue, each kilogram of lean tissue gain will require 10.3 MJ ME (10.7 MJ DE). Thus the energetic cost of depositing fatty tissue is five times the cost of depositing lean. Using these values it is possible to derive estimates of the energy requirements for pigs of different weights (Table 21.26). An important question that arises with modern genotypes is whether they actually have the capability to consume enough food to maximize their growth potential. Some estimates of the feed and energy intake that can be achieved under commercial conditions are given in Table 21.27. A comparison of the values for energy requirement in Table 21.26 with the values for predicted energy intake in Table 21.27 suggests that they are unlikely to be met unless diets have a very high energy density.

The other major component needed in the pig's diet is protein, or more specifically a supply of amino acids. The pig cannot elaborate protein without a supply of amino acids of which ten are termed essential. These amino acids must be supplied in the diet, as the pig cannot synthesize them from nitrogen sources or transaminate other amino acids to form them. In pig diets lysine is usually the first limiting amino acid; that is, the amino acid most likely to be in short supply and therefore determining the rate at which protein can be produced. For this reason the requirements of amino acids are usually expressed as a proportion of the requirement for lysine (Table 21.28). The proportions

Table 21.26 Factorial estimates of the daily energy requirements of superior genotypes[1]. (After Close, 1994)

	Body weight (kg)				
	20	*40*	*60*	*80*	*100*
Biological performance (g/d)					
Growth rate	800	1000	1200	1200	1200
Protein gain	137	171	205	205	205
Fat gain	141	178	215	215	215
Energy requirement (MJ/d)					
Maintenance	6.2	9.5	12.1	14.5	16.6
Protein gain	6.0	7.5	9.0	9.0	9.0
Fat gain	7.6	9.5	11.5	11.5	11.5
Total ME	19.8	26.5	32.6	35.0	37.1
Total DE[2]	20.8	27.8	34.2	36.8	38.9

[1] Assumes pigs capable of gaining 1.2 kg/d between 60 and 100 kg.
[2] ME × 1.05.

Table 21.27 Estimates of feed intake and digestible energy intake by pigs of different weights

Liveweight (kg)	*Feed intake*[1] *(kg/d)*	*DE intake*[2] *(MJ/d)*
10	0.56	9.21
15	0.76	11.90
20	0.95	14.26
25	1.12	16.41
30	1.28	18.41
35	1.44	20.29
40	1.59	22.07
45	1.74	23.77
50	1.88	25.40
55	2.02	26.97
60	2.16	28.49
65	2.29	29.96
70	2.42	31.40
75	2.55	32.79
80	2.67	34.15
85	2.80	35.48
90	2.92	36.78
95	3.04	38.06
100	3.16	39.31

[1] Feed intake (kg/d) estimated by the equation $0.10\,W^{0.75}$.
[2] DE intake (MJ/d) estimated by the equation $2.4\,W^{0.63}$ (less 10% to account for practical intakes).

will vary according to the purpose for which the protein is being used so, for example, the proportions will be different in a growing pig, in which the protein is being used for lean tissue production, than for a lactating sow,

Table 21.28 The optimum balance of essential amino acids in ideal protein (relative to lysine = 100). (After Close & Cole, 2000)

Amino acid	*Growing pig*	*Pregnant sow*	*Lactating sow*
Lysine	100	100	100
Methionine	28	28	28
Methionine + cystine	50–55	55	55
Tryptophan	18–20	20	17
Threonine	65–67	70	60
Leucine	112	100	112
Valine	78	78	78
Isoleucine	58	70	58
Phenylalanine	55	55	55
Phenylalanine + tyrosine	115	100	120
Histidine	33	33	33

in which most of the protein will be used to form milk protein.

When constructing diets it is usual to specify the protein requirement in terms of the ratio of lysine to digestible energy. Unfortunately, there can be no one definitive value as the protein requirement changes as the animal grows. The protein requirement is affected by sex (boars have a higher protein deposition rate than gilts) and by the genetic capability of the pig to deposit protein. An example of the way in which the lysine:DE changes during growth is given in Fig. 21.8.

It will be clear from the foregoing discussion that in order to meet the requirements of the pig precisely it would be necessary to change the diet on a daily basis. However, in practice most producers will only want to store and feed a limited number of diets. Many producers would feed three or four diets of different specification during the growing period (Table 21.29). This means that most pigs are being fed an excess of protein throughout their life. Not only does this represent a waste of valuable protein but also it means that pig effluent contains more nitrogen than it need. Unless slurry spreading is carefully managed this excess nitrogen can be a potential pollutant. On large units which can practise an all-in all-out policy it is possible to have a series of diets that better match the pig's requirement. These can be fed in sequence without increasing the cost of the feed delivery system. This system is known as 'step feeding'. An alternative approach which can be used in continuously stocked buildings is to use two diets at the extremes of the lysine:DE ratio and feed them in different proportions in order to match the requirement of the pig. This system, known as 'phase feeding', tends to

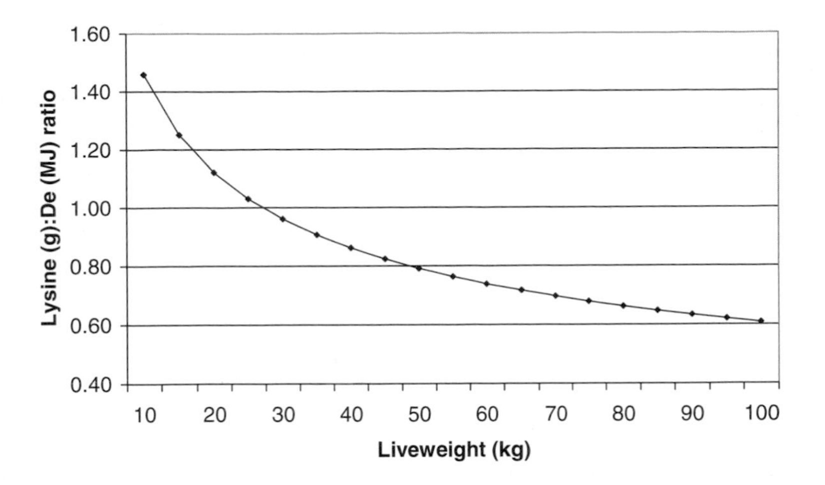

Figure 21.8 Phase feeding reduces the wastage of protein and reduces the nitrogen content of effluent.

Table 21.29 Suggested concentration of digestible energy (MJ/kg) and lysine (g/MJ DE) in diets for growing-finishing pigs of different weights

Liveweight (kg)	*Suggested DE concentration (MJ/kg)*	*Suggested lysine : DE ratio*
10–15	17.0	0.95
20–30	15.0	0.92
35–45	14.5	0.88
50–70	14.0	0.80
75–100	13.5	0.75

be expensive as the cost of producing the two extreme diets can be higher than the cost of producing individual diets. It is also expensive to equip buildings to feed in this way. Both step and phase feeding are most easily accomplished in units that liquid feed pigs and have computer-controlled feeding systems that both mix and deliver feed to the pigs. However, as environmental concerns continue to increase it may be necessary for pig producers to invest in more expensive feeding systems and engineer their diets to reduce environmental impact.

Feeds and feeding systems

The pig is an omnivore and as such is able to utilize a vast range of materials as food. In developed economies pigs are usually fed on a combination of cereal grains, oilseeds and their residues and high protein food by-products such as fishmeal and meat and bone meal. The cereal component varies with country. In the mid-west USA and in Brazil, and in countries importing feed grains from those countries, diet will be based on corn (maize) and soyabean meal. In Canada, Australia and the European countries diets will generally be based on wheat and barley with some of the protein being provided by rapeseed meal (canola in Canada). Some Far Eastern countries provide the carbohydrate portion of the diet from cassava (also known as manioc or tapioca) rather than cereals. Cassava has also been imported into Europe and has been used extensively in the Netherlands. Oilseed residues are widely used in pig diets and soyabean meal is ubiquitous in pig diets around the world. Although vegetable proteins are well utilized by the pig they are low in lysine and it is difficult to produce the whole range of diets needed by the pig from vegetable sources alone. Fishmeal and meat and bone meal both contain high levels of protein and are better suppliers of essential amino acids. Unfortunately, the availability and usage of both these valuable protein sources have declined in recent years. Fishmeal is produced as a by-product of fish manufacture for human consumption, but a considerable amount has been produced through the harvesting of non-edible species specifically to produce animal feed. There are concerns about the impact that this practice has on the aquatic food chain. This together with changes in traditional supplies due to a combination of overfishing and global warming has reduced supplies. A further trend putting pressure on available supplies is the rapid growth of aquaculture. Many farmed fish species will not grow adequately on diets based on vegetable proteins and need fishmeal as a component of diets. The fishmeal that is available to the livestock industry tends to be used in diets for young animals that have a very high requirement for protein and only limited appetite. Fishmeal is generally excluded from the diets of pigs from 50 kg liveweight as its inclusion can lead to fish taints in the fat of the slaughtered animal.

Meat and bone meal is a sterilized rendered by-product from the meat processing industry. Between 30 and 50% of the liveweight of slaughtered animals is material that does not have a market for human consumption. Until the emergence of Bovine Spongiform Encephalopathy (BSE) this material was processed and used in diets for pigs, poultry and ruminant animals. Initially in the UK, and subsequently in the EU, the emergence of BSE led to a ban on the use of meat and bone meal in farm animals. The ban is precautionary and consumer-led in the case of pigs and poultry, as there is no evidence that either of these species can contract transmissible encephalopathy by the consumption of prion-infected meat and bone meal. The ban has had two serious effects. First, it has removed a very valuable source of protein from pig and poultry diets and in so doing has probably increased the cost of diets by £5–10 per tonne. Second, it has had an adverse effect on the environment. Instead of the material being recycled it now has to be incinerated at high cost and with an adverse environmental impact. Meat and bone meal is still widely used in other countries and is likely to continue to be used unless those countries develop BSE.

In order to make good the deficiencies in vegetable protein, nutritionists have increased the use of synthetic amino acids. These amino acids are produced by micro-organisms in a fermentation process. Lysine and methionine are available at a competitive price and are widely used in diets for pigs and poultry. Trytophan and threonine are also available commercially but are less widely used partly because of their cost and partly because they are generally the third or fourth limiting amino acids and therefore less commonly deficient in diet formulations.

Diets are infinitely varied and so are the methods of presentation and feeding. It is beyond the scope of this chapter to consider the range of feeding systems used in commercial practice. Nevertheless, it is worth mentioning the four main options: presenting the feed as a meal or as pellets, in dry or liquid form. When food is prepared and mixed on the farm using a high proportion of homegrown cereals, it is generally fed in meal form. This is because pelleting the diet increases cost and slows throughput. It is not easy to produce good pellets and home-mixers often lack the appropriate expertise. Most diets supplied by feed compounders are in pelleted form. This is because pelleted diets are easy to handle, increase intake and reduce waste. Another reason why compounders pellet diets is that they wish to heat-condition diets in order to kill enteropathogens, such as *Salmonella*, which may contaminate raw materials.

If pigs are going to be fed dry diets then these are best fed in a pelleted form. Pellets reduce dust in buildings and reduce the wastage of feed from troughs and *ad libitum* feeders. This results in improved feed conversion and growth rate. The main alternative is to feed diets in liquid form. In Europe, around 30% of pigs are fed on liquid diets. These diets may comprise meal and water or may incorporate a wide range of liquid food-industry co-products like skim milk, whey, starch residues, potato steam peelings and vegetable processing residues. Where these materials are readily available they can produce significant reductions in cost. The Dutch make the best use of liquid co-products of any country in the world and estimate that the incorporation of two liquid co-products in a diet will reduce diet cost by 10–14%. As mentioned earlier, liquid feeding allows for much more sophisticated rationing of pigs and for the production of many more diets. Therefore, on a large unit the high initial capital costs can be justified by the improved cost per kilogram of gain achieved by using such systems and the ability they provide to control carcass composition to meet the specification set by processors.

Carcass quality and classification

The prime objectives in producing pigs for slaughter are to minimize cost of production (which can be measured in cost per kilogram of carcass gain) and maximize return from the carcass produced (measured in price paid per kilogram of carcass weight). In the past this presented the producer with a dilemma. Cost of production could be minimized by maximizing growth rate and thereby reducing the proportion of the food input used for maintenance. However, rapid growth used to be associated with excessive fat deposition in the later stages of growth and this in turn reduced the price paid for the carcass. To prevent pigs becoming overfat it was necessary to restrict feed intake quite considerably (up to 30%) in the later stages of growth. This increased the feed cost per kilo of gain and extended the growing period and hence the building and labour cost per pig. Over the last 30 years, improvements in the genotype of the pig coupled with improved nutrition have made it possible for many pigs to be taken to slaughter weight on an *ad libitum* feeding system. This simplifies management and minimizes the building and labour costs.

The market for pig meat has tended to become more uniform as well. In the past different breeds of pig were kept to different weights to satisfy the market demands for fresh pork, bacon and ham and pork for manufacturing into processed products. The needs of these different sectors can now be met by a single genotype and a single weight of pig. In the UK there has been a considerable reduction in the proportion of low liveweight pigs and an increase in the percentage of pigs in the 60–80 kg class (Table 21.30). There is increasing uniformity in the type of carcass required around the world. With the exception of some speciality markets, like that for Parma ham production, the demand is for a carcass of between 70 and 90 kg. Producing a heavier carcass has the advantage that it reduces the overhead cost per pig from the sow herd. For example, if a sow in the UK produced 20 pigs and these were slaughtered at 71 kg carcass weight the cost of maintaining the sow would be spread across 1420 kg pig meat. In comparison for a sow in Italy producing 20 pigs slaughtered at 112 kg the sow costs would be spread across 2240 kg pig meat. If we assume that the total variable costs incurred in producing the weaner pigs was £412 per year, the sow cost per kilogram of pork produced would be reduced from 29 to 18 p/kg. In the past the argument against taking pigs to heavier weights was that the carcass fat percentage invariably increased. With modern high-lean genotypes this argument no longer applies and the trend is for average carcass weight to increase to the optimum for lean meat production (around 120 kg liveweight or 90 kg deadweight).

Worldwide the requirement is for pigs with a high percentage of lean tissue and minimal fat. Consumers tend to have a more sedentary lifestyle than they did in the last century and are both better informed and more concerned about health issues and do not wish to purchase meat that is high in fat. Fortunately, the pig industry has been well placed to respond to the demands of the consumer. For many years pig geneticists have been reducing the fat content of the pig carcass and this has allowed pig meat to continue to compete in the market place. One advantage of pig meat over most other meats is that the level of intramuscular fat is minimal and for a given weight of lean tissue the fat content is significantly lower; although paradoxically there has been a move to use genotypes that include Duroc genes to increase the intramuscular fat content as this is seen to make the meat more succulent.

The value of the finished pig is determined by its weight and the fat content of the carcass and in particular the amount of subcutaneous back fat. In the UK, the Meat and Livestock Commission operates a national scheme for the classification of carcasses and in 1999 72% of the national kill (10.5 million pigs) was classified within this scheme.

Classification aims to describe carcasses objectively, in a manner that can be used by producers, meat wholesalers and retailers as a framework for payment. It is fortuitous that there is such a clear relationship between the depth of subcutaneous back fat on the pig and its saleable meat yield and lean content. This has meant that a relatively simple scheme of carcass measurements

Table 21.30 Percentage of pig carcasses in three weight ranges with average carcass weights and P_2 fat depths (1995–1999). (MLC, 2000)

weight		*1995*	*1996*	*1997*	*1998*	*1999*
<60 kg	% of total	16.6	11.1	9.2	8.9	7.8
	Average carcase weight (kg)	54.1	54.4	54.6	54.5	54.9
	P_2 (mm)	9.4	9.4	9.4	9.5	9.4
60–80 kg	% of total	78.5	83.1	83.3	82.5	84.4
	Average carcase weight (kg)	69.4	69.9	70.2	70.4	70.3
	P_2 (mm)	11.2	11.1	11.2	11.2	11.0
>80 kg	% of total	4.9	5.8	7.6	8.6	7.7
	Average carcase weight (kg)	83.6	83.4	83.7	83.7	83.5
	P_2 (mm)	13.5	13.1	13.2	13.1	12.9
All carcasses	Average carcase weight (kg)	67.6	68.9	69.8	70.1	70.1
	P_2 (mm)	11.0	11.0	11.2	11.2	11.0
Average lean meat %[1]		58.0	58.1	57.9	58.0	58.2

[1] Average predicted lean meat percentage based on the equation Lean meat % = $65.5 - 1.15 \times P_2$ (mm) + $0.076 \times$ carcass weight (kg).

gives easily defined grading schedules based on what the meat trade and consumers want. The three basic measurements of back fat used in the MLC scheme, known as the P_1, P_2 and P_3 measurements, are taken at the last rib 4.5, 6.0 and 8 cm respectively from the mid-line. The measurements are taken with an optical probe that is inserted in the carcass at the appropriate point(s). Lean percentage can then be predicted using the following equation:

$$\text{Lean meat \%} = 65.5 - 1.15 \times P_2\ (\text{mm}) + 0.076 \times \text{Carcass weight (kg)}$$

Each 1-mm increase in P_2 measurement between 9 and 13 mm will decrease the carcass lean content by 0.85 kg (1.15%).

Measurements taken at all three points are not necessarily used in any one grading scheme. For example, in method 1 the P_1 and P_3 measurements are taken and added together to give a joint score. In method 2 the measurement is taken at the P_2 position. In recent years a number of automatic recording probes such as the Hennessey Grading Probe and the Fat-O-Meater have been introduced. These are used in method 3. In addition to the P_2 thickness, these measure rib eye muscle depth and use both measurements to calculate carcass lean percentage. The Ultra Fat-O-Meater takes similar measurements but uses ultrasonics rather than an optical probe. The fourth method uses the CSB Ultra-Meater. This is also an ultrasonic technique and measures back fat and muscle depth 6 cm from the mid-line at the third/fourth last rib. The work involved in taking measurements can be automated, and computer-interfaced systems are available which can identify individual pigs, take and record the necessary information and process the data for grading the pig and paying the farmer.

Grading schemes are usually based on a combination of carcass weight and back fat thickness and will categorize pigs into three or more 'grades'. The categories are designed to encourage the producer to send the maximum number of pigs within the top grade, for which the best price is paid. A major advantage of such schemes is that pig breeding companies make the same measurements on live animals that are used for assessing payment on carcasses and use this information as a basis for selecting breeding stock for improved carcass characteristics.

Over the last 30 years the average back fat thickness has declined significantly. Despite the fact that there is considerable variation in the back fat scores of classified pigs, in general pigs are now leaner than ever before. Some wholesale meat companies have built into their grading schedules a penalty for pigs that are too lean. This is partly because some extremely lean pigs have gone beyond the threshold of consumer acceptability and partly because the texture of the meat is adversely affected and cutting becomes difficult. This seems a rather intractable dilemma and the producer is caught in the middle of consumers demanding increasing leanness and the meat trade seemingly reluctant to pay when very lean pigs are produced.

Within the EC, individual carcasses are designated into one of six EC grades, which are based on lean meat percentage (Table 21.31). As the data in Table 21.31 show, 87% of pigs classified in the UK fall in the top two grades.

A particular feature of the UK pig industry is the use of entire male pigs for meat. The entire boar has a higher lean growth rate and so converts food to meat more efficiently than the castrate. Although boar carcasses contain more lean meat they do have the disadvantage that the meat is not as tender. This is because the muscle fibres tend to be coarser in boars than castrates and the lean tissue somewhat drier. Another disadvantage of boars is that the carcasses can suffer from 'boar taint'. This results in them producing an unpleasant odour when cooked. This odour has been shown to have two components, one produced by androstenone (a breakdown product of the male hormone testosterone) and a second produced by skatole (produced by the microbial breakdown of the amino acid tryptophan in the lower gut). In fast-grown modern genotypes only a very small percentage of carcasses (probably less than 2%) may have detectable levels of taint. These can be identified at the meat processing plant and incorporated into cooked products where the taint will present no problems. There is considerable variation in the ability of consumers to smell boar taint, and the extent to which they consider it distasteful. For those consumers who can detect boar taint, cooking affected meat can be an unpleasant experience that makes them unlikely to purchase pork again for a long time. Thus, in order for the

Table 21.31 EC grade classification

Lean meat percentage	*EC grade*	*% UK pigs in classification (1999)*
60% and above	S	27.8
55–59%	E	59.4
50–54%	U	11.4
45–49%	R	1.10
40–44%	O	0.16
39% or less	P	0.14

industry to maintain repeat sales, it is important that the number of tainted carcasses reaching the market is minimized.

Quality Assurance schemes

Of all the changes that have occurred in the pig industry over the last 50 years the most significant and far reaching is the emergence of Quality Assurance schemes. Because of the need to produce a carefully specified product to meet the demands of the bacon market there has always been an articulation between producers and processors. The deadweight and grade payment system provided an important feedback to producers and ensured a dialogue between the producer and end user. Those linkages have now been extended down the production chain to include the retailer and the consumer and backwards to include the agricultural supply industry, the feed compounder and even the producer of raw materials used in diets. These changes have been driven by a number of different interests. The animal welfare movement sought to change the way in which pigs were housed and in particular to outlaw the use of sow stalls. Nutritionists wished to influence the 'healthiness' of food by encouraging lower fat content and an increase in the unsaturated to saturated fat ratio in meat. Health professionals wished to see reductions in antibiotic use because of the development of antibiotic resistance. Linking all these interests were the multiple retailers who wished to gain market share by reassuring the consuming public that their disparate interests were being protected. The net result has been the emergence of Quality Assurance schemes that go far beyond a desire to ensure that food is 'safe' and of consistent quality. The schemes that have been put in place seek to prescribe methods of production, and control every aspect of the food chain. While producers can see merit in many aspects of such schemes there is an inevitable and justifiable concern that compliance with such schemes has and will continue to increase the costs of production. Producers also fear that consumers are fickle and that expensive changes in response to short-term media attention or food faddism will neither secure their market nor increase the value of their products. As barriers to free trade are removed, national industries will become increasingly vulnerable to products that have been produced more cheaply in countries that impose lower standards of animal welfare, have less concern for human health and safety and less concern for the environment. Sadly, there is abundant evidence that in the final analysis consumers leave their ethics at the supermarket door and buy on price.

References and further reading

Agricultural Research Council (1981) *The Nutrient Requirement of Pigs.* Commonwealth Agricultural Bureau, Slough.

Boyd, J. (1999) Gilt edged strategy for lifetime performance. *The Pig Journal*, **43**, 110–21.

Close, W.H. (1994) Feeding new genotypes: establishing amino acid/energy requirements. In: Cole, D.J.A., Wiseman, J. & Varley, M.A. *Principles of Pig Science.* Nottingham University Press, Nottingham, pp. 123–40.

Close, W.H. & Cole, D.J.A. (2000) *Nutrition of Sows and Boars.* Nottingham University Press, Nottingham.

Cole, D.J.A. & Haresign, W. (eds) (1985) *Recent Developments in Pig Nutrition.* Butterworths, London.

Cole, D.J.A., Haresign, W. & Garnsworthy, P.C. (eds) (1993) *Recent Developments in Pig Nutrition 2.* Nottingham University Press, Nottingham.

Cole, D.J.A., Wiseman, J. & Varley, M.A. (1994) *Principles of Pig Science.* Nottingham University Press, Nottingham.

Darragh, A.J. & Moughan, P.J. (1998) The composition of colostrum and milk. In: Verstegen, M.W.A., Moughan, P.J. & Schrama, J.W. *The Lactating Sow.* Wageningen Pers, Wageningen, pp. 3–21.

DEFRA (2002) http://www.defra.gov.uk/esg/econfrm.htm.

English, P., Fowler, V.R., Baxter, S. & Smith, W. (1982a) *The Growing-Finishing Pig – Improving Efficiency.* Farming Press, Ipswich.

English, P., Smith, W. & Maclean, A. (1982b) *The Sow – Improving Her Efficiency*, 2nd edn. Farming Press, Ipswich.

Farm Assured British Pigs (1998) *Farm Standards Manual.* Farm Assured British Pigs, Stourport on Severn.

Food and Agriculture Organization (2001) http://apps.fao.org.

Freese, B. (2001) http://www.agriculture.com/sfonline/sf/2001/october/0111pork powerhouses.html.

Gill, B.P. (1998) *Phase Feeding: Effects on Production Efficiency and Meat Quality.* Meat and Livestock Commission, Milton Keynes.

Gordon, I. (1997) *Controlled Reproduction in Pigs.* CAB International, Wallingford.

HMSO (1998) *Codes of Recommendations for the Welfare of Livestock; Pigs.* HMSO, London.

Lyons, T.P. & Cole, D.J.A. (eds) (1999) *Concepts in Pig Science 1999.* Nottingham University Press, Nottingham.

Lyons, T.P. & Cole, D.J.A. (eds) (2001) *Concepts in Pig Science 2001.* Nottingham University Press, Nottingham.

MLC (1999) *Pig Yearbook 1999.* Meat and Livestock Commission, Milton Keynes.

MLC (2000) *Pig Yearbook 2000.* Meat and Livestock Commission, Milton Keynes.

MLC (2001) *Pig Yearbook 2001.* Meat and Livestock Commission, Milton Keynes.

National Research Council (1998) *Nutrient Requirements of Swine*, 10th revised edn. National Academy Press, Washington.

Piva, A., Bach Knudsen, K.E. & Lindberg, J.E. (eds) (2001) *Gut Environment of Pigs.* Nottingham University Press, Nottingham.

Standing Committee on Agriculture Pig Subcommittee (1987) *Feeding Standards for Australian Livestock. Pigs.* CSIRO, East Melbourne, Victoria.

Stark, B.A., Machin, D.H. & Wilkinson, J.M. (1990) *Outdoor Pigs, Principles and Practice.* Chalcombe Publications, Marlow.

Stotfold Pig Development Unit (2001) http://www.stotfoldpigs.co.uk/.

Thornton, K. (1988) *Outdoor Pig Production.* Farming Press, Ipswich.

Varley, M.A. (1995) *The Neonatal Pig: Development and Survival.* CAB International, Wallingford.

Verstegen, M.W.A., Moughan, P.J. & Schrama, J. (eds) (1998) *The Lactating Sow.* Wageningen Pers, Wageningen.

Whittemore, C. (1993) *The Science and Practice of Pig Production.* Longman Scientific and Technical, Harlow.

Useful websites

Statistics

http://apps.fao.org – Food and Agriculture Organization.

http://www.defra.gov.uk/esg/m_index.htm – Department for Environment, Food & Rural Affairs.

http://www.usda.gov/nass/ – US Department of Agriculture.

Organizations

http://www.danskeslagterier.dk/english/index.htm – Danish Bacon and Meat Council.

http://europa.eu.int/comm/food/fs/sc/scan/index_en.html – Scientific Committee on Animal Nutrition.

http://www.nppc.org/ – National Pork Producers Council.

http://www.pigsuk.com/option.html – Pigs UK.

http://www.stotfoldpigs.co.uk/ – Meat and Livestock Commission, Stotfold Pig Development Unit.

University and college pig sites

http://www.anr.ces.purdue.edu/anr/anr/swine/porkpage.htm – Purdue University Pork Pages.

http://www.bishopb-college.ac.uk/ncpit/ncpitf.htm – National Centre for Pig Industry Training.

http://cal.vet.upenn.edu/swine/index.html – University of Pennsylvania Swine Production Homepage.

http://www.greenmount.ac.uk/pigs/index.htm – Greenmount College Pig Pages.

http://mark.asci.ncsu.edu/ – North Carolina State University's Extension Swine Husbandry.

http://www.pighealth.com – Pig Disease Information Centre, University of Cambridge.

http://porkcentral.unl.edu/ – University of Nebraska, Pork Central.

On-line magazines

http://www.pigworld.org/ – Pig World.

Internet resources

(Sites with search engines and links to many other pig resources)

http://members.ozemail.com.au/~rroutley/ – The Pig Page (Australia).

http://www.oneglobe.com/agrifood/aginform/swine/idxswine.html – One Globe.

http://www.pighealth.com – Pig Disease Information Centre, University of Cambridge.

http://www.swine.net/swine.html – Swine Net.

http://www.thepigsite.com/ – The Pig Site.

22

Poultry

J. Portsmouth & T. Marangos

Introduction

The UK poultry business is part of an international production and marketing network. As a member of the European Union the UK industry produces poultry, meat and eggs under strict codes of welfare and hygiene regulations. At the same time it allows imports from non-regulated but never-the-less highly cost-efficient producing countries. This 'open marketing' policy has meant that UK producers are competing against countries as far afield as Brazil and the Far East, and unless they increase and maintain their competitiveness the future livelihood of many UK poultry producers is uncertain.

In all the major poultry countries chicken and turkey meat is produced from specific genetic material bred to produce high quality meat in a most efficient manner. Eating eggs are also produced from specifically selected stock bred to lay large numbers of high quality eggs whilst consuming low amounts of food.

World poultry information

Egg layers

There are 4.65 billion laying birds producing 820 000 billion eggs per annum: an average of 176 eggs per bird. China produces some 37% of all eggs produced, the USA 10% and the UK 1.34%. Consumption of eggs and egg products is 159 per capita in the UK compared to 251 in France, 242 in the USA and 368 in Japan, the world's highest consumer of eggs.

Broilers

In 1998 the world marketed 38 billion broilers, of which the USA produced 8.1 billion (21%), Brazil 3–4 billion (9%), Thailand 785 million (2%), China 5.87 billion (15.4%) and Europe 6.74 billion (18%), with the UK accounting for 11% of total European production. The largest broiler producer in the world is Tyson Foods with 26% of the USA total.

Average killing weight of broilers has risen from 2 kg in 1990 to 2.27 kg (5 lb) in 1997 in response to the increased demand for white meat, which yields higher at the larger liveweight. Cobb and Ross broiler strains account for approximately 50% of the world's production of broilers. Both these strains have excellent growth and feed efficiency and superior meat yield to other broiler strains available to the market.

Turkeys

World output is approximately 4.7 million tonnes per annum, of which the USA accounts for 50% and Europe 40%. Average killing weight of toms and hens in the USA is 13.5 and 7 kg respectively. The major turkey strains used in Europe and the USA are BUT (British United Turkeys) and Nicholas.

Meat production in the UK and Europe

Table 22.1 shows the growth of the UK poultry meat industry relative to the EU between the years 1989 and 1999. In this 10-year period broilers expanded by 51%

Table 22.1 The UK poultry meat industry (1989–1999). (Source: *Poultry World*, Reed Publications.)

	1989	*1994*	*1995*	*1996*	*1997*	*1998*	*1999*
Broilers – output ('000 tonnes)	753	993	1020	1075	1117	1132	1139
Slaughterings (million)	470	671	687	723	751	755	752
Consumption/head (kg)	17.1	19.6	19.8	20.6	20.4	21.8	22.6
Placings (millions)	593.5	719.5	730.2	765.1	792.2	793.4	787.7
Producer prices (p/lb)	25.9	27.2	25.8	28.8	26.5	23.4	22.3
Turkeys – output ('000 tonnes)	177	268	296	294	296	301	264
Slaughterings (million)	34	36	40	39	37	34	28
Consumption/head (kg)	2.6	4.6	5.0	5.5	5.2	5.1	4.8
Placings (million)	37.21	41.26	43.21	42.29	40.21	35.17	29.00
Ducks – output ('000 tonnes)	24	28	30	35	38	39	41
Slaughterings (million)	11	13	14	16	17	17	18
Geese – output ('000 tonnes)	3	3	2	3	3	3	3
Slaughterings (million)	6	1	1	1	1	1	1
Hens – output ('000 tonnes)	37	53	53	55	56	56	52
Slaughterings (million)	24	34	34	36	36	36	34
EU 15 poultry meat – output ('000 tonnes)	N/A	7771	8041	8353	8549	8789	8858
UK – output as percentage of EU 15	N/A	12.8	16.7	13.6	17.7	17.4	16.9

whilst turkey output rose by 49% and waterfowl by 63%, with ducks accounting for almost all of this. The UK meat output as a percentage of the EU output has risen from 13% to almost 17%, a growth of 32% or 5.3% per annum. UK poultry meat consumption has risen from 13.5 kg in 1981 to 29 kg/head/year in 1999 (Table 22.2). This is an annual increase of 6.4%. In the last decade the annual rate of increase has been approximately 2.5%. It is likely that this increase will be maintained and even improved upon as health-conscious consumers increase their consumption of poultry meat, particularly breast filets which are relatively low in saturated fat. Duck and goose meat is also set to rise as average consumer income and spending increase and duckling in particular gradually increases its competitiveness with other poultry meats.

Table 22.2 UK poultry meat consumption (kg/head/year) (1981–1999). (Source: *Poultry World*, Reed Publications.)

Year	*Total*	*Turkey meat*	*Broiler meat*
1981	13.5	2.2	10.0
1982	14.4	2.3	10.9
1983	14.5	2.5	11.3
1984	15.5	2.7	11.7
1985	16.0	3.0	11.9
1986	17.3	2.6	13.2
1987	18.4	2.6	14.1
1988	19.4	2.7	15.0
1989	20.9	2.6	17.1
1994	24.5	4.6	19.6
1995	25.0	5.0	19.8
1996	27.6	5.5	20.6
1997	27.1	5.2	20.4
1998	28.4	5.1	21.8
1999	29.0	4.8	22.6

The broiler industry

Since it began in the late 1950s, broiler production has been the most integrated form of livestock production. A mere handful of broiler integrations accounts for nearly 80% of total production. By 2010 the number of integrated companies will have decreased still further and whilst consumption per capita will continue to rise, an increasing proportion of this will be from imported broiler meat which is converted into a wide variety of ready-to-eat and ready-to-cook gourmet dishes by UK companies. Most of this will be imported from South and North America, Europe and the Far East.

A broiler is a young roasting chicken of either sex marketed between 1.75 and 3.50 kg liveweight (1.2–2.5 kg eviscerated weight). Age to marketing is between 35 and 56 days depending on the demand for whole birds, with heavier birds used for meat stripping.

Table 22.3 illustrates broiler performance, whilst Tables 22.4 and 22.5 provide recommendations on nutritional needs and feeding programmes to attain the required performance. Males are used to produce the heavier killing weights of 2.5–3.5 kg liveweight due to their superior feed efficiency and lower tendency to deposit unwanted carcass fat compared to females.

Anticoccidials and antimicrobials

An anticoccidial is a drug that inhibits the developments of the intestinal protozoa that cause the disease coccidiosis. The parasite is *Eimeria* and there are five major species: *tenella*, *acervulina*, *necatrix*, *brunetti* and *maxima*. The drugs used to control this important disease are either ionophores or chemicals and they can be used either as a single drug programme or in tandem, with one class being used in starter diets and the other in the finisher diets. These programmes are referred to as shuttle programmes.

Elancoban (monensin) is a major ionophore, whilst nicarbazin is one of the major chemical anticoccidials used. The future control of this important disease will be through vaccination as not only is this method likely to be more effective but also anti-antibiotic drug lobbies will lead to the demise of the established route of control for this disease.

Antibiotic growth enhancers, developed and used since the modern intensive poultry industry began in the 1950s, have been phased out. Used to enhance growth rate and improve feed efficiency through control of the gut microflora, many of them were alleged to have risked cross-resistance in humans. Two products remain on the approval list, Maxus (Avilamycin) and Flavomycin (Bambermycin). The demand for poultry meat produced from drug-free fed birds will continue and expand. Growth in organically produced meat and eggs from chickens housed and managed under semi-

Table 22.3 Broiler performance guide. (Source: JP Enterprises.)

Age at slaughter (days)	32	42	45	52	60
Average liveweight (kg)	1.7	2.2	2.6	3.1	3.5
FCR	1.6	1.8	1.9	2.0	2.2
Liveability (%)	97	96	96	95	92

Table 22.5 Broiler feed programmes (kg/1000 birds). (Source: JP Enterprises.)

	Age at slaughter (days)		
	35–42	*43–49*	*49–53*
Starter	250	250	250
Grower	1250	1250	1000
Finisher 1	To finish	2000	2250
Finisher 2	0	1000	1250

Table 22.4 Recommended broiler ration specifications. (Source: JP Enterprises.)

Nutrient	*Starter*	*Grower*	*Finisher 1*	*Finisher 2*
Protein (%)	23	21.5	20	18
Total Lysine (%)	1.40	1.28	1.17	1.05
Digestible Lysine (%)	1.20	1.10	1.00	0.90
Total Methionine (%)	0.62	0.58	0.54	0.48
Digestible Methionine (%)	0.55	0.51	0.48	0.42
Methionine + cystine (%)	1.00	0.95	0.90	0.84
Threonine (%)	0.94	0.88	0.78	0.70
ME (kcal/kg)	3035	3166	3266	3200
ME (MJ/kg)	12.70	13.25	13.50	13.38
Calcium (%)	0.90	1.00	1.00	1.00
Available phosphorus (%)	0.45	0.45	0.45	0.45
Sodium (%)	0.20	0.16	0.16	0.15
Linoleic acid (%)	1.25	1.25	1.25	1.25

and total extensive management systems is forecast to double within the next decade.

Neutraceuticals

Modifying and controlling the gut microflora will be accomplished by a new range of natural products such as herbs, herbal essential oils, organic acids and yeast, combinations of which are still in the embryonic stages of development. The oil of the herb oregano has been shown to exhibit a strong antibacterial and antiviral activity and organic acids such as formic and butyric lower the gut pH to the detriment of many pathogens such as *Salmonella* and clostridia. Certain yeast cultures such as *Saccharomyces cerevisiae* have been used to prevent an overgrowth of pathogenic bacteria. *S. bulardi* remains viable in the gastrointestinal tract and inhibits the growth of several bacterial species whilst improving the growth rate and feed efficiency in broilers and turkey poults.

International regulation of product quality, safety and efficacy of botanical feed additives has not yet approached the standard set by pharmaceutical companies. For continued expansion of this new area, higher standards of control will be demanded by the food industry. Table 22.6 provides a limited list of botanical feed additives.

The broiler breeding industry

The two major primary breeders in Europe and worldwide are Cobb and Ross. They account for some 50% of the world market in broiler breeding stock. Both strains are so-called heavy meat yielders because of their high meat to bone ratio and high level of breast meat and thigh meat.

Both strains perform according to the data in Table 22.3. The parent stock of the broiler produces between 115 and 130 first-quality chicks per breeder hen in a 60- to 62-week cycle. Very dedicated management and feeding practice is essential during both rearing and laying periods. Careful feed control is essential in order to follow the breeders' recommended growth curve during the rearing period. This is paramount and both breeder companies produce management and feed guides for the rearing and laying periods. An example of a feed control programme for the Cobb 500 pullet is shown in Table 22.7.

As feeds and feed quality vary in different parts of Europe and the world the guide must be used with caution and with technical advice from the specific breeding company. Special feed programmes must be carefully followed with emphasis on monitoring a pre-assigned body weight for age guide. At 20–21 weeks of age increases in both feed amounts and daylength are synchronized to bring the birds in lay without them becoming overweight. A well-managed and well-fed breeding flock will peak in lay at 80–84% and have good reproduction performance, averaging 82–85% hatch of all eggs incubated.

Commercial egg production

The UK laying flock continues to decline owing to the impact of social, environmental, economic and above all legislative influences and uncertainties. The fall in demand for eggs caused by controversial issues such as *Salmonella* and cholesterol has played a major role in this decline. Recent research, however, has shown that the type of fat that we eat rather than the cholesterol in our food affects the cholesterol in our body.

Improvements in egg production brought about by genetic developments of the commercial breeds listed in Table 22.8 have also increased supply throughout the world and with it lowered the price paid to the producer (Table 22.9). Within 10 years such genetic improvements will give rise to hens laying an egg a day for their productive laying cycle and achieve an increase of about 20 eggs over current acceptable performance. Assuming consumption does not alter, this could mean a further decrease in bird numbers of approximately 12.5%.

In June 1999, the UK laying flock, predominantly the brown varieties for the preferred brown eggs, numbered 30.6 million. By May 2000 this figure was 5.5% lower at 29.93 million according to MAFF. Numbers could fall to possibly 25 million by the year 2010 by extrapolating the effects of increased bird potential. In the UK these hens are housed in both conventional cage and non-cage (free-range and barn) systems of egg production.

Bird population and distribution

The total number of poultry farms with laying hens has fallen from 125 258 in 1971 to 26 500 in 1999, made up as follows:

Table 22.6 Plant extracts whose multiple activities are directly linked to active substance composition. (Source: *Feedmix*, vol. 8, no. 3, 2000.)

Vegetal form	*Utilized parts*	*Main compounds*	*Reported properties*
Aromatic spices			
Nutmeg	Seed	Sabinene	Digestion stimulant, antidiarrhoeic
Cinnamon	Bark	Cinnamaldehyde	Appetite and digestion stimulant, antiseptic
Clove	Cloves	Eugenol	Appetite and digestion stimulant, antiseptic
Cardamon	Seed	Cineole	Appetite and digestion stimulant
Coriander	Leaf, seed	Linalol	Digestion stimulant
Cumin	Seed	Cuminaldehyde	Digestive, carminatif, galactagogue
Anise	Fruit	Anethole	Digestion stimulant, galactagogue
Celery	Fruit, leaf	Phtalides	Appetite and digestion stimulant
Parsley	Leaf	Apiol	Appetite and digestion stimulant, antiseptic
Fenugreek	Seed	Trigoneiline	Appetite stimulant
Pungent spices			
Capsicum	Fruit	Capsaicin	Antidiarrhoeic, anti-inflammatory, stimulant, tonic
Pepper	Fruit	Piperine	Digestion stimulant
Horseradish	Root	Ally lisothiocyanate	Appetitie stimulant
Mustard	Seed	Ally lisothiocyanate	Digestion stimulant
Ginger	Rhizome	Zingerole	Gastric stimulant
Aromatic herbs and spices			
Garlic	Bulb	Allicin	Digestion stimulant, antiseptic
Rosemary	Leaf	Cineole	Digestion stimulant, antiseptic, antioxidant
Thyme	Whole plant	Thymol	Digestion stimulant, antiseptic, antioxidant
Sage	Leaf	Cineole	Digestion stimulant, antiseptic, carminatif
Bay laurel	Leaf	Cineole	Appetite and digestion stimulant, antiseptic
Peppermint	Leaf	Menthol	Appetite and digestion stimulant, antiseptic

- 25 600 with under 5000 laying birds;
- 600 with 5000–20 000 laying birds;
- 300 with 20 000 or more laying birds.

(Source: BEIS *Facts and Figures 1999/2000*.)

Flocks of 20 000 or more laying birds number 300, representing 78.5% of the UK laying flock. This trend is likely to reverse as welfare lobbies influence legislation and stocking densities decrease and non-cage units increase in numbers.

The market and marketing organizations

The decline in bird numbers and margins throughout the production chain combined with an increasing impact of cheaper imports is taking its toll on the major egg packers. Merges, acquisitions, reorganization and downsizing have occurred in the UK. Further rationalization is inevitable until and if the market gets into profitable balance.

During 1999, 78.5% of eggs were produced from birds housed in conventional cages. Free-range accounted for 15.5% and barn systems for 6%. The rise and fall of each sector over the last 40 years is shown in Table 22.10. This identities a trend of cage decline and a concomitant increase in alternative systems which is set to continue.

New niche egg lines and products are continuing to be developed in the free-range sector. These incorporate organic production and the trend towards natural, non-genetically modified raw materials, eggs from pure breeds, and also for the health-conscious consumers, functional food products incorporating omega-3 essential fatty acids such as DHA (docosahexaenoic acid).

Table 22.7 Cobb 500 breeder body weight targets and feed guide (female)

Age		*Body weight (g)*	*Feed*[1]	*Key points (Refer to* Cobb Breeder Nutrition Guide *for recommended nutrient levels)*
Days	*Weeks*			
	0–1			Feed type – Chick Starter
7	1–2	120		
14	2–3	260		
21	3–4	400	40–42	
28	4–5	520	44–46	
35	5–6	620	48–49	
42	6–7	720	50–52	
49	7–8	820	52–54	Feed type – Grower
56	8–9	920	54–56	
63	9–10	1020	56–58	
70	10–11	1120	57–59	
77	11–12	1220	58–60	
84	12–13	1300	60–63	
91	13–14	1380	62–64	
98	14–15	1440	64–65	
105	15–16	1520	66–69	
112	16–17	1600	69–73	
119	17–18	1700	74–79	
126	18–19	1820	80–86	Feed type – Pre-breeder
133	19–20	1960	87–93	
140	20–21	2160	95–102	
147	21–22	2320	103–109	
154	22–23	2500	111–113	Feed type – Breeder
161	23–24	2680	118–119	
168	24–25	2840	122–127	
175	25–26	2950	127–132	
182	26–27	3040	130–137	
189	27–28	3130		
196	28–29	3220		
203	29–30	3260		
210	30–31	3310		

[1] Feed amounts are only a guide. Ad libitum to maximum 40 g/bird/day.

Value of the UK market 1999

The value of egg output in the UK in 1999 was £359 million, down from £400 million the previous year, and the UK was 91% self-sufficient (Table 22.11). In terms of market volume approximately:

Table 22.8 Selection of egg-producing strains available in the UK. (Source: JP Enterprises and Nutrition Solutions.)

Commercial name	*Company*
Babcock 380	Hubbard ISA
Hisex Brown	Euribrid
Lohmann Brown	Lohmann Germany
Shaver 585	Hubbard ISA
Hy-Line Brown	Hy-Line International
Isa Brown	Hubbard ISA
Bovans	Joice and Hill

Table 22.10 Percentage of eggs produced under different systems. (Source: BEIC, 2000.)

Year	*Free-range*	*Barn*	*Cage*
1960/61	31	50	19
1970/71	6	9.5	84.5
1975/76	3	5	92
1987/88	7	1	92
1993/94	11	4	85
1998/99	15	5	80
1999	15.5	6	78.5

Table 22.9 Average producer prices

	1990	*1991*	*1992*	*1993*	*1994*	*1995*	*1996*	*1997*	*1998*	*1999*
Pence/dozen	53.5	48.6	46.0	52.2	52.4	49.5	58.4	52.2	48.1	46.4

Table 22.11 Volume and value of UK egg production. (Source: MAFF)

	1990	*1991*	*1992*	*1993*	*1994*	*1995*	*1996*	*1997*	*1998*	*1999*
All or total poultry and poulty meat (£ million)	1079	1079	1124	1271	1336	1333	1501	1482	1347	1247
Value (£ million)	439	410	397	427	432	399	467	428	400	359
As % of total	40.7	38.0	35.3	33.6	32.3	29.9	31.1	28.9	29.7	28.8
Production (million dozen)	804	818	806	791	787	774	775	794	792	740
Imports from (million dozen)										
the EU 14	73	50	41	47	65	64	70	69	63	64
the rest of the world	0	0	0	2	2	1	2	2	1	2
Exports to (million dozen)										
the EU 14	13	14	14	17	16	20	12	20	30	11
the rest of the world	1		1	2	1	2	2	6	7	8
Total imports as a % of production	9	6	5	6	9	9	9	9	8	9

Table 22.12 UK retail egg sales (expenditure) by outlet (year ending 2 April 2000)

	%
Multiple grocers	76
Co-ops	3
Independent/symbols	3
Butchers	3
Milkmen	4
Other outlets	11

(Source: BEIS Facts and Figures 1999/2000.)

- 60.5% of eggs are sold at retail level;
- 21.0% are purchased by the catering trade; and
- 18.5% are manufactured/processed into various egg products (excluding eggs for boiling).

The processing sector has increased dramatically in recent years as consumer preferences have moved to fast food and ready-made meals.

'Lion quality'

In 1995 the British Egg Industry Council (BEIC) developed and introduced a far-reaching Code of Practice for '*Lion Quality*' eggs, covering all stages of pullet production through to the ultimate consumer of eggs. This standard which includes compulsory vaccination against *Salmonella enteriditis* has more recently halted the decline in egg consumption and has seemingly protected sales to the multiple retailers against the threat of imports. The importance to and major influence on the egg sector of the retailers is shown in Table 22.12.

The '*Fast Food and Good for You*' campaign for Lion Eggs has gained considerable market recognition despite limited resources and sporadic advertising exposure.

Welfare legislation

New animal welfare legislation in the UK and EU such as the Council Directive 1999/74/EC will necessitate considerable structural change requiring significant capital investment in both new buildings and equipment. In addition the cost of producing eggs in all systems of production will increase as the industry is forced to move to less efficient systems of production.

The laying cage system

Laying cages still remain the most common method of commercial egg production in the UK, representing approximately 80% of eggs sold. Cages offer a high degree of control over the bird's environment. They also afford a good opportunity for mechanization of routine operations such as the provision of feed and water and removal of faeces.

Currently birds are kept in cages at a stocking density of 450 cm^2 per bird. The new Council Directive which will come into force on 1 January 2003 will require a reduction in stocking density to 550 cm^2 and from that date the only cages that can be brought into use will have to be 'enriched'.

Enriched cages

By 1 January 2012 conventional cages will be withdrawn from use. The only cages then allowed will be 'enriched'. The Directive requires such cages to include the following specifications:

- at least 750 cm^2 of cage area per bird, of which 600 cm^2/bird must be usable;
- the height of the cage must be at least 45 cm, except over the nesting area, where it must be at least 20 cm;
- perching space of at least 15 cm/bird;
- a suitable litter area and nesting space.

Enriched cages may compromise bird welfare. The recommendations were based on small-scale trials that were not scientifically sound.

Understandably there is a great deal of uncertainty in the UK egg industry, which is discouraging investment in both cages and alternative systems. Producers with birds in older conventional cages would have been reinvesting in new cages, yet due to the uncertainty will now operate old cages until the end of 2011. This is not in the interests of bird welfare.

The barn system

Approximately 5% of eggs sold in the UK are produced in the barn system. In this system, the hen house has a series of perches and feeders at different levels. The European Egg Marketing Regulations stipulate a maximum stocking density of 25 hens/m^2 of floor space. Perches for the birds must be installed to allow 15 cm of perch per hen. In the deep-litter system the birds are kept in hen houses in which all the floor area should be solid with a litter of straw, wood shavings, sand or turf. The maximum permissible stocking density for the deep litter system is 7 birds/m^2. In both systems one nest box per five birds or communal nest at the rate of 120 birds/m^2 of floor area of communal nests are provided. Natural lighting may be supplemented by electric lighting to provide an optimum daylength throughout the year.

The Lion Code of Practice stipulates the following additional standards for Lion Quality barn eggs: maximum flock size of 32 000 birds divided into colonies of 4000 where flock size is over 60 000 birds in total; maximum stocking density in-house of 11.7 birds/m^2 for part litter and slat systems and 15.5 birds/m^2 in multi-tier systems; scratching area for dust bathing; 5-cm linear length or 4-cm circular feeding and drinking space per hen.

The free-range system

The free-range system currently accounts for 15% of eggs sold in the UK and will increase in importance. The European Egg Marketing Regulations stipulate that for eggs to be termed 'free-range' hens must have continuous daytime access to runs which are mainly covered with vegetation and with a maximum stocking density of 1000 birds/ha 395 birds/acre. The hen house conditions must comply with the regulations for birds kept in perchery (barn) systems, or deep-litter stocked at 7 birds/m^2 when no perches are provided.

The Lion Code of Practice stipulates a maximum flock size of 16 000 birds and the same additional standards as for Lion Quality barn eggs, provision of outdoor shading in absence of a veranda and one pop-hole per 600 birds open for 8 hours daily to allow access to the outside.

The multiple retailers have also introduced their own far-reaching codes of practice which require the RSPCA Freedom Foods organization to police and audit their producers, affording them the recognized and consumer-accepted Freedom Foods logo.

For non-cage systems of production (deep litter, perchery/barn, free-range) the Directive requires an eventual maximum stocking density of 9 birds/m^2 usable area from 2012, i.e. a 23% reduction. New non-cage units are made uncompetitive by the requirement to comply at once with a stocking density of 9 birds/m^2, whereas existing units with a stocking density of up to 12 birds/m^2 are allowed to operate until 2012, so no new non-cage stems will be introduced.

The capital cost to the UK egg industry of implementing the Council Directive has been estimated by the BEIC as over £500 million (BEIC, 2000) An NFU-conducted survey of egg producers showed that over 60% of respondents do not expect to remain in egg production after 2012 (BEIC, 2000).

World trade considerations

The new Council Directive will raise the cost of production within the EU at a time when there is pressure in the World Trade Organization (WTO) for a reduction in import tariffs and increase in import access. This would leave the egg industry, and more particularly the egg processing industry, vulnerable to imports from less developed countries which have lower costs of production and lower welfare standards. The egg processing industry currently accounts for 25% of the EU market. The Commission estimates that if import duty were decreased by 33% in the years up to 2010 then the industry would no longer be competitive if costs increased by 10%.

The UK Government and Commission have indicated their wish to see animal welfare standards included in any future trade agreements and efforts are being made in discussions of the WTO. Nevertheless there seems little support for their inclusion in the next round of WTO talks.

Feeding programmes for laying birds

Laying hens are invariably fed *ad libitum*. The hen's adjustment of feed intake to change in ration energy level is not precise but sufficiently accurate to make *ad libitum* feeding the most economical system.

To obtain high and economic production certain minimum amounts of nutrients must be provided. These are shown in Table 22.13.

Table 22.13 Recommended intake of certain essential nutrients to produce 58 g egg output with a house temperature of 21°C. (Source: JP Enterprises and Nutrition Solutions.)

Nutrient	*Minimum daily requirement*	*Percentage nutrient in ration at specific average daily intake*		
		100 g	*110 g*	*120 g*
Protein	19.0 g	19.00	17.30	15.80
Lysine	950 mg	0.95	0.86	0.79
Methionine	450 mg	0.45	0.41	0.38
Methionine + cystine	700 mg	0.70	0.64	0.58
Calcium	4.20 g	4.20	3.80	3.50
Available phosphorus	0.33 g	0.33	0.30	0.28
Linoleic acid[1]	1.25 g	1.25	1.14	1.04
Xanthophylls[2]	14 mg/d	14.00	12.75	11.70
Energy (MJ/kg feed)		11.9	11.5	11.3

[1] The level of linoleic acid influences egg weight. To maximize egg weight, 3 g linoleic acid/bird/day is required.
[2] Recommended level of a blend of xanthophylls ensures a good yellow–orange colour (Roche fan 10–11).

Feed restriction for replacement pullets

Excess body fat adversely affects subsequent rate of lay. Commercial laying pullets are reared on an *ad libitum* feeding system as it brings birds into lay at an earlier age. Such birds should not, however, be fat and the balance between additional body weight and earlier sexual maturity is a delicate one as it is important to optimize not only egg production but also egg weight.

It is advisable that the breeder companies' recommendations for the growing period and body weight for age are closely followed.

Turkeys and water fowl

Both turkeys and ducklings are sold all the year round. Festive occasions such as Christmas, Easter and Independence Day (in the USA) still, however, account for a major part of overall sales. Changes in eating habits, working wives, etc. have led to turkey being increasingly consumed in ready-to-eat, quickly heated consumer-friendly dishes. A major pioneer in this market in the UK is Bernard Matthews Ltd. whose turkey and meat processing operation has been operating for about 50 years. This business is one of the most successful in the global poultry business.

Table 22.1 shows UK turkey meat supply balance. Over the past 10 years total output has increased by 49% whilst imports of whole birds has fallen by half. Imports as a percentage of off-take have changed only slightly and currently stand at 17.8%.

Table 22.14 Management standards: turkey growth rate and FCR (as hatched birds) of BUT Big 6 turkeys. (Source: British United Turkeys Ltd.)

Age (weeks)	*Live weight (kg)*	*Food consumed (cumulative) (kg)*	*FCR (cumulative)*
4	1.12	1.66	1.49
8	4.00	6.80	1.77
12	7.81	16.0	2.05
16	11.67	25.40	2.18
20	15.06	41.26	2.74
22[1]	20.10	55.47	2.76

[1] Stags only.

Table 22.15 Nutrient : energy ratios for growing turkeys. Requirement values for nutrients for growing turkeys expressed in relation to the metabolizable energy content of feed (g nutrient/MJ ME). These values apply in conditions where temperatures, stocking density and pellet quality do not depress feed intake. For hens, BUT research has not indicated a need to feed sexes differently up to 8 weeks of age. Thereafter, diets formulated to meet the recommended nutrient : energy ratios for males can be used for hens in a modified feed programme. (Source: British United Turkeys Ltd. 2000.)

Age (weeks)	*Nutrient (g/MJ ME)*										
	Lysine	*Methionine*	*Total sulphur amino acids*	*Tryptophan*	*Threonine*	*Arginine*	*Calcium*	*Available phosphorus*	*Sodium*	*Salt*	*Essential fatty acids*
0–4	1.57	0.57	1.02	0.27	1.00	1.69	1.14	0.64	0.13	0.25	1.27
4–8	1.34	0.53	0.94	0.23	0.86	1.46	1.04	0.58	0.13	0.25	1.09
8–12	1.10	0.46	0.83	0.19	0.75	1.21	0.94	0.53	0.13	0.25	
12–16	0.89	0.40	0.71	0.15	0.58	1.02	0.86	0.49	0.13	0.25	
16–20	0.75	0.36	0.64	0.13	0.48	0.88	0.78	0.45	0.13	0.25	
20–24	0.65	0.32	0.57	0.11	0.42	0.80	0.74	0.41	0.13	0.25	

Table 22.16 Comparative growth performance of broilers, ducks and turkeys at age 45 days. (Source: JP Enterprises.)

	Broiler	*Duck*	*Turkey*
Average liveweight (kg)	2.60	3.80	3.10
FCR	1.90	2.20	1.65

UK turkey poult numbers placed for meat production peaked at 43.2 million in 1995 but declined dramatically to 29 million in 1999, whilst total placing of poults in the EU stands at 241.3 million birds. Within the EU France produces in excess of 50% of total output and is also a major exporter of turkey meat. Turkey production continues to be dominated by large integrated producers. However, the last 5 years have seen a revival of extensive traditional turkey production using specific slower-growing dark-feathered varieties such as the Kelly Black, Kelly Bronze. This small breeder and producer in Essex, England, has dominated this section of the market, demonstrating how a niche market area can be productive when operated by first-class management and with first-class products.

Tables 22.14 and 22.15 provide information on turkey growth rates (Table 22.14) and recommended diet specification for feeding fattening turkeys to 24 weeks of age (Table 22.15), whilst Table 22.16 shows comparative growth rates of the turkey, broiler and duck to 45 days.

Ducks

Having remained static for a number of years this specialized area of the poultry meat market shows definite signs of growing – albeit slowly. Consumption of duck meat has increased 63% in the last 10 years. Cherry Valley, the UK market leader in breeding and fattening, has done much to promote duck meat and has made marked improvements in meat to bone ratios.

Geese

Statistics on goose production are poor compared to other meat sectors. The British Goose Association has been proactive in the past 5 years and consumption of goose meat at festive times is increasing.

Micro-nutrients

Good nutrition is not simply knowledge of the birds' requirements for protein (amino acids) and energy but also for a wide range of micro-nutrients such as vitamins and trace elements. Table 22.17 sets out the recommended levels of vitamins and trace elements to be added to the diet. This table covers all classes of poultry and is based on the use of wheat as the major cereal used in UK poultry rations.

Table 22.17 Recommended micro-nutrient levels for poultry fed wheat-based diets. (Source: JP Enterprises.)

Micro-nutrient	*Broilers*		*Chick/layer/breeder*			*Turkeys*			*Ducks*
	Starter	*Finisher*	*Breeder*	*Chick/ grower*	*Layer*	*Starter*	*Grower/ finisher*	*Breeder*	*Starter/ finisher*
Vitamin A (m iu)	12	12	13	10	8	15	10	15	12
Vitamin D_3 (m iu)	5	5	5	3	3	5	3	5	3
Vitamin E (k iu)	50	40	100	50	20	80	50	100	50
Vitamin K_3 (g)	10	10	12	8	6	15	10	20	5
Vitamin B_1 (g)	2	2	3	1	1	5	1	2	2
Vitamin B_2 (g)	8	7	14	8	4	8	6	20	12
Nicotinic acid (g)	70	50	50	20	10	75	50	70	80
Pantothenic acid (g)	12	12	20	10	8	25	15	25	15
Vitamin B_{12} (mg)	20	20	25	20	10	20	20	30	20
Vitamin B_6 (g)	3	2	7	3	1	7	5	5	4
Choline (g)	300	250	400	200	0	400	150	200	400
Folic acid (g)	2	1.5	3	2	0	3	2	3	2
Biotin (mg)	250	200	300	100	0	300	300	380	200
Manganese (g)	110	110	120	110	100	120	100	120	100
Zinc (g)	100	100	100	100	100	100	70	100	100
Copper (g)	20	20	20	20	10	20	20	20	15
Cobait (g)	0.40	0.40	0.50	0.50	0.40	0.40	0.40	0.50	0.50
Iron (g)	30	30	30	40	20	50	20	50	40
Selenium (g)	0.30	0.30	0.30	0.25	0.20	0.30	0.25	0.25	0.25
Iodine (g)	1.0	1.0	2.0	1.0	1.0	1.0	1.0	1.0	1.0

References

BEIC (2000) *The Welfare of Laying Hens – Overview of Current Issues*. The British Egg Industry Council, London.

Dennett, M. (1995) *Profitable Free Range Egg Production*. Crowood Press, Marlborough.

Leeson, S. & Summers, J.D. (1997) *Commerical Poultry Production*, 2nd edn. University Books, Ontario.

Leeson, S. & Summers, J.D. (2000) *Broiler Breeder Production*. University Books, Ontario.

Sainsbury, D. (1998) *Animal Health*, 2nd edn. Blackwell Science, Oxford.

Sources of further information

Aviagen Ltd., Newbridge, Midlothian, Scotland.

British Egg Information Service, Cromwell Road, London SW7 4ET.

British Goose Producers Association, London SW7 4ET.

British United Turkeys Ltd., Chester, CH4 0EW.

Cherry Valley Ducks, Market Rasen, LN7 6BP.

Cobb Europe, Midden Engweg 13, 3882 TS Putten, The Netherlands.

International Hatchery Practice (monthly magazine), Positive Action Publishing Ltd., Driffield.

Kelley Turkeys, Danbury, Essex, CM3 4EP.

Poultry World (quarterly magazine), Reed Business Publishers, Sutton, Surrey.

World Poultry (monthly magazine), Elsevier, Amsterdam.

23

Animal health

D. Sainsbury

Introduction

Good health is the birthright of every animal and it is the duty of the livestock keeper to do everything possible to ensure this. Considerable advances in the control of animal infections have led to the effective elimination of many of the more traditional causes of acute disease and it is no longer necessary to consider disease an inevitable part of a livestock enterprise. This situation has been achieved by a combination of good hygiene, the appropriate use of vaccines, good medicinal therapy, and the development of disease-free strains of livestock, but above all by improved husbandry and management. These advances have made it possible to keep animals in much larger groups and more densely housed than hitherto, but the results have by no means been a gradual disappearance of infectious and contagious diseases altogether.

On the contrary, a number of complex diseases have emerged, which are difficult to diagnose and induced by a multiplicity of pathogenic agents. Whilst these may cause an apparent or 'clinical' disease it is more likely that the effect will be less obvious and may only reduce the overall productivity of the livestock by, for example, slowing growth and reducing the feed conversion efficiency. Animals may not die or even show any symptoms at all, so that the farmers may be unaware of what is happening unless they keep very careful records and use them with more than the usual degree of skill. It is also a very common phenomenon that intensive production on a livestock unit may start efficiently and effectively, but deteriorates in time so gradually that it is not noticed until the consequences have become very serious and control becomes extremely difficult. The nature of these infections is of especial interest and concern because the environmental and housing conditions have a profound effect on their severity. In the field of contagious diseases the new problem of the 'viral strike' is emerging: many apparently new virus diseases, borne by wind or vectors, travel through areas causing devastating effects for a time, especially in the large livestock units. More virulent and mutating strains of pathogens appear to be produced the larger the 'biological mass' and the more vaccines are used to protect the animals from diseases.

A further major problem is that of the *metabolic diseases*. These are a group of diseases that are caused intrinsically by the animals being called upon to produce an end-product faster than the body can process its intake of feed. Enormous efforts are made to provide the right nutrients in an easily assimilated form, but as the metabolic disease is rather different from a deficiency disease, this does not necessarily work. In a way the metabolic diseases are the inevitable outcome of the success in conquering the acute virulent disease, together with the advances made in improved genetics, nutrition and housing. These improvements have led to much increased growth and productivity, but the capability of the animals to keep pace with these has outstripped the normal functioning process (or metabolism) of the body.

An example of this is provided by considering the growth of table chickens. When the industry started intensively it took about 13 weeks to produce a bird weighing 2 kg, at a feed conversion efficiency of approximately 3:1 (that is 3 kg of feed to produce 1 kg of liveweight). The hazards of growing the birds at that time were innumerable and were largely related to contagious and infectious diseases. Mortalities were high, often up to 30%, and the diseases played their part in slowing growth and damaging the feed conversion efficiency. Now some 40 years later, the position is quite different. It takes less than half the time for the birds to reach 2 kg: that is, about 5 weeks. Mortalities normally average 3–4%, and contagious and infectious diseases are controlled largely by good hygiene together with judicious use of vaccines and medicines. However, now we have a number of the so-called metabolic or production diseases: for example, birds are affected with excess fatty deposition, causing degeneration of the liver and kidney and heart attacks. The skeletal growth may not keep pace with the rate of muscle development so that many locomotor disorders occur, such as twisted, rubbery and broken bones, and slipped tendons. Not dissimilar conditions occur in other livestock subjected to intensive demands.

Table 23.1 The disease pattern on the farm

Poor husbandry + Primary disease agent (often a virus)
+ Secondary disease agents (often bacteria or parasites)
= Subclinical or overt disease

Examples of bad husbandry
Too many animals on a site
Overcrowding within the buildings
Mixed ages within a house or site
Excessive movement of animals
Poor ventilation and environmental control
Bad drainage and muck disposal
Insufficient bedding
Lack of thermal insulation in construction
Unhygienic or insufficient food and watering equipment
Absence of routine disinfection procedures
Faulty nutrition

Factors influencing the incidence of disease

An important factor influencing the incidence of disease in livestock today is the increasing immaturity of livestock. Improved performance has resulted in animals reaching market weight much earlier, whilst, for genetic reasons, breeding animals are also younger on average than previously. Thus, on a livestock farm, there is usually a high proportion of young animals that are in a state of susceptibility to infectious agents whilst they are still developing the ability to resist disease naturally or natural immunity to disease which will normally take place over a prolonged period. This difficult state of affairs is further exacerbated by the considerable size of many livestock units in which the young animals may have originated, from various parents of very different backgrounds. In many cases the young or growing stock have come from widely separated areas. They may have no resistance to local infections and will therefore be totally susceptible to them, and at the same time contribute a new burden of pathogenic micro-organisms to the unit they have entered. Altogether the modern livestock unit may present at any one time a confusing immunological state and the basic design and its management will influence the success or otherwise of disease control (see Table 23.1).

There are certain major groups of infections that account for most of these problems. The most widespread are probably respiratory diseases. Many of these are subclinical, have a pronounced debilitating effect on the animals and are caused by a large number of different infective agents even in any one disease incident. They may not respond satisfactorily to vaccines, antisera, antibiotics or any other medicines, so that the fundamental method of approach to their control is by managerial, environmental and hygienic measures. In the first place, considerable advances in the control of animal infections have led to the effective elimination of many of the traditional causes of acute disease. This has been achieved by a combination of appropriate vaccine usage, good veterinary medicines, continually improving hygiene measures and the development of disease-free strains of livestock. These advances have made it possible to keep animals in much larger groups and more densely housed than hitherto, but the results have by no means led to a gradual disappearance of infectious diseases altogether.

Another significant group of diseases influenced similarly are enteric infections. These also have many different primary causative agents, ranging from parasites to viruses and bacteria. The reasons for their increasingly harmful effects in recent years are not only those already listed above but also the general trend with certain livestock to eliminate the use of bedding such as straw. The harmful effects of this may often be corrected by the use of good pen design, especially by the use of slatted or slotted floors, but it is nevertheless more difficult to separate animals from their urine and faeces when the flooring is without litter, as the latter has a diluting and absorbent effect on the excreta.

There are, in addition, several important bacterial infections that have tended to increase in large intensive units. Examples of these are *Salmonella* and clostridia bacteria spores, together with *Escherichia coli, Pasteurella* and *Campylobacter*. Many forms of these organisms are normal inhabitants of the animals' intestines in small numbers, but excessive 'challenges' causing disease many build up under unhygienic intensive conditions, encouraged by buildings with poorly constructed surfaces which cannot be cleaned.

Livestock unit size

In addition to the risks of the gradual buildup of disease-causing agents within livestock buildings there are serious dangers of livestock enterprises becoming too large. In all parts of the world where the development of large units has taken place a number of diseases have emerged which tend to 'sweep' like a forest fire through areas with a large livestock population, sometimes leaving a trail of devastation. Often there is likely to be considerable loss over a concentrated period, after which the animal population may develop a natural immunity, at least for a time, or artificial immunities are promoted by the use of vaccines. It is now known that contagious virus and other particles can travel great distances from large infected sites. Certainly distances of 50 miles have been virtually proven, but they may well travel much further than this. If livestock enterprises continue to grow in size then the dangers in this respect can only become greater. At the present time it is impossible to give soundly based objective advice as to the optimal unit size, and in any event the factors that would lead to making proposals are highly complex. However, there is clear evidence that animals thrive less efficiently in large numbers even in the absence of clinical disease. Apart from this there is the difficulty of hygienic disposal of the dung when the unit is of excessive size.

It is normally possible to keep very much greater numbers of *adult* animals together than young stock, since there is nothing like the same number of contagious disease problems after the difficult growing stage and its immunological uncertainties are passed.

As to the maximum number of livestock that might be kept on a site, it will be appreciated that this will depend on a number of factors, apart from health considerations. Figures have been proposed that attempt to allow for all factors and those suggested are as follows:

Dairy cows	200
Beef cattle	1000
Breeding pigs	500
Fattening pigs	3000
Sheep	1500
Breeding poultry	3000
Commercial egg layers	60 000
Broiler chickens	150 000

Such figures can be no more than suggestions made in the light of practical experience. In due time more scientific evidence may be available to establish a better degree of accuracy. It is certainly likely that the figures will require constant adjustment in the light of new developments in husbandry, housing and disease control and especially with the anticipated trend towards the increasing provision of livestock free of specific disease.

The benefits of smaller livestock units are becoming appreciated more widely worldwide and there is evidence that 'small is beautiful' is becoming more fashionable.

Depopulation of buildings or site

A fundamental concept in maintaining the health of animals is to ensure the periodic depopulation of a building or a site. The benefits of eliminating the animal hosts to disease-causing agents are well understood and the virtue of being able to clean, disinfect and fumigate a building when the animals have been cleared is also accepted. Periodic depopulation is especially important for young animals but is less so for groups of older animals which have probably achieved an immunity to many contagious diseases. Much also depends on whether the herd or flock is a 'closed' one with few introductions of fresh animals or an 'open' one with a constant renewal of the animal population. If the latter is the case, then constant depopulation is of greater importance as there is little or no opportunity for natural immunities to develop and the regular removal of the 'buildup' of infection is of great assistance in ensuring the good health of the livestock.

Health status of stock

The policy will also depend on the health status of the stock. At one extreme there are the so-called minimal disease or specific pathogen-free herds which have been developed to be free of most of the common disease-causing agents of that species. Here, depopulation is less critical than the protection of the animals from outside infections. Since this danger is very serious in most localities it is important to subdivide the animals in a unit into smaller groups, lessening the likelihood of a breakdown and/or enabling isolation and elimination of a group that may become infected. At the other extreme there are those units that have a constant intake of new animals from outside and of a totally unknown health status. In this case there is a high risk and, more usually, a near certainty that some will be either clinically infected with, or carriers of, disease-causing organisms. Design specifications for such units should be quite different from those of the closed herd or flock, so that defined areas of the unit should have groups of animals put through them in batches after which the area can be cleared, cleaned and sterilized. Obviously, it is preferable if the whole unit can be so treated since it ensures an absolute break in the possible disease 'buildup' cycle.

Between these two extremes is the more usual case in which a herd or flock is of reasonable health status, though certainly not free of all the common diseases, and in which new livestock are added only occasionally. In such cases the precautions in the housing against disease buildup and spread of infection can be relaxed, but there should still be proper provision for the isolation of incoming and sick animals.

Group size

Animals thrive best in groups of minimal size. If groups are small, it is usually easier to match the animals in them for size, weight and age, and it is well established that growth under these circumstances is likely to be most even and economical. Behavioural abnormalities such as fighting and bullying are also kept to a minimum – indeed they may be prevented altogether.

Fighting amongst animals is a highly contagious condition and under the most intensive husbandry systems an almost casual accident that may draw some blood can escalate into a blood-bath. Pens that keep the animals in small groups will tend to reduce the occurrence of such disasters and indeed with good management the removal of an animal that has accidentally injured itself or is off-colour and therefore prone to being bullied may stop the trouble before it has had a chance to develop seriously.

There is yet another economic advantage in keeping animals in small groups. The farmer will achieve the best economic return if the stock are housed at the densest possible level for optimal productivity. If animals are kept in a house to allow a density such as this, it means that they should spread across the house evenly so they do in effect occupy and use this area. In practice, however, this is very difficult to achieve, especially when large numbers are housed together without any subdivision at all. Hence the birds, or any other livestock as may be at risk, crowd in certain parts of the buildings which can lead to grossly overstocked floor areas. If livestock crowd excessively in certain parts of a house, this area is likely to become more polluted with dung and respiratory exhalations to an abnormal and harmful degree; the humidity becomes high, proper air movement is impeded and the animals soon may become ill. Sick animals feeling cold tend to huddle together more, so the vicious circle is perpetuated and there is seemingly no end to it unless some measures are taken to ensure a better distribution of the stock. When this problem arises under practical conditions it may often be impossible to subdivide the animals at once. An immediate trend in the right direction, that is encouraging the animals to spread themselves more uniformly over the house, can often be achieved by introducing some artificial heat. There are excellent gas radiant heaters and oil-fired and electrical blower heaters available.

When animals are penned in large numbers, the effects of a fright caused by an unusual disturbance can be extremely serious. It is almost impossible to guard against all the extraneous sounds and sights that may affect the stock. The best safeguard, therefore, is to have the animals housed in small groups so that the effect of a panic movement will be more limited and will never build up into highly dangerous proportions.

Floors

The profound effect of the floor surface on the health and well-being of livestock is well established. When bedding was almost invariably used in animal accommodation there were relatively few problems related directly to the flooring. Now that the farmer must frequently use housing systems without bedding, often with slatted or other forms of perforated flooring, to produce economically a comfortable and clean environment for the animals, new problems have emerged. Though we are still far from being able to advise on the ideal floor, it is possible, by choosing a good combination of surface and bedding where used, to provide the animals with a comfortable, warm, hygienic and well-drained surface.

The best solid flooring is usually based on concrete because when properly made and laid it is hard wearing, hygienic and impervious to fluid. When an animal lies directly on the floor, without bedding, there should be an area of insulated concrete, most often incorporated by using 100–200 mm of lightweight or aerated concrete under the top screed and above a damp-proof course.

The surface of the concrete must not be so smooth that the animals slip or injure themselves; on the other hand, if it is too rough it can cause abrasions and injuries. A happy medium is not easy to find but is usually achieved by using a wood float finish or by tamping the floor lightly with a brush. Also, very rarely are the 'falls' of the floor correct; again, there must be a balance between being steep enough to drain away liquids yet not so sloping that the animals slip. With some animals the floor is made more comfortable by placing rubber or plastic mats on top; these are used with some success with cattle, calves and pigs. Some of the worst problems of injury, especially with cattle and pigs, have occurred with slatted or other forms of perforated floors. These have been particularly troublesome when the edges have been left too sharp or have worn, leaving injurious protrusions.

If bedding can be used for livestock it is usually better to do so. Good bedding is an insulator and warmer for animals and is probably the cheapest flooring that there is. The air temperature of buildings can more safely be lowered if bedding is used. When the bedding is straw it may form an acceptable, if not an essential, adjunct to the diet of the animal. The fibre may help digestion and create a sense of well-being. It keeps the animal occupied and assists in the prevention of vices.

Bedding also provides just about the safest flooring for the animal, so reducing the danger of injuries to the body, particularly the limbs. It is also much more difficult to keep the floor dry when no bedding is used, and if the animal's surface is kept wet there is a further chilling effect created by contact with a wet floor. It is also important to emphasize the health risk from undiluted muck which does not pass through slats but accumulates at some points, and the great risk to the respiratory system of humans and animals from gases rising from slurry tanks or channels below the slats. Not surprisingly, therefore, there has been a trend towards the advocacy of bedding for most animals.

Isolation facilities

There is an urgent need for better consideration to be given by the farmer to the isolation of sick animals. Isolation removes the dangers of contagion to the normal animals. It also makes it much more likely that the sick animal will recover without the unwelcome attentions of the other animals who will always tend to act as bullies. In nature the sick animal usually separates itself from the rest of the herd or flock, but it is generally impossible for the housed animal to do this. Also, under intensive management, more animals suffer from the aggression of their pen-mates and it is essential that such animals are removed to prevent them from suffering or being killed.

It is also easier, with an isolated animal, to give it such therapy as it requires and to give it any special environmental conditions, such as extra warmth. It is an interesting but important observation, known to most farmers, that if you take 'poor doers', perhaps animals suffering from subclinical disease, out of a group, and pen them with more space and extra comfort, they may thrive without any further attention. Nevertheless, with the addition of suitable therapy, the cure may be accelerated and completed.

The disposal of manure

The method used for manure disposal has potentially important effects on health, both human and animal. Whilst the smell from composted solid manure tends to be strong, it rarely travels far or creates a nuisance problem and there is little, if any, risk to the human or animal population from this form of manure under temperate climatic conditions.

Slurry, however, is quite a different matter. If placed straight on the land from the animal house or after holding in a tank anaerobically it has an extremely offensive smell. While masking agents are possible, they are too expensive at present to be considered economic. The worst smell comes from the pipeline and gun spreader, because the droplet size is small and light, and particles may carry for considerable distances. Less smell arises from a tanker spreader because the slurry is much thicker and not spread by aerially dispersed small droplets. Satisfactory ways of preventing the slurry from causing offence are to treat it aerobically in some way before spreading or inject it under the surface. Many human and animal health problems may arise from the spreading of slurry. In surveys it has been found that potentially pathogenic bacteria were able to survive for up to nearly 3 months in slurry kept under anaerobic conditions. Whilst the particular bacteria studied were *Salmonella* species and *E. coli*, there is little doubt that more resistant organisms, such as *Bacillius anthracis, Mycobacterium tuberculosis, Clostridium* species and *Leptospira* species, could survive as long or probably very much longer. If the slurry should enter a river or stream, the pollution may have far-reaching and infinitely more serious effects. It is thus essential for all enterprises with slurry as the disposal system that either the slurry is placed on land where it cannot be a nuisance or health risk or it is treated beforehand so that the risks are removed.

Dangers from gases in farm buildings

Hazards associated with gases in and around livestock farms have been highlighted recently, particularly in association with gas effusions from slurry channels under perforated floors. Numerous fatalities have been recorded of livestock, and even humans, due to gas intoxication. There is, in addition, mounting evidence that there may be concentration of gases in many livestock buildings that may affect production adversely by reducing feed consumption, lowering growth rates and the animals' susceptibility to invasion by pathogenic micro-organisms.

The most serious incidence of gas intoxication arises from areas of manure storage in slurry pits or channels under the stock, usually, but not always, associated with forms of perforated floors. The greatest risk arises when the manure is agitated for any reason, usually when it is removed. There is also an ever-present danger if a mechanical system of ventilation fails and this is the only method of moving air in the house. Several cases of poisoning have been reported when sluice gates are opened at the end of slurry channels and the movement of the liquid manure has forced gas up at one end into the building.

High concentrations of gases, chiefly ammonia, may also arise from built-up litter in animal housing. This is most likely in poultry housing, since the deep litter system is the most commonly used arrangement with broiler chicken and poultry breeders. The danger has undoubtedly been exacerbated within the past few years, owing to the necessity of maintaining relatively high ambient temperatures in order to reduce food costs, while at the same time there has been good evidence that higher temperatures than hitherto should be maintained for optimal productivity. Poultry farmers have often attempted to achieve such temperatures by restricting ventilation in the absence of good thermal insulation of the house surfaces, and the result can be generally harmful if not dangerous.

One of the most popular forms of heating is by gas radiant heaters which are suspended from the ceiling. Well over half of all poultry housing is heated in this way and because the cost of gas as a fuel appears in some respects to be improving in relation to that of other fuels, the system is being used more for pig housing and especially for piglets in the early weaning system. There is a risk that with inexpert use such heaters may be improperly serviced and the house insufficiently ventilated to give complete combustion and that toxic quantities of carbon monoxide may be produced.

Growth promoters

For many years the poultry industry has made use of antibiotics and other substances that, when incorporated in very small amounts in the ration, may increase the growth rate of livestock and improve food conversion efficiency. Since medicines as growth promoters are used at a much lower level than for therapeutic use, the incorporation of antibiotics at a low level in the feed is unacceptable since resistant organisms will soon emerge and not only will the antibiotics no longer be of use as growth promoters but they will also have lost their value for therapeutic purposes. Because of this there is now a clear separation, legally enforced, between the use of antibiotics for these two purposes. Those antibiotics that are used as growth promoters (e.g. Flavomycin) have no other use and are not used for any therapeutic purposes. The therapeutic antibiotics are used only for prevention and treatment of disease and are available only on veterinary prescription to be used under the control of the veterinarian.

This problem of resistance is not the only one in connection with the use of the feeding of growth-promoting antibiotics. These substances have their greatest effect when the hygiene or husbandry system is less than ideal. Under really good management practices they may do little good at all. Thus they tend to 'mask' bad husbandry and this is a dangerous characteristic since such an effect will not last for ever; the bad practices, however, cannot only last indefinitely but will tend to have a cumulative effect on results.

There is still no absolute certainty as to the means by which growth-promoting antibiotics exert their effects, but it seems most likely they are due to their favourable action on the balance of bacteria in the intestines, reducing the harmful ones and promoting the beneficial ones and thereby increasing the efficiency of food utilization and production.

The importance of disinfection

Livestock are generally housed much more intensively than hitherto. This involves not only the stocking of animals more densely but equally an increased size of unit and the more efficient use of building space. It is rare for a livestock pen to be without stock; if there is a break between batches then it is usually a short-lived one. In the days before the use of antibiotic therapy, intensive methods tended to collapse under the burden of disease. Now they are again under threat because an increasing number of disease-producing organisms are resistant to medicines and also the cost of antibiotic therapy is becoming an economically damaging burden to the stock keeper.

There is one viable answer and it is to institute a very efficient programme of cleaning and disinfection of the buildings. Let us be quite clear, however; disinfection needs to be done in a carefully organized way and must not be, as so often is the case, an application of any disinfectant chosen at random in the hope that it will kill the pathogens. First, it is imperative, at least with young livestock, to plan the enterprise in such a way that a building or even a site is totally depopulated of stock before the cleaning and disinfection takes place. This is necessary because the live animal, and it might be literally only one, is potentially a much greater reservoir of infection than the building itself. So the complete emptying of a house can be more important than the disinfection process itself. The disinfection process in sequence is as follows:

(1) Empty the pen and preferably the house of all livestock.
(2) Clean out all 'organic matter' – dung, bedding, old feed, any other material that could contain pathogenic micro-organisms. Remove totally from the area of the buildings.
(3) Remove all portable equipment for cleaning and disinfecting outside the building.
(4) Wash down with heavy-duty detergent-disinfectant. Use a power washer wherever possible.
(5) Apply a disinfectant appropriate to the types of infection being dealt with. Usually required is a mixture active against viruses, bacteria, parasites and insects which may carry infection from one batch of livestock to another.
(6) Wherever possible fumigate the house with a disinfectant in aerosol or spray form after the equipment has been reassembled.
(7) Dry out and rest for a day or two before restocking.

Disease and immunity

Organisms causing infections

There are different groups of organisms causing infectious disease in animals. The smallest of them are viruses, below 300 nm in size, but many are well below this. For example, the foot-and-mouth disease viruses are 10–27 nm (1 μm = 0.001 nm and 1 nm = 0.000 001 mm). Viruses cannot usually be seen under the ordinary microscope. They are simple organisms that can only multiply within living cells and this property distinguishes them from bacteria. Viruses are classified into a number of different groups, e.g. reoviruses, adenovirus and herpesviruses.

Bacteria are relatively large organisms compared to viruses and can multiply and grow outside living tissues. In size, for example, cocci (round bacteria) measure 0.8–1.2 μm, bacilli (rods) are 0.2–2 μm and the spiral-shaped spirella are up to 50 μm. They are visible under the ordinary microscope. Many types, once outside the animal body, may sporulate to form a protective coat so that they can live many years in buildings, soil or elsewhere and can still be capable of infecting animal life. For example, the spores of anthrax and *Clostridium* can live for 20 years or more under favourable circumstances.

Mycoplasma are smaller organisms than bacteria, being about 0.25–0.5 μm (or 250–500 nm). They are rather like oversize viruses, but, unlike viruses, they can be cultured on artificial media. Rickettsiae are of somewhat similar size to *Mycoplasma*, being 0.25–0.4 μm, but they cannot be grown in ordinary culture media and will only multiply intracellularly, like viruses.

All these organisms are classified as from the the vegetable kingdom and can be joined by the yeasts, moulds and other fungi which are larger than bacteria, several varieties of which cause diseases in animals and humans.

There are also many simple organisms from the animal kingdom, known as parasites, which cause disease in farm livestock. These range from the single-celled protozoa, such as the coccidial parasites, to parasites of ever-increasing size and complexity, culminating in such relatively complicated organisms as the roundworms (helminths), lice, flies and ticks.

Organisms from these groups are the main cause of infectious disease in farm livestock. The only other agents causing disease, and which are not infections, are the metabolic disorders, poisons, deficiencies, excesses and injuries, examples of which are found throughout this chapter.

Immunity

In our understanding of the natural mechanisms that maintain good health it is essential to know the principles of the normal animal's immunological system and the way in which we make use of such products as vaccines and antisera to boost it as necessary. This is best understood by studying the cycle of events in any animal's life.

When an animal is born it is relatively free of disease organisms, but it has what is known as *passive immunity*, which it receives from the mother whilst still *in utero*. If the mother has been vaccinated against a disease or has had an experience of the actual disease, then her immunological system will normally produce the antibodies that are capable of resisting the disease. These antibodies circulate in the blood and will also be transferred to the fetus. The antibodies themselves are effective only for about 2 or 3 weeks before they disappear.

The young animal can produce its own antibodies and thus develop what is known as *active immunity* only if it has experienced the disease organisms that cause the disease in question, or has been vaccinated against it. It should be stressed that an animal will have some passive immunity only to those particular infections experienced by the mother. These may be quite limited and 'local'. This is a good argument for rearing an animal in the environment in which it is conceived and a good reason for not transporting young animals too early in life before they have some better ability to resist a 'foreign' disease challenge. Passive immunity in mammals is fortified immediately after birth by the young drawing the first milk (or colostrum) from the dam. Colostrum is especially rich in antibodies and also nutrients.

As the passive immunity the young animal receives from its mother fades it may be replaced by an active immunity if the young animal is vaccinated or has some challenge from the organisms that can cause disease. Under practical conditions the aim is to make sure that any natural challenge of potentially pathogenic organisms is always mild or gradual, not massive and overwhelming, as in the latter case disease will arise.

Sometimes it is perfectly satisfactory to allow the animals to receive a 'natural' challenge from the environment, but this is at the best inexact and at its worst ineffective. Thus, various methods of induced immunity are given, either *passive* or *active*. The passive immunity is induced by injecting into the animal hyperimmune antiserum prepared from the serum of animals that have experienced the actual infection and have recovered. This serum may be used for treatment of the disease and for passive immunity but has only a transitory effect lasting about 3 weeks, similar to the natural immunity received by the young animal from its mother.

For an active immunity, vaccines are used. A vaccine is a way of activating the body's defence mechanism by challenging it with the pathogenic organisms so modified that they are just active enough to produce an immunity but not the disease itself. The modification is done in very many ways, by adding chemicals, by growth of the organism in special media or livestock of a different species or by irradiation. Sometimes no modification is required as an organism can be used which is sufficiently closely related to induce immunity but not disease. Occasionally, in actual farming practice, if no vaccine is available for a disease which is almost certainly likely to challenge the animals later in life, the animals are deliberately exposed to the disease-causing agent and at such a time when no actual disease will be caused. This is a venture that is hazardous and it is unwise to do it without proper veterinary supervision.

The production of vaccines is a very sophisticated process but can induce an enormously efficient protection for the animal. Vaccines must be used by skilled hands at the right time in the correct dose and repeated as necessary since the length and strength of the immunities developed vary according to the infection. It should be borne in mind that when a vaccine is given, the immunity takes time to develop, somewhere in the order of 3 weeks, so it may be necessary to protect an animal during that time by antisera, antibiotics or simply by isolation. In general, the live vaccine may give a stronger immunity more quickly, but it is not necessarily any longer lasting and it may induce more reaction. Dead vaccines given by infection in oil or other special bases can be very effective for long-term immunities and are increasingly widely used. It should be emphasized that when a vaccine is administered to an animal it causes a degree of illness and during the period of immunological development the animal is experiencing stress and is *more* susceptible to infection than at other times. Thus it is necessary to practise even better management at the time of vaccination. Very often this factor is not appreciated and a failure that is unjustifiably ascribed to the vaccine is essentially due to mismanagement of the animals.

Medicines

Vaccines and sera are used to protect or treat animals for specific diseases. They should be used wherever they can be in the prevention of infections, but in many cases no serum or vaccine exists and we must deal with illness using one or many of a wide choice of medicines. Also, many of the disease conditions we are dealing with are caused by a number of different organisms and the identity of some or all may not be known definitely. Frequently the urgent need for treatment demands that a medicine is used before laboratory examinations have elucidated the cause, and we must rely on the experience of the farmer and the clinical skill of the veterinarian to judge the correct medicine to use. When the results of the tests are known, an appropriate change in treatment may be indicated.

Medicines can be given to have an almost immediate action. One administration may have an effect for only a few hours or in some cases a few days, but usually, in order to keep the levels up within the bloodstream of the animals, there must be a constant boosting of levels. It cannot be emphasized too strongly that the full course of treatment prescribed should be completed and levels maintained correctly or the benefits may be lost and medicine-resistant organisms produced: this possibility is one of the worst outcomes of the use of medicines. Care must also be taken in the agricultural use of medicines to observe certain conditions of use. For example, with many medicines a compulsory withdrawal period is required before animals can be sent for slaughter, varying from about 5–10 days and in some cases even longer, and milk from cows being treated for mastitis must not be mixed with normal milk for specific periods. Withdrawal periods are supplied with the products and are essential to obey.

The majority of medicines used for the prevention and treatment of disease are chemotherapeutics, the term implying the use of chemicals that have been either synthesized, such as the sulphonamides, or produced by the action of living organisms, these being the antibiotics.

The routes by which medicine are given vary from injection by the intravenous, intramuscular or subcutaneous routes to oral dosing and incorporation in feed or drinking water. With so many medicines available and the routes and methods of administration being so variable it is perhaps inevitable that careless and abusive use occurs. One cannot do more than emphasize that the prescribing of medicines requires a high degree of skill and their administration an abundance of diligence and competence.

Signs of health and disease

Early recognition of illness is particularly vital because the risk of spread of anything contagious is very great. Early recognition also increases the opportunity for successful treatment, enables the affected stock to be isolated and speeds any laboratory diagnostic work. It also helps in the good welfare of the animals since sick animals are often maltreated by their neighbours. It will also generally reduce losses to a minimum and thus make sound economic sense.

The following are the most important signs to look for.

Appetite

One of the earliest and surest signs of illness is a lack of interest in feed (anorexia) or a capricious appetite, though on occasions it may be the feed itself which is at fault and requires investigation.

Separation from the group

Such animals will usually separate from the others and hide in a corner or bury themselves in bedding if available.

Excreta

This may be 'abnormal'. Sick animals often have scour (diarrhoea) or, more usually in the earliest stages of disease, they will be constipated, especially if there is a fever. In other circumstances the faeces may contain blood (dysentery) or may be of abnormal colour (which sometimes happens in cases of poisoning, for example).

Urine

This may show abnormalities such as the presence of blood or may be cloudy or yellow (jaundiced). Normal urine is pale and straw-coloured.

Posture

Sick animals may find it impossible to rise, or may hold themselves uncomfortably, and may be lame. The nature of the abnormality of posture may indicate the organ affected.

Appearance

Sick animals will often have a droopy head, with dull eyes and dry muzzle and possibly discharge from the nose and watery or catarrhal eyes. All mucous membranes may be discoloured.

Skin and coat

A healthy coat is clean and glossy. Abnormal coats may be 'hide-bound' which is when dehydration makes it difficult to move the skin over the underlying tissues. Sick animals may have bald, scurfy, staring, 'lousy', mangy or scabby skin, and the animals may scratch or rub. The skin, coat and feathers are very good indicators of disease.

Coughing

Most abnormalities of the respiratory system produce coughing in animals and the nature of the cough will often be diagnostic of certain conditions.

Pain

The presence of pain shows in a number of ways: grunts, groans, grinding of the teeth, squealing or crying, and arching of the back.

Mucous membranes

The normally moist membranous linings to the mouth, nose and other external orifices may show abnormalities such as discoloration, dehydration, discharge of a serous, mucous or purulent nature, or haemorrhages.

Temperature

An abnormally high temperature usually indicates an infection; an abnormally low one may be due to a metabolic defect, poisoning or simply the later or terminal stages of an infection. Normal temperatures and respirations are as follows:

	Temperatures		*Respirations/min*
	°C	°F	
Cattle	38.7	101.5	12–20
Sheep	39.4	103.0	12–30
Pig	39.2	102.5	10–18
Fowl	41.5	106.5	12–28

Respirations tend to be accelerated by infectious diseases and slowed by others though the nature of the breathing will also be vastly affected and may change from the deep to the shallow on the occurrence of illness.

Pulse rates

These are normally only studied by the veterinarian, but the normal rates are as follows:

	Pulse rate/min
Cattle	45–50
Sheep	70–90
Pig	70–80
Fowl	130–140

Animal welfare and animal health

A recent phenomenon has been the widespread questioning of the humanity of certain methods of rearing livestock. Under critical attack from so-called welfare groups have been systems of housing that greatly restrict the movement of animals, amongst these being cages for chickens, calves housed in crates for the production of veal and sows closely tethered in stalls during pregnancy. In the UK the rearing of calves in crates where there is no bedding and where they cannot turn round is now banned, as are sows kept in tethers or in stalls where there is also severe restriction on movement and usually no bedding. Few other countries are copying such welfare legislation and there is a danger that unilateral welfare measures will merely encourage livestock production to expand in countries without welfare rules and regulations.

Animal welfare is very pertinent to the question of health since there is no dispute that one of the essentials for the provision of good welfare is the maintenance of health in the animals. It has also become apparent from recent activities in the research field that the eventual goal of establishing what constitutes good welfare in a scientific way is going to be very difficult indeed, if not impossible. Thus it will be necessary to continue to rely, at least in the foreseeable future, on the overall knowledge, expertise and even instinct of the persons concerned with the management and care of the animals, these being the farmers, stock keepers and husbandry and veterinary advisers. It is essential to keep in the forefront of consideration the overriding importance of good welfare and humanity to animals on all occasions. Anyone who has closely observed sick animals can understand what constitutes 'misery' in the animal kingdom and contrast this with the alert, buoyant and inquisitive appearance of an animal in good health.

A feature of this controversial field which has been beneficial is the way in which it has forced us to enquire whether it is necessary to house animals using methods which involve a high degree of restriction on their movement. There has been a vigorous search for alternatives which has been fruitful as it appears there is less need for the extremes of intensification than has been alleged and indeed it may be economically preferable to have systems that are more in harmony with the overall agricultural scene.

Large and intensive livestock enterprises may be unsuccessful for three principal reasons. First, the management of large numbers of *living beings* is not easily organized on the large scale. Stockmanship is partly individual. If only one small item goes wrong or is overlooked, it can lead to a very serious chain of events on the large scale.

The second reason is that disease is much more likely to have a serious deleterious effect in the large unit in subclinical or clinical forms. The failure of many units can be ascribed to their inability to control disease so that the economic viability of the unit has been totally destroyed.

The third reason is the heavy cost of buying in or bringing in fodder and moving out the muck in such enterprises. It is common for the highly intensive unit to be totally dependent on food from outside the farm or the locality and to be faced with great difficulty in disposing of all the waste products. In smaller units the crops can be used to feed the livestock and the 'muck' they generate is of great value to the land; energy is conserved and this is agriculture of a totally balanced sort.

It may be stated also that the welfare standards that are gradually emerging across the world are based on evidence which, so far as is possible, is not in conflict with economics. For example, if animals are housed at concentrations and intensities above those advised, it is likely that their productivity will fall. Particularly this is so when it comes to stocking density and its relationship to growth, production and the efficiency of food utilization. Good welfare standards can therefore be used as husbandry standards with some confidence.

The key to good welfare is a high standard of stockmanship. If this is to function, good facilities must be provided for the stock person. Many systems that have been developed in recent years completely ignore this. The essentials are as follows:

- plenty of room, for the stock person to see the stock whenever he/she needs to;
- good lighting and no undue saving on passage space;
- facilities for isolation of sick stock;
- easy access by the stock person to the animals and, above all, a very high standard of handling facilities so that it is easy to medicate or treat the animals in any way necessary.

Disease legislation

In Great Britain, under the Diseases of Animals Act, there are a substantial number of Orders relating essentially to the state's interest in animal disease control. These regulations cover a wide spectrum of activity. For example, they deal with the importation of animals and animal by-products, the movement of animals and the inspection of livestock and vehicles, and allow for the Ministry of Agriculture, Fisheries and Food to institute a large number of Orders relating essentially to specific diseases.

Under the Diseases of Animals Act certain specified diseases are designated notifiable. Those of current importance are:

- Anthrax
- Foot-and-mouth
- Fowl pest
- Bovine tuberculosis
- Bovine spongiform encephalopathy ('mad cow disease')
- Scrapie
- Swine vesicular disease
- Aujeszky's disease
- Enzootic bovine leucosis
- Rabies
- Warble fly

There are also others, such as swine fever, that could well occur again as they are widespread in other parts of the world.

The most important feature about the notifiable diseases is that the owners or persons in charge of an animal suspected to be affected by a notifiable disease must report immediately his or her suspicion to a police officer or an inspector of the Ministry of Agriculture. Once the owner has notified the inspector or police, steps will be taken to diagnose the condition and measures may be taken to isolate the premises and treat, vaccinate or slaughter the stock. Different diseases have different procedures and these are dealt with later in the chapter under the respective diseases. Also, under the Zoonoses Order, the presence of conditions such as salmonellosis and brucellosis, which infect both animals and humans, must be notified.

The health of cattle

In this section health problems are dealt with on a life-cycle basis, dealing firstly with those health problems affecting the calf, continuing with growing stock and concluding with the mature beast and those diseases that have no special age incidence.

The health of the calf

Of the 5–10% mortality that occurs in calves, nearly two-thirds takes place in the first month of life and much of this is due to what is commonly called calf scours. The main cause is ultimately the ubiquitous bacterium *Escherichia coli* following initial environmental stress and usually a virus challenge.

Colostrum and protection against diseases in the calf

It is essential that a calf should have a good drink of colostrum very soon after birth. Colostrum contains the immunoglobulins that are the essential protection to infection. It also contains large quantities of vitamin A (six times as much as in ordinary milk). Immunoglobulins are the proteins that contain the antibodies which protect the calf from disease organisms, whilst vitamin A is the most important vitamin associated with the animal's resistance to infection.

Because calf scour is caused by a number of organisms, the use of specific vaccines and sera may have a limited chance of providing either good treatment or prevention, though sometimes this can be the case. More often the disease must be treated with antibiotics with a broad range of activity, such as the synthetic penicillins, together with replacement therapy for the enormous fluid depletion that takes place. This will include electrolytes, glucose, minerals and vitamins in easily assimilated liquid form.

Salmonellosis

The salmonella group of bacteria is a very large one with hundreds of different types, many potentially pathogenic to man and animals. Only relatively few are harmful, the majority being 'exotics' that come in via the feed and are quickly eliminated.

The two most important species in the condition in calves are *Salmonella typhimurium* and *S. dublin*. The symptoms with either of these infections or indeed with other virulent salmonellae can be largely similar. A percentage of calves may die so rapidly that no symptoms are seen; others scour profusely with dysentery, pneumonia, arthritis, jaundice and even nervous symptoms manifesting themselves in some cases. Urgent treatment will be essential to include antibiotics such as ampicillin, chemotherapeutics and fluid therapy, isolation and general nursing, especially warmth. Thorough disinfection of the quarters will be essential and preventive measures may include vaccination and possibly preventive medication. However, the use of the last measure is hazardous as it may generate drug resistance and encourage the establishment of the 'carrier' animal which can excrete salmonella organisms throughout its life.

Whilst salmonellosis is especially a disease of young stock, adult cattle do become infected and may not show symptoms, becoming carriers of great potential danger to man and other livestock. Abortions may be caused by *Salmonella*, and *Salmonella* can be a source of infection via milk. In the main the biggest problems from salmonellosis arise from farms where the calves are brought in from outside and infection can easily spread to the adults if they are on the same premises.

In Great Britain salmonellosis is a notifiable infection and official measures can be taken to control the disease as appropriate. In the event of an outbreak in adult cattle, measures that may be taken will include laboratory testing of the cattle to identify the carriers, treatment of animals with antibiotics, such as ampicillin, and appropriate care of the milk to remove the risk of human infection.

Calf diptheria

The bacterium associated with foul-in-the-foot in cattle, known as *Fusibacterium necrophorum*, is also capable of causing a serious diseases in calves known as calf diphtheria. The infection attacks very young calves, causing inflammation within the mouth, which leads to soreness, ulcers and ultimately necrosis. The calf eventually becomes seriously ill. The condition responds well to treatment. Sulphonamides are effective as are a wide range of broad-spectrum antibiotics. It is a disease that is associated with bad hygiene.

Navel-ill or joint-ill

These two terms describe the disease affecting calves where bacterial infection causes inflammation in the region of the joints. It usually occurs because the calves are managed under dirty conditions and are challenged by a serious burden of bacteria before the navel has healed.

Treatment may be undertaken successfully in the early stages of disease with sulphonamides or broad-spectrum antibiotics; local application of dressings to the navel will help to prevent infection.

Calf tetany

Many years ago it was established that calves could not be reared to maturity on milk alone because of the onset of deficiency conditions and in particular hypomagnesaemia or calf tetany during growth. However, the condition does also occur under other conditions of feeding and it is an increasing problem in beef calves, especially if they are grazing at the time of the most active growth of spring grass.

The symptoms in calves are similar to those in older animals: hyperexcitability, irritability and twitching of the muscles. The walk often becomes spastic, the feet being carried well above the usual height. Thereafter in the progress of the disease convulsions may follow and eventually death.

Treatment consists of immediate administration of suitable magnesium injections. Prevention is effected by administering about 15 g of calcined magnesite daily, or 30 g of magnesium carbonate daily.

The health of growing cattle

Internal parasites

There are three major groups of diseases in cattle caused by parasitic worms. There is 'husk' or 'hoose', which is due to worms which live in the lungs; liverfluke disease, caused by flatworms or flukes which live in the liver; and parasitic gastroenteritis, caused by roundworms which live in the abomasum and/or intestines of cattle.

Parasitic worms do not multiply within the body of the animal in which they live as adults. The adults produce millions of eggs but none reaches maturity without first passing out of the animal in the faeces. They must then have a period of time outside the animal before they become infective, and this period varies from a few days to many weeks. Thus every worm in the animal has been picked up in the herbage while the animal was grazing. The significance of this knowledge is that it provides for two ways of attack, one within the body and one without. Removal of contact of animals with their own faeces is a vital factor. If the husbandry is good, with a clean environment and no overcrowding, the infestation may be so slight as to be unimportant. Mixed grazing is also useful since the parasitic worms are host specific and infect only one species. It is also important to know that it is chiefly the young and growing stock that are vulnerable to infection. Adults should have a naturally produced immunity due to earlier contact with the parasites. Nevertheless, the adults often have a worm burden and will produce some eggs. Furthermore, even with adults they can become badly affected if they become stressed by poor conditions or nutrition. The worm eggs once voided can remain infective for up to or even more than a year.

Husk or hoose

This is an infection of the bronchial tubes and leads to the cough known as 'husk' or 'hoose'. It is caused by the lungworm *Dictyocaulus viviparus*. The white worms are threadlike, up to 75 mm long, and the females lay vast numbers of eggs. After hatching they produce minute larvae which are carried in the mucus up the windpipe and to the mouth. They are then swallowed and pass out onto the pasture with the dung. These larvae have to develop on the ground before they are able to infect cattle who take them in with the grazing. The development takes about 5 days in warm, moist weather and over a month when the weather is cold. Whilst developing in the dung the larvae go through a series of changes and moult their skin twice. The larvae have little resistance to drying and can die off rapidly under these circumstances.

When the infective larvae are swallowed by grazing cattle, they penetrate the intestinal wall and pass into the lymphatic system and by this route reach the bloodstream and then pass to the lungs. Here they develop into the adult worms and begin to lay eggs after 3 weeks.

Cattle that have been exposed to lungworm infestation will in due course acquire a resistance. If the level of pasture infestation rises at a moderate rate, the gradual and increasing resistance of the calf will enable it to reject the effect of the worms and no disease will actually occur. However, if the pasture infestation rises rapidly, as it may do if the climatic conditions are favourable, disease may occur.

High pasture infestation may occur in a number of circumstances. In early spring most pastures will be either clean or at least only very slightly infested. Contamination of the pasture by older animals, which have carried over some worms in their lungs from the previous season, may produce more significant levels of infestation on the herbage and may even give rise to dangerous infestation if the season and pasture conditions are suitable. Low levels of herbage infestation may be increased when susceptible calves become infected on the pasture and in consequence increased numbers of larvae are passed on to it in their dung.

Symptoms The disease may be recognized, or at least suspected, if there is an increase in the respiratory rate, coughing, especially when the animals are moved, and some loss in body condition. If cases become severe, the coughing becomes louder, the respirations shallow with clear pain and distress, and the animal stands uncomfortably with head and neck outstretched and the tongue protruding.

Control There are a number of useful medicines for lungworms, such as ivermectin and doramectin. Wherever possible it is best to remove the animals from the pasture and house them in good buildings where they can be well fed and warmly maintained.

Prevention It is wise to keep animals grazing for their first year off pasture that has had animals on it recently. Strict systems of rotational grazing can always be of help since they will prevent heavy buildup of disease. The calves should be moved twice a week onto ground which has not had calves on it for about 4 months, but even a short time would be of help in reducing infection. The vaccine is also a great help in preventing the disease, but it is an *adjunct* to good management and not a substitute.

Liverfluke disease

Liverfluke disease tends to go quite undetected or unsuspected until animals start dying. It is admittedly more a disease of sheep than cattle, but there is good evidence that liverfluke disease causes a very serious effect on growth and productivity. The liverfluke is a flat worm about the size and shape of a privet leaf, which lives and feeds on the liver of cattle. The flukes lay many thousands of eggs which pass on to the pasture in the dung. The eggs hatch and produce a very small swimming creature which has to enter a particular type of snail if it is to survive. The parasite grows in the liver of the snail, producing many young liverflukes. These look like the flukes that cause disease in the older stock except that they are much smaller. The young flukes leave the snail when the snail is in water and encyst on any herbage which they contact. They lose their tails and surround themselves with a protective covering. There they remain as cysts but are inactive for long periods, up to months on occasion, until eaten by the grazing animal. When eaten the protective coat disappears and the fluke emerges and makes its way into the liver. During the next 2–3 months the fluke grows to maturity and then lays eggs.

Symptoms There are two forms which are clinically differentiated. The acute disease occurs when the grazing animal picks up many flukes in a short time. These damage the liver seriously and this can lead to sudden death in the apparently healthy animal. The chronic form is most common. It is due to the presence of flukes in the liver over a period of months, and affected animals waste progressively.

Control Certain medicines now exist for treatment, namely oxyclozanide, nitroxynil and albendazole. Also recommended are rafoxanide, netobimin and triclabendazole. All will remove more than 90% of adult flukes from the bile ducts but have variable efficiencies against the immature stages migrating from the liver. Professional advice should generally be sought before use of these medicines in view of possible side effects, contraindications and withdrawal periods of milk for human consumption.

It is also effective to prevent the disease by controlling the secondary hosts, the snails. Infestation in flukey areas is worse in the months of October to December and hence fluke-free pastures should be grazed in this period wherever possible. Badly drained areas should be properly drained. The snails may also be dealt with by treating the pasture with either copper sulphate at 30 kg/ha or sodium pentaclorphenate at 10 kg/ha. As these are dangerous materials they should be applied with due caution.

Parasitic gastroenteritis

This is a disease, or infestation, caused by the presence in the abomasum or fourth stomach or the anterior part of the small intestine of a group of small roundworms. The most important species are *Ostertagia ostertagi* and *Trichostrongylus axei* in the abomasum; *Cooperia oncophora, C. punctata*, *T. vitrinus* and *T. colubriformis* in the small intestine and *Oesophagostomum radiatum* in the large intestine. They cause considerable economic loss from both poor productivity and overt disease in young animals.

The life-cycle of the worms is 'direct,' there being no intermediate host. Eggs are laid by the adult worms passing out with the dung on to the pasture where they can hatch within about 24 h, producing minute larvae which feed on bacteria in the faeces. Within the next few days they moult twice and only then are they capable of infesting stock who ingest them in the herbage. In order to make the ingestion more certain they leave the dung and pass on to the grass, climbing upwards so that they will readily be consumed by the animals. All this can take not more than 4 days in warm, damp weather but may take weeks in cool conditions.

A limited number of worms do little or no harm to the beasts, but a heavy infestation can be lethal. Thus infestation is assisted by overcrowding, moist warm conditions and long grass in which moisture more readily persists.

Symptoms These are nearly always indefinite and the usual signs are a progressive loss of condition possibly leading eventually to emaciation and death. There is usually a severe scour but the animals continue to eat voraciously. However, one should be aware of the fact that some animals do not scour and signs will be so indefinite as to go undetected. In every case it is wise to have positive identification by examination of the faeces to determine the degree of infection.

Control The first course of action is to consider medication, and there are a wide variety of mediciness that can be given; e.g. doramectin, ivermectin, benzimiadoles and levamisole. The medicines can be administered by injection or orally in several cases and a full course, as advised in the specific literature of the medicine, should be given.

Preventive medicine should consist of the following:

- Overcrowding of the young stock should be avoided.
- Grass should be kept short as larval worms tend to perish in such grass.
- Only adult stock should be put on badly infested land.
- Periodic treatment should be given to animals after laboratory examination of faeces has established the degree of infection.

Skin conditions and external parasites

Ringworm

This is an extremely common skin disease of cattle which tends to be a problem in winter when the animals are housed. It is caused by fungi belonging to the genus *Trichophyton,* two species of which occur in cattle in the UK and also attack horses, pigs, dogs and humans. *Trichophyton verrucosum* is the most common and is responsible for most cases of ringworm in cattle, farmers, stock keepers and their families.

The fungi that cause ringworm are capable of surviving for very long periods in the farmyard, certainly for over a year and probably much longer. The spores that are picked up by the animals find their way into cracks on the skin and germinate to produce fine filaments which flourish in the surface of skin and hair. These fungal threads grow downwards inside the hair keeping pace with its growth from the hair bulb and weakening it so that it snaps off at the surface of the skin, producing the characteristic bald areas. The skin reacts to the infection by inflammation followed by the formation of a grey to yellow–white crust. The lesions of ringworm are most often seen on the head and neck, especially in calves, but in adults patches occur on the flanks, back and rump, wherever rubbing takes place. Animals that are run down are likely to suffer a severe infestation.

Whilst ringworm will disappear spontaneously in the course of time, it is not advised to ignore treatment since the disease has a serious debilitating effect. Treatment may be effected by removing the crusts with a soft wire brush and applying one of several imidiazole preparations, for example 5% thisabedazole, handling with care, wearing rubber gloves and overalls. A much better treatment, though expensive, is an antibiotic griseofulvin. It is more vital to practise good hygiene measures by efficient cleansing and disinfection of the buildings and all objects that could harbour the fungal spores.

Mange

Certain forms of mites infest cattle. These are *Sarcoptes scabei, Chorioptes bovis, Psoroptes communis* and *Dermodex bovis.* All are extremely contagious conditions (sometimes called scabies). The result of the disease is very patchy coats and areas of bare skin with an extreme amount of irritation.

The parasitic mites spend their whole lifetime on the animals and only survive a few weeks off them. The symptoms are always at their worst during the final winter months. Diagnosis is made by microscopic examination of skin scrapings. The mites are visible to the naked eye.

Control Control is rather difficult and requires professional veterinary advice. Any incidence should firstly be identified and once this is done measures are advised to eradicate it from the herd. Armitraz and ivermectin, certain organophosphorus compounds and synthetic pyrethroids can be applied. All animals should be treated and in order to eradicate the condition it will be necessary to re-treat two or three times at 10-day intervals.

It is also necessary to keep the cattle away from those parts of the building that could harbour the mites and/or disinfect these areas as effectively as possible to prevent re-infestation.

Blackquarter (blackleg)

The disease of blackquarter is similar to so-called blackleg of sheep. It is caused by a bacterium *Clostridium chauvoei*, the spores of which are common inhabitants of the soil in areas where sheep and cattle have long been grazed.

The disease occurs because the organism is able to multiply in the muscles of the leg. The symptoms are stiffness and lameness with swellings appearing in areas such as the loin, buttocks or shoulder. The swellings are hot and painful at first but later become cold and painless. The skin over the affected muscles will become hard and stiff and if pressure is applied there is a papery effect, a dry crackling noise due to the stiffness of the skin and the movement of the gas underneath. In fatal cases the condition of the animal rapidly deteriorates and it will die quite quickly. Initially there is probably a bruise or injury to the muscles and it is in this area that the clostridia can multiply in the absence of oxygen. Then they produce the toxins which have the serious destructive effect on tissues.

Animals affected with the disease can be treated successfully with antibiotics, but only if the disease is recognized very early. Penicillin preparations are satisfactory. Prevention can be effected by using a vaccine, but the best measures of prevention rely on good pasture management, avoiding areas of known susceptibility and draining or cultivating those that must be used.

Coccidiosis

Coccidiosis is a form of dysentery that primarily affects young stock in the summer and autumn months. It is caused by a microscopic unicellular parasitic organism known as a coccidium. The coccidial parasite forms spores, known as oocysts, which can live outside the animal for many months. Adults are frequently carriers of the disease to younger animals.

The oocysts are very resistant to destruction and can often remain infective for a year, but if they are swallowed by a susceptible animal they hatch and the active forms of the coccidia are liberated. These penetrate the cells lining the intestine where they mature and multiply together with bleeding.

Animals acquire a resistance to coccidiosis by receiving a light infestation early in life and developing an immunity. However, if the intake of oocysts is sudden and too great at one time before a resistance has been produced, then disease occurs. This is most liable to happen when animals are densely stocked and the weather conditions are warm and wet. Dirty walls of pens of boxes, as well as the floor are sources of infection.

The disease is controlled in a number of ways. Affected animals should be treated with a suitable medicine, such as sulphonamide, plus plenty of really good nursing, especially fluids. Overcrowding must be avoided: good bedding and cleanliness are important and animals should be kept in thriving conditions so that they can readily resist such infection that occurs.

Bovine virus diarrhoea (mucosal disease)

Bovine virus diarrhoea (BVD) or mucosal disease (MD) is an infectious disease of cattle caused by a virus. It was first recognized in the USA, but in recent times has occurred in virtually all parts of the world. Originally the two diseases BVD and MD were distinguished, but it is now clear that they are caused by the same virus so the best term is probably the BVD–MD complex. The evidence is that this is an increasingly pathogenic condition and professional advice is needed to deal with it. This may involve blood testing the herd and removing those persistently excreting the virus.

Symptoms The disease causes ulceration of the mouth and lips accompanied by foul-smelling diarrhoea. There is often lameness and extreme scurfiness of the skin. By far the majority of cases occur in cattle between 4 and 18 months of age. Most cases can recover spontaneously, and only exceptionally, if there are many young animals affected, the death rate can be high.

Treatment and control There is no specific treatment, but supportive therapy in the form of intestinal astringents and electrolytes can be of value. Vaccines are now available to prevent the disease.

Respiratory diseases

The number of pathogenic organisms that can infect the respiratory system of cattle is enormous: there are some 20 bacteria, six fungi and four parasites that *commonly* infect them, not to specify the underlying primary damage that may be wrought by viruses. The wide involvement of so many agents has made the production of satisfactory vaccines most difficult and the use of medicines complicated and expensive. Thus the prime responsibility for dealing with such conditions rests with getting the environment correct. The general principles are given in Table 23.2.

Specific advice for cattle may be summarized as follows:

- Depopulate between batches and institute a rigid disinfection programme.
- Reduce numbers in one environment to a minimum.
- Keep different ages and sizes of animals separate.
- Ensure copious ventilation without draught as cattle are hardy animals and do not require cossetting.

Infectious bovine rhinotracheitis (IBR)

This is an acute respiratory disease of cattle caused by a virus and is spread by contact or by the airborne route.

Table 23.2 Essential action to counter respiratory disease in cattle

- Expert attention required for correct therapy.
- Improve ventilation – more air flow is usually needed with less draught.
- Check stocking density – are the cattle overcrowded?
- Is more bedding required?
- Separate and if possible isolate those cattle that are badly affected.
- Provide warmth for the very sick.
- Attempt separation of different ages.
- Reduce movement of animals to a minimum.
- Consider insulation and other essentials of construction relevant to the building.
- Can serum, vaccine or medicines be used to protect the affected animals?

It is a very contagious disease and can spread through a herd within 7–10 days.

Symptoms The symptoms are severe inflammation of the eyes and the upper respiratory tract. The disease can also cause a substantial fall in milk yield, abortion and infertility. There may be a fever, loss of appetite and drooling of saliva. Sometimes there is a pronounced cough.

Control This is not too serious a disease in Great Britain. IBR is treated with broad-spectrum antibiotics to help reduce pyrexia or the effect of secondary invaders. A vaccine is available to protect the herd.

Wooden tongue and lumpy jaw

These are two distinct diseases which have rather similar symptoms and are caused by unrelated organisms. They both run a slow course and are characterized by a steady enlargement of the tongue and swellings in various parts of the animal but especially on the head and neck. Their presence may generally interfere with the normal functions of the animal. They can also form abscesses or ulcerate, discharging into or outside the body. Untreated they cause a steady wasting of the animal so that it becomes unproductive and will eventually die.

The causes of the two diseases are quite different. Actinomycosis or lumpy jaw is caused by a ray fungus known as *Actinomycetes bovis*, whereas actinobacillus or wooden tongue is caused by a bacterium *Actinobacillus ligneresi*. The main difference between the two diseases is that actinobacillosis tends to affect the soft tissues, such as the tongue, and actinomycosis usually causes diseases of the bones, especially those of the jaw.

Symptoms The tongue is most commonly affected and becomes hard and rather rigid, hence the name 'wooden tongue'. There is a constant dribbling of saliva and food tends to be rejected. The disease will also affect the glands of the neck and swellings appear between the angles of the jaw. If this swelling is allowed to continue, it will affect breathing and swallowing also becomes difficult. The lesions may burst and release a characteristic yellow and granular pus. In addition the bones of the jaw are often affected. Pus forms within the bone and these disease areas can eventually break through the skin to form an unpleasant discharging fistula. They may also affect the internal organs and will then cause a progressive wasting.

Control and treatment Other than in advanced cases, treatment can be very successful. If the disease does seem to have got a firm hold on the animal which is fast losing condition, it is probably best to slaughter it. *Actinobacillus* responds quite well to iodides and sulphonamides, and actinomycosis can be treated with penicillin or broad-spectrum antibiotics.

Redwater

Redwater, also called bovine piroplasmosis or babesiosis, is a curious condition caused by the presence in the blood of minute parasites called piroplasms. These parasites live inside the red blood corpuscles where they can be seen in the early stages of the disease. It is a condition that is found in certain areas, for example in Wales, southwest England and East Anglia.

The organisms are especially associated with rough scrubland, heathland or woodlands where the ticks *Ixodes ricinus* are present as they transmit the organisms from animal to animal. The tick is infected by sucking the blood of an infected animal. Once the tick is engorged with blood it falls off and lies in the herbage. Later the tick may attach itself to other cattle and when it sucks from these it introduces the parasites into the blood. The two worst periods for redwater disease are from March to June and October to November.

Symptoms The main effects of the disease are to cause loss of condition and a considerable loss of milk yield in the dairy cow, together with blood pigments in the urine. Animals affected show a high fever and diarrhoea and later constipation. The animal also often becomes jaundiced, the mucous membranes becoming yellow. The heart beat becomes very loud and there are tremors of the muscles of the shoulder and legs. It should be noted that young cattle are less susceptible than adults which have not previously been infected. If contracted early in life, the disease is only mild and the animal is then immune against further infection. This indicates one of the great dangers; if adult cattle from a tick-free area are introduced into a redwater area, some of these may contract the infection and die quite rapidly without treatment.

Control Treatment of infected animals will be twofold. First, the animal will require an early injection with one of a number of chemotherapeutic agents that are lethal to the parasite. Second, careful nursing is perhaps even more important. Feed well, house comfortably, do not drive or excite for fear of heart failure, and keep the animal warm.

Prevention The disease could be eradicated if ticks were eliminated. The tick is extremely resistant to most measures of control. In addition it can feed off several species of animals and is not dependent on cattle alone.

Sometimes sheep are put on pasture as 'tick collectors'. By placing them alone on the pasture at the worst tick season of the year and then adopting a heavy dipping programme, this will reduce numbers considerably. The only effective way of eliminating ticks is to clear the rough ground by cultivation and normal crop rotation.

Wherever this procedure is adopted and a part of the farm is freed from ticks, the importance of exposing young animals to tick infestation must be considered otherwise they will be highly susceptible as adults to a first infection. Great care must be taken not to put them on to tick-infested ground for the first time in adulthood.

Diseases of mature cattle

Metabolic disorders

Bloat

In bloat, also known as blown, hoven or tympanitis, the rumen becomes distended with gas and the pressure from this on the diaphragm may lead to the animal dying from asphyxia or shock. Two types of bloat occur. The first is where the gases separate from the contents of the lower part of the rumen, and the second is where the gases remain as small bubbles mixed with the contents in a foamy mass, giving rise to what is known as 'frothy bloat'.

Bloat most often occurs after the grazing of lush pastures which contain a high proportion of clover. It may be due to the fact that some kinds of herbage contain substances that may cause paralysis of the rumen muscles and of other organs concerned in the process of rumination. Or it may be due to the fact that more gas is produced than can be got rid of by eructation. Bloat may also be due to the result of foaming in the rumen caused by saponin and protein substances obtained from plants.

Symptoms Symptoms are obvious distension of the abdomen which is particularly pronounced on the left flank between the last rib and the hip bone. In the early stages the animal becomes uneasy, moves from one foot to another, switches its tail, and occasionally kicks at the abdomen. Breathing becomes very distressed and rapid. If the attack is a severe one, any movement intensifies the discomfort and so the animal stands quietly with legs wide apart. If relief is not given very quickly, death from suffocation or exhaustion can follow rapidly.

Control It is advisable to summon a veterinary surgeon as soon as possible. If time allows, a stomach tube may be inserted into the rumen to allow the gas out. In acute cases the only remedy is to puncture the rumen on the left flank with the surgical apparatus the trocar and cannula. In an emergency, where death seems imminent and specialized assistance is not available, the only course of action is to puncture the area with a short pointed knife with a blade which is at least 6 inches long, the blade being plunged into the middle of the swelling on the left side and twisted at right angles to the cut to assist the gas to escape.

Although this treatment will serve to give immediate relief, more treatment is required if the case is one of frothy bloat. Drenching with one of the substances known as silicones, or with an ounce of oil of turpentine in a pint of linseed oil, often helps in the less severe cases and some gentle exercising may help to relieve the pressure of the gas in the rumen. There are also a number of drug treatments available, such as Avlinox, which is an ethylene oxide derivative of ricinoleic acid.

Prevention Efficient pasture management can help to avoid the trouble. After a pasture is closely grazed and then rested, the clover recovers more quickly than the grasses so that when pastures are used intensively the clover becomes dominant in the sward. However, it is generally considered that if there is less than 50% of clover present, it is usually safe. A variety of seed mixtures have been tried to provide a suitable balance between clover and grass: one that has been given promising results contains tall fescue. After it has been closely grazed this species of grass recovers almost as quickly as the clover and maintains a reasonably safe balance between grass and clover.

Another system that has been advocated is to allow the herd to graze a potentially dangerous herbage for a limited period and then turn it on to an old pasture for a time.

Controlled grazing can contribute to the prevention of bloat by ensuring that the grazing animal eats both leaf and stem, whether clover or grass, and that it cannot select clover or clover leaf which is thought to be more dangerous. If an electric fence is moved forward to that the cows have access to a succession of narrow strips of pasture, during the course of the day, bloat may be reduced or eliminated, but it may reduce the intake of the feed and hence milk yield.

The second method of controlling grazing is to mow strips of herbage, then allow it to wilt and let the herd graze it as it lies on the ground. Alternatively the wilted herbage can be carted to the cattle and fed indoors. The feeding of roughage in the form of hay or oat straw shortly before the grazing of dangerous pasture has also been practised for many years. It is only completely reliable if the roughage is fed overnight.

Milk fever

Milk fever occurs principally in cows after calving, usually in dairy breeds of high milk-yielding capacity. It is given various synonyms, such as hypocalcaemia, parturient hypocalcaemia, parturient paresis and parturient apoplexy. There is, in fact, no fever and indeed it is more usual for the cows to show subnormal temperatures.

Symptoms Usually within 12–72 h of calving there is a short period or uneasiness with paddling of the hind legs, swishing of the tail and convulsive movements, soon followed by depressed consciousness and paralysis. The animal usually falls on its side with its legs extended, rolling its eyes and breathing heavily. The characteristic picture is a twisting of the neck as it lies on one side, with spasm of the neck muscles, shallow laboured breathing, grunting, cessation of rumination and dryness of the muzzle. In straightforward milk fever the calcium and phosphate levels fall and the magnesium levels rise.

In practice, it is quite difficult to distinguish between the simple hypocalcaemia and the more complicated metabolic disorder where there are a number of deficiencies which have to be corrected.

Background to occurrence There does not appear to be any real breed incidence, but it is usually well-fed, high-yielding cows that are affected and the severity of the condition increases with age. The cause is usually ascribed to a temporary failure in the physiological mechanism which controls calcium levels in the blood. The stress of calving and the physical adjustment necessary for lactation will upset the balance of chemical regulators (hormones) produced by various glands. If the cow does not adjust itself quickly enough, it may suffer an attack of parturient hypocalcaemia.

Control The immediate administration of a solution of calcium borogluconate is effective in curing uncomplicated milk fever. It may be given intravenously for a quick response or subcutaneously for a slower effect. In addition, the cow should be propped up in the box by means of straw bales on each side to prevent it from becoming 'blown' and to prevent regurgitation of ruminant contents. Further injections of calcium may be required and in addition, in complicated cases, there will be the need for administration of phosphorous injections and even other minerals. Professional advice is certainly required for all but the most simple case of hypocalcaemia.

Injections of vitamin D increase the mobilization of calcium in the body and thereby may help to prevent levels falling too low.

Ketosis (acetonaemia)

Ketosis is a disease involving ketones which are chemical compounds of fat. This condition may occur when an animal's food intake is inadequate for its needs and it is drawing on its own reserves in considerable amounts to make good the deficiency. The usual form of ketosis is ketonaemia which includes an excess of ketones in the blood. Since acetone is the simplest and most characteristic of the ketones, the disease is generally termed acetonaemia. It is also called 'post-parturient dyspepsia' as it often occurs soon after calving and is associated with indigestion, It often follows mild attacks of milk fever and almost appears to be a subnormally slow recovery, hence the traditional term for the disease of 'slow fever'.

Symptoms The symptoms of ketosis are usually seen within 2 weeks of calving and in well-nourished stock of high milking capacity. However, it may also occur much later than this, even several months after calving. It is most common during the tail-end of winter. Affected cows appear listless and show a marked loss of condition, have a dull coat and there is a sickly sweet smell of acetone on the breath and also in the urine and milk. The milk production falls away quickly and the milk becomes tainted and undrinkable. Constipation with dark, mucous-covered dung is usual. Sometimes there is a licking mania, a rather wild-looking appearance and champing of the jaws and salivation, and there may also be hyperexcitability.

Control First, tests are carried out to confirm the presence of ketones in the blood. It will be usual for the blood sugar to be low and injections of glucose are an early procedure. Corticosteroids may be administered to give a longer-term beneficial effect. Other useful treatments are glycerine and preparations containing sodium propionate. Molasses and/or molassine meal are also helpful.

Prevention Prevention will depend principally on skilful management. The energy provided in the food must keep pace with the draining demands of production, especially early lactation. In late pregnancy the cow should be kept in good condition but not over-fat. The roughage in the ration should also be carefully selected and kept up to a good level or at least reduced slowly if the cow's appetite and enthusiasm for a balanced diet are to be maintained.

Hypomagnesaemia
The amount of magnesium in the blood of normal animals keeps within a fairly constant range. However, should the magnesium level of the blood drop below this range, then the condition develops known as hypomagnesaemia, or grass tetany, grass staggers, lactation tetany or Hereford disease, whilst in calves it is known as milk tetany or calf tetany.

Underlying causes The occurrence of this disease is in the period of sudden change of lactating cows from winter housed conditions to those of the rapidly sown spring grass. However, it is now known to occur in cattle under most feeding and management regimes, including winter feeding periods. It is not due to a simple deficiency of magnesium, for sometimes the condition occurs in cattle on pasture that has a very adequate magnesium content. Underfeeding appears to be a predisposing cause, especially in outwintered beef cattle, but overall the disease is due to physiological dysfunctioning that interferes with the absorption and utilization of the magnesium that is in the food.

Symptoms Symptoms vary according to the intensity of the attack. The first signs are often nervousness, restlessness, loss of appetite, twitching of the muscles, especially of the face and eyes, grinding of the teeth and, quite soon after, staggering. In less acute cases animals that are normally placid become nervous or even fierce. In severe cases paralysis and convulsions develop very soon after the onset of symptoms and if treatment is not given death can follow very quickly. In milking cows a sudden reduction in milk yield may occur just before an attack. Sometimes the attack comes on so rapidly that no symptoms are seen, merely a dead beast. There is also a chronic form of the disease in which cows show a gradual loss of condition although appetite and even milk yield do not show any drop. This chronic state can last several weeks and then develop into the more acute condition.

Control The veterinary surgeon will inject magnesium plus other solutions into the affected animal to cure the condition. To prevent the disease, supplements of magnesium will be used; 60 g of magnesium oxide/head/day has been used for many years successfully in dosing adults to prevent the condition. The magnesium oxide is usually given as calcined magnesite in a granular form. This must be given daily to cattle under risk or blood magnesium levels can fall quickly.

Prevention The use on pastures of magnesite-rich fertilizers has been quite successful in preventing the disease. On some soils top dressing in January or February with 1200 kg of calcined magnesite per hectare can be a perfect preventative. Even 600 kg may be sufficient. Another method of preventing the disease is the magnesium 'bullet' in the reticulum with its slow release of magnesium, and attempts have also been made to give multi-vitamin injections to increase the absorption and utilization of magnesium.

Mastitis

Mastitis means literally inflammation of the mamary gland or udder. It has always been an important disease of dairy cattle with potentially disastrous consequences, but with the advent of antibiotic treatment there was a general optimism that mastitis would be eliminated; however, organisms causing the disease have become increasingly resistant to the use or, perhaps more correctly, misuse and abuse of antibiotic therapy. Also there has been a growth of large dairy cattle units with the greater opportunity for disease buildup and cross infection. Cow cleanliness has become less satisfactory and some of the most modern practices lead to dirtier cows and udders, for example, with loose housing and cubicles. There has been the emergence of different forms of mastitis, called environmental because of their association with the newer forms of husbandry and the rather different balance of organisms involved.

Mastitis may affect as many as 50% of the cows in the national herd in one form or another, many cases being subclinical and so not easily recognizable. Since a cow with mastitis may *readily* lose a quarter of its milk output there is no doubt of the extreme economic importance of this condition. Whenever mastitis occurs it must be looked upon as a herd problem and should be dealt with urgently and thoroughly. Even a single cow may be a warning of a managemental error and should be carefully considered in this respect.

Subclinical mastitis means that the udder is affected mildly but the cow's health or even the milk yield are not obviously affected. It can be recognized by *cell counts*. If an inflammatory process is proceeding, then the milk will contain more white blood cells than normal. The best way for the farmer to deal with this is to arrange for a regular monitoring of the cells in the milk. The range of 'counts', with their significance, is shown in Table 23.3

The milking machine
A major predisposing cause of mastitis is the misapplication of milking machines, together with bad maintenance and a poor hygiene routine during milking. One of the most serious of these is overmilking, i.e. the milking machine is left on much longer that it should

Table 23.3 The significance of cell counts

Cell count ranges (cells/ml)	*Estimate of mastitis incidence in a herd*	*Estimate of milk production loss per cow (litres)*
Below 250 000	Negligible	—
250 000–499 000	Slight	200
500 000–749 000	Average	350
750 000–999 000	Bad	720
1 000 000 and over	Very bad	900

be, causing considerable damage to the delicate mucous membranes lining the teat canal, allowing pathogenic micro-organisms to invade so that mastitis results. Teat liners must be chosen of the right size and gentle fit. They must never be slack and should be carefully maintained to prevent them from causing damage. Pulsations of the machine must be maintained between 40 and 60/min; if the rate increases, squeezing may fail and mastitis is more likely to occur. The ratio between the release and squeeze phase should be between 2 : 1 and 1 : 1.

The pump needs to be maintained properly as it 'ages' or a surge of milk may result around the teats, such circumstances also being a predisposing cause of mastitis. The vacuum regulator prevents fluctuation in vacuum which can be very damaging to the udder and teats for, if it fails, a surge of milk results around the teat and clusters may fall off or, if it rises, it damages the teats causing extrusion of the lining of the teat canal.

Types of mastitis

The easily recognizable forms of clinical mastitis are the acute, sub-acute and chronic forms.

Acute mastitis The cow is obviously ill, feverish, fast breathing and depressed and there is no cudding. Examination of the udder will show that one or more quarters are tense, swollen, painful and possibly discoloured (blue). It may be possible to withdraw from the affected quarter(s) a small quantity of grossly abnormal fluid. Gangrene may set in causing the destruction of part of the udder. This 'acute' mastitis is often the same as 'summer mastitis' which, whilst it is most common in dry cows and heifers in the summer months, can occur in cows at other times. Such mastitis is caused primarily by a bacterium, *Corynebacterium pyogenes*, and must be treated vigorously with antibiotics and sulpha drugs. Because of a common association with 'dry' cows its incidence may be reduced by using long-acting antibiotic therapy as soon as the cow is dried off.

Sub-acute mastitis The symptoms are somewhat similar but milder and slower in their progress. The first sign of disease may be the appearance of small clots in the milk with some increased difficulty in extracting it. Pain also gradually increases and the affected quarter(s) swell. The milk eventually becomes yellowish and much decreased in amount.

Chronic mastitis In this form of mastitis there is no pain and any general sickness in the cow is unlikely, but there is a gradual hardening of the udder tissue and a decrease in the amount of milk.

Causes At one time the most frequent-occurring bacterium was *Streptococcus agalactie*, but this has generally decreased in incidence with the advent of successful antibiotic treatment. Other streptococci, e.g. *S. dysgalactia* and *S. uberi*, are common and also *Staphylococcus* and *Corynebacterium*. There are, however, some serious new forms of mastitis that are of comparatively recent origin and are most difficult to treat.

Escherichia coli is usually associated with the so-called environmental mastitis. Also *Pseudomonas* is a common bacterial cause of mastitis and *Corynebacterium bovis*. Mastitis may also be caused by micro-organisms other than bacteria, such as mycoplasma (*M. agalactie*) and moulds, the latter being termed mycotic mastitis.

Leptospira infections in cattle are also causing considerable concern, especially as such infections can cause serious illness in man. The most common organism is *Leptospria hardjo*, responsible for the so-called milk drop syndrome, and also a cause of abortion. The milk drop appears initially as a mastitis usually in all four quarters, falling almost to zero within 24 hours. The condition is accompanied by a high fever and the milk is thickened or clotted or even bloody. A very high percentage of cattle, as many as 50% of cows in a herd, may be affected.

Control and prevention The condition can be successfully treated by broad-spectrum antibiotics and can be prevented with a vaccine. The view is now taken that the control of mastitis must depend largely on the application of hygienic criteria of the highest standard. These methods are summarized in Table 23.4

Disease problems associated with breeding in cattle

Lowered fertility results in lowered production and causes great economic loss. However, infertility is not a disease as such; it is only the symptom of one. It may be due to any condition that prevents a vigorous sperm from fertilizing a healthy ovum to produce a robust calf.

Table 23.4 Summary of measures aimed at preventing mastitis in the dairy cow

- Basic design of cow accommodation must be right to ensure cow's bed is dry and clean and to minimize the chance of injury to udder.
- Replace fouled bedding frequently in stalls, yard or cubicles.
- Test milking machine regularly, check vacuum pressures, pulsation rates, air bleeds and liners daily.
- Monitor cell counts in milk. Keep detailed records for frequent reference.
- Teat-dip always to be used after milking. Use a sanitizing mixture with an emollient such as lanolin.
- Wash udders before milking with clean running water and dry with disposable towels. Milker should preferably wear smooth rubber gloves.
- Ensure early diagnosis by use of 'fore-milk' cup.
- Treat all cases of mastitis promptly.
- Cull chronic cases of mastitis.
- Correct laboratory diagnosis of causal organisms assists in taking proper remedial measures.
- Treat cows as they dry off with a long-acting antibiotic.

The process involved is complicated and the different causes of infertility are numerous. Consider the whole physiological process. The bull must produce healthy sperm only, but the cow must not only produce a healthy egg but also provide the ideal condition for the fertilization of the ovum by the sperm and the establishment of the ovum by the sperm and the establishment of the embryo in the uterus. The cow then has to provide the nutrients for the calf and then, when it is mature, she must expel the calf from the uterus.

Infertility has probably become more serious in recent years because it tends to be associated with cows giving an immensely good milk yield.

Causes of infertility

Fundamentally the breeding potential of an animal depends on certain inherited genetic tendencies and the presence of any inherited abnormalities. Such genes may cause the animals either to be infertile or to produce faulty calves. However, some of the most important causes of infertility are bacterial or other infectious agents. Such an agent may be a specific organism producing a condition of which infertility is only one symptom. An example of this is contagious abortion (brucellosis), a disease that has now been virtually eradicated from the UK. Another example is trichomoniasis, the causal organism being *Trichomonas foetus*, which is spread at service by infected bulls which themselves show no obvious infection. A third example, vibriosis, is due to the organism *Vibrio foetus* and is carried and transmitted by apparently normal bulls. Both trichomonad and vibrio organisms multiply in the uterus and cause the death and abortion of the fetus. Infertility may also be caused by non-specific infections of the genital tract.

Control and prevention When a cow or heifer fails to conceive to a second service, she should be examined by a veterinary surgeon who, by carrying out tests, can establish if there is an infectious cause. If there is, it may well be treatable. Prevention depends on a number of factors. The herd should be kept as self-contained as possible and only maiden heifers and unused bulls should come into the herd. There should be no exchange of breeding animals. The diet must be adequate and balanced.

Diseases affecting cattle of all ages

Foot-and-mouth disease

Foot-and-mouth disease is an acutely contagious disease which causes fever in cattle and all cloven-hoofed animals, followed by the development of blisters or vesicles, which arise chiefly on the mouth and feet. The disease is an infection, worldwide in incidence, caused by at least seven major variants of the foot-and-mouth virus. Each type produces the same symptoms and they can only be identified by specialist examination. The disease is notoriously contagious, perhaps more so than any other disease, and it is known that it can spread some 50 miles down-wind from one outbreak to another. It has a fairly short incubation period of about 3–6 days so it can spread rapidly throughout a susceptible population.

The disease does not usually cause deaths, except in the young, but it leads to damaging after-effects. It will seriously lower productivity in all animals, especially milking cows. Mastitis will possibly develop so milk production is permanently impaired and the infection of the hooves may cause lameness which leads to secondary infections.

Infection with the virus of foot-and-mouth disease can occur in many ways other than by wind-borne spread. People, lorries, wild birds and other animals, wild life, such as hedgehogs, markets and so on are just some of the ways in which the disease can spread. In addition, infective material can come in from abroad in imported substances such as hay and meat and bones which have not been sterilized.

In the UK foot-and-mouth disease is a notifiable disease. The notification of the disease and slaughter of infected and in-contact animals had been a successful policy and there were few outbreaks for many years. The last outbreak until the year 2001 was in 1967, with just two very small outbreaks since then, but in February 2001 the most widespread outbreak of all time occurred. It took over 6 months to control and led to the slaughter of millions of infected or contiguous herds of cattle and pigs and flocks of sheep at a cost of several billions of pounds. The worst affected areas were in Cumbria, Wales and the southwest of England and for a time the disease appeared to be out of control. After the 1967 outbreak a committee of enquiry advised that in any future outbreak consideration should be given to instituting vaccination in a ring around infected areas to prevent spread of the disease, combining this policy with the continuing slaughter of infected animals. This worked well in outbreaks that occurred on the continent of Europe at the same time and the disease was speedily eliminated.

Several enquiries into the 2001 outbreak are currently underway. The origin was almost certainly from imported infected meat which was inefficiently sterilized before feeding to pigs. From infected pigs the disease then spread to sheep, in both species of which the disease is not always immediately apparent, and after which it infected cattle with especially catastrophic results. The symptoms in cattle that would lead one to suspect foot-and-mouth disease are as follows.

Infected cattle suffer from fever and anorexia with a sudden drop in milk yield in the milking cow. There are blisters on the upper surface of the tongue and the balls of the heels. Animals prefer to lie down, but when they are forced to walk they move painfully, occasionally shaking their feet. The lameness then becomes worse so the animal can barely move. Blisters also develop on the teats so cows cannot be milked. Positive diagnoses are always dependent on specialist laboratory examinations.

It may be surmised that the government enquiries will advise that the following measures must be taken to prevent a future catastrophe on the lines of the 2001 outbreak: the control of imported products that may be infected; much speedier diagnosis and disposal of infected cattle; limited use of ring vaccination; much stricter regulation of animal movement and disinfection. Such measures administered stringently may prevent further disasters.

Rabies

Rabies is a notifiable disease which affects nearly all animals and is especially dangerous to humans. It causes progressive paralysis and madness in most animals. The virus is present in the saliva so that the animals affected may bite one another, or humans, and in this way cause infection to spread. It is nearly always fatal. The greatest danger presented by this disease is that infection becomes endemic – virtually permanent – in wild animals, particularly foxes but also badgers, deer and vampire bats, and these infect dogs, cats and farm animals. In this way is the particular danger to humans created.

The disease may be prevented in humans and animals by vaccination, but it should be emphasized that no vaccinations are 100% effective and it is difficult to exaggerate the ghastly effects of the disease in the human.

One of the features of this virus is that after an animal becomes infected, perhaps via the saliva by being bitten by another rabid animal, several months may elapse before any signs of the disease are seen.

Symptoms Symptoms in animals may show great variation, ranging from the classic mad, biting, salivating beast, to 'dumb' forms in which the animals are incoordinate, progressively paralysed and make no noise. Usually the 'dumb' form follows the mad stage.

Control The British Isles have been free of rabies outside quarantine for 60 years, other than two cases in dogs which were quickly dealt with. Our rigid regulations prohibiting the importation of dogs and cats and certain other mammals without a 6-month quarantine period which includes vaccination was thoroughly justified. The fear remains that animals incubating the disease may be smuggled in, or could jump off a boat, and thereby rabies could become endemic in the wild animal population. The danger has increased with the spread of rabies in areas of Europe near the UK. In those countries where the disease is endemic, great strides are now possible towards eradicating rabies with the aid of the greatly improved vaccines now available. In the UK the long-established quarantine policy has now been abandoned in favour of a vaccination policy for susceptible species.

Tuberculosis

Tuberculosis is a chronic infectious disease affecting virtually all species of animal and also humans and birds. It is due to a bacterium, *Mycobacterium tuberculosis*. There are three main strains of this organism. There is the human strain which affects primarily the human but can also affect cattle; the cattle (bovine) strain which is most prevalent in cattle but also affects humans, pigs and certain types of wild life; and finally we must be aware of the avian strains which primarily affect birds but also cattle, pigs and other animals at times. All types of cattle may be affected, but it is in the dairy cow that the main risk occurs. The milk from an infected dairy cow may contain the organism that causes infection and this may infect calves or human subjects if the milk is not pasteurized or sterilized. Also, in advanced cases of tuberculosis the uterus may become infected so that the calf is born with the disease.

Tuberculosis is essentially a slow, inflammatory action which produces almost anywhere in the organs of the body nodular swellings, known as tubercles, which are of fibrous tissue with a core of pus-like caseous (cheesy) material. Fortunately it is possible to test an animal for the presence of the disease long before it has reached the contagious stage.

Tuberculosis has been virtually eradicated from the UK though testing continues and some cases do occur. Recently there has been an increase in certain parts of the UK which is of real concern since the reason has not been established. Animals may be infected by man and wild life: there is, for example, much concern about the infection of cattle by badgers, which, in some areas, are serious reservoirs of the disease. Those same measures that have generally been successful in eliminating tuberculosis in the UK are also being applied in a similar fashion elsewhere.

Johne's disease (paratuberculosis)

Johne's disease is a chronic infectious enteritis causing progressive wasting and eventually death. The cause is a bacterium, *Mycobacterium johnei* (paratuberculosis). In the UK the incidence has generally declined in recent years. Infection occurs usually in young calves from their dams or in contact with the faeces of carrier cows. There is then a long incubation period of 2–6 years whilst the lesions develop in the small intestine, with the animal scouring profusely accompanied by a steady loss of weight.

Treatment is not advised as it is rarely effective. A vaccine is available to control infections on affected herds, but a programme of control and eradication should only be carried out with veterinary guidance.

Anthrax

Anthrax is a serious bacterial disease which infects cattle, other animals and humans. It is rare to see symptoms in cattle as infection is so acute that it usually causes sudden death. The cause of the disease is a bacterium, *Bacillus anthracis*, which may persist in the soil of a farm, once present, almost indefinitely. However, few cases occur in cattle in the UK, around 50 a year, because of the means taken to prevent its spread and re-infection.

Anthrax should be suspected if an animal is found dead or has a very high fever and swollen neck. At either point professional help should be sought. Diagnosis is made by the veterinary surgeon taking a very small quantity of blood and examining this under the microscope. If the animal died from anthrax, or is infected with it, the blood will be teeming with the rod-shaped organisms that are distinctive features of this disease.

A major warning: do not allow the carcass to be cut or air to reach the blood since it is then that the organisms form spores which can survive for many years.

Some blood may be oozing from parts of the animal and this must be treated with disinfectant. Anthrax is a notifiable disease and measures that must be taken are mandatory. The carcass must be burned or buried deeply so that it is removed from any opportunity to cause re-infection or for the organisms to form spores. The area around which the animal was found dead should be disinfected and isolated for a short while in case of spillage of other organisms. It is most likely that the source of infection is from a feedstuff, possibly meat or bone meal, which has come from imported infected material.

Enzootic bovine leucosis (EBL)

This is a comparative newcomer to the list of notifiable diseases. It is a virus disease, slow and insidious in its effect, which may have been imported with Canadian Holstein cattle. It causes multiple tumours, known as lymphosarcomas, and whilst it is certainly not widely present in the UK, its presence has been confirmed. The symptoms in the live and adult animal are chronic ill health, anaemia, weakness and inappetance, which are not diagnostic. Only special laboratory examinations will give a definite diagnosis.

Bovine spongiform encephalopathy (BSE) or 'mad cow disease'

Bovine spongiform encephalopathy is an extremely serious disease of cattle, believed to have originated from infected meat and bone meal in cattle feed concentrates, which was derived from a similar disease in sheep known as scrapie. The eradication of BSE is of vital importance to safeguard herds and hence the future supply of bovine meat and dairy products for human and animal food together with important bovine by-products. In addition it has been alleged that the consumption of infected bovine products may cause new variant Creutzfeld-Jakob disease (vCJD) in the human subject.

BSE is a fatal brain disease of cattle which has emerged in recent years, being first recognized in the

UK in 1986. The incubation period for BSE is very long, commonly 3–5 years, but the range can be considerably wider, from 30 months to 8 years and possibly even longer. Cases have been reported in many other countries but not in the same magnitude as in the UK. The symptoms in cattle are as follows: the affected cattle appear to become extremely nervous, although there is no conventional madness. The cattle lose weight, have difficulty in walking and the milk yield declines markedly. Confirmation of the disease is dependent on post-mortem examination of the brain tissue. There is no certain test to detect the disease in live animals.

Cause

In neurodegenerative disorders such as BSE in cattle, scrapie in sheep and CJD in man, referred to collectively as transmissable spongiform encephalopathies, the causative infective agents are distorted abnormal prions. A prion (PrP) is a small protein molecule found in the cell membrane but which has no associated nuclei and so it is not classified as a living cell. The term 'prion' is a generic one and different species of animals have brain cell prion proteins of different composition. For example, the amino-acid sequence of the human prion differs at more than 30 positions from that of the cattle prion, whereas sheep and cattle prions differ at only seven positions. Prions are extremely resistant to disinfection and temperatures greater than 120°C, over long periods, are needed to inactivate the finely divided material. The only chemical treatment so far found completely successful is sodium hypochlorite at 20 000 parts per million of available chlorine for 1 hour.

Origins of BSE in cattle

After the first cases of BSE appeared in cattle, an epidemiological study of the first cases indicated that the only factor in common had been commercial cattle feed concentrates which contained meat and bone (MBM) derived from sheep (and possibly cattle) presumed to have been infected with scrapie or BSE. This was coupled with a change in processing that eliminated a solvent extraction step and hence also eliminated a steam stripping treatment which had been needed to remove solvent residues. This led to the banning of ruminant-based feed for ruminants.

Some experts have reservations about the sheep MBM hypothesis, particularly with regard to how the sheep scrapie prion changed, during passage through cattle, to the BSE prion. It is possible that BSE developed in such large numbers as a result of a number of inciting factors coming into play in combination. There is currently a serious discussion among experts on the possible roles of exposure to organophosphorus compounds at sub-toxic levels, or of the possible role of hay mites as a vector. Nevertheless, the prohibition of ruminant material from animal feed and the successive dramatic reduction in new cases over recent years seem to indicate that the animal feed was a key factor. The downward trend from some 35 000 cases in 1992 to an anticipated virtual elimination in the year 2001 is evidence of the efficacy of the control methods. There is no evidence of vertical transmission in cattle and it is believed to be entirely by the oral route.

The risk to man

If there is a danger that BSE could be a risk to man, two situations must be considered: firstly that the disease can be transmitted from cattle to man, and secondly that parts of the diseased animals carrying the infective agents can enter the food chain.

On the first issue, there is no direct evidence to show that BSE can be transmitted from cattle to man. On the other hand the emergence in the UK in recent years of some cases of anomalous cases of CJD caused by a new pathogen, previously unseen, led to the view of experts that cases were most likely to have been caused by exposure of the affected people to infected brain or spinal cord before 1989 at which time they were banned from the food chain.

The second issue is whether parts of a BSE-infected animal can enter the human food chain. The evidence is that whilst the BSE infective agent can be found in the brain, spinal cord and also the retina of affected cattle, extensive tests have failed to detect it in the muscle, meat and milk of infected beasts. Mandatory measures have been taken to prevent all that might be infected from entering the food chain. Thus consumption of muscle and meat, milk, gelatin and tallow from cattle in the UK should be without risk.

Animal health controls

To summarize, the principal measures for the control of BSE in cattle are:

- to slaughter all animals clinically diagnosed on the farm;
- to prohibit the feeding of all material containing animal protein derived from ruminants to all farm animals;
- to incinerate all carcasses of cattle affected with BSE;
- the removal of all tissue from beef carcasses that could be infected with the prion ('specified risk material');
- efficient controls on all imported products that could be infected;

- OTMS Scheme (Over Thirty Month Scheme) – all cattle slaughtered that are over 30 months in age are incinerated or rendered for safe disposal;
- introduction of the Beef Assurance Scheme – meat from animals aged 30–42 months can be exempted from the 30-month rule and sold for food provided the herds meet stringent conditions;
- introduction of the Cattle Tracing System – a computerized cattle tracing system is now in hand in the UK which registers cattle and their movements from birth to death. All cattle have 'passports' to record all movement and other essential information.

External parasites

Warble fly

There are two species of cattle warble fly that cause trouble, *Hypoderma lineatum* and *H. bovis*. Both are rather like bumble bees in appearance as their dark bodies are covered with white, yellow and black hairs. They have no mouths and only live for a few days. They have been found almost everywhere in the British Isles and most other parts of the world.

Adult female flies cause substantial losses to the stock person, being responsible for 'gadding' during the spring and summer when the cattle can be so panicked by the approach of the flies that they may stampede in the fields. Such stampeding lowers milk yield and may cause serious injuries to the beasts and even broken limbs. There is also hide damage and this is by far the worst loss. The maggots of the fly make breathing holes and these can ruin parts of the hide, even after the maggot has left the hide and the hole has healed over with fibrous tissue: this remains a weak area in the leather so that it has a much reduced value. It also damages the meat itself as a yellowish jelly-like substance forms around each maggot and spoils the appearance of the meat.

The life-cycle of the warble fly is as follows. During the short life of the fly the female may lay some 50 or so eggs which are attached to hair on the legs and bellys of cattle, those of *H. bovis* being laid singly (like louse eggs) and those of *H. lineatum* in rows of up to about 20. In 4–5 days the eggs hatch and tiny maggots emerge, crawl down the hair and burrow through the skin of the beast by means of sharp mouth hooks.

The entry of the maggot causes some injury and shows as a small pin-pimple or scab. Under the skin the maggot moves along, feeding and growing. It takes 7–8 months to reach the skin of the back of the animal. When this point is reached the maggots make breathing holes and become isolated within a small abscess. They stay there for 5–7 weeks, casting their skin twice and growing to a length of about 1 inch. The full-grown maggot is dark brown, fleshy, barrel-shaped and covered with groups of tiny spines. When fully ripe they squeeze themselves out of their 'warbles' and fall to the ground. Here they slide into shallow crevices in the soil and pass underneath dead vegetation. The skin of the maggot darkens and it then becomes a puparium or chrysalis inside which the warble fly is formed. This usually takes about 4 weeks.

Control must be achieved by dealing with the maggot. These can be destroyed by treating the animal with one of the more recently produced systemic insecticides.

Warble fly is a notifiable disease. The disease had recently been eradicated from the UK but has been reintroduced in recent years due to the importation of infected cattle. Any suspicions must be reported to the divisional veterinary officer of the Ministry who will supervise treatment of all cattle over 12 weeks old on the premises. Great care is needed in the use of the systemic insecticides as they are organophosphorus compounds and are given by pouring on the backs of the animals or by injection. Sick animals must not be treated. Milking cows must be treated immediately after milking to allow at least 6 hours before the next milking. The approved 'withdrawal' period – which is on the label of the product used – will specify that no animal may be slaughtered for meat less than 14–21 days after treatment.

Lice

Lice infestation is a major problem only in cattle when they are housed. The thick winter coat, lack of sufficient air and sunlight and possibly poor nutrition can contribute to the occurrence of this problem. There are four species of lice infesting cattle, three being sucking lice and the fourth, which is also the most common, a biting louse. The sucking lice are *Haematopinus eurysternus, Linognathus vituli* and *Solenoptes capillatus*, and the biting louse is *Damalinia bovis*. The sucking lice cause the greater damage and irritation as their sharp mouthparts pierce the skin to draw blood on which they feed. The biting louse feeds on the scales of the skin and on the discharge from existing small wounds.

These four types of louse affect only cattle and cannot exist on other farm animals. The entire life-cycle takes place on the skin on the animals where the females lay eggs in small groups. They are attached to the base of the hair by a sticky secretion. There the eggs remain for up to 3 weeks after which the live young hatch out and very quickly start feeding and reproducing.

Reproduction lasts for about 5 weeks; each female is able to lay about 24 eggs and if conditions are favourable the number of lice can increase very rapidly. Lice cannot survive for more than 3–4 days away from cattle so that transfer of infections is largely by direct contact.

It is not too difficult to recognize an infestation with lice. They tend to congregate on the animal's shoulders, the base of the neck, and the head and root of the tail, and examination of these areas will show them to be extremely scurfy and the lice should be obvious. As the infestation worsens more of the body can become affected. The affected animals rub and scratch themselves, greatly adding to the damage. The hair goes altogether in some areas. Rubbing thickens the skin which often breaks, so that bacterial infection gets in. All the effects of a bad infestation are ultimately to cause much loss of condition and production and should not be tolerated.

Insecticides must be applied to kill the lice. Materials that are satisfactory include organophosphorus preparations. Liquid preparations can be applied either under pressure or with a scrubbing brush. As these preparations have a reasonable residual effect, it is feasible for one good application to kill the lice and the larvae that hatch out after the application.

After treatment, measures must be taken to prevent reinfestation by contact with untreated animals and also by putting in fresh bedding and probably cleaning equipment.

Never underestimate the damage that can be done by lice both as mechanical irritators and as vectors of any livestock disease.

Foul-in-the-foot

Foul-in-the-foot is an infection with *Fusibacterium necrophorum* between the claws of the feet of cattle, causing pus formation and necrosis. The organisms enter only through the broken skin, often caused by sharp stones. Cattle become lame and productivity can be severely affected. Treatment consists of trimming, cleaning and dressing the affected feet and giving an injection of a sulpha drug or antibiotic. As a routine preventive where the infection is widespread in a herd, the cattle may be routinely walked through a foot-bath containing 5–10% copper sulphate or 5% formalin.

A vaccine should soon be available for the prevention of foul-in-the-foot.

Cattle health standards

A new body has been formed in the UK, led by the industry, with a view to promoting and improving the health of cattle in the national herd. It is called Cattle Health Certification Standards (UK) (or CHCS for short). It is an umbrella organization for various cattle health schemes operating in the UK and to set common standards for testing for non-notifiable diseases. The organization is focusing initially on four diseases: bovine virus diarrhoea, infectious bovine rhinotracheitis, leptospirosis and Johne's disease.

The health of pigs

The sow and piglets around farrowing

Farrowing fever (metritis, mastitis, agalactia syndrome or MMA)

This is a complex condition that occurs in sows during the period of farrowing and which seriously affects the viability of the piglets and the health of the sows. The sow runs an elevated temperature (about 40.5–42°C) and the udder is often hard and tender. There will be little or no milk 'let-down' ('agalactia'), and there is often inflammation of the uterus ('metritis'), which may be associated with retained afterbirths.

It is usually possible to cure the disease by immediate administration of broad-spectrum antibiotic injections such as tetracyclines or ampicillin, and posterior pituitary hormone (oxytocin) to 'let-down' the milk.

Transmissable gastroenteritis (TGE)

This is an alarming disease of pigs caused by a virus which causes a severe diarrhoea, with many very young piglets rapidly dying of dehydration. Some piglets vomit; all tend to have a very inflamed stomach and intestines. The sows may also be ill and have little milk. Older piglets, above about 3–4 weeks, will also be affected to varying degrees and for a period will scour and make little progress but should recover. The only treatment is to administer fluids, provide good nursing and inject antibiotics to reduce the likelihood of secondary bacterial infections. (Also see later section for general advice against viral infections.)

Piglet anaemia

Every pig that is born in intensive housing requires some extra administration of iron to ensure that it does not become anaemic. The piglet is born with limited reserves of iron in its liver, iron being an essential constituent of haemoglobin, the oxygen-carrying element in the red blood cells. Some supplementary source of assimilable iron must be given. The preferred way of giving the iron is by injection of 200 mg of a soluble iron preparation at not more than 3 days of age. This is sufficient to carry the piglet through to the time it is eating solid food which should be liberally supplemented with iron.

Escherichia coli lifection

There are various ways in which this ubiqituous cause of trouble on the livestock farm can do damage. In the young piglet during the first few days of life, *E. coli* may be a cause of an acute septicaemic condition. Within only a few days of birth a number of pigs in the litter can become very ill with no particular symptoms except extreme illness leading to death.

Neonatal diarrhoea

Those pigs that are infected with *E. coli* but less seriously than are the piglets killed with the septicaemic form of the disease show a most serious profuse watery diarrhoea.

Milk scours and post-weaning diarrhoea

Another form of diarrhoea due to *E. coli* but occurring later is that known as 'milk scours'. A pale, greyish scour affects piglets in the period between 1 and 3 weeks of age.

There are many strains of *E. coli* that cause these symptoms and almost certainly many of these strains will be living in perfect harmony within the intestines of the pig and will generally be present around the pigs' quarters. The *E. coli* appear to be able to multiply if there are certain stresses in the management.

Treatment and prevention The treatment of *E. coli* infections is best achieved either by injections or by medication incorporated in the feed or water. There are many antibiotics that can be used for this purpose, such as chlortetracycline, oxytetracycline, streptomycin, amoxycillin, lincospectin or the chemotherapeutic trimethoprim sulphonamide. Orally, nitrofurans, framomycin or neomycin may be given. The recovery of pigs suffering with *E. coli* diarrhoea will also be helped by administering fluids fortified with minerals, electrolytes and vitamins.

Prevention may be achieved by using certain antibiotics in the feed. There are a number of alternatives to choose from including furazolidone, organic arsenicals, chlortetracycline and carbadox. An alternative approach is to attempt to create an active or passive immunity by the use of serum or vaccine. The former course makes use of *E. coli* hyperimmune sera. In the case of the vaccines, *E. coli* vaccine may be given to the sow during pregnancy.

Diseases of young and growing pigs

Bowel oedema

Also caused by *E. coli*, the condition invariably occurs about 10 days after weaning. The first indication of an outbreak is usually the appearance, in a pen of weaners, of one dead pig, often the largest of the bunch. Others may show 'nervous trouble', staggering about when roused. As the disease progresses in these pigs they will lie on their side and paddle their legs and they then go into a coma and die within 1 day. In bowel oedema such pigs usually show swollen (oedematous) eyelids, nose and ears and may have a moist squeal resembling a gurgle. A post mortem, if carried out speedily, shows oedema of the stomach and in the folds of the colon and in the larynx.

The occurrence of this disease seems to be associated with the stress of weaning, with the addition of *ad libitum* dry feeding.

Treatment is rarely successful. To prevent the spread of the disease all dry foods should be withdrawn and replaced with a limited wet diet.

Streptococcal infections

In recent years streptococcal infections of young piglets have become very common. *Streptococcus suis* Type 1 infects the sow and she then transfers the infection to the piglets soon after birth. The organisms, after getting into the piglet, may cause a general infection.

The disease usually occurs at about 10 days of age. Affected pigs run a temperature, show painful arthritis, have muscular tremors, appear blind and cannot coordinate their movements. Some die suddenly with heart inflammation. Typically, some 20–30% of a litter is affected.

Affected pigs may be treated successfully with injections of suitable antibiotics such as penicillin or with sulphonamides such as trimethoprim.

The same organism, *Streptococcus suis* Type 1, can also cause serious disease in older pigs. Here, what seems to happen is that young pigs, already carriers of this disease, are mixed with older pigs around the early fattening stages, say 8–12 weeks. The carriers infect the pigs they are mixed with and after a few days of incubation the affected pigs show symptoms similar to those displayed by the younger pigs.

Clostridial infections

Whilst the effects of clostridial infections on pigs are in no way comparable with those that affect sheep or even cattle, there are problems from time to time, particularly in outdoor-reared pigs on 'pig-sick' land. For example,

Clostridium welchii (perfringens) can produce a fatal haemorrhagic enteritis in pigs up to about a week old. The disease has a very dramatic effect, with profuse dysentery, becoming dark from the profusion of blood within it. Usually pigs so affected die very rapidly though a few take a more chronic turn. Most die within 24 hours of the symptoms being noted. The litter affected should be treated with an antibiotic such as ampicillin, and those under risk of infection can be given antiserum. A more permanent answer may be needed and the sows can be given a vaccine for the infection, this being administered twice during pregnancy.

Piglets are also infected with *Clostridium tetani* in the same way as lambs by penetration of the organism from dirty conditions, possibly through the navel or by the injury caused during castration.

Greasy pig disease (exudative epidermititis; Marmite disease)

This is an acute skin inflammation of pigs 2–8 weeks old, with the production of excessive amounts of sebaceous secretion and exudation resembling the product Marmite in appearance and consistency and which does not cause irritation. It tends to be most common in hot weather.

For treatment inject a broad-spectrum antibiotic such as ampicillin or lincomycin. It is vital then to make sure that the cause is removed; in particular the best procedure is to provide plenty of soft bedding.

Swine influenza

The swine influenza virus causes serious losses amongst pigs in all parts of the world. In the normal way, if the animals are well housed and are not infected with other respiratory invaders, the course of the disease is acute but the animals quickly recover.

Atrophic rhinitis

Atrophic rhinitis is a respiratory disease caused principally by a bacterium, *Bordetella bronchiseptica*, and other organisms, notably a virus – the inclusion body rhinitis virus. Atrophic rhinitis affects primarily the membranes lining the delicate bones of the snout. This does profound damage to the growth of the pig and it also makes the animal more susceptible to enzootic pneumonia and any infections of the respiratory tract.

In order to prevent the condition a vaccine may be used or continuous preventive medication may be instituted with a suitable antibiotic such an Tylan or chemotherapeutic medicine such as a sulphonamide. It is better to improve the environment and especially ventilation and management than to rely on measures of vaccination and medication.

Porcine dermatitis nephropathy syndrome (PDNS) and post-weaning multi-systemic wasting syndrome (PMWS)

PDNS and PMWS are two conditions that have emerged in pigs worldwide in very recent years. The clinical signs of both conditions can be similar to those of classical swine fever and African swine fever, presenting a diagnostic challenge to veterinarians. The causal agents of both of these conditions are porcine circoviruses. Their effect is to cause persistent diarrhoea, wasting and an increased mortality. There are also oval red skin lesions and a severe pallor. Specialist diagnosis is required as there are unique and distinctive post-mortem signs.

No vaccines exist to protect herds. Herds may be found to carry the circoviruses that are believed to cause the disease yet may show no symptoms. The only treatment is to give anti-diarrhoea medicines, antibiotics to reduce secondary bacterial infections and, above all, reduce stress by ensuring plenty of space and bedding and clean and hygienic quarters.

Diseases largely of fatteners

Salmonellosis

All forms of salmonellosis can occur in pigs. Symptoms vary tremendously depending on the type and virulence of the *Salmonellae*. The worst outbreaks are usually after weaning in large groups of 'stores'. This often shows itself in the septicaemic form, causing sudden death or acute fever with blue (cyanotic) discoloration of the ears and limbs. Another form of disease shows itself as an acute diarrhoea, also with a fever, but these are not all the signs. Pneumonia may occur with abnormal breathing, and also nervous abnormalities, such as incoordination and paralysis. Also the skin is often affected and parts may even 'slough-off' later in the disease.

Treatment will require the use of antibiotics with known activity against *Salmonella*. Salmonellosis is a notifiable disease under the Zoonoses Order (see the earlier section headed 'Disease legislation').

The only form of *Salmonella* that can be prevented by vaccination is *S. cholerae suis*.

Enzootic pneumonia

One of the most serious scourges and causes of economic loss on the pig farm is the disease of pneumonia caused by a species of mycoplasma, *M. hyopneumoniae.* Though this is recognized as the principal cause of the disease, other organisms are common secondary invaders, including other types of mycoplasma and bacteria, such as *Pasteurella multocida* and *Bordetella bronchiseptica*. The lesions of enzootic pneumonia tend to be in the anterior lobe of the lungs which become consolidated (solid) and grey in colour.

Treatment should be instituted in pigs showing serious clinical signs. Broad-spectrum antibiotics, such as chlortetracycline and oxytetracycline, may be injected but may not have a permanent beneficial effect as no immunity follows an attack.

Whilst the disease may kill a proportion of pigs its most serious economic effect is to lower the growth rate and food conversion efficiency of nearly all the animals in the building. The longer-term measures to deal with the disease are to improve the housing and environment and/or to consider restocking with enzootic-pneumonia-free pigs.

Swine dysentery ('bloody scours')

This is a disease largely of fattening pigs and the infection causes inflammation of the intestines which leads to dysentery (diarrhoea together with some blood effusions), and a general debilitation of the pigs. It is highly contagious, being spread via contact with infected faeces.

The main causal organism involved in swine dysentery is a bacterium, *Treponema hyodysenteriae*, though it is not the only one involved and others such as *Campylobacter, E. coli* and *Salmonellae* of various species may always be present as secondaries.

Treatment A number of medicines may be given via the food or water, such as the macralides, in the form of tylosin, erythromycin or spiramycin, or lincomycin, which may be used as a feed additive or in combination with spectinomycin may be given in water. Further effective medicines are tiamulin and dimetridiazole.

Diseases of pigs largely without age incidence

Swine erysipelas ('the diamonds')

This disease is caused by a bacterium known as *Erysipelothrix insidiosa* or *E. rhusiopathiae*. Most pigs can live in risk of this disease because the organism can live in and around piggeries and 'pig-sick' land for very many years in the sporulated form. In the peracute form there may be hardly any signs at all, and the pig may be found dead after a very brief period of severe illness. In the acute form the pig is also ill, has a high temperature and the skin is inflamed: it will not eat. Then there is a sub-acute form which is the more typical disease, showing characteristic discoloration of the skin with raised purplish areas said to be roughly diamond-shaped throughout the back and on the flanks and belly. Finally, there are two 'chronic' forms of the disease. In one the joints are affected which causes the pig severe lameness, whilst in the other the organism causes erosion of the valves of the heart, leading to signs of heart dysfunction or even heart failure and death. A feature of swine erysipelas is that it tends to affect pigs during hot and muggy weather conditions. The condition is effectively treated with penicillin or broad-spectrum antibiotics and it is prevented with an annual vaccination.

Intestinal haemorrhage syndrome ('bloody gut')

This condition affects fatteners in the latter stages of growth, and, after a short period of depression, with an appearance of paleness and an enlarged abdomen, they die rapidly. Post-mortem shows the small intestine full of blood-stained fluid and also gas in the large intestine. There may also be a twisting of the gut. The condition is more particularly common in whey-fed pigs, but not exclusively so, yet its cause remains a mystery.

Mulberry heart disease

This is a curious disease of fatteners causing sudden death or the pigs become depressed and weak, collapse and die within about 24 hours of the onset of the disease. Few recover. On post-mortem examination the signs are an enlarged and mottled liver and the surface of the heart is streaked with haemorrhages running longitudinally and also occurring in the endocardium. The cause is not really known, but there is a suggestion that a deficiency of vitamin E or selenium may be involved. There is little evidence that treatment does much good, but the use of multi-vitamin injections and broad-spectrum antibiotics may be of some benefit.

Swine fever (hog cholera)

This is a very serious virus disease which had been eliminated from the UK. Affected pigs show many symptoms especially a high fever, depression and discoloration of the skin and usually diarrhoea of a particularly foetid type. Pigs also have a depraved appetite and thirst. There is also a highly virulent disease known as African Swine Fever with rather similar symptoms but the causal agent is a different virus. It has a devastating effect and may kill 90% of the pigs in a herd. It has occurred not only in Africa but recently in parts of southern Europe.

Recently there have been serious outbreaks of swine fever (not African but the classical form of swine fever and designated CSF or classical swine fever) in the east of England. They have been dealt with by slaughtering infected herds, creating areas of isolation around the infected farms and restricting the movement of pigs. CSF has now been eliminated from the UK but outbreaks are still prevalent in many other parts of the world, including continental Europe.

Aujeszky's disease

This is a herpes virus which leads to nervous and respiratory symptoms with a fever. Mortality can be high in young pigs. It is a notifiable disease in the UK and affected herds are slaughtered.

Tuberculosis

Three forms of tuberculosis can infect the pig: the bovine, avian and human forms. Because of the much reduced incidence of tuberculosis in cattle and man, there is now very little in pigs, but it does still exist as a clinical entity. Of the cases that occur, most are of the avian type doubtless infected from wild birds since tuberculosis is virtually non-existent in the domestic fowl.

The symptoms are rather non-specific and would rarely be suspected by the pigperson. Loss of weight, coughing and discharge from the nose will hardly lead to suspicion so that it is unusual for the disease to be recognized before it is seen in the slaughterhouse by the occurrence of swollen and infected lymph nodes in the throat, chest and abdomen. If samples are taken the disease can then be diagnosed positively.

Treatment is rarely called for, but careful note should be given to any lessons there may be in isolating stock from sources of infection and also in improving the hygiene of the premises. The organisms that cause tuberculosis in all species are very persistent and resist destruction by most means; it is therefore very important to take such measures that eliminate them. It is also pertinent to stress the risk to the human population and whilst it is not a highly contagious agent of disease, it could represent a very undesirable challenge to man if the infection became widespread in a pig herd.

A group of important virus infections affecting pigs of various ages

Pig enterovirus infections including 'SMEDI'

A number of enteroviruses cause stillbirths, mummification, embryonic death and infertility – the initials SMEDI – and hence this name is given to this group of diseases.

Pig enteroviruses may also be associated with nervous symptoms such as incoordinatioon, followed by stiffness, tremors and convulsions. The acute form of this disease, known as Teschen disease, is probably not present in the UK, but it is a notifiable disease and would be dealt with by a slaughter policy to attempt to eradicate it. The mild form, known as Talfan disease, is certainly present in the UK, but no special control policies are instituted and reliance is made on the pigs developing a natural immunity, as with many other virus diseases of rather mild and uncertain symptoms.

Rotavirus infection

This causes very severe scouring in piglets which can lead to quite a heavy mortality. The symptoms are anorexia, vomiting, and diarrhoea, yellow or dark grey in colour, and very profuse. The diarrhoea causes rapid dehydration, which can kill, otherwise the pigs can recover in about 7–10 days.

These clinical signs are very similar to other conditions of young pigs causing profuse diarrhoea, such as TGE and *E. coli* infection, and only a laboratory diagnosis which identifies the virus will confirm the cause.

Vomiting and wasting disease

This is a disease of the newborn pig caused by a coronavirus similar to the TGE virus. Affected pigs firstly vomit, huddle together with general illness, are depressed, run a high temperature and show little interest in suckling. Only a proportion of piglets in a litter are affected and only a proportion of all litters may be affected at all.

Epidemic diarrhoea

This is a very contagious disease of pigs caused by a virus which produces a profuse diarrhoea, vomiting, wasting and inappetance. The disease is similar to TGE but affects largely the older pigs; younger pigs are affected very much less or not at all.

Parvovirus infection

Parvovirus infection of pigs is a cause of infertility, stillbirths, small litters and mummification. It is a relatively new condition that has been recognized in a number of large units. It tends to be most commonly seen in young gilts or newly introduced pigs which have no resistance to the 'local' infection. It may also infect boars which then act as spreaders of infection. It appears to be far more of a problem in those housing systems where the sows are kept as individuals (as in tethering or sow stalls) and where the lack of contact fails to produce a passing infection and then resistance. In this type of condition there is no treatment that is of any real use. Licensed vaccines have been produced in the UK. An alternative procedure is to 'infect' a young gilt before service and the most effective way is to 'feed' homogenized placenta from known infected cases. This is, however, a crude method of 'vaccination' and has great dangers of spreading other infections indiscriminately.

Porcine reproductive and respiratory syndrome (PRRS or 'blue-eared' pig disease)

This condition has already been mentioned as a good example of a recent viral condition that has swept through heavily populated pig areas. Attempts to limit its spread by notification, isolation and slaughter have completely failed and it is now accepted as a condition that has to be lived with and which appears to be due to one or more viral infections.

Symptoms The clinical signs are very variable but the following usually occur when the disease first strikes: inappetance and listlessness; abortions; embryonic death leading to a substantial fall in total births; irregular oestrus or anoestrus; skin changes including blue coloration; poor milking; poor piglets; lameness; and puffy eyes. A general increase in diseases occurs due to the immunosuppressive effect of the PRRS virus, and especially respiratory disease. One of the biggest problems that has occurred in the wake of PRRS has been infection in pig herds with the swine influenza viruses. Their incidence and effect appear to become much greater after a primary outbreak of PRRS. When a herd is affected with both PRRS and one or more of the strains of swine influenza virus, the effect can be quite devastating. This effect is especially serious in intensive units, and some farmers have reverted to less intensive methods in their determination to eliminate the effects of this range of infection.

General control measures for virus infections when no vaccines are available

The upsurge in the number of virus infections in pig herds in recent years is worrying and provides us with lessons of great importance. They appear to be very similar to those conditions that have affected poultry, but the pig industry is much more vulnerable. It tends to be less enthusiastic about hygiene, has no general policy of depopulation of sites and moves pigs around during their lifetime much more generally than is the case with poultry.

To minimize the effects, the following preventive measures should be considered:

- Introduce a minimum of new stock from outside the unit.
- Depopulate the housing of young pigs as frequently as possible.
- Keep adult breeders closely but cleanly housed.
- Do not use sow stalls or tethers but kennels or yards which group the gilts, sows and boars.
- Limit the size of each self-contained unit to a minimum. This will ensure that the effects of any virulent virus take place as rapidly as possible and immunity may develop quickly and uniformly.
- Practise careful isolation of the site in terms of feed deliveries, visitors, collecting lorries for fat pigs and any other potential danger.

External parasites

Pigs can be quite badly infected with lice (*Haematopinus suis*), the mange mite (*Sarcoptes scabei* var. *suis*) and the stable fly (*Stomoxys calcitrans*).

Lice

Lice tend to be most common on the folds of the skin of the neck, around the base of the ears, on the insides of the legs and on the flanks. The constant irritation causes the pig to rub and scratch and this reduces growth and food conversion efficiency. Also the lice may be vectors of other disease agents.

Mange

Mange is an infinitely more worrying problem than lice. The parasite which is most common in the pig (*Sarcoptes scabei* var. *suis*) burrows in the skin and lives in so-called galleries. The mange parasites are about 0.5 mm long and lay their eggs in the galleries and develop through larval and two nymphal stages to adults within a period of about 15 days. Whilst the parasites can only multiply on the pig, the mites can survive up to 2–3 weeks in piggeries. The disease causes considerable damage to the skin and intense irritation and in the final stages of the disease the skin looks more like that of an elephant than a pig. To eliminate this disease suitable medicines are phosnet, diazinon or bromocyclen. Two applications at 10-day intervals should be fully successful.

Internal parasites

Ascaris lumbricoides is the common large roundworm that lives in the small intestine. Some can achieve a length of up to 45 cm. Infection can be so bad that the intestines are literally blocked with large numbers of these worms. The life-cycle is direct and after the eggs are laid and ingested by pigs kept under unhygienic conditions, the eggs hatch out. The larvae, in their development to the adult stage (which is always in the intestine), actually pass through the lungs and liver. Coughing and pneumonia result from the migration of the larvae in the lungs, and tracking of the larvae through the liver also causes damage and leaves small white areas known as 'milk spots'. Treatment of the pigs with anthelmintics will eliminate the worms and improved hygiene can destroy the eggs.

Other intestinal worms of pigs occur in various parts of the intestine. There are the two stomach worms *Ostertagia* and *Hyostrongylus rubidus* and the worm *Oesophagostomum*, which lives in the caecum and colon.

Finally there is a lungworm of importance known as *Metastrongylus*. The adult worm lives in the bronchioles of the lungs. In this case the life-cycle is indirect, infection being caused by ingestion of the earthworm which is the intermediate host of the larvae of the *Metastrongylus*. The lungworms undoubtedly cause damage to the lungs and pneumonia, but it is chiefly in younger pigs that the symptoms are serious. Older pigs are not usually adversely affected but do remain as carriers of the infection.

Treatments can be carried out by in-feed medication or by infections using products such as fenbendazole, tetramisole, thiabendazole, piperazine and dichlorvos.

Other pig diseases

Pigs can be affected with anthrax; foot-and-mouth disease; a mild condition but rather similar to foot-and-mouth disease known as swine vesicular disease; and rabies, but all are only very rare occurrences if occurring at all in the UK. All, however, are notifiable diseases.

There are certain other viral and bacterial infections that the pig industry is wrestling with. For example, post-weaning multi-systemic wasting syndrome (PMWS) is present in most of the intensive pig-rearing areas in East Anglia and Yorkshire, as also is porcine dermatitis nephropathy syndrome (PDNS). These last two conditions are both viral infections. Flank biting and necrotic ear syndrome is caused by *Staphylococcus hyicus* and porcine proliferative enteropathy or ileitis is caused by *Lawsonia intracellularis*. The names of these conditions indicate the symptoms, but proper diagnosis can only be carried out by a veterinary laboratory and treatments thereby organized, mostly using antibiotic mixtures. It is a characteristic of the large intensive and extensive units that such diseases emerge from apparently nowhere and cause a variable amount of morbidity and mortality.

The health of sheep

Health problems at or near lambing

The first point to stress is the essential need for the highest standard of hygiene at all times. Lambing is the period of greatest risk from infection. Ewes must be carefully observed during lambing to reduce losses due to difficulties in the birth of the lamb (dystocia). If no progress is made within 3 hours of the start of lambing, the competent shepherd should carry out an examination. The shepherd may find he/she can readily correct any malpresentation or may decide to call in a veterinary surgeon, according to the judgement of the position.

E. coli infection (colibacillosis)

There are two types of *E. coli* infection: the enteric and the septicaemic forms. Symptoms in the enteric form usually manifest themselves at 1–4 days of age and the lamb becomes depressed, shows profuse diarrhoea or dysentery and dies, usually within 24–36 hours of the onset of symptoms. Those affected with the septicaemic form are usually 2–6 weeks old. Affected lambs have a fever and become stiff and uncoordinated in their movements; later they lie down, paddle with their legs and become comatose.

To prevent this disease the ewes should be vaccinated during pregnancy. Treatment is by antiserum or broad-spectrum antibiotics.

There is a further condition affecting young lambs and also caused by *E. coli*. This is watery mouth or rattle belly. Lambs have cold wet mouth with drooling of saliva – or even abomasal contents. Such lambs are depressed, usually scour and have an elevated temperature. Treatment is by nursing, warmth, antibiotics and vaccination for prevention of coliform infection in general.

Navel-ill

This occurs in lambs for the same reason as in calves (see 'The health of cattle') and requires identical attention.

Mastitis

Mastitis in the ewe is a very damaging disease which often leads to the complete destruction of the parts affected. Immediate treatment with a broad-spectrum antibiotic, given both locally into the teat and by injection, may save the udder and the ewe.

The clostridial diseases

The group of bacteria known as clostridia are spore-forming organisms found universally wherever sheep are kept. As soon as the bacteria pass out of the animal's body they form tough spores (or capsules) which make them extremely resistant to destruction. Their mode of action is to manufacture toxins which make the animal ill or kill it. Clostridia will multiply only in areas where there is no oxygen, and they do this in wounds, especially deep ones, in the intestines of animals, or in organs within the body such as the liver. Clostridial diseases tend to affect animals in very good, thriving conditions. The main clostridial diseases are described below.

Lamb dysentery

Lamb dysentery occurs in lambs during the first few weeks of life and is caused by the organism *Clostridium perfringens* Type B. The principal symptoms are a bloody diarrhoea (dysentery). There may be many deaths – up to 30% of the flock if it is unchecked.

Immediate treatment or prevention is by the use of antiserum. To protect from future attacks, the ewe should be given two vaccinations, the last one being about 1 month before lambing, thus transferring a strong immunity to the lambs. In succeeding years ewes will only need one additional 'booster' injection.

Pulpy kidney disease

This is caused by *Clostridium perfringens* Type D. The first sign is often the sudden occurrence of a number of deaths in really good lambs at 2–3 months of age. The name 'pulpy kidney' is given because on post-mortem the kidneys show a high degree of destruction.

Antiserum or vaccination procedures will be similar to those that apply to lamb dysentery. A major predisposing cause is when sheep are placed on rather too good a pasture, and much can be done to prevent trouble by watching their condition and regulating their nutrition.

Blackleg

Caused by *Clostridium chauvoei*, blackleg is similar to blackquarter in cattle which has been described earlier.

Braxy

Caused by *Clostridium septicum*, usually the first sign is the sudden death of some young sheep in good condition on frosty autumn mornings. The predisposing cause is the eating of frosty food which damages the wall of the abomasum and allows the invasion of the clostridial organisms. Vaccination is totally effective as a preventive.

Black disease

The symptoms are sudden death, usually in adult sheep. The organism *Clostridium oedematiens* invades the liver after damage by the liverfluke. It usually occurs in the autumn and early winter when flukes migrate in their largest numbers from the intestine to the liver.

The same measures of control are used as in other clostridial diseases. It will also help to prevent black disease if the fluke infestation is controlled.

Tetanus (lockjaw)

Tetanus is caused by the organism *Clostridium tetani* which finds entry through a wound which heals and the organism multiplies within this anaerobic atmosphere, producing the toxin which causes nervous symptoms, spasms of muscular contraction, stiffening of the limbs and eventually death.

Control and prevention is on the same lines as other clostridial diseases. Proper treatment of wounds will greatly reduce the likelihood of tetanus.

The use of vaccines and sera with clostridial infections

All the clostridial diseases can be countered by the use of sera and vaccines – sera for immediate prevention or treatment or vaccines to build up an immunity over a period of weeks. Preparations combining several vaccines together (up to at at least eight) covering all the common sheep diseases are now available.

Pneumonia

Acute pneumonia infections are especially serious in winter and autumn and are caused by infection with *Pasteurella* bacteria. The predisposing causes of infection can be intensification, harmful weather conditions, transportation under less than ideal conditions and infestation of the lungs with parasites.

Treatment is by appropriate antibiotics, and prevention by vaccination.

Orf (contagious pustular dermatitis)

This is a virus infection of sheep causing vesicles and scabs on the skin and especially over the mouth and legs. It can also cause an unpleasant disease to sheep handlers. After infection older animals develop an immunity. An effective vaccine is available for prophylaxis.

Deficiency conditions

Swayback (enzootic ataxia)

Swayback affects the nervous system of young lambs causing incoordination and paralysis of the limbs of the body due to a degeneration of the nerve cells in the brain and spinal column; it is associated with a low level of copper in the blood and tissues of the lambs and ewes. The basic cause of the disease is a deficiency of copper, this element being essential in the formation of enzymes for the construction of nervous tissue. The critical level of copper necessary for the proper development of nervous tissue is 5 ppm and in swayback areas levels of copper are generally well below this. To prevent the disease it is necessary to supplement the copper intake of the pregnant ewe. Adult ewes require an average daily intake of 5–10 mg of copper as an adequate allowance. Mineral block mixtures containing 0.5% copper sulphate are satisfactory, or diet supplementation with 0.2 g of copper sulphate fed weekly to each sheep would be adequate.

Pine (pining)

Pine is a disease of sheep which leads to wasting. It is caused by a deficiency of cobalamin (vitamin B_{12}) due to a lack of sufficient cobalt in the diet.

A simple way of preventing pine is to add a supplement that has been enriched with cobalt to the normal concentrate feed. Animals may also be dosed individually, the cobalt 'bullet' being the best treatment since it is released gradually over a long period. The pasture may also be treated with 5 kg of cobalt sulphate per hectare every 5 years.

Rickets

Rickets occurs in lambs and is a condition of poor bone construction when the calcification fails to take place efficiently. Proper formation of bones is due to a sufficient quantity and balance of the minerals calcium and phosphorus, together with usually vitamin D. In practice injections of vitamin D are successful in arresting the problem.

Internal parasites

The three main types of internal parasite that infest sheep are tapeworms, flatworms represented by the liverfluke, and the roundworms, which infest both the intestines and the lungs.

Tapeworms

Tapeworms require an intermediate host, as well as sheep, for their existence. The adult tapeworm lives in the sheep's intestines, and lays eggs which pass out in the dung on to the pasture. Further development occurs in the intermediate host, which is a small pasture mite; lambs become infested when they consume the mites accidentally when grazing.

Sheep may also be infested with a number of bladder worms, infestation being picked up on grazing contaminated with the droppings of dogs and foxes which carry the adult tapeworms in their intestines. The bladder worms occur as thinly walled, fluid-filled bladders among the intestines, in the lung and liver and in the nervous system.

Liverfluke

Liverfluke can be a serious problem in wet years and has been fully described earlier in connection with cattle. The snail can exist only in areas where the soil is saturated with moisture for considerable periods, though it is not found naturally in running water or ponds. In dry periods or in winter, it tends to burrow into the mud and can survive in the inactive state for a considerable time.

Roundworms

At least ten species of roundworms (nematodes) cause parasitic gastroenteritis. Heavy infestations are disastrous but even light ones can cause great loss of productivity. Principal worms are: *Haemonchus contortus; Ostertagia circumincta; Trichostrongylus axei; Nematodirus spathiger; Trichuris ovis*; and *Oesophagostomum columbianum*.

Control is tackled by treatment of the animal to eliminate the worm burden and by pasture and flock management to minimize infection. Medicines used to control the infection are benzimiadoles, imidazothiazoles, tetrahydropyrimidines and organophosphorus preparation. Suitable pasture management is based on rotational grazing, allowing up to a week on pasture then closing it for 3 weeks.

Sheep may also be affected by lungworms. The reader is referred to the description of the similar condition in cattle.

External parasites

These include insects, mites and ticks. The easiest to kill are those that spend their whole time on the sheep – the mites, lice and keds. This group is effectively controlled by a single whole-body treatment with a preparation which is lethal to adults and young stages and persistent enough to kill the larvae as they hatch from resistant eggs, or by repeated use of a material that kills the adult only.

The blowfly, in comparison with the other external parasites, completes only one fairly short state of its development on the sheep and is thus open to attack by an insecticide for only a short time. Larvae hatch rapidly and feed on the skin and tissues for several days, causing great damage. They fall on to the herbage and form the resting stage, known as the pupa, from which the adult fly emerges several weeks later.

The problem of fly-strike requires additional measures to be effective. All cases of scour should be treated and wounds should be dressed immediately.

In the north of England and in Scotland a serious problem has recently emerged – the *headfly*. This fly feeds on the fluid from the eyes, nose and mouth of sheep. As a result of the irritation the sheep rub their heads against fences and walls causing wounds which make them even more attractive to the unwelcome attention of the headfly. The result of this disturbance is a very restless animal and considerable economic loss.

The *sheep tick* attaches itself for only a relatively few days during its 3-year cycle of development. The preparation that is used must therefore be highly active against all the developmental stages and should protect the sheep over a period of about 2 months.

Sheep scab, the commonly used name for a mange in sheep caused by the psoroptic mange parasite, was until recently a notifiable disease in the UK but has now been deregulated. The mange parasite can invade all parts of the body that are covered with wool and also the ears. The total effect is extremely damaging, which is the reason why the disease was notifiable. Although at one time the disease was eliminated from sheep in the UK it has now returned to quite a worrying extent. Treatment and prevention are instituted by a dipping of the sheep with approved materials according to a planned programme. It is also possible to deal with the condition by double injections of ivermectin, an organophophorus compound.

Infections of mature sheep

Scrapie

Scrapie is caused by an agent that acts like a virus but does not have the physical characteristics of one. The term 'prion' has been proposed. The disease affects the nervous system and the sheep becomes uncoordinated, develops a severe itchiness and rubs furiously against posts. No treatment is of any use and all sheep that are infected should be slaughtered.

The causal agent is believed to be the same as, or similar to, that which causes bovine spongiform encephalopathy (see section on this discease earlier in this chapter). It is believed that scrapie is transmitted vertically from ewe to lamb. It is also believed that some breed lines have a built-in susceptibility, so that where such lines are identified they should be culled from the breeding programme.

Foot-rot

This is one of the biggest causes of serious economic loss in sheep everywhere. It is primarily caused by a bacterium *Fusiformis nodosus* and spreads from sheep to sheep via infected soil. The harbourers of infection are in fact the sheep with bad feet in which the organism can survive indefinitely. The conditions that favour the spread of infection are wet weather and soil when the feet are softened and the organisms are released to invade other feet through any injured point.

Treatment and control have been described in the section on foul-in-the-foot in cattle. An effective vaccine is available.

Metabolic diseases of sheep

There are three most important metabolic diseases of sheep: pregnancy toxaemia (or twin lamb disease), milk fever (lambing sickness) and hypomagnesaemia (grass staggers).

Pregnacy toxaemia (twin lamb disease)

This is a disease of ewes which affects them in the last few weeks of pregnancy and nearly always occurs in ewes that are carrying more than one lamb. Affected ewes are incoordinate and then totally recumbent and comatose, and will invariably die unless treated.

To some extent the condition is due to excessive demands being made on the ewe's metabolism when associated with insufficient nutrition. Affected ewes may have injections of intravenous glucose solutions together with dosing with glycerine and/or glucose solutions. However, most of the effort should go into prevention. The important and indeed essential requirement is that during the last 6 weeks of pregnancy the ewe is given a diet that is low in fibre, nutritious and easily digested.

Milk fever and hypomagnesaemia

Milk fever (lambing sickness) and hypomagnesaemia (grass staggers) both occur in sheep and are similar to the condition in cattle (see earlier section).

Diseases causing abortion

Enzootic abortion

This is caused by a chlamydial organism and tends to be common in certain well-defined areas; in the UK it is especially prevalent in NE England and SE Scotland. The infection is usually introduced into a flock by a 'carrier' which releases a great amount of infected material when it aborts. This infects the ewe lambs which will not abort the first season but will abort in the second. After abortion they are then immune but young sheep and brought-in-animals will continue to become infected.

There is no specific treatment for infected animals, but it is good policy to isolate ewes that have aborted for the period when they may be discharging infected material. Thereafter a policy of vaccination may need to be instituted.

Salmonellosis

Salmonella abortion is an acute and contagious condition caused by the bacteria *Salmonella abortus ovis*, and can also be caused by the non-specific salmonellae, such as *S. typhimurium* and *S. dublin*.

Control can be instituted, including antibiotics or chemotherapy, vaccination and hygienic measures.

Vibriosis

Vibrionic abortion is caused by the bacterium *Vibrium foetis* var. *intestinalis*. Like salmonella infections, this causes late abortions. There is no specific treatment, but vaccines are available to control the condition.

Brucella

The *Brucella* species of bacteria can also cause abortion in sheep, the effects being the same as in cattle. *Brucella abortus*, *B. mellitensis* and *B. ovis* are all capable of causing the problem. After positive diagnosis, control relies on hygiene and vaccination.

Toxoplasmosis

There is also a parasitic cause of abortion in sheep which is due to a toxoplasm, *Toxoplasma gondii*. The best procedure in an infected flock is to mix infected ewes with non-pregnant ones to stimulate natural immunity before pregnancy.

Other causes of abortion

Other causes of abortion are *Listeria homocytogenes*, *Cornilla burnetic* and the viral infection border disease. Control of many of the causes is by vaccines, where available, and strict hygiene and isolation standards.

Other diseases

Maedi-visna

This is a 'slow' virus infection, also known as ovine progressive pneumonia and characterized by an insidious incurable respiratory disease. It is not a notifiable disease but there is an official M-V Accredited Flocks Scheme which enables farms to set up and maintain sheep known to be free after appropriate blood tests which are continued for member flocks.

Notifiable diseases of sheep

The notifiable diseases of anthrax and foot-and-mouth disease also occur in sheep and the reader is referred to the section covering the diseases affecting cattle.

Further reading

Andrews, A.H., Blowey, R.W., Boyd, H. & Eddy, R.G. (eds) (1991) *Bovine Medicine.* Blackwell Science, Oxford.

Blood, D.C. & Studdert, V.P. (1996) *Baillière's Comprehensive Veterinary Dictionary*. Baillière Tindall, London.

Hill, J. & Sainsbury, D. (1995) *The Health of Pigs: Nutrition, Housing and Disease Prevention.* Blackwell Science, Oxford.

The Merck's Veterinary Manual (1996) Merck & Co Inc., New Jersey.

Moss, R. (ed) (1992) *Livestock health and Welfare.* Blackwell Science, Oxford.

Sainsbury, D. (1998) *Animal Health.* Blackwell Science, Oxford.

West, G.P. (ed) (1994) *Black's Veterinary Dictionary.* A & C Black, London.

Part 4

Farm Equipment

24

Farm machinery

P.H. Bomford & A. Langley

Introduction

Farm machinery allows the farmer to carry out farming operations safely, economically and within the time available. Successful management of machinery involves the selection of equipment having the correct function and capacity, and the supervision of its efficient and safe operation to achieve quality work at the lowest possible cost.

The agricultural tractor

The tractor provides power for most mobile operations on the farm and needs to be flexible to carry out tasks ranging from heavy pulling to transport to power take-off (pto) work to loading duties. Power for the tractor is produced by its engine, and this power is used in three forms:

(1) as pulling power at the drawbar, to operate trailed equipment;
(2) as rotary power at the power take-off (pto) shaft;
(3) as hydraulic power, to operate hydraulic rams and motors.

The 'size' of a tractor is generally described by quoting the power produced by its engine. Since the full power of the engine is not available to do work outside the tractor, power that is available at the pto is a more useful indication of the work a tractor may be able to carry out.

The engine

The tractor's engine converts chemical energy contained in diesel fuel into rotary power at the flywheel. Two-thirds or more of the energy value of the fuel is lost as waste heat via the exhaust and cooling systems. Flywheel power, often called brake power because it is measured by applying a braking load at the engine's flywheel, is the product of the engine speed (N) in revolutions per minute (rpm) and the torque (T) or twisting effort, in Newton metres (Nm), that the engine can maintain at that speed. The relationship between these factors is:

$$\text{Power}(\text{kW}) = \frac{2\pi N(\text{rev/min})\,\text{T}(\text{Nm})}{60000}$$

A typical tractor engine produces little torque below 500 rpm. Maximum torque is reached at about 1400 rpm, and torque then decreases with increasing speed to 90% or less of maximum at full speed, which is 2000–2800 rpm. This torque reduction is called 'torque backup'. A large torque backup indicates a flexible engine with good 'lugging ability', and a reduced need for gear changing during work. Brake power, however, increases almost linearly with engine speed, and maximum power is produced only at maximum rated engine speed.

The conventional engine is the best mobile power source at present available in terms of efficiency, weight, utilizing available fuel and cost. However, it is by no means perfect. It is noisy, it vibrates and it produces exhaust gases which pollute the atmosphere. It is made up of many hundreds of individual components and has many points of wear. The working life of a well-maintained tractor engine, before it needs a major overhaul to renew worn parts, is between 4000 and 7000 hours. This compares with 200 hours for small air-cooled engines, 1500–2000 hours for a car engine and 12 000 hours for heavy-duty industrial engines.

The faster an engine rotates, the greater the amount of power (and fuel) that is consumed in just keeping the engine turning at that speed, in relation to the power that is available at the flywheel. If it is not necessary to run an engine at full speed, because maximum power is not needed, operating the engine more slowly will save fuel and prolongs its working life.

Many diesel engines are fitted with turbochargers, to increase power output. A turbocharger is an exhaust-driven rotary compressor which forces more air into the engine. This allows more fuel to be burnt, releasing more energy and producing more power. A power increase of 30% or more may be achieved, but this will subject the engine to greater thermal and mechanical stresses. Mechanical components, and cooling and lubricating systems, must be upgraded accordingly. Since the increase in power is achieved with no increase in engine speed or size, frictional losses do not increase in proportion, and the increase in power also results in an increase in engine efficiency.

The transmission

In order to deliver engine power at a wide range of forward speeds, the engine is connected to the wheels by a transmission offering from 6 to over 30 gear ratios. With so many ratios available it is necessary for the driver to change gear often in order to match power and forward speed to changing conditions. Various devices are provided to make this task easier. Synchromesh transmissions match the speeds of rotating parts, to allow quiet gear-changing on the move. Semi-automatic transmissions change gear hydraulically by releasing one clutch and engaging another. The most common application of a semi-automatic transmission is in a 'high–low' change which inserts an extra ratio between each pair of existing gears; complete transmissions can operate this way, providing up to 15 forward and four reverse ratios. Semi-automatic transmissions deliver less of the engine's power to the wheels than conventional systems because power is lost due to friction and in operating the hydraulic control system of the transmission itself (Renius, 1992).

Some tractors are fitted with hydrostatic transmissions which provide an infinitely variable range of ratios with a stepless single-lever change, even from forward to reverse. This system is ideal for the operation of trailed pto-driven machines such as balers and forage harvesters. Forward speed can be continuously adjusted to match crop and ground conditions while maintaining a constant engine (and pto) speed. For the same reason, hydrostatic transmissions are fitted to many combines and self-propelled forage harvesters. Where a major proportion of the tractor's power is to be used in traction, the lower efficiency of the hydrostatic transmission means that less power is available at the wheels from a given engine power (Tinker, 1993).

Drawbar power

Work is done or energy is expended when a force acts through a distance (Equation 1) and power is defined as the rate of doing work (Equation 2). The unit of work is the Newton metre (Nm) or the Joule (J) and power is measured in Nm/s or J/s or watts (W). A rate of work of 1000 Newton metres per second is 1 kilowatt (kW).

$$\text{Work}\left(\text{Nm}\right) = \text{force}\left(\text{N}\right) \times \text{distance}\left(\text{m}\right) \quad \text{(Equation 1)}$$

$$\text{Power}\left(\text{W}\right) = \text{work}\left(\text{Nm}\right)/\text{time}\left(\text{s}\right) \quad \text{(Equation 2)}$$

Note that Equation 2 can be manipulated to show that:

$$\text{Power}\left(\text{W}\right) = \text{force}\left(\text{N}\right) \times \text{speed}\left(\text{m/s}\right)$$

The drawbar pull developed by a tractor is dependent on the gross tractive effort developed between its drive wheels and the soil less the force required to overcome rolling resistance of the tractor's wheels, as follows:

$$\text{Drawbar pull} = \text{gross tractive effort} - \text{rolling resistance}$$

Maximum drawbar pull is achieved both by maximizing the gross tractive effort and by minimizing the 'parasitic' effect of rolling resistance.

Drawbar power is the product of drawbar pull and forward speed, as expressed by the equation:

$$\text{Drawbar power}\left(\text{kW}\right) = \frac{\text{Drawbar pull}\left(\text{kN}\right) \times \text{speed}\left(\text{km/h}\right)}{3.6}$$

Factors affecting gross tractive effort

When the lugs of a drive wheel or track bite into the soil, the rearwards thrust of the lugs tends to push the soil back. As the soil trapped between the lugs is sheared from the underlying soil, a horizontal force is developed. The further the soil is displaced, the greater is this force.

As the tractor moves forward, exerting a pull on some following attachment, the soil is pushed backwards a little. The percentage the soil is moved back, in relation to the distance the tractor would move forward on a rigid surface with no drawbar load, is called the slip. A wheeled tractor develops its maximum drawbar pull at 20–25% slip, but maximum tractive efficiency occurs at about 10–12% slip. Any slip at all means that some of the tractor's power is being lost in pushing soil backwards instead of pushing the tractor forwards, but since no pull can be generated without some slip, this must be accepted.

The shear strength of the soil under the tractor's wheel or track depends to a small degree on the rate of shearing; a tractor operating at a higher speed can generate a slightly greater pull because of this property. The major factors affecting soil strength are its coefficient of internal friction and its cohesion.

The more weight that is applied to the soil under the driving wheel, the greater will be its frictional strength and the greater will be the tractive effort generated at a particular level of slip. Coefficients of internal friction range from below 0.2 for a plastic clay to over 0.8 for a coarse sandy soil. An 'average' figure is 0.6, which means that 60% of the vertical force applied to the soil by the driving wheels would be available as gross tractive effort. Excess water acts as a lubricant between the soil particles, and can reduce the coefficient of friction almost to zero.

Frictional tractive effort is increased by adding weight (ballast) to the driving wheels or tracks. The loading may be by means of iron weights or water ballast in the tyres, or the tractor can be made to carry part of the weight of the implement it is pulling. This last approach has the advantage that when the implement is detached from the tractor, so is the extra weight.

As the tractor pulls a load, the resistance of the load tends to tip the tractor backward about a point on the ground beneath the rear axle. This has the effect of transferring weight from the front to the rear wheels, and thus increasing the available tractive effort. The height of the hitch point must be kept low enough to eliminate any risk of overturning the tractor. Front-end weights may be fitted to ensure that at least 20% of the tractor's weight remains on the front axle to give steering control. Four-wheel-drive tractors and crawlers gain no benefit from weight transfer, since all the weight of the machine is carried on the driving members. On unequal-wheel four-wheel-drive versions of two-wheel-drive tractors, weight transfer can remove much of the weight from the front axle under good tractive conditions so that the powered front axle contributes little to traction unless it is ballasted by front-end weights or front-mounted implements.

The cohesive strength of the soil is determined by the degree to which the soil particles cling together, even when no weight is applied to the soil. A typical value for a friable soil is 30 kN/m^2, with a range from zero for very coarse-textured soils to a maximum of 60 kN/m^2 in some clay soils. The greater the area of cohesive soil that is put in shear, the greater will be the tractive force generated.

Cohesive tractive effort is increased by increasing the contact area between drive member and soil. The fitting of larger section rear tyres or dual wheels, or the use of four-wheel drive or crawler tracks all have this effect. Additional soil may be put in shear by the use of grousers, strakes or spade lugs, which can penetrate through a slimy surface layer into stronger underlying soil. However, traction in this case is increased at the expense of reduced tractive efficiency since power is lost in digging into the soil. The use of tyres at no more than the recommended inflation pressure for the load carried will ensure the maximum safe contact area between tyre and soil.

Compacting a loose soil, for example by running a wheel over it, will increase both the frictional and cohesive strength so that a following wheel can generate a greater tractive effort than it could if running on uncompacted soil.

Ballasting

Correct ballasting is essential if a tractor's maximum tractive power output is to be realized on typical frictional-cohesive agricultural soils. Research (Dwyer & Dawson, 1984) has shown that the total weight to be carried on the driving wheels (w) is found where

$$w\,(\mathrm{kg}) = \frac{650 \times \text{pto power of tractor } (\mathrm{kW})}{\text{working speed } (\mathrm{km/h})}$$

Much of this weight will be provided by the tractor itself and by any mounted implements that it carries, but additional iron weights and water-ballasting of tyres are normally required. In the case of four-wheel-drive tractors, front to rear weight distribution should be in proportion to the tyre-maker's recommended carrying capacity of front and rear tyres at equal inflation pressures. If ballasting is correct with the tractor stationary, the effect of weight transfer will not be large enough to reduce tractive performance.

When the correct ballasting has been established, the tractor must be equipped with drive wheels and tyres large enough to carry the necessary weight at a low inflation pressure (preferably no more than 1.5 bar). In work the tractive load of the tractor should be adjusted so that wheel slip is 10–12%. This is not easy to judge, but it can be measured, and tractor-mounted slip indicators and slip control systems are available for this purpose.

Factors affecting rolling resistance

In order to roll a wheel along a surface, a force must be applied to overcome its rolling resistance. Rolling resistance increases with the load carried by the wheel. In agricultural conditions the force ranges from 5% of the weight carried by the wheel when travelling over a dry field after a cut of silage, to 20% or more when harvesting root crops.

Rolling resistance is minimized by using the largest available tyre size inflated to the minimum pressure recommended for the load being carried, and by using more wheels to support the load in dual or tandem formation. Where radial tyres are available, a 5% reduction in rolling resistance can be attained by their use.

Tyres

Most tractors rely on rubber tyres to transmit the power of the engine to the soil and to generate tractive effort. So long as lug height is not less than 20 mm, tread pattern has little effect on overall tractive performance. There is negligible advantage in increasing lug height above this value, as taller lugs will deform under load and lose their bite into the soil.

The familiar chevron tread pattern has the advantage that it is self-cleaning under quite sticky conditions so that the tread bars can continue to bite into the soil when other tread patterns would become completely clogged with mud. Because the tread is more rigidly supported, and the side walls are more flexible, radial tyres have less rolling resistance, give 5–15% better tractive performance, last longer and produce lower peak soil pressures than similarly loaded cross-ply tyres.

The size of a tractor tyre is described by two dimensions (usually in inches). The first of these gives the maximum width of the tyre section, and the second indicates the diameter of the wheel rim on which the tyre is mounted. A 50-kW tractor can be fitted with 13.6 × 38, or 18.4 × 30 rear tyres, which are similar in overall diameter. However, the wider 18.4 × 30 tyre has a larger ground-contact area and a greater carrying capacity, and when ballasted to take advantage of this will give a 5–15% increase in tractive performance. It is not possible to use wide section tyres for row-crop work, but for most tillage and haulage operations there is no restriction on tyre width.

Tyres and compaction

Soil compaction by heavy machines is a problem not only of rutting the soil surface, but also of compressing the soil itself which reduces pore space, inhibits water movement and increases the formation of clods. Compaction, under any particular combination of soil conditions, is largely a function of applied ground pressure and number of passes; the higher the pressure, the more dense the soil becomes, and the greater the depth to which compaction occurs. Ground pressure is reduced by spreading the weight of the machine over a greater ground contact area by using wider section tyres or dual wheels. Although the degree and depth of compaction are reduced, more soil is compacted by the wider contact surface. Minimum degree, depth and volume of compaction are achieved by a long, narrow contact patch, or by the use of wheels in tandem.

Compaction is also increased under a wheel that is operating at high slip, the maximum effect occurring at a slip range of 15–25%. Since this is above the level at which maximum tractive efficiency occurs, it is advisable, for reasons of efficiency as well as reducing compaction, to operate a tractor at loads that only require a slip of 10–14%. Working rate is maintained by pulling these lighter loads at higher speeds.

The most effective method of controlling compaction is to use the lightest machines that will do the job, keep off the soil when it is wet and reduce the number of passes over the field to the minimum. To eliminate compaction of the cropped soil entirely, the tractor can be run in the same wheel-tracks for all operations, while the crop is grown in beds between the wheel-tracks. This system is known as 'controlled traffic' or 'zero compaction'.

Four-wheel-drive tractors

The majority of new tractors sold in Europe are fitted with four-wheel drive (Fig. 24.1). Driving all four wheels of a tractor offers a number of benefits:

- Greater soil contact area produces a greater cohesive pull, which is particularly advantageous under wet conditions when friction is low. When tractive conditions are good, the advantage is small.
- The pull is shared between four drive wheels, reducing each wheel's slip, compaction and sinkage. Ballast is also spread between all wheels, reducing total weight for a given tractor power. Two-wheel-drive tractors above 50–60 kW cannot be ballasted sufficiently for maximum performance because of tyre limitations.
- The powered front wheels give improved steering control in wet conditions.
- The tractor's brakes are effective on all four wheels. Some tractors have front brakes, but normally only rear brakes are fitted, and front-wheel braking is only effective when front-wheel drive is engaged.

Figure 24.1 A105 kW four-wheel-drive tractor pulling a five furrow reversible plough. (Courtesy of John Deere Ltd.)

Systems are available that automatically engage front-wheel drive (if not already engaged) when the brakes are applied.

Unequal-wheel four-wheel-drive systems are common on medium and large sized tractors. Some very large sized tractors have drive systems where all four wheels are the same size. This configuration produces the best tractive performance. The relatively large turning-circle problem can partly be overcome with centre-pivot or four-wheel steering, which also allows front and rear wheels to run in the same tracks during turns.

If front wheels are to contribute fully to tractive performance, they must be correctly ballasted.

The power take-off (pto)

Rotary power is available at the rear, and on some tractors at the front, by attaching a drive to a splined shaft. There are two internationally standardized shaft sizes and speeds – a six-spline shaft turning at 540 rpm and a 21-spline shaft turning at 1000 rpm. Rotation is clockwise when viewed from the rear of the tractor. Many tractors have dual pto systems to accommodate all types of machine. As rotary power is the product of torque and speed, it can be seen that the higher pto speed can transmit almost twice the power at any given torque.

Whereas only 50–70% of the power of a tractor's engine is available through the wheels under average tractive conditions, 80–94% is available through the pto shaft at standard speed. The difference is mainly due to wheelslip and rolling resistance losses.

A fixed ratio between the engine and the pto sometimes allows the engine to develop its maximum power at the standard pto speed. Where this is not so, it is generally possible to over-speed the pto so that engine power can be maximized.

The pto is connected to the engine by its own clutch. If this can be operated quite separately from the transmission clutch, the system is called an 'independent' pto. A two-stage clutch pedal controlling the transmission at half depression and disconnecting the pto at full depression gives a 'live' pto.

Hydraulic systems

Approximately 15–30% of a tractor's power can be available through the hydraulic system. This is adequate for light work such as the operation of the three-point linkage, tipping trailers, most front-end loaders and a few light duty excavator attachments. Larger hydraulically operated machines such as high-lift loaders or hedge cutters must have their own hydraulic power units, driven by the tractor's pto.

Hydraulic power is a function of fluid flow rate and pressure, and is represented by the expression:

$$\frac{\text{flow rate (litre/min)} \times \text{pressure (bar)}}{600} = \text{power (kW)}$$

Tractor hydraulic systems operate at pressures of 140–200 bar, with flow rates of 20–100 litre/min.

The tractor's three-point linkage is of standard dimensions and pin sizes (category I, II or III or combinations according to tractor size). It is able to carry mounted equipment for transport and to control the working depth of many types of soil-engaging implements. Two alternative control systems are usually available, draught control and position control.

- Draught control – this system adjusts the working depth of ploughs or other high-draught implements to maintain a constant draught or tractive load. Draught is sensed by springs or pins in the upper or lower linkage; changes in draught cause the hydraulic system to raise or drop the linkage. A 'response' adjustment controls the rate at which these hydraulic corrections are made; slow response gives the smoothest work, but fast response may be needed if the ground is uneven (Hesse & Withington, 1993).
- Position control – this system will hold the linkage at a constant height relative to the tractor. This is generally used for transporting equipment in a raised position but may also be used to control the working position of some machines such as light cultivators or weeders.

Front three-point linkages are available for many tractor models. The use of front-mounted implements can apply ballast to driven front axles, and can allow two light operations to be carried out simultaneously, saving time and labour.

Most tractors have available two or more pairs of external hydraulic couplings, controlled by separate double-acting control valves, and a coupling for the automatic operation of hydraulic trailer brakes. The valves may be used to control a front-end loader, rear fork-lift or other accessory, saving the cost of purchasing separate control valves for each attachment. External hydraulic hoses are usually connected to the tractor by way of snap-on couplings, which will pull out and seal themselves if the machine becomes detached from the tractor. All couplings are a source of contamination to the tractor's hydraulic system, and care must be taken to avoid the entry of dirt when attaching and storing hydraulic accessories.

Health and safety for the tractor driver

Hearing loss from noise exposure and injury from tractor overturning accidents are the major risks to which a tractor driver is subjected.

Current safety cabs must reduce the noise level inside the cab to no more than 85 dB(A). Since much of the noise from a tractor is airborne, most of the soundproofing of the cab is lost if it is necessary to leave a door or window open for access to implement controls or for ventilation in hot weather. Remote controls eliminate the first problem and an air-conditioning system does away with the second.

Tractor overturns most commonly occur sideways, when operating on steep slopes or driving too close to gulleys, ditches or steep banks. Trailers, slurry tankers or other heavy, unbraked machines can push the tractor down hills. Backward overturns occur less frequently, usually from attempting to pull from a high hitch point (Hunter, 1992).

Since the introduction of BS-approved safety cabs on all new tractors, the number of deaths from tractor overturns has diminished dramatically. In an overturning incident, it is most important to hang on and stay inside the cab until the tractor has come to rest completely.

Cultivation equipment

Soil working usually is required to convert a field carrying the remains of a previous crop into an environment suitable for the establishment and the continued growth of the next crop. Burying trash, reducing compaction, producing a seedbed, shaping the soil for a particular crop and improving harvesting conditions are examples of operations involving soil disturbance. A detailed understanding of the actions and interactions of cultivation equipment, under a full range of soil conditions, is essential if the aim is to achieve the desired result at least cost.

The major factor influencing the effect that a cultivator has on the soil is the rake angle of the tine in contact with the soil. The rake angle is defined as the angle between the tine and the soil surface in the direction of travel of the tine and can range from 15° to almost 180°.

The actions of cultivation equipment on the soil fall into the following categories:

- loosening – reducing the soil bulk density – achieved by rake angles less than 45° (forward raked);
- clod disintegration and compaction – achieved by rake angles greater than 135° (backward raked);
- consolidation – increasing soil bulk density by arranging soil particles – achieved by rake angles around 90°;
- cutting – achieved with sharp-edged backward raked tines or discs.

Primary cultivations are those involved in breaking up the soil initially. Secondary cultivations refine the soil to produce the required conditions in terms of particle size, firmness and surface shape for subsequent seed germination and plant growth.

Subsoilers

Repetitive passes of tractors and equipment, especially in wet conditions, can increase soil density beyond the topsoil layer. Cultivation 'pans' – localized dense soil layers – also can be created by the smearing of soil by wheels or tines. Both occurrences result in poorer drainage and root penetration. Subsoilers are used to loosen the soil to a greater depth than that reached by routine cultivations.

The subsoiler consists of one or more heavy vertical tines, each fitted with a replaceable point or foot and can operate up to 600 mm deep. Feet can be fitted with horizontal wings, about 300 mm wide, which considerably increases the width of soil below ploughing depth loosened by the subsoiler.

Optimum subsoiling (i.e. soil disturbance) will be achieved if both surface (for good traction) and subsurface conditions are reasonably dry. Wet conditions at the base of the subsoil tine can lead to soil smear and the production of channels rather than loosened soil. The depth at which soil loosening ceases and smearing occurs is known as the 'critical depth' for the subsoiler and will depend on the soil moisture content and its density. As soil moisture content increases, so the 'critical depth' will decrease.

A pair of shallow leading tines, spaced at 0.5 m to either side of the main winged tine, loosen the upper soil layers, decrease the forces on the main subsoil leg and result in a large increase in soil disturbance with negligible increase in draft.

Subsoiling is a slow operation and requires high power input. Test digging after a short run should be carried out to determine if soil shatter is being achieved and to identify the depth and spacing of work required.

Ploughs

Types of ploughs

The mouldboard plough is the standard primary cultivation implement and has been developed to keep pace with larger tractor power sizes. 'Conventional' or 'one-way' ploughs have been superseded by reversible ploughs which avoid the necessity of extensive field marking and leave a level finish. Reversible ploughs carry two sets of mouldboards, one of which moves soil to the right, the other to the left. This arrangement allows continuous ploughing to be carried out from one side of a field to the other. When using a 'conventional' plough, accurate field marking is required and ploughing is carried out in sections ('lands') to avoid excessive headland travel.

Ploughs are normally fully mounted on the tractor's three-point linkage. Larger ploughs can be semi-mounted. Attached to the tractor's lower links and having a rear-mounted plough wheel, these ploughs are easier to lift but less weight transfer (for improved traction) is available.

Swing-beam ('square') and disc ploughs were recently promoted in Britain on the basis of high-speed operation and their ability to deal with chopped straw. These are important facts when considering methods of reducing the autumn workload bottle-neck. However, their reduced performance when operating in wet soil conditions has prevented their popularity increasing.

The disc plough resembles a mouldboard plough in layout, but the bodies have been replaced by angled, inclined, free-turning concave discs. Large stationary scrapers are fitted to prevent soil buildup on the discs and to increase the turning action on the soil. The soil is loosened and mixed, rather than inverted; where erosion is a problem, partially buried crop residues can be very effective in binding and stabilizing the soil. The machine is difficult to adjust, and the work produced does not look like conventional ploughing. Where obstacles such as roots or rocks abound, the discs avoid damage by rolling over the obstructions rather than catching under them. Although not much used in this country, the disc plough and its derivatives are widely accepted in areas of the world where their special properties can be used advantageously.

Mouldboard ploughs

Soil loosening, disintegrating large clods, and inverting the topsoil to bury weeds, trash and potential disease are carried out during ploughing. The soil-engaging components of the mouldboard plough consist of:

- horizontal (the share) and vertical (the shin or breast) cutting elements;
- a forward-raked point to aid penetration;
- a soil inversion plate (mouldboard);
- a tailpiece to help position the furrow;
- and a vertical plate (landside) to resist the horizontal thrust of the turning furrow.

All these components are bolted to a frame (frog), which, via a leg, is attached to the rigid plough frame. On reversible ploughs, the frame is connected to the headstock, which carries the tractor mounting brackets.

The frame rotates through 180° relative to the headstock to allow the second set of furrows to engage the soil at the start of a new run.

Mouldboards are available that can produce a range of effects on the soil and are characterized by their length and the degree of curvature. Semi-digger boards are most widely used and leave a broken furrow. Digger bodies are shorter and sharply curved to produce even more furrow disintegration. Ley or general purpose bodies are long and gently curving and leave an unbroken furrow. The various effects produced by different boards can be modified by operating at different speeds.

Coulters are often used instead of the shin to create the vertical cut. Disc coulters cut through surface trash but may have difficulty in penetrating the soil under hard conditions. Their physical size also requires the plough to be longer and therefore more difficult to lift. On some ploughs, a disc coulter is fitted to the final furrow only to leave a neat edge for the next run. Knife coulters are simpler, smaller and penetrate more readily, but can block easily in trashy or matted turf conditions. Sword landsides can be fitted to the frog adjacent to the point to provide an upward vertical cut where surface trash or stones are a problem. Skim coulters are used to help ensure complete burial of trash by removing a small section of the furrow and dropping it into the furrow bottom. Skimmers can operate independently or can be attached to a disc coulter.

A standard plough furrow width is 350 mm and there can be as many as 12 mouldboards on one plough, giving a working width of 4.2 m. Furrow width on some ploughs can be adjusted manually by +/−50 mm, whereas on others the width can be varied hydraulically from 300–500 mm depending on the soil conditions, to allow full tractor power to be utilized. Four- and five-furrow ploughs are common.

Ploughs can be fitted with a safety release mechanism on each leg (or pair of legs on a reversible plough) which allows the whole leg to fold back if it strikes an obstruction. This is particularly valuable where large ploughs are operated at high speeds and/or in stony conditions. A small investment in protective devices can reduce the risk of long and costly delays and expensive repairs at a busy time.

Good ploughing (i.e. level, uniform furrows) can only be achieved if the plough components are uniform relative to one another. This can be checked with a straight-edge. Once in the field, the following adjustments may be required:

- matching the front furrow width to the remainder of the (fixed) furrows;
- levelling the plough from front to back – achieved by the tractor's top-link;
- levelling the plough from side to side.

Ploughing depth, which should not exceed two-thirds of the width of the furrow slice if the slices are to turn satisfactorily, may be set by means of an adjustable depth-wheel. More commonly, the tractor's hydraulic draught control system is used. The use of a depth wheel gives more accurate control and is not affected by changes in soil conditions, but the use of draught control without a depth wheel puts more weight on the tractor's rear wheels. This improves traction and also eliminates the extra rolling resistance of the trailed wheel. Long ploughs, of four furrows or more, often have a depth wheel at the rear to supplement the draught control system.

Rates of work and power requirement are dependent on soil conditions and working depth. A typical value for an 80 kW tractor and four-furrow plough would be 0.6 ha/h. The plough and tractor should be matched so that the power of the engine is fully utilized at a speed of 5–8 km/h. Fully loading the tractor at slow speeds causes high losses due to wheelslip. Operating at very high speeds puts great strain on the plough and leads to a high draft force, as draft increases with forward speed.

Tine cultivators

Tine cultivators can be fitted with either rigid or spring-tines and, depending on the strength of the tines fitted, can be used for either primary or secondary cultivations. The tines have a forward rake and loosen the soil while breaking down weak clods. A frame carries a number of tines uniformly distributed so that each tine passes through a 100- to 200-mm strip of soil. Replaceable points of various widths to suit the desired cultivation effect are available. The machine is normally tractor-mounted, but is carried in work by adjustable depth control wheels. Machines above 3 m in width are made up of a central section with two hinged wings which fold for transport.

Rigid tined implements are used for primary cultivation in place of ploughing and two passes of the field at different angles are usually required to obtain sufficient soil disturbance. Rigid tines have shear bolts to protect them from damage by large stones or rocky outcrops.

Spring tines vibrate as the machine moves forward and this movement is very effective in both shattering weak clods and promoting self-cleaning of the tines when working in trashy conditions. In work, the tines have a rake angle of about 90° and trash and stronger clods are brought to the surface. The machine leaves the soil furrowed corresponding with the spacing of the last row of tines. Crumbler rollers or light harrows, available as extras to fit on the rear of the machine, will produce a level surface finish.

The spring-tine cultivator has a good mixing action at speeds above 7 km/h and may be used for the incorporation of soil chemicals. Two passes are recommended, at 45° to each other.

Disc harrows

A disc harrow is made up of two or four adjustable axles, each axle having a number of concave discs mounted along its length (Fig. 24.2). The axles are angled to the direction of travel and the discs are mounted in such a way that the front discs throw soil outwards, while the rear discs throw soil towards the centreline of the implement. Units having four axles are termed 'X' or 'H' framed harrows: two axle units are referred to as 'A' framed or offset harrows. The front axles are often fitted with scalloped discs, which penetrate firm soil and cut through surface trash better than plain discs. The rear gangs use plain discs, which last longer and move more soil.

The action of the disc harrow on the soil depends on the angle of the axles. With the axles set almost at right-angles to the direction of travel, the discs cut into the soil, resulting in a breakdown of clods and a minimal degree of soil throw and mixing. By increasing the axle angle, the discs penetrate the soil, resulting in a greater degree of soil loosening and mixing. The penetration of the discs can be improved by adding weights to the frame of the unit.

Tractor-mounted disc harrows are available, but, since much of the penetrating effect depends on weight, heavier trailed machines are better able to deal with tough or hard soil conditions. Transport from field to field is achieved by hydraulically operated wheels.

When used for straw incorporation – a primary cultivation designed to initiate the breakdown of straw and stubble by mixing with soil – it is usual to pull a roller behind the discs to consolidate the soil to prevent excess soil moisture loss.

Disc harrows are particularly suited for secondary cultivations where heavy soils have dried and a large downward force is required to break clods.

Vertical tine harrows

A harrow consists of a large number of small vertical tines or spikes carried on rigid or flexible (chain) frames. As they are dragged through the soil, the action of the tines is to disintegrate weak clods and sort, level and compact the seedbed. The degree of soil penetration depends on the weight of the frame and the size of the spikes. Harrows are also used after drilling to ensure seed is covered. Chain harrows may be used on grass-

Figure 24.2 A set of disc harrows and a following press operating in stubble. (Courtesy of Simba Ltd.)

land to break up matted swards or to spread cow pats after grazing.

The power requirement of harrows is low, so it is uncommon for any tractor to be fully loaded by a set of harrows of normal width. Harrows may be pulled behind other machines (spring-tine cultivators, seed drills) to achieve two operations in one pass.

Pto-driven cultivators

By applying power to the soil-engaging parts of a cultivator via the tractor's pto shaft, the resulting effect on the soil is independent of forward speed and the draught requirement is reduced. The cultivating effect can be manipulated by varying the speed of the tines relative to forward speed. There are two main pto-driven cultivators – the rotary cultivator and the power harrow. Other machines are the spading machine, used for horticultural work, and the reciprocating harrow.

The rotary cultivator

The rotary cultivator may be used for primary or secondary cultivation work. A horizontal rotor, driven by the tractor's pto via a gearbox and chain drive, carries either spikes or L-shaped blades. The spikes or blades rotate at 120–270 rev/min to produce either a coarse or a fine tilth. A rear hood may be raised to allow clods to be thrown out, or lowered to give a further shattering effect as the clods strike the inner surface of the hood. Working depth, to 200 mm, is controlled by a land wheel or a rear crumbler roller. Machines fitted with spikes have generally replaced L-bladed units in agriculture and are used for straw incorporation, chemical incorporation and cultivation for both cereals and root crops. L-bladed machines have a better chopping action than spikes but can cause soil smearing at cultivation depth if the soil is moist. The usual arrangement is for the rotor to rotate in the same direction as the tractor wheels resulting in the machine 'pushing' the tractor forward.

The power harrow

The cultivating elements of the power harrow consist of pairs of almost vertical spiked tines rotating about vertical axes. Each pair of tines is attached to a gear which drives, or is driven by, the adjacent gear. This results in neighbouring sets of tines contrarotating. The speed of the tines is much greater than the forward speed of the tractor and their shattering action is thus more effective than that of rigid tines being pulled through the soil. The action of the tines also levels and compacts the seedbed, without raking up much buried trash or subsoil. Seed drills are often used in conjunction with these machines to combine the operations of seedbed preparation and drilling.

Rollers

Rollers have a similar effect as backward raked tines in that they disintegrate clods and compact the soil. They range from plain through Cambridge, crosskill, flexicoil and crumbler rollers to furrow presses. The effect of a roller depends on its weight and soil contact area and decreases with increasing roller diameter and with increasing forward speed.

Lightweight plain rollers are commonly used after grain drills in spring to firm the soil and ensure good seed/soil contact. Heavier units are used for levelling grassland and pressing-in stones in spring to prepare for hay or silage harvesting later in the season. By adding ballast to the hollow cylindrical rollers, weights from 0.5–1.3 t/m width can be applied. Care must be exercised in matching the grass roller to a tractor of adequate weight. A roller weighing up to 5 t or more can sometimes take control on steep ground when moving downhill, resulting in tractor control loss.

Cambridge and crosskill rollers, made up of a number of ribbed cast iron wheels on an axle, are often used in seedbed preparation to crush clods, to compact the soil and to leave a smooth finish. The point contact of the 'wheels' results in greater disintegrating and compacting effects compared to a plain roller. Cambridge rollers are available in widths to more than 7 m, the larger sizes being made up of a central section and two wings which can be folded hydraulically for transport.

Some secondary cultivation machines may be fitted with a crumbler roller. This is an open-cage roller, with a surface of spaced straight steel rods held in position by two or more integral flanges. The whole unit is free to turn in bearings at either end. As the roller moves forward, its weight is concentrated successively on each steel rod, which is quite effective in crushing clods at that point. There is also a good levelling effect. A heavier version is the packer roller.

The purpose of the furrow press is to compact soil and disintegrate clods on ploughed ground. It consists of a number of large (700 mm) diameter cast iron wheels mounted on an axle or axles. These units can operate independently, mounted either on the front or on the rear of a tractor, but are usually used in conjunction with a reversible plough. An arm is secured to the plough which connects to the drawbar of the furrow press. As the plough travels from one end of the field to the other, the furrow press is dragged behind and operates on the preceding ploughed furrows. At the headland, the plough is raised out of work, the furrow press is disengaged, the plough is turned over, begins its return across the field and the furrow press is automatically re-engaged.

Combination seedbed-preparation equipment

Since many seedbed-finishing operations demand little power, implements have been developed that combine several operations and can thus utilize a tractor's power more completely. Such items as ridged rollers, rigid or spring tines, crumblers or toothed rollers may be combined in such a way to progressively level and firm the soil, disintegrate clods, lift and expose large clods to the surface, attempt to disintegrate these clods and finally leave a fine and level seedbed. This can be achieved, for example, with a front roller, rows of vertical tines, a row of backward raked tines and a rear roller. In addition to firming and breaking clods, the front and rear rollers act as the depth control mechanism for the implement.

Reduced or minimal cultivations

In situations where the previous crop has been harvested in good conditions and the subsequent crop requires a shallow seedbed (a cereal crop, for example), there can be savings in energy and increases in work rate if the soil is only cultivated to a depth of about 100 mm. This system has been used successfully, even in heavy soils, for cereal production, without loss of yield. In many situations this practice has also led to long-term improvements in the structure of the upper soil layers. Problems with the buildup of annual grass-weed populations can occur unless this is controlled by herbicide applications or by occasional ploughing.

Discs or heavy spring-tined or rigid-tined cultivators, with tines spaced at 200 mm overall, are suitable for the task. At least two passes usually are needed before drilling in heavy soils. Combination machines, generally using gangs of discs in conjunction with banks of heavy spring tines, can reduce the number of passes needed to produce a satisfactory seedbed. One example, 3 m wide, requires a tractor of 75–105 kW, and can cultivate up to 2.5 ha/h.

Fertilizer application equipment

The majority of fertilizer is applied as a solid in the form of granular, prilled, crystalline or powdered materials. Liquid fertilizer solutions are available and have the advantage of being handled easily by pumps. Application is achieved using machines similar to those described in the later section 'Crop sprayers'. Liquid fertilizer is generally less concentrated than solid and thus more material must be handled to apply a given weight of nutrients. Also, whereas solid fertilizer can be stored safely in a general-purpose shed, rigid storage tanks must be available on the farm to hold at least part of a year's supply of fertilizer.

Fertilizer can be applied;

- prior to sowing or planting a crop (e.g. on ploughed ground);
- during the sowing/planting operation (e.g. with a combine drill);
- or during the growing season (e.g. top dressing).

Application of solid fertilizer is achieved with:

- broadcasters – single disc, twin disc and oscillating spout machines;
- full width distributors – as part of a grain drill, pneumatic fertilizer spreader.

Broadcasters

The majority of fertilizer is applied by broadcasters. Fertilizer flows from a hopper via a metering device onto a pto-driven spreading mechanism comprising a single disc, twin discs (Fig. 24.3) or an oscillating spout. The material is thrown from the disc(s) or spout behind and to both sides of the spreader, the majority of the fertilizer being thrown behind the spreader. Uniform

Figure 24.3 A twin disc fertilizer broadcaster top-dressing a crop of wheat. (Courtesy of Amazone Ltd.)

application is achieved by spreading fertilizer during the adjacent run at a suitable spacing (e.g. 12, 18 or 24 m) from the previous run to produce an even application across the width. Mounted and trailed machines are available with hopper capacities up to 1200 kg on mounted units and up to 8000 kg on trailed models.

Full width distributors

Rather than throw fertilizer from a point source, these units apply fertilizer across the full width of the machine. This approach has the benefit of reducing the influence of external factors on spread uniformity.

Pneumatic spreaders use an airstream to convey the fertilizer. They consist of a central hopper, a fertilizer metering mechanism and a boom along which a number of pipes convey fertilizer to outlets spaced about 600 mm apart. The fertilizer, emanating from an outlet, strikes a distribution plate, which spreads the fertilizer across a 120-mm width, which, in conjunction with the neighbouring outlets, creates a uniform overlapped pattern. The hopper base has a number of metering units corresponding to the number of outlets. Fertilizer flowing through a metering unit is introduced to its conveying pipe by means of a venturi. The fertilizer-carrying airflow is generated by a pto-driven fan. Mounted and trailed machines are available with hopper capacities ranging from 1–5 t and having folding booms up to 30 m wide. The metering unit's fluted rollers are land-wheel driven to ensure accurate application rate independent of forward speed.

Pneumatic units also are found on combine grain drills. The fertilizer metering and conveying system on some machines is similar to the unit described above. On other units, one metering unit is employed at the base of the hopper, the fertilizer is introduced to the airflow and is blown to a distribution head which splits the fertilizer flow into pipes which carry the material to the soil openers.

On gravity-fed grain drills (combine drills), a secondary hopper can be mounted behind or incorporated with the grain hopper. Along the base of this hopper are a number of outlets, each fitted with a metering device. Fertilizer flows by gravity from the hopper, through the metering device to the soil via tubes and soil openers. Combine grain drills have diminished in popularity due to both the swing to autumn cereals with the reduced need to sow grain and apply fertilizer simultaneously and the higher output achieved by grain-only drills.

Calibration

The calibration of a spreader is carried out to check and adjust, if necessary, the application rate.

On most tractor-mounted broadcasters, the flow of fertilizer to the disc/spout is controlled by the aperture opening(s) on the hopper bottom or by belts running under the hopper. On trailed units, belts or chains run in the hopper base and are land-wheel driven.

The fertilizer flow rate (kg/s) onto the discs/spout, tractor forward speed (m/s) and spread bout width (m) are required to calculate the application rate (in kg/ha) as follows:

$$\text{Application rate} = (\text{flow rate} \times 10000) / (\text{forward speed} \times \text{bout width})$$

The application rate can be checked in the field by calculating the area covered when spreading a known quantity of fertilizer. Stationary calibration involves removing or bypassing the disc(s) and weighing the quantity of fertilizer collected over a measured time period. The time needed to cover 1 ha at a known speed and spreading width (Table 24.1) can be used to convert the delivery rate/min to rate/ha. To adjust the application rate to that required, the aperture openings can be opened or closed.

Where the metering device is land-wheel driven, the machine may be calibrated by rotating the mechanism by hand for a specified number of turns (which relates to a distance travelled) and collecting the fertilizer in a tray provided. The application rate can then be calculated by dividing the weight collected by the area covered (i.e. the distance travelled multiplied by the spread width). Adjustment of the application rate is achieved by changing, usually by means of a gearbox, the speed of the metering wheels relative to the ground-wheel speed.

Spread accuracy

Having adjusted the machine to apply the required quantity of fertilizer, it is important to ensure that the fertilizer is applied uniformly so that problems such as lodging, striping, green grains, variable ripening and yield loss are avoided. Factors influencing the uniformity of application are machine maintenance, machine settings, operation, fertilizer quality and operating conditions.

Fertilizer, particularly in combination with atmospheric moisture, is very corrosive to mild steel, iron and

Table 24.1 Time (minutes) taken to cover 1 ha when working at various widths and forward speeds. Non-productive time for such operations as filling hoppers, adjustment or turning is not included. Combination of width and speed not shown in the table may be evaluated by means of the formula $t = 600/(w \times s)$, where t is the time (in minutes) to cover 1 ha, w is the working width (in m) and s is the speed (in km/h)

Effective width (m)	*Speed (km/h)*								
	5	*6*	*7*	*8*	*9*	*10*	*12*	*14*	*16*
3	40	33.3	28.6	25	22.2	20.0	16.7	14.3	12.5
4	30	25	21.4	18.8	16.7	15.0	12.5	10.7	9.38
5	24	20	17.1	15.0	13.3	12.0	10.0	8.57	7.50
6	20	16.7	14.3	12.6	11.1	10.0	8.33	7.14	6.25
7	17.1	14.3	12.2	10.7	9.52	8.57	7.14	6.12	5.36
8	15	12.5	10.7	9.38	8.33	7.50	6.25	5.36	4.69
10	12	10	8.57	7.5	6.67	6.00	5.00	4.29	3.75
12	10	8.3	7.14	6.25	5.56	5.00	4.17	3.57	3.13

aluminium components. It should not be left in a spreader overnight. Manufacturers use corrosion-resistant materials such as plastics or stainless steel for many parts of the broadcaster, and it is important also that the machine is easy to dismantle and clean. Fertilizer is abrasive and causes wear to components. Worn or damaged discs, vanes, spouts and deflectors cannot be expected to spread fertilizer accurately and should be replaced regularly. Vulnerable components can be brushed or washed clean, and coated with oil to protect and keep them in good condition for the following season.

Prior to spreading, broadcasters should be levelled transversely and the longitudinal angle of the discs should be set according to the instruction book. Spread uniformity will also be promoted if the discs are set at the correct height above the target. Mounted broadcasters, having a gravity fertilizer flow system, require to be driven at a constant forward speed and with a constant pto speed.

Accurate driving is essential. In cereal crops, tramlines promote constant spacing between runs across the field. In grass crops, however, some form of marking system, such as foam, is useful in maintaining regular bout widths. Full width distributors are particularly susceptible to inaccurate bout width matching, which leads to either double-dosing or complete missing. The spread pattern from broadcasters is more forgiving in that a small inaccuracy in bout width matching will not have much effect.

The fertilizer being spread also contributes to the application accuracy. The particle density, shape, frictional characteristics and size range will affect the distance the fertilizer travels. Machine manufacturers provide setting recommendations for different types of fertilizer. Fertilizer manufacturers have introduced fertilizer 'spread ratings' which, based on the physical characteristics of the material, indicate the uniformity of application likely to be achieved.

Operating conditions (e.g. field slope and direction of work) will influence spread accuracy. Forward speed can change due to wheel slip when operating either up or down slopes and will be influenced by the quantity of fertilizer in the hopper. The application rate of broadcasters can be controlled by a tractor-mounted forward speed sensor, which can adjust the fertilizer flow rate mechanism to ensure uniform application rate as forward speed changes. Wind and rain can also contribute to spreader application inaccuracy.

Spreading accuracy of distributors can be checked by weighing the fertilizer from each outlet. For broadcasters, two checks can be made. Firstly, the broadcaster can be operated indoors in a wide shed. After having run the machine for a short period, the fertilizer thrown to each side of the machine is brushed up and weighed. If the two weighings are within 5%, the balance of spread can be assumed to be satisfactory. For a more accurate check, a set of trays covering the full width of spread, can be laid out across the direction of travel of the spreader. The fertilizer collected in each of the trays can be weighed or poured into tubes to give a visual representation of the spread pattern.

Care must be taken to avoid indiscriminate spreading near watercourses and hedgerows. Fertilizer's polluting effect and its influence on biodiversity can be minimized by using deflectors or tilt mechanisms to reduce the spread width to one side of the spreader when operating near field boundaries.

Sowing and planting machines

The range of sowing and planting equipment reflects the various characteristics and requirements of current crops. Many crops are grown from true seed (e.g. sugar beet, swedes) that is relatively tough but may require individual placement in the soil to optimize growth potential. Others, such as cereals, can be sown randomly due to their tillering action. Potatoes are grown from easily damaged tubers ('seed') and benefit from regular spacing in the row. Transplanting is carried out to remove the uncertainty of seed germination where individual plant growth is important. Whether seeds are sown or plants are transplanted, machines must satisfy the following:

- Accurate metering is necessary to ensure the required plant population is achieved.
- Material (seed or plants) can be sown into a range of soil conditions.
- A range of seed/plant sizes can be handled (e.g. compare seeds of OSR and peas).
- A range of sowing rates are available to suit a range of crops.
- Sowing rate is easily adjusted.
- Soil-engaging elements of the machine cannot block easily.
- Maintenance is simple.
- Depth of sowing/planting is uniform.
- There is sufficient hopper capacity to maintain output.
- The machine is easily cleaned.

Grain drills

Grain drills can sow a range of seed sizes from grass or clover to beans. Seed is discharged in rows, in an even trickle. Most crops are sown into a prepared seedbed, but other techniques exist:

- Seed can be sown into soil, which has been 'minimally cultivated'.
- Seed is sown directly into the previous crop's residue.
- Seed is broadcast onto lightly cultivated stubble.
- Seed is broadcast on the soil surface and ploughed in (e.g. beans).

Drills are available to suit a wide range of farming conditions and requirements. Factors such as soil type, soil fertility, cropping area, available sowing 'windows', tractor power and labour availability result in farmers requiring specific equipment to suit their circumstances. Consequently, drills have developed to cater for these requirements and four main categories can be identified;

(1) drills that sow grain only;
(2) drills that sow both grain and fertilizer;
(3) drills that sow grain (+/− fertilizer) and have a cultivating effect;
(4) drills that sow grain directly into uncultivated soil.

The main components of a grain drill are a seed hopper, a seed metering device and a soil opening device (coulter) to allow the grain to be introduced to the soil.

Gravity-fed grain drills have a hopper which extends across the full width of the machine. There is a metering mechanism located in the base of the hopper for each seed row. The metering mechanisms consist of either studded or fluted rollers and are driven from a ground wheel and a common drive shaft. As the rollers rotate, grain is moved from the hopper to seed tubes. The flow of seed is regulated by the rotational speed, in the case of the studded roller, and for the fluted roller, by the width of the fluted section exposed in the hopper. Different sizes of fluted sections are available to cope with large- and small-seeded crops.

Individual rows can be shut off by slides in the bottom of the hopper. Automatic or semi-automatic 'tramlining' attachments can be fitted to close off certain pre-selected rows on the drill's second, third or fourth pass across the field. As the crop grows, a regular pattern of marks ('tramlines') develop which can be used by subsequent equipment, such as sprayers and fertilizer spreaders, to promote timely and accurate application of material. For example, shutting off rows during every fourth pass of a 3-m-wide drill results in tramlines being produced at 12-m centres. Other standard tramline widths are 16, 18, 21 and 24 m.

Pneumatic grain drills (see Fig. 24.4) have a central hopper which allows easier filling, especially when using 1-t seed bags. After metering the grain, a fan is used to blow seeds along tubes to the coulters. A further advantage of this type of drill is that the frame carrying the coulters can be folded to allow rapid movement from field to field. This is in contrast to wide gravity-fed drills which require to be towed end-on during transport.

One type of pneumatic drill meters the seed in bulk from the hopper using a fluted roller. Seeds are introduced to the air-conveying system by means of a 'venturi' and are blown to a distribution head which divides the bulk of the grain evenly into the seed tubes for delivery to the coulters. Another pneumatic drill meters grain for each row individually from the hopper (similar to a gravity-fed drill). Grain is then introduced

Figure 24.4 A combination cultivator/drill unit comprising a power harrow and a pneumatic grain drill. (Courtesy of Lely Ltd.)

to the air-conveying system and blown to each of the coulters. The fan is pto-driven, the metering mechanism is driven from a land-wheel or 'spider' wheel and the tramlining mechanism is usually electrically operated via solenoids.

The seed is placed in the ground by the coulter. Ideally, each seed will be planted at the same depth, and surrounded by firm, moist, warm soil. Suffolk coulters and single-disc coulters are standard units. The Suffolk coulter is relatively inexpensive and is a backward rake blade which presses a groove in the soil, into which the seed falls. It operates well in good conditions, but in hard or cloddy conditions, poor penetration and uneven drilling usually result. The single disc coulter comprises a curved disc with the seed tube connected to the concave side. The disc can penetrate hard soils, break hard clods and cut through soft clods and rubbish. It can therefore operate more satisfactorily in unfavourable conditions compared to the Suffolk coulter.

Coulters are spaced at between 100 and 175 mm apart. On a 3-m drill, for example, there will be 24 coulters spaced at 125 mm apart. To avoid the chance of blockages due to surface trash, alternate coulters are mounted on separate staggered rows. Grass seed drills have coulters spaced at 75 mm centres to promote crop cover.

Both types of coulter are carried on spring-loaded arms; increasing spring tension will increase coulter penetration and planting depth. Individual coulters can move vertically to follow uneven ground. A following vertical-tine harrow can improve seed cover. To match up subsequent passes of the drill, disc markers are used which are carried on arms and extend on either side of the drill. The disc produces a small furrow that can be followed by the front wheel of the tractor at the next pass across the field.

When operating in harsher conditions, such as drilling into stubble, penetration ability together with imparting a degree of cultivation are desirable coulter characteristics. Double and treble disc coulters and spring-tine hoe coulters are available for these circumstances.

Manufacturers give application rate guide settings for their machines. However, grain drills meter grain on a volume basis whereas application rates are based on weight. Grain varieties have differing bulk densities and to ensure that the correct seed application rate is achieved, the drill must be checked, and adjusted if necessary, prior to drilling. Part of the metering device drive system will accept a handle which can be turned manually. Grain emanating from the metering units can be diverted onto a tray. The handle is operated for a set number of turns to represent an area covered by the drill (say 0.1 ha) and this figure used in conjunction with the weight of grain collected in the tray can determine the application rate (in kg/ha). Static calibration does not, however, take into account the effect of forward speed and wheelslip which can vary from 0.5–15% according to soil and tyre conditions. A final calibration where the drill is run at full planting speed over a measured area (see Table 24.2) can ensure that the amount of seed delivered is correct under field conditions.

Simultaneous application of fertilizer and seed can be beneficial for spring crops, both to maximize crop growth and overcome deficiency problems (e.g. manganese). Gravity-fed combine drills have a combined grain and fertilizer hopper with a central divider, which can be reversed to give ratios of either 1:1 or 1:2 in weight of grain to weight of fertilizer. Removal of the divider allows the full hopper to be used for grain. The fertilizer is metered similarly to that described above for grain and is delivered through fertilizer tubes either to the coulters which introduce the grain to the soil or to separate coulters spaced between the grain coulters.

Combine pneumatic drills have either a rear-mounted split hopper or twin hoppers, or have a front-mounted hopper for grain and a rear-mounted hopper for fertilizer. The grain and fertilizer are independently blown to common coulters.

Combined soil cultivation and grain drilling is commonplace and reflects the needs to save labour by reducing the number of passes across the ground. A power harrow with a mounted drill requires cultivated ground, whereas many units, utilizing either draught or pto-driven rotary cultivators, can be used on stubble.

Seed population can be further checked after drilling, and plant population after emergence, with reference to Table 24.3.

Precision seeders (unit spacing drills)

Where crops are grown as individual, spaced plants, it is necessary, in addition to sowing the correct quantity, to plant seeds singly rather than in a stream. These two functions are carried out most commonly by cell metering devices.

Cells (holes or indentations) of the correct size to match the seed (which should itself be graded or pelleted to improve uniformity and accuracy) are carried in the outer rim of a wheel, or as perforations in a flexible belt. As the wheel or belt passes under the seed lying in a small hopper, a seed drops into each cell by gravity. The seeds are removed from the hopper by the wheel or belt and fall from the cell to the ground by gravity, usually aided by an ejector. The seeds are directed rearwards as they leave the seeder to counteract the machine's forward speed and reduce the tendency of the seeds to roll or bounce along the ground.

The most satisfactory results are achieved with uniform graded seed and at low speeds (3–5 km/h), when sufficient time is available for the seeds to find their way into the cells. With increasing speed, more and more cells remain unfilled, and the number of gaps in the row of seeds increases.

Precision seeders, using gravity to fill cells, operate well with spherical seeds. For seeds that are not spherical (e.g. lettuce) and do not readily flow, a more positive seed selection method is desirable. Vacuum seeders, the vacuum being provided by a pto-driven fan, suck seeds onto a ring of small, perforated depressions on one side of a thin circular disc, which rotates vertically through a seed hopper. As the disc rotates, the seeds are retained and lifted out of the hopper. An adjustable finger displaces any doubles. When the seed is directly above the coulter, the vacuum is cut off and the seed drops to the ground, again with a rearwards impetus in many cases. Some vacuum machines can plant seeds accurately at 8–10 km/h. One size of disc can meter a considerable range of seed sizes, so the time and cost normally involved in changing from one set of cells to another between crops is much reduced.

Table 24.2 Calibration distances for machines of different widths

Working width (m)	*Distance (m) to cover*	
	1/10 ha	*1/25 ha*
3	333	133
4	250	100
6	167	66.7
8	125	50.0
9	111	44.4
12	83.3	33.3

Table 24.3 The number of seeds (or plants)/m^2 according to row width and spacing within the row

Within-row spacing (mm)	*Row width (mm)*				
	120	*125*	*140*	*170*	*190*
10	833	800	714	588	526
15	556	533	476	392	351
20	417	400	357	294	263
25	333	320	286	235	211
30	278	267	238	196	175
35	238	229	204	168	150
40	208	200	179	147	132
Metres of row/m^2	8.33	8.0	7.14	5.88	5.26

Precision seeders are made up of a number of single-row units, each with its own seed hopper, attached by flexible links to a tractor-mounted toolbar. By moving the units along the toolbar, row widths down to 200 mm are possible on some machines. Closer spacing is possible by staggering units in two or three transverse rows.

Each single-row unit comprises the selecting/metering device, front and rear depth control wheels and a backward raked coulter which forms a groove in the soil. The coulter can be lowered relative to the depth wheels to increase planting depth. A covering device is located behind the coulter.

As the seeds are small, a fine tilth is required to ensure good seed/soil contact. Also, as there is little ground clearance, the seedbed must be smooth and trash free. Accessories are available for some machines to push aside loose dry clods, or to level and compact an uneven tilth. Angled discs or blades can draw soil over the seed, and heavy narrow press wheels can improve compaction round the seed. Use of the correct accessories can ensure fast, even germination under a range of soil conditions, but care must be taken not to disturb the seed spacing with any following treatment.

Additional equipment can be mounted on the precision seeder frame for the simultaneous application of liquid or granular chemicals.

A seeder unit is very compact and close to the ground, so that it is difficult for the operator to see if it is working satisfactorily. Some machines can be fitted with simple electric monitors to show whether the mechanism of each unit is turning, or whether the seed hoppers are empty. A more sophisticated device causes a light beam to be broken by each seed as it is released. Any interruption in seed flow sets off an alarm, so that the fault can be identified and corrected immediately.

The number of rows of the seeder should be a multiple of the number of rows to be harvested simultaneously. For example, a crop to be harvested by a three-row harvester should be planted by a 6, 9, 12, 18 or 24 row planter. This avoids the risk of misalignment which might occur if the harvester had to overlap two passes of the seeder.

The metering mechanism is driven either from a ground wheel on each seeder unit or from a master landwheel which drives all units. Changing the drive ratio, by means of stepped pulleys or gears, changes the rotational speed of the metering mechanism relative to the ground speed, and thus the seed spacing. On most machines, cell wheels or belts can be changed for ones with space for more or less seeds per turn, which will also change the spacing.

Once the machine has been prepared for work, by matching cells to seed size and by adjusting drive ratios, it is important to check performance in the field, at normal planting speed. A portion of every row should be uncovered so that spacing and evenness may be assessed, and corrected if necessary. Table 24.4 shows the relationship between spacing, row width and seed population. Seed population must not be confused with plant population, which will be lower in proportion to the emergence percentage of the crop involved. The figure for sugar beet is 60–65%; extra seeds must be planted to allow for this loss.

Potato planting and establishment equipment

Potato establishment equipment

To increase harvester workrate and to decrease tuber damage at harvest, a clod- and stone-free tilth is desirable. Deeper cultivation than that for cereals is required to allow ridges to be formed. A standard sequence of events is to form 1.8-m beds, using a deep ridger or bed maker. This implement consists of ridging bodies mounted behind the tractor wheels. A pto-driven rotary cultivator, fitted with ridging bodies, can be used on soils

Table 24.4 Seed population (10^3/ha) according to row width and spacing

Spacing (mm)	*Row width (mm)*					
	400	*500*	*600*	*700*	*800*	*900*
100	250	200	167	143	125	111
125	200	160	133	114	100	88.8
150	167	133	111	95.2	83.3	74.1
175	143	114	95.2	81.6	71.4	63.5
200	125	100	83.3	71.4	62.5	55.6
250	100	80.0	66.7	57.1	50.0	44.4
300	83.3	66.7	55.5	47.6	41.7	37.0
400	62.5	50.0	41.7	35.7	31.3	27.8
Kilometres of row/ha	25.0	20.0	16.7	14.3	12.5	11.1

that have numerous clods to prepare a fine tilth. A stone and clod separator is then used to remove stones and any remaining clods from the bed and deposit them in an adjacent furrow bottom. The separator consists of a share, which lifts the bed onto a series of pto-driven rod conveyors (webs). As the bed passes over the webs, soil and small material fall through the webs to create a stone- and clod-free tilth. Strong clods (which are not disintegrated during their passage over the webs) and stones, that are too large to pass through the web fall onto a cross conveyor and are deposited in an adjacent furrow bottom. Planting should take place immediately after stone separation, as the finely cultivated beds do not dry quickly after rain. Figure 24.5 shows the sequence of cultivation events prior to potato planting.

On light soils and where stone populations are small, the above sequence of events can be simplified. For example, it is possible to cultivate ploughed ground to form a tilth and then plant and form ridges simultaneously.

Potato planters

A potato planter must be capable of dealing with a large amount of 'seed' material, typically 40 000 tubers when planted at a rate of 2.5 t/ha. Gentle handling is essential and especially more so if the seed has been artificially sprouted.

Although manual planters are still used, the majority of potatoes are planted by automatic planters. Manual machines usually require operators to remove tubers from a hopper and fill a series of cups mounted radially on a horizontal shaft. The cups rotate and release tubers at their lowest point. Regular spacing is achieved and alterations to the spacing are effected by changing the sprockets in the ground-drive system. One operator per row is required.

Automatic planters

Automatic planters are of two types: cup-fed and belt-fed.

Cup-fed planters A cup-fed planter (Fig. 24.6) consists of a hopper, a tuber selection and metering unit, a soil opener and a tuber-covering element. Tuber selection and metering are achieved by a double row of cups, mounted on a vertical endless belt, which pass upwards through a hopper. Tubers are removed from the hopper

Figure 24.5 Bed-forming (foreground), bed-tilling and stone/clod separating (background) operations designed to promote potato damage reduction and high work rates at harvest. (Courtesy of Reekie Netagco Ltd.)

Figure 24.6 A two-row cup-fed potato planter. (Courtesy of Reekie Netagco Ltd.)

by the cups and are carried over the top pulley and downwards towards the soil opener. As the tubers reach the bottom pulley, they fall from the cups into the slot created by the V-shaped opener. The ridge is reformed and the tubers covered by a set of discs or mouldboard ridgers.

Two-row planters are standard (four-row versions are also available) and have a hopper capacity of 500–1000 kg. Row widths can be varied between 750 and 920 mm. Planters are available that can plant three rows of potatoes in beds.

The planting rate (kg/ha) is adjusted on ground-wheel-driven units by changing gears or sprockets in the driveline to alter the speed of the cup-carrying belts relative to forward speed. Uniform spacing is achieved by the cups releasing tubers at regular intervals. It is important that each cup carries only one tuber so that misses and doubles are avoided. This can be achieved by ensuring that the tubers are closely graded, the cup size matches the tuber size, the speed of the belts through the hoppers is not excessive and the belt agitation system (to dislodge extra tubers in a cup) is set correctly. Tubers are disturbed as the cups enter the hopper and long sprouts can be broken off. It is important to produce mini-sprouts when using a cup-fed planter with chitted seed.

Where it is important to accurately space tubers at spacings that are not catered for by the sprockets and gears available, planters can be used that allow adjustment of spacing from the tractor seat. The belts carrying the cups are driven by an hydraulic motor, and planter forward speed is monitored. A microprocessor is used to control the hydraulic motor speed to ensure that the required tuber spacing is achieved irrespective of forward speed.

Belt-fed planters Belt-fed planters have similar components to cup-fed units except in the method used to select and meter the tubers. Tubers flow from the hopper and are arranged by means of multi-belts which gradually discard tubers to produce a line for planting. The tubers are not planted discretely but 'flow' from the planter to the soil via the furrow opener. Uniform grading will promote regular spacing in the row. Being land-wheel driven, the application rate is adjusted by changing gears or sprockets to alter the belt speeds relative to forward speed. The 'flow' characteristic of belt-fed planters results in less damage to sprouted tubers.

Transplanters

Many vegetable and nursery crops are planted out as transplants both to reduce the risk of seed establishment failure and to optimize the growing season. The standard transplanter is tractor-mounted and semi-automatic. It consists of a plant tray holder, a plant feeding and metering device and a soil opener. An operator removes plugged plants from a tray and drops individual plants into one of four (or six, depending on model) rotating cups. The base of the cups is designed to open by means of a cam. As a cup rotates above the soil opener, the base opens and the plant drops into the soil where it is firmed into the soil by two angled press wheels. The cups are driven by the press wheels and plant spacing is adjusted by changing the gearing in the drive mechanism.

Crop sprayers

Most chemicals are applied as water-based solutions or suspensions in the form of droplets to give an even crop cover. The required droplet size depends on the chemical being applied and its target. Very fine droplets are required for an even crop cover (e.g. when spraying against insects and fungi), whereas larger droplets will suffice if the chemical is soil-acting and will subsequently be mixed in. Producing the wrong droplet size or indiscriminate application can have consequences in efficacy, economic and environmental terms. Large droplets can roll or bounce off leaves whereas very fine droplets are subject to wind effects and may not reach the target and drift away (Table 24.5). Spray drift has been measured as far as 6 km from the spraying site. The problem of drift is increased by the fact that water evaporates from droplets in dry weather, making them lighter and more drift prone.

A typical field volume rate when using a standard sprayer is 200 l/ha. One method of reducing this water volume (yet still applying the same quantity of chemical) and allowing a tankful of spray to treat a greater area is to reduce the spray droplet size. For example, halving the spray droplet diameter results in the same number of droplets being produced in an eighth of the volume. Table 24.6 demonstrates this effect.

Table 24.5 Effect of droplet size on spray drift (Spillman 1984)

Diameter of droplet (μm)	*Horizontal drift (m) when released from a height of 0.6 m*		
	Wind 0.1 m/s	*Wind 1 m/s*	*Wind 5 m/s*
10	19.1	191	859
50	0.84	8.4	42
100	0.24	2.4	12.3
300	0.05	0.5	2.5
1000	0.01	0.1	0.6

Sprayer components

A standard tractor-mounted sprayer is shown in Fig. 24.7. The main components of the sprayer are the tank, pump, spray boom and spray nozzles. The tank, formed usually from a plastic material to resist corrosion, has rounded corners to promote easy internal cleaning and ranges in capacity from 200–1000 litres on mounted machines and up to 3500 litres on trailed models. Some form of tank content agitation, usually by a pressure jet return, is necessary to ensure the spray material maintains a uniform mix. Agitation is especially important where suspensions are applied. The tank has a filler

Table 24.6 Effect of droplet size on closeness of spray cover (Spillman, 1984)

Diameter of droplet (μm)	*Number of droplets/cm² at 45 l/ha application rate*
400	13
250	55
150	254
50	1750

Figure 24.7 A tractor-mounted hydraulic sprayer. (Courtesy of Amazone Ltd.)

opening and strainer on top and a fine strainer at its outlet to protect other components from blockage. Since the top of the tank may be quite high with difficult access, machines are fitted with a low-level filling hopper. This induction bowl allows the operator to add chemical to the sprayer safely at ground level. The chemical is flushed into the main tank by the pump. Most sprayers transport the spray mix as a dilute water-based solution. Some sprayers have separate containers for water and for chemicals, which are metered separately and mixed continuously as the machine crosses the field. The development of this direct injection principle minimizes the need to handle the chemical concentrate and avoids the problem of disposing of dilute chemical should spraying have to stop unexpectedly (Landers, 1992).

The main function of the pump is to deliver the spray material from the tank to the nozzles. The design of the sprayer pipework also allows it to be involved in filling the sprayer (with water and chemical), in maintaining agitation of the tank contents, in moving and mixing concentrate chemical from the induction bowl to the main tank and in washing the tank. The pumps are of the positive displacement type, delivering 0.5–2.5 l/s at 540 rpm, at pressures of 2–4 bar. Roller-vane pumps may be used on small sprayers, but diaphragm pumps are used on the majority of machines. Piston pumps are available and can operate to pressures of 40 bar.

A pressure relief valve is an essential safety component when using a positive displacement pump. This can be incorporated into an adjustable pressure regulator that limits pressure fluctuations, or in more complex forms, keeps pressure constant. A constant pressure system allows the driver to vary gear selection to maintain a steady forward speed. Automatic volume control systems contain a forward-speed measuring device which increases or decreases nozzle flow to compensate for changes in forward speed. Within a limited speed range, errors in application rate because of speed variations are eliminated.

The boom, which carries the spray nozzles, can be from 6–36 m in width, mostly in the range 18–24 m, the choice depending on the tramline system. It is generally made in sections that can be folded horizontally for transport. Each section can be shut off independently. The function of the boom is to carry the nozzles at a set height above the target with minimal vertical or longitudinal movement. Boom height can be adjusted to maintain the desired clearance above the soil or growing crop. Any fore and aft (yaw) or vertical movement (bounce) of the boom will affect the application rate. These movements will be accentuated when travelling at speed and over rough ground and will be greatest at the extremities of the boom. The ground speed of the outer nozzles may vary from backwards to forwards at 1.8 times tractor speed. Bounce and yaw can be reduced by mounting the boom on a damped, flexible linkage to isolate it from the movement of the tractor and spray tank.

Nozzles are mounted along the boom, usually at intervals of 0.5 m, and must be held at the recommended height above the target if application is to be uniform. The nozzle breaks up the stream of spray solution into droplets and discharges them in a triangular pattern towards the target. The orifice of the nozzle meters the flow of spray solution according to fluid pressure; an increase in pressure gives an increase in flow rate through a particular nozzle. The pattern developed by the nozzle will depend on the type of nozzle used. The three nozzle types are the flat fan, the hollow cone and the deflector (anvil). The flat fan nozzle is used in agriculture for the application of herbicides and fungicides. Hollow cone nozzles are used when small droplets are necessary. Deflector nozzles tend to produce larger droplets than flat fans and are used for the application of soil-acting herbicides and liquid fertilisers. Nozzles are classified according to their pattern, output and rated pressure. The classification F/80/1.2/3, for example, denotes a fan spray nozzle having a spread angle of 80° and producing an output of 1.2 l/min at its rated pressure of 3 bar (British Crop Protection Council, 1986).

The size of droplets produced (i.e. the spray quality) depends on spray pressure and nozzle size, but nozzle size has the greatest effect within the usual 2–4 bar pressure range. The droplet size is not uniform, but the spray quality can be classified as being very fine, fine, medium, coarse or very coarse. Some nozzles can produce more than one spray quality and manufacturers produce charts showing the pressures at which one or other grade is produced by a particular nozzle. Chemical manufacturers specify the spray quality that should be used for each product, and this is shown on the product label. Fine spray is used for foliar-acting weed control in cereals and for contact-acting fungicides and insecticides. Because of the risk of drift, fine sprays must not be used for products labelled toxic. Medium sprays are suitable for most purposes, while coarse are used for soil-incorporated materials. Very fine sprays are used for fogging, and very coarse are used for liquid fertilizer application.

Nozzle tips may be made of plastic, stainless steel or ceramics; durability of these materials increases in the same order. During operation, the nozzle orifice widens, resulting in a narrower spray pattern and a concentration of spray directly in line with the axis of discharge. The cost of a set of nozzles is cheap relative to the chemicals applied and nozzles should be replaced at least annually to ensure accurate application. Nozzle manufacturers use an international standard colour code to identify nozzle sizes and thereby avoid errors when changing nozzles.

Booms can be fitted with rotating multi-nozzle holders that allow easy change from one application to the next. Nozzle holders are equipped with check valves which prevent flow of spray when pressure drops to 0.2–0.3 bar. This prevents dribble from the nozzles after the machine has been shut off at the end of a run. Corrosion-proof strainers may be fitted to protect the nozzle from blockage. It is important to match the screen size of the strainer to the aperture of the nozzle; if the strainer is too coarse, the nozzle is not protected; if too fine, the screen may clog up with small particles which could otherwise pass freely through the nozzle.

To increase the chance of droplets reaching their intended target and therefore reducing drift problems, a number of developments has occurred:

- Compressed air is mixed with the droplets to help propel them towards the target.
- An air curtain is used to influence the direction of the droplets. A fan inflates an air bag that is mounted on the boom. Air escapes from orifices on the underside of the bag and entrains the nozzle output and directs the droplets towards the crop.
- An electrostatic charge is applied to the droplets as they leave the nozzle. The charged droplets are attracted onto the surface of the crop, ensuring that a large proportion of the spray actually ends up on the crop.

Conventional hydraulic nozzles produce a wide spread of droplet sizes, from those that are too heavy to stick to a leaf, to those that are light enough to be carried away by a breeze. Rotary atomisers produce a very narrow spectrum of droplet sizes as the liquid spins off the edge of a rotating disc. Controlled droplet application (CDA) sprayers produce droplets in the range of 250–300 μm. These resist drift and give good cover using less spray. Spray volumes of 20–40 l/ha are recommended, allowing a very large area to be treated by one tankful.

The ultra low volume (ULV) system generates a droplet size of 70 μm and uses very low spray volumes, typically applying neat chemical. The droplets are carried by the wind to give a very thorough cover of the crop with excellent penetration into dense foliage. Herbicides and poisonous materials are not applied, because of the risk of drift, but satisfactory results can be achieved with approved fungicides and insecticides. Special oil-based formulations are used to reduce the risk of evaporation while the tiny droplets are in the air. Only chemicals approved for this system of application may be used, and manufacturers' recommendations must be followed. Due to the high risk of operator contamination and drift outside the treatment area, few if any products are approved for application by ULV in the UK.

Sprayer calibration and maintenance

The value of spray chemicals handled by a sprayer in a year can be far greater than the value of the sprayer itself. Time spent calibrating and checking the machine can yield large dividends in terms of more efficient use of chemicals.

As nozzles wear with use, both the quantity output and the spray pattern uniformity will alter. A visual check on the pattern of each nozzle should be undertaken regularly – if the pattern is not uniform, the nozzle should be replaced. The output of a nozzle can be measured using a special meter, or by collecting the liquid in a graduated 2-litre container over a period of 1 min. If a nozzle is found to be delivering more than 5% in excess of the rated output, it should be replaced.

Checking the sprayer's application rate can be achieved by firstly measuring the mean nozzle output (litres/min) using at least four nozzles, at least one from each section of the boom. From Table 24.1, the time taken to spray 1 ha at a given forward speed can be determined, and the application rate (litres/ha) of the sprayer can be calculated and compared with the desired application rate.

Correct forward speed is essential for precision spraying; tractor speedometers, if fitted, must be calibrated to take account of alternative tyre sizes and field conditions. Speed monitors, using independent ground wheels or radar sensors, are available, and some monitors will also display flow rates and application rate per hectare. Control systems known as 'spatially variable technology' are capable of adjusting the sprayer's output of chemicals as conditions vary across the field. This will give greater precision in husbandry and better efficiency of pesticide use (Paice, 1993).

Sprayer workrate

Although the field sprayer is a wide machine and high rates of work are possible, actual workrates can be disappointing. Tank filling and travelling to and from the water supply can use up much of the available time. Where only a few days are available for the application of a particular chemical, maximizing output can be valuable. Fast refilling can be achieved by providing a water tanker on the headland, fitted with a high capacity pump. This cuts down travelling time almost to zero and reduces filling-up time in comparison to the time taken to fill the tank via a hose pipe from the mains, for example.

Sprayer safety

The use of spraying equipment and spray chemicals is potentially hazardous, both to the operator and to members of the public, livestock, pets, beneficial insects and non-target crops. The Food and Environment Protection Act 1985 (FEPA) and the Control of Substances Hazardous to Health Regulations control all aspects of the importation, advertisement, sale, supply, storage, handling and application of pesticides. An appropriate recognized certificate of competence is required by any person who sells or supplies pesticides approved for agricultural, horticultural, forestry or amenity uses. Anyone applying these pesticides as a contracting service, or who was born after 31 December 1964, must also hold a recognized certificate of competence unless directly supervised by a certificate holder. Only approved products may be advertised and sold, and users must comply with the conditions of approval relating to use.

The test for the certificate of competence in pesticides application (mandatory from 1 January 1989) is administered by the National Proficiency Tests Council (NPTC), and in Scotland by the Scottish Skills Testing Service (SSTS). Training is provided by Lantra, some colleges and various private training organizations. The following areas of expertise are included:

- legal requirements, including the importance of understanding the conditions attached to product approvals;
- hazards and risks of pesticides, including hazards to humans, animals, crops, insects, other wildlife and the environment;
- the nature and degree of potential health risks, and the likely ill-effects, of exposure to pesticides;
- recommended ways of storing, handling and mixing pesticides, and of disposing of empty containers, contaminated materials, and surplus pesticide whether in concentrate or dilute form;
- procedures for the preparation and calibration of application equipment, without risks to the user or other people, plants, animals, beneficial insects or the environment;
- the control measures necessary to protect the user and others, the reasons for them and how to use them. This includes the reasons for using suitable personal protective equipment (PPE), including protective clothing, the jobs where it is necessary, and the correct methods for cleaning, storing or disposing of contaminated items;
- safety procedures to be followed before, during and after application. Recognition of circumstances where risks can arise to people, animals, beneficial insects or the environment including water, and the need to avoid use in such circumstances;
- emergency action necessary in the event of a spillage of concentrate or dilute pesticide;
- emergency action to decontaminate people, the signs and symptoms of pesticide poisoning, and the action to be taken and details of how to obtain medical advise in case of accidental exposure;
- when to use monitoring and health surveillance;
- the need to make, keep and provide access to records to meet statutory obligations;
- determination of safe wind and weather conditions for spraying in order to minimize the risk of spray drift.

Irrigation

Field irrigation enables the farmer to optimize crop production by avoiding moisture stress at any time during the growing season. Benefits include good early establishment, disease control, uniform crop development and increased crop value through product uniformity and enhanced yield and quality.

Other uses for irrigation equipment include softening the ground to aid seedbed preparation or root harvesting, the application or washing-in of fertilizers and other chemicals, the prevention of wind erosion and the protection of sensitive crops from frost. Irrigation systems are also used for the application of livestock wastes, especially dirty water.

Water supply

A reliable water supply of a quality appropriate to the crops being treated is required and the appropriate water control authority must be consulted before abstracting water.

Direct abstraction

The cheapest water source for irrigation is water from a river or stream which is pumped directly into the irrigation system. In areas where underlying geological structures are suitable, boreholes may offer a source from which water may be pumped directly.

Water storage

If direct abstraction of water is not possible due to either legislation or low summer flow rates, a reservoir can provide storage for water collected over winter for summer use. The reservoir may be constructed by enlarging a hollow or closing off a valley. Being part of the river flow system requires that adequate spillway facilities and a system of cleaning to remove sedimentation have to be considered. In flat country, off-stream storage involves the construction of a rectangular bank from the soil excavated from the area enclosed by the bank. Field drainage systems and any porous subsoil layers must be removed. If suitable clay is available, preferably from the excavation itself, this may be used to line the reservoir; if not, a lining of flexible sheet material will be required. A lining of this type can double the cost of a reservoir.

Banks must be protected from the effects of weathering and from wave action at the waterline. Points of water entry, and spillways to allow for overflow, must be reinforced to resist the erosive effects of concentrated water flow. The advice of a chartered civil engineer should be sought in the design of reservoirs and associated works, especially where reservoir capacity exceeds 20 000 m^3, or banks are higher than 5 m.

Ideally, the reservoir will be filled by gravity by a pipe from the adjacent stream, with flow being controlled by a sluice at the pipe inlet. Where this is not feasible, the reservoir can be filled by a low-lift, high-volume pump positioned as near as possible to the supply water level.

The required storage capacity will depend on likely application quantities, which will range from 100–175 mm/ha over the growing season according to crop. An additional quantity, normally equivalent to 500 mm depth in the reservoir, is allowed for seepage and evaporation from the reservoir.

Irrigation systems

The basic components of an irrigation system are a pump, a water distribution system and an application system.

Irrigation nozzles

In the UK, most application systems use nozzles to distribute the water on to the crop. When water is forced out through a nozzle, the stream breaks up into droplets in a range of sizes. The degree of atomization depends on the pressure at the nozzle. At low pressure, there is little atomization, and a solid stream of water may result, while at high pressure there is very thorough atomization, resulting in a wide, feathery spray which does not carry far and is easily deflected by the wind. Between these extremes, there is an optimum range of pressures at which droplet size range and throw are satisfactory, and flow rate is sufficient. It is unusual for excessive operating pressures to be a problem, but many irrigation systems suffer from lack of pressure, resulting in poor distribution from nozzles, reduced water application and soil damage from the larger droplets that are produced.

Large droplets are produced by large nozzles, even when operated at the recommended pressure. The use of such nozzles carries with it the risk of soil damage from impact by droplets larger than 3–5 mm, especially when the soil is not fully protected by a growing crop. For this reason, 22 mm is the largest nozzle diameter recommended for use in beet or potato crops, while 28- or 30-mm nozzles can be used on grassland.

Nozzles rotate in use, either through 360° or through a segment of a circle, depending on the application system used. Sprinklers, incorporating one or two nozzles, have a discharge rate of 0.5–5 m^3/h and are used at spacing of 12–24 m (Fig. 24.8). Rainguns, generally with a single large nozzle, have discharge rates of up to 150 m^3/h or more and may be spaced up to 140 m apart and irrigate over 1 ha from a single point (Fig. 24.9).

Application systems

Mobile irrigators

The popularity of mobile irrigators is due to their reduced labour requirement compared with traditional sprinkler systems (see Table 24.7). The machine covers

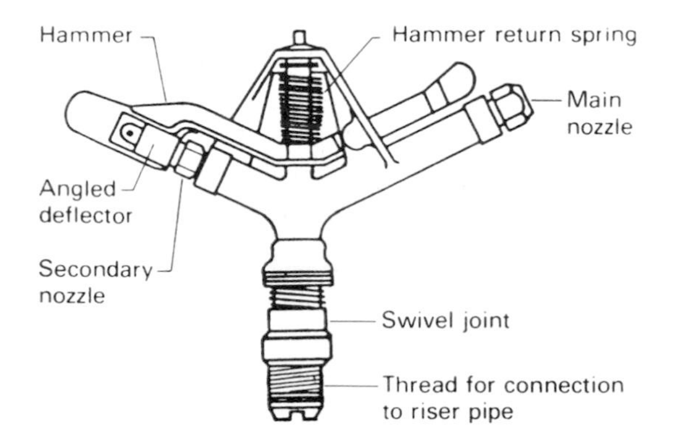

Figure 24.8 A sprinkler. (Courtesy of Wright Rain Ltd.)

a large area per setting unattended, before being moved to a new position. Models are available to cover a wide range of areas per setting, from less than 1 to more than 400 ha.

Because mobile irrigators are constantly on the move, they only actually apply water to a fraction of the set area at any one time. This fraction receives the full output of the machine, resulting in an extremely high spot application rate. Typical spot rates for hosereel boom machines are 60–80 mm/h, while the outer regions of centre-pivot systems can deliver more than 300 mm/h. Such high application rates will cause ponding if the terrain is fairly flat, while runoff and possibly erosion can occur on sloping ground.

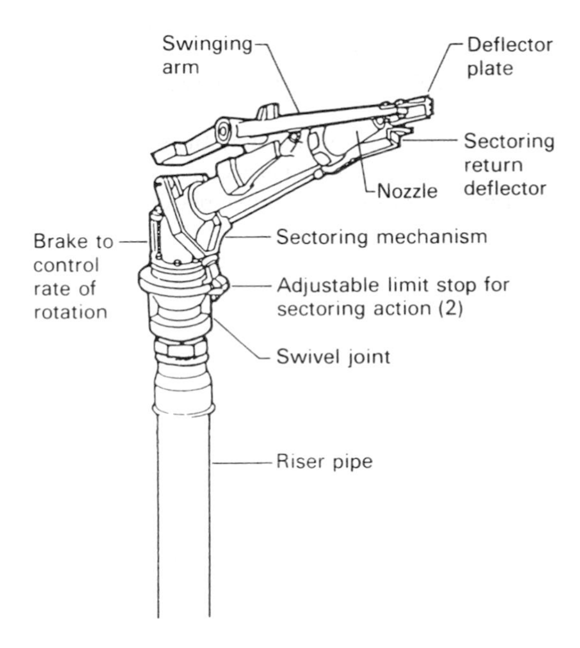

Figure 24.9 A raingun. (Courtesy of Wright Rain Ltd.)

Hosereel irrigators Most agricultural irrigation systems in the UK use hosereel machines (Stansfield, 1992). The main component is a large reel, carrying up to 400 m of polyethylene hose. One end of the hose is connected to the water supply inlet; the other end is attached to a raingun mounted on a wheeled trolley (Fig. 24.10). To prepare the machine for work, the reel is positioned at the headland of the field and the trolley is pulled across the field by a tractor. In doing so, the hose unwinds from the reel. The limit of operation is either the length of the hose or the width of the field.

Table 24.7 A relative comparison of irrigation methods (after Curtis, 1980)

Type of system	*Cost*	*Labour*	*Energy*
Hand-moved sprinklers	100	100	100
Solid-set sprinklers	400	20	100
Mobile hosereel irrigator	170	20	190
Mobile centre pivot (low-pressure nozzles)	250	15	75

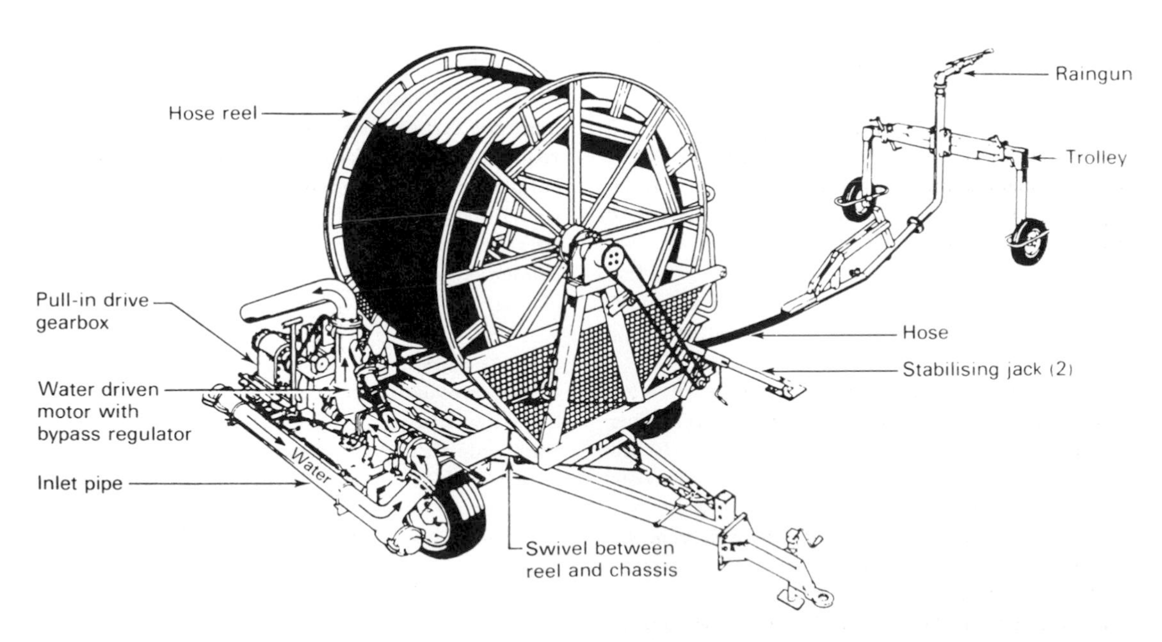

Figure 24.10 A hosereel mobile irrigator. (Courtesy of Wright Rain Ltd.)

When the water supply is activated, the raingun irrigates a sector of ground facing away from the reel, and, after a delay, a water-powered motor begins to turn the reel, rewinding the hose and drawing in the trolley and raingun. Application rate depends on nozzle output and rate of travel. Most machines have compensators to allow for the fact that the reel increases in diameter as the hose is wound on to it, but it is advisable to check the accuracy and consistency of speed settings from time to time.

When the trolley reaches the reel, the trolley contacts a bar that is linked to a valve, which stops reel rotation. Irrigation continues for a further pre-set period before the pump is shut down. The machine is then repositioned for another period of unsupervised work.

Some machines are fitted with a turntable so that the reel can be swung round to face in any direction without moving the chassis. This allows precise alignment with crop rows where these are not exactly at right angles to the headland, and it also facilitates irrigation to either side of a central laneway or 'midrig' where a field is too wide to be irrigated fully from one side headland.

Where delicate crops are to be irrigated or where poorly structured soils are not fully protected by crop cover, the raingun may be replaced by a wide boom carrying several sprinklers which will produce fine droplets and minimize impact damage.

Because of the long runs possible, reel irrigators can be made to operate for almost 24 hours per day. Larger machines can work two 11-hour runs, while smaller machines can be set for a single 22-hour run. Where awkward fields do not allow this, short runs can be worked in the daytime, and long runs saved for night operation.

Centre pivot and linear-move irrigators These machines allow a uniform application of water to be applied automatically to a large area of land. Coefficient of uniformity for centre-pivot machines is 90–95%, compared with 50–60% for hosereel machines, and this allows more precise control of water use. Power requirement is also less, although capital cost is greater than for hosereel machines.

Both types of irrigator consist of a long boom, often several hundred metres in length, made up of a number of segments, each approximately 30 m long. Wheeled towers at the outer end of each segment carry the boom above crop height. Sprinklers or downward-facing low-pressure nozzles are spaced along the length of the boom.

Centre-pivot machines are anchored to a central mast, through which water and electrical power (for the propulsion system) are fed in. Driven by the wheeled towers, the boom rotates around the mast, covering a circular area of land, ranging from 15–400 ha. Rotational rates of between 1 and 100 hours are controlled by the travel speed of the outermost drive unit; sensors at each tower ensure that the boom is kept straight. Most models can be fitted with a hinged outer extension, guided by a low-voltage cable buried in the soil below plough depth, which will unfold to reach into the corners of a square area.

Flat ground is ideal, both for the drive system and because of the high application rate, but the drive system can negotiate undulating ground if necessary. Circular tracks are left by the wheels; hedges and banks must be opened up to allow the towers to pass. The area lost to wheel tracks is less than 1% of the whole. The size and spacing of output nozzles along the boom vary so that all parts of the cropped area receive the same application of water.

The linear-move irrigator follows a straight path, covering a rectangular area. Water enters the boom at one end, either from a trailing or reeled pipe or pumped directly from a supply ditch running down the field. At the end of a run, the machine can be turned through 180° to irrigate a fresh strip on the return journey. A semicircular area can also be irrigated at each end of the run.

Fixed sprinkler irrigation systems

Sprinklers and laterals A number of sprinklers, spaced along 75-mm-diameter pipes, are laid out to cover, typically, an area of 0.4 ha. The pipes, or laterals, are available in 6- and 9-m lengths to allow for different spacings, and sprinklers are normally held above crop height by vertical risers every 12–18 m. An application of 25 mm of water is delivered in 2–4 hours, after which irrigation must stop and two or more people move the lateral to its next position. The work in wet conditions after irrigation is not pleasant, and involves the interruption of other activities at what is often a busy time of year.

Solid-set sprinklers The chore of moving sprinkler laterals three or four times a day can be eliminated where there are sufficient sprinklers and laterals to cover the entire crop area. The system is set out at the beginning of the life of the crop, and remains in place until no more irrigation is required.

Semi-solid-set systems are also in use. Here, an extended area is laid out with pipes, which offer many quick-coupling sockets into which standpipes and sprinklers can be inserted. Sockets not in use are sealed automatically. A 0.4-ha set of sprinklers can be moved from socket to socket across the area as irrigation proceeds, without the need to reposition the pipes. However, as with conventional sprinkler systems, attention is needed three or four times a day.

Trickle irrigation system

Trickle irrigation systems use a water application system consisting of small-diameter plastic irrigation pipe or tape, fitted with outlets, which is positioned along each crop row, either on or below the soil surface. The system is suited to crops such as field-scale raspberries that are cropped over a 7- to 10- year period. The merits of the system are:

- Water use efficiency is high – water is delivered to where it is needed.
- The system can be automated – soil moisture sensors within the crop can control the pump and distribution system.
- Labour input is minimal.
- Fertilizer can be applied along with the irrigation water.

A major disadvantage for annual crops (e.g. potatoes, vegetables) is the capital outlay of the plastic pipe and the problems of retrieving it and reusing it in subsequent seasons.

Distribution of water

Where appropriate, water is distributed from the pump via a 'skeleton' of permanent underground 'mains', of at least 150 mm diameter, terminating in hydrants. At the hydrants, portable aluminium or steel mains, in 9-m lengths and a minimum diameter of 100 mm, are attached to carry the water overground to the area to be irrigated, with further hydrants at spacings to suit the particular method of irrigation.

Above-ground irrigation pipes have been responsible for many serious accidents where pipes, being carried upright, have contacted overhead electricity cables. The pipes are good conductors of electricity and care should be taken when irrigating near overhead cables, both to avoid direct contact by pipes and to avoid contact by solid water jets from large rain guns, which can also conduct electricity.

Pumping

Because of their simplicity and reliability, centrifugal pumps are almost universally used in clean water irrigation systems. One or more vaned rotors spin the water to build up its kinetic energy, and this is converted to pressure energy due to the internal shape of the pump body. This pressure then causes the water to flow through the system. Increasing the rotational speed of the pump increases pressure and flow. Typical rotor speed is 1800–2800 rev/min; this allows direct drive from diesel engines or electric motors, but requires gearing-up if the pump is to be driven by a tractor pto. The maximum efficiency of a centrifugal pump is 65–80%, and the peak is at a specific combination of flow and pressure. The pump for any particular irrigation system should be selected and then operated so that it will work under conditions at which its greatest efficiency will be realized.

Compared to its maximum delivery pressure of 9–15 bar, the inlet suction or lift capacity of a centrifugal pump is small. The ideal position for a pump is on a floating raft, as close as possible to the water level. Where this is not convenient, the pump should be located on the bank, again as close as possible to the water. Pumps are fitted with a hand-operated priming pump so that they can be filled with water before start-up and a non-return foot valve prevents water from escaping from the pipework when the pump is stopped.

In addition to manual start and stop controls, irrigation pumps are normally fitted with safety controls which will shut down the power supply system if the motor becomes overloaded (electric drive) or if the engine temperature rises or oil pressure falls beyond a pre-set value (diesel drive). The system will also shut down if water pressure falls, indicating a loss of inlet prime.

To allow adjustments to be made to application equipment without a double visit to the pump, high- and low-level pressure switches can stop the motor when hydrants are closed prior to an adjustment, and start it up again once the applicators are re-set and the hydrant is reopened. Timers can be used to set a given period of operation, allowing the final irrigation setting of the day to finish without supervision.

When selecting an irrigation pump, the required water flowrate (m^3/s) and the operating pressure (bar) must be specified. The components of total operating pressure are:

- the suction head – the vertical distance between the water surface level and the centreline of the pump;
- the delivery head – the vertical distance between the centreline of the pump and the nozzle;
- the friction head – the pressure required to overcome friction in the pipework and losses at corners and changes in pipe diameters;
- the operating head – the required pressure at the nozzle or the inlet to the hosereel irrigator.

Planning the system

Each irrigation treatment will involve the application of between 12 and 25 mm of water. When considering the required capacity of the system, the aim should be to cover the irrigated area within 6–8 days to allow for a peak depletion rate of over 3 mm/day. With careful management, a modern irrigation system can be kept in operation for 20–22 hours daily. The application of 1 mm of water to 1 ha (1 ha mm) represents 10 m^3 of water.

For a hosereel unit applying 25 mm to 6 ha in a 22-hour day, the water flowrate requires to be 68.2 m^3/h and the unit has the capacity to cover 48 ha in 8 working days. A traditional lateral and sprinkler system, having a '1 acre setting' and applying 25 mm to 0.4 ha in 3 hours, will require a water flowrate of 33.3 m^3/h. If the system is moved four times each day, 12.8 ha can be irrigated in 8 working days.

Irrigation scheduling

During the growing season, the soil is depleted of water, the resulting deficit depending on whether the amount of water taken out by the crop (transpiration) exceeds the amount returned by rainfall. This deficit is expressed either as the amount of rainfall or irrigation that would return the soil to field capacity (in millimetres) or the tension with which the remaining water is held in the soil.

As water is depleted, crop growth begins to be affected. The tension or soil moisture deficit (SMD) at which this takes place is known as the critical SMD, and will depend on the crop itself (see Table 24.8), its rooting depth and the available water storage capacity (AWC) of the soil in which it is growing (see Table 24.9). The scheduling of the timing and size of irrigation treatments must be based on an understanding of these variables. A deep-rooting crop on a soil with a high AWC and a high infiltration rate (see Table 24.10) can be given heavy, infrequent water applications which are only necessary when a large SMD has developed, while a shallow-rooted crop on a soil of low AWC must be irrigated frequently with smaller applications of water, to maintain a small SMD. Some crops are particularly sensitive to soil water conditions at certain stages of their growth, and this must also be taken into account.

Soil moisture tension may be measured directly by means of tensiometers buried in the ground. This system has the great attraction that no other measure-

Table 24.8 Soil moisture deficit criteria for the major irrigated crops (MAFF, 1984)

Crops to be irrigated	*Recommended water per application (mm)*			*Critical SMD (mm)*			*Start SMD[1] (mm)*		
	Soil class[3]			*Soil class[3]*			*Soil class[3]*		
	A	*B*	*C*	*A*	*B*	*C*	*A*	*B*	*C*
Potatoes									
Early and canning: emergence onwards	25	25	25	25	30	30	20	25	25
Second early from tuber size 6 mm	25	25	25	25	30	30	20	25	25
Main crop from tuber size 10–20 mm	25	30	30	35	50	50	20	30	30
Sugar beet									
May, 75% crop cover	25	25	25	25	30	50	20	25	25
June	25	25	25	35	35	50	20	30	30
July	25	40	50	45	50	100	35	40	90
August	25	50	50	55	75	125	50	60	100
Cereals	25	—	—	50[2]	—	—	35	—	—
Grass									
Grazing (May onwards)	25	25	25	25	35	35	20	30	30
Conservation (May onwards)	35	50	50	35	50	50	30	40	40

[1] Start SMD is on assumption that it will take 2–4 days to irrigate an area and one has to anticipate the deficit building up.
[2] Up to anthesis.
[3] See Table 24.9.

Table 24.9 Classification of the available water capacities (AWC) on common soil textures in approximately increasing order (MAFF, 1982)

A. *Low available water storage capacity*	
AWC <12.5% by volume (<60 mm/500 mm soil depth)	Coarse sand Loamy coarse sand Coarse sandy loam
B. *Medium available water storage capacity*	
AWC between 12.5 and 20% by volume (60–100 mm/500 mm soil depth)	Sand Loamy sand Fine sand Loamy fine sand Clay Sandy clay Silty clay Clay loam Sandy loam Sandy clay loam Silty clay loam Fine sandy loam Loam
C. *High available water storage capacity*	
AWC >20% by volume (>200 mm/500 mm soil depth)	Very fine sand Loamy very fine sand Very fine sandy loam Silt loam Silty loam Peaty soils

Table 24.10 Infiltration capacities associated with some common soil textures (MAFF, 1982)

Category	*Equilibrium infiltration capacity range (mm/h)*	*Soil textures*
Very high	>100	Coarse sands, sands, loamy coarse sands, loamy sands
High	20–100	Sandy loams, fine and very fine sandy loams, loamy fine sands and loamy very fine sands
Moderate	5–20	Loams, silt loams, silty loams, clay loams
Low	<5	Clays, silty clays, sandy clays

ment and no calculation is necessary. Some irrigation systems are controlled automatically on the basis of tensiometer measurements. However, several tensiometers must be distributed in each field to give accurate results, and much care is necessary in their use.

More commonly, a running balance sheet of SMD is kept, either paper- or computer-based. Potential transpiration data are available, based on regional weather measurements, and these figures are combined with local rainfall, crop cover and irrigation records to arrive at the weekly SMD for each field. Soil information for each field is needed to indicate the critical SMD and the point at which irrigation should commence. Irrigation plans for the coming week are based on the assumption that no rain will fall. Many consultancy services are available to assist the farmer in scheduling irrigation applications.

Organic irrigation

Irrigation equipment may be used to transport and spread livestock wastes and 'dirty water' on the land. Labour required is minimized and the movement of heavy tankers on the land with attendant risks of soil damage and safety hazards on slopes is avoided.

Because of the fibrous nature of manure solids and the probability of hay, silage or straw being present, special pumps, often fitted with chopping attachments, are specified. Also to avoid blockage, rainguns with large nozzles are used, either static or as part of a hosereel irrigator. Only machines recommended for organic irrigation should be used, as liquid manure can affect the water motor and valves of conventional irrigators.

The suitability of liquid manure for handling in this way depends very much on its solids content: generally, slurries containing up to 8–10% DM are suitable for organic irrigation (Schofield, 1984). This corresponds to the dilution of faeces and urine combined with an equal quantity of water, or to the liquid fraction produced by a slurry separator.

Umbilical systems, in which a mobile raingun is attached to a tanker on the headland by means of a flexible pipe, are used for the distribution of both livestock slurry and sewage sludge. Dirty water systems, used for the disposal of rain and wash water from dairy parlours and scraped concrete yards, use a raingun mounted on a carriage, which moves along the field by means of an anchored wire and an on-board winch, powered by the rotation of the raingun.

When planning and using organic irrigation, it is essential to follow the guidelines set out in the Code of Good Agricultural Practice for the Protection of Water (MAFF, 1998).

Forage harvesting

Engineering aspects of forage conservation

Whether grass is to be conserved as hay or as silage, the majority of crops are first cut and then allowed to wilt or dry. The length of this drying period will vary from 12–36 hours in the case of silage to several days when making hay. Crop drying rate is influenced by solar energy input (sunshine), air relative humidity, wind speed, swath shape, swath density, degree of leaf and stem abrasion and stubble length.

On a hot, breezy day, sun and wind can remove more than 30 t/ha from the leaves of a growing crop. If this moisture removal rate could continue after the crop was cut, hay would be made in a single day! As it is, the crop resists water loss, and conditions in the swath do not allow sun and wind to reach all parts of the crop.

The objectives of mowing and swath treatment machines are to cut the crop, to produce material that will field-dry quickly and to avoid contamination with soil and stones. These objectives can be met by:

(1) ensuring the field has been rolled in the spring to flatten molehills and press stones into the soil;
(2a) producing a fluffy, durable swath for maximum wind penetration, or
(2b) producing a layer of uniformly spread grass having maximum surface area;
(3) mixing the crop to expose wetter material to wind and sun;
(4) abrading ('conditioning') the waxy outer cuticle of the crop to allow water vapour to escape once stomata have closed. Ideally, this surface treatment should be adjustable in its severity to match crop requirements and should be concentrated more on the thicker stems than on the more easily dried and fragile leaves. This 'conditioning' must be achieved without excessive fragmentation and leaf loss, which can result from overvigorous treatment. Losses increase with the length of time that the crop is drying in the field, so a treatment that gives a rapid drying rate for a 24-hour wilted silage crop may well be too severe for a 5-day hay crop;
(5) leaving stubble lengths of 50 mm or more to allow air to circulate under the crop. The risks of contaminating the crop with soil, and also of damaging the machine, are reduced considerably by leaving a long stubble.

Silage production requires anaerobic conditions to achieve good fermentation. Short chopping of grass when forage harvesting increases the chance of good fermentation conditions by exposing more 'sites' for bacteria and by improving the ease of consolidation.

Mowers

Drum and disc mowers have superseded cutterbar and flail mowers as the standard units for cutting grass for hay or silage. The basic distinction between drum and disc mowers is in their drive mechanisms. Drum mowers consist of two or more drum-shaped rotor assemblies driven from above by either belts or shafts. The cut material passes back between adjacent pairs of drums. If the cut material does not pass quickly back off the machine there is a risk of double-cutting, where small fragments are severed from the butt end of the crop and are subsequently lost in the stubble. The rotors of a disc machine are driven by a gear train from below and the cut material passes over them.

Cutting is carried out by two or three small, free-swinging blades mounted on the periphery of each drum/disc. The blades move at 50–90 m/s and the crop is cut on impact. Blunt blades may absorb 50% more power than sharp ones and power requirement is at its highest when mature, stemmy crops are cut. There is danger both to the machine and to the user if a blade strikes a stone or other obstacle. Blades can be broken and fragments can be thrown out. Guards and shields must be kept in place.

High rates of work (0.5–1.5 ha/h/m) are possible, machines require little maintenance or downtime and are unaffected by the condition of the crop. Power requirement is 35 kW for a 1.6-m drum machine, with a 15% reduction for disc machines. Capital cost increases with width, with a large increase for trailed machines. For this reason, and as a means of avoiding the reduction in drying rate that is experienced in swaths produced by large machines, the use of twin smaller machines, one front and one rear mounted on the same tractor, may be considered.

Developments in trailed machines have resulted in units whose driveline is connected to the tractor by a right-angled gearing. This arrangement allows the tractor to turn through 90° relative to the mower and results in the production of 'square' corners when cutting the field in a round-and-round pattern. Another development is the provision of a central drawbar and driveline which can be moved hydraulically either to the right or left relative to the cutterbar. The mower can cut a field off one side rather than round and round or in 'lands'.

Mower-conditioners

The swath produced by drum and disc mowers is not 'set up' for rapid drying without further treatment. The combination of a mower with a 'conditioner' can leave the swath in an ideal state for fast drying. A mower-conditioner requires 35% more power than the equivalent mower and may cost 50% more to purchase.

Two types of conditioner are available. Roller machines have a pair of rubber-covered interlocking horizontal rollers which bend and crack the crop before forming it into a loose swath. These machines are best suited to stemmy crops such as lucerne, but deal less well with heavy, leafy, grass crops. Flail conditioners scuff the crop using rotating rows of narrow fixed or hinged flails. This type of machine deals more thoroughly with thick, leafy crops, where the rigidly mounted resilient flails are able to penetrate throughout the material. The severity of treatment can be easily adjusted by moving a baffle which can delay the passage of the crop so that the flails apply more impacts to it.

The use of any steel-tined machine carries with it the risk that a broken tine may occasionally be left in the swath and cause serious damage to following machinery. Designs of conditioner which use plastic bristles eliminate this problem, as well as offering greater improvements in crop drying rates.

A mower-conditioner development is the addition of a belt conveyor behind the conditioning unit. The belt can be swung in and out of work by an hydraulic ram and the system allows adjacent bouts of grass to be positioned next to each other for pick up by the forage harvester (Fig. 24.11).

Swath treatment machines

Further treatment to assist drying after mowing and conditioning requires the swath to be fluffed up (tedded), moved laterally onto dry ground, inverted to expose damp material or spread out. Windrowing is the process of reforming the swath to match the next stage of the harvest operation or to reduce the risk of spoilage if rain is expected. The crop should be tedded at least daily in good weather.

Spreading the swath produced by a mower or mower-conditioner allows the cut material to intercept most of the solar energy falling on the field, rather than half or less when gathered into rows. Drying breezes also have better access to all parts of the crop. Low drying rates in swaths produced by wide mowers can be avoided if the crop is spread immediately after cutting. A Dutch system involving early-morning cutting with a mower-conditioner at a stubble height of 75 mm, spreading immediately and spreading again after 2 hours has been shown to produce wilted material of 35% DM or more by the end of the day of cutting. Where spreading is

Figure 24.11 A silage crop is cut by front- and rear-mounted mower-conditioners. The rear 'grouper' attachment places two swaths close together to suit the forage harvester. (Courtesy of John Deere Ltd.)

used, crop losses are increased both by the scattering action and because a tractor must run over part of the crop in order to gather it back into windrows.

Most swath treatment machines use spring-steel tines to move the crop. If these are set too close to the ground the risk of breakage is increased. Farmers often modify their machines by adding small clips or ties so that broken tines are retained rather than being allowed to fall into the windrow. Machine designers have addressed this problem by the use of plastic or rubber crop-handling elements.

Single purpose swathing and tedding machines tend to be gentle on the crop by having relatively low rotor speeds. Careful setting of the machines is necessary to avoid contaminating the swath with soil.

Tedding machines

Tedders either fluff up a swath in situ or spread a swath. Units that maintain the swath's shape consist of a rotor on which is mounted four rows of tines. The tines, rotating in the opposite direction to the tractor wheels, pick up the swath and in doing so loosen the swath. Spreading machines use four or six contra-rotating pto-driven tined rotors. The rotors are angled slightly to the horizontal and the tines throw the swath upwards and backwards. Wide machines are articulated to follow ground contours.

Swathing machines

Swathing machines have one or two rotors carrying radially mounted groups of vertical tines. A strip of spread material is gathered into a windrow by the tines as they sweep the ground when passing across the front of the machine. A cam operates the radial shafts which cause the tines to fold back and to release the crop and leave it in a windrow. On single rotor machines, a vertical guide, of canvas or steel rods, assists in the formation of an even swath. With twin rotor units, the rotors either operate in tandem to leave a windrow to the side of the machine, or contra-rotate and the windrow is produced between the two rotors.

Dual-purpose tedders and swathers

These machines are able to perform the functions of tedding, turning and windrowing. Covering a 3- to 5-m width, they usually consist of two pto-driven tined rotors mounted on vertical spindles. The whole machine is tilted forwards to ted or scatter the crop, or levelled to produce a windrow. A rear cage is used when swathing to allow the material passing between the rotors to be formed into a windrow.

Baling and bale handling

The development of balers and bale handling equipment has reflected the increase of materials handling ability on farms. Small bales, convenient for moving by hand, have largely been replaced by round bales, and in cases where bales are to be transported, large 'square' bales are used. Straw, hay and grass (for silage) can be packaged into bales. Small bales are generally used for hay and straw; round bales for straw, hay and grass and 'square' bales for straw and grass. In all cases, mechanical handling systems have been developed for field clearing and stacking.

The conventional (small bale) pickup baler

One machine produces bales having a 360 × 480 mm cross section and variable length up to 1.2 m. Bale density is adjustable and bale weights range from 15–40 kg.

The spring-tined pickup, its height controlled by skids or small wheels at either side, can be offset or centrally located, and lifts the crop into the machine. Machines with an offset pickup have a crossfeed, consisting of an auger or packing tines that move wads of material from behind the pickup and feed them laterally into the bale chamber each time the ram draws back. The heavy ram slices off each wad from any trailing material, and forces it down the bale chamber towards the rear of the machine. As the compressed material moves along, driven by the ram working at 60–95 strokes/min, it carries with it two loops of string which will eventually form the securing bands of the bale.

When sufficient material has passed the bale length sensor, the tying cycle is initiated. This is timed to occur when the ram is forward. The needles bring up the ends of the strings to encircle the bale, place them in the knotters and withdraw, trailing out a new string ready for the next bale. The knotters tie each bale string by forming a loop and pulling the two ends through. The strings are cut at the knotters to release the bale, which continues rearward past the adjustable restrictor that controls bale density, and is finally discharged via the bale chute.

Many balers have potential work rates of 15 t/h or more; typical field performance is 6–8 t/h. Large, even windrows allow the machine to operate steadily at high throughput rates.

Round balers

Round balers produce a cylindrical bale, either 1.5 m wide × up to 1.8 m diameter or 1.2 m wide × up to 1.5 m diameter. Bale weights range from 150–800 kg, with densities from 90–140 kg/m^3. A spring-tine pickup

Figure 24.12 A round baler discharging a 1.2-m diameter × 1.2-m wide bale. (Courtesy of Claas UK Ltd.)

Figure 24.13 This high-density baler produces bales 0.5 × 0.8 × up to 2.4 m long. A silage bale 1.5 m long will weigh up to 0.5 t. (Courtesy of Claas UK Ltd.)

carries material to the cylindrical bale chamber. The machine has no crossfeed, so careful swath building and care in aligning windrow and pickup are essential for the production of good-shaped bales.

Fixed-chamber balers (Fig. 24.12), producing standard 1.2-m diameter × 1.2-m wide bales, accumulate material until enough has entered to cause the material to rotate and commence forming the bale. Variable-chamber balers can produce bales of different diameters, the bale chamber expanding as more material is fed in. When the bale chamber is sufficiently full, the operator stops forward travel and initiates the wrapping of six or more turns of string round the bale as the baler continues to turn. The string is cut, but not knotted, and the bale is discharged through a rear gate. The bale rolls down a ramp and away from the baler to give room for the rear gate to close. The alternative to string is self-clinging plastic net. Although more expensive, net-wrap saves time and provides rain protection for the bale. On some balers, continuous baling, with no stops for wrapping, is achieved by the use of a 'feed chamber' where picked-up material can build up while wrapping and ejection are taking place.

The increasing popularity of baled silage has prompted the introduction of chopping mechanisms at the entrance to the bale chamber. Grass is pushed through a set of fixed knives to reduce chop length, increase bale density and promote good fermentation.

'Square' balers

Bales weighing up to 0.6 t are produced by these large machines which require a tractor of at least 75 kW (Fig. 24.13). Bale dimensions vary depending on the make and model. For example, Table 24.11 gives examples of the three models, each having a different bale cross section, offered by one manufacturer.

The baler has a central tine pickup and feeds material into a precompaction chamber before the main chamber. From this chamber, wads of material are pushed into the main chamber to form a bale which falls apart easily for feeding or bedding once the strings (four or five depending on baler model) have been cut. Chopping mechanisms for use when baling grass for silage can be fitted.

This system is suitable for handling very dry hay, straw or silage crops.

Table 24.11 Characteristics of square balers from one manufacturer

Bale cross section (mm × mm)	*Bale length (m)*	*Material*	*Bale density (kg/m³ max.)*	*Bale weight at length (kg, m)*
800 × 700	1–2.5	Straw	180	240, 2.4
		Hay	220	245, 2.0
		Grass	420	375, 1.6
1200 × 700	1–2.5	Straw	180	365, 2.4
		Hay	220	370, 2.0
		Grass	420	565, 1.6
1200 × 850	1–2.5	Straw	160	390, 2.4
		Hay	220	450, 2.0
		Grass	380	620, 1.6

Bale handling

Table 24.12 gives an indication of the relative numbers of bales produced by different balers. Small bales are of a suitable size for manhandling, and are convenient for feeding and bedding livestock in small pens and in buildings where tractor access is restricted. However, as is shown in Table 24.12, the numbers produced necessitate some form of mechanical handling system.

A common method of handling small bales uses the 'flat eight' system. A sledge is pulled behind the baler and automatically arranges the bales into two rows of four. On completion of the 'flat eight', the bales are released. The 'flat eight' is used to form part of a larger unit load comprising 40 bales – eight layers arranged alternately to provide a degree of interlocking stability. The stack is formed by a grab mounted on a tractor or material handler. The grab lifts the 'flat eight' unit either by squeezing two opposing sides or by impaling the bales with hydraulically operated spikes from above. A specialized trailer lifts and transports the stack to the storage area, where a grab can stack the bales to heights greater than eight bales high if required.

Many other systems of handling small bales are used and have varying degrees of labour input. The systems range from manual stacking of bales into groups of 20 which are subsequently removed from the field by tractor rear-mounted transporters, to self-propelled trailers which pick up the 144 bales, arrange them into a stack, transport the stack to the storage area and deposit the 12-bale-high stack in its final storage position.

The simplest method of handling round bales is with a spike. Tractor-mounted spike attachments capable of handling one, two and four bales are available, as are specialist trailers which can lift and transport up to 16 straw bales. 'Square' bales are usually handled with grapple or squeeze-loader attachments.

Barn hay drying

Field drying or 'curing' of hay usually takes in excess of 5 days and requires an unbroken sequence of dry weather. To reduce the moisture content to below 20% requires ambient air of low relative humidity (RH), as shown in Table 24.13.

Once the crop has reached 30% moisture content, further drying can be a long-drawn-out process depending on the weather conditions. This final stage can be speeded up, and exposure time in the field reduced by 1 or 2 days, if the hay is removed from the field at moisture contents of 50–35%, and dried with forced air in a barn or stack. The result will be a better quality product with reduced field losses and weather dependence, although it is produced at a higher cost than field-cured hay.

Small rectangular bales are generally dried in the storage area, which they will occupy until used. Batch systems involve a second move of the crop which increases cost. The air is blown into the stack by a fan, via a false floor or large duct under the stack, or a vertical duct up the centre of the square stack. Mesh floor systems require an airflow of $0.25\,m^3/s/m^2$ of floor area, while the vertical duct system demands $0.005\,m^3/s$/bale, at a pressure of 60–85 mm water gauge. Where the air temperature can be raised by 5°C above ambient, drying can be continuous. Without heat, drying is restricted to these periods when the air relative humidity is below the crop's equilibrium value. Constant monitoring of the relative humidity is required and the drying process will be extended over a longer period.

Drying rate will be influenced by the ease of air movement through the bale. Small bales should be formed so that a hand can be pushed between wads of hay; fixed-chamber balers are more suitable than variable chamber; 'square' balers are unsuitable due to the high hay density. Unless the hay is drier than 20% moisture content or treated with a preservative to prevent moulding, round and 'square' bales must be dried with caution.

Baled silage

Increasing quantities of grass are being conserved as baled silage. Both round and 'square' balers are used and

Table 24.12 Number of bales/ha produced from a cereal crop yielding 3 t straw/ha

Baler type	*Bale dimensions (m)*	*Bale density (kg/m^3)*	*Bale number*
Small	0.4 × 0.45 × 1.0	100	167
Round	1.2 × 1.2 diameter	100	22
Round	1.2 × 1.5 diameter	100	14
'Square'	1.2 × 0.7 × 2.0	150	12

Table 24.13 The equilibrium relationship between air relative humidity and hay moisture content

Air relative humidity (%)	*Hay equilibrium moisture content (%)*
95	35
90	30
80	21.5
77	20
70	16
60	12.5

crop chopping devices can be added to reduce the chop length to promote good fermentation.

The grass should be wilted to at least 25% dry matter. As soon as possible after baling, bales should be wrapped in a minimum of four layers of polythene to promote anaerobic conditions. Round bale wrapping machines can be stationary units, mounted or trailed behind a tractor, or trailed behind or as an integral part of a baler. Trailed units have a hydraulically operated bale pickup arm; stationary models are loaded by a material handler. The bale is loaded onto a turntable having a base of driven belts. The bale is rotated about two axes while pulling 500- or 750-mm-wide polythene film from a dispenser. As the bale rotates it is wrapped. On completion of the wrapping operation, the film is cut as the turntable is tipped and the bale is lowered onto the ground. Wrappers used for 'square' bales rotate the bale about a horizontal axis while one or two bale dispensers rotate around the bale in a vertical axis.

Once wrapped, bales need to be handled carefully to avoid damaging the film. Cradles, formed from two hydraulically closing forks, and squeeze handlers are two pieces of equipment that are suitable. To avoid film damage and misshapen bales, stacking height should take grass moisture content into account. Bales having a low moisture content are more stable and can be stacked three high. Siting of the stack is important to avoid any risk of polluting a watercourse with escaping effluent.

Forage harvesters

The role of the forage harvester is to pick up the swath, to reduce the chop length of the grass and to deliver the material into a trailer. The chop length is dependent on the quality of silage required, grass moisture content, the conservation method and the feeding system.

Short chopping of grass has the following features:

- It improves the chances of a good fermentation – more sites are available for bacteria.
- It aids consolidation – short material packs better.
- It allows easier extraction of silage from the clamp.
- It increases animal intake.
- It increases the trailer payload.
- It requires more power.

For satisfactory compaction and fermentation in a clamp silo, the following chop lengths are recommended:

Grass dry matter (%)	*Recommended chop length (mm)*
Up to 20	200
20–25	130
25–30	80
Above 25	25

A 60% increase in payload can be achieved if the chop length is reduced from 220 to 25 mm.

Precision-chop forage harvesters

Historically, a range of machines was used for harvesting grass and other materials that were to be ensiled. Precision-chop harvesters have largely superseded flail and double-chop forage machines. The harvester is fitted with a tined pickup for windrowed crops. Alternative headers for direct cutting and for harvesting maize are available. Self-loading forage wagons are still found on the European mainland.

'Precision-chopping' is achieved by controlling the feed of the crop into the cutting mechanism so that the crop moves forward a set distance between each cut of the blades (Fig. 24.14). This distance is known as the 'theoretical chop length' and can normally be adjusted between 5 and 50 mm. The average length of the chopped material will be greater than the set value, because of the way the crop is arranged in the swath. Chop length is adjusted by changing the rate of crop feed, controlled by the speed of one or two pair(s) of feed rollers, or by varying the number of knives on the rotary cutter. The shorter the chop length, the greater is the power consumption, or the lower the output of the machine.

Cutting is achieved by the shearing action of sharp, fast-moving knives against a solid square-edged shear bar. Dull knives or excessive clearance between knife and shear bar waste power and reduce work rate. Machines are fitted with knife sharpeners and these should be used at least daily, followed by readjustment of the shear bar:knife clearance if necessary.

The cutting mechanism is particularly vulnerable to damage from stones or metal objects picked up in the windrow. Damage-resistant designs of cutter have been developed by several makers, and electronic metal detectors can stop the crop intake mechanism instantly if a piece of ferrous metal is detected. The particle of metal has to be picked out before harvesting continues. This device protects the machine from expensive breakdowns and also prevents small metal fragments from being incorporated into the silage and later fed to livestock. A mechanism is used to reverse the direction of the feed rollers to allow the metal to be removed.

Many harvesters have cylinder choppers (Fig. 24.14) which are similar to the mechanism of a traditional lawn mower. Up to 12 knives, arranged round the periphery of a cylinder, slice the crop against the shear bar. Each knife is sectioned so that in the event of damage, only a small part of the blade needs to be replaced. The direction of cut is normally downwards, and the cut material is dragged round under the cutting cylinder. The cylinder, acting as a blower, propels the chopped grass into

Figure 24.14 Flow of material through a self-propelled forage harvester, showing feed rolls, chopping cylinder, corn cracker (for maize) and material accelerator. (Courtesy of Claas UK Ltd.)

the trailer. Upwards-cutting systems throw the cut crop directly upwards, eliminating drag and releasing more power for cutting and throwing. The cutting device on flywheel choppers comprises a number of knives radiating from a hub. The knives rotate in a vertical plane and grass is cut between a knife and the shear bar. The higher inertia of the flywheel chopper gives it a better ability to handle 'lumpy' swaths.

A trailer can be towed behind the forager but usually it is towed alongside by another tractor. The latter system is both faster (quicker trailer turnaround) and safer (improved traction; safer turning) but requires a greater tractor and labour input. Outputs range from 20–50 t/h, dependent on available power, crop density, chop length and field organization. Self-propelled harvesters (Fig. 24.15) are fitted with engines up to 375 kW. Outputs of over 100 t/h are possible, but this depends very much on the availability of large, even windrows and empty trailers.

Figure 24.15 A self-propelled forage harvester can achieve spot work rates of over 100 t/h. (Courtesy of Claas UK Ltd.)

Silo filling

Equipment for filling the silage pit or tower silo must keep pace with harvesting equipment. The standard filling equipment at the clamp is the push-off buckrake, a unit comprising a number of tines which lift a volume of grass and spread it over the clamp by means of the hydraulically operated backplate. The aim of the operation is to spread the grass uniformly in the clamp to promote even consolidation and the removal of air when rolling. A tractor with rear push-off buckrake can handle up to 30 t/h. Increased outputs are possible with a front-mounted buckrake, and more still from four-wheel-drive handling vehicles with fork capacities of 1–2 t.

The storage of grass in towers allows every stage of silage handling, from field to feeding, to be mechanized. Tipping trailers deliver precision-chopped forage into a stationary dump box, which feeds it at up to 30 t/h into a forage blower driven by a tractor of 40–80 kW. The forage is blown up a vertical pipe on the outside of the tower and against a spreading device in the centre of the roof within the silo. Spreading is essential if the silo is to be filled evenly and its potential storage capacity is to be fully utilized. A minimum filling rate of 2 m depth per day ensures satisfactory fermentation.

Grain harvesting

The combine harvester

Combine harvesters (Fig. 24.16) are used universally for the harvest of grain crops (and also pulses, oil seeds and other seed crops). The machine carries out four major functions: it cuts and gathers the crop, separates (threshes) the seed from the ear, removes loose seed from straw and cleans the seed of straw, chaff, weed seeds, other rubbish and dust. After temporary storage in a hopper on the machine, the grain is delivered into a bulk trailer or lorry. Most machines are self-propelled, with cutting widths ranging from 2.1–9.0 m. Engine sizes up to 240 kW are available.

The rate of work achieved by a combine depends on the physical aspects of the machine (cutterbar width, drum width and diameter, type and size of the straw separation area, tank capacity, engine size), the type, variety, maturity and condition of the crop and the harvesting conditions. Under good working conditions, the factor that normally limits a combine's rate of work is its ability to shake out the grain from the remainder of the 'material other than grain' (MOG) in the separating mechanism and manufacturers have addressed this problem by replacing gravity separation systems with centrifugal units.

Figure 24.16 Side exposure of a combine harvester, showing the threshing drum, straw walkers, fan and cleaning sieves. (Courtesy of John Deere Ltd.)

A typical grain harvesting season contains 12–20 dry working days (up to 200 working hours), according to geographical location. While it is common to select a combine size that will provide approximately 1 m of cutting width for every 35 ha of grain (up to 60 ha in a few cases), an alternative approach is to specify a harvesting capacity that can deal with an expected crop within the working time available in the region. As explained earlier, no firm output figures can be given, but the following approximations are satisfactory for planning purposes; machines of 60–80 kW can harvest 6–10 t/h; 80–100 kW, 8–12 t/h; more than 100 kW, 11–16 t/h. The largest 'rotary' combines can attain rates in excess of 24 t/h in good conditions.

Inadequate combine capacity leads to an extended harvest period with greater weather risk and rapidly rising levels of crop loss from shedding, both in the field and at the combine intake. A doubling of pre-harvest losses can occur after a delay of only 2 weeks in some crops.

Cutting and gathering the crop

The crop is gathered into the machine at the 'table' or 'header'. Most tables have a cutterbar incorporated, but conveyor pickups are used in swathed crops (e.g. oilseed rape).

The crop is cut by a reciprocating cutterbar, which comprises a number of triangular-shaped knives. Two sides of the triangle have serrated blades, which grip and cut the hard stems of the crop. Control of the height of cut, which is adjusted hydraulically, is the most demanding task facing the combine operator. Many larger machines have automatic table-height controllers, and other machines have flotation devices, which allow the table to follow ground contours.

To help with cutting and feeding the crop, a reel is mounted above the cutterbar. The reel rotates and has a number (usually six) of bars on which tines are mounted. It can be moved forward and backward, up and down, its rotational speed can be adjusted and the tine angle can be changed. The ability to change settings is important given the range of conditions in which the combine will have to work – short and tall crops, leaning crops and laid or lodged crops. Inappropriate settings can lead to crop losses. Crop dividers, mounted at each end of the cutterbar, separate the crop to be cut. A vertical knife can be fitted at one end of the cutterbar to cut through tangled crops such as oilseed rape. Crop lifters can be fitted along the cutterbar to slide the crop gently up on to the table; these are especially useful in lodged crops.

The cut crop falls onto the table and is drawn towards its centre by a double flight auger. Loss of grain from the table itself is minimized by the length and profile of the deck, which is shaped to retain loose grain should any be threshed out of the ears by the reel or the auger. At the centre of the auger, the crop is propelled into a chain and flight conveyor by retractable tines. The conveyor width matches the width of subsequent mechanisms in the machine. The cutterbar, reel, table and auger form the combine header. Most headers can be detached quickly for transport or storage and are trailed behind the combine when moving from field to field.

Alternative gathering equipment is available for such crops as maize, windrowed crops, sunflowers or edible beans.

A 'rotary stripper', comprising a number of bars with keyhole-shaped holes, which removes the ears from the standing straw, has been shown to be more effective than a conventional header in gathering-up lodged crops. In addition, the amount of crop material entering the combine is halved, allowing faster working rates. A ban on straw burning and the move to crop establishment by minimal cultivation techniques have not allowed this system to be fully developed.

Threshing

On leaving the conveyor, the crop passes over a stone trap into which collects, by gravity, material (stones, clods, soil, etc.) that could damage the drum and concave threshing mechanism. The drum is cylinder-shaped and fitted with 10–12 rasp bars. The concave comprises an open grate and matches the drum's curvature for almost one-third of its circumference. The crop is introduced between the drum and concave and the impact of the rough-faced rasp bars on the crop at up to 30 m/s, and of the moving crop on the stationary concave, rubs the seeds loose from the head. The majority (90%) of the loose grain and other small rubbish fall through the concave on to the preparation pan, while the longer material passes between the drum and concave towards the rear of the machine. The severity of threshing is adjusted to match the type and condition of the crop by altering cylinder speed or varying the gap between concave and cylinder. Too high a drum speed and/or too narrow a gap will scuff and split grains and cause excessive breakup of straw and chaff. A slow drum speed and/or wide drum/concave setting can leave grains attached to the ears which can be discharged with the straw and lost.

Removing loose grain from straw

Removing the remaining threshed grain from the straw is the function of the straw walkers. Four, five or six walkers, depending on the combine make and model, are positioned behind the threshing mechanism and their purpose is to agitate the straw mass to shake out the grain. The agitation process throws the straw upwards and rearwards until finally it is discharged on to the ground, either in the form of a swath to be baled, or passed through a chopping device and spread across the width of the header. The grain falls through the open bottom of the straw walkers and moves to the preparation pan.

A thick mat of straw on the straw walkers reduces separation efficiency. The situation is exacerbated if the straw is damp or has been excessively broken during threshing. Beyond a certain straw throughput, different for every machine, satisfactory separation cannot be completed within the length of the straw walkers and grain is carried over the back of the machine. As straw throughput increases, grain losses increase. Electronic loss monitors can be fitted at the rear of the straw walkers and measure the rate of grain loss; with their aid (if regularly calibrated), the forward speed (and hence throughput) of the combine can be adjusted to maintain an acceptable level of loss at this point.

To overcome the limitations of straw walkers at high straw throughputs, manufacturers have developed systems that use centrifugal force to separate the grain from the straw. The aim is to create a thin layer of straw which is rotated against a grating. The separation units can be either transverse or longitudinal.

Axial flow combines unite the threshing action and grain removal from straw operations in one unit. A rotor, fitted with rasp bars and guide vanes, is surrounded by a concave along its length. The rotor is positioned behind the feed conveyor and is longitudinal to the direction of travel. The crop is fed between the rotor and concave and threshed grain and small material falls through the concave to be moved by augers to the cleaning system.

Cleaning

Impurities are removed from the grain by a combination of size and density separation. The preparation pan has a reciprocating motion and moves grain and unwanted material to the separation area. The preparation pan carries out an initial separation of material, resulting in low-density material occupying the top layer while high-density material makes up the lower layers. At the end of the preparation pan, material falls onto the topmost sieve of a bank of two. The combination of the reciprocating action of the sieves and an air blast from a cleaning fan moves both low-density and large material to the rear of the sieves and out of the combine, while smaller dense material (including grain) passes through the sieves. Material passing through both sieves is conveyed to the grain tank. Material that has not passed through the second sieve may contain unthreshed pieces of head and is either returned to the main drum and concave or passed through a rethreshing unit. Adjustable sieves are convenient, but interchangeable fixed-size sieves are more precise in their dimensions, and also permit greater throughputs of smaller-sized seeds.

Losses over the sieves can be detected by a grain-loss monitor in the same way as straw walker losses.

The combine works best on level ground, and any inclination affects the operation of the straw walkers and, in particular, the sieves. Combining up and down the slope can result in excess grain losses on the uphill run as material is not retained on the sieves and blockages on the downward run and a subsequent loss as the combine turns at the end of the run. Working across a slope can result in material moving to the downward side of the sieves and allowing cleaning air to escape at the topmost side. Insufficient cleaning with good grain being lost over the back of the combine is the result. Sieves are fitted, as standard, with intermediate vertical sections along their length to prevent mass sideways movement of material. Manufacturers offer self-levelling kits for the cleaning sieves so that the sieves remain horizontal irrespective of the ground slope.

Combines are available that automatically level the body of the machine when working on slopes. Some machines only level in a fore and aft direction to allow combining up and down a slope. Others can level the combine body relative to the header to allow working across a slope.

Monitoring combine operation and performance

Combines are equipped with sealed air-conditioned, sound-proof cabs. Even without this barrier it is very difficult for the operator to keep in touch with all the processes and components of a complex machine. In addition to loss monitors, most of the larger machines have as standard, or can be fitted with, a range of 'function monitors' which detect blockages or malfunctions and alert the operator before a major breakdown can occur.

Yield mapping and precision farming

Combines can be fitted with instrumentation that can calculate continuously the weight of grain entering the grain tank. This information can be combined with a ground positioning system to allow the production of a detailed yield map of the field. The yield map is one part of the precision farming jigsaw. Building up year-on-year information for a field and the use of a ground positioning system for other tasks such as soil sampling (for pH, P, K) allow very precise control of subsequent

Table 24.14 Grain losses at harvest

Type of loss	*Reason for loss*	*Identifying the loss*
Pre-harvest loss	Wind, rain, birds, lodging, combine capacity too small	Loose grains in uncut crop
Cutterbar loss	Reel speed too high, reel incorrectly positioned, knife too high, inaccurate driving	Loose grains across full width of stubble
Threshing loss	Drum too slow, drum:concave clearance too large	Unthreshed heads in straw swath
	Drum too fast, drum:concave clearance too narrow	Damaged grains in tank
Straw walker loss	Forward speed too high, large straw throughput, moist straw	Loose grains under straw swath
Sieve loss	Air blast too high, air blast too low, sieve openings too small, drum too fast, drum:concave clearance too narrow	Loose grains under straw swath

husbandry inputs to optimize production with a minimum use of resources. Variable rate application of lime and fertilizer is possible with machines being controlled by computer from information generated at harvest time.

Grain losses

Grain losses at harvest time are caused by a number of factors and occur at different locations within the combine. Table 24.14 shows types of losses that occur, why they occur and how they can be identified. Acceptable loss levels vary with geography and weather conditions. The magnitude of acceptable loss when harvesting winter barley in July may appear ridiculously low when trying to harvest winter wheat in early November.

Table 24.15 shows an example of loss levels. In some cases, a greater straw walker loss may be acceptable, as this will allow harvesting to progress at a faster rate. Grain losses are measured in the field by collecting and quantifying the grain from a known area. If the combine is stopped when harvesting an 'average' part of the field, losses may be assessed in the standing crop in front of the combine to give pre-harvest losses. After reversing the combine a short distance, lost grain from beneath the centre of the machine in its initial position is the sum of preharvest and gathering losses. Behind the machine lies the straw, with unthreshed grains still attached, and the sum of all types of loss. By subtraction, the extent of each individual type of loss may be determined, and the appropriate corrections can be made. Note that the straw walker, sieve and threshing losses, although found in or under the swath, have actually emanated from the full header width.

A convenient unit area is 1 m^2, which should extend across the full width of cut and can be marked out with a string and four pegs. For a 4-m cutting width, the length under investigation would be 250 mm. All lost grain within this area is then gathered up. Table 24.16 shows the relationship between the number of grains found and the total loss of grain per hectare.

Table 24.15 Acceptable grain losses from a combine harvester in good harvesting conditions

	Acceptable level of grain loss (%)	*Loss (kg/ha) in a 7-t/ha crop*
Gathering loss	0.2	14
Threshing loss	0.1	7
Straw walker loss	1.0	70
Sieve loss	0.2	14

Table 24.16 Assessing grain losses

Crop	*Number of seeds/m^2 equivalent to 1 kg/ha*
Barley	2
Maize	0.2
Oats	2.5
Rape	20
Wheat	2

Root harvesting

Potatoes

Prior to harvesting, the vegetative growth of the potato plant needs to be killed or removed to ease the harvesting operation. The exception to this practice is when lifting early potatoes where maximum financial returns depend on rapid lifting and delivery to retailers. The standard practice of using sulphuric acid as a desiccant has environmental implications, and mechanical methods, such as pulverization and haulm pulling, and thermal desiccation may become prevalent.

Historically, hand-lifting was the universal method of harvesting potatoes. For small-scale operations, potatoes still are hand-lifted into baskets after the potatoes have been exposed on the soil surface. A tractor-mounted elevator digger undercuts the potato ridge with a flat share and the ridge contents are conveyed up and back by an elevator consisting of round steel crossbars or 'webs'. Loose soil falls between the webs and the tubers are carried over the back to fall on top of the sifted soil. The difficulty of locating and organizing labour for hand-lifting has resulted in the development of complete harvesters, which can deliver potatoes directly into sacks, boxes or bulk trailers.

Although self-propelled and single-row machines are available, potatoes are commonly harvested with two-row trailed units (Fig. 24.17). Their main components are:

- a share to undercut the ridge and lift it on to the main elevator;
- vertical side discs to cut through haulm;
- a main elevator (main web) which removes loose soil and small stones and clods;
- a haulm-removing device;
- a stone- and clod-removing device;
- an elevator to deliver potatoes into boxes or trailers.

The share and discs separate the ridge from the surrounding soil. Depth control is achieved by a diabolo roller running on top of the ridge that maintains a fixed vertical distance relative to the share. The main web's action of sifting out small material can be enhanced by agitation using rocker arms under the web. In addition to the main web, most harvesters have a second (and sometimes a third) web. At the end of each web, there is a haulm roller, located below the web pulley and rotating in the opposite direction. Guide fingers direct haulm towards this roller where it is caught between the two contra-rotating units and ejected to the ground. A range of mechanical devices are used to remove stones and clods from potatoes, and include star-wheels rollers and axial-flow rollers. In situations where poor harvesting

Figure 24.17 Potato harvester working in 1.8-m-wide beds. (Courtesy of Reekie Netagco Ltd.)

conditions can be expected, harvesters with inspection conveyors that allow up to four pickers to remove stones, clods, damaged tubers and other rubbish are used. The delivery elevator's position and drive are hydraulically controlled to allow the swan-neck section of the unit to be positioned near the base of boxes or trailers to minimize damage. During changeover of the transport vehicle, the discharge conveyor can be stopped and will act as a temporary hopper to allow harvesting to proceed.

The potato market, both for seed and ware, requires a disease- and damage-free sample. Damage sites are a source of infection and scuffed, spilt, cracked and bruised tubers will be difficult to place in the premium markets of pre-packs, crisps and french fries. Potatoes are easily damaged, with most damage occurring at harvest. Grading lines can be another major source of damage. Tuber damage can be reduced by harvesting under good conditions – not always possible – and by preparing a ridge that does not contain clods and stones. Stone windrowing dramatically reduces potato damage during harvesting and increases workrates. During harvesting, tuber damage can occur at various points on the harvester and for the following reasons:

Damage	*Reason*
Slicing	Share set too shallow; discs set too narrow
Scuffing	Immature crop, insufficient interval between haulm treatment and lifting; excessive movement between tubers and other material; excessive agitation on main web; insufficient soil on the main web (aim for a soil cushion for two-thirds of the web's length); dry conditions and high-speed rubber rollers on cleaning units
Cracking	Immature crop, haulm firmly attached to tuber and pulled off at haulm rollers; excessive drop height into box or trailer; harvesting in cold conditions
Bruising	Excessive drop height into box or trailer (aim for less than 200 mm); excessive agitation on main web; insufficient soil on main web

Sugar beet

Sugar beet is harvested by machine. The harvesting process involves: removal of the tops, which may be conserved for livestock feeding but are more usually discharged back on to the ground; lifting of the roots from the soil – options are (1) a pair of inclined, sharp-edged, spoked wheels which cut into the soil on either side of the beet, (2) a pair of parallel shares which reciprocate transversely and gently raise and remove the beet from the soil and (3) a pair of belts which grasp the unremoved green top of the beet and pull it form the ground. A minority of machines use fixed shares; the beet is then cleaned of loose soil by web conveyors or rotary units, before being discharged into an integral hopper or into a trailer running alongside.

Harvesters range from one-row trailed models to six-row self-propelled units. Rates of work of 0.125–0.2 ha/h/row can be expected.

Crop losses result from incorrect topping, from leaving the lower parts of roots in the ground and from failing to gather whole roots into the machine. One 0.6-kg root lying on the surface in a 20-m length of row represents a loss of 2 t/ha.

Harvested beet are often piled at the roadside for considerable periods before being transported to the factory. Losses are minimized if a concrete area is available for storing and loading the crop. The roots are removed from the pile by front-end loaders; industrial loaders can move up to 90 t/h; rough-terrain fork-lifts 60 t/h; tractor front-end loaders 20–30 t/h. Slatted buckets allow some soil to escape, and this process can be continued by passing the beet through a cleaner-loader on its way to the lorry.

Feed preparation and feeding equipment

Animals are fed products ranging from concentrates to bulky fodder crops. The requirements of any feed preparation and feeding system are that:

- it is reliable, robust and simple;
- it avoids feed wastage, deterioration and contamination;
- it is fast, accurate and either easy to clean or self-cleaning;
- it retains the uniformity of the mix and the physical condition of the material;
- it is cost effective, safe and provides an equal opportunity for all animals.

On-farm feed production usually benefits from cost savings on raw materials compared to finished feed products. However, there is no point in saving on feed costs if animal performance suffers due to other aspects of the system such as incorrect feed formulations, insufficient ingredient analysis and inaccurate ingredient weighing.

Concentrate feed preparation

Grinding

Cereals and other ingredients of livestock feeds are ground into smaller particles for a number of reasons:

- to improve digestion by breaking up the protective outer skin and increasing the surface area of the material;
- to increase intake;
- to aid mixing of ingredients into homogeneous final diets;
- to allow the preparation of stable pellets, cubes and nuts.

Desirable features of a grinding machine include:

- the ability to produce an end-product of uniform particle size, with a minimum of dust;
- the ability to adjust the degree of product comminution to suit different classes of livestock;
- a minimum power requirement in relation to output;
- the ability to operate reliably without supervision.

Hammer mills

Hammer mills are used for processing grain for a range of livestock and are particularly used for pigs and poultry and for feeds that will be cubed or pelleted.

The mill consists of a number of rotating beaters (hammers) mounted on a shaft. A perforated screen surrounds the hammers. Grain flows into the centre of the mill where it is sheared by the hammers rotating at up to 100 m/s. The resulting particles are subjected to this action until they are small enough to pass through the screen perforations. A range of screens having 1.0- to 9.5-mm-diameter holes cater for all requirements. An impeller can be mounted on the power shaft to convey the meal emanating from the mill into a storage hopper or mixer. This pneumatic conveying system allows the mill to be sited remotely from other components. A lower power input option is to move the meal with an auger. The mill's throughput decreases when used with a small screen hole diameter and when processing high moisture content grain; 14–16% mc is considered ideal. Edges of the rotor and the screen become blunt with wear, and must be turned (an intermediate option for hammers) or renewed to restore performance. A typical output would be 45–85 kg/h/kW of power available; electric motor sizes are typically in the range of 2.2–37 kW, those up to 7.5 kW being the most popular on farms.

Hammer mills are vulnerable to damage by dense materials, and most machines are fitted, near the inlet, with a magnet (to retain ferrous metal) and a stone trap. A considerable amount of dust is generated, and cyclones and dust socks are necessary to safeguard operators' health.

Hammer mills can operate unattended; many small machines are run at night, drawing grain from an overhead hopper. When the hopper is empty, a pressure-sensitive microswitch is activated and the mill's electric motor is switched off.

Plate mills

The plate mill consists of a pair of cast-iron discs, each having serrated surfaces. Grain is ground between the discs, as one rotates as it is pressed against the other. Fineness of grinding depends on the pressure setting between the discs and grain of up to 18% moisture content can be accommodated. A more uniform meal is produced, with less dust than the hammer mill. Discharge is by gravity only, so a conveying system must be added unless the mill is positioned directly above a receptacle for its output. The specific output of a plate mill is 60–110 kg/h/kW of power available.

Roller mills (bruisers)

A roller mill crushes grain between a pair of flat or serrated cylindrical rollers. The rollers are spring-loaded against one another, and spring tension can be adjusted. Dry grain is cracked to produce a dusty, coarse meal, while grain of higher moisture content is flattened and crushed into flakes. A moisture content of 18% or more is recommended and the machine lends itself well to a system of high-moisture grain storage. Dry grain can be moistened to improve its rolling characteristics.

The meal (using grain >18% mc) is almost dust-free and hence more palatable to livestock. Breakup of the fibrous portion of the grain can be minimized, making a suitable product for feeding to ruminants.

A specific output of 90–120 kg/h/kW of power available can be expected, and machines are available in sizes up to 7.5 kW with electric drive, as well as larger pto-driven models. Discharge is by gravity, so that further handling or receiving equipment must be installed. Rolled material is fragile; handling and mixing equipment must be selected and operated with this in mind if the product is to be preserved in this desirable form.

Mixing

On-farm mixing is generally a batch process, producing 0.5–2.0 t per batch. Some ingredients are weighed into the mixer, but most ingredients are added volumetrically – a hopper is filled to a level that represents the required weight. Since ingredients have different bulk densities, the container must be calibrated with care if the final diet is to be as required. The bulk densities of various ingredients are shown in Table 24.17 (see also Table 24.18).

Continuous-flow mixing, or blending, is possible and may be a preferable option to larger operators.

Vertical mixers
The vertical mixer is an upright cylinder, narrowing to a funnel-shaped base. A central auger circulates and mixes the ingredients. Capacities of 0.5, 1 and 2 t are available. The mixer is often used in conjunction with a pneumatic hammer mill. The mill blows meal tangentially into the top of the mixer. The mixer acts as a cyclone to separate the conveying air from the meal. A vent pipe in the top of the mixer allows air to escape and directs it to a filtration unit where any dust particles are removed. The meal falls into the body of the mixer. Other ingredients can be added using a small hopper near the mixer's base. The mix is discharged from an outlet or sacking-off spout just above the base of the machine.

Adherence to the recommended mixing period of 12–20 min is essential; too short a period results in incomplete mixing, while an excessive time may result in separation of some ingredients. Automatic timers are available. The aggressive nature of this type of mixer makes it unsuitable for rolled products.

Table 24.17 Bulk densities of various livestock feedstuffs

	Bulk density (kg/m³)
Wheat	1.29
Oats	1.96
Barley	1.43
Beans	1.21
Wheat meal	2.13
Oat meal	2.38
Barley meal	1.96
Bean meal	1.85
Grass meal	2.85–3.90
Dry beet pulp	5.67–3.90
Fish meal	1.83–2.08
Soyabean meal	1.48–1.83
Pelleted ration	1.60–1.69
Crumbed ration	1.83

Table 24.18 Typical annual concentrate consumption (t) by various types of livestock

	Annual consumption (t)
Dairy cow	1.25–2.0
Sow or gilt	1.0–1.8
Baconers (from 8 weeks)	0.7–1.0
Porkers (from 8 weeks)	0.5–0.7
1000 laying hens	35–45
1000 broilers	15–20

Chain and slat mixers
The chain and slat mixer comprises a rectangular box with a sloping bottom. The ingredients are mixed by a chain and slat conveyor moving up the sloping bottom of the mixer. Material tumbles and cascades down the slope and is mixed. A cross-auger ensures that lateral mixing is achieved. Bulky material is added through a hole in the top of the mixer – a roller mill is often mounted at this point – and other ingredients are added through a hopper at the low end of the slope. The final mix is discharge through spouts at the opposite end.

Mixing time is short (2–3 min) and the gentle action is very suitable for rations containing rolled ingredients.

Blenders
A blender prepares the diet as a continuous process. All ingredients are metered out simultaneously in a stream, which is mixed as it is carried away by an auger conveyor. The metering devices are generally short auger conveyors, the speeds of which are adjusted to give the desired flow rate of each ingredient. Since the metering is in fact volumetric, each auger must be carefully and regularly calibrated to ensure that the proportions, by weight, of each ingredient in the mix remain correct.

Mobile mill and mix units
An alternative to fixed equipment that still permits the farmer to include his/her own grain in his/her livestock rations is to use a contractor-operated mobile mill and mix unit. The unit, mounted on a lorry chassis, has weighing, hammer milling, roller milling and mixing equipment. Ingredients, such as straw and molasses, can be processed and incorporated in the final mix. The unit is required to visit the farm at regular intervals over the feeding period to ensure the feed mix is fresh.

Cubers

The conversion of meal to dense cubes ('cubes', 'crumbs', 'pellets', 'nuts', 'pencils' or 'cobs') has a number of advantages:

- The mix cannot segregate during transport.
- Selective feeding (of individual ingredients) is avoided.
- The cubed material 'flows' better.
- Less storage space is required.
- Animals' intake of feed is increased.

- The product is dust-free.
- The product may be more palatable.

The cubes are produced by forcing the meal, from a central chamber, into radial holes in a steel ring ('die'). Steam, water or molasses are used as binding agents. High temperatures are generated during the cubing process and cubes must be cooled before being stored.

Farm-made cubes require gentle handling to avoid breakages. Cubing is a slow, power-consuming process and requires careful consideration before being added to a farm-scale feed processing plant.

Feeding equipment

Non-ruminants (pigs and poultry) are fed concentrates. Most ruminants will be fed concentrates and fodder crops, either separately or as a mix.

Concentrate feeding equipment

Pigs and poultry
Concentrate feed is usually dispensed from hoppers, filled either automatically from a feed storage hopper by a pneumatic or mechanical feeding system or manually. Automatic feeding is common.

Sheep and cattle
Sheep and cattle are fed from manually filled outdoor hoppers. Each feeding point can be part of a stall to prevent bullying. 'Creep' feeders prevent older stock from gaining access to calf feed. Alternatively, troughs can be used, either in- or out-doors, filled from bags or sacks by hand or from a tractor-mounted hopper and unloading auger or from a self-loading tractor-mounted bucket with an integral unloading auger (can also be used for root crops such as turnips, swedes and potatoes).

Dairy cattle
In-parlour feeding occupies the cows during milking and also allows individual rationing of each cow. Automatic dispensing of the correct quantity for each cow is possible once the operator has manually inputed the cow's identifier or the cow has been automatically identified by a coded transponder worn around its neck.

Out-of-parlour feeders overcome the problem of slow concentrate intake in the parlour causing a bottleneck to the milking routine. The feeder(s) is sited in the cattle court or cubicle house and each cow is identified by a coded neck transponder as it approaches the feeder. Only part of the daily ration (0.5–1 kg) is delivered (in case the cow is bullied or is frightened off). The system records the number of visits and controls the quantity of concentrate delivered.

Fodder feeding equipment plus/minus concentrates

Hay, straw, silage and roots are the main fodder crops and are fed using a range of equipment. Loose hay and straw can be fed in hand-filled racks, the racks being at or above stock head height. Round bales of hay, straw and silage are fed in circular feeders (one bale capacity) within the cattle court and in bunkers, usually sited in a doorway of the court. For outdoor feeding, circular roofed feeders and trailers with diagonal feed barriers are used. Round and 'square' bales can be shredded and delivered into a trough using a tractor-mounted or trailed unit. A bale is placed in an angled drum that forces the bale against rotating knives. The shredded material is thrown from the side of the machine by an impeller into the trough.

Clamp silage can be self-fed. The clamp should be sited next to the stock housing so cows can have easy and unlimited access. The system is cheap and simple, but its popularity is limited due to:

- the clamp width restricting stock numbers;
- cow safety and reach restricting silage height;
- the inability to ration silage;
- stock having difficulty in extracting silage and rejecting less palatable layers.

For these and other practical reasons, the majority of clamp silage is extracted and fed mechanically.

Silage extraction is commonly achieved with a tined front loader and crowd grab. The horizontal tines are pushed into the clamp and in conjunction with the curved crowd tines remove a block of silage. Shear grabs take out a regular block of silage using horizontal tines and a hydraulically operated three-sided knife which is forced down into the silage. The clean face left by the shear grab helps to prevent secondary oxidation of the clamp material. Once extracted, the loader can deliver the silage to a bunker feeder or to a feed pass where it can be manually forked into a trough.

The two machines used for mechanically delivering silage to a trough are the forage box and the mixer wagon. The forage box comprises a trailer fitted with a chain and slat floor conveyor. The conveyor moves the silage to the front of the unit where two horizontal beaters shred the silage and deliver it to a belt cross-conveyor or an auger. Belt conveyors can be reversed to allow delivery of silage to either side of the forage box. Some units have a split belt, allowing silage to be fed to troughs on both sides of the machine simultaneously. Forage boxes are designed for use with silage. If other products are to be fed with the silage, such as potatoes or concentrates, limited mixing is possible by loading alternate layers of feed.

Mixer wagons, as their name suggests, are designed to mix intimately and deliver a range of feedstuffs, including concentrates and fodder. The main components are a hopper and a mixing device. Two, three or four augers (Fig. 24.18) or a single paddle are used for mixing the ingredients. The majority of machines have cylindrical hoppers and mixing devices that rotate on a horizontal axis, but vertical rotating mixers are also available. The hopper is mounted on load cells and the operator can set the weighing unit for each ingredient. When the correct weight is added, an alarm sounds to alert the operator to begin adding the next ingredient. At the trough, a door is opened on the side of the hopper and the feed is delivered to the trough via a cross-conveyor. The aim is to produce a uniform mix which can be fed to a group of stock as a base diet. Dairy herds may be split into three groups – low, medium and high yielders – with the ingredients used in the mixer wagon being adjusted to meet each group's dietary need.

Dung (FYM) spreaders can also be converted to feed silage. A popular unit is the side delivery spreader, which, with the addition of a deflector, can deliver silage into a trough. Other machines can both extract and deliver silage to a trough. Mounted or trailed, these machines have a hopper, the rear door of which doubles as a rake and is hydraulically operated. The rake is used to 'comb' the silage face and fill the hopper. During delivery to the trough, a moving floor conveys the silage to a cutting rotor and impeller, where the silage is shred and is thrown into the trough.

Figure 24.18 A twin auger mixer wagon. (Courtesy of Kverneland Ltd.)

Milking equipment

Milking parlours

The milking operation can be either a batch process in which groups of cows enter the milking unit (parlour), are milked and exit simultaneously, or a continuous flow system in which cows enter and exit the unit individually.

The most popular form of batch process uses a 'herringbone' parlour, where a row of standings is constructed on each side of a sunken operator's pit. The cows face slightly outwards, so that their rear ends are in towards the operator. Feeding may be provided for each cow to allow individual rationing. Milking units and milk handling pipelines are centrally installed. In many cases, only one line of cows can be milked at a time, while the other batch is being installed, fed and washed. The provision of a second set of milking units allows milking to begin on all cows as soon as udder washing is complete, regardless of the progress of the previous batch, and can increase the throughput of cows. Herringbone parlours are described by quoting the number of standings and the number of milking units. For example, a 16/8 unit comprises two rows of eight standings and one set of eight milking units. A 20/20 unit has 20 standings and 20 milking units.

To increase cow throughput, herringbone parlours have been modified to increase the number of batches being milked at any time. Triangular and diamond-shaped configurations increase the batches to three and four respectively.

Continuous flow parlours use a rotating milking platform. Various designs exist, the common factor being that the cows are milked as the platform moves through 360° from a cow entrance point to an adjacent exit. Cows can be positioned in tandem, angled (herringbone) or facing radially inwards or outwards. The operator's location can be either inside or outside the platform depending on the parlour design. Higher cow throughputs can be achieved with rotary parlours and they are generally associated with larger herds.

Milking machines

Milk is extracted from the udder by applying a vacuum of 50 kPa to the teats. A higher vacuum will extract the milk faster, but the risk of teat damage is also increased. Although a constant vacuum is applied to the interior of the flexible teat-cup liner, it is only applied to the teat for two-thirds or half of the time. The alternation of vacuum with atmospheric pressure in the space between the liner and the teat-cup, controlled by the pulsator, allows the liner to collapse round the teat and shield it from the vacuum for 0.33 s, at 1-s intervals. This allows blood to circulate more freely, and maintains udder health.

The pulsation ratio, the ratio of the time the teats are exposed to vacuum to the time the liner is collapsed, is typically 2 : 1 as described above, although some equipment operates at 1 : 1. Pulsation rate is normally 60 cycles/min. Milking rate can be increased to some extent by widening the pulsation ratio or by increasing the pulsation rate.

The four teat-cups are connected to the claw, which provides pulsation and vacuum connections to each teat-cup. This whole assembly is the cluster. Some claws have a built-in shut-off valve to prevent air entry when the cluster is removed from the cow or if it falls off during milking. Alternatively, a pinch valve on the long milk tube can perform the first function. A small hole in the claw allows air bubbles to enter the milk, making it easier for the vacuum to raise it up the long milk tube.

Most parlours are provided with graduated glass recorder jars of 23 litres capacity, either at operator's head level or below the cows, where low-level pipelines are used. The milk from each cow accumulates in the corresponding jar, so that yield can be recorded or a sample taken. An individual batch of milk can be rejected, if necessary, without contaminating the entire system.

The recorder jar has in some cases been replaced by a milk meter. When installed in the long milk tube, the milk meter indicates milk yield and collects a representative sample of the milk for butterfat analysis. When yields are not being recorded, the meters are not installed and milk passes directly into the pipeline system. A milk flow indicator may be used to let the operator know when a cow has finished milking, or a transparent section of tube or a transparent milk filter may be included in the long milk tube.

A further refinement is to couple the automatic detection of the end of milk flow with the automatic cluster removal (ACR) from the cow. The cluster is raised by a cord powered by a vacuum-operated piston, while the milking vacuum is shut off to release the udder and prevent air entry. If an individual cow has an erratic milk output, ACR can terminate milking before all her milk has been removed.

Milk enters the pipeline either from a transfer valve at the base of the recorder jar, or directly from the milk meter or the long milk tube. An air bleed into the claw or the transfer valve helps the milk to flow towards the source of vacuum until it reaches the releaser vessel (capacity 25 litres) which is normally mounted on the milk room wall. An electric milk pump is actuated by a high level of milk in the vessel to withdraw milk against the vacuum and discharge it through a filter into the bulk tank.

Pipeline milking systems and all their components are designed to be cleaned and sterilized in place. Circulation cleaning is a routine by which the system is first rinsed through with hot water, then a hot solution of detergent and sodium hypochlorite is circulated through the system for 5–10 min, followed by a final water or hypochlorite rinse. The various liquids are drawn from a trough or tank into the wash line, and enter each cluster via a set of four jetters which fit over the ends of the teat-cups. From here, the liquid follows the route of the milk described above, until it reaches the releaser pump which either returns it to the trough for recirculation or discharges it from the system.

A saving in chemicals, but an increased consumption of electricity for water heating, is achieved by the process of acidified boiling water (ABW) cleaning. Sterilization is mainly by heat, the entire system being raised to 77°C for at least 2 min. For each unit, 14–18 litres/unit of near-boiling water (96°C), acidified by a small quantity of sulphamic or nitric acid, is drawn once through the system at a rate controlled by an inlet orifice, and discharged by the releaser pump. In addition to a mild sterilizing action, the acid helps to keep the glass tubes clear of hard-water salts.

The milking plant is used 365 days of the year; regular maintenance is necessary to ensure that rubber parts are in good condition, pulsators and vacuum regulator are operating correctly and the vacuum pump is lubricated. There should be an annual machine test by the machine manufacturer's agent. Such care will help to ensure that the installation gives reliable, efficient milking, good milk hygiene and minimum machine-induced mastitis.

Robotic milking

Systems are currently available for the robotic milking of cows. The milking machine is available round the clock, and the animals may present themselves for milking at any time (Street, 1993). The unit automatically carries out all the necessary operations: identifying the cow, putting on the teat-cups, milking, recording the yield, removal and cleaning of teat-cups, and udder disinfection. Up to 70 cows can be milked using one robotic milking unit depending on the herd management system in place.

Benefits of the system include:

- Welfare – cows choose when and how often to be milked.
- Yield – more frequent milking generally results in increased milk yield per cow.
- Better working conditions – the robot can perform tasks that are unpleasant and monotonous for workers (Mottram, 1992, 1993).

Milk cooling and storage

Milk is normally cooled and stored prior to collection in a stainless-steel bulk tank of 600–4000 litre capacity. The milk leaves the cow at about 39°C and has to be cooled to around 4°C. Refrigerator capacity will depend on the cooling rate required and the quantity of milk involved. The milk is cooled by contact with the chilled inside surface of the tank; an agitator paddle keeps the milk in circulation to assist in this process. The agitator also has a homogenizing function so that a uniform sample is produced. The tank is insulated so that the chilled milk will not rise in temperature by more than 1.7°C over an 8-hour period when the ambient temperature is 32.2°C.

Most bulk tanks are cooled by an ice bank to avoid using a large-capacity refrigerator unit for direct cooling of the milk. The evaporator coils are submerged in water in the space between the inner and outer walls of the tank, and the refrigerator operates over a long period to build up an ice bank. A control will stop the refrigerator when the ice bank is sufficiently large to cope with the expected milk yield and ambient temperature. During tank filling, the circulating warm milk inside the tank gives up its heat, which is absorbed as latent heat by the melting ice.

Where very large quantities of milk are handled, it may be more economical to cool the milk in a separate plate-type heat exchanger before it is stored in an insulated tank. The conventional bulk tank will cool about 45 litres of milk for each kilowatt hour of energy consumed. The addition of a heat exchanger can double this figure. A heat pump may also be used to recover some of the energy from the milk and use it to heat washing water.

Bulk tanks are always cold, so that heat sterilization is impossible; chemical sterilization must be used. Most tanks are equipped with a programmed cleaning unit which, by means of a spray head, rinses, washes, sterilizes and finally rinses the tank. This cycle is set in motion by the tanker driver after he has collected the milk.

Information technology in the milking parlour

In addition to the identification of individuals, the recording of milk yields and the dispensing of feed, detectors and automatic recorders can gather and display valuable herd management data such as milk conductivity (indicator of mastitis), milk temperature (indicator of ill health or 'heat'), activity of the animal (indicator of 'heat') and animal weight (indicator of condition) (Peiper *et al.*, 1993). Such measurements can be made at every milking, and are immediately available on-screen to the herd manager. Any necessary medication, nutritional change or other action can be implemented at once.

Manure handling and spreading

Waste from animal housing can take three forms: solid [farmyard manure (FYM), dung], liquid (slurry) and dirty water. Quantities of excreta produced by livestock are shown in Table 24.19. The quantities of waste material to be handled depend on the degree of admixture and housing system. By storing and subsequently spreading organic waste at an appropriate time, best use can be made of its crop nutrient value and pollution problems can be avoided.

Table 24.19 Daily production of excreta by livestock. (Source: PEPFAA Code, 1999)

	Daily excreta production (litres)
Dairy cow	35–57
Beef cow	30
Store cattle	14
Finishing cattle	21
Fattening pigs (dry meal)	5
Fattening pigs (wet feed)	8
Sow and litter	15
Fattening lamb	2
1000 laying hens	115

Farmyard manure

Straw or other absorbent material is used as bedding and combines with faeces and urine to produce FYM. Approximately 0.5 t of straw will be used over winter as bedding for dairy and beef cattle. If there is sufficient capacity within the cattle court, the FYM may be stored in place until it can be spread. However, as the FYM depth increases, there can be problems with feeding and fences and double-handling is usually involved. The FYM is removed from the courts to a temporary storage site prior to spreading onto the land, usually stubble, at a convenient time of year. Dairy cows or pigs housed in straw-bedded cubicles/pens can have FYM scraped into a concrete area where it can be stored until spread.

Using a tractor-mounted front-end loader, a rough-terrain forklift with manure bucket or an industrial loader similarly equipped, FYM can be loaded into trailers or spreaders at a rate of 20–60 t/h. Spreading on the land is by rear- or side-discharge spreaders, having carrying capacities of 4–18 t.

Rear-discharge machines resemble trailers. They are equipped with a slatted floor conveyor to move the load rearwards to two vertical or horizontal rotating beaters. These shred the FYM and distribute it in a swath behind and to either side of the machine. Some spreaders can be adapted to handle liquid manure by the addition of a hydraulically operated rear door.

One type of side-discharge spreader is cylindrical in shape, with an opening along the top portion for loading and discharge. A shaft through the centre of the cylinder carries numerous chain flails, and a rigid flail at either end. As the shaft rotates, manure is discharged first from the ends of the load, and then progressively towards the centre of the load. Spreading is to one side of the machine only; a wide range of materials from FYM to semi-liquids can be carried and spread. Side-discharge spreaders are available that have an auger running in the base of a V-shaped hopper. The auger moves the FYM to an impeller that throws the material in an arc to the side of the machine.

Slurry

In livestock areas where straw is not readily available, housing systems involving slurry predominate. The manure consists only of faeces and urine, contaminated with waste food. It is removed from contact with the animals by falling or being trampled through a slatted floor into a chamber below. In the case of cubicle housing for dairy cows, the passageways between rows of cubicles are scraped regularly either by a tractor rear-mounted scraper or an automatic scraping system.

Weather conditions render it impossible to spread manure daily, so a storage tank must be constructed to hold at least 6 months' production. This volume can be reduced if a waste management plan can demonstrate that a shorter storage period will not give rise to pollution problems.

Calculating store capacity

The store's capacity will depend on the type of livestock and the slurry storage period. The final capacity will need to account for additional liquids such as spilt drinking water and washing water. It is important to prevent clean rainwater (from roofs) which could be discharged to a stream or river from entering the store as this increases the storage requirement and cost.

Types of slurry stores

Slurry stores can be constructed as an integral part of the livestock house, sited directly below the stock. Storage capacity is often limited, necessitating the tank to be emptied during the housing period. Animals should be removed from the house to avoid Hydrogen Sulphide (H_2S) poisoning when agitating and homogenizing the tank's contents are carried out. Stores adjacent to the livestock house are generally above-ground structures.

Concrete rectangular structures have slurry pushed in from ground-level and may be entered by a ramp down from ground level.

A common cylindrical store is one mounted on a ground-level concrete base and constructed of glass lined steel panels, which are bolted together. Stores are available in diameters of 4.5–25 m, heights of 1.2–6 m and capacities up to 3000 m^3. When planning a manure handling system, provision must always be made for increasing its capacity, even if no immediate increase in livestock numbers is envisaged. An increase in store height to give greater capacity is only feasible if the original store was designed to accept this increase.

The storage tank may be below the level of the yard, if topography allows, and can be filled directly by the scraping system. Alternatively, a small below-ground sump, covered by a heavy grid, receives the scraped slurry from the buildings and yard. This is transferred into the adjacent main tank at intervals by a pump, fitted with a fibre shredding device at its inlet.

Successful operation of a slurry system demands care and good management at all stages and should include the following points:

- Any fibrous material (hay, silage and straw) which can block pumps and contribute to surface crust formation should be excluded.
- Pumping of the manure can be facilitated by the addition of water to the dung and urine, although this also increases the volume of material to be handled. Dung and urine together form a slurry at about 12–15% solids content which is barely pumpable; adding water to produce a less viscous consistency at 6–7.5% solids makes the resultant slurry very easy to pump. This additional water could be wash water from a dairy parlour.
- To ensure trouble-free emptying of the tank, the contents should be thoroughly agitated at least fortnightly to prevent the formation of surface crusts in the case of cattle manure or sludge deposits in the case of pig manure. Recirculation of the contents of the store via the filling pump is one common option. A jetting nozzle can be attached to the pump's pipework outlet, the jet being used to break up the crust, dislodge sludge deposits and mix the contents. Propeller agitators can be used to homogenize the tank's contents and are especially useful where regular mixing has been neglected or mixing power is inadequate and crusts or deposits have built up. Aeration of the tank's contents by surface-mounted axial pumps is possible to encourage the oxidation of organic products.

Trailed tankers with cylindrical bodies distribute slurry in a wide arc in the field from an outlet at the rear of the unit. A 'gravity' tanker is filled from the sump by the store-filling pump. Slurry is spread by a pto-driven pump and nozzle. 'Vacuum' tankers have a pto-driven compressor which reduces the pressure in the tanker's chamber. A pipe is connected to the inlet of the tanker and placed in the sump. On opening the tanker's inlet valve, slurry is 'sucked' into the chamber. Emptying is achieved by altering the plumbing on the tanker so that the chamber is pressurized and the slurry is ejected from the tanker through a distribution nozzle.

Considerable odour may be produced, particularly if the liquid manure is thrown high into the air. Tankers can be fitted with heavy tine 'injectors' to place the manure below the soil surface to minimize odour and reduce run-off.

Care must be exercised when operating tankers across slopes, particularly when partly full. The load shifts to the downhill side, moving the centre of gravity nearer to the downhill wheel. Overturns can occur on side slopes in situations where a full tanker, or a trailer, could be used safely.

Dirty water

Dirty water contains small proportions of organic waste and requires to be stored and land spread. Examples are wash water from dairies and water collected from concrete yards. Often, the material is added to the slurry tank, but it can be stored and treated as a separate material. Its low pollution potential may allow it to be spread at regular intervals by an irrigation system (see earlier section on Irrigation').

Slurry separation

Slurry is separated into solid (30–60% DM) and liquid (1.5–8% DM) fractions, generally by roller pressure against a perforated screen, but sometimes by gravity or vacuum. The solid fraction may be stored for further processing (e.g. composting) or subsequent spreading. The liquid fraction is retained in a tank (see above), and applied to the land by means of an organic irrigation system when conditions permit.

References

British Crop Protection Council (1986) *Nozzle Selection Handbook*. BCPC.

Curtis, G.C. (1980) Application equipment – present and future. *Soil and Water*, **8**, 17–18.

Dwyer, M.J. & Dawson, J. (1984) Tyres for driving wheels. *Proc. BSRAE Assoc.* Members Day, 4 Oct. 1984.

Hesse, H. & Withington, G. (1993) Digital electronic hitch control for tractors. *Agricultural Engineering*, **48**, 1–17.

Hunter, A.G.M. (1992) A review of research into machine stability on slopes. *Agricultural Engineering*, **47**, 49–53.

Landers, A. (1992) Direct injection systems on crop sprayers. *Agricultural Engineering*, **47**, 9–12.

MAFF (1998) *Code of Good Agricultural Practice for the Protection of Water*. MAFF Publications, London.

MAFF (1984) Daily calculation of irrigation need. Booklet 2396, MAFF Publications, Alnwick.

MAFF (1982) Irrigation. Reference Book 138. HMSO, London.

Mottram, T.T. (1992) The design and management of automatic milking systems. *Agricultural Engineering*, **47**, 87–90, 115–18.

Mottram, T.T. (1993) The design and management of automatic milking systems. *Agricultural Engineering*, **48**, 6–12.

Paice, M. (1993) Patch spraying. *Agricultural Engineering*, **48**, 24–5.

Peiper, U.M., Edan, Y., Devir, S., Barak, M. & Maltz, E. (1993) Automatic weighing of dairy cows. *Journal of Agricultural Engineering Research*, **56**, 12–24.

PEPFAA (1999) *Code of Prevention of Environmental Pollution from Agricultural Activity. Scottish Office*, PEPFAA, Scotland.

Renius, K.T. (1992) Developments in tractor transmissions. *Agricultural Engineering*, **47**, 44–8.

Schofield, C.P. (1984) A review of the handling characteristics of agricultural slurries, *Journal of Agricultural Engineering Research*, **30**, 101–9.

Spillman, J.J. (1984) Spray impaction, retention and adhesion: an introduction to basic characteristics. *Pesticide Science*, **15**, 97–106.

Stansfield, C.B. (1992) Irrigation surveys by ADAS. *Agricultural Engineering*, **47**, 6–8.

Street, M.J. (1993) Robotic milking. *Agricultural Engineering*, **48**, 24–25.

Tinker, D.B. (1993) Integration of tractor engine, transmission and implement depth controls. Part 1, Transmissions. *Journal of Agricultural Engineering Research*, **54**, 1–27.

Further reading

Culpin, C. (1992) *Farm Machinery*, 12th edn. Blackwell Science, Oxford.

DOE (1996) *Code of Practice for Agricultural Use of Sewage Sludge. DOE*, London.

Journal of Agricultural Engineering Research.

Journal of the American Society of Agricultural Engineers.

Landwards, journal of the Institution of Agricultural Engineers.

MAFF (1995) *Farm Waste Management Plans: Making Better Use of Nutrients While Minimising the Risks of Water Pollution*. MAFF Publications, London.

MAFF (1998) *Code of Good Agricultural Practice for the Protection of Air (The Air Code). Revised 1998.* MAFF Publications, London.

MAFF (1998) *Code of Good Agricultural Practice for the Protection of Soil (The Soil Code). Revised 1998.* MAFF Publications, London.

MAFF (1998) *Codes of Good Agricultural Practice for the Protection of Water, Air and Soil: Summary.* MAFF Publications, London.

Witney, B.D. (1988) *Choosing and Using Farm Machines.* Longmans, London.

Useful websites

http://www.alvanblanch.co.uk
http://www.amazone.co.uk
http://www.casecorp.com
http://www.claas.com
http://www.deere.com
http://www.fraser.co.uk
http://www.horsch.co.uk
http://www.jcb.co.uk
http://www.keenan.ie
http://www.kvernelandgroup.com
http://www.lemken.com
http://www.lely.com
http://www.masseyferguson.com
http://www.mchale.net
http://www.newholland.com
http://www.opico.co.uk
http://www.reekie-mfg.co.uk
http://www.ritchie-d.co.uk
http://www.stanhay.com
http://www.teagle.co.uk
http://www.wrain.co.uk

25

Farm buildings

R. Coates

Design criteria (BS 5502, Part 21, refers)

Introduction: quality control

The great variety of purpose and intensity of use, together with alternative construction methods, results in considerable differences in building investment on individual farms. Even when there are similarities of production, it may range from basic weather protection for stock up to elaborate environmental control systems, producing variations from 10 to 90% of a farm's capital. Bearing this potentially heavy, possibly crucial, expenditure in mind it is imperative that construction projects are well conceived, thoroughly thought out, and carefully managed to a satisfactory completion for use.

The extent and scale of change is accelerated by the increasing difficulty in making a reasonable living from a family farm combined with the high value of the farmstead – especially where there is a prospect of conversion. New farms are being created as a result, requiring large-scale and well-designed production units.

This implies that it is more important than ever that the farmer/investor should ensure optimum value for money from any farm building installation. This requires, in turn, that a greater degree of professionalism be used in all phases of the design and construction process. Although, clearly, a hired specialist designer, whether an architect, surveyor or other expert, should be in a position to best advise, they are relatively rarely employed to do so, partly due to their 'added cost' but also because there are comparatively few of them. However, to be of maximum use they must be consulted as early in the design process as is possible – before mistakes are made, rather than after. (Such consultants can be found through the Rural Design and Building Association – see also their journal, *Countryside Building*.)

The design quality factors fall into three main headings:

(1) Function
(2) Construction
(3) Appearance

Function quality

The industry has been traditionally adept at buying a building and then fitting it out to suit the functional requirements! The flexibility of practice has gone with the tight margins, pollution control and quality assurance schemes. A simple example is using the wrong height of building for livestock housing, resulting in natural ventilation failure, which could seriously affect the health of the stock.

Construction quality

Agricultural buildings are one of the few structures exempt from Building Regulation control – "providing they are sited at a distance of not less than one-and-a-half times its own height from any building containing sleeping accommodation; *and*, is provided with a fire exit not more than 30 metres from any point within the building" (Sch. 2 Class III. Building Regulations 1991). This has had unfortunate consequences on maintaining construction quality standards in this highly competitive market, with no other mandatory quality standards and a reluctance to pay for advice.

Appearance quality

Although some form of planning consent is now mandatory in all cases (see Chapter 14 on Agricultural Law) such control is very limited and mainly concerned with siting. Planning authorities may have difficulty in understanding the agricultural justification of the design so appearance enhancement tends to be neglected. The extra cost of good design detail is usually minimal and yet such design has a substantial effect on both the lifespan and user 'pride factor' – influencing the degree of care exercised.

Design control factors

Despite the lack of overall control, agricultural building design is currently under considerable pressure to improve quality from a rapidly awakening product, health and safety, and environmental consciousness, brought about by general market and legislative forces:

- Farm construction has a British Standard (BS 5502) to guide its construction and functional design. Its application is voluntary unless the relevant sections are specified either by the purchaser or as part of environmental or quality assurance control. It may also be a planning approval condition (usually using a Class II classification: 20 year minimum life). Some of the requirements are dated and should be checked with more recent sources. (See also references throughout this chapter.)
- The Construction (Design Management) Regulations 1994 (CDM) apply to all farm building sites where there is any demolition or construction work where the complete works (including fitting out, etc.) will last more than 30 days, or involve five people or more at any one time. However, the designer's responsibility applies on all sites regardless of size. This is now being enforced by site inspections by the HSE. (See *HSE Construction Sheets* nos. 40–44 inc. and *Countryside Building Journal*, vol. 1, issue 2.)
- The Code of Good Agricultural Practice for the Protection of Water – 1998 Revision (the Water Code) – and to a much lesser extent the Air Code – imposes both legal conditions under the Control of Pollution (Silage, Slurry and Agricultural Fuel Oil) Regulations 1991 as amended in 1997 (affecting field silage only) and non-statuary recommendations within the Code. (See section on 'Water pollution control'.)
- Construction standards are also partially controlled by specific material British Standards Institution (BSI) or Agreement certification or kitemark, sometimes backed up by manufacturers' performance guarantees.
- All contracts involving mains water installations are required to adhere to the Water Supply (Water Fittings) Regulations 1999 (Water Byelaws 2000, Scotland) – which includes a system of prior notification and inspection.
- Farm Product Assurance Schemes now cover most of the main produce areas and set wide-ranging standards in which buildings and their fittings usually play a large part. Voluntary codes of practice are specified as minimum standards – whether it is for vermin-proof stores or lying space for cows. Functional design standards are being scrutinized under a system of inspection. It follows that it would be very unwise to embark on any proposed development without ensuring conformation of the whole detail of the concept.
- Proposed buildings in land-classification areas such as AONBs, EASs, etc. are coming under increasingly sensitive scrutiny by the planning authority concerned. The same applies to buildings in Conservation Areas or in close proximity to listed buildings.
- Many rural planning authorities are now producing building design guidance booklets which are usually adopted into their local plans and can be helpful in illustrating acceptable design concepts. [See section on 'Appearance Design (aesthetic quality)'.]
- The last influence is the concept of sustainable development now incorporated in the planning guidance notes (PPG7, etc.), which incorporates all three of the above design quality principles as well as those more difficult to define. These might include: improving the viability of the holding, energy efficiency, both in construction and operation – including sourcing materials locally (e.g. using home-grown timber cladding), enhancing the quality of the environment (e.g. reducing pollution risk), respecting the wild life (e.g. providing barn owl and bat boxes), reusing existing buildings or building materials if possible, and flexibility of use – ensuring a long life of benefit to the holding.

Note: there are many valuable sources of research and advisory information on topics included in this section. The most relevant are summarized later. See also 'Further reading'. These are mainly of a 'general text' nature, but will provide pointers to particular further detailed references.

Function design

Specific function design is dealt with under the appropriate use heading, but despite the great variation implied in the Introduction above, derived from type and scale of farm operations, there are some considerations that are common to the planning processes of all farm buildings. These may be summarized as follows:

- The functional design brief has to come first in which all capacity requirements are assessed in relation to a time period to ascertain the maximum – using bar charts for the year and then testing against possible changes – e.g. calving pattern/index, or increase in grain variety separation. Include inputs (food, bedding, etc.) and outputs (waste, etc.).
- Capacities must then be converted into spatial requirements with the appropriate ratios, e.g. feed space in proportion to bedded space. Calculate *areas* required (of floor, pens, circulation, access, slats, handling equipment, feeders, fixed equipment, etc.). Arrange the *interior layout* for stock control, materials movement, equipment installation and environmental effects.
- *Heights* (of ceilings, doorways, walls, mechanical equipment, services, etc.). Remember to allow for essential clearances, e.g. trailer tipping, pig reach, as well as the environmental considerations.
- Derive the *external shape* from the above taking account of weather resistance (roof shape and wall presentation) and impact on the surrounding rural environment, including any adjacent buildings (odours, contrasts, textures, etc.).
- Most of the minimum requirements can be sourced under the appropriate Part Number under BS 5502, but care and experience are essential in interpretation. The date of the last time each Part has been revised will remind users to check subsequent developments and examples of good practice.
- Appropriate cost, relative to permanence and maintenance and according to finance available.
- Flexibility of design to allow for: expansion, produce and method of production changes, alternative (or alternate) use.
- Efficiency of design to minimize labour and energy requirements.
- Risk assessment for health of crop, stock (disease separation) or manpower, and pollution. Check and include Health and Safety elements relating to function design, which are intended to limit hazards (to operatives, stock and goods) from fire, structure failure, poor atmospheric composition and themselves, e.g. separating the high fire risk element such as straw storage, ensuring stock can be quickly evacuated from a building, etc.
- Utilization of common/standard/local materials (with account taken of DIY element).
- Appropriate use of site topography, e.g. for aspect, foul and surface drainage, shelter and gravity movement of materials.
- Compliance with legislation, codes of practice and all relative assurance schemes. This requires a comprehensive list of all the relevant criteria to achieve full compliance in both basic spatial requirements and attention to design detail, e.g. making all food storage facilities vermin proof. Minimal quality standards are no longer an option as soon as any product is to be marketed under an assurance scheme.
- Potential for modification or extension of existing buildings – allowing for age and technological differences – for all or part of enterprise.
- Effect of proposed enterprise on existing enterprises: all aspects – but as far as this chapter is concerned it is usually the covered spatial requirement that is affected. For instance, an enlarged livestock enterprise will increase storage requirements for fodder, bedding and waste.

Construction design

It will be understood that the lack of mandatory control makes the industry vulnerable to abuse. So it should be everyone's interest that the minimum standards laid down by BS 5502 are universally adopted. This standard is intended to embrace all agricultural structures and impose upon them precise structural and environmental requirements which will result in satisfactory operation for a predicted and assured lifetime whilst providing safe working conditions during that lifetime. All buildings conforming to the BS should carry a plate showing their classification, manufacturer and date. Classification (see Table 25.1) relates to occupancy, predicted safe lifespan, type of activity intended for, and the construction material's safe stressing limitations together with its general suitability for the purpose and use.

Appearance design (aesthetic quality)

There is an increasing awareness of the need for thought to be given to the quality of design in the countryside in order to give buildings character and local distinctiveness. However, in view of the infinite variety of sites and circumstances, the following ground rules can only be a general guide to stimulate quality.

Table 25.1 Summary of provisions for BS 5502 Part 22 classification

Class (and potential example)	*Expected structure life (minimum) – reflecting strength of components (years)*	*Human occupancy (hours/day) at population density (persons/m²)*	*Situation (m minimum from classified road or non-owned dwelling place)*
1 (vegetable processing shed)	50	Unrestricted	Unrestricted
2 (milking parlour)	20	6 at 2	10
3 (pig fattening)	10	2 at 1	20
4 (plastic envelope sheep house)	Temporary	1 at 1	30

Siting (Fig. 25.1)

- Reduce impact by letting into slope – avoid made-up ground especially for loadbearing structures.
- Avoid skyline – unless integrated with existing buildings.
- Position to relate to existing buildings – consider access, security, circulation (keeping clean and dirty separate) and future development.
- If existing range has potential for diversification look at forming second farmstead, especially with unspoilt traditional buildings of character.
- Utilize existing landscape features to provide backcloth: retain as much as possible when preparing site; especially old walls which can link new and old buildings.
- Allocate space for additional landscaping remembering a natural shelter belt is the best wind filter. Consider using excavated spoil sensitively to create natural windbreak or flash flood diversion bank.
- Try to avoid 'all under one roof'. (Will probably improve access and ventilation, and create fire break in doing so.)
- Step floor and roof levels down steep site.
- Consider 'near' and 'far' viewpoints, especially from the farmhouse, approach and nearest highway.
- Orientate to suit use in relation to sun/prevailing wind.
- Consider breaking up building to create second structure at right angles (instead of lean-to) to provide shelter and create courtyard with other buildings.
- Utilize tower silos, elevator penthouses, etc. to create focal point – but ensure non-reflective and harmonize colour.

Roof

- Colour (new buildings) should be dark and non-reflective, and relate to existing roof colour – slate or tile, etc. ('Plastisol' coated steel sheet is not recommended for roofs in dark colours.) Use Farmscape/Agriscape matt range for fibre cement. (Remember fibre cement roofs should be used for livestock buildings because of the condensation problem associated with steel.)
- Relate the facing roof of new buildings to those adjacent by creating the same roof height, length and pitch by offsetting ridge from centre, and then clad in the same material.
- Break up large areas of roof, vary pitch, offset ridge from centre, or turn at right angles.
- Extend verges by say 300–400 mm not only to give strong shadow line but also to reduce cladding staining.
- Extend eaves by say 1.2–1.8 m of open-sided buildings to protect troughs and/or bedding, implements, etc., or 300–400 mm as above if clad.
- Interrupt long lengths with dormer over large doors to give tithe barn effect – but follow local tradition.
- Avoid all monopitch roofs and pitches below 10 degrees (causes wind/snow load problems). Minimum 15 degrees advised.
- Space rooflights evenly (quantity usually 10%), or omit (especially on south face), or use 5% if spaced roof used using only heavy-duty fibreglass or polycarbonate.

Side cladding

- Colour and texture must contrast, and normally be a lighter shade than roof, but follow local custom.
- Vertical home-grown treated sawn board, usually 150 × 25 mm, is both the most economic and harmonious, as spaced board, lapped or with tempered 6-mm hardboard backing. Stain/paint black to match traditional farmstead buildings. Horizontal lapped boarding is traditional and more appropriate for smaller buildings. Featheredge is cheaper and effective for domestic size buildings.

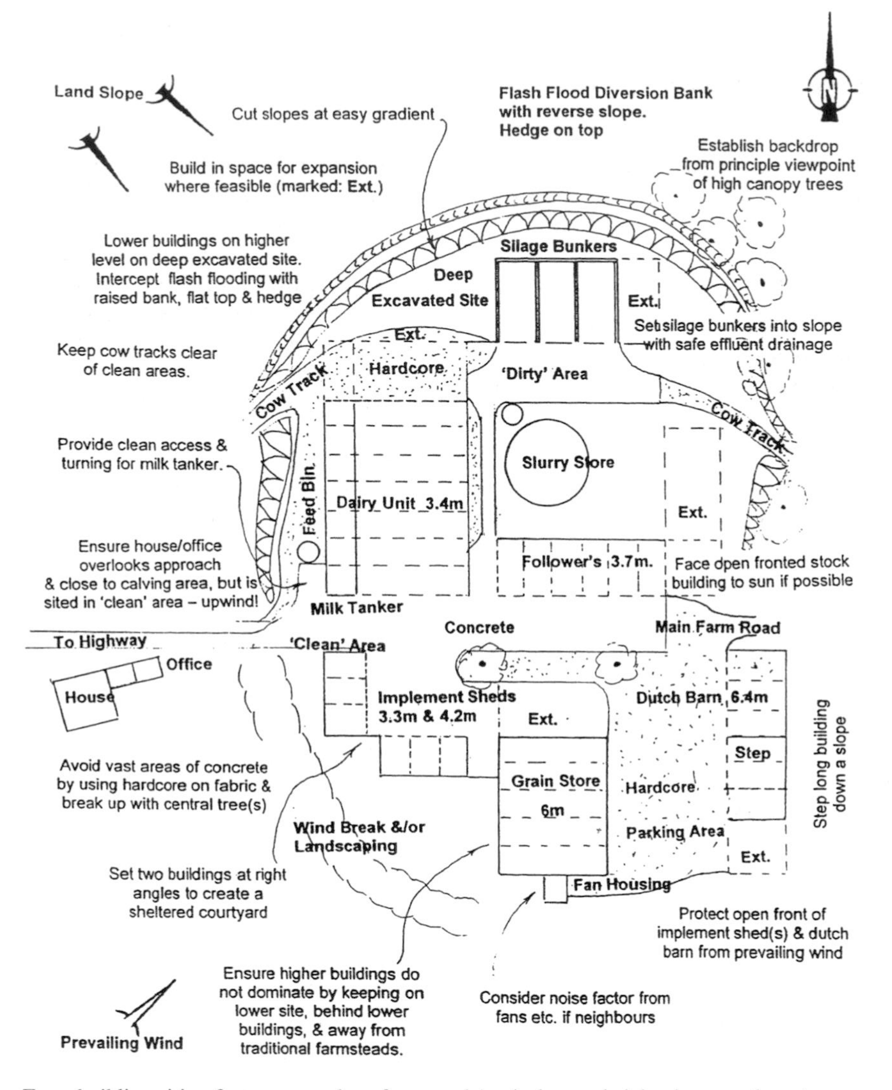

Figure 25.1 Farm building siting factors: a modern farmstead (typical eaves heights in parentheses).

- Copy local practice in provision of a focal point in the gable peak apex, e.g. name plate, hay loft door, e.g. barn owl box access, louvered ventilator, e.g. weather vane.
- It is usually better to clad over stanchions.
- Arrange cladding junction with wall using local practice, e.g. with a cill, but in all cases try to ensure overhang is projecting to reduce wall staining and create shadow.

Walls

- Contrast with cladding in colour, texture and alignment.
- Incorporate old wall if at all possible.
- Pay special attention to elevations facing existing buildings, or in a prominent position, by incorporating local materials and detail of design.
- Feature stanchions to break up long wall.
- Where concrete blocks are to be used specify colour, finish, mortar and pointing. Choose fair faced if possible – better finish and colour.
- If piers used, cap with local slate or tile.
- 'Frame' door openings with contrasting local detail.

Rainwater drainage systems

- Match roof colour at eaves.

- Avoid standard half round gutter with many downpipes.
- Site downpipes with care and protect from impact.
- Consider use of large boundary gutter instead of overhanging eaves to create shadow line, or leaving out gutter if good overhang.

Doors and windows

- Give thought to siting, size and type. Use wood? (Douglas fir is preferable.)
- May be appropriate to make main entrance focal point.
- Conceal/protect sliding door track by recessing behind cladding above. Paint sheet steel once 'matured'.
- Copy style and line of doors/windows in nearby buildings. Copy local detailing in finish, provision of fastenings (including cabin hooks, etc.). Protect from impact.

Fences, walls and hedges

- Link to existing buildings and surrounding land (Visually very important as well as practical).
- Plant 'waste' areas with indigenous shrubs and trees of size to suit location. (No high canopy trees within say 20 m of buildings.)
- Consider forming tree backcloth if none present.
- Follow local custom with gate and fence design.

Quality control: concluding thoughts

These parameters need skill and experience in interpretation in a field that requires good agricultural knowledge and the ability to convert that knowledge into a design brief that will give lasting satisfaction in terms of the functional suitability, constructional life and a pleasing appearance. For such a process to be universally adopted needs both carrot and stick – supported with an underlying increase in education – probably best done through the medium of product assurance scheme training and support. Such assurance schemes are likely to make BS 5502 and its associated codes of practice compulsory in the fairly near future. Building Regulations are kept constantly under review. If it is not to be extended to farm buildings then it is essential that BS 5502 is treated with the same respect by being reviewed and updated on a regular basis.

Planning and construction design

Introduction

Whilst the general points above must be borne in mind at all times in any building development, specific use structures will have particular requirements concerned with their productive functions, e.g. precise environmental control. The route by which they may be included in the overall design is indicated below in the scheme of work represented by steps A–F and the chart in Fig. 25.2.

BS 5502 incorporates animal welfare guidance and lays down environmental standards desirable for healthy livestock in humane production systems. It does this by recommending space allocations, target temperatures, ventilation rates, confinement methods, etc. Another important function of the standard is to ensure that the fabric and equipment, as well as the layout, of farm buildings are adequate for their purpose and use. It thus lays down rules for types, qualities, installation methods and required long-term abilities of material to be used in the construction of all elements of a farm structure.

All parts of this section of the chapter should be read with the guidelines laid down in BS 5502 in mind. Where appropriate, reference is made to specific parts of the Standard as and when they are relevant to the text content.

The building process

A. Initial preparation

(Confer at all appropriate stages with the appropriate advisory organization.)

(1) Establish the business motives for the building investment and the level of acceptable cost relative to the return on the capital expended.
(2) Consider alternative systems to achieve similar results and consult other existing users (if any) who have experience of them.
(3) Estimate the overall size of the completed structures and ancillaries required and hence assess the approximate overall cost by reference to the current *SAC Farm Building Cost Guide* and comparison with costs of similar structures. Include:
    - the substantial planning application fees for a large building: £190 for first 540 m^2 and

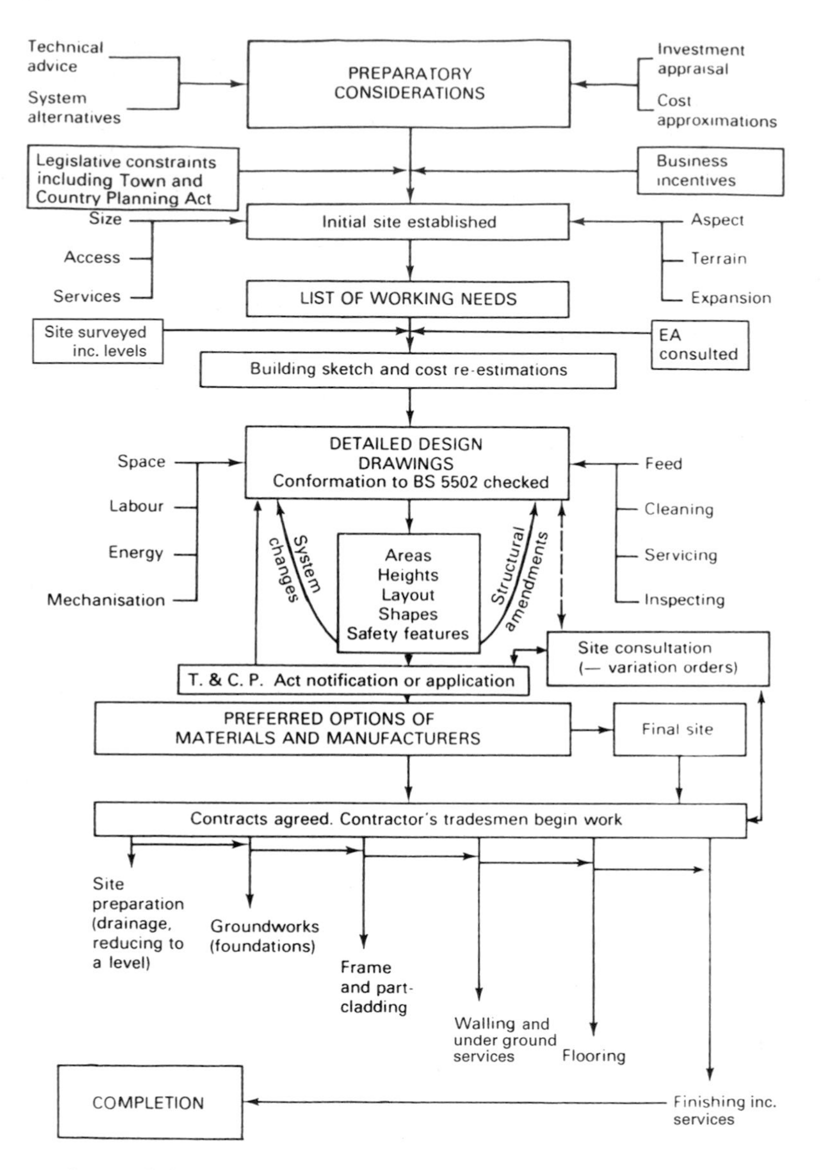

Figure 25.2 Steps to the completion of a farm building.

£190 for every 75 m^2 thereafter, as from April 2002;

- designer and supervisor fees;
- site works, before and after (excavation, roads/aprons, landscaping, etc.);
- services from/to building including reinforcement if necessary;
- other costs and contingencies including relocating displaced enterprises, ancillary structures, etc.
- VAT if it is not funded by registered business.

(4) Obtain information on:
- grant aid potential, tax concessions, etc.;
- design control factors as previously listed.

(5) Examine possible sites for the building(s), considering their interactive effects on the farm business and the rural environment both during and after construction. Include the following aspects in a feasibility study:
- size of building and ancillary works (area and height – note planning permission restrictions);

- access (for vehicles, animals, operatives, building and other materials handling);
- adequate soil strength – by classification and avoidance of made-up ground;
- avoidance of both flood plains and flash flood routes – or make provision to divert;
- avoidance of sites of substantial trees that would have to be felled (to avoid soil heave; it takes several years for such sites to stabilize);
- avoidance of overhead or underground services or planned rerouting. A location plan should be submitted at an early stage to all the service authorities for checking;
- circulation space (and expansion/change potential);
- drainage (within, from and around building, note anti-pollution requirements);
- aspect (climate and its environmental effects, note also visual effects in areas of particular sensitivity);
- availability of services (provision in remote areas may be very costly);
- utilization of slopes (gravitational movement of materials, use of storage space and drainage);
- space provision for landscape planting, cut slope batters (usually maximum 45 degrees), and flash flood diversion banks on top of slopes;
- potential for expansion or change (space of surroundings, layout of services).

(6) Confirm the building concept by sketching and listing the requirements of the building and impact on other buildings and enterprises. Include provision for all necessary activities within and surrounding it.

(7) Re-estimate the cost in the light of decisions taken about siting and building system/methods, which may modify the original concepts or facilities required.

B. Design criteria establishment

(1) Consider the space, labour, mechanization and energy inputs required to operate the building system within and around the building. Note and list requirements for activities such as feeding, cleaning, handling, emptying, filling, heating, cooling, weighing, lighting.

(2) Draw up a list of relevant criteria to be met such as living or storage space requirement, insulation standards, slurry capacity, mechanical handling clearances, ventilation rates, suitability of materials, circulation areas, etc.

C. Dimensional formulation

Dimensional formulation of plans should now be possible. This entails accurate drafting, preferably to a standard scale (1 : 10, 20, 50, 100, 200 as appropriate), of all the external and internal aspects, in detail, which will enable the builder to construct without further instruction. At this point it may be advisable to employ specialist help with knowledge of drawing interpretation. However, it should always be remembered that experts in the building trade are not experts in agriculture, and so thoroughgoing instructions should be given wherever possible.

(1) Drawings (see Fig. 25.3) should be prepared to scales of 1 : 100 or 1 : 200 showing:
   - the layout in plan view indicating position of doors, barriers, feeders, circulation space, drainage, water and electricity supplies in relation to frame members, walls and other structural features at or above ground level. (Some of these might be more clearly shown on separate additional drawings to avoid confusion of lines and shading);
   - the building in all four elevations to demonstrate shape, volume, cladding materials and external features likely to be obtrusive and relationship to ground levels;
   - Cross sections of the structure demonstrating material details and thickness, positions of service fixings and the nature of space usage in the building. Scale usually 1 : 50.

(2) Prepare a block plan to the scale of 1 : 500 showing the existing and proposed excavation site levels and the location in outline of the proposed building – if the site is to be excavated. A site survey is therefore necessary before any scheme is submitted for full planning permission which must be related to a clearly identified datum or temporary bench mark (TBM).

(3) Prepare a block plan to the scale of 1 : 500 showing the proposed building, aprons/roads, services to/from the building, fencing and landscaping. The plan should show proposed finished floor and ground levels with slopes of aprons/roads in relation to surrounding ground levels and TBM.

(4) A scale map of 1 : 1250 or 1 : 2500 should be used to show the location and general arrangement of the building relative to its surroundings. (Permission may be granted for a copy to be made for personal use from an Ordnance Survey map of this

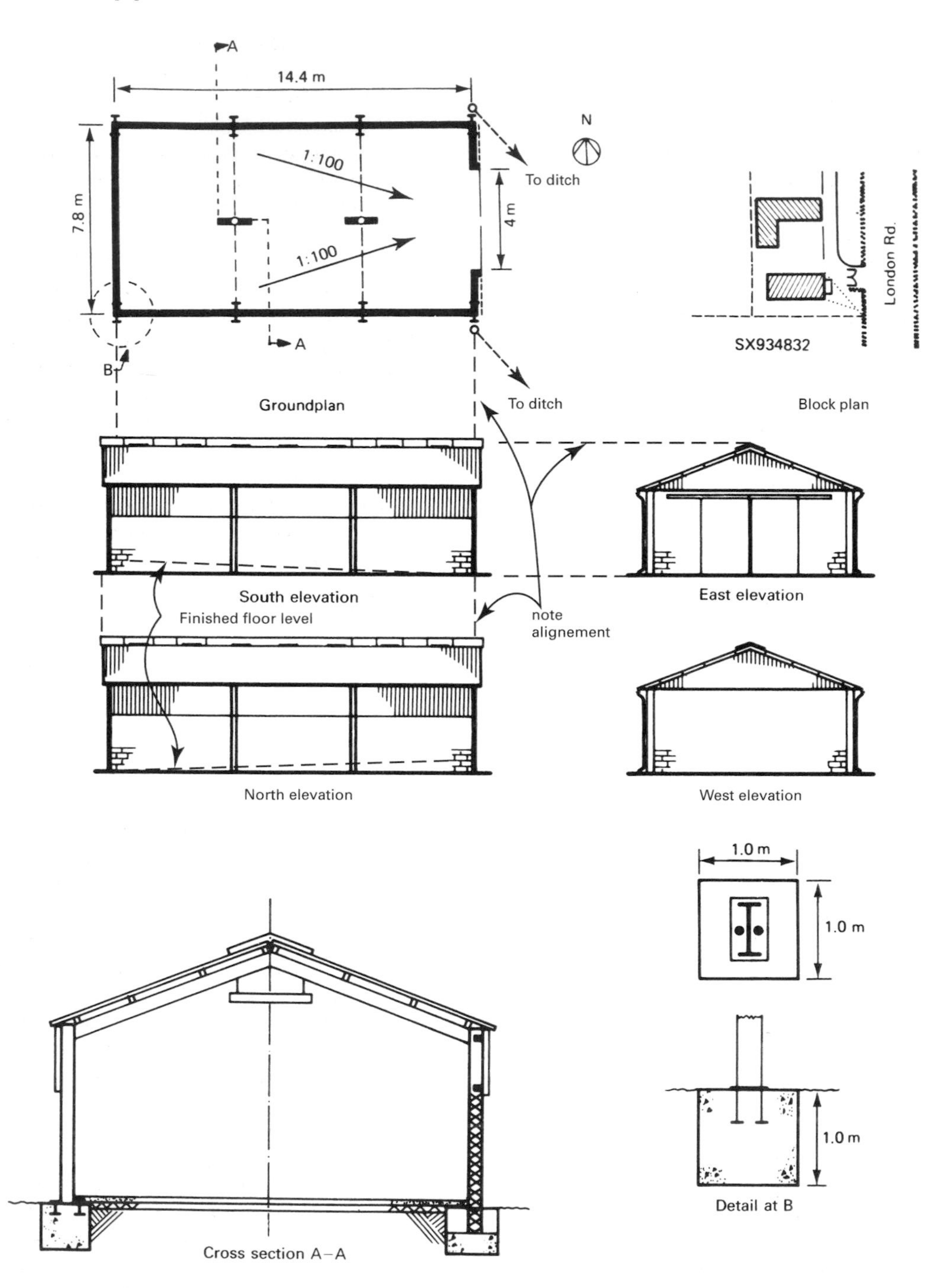

Figure 25.3 Building drawings.

scale.) This is also an essential requirement for any planning application coloured in accordance with their requirements.

(5) Further drawings to scales of 1:20 and 1:50 may be prepared to show constructional features such as foundations, framing and fixings, drainage schemes, damp and vapour proofing, insulation methods and mechanical services. These are generally the province of professional experts, surveyors and architects or may be provided by commercial building firms as part of a sales service.

D. Site survey

This is a three stage process:

(1) The initial site survey has to be carried out before working drawings are prepared, unless it is a simple building on a near-level site, in order to prepare the block plan defined above in (2). Having established the overall dimensions of the proposed building and its approximate position the location can be pegged out on the ground. The longest side of the building can then be used to set out a grid of say 12-m squares, or 24 m if the site is fairly level. Ground levels are then taken and related to a fixed point: the temporary bench mark. Any fixed features and services on the site are plotted in relation to the grid.

(2) The proposed excavated and finished floor levels (ffl) are then superimposed on the existing plan.

(3) From the results of this survey estimates may be made of soil quantities to be removed and the problems, if any, that might arise in the installation or relocation of services on the site. Once excavated the precise position of the building outline can be marked by pegs and string or chalk lines (described in the next section) as a first step in the construction process.

E. Contracting out options

(1) Design and Build – a single fixed price contract under which the contractor is responsible for the whole design and construction process. A fairly expensive option but the safest on cost control and the integration of the stages of the work. The design and attention to detail can be very good using a specialist highly reputable contractor. It is, however, essential to set out a very detailed design brief to ensure there is no misunderstanding. It is normal to negotiate with a single contractor. However, if the contract is very substantial in scale a tender stage can be introduced providing there is professional supervision.

(2) Separating the design stage – all other options require a design to be produced preferably in the form of working drawings which are required anyway for a planning application. Such drawings should be completed by a designer who is competent in interpreting both the constructional and functional design brief. Smaller contracts may contain all the necessary information on the drawing, but it is usual and preferable for the drawings to be accompanied by a specification or at least a complete schedule of works. Large complex contracts requiring a high level of quality and cost control should also be accompanied by a bill of quantities.

(3) Split contract – the most common option often split as follows, but note the need to ensure the clear responsibility of the different parties under the CDM regulations:
- Excavation – direct contract with plant operator on daywork or fixed price tender, or could be included in the infill works contract.
- Framed building – usually a fixed price contract as a result of a competitive tender to supply and erect the frame and cladding. The contractor taking responsibility for the structural design and ensuring conformation with the specified BS 5502 classification.
- Infill works – usually a fixed price contract as above after a competitive tender but could also be carried out at a fixed price as a result of a negotiated contract with a favoured contractor. This has the advantage of a much earlier commitment by the contractor concerned and of the work being carried out by builders who may be familiar with the site. In both cases there will be a known final cost plus a contingency allowance (allow minimum of 5%). If the work is required to start at very short notice or the precise extent of the work is not known and a well-trusted contractor is employed, all or part of the contract can be carried out on daywork on a cost plus basis (cost of materials and labour plus a reasonable addition for expenses and profit). This method also has the advantage of farm staff carrying out certain works, but health and safety implications may prohibit some joint enterprises. The obvious drawback is the lack of cost control.

F. Construction supervision
Construction must be supervised to ensure the quality of construction is maintained and the appropriate health and safety requirements are observed in accordance with CDM regulations (see 'Design control factors' earlier). This will include ensuring the erection of safety nets and edge protection, working off platforms not ladders, etc.

Although the above stages in the design have been taken in logical order, one by one, it is likely that there will be a continual need throughout to readjust or refine ideas and objectives in attempts to match conflicting (even contradictory) requirements. Compromise between farm business aims and building techniques is often unavoidable, as for instance in the situation where the drainage of slurry may create foundation design problems. Even when construction has started, cases arise where the three-dimensional nature of the building work demonstrates that the two-dimensional planning process is inadequate and necessitates changes in the instructions to the builder. As this will increase the cost of the work significantly, a contingency sum should be set aside to cover these eventualities, but they should be avoided wherever possible.

Stages in constructing a building (BS 5502, Parts 21 and 22, refer)

Preparation – setting out a simple rectangular building

(1) Strip the turf and surface soil from the site, along with any vegetation, complete with roots, if possible. The subsoil thus exposed is 'levelled' over an area exceeding that of the building. If the site slopes, then the surface must be 'reduced' to a satisfactory gradient and banks formed as required by use of a surveyor's levels. Maintain a slight slope in initial excavation even if floor to be finished level to disperse surface water.
(2) Divert existing land drainage drains which cross the site of proposed works. Intercept surface water with a 'barrier' drain trench upslope from the site (this latter is sometimes known as a 'French' drain). It is a legal requirement that any such drain is positioned at least 10 m away from sources of pollution.
(3) On the prepared subsoil surface stretch a line between pegs to represent the position of the longest side of the building but much greater in length such that the pegs are well clear of the proposed corners as at (a) in Fig. 25.4.
(4) Mark one of the corner positions with another peg (b, Fig. 25.4) and at right angles from the first line stretch another to mark a second building side (c, Fig. 25.4). An accurate right angle may be made by using a builder's square, a surveyor's site square, or the method of Pythagoras (3 : 4 : 5 triangulation giving 6 m along one side, 8 m along a second and 10 m across the measured hypotenuse).
(5) In a similar manner mark the third and fourth sides (d, Fig. 25.4). The dimensions and position of the building are now shown in plan by the lines and corner intersections. It is usual for these to be the outside edges of the building above ground, but they may represent centrelines or inside surfaces. It is important to plan according to fixed reference surfaces and levels throughout the construction and to avoid changing them unless it is absolutely necessary.
(6) Check the diagonals (e, Fig. 25.4) of the building outline: they should be the same (for a rectangular building).
(7) Profile boards (Figs 25.4 and 25.5) are now sited beyond each corner, well clear of the building area. These are pieces of timber which have permanent markers (nails or sawcuts) indicating the position of one or more aspects of construction, for example the edges of the foundation trench, the position of the wall to be built on the foundation and internal features such as doorways and walls (h, Fig. 25.4).
(8) Lines (g, Fig. 25.4) stretched from marker to marker on the profile boards provide an accurate facsimile of the ground plan full size on the surface to be worked on. Before construction begins, they are replaced by trickled silver sand, chalk dust, paint lines or other easy to follow methods of ground marking. The boards, however, are retained until such time as the reference marks on them have no value.

The heights and levels over the site should be set out and checked against a vertical reference stake driven firmly into the ground, clear of all site activity. Marks on the stake should indicate ground level when the stake was driven in, a general reference height, say 1 m, above ground level known as the temporary bench mark. (Clearly visible reference lines painted on a nearby permanent structure might be preferable in some cases.)

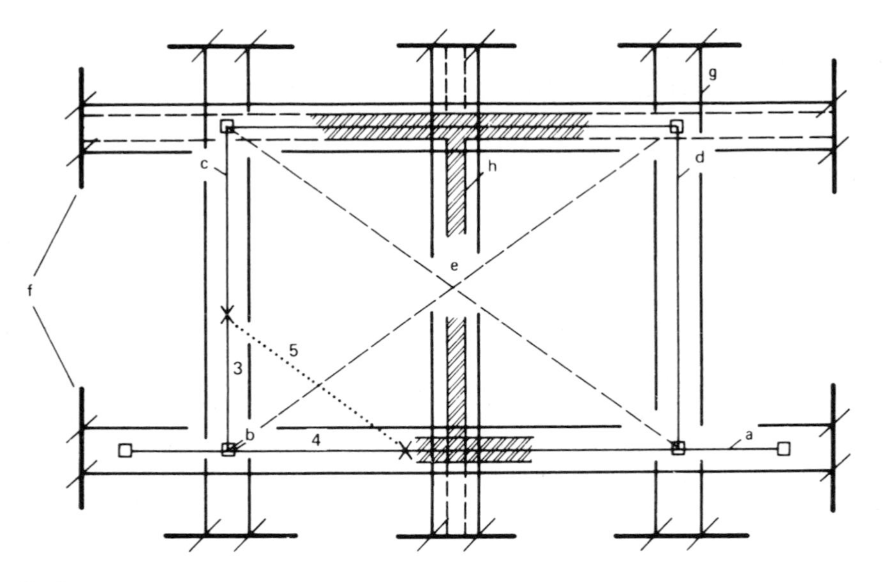

Figure 25.4 Setting out.

Construction – frame (see Figs 25.6–25.8)

(1) Steel, concrete and timber frames have different shapes and material characteristics which require different assembly and fixing techniques (see Fig. 25.7). The manufacturer of the frame will give specific guidance on the fixing and foundation requirements of particular structures.

(2) Each foundation pad supports and locates a frame member. The location system, of which there are many, using sockets and bolt assemblies, should be positioned exactly so as to receive the stanchion of the frame without distortion. Most frame erectors will cast pad foundations themselves to avoid potential site difficulties.

(3) The frame members will arrive on site, sectionalized, for assembly prior to lifting into position by mobile crane. The span sections are loosely bolted in place and stabilized spatially by use of longitudinal purlin and wind-bracing members until the whole structure is complete at which time the joints are tightened up and the structure becomes rigid. Once the frame is secured, roof cladding is fixed, starting from one (usually leeward) gable end. Care is needed to align the corrugated sheets squarely with the verge and eaves of the building. Special features like translucent sheeting, eaves fillers, ventilation ducting and insulation materials may be incorporated as the roof is clad.

(4) After both slopes are covered, the ridge detail is fixed in place. The overlap on both sheets and

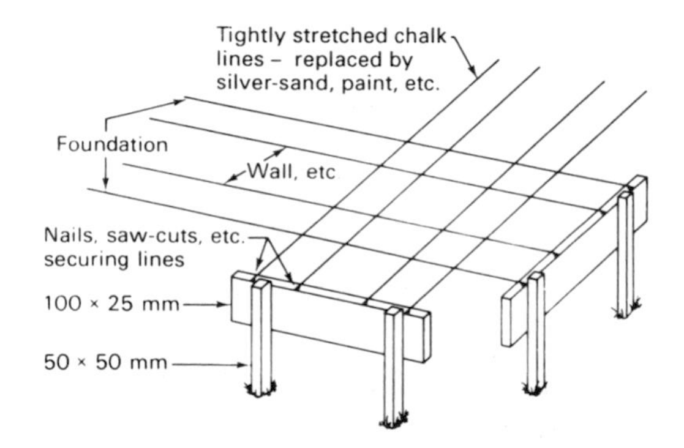

Figure 25.5 Use of profile boards in marking out.

capping must be in accordance with the manufacturer's requirements including sealing. The gable ends and side cladding are then fixed before the bargeboards and guttering, at a suitable slope, and the 'down' pipes to channel the surface water to ground level.

Construction – ground works (BS 5502, Part 22, refers)

(See sections on Foundations and Drainage.)

(1) Trenches and/or pad support areas are dug out to profiled width and correct depth with vertical sides

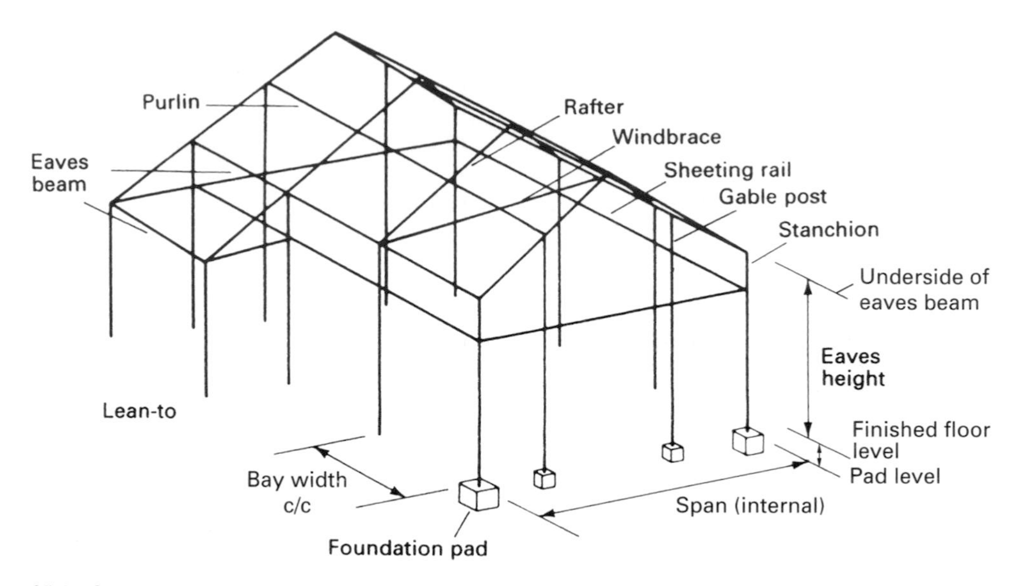

Figure 25.6 Structure elements.

and firm bottom. Any soft areas should be stabilized with stone ballast or concrete and rammed down until firm. Water and water-softened soil should be removed prior to the foundation concrete being poured. In very soft soil conditions it may be necessary to support the trench sides with struts and board temporarily, not least to ensure safety of workers.

(2) Foundations should always be horizontal; on sloping sites a 'stepped' foundation is required. The depth of each step should, for convenience, be a multiple of whole building units, e.g. block(s) or brick(s). Thus:

$$\text{Distance between steps} = \frac{\text{Depth of step} \times 100}{\text{Slope \%}}$$

Each step must project over the slab below substantially more than the step 'depth'.

(3) To ensure horizontal levelling and the correct thickness of materials used, drive pegs into the trench bottom so that the tops represent the finished foundation surface. Use a spirit level and a straight board over short distances, but a surveyor's level is normally essential.

(4) The concrete mixture used for most farm works will normally be the designated mixes specified in BS 5328: 1991 (e.g. as type RC 35) as described later. Note: liquid-retaining structures are also covered by BS 8007, which normally specifies type RC 45 for that purpose. Refer to BCA Farm Concrete Facts No.16, *Concrete Mixes for the Farm*. (Reference may be made by contractors and others to the use of mixes described as 1:2:4, 1:2 2:4, 1:3:6. These outmoded descriptions relating to volumetric proportions do not imply satisfactory quality standards and should be avoided. Concrete mixes may be described by the also outmoded designations of C20P, C7.5P, etc. These are more reliable descriptive guides to concrete quality, but still may mislead the unwary and are not as defined as those under BS 5328. They will not be used in specifications henceforth.)

(5) The top surface of the foundation concrete must be level, smooth and horizontal to facilitate the laying of brick, block or framework.

(6) Drainage trenches are dug out and pipes laid where appropriate, at the same time as the foundations are excavated. Any work that affects both is carried out first. For instance any pipes to be positioned under or through strip foundations should be placed and fixed by, for instance, encasing in 150 mm of concrete (to BS 5238) before the rest of the mass is poured around the pipework.

(7) Services are laid onto the site and led to the main distribution points for connection on a temporary basis during construction.

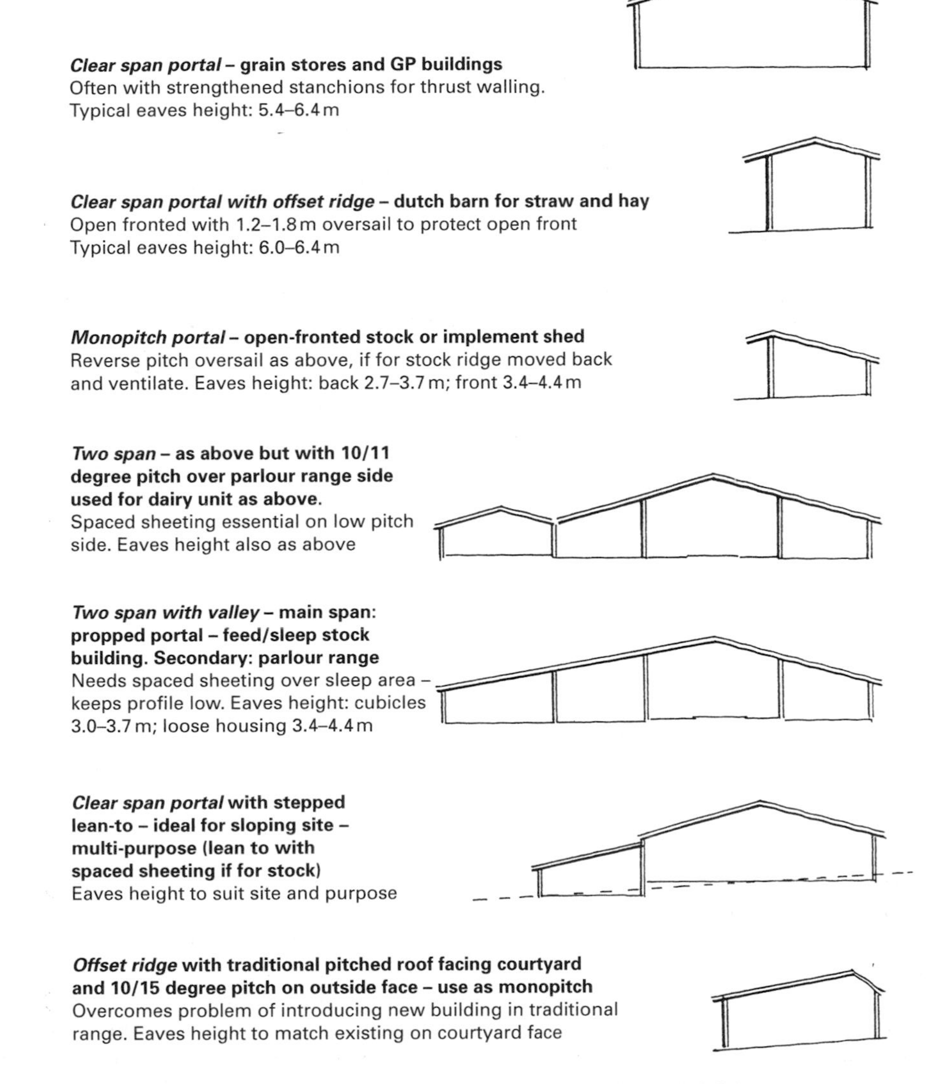

Figure 25.7 Examples of frame systems (giving potential uses and typical eaves height).

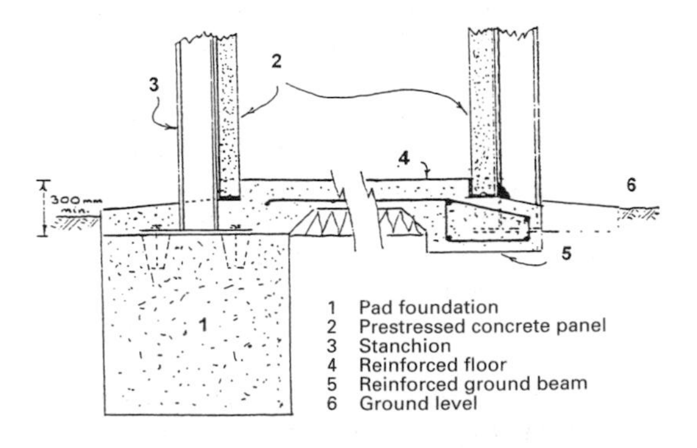

Figure 25.8 Stanchion pad: with panel wall and ground beam.

Construction – masonry cladding and floor laying

(1) With the frame in place, blocks may be laid on 'strip' foundations between stanchion members up to 'damp-course' (DPC) level which should be at least 150 mm above the outside ground surface. Note: no DPC may be used where there is any possibility of side thrust being applied to a wall. They are not therefore commonly used in agricultural situations, even though they are mandatory in other spheres.

(2) Doorways and other gaps are left in the wall, as appropriate, with timber 'liners' or with fixing blocks left in the masonry for subsequent connections.
(3) The walls will be built according to the design specification and may incorporate support piers and/or other reinforcements, insulation, waterproofing or a cleanable surface. Blocks should normally be 7N dense concrete to BS 5628 laid flat to give 200 mm thickness if there is any chance of impact, otherwise 150 mm is sufficient. Blocks should not be used for load-bearing structures.
(4) When the walls have reached damp-course level (see 1 above), it is often convenient to lay the concrete floor. (For lightweight agricultural buildings the method of construction sometimes used is the 'ring-beam' foundation system, in which the floor is cast at the same time as the frame and wall supports with reinforcement incorporated at the edges.) The ground surface should already be level and free of topsoil so pegs are placed to indicate the finished floor surface. Hardcore is laid on to the soil to the required thickness (normally minimum 150 mm) and vibration-rolled into a hard, dense mass. The top of this is then 'blinded' by sand to fill all surface cracks and provide a smooth plane on which to place the damp-proof membrane. Polythene sheet, 125 μm thick, is laid, with a minimum overlap of 150 mm at the sheet edges, to cover the floor and is bonded to any adjacent vertical wall and its DPC by bitumen-based paint or mastic. (The sheeting also serves as a 'slip' plane for the concrete mass and improves the quality of the curing process.)
(5) The floor concrete (see 'Mass concrete' later) should be poured in convenient shuttered bays, not exceeding 10 m × 3 m in area, for manageability and timeliness. It should be to BS 5328 specification, normally a 75-mm slump and of a suitable thickness (minimum 100 mm). It must be tamped and surfaced by floating or other method, preferably after the use of vibration equipment for consolidation. It should be left to 'cure' for at least 7 days and preferably 14 days before it is required to bear a load. If the curing period is very warm or dry, then a gentle rewetting of the surface or spraying with resin sealant is recommended to preserve the mix moisture level. Special admixtures must be used for specialist applications or conditions, such as plasticizers to increase workability. Pulverized fuel ash cement should be specified for all areas used by livestock.
(6) An additional surface may be called for in some buildings, for instance in a farm dairy where a granolithic 'screed' may be necessary to provide anti-slip characteristics. This is a thin (40–50 mm) top layer of fine granite aggregate concrete mix, applied to the base mix before curing has finished, and preferably within 72 hours of laying. This is a skilled job, usually best left to specialist tradesmen. Ordinary 1 : 4 screeds should be dusted with carborundum to improve non-slip/wear characteristics.

Construction – internal finishing and services

The internal fittings such as water supplies, electrical installations, joinery work and masonry finishing (plastering, etc.) are normally the province of professional tradesmen, whose skill can make or mar a building. In this work there is a flexible element of design, which is not possible in other areas of construction. 'On-site' decisions are quite usual and in any case craftsmen would be expected to think for themselves in respect of details like pipe and wire runs and fixing methods on all but the largest farm construction. Compliance with the relevant regulations is essential including the 16th Edition of the IEE Regulations for the electrical installation and the Water Supply (Water Fittings) Regulations 1999 (Water Byelaws 2000, Scotland). The water regulations are substantially different from accepted practice in the past.

The complexity of the work to be done and the need for a careful order of procedure to be followed means that the timeliness of the attendance of the specialists is often important to the smooth transition in the stages of building construction to avoid time wasting and mistakes. The employment of an experienced professional to provide overall supervision may be crucial to the successful completion of a complicated building.

Building materials and techniques (BS 5502, Part 21, refers)

Mass concrete

There are various types of concrete in general use but that for farm construction is usually a mixture composed of ordinary Portland cement (as opposed to rapid-hardening Portland cement and other types), fine aggregate, coarse aggregate and water. Aggregate may consist of gravel or crushed stone, of a light or heavy nature, and may be round or sharp (angular). The fine aggregate (or sand) fills in the voids between the coarse particles (stone, gravel), the cement coating adhering to both. This should produce an almost solid matrix when set if air is expelled by good mixing and consolidation by vibration and tamping. All aggregates must be clean and free from vegetable matter, salt and clay. They must be dry (for accurate volume or weight measurements to be made prior to mixing). 'All-in' or 'as-dug' aggregates should not be used. Always use clean water, preferably from a drinking source. Although it is possible to mix concrete by hand in small batches, it is not recommended practice. The most reliable way of apportioning mix ingredients is by 'weigh batching', that is, weighing out enough materials to mix with 1 bag (50 kg) of cement, or a multiple number of bags, and to load the quantities into a suitably sized mixing machine in one whole batch.

The time of mixing should be long enough to ensure that all the aggregate particles are coated evenly with cement to give a consistent greyish coloration, for which, normally, a minimum of 2 min is required after the last ingredient has been placed in the drum. Most mixers work best when the coarse aggregate is placed in the drum first, followed by the water, then half the sand, then the cement and finally the second half of the sand. Some 'sticky' aggregates may require that this order be varied to achieve the best possible mixture.

The quantities of cement, aggregate and water will vary according to the job to be done. Table 25.2 shows the normal proportions by volume for the work loads expected in three types of circumstance, on farm sites. (A recent reclassification of concrete mixes under BS 5328: 1991 has altered the designated coding for all uses: see Table 25.3.)

The size of aggregate must be taken into account when selecting an appropriate mix. Thin section or reinforced concrete requires stone that is relatively small in size, graded from 6 to 20 mm, whereas large mass concrete can satisfactorily enclose aggregate of up to 40 mm size grade. Fine aggregate is 5 mm or less in size usually, but must not contain too much dust.

Water quantities are critical – almost as important to the ultimate strength of the concrete as the cement

Table 25.3 Designated mixes for premixed concrete (BS 5328: 1991). Sample of types and applications (see British Concrete Association Fact Sheet No. 16 for more information)

Typical application	*Designated mix*
Concrete blinding under floor	GEN 1
Oversite concrete to be finished with screed, mass concrete	GEN 2
Unreinforced foundations – strip, etc.	GEN 3
Reinforced foundations, stanchion bases	RC 30
Crop store and livestock floors	RC 35
Slurry floors (and walls), workshop and sugar beet floors	RC 40[1]
Silage floors (and walls), stable floors, brewers' grain stores	RC 45[1]
Parlour and dairy floors (50 mm slump plus carborundum dusting)	RC 45[1]
External yards and roads – air entrained mix	PAV 1

[1] Equivalent to C35A.

Table 25.2 Concrete mixes for various purposes proportional by volume

Purpose and use	*Large bulk volumes, e.g. deep foundations*		*Average use, floors, roadways*		*Heavy use and thin sections*	
Cement	2 vol.	1 bag	3 vol.	1 bag	3 vol.	1 bag
Fine aggregate (damp)	5 vol.	190 kg	5 vol.	115 kg	4 vol.	90 kg
Coarse aggregate	8 vol.	270 kg	9 vol.	195 kg	7 vol.	170 kg
Water	1–1.25 vol.	20 litres	1.25–1.5 vol.	23 litres	1 vol.	17 litres
Approx. yield	7–8 vol.	200 litres	9–10 vol.	170 litres	7–8 vol.	140 litres
Previous designations	1:3:6/C7.5P		1:2:4/C20P		1:1.5:3/C25P	
BS 5328 Code	GEN 1/2		ST4		ST5	
Table 25.3 near practical equivalent			GEN 3/PAV 1		RC 35	

itself. Too much weakens the concrete by leaving cracks when it dries out, too little will give ineffective bonding due to lack of chemical reaction. Workability is also governed by the water content of a mix and as a consequence very often concrete is mixed and laid much too wet in order to increase the facility of placement. Increased workability can only be achieved by adding a superplasticizer (see 'Admixtures'). The moisture content can be checked by use of the 'slump' test. Special testing cones can be purchased or hired to measure the required 75-mm slump, but a steel bucket may be used as follows instead to provide a rough guide. The bucket is filled with concrete and inverted on to a flat surface (steel/wood plate). It is then drawn gently away from the contained concrete leaving a cone-shaped heap, the height of which has 'slumped'. A slump of more than 25% of the height of the bucket shows the mix is too wet.

All concrete should be laid as soon as possible after mixing and always before 1 hour has elapsed. It should be carefully positioned from a low height avoiding any movement that might cause desegregation of the constituents, for instance rough barrowing. There is a brief period of 'setting' (about 30 mins to 2½ hours) followed by 'hardening' during which it gradually achieves greater strength. During this latter 'curing' period, which is indefinite but for practical purposes is said to be 28 days, the concrete should be undisturbed, kept from drying out and in a frost-free environment. Under no circumstances should concrete be mixed when the aggregates are frosted, but in very cold conditions (less than 4°C) use accelerator (see 'Admixtures for concrete' below).

Note: pulverized fuel ash (pfa) cement should be specified where alkali hoof erosion is potentially possible on new concrete. Ensure that no limestone aggregate is used in silage bunker floors.

Admixtures for concrete

- Plasticizers or superplasticizers increase workability, enabling easier placing and compaction. Important to use in difficult situations. The super grade is extra effective – for instance for pouring or pumping. Note remark above: never add extra water to try to achieve same result.
- Accelerators increase the rate of gain of early strength. Useful for setting in tanks, etc. in high water table situations, etc. Also suitable to compensate for delayed cold-weather setting. Not suitable for high strength situations or (depending on type) in reinforced concrete.
- Retardants slow down setting of concrete, prolonging time it remains workable. Particularly useful for premixed loads on restricted sites with time-consuming placement.
- Air-entraining produces a controlled amount of very small air bubbles which reduce both durability and workability. As it is particularly effective in reducing the corrosive effect of salt it is always specified for roads and aprons (designated mix code: PAV 1).
- Waterproofing reduces the permeability of concrete. Especially useful for screeds and renders. Not a substitute for using right mix and joints under BS 8007. For instance 150-mm thickness of RC 45 is not permeable.

Reinforced concrete

Reinforced concrete contains steel reinforcement at a position determined by the need to carry tensile loads, for which concrete is relatively weak, having only one-tenth of its capacity to resist compression. It is required in all load-bearing floors, roads and aprons. The steel is usually below the centre line, towards the bottom surface in a mesh form; sized according to load; the standard minimum being A 142 fabric. Secure placement is essential to maintain the design strength and ensure adequate cover. Tamping by beam is no longer sufficient so a vibrating poker must be used – with great care to ensure mesh is not displaced. The design of more complex structures such as ground beams and cast-in-situ walling should only be carried out by experienced installers to an engineer's specification.

Note: in the case of liquid-retaining structures see also Mason (1992).

Concrete blocks

These are supplied in various common forms, standard sizes (see Fig. 25.9 for a typical range) and aggregate materials to suit the requirements of the structure. The majority of those used in farm construction work are of 'dense' aggregate, although 'cellular' blocks are used from time to time for insulation purposes. These latter have a bath sponge appearance, due to the foaming method of manufacture. Although of low heat conductivity they are relatively weak and are porous to water passage. (Note, however, that all concrete blocks are poor moisture barriers compared with bricks.)

When calculating quantities to be purchased it should be noted that the actual sizes of the blocks may be less

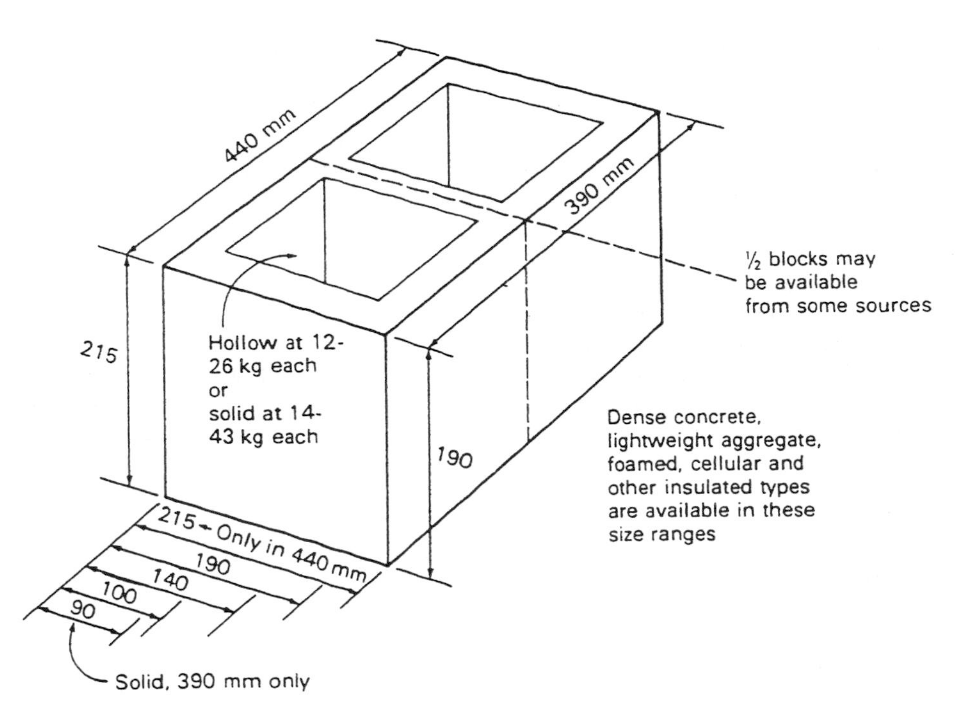

Figure 25.9 Alternative concrete block dimensions.

than the 'nominal' size by 10 mm, the mortar gap allowance.

Concrete block walling

Most walls constructed in farm buildings use 390 × 190 × 140 (or 190) mm 7N solid dense block, (laid upright to give 140-mm thickness or laid flat to give 190 mm), or 440 × 215 × 215 mm wide hollow blocks laid in 'stretcher' bond pattern (see Fig. 25.10). 'Cavity' walls should not be used in agricultural buildings, although they are good insulators, as they are complicated to build, are weak under lateral pressures and can be expensive. If insulation is required use a 190-mm insulated block protected on both sides with rendering or cladding.

For some purposes, however, the wall may require strengthening. This is carried out, traditionally, by either the incorporation of piers or the insertion of steel rods or netting into the bond pattern of the wall. Piers (see Fig. 25.10) are part of the wall and are thickened by the inclusion of blocks such that a vertical 'ridge' is formed, normally one block wide up the wall. The various forms of steel reinforcement provide strength in tension which the concrete block-work requires but lacks under bending loads. The steel must be bonded into the wall such that it all behaves as one composite material and yet is correctly positioned to resist the tension forces. In practice, piers and reinforcement are used together to achieve maximum strength, one method supplementing the other. Horizontal reinforcement is bedded in the mortar joint using 3.5 mm stainless-steel wire mesh such as BRC Wall force. Vertical reinforcement is only possible in hollow blocks using mild steel rods with looped ends.

See Table 25.4 regarding wall stability and 'rule of thumb' relationships of wall width, strength and use. Note: many agricultural buildings carry wall loads that are lateral and quite severe. 'Rules of thumb' are not recommended for these and expert advice should be sought for their design. In such severe load situations blockwork is not recommended if prestressed concrete panels can be used.

Where the length of a wall exceeds its height by 1.5–2 times or where there is no natural bond break within 6 m, an expansion break must be included to accommodate thermal and moisture movement to which concrete blocks are particularly prone. This may be provided by a slot cut in the outer block surface at least 50 mm deep or, possibly, completely through the wall. In the latter case the stability is retained by the use of steel dowels

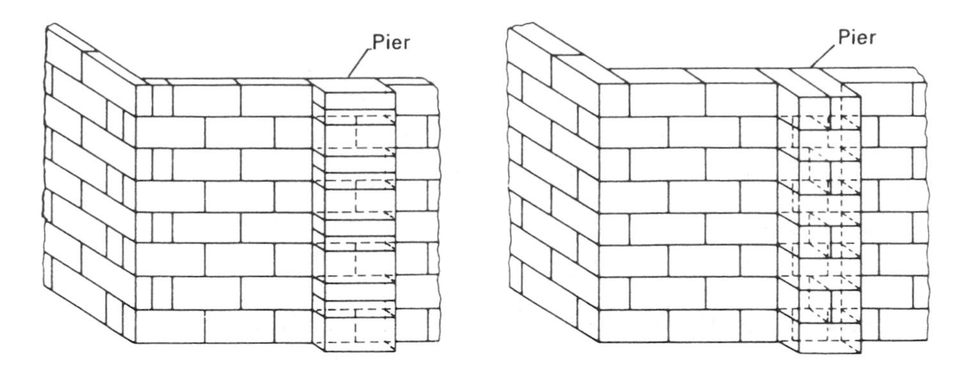

Figure 25.10 Alternative wall bonding (100–150 mm on edge and 190–215 mm laid flat).

Table 25.4 Block thickness (m) for unreinforced walls (but which may include piers). [Note: walls designed to withstand wind and stock pressures; not to retain stored grain or dung. For larger stock use the large block sizes.]

		Block size (mm)		
		100	*140*	*215*
Walls lacking any support at top	Max. length	3.6	4.5	9
	Normal max. related height	1.8	3	3.6
With continuous top support, e.g. roof framing, wall-plates, eaves beams	Max. length	4.5	6	7.5
	Normal max. related height	4.5	6	7.5

set in the mortar on one side of the break and greased before laying in the other side. The slot should be filled with mastic at the outer surface to provide a weather-seal.

The mortar used in concrete block laying should be in the volume proportions of 1:2:9 or 1:1:6 mixed from cement, lime and builder's sharp sand. The lime may be replaced with proprietary ingredients known as plasticizers, which make the mortar 'fatty', that is, slip easily into place between the blocks. These mixes are weak, relative to block strength, so that if there is a tendency to crack it will appear less unsightly and be of less significance to wall stability when it is confined to the mortar gaps.

Concrete block walls may, depending on their purpose, require waterproofing. This is primarily done by use of render coatings of mortar-like material, usually in two coats; the first, at 10 mm thick of 1:1:6 mix as described before and the second at 6 mm thick of 1:4:3 mix. Alternatively, the block pores can be filled in and the mortar gaps flush pointed to give a flat surface which can be painted with chlorinated rubber, epoxy-resin or acrylic-resin bonding paints. This is not sufficient for a parlour or dairy which requires a single render coat followed by a long-life coating such as fibreglass.

A new approach to block laying combines the simplicity of 'dry-laying' (with no mortar) and render-coating for waterproofing, by the use, as a wall surface fixative, of a mix of cement, sand, specially prepared glass fibre and bonding chemical laid on to produce a hard, durable coat which stabilizes and creates the strength of the wall from the outside surfaces. Such a coating is an acceptable finish for parlours and dairies.

Timber

Timber used in construction should be well seasoned, pressure treated with biocides and fire retardants, and free from warping, knots, sapwood, bark and shakes (splits). Softwood (see Fig. 25.11 for available size ranges) is used for nearly all work other than decorative or highly specialized functions due to the high cost and difficult working of most hardwoods. Seasoning is not required for cladding timber; treated home-grown timber is ideal especially larch or Douglas fir.

Lengths of timber are generally purchased in a 300-mm module, but for large quantities it may be sold by the cubic metre with additional sawing charges. Pieces in excess of 5.4 m long may exact a premium price. Sawing and planing 'allowances' must be taken into account when sizing timber sections as these may reduce the dimensions by up to 2 mm.

Probably the biggest drawback to the use of timber in construction is biotic decay. In the UK timber is affected by three main agents, 'woodworm' (actually the larvae

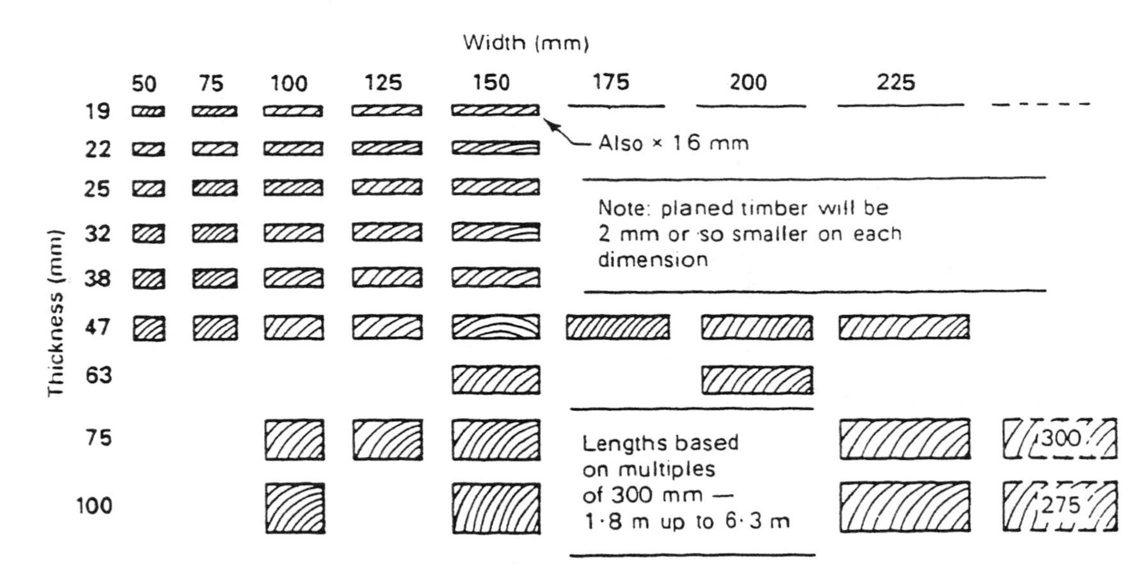

Figure 25.11 Sawn softwood standard sizes.

of a beetle, *Anobium punctatum*) 'dry rot' (a fungus, *Merulius lacrymans*) and 'wet rot', which has several possible causal fungi. All these problem pests can be prevented from doing damage by treating the timber initially with a cocktail of potent chemical biocides, but the life-span of timber can, in any case, be prolonged by the avoidance of poor climatic or environmental conditions. Repair of timber that is already infected is difficult and sometimes, as in the case of dry rot, almost impossible without wholesale replacement.

Recently the use of 'stress-graded' timber has become a requirement in all types of load-bearing structure. This is timber that has been inspected visually for flaws such as knots, size imperfections, shakes and other defects and is stamped with a grade category which indicates the limitations to its use in structural work. GS (general structural) has been passed as acceptable for general use, for example as floor joisting; SS (special structural) implies a more particular usage (see BS 4978 on stress grading of timber).

For information about screw and nail sizes for timber work see BS 1210 and 1202. Recently there have been a number of new types of fixing method devised, most of which combine the driving ease of the nail with the firm hold of the screw. A review of these innovations is recommended before extensive timberworks are carried out.

There are two further general design concerns expressed about structural timber. The first is its apparent susceptibility to fire damage which, owing to its characteristic of charring rather than burning, is not as threatening as might at first be thought. If proper safety factors are used in design, considerable parts of the cross section of timber beams must be burnt away before failure occurs. The second concern relates to its flexibility. Timber lacks 'stiffness' compared with structural concrete and steel so its use is restricted in simple beam or lintel framing to short spans. In larger spans the timber must be used in a lattice girder or similar pattern, to achieve the required strength and rigidity.

Cladding options

Timber

The structural framework used for mounting these boards (usually supplied in 2.44 × 1.22 m sheets) is generally based on the traditional 'studwork' used for timber planking (see Fig. 25.12a), but they are sometimes supported on rails in a similar manner to corrugated sheeting (see Fig. 25.12b). Options include:

- External grade (WBP, weather and boil proof) plywood sheets for walls and roofs (the latter with a felt covering) though care must be taken when fixing if warps are to be avoided. Note also that unprotected ply will tend to delaminate over a period of time in strong sun. All edges should be secured with galvanized nails or screws, with intermediate nailing for hardboard (or other types of 'particle' board) at

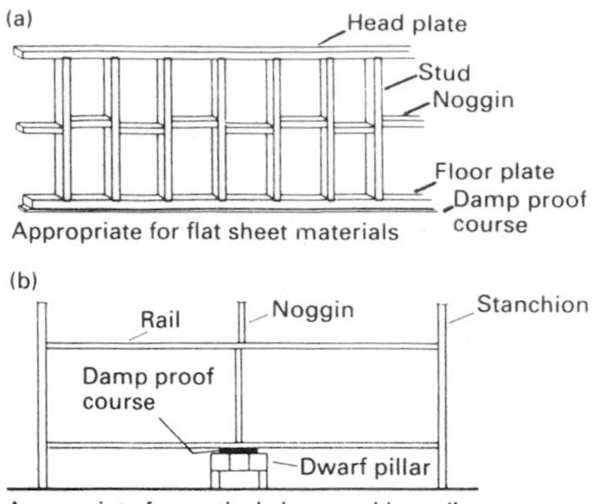

Figure 25.12 Framing for side cladding.

300-mm centres. Generally, the main studs should be at 1.2-m centres, equal to the sheet widths with noggins at 600-mm vertical centres.

- Oil-tempered hardboard, type TN or TG, usually 6 mm, an inferior but cheap alternative, should be nailed every 150 mm to a frame specially provided with studs and noggins to accommodate it. Useful also as permanent backing sheet. Form drip on bottom edges for long life.
- Chip (particle) board, usually 12 mm T&G, should not be used externally and only in the water-resistant grade, V313, internally.
- Traditional sawn timber boards are used as an outside wall cladding:
  - Horizontal dung walling, 50 mm thick, T&VG, ex 175 mm to give 150 mm c/c with posts at 1.2 m c/c.
  - Vertical board 18 or 25 mm thick, 100 or 150 mm wide, spaced, butt jointed or lapped using 150 mm backing and 100 mm on front with 25-mm laps with/without oil tempered hardboard backing.
  - Horizontal weatherboard, 25 mm thick, 150 or 175 mm or wavy edged, with/without micro porous building paper backing.
  - Featheredged 18–7 mm, 175 mm fixed horizontally with 25 mm lap. Sometimes available in home-grown cedar. Backing as above.
  - Shiplap, prepared ex 18 mm in various propriety forms for domestic buildings.

Fibre cement

Fibre cement replaces asbestos cement – now banned for use. (Existing asbestos roofs do not have to be replaced, but there are very strict legal requirements relating to the disposal of asbestos sheeting.) The old 6-inch profile is the normal one for farm buildings. It should always be specified with strip reinforcement. It is still classified as a fragile under HSE 33 and so all building with fibre cement must be labelled accordingly. Its lack of resistance to impact makes it less suitable than steel or timber for side cladding, but it is the preferred option for most roofs. The advantages over steel are the vapour permeability significantly reducing condensation, and the high resistance to corrosion giving a long maintenance-free life expectancy in the aggressive environment of agricultural use – especially livestock buildings. The better acoustic properties are also useful in some circumstances. If it is to be used for spaced roofing it must be ordered pre-cut to suit (sheet width reduced to 1000 mm from 1016 mm). High specification fixing should be designed to match the material life expectancy, and all fastener heads should be capped. Sheeting laps should be sealed on exposed sites (more than half of the UK). The new matt colours are specifically designed for the agricultural industry and are available at the same cost as natural sheeting. The wider range of acrylic colours is highly reflective and so less suitable in the rural environment.

Metal

There are two preferred choices in steel cladding: Colorfarm AP has been specially designed for the market with a choice of colours and the same protective treatment of co-polymer on the underside – to help resist the aggressive environment. Plastisol does not have the same underside protection but is coated with a textured bonded plastic upper surface which is available in a range of colours. The texture greatly reduces the reflectivity to give a pleasing effect. Plastisol coating is not recommended for roof covering in dark colours. Both normally use a Galvatite 0.5-mm substrate and are available in much longer lengths than fibre cement to a range of profiles. Both should be separated from the purlin with a DPC material and fastened as above. The cheap alternative known as Ruralclad is not suitable for BS 5502 Class II buildings. Aluminium and uncoated steel are too intrusive in appearance to be used in the countryside.

Rooflights

PVC rooflights are no longer recommended due to problems of compliance with HSE 33 and life expectancy. Existing PVC rooflights are also highly susceptible to gale damage if not stitch bolted. Glass reinforced polyester (GRP) roof lights on the other hand have increased in choice to give the extra strength required with such products as 'Supasafe' – given a non-fragile roof rating. Many prefer the more diffuse light offered. The drawback is that the surface becomes textured over a period of time – good news as it reduces reflection but bad news owing to the accumulating lichen reducing the light.

Polycarbonate has only just appeared in the roof light market and is more expensive than PVC and is as yet untried in this format. It gives a clear shaft of light like PVC.

Frames and simple roof support structures

The frequently used term 'portal' should specifically be applied to support frames providing clear, unimpeded headroom to the ridge inside the building, but it is often corrupted to mean any factory-made, site-assembled, large-span system (Fig. 25.7). See Fig. 25.6 for terminology applied to the common components.

The common and 'preferred' size ranges for major frame dimensions to which most manufacturers adhere are shown in Fig. 25.13. Eaves height is comparatively simple to adjust during manufacture or assembly and so provides many options. A sloping floor building will have different heights for each stanchion. As a general rule wider buildings have shallower slopes to their roofs to reduce overall height, but the framework elements must become stiffer to compensate. Thus it is more economical to use a propped portal frame to create very wide structures or, alternatively, multi-span arrangements with roof valleys between slopes.

Purlins are usually Zed steel, although aluminium box section are now available and treated timber should be on livestock buildings to overcome the problem of dripping condensation. A DPC type material should be used to separate metal sheeting from the purlin to prevent chemical or electrolytic reaction. Centres must follow the requirement of the sheet length and type used: 1.145 m for 2.44-m sheets and 1.375 m for 2.9-m are typical for fibre cement.

Foundations

Foundations are intended to sustain building structures against three main loading problems in agriculture (note: reference to 'soil' implies subsoil, i.e. below the agricultural soil horizon):

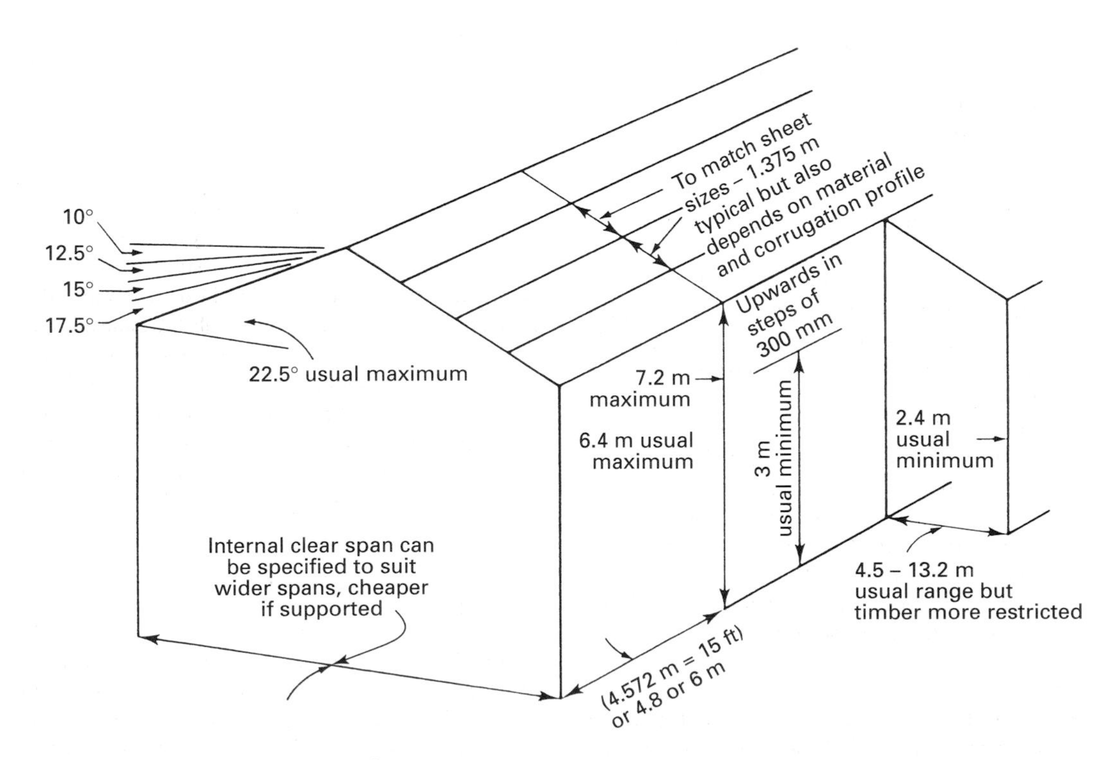

Figure 25.13 Common frame dimensions.

- Sinkage – when the building weight (its own structure load and that imposed upon it by contents) exceeds the support capacity of the soil beneath, downward or sideward movement may occur. This can be differential in nature, varying from place to place, creating bending and twisting loads on the structure. Made-up ground is especially susceptible unless made up in 150-mm consolidated layers to an engineer's specification to suit the soil type.
- Water effects include the following:
  - volume changes due to soil moisture content variation (sand soils best, clay and peat worst) cause structure stress and movement by soil 'heave';
  - frost action giving volume and structure changes in soil (minimum depth for foundations related to climatic considerations), by expansion/contraction;
  - leaching and erosion undermining the base for support, caused by poor drainage (from soil or inadequate piping configurations);
  - tree root influences, creating changing moisture levels in the soil horizon, especially after felling.
- Cart wheeling – where lateral (side) loads applied by building contents, e.g. stored grain, turn the structure base through 90 degrees.
- Foundations must resist other forces such as vibration and load cycling, but fortunately these are comparatively rare in agriculture and can be ignored for practical design purposes. (They must also resist chemical attack from soil acids and sulphates – but these are not directly structural matters, although important.)

Thus foundations must be:

- correct in area (load imposed = soil bearing capacity × safety factor) which sometimes, for simplicity, results in a standard recommended size;
- at a correct depth (for example, 1000 mm below excavated ground level according to load, soil and climate, see water effects above);
- of a correct width to prevent cart wheeling and provide for easy construction practices, e.g. trench and hole sizes are dictated partly by the skill and ease of digging; also masonry units require a margin of measurement accommodation;
- of a thickness to be strong enough to resist failure in compression, i.e. to avoid a hole being punched through by the load applied (it may be necessary to reinforce with steel to achieve this);
- of sufficient material quality to resist failure by chemical reaction and premature loading during curing;
- below any 'fill' material on 'made-up' ground area.

The actual size and depth of foundations are therefore inherently related to particular sites and building circumstances, and there can be no safe 'rules of thumb'.

Note: foundation design advice should always be obtained from an expert source familiar with the site and the proposed building before commitment to a type of construction is made. The practice of assuming a soil-bearing capacity at tender stage and failing to investigate before construction is still occurring in the farm building industry. Beware: there is no Building Regulations inspection to check!

Strip, trench fill and ground beam foundations (Fig. 25.14)

Typical dimensions for a 2.1-m-high, 200-m-thick dense concrete block wall are:

Depth: 1 m
Width: 600 mm
Thickness: 300 mm or full depth less 150 mm for trench fill

Lightweight walls or concrete panels can be supported on a ground beam: a thickened and reinforced floor slab – providing the base is at least 300 mm below ground level. The reinforcing must tie back into the floor slab for at least 1.2 m. Such an arrangement may be necessary anyway to strengthen an exposed floor edge with a reduced thickness – 300 mm will usually suffice.

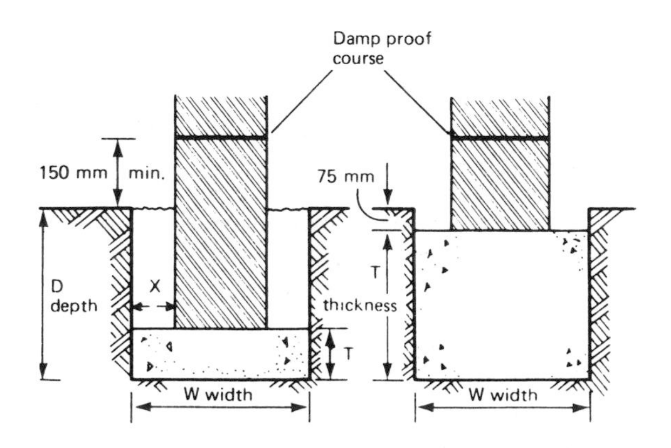

Figure 25.14 Strip and trench fill foundations.

Pad foundations

These provide support for frame uprights and locate them laterally. Manufacturers of frames will provide detailed drawings of the socket, bolt or dowel fixings with their positions, in pads of recommended size. It is not possible to give guidelines for these as they vary tremendously with shape, dimension and type of frame and the properties of the soil on which they rest.

Drainage

Categories

Two categories of drainage must be considered (and treated) separately in practice (see 'Water pollution control' later in this chapter for further information). They are:

- Surface water (SW), that is, water from rainfall run-off, which should be 'clean' and may be discharged directly into a watercourse for disposal. In effect this implies water from roofs and field drainage only. The volumes usually mean it cannot be discharged into a soakaway so adequate conveyance by pipe or ditch to a watercourse is essential. Old conveyance systems often used clay land drains in close proximity to buildings (within 10 m under the Water Code). Such arrangements must be replaced with a sealed system if there is any risk of pollution.
- Dirty water or foul water (FW) contains less than 4% dry matter; if above this figure it is treated as slurry and cannot be conveyed in drainage pipes (see Water Code and sections on 'Water pollution control' and 'Slurry storage'). Dirty water includes all water used for washing, any spilt effluent, and the rainfall from any open areas in which livestock is held. This category must in no circumstances be discharged into watercourses.

Choice of materials

- *Vitrified clay* – manufactured to BSEN295:1991 is ideal for the aggressive environment of a farmyard where crushing is a risk and available in the three normal sizes of 100, 150 and 225 mm. Larger sizes are also available. Can be laid without bedding material in certain circumstances.
- *Ribbed uPVC* – available as above. Plain pipe, without ribs is unsuitable, not having the crushing strength required. Always needs bed of granular material.
- *Concrete* – only suitable for SW in areas of no pollution risk. Push-fit joints which should always be taped to prevent access of roots. Available in 150 and 225 mm and larger. Does not require bed and is cheaper.
- *Clay or uPVC land drains* – not suitable in the farmyard environment for the fast heavy flows required, the pollution risk and crushing strength. In French drains use perforated pipe – not land drains.
- *Pitch fibre* – is never used in new work. Existing installations usually have to be replaced as pipe will have wholly or partially collapsed.

Installation

- Flow rates – see Fig. 25.15.
- Pipe protection – lay in granular bed if required to give minimum backfill coverage of 600 mm in no traffic areas or 900 mm with traffic. If it has to be laid at less (for instance close to downpipe) provide 100-mm cushion over pipe, then step trench cast 100-mm concrete 'raft lintel' over (see Fig. 25.16).
- Access – provide access on all FW bends or junctions, and at least every other SW bend or junction. Provide access for all strait runs at 24 m c/c FW or 36 m c/c SW. Access to comprise of an inspection chamber. A rodding eye is sufficient in some circumstances. Downpipes should be fitted with easy bend strait junction with rodding eye – not a trapped gully which is liable to pollute. Inspection chamber must be designed to take full traffic loads using a Class B cover. Pressed steel covers (even so-called 15-tonne loading) are not suitable except in non-traffic areas and should be galvanized. The construction should be in 200-mm-thick semi-engineering brick or heavy-duty concrete rings if deeper. A catch pit is often used with a 'T' pipe connection on the outfall, e.g. for silage effluent before discharging into tank.
- French drain – use a perforated vitrified clay pipe with perforations uppermost in granular bed and surround, backfilled to the surface with round stone. In most soils a geotextile fabric should be used as trench lining (pipe reversed for septic tank type soakaway).

Water supply

Legislation

Water supply is controlled by the Water Supply (Water Fittings) Regulations 1999 and Water Byelaws 2000 Scotland (see DETR website: www.environment.detr.gov.uk/wsregs99/waterfit/index.htm). These regulations are an update on the previous Byelaws and are

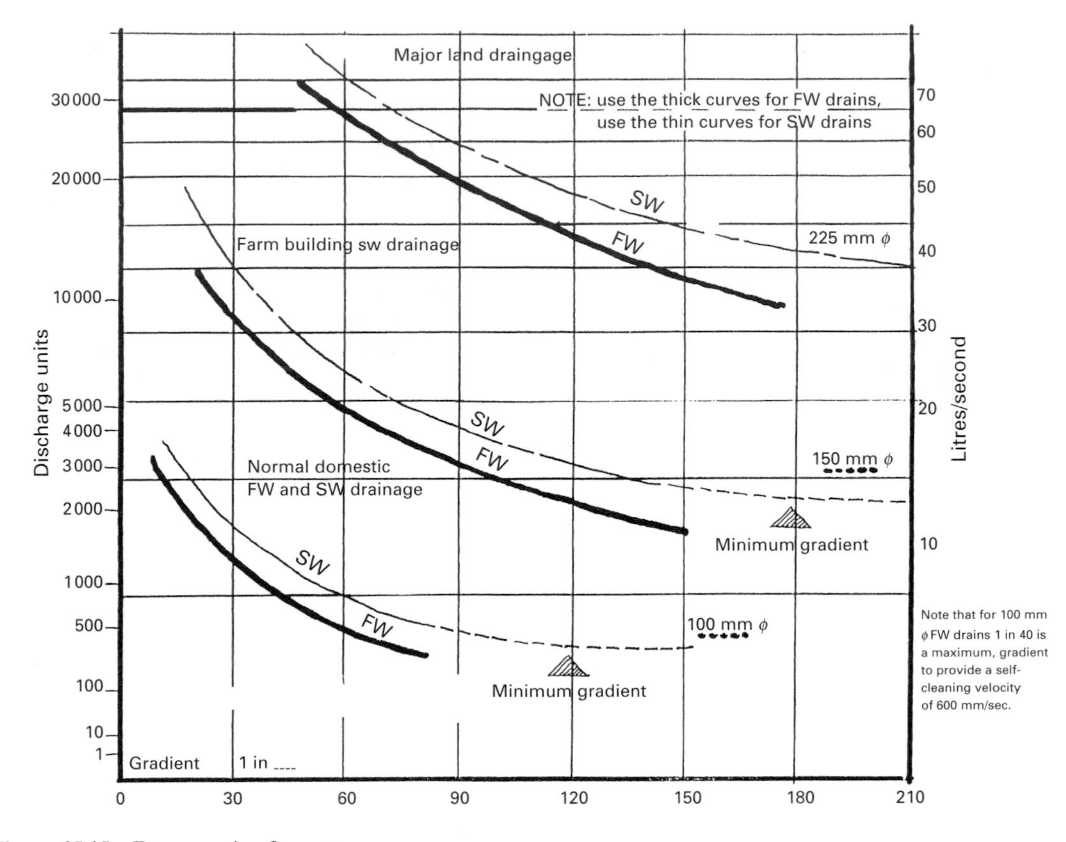

Figure 25.15 Dramage pipe flow rates.

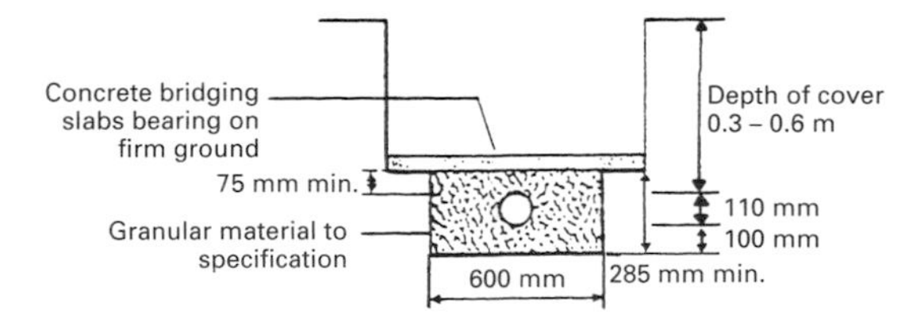

Figure 25.16 Shallow pipe protection.

as complex as the website address suggests! They are tailored to an industrial and domestic situation with insufficient consideration of the farm situation – for instance, the inspection traps required are likely to fill up with effluent. The requirements substantially differ from accepted good practice in the past. It is therefore essential that any work is carried out with a full understanding and the Water Authority Inspector notified and consulted well in advance. Work cannot start without prior consent and continue without stage inspections.

Material and pipe sizes

The only acceptable material is medium density polyethylene (MDPE) colour-coded blue which is ultraviolet unstable so it cannot be used uncovered above ground (black pipe is still available for dirty water irrigation). The pipe sizes are now given in external diameter. See Table 25.5 for comparative internal diameter. The flow rates achieved increase out of all proportion to the expected difference in pipe size (see Fig. 25.17). It is therefore essential that all distributor pipes covering long distances or services to a livestock unit – especially dairy – are installed in 63-mm MDPE.

Installation

It is not possible to describe more than a few points required under the Water Supply Regulations:

- All pipes shall have at least 750 mm cover (maximum 1350 mm unless ducted), be bedded in sand (so no moling in) or be ducted if under concrete or inside

Table 25.5 Recommended MDPE pipe sizes: agricultural use

MDPE External diameter	*Previous equivalent Internal diameter*	*Typical use*
20 mm	0.5 inch (ap. 12 mm)	Water bowl connections
25 mm	0.75 inch (ap. 18 mm)	Troughs, standpipes, most fittings
32 mm	1 inch (ap. 25 mm)	High flow-rate fittings, tanks
50 mm	1.5 inch (ap. 38 mm)	Local distribution
63 mm	2 inch (ap. 50 mm)	Main and long-distance distribution

building. No joint is allowed unless it is accessible. If above 800 m it shall be protected from frost, impact or stock damage, or vandalism, and be capable of draining down.

- All services shall be separated and capable of isolation, so each trough or bowl must either be serviced by a separate pipe or be on a separate branch with an individual stop valve at both ends.
- The whole system must be protected from backflow or siphonage in the construction of all fittings and cisterns, use of double check taps, etc.
- Despite the drawbacks it does mean that most of the farm installation, especially water bowls, stand pipes, etc., has to be via cisterns.

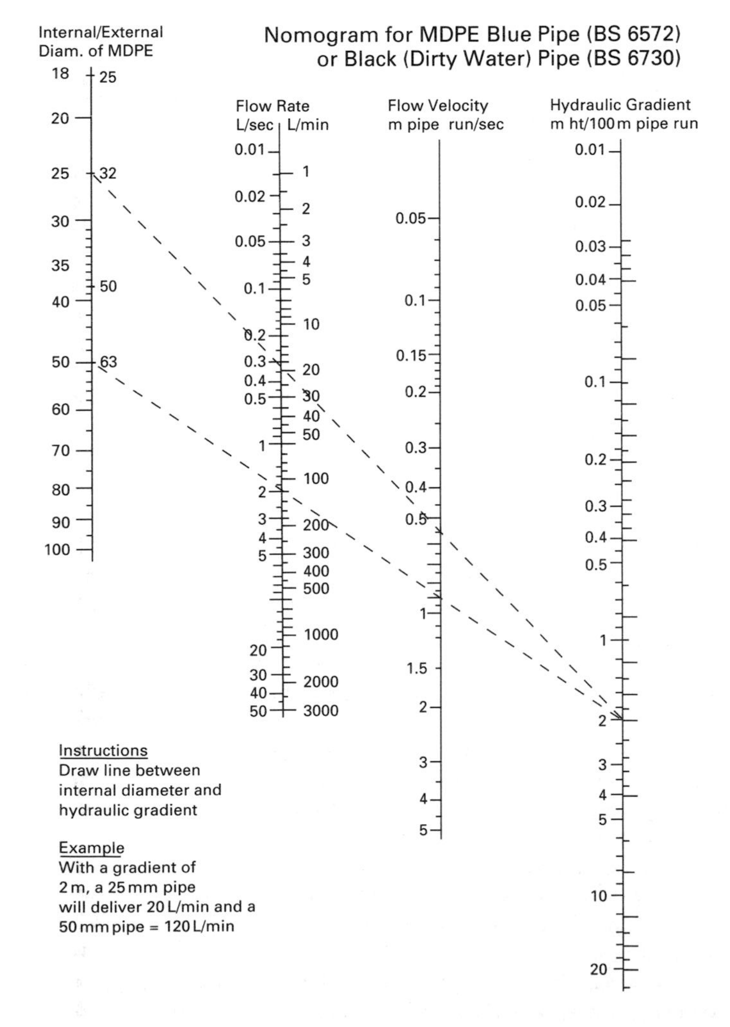

Figure 25.17 Water pipe flow rates.

Electrical supply

Legislation

The Institute of Electrical Engineers (IEE) produces a detailed code of practice now in its 16th Edition, which all practising electrical engineers are required to observe. In addition, the Health and Safety Executive impose conditions and any new installation is tested by the authority concerned before connection.

Power supply

All cabling has to be in impact-resistant conduit to suit the application for either 3 phase (415 v) or single phase (220 v) sockets. Three phase should always be chosen if available for installations requiring motors and high power. The motors are cheaper and much more power is available. The incoming supply is likely to have a protected multiple earth (PME) – look for the label PME on the transformer pole – but additional protection is essential in the damp environment of farm operations. All motors should have an isolator switch within reach of the operator.

Surprisingly it is not compulsory to use only 110-v power tools through a transformer for outside or farm building use, though it is strongly advised. All power outputs should be protected with a circuit breaker. All metal fittings such as in parlours and dairies must be well earthed.

Back-up generator power is essential for critical installations like parlours and environment-controlled housing. To avoid overload the critical parts of a system need to be wired independently to a change-over switch rather than trying to serve the whole complex.

Lighting

As above, the installation and fittings must be protected and suitable for the damp environment. The choice of lamps and fittings is wide. Fittings must be positioned to ensure a reasonable spread of light but also to be safely accessible for maintenance. The choice of lamps is dictated by efficiency and life. So the sodium lights are ideal for background lighting which can be connected to a light-sensitive switch in stock buildings. Fluorescent lighting also has a long life and economic output. Double tubes in enclosed fittings are the best option for a light that is being switched on and off. Halogen flood lights are more expensive but ideal in conjunction with infra-red switched security lighting.

The level of artificial light provided is rated by the output in terms of lux. The design of a lighting system is therefore dependent on the lux rating (see Table 25.6). Care must be exercised, however, in matching the positioning of the fittings with accessibility for maintenance.

Table 25.6 Recommended lux rating

Lux	*Application*
500+	Extending daylight hours in specialist stock units
300	Over parlour pit, machine working areas and other local tasks
150–100	Parlour (overall), dairy, office, workshop (overall) and general working areas
50–20	Background lighting for grain stores, all night stock yard lighting

Highway access road

The junction with the highway needs special attention to cater for the speed of rural traffic and the size of delivery vehicles; 60-m visibility from a position 2.4 m back is required in both directions if possible especially on the near side. The exit must be at least 4.8 m wide for a depth of not less than 6 m, with 6-m radius splays both sides. Additional turning space may be required if the highway is very narrow. The construction must be approved as to the required standard by the highway authority. Planning permission is required for any alteration to any access (including fields) to the highway unless the highway is unclassified. Security is improved with gates or grids set at least 6 m in from the highway.

The main incoming road may be single: 3 m width with passing places at say 150-m centres, or double: 4.8 m width. In both cases a verge of 1.5 m minimum should be provided. This should be increased considerably if the verge is also to be used as a cow track, or trees are to be planted, or to provide for a ditch. Provision must also be made for agricultural machinery to turn into gateways: either by increasing the width of the gateway or by setting it further away. Roads should be built with a cross fall of say 50 mm (or occasionally with a 4.8-m road with a camber) to discharge surface water into a drainage system which might comprise a ditch or French drain.

Construction can be either concrete or flexible, bitmac or asphalt. Asphalt should always be used rather than bitmac if the road also serves livestock. The dung will tend to lift bitmac. Both types require a good hardcore base of not less than 150 mm consolidated thickness. A rubble base is very good providing allowance is made for it to be laid with a Traxcavator at say 300-mm thickness and blinded.

PAV 1 concrete is laid at a thickness of 150 mm usually with A142 fabric on a DPM to provide slip movement with expansion joints at 6 mc/c. The surface should be retamped 30 min after laying to provide a good coarse texture. Passing places may be a hardcore finish as specified below for farm tracks.

Bitmac is laid in two layers with a base course of 75 mm laid on a rolled scalpings bed and wearing course of 40–50 mm, the wearing course to be asphalt if used by stock. Edge protection will determine the life of the surface. Curbing may be necessary on the inside of sharp bends, etc. but is expensive for the whole road. A dished concrete channel is very effective. As a minimum the edge should be backfilled with a width of 300 mm of rammed 75 mm/dust scalpings sloping down from the edge and laid on a strip of geotextile fabric. This should be increased at gateway junctions.

Farm tracks

Farm tracks may be mainly for vehicles or mainly for dairy cows. There are some excellent propriety cow track surfaces now available using geotextile fabric in conjunction with a mix of bark strips or the like. The advantage of such tracks must be measured against the maintenance and the amount of vehicle movement. Great care is also necessary to ensure the material does not wash out in adverse weather. The traditional solution of a hardcore track has been transformed by the use of geotextile fabric such as 'Terram'. By laying such fabric on the excavated and graded surface – and excavating and filling any soft spots with rammed large stone – the road should remain maintenance-free for several years. The fabric prevents the mud oozing up through the hardcore. The topping surface will depend on what is available locally. Perhaps the best result is achieved with a minimum consolidated thickness of 75 mm/dust of damp scalpings rolled in with a 5-tonne vibrating roller. Note that the material should be laid as it arrives and not left in a heap for later use. As above, a cross fall, adequate drainage and verges are equally important.

Environmental control

The term 'environmental control' in agriculture is generally understood to mean the manipulation of temperatures (and occasionally humidities) by ventilation and heating. Its main use is in the optimization of livestock housing climates, but in limited form 'conditioning' in crop stores may also be described as such. In horticulture, the term has a much wider implication and may imply, among other things, adjustment of gas levels (CO_2, O_2, N_2), supplementation of light and 'natural' control of pests. These activities call for a precision of monitoring and control which is not usually justifiable economically for less-sensitive agricultural produce such as potatoes and pig-meat. However, environmental control is still appropriate to these latter, though on a lesser scale, because it is only by the achievement of a reasonable energy flow relationship between input and output that satisfactory profits can be made.

The major elements in the balancing act between input and output of energy, which is said to be 'environmental control' in livestock buildings, are represented in Fig. 25.18. As can be seen, some of these factors are the result of management decisions, whilst others are established by the nature of the building's design. However, by far the biggest influences brought to bear are those of the weather. Thus, whereas we loosely use the word 'control' to denote our ability to change conditions within the environment, often our attempts to do so are rendered difficult or even ineffective by the rapidly altering climate outside. Nevertheless, by manipulating the two 'mechanical' pieces of the equation, ventilation and heating (or cooling), some semblance of temperature (and humidity) control may be achieved. The process is made much easier, and less costly, if there is a 'buffer' between the outside elements and the internal environment. This takes the form of an insulative envelope, which can be simple as in the case of 'Yorkshire board', or complicated like the composite cladding around a pig fattening house.

Any of these systems has the objective of providing initial weather protection. Therefore, it is obvious that the starting point for all environmental control systems should be consideration of the standard of insulation to be used. In addition, it is now being realized that animal welfare is often prejudiced by inadequate insulation, particularly in flooring. This reinforces the need to review this factor first.

Insulating methods

Virtually any structure that encloses an environment is acting as an 'insulator', but the term is generally reserved for materials that have the specific property of resisting, or opposing, heat transfer. This implies that they are capable of preserving a high (relative) temperature difference between external and internal environments, irrespective of which of these is the higher. Some insulators are clearly better than others (e.g. timber as opposed to dense concrete), but the great majority of specific-purpose materials cannot be judged by instinct or on 'common knowledge' grounds. They need, therefore, to be compared by numerical values, of which there are several standard alternatives. The most frequently used of these is the 'U' value, a measure of the rate of

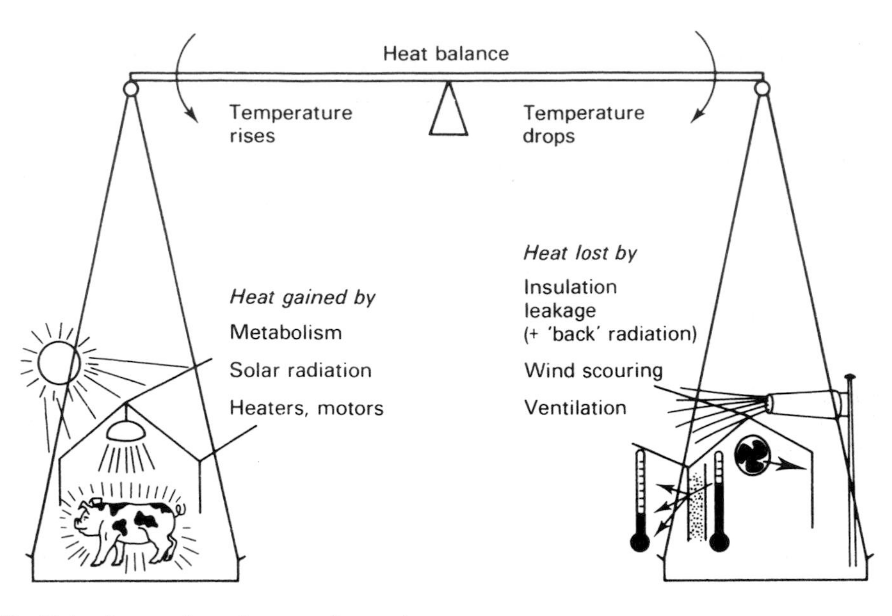

Figure 25.18 Major elements in environmental control.

heat flow expected through a structure under given temperature conditions; better insulators are associated with lower values. Table 25.7 gives some examples. Perhaps more logically, the 'R' value – for resistivity – is quoted for some materials. This is the inverse of 'U' and as such gives larger numbers to represent greater insulating effect. Both 'U' and 'R' are affected by their working 'conditions'. For instance, on exposed sites the 'U' will be greater and the 'R' smaller than normal, due to heat being scoured away from the insulating surfaces, creating more rapid transmission effects. In still air, however, the values will be appreciably better. Thus, they reflect the insulation properties for typical situations of material, climate and aspect rather than indicating absolute values which hold good at all times. Care should therefore be exercised in their interpretation in unusual situations.

Two further factors should be accounted for in the practical use of insulators. First, the majority of the better ones contain large quantities of air, usually in tiny bubble formations; these cavities should not be allowed to fill with water by condensation or wetting otherwise their insulating properties will be lost. Second, radiant heat (positive solar gain and negative 'back' radiation loss to sky) can further influence the insulation benefits, which is difficult to counteract. Insulators that inherently have antiradiation properties, being light in colour or having silvered surfaces, will help to limit heat transfer from this cause.

A good insulator, therefore, usually takes the form of a lightweight, hydrocarbon-plastic mass, filled with minute, discrete air bubbles and covered with a shiny, preferably silvered, surface which is also a water-vapour proofing coat. Care should be taken, however, not to compromise the fire flashover potential by using rockwool or glasswool insulation.

Summary check list of criteria

- 'U' value minimum $0.45\,W/°C/m^2$ but aim for 0.25 for ceilings/roof.
- Try to create a microclimate in local environment if not possible for whole building.
- Ensure insulation is protected with internal and external vapour check and air-tight joints.
- Ensure not broken by cold bridging causing condensation.
- Protect from impact, stock and rodent damage.
- Design for long maintenance-free life – minimum 20 years.

Ventilation methods (Fig. 25.19)

Whereas insulation is normally an inseparable part of a building, ventilation systems are often 'add-ons' or at least intrusions (e.g. ridge vents) into the structural envelope. They are therefore much more variable in pattern, but can broadly be divided into two types, although there is overlap between them. The simplest of

Table 25.7 'U' and 'R' (= 1/U) values of common insulators

	'U' ($W/°C/m^2$)	'R' ($W/°C/m^2$)
Wall		
215-mm dense, hollow, concrete blockwork	2.05	(0.49)
As above + outside rendering + 12 mm expanded polystyrene slab	1.14	(0.88)
275-mm 'cavity wall' of 100-mm dense concrete blocks + 25 mm expanded polystyrene slab	0.68	(1.47)
215-mm solid wall of foamed blockwork + rendering both sides	0.45	(2.22)
6-mm cement-fibre sandwich on 50 × 50 mm timber frame with expanded polystyrene core	0.45	(2.22)
Roof		
Corrugated cement-fibre sheet	6.53	(0.15)
Corrugated sheet steel + fibre 'insulation' board	4.82	(0.21)
Corrugated cement-fibre sheet + 80 mm mineral wool (+ vapour barrier) + 4.5 mm cement-fibre lining board	0.45	(1.67)
Corrugated cement-fibre sheet + 40 mm foil-faced polyurethane board	0.45	(1.69)

Note: mandatory minimum requirement for 'U' of industrial walls and roof is 0.45. Dwellings require 0.45 (walls) and 0.25 (roof).

Floor	'R_{f45}'
Dense concrete floor	0.042
Concrete slatted panel	0.086
18-mm screed on 150 mm lightweight aggregate	0.17

Note that R_{f45} is not directly equivalent to R(= 1/U) above as it takes into account sideways heat movement from 'point' body contacts, etc.

these from the point of view of equipment and installation is 'natural' ventilation, the power source of which is derived from wind force or warm air buoyancy (or more generally a combination of the two). 'Forced' ventilation, the alternative, is more complicated, needs mechanical/electrical power to operate and may involve the use of fans, ducts, diffusers, filters and controllers. Its advantage in contrast to 'natural' ventilation systems, which may outweigh these complications, is that it provides positive air movement and exchange irrespective of climatic or housing conditions. In livestock housing each of these systems has its appropriate place directly related to economic levels of investment, whereas in crop storage only the more positive powered methods are likely to be successful.

Natural ventilation

Natural ventilation (Fig. 25.20), created from 'stack' effects (buoyancy) and wind flow, is generally used for housing animals that have low 'critical' temperatures and where stocking rates are sufficient to produce satisfactory flows/exchange rates. Two general problems arise from the use of this system. The first occurs where there are no wind flows of any consequence, as for instance in anticyclonic weather conditions. The second is where the building is so wide from eaves to ridge that air flow patterns through the building are sluggish due to inertial effects. Both of these situations result in stagnant air masses which, once established, are difficult to move or exchange, and thus create health and structure hazards from condensation and general dampness. The problem is made worse by a reluctance to provide an adequate width of opening at the ridge, mainly due to the apparent risk of weather penetration. A protected/unprotected ridge (see Fig. 25.21) which is open overcomes this problem.

It is acknowledged that natural ventilation is at best unpredictable and at worst impossible to control, the situation not being improved by common variables such as building shape and size. However, Fig. 25.21 attempts to do this using a formula delevoped by MAFF many years ago that gives a satisfactory result in cattle and sheep buildings.

Many structures, despite having an open ridge, are underventilated much of the time. In recognition of this there has been a resurgence of interest in spaced sheet roofing for large livestock buildings – especially for those with no livestock in the centre area (e.g. tractor passage). A simple way of creating this effect is to leave gaps of 12–18 mm (usually 16 mm) between each line of sheets

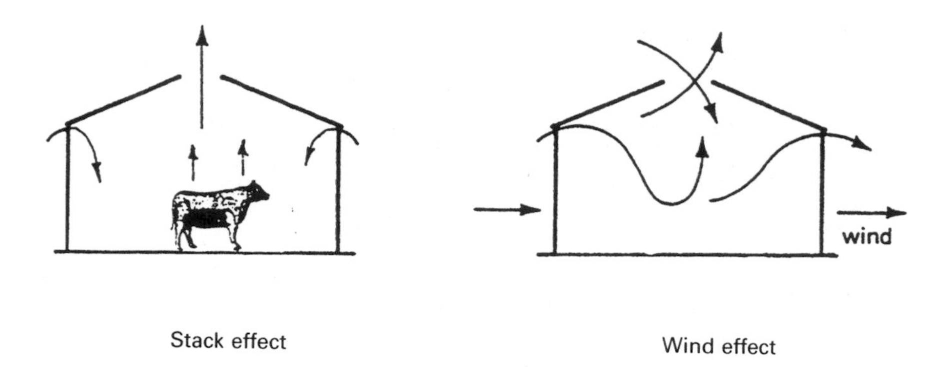

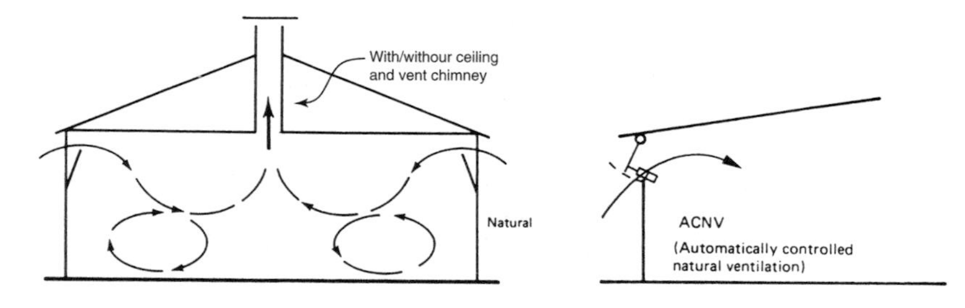

Natural ventilation

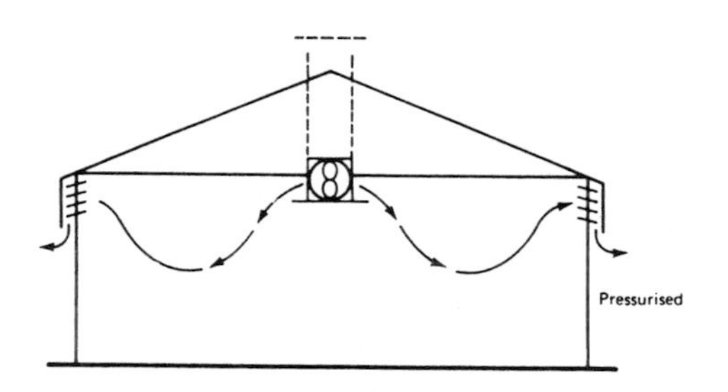

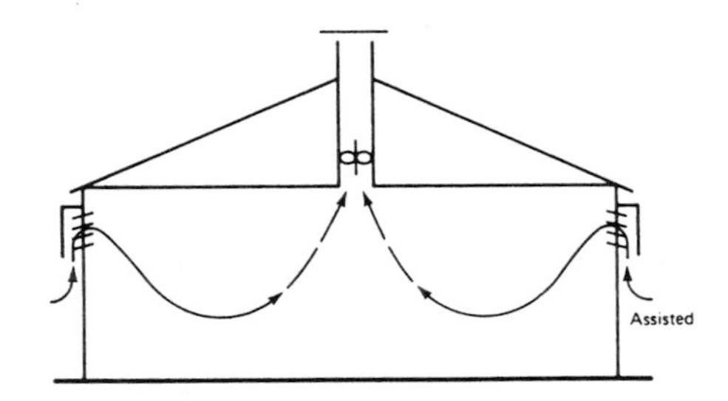

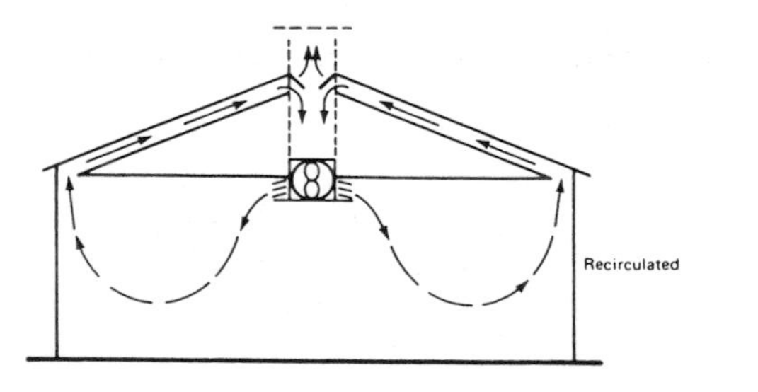

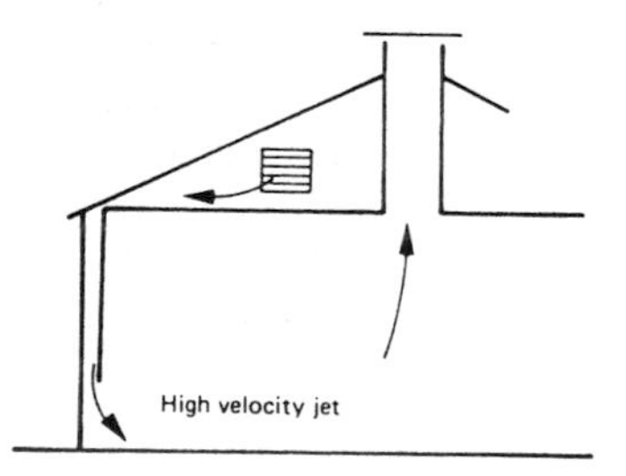

Forced ventilation

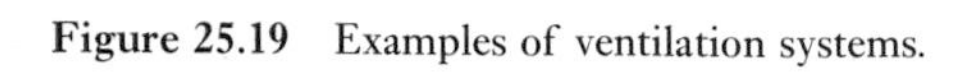

Figure 25.19 Examples of ventilation systems.

YOU NEED TO KNOW :–

BUILDING SPAN [ ] m   BUILDING LENGTH [ ] m   ROOF PITCH [ ]

No. OF STOCK [ ]   WEIGHT OF STOCK [ ] Kg.

Stocking density relates only to bedded and loafing areas – not tractor passages.
Ensure stocking density and feed space length relates to BS 5502 for final weight.
This calculation assumes reasonable eaves height of normally 3.2 m for cubicle buildings to 3.9 m for strawed yards and assumes a fibre cement roof.

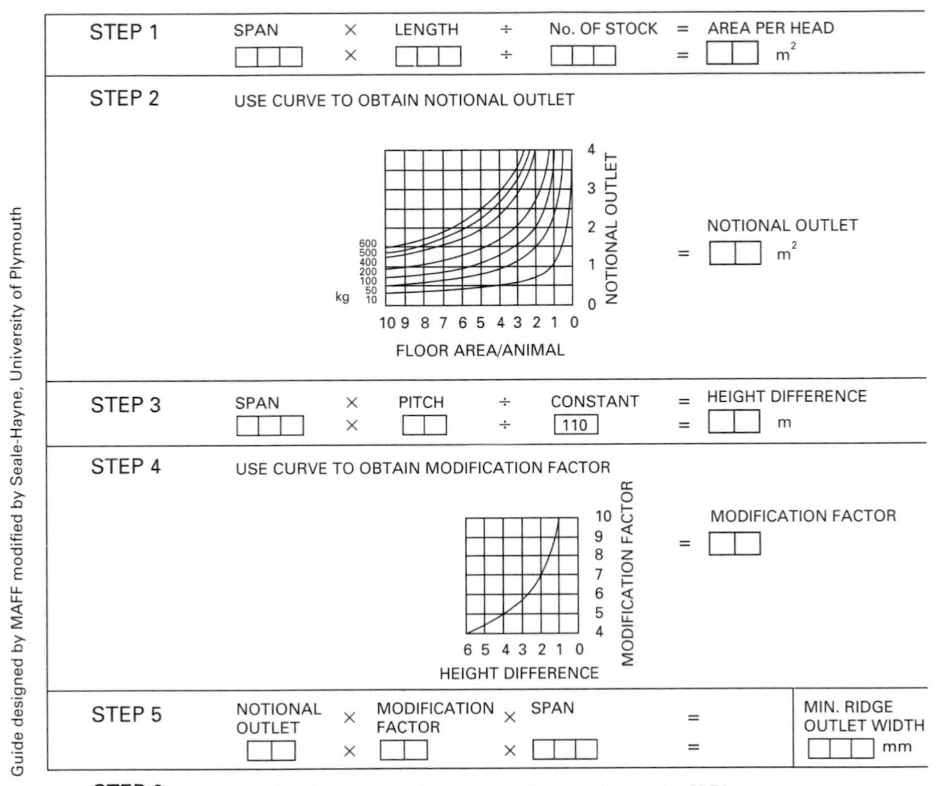

STEP 6: For extra wide buidings especially with centre tractor passages use 50/75 mm unprotected ridge gap and use spaced roof sheeting over livestock areas only. To calculate spaced roof gap establish total outlet ventilation area (TOVA) = ridge length x outlet width.
EITHER: TOVA:- number of sheet joints along area to be ventilated
(usually bld. length ÷ 1 m) multiplied by the length of the joints. = SPACED ROOF GAP [ ] mm
OR: use 16 mm gap over livestock areas only - (lapped over remainder)
FOR BOTH: specify fibre cement sheeting cut to 1 m. REDUCE GAP IN EXPOSED SITUATIONS OR WINDWARD SIDE

Complete LH graph before reading across to board width to establish gap

STEP 7

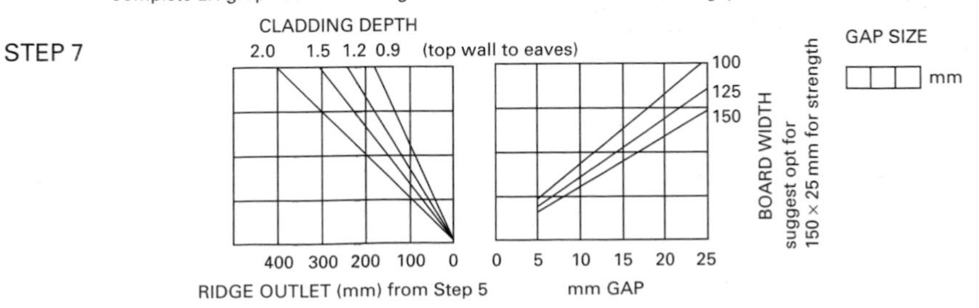

GAP SIZE [ ] mm

Figure 25.20 Natural ventilation design guide (cattle/sheep).

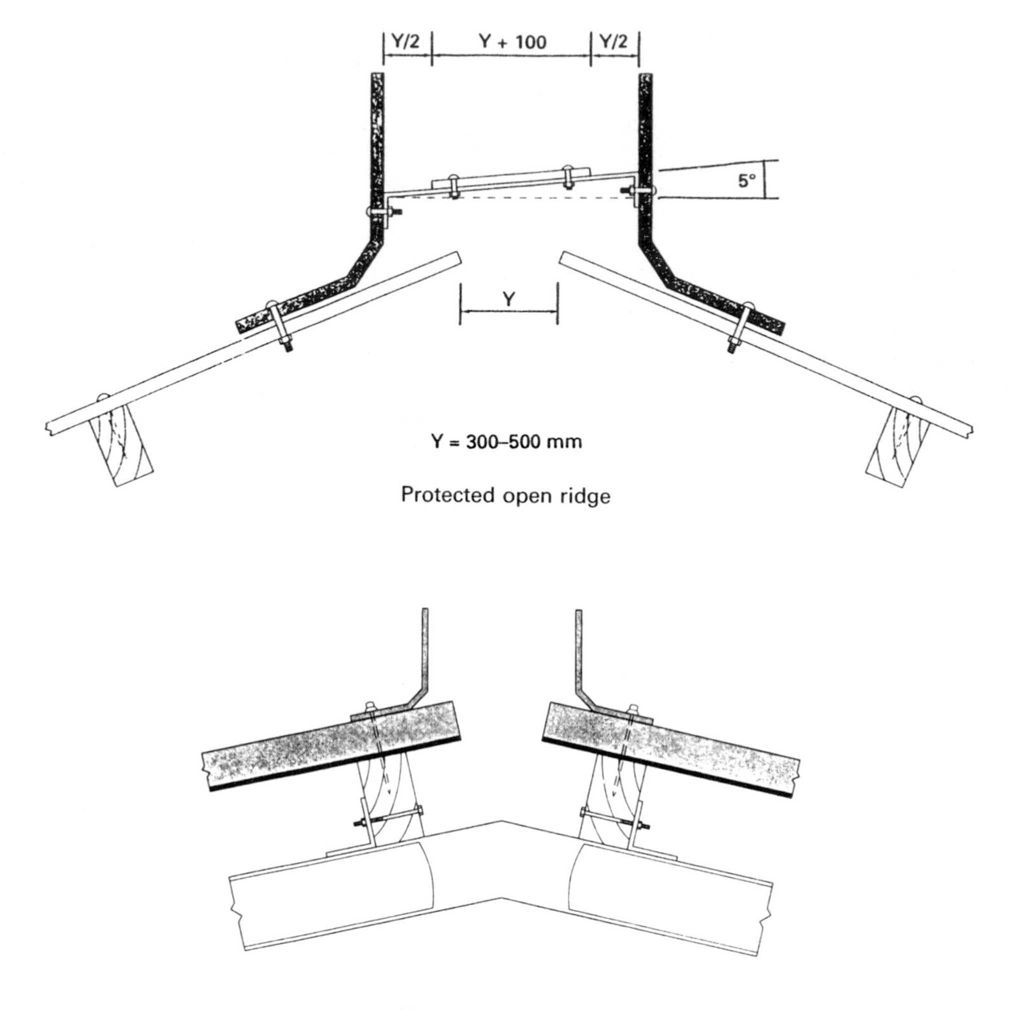

Figure 25.21 Ridge ventilation [taken from Eternit (2001)].

when covering the roof (but the sheets have to be specially ordered as they are cut to ensure the corrugations conduct rain satisfactorily to the gutters!). This provides good ventilation rates and patterns, balancing adverse wind forces on the roof, and has the bonus of providing 100% natural light in the gaps, but nevertheless still lacks the essential element of controllability.

Automatically controlled natural ventilation (ACNV), a fairly recent development based on improved electrical-control equipment (electronics), is now commonly used for small volume, high stocking density situations such as weaner housing, where expensive external heat sources are often required. This system uses a combination of wind flow and stack effect to achieve good air flow rates (and control) by mechanically actuated ventilation flaps arranged in optimal structure positions. It may also be used in conjunction with fan power assistance, as a backup, when the wind flow fails. This circumvents some of the drawbacks inherent in a purely 'natural' system.

Forced ventilation

Forced ventilation which relies solely on fan power for air exchange and distribution can be divided into 'suck' (exhaust) and 'blow' (pressurizing) methods. Each has advantages for specific types of building but generally those that exhaust are simpler and cheaper but less effective, particularly in achieving good air distribution. In contrast, pressurizing methods will exchange air efficiently, enabling better scouring and control. They also offer the possibility of air recirculation, filtration and silencing but almost invariably cost more. The potential for creating draughts will also be less with pressurized air flow, so long as it is conducted in properly designed ducting at appropriate (low) velocities. Figure 25.19 demonstrates some of the air-flow pattern variants which are possible with both these systems.

Ventilation rates (see specific 'functional design of purpose buildings')

The need to achieve a reasonable temperature stability in a building at a set level can result in very different (and variable) air flow needs, possibly requiring alteration from the minimum to remove noxious gases in winter up to a high rate sufficient to reduce heat build-up in still-air in summertime conditions. This ratio of demand may be as much as 30 or 40 : 1, making successful ventilation very difficult to achieve, even without interference from adverse wind flows or the inadequacies of control equipment.

The actual air-flow rates required can be calculated on two bases: either whole building exchange (house volumes/hour) or the needs of the particular species or crop concerned (m/kg liveweight, m^3/t, litres/bird). Both these methods have their merits, but recently the latter approach has been more popular.

(See also BS 5502, Part 52, for alarms and emergency ventilation.)

Fan-assisted airflow criteria check list

- Ensure optimum insulation first, and for most systems the building is sealed.
- Design for all seasons and stock size/age range.
- Design for rotary motion preferably finishing over dunging area, but retain option to recirculate in very low external temperatures.
- Ensure circulation modification by internal obstructions is taken into account. Smoke test neighbour's installation!
- Strong winds may have to be baffled in exposed situations before reaching inlet.
- Programme automatic controls for multiple sensing of both variable temperature and minimum ventilation. Remember it is expensive to extract warm air on a cold day, but overheating must be controlled by ventilation and reduced by insulation.
- All systems must be protected with alarm, manual back-up, and generator capable of maintaining metabolic heat to not more than 5° above external temperature.

Humidity control

This has not hitherto figured very largely in environmental control design, due mainly to the difficulty of sensing any change in conditions with reasonable precision. Added to this, the reaction of stock people to humidity tends to be anthropomorphic and so controllers are often overridden when the atmosphere is felt by them to be 'uncomfortable'. However, as there is a growing awareness of the effects of 'damp' conditions in disease transmission and general animal welfare, more emphasis is bound to be placed on this in future. One particular problem of high humidities lies in the implied risk to structures from condensation effects, both on cold structure surfaces and within insulation under 'dew point' conditions. These situations if permitted over long periods will depreciate building materials and reduce their effectiveness, ultimately affecting stock and crop health.

In crop drying or ventilating humidity is an equally important factor. An important energy saving is made by incorporating a dehumidifier at the air intake of a grain drying plant. In ventilating crops systems will often incorporate a measuring device to assess the humidity of the ambient air. If this is greater than the internal air a parasite fan is brought into operation to recirculate the internal air instead.

Water pollution control (BS 5502, Part 50, refers)

The problem of watercourse pollution from farms is one of the main issues on the environmental agenda. It has been tackled by means of a comprehensive Code of Practice, backed up by specific legislation that affects every holding. Refer to sections on slurry and farmyard manure storage and silage bunkers for detailed information. Enforcement and monitoring are the responsibility of the Environment Agency (EA).

Code of practice

No installation of control should be attempted without knowledge of the requirements laid down in MAFF/WOAD (1998a) *The Water Code (Code of Good Agricultural Practice for the Protection of Water)*, and the detailed CIRIA Report 126, *Farm Waste Storage: Guidelines for Construction* (Mason, 1992). The MAFF/WOAD (1998b) *Farm Waste Management Plan: A Step by Step Guide for Farmers* is also useful (see 'Further reading').

The requirements summarized here are mandatory for England and Wales. Scotland and Northern Ireland have equivalent regulations and codes which require the same high standards, but there are important differences in interpretation so the correct Codes for the location should be applied.

Legal requirements

Adherence to the above Code and Guidelines is not legally binding. The legal requirements are set out in Statutory Instrument No. 324 (1991), The Control of Pollution (Silage, Slurry and Agricultural Fuel Oil) Regulations 1991 (under The Water Act 1989) as amended. Contravention is a criminal offence.

Air pollution must also be considered: now covered by MAFF: Code of Good Agricultural Practice for the Protection of Air.

See also relevant sections in *The Agricultural Notebook* on health and safety (Chapter 15) (Health and Safety at Work Act 1974) and animal health (Chapter 23) [Code of Recommendations for the Welfare of Livestock and the supporting regulations (1990)]. This will include aspects on child-proof fencing of lagoons, etc.

Planning permission and building regulations

All structures or engineering works will require either full planning permission or a planning notice, although the extent of applicable works may vary from one authority to another. There should be no problem, however, with complying with the exemption requirements as prescribed for Class III of the Building Regulations 1985. See under 'Design criteria'.

Consultation

Prior consultation with the EA is always advisable. Free general advice on Farm Waste Management Plans may be available from ADAS in specific catchment areas. Neither of the above consultations will be sufficient to draw up a scheme, which should be done by ADAS, consultants, equipment suppliers, or a combination.

Notice of construction

It is a legal requirement that notice must be served on the EA of the relevant works at least 14 days before use. Having served the notice you can proceed with the use after the 14 days have elapsed. There is no need to wait for an inspection. If it has to be brought into use, even in unfinished state, notice should still be served in time with an explanation.

Construction standards (see 'functional design of specific purpose buildings')

Many pollution incidents are the result of substandard construction. It is therefore essential for farmers to stipulate that the installation must be constructed to:

- BS 5502 Class I (50-year life) or Class II (20-year life). The regulations stipulate a minimum 20-year design life (maintenance-free for underground works);
- the relevant part of BS 5502 for that structure, e.g. Part 50:1989 for slurry storage;
- the relevant BS and manufacturer's instructions for each material, e.g. BS 8007 for concrete.

Applicable works (summary only)

(1) All forms of silage storage except:
- baled silage or Ag.Bags in sealed bags at least 10 m from any unsealed water, e.g. land drain, ditch, pond – so bags can be stored in a field or on hardcore base subject to risk assessement;
- field silage subject to all other restrictions and in position agreed with EA. Minimum 14 days notice before use, but cannot be used until agreed by EA – unlike permanent structures;
- exempt if built before 1 March 1991 or contract for construction placed before 1 March 1991 and completed by 1 September 1991 – providing prior notice of intention to continue to use was served before 1 September 1991, and the above 10-m rule applies. If an exempted structure is substantially altered or extended, then it loses the exemption.

(2) All forms of slurry (which includes water contaminated with excreta) storage except:
- slurry temporarily stored in mobile tanker, maximum capacity of 18 000 litres;
- exempt structures as above;
- solids that cannot be pumped, i.e. farmyard and poultry manure, and solids from a separator.

(3) All forms of fuel oil storage on farms except:
- fuel temporarily stored in a mobile tanker;
- fuel stored in an underground tank;
- exempt structures as above;
- total quantity on farm does not exceed 1500 litres;
- fuel oil used exclusively for heating a dwelling.

Notice requiring works

Where the EA considers there is a significant risk of pollution from an existing installation, it may serve notice on the person having custody or control of the relevant substance requiring him/her to carry out such works as it considers appropriate to reduce that risk to a minimum within a reasonable period (minimum 28 days). There is also provision for negotiation and appeal.

General design calculations

(1) Calculate total livestock excreta in each age group held in buildings/yards and period held. Assess maximum daily flow and minimum total storage requirement for each category.
(2) Assess total silage effluent and maximum flow rate if combined in open storage (dangerous to mix in enclosed) and add to the above if overlapping in time scale.
(3) Add total wash-down water and any other effluent. Assess maximum daily flow.
(4) Add rainfall – a report from the Meteorological Office is required (1999, cost £65) giving the average maximum for 1–3 hours, 24 hours and 4 months on a 5-year return period. (The four winter months figure will not be one third but more like two thirds.) The Met. Office figure should be multiplied by the following areas:
   - any external uncovered area where livestock is held, or areas polluted from such holding locations, or polluted by silage effluent (it is reasonable to exclude roads and tracks where dairy stock is passing through but not held);
   - any roof area discharging on to first bullet point; so it is important to eliminate if possible;
   - any surface water discharging on to first bullet point from fields or roads; – as above;
   - the total surface area of the uncovered storage facilities (in the case of a slurry lagoon this could double the required storage capacity);
   - addition for future expansion and/or contingencies.
(5) Calculate total low-risk land available for effluent distribution, after taking out land within 10 m of any watercourse – increase the 10 m according to risk (e.g. sloping towards watercourse and to 50 m from a spring, well or borehole).
(6) Consider options for spreading and relate to spreading rates (maximum rate 50 m^3/ha per application) and maximum available nitrogen, and other limiting factors (see Code). Relate to volume above.
(7) It is likely that the main problem will be excess surface water especially in high rainfall areas. It is therefore essential that a great deal of thought be put into the design of the system to eliminate pollution of 'clean' areas and separation/diversion of surface water running on to the collection areas or into the drains.
(8) The minimum total capacity of the system is laid down in the regulations as set out under 'functional design of specific purpose buildings' with one important exception. The pumping chamber in a low volume irrigation system merely states that the pump should be able to cope with a maximum 24-hour volume, whereas it is also essential that at least 24/48 hours' storage is available to allow for breakdown, freeze up, etc. (with back-up such as a tanker spreader) or 48/72 hours without. In many situations, especially higher altitudes, it will also be necessary to provide for emergency additional storage for much longer periods of freezing and/or snow-covered conditions: use adjusted figures omitting most of the rain water.

Special areas

If the farm is sited in a National Park, Environmentally Sensitive Area, Nitrate Vulnerable Zone or Site of Special Scientific Interest, you must approach the appropriate authority first before designing any scheme as the regulations are different especially for the last three. Some restrictions may also be imposed by the EA on aquifer zones.

Site ground conditions

As with all farm buildings it is essential to know the soil-bearing capacity. Ignoring it in the design not only contravenes the British Standards but is courting financial disaster and prosecution. Likewise in the construction of any earth-sided bunker (not allowed for silage in England and Wales) or lagoon, a soil analysis is required unless a lining is to be used (see the excellent guidance in CIRIA Report 126).

Environmental statement

This may be required to support a planning application on highly sensitive sites and large-scale pig/poultry enterprises especially with low areas of land.

Functional design of specific purpose buildings

Cattle accommodation (BS 5502, Part 40, refers)

Dairy cows

Space for loose housing Cost range: £140–£180/m^2	Between 5.0 m^2/small animal and 8.0 m^2/large animal (of which 75% strawed) + 1.0–3.0 m^2 when housing not completely covered
Space for cattle on slats Cost range: £180–£200/m^2	3.0–4.0 m^2. Slats 125 mm wide with 40-mm gap (see Fig. 25.22)
Cubicle dimensions (see Figs 25.23–25.26)	1.1–1.3 m wide c/c × 2.3–2.5 m deep (for small and large cows) + passageway (2.7–3.0 m)
Feeding and loafing space	Cubicles: 150% of cubicle area Loose housing: 50% of bedded area
Minimum height to eaves	2.4 m: from cubicle bed (for cubicle systems) 3.3 m: loose housing plus max. bedding depth; say 3.9 m overall
Temperature and insulation	Not important down to 5°C, no control of environment usually. 2°C is 'lower critical' temperature of small (50-kg) animal. Insulation limited to bedding or, occasionally, flooring.
Ventilation	Natural; by open ridge or slotted/spaced roof and spaced boards in upper wall area. Usually underestimated, with too little ridge vent provided. Need to reduce gable end draughts. (See Fig. 25.20) for calculations
Bedding	Straw 1–3 tonnes/cow/winter for loose housing, 200–300 kg/cow/winter chopped straw for cubicle building Wood shavings 400–600 kg/cow/winter
Water	40–70 litres/cow/day (from trough with allowance of 0.03 m^2/cow or one drinking bowl/10 cows)
Effluent output and drainage	Urine 25–60 litres/cow/day – depends on feed and housing system Dung 10–35 litres/cow/day – depends on feed and housing system 53 litres/day excreta adopted by MAFF for pollution potential calculations. Allow 27 litres on scraped areas if loose housing Drainage slopes on scraped floors: 1 : 100/200 (near level to retain liquid) Slats may be part or whole of floor, may have drainage channel or tank under, to contain 4–6 months' accumulation, may be raised 2+ m to provide total storage for permanently housed stock. Slats may be of timber, concrete or steel with general dimensions of 125 mm wide with 38 mm gap, 150 mm deep and 25 mm taper (see Fig. 25.22)

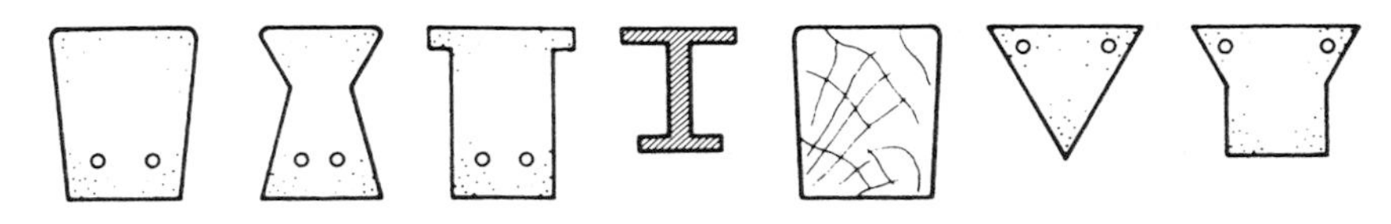

Figure 25.22 Slats: some alternative sections (BS 5502, Part 51, refers).

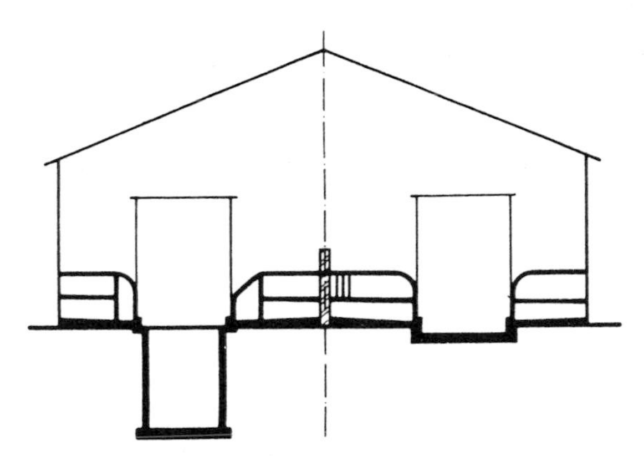

Figure 25.23 Section: slurry cellar/solid floor cubicle building.

Figure 25.24 Section: typical kennel-type cubicle building.

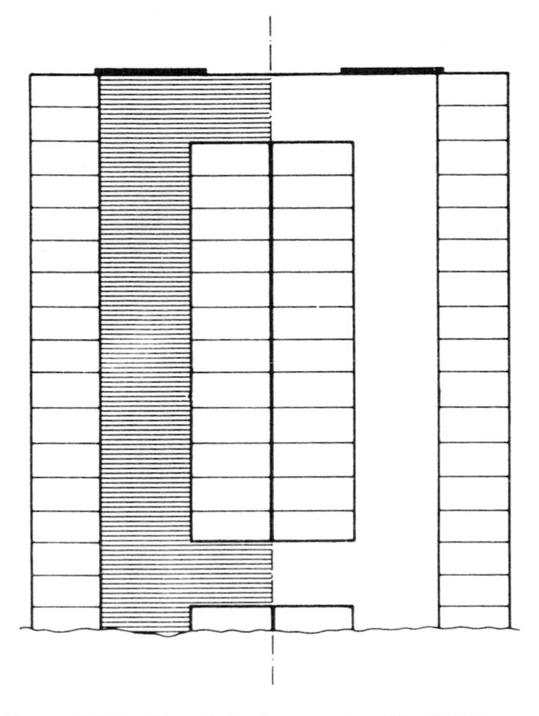

Figure 25.25 Plan of the above section (Fig. 22.23).

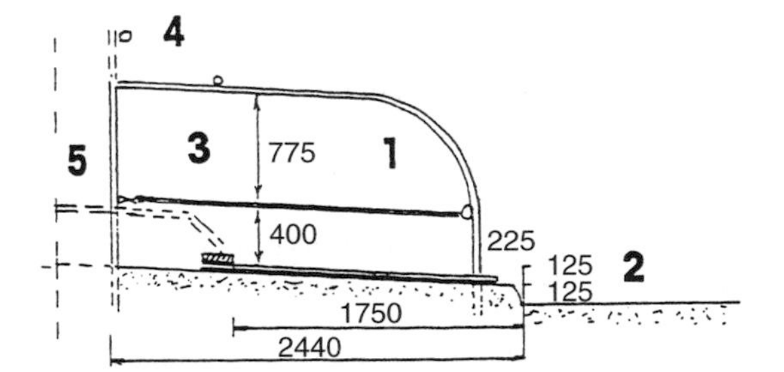

Figure 25.26 Section of typical cubicle.

1. 48 × 4 mm tube with twisted rope bottom rail
2. Bullnosed 125 mm step. 125 mm fall to step
3. 150 × 75 mm oak curb bolted to floor securing mat (option to build brisket board over)
4. Adjustable headrail (raised if brisket board)
5. Option of additional lunging space

Feed space	600–760 mm trough length/cow 150–200 mm self-feed silage face/cow with unrestricted access 300–400 mm self-feed silage face/cow with limited access
Collecting area Cost range: £150–£180/m^2	1.4–1.6 m^2/cow
Other considerations	Races (800 mm wide, internal), and handling equipment should be adjacent to and integrated with the housing Floors: all non-slip: wet or dry grooved, tamped finish not good enough Avoid steps if possible. If essential, rise max. 250 mm/step and 'going' or tread min. 760 mm Need an isolation box separated from rest of buildings Provide calving boxes (min. 15 m^2 each), number to suit calving pattern and herd size (e.g. typical min. 1 box/30 cows)
General	Consider cow management and handling, effluent problems, feed store siting, cow comfort, access for mechanical equipment

Calves

Space for individual pen housing up to 4 weeks old Cost range: £130–£160/m^2	1.5–1.8 m^2 in single pens of approximately 1 × 1.5 m (minimum depth) for small calves
Group housing space, up to 3 months	1.5 m^2 increasing to 2.5 m^2
Space in open-fronted housing, 3–6 months	2.5–3.6 m^2 with three or four per pen (see Fig. 25.27)
Height recommendations	2.5 m to ceiling, minimum of 6 m^3/calf. Barrier height 1.0 m minimum, up to 1.4 m for larger calves to provide isolation between pens
Temperature	20°C for young calves, 15°C at 1 month old and 12°C at 3 months. Note that lower critical temperature (LCT) varies greatly with bedding type and dryness.

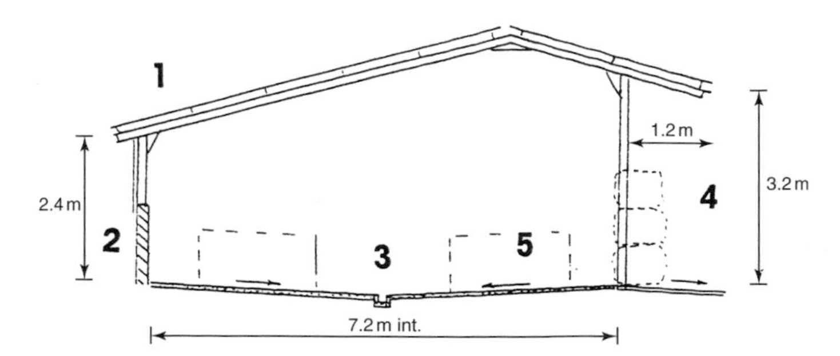

Figure 25.27 Section through a typical open calf shed.
1. Spaced sheet roof, 10% roof lights, butt jointed ridge
2. Block rear wall, space boarding over
3. Conc. floor sloping to channel or porcupipe drain
4. Open Front, feed fence or bale wall for youngest group
5. Single pens for youngest group (with kennel roof in severe weather)

	Thus, if kept in 'non-controlled' accommodation may be very close to LCT, particularly if draught-prone position. Local/individual snug areas can assist
Environment controlled	Such housing is no longer recommended on grounds of both cost and health problems caused by the lack of sufficient response to environmental changes in all but the best designed housing
Ventilation	Adequate, draught-free ventilation essential for good health. Calves are highly susceptible to disease in poor conditions
Feed and water	Feed from bucket or machine at pen front, or trough for older calves (350 mm/animal), all near passageway. Water: from troughs in groups or bucket near passage for pens
Effluent output	10–30 litres/day depending on food, size and age
Flooring	0–4 weeks: hard, cleanable, durable concrete, resistant to disinfectant. Surface with wood float. Drain to 'centre dome' passageway at 1 : 20, and via step of 100 mm at edge of pen to drain channel at passage-edge
	4 weeks to 3 months: hardcore floor also acceptable
	Above 3 months: slats also acceptable; 100 mm wide, 30 mm gap
Walling	As for floor with hard cement render and waterproofing additive. Form curved junction (coving) with floor
Other considerations	Solid or 'see-through' pen divisions? Opinions vary as to value of each because of hygiene and social conflict
	Care with surfaces as calves chew and lick everything
	Drainage must never interconnect pen to pen
	Relative humidity should be kept down to 70% or less. (May be more easily achieved with 'open' than 'controlled environment' accommodation)

Beef animals

Space on straw (see Fig. 25.28)	2.0–3.5 m^2 per animal at 12 months, 4.5–6.0 m^2 per 2-year-old bedded animals (4.0 m^2 with self-feed silage systems)

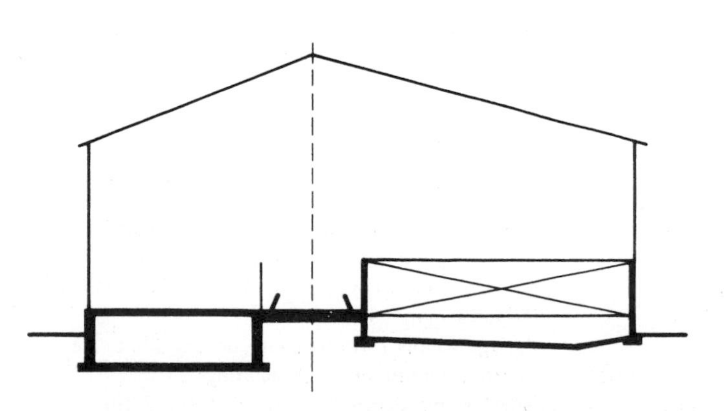

Figure 25.28 Slatted floor/deep straw beef housing.

Cost range: £120–£160/m^2	10–20 in group to one court or pen
Space for animals on 'sloped' ('Orkney') lightly littered floors	2.5–3.5 m^2 per animal 12–24 months old (sloped area at 1 : 16)
Height to eaves	3.0 m minimum (plus extra for over-winter dung build-up of 1.2 m in strawed yards, or slurry depth under slatted floor); 4.2 m max.
Temperature and insulation	Not critical, within range 2–20°C Insulation not required
Ventilation	Natural, with space-boarded upper walls and ventilated ridge. Avoid draughts (slotted roofing appropriate)
Bedding consumption	Straw 1–2 tonnes/beast/winter
Trough space	500 mm length/9- to 12-month-old beast; 600 mm length/18-month-old beast; 700 mm length/2-year-old beast; 760 mm/mature bull beef animal; 150 mm/animal for *ad libitum*-fed young stock
Water consumption	25–45 litres/head/day Allow 0.025 m^2 trough per head or one drinking bowl per 10 animals
Other considerations	Kennels and cubicles may be appropriate (see 'Dairy cows') especially for training dairy herd followers Safety of personnel especially with bull beef animals: requirement for sturdy barriers and doors.
Bull beef	Vertical rail partitions and gates at 150-mm centres. Feed barrier at 380-mm centres. (Normal beef centre barrier at 300 mm.) Heights to be a minimum of 1.3 m (normal 1.2 m)

Milk production buildings [BS 5502, Part 49, and Dairy Products (Hygiene) Regulations]

Cost range: too many variables to cost on m^2. Allow £240–£300 for building shell

Abreast parlour (six standings) (see Fig. 25.29)	Overall depth 4.8–5.8 m Overall width 6.9 m Side exit passage 0.3 m for personnel Minimum headroom over standings 2.0 m (plus 0.45 m step to standings gives total height minimum 2.45 m)
Tandem parlour (six standings) (see Fig. 25.30)	Overall width 5.1 m (plus 2 × 100 mm for rubbing rail) Overall length 9.3 m Passage widths 1.0 m (one end, across, and two exit passages) Minimum headroom over standings 2.0 m Operator pit: depth 0.8 m, width 1.8 m/2.3 m

Herringbone parlour (eight standings) excluding 'quick-exit' parlours (see Fig. 25.31)	Overall width 4.8/6.0 m Overall length 6.9 m (Allow for outside hoppers if applicable) Operator pit: as tandem Minimum headroom over standing 2.0 m Width of cow standing 1.5–2.0 m each side May be open at entry end
Chute parlour (Fig. 25.30)	As tandem, less end and exit side passages and with narrower pit. Six standing measures 7.4 × 3.1 m overall
Rotary, tandem (Fig. 25.32)	Space occupied overall from 5.5 × 5.5 m for six milking points up to 15.5 × 15.5 m for 18 points
Rotary, abreast	Space occupied overall from 8.1 × 8.1 m for 12 milking points up to 14.0 × 14.0 m for 30 points
Rotary, herringbone	Space occupied overall from 7.0 × 7.0 m for 12 milking points up to 14.0 × 14.0 m for 28 points
Polygon parlours (see Fig. 25.33)	Four ranks (or three in 'trigon') of standings with up to eight cows per rank, in flattened diamond or triangle shape arrangement, 16–40 cows at one time in parlour, 16–18 m wide × 18–33 mm long, with 7–9 m at pit centre
Collecting yards and circulation space	Circular or rectangular, open or covered (in which case ventilation should not be through the parlour) allowance of 1.5/1.7 m^2 per cow. Taper at parlour entrance. Consider backing gates.
Drainage	Smooth floated channels or vitrified pipe. Overall falls 1 : 100 or 150 over lengths of buildings. 1 : 25 for standings and falls to outlets and gullies. Channels to have 50-mm-high minimum rise at edges
Water supplies	0.5–1.0 litres/cow/day udder wash allowance Cleaning water without power hose: 14–22 l/cow/day (18 av?) Cleaning water with power hose: 27–45 l/cow/day (35 av?)
Services	Pipes and fittings required for vacuum, milk transfer, cleaning systems; storage facilities for hot water, warm water, milk All accessible electricity supplies and fittings should be low voltage (110, 24 or 12 V) or leakage protected. All metal fittings to be 'earth' bonded. Consider heating installations against freezing of pipes and for operator comfort
Light levels and sources	Natural preferred from above for general illumination. Local high-intensity artificial light at udder level. 150 lux overall. Visual contact to be maintained between parlour, dairy (window), collecting yard and shedding gate exit
Flooring	Hard, durable, resistant to detergents and disinfectants with non-slip surface (granite chippings or carborundum dust or epoxy) which is cleanable by pressure hose. Must not puddle
Walls	Rendered with waterproof, cleanable, smooth surface to a height of 1.5 m minimum (1.8 m preferable), painted with epoxy resin or other acceptable finish. Fibreglass particularly effective. No insulation provided. Curved filler in render at junction between floor and walls

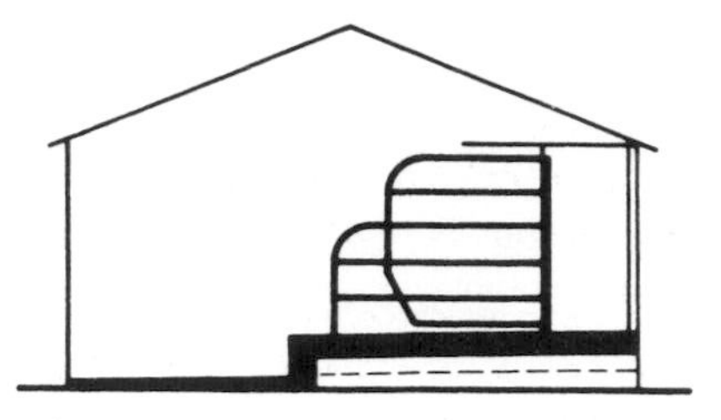

Figure 25.29 Section: abreast.

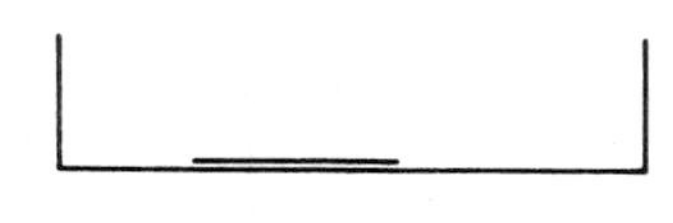

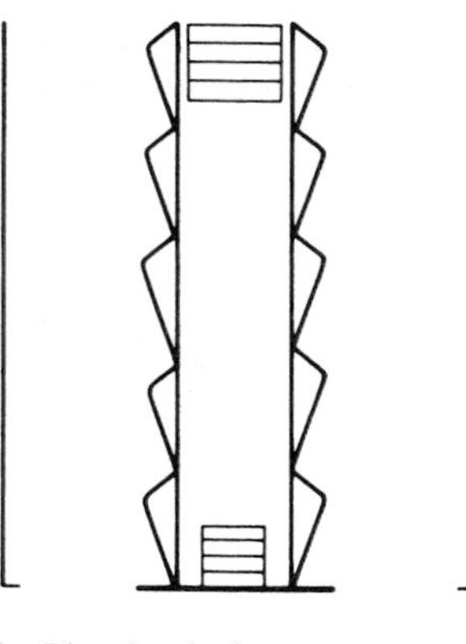

Figure 25.31 Plan: herringbone.
Feed hoppers normally outside if required.

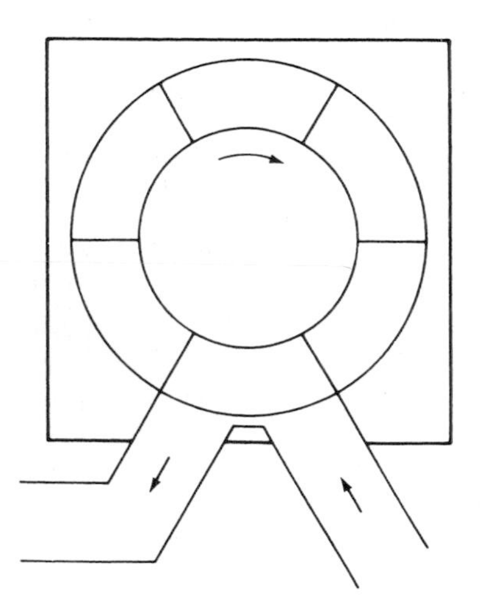

Figure 25.32 Plan: rotary tandem (simplified).

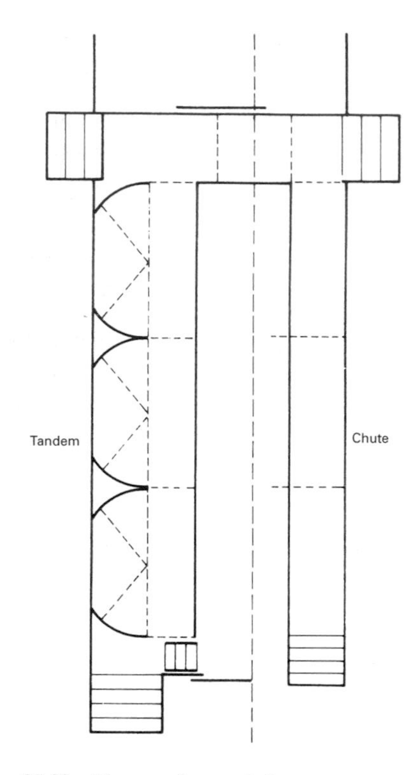

Figure 25.30 Plan: tandem and chute.

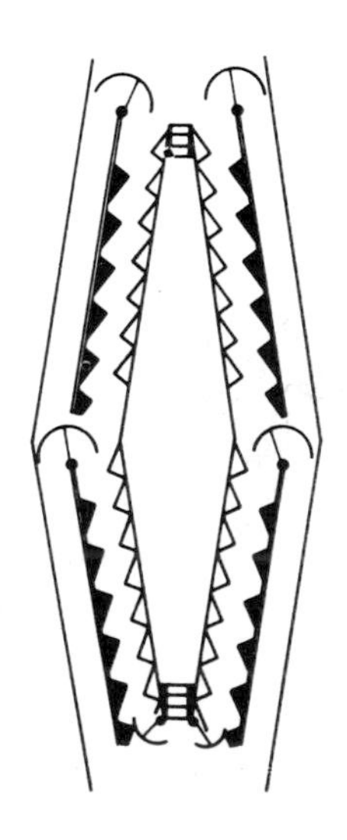

Figure 25.33 Plan: polygon.

Figures 25.29–25.33 Some parlour options.

Other considerations

Consider implications of open or enclosed parlour. This country is moving towards enclosed with vermin-proof doors

Recommend insulated ceiling or roof to reduce temperature extremes especially on hot summer days with controllable mechanical ventilation or opening skylight

Step into/out of parlour for cows should be non-slip with 150- to 250-mm 'rise'. Heel stones on pit edges should be a minimum of 100 mm above standing level. Steps for operatives should not exceed 220 mm rise and should not be less than 250 mm going

Sliding doors or rubber/plastic flap doors required with automatic opening and closing as appropriate

Dairy

See Fig. 25.34. Wall: washable wall surfaces to 1.8 m (e.g. fibreglass), easy clean above. Coved junction with floor. Roof: dust proof and easy clean. (Recommend insulated to reduce condensation.) Floor: easy clean screed, free draining with fall to external drain or trapped gully. No slip finish (carborundum dust). Vermin proof. Secure: door kept locked. Adequate natural (10% floor area) and artificial (150 lux) light. Adequate natural or artificial ventilation (suggest hopper type window). Any timber must be easy clean – so inside wood doors must be clad with sheet steel

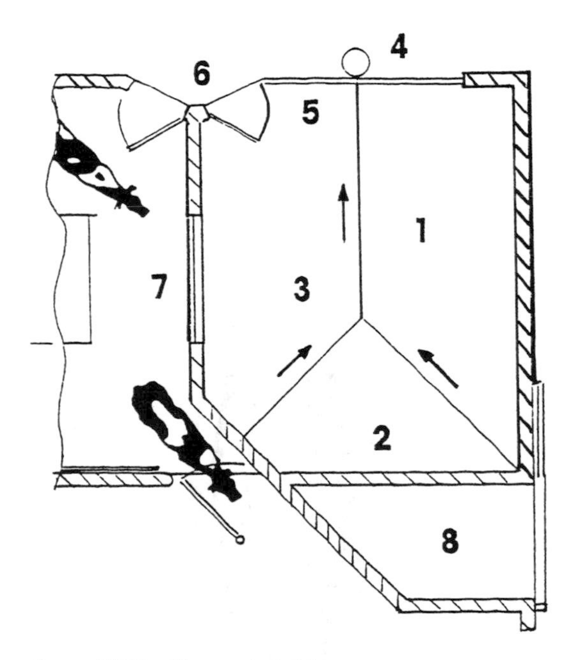

Figure 25.34 Plan: typical dairy.
1. Insulated roof with 10% rooflight area
2. Walls rendered with fibreglass up to 2 m
3. Sloping screed floor with carborundum dust
4. Ext. trapped FW drain via vermin proof slot
5. Demountable section to extract tank with opening or louvred window with fly screed
6. Locked door – not direct from parlour
7. Protected viewing window – acrylic glazing
8. Ventilated machine area with tank loft over

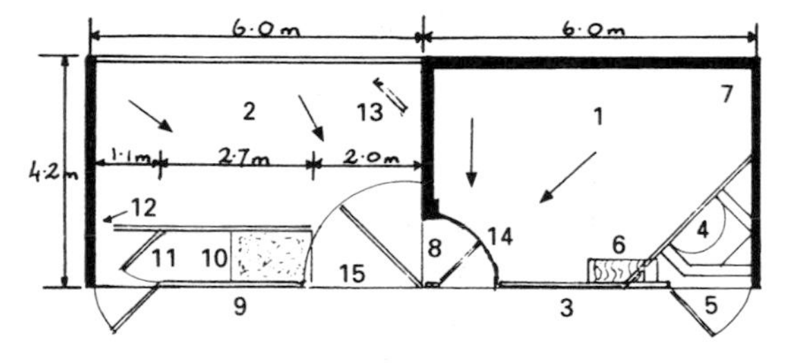

Figure 25.35 Bull box and exercise yard.
1 Box – covered (insulated or heated floor?)
2 Yard – covered or open (both with non-slip floors)
3 Dairy collecting or dispersal yards (usual site in relation to)
4 Large manger for silage with yoke
5 Childproof access door – width to suit feed system
6 Water trough – protected service box and rails over
7 Reinforced solid walls – rendered if concrete block
8 Can add further gate or slide in rails
9 Vertical rails at 150 mm centres to ground level
10 Service pen – 1-m internal width with sand bed
11 Cow yoked gate with feed area and exit gate
12 Squeeze gap, 300 mm wide – must not be accessible to children
13 Refuge with vertical ladder rails – 300 mm gap either side
14 All gates double latched – capable of adult operation only
15 Steps to suit, especially if outside pen area is scraped

Bull pens Cost range: £8000–12 000 each	See Fig. 25.35. Housing elements designed to resist aggressive behaviour with escape routes from areas used by operatives. Ensure specific Health and Safety requirements observed including exclusion of children – making squeeze gaps to outside illegal. Note also that any blockwork must be reinforced. The bull pen must adjoin and overlook the cow circulation areas

Pig housing (BS 5502, Part 42, refers)

The last few years have seen very rapid changes in both the structure and the technology of pig-producing systems in Europe. Among these changes has been a major move towards freedom of movement and away from restrictive enclosure or tethering. Additionally, there has been a trend towards liquid and/or automatic feed dispensing systems which have had a radical effect on building design. Straw has been deemed to be the most appropriate bedding method for pigs long term, and ways have been sought to incorporate the material in housing methods that do not bedevil the problems of manure handling and labour use. A further factor has been the advancement of group (family) housing theories among the leading exponents of 'natural' pig production.

All these causes have allied with the recent financial problems of the industry to ensure that pig housing design was, and is, subject to considerable reversionary influences. Some of the text below describing housing types therefore refers to systems as they are, rather than 'state of the art' techniques as they should be if the world was perfect. The above remarks should be borne in mind and appropriate advice sought so that intended new systems do conform to the latest strictures on the health and welfare of pigs as the descriptions below cannot encompass all the potential variations – or pitfalls for the unwary!

Pig fattening accommodation

Space within controlled environment housing
Cost range: £280–£360/m²

Absolute minimum of 0.5 m²/'grower' pig place

	Total area (m²)	Dunging area (m²)	Alternate slatted (m²)
Porkers	0.6–0.7	0.25	0.12
Baconers	0.75–0.9	0.3	0.15
Heavy hogs	0.9–1.0	0.35	0.2

(May have slightly less than above with some straw-based systems where there is a definite strawed sleeping area)

10–20 pigs/pen normal; some systems 30–40

Pen dimensions and arrangement (Figs 25.36–25.39)	Optimum depth of lying area 1.8 m. Trough width 300–400 mm. Trough length/pig 250, 300, 350 mm porkers, baconers and heavy hogs respectively. *Ad libitum* feeding systems only require 225 mm for all classes of fattening pig. Some systems have no trough allowance – floor feeding, etc. (Trough length usually governs length and shape of pen and thus the house) Square (or nearly square) pens are now favoured, particularly for larger groups
Height of building	As low as possible. Passageways for operatives give minima; 2.15 m normal, 2.7 m minimum with scraping, 3.0 m with catwalk over pen walls for feeding, etc. Eaves height 1.7 m over slatted areas or 1.1 m with strawed/open yards and 'pop-hole' access. (If liquid feed systems are to be used, the height must provide for convenient siting and fixing of pipes and controls to give sufficient drop out of the pigs' reach)
Passage widths	Dung passages 1.05–1.35 m slatted outer, 1.15–1.65 m slatted central, 1.15–1.5 m concrete solid outer, 2.3 m minimum for tractor. Feed passages 1.0–1.2 m

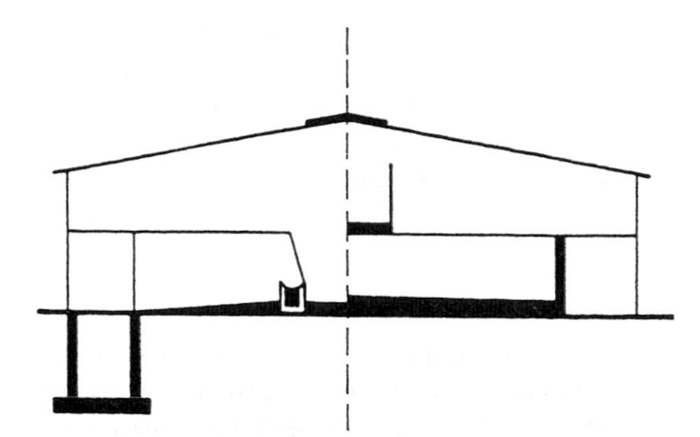

Figure 25.36 Section: feed trough and cellar/platform and scraped passage.

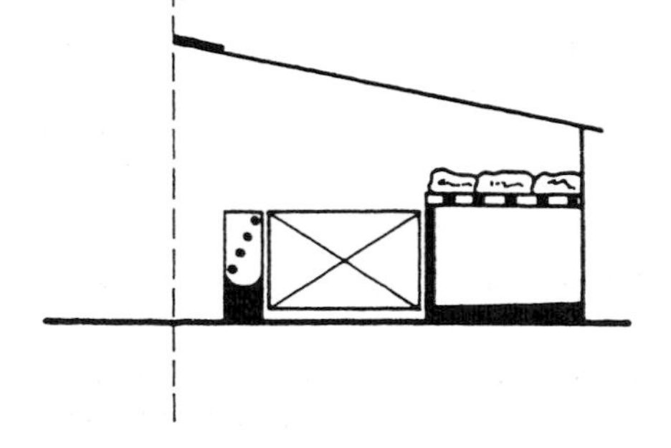

Figure 25.38 Section: kennel type for uninsulated housing.

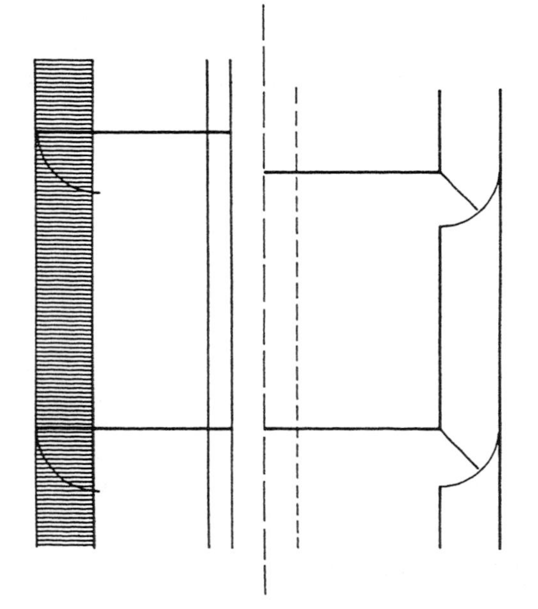

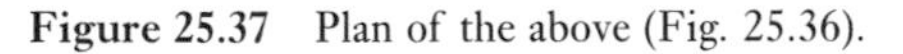

Figure 25.37 Plan of the above (Fig. 25.36).

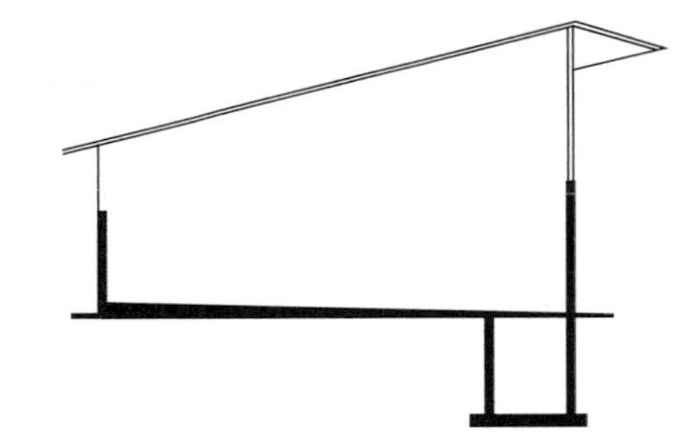

Figure 25.39 Section: open fronted with cellar.

Figures 25.36–25.39 Some fattening house options.

Temperature and insulation	Minimum of 21°C mean throughout environment. Note: Nurtinger kennels described later under 'Weaner housing', measuring up to approx. 2.4 × 1 × 0.9 m. Straw bedding affects lower critical temperature, to reduce thermal risks. 2°C less, generally, for pigs over 4 months old. Roof and walls should have 'U' value of not more than 0.45 W/m^{2}K. Floor in lying area should be insulated and vapour proofed to a particularly high standard
Ventilation	2.5 litres/s/50 kg liveweight in winter 12.5 litres/s/50 kg liveweight in summer
Feed and water consumption	2–4 kg/pig/day of meal
	4.5–9 litres/pig/day through trough with food, + bowls or drinkersin dunging area or over trough at one per six pigs. (Minimum of two drinkers per pen)

	Pipe-line feeders supply through delivery pipe at high level down to troughs in conventional position
Effluent and drainage	7 litres/day from baconer on dry feed. 14 litres/day from baconer on *ad libitum* wet feed. Floor falls 1:15 in lying area; 1:50 to 1:100 for dung passages or channels
	Slatted area – part or whole. Slats-concrete, 75–100mm top width + 15- to 20-mm gaps. Steel T bars, 35–50mm top width + 10- to 12-mm gaps. Step of 50mm to slatted area sometimes
	'Straw-flow' systems create a natural movement of straw from the back of the pen to the front and to waste clearance through a much more pronounced floor slope
Other	Chew-proof doors and partitions with secure latching and fixings considerations

Farrowing accommodation

Cost range: very variable but likely to be £290–£380/m^2

Space for crate (see Fig. 25.40)	2.4–2.95m long × 0.6m wide (+2 × 400mm side clearance) × 1.4m high. (Plinth may be provided 200–300mm high.) Crates vary considerably both in their construction and ultimate effects on the pigs, so it is important that a thorough review of current options is carried out before any contemplated investment is made. In general they are tending to provide greater and even unrestricted movement, positioning the sow according to her natural tendencies rather than forcing her into an unnatural posture
Space for circulation space round crate	Passage at rear minimum of 1.4m. Feed passage at head end minimum of 0.8 m. Eaves height 2.4m from passage floor
Space for creeps surrounding crate	Up to 0.1m^2 per piglet allowance. Maximum is likely to be 0.4–0.5m wide × 1.4m long (length of crate). Covered with lid and heat source above piglets below lid. Bottom rail or pop hole access for piglets. Rail height 250–300mm or three to four pop holes along creep base
Traditional farrowing places/pens with weaning space and creep (see Fig. 25.41)	2.4m deep × 3.0m long (pens either side of feed passage). Eaves height 2.4m. Creep rails on rear and side wall nearest creep, set 250mm out from walls and 250mm above floor
	Note: the development of 'family' or other 'natural' group sow (and other pig class) housing in batches of 5–10 occupying one very large space
	Creep at pen front, 0.9 × 0.8m area with access from pen via pop holes and creep rail. Lid over, with heat source below
Open front pen (see Fig. 25.42)	1.5m wide × 4.5–5.5m deep × 1–1.5m at rear eaves (mono-pitch with front eaves of 2.1–2.4m), against rear wall. Removable farrowing crate rails linked to creep, arranged along the pen at centre rear
Temperature and insulation	General areas 16–20°C. Creep at 30°C initially, reducing to 24°C over 3–4 weeks from birth. 'U' values of not more than 0.45 /m^{2}K. Crate and creep

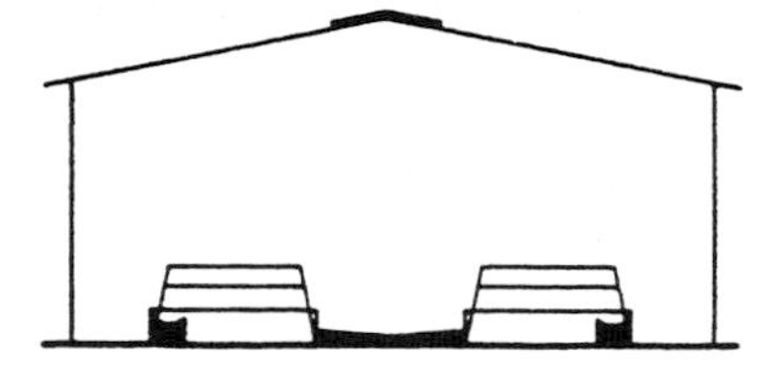
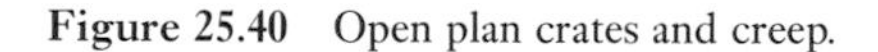

Figure 25.40 Open plan crates and creep.

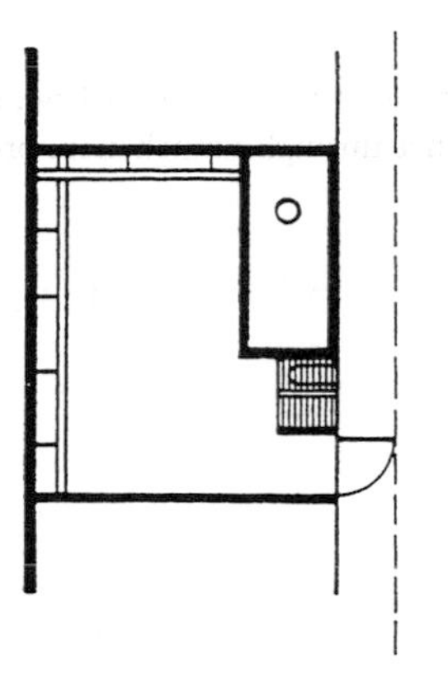

Figure 25.41 Traditional pen.

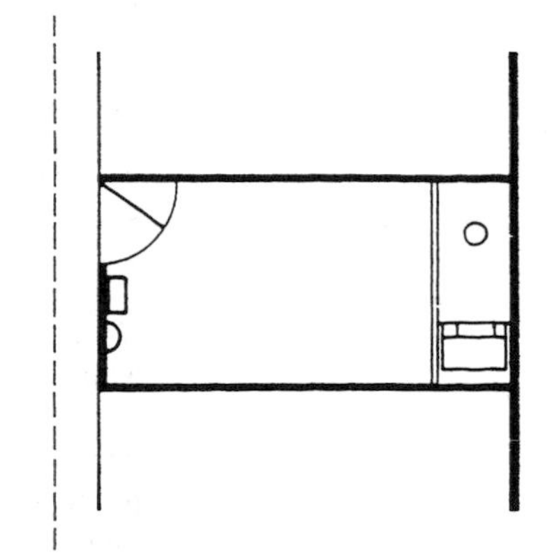

Figure 25.42 Open front pen.

Figures 25.40–25.42 Some farrowing house options.

areas to be well-insulated concrete, may have in- or under-floor heating. Note: Nurtinger kennels described later under 'Weaner housing'

Ventilation

Natural in open systems and in small, pen-type housing (up to five sows). 5 litres/s per sow in winter and 15 litres/s per sow in summer in controlled environment houses

Feed and water consumption

5–8 kg/lactating sow/day. 0.6 m long trough with drinker over. 250 mm above floor level. 18–25 litres/lactating sow. Supplementary water for piglets provided by 'tube' drinkers or trough away from creep feeder, over slats if present. Use of liquid feed is reducing the need for service passageways at the sides of houses, etc.

Effluent and drainage

20–25 litres/sow/day

Drainage falls in lying and creep areas to be 1 : 20, 1 : 40 in general areas. Steps to drain or dung areas not to exceed 50 mm rise. Slats or mesh area in rear half of crate. Slats consist of slabs of concrete with 10-mm slots and 75- to 100-mm slat tops. Steel tube or cast iron, 15–50 mm with 10- to 12-mm gaps. Meshes of steel, plastic, plastic-coated steel of various conformations and aptitudes, care needed for foot comfort of piglets. Crate raised 400–500 mm above surrounding levels at rear to accommodate slurry buildup and drainage flows.

Materials of walls and floor

Easily cleaned and resistant to strong disinfectants. Floor flat but not smooth trowelled – no tamping grooves to impede drainage. Dust with carborundum to improve wearing and non-slip quality.

Other considerations	Robust fittings, including water bowls required. Strong post and frame fixings. May have crates in an 'open' arrangement with or without low partitioning between each, situated on a plinth or on a well-sloped floor. All crates should be removable for easy cleaning of crate sections and floor.

Dry sows and boar accommodation

Space allowances in open yards Cost range: variable, £130–£160/m^2	2.5–3.7 m^2/sow of which 1.2–1.5 m^2 is covered and possibly strawed. 65% lying/feeding and 35% dunging area. Group size normally up to 20–40/yard. feeders require 4–5+ m^2
Sows in pens Cost range: £240–£280/m^2	2.1 × 1.5 m/individual sow. Trickle feeder space 400+ mm width
In boar pens	4.5 × 3.0 m pen of which 4.5 m^2 is lying area (2.4 × 1.8 m minimum)
Temperature and insulation	21°C maximum, 14°C minimum. Roof and walls of 0.5 and 1.0 W/°C/m^2 'U' value respectively
Ventilation	Natural usually adequate but forced ventilation in some totally enclosed houses at 50 litres/s/pig maximum
Feed and water consumption	2–3 kg/day per pregnant sow in individual feeders 3–5 kg/day per boar from a trough of 0.6 m length 9–18 litres/day per sow or boar
Effluent and drainage	20–25 litres/day per animal. Floor falls laid to 1:40 generally; 1:20 to discharge into open gullies. House end falls 1:50–100. (Slats may be used, see 'Farrowing accommodation' earlier)
Other considerations	Boars kept within visual contact of sows in extensive houses. Sow restraint may reduce bullying, can produce sores

Weaner housing

Space requirements Cost for veranda type: £240–£290/m^2	0.5–0.6 m^2/pig in strawed yards of which up to 0.3 m^2 is lying area if open fronted or 'veranda' type (see Fig. 25.43 for slatted uncovered veranda version). Eaves height 0.9–2 m depending on whether kennel or framed type. Groups of 10–20/pen are usual
Temperature and insulation	23–27°C inside kennels and in controlled environment housing. Roofing or kennel lid 'U' value should be 0.35–0.45 W/K/m^2. Wall 'U' value should be 0.45 W/K/m^2. Well-insulated floors with in-floor heating using electric cable or hot water piping are strongly recommended. Extra heat may be needed in winter – gas heaters common for air flow systems Nurtinger kennels may be used to provide 'an environment within an environment', to give a near-natural pig microclimate. This may be an asset for all pig-producing systems, but it is particularly useful for weaners. The kennels take the form of small-section but relatively long 'hutches' with one long side curtained in flexible plastic. Their dimensions vary widely according to the class of pig accommodated, but the range is: length 1.5–2.4 m, width 0.35–1 m, height 0.45–0.9 m

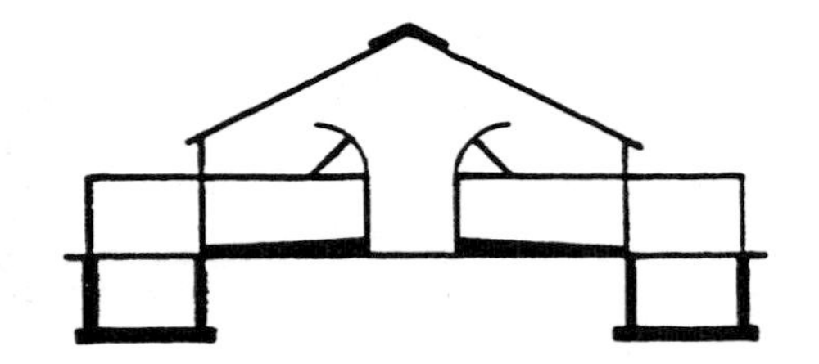

Figure 25.43 Section: weaner veranda house.

Ventilation	2.5 litres/s/50 kg liveweight in winter 15.0 litres/s/50 kg liveweight in summer
	Take care to ventilate, avoiding passage of noxious gases from dung through house in total control systems
Feed and water	*Ad libitum* feeding from 50-kg-capacity hoppers with 50 mm/pig access. Some rationing systems use troughs at 75–100 mm/pig; should be spill proof. 1–2 litres/pig/day from nozzle-drinkers (two/pen minimum) over well-drained area. Very young pigs might need temporary ramp, etc. to reach drinkers
Effluent and drainage	1–2 litres/pig/day. Either open dunging area outside kennel or weldmesh/expanded metal/plastic mesh part or whole floor, e.g. Weldmesh is 76 mm × 13 mm × 8 swg (10 swg for smaller) over slurry pit 1–2 m deep or raised 0.6 m in total control housing. Many alternatives available
Materials	Prefabricated buildings with easy erect, easy clean and control facilities

Sheep accommodation (BS 5502, Part 41, refers)

Sheep housing (Fig. 25.44)

Space requirements Cost range: £70–£100/m^2	(See Table 25.8)
Temperature and insulation	Ambient of surroundings, down to 2–5°C. No insulation necessary
Ventilation	Adequate air exchange essential to maintain health of stock
	Natural from space boarding walls or netting and ridge ventilation (open 300–600 mm, relative to building span) or spaced roofs. Avoid low-level draughts with timber dung walling 1.2/5 height. Building may be open at one side on sheltered aspect
Bedding, feed and water consumption	0.8–1.5 kg/straw/ewe/day in yards. 0.25–1.0 kg/concentrate/ewe/day. 1–1.5 kg/hay/ewe/day. Up to 5 litres/animal/day from a water trough 0.3 × 0.6 m in each pen or one bowl to 50 ewes. (Running water to overflow desirable for hill breeds)
Floor and drainage	Excreta: 4.1 litres/adult animal/day. Floor usually chalk or hardcore of 150-mm stone scalpings (graded 50 mm to dust) on geotextile fabric, vibration roll consolidated. Slats are usually hardwood, 38–60 mm top width × 25–30 mm

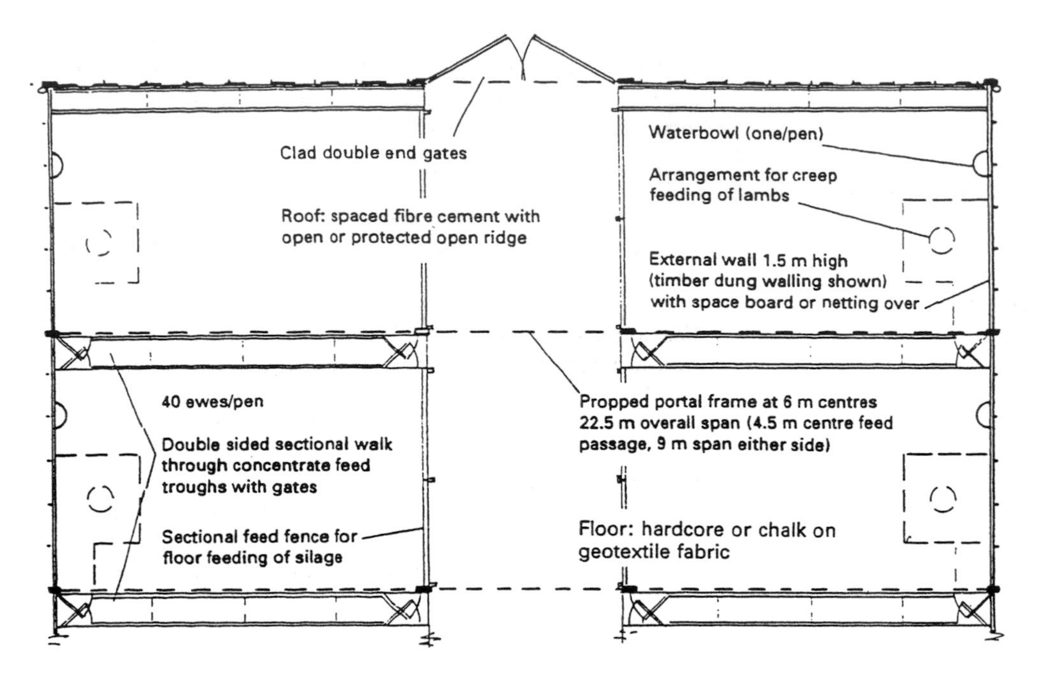

Figure 25.44 Plan of end two bays of typical sheep housing.

Table 25.8 Sheep housing: special requirements. Note: for clipped ewes the figures may be reduced by 10%. [From Astley Cooper *et al.* (1991), with permission]

Type of housing	*Area on straw (m^2)*	*Area on slats (m^2)*	*Trough space (access from outside pen)*	
			Concentrates (mm)	*Ad-libitum hay or silage (mm)*
Large ewe 60 kg up to 90 kg in-lamb	1.2–1.4	0.9–1.1	450–500	200–225
Large ewe 60 kg up to 90 kg + lamb	1.4–1.8	1.2–1.7	450–500	200–225
Small ewe 45 kg up to 60 kg in-lamb	1.0–1.3	0.7–0.9	400–450	175–200
Small ewe 45 kg up to 60 kg + lamb	1.3–1.7	1.0–1.4	400–450	175–200
Hoggs 32 kg up to 45 kg	0.7–0.9	0.5–0.7	350–400	150–175
Hoggs 23 kg up to 32 kg	0.15	0.4–0.5	300–350	125–150
Lamb creep (2 weeks)	0.4	—	—	—
Lamb creep (6 weeks)	2.2	—	—	—
Ewe plus lamb in pen	1.5	—	—	—
Lambing pen		—	—	—

	thickness, with 12- to 20-mm gaps according to size of breed, etc. Leave 0.65 m clearance underneath for one season's holding capacity, 0.2–1.0 m common. If concrete laid must fall away from feed fence: 1/100
Special areas (lambing, shearing)	Clean area, isolated from general holding area Adequate services. Pens of 0.75–1.2 × 1.2–1.8 m 0.5–0.6 m^2/animal holding area. 1.5 m^2/shearer

Other considerations	Polythene or polyester 'tunnels' provide low-cost short-term accommodation, but be careful to ensure good ventilation. Provide races and satisfactory entry and exits to dipping and handling areas. Ramps to and from house to be 1 : 10 maximum

Sheep handling areas

Space allocations Cost range: £40–£50/m^2 for concreted pen areas	Gathering and dispersal pens to contain groups of up to 100 sheep at 0.5–0.6 m^2/animal
	Catching pens to contain 10–15 sheep at 0.3–0.4 m^2/animal; may be circular with centre pivot gate to crowd into race, dip, etc., or rectangular
	Central races 3.0–6.0 mm long × 0.45 m wide × 0.8–1.1 m high
	Footbath and race 3 m long minimum × 0.25 m wide at bath (0.15 deep with 0.1 m fluid depth) and 0.45 m wide at top of boards, about 1.0 m high
	Dip tank capacity from 0.75 m^3 for 300 ewe flock to 2.5 m^3 for flocks of 1000+ (see below for general advice)
	Ancillary holding pens for up to 50 sheep, short term, at 0.4 m^2/animal
Other considerations	Allow 2.5 litres/sheep (minimum 350 litres, then 1000 litres/500 ewes up to 4000 litres) of dip fluid. Circular dip best for large flocks. Floors should be roughened concrete with slope to drain channels of 1 : 50–100, or may be hard, dry soil in well-drained areas, or of introduced stone. Slope from bath should be 30°
	Drain baths tanks to an approved holding or disposal point. (Care to avoid pollution. See Water Code)
	Pen posts for rails should be away from sheep, either on non-sheep side or with shoulder protector boards

Crop storage (BS 5502, Parts 70 and 74, refer)

Grain storage (Table 25.9)

Cost range of four types:

- automated bin and dry: £200–£260/t
- towers with LV ventilation: £120–£150/t
- on-the-floor and dry: £100–£130/t
- on-the-floor with LV ventilation: £80–£100/t

Store types (see also Chapter 7 on grain storage)	Circular bins 2–6 m in diameter, up to 6 m high (height not to exceed twice the diameter) (see Figs 25.45 and 25.46). Materials may be hessian and wire mesh, oil-tempered hardboard, galvanized corrugated steel, galvanized slotted steel, ply board, welded steel plate, glass- or enamel-surfaced steel, or concrete stave with steel hoops. Floor may be plain concrete, aerated mesh or sloped steel (45°). Emptied by 'air-sweep', slope, or auger conveyor

Table 25.9 Dry crops spatial requirements. Note: further moisture adds weight/m^3 at 1–1.5% extra per 1% addition of moisture over standard conventional.

		Barley	*Beans*	*Herbage seeds*	*Linseed*	*Maize*	*Oats*	*Peas*	*Rape*	*Rye*	*Wheat*
Space requirements (based on conventional storage MC and 20–30° repose angle)	(m^3/t)	1.45	1.2	3.0–5.0	1.45	1.35	2.0	1.3	1.6	1.45	1.3
	(kg/m^3)	705	833	200–500	705	740	513	785	625	705	785
Capacity of level, 2-m fill, of loose grain, per 4.8-m bay and 6-m span		42	51		42	47	29	44	40	42	44
Capacity of fill when heaped above 2 m, per 4.8-m bay and 6-m span		57				64	39				
Tonnes of product 2 m high, level fill/m^2 of floor area		1.34	1.67	0.4–0.6	1.34	1.48	1.0	1.54	1.33	1.34	1.54
Capacity of 4.5-m-diameter silo 9 m high		102					76	220		98	110

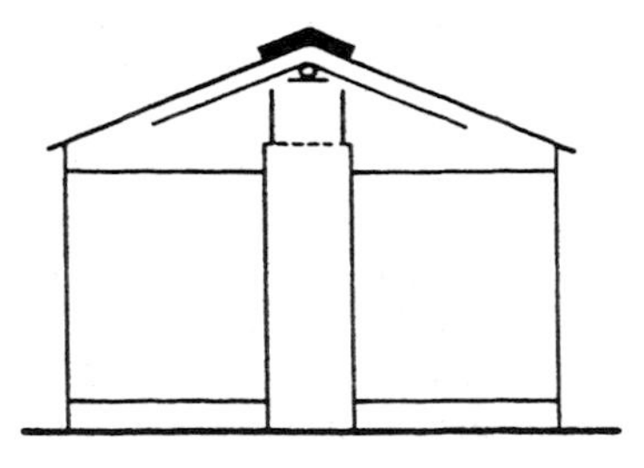

Figure 25.45 Section: automated square/circular bin store.

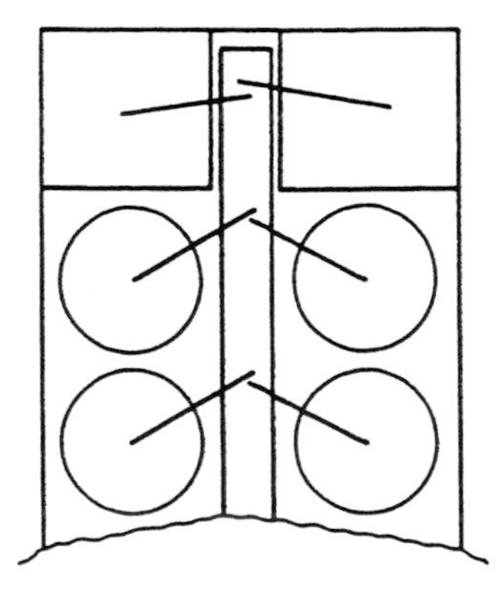

Figure 25.46 Plan of section (Fig. 25.45).

Square or rectangular bins 2.4–4.5 m with heights up to 9 m (6 m commonly for drying purposes). Materials may be corrugated steel sheet (square profile), galvanized steel pressed panels. Flooring will be plain, ventilated or sloped as with circular type

Loose bulk ('on-the-floor') storage uses standard building types and frames; with or without simple retaining walls for containment at the edges or at dividing partitions (see Figs 25.47 and 25.48). Variants include systems for air passage through grain from main ducts and laterals, above- or under-floor ducting, drive on perforated floor, and radial distributors buried from top (see Chapter 7 on grain storage). Duct/outlet type and positioning are dictated by grain type, moisture levels, suck or blow system, high or low ventilation rates. The building size and shape are also critical to the system used. Refer to specialist texts for details. It is important to appreciate that the building width in particular is critically dependent on the duct system chosen

Wet grain storage, in sealed silos (or with non-sealed roofing), using circular, galvanized, glass fused or enamelled steel. (Allow for density of product at 14% MC plus 5–10%.) Danger from CO_2 buildup

Wet grain heaped on-the-floor in clear span framed buildings, preserved by propionic acid application. (Density as above)

Refrigerated grain stored at 18–22% MC in bins at a density of 3–5% more than at 14% MC is not viable in this country at present

Floor

Hard, durable, smooth, dust-free and dry. 150 mm of RC 35 concrete, reinforced on soft, difficult soils and at edges when supporting bins; steel floated and incorporating a damp-proof membrane. May have to accommodate air ducts or conveyors

Walls

Thrust resistant and moisture/air tight. Must be designed with adequate foundations. Building frame may support thrust if designed to do so. Maximum economic height for on-the-floor drying is 3 m of level grain. Purpose-designed stores for pre-dried grain may increase this up to 6 m with an LV ventilation system. Surcharged grain is not feasible if the grain is to be dried in situ or if the wall and structure was not designed to take the extra load. Height of eaves above grain will depend on system. On-the-floor loading is likely to require twice grain height – so 6 m usual. Conveyor systems buildings will depend on width of building to obtain an even fill

Reception pit

Reception pit needed for bin or pre-drying systems, with a sloping delivery floor at 60° to the horizontal. Essential to build up with upstand and easy slope ramp tipping platform to ensure capacity sufficient for largest delivery trailer. Conveyor pit attached to reception pit must be firmly anchored with exterior

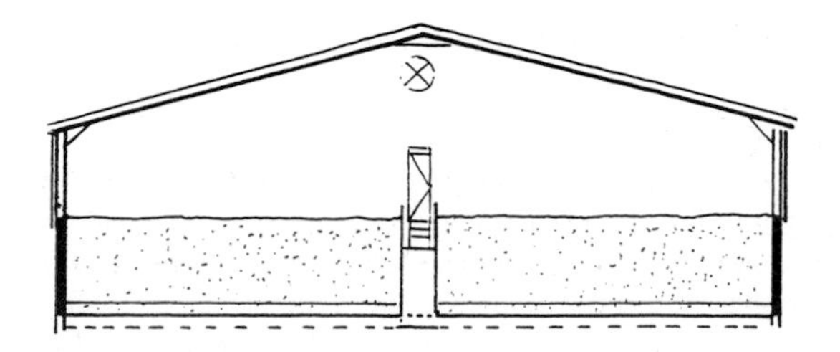

Figure 25.47 Section: on-the-floor store showing above/below ground laterals).

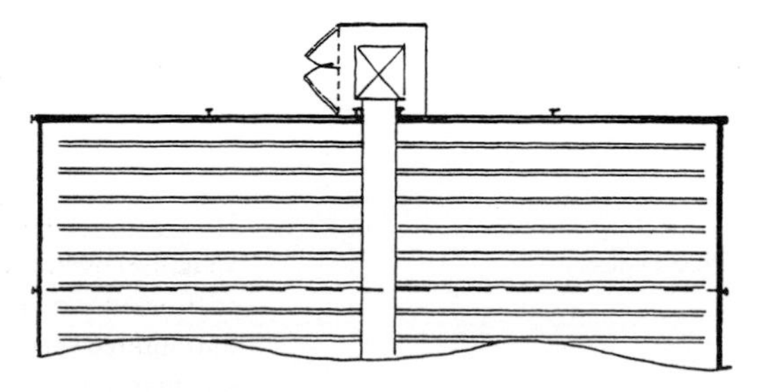

Figure 25.48 Plan of the above (Fig. 25.47) (end bay, also showing fan housing).

lugs in concrete surround to prevent water table lifting. Note all Health and Safety requirements especially in covering all pits – reception with heavy-duty weldmesh which grain can be tipped through. It must never be lifted if there is grain in the pit

Other considerations

High power consumption (e.g. 30–60 kW) likely in some stores. Lighting need only be background floodlight to provide min. 50 lux, positioned on gable ends and eaves for easy access with 150 lux over control panel. Provide well glass lighting for catwalks and conveyor pit ladders. Do not use pendant lights

Bin stores need overhead hopper to fill lorries rapidly, with high delivery rate conveying systems to match. On-the-floor stores need good access for telescopic handler. Large headroom and strong portal framing required with no overhead obstruction

Must be vermin proof (mandatory control legislation against *Salmonella*), thus totally enclosed and no natural light – to discourage birds (see BS 5502, Part 30, for general guide to infestation proofing)

Doors should provide clearance for tipped trailers (up to 6 m). Provision should be made for heavy vehicles in the flooring/aprons in the vicinity of grain stores. Special consideration should be given to fire hazards in grain drying areas. Deep-pit elevator equipment and temporary storage prior to drying may be required. (Service access needed below ground level)

Hay, straw, grass (see next section for silage)

		kg/m³
Space required for bulky, low-density materials	Straw, high density	120–160
	Hay, medium-density baled	160–210
	Straw, small bale	80–130
('average' figures provided: there will be considerable variation from sample to sample, dependent on moisture content, species, etc.)	Straw, large round bale	80–130
	Round bale bagged silage	300–500

Cost range: (Dutch barn) £70–£90/m²

Volumes and tonnages	A 4.8-m bay of a building will contain:	35 t of baled hay at a height of 4.8 m
		20 t of medium-density baled barley straw at a height of 5.4 m

Store types

Usually known as Dutch barn: a part or wholly open sided pole barn, curved roof, steel frame or portal frame with open or closed sides – usually open or butt jointed space boarding – with strengthened frame if round bales are stored on their side. No wall or simple timber dung walling dwarf wall. (See Fig. 25.49 or drying floor type.)

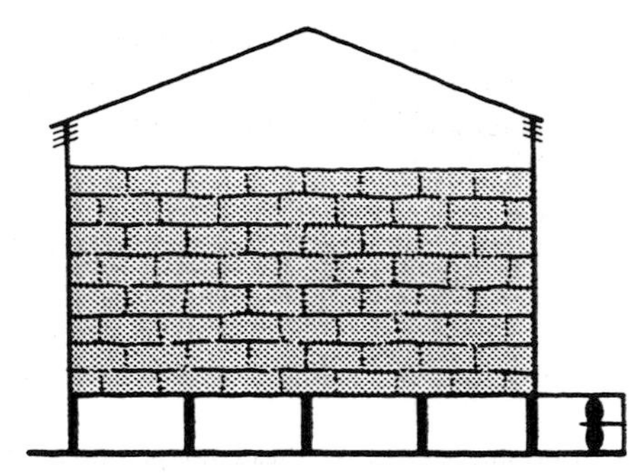

Figure 25.49 Section: hay drying barn.

Flooring	Earth or rammed hardcore over Terram fabric sloping to drain off driving rain, for simple storage of baled hay and straw. Can have raised weldmesh floor to provide plenum chamber for drying of hay in batch or storage driers. (See also Chapter 24, 'Barn hay drying')
Walls	Hay barn walls may be removable in the form of wooden baulks or steel sheet or they may, in the case of loose material, be of wire netting. Some types of walling must be air tight in storage drying. Top of wall near eaves should be slatted for up to 1 m to allow air movement during drying
Roofing	Hay barn roofing should be ventilated and of non-condensation-drip-forming type – fibre cement sheeting with timber purlins
Other considerations	Fire risk should be minimized. Ease of access to farm roads and livestock buildings. Power requirement may be high in storage driers or tower silo systems (20–100 kW). Drying fans tend to be noisy in operation

Silage bunkers (see Fig. 25.50)

Cost for 1000 tonnes: £40–£50/t

Walls	BS 5502, Part 22:1987 (legal requirement)
Concrete floor	BS 8007:1987 (to comply with legal requirement to be impermeable)
Space	1.4 m^3/tonne grass silage @ 20% dry matter (minimum) 1.5 m^3/tonne grass silage @ 25% dry matter (optimum) 1.5 m^3/tonne maize silage @ 30% dry matter (maximum)
Depth	Optimum 2.4 or 3 m. Maximum height for self feed 1.8 m unless top removed daily. Must never increase height by surcharging – consolidated height must not project above top of wall for at least 600 mm from edge and only slightly domed in centre
Width	Optimum face, for grass, to prevent spoiling, 7.2–9.0 m or up to 12 m if can use fast enough. Conflict if self-feeding as must allow min. 150/200 mm of face width/cow (or large beef) for 24-hour access

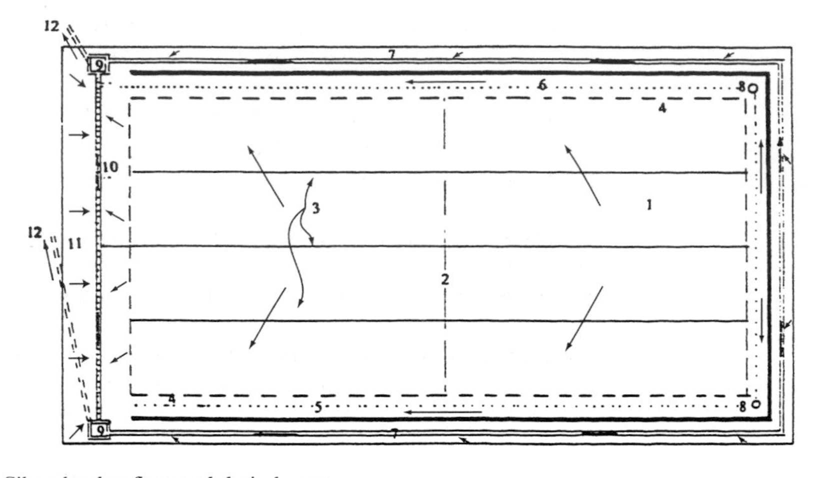

Figure 25.50 Silage bunker floor and drain layout.
1 Concrete floor to BS 8007 (very severe rating)
2 Contraction joint (see Fig. 25.51)
3 Construction joints (see Fig. 25.52)
4 Expansion joints (see Fig. 25.53)
5 Perimeter concrete (to design for wall)
6 Internal drainage (e.g. Porcupipe)
7 External drainage (e.g. concrete channel)
8 Rodding chamber
9 Catch pit
10 Front drainage (e.g. polymer concrete with grating)
11 Front apron with reverse fall to drain
12 To effluent tank

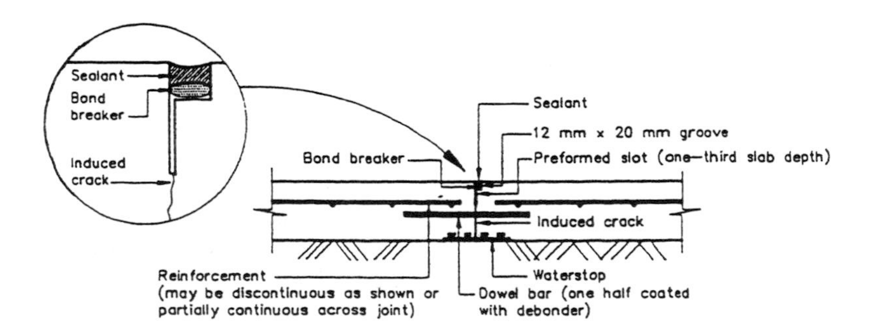

Figure 25.51 Contraction joint detail. (Adapted from BCA *Farm Concrete Fact Sheet* and CIRIA Report 126.)

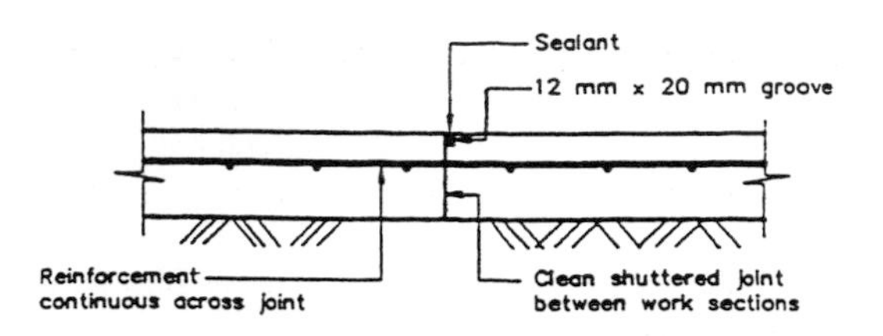

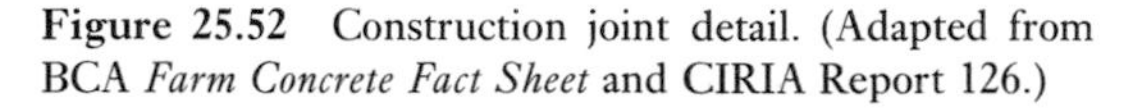

Figure 25.52 Construction joint detail. (Adapted from BCA *Farm Concrete Fact Sheet* and CIRIA Report 126.)

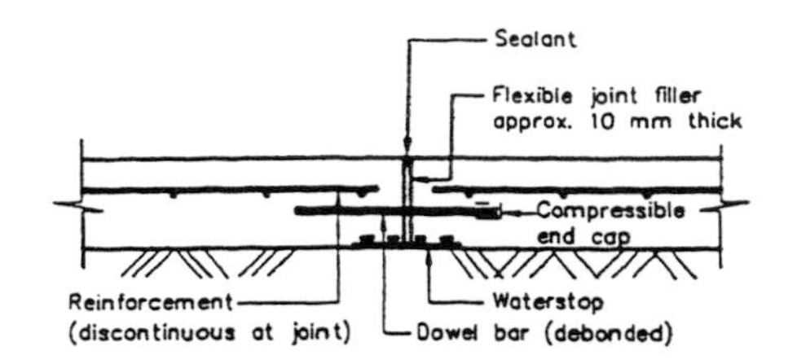

Figure 25.53 Expansion joint detail. (Adapted from BCA *Farm Concrete Fact Sheet* and CIRIA Report 126.)

Length	No limit except practical considerations. Long bunkers much more flexible if open both ends – if site permits. Avoid demountable front walls if possible
Falls	From back to front: 1 in 100 (CIRIA) or 1 in 75 (Code) From centre to sides: 1 in 75 (CIRIA) or 1 in 50 (Code) This means that the walls must either be stepped or sloping – if sloping they will no longer be in the vertical plane. Confirmation that this is acceptable within the design of prefabricated panels must be obtained
Roof	Not justified on economic grounds, but does eliminate rainwater from effluent system and provides for straw storage over, cover for self feed and scope for other uses
Walls	Options (all of which must have an engineer's certificate confirming that the wall type and foundation requirements conform to BS 5502): • No walls (still needs impermeable floor with upstanding curb and outside channel unless only used for bagged silage) • Stub or building stanchions with horizontal rail(s) and: – concrete or timber (usually external ply) vertical panels – sleepers (vertical) • Stub or building stanchions and: – concrete horizontal prestressed panels – cast in situ concrete walling – sleepers (horizontal) would need stanchions at max. 2.6 m c/c • Cantilevered vertical prestressed concrete panel • Cantilevered cast in situ concrete (with engineer's supervision) • Precast concrete 'L' shape retaining wall sections (height up to 3.6 m) Note: reinforced hollow concrete block walls no longer recommended
Floor	Concrete floors must be laid under expert supervision with a working knowledge of the very high standards of the 'very severe' category of BS 8007, which could include the following: Mix: RC 45 (previously C35A) 0.55 water/cement ratio and 325 kg/m^3 minimum cement content, 75 mm slump, use PFA (pulverized fuel ash) cement if available; never use limestone aggregate Thickness: 150–250 mm depending on ground bearing and loading factors, with min. 50 mm cover over reinforcement Reinforcement: Weldmesh fabric to suit jointing and bay size; min. A142 Membrane: 1200 gauge polythene (used mainly as slip surface for contraction, but also additional pollution protection) Joints: see Fig. 25.50, or lay single slab with pumped concrete

Finish: retamp approx. 20 min after laying across normal traffic direction

Temperature: do not lay in extremes of temperature – hot or cold

Surface coating: use purpose-designed acid-resisting coating – also required for inside face of walls (two coats at base)

Joint sealant: use only purpose-designed one- or two-part polysulphide or polyurethane – not bitumen. See CIRIA Report section 7.4.4 about the precautions necessary

Hot rolled asphalt (HRA)

Although offered in the Code as an alternative with a bitmac base, it is not recommended at present until the problem of the expansion joint with the perimeter concrete has been overcome. It does, however, have a very useful role to play in upgrading existing concrete floors (and for surfacing new floors providing there is sufficient time lapse), using 50 mm hot rolled asphalt laid by mechanical paver and 8-tonne roller using 40% hardstone type aggregate – not limestone – to BS 594, column 21, Table 5 modified. Preparation is the secret to success. The tack coat must be applied to a very clean and textured surface

Channel external

There is a legal requirement to install an impermeable channel around the outside with adequate fall (min. 1/100) in an extension of the base of the silo to discharge into the effluent system; usually formed by insetting half-round salt-glazed or polymer pipe or dishing the concrete

Drain internal

The design of the walls will be dependent on the provision of an internal drain within 500 mm of the inside face to relieve the hydraulic pressure. This can be omitted if the wall design allows for the effluent to pass through or underneath. There are three options:

(1) temporary, flexible, 75-mm, perforated, PVC land drain – often tied to 75-mm timber batten to prevent crushing
(2) permanent 100-mm drain – slot or Porcupipe with rodding access points
(3) permanent channel – with engineering brick or timber infill or heavy duty narrow slot grating

Note: with the second and third options thicken concrete accordingly and maintain 50-mm cover over mesh

Front drain

Must be set approx. 1 m beyond front edge of retaining wall and have further 1 m of apron reverse falling towards bunker so that even if drain is blocked the effluent will pond. Use same drainage system as internal with letterbox slot and sump at end to collect ponding effluent.

Effluent tank legal requirement to provide to following capacity:

up to 1500 m^3 = 20 litres/m^3
over 1500 m^3 = 30 m^3 + 6.7 litres/m^3 over 1500 m^3
It must also have a maintenance-free life expectancy of 20 years. The alternative is to discharge direct into the dirty water system providing that system conforms and precautions are taken (and notices erected) about the danger of the production of highly toxic hydrogen sulphide (bad-egg smell). Both the Code and CIRIA recommend GRP bottles or cylinders with the special silage coating may be used, but access to remove heavy sand deposits which cannot be pumped is difficult. (They must not be entered without full breathing apparatus and safety harness, etc.)

Precast concrete tanks are offered by several manufacturers which overcome the access problem with slat covers, but must be designed and installed with instructions certified by an engineer, and tested on completion. (See Box 51, CIRIA Report)

Direct engineer's supervision is required for a cast in situ tank, and blockwork is not suitable

Notice

Legal requirement to display notice showing the following:

- moisture content after rolling not to exceed . . . %
- internal drainage provided
- gross vehicle not to exceed . . . (usually 8 t)
- maximum depth of silage when rolled not to exceed . . .
- minimum distance of centre of wheel of rolling . . .
- vehicle to edge of bunker 0.6 m
- (usually added) complies with BS 5502, Part 22: (and date of construction)

Concentrate and root crop feed material

Space for 'concentrates' ('average' values provided – there will be considerable variation from sample to sample, dependent on moisture content and consolidation)
Cost range: £70–£120/m^2

Meal, loose, 2.0 m^3/t or 500 kg/m^3

Pellets, loose, 1.7 m^3/t or 600 kg/m^3

Nuts, loose, 1.4 m^3/t or 700 kg/m^3 (50-kg bags stack 2 high, occupy 2.0 m^3/t or 500 kg/m^3)

Beet pulp 1.5 m^3/t or 650 kg/m^3

Fodder beet 1.7 m^3/t or 600 kg/m^3

Turnips 1.8 m^3/t or 550 kg/m^3

Brewers' grains (dry) 2.0 m^3/t or 500 kg/m^3

Space for equipment used in milling and mixing

Horizontal mixer 4–10 m^2 × 2–3 m high

Vertical mixer 3–5 m^2 × 3–6 m high

Crushing mill 2–3 m^2 × 1–1.5 m high

Hammer mill 1–6 m^2 × 1–1.5 m high (5 m with cyclone above)

Allow 1–1.5 m headroom for overhead conveying

Other considerations

All feed storage on farm must be vermin proof under the Code of Practice for the Control of *Salmonella*. This includes straits storage used in complete diet mixes, raising the cost considerably. It is obviously not possible for silage bunkers to be vermin proof nor the troughs in which stock are fed. Since the milk processing treatment eliminates *Salmonella* anyway this regulation does seem a bit excessive

Self-emptying bins should have floors at 60° to the horizontal. Noise and dust are potential health hazards. Risk of explosion in a dusty atmosphere. Very high power requirement by most barn machinery (5–40 kW). Mechanical handling requires smooth running, hard flooring. Use of gravity desirable in conveying – overhead hoppers for lorry and trailer filling with 10–20 t capacity for big units. High clearance needed (up to 6 m)

Vegetable storage (BS 5502, Part 71, refers)

Space requirements and environment		*m^3/t*	*Mean temperature to store long term*	*Maximum height*
Cost range: insulated + vent, £210–£250/m^2; non-insulated, £150–£180/m^2	Potatoes, bulk (see Fig. 25.54)	1.5–1.6	Ware 4–5°C	4.0 m (6 m if air flows can reduce temperature gradient)
	Potatoes, in pallet boxes (1.2 × 1.4 × 0.8–1.0 m) (see Figs 25.55 and 25.56)	2.4	Ware – as above. Crisping and processing, 6–8°C	4–6 boxes (4.0–5 m)
	Onions, dry (see Fig. 25.54)	1.5–2.0	2–3°C	4–5 m
	Onions, green topped	3.0	Ambient temperature −3 to 6°C until dry	2.5 m
	Onions, refrigerated	2.0	0.5–1.5°C	3.0 m
	Onions, dry, in boxes (1.1 × 1.3 × 1.05 m high)	3.0	Ambient temperature +3 to 6°C	Boxes stacked to eaves

Store types

Standard, clear span frames

Floor

As under 'Grain storage' plus smooth-running heavy-duty reinforced concrete floor for pallet handling equipment

Walls

Thrust resistant (see under 'Grain storage') – angle of repose for potatoes is 30–40°). Differences include the ability to store up to 4 m for on-the-floor ventilation and in pallet box stores cheaper walling may be used which is not thrust resistant. The extra cost, however, is in the need to insulate: a 'U' value of 1.0 W/°C/m^2 max. for walls (and roof) is necessary to protect potatoes from frost. Care must be taken to avoid 'cold bridging' across the walls via reinforcing or framing systems. Walls should also be smooth and cleanable. Grain stores temporarily used for potatoes can be insulated with straw bales around the inside perimeter

Ventilation

Main purpose is to control temperature rather than moisture. Insulation of structure greatly assists but does not remove need for ventilation except in short-term box storage or shallow (max. 1.8 m) on-the-floor storage. Even then a low volume ventilation system is desirable to dissipate the heat and moisture given off by the crop. The systems illustrated above are usually controlled by automatic heat and humidity sensors placed in the crop and measuring the ambient air temperature and humidity. This gives the option to recirculate the

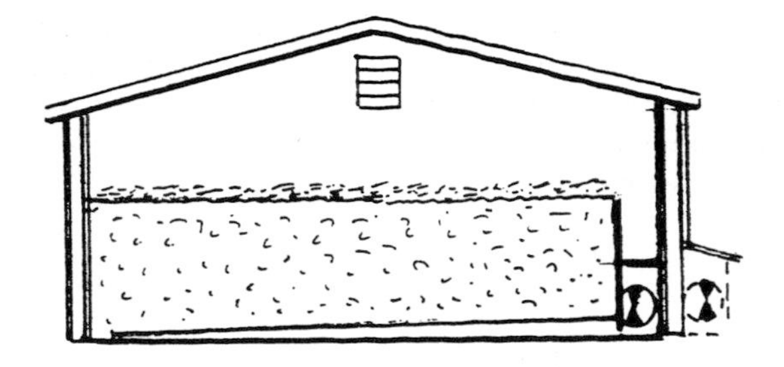

Figure 25.54 Section: bulk potato store with side main duct (inside/outside), tapered laterals, 16-m internal span (see centre duct in Figs 25.47 and 25.48, 23-m internal span).

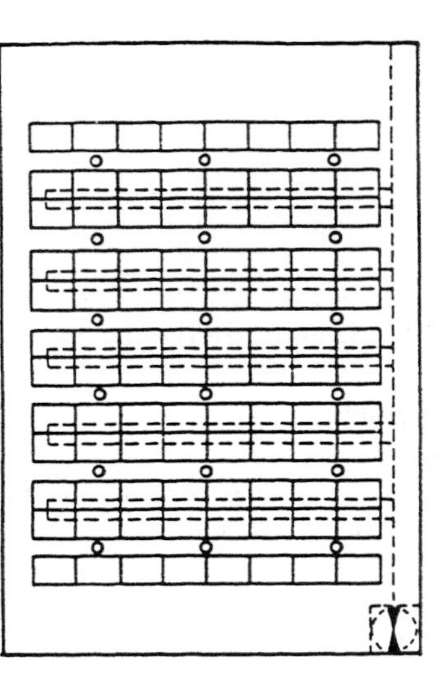

Figure 25.55 Plan: potato chitting house.

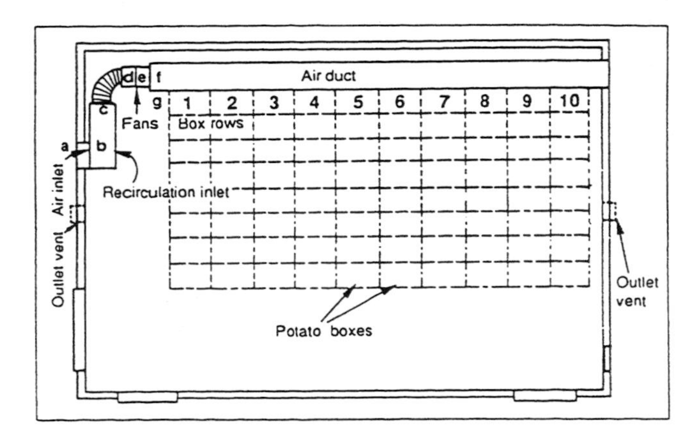

Figure 25.56 Plan: potato pallet box store.

Figures 25.54–25.56 Some potato storage options.

air if the need arises. Note also the laterals are wider spaced (usually 1.8 m c/c) than for grain storage and can exceed 10.5 m in length if tapered. Essential to provide pressure release flaps in a controlled environment with input fans

Other considerations

Avoid materials in structure with a high thermal conductivity. Use thick bed of loose straw as light excluder, insulation and condensation collecting layer on top of stack

Porch desirable to enable store to be filled to end wall, potatoes rest on removable boards supported on porch side walls which carry stack weight as it is built up

If only used for storage exclude natural light, but if building used for sorting and grading, roof lights preferable – keeping crop covered with straw. Strong artificial light, 300 lux, is required in all instances over working machinery, otherwise background flood lights as grain store

Chitting houses for potatoes (see Fig. 25.55) (BS 5502, Part 66, refers)	Potatoes stacked in trays, with access to illumination from movable lighting columns suspended from ceilings. Trays are 750 × 450 × 150 mm on pallets 1.5 × 0.9 m in area, stacked to depth of 3.5 m max. Alleyways 0.5 m min. between rows. 1-m side alleys. Generous end space in building. Should be frost proof and force ventilated, with heating equipment available when necessary

Other buildings and structures

Fertilizer storage

Space required Cost range: GP building: £140–£170/m^2	Loose, bulk, 0.9–1.1 m^3/t average 50-kg bags, stacked 1.0–1.2 m^3/t (at 180 mm thickness per bag, with maximum of 10 high)
	'Big' bags containing 0.5 t, stack two or three high; are nearly equivalent to loose
Other considerations	Stored in standard frame building with weather protection and anticondensation provisions to prevent dripping on to bags. Thrust-resistant walls for bulk fertilizer with corrosion-resistant surfacing. Ends and top of heap covered with plastic sheeting. Walls and floor clearance of 0.5 and 0.15 m respectively for bags (on pallets or boards). Provide temporary divisions to isolate different fertilizer types.

Slurry and farmyard manure storage

BS 5502: Part 50:1989 (legal requirement) and BS 8007:1987 for impermeable concrete floors where applicable

Space required	Dirty water [less than 4% dry matter (DM)]: 1 m^3/tonne
	Thin slurry (3–10% DM) in above-ground tanks: 1–1.2 m^3/tonne
	Thick slurry (10–15% DM) in weeping wall stores or pumped earth compounds: 1.2–1.4 m^3/tonne Farmyard manure: 2.0–4.0 m^3/tonne (allow higher figure for unrotted FYM with high volume of straw)
Cost range: £10/m^3 (earth lagoon); £25/m^3 (steel store); £40/m^3 (weeping wall excluding LV irrigation system)	
Minimum capacities	*Dirty water*: *either* total containment – min. 4 months (legal) – *Or* with distribution system (usually low volume irrigation) and waste management plan: no specific obligation except to be adequate. Recommend three-chamber with 'H' pipes (see Fig. 25.57). Min. 48 hours in pump chamber
	Slurry: *either* unseparated total containment – 4 months plus 300 mm min. 'freeboard' for above-ground store or 750 'freeboard' for earth bank lagoons (legal) –

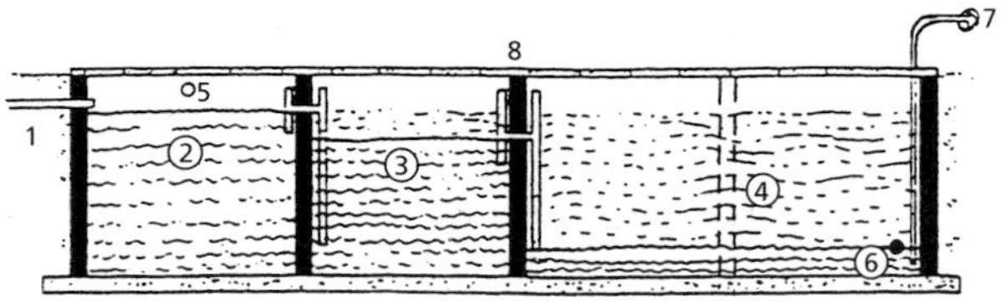

Figure 25.57 Dirty water settlement tank.
1 Inlet (to be kept as high as possible)
2 Primary settlement tank (usual actual min. 7500 litres*)
3 Secondary settlement tank (usual actual min. 5000 litres*)
4 Pumping tank [with/without intermediate wall, including top section of secondary settlement tank and excluding level below float switch (300 mm) – usual minimum capacity 24 hours *without pump functioning* and providing emergency overflow arrangements – subject to EA agreement (if no overflow, recommend min. 48 hours' capacity]
5 Overflow as above
6 Pump intake min. 150 mm above base and float switch 150 min. above intake
7 To low-volume irrigation system. (Pump kept well above tank in case of flooding and provided with frost stat control heaters with warning light – usually on adjacent buildings)
8 Construction to regulations – need easy access for desludging – slats shown
* Will not be sufficient for units with large open yards.

Or separated by weeping wall store, retaining all slurry less 10%. Recommend 12 months' capacity to ensure distribution when fully dried out. Has to be with dirty water distribution system on waste management plan

Or mechanically separated by extracting up to 20% solids converting remainder into dirty water as above. Extracted solids treated as FYM

Slurry reception pit: min. 48 hours (legal)

Farmyard Manure (FYM): no minimum requirements. Permanent stores should have impermeable floor. Can be temporarily stored in field providing does not cause pollution from run-off

Options for storage

- Under slat storage cellars or channels leading to other storage. Floors: min. 150 mm RC 40 reinforced with membrane laid horizontal with 100/150 mm min. water containment so dung solids will float. Walls: impermeable and double load bearing (for outside earth thrust) in reinforced concrete (BS 8007) or masonry (BS 5628) waterproofed inside and outside. Cellars with ramp or reinforced panel access for emptying. Channels with weir every 24 m; 150 mm high, usually 1.2 m wide
- Above-ground circular store: proprietary steel or concrete on concrete base and ring beam to manufacturer's requirements up to 33 m diameter and 7.2 m high with access platform and steps, mixing system. Used in conjunction with reception pit (minimum capacity 48 hours with two locked sluice valves – both legal requirement) and tanker filling pump. Covers now available if rainfall or smell problem
- Dirty water below-ground tanks for low-volume irrigation (see Fig. 25.57): three chambers as above usually prefabricated concrete panels or rings, but can be constructed as underground cellars above in purpose-designed cast in situ or masonry. GRP not suitable. Must also have adequate locked access for pumping out, preferably with removable cover on settlement chamber

for mechanical removal of silt. Accessible 'H' pipes between chambers. Can use earth lagoon (see below) but need extra filtration for pump

- Weeping wall store – used with a dirty water system. Floor construction as option 1 with level floor with 150 mm water-holding upstand, or with floor level around perimeter, but with slight fall towards fill area. Outside the walls the floor must be extended by 300 mm incorporating a channel as per silage (legal requirement). Walls: 2.4–3 m high with 25- to 30-mm gaps at 200- to 450-mm centres on some or all faces using concrete panels or wooden sleepers in steel uprights. Filled by channel or ramp. Emptied via ramp or removable panels
- Impermeable wall store – usually used as above or with slurry pump. Construction of floor as above except that the external channel is not required, and walls as per silage bunker. Usually too expensive for 4- month storage. Can be used with internal weeping wall strainer used in conjunction with drain or pump to remove dirty water
- Earth bank lagoon (sometimes called compound when used for slurry). Suitable for either slurry or dirty water, short term, emergency overflow or 4-month minimum. The 'freeboard' minimum goes up to 750 mm and the soil must be suitable (legal requirement). Criteria for the construction are set out in CIRIA Report 126. (Note that the Code does not give sufficient guidance.) In very simple terms, providing the soil samples taken have passed a detailed analysis, it can be used without lining, as shown in Fig. 25.58. It must also be constructed in accordance with the standards laid down including compaction in layers to the table provided, e.g. 150 mm using eight passes of a vibratory roller between 1300 and 1800 kg. If there is a shortage of suitable material, it is possible to use an impermeable core, or, if the material is not suitable, then a liner can be used in one of the approved materials – usually with a guarantee, but a leakage monitoring system is required, so it is a last resort. All lagoons must be protected with child-proof fencing, locked gates and a hazard warning sign. Earth-bank lagoons can be used with low-volume irrigation systems with separating banks to subdivide the three tanks with T-pipe overflows between each

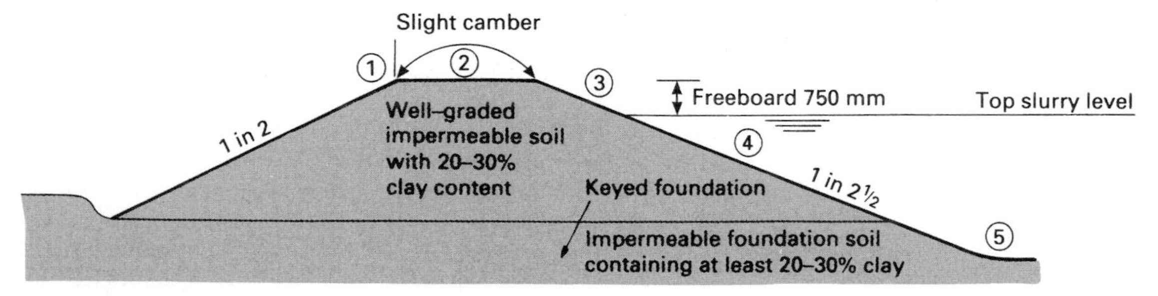

Figure 25.58 Lagoon embankment on impermeable soil (one of four options from CIRIA Report 126).
1 Suggested position of child-proof fence
2 Top width varies with depth but allow minimum 3 m for vehicle access, 4 m if emptied from top
3 Seed exposed faces to stabilize
4 Place 'ladders' around bank – often made out of tyres
5 Floor of lagoon should have shallow fall towards fill position

(usually two to each). Large lagoons must be designed to be emptied economically – preferably with a ramp access. If to be emptied from the top by swing shovel, the width must be restricted to suit the machine with a 4-m top access strip. A strainer box to ease the extraction of liquid is also essential with provision for operator access.

Machinery storage

(BS 5502, Part 80, refers)

General

Site building near good hard road, central to farm operations, adjacent to work shop and cleaning facilities. Enclosed roof with 10/15% roof lights. Open fronted with deep oversailing eaves facing sheltered aspect, or enclosed with doors if security problem. Eaves height at access usually 4.2 m for larger machinery or 3.7 m. Dwarf walls and enclosed cladding on other sides. Hardcore floor (preferably concrete at entrance). Need good lighting for maintainence: 150 lux artificial light plus sockets (preferably low voltage) for power tools

Cost range: £80 (open) to 140/m^2 (enclosed)

Clearance requirements for tractors and implements

Medium size tractor 3.0 m high × 1.85 m wide (may be extended to 2.5 m wheel track), area 6.75 m^2, with a turning circle of 7.5 m diameter (exhaust should have 0.5 m minimum height clearance from structure). Medium size tractor plus loader, turning circle 10 m diameter

Combine harvester 3.0–6.0 m wide × 7.5–10 m long × 3.0–4.0 m high

Trailers 2.0–2.43 m wide × 4.0–5.5 m long × up to 4.5 m high (tipped)

Lorries 2.43 m wide × 4.5–9.1 m long (rigid chassis) or 10.0–12.5 m long (articulated) × up to 4 m high (unusually 5 m high by 16 m long)

Plough, reversible, three-furrow, 1.5 m wide × 2.5 m long. Disc harrow 2.5–5 m wide × 1.9 m long. Seed and fertilizer drills 2.5–4 m wide (trailed) or 2.5–6.5 m wide (mounted) × 2.0–3.0 m long

Balers, medium density, 2.5–3.0 m wide × 4.5–5.5 m long

Beet harvesters 2.5–4 m wide × 5.5–7 m long × 3.5 m high

Fuel storage

Diesel store only (petrol storage restricted and licensed)

The Water Code refers to bunding requirement (legal)

Typical tank size (volumes approximate):

Steel
1 × 1 × 1 m (1000 litres); 2 × 1.25 × 1.25 m (3125 litres)

Plastic
1.15 m diam. × 1.3 m (1000 litres); 1.60 m diam. × 1.32 m (2000 litres)

Steel may be supported on walls; plastic on a 'solid' base (paving slabs on concrete lintels). All high enough to discharge to highest tank filler on vehicles when almost empty. Bottom slope from delivery pipe to sludge drain cock. Dipstick and ladder access or sight-gauge fitted. Only approved materials to be used. If 1500 litres or more is stored, then tank should be mounted in enclosing 'bund' walling to hold 110% of tank fill. (Tanks now available with integral bunding so no rainwater problems.) Flexible hose should have auto-close valve. All valves kept locked until use. Tanks are not installed within 10 m of watercourse. Minimum life of 20 years

Workshop accommodation

(BS 5502, Part 80, and Building Regulations, refers)

General	BS 5502 Class I if no mechanic employed, but since situation may change it is advisable to obtain full Building Regulations consent. In either case the Health and Safety requirements must be observed.
Area required and facilities	One or two bays of standard frame building of 9–15 m span insulated throughout to U-value of 0.45 W/K/m^2. Double skin roof lights min. 10%. Pit 2.0 long m × 0.8–1.0 m wide × 1.7 m deep with steps at each end, waterproofed, with drain sump: placed centrally in one of the bays with access from main door to workshop. Insulated walls with DPC to 1.8 m. Clear wall length of 5 m for benching with electric power (110 V). Good overall lighting to min. 150 lux and spot lighting to 400 lux. Concrete area of 50 m^2 min. outside main workshop door with good drainage falls to trapped sump with fuel separator. Coated non-slip dust-free floor inside of 150 mm RC40 concrete on DPM, reinforced with A142 fabric. Overhead gantry for crane desirable, but frame must be designed for load. Space heating and washing facilities. Fire precautions. High security

Farm office accommodation

Building Regulations and Offices, Shops and Railway Premises Act, 1973, refer	10 m^2 minimum, 15–20 m^2 preferred, with washroom area, kitchenette, secure veterinary store, and WC adjacent, all served by porch lobby of 3 m^2 min. Ceiling 2.5 m high. Double glazed window (min. 10% floor area) preferably with view of main farm access road. Electric power, water and telephone services essential. Insulate and ventilate to Building Regulations depending on situation. Large area of clear wall space desirable. Full or background space heating essential for computer equipment

Further reading

General

Astley Cooper, Sir P. (ed), *et al.* (1991) *Farm and Rural Buildings Pocketbook*. Farm and Rural Buildings Centre, NAC, Stoneleigh.

Barnes, M. & Mander, C. (1992) *Farm Building Construction: The Farmer's Guide*. 2nd ed. Farming Press, Ipswich.

BSI (1989–94) BS 5502 *Buildings and Structures for Agriculture*. Parts 10–19, General data series. Parts 20–39, General design series. Parts 40–59, Livestock building series. Parts 60–79, Crop building series. Parts 80–99, Ancillary buildings. BSI, London.

Eternit (2001) Profiled Sheeting Compendium. Eternit UK Ltd., Meldreth.

MacCormack, J. (ed) (2001) *Farm Buildings Cost Guide, 2001*. Centre for Rural Building, Aberdeen.

Noton, N.H. (1982) *Farm Buildings*. College of Estate Management, Reading, distributed by Spon, London.

Livestock

Baxter, S. (1984) *Intensive Pig Production, Environment Management and Design*. Granada, London.

Brent, G. (1986) *Housing the Pig*. Farming Press, Ipswich.

British Veterinary Association (1987) *Farm Animal Housing*. BVA, London.

Bryson, T. (1984) *The Sheep Housing Handbook*. Farming Press, Ipswich.

Farm Electric Centre (1990) *Controlled Environments for Livestock*, EA 1011. NAC, Stoneleigh.

Farm and Rural Buildings Centre (1992) *Planning Sheep Handling Units*. NAC, Stoneleigh.

Farm and Rural Buildings Centre (1992) *Planning Sheep Housing*. NAC, Stoneleigh,

Loynes, R. (ed) (1992) *Planning Dairy Units*. Farm and Rural Buildings Centre, NAC, Stoneleigh.

MAFF (1986) *Guidelines for Housed Livestock*. HMSO, London.

Meadowcroft, S. & Hardy, R. (1990) *Indoor Beef Production*. Farming Press, Ipswich.

Mitchell, D. (1976) *Calf Housing Handbook*. Farming Press, Ipswich.

Sainsbury, D. & Sainsbury, P. (1988) *Livestock Health and Housing*. Baillière Tindall, London.

Wathes, C.M. & Charles, D.R. (1994) *Livestock Housing*. CAB International, Slough.

Crop storage and effluent control

Farm Electric Centre (1985) *Hay Drying*, EC4744. NAC, Stoneleigh.

Farm Electric Centre (1985) *Potato Storage*, EC4457. NAC, Stoneleigh.

Farm Electric Centre (1990) *Bulk Grain Drying and Conditioning*, EA1032. NAC, Stoneleigh.

MAFF/WOAD (1998a) *The Water Code (Code of Good Agricultural Practice for the Protection of Water)*. MAFF Publications, London.

MAFF/WOAD (1998b) *Farm Waste Management Plan: A Step by Step Guide for Farmers*. MAFF Publications, London.

Mason, P.A. (1992) *Farm Waste Storage: Guidelines for Construction*. CIRIA Report No. 126. Construction Industry Research and Information Association, London.

McLean, K.A. (1989) *Drying and Storing Combinable Crops*, 2nd edn. Farming Press, Ipswich.

Building and surveying

Barry, R. (1992–1998) *The Construction of Buildings*, vols. 1–5. Blackwell Science, Oxford.

Powell-Smith V. & Billington, M.J. (1995) *The Building Regulations Explained and Illustrated*, 10th edn. Blackwell Science, Oxford.

Useful websites

www.agrifor.ac.uk – Agricultural (food & forestry) web catalogue

www.potato.org.uk – British Potato Council

www.bsi-global.com – British Standards Institution

www.countryside.gov.uk – Countryside Agency

www.defra.gov.uk – Department for Environment, Food & Rural Affairs

www.environment-agency.gov.uk – Environment Agency

www.hse.gov.uk – Health & Safety Executive

www.hgca.com – Home Grown Cereals Authority

www.ndfas.org.uk – National Dairy Farm Assurance Scheme

www.nfu.org.uk & www.nfucountryside.org.uk – National Farmers Union

www.nics.gov.uk – Northern Ireland DoE

www.rics.org/rural – Royal Institution of Chartered Surveyors

www.rdba.org.uk – Rural Design & Building Association

www.rspca.org.uk – RSPCA

www.soilassociation.org.uk – Soil Association

www.scotland.gov.uk/pages – Scottish Executive

www.hmso.gov.uk – The Stationary Office

www.ukfpa.co.uk – UK Forest Products Association

www.defra.gov.uk/environment/index.htm – Water Regulations Advisory Service

Index